AF361766

Targeting STAT3 and STAT5 in Cancer

Targeting STAT3 and STAT5 in Cancer

Editors

Richard Moriggl
Patrick T. Gunning
György Miklós Keserü

MDPI • Basel • Beijing • Wuhan • Barcelona • Belgrade • Manchester • Tokyo • Cluj • Tianjin

Editors
Richard Moriggl
University for Veterinary Medicine Vienna
Austria

Patrick T. Gunning
University of Toronto Mississauga
Canada

György Miklós Keserü
Medicinal Chemistry Research Group
Hungary

Editorial Office
MDPI
St. Alban-Anlage 66
4052 Basel, Switzerland

This is a reprint of articles from the Special Issue published online in the open access journal *Cancers* (ISSN 2072-6694) (available at: https://www.mdpi.com/journal/cancers/special_issues/STAT_cancers).

For citation purposes, cite each article independently as indicated on the article page online and as indicated below:

LastName, A.A.; LastName, B.B.; LastName, C.C. Article Title. *Journal Name* **Year**, *Article Number, Page Range*.

ISBN 978-3-03943-036-9 (Hbk)
ISBN 978-3-03943-037-6 (PDF)

Cover image courtesy of Richard Moriggl.

Contents

About the Editors

Richard Moriggl studied biotechnology in Germany and completed his Ph.D. in 1997 working on cytokine signaling with Prof. Groner at the Friedrich Miescher Institute, Basel, Switzerland and the Institute of Experimental Cancer Research, Freiburg, Germany. After a postdoctoral fellowship with Prof. Ihle in the area of immunology at St. Jude Children's Research Hospital in Memphis, USA, he undertook a second postdoc in 2000 to work on leukemia with Prof. Beug at the Institute for Molecular Pathology, Vienna. Dr. Moriggl served for 14 years as director of the Ludwig Boltzmann Institute for Cancer Research, heading six research groups that all progressed to tenure tracked positions and integrated into medical research performing universities in Austria. The Moriggl lab explores targeting of STAT transcription factors and performs basic and translational cancer research with a main focus on generation and utilization of gene-targeted mouse models. Established cellular systems are used to investigate neoplastic T-cell development in PTCL and how these processes can be blocked. The Moriggl lab contributes to the understanding of gene regulatory and chromatin landscape consequences in cancer upon hyperactivation of JAK-STAT3/5 action. Unpublished work investigates T-cell development and targeting STAT5 to understand cancer progression in association with fatal clinical problems.

Patrick T. Gunning is a full professor of chemistry and Canada Research Chair in Medicinal Chemistry at the University of Toronto. Dr. Gunning received his BSc (2001) and Ph.D. (2005) from the University of Glasgow before pursuing postdoctoral studies with Professor Andrew D. Hamilton (Yale University). His research group is focussed on the development of small molecule inhibitors of protein-protein interactions, namely disruptors of STAT3 and STAT5 proteins, HDAC inhibitors, and monovalent degradation. In this field, Dr. Gunning has co-authored >100 research publications and received numerous awards and honours. These include the 2015 Rose Winer Levin Lectureship by Dana-Farber Cancer Institute, the 2012 RSC MedChemComm Emerging Investigator Lectureship by the Royal Society for Chemistry (UK), the 2016 Canadian Society for Chemistry's Bernard Belleau award (Canada) and named in Canada's Top 40 under 40 awardees.

György Miklós Keserü obtained his Ph.D. at Budapest, Hungary. He worked for Sanofi and in 1999 moved to Gedeon Richter. Since 2007 he was appointed as the Head of Discovery Chemistry at Gedeon Richter. He was involved in the discovery of 10 clinical candidates (41 patent applications) entered into clinical development in neurological and psychiatric indications. He contributed to the discovery of the antipsychotic Vraylar® (US)/ Reagila® (EU) that has been approved by the US FDA in 2015 and European authorities in 2017. He served as a director general of the Research Centre for Natural Sciences (RCNS) at the Hungarian Academy of Sciences. He played a significant role in the development of physicochemical profiling in drug discovery, the concept of compound quality and the improvement of optimization strategies in medicinal chemistry. His results has been acknowledged by the George A. Olah Award and the Prous Institute Overton and Meyer Award of the European Federation of Medicinal Chemistry. From 2015 he is a full professor at the Budapest University of Technology and Economics and heading the Medicinal Chemistry Research Group at RCNS. In 2019 he has been elected as corresponding member of the Hungarian Academy of Sciences. His research interests include medicinal chemistry and drug design. He has published over 200 papers and more than 10 books and book chapters.

Preface to "Targeting STAT3 and STAT5 in Cancer"

Cancer is often driven by hyperactivation of the JAK–STAT core cancer pathway, which is associated with inappropriate functions of normal signaling pathways involving cytokine, growth factor and hormone action. The successful implementation of clinically approved JAK kinase inhibitors (baracitinib, ruxolitinib and tofacitinib) validates that JAK–STAT targeting is beneficial. Consequently, targeting oncogenic transcription factors of the STAT family, namely STAT3, STAT5A, and STAT5B, as major funnels for gene regulatory processes including chromatin remodeling has clear therapeutic power. STAT3/5 proteins are attractive targets for drug discovery since they steer cell proliferation, survival and metabolism. Targeting protein-protein interactions, however, is challenging since the interacting surfaces are typically large and flat, having non-contiguous binding hot spots and a lack of deep tractable pockets. Furthermore, transcription factors were long considered "undruggable". Emerging drug design strategies and medicinal chemistry approaches, including methods to impair function, to destabilize or degrade transcription factors or to interfere with interaction partners and cofactors, initiated new concepts and changed this paradigm. Here, we summarize approaches for targeting STAT3/5, where the field moves into clinical application. We cover the design paradigms and medicinal chemistry approaches to illuminate limitations in specificity, potency, and in vivo bioavailability, necessitating new approaches and further developments.

Richard Moriggl , Patrick T. Gunning, György Miklós Keserü
Editors

Editorial

Targeting STAT3 and STAT5 in Cancer

Elvin D. de Araujo [1,2], György M. Keserű [3], Patrick T. Gunning [1,2] and Richard Moriggl [4,*]

1   Centre for Medicinal Chemistry, University of Toronto at Mississauga, Mississauga, ON L5L 1C6, Canada; e.dearaujo@mail.utoronto.ca (E.D.d.A.); patrick.gunning@utoronto.ca (P.T.G.)
2   Department of Chemical and Physical Sciences, University of Toronto at Mississauga, Mississauga, ON L5L 1C6, Canada
3   Medicinal Chemistry, Research Center for Natural Sciences, 1117 Budapest, Hungary; keseru.gyorgy@ttk.mta.hu
4   Institute of Animal Breeding and Genetics, University of Veterinary Medicine, A-1210 Vienna, Austria
*   Correspondence: richard.moriggl@vetmeduni.ac.at

Received: 15 July 2020; Accepted: 16 July 2020; Published: 22 July 2020

1. Introduction

Insights into the mutational landscape of the human cancer genome coding regions defined about 140 distinct cancer driver genes in 2013, which approximately doubled to 300 in 2018 following advances in systems cancer biology studies [1,2]. The rapid growth and understanding led to taxonomical organization of known oncogenes into 12 core cancer pathways that regulate cellular differentiation, survival, and genome maintenance [1,3]. Despite significant advances in functional cancer genomics, only a subset of these onco-targets has been successfully tackled in clinical settings, with a predominant emphasis on small molecule kinase inhibitors. Contrastingly, targeting transcription factors has been challenging due to shallow binding sites and non-contiguous surfaces. However, several signaling networks converge upon the Janus Kinase-Signal Transducer and Activator of Transcription (JAK-STAT) pathway and are explored in this special issue (Figure 1). Herein, we describe current concepts and paradigms of oncogenic STAT3/5 targeting.

The JAK-STAT pathway controls cell survival, differentiation, and metabolism, with critical roles in shaping the chromatin landscape, immune response, and mitochondrial function (Figure 1) [3]. To add another layer of complexity, current STAT studies are focused on in vitro model systems, which may not fully represent the intricate complexities of stimulatory or inhibitory cancer cell-stroma interactions as well as paracrine and endocrine cellular signaling. Additionally, the pleiotropic action of the JAK-STAT pathway in different cell types is complex. It is well known that JAK-STAT3/5 hyperactivation promotes tumorigenesis, both in solid tumors as well as blood cancers. As such, we have focused the themes of this special issue on targeting STAT3 and STAT5, while also highlighting the effects of changes to the collective expression and activity profiles. This is critical as the entire signaling cascade is interconnected with iterative response profiles, best illustrated by the tumor suppressive action of STAT family members, STAT1/2.

The first chapter explores targeting concepts of STAT3/5, with an emphasis on hematopoietic cancers and disease-associated mutations, providing both overview articles as well as original work. Chapter 2 discusses the role of STAT3 and STAT5 across a wide range of solid cancers. Chapter 3 highlights emerging concepts influencing upstream or downstream regulatory mechanisms for targeting oncogenic STAT3/5 action. Each chapter is prefaced with an overview article to appropriately outline and introduce the disease topics for targeting concepts.

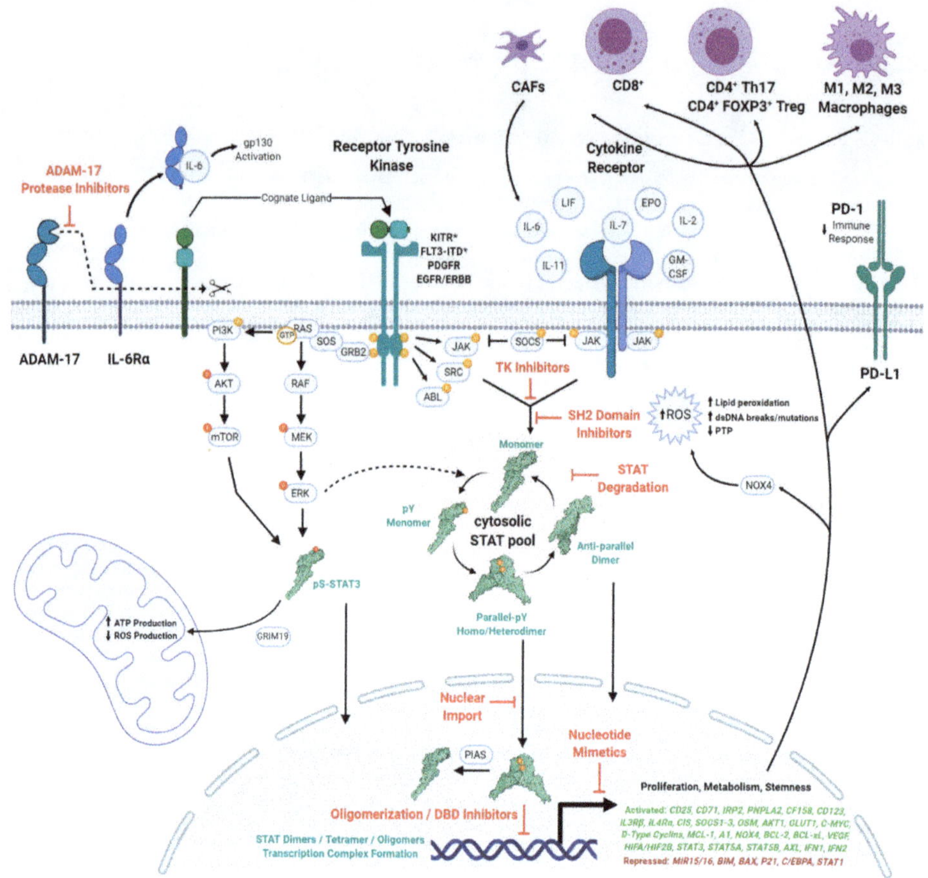

Figure 1. STAT activation pathway with opportunities for inhibition.

An oversimplification of the definition of core cancer pathways is helpful for pharmacologic targeting concepts, since one can aim to target single or multiple core cancer pathways with pathway blocker(s). The JAK-STAT signaling pathway plays a dominant role interconnecting and driving several core cancer networks. Hyperactivated JAK-STAT activity is associated with aberrant functioning of cytokine, growth factor, chemokine, and hormone signaling pathways, ultimately transforming chromatin landscapes and gene reprogramming [3]. Stimulating ligands normally trigger JAK-STAT action in a transient or gradient-like manner. This is well documented for Wnt signaling in crypt villi or upon expansion of the immune system during infection, where a cytokine storm is followed by down-regulation to homeostatic conditions and memory response [4–6]. However, the situation in the cancer phenotype is very different and where age dominant clones can originate. For example, 10–15% of individuals over the age of 70 develop Clonal Hematopoiesis of Indeterminate Potential (CHIP), whereby small imbalances of external signals are known to promote acute leukemia and lymphoma outbreak [7,8].

Loss of control over cytokine signaling, as observed with gain-of-function mutations in oncogenic JAK-STAT components, directly promotes neoplastic outgrowth [9–11]. Moreover, in recent years, advancements in understanding processes, such as unphosphorylated uSTAT action [12] as well as STAT involvement in mitochondrial metabolism [13], have transitioned the consensus away from the idea of a linear signaling JAK-STAT cascade. Several negatively acting controls are also implemented on JAK-STAT signaling to suppress transformation [10,11]. However, cancer is a multi-genetic disease, and lost or lowered tumor suppressor protein function associated with global methylation changes [14] decreased proteolysis of receptor-kinase components (e.g., lost or methylated SOCS family members) [15] and/or inhibited phosphatase action [16] can all provoke hyperactivated JAK-STAT action, culminating into cancer initiation and progression. Within the scope of this special issue, several articles also focused on STAT3/5 hyperactivity in the microenvironment for complex immune cell function or interplay with cancer cells or other stromal components. Overall, aberrant STAT3/5 functioning extends beyond cancer phenotypes into additional malignances, including autoimmunity, chronic inflammation, and infectious disease.

The successful implementation of clinically approved JAK kinase inhibitors (ruxolitinib, tofacitinib, baracitinib, and ifacitinib) validates the therapeutic utility of JAK-STAT targeting in a wide variety of cancers and chronic inflammatory and autoimmune diseases. Current JAK kinase inhibitors are ATP competitive analogues, with limited subtype selectivity. Some selectivity was achieved by covalently targeting a Cys residue in close proximity to the orthosteric pocket of JAK3 [17]. JAK kinase inhibitors have recently been employed for autoimmune-driven diseases including rheumatoid arthritis and exploration for basket trials in cancer, such as those performed for BRAF or HER2 kinase inhibitors, and could further propel clinical use [18]. However, there are significant challenges to overcome in cancer targeting, and several of the proposed inhibitors in literature could be enhanced through deeper understanding of the direct biophysical and target engagement profiles, in conjunction with phenotypic screening.

Here, we summarize targeting approaches on STAT3/5, where we combine review articles from experts in the field as well as original articles from multiple groups employing different targeting strategies to advance clinical development. Inhibitor designs and medicinal chemistry approaches are covered, highlighting limitations in the current modalities, such as selectivity, potency, and in vivo efficacy for targeting hematopoietic and solid cancers. We have structured this special issue into three main chapters.

Chapter 1: Targeting STAT3/5 in Hematopoietic Cancers

This chapter examines the drug targeting concept of STAT3/5 in hematopoietic cancers as well as exploring less-studied STAT-driven indications. The chapter is prefaced by a review outlining a brief history and identification of the involvement of STAT3 and STAT5 gene products and their current mutational landscape. The major gain-of-function driver mutations in hematologic cancers are highlighted, as well as detailed reports of indirect and direct inhibitors of STAT3 and STAT5 and their current stage of clinical development [19]. These themes are further developed by Orlova et al., where the current paradigms in direct small molecule targeting of STAT3 and STAT5 are explored in the context of unphosphorylated STATs, as well as chromatin remodeling [20]. Furthermore, the involvement of STAT signaling in the immune system and T-cell development have also led to proposed studies for hijacking and re-wiring the immune suppressive tumor-associated macrophages in the context of STAT3/5 [21]. The central ideas are expanded upon by Rébé et al., where they examine the critical role of STAT3 in the immune system and T-cell differentiation as well as checkpoint inhibitors [22]. As seen above, several STAT3/5 targeting options were explored for hematopoietic cancers. However, the role of STAT3 and STAT5 in specific indications is still being established and is often revealed through somatic mutations identified from patient tumors. The molecular-level functional effects of specific disease-associated SH2 domain mutations are highlighted in the context of the current protein crystal structures [23]. Additional studies have explored the role of STATs in hematological malignancies. Phospho-STAT5 was shown to play a key role in CD34+/CD38− myeloproliferative neoplasms with downregulation by pharmacologic inhibition of JAK and STAT [24]. JAK/STAT mutations were also identified in patients with the rare and aggressive T-cell prolymphocytic leukemia (T-PLL), pointing towards a possible mode of transformation [25]. Additionally, STAT3 (and not

STAT5B) mutations were identified in multiple peripheral T-cell lymphomas (PTCL), as well as high pY-STAT3 expression, particularly in angioimmunoblastic T-cell lymphoma (AITL) and anaplastic large cell lymphoma (ALCL) patient samples, two different PTCL cancer types [26]. From an inhibition perspective, a derivatized indole was shown to be effective in both chronic myeloid leukemia (CML) and acute myeloid leukemia (AML), as well as aggressive STAT5-N642H tumors [27]. Collectively, these studies shed light on the importance of STAT3/5 in several hematopoietic cancers, potentially offering disease biomarkers and hallmarks across several different indications.

Chapter 2: Targeting STAT3/5 in Solid Cancers

This chapter highlights the role of STAT3/5 across a range of solid cancer models and is prefaced by a thorough review highlighting the importance of STAT3 in inflammation, stemness, and mitochondrial functions. It also depicts the mutational landscape with recurrent hotspot mutations in the three different STAT3/5 gene products throughout these solid cancers [28]. The detailed mutational maps, related to the oncogenic STAT3/5 proteins, indicate key functional amino acid residues, and their appearance highlights where they might play critical cell-type specific roles [28]. Successively, Polak et al. highlight the importance of leveraging selective STAT-targeting, arising from the STAT3-driven cancer-stem cell properties and STAT3/5-enhanced immunosuppression, in contrast to the tumor suppression activity of STAT1/2 [29]. These concepts form the foundation for the discussion of STAT3/5 across multiple solid cancers and indications. Notably, aberrant STAT3/5 function alters tumorigenic properties across multiple organs, further highlighting their critical role as cancer drivers. Herein, we briefly outline the different indications explored within the chapter.

Melanoma and causatively linked autoimmune diseases are directly tied to hyperactivation of STAT3/5 cascades, and dual/simultaneous targeting could reduce the overall disease burden [30]. For colon cancer-based therapies, upstream targeting of IL-6 trans-signalling was explored in the context of the ligand-releasing protease, ADAM17 [31]. Aggressive epithelial ovarian cancer was discussed relative to the STAT3/5 activating tumor microenvironment and their correlation to the persistence of recurrent neoplasms [32]. In prolactin-driven prostate cancer mouse models, STAT5A/B deletions were found to significantly delay the onset of tumorigenesis, further highlighting a potential targeting opportunity [33]. For breast cancer-based models, STAT3 was identified to promote PDL-1 expression, leading to reduced immune responses, which was validated through gene silencing and pharmacologic inhibition [34]. Direct targeting nanomedicine-based approaches were also applied in syngeneic glioblastoma mouse models, where lipid-formulated siRNA was capable of suppressing STAT3 activity and tumor growth [35]. STAT3 activation was also explored in hepatocellular carcinoma (HCC) arising from *hepatitis C* infections. Aydin et al. identified that HCC progression was mediated through NRF2, leading to STAT3 activation and suppression of *miR-122* [36]. Finally, nano-formulated STAT3 inhibitors were also shown to be effective in myeloma xenograft models, enhancing the therapeutic properties of the inhibitor alone [37].

Collectively, the range of indications examined in this chapter underscore the importance of STAT3/5 signalling in transformation, but also the potential for cross-indication utility for prospective therapies.

Chapter 3: General Targeting Aspects to Block the JAK1/2/3/TYK2-STAT3/5 Core Cancer Pathway

This chapter broadens the scope of current STAT targeting strategies. This has led to studies of STAT within companion animals, particularly canine species, that offer advantages over current pre-clinical models which are also a major hurdle in drug development safety/toxicity studies. Moreover, companion animals share the same environment as their owners and similarities in cancers, e.g., driver mutation conservation for key STAT domains could be important for understanding the mechanistic detail within comparative pathology studies. Different animal species can also provide additional mechanistic insights into STAT evolution and the signaling cascades [38]. For instance, understanding STAT activation in the context of heat shock proteins (HSPs) is critical for their involvement in cellular stress and the immune response which was explored by Jego et al [39].

Alternative strategies for STAT targeting were also discussed within this chapter. Nucleotide therapeutics, which were directed towards the STAT3 DNA-binding domain, were also employed in clinical trials and represent a promising avenue in STAT inhibition that was distinct from small molecules [40]. Alternatively, targeting the specific nucleocytoplasmic STAT transport shuttles offers a creative mechanism for gaining STAT-inhibition selectivity [41]. The role and importance of targeting TYK2, in addition to the JAK1-3, in the context of inducing STAT activation, was also thoroughly discussed [42]. Similarly, the mTOR-STAT3 axis was investigated in Shwachman-Diamond syndrome as a potential kinase target in blockading persistent STAT3 activation [43]. Collectively, these strategies offer non-conventional approaches to the JAK-STAT targeting regimens.

2. Conclusions

In summary, this special issue offers an overview of different STAT3/5 targeting approaches in the context of multiple disease indications. Emerging drug design strategies and medicinal chemistry approaches, including methods to impair function [44] destabilize or degrade transcription factors [45–47], interfere with interaction partners and cofactors [48], block DNA binding [49], nuclear shuttling [41], or target specific subsets of cell types/microenvironment [50], initiated new concepts for broad views on targeting. Consequently, targeting oncogenic transcription factors of the STAT family, namely STAT3, STAT5A, and STAT5B as three distinct gene products are major funnels for gene regulatory processes, including chromatin remodeling. Targeting these STAT gene products has therapeutic power, as demonstrated with proof-of-concept studies, although advanced clinical trials are not available. Thus, greater efforts need to be undertaken to identify new targets and develop selective and less toxic clinical-grade inhibitors.

From a contemporary standpoint, the COVID-19 pandemic has highlighted the vulnerability of cancer patients to infection [51]. We want to conclude that research on STAT3/5 targeting molecules and concepts are broadly underexplored, as the 2020 SARS-CoV-2 pandemic revealed. SARS-CoV-2 infection accounts for ≈28% mortality in New York cancer patients, from a cohort of 218 infected individuals. Even therapeutic success will leave potential comorbidities that can flip the coin between life and death [52]. SARS-CoV-2 infection rates were also associated with a 37% higher mortality in hematologic malignancies predominantly associated with CHIP, pointing to the importance of regulation of the blood and immune system, which is under the control of external stimuli, i.e., the JAK-STAT pathway. Currently, mortality of SARS-CoV-2 infections can be reduced by blocking the cytokine storm paradoxically with either tocilizumab or dexamethasone [53,54]. These drugs display well known immunosuppressive action aside from other more systemic impacts on metabolism. STAT3/5 are well known to interact with the glucocorticoid receptor. They can promote cytokine and growth factor feed forward loops, and are critically involved in cytokine or growth factor, as well as hormone signalling. Whether or not STAT3/5 inhibition is beneficial to treatment of COVID-19 infections is unclear, illustrating the gaps in the research ahead of us, as well as the many opportunities, options, and value in targeting STAT3/5.

Funding: R.M. was supported by the Austrian Science Fund (FWF; SFB-F04707, SFB-F06105, and under the frame of ERA ERA-NET (I 4157-B)). G.M.K. was supported by OTKA K 116904 (National Research, Development and Innovation Office, Hungary). P.T.G was supported by research grants from NSERC (RGPIN-2014-05767), CIHR (MOP-130424, MOP-137036), Canada Research Chair (950-232042), Canadian Cancer Society (703963), and Canadian Breast Cancer Foundation (705456); and infrastructure grants from CFI (33536) and the Ontario Research Fund (34876).

Acknowledgments: All figures were created with https://biorender.com/.

Conflicts of Interest: The authors declare no conflict of interest.

References

1. Vogelstein, B.; Papadopoulos, N.; Velculescu, V.E.; Zhou, S.; Diaz, L.A.; Kinzler, K.W. Cancer genome landscapes. *Science* **2013**. [CrossRef] [PubMed]
2. Crescenzo, R.; Abate, F.; Lasorsa, E.; Tabbo', F.; Gaudiano, M.; Chiesa, N.; Di Giacomo, F.; Spaccarotella, E.; Barbarossa, L.; Ercole, E.; et al. Convergent mutations and kinase fusions lead to oncogenic STAT3 activation in anaplastic large cell lymphoma. *Cancer Cell* **2015**, *27*, 516–532. [CrossRef] [PubMed]
3. Wingelhofer, B.; Neubauer, H.A.; Valent, P.; Han, X.; Constantinescu, S.N.; Gunning, P.T.; Müller, M.; Moriggl, R. Implications of STAT3 and STAT5 signaling on gene regulation and chromatin remodeling in hematopoietic cancer. *Leukemia* **2018**, *32*, 1713–1726. [CrossRef] [PubMed]
4. Xu, N.; Wang, S.Q.; Tan, D.; Gao, Y.; Lin, G.; Xi, R. EGFR, Wingless and JAK/STAT signaling cooperatively maintain Drosophila intestinal stem cells. *Dev. Biol.* **2011**, *354*, 31–43. [CrossRef]
5. Zou, W.Y.; Blutt, S.E.; Zeng, X.L.; Chen, M.S.; Lo, Y.H.; Castillo-Azofeifa, D.; Klein, O.D.; Shroyer, N.F.; Donowitz, M.; Estes, M.K. Epithelial WNT Ligands Are Essential Drivers of Intestinal Stem Cell Activation. *Cell Rep.* **2018**, *22*, 1003–1015. [CrossRef] [PubMed]
6. Villarino, A.V.; Kanno, Y.; O'Shea, J.J. Mechanisms and consequences of Jak-STAT signaling in the immune system. *Nat. Immunol.* **2017**. [CrossRef]
7. Jaiswal, S.; Libby, P. Clonal haematopoiesis: Connecting ageing and inflammation in cardiovascular disease. *Nat. Rev. Cardiol.* **2020**, *17*, 137–144. [CrossRef]
8. Libby, P.; Sidlow, R.; Lin, A.E.; Gupta, D.; Jones, L.W.; Moslehi, J.; Zeiher, A.; Jaiswal, S.; Schulz, C.; Blankstein, R.; et al. Clonal Hematopoiesis: Crossroads of Aging, Cardiovascular Disease, and Cancer: JACC Review Topic of the Week. *J. Am. Coll. Cardiol.* **2019**, *74*, 567–577. [CrossRef]
9. de Araujo, E.D.; Erdogan, F.; Neubauer, H.A.; Meneksedag-Erol, D.; Manaswiyoungkul, P.; Eram, M.S.; Seo, H.S.; Qadree, A.K.; Israelian, J.; Orlova, A.; et al. Structural and functional consequences of the STAT5BN642H driver mutation. *Nat. Commun.* **2019**, *10*. [CrossRef]
10. O'Shea, J.J.; Schwartz, D.M.; Villarino, A.V.; Gadina, M.; McInnes, I.B.; Laurence, A. The JAK-STAT Pathway: Impact on Human Disease and Therapeutic Intervention. *Annu. Rev. Med.* **2015**, *66*, 311–328. [CrossRef]
11. Kiu, H.; Nicholson, S.E. Biology and significance of the JAK/STAT signalling pathways. *Growth Factors* **2012**, *30*, 88–106. [CrossRef] [PubMed]
12. Timofeeva, O.A.; Chasovskikh, S.; Lonskaya, I.; Tarasova, N.I.; Khavrutskii, L.; Tarasov, S.G.; Zhang, X.; Korostyshevskiy, V.R.; Cheema, A.; Zhang, L.; et al. Mechanisms of unphosphorylated STAT3 transcription factor binding to DNA. *J. Biol. Chem.* **2012**, *287*, 14192–14200. [CrossRef] [PubMed]
13. Rincon, M.; Pereira, F.V. A new perspective: Mitochondrial stat3 as a regulator for lymphocyte function. *Int. J. Mol. Sci.* **2018**, *19*, 1656. [CrossRef] [PubMed]
14. Hernandez-Vargas, H.; Ouzounova, M.; le Calvez-Kelm, F.; Lambert, M.P.; McKay-Chopin, S.; Tavtigian, S.V.; Puisieux, A.; Matar, C.; Herceg, Z. Methylome analysis reveals Jak-STAT pathway deregulation in putative breast cancer stem cells. *Epigenetics* **2011**, *6*, 428–439. [CrossRef] [PubMed]
15. Seif, F.; Khoshmirsafa, M.; Aazami, H.; Mohsenzadegan, M.; Sedighi, G.; Bahar, M. The role of JAK-STAT signaling pathway and its regulators in the fate of T helper cells. *Cell Commun. Signal.* **2017**, *15*. [CrossRef]
16. Kim, M.; Morales, L.D.; Jang, I.S.; Cho, Y.Y.; Kim, D.J. Protein Tyrosine Phosphatases as Potential Regulators of STAT3 Signaling. *Int. J. Mol. Sci.* **2018**, *19*, 2708. [CrossRef]
17. Abdeldayem, A.; Raouf, Y.S.; Constantinescu, S.N.; Moriggl, R.; Gunning, P.T. Advances in covalent kinase inhibitors. *Chem. Soc. Rev.* **2020**, *49*, 2617–2687. [CrossRef]
18. Ribas, A.; Lo, R.S. Trying for a BRAF Slam Dunk. *Cancer Discov.* **2020**, *10*, 640–642. [CrossRef]
19. Brachet-Botineau, M.; Polomski, M.; Neubauer, H.A.; Juen, L.; Hédou, D.; Viaud-Massuard, M.C.; Prié, G.; Gouilleux, F. Pharmacological inhibition of oncogenic STAT3 and STAT5 signaling in hematopoietic cancers. *Cancers* **2020**, *12*, 240. [CrossRef]
20. Orlova, A.; Wagner, C.; De Araujo, E.D.; Bajusz, D.; Neubauer, H.A.; Herling, M.; Gunning, P.T.; Keserü, G.M.; Moriggl, R. Direct targeting options for STAT3 and STAT5 in cancer. *Cancers* **2019**, *11*, 1930. [CrossRef]
21. Verdeil, G.; Lawrence, T.; Schmitt-Verhulst, A.M.; Auphan-Anezin, N. Targeting stat3 and stat5 in tumor-associated immune cells to improve immunotherapy. *Cancers* **2019**, *11*, 1832. [CrossRef] [PubMed]
22. Rébé, C.; Ghiringhelli, F. STAT3, a master regulator of anti-tumor immune response. *Cancers* **2019**, *11*, 1280. [CrossRef] [PubMed]

23. de Araujo, E.D.; Orlova, A.; Neubauer, H.A.; Bajusz, D.; Seo, H.S.; Dhe-Paganon, S.; Keserű, G.M.; Moriggl, R.; Gunning, P.T. Structural implications of stat3 and stat5 sh2 domain mutations. *Cancers* **2019**, *11*, 1757. [CrossRef]

24. Hadzijusufovic, E.; Keller, A.; Berger, D.; Greiner, G.; Wingelhofer, B.; Witzeneder, N.; Ivanov, D.; Pecnard, E.; Nivarthi, H.; Schur, F.K.M.; et al. STAT5 is expressed in CD34+/CD38– stem cells and serves as a potential molecular target in ph-negative myeloproliferative neoplasms. *Cancers* **2020**, *12*, 1021. [CrossRef]

25. Wahnschaffe, L.; Braun, T.; Timonen, S.; Giri, A.K.; Schrader, A.; Wagle, P.; Almusa, H.; Johansson, P.; Bellanger, D.; López, C.; et al. Jak/stat-activating genomic alterations are a hallmark of t-pll. *Cancers* **2019**, *11*, 1833. [CrossRef] [PubMed]

26. Andersson, E.I.; Brück, O.; Braun, T.; Mannisto, S.; Saikko, L.; Lagström, S.; Ellonen, P.; Leppä, S.; Herling, M.; Kovanen, P.E.; et al. STAT3 mutation is associated with STAT3 activation in CD30+ ALK– ALCL. *Cancers* **2020**, *12*, 702. [CrossRef] [PubMed]

27. Brachet-Botineau, M.; Deynoux, M.; Vallet, N.; Polomski, M.; Juen, L.; Hérault, O.; Mazurier, F.; Viaud-Massuard, M.C.; Prié, G.; Gouilleux, F. A novel inhibitor of stat5 signaling overcomes chemotherapy resistance in myeloid leukemia cells. *Cancers* **2019**, *11*, 2043. [CrossRef]

28. Igelmann, S.; Neubauer, H.A.; Ferbeyre, G. STAT3 and STAT5 activation in solid cancers. *Cancers* **2019**, *11*, 1428. [CrossRef]

29. Polak, K.L.; Chernosky, N.M.; Smigiel, J.M.; Tamagno, I.; Wjackson, M. Balancing STAT activity as a therapeutic strategy. *Cancers* **2019**, *11*, 1716. [CrossRef]

30. Logotheti, S.; Pützer, B.M. STAT3 and STAT5 targeting for simultaneous management of melanoma and autoimmune diseases. *Cancers* **2019**, *11*, 1448. [CrossRef]

31. Schumacher, N.; Rose-John, S. Adam17 activity and il-6 trans-signaling in inflammation and cancer. *Cancers* **2019**, *11*, 1736. [CrossRef] [PubMed]

32. Wu, C.J.; Sundararajan, V.; Sheu, B.C.; Huang, R.Y.J.; Wei, L.H. Activation of STAT3 and STAT5 signaling in epithelial ovarian cancer progression: Mechanism and therapeutic opportunity. *Cancers* **2020**, *12*, 24. [CrossRef] [PubMed]

33. Boutillon, F.; Pigat, N.; Sala, L.S.; Reyes-Gomez, E.; Moriggl, R.; Guidotti, J.E.; Goffin, V. STAT5a/b deficiency delays, but does not prevent, prolactin-driven prostate tumorigenesis in mice. *Cancers* **2019**, *11*, 929. [CrossRef] [PubMed]

34. Zerdes, I.; Wallerius, M.; Sifakis, E.G.; Wallmann, T.; Betts, S.; Bartish, M.; Tsesmetzis, N.; Tobin, N.P.; Coucoravas, C.; Bergh, J.; et al. STAT3 activity promotes programmed-death ligand 1 expression and suppresses immune responses in breast cancer. *Cancers* **2019**, *11*, 1479. [CrossRef] [PubMed]

35. Linder, B.; Weirauch, U.; Ewe, A.; Uhmann, A.; Seifert, V.; Mittelbronn, M.; Harter, P.N.; Aigner, A.; Kögel, D. Therapeutic targeting of stat3 using lipopolyplex nanoparticle-formulated sirna in a syngeneic orthotopic mouse glioma model. *Cancers* **2019**, *11*, 333. [CrossRef]

36. Aydin, Y.; Kurt, R.; Song, K.; Lin, D.; Osman, H.; Youngquist, B.; Scott, J.W.; Shores, N.J.; Thevenot, P.; Cohen, A.; et al. Hepatic stress response in HCV infection promotes STAT3-mediated inhibition of HNF4A-miR-122 feedback loop in liver fibrosis and cancer progression. *Cancers* **2019**, *11*, 1407. [CrossRef]

37. Huang, Y.H.; Vakili, M.R.; Molavi, O.; Morrissey, Y.; Wu, C.; Paiva, I.; Soleimani, A.H.; Sanaee, F.; Lavasanifar, A.; Lai, R. Decoration of anti-CD38 on nanoparticles carrying a STAT3 inhibitor can improve the therapeutic efficacy against myeloma. *Cancers* **2019**, *11*, 248. [CrossRef]

38. Kieslinger, M.; Swoboda, A.; Kramer, N.; Pratscher, B.; Wolfesberger, B.; Burgener, I.A. Companion animals as models for inhibition of STAT3 and STAT5. *Cancers* **2019**, *11*, 2035. [CrossRef]

39. Jego, G.; Hermetet, F.; Girodon, F.; Garrido, C. Chaperoning STAT3/5 by heat shock proteins: Interest of their targeting in cancer therapy. *Cancers* **2020**, *12*, 21. [CrossRef]

40. Lau, Y.T.K.; Ramaiyer, M.; Johnson, D.E.; Grandis, J.R. Targeting STAT3 in cancer with nucleotide therapeutics. *Cancers* **2019**, *11*, 1681. [CrossRef]

41. Ernst, S.; Müller-Newen, G. Nucleocytoplasmic shuttling of stats. A target for intervention? *Cancers* **2019**, *11*, 1815. [CrossRef] [PubMed]

42. Wöss, K.; Simonović, N.; Strobl, B.; Macho-Maschler, S.; Müller, M. Tyk2: An upstream kinase of stats in cancer. *Cancers* **2019**, *11*, 1728. [CrossRef] [PubMed]

43. Vella, A.; D'aversa, E.; Api, M.; Breveglieri, G.; Allegri, M.; Giacomazzi, A.; Busilacchi, E.M.; Fabrizzi, B.; Cestari, T.; Sorio, C.; et al. MTOR and STAT3 pathway hyper-activation is associated with elevated interleukin-6 levels in patients with shwachman-diamond syndrome: Further evidence of lymphoid lineage impairment. *Cancers* **2020**, *12*, 597. [CrossRef] [PubMed]

44. Huang, Q.; Zhong, Y.; Dong, H.; Zheng, Q.; Shi, S.; Zhu, K.; Qu, X.; Hu, W.; Zhang, X.; Wang, Y. Revisiting signal transducer and activator of transcription 3 (STAT3) as an anticancer target and its inhibitor discovery: Where are we and where should we go? *Eur. J. Med. Chem.* **2020**, *187*. [CrossRef] [PubMed]

45. Zhou, H.; Bai, L.; Xu, R.; Zhao, Y.; Chen, J.; McEachern, D.; Chinnaswamy, K.; Wen, B.; Dai, L.; Kumar, P.; et al. Structure-Based Discovery of SD-36 as a Potent, Selective, and Efficacious PROTAC Degrader of STAT3 Protein. *J. Med. Chem.* **2019**, *62*, 11280–11300. [CrossRef] [PubMed]

46. Genini, D.; Brambilla, L.; Laurini, E.; Merulla, J.; Civenni, G.; Pandit, S.; D'Antuono, R.; Perez, L.; Levy, D.E.; Pricl, S.; et al. Mitochondrial dysfunction induced by a SH2 domain-Targeting STAT3 inhibitor leads to metabolic synthetic lethality in cancer cells. *Proc. Natl. Acad. Sci. USA* **2017**, *114*, E4924–E4933. [CrossRef]

47. Bai, L.; Zhou, H.; Xu, R.; Zhao, Y.; Chinnaswamy, K.; McEachern, D.; Chen, J.; Yang, C.Y.; Liu, Z.; Wang, M.; et al. A Potent and Selective Small-Molecule Degrader of STAT3 Achieves Complete Tumor Regression In Vivo. *Cancer Cell* **2019**, *36*, 498–511.e17. [CrossRef]

48. de Araujo, E.D.; Manaswiyoungkul, P.; Erdogan, F.; Qadree, A.K.; Sina, D.; Tin, G.; Toutah, K.; Yuen, K.; Gunning, P.T. A functional in vitro assay for screening inhibitors of STAT5B phosphorylation. *J. Pharm. Biomed. Anal.* **2019**, *162*, 60–65. [CrossRef]

49. Hong, D.; Kurzrock, R.; Kim, Y.; Woessner, R.; Younes, A.; Nemunaitis, J.; Fowler, N.; Zhou, T.; Schmidt, J.; Jo, M.; et al. AZD9150, a next-generation antisense oligonucleotide inhibitor of STAT3 with early evidence of clinical activity in lymphoma and lung cancer. *Sci. Transl. Med.* **2015**, *7*. [CrossRef]

50. Bournazou, E.; Bromberg, J. Targeting the tumor microenvironment. *Jak-Stat* **2013**, *2*, e23828. [CrossRef]

51. Mehta, V.; Goel, S.; Kabarriti, R.; Cole, D.; Goldfinger, M.; Acuna-Villaorduna, A.; Pradhan, K.; Thota, R.; Reissman, S.; Sparano, J.A.; et al. Case Fatality Rate of Cancer Patients with COVID-19 in a New York Hospital System. *Cancer Discov.* **2020**. [CrossRef]

52. COVID-19 More Frequent, Severe in Cancer Patients. *Cancer Discov.* **2020**. [CrossRef]

53. Guaraldi, G.; Meschiari, M.; Cozzi-Lepri, A.; Milic, J.; Tonelli, R.; Menozzi, M.; Franceschini, E.; Cuomo, G.; Orlando, G.; Borghi, V.; et al. Tocilizumab in patients with severe COVID-19: A retrospective cohort study. *Lancet Rheumatol.* **2020**. [CrossRef]

54. Horby, P.; Lim, W.S.; Emberson, J.; Mafham, M.; Bell, J.; Linsell, L.; Staplin, N.; Brightling, C.; Ustianowski, A.; Elmahi, E.; et al. Effect of Dexamethasone in Hospitalized Patients with COVID-19: Preliminary Report. *MedRxiv* **2020**, 2020.06.22.20137273. [CrossRef]

Review

Pharmacological Inhibition of Oncogenic STAT3 and STAT5 Signaling in Hematopoietic Cancers

Marie Brachet-Botineau [1], Marion Polomski [2], Heidi A. Neubauer [3], Ludovic Juen [2], Damien Hédou [2], Marie-Claude Viaud-Massuard [2], Gildas Prié [2] and Fabrice Gouilleux [1,*]

[1] Leukemic Niche and Oxidative metabolism (LNOx), CNRS ERL 7001, University of Tours, 37000 Tours, France; marie.brachet@univ-tours.fr

[2] Innovation Moléculaire et Thérapeutique (IMT), EA 7501, University of Tours, 37000 Tours, France; marion.polomski@etu.univ-tours.fr (M.P.); ludovic.juen@mcsaf.fr (L.J.); damien.hedou@univ-tours.fr (D.H.); marie-claude.viaud-massuard@univ-tours.fr (M.-C.V.-M.); gildas.prie@univ-tours.fr (G.P.)

[3] Institute of Animal Breeding and Genetics, University of Veterinary Medicine Vienna, A-1210 Vienna, Austria; heidi.neubauer@vetmeduni.ac.at

* Correspondence: fabrice.gouilleux@univ-tours.fr

Received: 6 December 2019; Accepted: 13 January 2020; Published: 18 January 2020

Abstract: Signal Transducer and Activator of Transcription (STAT) 3 and 5 are important effectors of cellular transformation, and aberrant STAT3 and STAT5 signaling have been demonstrated in hematopoietic cancers. STAT3 and STAT5 are common targets for different tyrosine kinase oncogenes (TKOs). In addition, STAT3 and STAT5 proteins were shown to contain activating mutations in some rare but aggressive leukemias/lymphomas. Both proteins also contribute to drug resistance in hematopoietic malignancies and are now well recognized as major targets in cancer treatment. The development of inhibitors targeting STAT3 and STAT5 has been the subject of intense investigations during the last decade. This review summarizes the current knowledge of oncogenic STAT3 and STAT5 functions in hematopoietic cancers as well as advances in preclinical and clinical development of pharmacological inhibitors.

Keywords: STAT3; STAT5; hematopoietic cancers; therapeutic targeting; pharmacological inhibitors

---

1. Introduction

Signal Transducer and Activator of Transcription (STAT) proteins are a seven-member family of cytoplasmic transcription factors that relay signals emanating from cell-surface cytokine and growth factor receptors to the nucleus [1,2]. STAT proteins control fundamental cellular processes, including survival, proliferation, differentiation, and immune responses [3]. It is now well-established that three of these members, STAT3 and the closely related STAT5A and STAT5B proteins (Figure 1) are also important effectors of cellular transformation. Aberrant STAT3, STAT5A, and STAT5B signaling have been described in different solid tumors such as prostate, breast, colon, gliomas, head and neck cancer, melanoma, and in hematopoietic malignancies [4–7] (see also [8] in this issue). Historically, persistent activation of these transcription factors was frequently found in many tumor cells as a consequence of deregulated tyrosine kinase activity. STAT5A/5B and/or STAT3 are downstream effectors of various tyrosine kinase oncogenes (TKOs) such as TEL-JAK2, JAK2^{V617F}, SRC, TEL-ABL, BCR-ABL, TEL-SYK, NPM-ALK, TEL-PDGFR, and mutated forms of FLT3 and KIT receptors [9–19].

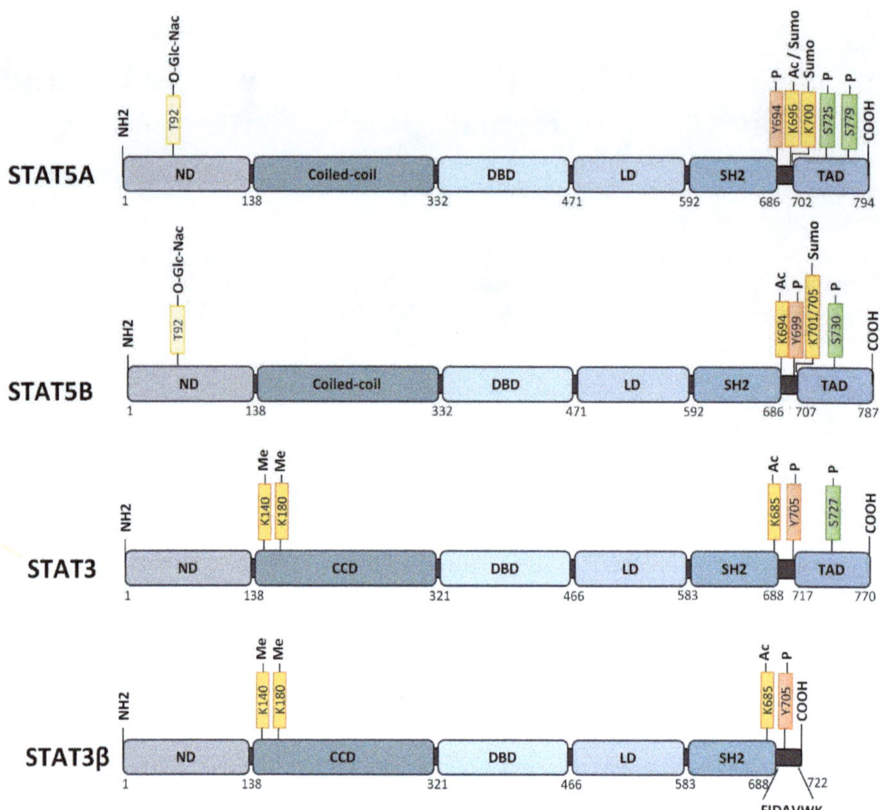

Figure 1. Structure of Signal Transducer and Activator of Transcription (STAT)5A, STAT5B, STAT3, and the spliced isoform STAT3β proteins. Functional domains: ND, NH2-terminal domain; CCD, coiled-coil domain; DBD, DNA binding domain; LD, linker domain; SH2, Src homology 2 domain; TAD, transactivation domain. Post-translational modifications: Ac, acetylation; Me, methylation; O-Glc-Nac, O-GlcNacylation; P, phosphorylation; Sumo, sumoylation.

Inhibition of STAT3/STAT5 signaling has a negative impact on the transforming potential of these tyrosine kinases in vitro and in vivo. Evidence for a direct role of STAT3 or STAT5A/5B in cell transformation was provided by the use of constitutively active variants. These proteins, designated STAT3C, STAT51*6 or cS5^F, are able to induce cell transformation in vitro and solid tumors or leukemias in vivo [20–23]. More recently, gain-of-function (GOF) mutations in STAT5B and STAT3 have been found in patients with leukemias and lymphomas (Figure 2) [24–27]. Occurring primarily within the SH2 domain, these mutations confer persistent and prolonged signaling and have been linked to poorer prognosis and relapse in patients [28–30]. Collectively, these data would undoubtedly define STAT3 and STAT5A/5B as important therapeutic targets in hematologic cancers. Nevertheless, STAT3 and STAT5 also behave as tumor suppressors in other tissues and regulate the antitumoral response of immune cells [31–37]. Thus, the respective and specific functions of STAT3 and STAT5, as well as their interactions in hematopoietic cancers, still need to be refined to develop therapeutic strategies that selectively block STAT3 and/or STAT5 activity in these diseases. In this review, we will first summarize the respective contribution of STAT3 and STAT5 in hematologic cancers as well as the canonical and non-canonical oncogenic properties of STAT3/STAT5. Finally, we will describe the different strategies used to target STAT3 or STAT5 and will discuss the potential future development of single or combined therapies to block STAT3 and/or STAT5A/5B activity/expression in hematologic cancers.

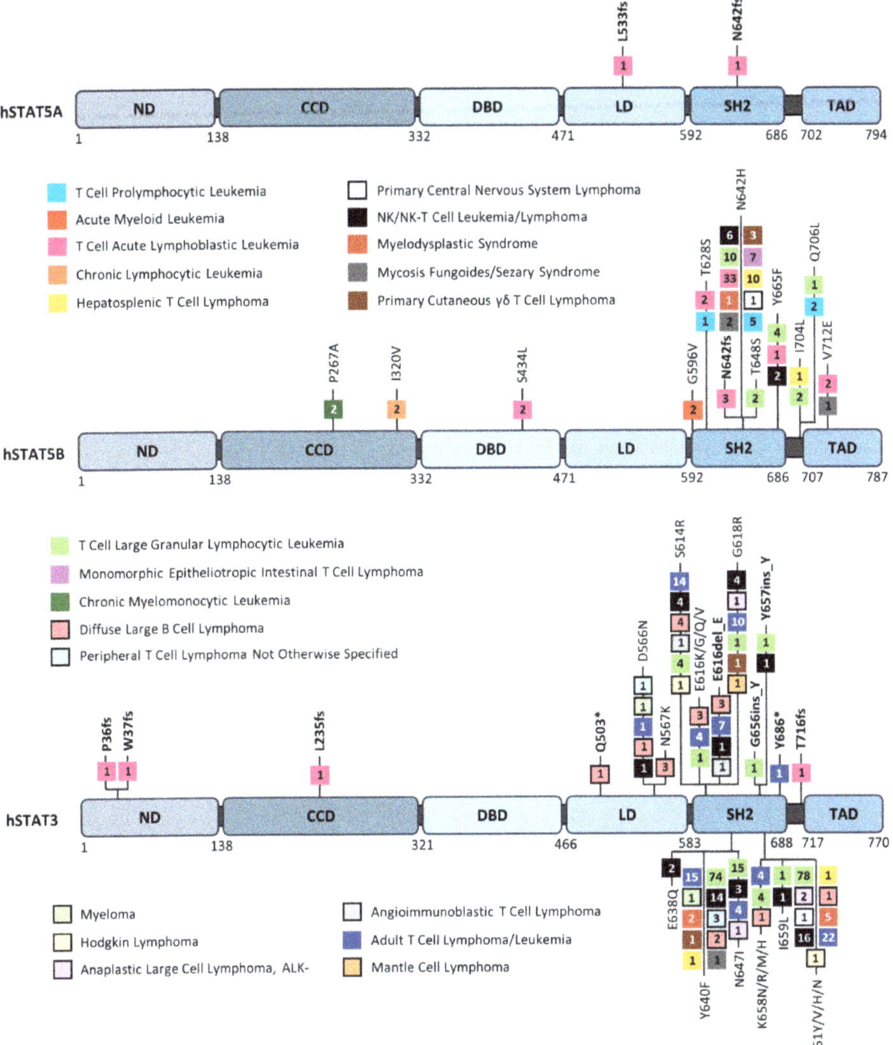

Figure 2. Map of somatic mutations detected in human STAT5A, STAT5B, and STAT3 in patients with hematologic cancers. Individual missense mutations found in at least two patients, as well as all reported nonsense (*) and frameshift (fs) mutations (bold), are depicted. The numbers in each box represent the number of cases reported for each mutation. Data were mined from the Catalogue of Somatic Mutations in Cancer (COSMIC) database. ND, NH2-terminal domain; CCD, coiled-coil domain; DBD, DNA binding domain; LD, linker domain; SH2, Src homology 2 domain; TAD, transactivation domain.

2. STAT3 and STAT5A/5B in Hematopoietic Cancers

2.1. STAT3/STAT5 in Hematopoietic Cancers: An Amazing 23-Year-Old Story

In 1996, pioneering works demonstrated that STAT3 and/or STAT5 are constitutively activated in leukemic cells from patients with acute myeloid leukemia (AML) or acute lymphoblastic leukemia (ALL) [38]. Surprising was the high constitutive STAT5 DNA binding activity detected in leukemic cells from Philadelphia chromosome-positive ALL (Ph$^+$ ALL) patients. These original findings already pointed to STAT5 as a potential effector of the BCR-ABL tyrosine kinase fusion protein, the transforming agent in Ph$^+$ ALL and chronic myeloid leukemia (CML), and were confirmed a few months later by other groups [15,16]. 23 years later with more than 4000 publications in the field, it is now clearly established that STAT3/STAT5A/STAT5B are essential and/or contribute to the development of hematopoietic malignancies, affecting both myeloid and lymphoid compartments (Table 1). Data also showed that deregulated STAT3 and STAT5 activity promotes drug resistance in leukemias/lymphomas/myelomas, highlighting the crucial interest to develop pharmacological molecules that selectively target STAT3 and/or STAT5 in hematologic cancers.

Table 1. STAT3 and STAT5 in hematologic cancers.

Hematopoietic Lineage	Hematologic Malignancies		Contribution of STAT3 and/or STAT5A/5B		
Myeloid compartment	Ph$^+$ MPN	**CML (BCR-ABL)**	STAT3	STAT5A	STAT5B
	Ph$^-$ MPN	PV JAK2^{V617F}	STAT3	STAT5	
		ET JAK2^{V617F}			
		PMF JAK2^{V617F}			
		SM KIT^{D816V}			
	AML	Flt3-ITD		STAT5	
		CBF-AML Kit^{D816V}	STAT3	STAT5	
		APL	STAT3	STAT5	
		t (15;17)			
Lymphoid compartment	ALL	Pre-B-ALL	STAT3	STAT5	
		B-ALL			
		T-ALL	STAT3	STAT5A	STAT5B
	T-LGL Leukemias		STAT3	STAT5A	STAT5B
	Lymphomas	CLL	STAT3		
		HL	STAT3	STAT5	
		DLBCL	STAT3	STAT5	
		ALCL	STAT3		STAT5B
		γδ-T cell Lymphomas	STAT3		STAT5B
		NK-T cell Lymphomas	STAT3		STAT5B
	Multiple Myelomas		STAT3		

Changes in font size and bold text refer to the level of contribution of each STAT protein in the disease. ALCL (anaplastic large cell lymphoma), AML (acute myeloid leukemia), APL (acute promyelocytic leukemia), B- or T-ALL (B- or T- acute lymphoblastic leukemia), CBF-AML (core binding factor-acute myeloid leukemia), CLL (chronic lymphocytic leukemia), CML (chronic myeloid leukemia), DLBCL (diffuse large B cell lymphoma), ET (essential thrombocythemia), Flt3-ITD (FMS-like tyrosine kinase 3-internal tandem duplication), HL (Hodgkin lymphoma), Ph$^+$MPN and Ph$^-$MPN (Philadelphia chromosome-positive and Philadelphia chromosome-negative myeloproliferative neoplasm), NK (natural killer cell), PMF (primary myelofibrosis), PV (polycythemia vera), SM (systemic mastocytosis), and T-LGL (T cell large granular lymphocytic) leukemia.

2.2. STAT3/5 in Myeloproliferative Neoplasms (MPNs)

Myeloproliferative neoplasms (MPNs) are hematologic diseases characterized by abnormal proliferation and accumulation of mature myeloid cells in the bone marrow and peripheral blood [39]. An increased risk of developing acute myeloid leukemia is also associated with MPNs. MPNs are classified as *BCR-ABL*-positive (or Ph$^+$) and *BCR-ABL*-negative (Ph$^-$) MPNs [40]. Both Ph$^+$ and Ph$^-$ MPNs are clonal disorders that result from the transformation of hematopoietic stem cells (HSCs). While the BCR-ABL fusion protein is the transforming agent in CML, driver mutations in *JAK2*, *CALR*, and *MPL* genes are variably present and are mostly mutually exclusive in Ph$^-$ MPNs, which include essential thrombocythemia (ET), polycythemia vera (PV), and primary myelofibrosis (MF) [41]. The JAK2 GOF mutation (JAK2^{V617F}) has been identified in 95% to 97% of PV patients [42,43]. This mutation, located in the pseudokinase domain of the JAK2 protein, constitutively activates the kinase. JAK2, MPL, and CALR mutants have been functionally validated and are sufficient to induce MPNs in mice [41]. Systemic mastocytosis (SM), a subcategory of MPNs, is a heterogeneous clonal disorder characterized by an accumulation of mast cells in various organs [44]. The GOF mutation in KIT (KIT^{D816V}) causing activation of the KIT receptor tyrosine kinase was found in 80–95% of patients with SM. Studies with transgenic mice suggested that this mutation alone is sufficient to cause SM [45]. The KIT^{D816V} mutant has also been detected in leukemic cells from AML patients [46]. The presence of KIT^{D816V} in AML is highly associated with co-existing SM [47]. Activation of STAT3 and/or STAT5 by BCR-ABL, JAK2^{V617F}, and KIT^{D816V} has been abundantly documented in the literature. However, conflicting results (cell lines vs. primary cells and/or human vs. murine leukemic cells) have emerged from these studies. For instance, tyrosine phosphorylation of STAT3 (Y^{705}) was observed in murine BCR-ABL$^+$ cells but barely detected in human BCR-ABL$^+$ cells [16,48]. Using *stat3*- or *stat5a/5b*-deficient mice, previous studies demonstrated that while STAT3 and STAT5 are required for the initial step of BCR-ABL-dependent cell transformation, only STAT5 is necessary for the maintenance of BCR-ABL-induced leukemia [49]. The effective role of STAT3 and STAT5 in maintenance, self-renewal, or transformation of normal HSCs might explain why both proteins are required in the BCR-ABL-dependent leukemia-initiating stem cell population [50–53]. More recently, dissection of the respective contributions of STAT5A and STAT5B in BCR-ABL-dependent transformation revealed that STAT5B, but not STAT5A, is a critical effector of BCR-ABL-driven leukemia development [54]. Besides survival- and growth-promoting effects, STAT5B facilitates leukemogenesis by suppressing IFN-α and IFN-γ signaling in these animal models. The development of tyrosine kinase inhibitors (TKIs) targeting BCR-ABL, such as imatinib mesylate (IM), has revolutionized the treatment of CML. However, imatinib mesylate (IM) is not totally curative and approximately 50% of patients remain therapy-free after IM discontinuation. The inability of IM to completely eradicate quiescent leukemic stem cells (LSCs) is probably responsible for the relapse of CML patients [55]. Moreover, the occurrence of BCR-ABL mutations in progressive or relapsed diseases promotes IM resistance of CML cells [56]. Studies indicated that high levels of phosphorylated STAT5 enhance the resistance of CML cells to TKIs but also triggers BCR-ABL mutations by inducing the production of reactive oxygen species (ROS) responsible for DNA damage [57–59]. Moreover, STAT5 was shown to play a key role in the maintenance of IM-resistant LSCs from CML patients [60,61].

The contribution of STAT3 in IM resistance was also demonstrated using an in vitro model of cocultures mimicking the bone marrow microenvironment of CML cells. In this model, bone marrow stromal cells stimulated the phosphorylation of STAT3 in CML cells in a BCR-ABL-independent manner and promoted the resistance of these leukemic cells to IM [62]. Interestingly, combined targeting of STAT3 and STAT5 has been proposed to overcome drug resistance in CML cells [63].

While STAT5 plays a critical role in JAK2^{V617F}-driven mouse models of MPN, studies have shown that STAT3 is not required for myeloid expansion induced by this TKO [64–67]. Moreover, activation of STAT3 negatively regulates JAK2^{V617F}-driven MPN in mice by enhancing thrombocytosis and shortening overall survival [67]. Further studies are therefore required to determine if STAT3 plays a pathogenic role in human MPNs.

STAT3, STAT5, and also STAT1 are activated by the KIT^{D816V} mutant in transformed mast cells, but only STAT5 seems to be transcriptionally active in these cells [68]. Phosphorylation of STAT5 has been detected in mast cells from SM patients. shRNA-mediated knockdown, dominant-negative mutant and pharmacological molecules targeting STAT5 all abrogate the growth of human neoplastic mast cells in vitro, indicating that STAT5 is a critical effector of KIT^{D816V} in human neoplastic mast cells [12].

Collectively, all data indicate that STAT5A/5B proteins are particularly relevant therapeutic targets in Ph$^+$ MPN and Ph$^-$ MPN and in the resistance to TKIs. Although the pathogenic role of STAT3 in human MPN still remains questionable, the contribution of STAT3 to TKI resistance elicited by the leukemic microenvironment would suggest that combination therapy or dual molecules targeting STAT3 and STAT5 might help to eradicate resistant leukemic cells in their "niche."

2.3. STAT3/5 in Acute Myeloid Leukemia (AML)

AML is a heterogeneous clonal disorder characterized by immature myeloid cell proliferation and bone marrow failure. A two-hit model has been suggested as the probable mechanism in the pathogenesis of AML [69]. In this model, gene mutations that give a growth advantage and block normal hematopoietic differentiation are responsible for AML development. For instance, activating mutations in FLT3 (FMS-related tyrosine kinase 3) and KIT receptors promote proliferation and survival, while mutations affecting the transcription factor CEBPα inhibit myeloid differentiation. However, there are several other gene classes such as those involved in epigenetic regulation or metabolism that are mutated in AML [70].

Analyses of primary peripheral blood and bone marrow specimens have demonstrated constitutive activation of STAT3 and/or STAT5 in AML [6,37]. Importantly, constitutive activation of STAT3 and STAT5 has been linked with disease outcomes in AML. Bone marrow evaluation of AML patients revealed that the activation of STAT3 was significantly associated with poor overall survival and reduced progression-free survival [71]. In sharp contrast, the spliced STAT3β isoform was shown to have a suppressor function in AML [33]. A higher STAT3β/α mRNA ratio was found in AML cells and correlated with a favorable prognosis and increased overall survival. Stat3β expression in mouse models of AML resulted in decelerated disease progression and extended survival. It is, however, unclear whether tyrosine phosphorylation and dimerization are required for the tumor suppressor activity of STAT3β in these animal models. The contribution of STAT3 in AML may not only depend on the STAT3β/α ratio but also on the subtype of AML cells or acquired mutations in this disease. For instance, KIT^{D816V}, which is frequently found in core-binding factor (CBF)-AML leukemias, stimulates autophagy through activation of STAT3 [72]. Inhibition of STAT3 blocked autophagy and reduced tumor growth in mouse xenograft models. Importantly, inhibition of STAT3 also stimulates the antitumoral immune response in animal models of AML, indicating that targeting STAT3 would not only block the growth and survival of AML but also AML-induced immune evasion [73,74].

Mutations in FLT3, either involving internal tandem duplications (FLT3-ITD) or point mutations in the activating loop of the tyrosine kinase domain (FLT3-KD), were observed in approximately 30% of AML patients and are associated with poor prognosis [75]. Although FLT3 mutants activate both STAT3 and STAT5 in cell lines, results showed that tyrosine-phosphorylated STAT5 is selectively associated with expression of FLT3-ITD in primary blasts from AML patients. Activation of STAT5 by FLT3-ITD is required to induce primary cell survival in vitro and leukemia in vivo [76–80]. Pharmacological inhibition of STAT5 also blocks FLT3-ITD-driven leukemias in mouse xenograft models [81]. FLT3-ITD promotes genomic instability by increasing ROS production via activation and association of STAT5 with the GTPase Rac1, which is an essential component of certain NADPH oxidases such as Nox2 [82]. Conversely, ROS production and p22phox, a membrane subunit of NADPH oxidase, are required for FLT3-ITD-induced STAT5 phosphorylation in the endoplasmic reticulum (ER) [83]. Importantly, recent works demonstrated that FLT3-ITD-independent activation of STAT5 induced by the leukemic microenvironment promotes resistance of FLT3-ITD$^+$ AML cells to quizartinib, an FLT3 inhibitor that is now in phase 3 clinical trial [84].

Lastly, we should also mention the identification of a fusion between *STAT5B* and *Retinoic Acid Receptor (RAR)α* resulting from an interstitial deletion on chromosome 17 in acute promyelocytic leukemia (APL) [85]. The corresponding fusion protein enhances STAT3 signaling and blocks myeloid maturation by inhibiting RARα/retinoid X receptor (RXR)α transcriptional activity [86].

2.4. STAT3/5 in Acute Lymphoblastic Leukemia (ALL)

ALL is the most common form of cancer in children and predominantly arises from the transformation of B cell progenitors (80–85% of cases) [87]. Mouse studies suggest that STAT5 is functionally important in certain types of B-ALL [88]. Transgenic overexpression of a constitutively active STAT5A mutant (cS5^F) cooperates with p53 deficiency to promote B-ALL in mice [89]. Genetic or pharmacological targeting of STAT5 suppresses human Ph+ ALL cell growth and leukemia development in mouse xenograft models [90]. Deregulation of precursor B cell antigen receptor (pre-BCR) signaling has been shown to be important in the development of B-ALL, and constitutive activation of STAT5B cooperates with defects in pre-BCR signaling components to initiate B-ALL [91]. Similarly, haploinsufficiency of B cell-specific transcription factors such as EBF1 or PAX5 synergizes with activated STAT5 in ALL [92]. Despite strong evidence for the oncogenic activity of STAT5 in TKO-driven B-ALL, the role of STAT5 appears to be context-dependent. For example, the deletion of STAT5 accelerates the development of B-ALL induced by c-myc in mouse models [93]. Activating mutations in *STAT5B* have been found in T-ALL [24,28]. The amino acid substitution N642H in the phosphotyrosine binding pocket of the SH2 domain promotes the constitutive activation of STAT5B and the capacity to induce T cell neoplasia in transgenic mice [29,30]. The role of STAT3 in ALL is poorly documented. However, data indicated that blockade of STAT3 signaling compromises the growth of B-ALL cells overexpressing the high mobility group A1 (HMGA1)-STAT3 pathway [94]. Unlike STAT5B, there are no recurrent STAT3 mutations detected in T-ALL and, in fact, only single frameshift mutations are reported (Figure 2).

2.5. STAT3/5 in T Cell Large Granular Lymphocytic (T-LGL) Leukemia

Activating mutations in the SH2 domain of STAT3 (Y640F, D661Y/V) and STAT5B (N642H) were also described in T-LGL leukemia which is a chronic lymphoproliferative disorder characterized by the expansion of some cytotoxic T cell or NK cell populations (Figure 2) [95–97]. *STAT3* mutations have been described in 30–40% of T-LGL leukemia patients while *STAT5B* mutations were found in rare but typical CD4$^+$ T-LGL leukemia cases. However, *STAT5B* mutations were more frequently detected in patients with a severe clinical course. In all cases, mutations were shown to increase the transcriptional activity of both STAT3 and STAT5B proteins, but only the STAT5B^{N642H} mutation was demonstrated to drive T-LGL leukemias in mouse models [98,99].

2.6. STAT3/5 in Chronic Lymphocytic Leukemias (CLL)

CLL is characterized by the accumulation of mature clonal B cells in peripheral blood, bone marrow, and lymphoid tissues. These cells are characterized by an extended lifespan due to intrinsic defects in apoptosis [100]. Increasing STAT3 phosphorylation on S^{727} but not on Y^{705} is believed to be a hallmark of CLL progression [101]. Phosphorylation of S^{727} regulates the transcriptional activity of the STAT3 protein but it is also involved in the mitochondrial localization of STAT3 in primary cells from CLL patients [102]. Cytokines such as interleukin (IL)-15 secreted by the microenvironment contribute to the survival of CLL cells through JAK-mediated tyrosine phosphorylation of STAT5 [103].

2.7. STAT3/5 in Lymphomas

Lymphomas are cancers of the lymphatic system. They are divided into two categories: Hodgkin lymphoma (HL) and non-Hodgkin lymphoma (NHL). Data from the literature underscored the important contribution of STAT3 and STAT5 in the proliferation and/or survival of HL cells [104–106]. Most NHLs are B cell lymphomas. Diffuse large B cell lymphoma (DLBCL) is the most common subtype of NHL and consists of at least two phenotypic subtypes: the germinal center B cell-like (GCB-DLBCL) and the activated B cell-like (ABC-DLBCL) groups. High-level STAT3 expression and activation are preferentially detected in ABC-DLBCL, which is associated with poor outcomes [107,108]. Inhibition of STAT3 expression/activity in ABC-DLBCL cells abrogates lymphoma cell growth and triggers apoptosis. Moreover, STAT3 coordinates migration to facilitate the dissemination of DLBCL [109]. Among NHL subtypes, peripheral T cell lymphoma (PTCL) and natural killer (NK)/T cell lymphoma (NKTL) represent a heterogeneous group of diseases with varied clinical features, prognosis and response to treatment. PTCL has been categorized into several subtypes including PTCL-not otherwise specified (PTCL-NOS), angioimmunoblastic TCL (AITL), anaplastic large cell lymphoma (ALCL), and the predominant subsets of cutaneous TCL (CTCL) [110]. Deregulation of STAT3/STAT5 activity was shown to be important for CTCL pathogenesis and cancer progression [111]. ALCL can be divided into anaplastic lymphoma kinase (ALK) positive and ALK negative subgroups, based on ALK gene rearrangements. The nucleophosmin-anaplastic lymphoma kinase (NPM-ALK) fusion protein is the major oncogenic driver in ALK+ ALCL and it activates STAT3 and STAT5 [17,18]. While STAT3 is required for NPM-ALK-induced cell transformation and B cell lymphomagenesis, the contribution of STAT5 is still unclear [112]. Studies indicated that STAT5A, but not STAT5B, is epigenetically silenced in NPM-ALK tumors and behaves as a tumor suppressor when reactivated to suppress NPM-ALK expression [34]. Activating mutations of STAT3 and STAT5B have been found in NKTLs and γδ-T cell lymphomas [25,27]. Recurrent hot spot mutations Y640F and D661Y/V/H/N in the SH2 domain of STAT3 have been identified most commonly in T-LGL leukemia, NK, NK/T, and adult T cell leukemia/lymphoma (ATLL) patients, while the aggressive STAT5B^{N642H} mutation has been predominantly found in T-ALL and γδ-T cell lymphomas, such as hepatosplenic TCL, monomorphic epitheliotropic intestinal TCL, and primary CTCL (Figure 2) [113]. These mutations were shown to be associated with increased phosphorylated STAT3 and STAT5B proteins and to confer a growth advantage to transduced cell lines or normal NK cells [27]. Activating STAT3 mutations E616G and E616K in the SH2 domain found in NK/T cell lymphoma patients were shown to increase the phosphorylation and transcriptional activity of STAT3 [114]. Interestingly, programmed cell death-ligand 1 (PD-L1) was overexpressed in NK/T cells harboring hotspot STAT3 mutations, and overexpression of a STAT3^{E616K} or STAT3^{E616G} mutant was sufficient to enhance PD-L1 expression. These data are consistent with the role of STAT3 as an important regulator of tumor immune evasion [115]. In summary, STAT3 and STAT5B, but not STAT5A, are relevant therapeutic targets in the treatment of lymphomas. All recurrent STAT3 and STAT5B mutations in hematopoietic cancers are summarized in Figure 2.

2.8. STAT3/5 in Myelomas

Multiple myeloma (MM) is a B cell malignancy characterized by the proliferation of clonal plasma cells in the bone marrow accompanied by secretion of monoclonal immunoglobulin [116]. Constitutive tyrosine phosphorylation of STAT3 has been evidenced in MM cell lines and primary CD138$^+$ cells from MM patients [117,118]. Patients with STAT3 activation were found to have significantly shorter progression-free and overall survival [119]. STAT3 activation has been reported to contribute to MM progression both directly, by upregulating survival and anti-apoptotic target genes, as well as indirectly by activating myeloid-derived suppressor-cells (MDSCs) in the bone marrow microenvironment, which facilitates tumor progression [120,121]. Myeloma cells rely on the pleiotropic cytokine IL-6, which activates the JAK/STAT3 pathway. IL-6 signaling is tightly regulated by tyrosine phosphatases SHP-1 and SHP-2, and suppressor of cytokine signaling 1 (SOCS1). The disruption of this negative feedback loop results in the constitutive activation of STAT3 [122]. Accordingly, *SHP-1* and *SOCS1* genes were found to be silenced by hypermethylation in MM patients [123]. In addition to the induction of anti-apoptotic gene expression in MM cells, STAT3 also regulates the expression of microRNA-21 with strong anti-apoptotic potential, suggesting that noncoding RNAs have an impact on the pathogenesis of human MM [124]. Moreover, the expression of five long noncoding RNAs (lncRNAs) was found to be regulated by STAT3 in MM cells [125]. However, the role of these lncRNAs in the progression of MM remains to be elucidated. The role of STAT5 seems to be marginal and restricted to immunoglobulin production in MM [126].

3. Canonical and Non-Canonical Roles of STAT3/STAT5 in Hematopoietic Cancers

It was presumed that most, if not all, of the oncogenic activities of STAT3/STAT5 are due to canonical functions. However, STAT3/5 proteins are also active through alternate non-canonical pathways impacting cell transformation. Post-translational modifications affecting canonical and non-canonical roles of STAT3 and/or STAT5 have been extensively reviewed in the literature. To be in line with this review, we will focus on canonical and non-canonical activities of STAT3/5 that have been described in cell transformation (Figure 3).

3.1. Canonical Function of STAT3/STAT5

In the canonical model, STAT3, STAT5A, and STAT5B are primarily activated by phosphorylation on tyrosine residues Y705, Y694, and Y699, respectively (Figure 1). In a physiological situation, STAT3 and STAT5 are transiently phosphorylated by JAK in response to cytokine receptor signaling, and contribute to hematopoietic cell proliferation and differentiation. In contrast, persistent tyrosine phosphorylation of STAT3/5 induced by TKOs promotes hematopoietic cell transformation [6,7]. Phosphorylation of additional tyrosine residues in STAT5A and STAT5B proteins in TKO-transformed cells has also been described in the literature, but the functional and physiological meaning of such phosphorylations remains unclear [127,128]. P-Y^{705}-STAT3 (herein referred to in this review as P-Y-STAT3), P-Y^{694}-STAT5A (P-Y-STAT5A), and P-Y^{699}-STAT5B (P-Y-STAT5B) proteins dimerize through reciprocal interactions between the SH2 domain of one monomer and the phospho-tyrosine residue of the other. P-Y-STAT3, P-Y-STAT5A, and P-Y-STAT5B form homodimers but P-Y-STAT5A can also dimerize with P-Y-STAT5B. The formation of STAT5A/5B heterodimers is not only dependent on the relative abundance of each protein but can also be differentially affected by upstream signaling events. For instance, BCR-ABL is less efficient in inducing STAT5A/5B heterodimerization than IL-3 [128]. Heterodimers between P-Y-STAT3 and P-Y-STAT5A or P-Y-STAT5B have never been reported despite the capacity of some TKOs to simultaneously activate these proteins. After dimerization, STAT3, STAT5A, and STAT5B are translocated into the nucleus. Nuclear import is dependent on a nuclear localization signal present in the coiled-coil domain [129]. The nuclear import of STAT3/STAT5 occurs independently of their tyrosine phosphorylation and is mediated by the importin-$\alpha 3/\beta 1$ system coupled to Ras-related nuclear (Ran) proteins bound to GDP or GTP [129]. Molecules other than importins and Ran also participate in the regulation of the nuclear translocation of STAT3/5. The small GTPase Rac1 and the GTPase-activating protein MgcRacGAP form a ternary complex with P-Y-STAT3 or P-Y-STAT5 to induce their translocation via the importin α/β pathway [130]. The MgcRacGAP/Rac1 complex also regulates the tyrosine phosphorylation of STAT3/5 induced by cytokines [131,132]. Interestingly, p21-activated kinases (PAK1 and PAK2), which are important effectors of Rac1, were shown to induce phosphorylation of the S779 residue and nuclear translocation of STAT5A in BCR-ABL-expressing cells [133]. In a similar vein, activation of focal adhesion kinase (FAK) by FLT3-ITD or KIT^{D816V} in AML cells was demonstrated to induce the nuclear translocation of STAT5 via the Rac1/PAK1 pathway [134]. In sharp contrast, several reports indicated that P-Y-STAT5 is abundantly detected in the cytoplasm of BCR-ABL$^+$ cells and that Src kinases might be responsible for the cytoplasmic retention of P-Y-STAT5A [135,136]. The reasons for these apparent controversial data remain unclear but might be related to the cellular context.

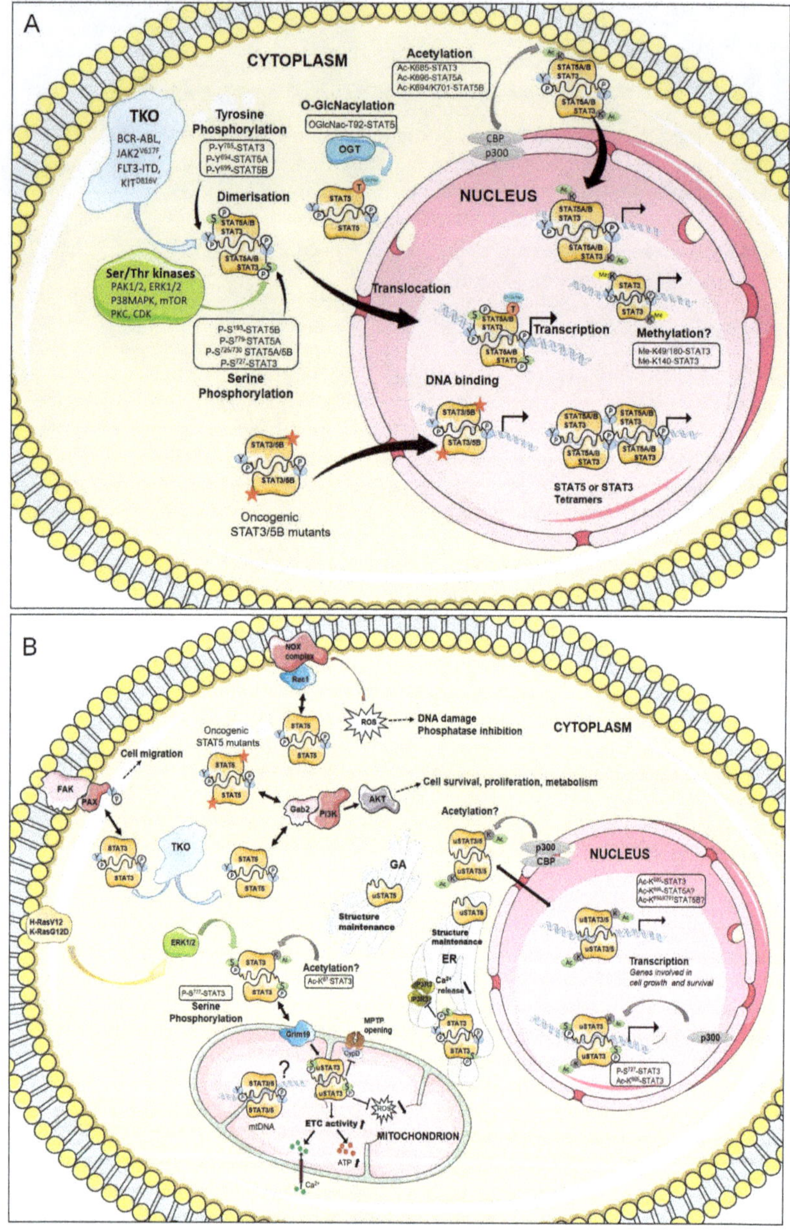

Figure 3. STAT3 and STAT5 signaling in hematologic malignancies. (**A**) Canonical function. In the canonical model, TKOs, or GOF mutations induce persistent activation (e.g., tyrosine phosphorylation) and nuclear translocation of STAT3 and STAT5, which then bind as dimers or tetramers to specific promoter sequences to regulate transcription of genes involved in cell growth, apoptosis, angiogenesis and immune response. Other post-translational modifications, as indicated, regulate the oncogenic activity of STAT3/STAT5. OGT (O-GlcNAc transferase). (**B**) Non-canonical function. uSTAT3 and uSTAT5 (non-tyrosine phosphorylated STAT3/5) proteins also have an active role in the nucleus, mitochondrion, ER (endoplasmic reticulum), GA (golgi apparatus), ETC (electron transport chain). Besides their transcriptional activity, P-Y-STAT3 and P-Y-STAT5 interact with different signaling pathways in the cytoplasm and ER.

Tyrosine phosphorylated STAT3/STAT5 dimers bind to specific DNA elements found in promoters, enhancers, and the first intron of target genes. These binding sites are characterized by clusters of conserved motifs with an interferon gamma-activated site (GAS)-like core sequence (TTCT/CNA/GGAA). STAT5 was shown to bind as a tetramer on adjacent GAS sequences with high or weak affinity in target gene promoters [137,138]. The tetramerization domain located in the NH2-terminal region of STAT5 was found to promote constitutively active, STAT5A mutant (cS5^F)-induced leukemia in mice [23]. The NH2-terminal domain of STAT3 is also required for oligomerization and IL-6-dependent transcriptional regulation [139]. This domain is involved in STAT3-mediated survival of solid tumor cells but its role in hematologic cancers is currently unknown [140].

TKO-mediated activation of STAT3 and/or STAT5 regulates expression of common as well as specific genes involved in hematopoietic cell survival, proliferation, metabolism, hypoxia, autophagy, migration and tumor immune evasion. BCL2-family members, cell cycle-regulated genes, proto-oncogenes such as PIM1, c-MYC, BCL6, and genes involved in cytokine receptor signaling or immune response are often targets of STAT3/STAT5-mediated transcription in hematopoietic cancers [141]. Transcription of these genes is also dependent on P-Y-STAT3/5 levels in the nucleus. For instance, previous works demonstrated that nuclear P-Y-STAT5 proteins at low, intermediate, and high levels differentially affect self-renewal, proliferation, and differentiation of leukemic stem cells by regulating the expression of distinct sets of genes [142]. Accordingly, P-Y-STAT5 regulates D-type cyclins and c-MYC expression at intermediate levels to promote proliferation but upregulates expression of the cyclin-dependent kinase (CDK) inhibitor p21waf1 at high levels to induce growth arrest and differentiation. Similarly, while constitutively active STAT3 was shown to protect CLL cells from apoptosis, at high levels it induced apoptosis and caspase-3 expression [143]. These data emphasize that increasing nuclear levels of P-Y-STAT3/5 might confer a growth disadvantage to leukemic cells [143,144]. Protein–protein interactions also drive the regulation of STAT3/STAT5-dependent gene expression. STAT3 and STAT5 interact with many transcription factors, co-factors, and/or chromatin remodeling proteins such as enhancer of zeste homolog 2 (EZH2), CREB-binding protein (CBP)/p300, ten-eleven translocation 2 (TET2), and DNA methyltransferase 1 (DNMT1), which largely impact hematopoiesis or leukemogenesis. Blocking these proteins also interferes with STAT3/STAT5-mediated transcription and cancer cell growth. Most of these interacting partners were exhaustively described in a recent review and will not be discussed here [141]. In addition, serine phosphorylation plays an important role in STAT3/5-dependent cell transformation. Constitutive phosphorylation of S727, which is located in the COOH-terminal transactivation domain of STAT3 (Figure 1), was shown to increase the transcriptional activity of this protein in CLL cells [101]. The oncogenic activity of STAT5A is highly dependent on S725 and S779 phosphorylation, and both residues (S726 and S780 in humans) were found to be constitutively phosphorylated in AML, ALL, and CML cells [145]. Importantly, PAK-dependent S779 phosphorylation of STAT5A is required for BCR-ABL-induced leukemogenesis but is not affected by BCR-ABL tyrosine kinase inhibitor treatment, indicating that oncogenic activation of STAT5A is mediated by independent pathways in BCR-ABL-expressing cells [133]. Phosphorylation of S779/780 on STAT5A also promotes the expansion and transformation of human hematopoietic stem/progenitor cells (HSCs/HPCs) [53]. Constitutive phosphorylation of S193 on STAT5B has been detected in various lymphoid tumor cell lines as well as in primary cells from leukemia or lymphoma patients [146]. Phosphorylation of S193 was found to be sensitive to inhibitors of mammalian targets of rapamycin (mTOR) and to play an important role in the DNA binding and transcriptional activity of STAT5B. Other post-translational modifications such as glycosylation affect the oncogenic functions of STAT5. O-GlcNAcylation at threonine 92 in STAT5A and STAT5B was shown to regulate tyrosine phosphorylation and transcriptional activity of oncogenic STAT5 in leukemic cells [147].

STAT3 and STAT5 also undergo acetylation on multiple lysine residues by the CBP/p300 histone acetyltransferase [148]. Acetylation of K685 located in the SH2 domain of STAT3 can enhance transcriptional activity by increasing dimer stability. Mutation of K685 affects dimerization but not tyrosine or serine phosphorylation of STAT3, and acetylation of STAT3 can occur in the absence of tyrosine phosphorylation [149,150]. AcK^{685}-STAT3 also recruits DNMT1 to induce the epigenetic silencing of tumor suppressor genes [151]. Acetylation of STAT5A and STAT5B at K696 and K694/K701 residues, respectively, is also required for STAT5 dimerization [152]. The contribution of acetylation in the canonical functions of oncogenic STAT3/5 on one hand, and in hematopoietic cancers, on the other hand, remains very unclear. Previous work indicated that acetylation of K685 does not play an essential role in the expression of the great majority of P-Y-STAT3-dependent genes, suggesting that acetylation might be more important for the non-canonical functions of STAT3/5 [153].

Methylation of nuclear STAT3 on K140 or K49/K180 by the histone-modifying enzymes SET domain-containing lysine methyltransferase 9 (SET9) or EZH2, respectively, have been reported to mediate opposing effects on STAT3-dependent transcriptional activity [154–156]. STAT5 also interacts with EZH2 in B cells to repress the Igκ locus [157]. The methylation status of STAT3 and STAT5 in hematopoietic cancers has yet to be investigated. Similarly, other post-translational modifications such as oxidation, glutathionylation, or sumoylation were shown to negatively regulate STAT3/5 activity, but their impact on STAT3/5-driven hematologic malignancies are still unknown [158–162].

The concomitant activation of STAT3/5 by TKOs is probably the most intriguing event in leukemogenesis as both proteins in cancer cells have compensatory and/or opposing effects on gene expression and cell fate [163]. For instance, the competitive binding of STAT3 and STAT5 to the regulatory loci of BCL6 and IL-17 has been shown to modify gene expression and cell phenotype [164,165]. The differential recruitment of co-activators and/or co-repressors might explain these opposing effects on gene expression. Activation status of both STAT5 and STAT3 might, therefore, provide important diagnostic and prognostic information in hematologic cancers. Notably, JAK2^{V617F} activates both STAT3 and STAT5, but only STAT3 negatively regulates JAK2^{V617F}-dependent MPN development [66,67].

3.2. Non-Canonical Functions of STAT3/STAT5

Subcellular localization of P-Y-STAT3/5 is not restricted to the nucleus and previously published data highlighted important functions of P-Y-STAT3/5 in the cytoplasm and/or other subcellular organelles of cancer cells. Cytoplasmic localization of P-Y-STAT5 was abundantly found in leukemic cells expressing BCR-ABL, JAK2^{V617F}, or KIT^{D816V}. Here, it was shown to interact with the scaffolding adapter Gab2 to favor the activation of the phosphatidylinositol-3-kinase (PI3K)/AKT pathway and leukemic cell survival [12,135,166]. The deletion of Gab2 attenuated the transforming potential of an oncogenic STAT5 mutant in mouse models [167]. Interaction of P-Y-STAT5 with Rac1, which is also an important component of certain membrane-bound NADPH-oxidase (NOX) enzymatic complexes, promotes ROS production in FLT3-ITD-expressing AML cells thereby increasing cell growth concomitantly with DNA damage [82]. BCR-ABL- and JAK2^{V617F}-induced ROS generation is mediated by P-Y-STAT5, and binding of P-Y-STAT5 to the NOX2 complex in BCR-ABL$^+$ cells also requires active Rac1 (unpublished data) [58,59,168]. P-Y-STAT3 has been detected in focal adhesions of cancer cells where it interacts with phosphorylated paxillin and FAK, thereby regulating cell migration [169]. Constitutively active STAT3 also controls Ca2$^+$ release in the ER by interacting with inositol 1,4,5-triphosphate receptors (IP3R3), facilitating its proteasomal degradation [170]. Accordingly, the binding of STAT3 to IP3R3 protects cells from oxidative/ER stress and apoptosis. Finally, the mitochondrial localization and the potential role of P-Y-STAT3 and P-Y-STAT5 in regulating the mitochondrial genome have also been reported [171–173]. Collectively, these data suggest that P-Y-STAT3/5 distribution in subcellular organelles or cytoplasmic structures may directly impact cancer cell growth and survival independently of their transcriptional activity in the nucleus. It would then be relevant to explore in further detail the subcellular localization and function of P-Y-STAT3/5 in hematologic malignancies.

Constitutive phosphorylation of STAT3 on residue S727, but not on Y705, plays a role in the pathogenesis of CLL by regulating STAT3-dependent expression of genes associated with cell growth and survival [101]. DNA binding and transcriptional activity of P-S^{727}-STAT3 probably require acetylation of K685 because (1) persistent acetylation of K685 on STAT3 has been observed in CLL cells, (2) acetylation of K685 promotes dimer formation, DNA binding and activation of transcription, and (3) K685 acetylation provides CLL cells with survival advantages [174]. In addition to its nuclear localization, P-S^{727}-STAT3 was also found in mitochondria where it contributes to the viability of CLL cells and protects against oxidative stress [102]. The oncogenic activity of mitochondrial STAT3 was first demonstrated in H-Ras^{V12}-transformed cells where it promotes anchorage-independent cell growth and tumor induction in mice [175]. Phosphorylation of S727 is carried out by the MAPK/ERK kinase (MEK)/Extracellular signal-regulated kinases (ERK) pathway and is necessary for Ras-mediated cell transformation [176]. Mitochondrial P-S^{727}-STAT3 also supports K-Ras^{G12D}-driven hematologic neoplasms [177]. Retinoid interferon-induced cell mortality (GRIM)19, a component of the electron transport chain (ETC) complex 1 and binding partner of STAT3, was shown to mediate STAT3 uptake into mitochondria [178]. However, other proteins such as the chaperone TOM20 have also been suggested as mitochondrial carriers of STAT3 [179]. Phosphorylation of S727 is important for mitochondrial STAT3 to upregulate ETC activity and ATP production, and mutation of S727 decreases the mitochondrial translocation of STAT3 [178]. In addition, CBP-dependent acetylation of K87 was also found to promote the mitochondrial localization of STAT3 [180]. Hematopoietic cell-targeted deletion of the *stat3* gene demonstrated the critical role of mitochondrial STAT3 in HSC/HPC function. *stat3*$^{-/-}$ mice have a blood phenotype with similarities to human diseases of myelodysplastic syndrome (MDS) and MPN, supporting, as mentioned above, a negative regulatory role of STAT3 in MPN development [181]. These data also suggest that targeting mitochondrial STAT3 in K-Ras-induced hematopoietic malignancies might have adverse effects on normal HSCs. In addition to mitochondrial functions, non-tyrosine phosphorylated STAT3 and STAT5 proteins (uSTAT3/5) also have important roles in the nucleus, ER, and Golgi apparatus (GA). The knockdown of STAT5A/5B in human pulmonary arterial endothelial and smooth muscle cells resulted in a loss of uSTAT5A and uSTAT5B associated with the GA, leading to a dramatic destabilization of the ER and GA. This was accompanied by the activation of ER stress pathways, indicating a potential role of uSTAT5 in maintaining the structure and function of these organelles [182]. Inhibition of uSTAT5 activity/expression induced ROS production and apoptosis in pre-B leukemic cells, indicating that uSTAT5 may also provide protection against oxidative stress [183] (unpublished data).

uSTAT3/5 are also involved in the regulation of transcription and chromatin remodeling as preformed anti-parallel dimers [129,141]. The NH2-terminal domain mediates the dimerization of uSTAT3 and is essential for its nuclear accumulation, DNA binding, chromatin remodeling, and regulation of gene expression [184,185]. Acetylation of K685 was shown to be crucial for uSTAT3 to form stable dimers and regulate gene transcription [149]. Accordingly, the majority of uSTAT3-mediated gene expression depends on the ability of K685 to become acetylated [153]. Interestingly, nuclear uSTAT3 regulates the expression of well-known genes in cancer, suggesting that uSTAT3 might contribute to oncogenesis [186]. The role of uSTAT5 is less documented but data from the literature indicates that nuclear uSTAT5 behaves as a partial antagonist of P-Y-STAT5 and acts as a repressor to maintain self-renewal of hematopoietic cells and to block differentiation [187]. The role of uSTAT3/5 in the initiation, emergence and/or progression of hematologic cancers has yet to be determined. However, previous studies suggested that uSTAT5 provides preferential and critical cell survival signals in lymphoid tumor cells, indicating that uSTAT3/5 should also be considered as therapeutic targets in certain hematologic malignancies [188]. Our understanding of the molecular mechanisms involved in non-canonical functions of STAT3/5 is still incomplete. Most of these activities contribute to the known roles of STAT3/5 in hematopoiesis and hematopoietic neoplasms, and this knowledge complicates the already difficult task of targeting STAT3/5 for therapeutic purposes.

4. Pharmacological Inhibitors of STAT3 and STAT5

4.1. Introduction

Modalities for targeting STAT3 and STAT5 in hematologic cancers can be classified into direct and indirect approaches. Compounds directly targeting STAT3 or STAT5 canonical functions may either inhibit dimerization, DNA binding, or transcriptional activity. Indirect approaches include preventing ligands binding to growth factor or cytokine receptors, inhibiting upstream tyrosine kinases such as TKOs, targeting the nucleocytoplasmic shuttling of STAT3/5, or activating negative regulators of STAT3/5 such as the tyrosine phosphatases SOCS or PIAS [189–191]. Both direct or indirect approaches might also be applied to non-canonical functions. Molecules targeting STAT3/5 protein expression/stability or disrupting interactions with partners that play critical roles in oncogenic STAT3/5 activity should also be considered.

Historically, abnormal activation of STAT3 in solid tumors was first appreciated before STAT5 was implicated, and as such, various therapeutics to target STAT3 were first developed based on one or more of these strategies described above. Pioneering works using phospho-Y-peptides to compete with P-Y^{705} for binding to the SH2 domain of STAT3 and peptidomimetic derivatives provided the proof of concept that disrupting the P-Y^{705}/SH2 domain interaction could be an efficient strategy to block oncogenic STAT3 signaling [192,193]. However, peptides and peptidomimetics have certain limitations in line with their in vivo instability and poor membrane permeability [194]. Nevertheless, these limitations provided the necessary impetus in many programs to design small molecules with greater and more specific inhibitory effects. Blocking DNA binding using decoy oligonucleotides (Duplex ODN), G-quartet oligonucleotides (GQ-ODN), or DBD peptide aptamers was another strategy to suppress canonical functions of STAT3 and STAT5. Treatment with STAT3 or STAT5 ODNs inhibits cell growth and/or induces apoptosis by preventing nuclear translocation of STAT3/5 in cancer cells [195]. RNA interference (RNAi) or antisense oligonucleotides (ASO) were also employed to target STAT3/5 mRNA in leukemia or lymphoma cells [195]. Some of these tools have been promising in modulating STAT3/5 signaling and inhibiting tumor cell growth. However, the therapeutic success of these different approaches relies on the effective entry and stability of the oligonucleotides in the cells and therefore requires chemical modifications for this purpose. Finally, cell-based screening with chemical compound libraries has allowed the identification of natural, synthetic, or clinically used molecules that inhibit STAT3/5 transcriptional activity [196,197]. However, hits derived from these assays have indirect effects that are challenging to determine. Due to an ever-growing list of STAT3/5 inhibitors and reviews in the field, we will focus on those that have been tested in hematopoietic cancers. We will discuss the limitations of STAT3/5 inhibitors in the treatment of these diseases but also promising outcomes when combined with other pharmacological compounds.

4.2. Indirect Inhibitors of STAT3 and STAT5 Signaling in Hematopoietic Cancers

4.2.1. Targeting Upstream Tyrosine Kinases

Most agents that are described as indirect STAT3/5 inhibitors actually target upstream kinases such as JAK, Src, BCR-ABL, FLT3, or KIT receptors. Activation of STAT3 and/or STAT5 is dependent on the leukemic cell type in which the kinase is active. The development of IM and related BCR-ABL kinase inhibitors such as nilotinib, dasatinib, bosutinib, and ponatinib has made a major breakthrough in targeted cancer therapy, CML and Ph$^+$ALL treatment [198–204]. IM leads to complete inhibition of BCR-ABL-dependent STAT5 activation and this is likely an important part of the effectiveness of this molecule [49,90]. However, some patients cannot tolerate the side effects of IM and related TKIs. In addition, the development of resistance to TKIs is a significant clinical problem, and is due, in part, to acquired point mutations in BCR-ABL such as T315I [205]. Although second- and third-generation TKIs were found to be effective against some BCR-ABL mutants, TKIs alone do not eradicate LSCs and de novo resistance of CML cells [55,56,62]. Targeting the kinase activity of FLT3 mutants has also been adopted to inhibit aberrant signaling in AML [206]. Among the agents that have been or are being evaluated in preclinical studies and/or clinical trials are multi-kinase inhibitors midostaurin (PKC412), sorafenib (BAY 43-9006), lestaurtinib (CEP701), KW-2449 (which also targets T315I-mutated BCR-ABL), sunitinib (SU11248), tandutinib (MLN518), and quizartinib (AC220), as well as compounds more specific for FLT3 mutants, including crenolanib (CP-868596) and gilteritinib (ASP2215) (Table 2) [207–219]. Few of them were approved for the treatment of FLT3-mutated AML as a single-agent or in combination with other therapeutic drugs. Although the effectiveness of these inhibitors was shown in preclinical studies, mixed results have been observed in clinical trials. Clinical activity of some of these molecules was evidenced in patients with FLT3-mutated AML but was often transient and relapse eventually occurred [206]. Secondary mutations in FLT3 and/or mutations associated with epigenetic regulators or transcription factors are responsible for the loss of therapeutic response to FLT3 inhibitors [206,220]. In addition, the bone marrow microenvironment provides protection against FLT3-TKIs [221]. Midostaurin, crenolanib, and tandutinib were also employed to inhibit KIT^{D816V} activity in mast cells from SM patients, while lestaurtinib was tested in two phase 2 trials for MPN treatment (Table 2) [222–226]. Other potent and selective KIT^{D816V} inhibitors including avapritinib (BLU-285) and DCC-2618 have entered clinical trials with promising results for the treatment of SM [222].

A breakthrough in understanding myeloproliferative diseases occurred after the discovery of GOF mutations in JAK2, leading to the development of small-molecule inhibitors of JAK2 for the treatment of MPNs [227]. In clinical trials, responses obtained with JAK inhibitors are independent of the driver mutations. However, treatment with JAK inhibitors was shown to have some limitations, partly because the targeted pathway is required for normal hematopoiesis and because specific inhibitors targeting the $JAK2^{V617F}$ mutant are yet to be developed. Consequently, JAK2 inhibitors have been disappointing in their ability to induce molecular remissions in MPN patients, indicating that JAK2 inhibitors do not preferentially target MPN cells over normal cells [228]. In addition, some of the JAK inhibitors that entered clinical trials were discontinued due to their adverse effects. A crystal structure and biochemical properties of the pseudokinase domain of JAK2 will certainly assist in developing $JAK2^{V617F}$-specific inhibitors in the future. Selective JAK and pan-JAK inhibitors that variably affect P-Y-STAT3/5 levels in hematologic neoplasms are presented in Table 2. Ruxolitinib is the first clinically-approved JAK1/2 inhibitor for PV and MF treatment and is also in clinical trials either alone or in combination with other pharmacological agents or TKIs for HL, MM, or CML treatment (Table 2) [227–232]. Ruxolitinib and most of the JAK inhibitors that are in clinical trials are type I inhibitors, which means that they block the ATP-binding site of JAKs under the active conformation of the kinase domain [227,233–237]. Type II inhibitors bind to the ATP-binding pocket of the JAK2 kinase domain in the inactive conformation, while allosteric inhibitors interact with other sites in the JAK2 protein [227]. Importantly, JAK2 target inhibition in MPN can be improved with the type II inhibitor NVP-CHZ868 offering increased therapeutic efficacy [238]. NVP-CHZ868 was also shown to act synergistically with dexamethasone in suppressing the growth and survival of human B-ALL cells in PDX models [239]. JAK2-specific inhibitors such as pacritinib are in the final stages of clinical trials for primary and secondary MF, and display increased potency compared to currently available JAK inhibitors [233]. Gandotinib, which is in a phase 2 study for $JAK2^{V617F}$-mutated MPN treatment, showed an increased potency for the $JAK2^{V617F}$ mutant [240]. Another example of a JAK2 inhibitor, fedratinib, which was previously burdened with a clinical hold in 2013, was recently FDA-approved for the treatment of MPN patients who have failed therapy with ruxolitinib [241]. Fedratinib also blocks the growth of HL and Mediastinal Large B-cell Lymphoma (MLBCL) in vitro and in vivo as demonstrated in preclinical studies [242]. Many JAK inhibitors with different selectivity and/or mechanisms of action have been tested in leukemias, lymphomas and MM (Table 2) [243–268]. For instance, AZD1480, INCB20, INCB16562, NS-018 and momelotinib (CYT-387) showed promising in vitro and/or in vivo efficacy against MM cells (Table 2) [246–248,251,261]. In all cases, P-Y-STAT3 was markedly reduced in

MM cells treated with these inhibitors. Collectively, these data strongly support that JAK inhibition has significant potential as a therapeutic strategy in MM.

4.2.2. Natural and Synthetic Molecules

Historically, natural compounds have been successfully used in the management of various human diseases. Natural products may also serve as a basis for the synthesis of derivatives aiming to increase their efficacy. There are several compounds that are known to exert anti-tumor effects through their indirect or direct action on STAT3 and/or STAT5 signaling. These natural molecules have a low toxicity profile and can act synergistically with other pharmacological agents to reverse chemoresistance. A number of plant-derived molecules such as avicin D, curcubitacin I, butein, honokiol, capsaicin, celestrol, and piperlongumine have been reported to inhibit growth or survival of leukemia, lymphoma, or myeloma cells in preclinical studies (Table 2) [269–277]. However, the mechanistic basis of their effects on STAT3/STAT5 signaling is still unknown. Inhibition of upstream kinases JAK1/2 and Src, and upregulation and/or activation of SHP1 or other protein tyrosine phosphatases appear to be a common feature of these compounds. Curcumin, a naturally derived phytochemical from plants such as turmeric (*Curcuma longa*), has been extensively investigated for its anti-tumor effects [278]. Curcumin was shown to block STAT3 and/or STAT5 phosphorylation in leukemia, lymphoma, and myeloma cells (Table 2) [279–283]. Although the administration of curcumin has been shown to be safe in humans, its clinical utility is somewhat limited due to the poor bioavailability and target selectivity. Therefore, efforts were made to design and synthesize novel curcumin analogs. FLLL32, one of these analogs, was shown to inhibit P-Y-STAT3 and growth of MM cells with greater efficacy but, again, target selectivity and mechanisms of action remained poorly defined [284]. The synthetic triterpenoid, CDDO-Imidazolide, which acts as an anti-inflammatory and anti-cancer drug, was demonstrated to suppress STAT3 and STAT5 phosphorylation and to induce apoptosis in MM cells [285]. EC804, a synthetic derivative of indirubin, an active component in a traditional Chinese medicine formulation, was reported to inhibit STAT3 and/or STAT5 phosphorylation as well as the growth of sensitive or TKI-resistant CML and AML cells [286,287]. This compound also blocks STAT3 activity in solid tumors [288]. Natural and synthetic compounds such as sulforaphane and the BET inhibitor JQ1 were found to inhibit STAT5-mediated transcription in CML and T-ALL cells, probably via epigenetic mechanisms [289,290]. Naphthoquinone (NPQ)-based derivatives could be also mentioned as indirect inhibitors of STAT5 probably through their multikinase modulatory effects in leukemias [291]. Research in our laboratory is focused on the synthesis and development of small-molecule inhibitors of STAT5. We recently identified 17f as a compound that selectively inhibits STAT5 phosphorylation and expression in AML and CML cells [292]. Moreover, we found that 17f overcomes the resistance of CML and AML cells to IM and Ara-C, a conventional therapeutic agent used in AML treatment [293]. We also found that 17f, when associated with IM or Ara-C, inhibits expression of STAT5B but not STAT5A in resistant CML and AML cells via translational or post-translational mechanisms. The mechanistic basis of this inhibitory effect is currently under investigation.

4.2.3. Drug Repositioning

Cell-based assays for high-throughput screening were employed to identify compounds that specifically block the transcriptional activity of STAT3/5 [196,197]. This type of strategy utilizes cells that are stably transfected with a construct containing the luciferase reporter gene under the control of a specific and high-affinity STAT3 or STAT5 responsive promoter. Chemical libraries of compounds biased toward bioactives and drugs known to be safe in humans were used in the screening. Using this approach, nifuroxazide, niclosamide, and pyrimethamine were identified as specific inhibitors of STAT3, while pimozide was found to inhibit STAT5 activity (Table 2). Nifuroxazide, an antidiarrheic agent, was shown to decrease STAT3 tyrosine phosphorylation, most probably via inhibition of JAK kinase activity, and to reduce the viability of MM cells [294]. Niclosamide, an antiparasitic drug, blocks P-Y-STAT3 and myeloma cell growth via unknown mechanisms. Niclosamide lacks selectivity because it also inhibits

NFkB activity in MM cells [295]. The antiparasitic and antifolate drug pyrimethamine also displays significant activity in vitro against MM cell lines harboring P-Y-STAT3 [196]. Pyrimethamine inhibited P-Y-STAT3 and transcriptional activity without affecting its upstream kinase JAK2. Three-dimensional modeling studies indicated that pyrimethamine binds to the SH2 domain, suggesting that it might be a direct inhibitor of STAT3, but this interaction needs to be biophysically demonstrated. Pyrimethamine was also found to be a potent inducer of apoptosis in AML cells [296]. However, it is still unclear whether the antitumor activity of pyrimethamine is due to its inhibitory effect on STAT3 and/or on folic acid metabolism [297]. Pyrimethamine is currently in a phase 1 clinical trial for the treatment of high-risk MDS and in phase $\frac{1}{2}$ trials for CLL and small lymphocytic lymphoma (SLL) treatment.

Pimozide was identified by high-throughput drug screening as a potent inhibitor of STAT5. Pimozide was shown to inhibit P-Y-STAT5 and survival of CML and MPN cells without affecting the kinase activity of BCR-ABL, JAK2 or Src [298,299]. Pimozide shows synergistic effects with IM/nilotinib in killing CML cells and overcomes TKI resistance in BCR-ABL^{T315I} mutant cells [298]. The effects of pimozide are not limited to CML and MPN cells, and the efficacy of this drug was also demonstrated in AML. Pimozide can also inhibit P-Y-STAT5 and STAT5-dependent gene expression in AML cells expressing FLT3-ITD, and it acts synergistically with FLT3 inhibitors to induce apoptosis in these leukemic cells [79]. The mechanisms involved in pimozide-mediated inhibition of P-Y-STAT5 is currently not known. Pimozide also inhibits P-Y-STAT3 in myeloma cells indicating that it is not a selective STAT5 inhibitor [196]. Antidiabetic drugs such as pioglitazone and rosiglitazone were shown to have antileukemic activity [300]. Both synthetic compounds belong to the thiazolidinedione (TZD) class of ligands that bind to the nuclear receptor PPARγ. Activation of PPARγ by pioglitazone not only inhibits the growth of CML cells but also reduces the expression of *STAT5* genes [60]. Quiescent CML stem cells, which are known to be resistant to TKI treatment, are highly sensitive to pioglitazone when combined with IM. This suggests that besides phosphorylation, targeting STAT5 expression might be important for eradicating resistant CML stem cells. Whether PPARγ directly regulates STAT5A and STAT5B gene promoter activity remains to be investigated. Pioglitazone combined with IM are now in a phase 2 trial to evaluate the impact of this combination therapy on CML residual disease [301]. Activation of PPARγ was also shown to induce apoptosis in Ph$^+$ ALL cells [302]. Finally, PPARγ agonists were found to inhibit the transcriptional activity of STAT3 in MM cells. In that case, it has been proposed that PPARγ and STAT3 may compete for binding to nuclear co-factors [303].

Table 2. Indirect inhibitors of STAT3 and STAT5 that have been tested in hematologic cancers.

Drug	Target(s) and Mode of Inhibition	Hematologic Malignancy	Stage of Clinical Development	References
		Upstream Kinases		
Atiprimod	JAK2/JAK3 Inhibits kinase activity	AML JAK2^{V617F}-MPN	Preclinical (cell lines, primary cells) Preclinical (cell lines, primary cells)	[249,250]
AZ960	JAK2/JAK3 Type I inhibitor	T cell leukemia/ lymphoma	Preclinical (cell lines)	[251]
AZD1480	JAK2/JAK1/Aurora/FGFR1/FLT4 Type I inhibitor	MPN B-ALL Lymphomas MM	Phase 1 (terminated) Preclinical (xenografts) Preclinical (cell lines) Preclinical (cell lines, xenografts)	[243–246]
Bosutinib (SKI-606)	BCR-ABL/Src Type I inhibitor	CML	Market authorization (Bosulif®, Pfizer)	[201,202]
Crenolanib (CP-868596)	FLT3/PDGFR/KIT Type I inhibitor	FLT3-mutated AML SM, AML	Phase 2 Phase 3 (CT) Preclinical (cell lines, primary cells)	[218,223]
Dasatinib (BMS-354825)	BCR-ABL/Src. Inhibits kinase activity Type I inhibitor	CML, Ph$^+$ ALL	Market authorization (Sprycel®, BMS)	[200]
Fedratinib (TG101348)	JAK2/FLT3/BRD4 Type I inhibitor	MPN HL, MLBCL	Market authorization (Inrebic®, Celgene) Preclinical (cell lines, xenografts)	[241,242]
Gandotinib (LY2784544)	JAK2/JAK1/JAK3 Type I inhibitor	JAK2^{V617F}-MPN	Phase 2	[240]
Gilteritinib (ASP2215)	FLT3/AXL Type I inhibitor	FLT3-mutated AML	Market authorization (Xospata®, Astellas)	[219]
Imatinib (STI-571)	BCR-ABL. Type II inhibitor	CML, Ph$^+$ALL	Market authorization (Glivec®, Novartis)	[198,199]
INCB20	Pan JAK	MM	Preclinical (cell lines, primary cells, mouse models)	[247]
INCB16562	JAK1/JAK2. Type I inhibitor	MM	Preclinical (cell lines, primary cells)	[248]
Itacitinib (INCB-039110)	JAK1 Type I inhibitor	MF B-cell lymphoma	Phase 2 Phase 1/2	[252,253]

Table 2. *Cont.*

Drug	Target(s) and Mode of Inhibition	Hematologic Malignancy	Stage of Clinical Development	References
KW-2449	FLT3/BCR-ABL/ BCR-ABL315I/Aurora Inhibits FLT3 and BCR-ABL phosphorylation	CML, AML MDS, AML, ALL, CML	Preclinical (cell lines, primary cells, xenografts) Phase 1	[212,213]
Lestaurtinib (CEP-701)	JAK2/FLT3/TrkA/ Aurora kinase Type I inhibitor	AML MF PV, ET	Phase 2 Phase 2 Phase 2	[211,225,226]
LS104 (AG490 analog)	JAK2/BCR-ABL Allosteric inhibitor	JAK2^{V617F}-MPN AML Ph$^+$ and Ph$^-$ ALL	Preclinical (cell lines, primary cells) Preclinical (cell lines, primary cells) Preclinical (cell lines, primary cells, xenografts)	[262–264]
Midostaurin (PKC412)	FLT3-ITD/KIT Type I inhibitor	FLT3 mutated AML, SM	Market authorization (Rydapt®, Novartis)	[207,222]
Momelotinib (CYT-387)	JAK2/JAK1/JAK3/ALK2 Type I inhibitor	MPN MM	2 Phase 3 Preclinical (cell lines, primary cells)	[259,260]
Nilotinib (AMN107)	BCR-ABL Type II inhibitor	CML, Ph$^+$ ALL	Market authorization (Tasigna®, Novartis)	[203]
NS-018	JAK2/Src Type I inhibitor	MPN MM	Phase 1-2 Preclinical (cell lines, mouse model)	[254,255]
NVP-BSK805	JAK2 (JAK2^{V617F}) Type I inhibitor	JAK2^{V617F}-MPN	Preclinical (cell lines, mouse model)	[256]
NVP-BVB808	JAK2 Type I inhibitor	MPN	Preclinical (cell lines)	[257]
NVP-CHZ868	JAK2/TYK2/KIT/PDGF/VEGFR Type II inhibitor	MPN B-ALL	Preclinical (cell lines, primary cells, mouse models) Preclinical (cell lines, xenografts, PDX)	[238,239]
ON044580	JAK2/BCR-ABL Allosteric inhibitor	JAK2^{V617F}-MPN BCR-ABL^{T315I}-CML	Preclinical (cell lines, primary cells)	[258]
Pacritinib (SB1518)	JAK2/FLT3/IRAK1 Type I inhibitor	MF Lymphomas, CLL, LPD	Phase 3 Phase 1	[233,234]
Ponatinib (AP24534)	BCR-ABL and BCR-ABL mutants FGFR/PDGFR/Src/RET/KIT/FLT1 Type II inhibitor	CML, Ph+ ALL	Market authorization (Iclusig®, Ariad)	[204]

Table 2. *Cont.*

Drug	Target(s) and Mode of Inhibition	Hematologic Malignancy	Stage of Clinical Development	References
Ruxolitinib (INCB18424)	JAK2/JAK1/JAK3 Type I Inhibitor	MPN CML HL MM AML, ALL	Market authorization (Jakavi®, Novartis) Phase 1 (CT) Phase 2 Phase 1 (CT) Phase1/2 (terminated)	[227–232]
Quizartinib (AC220)	FLT3 Type II inhibitor	FLT3-mutated AML	Market authorization (Japan) (Vanflyta®, Daiichi-Sankyo)	[217,218]
Sorafenib (BAY43-9006)	FLT3/PDGFR/ VEGFR/c-KIT/RAF Type II inhibitor	FLT3-mutated AML ALL, AML, MDS	Phase 1 Phase $\frac{1}{2}$ (CT) Phase 1	[208–210]
Sunitinib (SU11248)	PDGFR/ VEGFR/c-KIT/FLT3 Type I Inhibitor	FLT3-mutated AML	Phase 1 Phase $\frac{1}{2}$ (CT)	[214,215]
Tandutinib (MLN518)	FLT3/PDGFR/KIT Inhibits type III receptor kinases	AML/MDS AML SM	Phase $\frac{1}{2}$ Phase $\frac{1}{2}$ (CT) Preclinical (cell lines)	[216,224]
TG101209	JAK2/FLT3 Type I inhibitor	MPN MM T-ALL	Preclinical (cell lines, primary cells, mouse models) Preclinical (cell lines) Preclinical (cell lines, primary cells)	[235–237]
WP1066 and WP1034 (AG490 analogs)	JAK2 Phosphorylation and/or degradation	JAK2[V617F]-mutated MPN AML	Preclinical (cell lines, primary cells) Preclinical (cell lines, primary cells)	[265–267]
XL019	Pan-JAK. Type I inhibitor	MPN	Phase 1 (terminated)	[268]
Natural compounds				
Avicin D	JAK/STAT3 Inhibits P-Y/JAK1/2 phosphorylation Unknown mechanism	MM CTCL	Preclinical (cell lines) Preclinical (cell lines, primary cells)	[269,270]
Butein	STAT3. Inhibits P-Y Activates tyrosine phosphatase SHP1 Unknown mechanism	MM	Preclinical (cell lines)	[271]

Table 2. *Cont.*

Drug	Target(s) and Mode of Inhibition	Hematologic Malignancy	Stage of Clinical Development	References
Capsaicin	JAK1/Src/ STAT3. Inhibits P-Y/ Inhbits kinase activity Unknown mechanism	MM	Preclinical (cell lines, xenografts)	[272]
Celastrol	JAK/Src/STAT3. Inhibits P-Y/Kinase activity Unknown mechanism	MM	Preclinical (cell lines)	[273]
Curcubitacin I (JSI-124)	STAT3. Inhibits P-Y/P-S of STAT3 Unknown mechanism	B-ALL, CLL, lymphomas	Preclinical (cell lines, primary cells)	[274,275]
Curcumin diferuloylmethane	JAK/STAT3/STAT5. Inhibits P-Y/kinase activity Unknown mechanism	CML T-cell leukemia T-cell lymphoma HL MM	Preclinical (cell lines) Preclinical (cell lines) Preclinical (cell lines, primary cells) Preclinical (cell lines) Preclinical (cell lines)	[279–283]
Honokiol (HNK)	STAT3. Inhibits P-Y Activate SHP1 Unknown mechanism	AML	Preclinical (cell lines, primary cells)	[276]
Piperlongumine	STAT3. Inhibits P-Y and other pathways Binds to cysteine 712 of STAT3 Unknown mechanism	MM	Preclinical (cell lines, xenografts)	[277]
Sulforaphane	STAT5. Inhibits transcriptional activity. Deacetylase inhibitor Unknown mechanism	CML	Preclinical (cell lines)	[289]
Synthetic molecules				
CDDO (Imidazolide)	STAT3/ STAT5. Inhibits P-Y Unknown mechanism	MM	Preclinical (cell lines)	[285]
E804 (Indirubin derivative)	SFK/STAT5 Inhibits P-Y/ Inhibits kinase activity Unknown mechanism	CML AML	Preclinical (cell lines, primary cells) Preclinical (cell lines, xenografts)	[286,287]
FLLL3 (Curcumin derivative)	JAK/STAT3. Inhibits P-Y Unknown mechanism	MM	Preclinical (cell lines)	[284]

Table 2. *Cont.*

Drug	Target(s) and Mode of Inhibition	Hematologic Malignancy	Stage of Clinical Development	References
JQ1 (Thienodiazepine derivative)	STAT5. Inhibits transcriptional activity.BET inhibitor. Unknown mechanism	T-ALL	Preclinical (cell lines, primary cells)	[290]
17f	STAT5. Inhibits P-Y and expression. Unknown Mechanism	AML, CML	Preclinical (cell lines)	[292]
Naphtoquinone (NPQ) derivative	STAT5. Inhibits P-Y/ Unknown mechanism	CML	Preclinical (cell lines)	[291]
Drug repositionning				
Niclosamide (Antiparasitic)	STAT3. Inhibits P-Y Unknown mechanism	MM	Preclinical (cell lines, primary cells)	[295]
Nifuroxazide (Antidiarrheic)	JAK/STAT3. Inhibits P-Y Unknown mechanism	MM	Preclinical (cell lines, primary cells)	[294]
Pimozide (Antipsychotic)	STAT3/STAT5. Inhibits P-Y Unknown mechanism	CML MPN FLT3-mutated AML MM	Preclinical (cell lines, primary cells) Preclinical (cell lines) (CT) Preclinical (cell lines, mouse model) Preclinical (cell lines)	[79,196,298, 299]
PPARγ ligands (Antidiabetic) (Thiazolinediones)	STAT5. Inhibits *STAT5A* and *STAT5B* gene expression STAT3. Inhibits transcriptional activity	CM, Ph$^+$ ALL MM	Preclinical (cell lines, primary cells) Phase 2 (CT) Preclinical (cell lines, primary cells)	[60,301–303]
Pyrimethamine (Antiparasitic) (Antifolate)	STAT3. Inhibits P-Y Unknown mechanism	AML CLL/SLL MM	Preclinical (cell lines, primary cells, xenografts) Phase $\frac{1}{2}$ Preclinical (cell lines)	[196,296]

AML (acute myeloid leukemia), B- and T-ALL (B- and T-acute lymphoblastic leukemia), CLL (chronic lymphocytic leukemia), CTCL (cutaneous T cell lymphoma), CML (chronic myeloid leukemia), CTCL (cutaneous T cell lymphoma), ET (essential thrombocythemia), HL (Hodgkin's lymphoma), LPD (lymphoproliferative disorders), myelodysplastic syndromes (MDS), MF (myelofibrosis), MCL (mantle cell lymphoma), MLBCL (mediastinal large B cell lymphoma), MM (multiple myeloma), MPN (myeloproliferative neoplasm), Ph$^+$ and Ph$^-$ ALL (Philadelphia chromosome-positive and -negative acute lymphoblastic leukemia); PV (polycythemia vera), SLL (small lymphocytic lymphoma), SM (systemic mastocytosis), CT (combination therapy).

4.3. Direct Inhibitors of STAT3 and STAT5 in Hematopoietic Cancers (Table 3)

The mechanisms of direct STAT3/5 inhibition include disruption of tyrosine phosphorylation, dimerization, nuclear translocation and/or DNA binding. Targeting the SH2 domain was, therefore, the main focus for the design and identification of selective inhibitors.

4.3.1. Inhibitors Targeting the SH2 Domain

Dimerization is an essential step in the canonical functions of STAT3 and STAT5. Blocking this process seems to be an optimal solution to directly inhibit aberrant STAT3/5 signaling in hematopoietic cancers. The SH2 domain is not only required for dimer formation but also in the recruitment of STAT3/5 to tyrosine-phosphorylated receptor complexes. Therefore, initial strategies aiming to identify phosphopeptide (P-Y-peptide) inhibitors able to block SH2/P-Y interactions were pursued by several teams. This approach was first attempted by Turkson et al. who demonstrated that the peptide PY*LKTK (where Y* represents P-Y^{705}) inhibits STAT3 dimerization and tumor cell growth [192]. P-Y-peptides derived from STAT3 docking sites of gp130 and other cytokine or growth factors receptors were also used to identify the P-Y-peptide YLPQTV as a potent blocker of STAT3 dimerization and DNA binding [304,305]. Although highly specific, peptides usually have poor membrane permeability, low stability, and consequently low biological activities. This prompted investigators to develop peptidomimetics using the tripeptide PY*L, the minimum P-Y-peptide sequence required for STAT3 inhibition. One of these peptidomimetics, ISS610 was shown to be more potent in disrupting STAT3 dimerization and DNA binding but still had poor membrane permeability [193]. Structural and computational analysis of the interaction between ISS610 and the STAT3 SH2 domain led to the development of the peptidomimetic molecule, S3I-M2001, having increased membrane permeability but similar capacity in inhibiting STAT3 DNA binding [306]. Despite hard work in developing peptides or peptidomimetics with potent STAT3-inhibitory activity, poor permeability and metabolic stability have precluded their clinical testing. For the same reason, no peptide/peptidomimetic inhibitors of STAT5 have been developed. Based on the proof of concept that the STAT3-SH2/pY-peptide interaction was amenable to targeting, small nonpeptidic compounds that could specifically bind to the SH2 domains of STAT3 or STAT5 have become attractive candidates. Here, we will discuss the SH2 domain inhibitors that have been used in preclinical studies and clinical trials for the treatment of hematologic malignancies.

Most of the small molecules targeting the SH2 domain of STAT3 were identified by structure-based high-throughput virtual screening. The available X-ray crystallographic data of both the monomer and dimerized STAT3 bound to DNA were essential for these in silico studies [307]. Among these molecules, S3I-201, STA-21, C188, and STX-0119 were shown to block the phosphorylation, dimerization, DNA binding, and transcriptional activity of STAT3 as well as growth and survival of cancer cells [308–311]. In modeling studies, S3I-201 was shown to bind to the STAT3-SH2 domain through its salicylic acid moiety, which is known as a P-Y mimetic [308]. However, Genetic Optimization for Ligand Docking (GOLD) studies indicated that binding of S3I-201 was suboptimal, and structure–activity relationship (SAR) analyses based on this "hit" compound were conducted to derive analogs with improved STAT3 inhibitory activity. One of these analogs, S3I-1757, bound to the SH2 domain of STAT3 but still had a modest potency for decreasing STAT3 DNA binding [312]. However, S3I-1757 embedded in nanoparticles that had been conjugated with monoclonal antibodies against CD38 (denoted as CD38-S3I-NP) was demonstrated to have some increased efficacy in suppressing P-Y-STAT3 and tumor growth in xenograft models of MM. Data describing this compound are presented in this Cancers issue [313]. Among various S3I-201 derivatives, SF-1-066 was found to be the most potent. SF-1-066 was shown to directly bind STAT3 and to inhibit P-Y-STAT3 and growth of AML and MM cell lines with greater potency than S3I-201 [314,315]. A family of sixteen sulfonamide analogs of SF-1-066 was synthetized and BP-1-102 (17o) was proved to be the most active compound in this series. BP-1-102 exhibited improved inhibition of P-Y-STAT3 in MM cell lines [316]. In addition, BP-1-102 blocked P-Y-STAT3 in B-ALL cells overexpressing HMGA1 chromatin remodeling proteins and suppressed their growth [94]. Growth inhibition induced by this salicylic acid-based inhibitor was also reported for T-ALL cell lines [94]. N-alkylated derivatives of both SF-1-066 and BP-1-102, such as 16i and 21h molecules, were demonstrated to inhibit STAT3 function and to induce MM cell apoptosis, but these molecules also lost selectivity toward STAT3 [317]. BP-5-087, another analog of SF-1-066, was identified by SAR-based drug design and compound library screening. BP-5-087 was shown to be a more potent salicylic acid-based STAT3 inhibitor than SF-1-066 and to overcome IM resistance in CML stem cells [318].

A library of phosphopeptide mimicking salicylic acid-based molecules targeting the STAT3-SH2 domain was also employed to identify potent and STAT5-selective inhibitors. Among compounds that were selected in the screening via fluorescence polarization (FP) assays, BP-1-108 was found to be the more potent STAT5 inhibitor [319]. BP-1-108 significantly reduced P-Y-STAT5 and exhibited growth inhibitory properties in CML and AML cell lines. In silico-based studies using BP-1-108 as a scaffold identified compound 13a (or AC-4-130) as a selective and first nanomolar inhibitor of the STAT5B-SH2 domain [320]. AC-4-130 did not exhibit off-target kinase activity and suppressed P-Y-STAT5 at low micromolar concentrations in CML and AML cell lines, as well as in primary AML cells [81]. AC-4-130 inhibited the growth of AML cells in vitro and tumor formation in xenograft models of AML. Furthermore, AC-4-130 was also shown to synergistically increase the cytotoxic effect of the JAK1/2 inhibitor ruxolitinib and the p300/pCAF inhibitor garcinol in FLT3-ITD$^+$ AML cells [81]. IST5-002, another STAT5 inhibitor, was identified using in silico structure-based screening and medicinal chemistry. IST5-002 was demonstrated to bind to the SH2 domain of STAT5B, to inhibit P-Y-STAT5 and to reduce the growth and viability of sensitive and IM-resistant CML and Ph$^+$ALL cell lines [321]. Patient-derived CML or newly diagnosed and relapsed/TKI-resistant Ph$^+$ALL cells were also sensitive to IST5-002 treatment. Moreover, IST5-002 was found to reduce leukemia development in PDX models of Ph$^+$ALL [90]. A chromone-derived acyl hydrazine compound, identified through a high-throughput FP screen, was the first inhibitor to directly target the SH2 domain of STAT5B. However, high concentrations of this compound was required to disrupt STAT5 DNA binding and P-Y-STAT5 in CML and Burkitt's lymphoma (BL) cell lines, respectively [322]. The same group next identified catechol-biphosphate as a selective inhibitor of the STAT5B SH2 domain. One of the catechol-biphosphate derivatives, stafib-1, was shown to be a nanomolar inhibitor of the STAT5B SH2 domain with more than 50-fold selectivity over the STAT5A-SH2 domain [323]. SAR studies based on the Stafib-1 compound and rational design led to the development of Stafib-2 with improved and high selective activity against STAT5B in CML cells [324].

Structure-based high-throughput virtual screening of four different chemical libraries followed by cell-based reporter assays identified STA-21 (or deoxytetrangomycin) as a new SH2 domain-targeting inhibitor of STAT3 [309]. STA-21 induced apoptosis of primary cells from chronic lymphoproliferative disorders of natural killer cells (CLPD-NK) and T-LGL patients and DLBCL cells when associated with YM155, a survivin suppressant [325,326]. A structural analog of STA-21, LLL-3, was developed to improve the cellular permeability of STA-21. LLL-3 was shown to promote apoptosis of CML cell lines and to synergistically act with IM to suppress CML cell growth and survival [327]. LLL-3 was further optimized by replacing its acetyl group with a sulfonamide to produce LLL-12, which prevented the phosphorylation of STAT3 and induced apoptosis in MM cell lines and primary MM cells in vitro, even in samples from patients with relapsed/refractory MM. LLL-12 also suppressed tumor formation in xenograft models of MM [328]. However, LLL-12 was required in high concentrations for in vivo studies due to low bioavailability. C188 compound was identified by virtual ligand screening of eight chemical libraries [310]. C188 was shown to inhibit P-Y-peptide binding to the SH2 domain of STAT3 by surface plasmon resonance (SPR) assays. Using C188 as the scaffold, C188-9 was identified in a hit-to-lead program and was shown to bind to the STAT3-SH2 domain with greater potency than the C188 compound. C188-9 effectively suppressed G-CSF-induced P-Y-STAT3 in AML cell lines and patient samples without affecting JAK or Src kinases. C188-9 also induced apoptosis and blocked the clonogenic growth of AML cells [329]. In silico screening studies followed by STAT3-dependent luciferase reporter gene assays led to the identification of STX-0119, another STAT3 inhibitor impacting the growth of lymphoma cell lines. Oral administration of STX-0119 was shown to reduce STAT3 target gene expression and to induce apoptosis in a xenograft model of lymphoma [311]. The STAT3-SH2 domain inhibitor STATTIC was identified by screening chemical libraries in an FP-based binding assay. This compound was demonstrated to selectively inhibit the dimerization and transcriptional activity of STAT3, although a report indicated that it also blocks STAT1 phosphorylation [330]. STATTIC reduced P-Y-STAT3 and viability of MM cells in three-dimensional (3D) culture and sensitized them to bortezomib, a proteasome inhibitor that is clinically used in MM treatment (this issue) [331]. Moreover, STATTIC suppressed growth and survival of NK/T cell lymphoma cell lines expressing the STAT3^{Y640F} mutant [332]. Among the STAT3-SH2 domain targeting inhibitors, OPB-31121 and OPB-51602 were the only compounds to have reached early phase clinical trials for treatment of hematopoietic malignancies [333]. OPB-31121 was shown to inhibit P-Y-STAT3 and P-Y-STAT5 without affecting upstream kinase activity in various myeloid leukemia cell lines expressing BCR-ABL, FLT3-ITD, or JAK2^{V617F}, as well as in BL and MM cell lines [334]. Treatment with OPB-31121 also induced growth inhibition of these hematopoietic malignant cells and reduced tumor formation in PDX models of ALL, CML, and AML. Computational docking and molecular dynamics simulations (MDS) demonstrated that OPB-31121 bound with high affinity to the SH2 domain of STAT3 [335]. In the same way, OPB-51602 was also shown to interact with high affinity with the STAT3 SH2 domain to inhibit P-Y-STAT3 and to be effective against MM, BL, and AML cells in preclinical in vitro and in vivo studies [336]. Although promising data were obtained from these preclinical studies, clinical trials were terminated for both of these inhibitors due to poor pharmacokinetic properties, significant toxicity and lack of antitumor activity. Very recently, SD-36, a proteolysis targeting chimera (PROTAC) that selectively degrades STAT3 protein has been developed. This molecule consists of the cell-permeable STAT3-SH2 domain-targeting inhibitor, SI-109, conjugated to a ligand of Cereblon, an important component of the E3 ubiquitin ligase complex that is involved in ubiquitination and degradation of multiple cellular proteins. The resulting PROTAC binds to and recruits both STAT3 and the E3 ligase to form a productive ternary complex for ubiquitination and degradation. SD-36 was shown to induce STAT3 degradation, cell growth arrest, and apoptosis in lymphoma and leukemia cells lines as well as tumor regression in multiple xenograft mouse models of leukemias and lymphomas [337]. Lastly, several plant-derived molecules were also demonstrated to be direct inhibitors of STAT3. Using computational modeling and docking simulations, Withaferin A, a natural product isolated from the medicinal plant *Withania somnifera*, was found to interact with the SH2 domain of STAT3. Consequently, Withaferin A prevented IL-6–mediated or persistently activated P-Y-STAT3 and induced apoptosis in MM cells [338]. In a similar vein, YL064, a derivative of Sinomenine, a plant component that has been used to treat rheumatic diseases, was shown to target the SH2 domain of STAT3 and to induce MM cell death [339].

4.3.2. Inhibitors Targeting the DNA Binding Domain (DBD)

STAT3/5 bind to specific DNA response elements within promoters to mediate transcriptional activation of target genes. Concerted efforts were therefore made to identify specific inhibitors targeting the DBD of STAT proteins. Most of the DBD inhibitors presented in this section were historically used against STAT3 and proved to be effective in some hematopoietic cancers.

Platinum (IV) compounds such as CPA-1, CPA-7, and platinum (IV) tetrachloride, were shown to inhibit STAT3 DNA binding activity [340]. IS3 295, a member of this class of molecules, was identified by screening of the NCI 2000 diversity set using electrophoretic mobility shift assays (EMSA). IS3 295 was demonstrated to irreversibly bind to the DBD of STAT3 in its active and inactive form and to prevent its interaction with specific response elements. This compound inhibited the constitutive DNA binding of STAT3 and induced apoptosis in MM cells [341]. Peptide aptamers as inhibitors of STAT3 represent one of the effective approaches to disrupt STAT3 DNA binding. The DBD-directed peptide aptamer DBD-1 was identified in yeast two-hybrid screenings, and its cell-penetrating form DBD-1-9R was shown to inhibit STAT3 DNA binding in EMSAs and to induce growth inhibition and apoptosis in MM cells [342].

Another class of STAT DBD inhibitors being used are duplex ODN (decoy oligonucleotides or dODN) and GQ-ODN. dODNs targeting STAT proteins are double-stranded DNA molecules mimicking the consensus STAT DNA binding sequence. These duplex ODNs act by competitively inhibiting the DNA binding of STAT proteins to their endogenous promoter elements thereby preventing their nuclear function [195]. For instance, a STAT5 dODN was shown to block DNA binding and transcriptional activity of STAT5 and to induce growth arrest and apoptosis in CML cell lines [343]. STAT3 dODN, and the second-generation cyclic STAT3 decoy (CS3D) with a longer half-life, were only tested in solid tumors and will not be discussed here. However, a dual-function molecule CpG-STAT3 dODN, which consists of a STAT3 dODN fused to the TLR9 agonist cytosine guanine dinucleotide (CpG), was reported to induce growth-inhibitory and immune-mediated effects against AML and DLBCL [74,344]. These compounds are rapidly internalized by TLR9$^+$ immune and malignant cells to block the oncogenic activity of STAT3 and to promote an antitumor immune response. Both TLR9 stimulation and concurrent STAT3 inhibition were critical for immune-mediated therapeutic effects. GQ-ODN are G-rich oligodeoxynucleotides that form intra- and inter-molecular four-stranded structures [195]. For example, the GQ-ODN T40214 was shown to inhibit IL-6–induced P-Y-STAT3 and STAT3-mediated transcription [345]. Computer-simulated docking studies indicated that GQ-ODNs interact mainly with the SH2 domain of STAT3 and are able to disrupt STAT3 dimers bound to DNA. Blocking STAT3 with the GQ-ODN T40214 loaded into nanoparticles was shown to be effective in a mouse model of T-ALL [346]. However, the use of GQ-ODN remains problematic due to the large size and potassium-dependence of this molecular probe.

Table 3. Direct inhibitors of STAT3 and STAT5 that have been tested in hematologic cancers.

Target	Drugs	Hematologic Malignancy	Stage of Clinical Development	References
		Direct STAT3 inhibitors		
mRNA	AZD9150 (IONIS-STAT3Rx)	AML, MDS DLBCL, HL, NHL	Preclinical (cell lines, primary cells PDX) Phase 1	[347–349]
	CpG-STAT3-siRNA	AML, MM	Preclinical (cell lines, primary cells, mouse model of AML, xenografts)	[73,350]
SH2 domain	BP-5-087, BP-1-102/17o (S3I-201 derivatives)	CML B- ALL, T-ALL MM	Preclinical (cell lines, primary cells) Preclinical (cell lines, xenografts) Preclinical (cell lines)	[94,316–318]
	CD38-S3I-NP (S3I-1757)	MM	Preclinical (cell lines, xenografts)	[313]
	C188-9	AML	Preclinical (cell lines, primary cells)	[329]
	LLL-3	CML	Preclinical (cell lines)	[327]
	LLL-12	MM	Preclinical (cell lines, primary cells, xenografts)	[328]
	OPB-51602	MM, NHL, AML, CML	Phase 1 (terminated)	[336]
	SF-1-066 (S3I-201 derivative)	AML, MM	Preclinical (cell lines)	[314,315]
	STA-21	CLPD-NKs T-LGL DLBCL	Preclinical (primary cells) Preclinical (primary cells) Preclinical (cell lines)	[325,326]
	STATTIC	MM NK lymphomas T-cell lymphomas	Preclinical (cell lines) Preclinical (primary cells) Preclinical (primary cells)	[331,332]
	STX-0119	NHL	Preclinical (cell lines, xenografts)	[311]
	SD-36	AML, ALCL	Preclinical (cell lines, xenografts)	[337]
	YL064	MM	Preclinical (cell lines, primary cells)	[339]
	Withaferin A	MM	Preclinical (cell lines)	[338]

Table 3. *Cont.*

Target	Drugs	Hematologic Malignancy	Stage of Clinical Development	References
DBD	CpG-STAT3dODN (decoy oligonucleotides)	AML DLBCL	Preclinical (cell lines, xenografts, mouse models) Preclinical (cell lines, xenografts, mouse models)	[74,344]
	DBD-1-9R (peptide aptamer)	MM	Preclinical (cell lines)	[342]
	IS3-295 (Platinum IV)	MM	Preclinical (cell lines)	[341]
	T40214 (GQ-ODN)	T-ALL	Preclinical (mouse model)	[346]
Direct STAT5 inhibitors				
SH2 domain	AC-4-130	AML, CML	Preclinical (cell lines, primary cells, xenografts)	[81,320]
	BP-1-107, BP-1-108	AML, CML	Preclinical (cell lines)	[319]
	IST5-002	CML Ph+ ALL	Preclinical (cell lines, primary cells) Preclinical (cell lines, primary cells, PDX)	[90,321]
	Stafib-1	CML	Preclinical (cell lines)	[323]
	Stafib-2	CML	Preclinical (cell lines)	[324]
DBD	STAT5 dODN	CML	Preclinical (cell lines)	[343]
Dual STAT3/STAT5 inhibitors				
SH2 domain	OPB-31121	CML, AML, ALL, BL, MM, NHL, MM	Preclinical (cell lines, xenografts, PDX) Phase 1 (terminated)	[334]

AML (acute myeloid leukemia), ALCL (anaplastic large cell lymphoma), B-ALL/T-ALL (B- or T- acute lymphoblastic leukemia), Ph$^+$ ALL (Philadelphia chromosome-positive acute lymphoblastic leukemia), BL (Burkitt's lymphoma), CLPD-NKs (chronic lymphoproliferative disorders–natural killer cells), CML (chronic myeloid leukemia), DLBCL (diffuse large B cell lymphoma), HL (Hodgkin's lymphoma), MDS (myelodysplastic syndrome), MM (multiple myeloma), NHL (non-Hodgkin's lymphoma), T-LGL (T cell large granular lymphocytic) leukemia.

4.3.3. Inhibitors Targeting STAT3/5 mRNAs

Oligonucleotide-based inhibition of STAT3 and STAT5 has also been achieved by using antisense oligonucleotides (ASO) or mRNA knockdown (siRNA) [195]. Although siRNAs and ASOs were widely used to illuminate the oncogenic activity of STAT3 and STAT5 in hematopoietic malignant cells, few of these molecules were developed as anticancer therapeutic agents. We will discuss here two examples of ASOs and siRNAs that gave promising results in preclinical studies and clinical trials. The first example is the chemically modified ASO, AZD9150, which specifically targets STAT3 mRNA and is now in phase 1/2 clinical trials. AZD9150, without any delivery agent, was shown to inhibit STAT3 expression in primary AML/MDS leukemia stem cells and to inhibit leukemic cell growth in vitro and in vivo using PDX models of AML/MDS [347]. Inhibitory effects of AZD9150 on STAT3 expression and cell growth were also demonstrated in preclinical models of lymphomas [348]. Furthermore, AZD9150 was well tolerated and demonstrated efficacy in patients with highly treatment-refractory lymphoma [349]. The second example is a STAT3 siRNA conjugated to the TLR9 ligand CpG described above. STAT3 silencing mediated by this CpG-siRNA inhibited tumor growth in xenograft models of AML or MM [350]. In an elegant study, Hossain et al. demonstrated that the CpG-STAT3 siRNA conjugate stimulates systemic antitumor immunity and antigen-specific activation of CD8$^+$ T cells in a mouse model of AML [73]. Intravenous administration of CpG-STAT3 siRNA showed a direct immunogenic effect on leukemic cells indicating that targeted STAT3 inhibition and TLR9 triggering blocks leukemia cell growth by promoting antitumor immunity rather than direct tumor cell killing. To date, no inhibitors targeting STAT5A/STAT5B mRNAs were developed for therapeutic purposes.

5. STAT3/5 Inhibitors in Hematopoietic Cancers: Inherent Limitations and Future Challenges

Despite intensive efforts made during the last 15 years, clinically applicable STAT3/5 inhibitors still remain elusive. Toxicity, poor tumor targeting, lack of cell penetrance and rapid degradation are some examples of the problems to be considered and addressed. There are also several other reasons that could hinder the development of such inhibitors. First, STAT3 and STAT5 display multifaceted activity due to their canonical and non-canonical functions in cancer cells. The definition of constitutively active STAT3/5 only based on P-Y-STAT3 and/or P-Y-STAT5 levels as described in chapter 3 may not be broadly representative. For example, the use of decoy ODNs or SH2 domain-targeting inhibitors would not be effective in CLL cells, in which a predominant mitochondrial localization and function of P-S^{727}-STAT3 has been evidenced [102]. Furthermore, it has been shown that nuclear translocation and DNA binding of STAT3 or STAT5 can occur independently of their P-Y status. These observations indicate that SH2 domain-targeting inhibitors may not be sufficient to fully abrogate STAT3/5 oncogenic functions, which may contribute to the limited success of these compounds. Second, extensive crosstalk and alternative signaling pathways present in TKO-driven hematopoietic malignancies probably render single-agent STAT3/5 inhibition less effective. Third, adverse effects of STAT3/5 inhibitors may be associated with the physiological function of these proteins in normal tissues including hematopoietic and lymphoid tissues, resulting in STAT3- or STAT5-specific toxicity. Previous studies indicated that STAT5-deficient NK cells induce angiogenesis and promote tumor progression, highlighting the potentially detrimental effects of STAT5 inhibitors on NK cell-mediated tumor immune surveillance [37]. Moreover, STAT3 and STAT5 were also shown to activate tumor suppressor pathways, strengthening the possible adverse effects of STAT3/5 inhibitors. Most of these tumor suppressor functions were described in different reviews including one published as part of this special issue [8,32]. One of the tumor suppressor activities of STAT3 is linked to the level of STAT3β expression, which could explain why different AML patient subsets, based on the STAT3β/STAT3α mRNA ratio, are more or less sensitive to STAT3 inhibitors [34]. In such a case, STAT3 direct inhibitors would also target STAT3β which is unwanted for AML treatment. Despite these limitations, optimization of currently available STAT3/5 inhibitors and discovery of new compounds that specifically target other functional domains such as the NH2-terminal domain, domains involved in the subcellular localization of STAT3/5 or STAT3/5/Rac1 interacting domains, as well as drugs activating negative regulators of STAT3/5 such as SOCS and PIAS proteins, still have to be considered in the future. Although most of the direct inhibitors have shown a moderate efficacy in treating hematologic cancers as single agents, they act synergistically with clinically used chemotherapeutic drugs or TKIs. Importantly, targeting STAT3 and/or STAT5 signaling was demonstrated to overcome drug resistance in hematopoietic malignant cells, suggesting that combination therapy using STAT3/5 inhibitors is the most attractive approach to fight against relapsed hematopoietic malignancies. For example, STAT3 and STAT5 were both recognized as important effectors of de novo and acquired IM-resistance in CML cells [57,60,62]. In fact, activation of STAT3 was demonstrated to be an important positive autocrine-paracrine feedback loop in the therapeutic treatment of oncogene-addicted cancer cells [351]. Activation of STAT3/5 elicited by bone marrow microenvironment-derived signals was found to mediate resistance of CML cells to IM treatment, and inhibition of STAT3/5 suppressed IM-resistant CML cells in the niche. In both cases, extrinsic activation of STAT3 or STAT5 was induced by JAK kinases, and the JAK1/2 inhibitor ruxolitinib was shown to act synergistically with IM in killing resistant CML cells [62,229]. In a similar vein, the STAT5-SH2 domain inhibitor AC-4-130 increased the cytotoxicity of ruxolitinib in AML cells [81]. Combined targeting of STAT3 and STAT5 is another approach to overcome TKI resistance in CML cells. For example, the triterpenoid CDDO-Me (bardoxolone methyl) was shown to act synergistically with IM to kill resistant BCR-ABL-expressing cells [63]. In addition, blocking STAT3 and/or STAT5 with natural compounds or drugs that have been proven to be safe in humans might help to reduce the side effects of combination therapies. For example, piperlongumine (Table 2) enhanced the anti-MM effect of the proteasome inhibitor bortezomib, while pimozide increased the cytotoxic effects of TKIs in myeloid leukemias [277,298]. Besides its oncogenic activity, STAT3 is a central immune checkpoint regulator in cancer cells and tumor-associated immune cells. Oncogenic STAT3 drives PD-L1 expression in lymphoma and AML cells and may promote tumor immune evasion by inducing an immunosuppressive microenvironment [73,114]. In this context, the combination of STAT3 inhibitors with anti-checkpoint antibodies blocking PD-L1/programmed cell death 1 (PD-1) interactions might be a promising therapeutic approach for lymphomas and leukemias. Collectively, these data indicate that combination therapies using inhibitors that indirectly or directly target STAT3 and/or STAT5 might provide new therapeutic opportunities for relapsed or refractory hematopoietic malignancies.

6. Conclusions

The design and development of STAT3/5 inhibitors evolved rapidly during the last 10 years. Computer simulations and other in silico studies such as high-throughput virtual screening, cell-based assays, biophysical and biochemical approaches led to the identification of selective STAT3/5 inhibitors with potent anticancer effects. STAT3/5 inhibitors are likely to become a valuable addition to the expanding arsenal of drugs against hematopoietic cancers and solid tumors. However, there are some limitations to this relative success story that must be taken into account. Optimization of STAT3/STAT5 inhibitors by chemical modifications and drug delivery systems are both required for resolving issues linked to stability, cell permeability, and targeted delivery to tumor cells. The development of antibodies conjugated to STAT3/5 inhibitors or antibody-conjugated nanoparticles harboring STAT3/5 inhibitors might be promising but remains challenging. Identification of new inhibitors targeting the non-canonical functions of STAT3/5 is also desirable in treating some hematologic malignancies. Last but not least, the optimization of combination therapies using STAT3/5 inhibitors with molecules targeting tyrosine kinases or other key players in cancer will be required for finding the right combination that safely unlocks drug resistance in hematologic cancers.

Author Contributions: M.B.-B., M.P., H.A.N., L.J. and D.H. performed the literature search, designed the draft of the review and drew all figures and tables. M.-C.V.-M., G.P. and F.G. contributed to the editing and critical reading of the manuscript and provided the final version of the manuscript. All authors have read and agreed to the published version of the manuscript.

Funding: This research was funded by CNRS, Fondation pour la Recherche Médicale (Grant number: DCM20181039564), Ligue Contre le Cancer and University of Tours. M.B.-B. and D.H. are supported by FRM. H.A.N. is supported by the Austrian Science Fund (FWF), under the frame of ERA PerMed (I 4218-B). H.A.N. is also generously supported by a private donation from Liechtenstein.

Acknowledgments: Open Access Funding by the Austrian Science Fund (FWF).

Conflicts of Interest: The authors declare no conflict of interest.

References

1. Leonard, W.J.; O'Shea, J.J. JAKs and STATs: Biological implications. *Annu. Rev. Immunol.* **1998**, *16*, 293–322. [CrossRef]
2. Leaman, D.W.; Leung, S.; Li, X.; Stark, G.R. Regulation of STAT-dependent pathways by growth factors and cytokines. *FASEB J.* **1996**, *10*, 1578–1588. [CrossRef] [PubMed]
3. Kiu, H.; Nicholson, S.E. Biology and significance of the JAK/STAT signalling pathways. *Growth Factors* **2012**, *30*, 88–106. [CrossRef] [PubMed]
4. Yu, H.; Jove, R. The STATs of cancer—New molecular targets come of age. *Nat. Rev. Cancer* **2004**, *4*, 97–105. [CrossRef] [PubMed]
5. Liao, Z.; Nevalainen, M.T. Targeting transcription factor Stat5a/b as a therapeutic strategy for prostate cancer. *Am. J. Transl. Res.* **2011**, *3*, 133–138.
6. Benekli, M.; Baer, M.R.; Baumann, H.; Wetzler, M. Signal transducer and activator of transcription proteins in leukemias. *Blood* **2003**, *101*, 2940–2954. [CrossRef]
7. Dorritie, K.A.; McCubrey, J.A.; Johnson, D.E. STAT transcription factors in hematopoiesis and leukemogenesis: Opportunities for therapeutic intervention. *Leukemia* **2014**, *28*, 248–257. [CrossRef]
8. Igelmann, S.; Neubauer, H.A.; Ferbeyre, G. STAT3 and STAT5 activation in solid cancers. *Cancers* **2019**, *11*, 1428. [CrossRef]

9.   Vainchenker, W.; Constantinescu, S.N. JAK/STAT signaling in hematological malignancies. *Oncogene* **2013**, *32*, 2601–2613. [CrossRef]

10.   Yu, C.L.; Meyer, D.J.; Campbell, G.S.; Larner, A.C.; Carter-Su, C.; Schwartz, J.; Jove, R. Enhance DNA-binding activity of a Stat3-related protein in cells transformed by the Src oncoprotein. *Science* **1995**, *269*, 81–83. [CrossRef]

11.   Lacronique, V.; Boureux, A.; Monni, R.; Dumon, S.; Mauchauffe, M.; Mayeux, P.; Gouilleux, F.; Berger, R.; Gisselbrecht, S.; Ghysdael, J.; et al. Transforming properties of chimeric TEL-JAK proteins in Ba/F3 cells. *Blood* **2000**, *95*, 2076–2083. [CrossRef] [PubMed]

12.   Harir, N.; Boudot, C.; Friedbichler, K.; Sonneck, K.; Kondo, R.; Martin-Lanneree, S.; Kenner, L.; Kerenyi, M.; Yahiaoui, S.; Gouilleux-Gruart, V.; et al. Oncogenic Kit controls neoplastic mast cell growth through a Stat5/PI3-kinase signaling cascade. *Blood* **2008**, *112*, 2463–2473. [CrossRef] [PubMed]

13.   Kanie, T.; Abe, A.; Matsuda, T.; Kuno, Y.; Towatari, M.; Yamamoto, T.; Saito, H.; Emi, N.; Naoe, T. TEL-Syk fusion constitutively activates PI3-K/Akt, MAPK and JAK2-independent STAT5 signal pathways. *Leukemia* **2004**, *18*, 548–555. [CrossRef] [PubMed]

14.   Mizuki, M.; Fenski, R.; Halfter, H.; Matsumura, I.; Schmidt, R.; Müller, C.; Grüning, W.; Kratz-Albers, K.; Serve, S.; Steur, C.; et al. Flt3 mutations from patients with acute myeloid leukemia induce transformation of 32D cells mediated by the Ras and STAT5 pathways. *Blood* **2000**, *96*, 3907–3914. [CrossRef]

15.   Shuai, K.; Halpern, J.; ten Hoeve, J.; Rao, X.; Sawyers, C.L. Constitutive activation of STAT5 by the BCR-ABL oncogene in chronic myelogenous leukemia. *Oncogene* **1996**, *13*, 247–254.

16.   Carlesso, N.; Frank, D.A.; Griffin, J.D. Tyrosyl phosphorylation and DNA binding activity of signal transducers and activators of transcription (STAT) proteins in hematopoietic cell lines transformed by Bcr-Abl. *J. Exp. Med.* **1996**, *183*, 811–820. [CrossRef]

17.   Nieborowska-Skorska, M.; Slupianek, A.; Xue, L.; Zhang, Q.; Raghunath, P.N.; Hoser, G.; Wasik, M.A.; Morris, S.W.; Skorski, T. Role of signal transducer and activator of transcription 5 in nucleophosmin/ anaplastic lymphoma kinase-mediated malignant transformation of lymphoid cells. *Cancer Res.* **2001**, *61*, 6517–6523.

18.   Zamo, A.; Chiarle, R.; Piva, R.; Howes, J.; Fan, Y.; Chilosi, M.; Levy, D.E.; Inghirami, G. Anaplastic lymphoma kinase (ALK) activates Stat3 and protects hematopoietic cells from cell death. *Oncogene* **2002**, *21*, 1038–1047. [CrossRef]

19.   Wilbanks, A.M.; Mahajan, S.; Frank, D.A.; Druker, B.J.; Gilliland, D.G.; Carroll, M. TEL/PDGFbetaR fusion protein activates STAT1 and STAT5: A common mechanism for transformation by tyrosine kinase fusion proteins. *Exp. Hematol.* **2000**, *28*, 584–593. [CrossRef]

20.   Bromberg, J.F.; Wrzeszczynska, M.H.; Devgan, G.; Zhao, Y.; Pestell, R.G.; Albanese, C.; Darnell, J.E., Jr. Stat3 as an oncogene. *Cell* **1999**, *98*, 295–303. [CrossRef]

21.   Schwaller, J.; Parganas, E.; Wang, D.; Cain, D.; Aster, J.C.; Williams, I.R.; Lee, C.K.; Gerthner, R.; Kitamura, T.; Frantsve, J.; et al. Stat5 is essential for the myelo- and lymphoproliferative disease induced by TEL/JAK2. *Mol. Cell* **2000**, *6*, 693–704. [CrossRef]

22.   Onishi, M.; Nosaka, T.; Misawa, K.; Mui, A.L.F.; Gorman, D.; McMahon, M.; Miyajima, A.; Kitamura, T. Identification and Characterization of a Constitutively Active STAT5 Mutant That Promotes Cell Proliferation. *Mol. Cell. Biol.* **1998**, *18*, 3871–3879. [CrossRef] [PubMed]

23.   Moriggl, R.; Sexl, V.; Kenner, L.; Duntsch, C.; Stangl, K.; Gingras, S.; Hoffmeyer, A.; Bauer, A.; Piekorz, R.; Wang, D.; et al. Stat5 tetramer formation is associated with leukemogenesis. *Cancer Cell* **2005**, *7*, 87–99. [CrossRef] [PubMed]

24.   Kontro, M.; Kuusanmäki, H.; Eldfors, S.; Burmeister, T.; Andersson, E.I.; Bruserud, O.; Brümmendorf, T.H.; Edgren, H.; Gjertsen, B.T.; Itälä-Remes, M.; et al. Novel activating STAT5B mutations as putative drivers of T-cell acute lymphoblastic leukemia. *Leukemia* **2014**, *28*, 1738–1742. [CrossRef] [PubMed]

25.   Nicolae, A.; Xi, L.; Pittaluga, S.; Abdullaev, Z.; Pack, S.D.; Chen, J.; Waldmann, T.A.; Jaffe, E.S.; Raffeld, M. Frequent STAT5B mutations in γδ hepatosplenic T-cell lymphomas. *Leukemia* **2014**, *28*, 2244–2248. [CrossRef] [PubMed]

26.   Koskela, H.L.; Eldfors, S.; Ellonen, P.; van Adrichem, A.J.; Kuusanmaki, H.; Andersson, E.I.; Lagström, S.; Clemente, M.J.; Olson, T.; Jalkanen, S.E.; et al. Somatic STAT3 mutations in large granular lymphocytic leukemia. *N. Engl. J. Med.* **2012**, *366*, 1905–1913. [CrossRef]

27. Küçük, C.; Jiang, B.; Hu, X.; Zhang, W.; Chan, J.K.; Xiao, W.; Lack, N.; Alkan, C.; Williams, J.C.; Avery, K.N.; et al. Activating mutations of STAT5B and STAT3 in lymphomas derived from γδ-T or NK cells. *Nat. Commun.* **2015**, *6*, 6025. [CrossRef]

28. Bandapalli, O.R.; Schuessele, S.; Kunz, J.B.; Rausch, T.; Stütz, A.M.; Tal, N.; Geron, I.; Gershman, N.; Izraeli, S.; Eilers, J.; et al. The activating STAT5B N642H mutation is a common abnormality in pediatric T-cell acute lymphoblastic leukemia and confers a higher risk of relapse. *Haematologica* **2014**, *99*, 188–192. [CrossRef]

29. Pham, H.T.T.; Maurer, B.; Prchal-Murphy, M.; Grausenburger, R.; Grundschober, E.; Javaheri, T.; Nivarthi, H.; Boersma, A.; Kolbe, T.; Elabd, M.; et al. STAT5BN642H is a driver mutation for T cell neoplasia. *J. Clin. Investig.* **2018**, *128*, 387–401. [CrossRef]

30. de Araujo, E.D.; Erdogan, F.; Neubauer, H.A.; Meneksedag-Erol, D.; Manaswiyoungkul, P.; Eram, M.S.; Seo, H.S.; Qadree, A.K.; Israelian, J.; Orlova, A.; et al. Structural and functional consequences of the STAT5BN642H driver mutation. *Nat. Commun.* **2019**, *10*, 2517. [CrossRef]

31. Hosui, A.; Kimura, A.; Yamaji, D.; Zhu, B.M.; Na, R.; Hennighausen, L. Loss of STAT5 causes liver fibrosis and cancer development through increased TGF-beta and STAT3 activation. *J. Exp. Med.* **2009**, *206*, 819–831. [CrossRef] [PubMed]

32. Ferbeyre, G.; Moriggl, R. The role of Stat5 transcription factors as tumor suppressors or oncogenes. *Biochim. Biophys. Acta* **2011**, *1815*, 104–114. [CrossRef] [PubMed]

33. Zhang, H.F.; Lai, R. STAT3 in Cancer—Friend or Foe. *Cancers* **2014**, *6*, 1408–1440. [CrossRef] [PubMed]

34. Aigner, P.; Mizutani, T.; Horvath, J.; Eder, T.; Heber, S.; Lind, K.; Just, V.; Moll, H.P.; Yeroslaviz, A.; Fischer, M.J.M.; et al. STAT3β is a tumor suppressor in acute myeloid leukemia. *Blood Adv.* **2019**, *3*, 1989–2002. [CrossRef] [PubMed]

35. Zhang, Q.; Wang, H.Y.; Liu, X.; Wasik, M.A. STAT5A is epigenetically silenced by the tyrosine kinase NPM1-ALK and acts as a tumor suppressor by reciprocally inhibiting NPM1-ALK expression. *Nat. Med.* **2007**, *13*, 1341–1348. [CrossRef] [PubMed]

36. Yu, H.; Kortylewski, M.; Pardoll, D. Crosstalk between cancer and immune cells: Role of STAT3 in the tumour microenvironment. *Nat. Rev. Immunol.* **2007**, *7*, 41–51. [CrossRef]

37. Gotthardt, D.; Putz, E.M.; Grundschober, E.; Prchal-Murphy, M.; Straka, E.; Kudweis, P.; Heller, G.; Bago-Horvath, Z.; Witalisz-Siepracka, A.; Cumaraswamy, A.A.; et al. STAT5 Is a Key Regulator in NK Cells and Acts as a Molecular Switch from Tumor Surveillance to Tumor Promotion. *Cancer Discov.* **2016**, *6*, 414–429. [CrossRef]

38. Gouilleux-Gruart, V.; Gouilleux, F.; Desaint, C.; Claisse, J.F.; Capiod, J.C.; Delobel, J.; Weber-Nordt, R.; Dusanter-Fourt, I.; Dreyfus, F.; Groner, B.; et al. STAT related transcription factors are constitutively activated in peripheral blood cells from acute leukemia patients. *Blood* **1996**, *87*, 1692–1697. [CrossRef]

39. Spivak, J.L. The chronic myeloproliferative disorders: Clonality and clinical heterogeneity. *Semin. Hematol.* **2004**, *41*, 1–5. [CrossRef]

40. Van Etten, R.A.; Koschmieder, S.; Delhommeau, F.; Perrotti, D.; Holyoake, T.; Pardanani, A.; Mesa, R.; Green, T.; Ibrahim, A.R.; Mughal, T.; et al. The Ph-positive and Ph-negative myeloproliferative neoplasms: Some topical pre-clinical and clinical issues. *Haematologica* **2011**, *96*, 590–601. [CrossRef]

41. Vainchenker, W.; Kralovics, R. Genetic basis and molecular pathophysiology of classical myeloproliferative neoplasms. *Blood* **2017**, *129*, 667–679. [CrossRef] [PubMed]

42. James, C.; Ugo, V.; Le Couedic, J.P.; Staerk, J.; Delhommeau, F.; Lacout, C.; Garçon, L.; Raslova, H.; Berger, R.; Bennaceur-Griscelli, A.; et al. A unique clonal JAK2 mutation leading to constitutive signalling causes polycythaemia vera. *Nature* **2005**, *434*, 1144–1148. [CrossRef] [PubMed]

43. Kralovics, R.; Passamonti, F.; Busers, A.S.; Teo, S.S.; Tiedt, R.; Passweg, J.R.; Tichelli, A.; Cazzola, M.; Skoda, R.C. A gain-of-function mutation of JAK2 in myeloproliferative disorders. *N. Engl. J. Med.* **2005**, *352*, 1779–1790. [CrossRef] [PubMed]

44. Valent, P.; Horny, H.P.; Escribano, L.; Longley, B.J.; Li, C.Y.; Schwartz, L.B.; Marone, G.; Nuñez, R.; Akin, C.; Sotlar, K.; et al. Diagnostic criteria and classification of mastocytosis: A consensus proposal. *Leuk. Res.* **2001**, *25*, 603–625. [CrossRef]

45. Lim, K.H.; Pardanani, A.; Tefferi, A. KIT and mastocytosis. *Acta Haematol.* **2008**, *119*, 194–198. [CrossRef]

46. Goemans, B.F.; Zwaan, C.M.; Miller, M.; Zimmermann, M.; Harlow, A.; Meshinchi, S.; Loonen, A.H.; Hählen, K.; Reinhardt, D.; Creutzig, U.; et al. Mutations in KIT and RAS are frequent events in pediatric core-binding factor acute myeloid leukemia. *Leukemia* **2005**, *19*, 1536–1542. [CrossRef]

47. Fritsche-Polanz, R.; Fritz, M.; Huber, A.; Sotlar, K.; Sperr, W.R.; Mannhalter, C.; Födinger, M.; Valent, P. High frequency of concomitant mastocytosis in patients with acute myeloid leukemia exhibiting the transforming KIT mutation D816V. *Mol. Oncol.* **2010**, *4*, 335–346. [CrossRef]

48. Coppo, P.; Flamant, S.; De Mas, V.; Jarrier, P.; Guillier, M.; Bonnet, M.L.; Lacout, C.; Guilhot, F.; Vainchenker, W.; Turhan, A.G. BCR-ABL activates STAT3 via JAK and MEK pathways in human cells. *Br. J. Haematol.* **2006**, *134*, 171–179. [CrossRef]

49. Hoelbl, A.; Schuster, C.; Kovacic, B.; Zhu, B.; Wickre, M.; Hoelzl, M.A.; Fajmann, S.; Grebien, F.; Warsch, W.; Stengl, G.; et al. Stat5 is indispensable for the maintenance of Bcr/abl-positive leukemia. *EMBO Mol. Med.* **2010**, *2*, 98–110. [CrossRef]

50. Chung, Y.J.; Park, B.B.; Kang, Y.J.; Kim, T.M.; Eaves, C.J.; Oh, I.H. Unique effects of Stat3 on the early phase of hematopoietic stem cell regeneration. *Blood* **2006**, *108*, 1208–1215. [CrossRef]

51. Schepers, H.; van Gosliga, D.; Wierenga, A.T.; Eggen, B.J.; Schuringa, J.J.; Vellenga, E. STAT5 is required for long-term maintenance of normal and leukemic human stem/progenitor cells. *Blood* **2007**, *110*, 2880–2888. [CrossRef] [PubMed]

52. Kato, Y.; Iwama, A.; Tadokoro, Y.; Shimoda, K.; Minoguchi, M.; Akira, S.; Tanaka, M.; Miyajima, A.; Kitamura, T.; Nakauchi, H. Selective activation of STAT5 unveils its role in stem cell self-renewal in normal and leukemic hematopoiesis. *J. Exp. Med.* **2005**, *202*, 169–179. [CrossRef] [PubMed]

53. Ghanem, S.; Friedbichler, K.; Boudot, C.; Bourgeais, J.; Gouilleux-Gruart, V.; Régnier, A.; Herault, O.; Moriggl, R.; Gouilleux, F. STAT5A/5B-specific expansion and transformation of hematopoietic stem cells. *Blood Cancer J.* **2017**, *7*, 514. [CrossRef] [PubMed]

54. Kollmann, S.; Grundschober, E.; Maurer, B.; Warsch, W.; Grausenburger, R.; Edlinger, L.; Huuhtanen, J.; Lagger, S.; Hennighausen, L.; Valent, P.; et al. Twins with different personalities: STAT5B-but not STAT5A-has a key role in BCR/ABL-induced leukemia. *Leukemia* **2019**, *33*, 1583–1597. [CrossRef] [PubMed]

55. Chomel, J.C.; Bonnet, M.L.; Sorel, N.; Sloma, I.; Bennaceur-Griscelli, A.; Rea, D.; Legros, L.; Marfaing-Koka, A.; Bourhis, J.H.; Ame, S.; et al. Leukemic stem cell persistence in chronic myeloid leukemia patients in deep molecular response induced by tyrosine kinase inhibitors and the impact of therapy discontinuation. *Oncotarget* **2016**, *7*, 35293–35301. [CrossRef] [PubMed]

56. Gambacorti-Passerini, C.B.; Gunby, R.H.; Piazza, R.; Galietta, A.; Rostagno, R.; Scapozza, L. Molecular mechanisms of resistance to imatinib in Philadelphia-chromosome-positive leukaemias. *Lancet Oncol.* **2003**, *4*, 75–85. [CrossRef]

57. Warsch, W.; Kollmann, K.; Eckelhart, E.; Fajmann, S.; Cerny-Reiterer, S.; Hölbl, A.; Gleixner, K.V.; Dworzak, M.; Mayerhofer, M.; Hoermann, G.; et al. High STAT5 levels mediate imatinib resistance and indicate disease progression in chronic myeloid leukemia. *Blood* **2011**, *117*, 3409–3420. [CrossRef]

58. Warsch, W.; Grundschober, E.; Berger, A.; Gille, L.; Cerny-Reiterer, S.; Tigan, A.S.; Hoelbl-Kovacic, A.; Valent, P.; Moriggl, R.; Sexl, V. STAT5 triggers BCR-ABL1 mutation by mediating ROS production in chronic myeloid leukemia. *Oncotarget* **2012**, *3*, 1669–1687. [CrossRef]

59. Bourgeais, J.; Ishac, N.; Medrzycki, M.; Brachet-Botineau, M.; Desbourdes, L.; Gouilleux-Gruart, V.; Pecnard, E.; Rouleux-Bonnin, F.; Gyan, E.; Domenech, J.; et al. Oncogenic STAT5 signaling promotes oxidative stress in chronic myeloid leukemia cells by repressing antioxidant defenses. *Oncotarget* **2017**, *8*, 41876–41889. [CrossRef]

60. Prost, S.; Relouzat, F.; Spentchian, M.; Ouzegdouh, Y.; Saliba, J.; Massonnet, G.; Beressi, J.P.; Verhoeyen, E.; Raggueneau, V.; Maneglier, B.; et al. Erosion of the chronic myeloid leukemia stem cell pool by PPARγ agonists. *Nature* **2015**, *525*, 380–383. [CrossRef]

61. Casetti, L.; Martin-Lannerée, S.; Najjar, I.; Plo, I.; Augé, S.; Roy, L.; Chomel, J.C.; Lauret, E.; Turhan, A.G.; Dusanter-Fourt, I. Differential contributions of STAT5A and STAT5B to stress protection and tyrosine kinase inhibitor resistance of chronic myeloid leukemia stem/progenitor cells. *Cancer Res.* **2013**, *73*, 2052–2058. [CrossRef] [PubMed]

62. Bewry, N.N.; Nair, R.R.; Emmons, M.; Boulware, D.; Pinilla-Ibarz, J.; Hazlehurst, L.A. Stat3 contributes to resistance toward BCR-ABL inhibitors in a bone marrow microenvironment model of drug resistance. *Mol. Cancer Ther.* **2008**, *7*, 3169–3175. [CrossRef] [PubMed]

63. Gleixner, K.V.; Schneeweiss, M.; Eisenwort, G.; Berger, D.; Herrmann, H.; Blatt, K.; Greiner, G.; Byrgazov, K.; Hoermann, G.; Konopleva, M.; et al. Combined targeting of STAT3 and STAT5: A novel approach to overcome drug resistance in chronic myeloid leukemia. *Haematologica* **2017**, *102*, 1519–1529. [CrossRef] [PubMed]

64. Walz, C.; Ahmed, W.; Lazarides, K.; Betancur, M.; Patel, N.; Hennighausen, L.; Zaleskas, V.M.; Van Etten, R.A. Essential role for Stat5a/b in myeloproliferative neoplasms induced by BCR-ABL1 and JAK2(V617F) in mice. *Blood* **2012**, *119*, 3550–3560. [CrossRef]

65. Yan, D.; Hutchison, R.E.; Mohi, G. Critical requirement for Stat5 in a mouse model of polycythemia vera. *Blood* **2012**, *119*, 3539–3549. [CrossRef]

66. Yan, D.; Jobe, F.; Hutchison, R.E.; Mohi, G. Deletion of Stat3 enhances myeloid cell expansion and increases the severity of myeloproliferative neoplasms in Jak2V617F knock-in mice. *Leukemia* **2015**, *29*, 2050–2061. [CrossRef]

67. Grisouard, J.; Shimizu, T.; Duek, A.; Kubovcakova, L.; Hao-Shen, H.; Dirnhofer, S.; Skoda, R.C. Deletion of Stat3 in hematopoietic cells enhances thrombocytosis and shortens survival in a JAK2-V617F mouse model of MPN. *Blood* **2015**, *125*, 2131–2140. [CrossRef]

68. Chaix, A.; Lopez, S.; Voisset, E.; Gros, L.; Dubreuil, P.; De Sepulveda, P. Mechanisms of STAT protein activation by oncogenic KIT mutants in neoplastic mast cells. *J. Biol. Chem.* **2011**, *286*, 5956–5966. [CrossRef]

69. Gilliland, D.G. Molecular genetics of human leukemias: New insights into therapy. *Semin. Hematol.* **2002**, *39*, 6–11. [CrossRef]

70. Martignoles, J.A.; Delhommeau, F.; Hirsch, P. Genetic Hierarchy of Acute Myeloid Leukemia: From Clonal Hematopoiesis to Molecular Residual Disease. *Int. J. Mol. Sci.* **2018**, *19*, 3850. [CrossRef]

71. Benekli, M.; Xia, Z.; Donohue, K.A.; Ford, L.A.; Pixley, L.A.; Baer, M.R.; Baumann, H.; Wetzler, M. Constitutive activity of signal transducer and activator of transcription 3 protein in acute myeloid leukemia blasts is associated with short disease-free survival. *Blood* **2002**, *99*, 252–257. [CrossRef] [PubMed]

72. Larrue, C.; Heydt, Q.; Saland, E.; Boutzen, H.; Kaoma, T.; Sarry, J.E.; Joffre, C.; Récher, C. Oncogenic KIT mutations induce STAT3-dependent autophagy to support cell proliferation in acute myeloid leukemia. *Oncogenesis* **2019**, *8*, 39. [CrossRef] [PubMed]

73. Hossain, D.M.; Dos Santos, C.; Zhang, Q.; Kozlowska, A.; Liu, H.; Gao, C.; Moreira, D.; Swiderski, P.; Jozwiak, A.; Kline, J.; et al. Leukemia cell-targeted STAT3 silencing and TLR9 triggering generate systemic antitumor immunity. *Blood* **2014**, *123*, 15–25. [CrossRef] [PubMed]

74. Zhang, Q.; Hossain, D.M.; Duttagupta, P.; Moreira, D.; Zhao, X.; Won, H.; Buettner, R.; Nechaev, S.; Majka, M.; Zhang, B.; et al. Serum-resistant CpG-STAT3 decoy for targeting survival and immune checkpoint signaling in acute myeloid leukemia. *Blood* **2016**, *127*, 1687–1700. [CrossRef]

75. Kiyoi, H.; Naoe, T. FLT3 in human hematologic malignancies. *Leuk. Lymphoma* **2002**, *43*, 1541–1547. [CrossRef]

76. Spiekermann, K.; Bagrintseva, K.; Schwab, R.; Schmieja, K.; Hiddemann, W. Overexpression and constitutive activation of FLT3 induces STAT5 activation in primary acute myeloid leukemia blast cells. *Clin. Cancer Res.* **2003**, *9*, 2140–2150.

77. Obermann, E.C.; Arber, C.; Jotterand, M.; Tichelli, A.; Hirschmann, P.; Tzankov, A. Expression of pSTAT5 predicts FLT3 internal tandem duplications in acute myeloid leukemia. *Ann. Hematol.* **2010**, *89*, 663–669. [CrossRef]

78. Rocnik, J.L.; Okabe, R.; Yu, J.C.; Lee, B.H.; Giese, N.; Schenkein, D.P.; Gilliland, D.G. Roles of tyrosine 589 and 591 in STAT5 activation and transformation mediated by FLT3-ITD. *Blood* **2006**, *108*, 339–345. [CrossRef]

79. Nelson, E.A.; Walker, S.R.; Xiang, M.; Weisberg, E.; Bar-Natan, M.; Barrett, R.; Liu, S.; Kharbanda, S.; Christie, A.L.; Nicolais, M.; et al. The STAT5 Inhibitor Pimozide Displays Efficacy in Models of Acute Myelogenous Leukemia Driven by FLT3 Mutations. *Genes Cancer* **2012**, *3*, 503–511. [CrossRef]

80. Müller, T.A.; Grundler, R.; Istvanffy, R.; Rudelius, M.; Hennighausen, L.; Illert, A.L.; Duyster, J. Lineage-specific STAT5 target gene activation in hematopoietic progenitor cells predicts the FLT3(+) -mediated leukemic phenotype. *Leukemia* **2016**, *30*, 1725–1733. [CrossRef]

81. Wingelhofer, B.; Maurer, B.; Heyes, E.C.; Cumaraswamy, A.A.; Berger-Becvar, A.; de Araujo, E.D.; Orlova, A.; Freund, P.; Ruge, F.; Park, J.; et al. Pharmacologic inhibition of STAT5 in acute myeloid leukemia. *Leukemia* **2018**, *32*, 1135–1146. [CrossRef] [PubMed]

82. Sallmyr, A.; Fan, J.; Datta, K.; Kim, K.T.; Grosu, D.; Shapiro, P.; Small, D.; Rassool, F. Internal tandem duplication of FLT3 (FLT3/ITD) induces increased ROS production, DNA damage, and misrepair: Implications for poor prognosis in AML. *Blood* **2008**, *111*, 3173–3182. [CrossRef] [PubMed]

83. Woolley, J.F.; Naughton, R.; Stanicka, J.; Gough, D.R.; Bhatt, L.; Dickinson, B.C.; Chang, C.J.; Cotter, T.G. H_2O_2 production downstream of FLT3 is mediated by p22phox in the endoplasmic reticulum and is required for STAT5 signalling. *PLoS ONE* **2012**, *7*, e34050. [CrossRef] [PubMed]

84. Dumas, P.Y.; Naudin, C.; Martin-Lannerée, S.; Izac, B.; Casetti, L.; Mansier, O.; Rousseau, B.; Artus, A.; Dufossée, M.; Giese, A.; et al. Hematopoietic niche drives FLT3-ITD acute myeloid leukemia resistance to quizartinib via STAT5-and hypoxia-dependent upregulation of AXL. *Haematologica* **2019**, *104*, 2017–2027. [CrossRef] [PubMed]

85. Arnould, C.; Philippe, C.; Bourdon, V.; Gregoire, M.J.; Berger, R.; Jonveaux, P. The signal transducer and activator of transcription STAT5b gene is a new partner of retinoic acid receptor alpha in acute promyelocytic-like leukaemia. *Hum. Mol. Genet.* **1999**, *8*, 1741–1749. [CrossRef] [PubMed]

86. Dong, S.; Tweardy, D.J. Interactions of STAT5b-RARalpha, a novel acute promyelocytic leukemia fusion protein, with retinoic acid receptor and STAT3 signaling pathways. *Blood* **2002**, *99*, 2637–2646. [CrossRef] [PubMed]

87. Pui, C.H.; Gajjar, A.J.; Kane, J.R.; Qaddoumi, I.A.; Pappo, A.S. Challenging issues in pediatric oncology. *Nat. Rev. Clin. Oncol.* **2011**, *8*, 540–549. [CrossRef]

88. Hoelbl, A.; Kovacic, B.; Kerenyi, M.A.; Simma, O.; Warsch, W.; Cui, Y.; Beug, H.; Hennighausen, L.; Moriggl, R.; Sexl, V. Clarifying the role of Stat5 in lymphoid development and Abelson induced transformation. *Blood* **2006**, *107*, 4898–4906. [CrossRef]

89. Joliot, V.; Cormier, F.; Medyouf, H.; Alcalde, H.; Ghysdael, J. Constitutive STAT5 activation specifically cooperates with the loss of p53 function in B-cell lymphomagenesis. *Oncogene* **2006**, *25*, 4573–4584. [CrossRef]

90. Minieri, V.; De Dominici, M.; Porazzi, P.; Mariani, S.A.; Spinelli, O.; Rambaldi, A.; Peterson, L.F.; Porcu, P.; Nevalainen, M.T.; Calabretta, B. Targeting STAT5 or STAT5-Regulated Pathways Suppresses Leukemogenesis of Ph+ Acute Lymphoblastic Leukemia. *Cancer Res.* **2018**, *78*, 5793–5807. [CrossRef]

91. Katerndahl, C.D.S.; Heltemes-Harris, L.M.; Willette, M.J.L.; Henzler, C.M.; Frietze, S.; Yang, R.; Schjerven, H.; Silverstein, K.A.T.; Ramsey, L.B.; Hubbard, G.; et al. Antagonism of B cell enhancer networks by STAT5 drives leukemia and poor patient survival. *Nat. Immunol.* **2017**, *18*, 694–704. [CrossRef] [PubMed]

92. Heltemes-Harris, L.M.; Willette, M.J.; Ramsey, L.B.; Qiu, Y.H.; Neeley, E.S.; Zhang, N.; Thomas, D.A.; Koeuth, T.; Baechler, E.C.; Kornblau, S.M.; et al. Ebf1 or Pax5 haploinsufficiency synergizes with STAT5 activation to initiate acute lymphoblastic leukemia. *J. Exp. Med.* **2011**, *208*, 1135–1149. [CrossRef] [PubMed]

93. Wang, Z.; Medrzycki, M.; Bunting, S.T.; Bunting, K.D. Stat5-deficient hematopoiesis is permissive for Myc-induced B-cell leukemogenesis. *Oncotarget* **2015**, *6*, 28961–28972. [CrossRef] [PubMed]

94. Belton, A.; Xian, L.; Huso, T.; Koo, M.; Luo, L.Z.; Turkson, J.; Page, B.D.; Gunning, P.T.; Liu, G.; Huso, D.L.; et al. STAT3 inhibitor has potent antitumor activity in B-lineage acute lymphoblastic leukemia cells overexpressing the high mobility group A1 (HMGA1)-STAT3 pathway. *Leuk. Lymphoma* **2016**, *57*, 2681–2684. [CrossRef]

95. Fasan, A.; Kern, W.; Grossmann, V.; Haferlach, C.; Haferlach, T.; Schnittger, S. STAT3 mutations are highly specific for large granular lymphocytic leukemia. *Leukemia* **2013**, *27*, 1598–1600. [CrossRef]

96. Rajala, H.L.; Porkka, K.; Maciejewski, J.P.; Loughran, T.P., Jr.; Mustjoki, S. Uncovering the pathogenesis of large granular lymphocytic leukemia-novel STAT3 and STAT5b mutations. *Ann. Med.* **2014**, *46*, 114–122. [CrossRef]

97. Rajala, H.L.; Eldfors, S.; Kuusanmäki, H.; van Adrichem, A.J.; Olson, T.; Lagström, S.; Andersson, E.I.; Jerez, A.; Clemente, M.J.; Yan, Y.; et al. Discovery of somatic STAT5b mutations in large granular lymphocytic leukemia. *Blood* **2013**, *121*, 4541–4550. [CrossRef]

98. Dutta, A.; Yan, D.; Hutchison, R.E.; Mohi, G. STAT3 mutations are not sufficient to induce large granular lymphocytic leukaemia in mice. *Br. J. Haematol.* **2018**, *180*, 911–915. [CrossRef]

99. Klein, K.; Witalisz-Siepracka, A.; Maurer, B.; Prinz, D.; Heller, G.; Leidenfrost, N.; Prchal-Murphy, M.; Suske, T.; Moriggl, R.; Sexl, V. STAT5BN642H drives transformation of NKT cells: A novel mouse model for CD56+ T-LGL leukemia. *Leukemia* **2019**, *33*, 2336–2340. [CrossRef]

100. Chiorazzi, N.; Rai, K.R.; Ferrarini, M. Chronic lymphocytic leukemia. *N. Engl. J. Med.* **2005**, *352*, 804–815. [CrossRef]

101. Hazan-Halevy, I.; Harris, D.; Liu, Z.; Liu, J.; Li, P.; Chen, X.; Shanker, S.; Ferrajoli, A.; Keating, M.J.; Estrov, Z. STAT3 is constitutively phosphorylated on serine 727 residues, binds DNA, and activates transcription in CLL cells. *Blood* **2010**, *115*, 2852–2863. [CrossRef] [PubMed]

102. Capron, C.; Jondeau, K.; Casetti, L.; Jalbert, V.; Costa, C.; Verhoeyen, E.; Massé, J.M.; Coppo, P.; Béné, M.C.; Bourdoncle, P.; et al. Viability and stress protection of chronic lymphoid leukemia cells involves overactivation of mitochondrial phosphoSTAT3Ser727. *Cell Death Dis.* **2014**, *5*, e1451. [CrossRef] [PubMed]

103. de Totero, D.; Meazza, R.; Capaia, M.; Fabbi, M.; Azzarone, B.; Balleari, E.; Gobbi, M.; Cutrona, G.; Ferrarini, M.; Ferrini, S. The opposite effects of IL-15 and IL-21 on CLL B cells correlate with differential activation of the JAK/STAT and ERK1/2 pathways. *Blood* **2008**, *111*, 517–524. [CrossRef] [PubMed]

104. Kube, D.; Holtick, U.; Vockerodt, M.; Ahmadi, T.; Haier, B.; Behrmann, I.; Heinrich, P.C.; Diehl, V.; Tesch, H. STAT3 is constitutively activated in Hodgkin cell lines. *Blood* **2001**, *98*, 762–770. [CrossRef]

105. Holtick, U.; Vockerodt, M.; Pinkert, D.; Schoof, N.; Stürzenhofecker, B.; Kussebi, N.; Lauber, K.; Wesselborg, S.; Löffler, D.; Horn, F.; et al. STAT3 is essential for Hodgkin lymphoma cell proliferation and is a target of tyrphostin AG17 which confers sensitization for apoptosis. *Leukemia* **2005**, *19*, 936–944. [CrossRef]

106. Scheeren, F.A.; Diehl, S.A.; Smit, L.A.; Beaumont, T.; Naspetti, M.; Bende, R.J.; Blom, B.; Karube, K.; Ohshima, K.; van Noesel, C.J.; et al. IL-21 is expressed in Hodgkin lymphoma and activates STAT5: Evidence that activated STAT5 is required for Hodgkin lymphomagenesis. *Blood* **2008**, *111*, 4706–4715. [CrossRef]

107. Ding, B.B.; Yu, J.J.; Yu, R.Y.; Mendez, L.M.; Shaknovich, R.; Zhang, Y.; Cattoretti, G.; Ye, B.H. Constitutively activated STAT3 promotes cell proliferation and survival in the activated B-cell subtype of diffuse large B-cell lymphomas. *Blood* **2008**, *111*, 1515–1523. [CrossRef]

108. Lam, L.T.; Wright, G.; Davis, R.E.; Lenz, G.; Farinha, P.; Dang, L.; Chan, J.W.; Rosenwald, A.; Gascoyne, R.D.; Staudt, L.M. Cooperative signaling through the signal transducer and activator of transcription 3 and nuclear factor-{kappa}B pathways in subtypes of diffuse large B-cell lymphoma. *Blood* **2008**, *111*, 3701–3713. [CrossRef]

109. Pan, Y.R.; Chen, C.C.; Chan, Y.T.; Wang, H.J.; Chien, F.T.; Chen, Y.L.; Liu, J.L.; Yang, M.H. STAT3-coordinated migration facilitates the dissemination of diffuse large B-cell lymphomas. *Nat. Commun.* **2018**, *9*, 3696. [CrossRef]

110. Skarbnik, A.P.; Burki, M.; Pro, B. Peripheral T-cell lymphomas: A review of current approaches and hopes for the future. *Front Oncol.* **2013**, *3*, 138. [CrossRef]

111. Netchiporouk, E.; Litvinov, I.V.; Moreau, L.; Gilbert, M.; Sasseville, D.; Duvic, M. Deregulation in STAT signaling is important for cutaneous T-cell lymphoma (CTCL) pathogenesis and cancer progression. *Cell Cycle* **2014**, *13*, 3331–3335. [CrossRef] [PubMed]

112. Chiarle, R.; Simmons, W.J.; Cai, H.; Dhall, G.; Zamo, A.; Raz, R.; Karras, J.G.; Levy, D.E.; Inghirami, G. Stat3 is required for ALK-mediated lymphomagenesis and provides a possible therapeutic target. *Nat. Med.* **2005**, *11*, 623–629. [CrossRef] [PubMed]

113. Waldmann, T.A.; Chen, J. Disorders of the JAK/STAT Pathway in T Cell Lymphoma Pathogenesis: Implications for Immunotherapy. *Annu. Rev. Immunol.* **2017**, *35*, 533–550. [CrossRef] [PubMed]

114. Song, T.L.; Nairismägi, M.L.; Laurensia, Y.; Lim, J.Q.; Tan, J.; Li, Z.M.; Pang, W.L.; Kizhakeyil, A.; Wijaya, G.C.; Huang, D.C.; et al. Oncogenic activation of the STAT3 pathway drives PD-L1 expression in natural killer/T-cell lymphoma. *Blood* **2018**, *132*, 1146–1158. [CrossRef] [PubMed]

115. Kortylewski, M.; Yu, H. Role of Stat3 in suppressing anti-tumor immunity. *Curr. Opin. Immunol.* **2008**, *20*, 228–233. [CrossRef]

116. Roodman, G.D. Pathogenesis of myeloma bone disease. *Leukemia* **2009**, *23*, 435–441. [CrossRef]

117. Catlett-Falcone, R.; Landowski, T.H.; Oshiro, M.M.; Turkson, J.; Levitzki, A.; Savino, R.; Ciliberto, G.; Moscinski, L.; Fernández-Luna, J.L.; Núñez, G.; et al. Constitutive Activation of Stat3 Signaling Confers Resistance to Apoptosis in Human U266 Myeloma Cells. *Immunity* **1999**, *10*, 105–115. [CrossRef]

118. Bharti, A.C.; Shishodia, S.; Reuben, J.M.; Weber, D.; Alexanian, R.; Raj-Vadhan, S.; Estrov, Z.; Talpaz, M.; Aggarwal, B.B. Nuclear factor-kappaB and STAT3 are constitutively active in CD138+ cells derived from multiple myeloma patients, and suppression of these transcription factors leads to apoptosis. *Blood* **2004**, *103*, 3175–3184. [CrossRef]

119. Jung, S.H.; Ahn, S.Y.; Choi, H.W.; Shin, M.G.; Lee, S.S.; Yang, D.H.; Ahn, J.S.; Kim, Y.K.; Kim, H.J.; Lee, J.J. STAT3 expression is associated with poor survival in non-elderly adult patients with newly diagnosed multiple myeloma. *Blood Res.* **2017**, *52*, 293–299. [CrossRef]

120. Chong, P.S.Y.; Chng, W.J.; de Mel, S. STAT3: A Promising Therapeutic Target in Multiple Myeloma. *Cancers* **2019**, *11*, 731. [CrossRef]

121. Wang, J.; De Veirman, K.; Faict, S.; Frassanito Maria, A.; Ribatti, D.; Vacca, A.; Menu, E. Multiple myeloma exosomes establish a favourable bone marrow microenvironment with enhanced angiogenesis and immunosuppression. *J. Pathol.* **2016**, *239*, 162–173. [CrossRef] [PubMed]

122. Beldi-Ferchiou, A.; Skouri, N.; Ben Ali, C.; Safra, I.; Abdelkefi, A.; Ladeb, S.; Mrad, K.; Ben Othman, T.; Ben Ahmed, M. Abnormal repression of SHP-1, SHP-2 and SOCS-1 transcription sustains the activation of the JAK/STAT3 pathway and the progression of the disease in multiple myeloma. *PLoS ONE* **2017**, *12*, e0174835. [CrossRef] [PubMed]

123. Chim, C.S.; Fung, T.K.; Cheung, W.C.; Liang, R.; Kwong, Y.L. SOCS1 and SHP1 hypermethylation in multiple myeloma: Implications for epigenetic activation of the Jak/STAT pathway. *Blood* **2004**, *103*, 4630–4635. [CrossRef] [PubMed]

124. Löffler, D.; Brocke-Heidrich, K.; Pfeifer, G.; Stocsits, C.; Hackermüller, J.; Kretzschmar, A.K.; Burger, R.; Gramatzki, M.; Blumert, C.; Bauer, K.; et al. Interleukin-6 dependent survival of multiple myeloma cells involves the Stat3-mediated induction of microRNA-21 through a highly conserved enhancer. *Blood* **2007**, *110*, 1330–1333. [CrossRef] [PubMed]

125. Binder, S.; Hösler, N.; Riedel, D.; Zipfel, I.; Buschmann, T.; Kämpf, C.; Reiche, K.; Burger, R.; Gramatzki, M.; Hackermüller, J.; et al. STAT3-induced long noncoding RNAs in multiple myeloma cells display different properties in cancer. *Sci. Rep.* **2017**, *7*, 7976. [CrossRef] [PubMed]

126. Brown, R.; Yang, S.; Weatherburn, C.; Gibson, J.; Ho, P.J.; Suen, H.; Hart, D.; Joshua, D. Phospho-flow detection of constitutive and cytokine-induced pSTAT3/5, pAKT and pERK expression highlights novel prognostic biomarkers for patients with multiple myeloma. *Leukemia* **2015**, *29*, 483–490. [CrossRef]

127. Kloth, M.T.; Catling, A.D.; Silva, C.M. Novel activation of STAT5b in response to epidermal growth factor. *J. Biol. Chem.* **2002**, *277*, 8693–8701. [CrossRef]

128. Schaller-Schönitz, M.; Barzan, D.; Williamson, A.J.; Griffiths, J.R.; Dallmann, I.; Battmer, K.; Ganser, A.; Whetton, A.D.; Scherr, M.; Eder, M. BCR-ABL affects STAT5A and STAT5B differentially. *PLoS ONE* **2014**, *9*, e97243. [CrossRef]

129. Reich, N.C. STATs get their move on. *JAK-STAT* **2013**, *2*, e27080. [CrossRef]

130. Kawashima, T.; Bao, Y.C.; Nomura, Y.; Moon, Y.; Tonozuka, Y.; Minoshima, Y.; Hatori, T.; Tsuchiya, A.; Kiyono, M.; Nosaka, T.; et al. Rac1 and a GTPase-activating protein, MgcRacGAP, are required for nuclear translocation of STAT transcription factors. *J. Cell Biol.* **2006**, *175*, 937–946. [CrossRef]

131. Kawashima, T.; Bao, Y.C.; Minoshima, Y.; Nomura, Y.; Hatori, T.; Hori, T.; Fukagawa, T.; Fukada, T.; Takahashi, N.; Nosaka, T.; et al. A Rac GTPase-activating protein, MgcRacGAP, is a nuclear localizing signal-containing nuclear chaperone in the activation of STAT transcription factors. *Mol. Cell. Biol.* **2009**, *29*, 1796–1813. [CrossRef] [PubMed]

132. Tonozuka, Y.; Minoshima, Y.; Bao, Y.C.; Moon, Y.; Tsubono, Y.; Hatori, T.; Nakajima, H.; Nosaka, T.; Kawashima, T.; Kitamura, T. A GTPase-activating protein binds STAT3 and is required for IL-6-induced STAT3 activation and for differentiation of a leukemic cell line. *Blood* **2004**, *104*, 3550–3557. [CrossRef] [PubMed]

133. Berger, A.; Hoelbl-Kovacic, A.; Bourgeais, J.; Hoefling, L.; Warsch, W.; Grundschober, E.; Uras, I.Z.; Menzl, I.; Putz, E.M.; Hoermann, G.; et al. Pak-dependent stat5 serine phosphorylation is required for bcr-abl-induced leukemogenesis. *Leukemia* **2014**, *28*, 629–641. [CrossRef] [PubMed]

134. Chatterjee, A.; Ghosh, J.; Ramdas, B.; Mali, R.S.; Martin, H.; Kobayashi, M.; Vemula, S.; Canela, V.H.; Waskow, E.R.; Visconte, V.; et al. Regulation of Stat5 by FAK and PAK1 in Oncogenic FLT3- and KIT-Driven Leukemogenesis. *Cell Rep.* **2014**, *9*, 1333–1348. [CrossRef]

135. Harir, N.; Pecquet, C.; Kerenyi, M.; Sonneck, K.; Kovacic, B.; Nyga, R.; Brevet, M.; Dhennin, I.; Gouilleux-Gruart, V.; Beug, H.; et al. Constitutive activation of Stat5 promotes its cytoplasmic localization and association with PI3-kinase in myeloid leukemias. *Blood* **2007**, *109*, 1678–1686. [CrossRef]

136. Chatain, N.; Ziegler, P.; Fahrenkamp, D.; Jost, E.; Moriggl, R.; Schmitz-Van de Leur, H.; Muller-Newen, G. Src family kinases mediate cytoplasmic retention of activated stat5 in bcr-abl-positive cells. *Oncogene* **2013**, *32*, 3587–3597. [CrossRef]

137. Meyer, W.K.; Reichenbach, P.; Schindler, U.; Soldaini, E.; Nabholz, M. Interaction of STAT5 dimers on two low affinity binding sites mediates interleukin 2 (IL-2) stimulation of IL-2 receptor alpha gene transcription. *J. Biol. Chem.* **1997**, *272*, 31821–31828. [CrossRef]

138. Soldaini, E.; John, S.; Moro, S.; Bollenbacher, J.; Schindler, U.; Leonard, W.J. DNA binding site selection of dimeric and tetrameric Stat5 proteins reveals a large repertoire of divergent tetrameric Stat5a binding sites. *Mol. Cell. Biol.* **2000**, *20*, 389–401. [CrossRef]

139. Hu, T.; Yeh, J.E.; Pinello, L.; Jacob, J.; Chakravarthy, S.; Yuan, G.C.; Chopra, R.; Frank, D.A. Impact of the N-Terminal Domain of STAT3 in STAT3-Dependent Transcriptional Activity. *Mol. Cell. Biol.* **2015**, *35*, 3284–3300. [CrossRef]

140. Timofeeva, O.A.; Tarasova, N.I.; Zhang, X.; Chasovskikh, S.; Cheema, A.K.; Wang, H.; Brown, M.L.; Dritschilo, A. STAT3 suppresses transcription of proapoptotic genes in cancer cells with the involvement of its N-terminal domain. *Proc. Natl. Acad. Sci. USA* **2013**, *110*, 1267–1272. [CrossRef]

141. Wingelhofer, B.; Neubauer, H.A.; Valent, P.; Han, X.; Constantinescu, S.N.; Gunning, P.T.; Müller, M.; Moriggl, R. Implications of STAT3 and STAT5 signaling on gene regulation and chromatin remodeling in hematopoietic cancer. *Leukemia* **2018**, *32*, 1713–1726. [CrossRef] [PubMed]

142. Wierenga, A.T.; Vellenga, E.; Schuringa, J.J. Maximal STAT5-induced proliferation and self-renewal at intermediate STAT5 activity levels. *Mol. Cell. Biol.* **2008**, *28*, 6668–6680. [CrossRef] [PubMed]

143. Rozovski, U.; Harris, D.M.; Li, P.; Liu, Z.; Wu, J.Y.; Grgurevic, S.; Faderl, S.; Ferrajoli, A.; Wierda, W.G.; Martinez, M.; et al. At High Levels, Constitutively Activated STAT3 Induces Apoptosis of Chronic Lymphocytic Leukemia Cells. *J. Immunol.* **2016**, *196*, 4400–4409. [CrossRef] [PubMed]

144. Nosaka, T.; Kawashima, T.; Misawa, K.; Ikuta, K.; Mui, A.L.; Kitamura, T. STAT5 as a molecular regulator of proliferation, differentiation and apoptosis in hematopoietic cells. *EMBO J.* **1999**, *18*, 4754–4765. [CrossRef] [PubMed]

145. Friedbichler, K.; Kerenyi, M.A.; Kovacic, B.; Li, G.; Hoelbl, A.; Yahiaoui, S.; Sexl, V.; Mullner, E.W.; Fajmann, S.; Cerny-Reiterer, S.; et al. Stat5a serine 725 and 779 phosphorylation is a prerequisite for hematopoietic transformation. *Blood* **2010**, *116*, 1548–1558. [CrossRef]

146. Mitra, A.; Ross, J.A.; Rodriguez, G.; Nagy, Z.S.; Wilson, H.L.; Kirken, R.A. Signal transducer and activator of transcription 5b (Stat5b) serine 193 is a novel cytokine-induced phospho-regulatory site that is constitutively activated in primary hematopoietic malignancies. *J. Biol. Chem.* **2012**, *287*, 16596–16608. [CrossRef]

147. Freund, P.; Kerenyi, M.A.; Hager, M.; Wagner, T.; Wingelhofer, B.; Pham, H.T.T.; Elabd, M.; Han, X.; Valent, P.; Gouilleux, F.; et al. O-GlcNAcylation of STAT5 controls tyrosine phosphorylation and oncogenic transcription in STAT5-dependent malignancies. *Leukemia* **2017**, *31*, 2132–2142. [CrossRef]

148. Zhuang, S. Regulation of STAT Signaling by Acetylation. *Cell Signal.* **2013**, *25*, 1924–1931. [CrossRef]

149. Yuan, Z.L.; Guan, Y.J.; Chatterjee, D.; Chin, Y.E. Stat3 dimerization regulated by reversible acetylation of a single lysine residue. *Science* **2005**, *307*, 269–273. [CrossRef]

150. Wang, R.; Cherukuri, P.; Luo, J. Activation of Stat3 sequence-specific DNA binding and transcription by p300/CREB-binding protein-mediated acetylation. *J. Biol. Chem.* **2005**, *280*, 11528–11534. [CrossRef]

151. Lee, H.; Zhang, P.; Herrmann, A.; Yang, C.; Xin, H.; Wang, Z.; Hoon, D.S.; Forman, S.J.; Jove, R.; Riggs, A.D. Acetylated STAT3 is crucial for methylation of tumor-suppressor gene promoters and inhibition by resveratrol results in demethylation. *Proc. Natl. Acad. Sci. USA* **2012**, *109*, 7765–7769. [CrossRef] [PubMed]

152. Kosan, C.; Ginter, T.; Heinzel, T.; Krämer, O.H. STAT5 acetylation: Mechanisms and consequences for immunological control and leukemogenesis. *JAKSTAT* **2013**, *2*, e26102. [CrossRef] [PubMed]

153. Dasgupta, M.; Unal, H.; Willard, B.; Yang, J.; Karnik, S.S.; Stark, G.R. Critical role for lysine 685 in gene expression mediated by transcription factor unphosphorylated STAT3. *J. Biol. Chem.* **2014**, *289*, 30763–30771. [CrossRef] [PubMed]

154. Yang, J.; Huang, J.; Dasgupta, M.; Sears, N.; Miyagi, M.; Wang, B.; Chance, M.R.; Chen, X.; Du, Y.; Wang, Y.; et al. Reversible methylation of promoter-bound stat3 by histone-modifying enzymes. *Proc. Natl. Acad. Sci. USA* **2010**, *107*, 21499–21504. [CrossRef]

155. Kim, E.; Kim, M.; Woo, D.H.; Shin, Y.; Shin, J.; Chang, N.; Oh, Y.T.; Kim, H.; Rheey, J.; Nakano, I.; et al. Phosphorylation of ezh2 activates stat3 signaling via stat3 methylation and promotes tumorigenicity of glioblastoma stem-like cells. *Cancer Cell* **2013**, *23*, 839–852. [CrossRef]

156. Dasgupta, M.; Dermawan, J.K.; Willard, B.; Stark, G.R. STAT3-driven transcription depends upon the dimethylation of K49 by EZH2. *Proc. Natl. Acad. Sci. USA* **2015**, *112*, 3985–3990. [CrossRef]

157. Mandal, M.; Powers, S.E.; Maienschein-Cline, M.; Bartom, E.T.; Hamel, K.M.; Kee, B.L.; Dinner, A.R.; Clark, M.R. Epigenetic repression of the Igk locus by STAT5-mediated recruitment of the histone methyltransferase Ezh2. *Nat. Immunol.* **2011**, *12*, 1212–1220. [CrossRef]

158. Butturini, E.; Darra, E.; Chiavegato, G.; Cellini, B.; Cozzolino, F.; Monti, M.; Pucci, P.; Dell'Orco, D.; Mariotto, S. S-Glutathionylation at Cys328 and Cys542 impairs STAT3 phosphorylation. *ACS Chem. Biol.* **2014**, *9*, 1885–1893. [CrossRef]

159. Li, L.; Cheung, S.H.; Evans, E.L.; Shaw, P.E. Modulation of gene expression and tumor cell growth by redox modification of STAT3. *Cancer Res.* **2010**, *70*, 8222–8232. [CrossRef]

160. Sebastián, C.; Herrero, C.; Serra, M.; Lloberas, J.; Blasco, M.A.; Celada, A. Telomere shortening and oxidative stress in aged macrophages results in impaired STAT5a phosphorylation. *J. Immunol.* **2009**, *183*, 2356–2364. [CrossRef]

161. Krämer, O.H.; Moriggl, R. Acetylation and sumoylation control STAT5 activation antagonistically. *JAKSTAT* **2012**, *1*, 203–207. [CrossRef] [PubMed]

162. Zhou, Z.; Wang, M.; Li, J.; Xiao, M.; Chin, Y.E.; Cheng, J.; Yeh, E.T.; Yang, J.; Yi, J. SUMOylation and SENP3 regulate STAT3 activation in head and neck cancer. *Oncogene* **2016**, *35*, 5826–5838. [CrossRef] [PubMed]

163. Walker, S.R.; Xiang, M.; Frank, D.A. Distinct roles of STAT3 and STAT5 in the pathogenesis and targeted therapy of breast cancer. *Mol. Cell. Endocrinol.* **2014**, *382*, 616–621. [CrossRef] [PubMed]

164. Walker, S.R.; Nelson, E.A.; Yeh, J.E.; Pinello, L.; Yuan, G.C.; Frank, D.A. STAT5 outcompetes STAT3 to regulate the expression of the oncogenic transcriptional modulator BCL6. *Mol. Cell. Biol.* **2013**, *33*, 2879–2890. [CrossRef] [PubMed]

165. Yang, X.P.; Ghoreschi, K.; Steward-Tharp, S.M.; Rodriguez-Canales, J.; Zhu, J.; Grainger, J.R.; Hirahara, K.; Sun, H.W.; Wie, L.; Vahedi, G.; et al. Opposing regulation of the locus encoding IL-17 through direct, reciprocal actions of STAT3 and STAT5. *Nat. Immunol.* **2011**, *12*, 247–254. [CrossRef] [PubMed]

166. Weber, A.; Borghouts, C.; Brendel, C.; Moriggl, R.; Delis, N.; Brill, B.; Vafaizadeh, V.; Groner, B. Stat5 Exerts Distinct, Vital Functions in the Cytoplasm and Nucleus of Bcr-Abl+ K562 and Jak2(V617F)+ HEL Leukemia Cells. *Cancers* **2015**, *7*, 503–537. [CrossRef]

167. Li, G.; Miskimen, K.L.; Wang, Z.; Xie, X.Y.; Tse, W.; Gouilleux, F.; Moriggl, R.; Bunting, K.D. Effective targeting of stat5-mediated survival in myeloproliferative neoplasms using abt-737 combined with rapamycin. *Leukemia* **2010**, *24*, 1397–1405. [CrossRef]

168. Marty, C.; Lacout, C.; Droin, N.; Le Couédic, J.P.; Ribrag, V.; Solary, E.; Vainchenker, W.; Villeval, J.L.; Plo, I. A role for reactive oxygen species in JAK2 V617F myeloproliferative neoplasm progression. *Leukemia* **2013**, *27*, 2187–2195. [CrossRef]

169. Silver, D.L.; Naora, H.; Liu, J.; Cheng, W.; Montell, D.J. Activated signal transducer and activator of transcription (stat) 3: Localization in focal adhesions and function in ovarian cancer cell motility. *Cancer Res.* **2004**, *64*, 3550–3558. [CrossRef]

170. Avalle, L.; Camporeale, A.; Morciano, G.; Caroccia, N.; Ghetti, E.; Orecchia, V.; Viavattene, D.; Giorgi, C.; Pinton, P.; Poli, V. Stat3 localizes to the ER, acting as a gatekeeper for er-mitochondrion ca (2+) fluxes and apoptotic responses. *Cell Death Differ.* **2019**, *26*, 932–942. [CrossRef]

171. Yang, R.; Lirussi, D.; Thornton, T.M.; Jelley-Gibbs, D.M.; Diehl, S.A.; Case, L.K.; Madesh, M.; Taatjes, D.J.; Teuscher, C.; Haynes, L. Mitochondrial Ca^{2+} and membrane potential, an alternative pathway for Interleukin 6 to regulate CD4 cell effector function. *Elife* **2015**, *4*, e06376. [CrossRef] [PubMed]

172. Carbognin, E.; Betto, R.M.; Soriano, M.E.; Smith, A.G.; Martello, G. Stat3 promotes mitochondrial transcription and oxidative respiration during maintenance and induction of naive pluripotency. *EMBO J.* **2016**, *35*, 618–634. [CrossRef] [PubMed]

173. Chueh, F.Y.; Leong, K.F.; Yu, C.L. Mitochondrial translocation of signal transducer and activator of transcription 5 (STAT5) in leukemic T cells and cytokine-stimulated cells. *Biochem. Biophys. Res. Commun.* **2010**, *402*, 778–783. [CrossRef] [PubMed]

174. Rozovski, U.; Harris, D.M.; Li, P.; Liu, Z.; Jain, P.; Ferrajoli, A.; Burger, J.; Thompson, P.; Jain, N.; Wierda, W.; et al. STAT3 is constitutively acetylated on lysine 685 residues in chronic lymphocytic leukemia cells. *Oncotarget* **2018**, *9*, 33710–33718. [CrossRef] [PubMed]

175. Gough, D.J.; Corlett, A.; Schlessinger, K.; Wegrzyn, J.; Larner, A.C.; Levy, D.E. Mitochondrial STAT3 supports Ras-dependent oncogenic transformation. *Science* **2009**, *324*, 1713–1716. [CrossRef] [PubMed]

176. Gough, D.J.; Koetz, L.; Levy, D.E. The MEK-ERK pathway is necessary for serine phosphorylation of mitochondrial STAT3 and Ras-mediated transformation. *PLoS ONE* **2013**, *8*, e83395. [CrossRef]

177. Gough, D.J.; Marié, I.J.; Lobry, C.; Aifantis, I.; Levy, D.E. STAT3 supports experimental K-RasG12D-induced murine myeloproliferative neoplasms dependent on serine phosphorylation. *Blood* **2014**, *124*, 2252–2261. [CrossRef]

178. Tammineni, P.; Anugula, C.; Mohammed, F.; Anjaneyulu, M.; Larner, A.C.; Sepuri, N.B. The import of the transcription factor stat3 into mitochondria depends on grim-19, a component of the electron transport chain. *J. Biol. Chem.* **2013**, *288*, 4723–4732. [CrossRef]

179. Avalle, L.; Poli, V. Nucleus, Mitochondrion, or Reticulum? STAT3 à La Carte. *Int. J. Mol. Sci.* **2018**, *19*, 2820. [CrossRef]

180. Xu, Y.S.; Liang, J.J.; Wang, Y.; Zhao, X.J.; Xu, L.; Xu, Y.Y.; Zou, Q.C.; Zhang, J.M.; Tu, C.E.; Cui, Y.G.; et al. STAT3 Undergoes Acetylation-dependent Mitochondrial Translocation to Regulate Pyruvate Metabolism. *Sci. Rep.* **2016**, *6*, 39517. [CrossRef]

181. Mantel, C.; Messina-Graham, S.; Moh, A.; Cooper, S.; Hangoc, G.; Fu, X.Y.; Broxmeyer, H.E. Mouse hematopoietic cell-targeted STAT3 deletion: Stem/progenitor cell defects, mitochondrial dysfunction, ROS overproduction, and a rapid aging-like phenotype. *Blood* **2012**, *120*, 2589–2599. [CrossRef] [PubMed]

182. Lee, J.E.; Yang, Y.M.; Liang, F.X.; Gough, D.J.; Levy, D.E.; Sehgal, P.B. Nongenomic stat5-dependent effects on golgi apparatus and endoplasmic reticulum structure and function. *Am. J. Physiol. Cell Physiol.* **2012**, *302*, 804–820. [CrossRef] [PubMed]

183. Cholez, E.; Debuysscher, V.; Bourgeais, J.; Boudot, C.; Leprince, J.; Tron, F.; Brassart, B.; Regnier, A.; Bissac, E.; Pecnard, E.; et al. Evidence for a protective role of the STAT5 transcription factor against oxidative stress in human leukemic pre-B cells. *Leukemia* **2012**, *26*, 2390–2397. [CrossRef] [PubMed]

184. Vogt, M.; Domoszlai, T.; Kleshchanok, D.; Lehmann, S.; Schmitt, A.; Poli, V.; Richtering, W.; Muller-Newen, G. The role of the N-terminal domain in dimerization and nucleocytoplasmic shuttling of latent STAT3. *J. Cell Sci.* **2011**, *124*, 900–909. [CrossRef] [PubMed]

185. Timofeeva, O.A.; Chasovskikh, S.; Lonskaya, I.; Tarasova, N.I.; Khavrutskii, L.; Tarasov, S.G.; Zhang, X.; Korostyshevskiy, V.R.; Cheema, A.; Zhang, L.; et al. Mechanisms of unphosphorylated STAT3 transcription factor binding to DNA. *J. Biol. Chem.* **2012**, *287*, 14192–14200. [CrossRef] [PubMed]

186. Yang, J.; Chatterjee-Kishore, M.; Staugaitis, S.M.; Nguyen, H.; Schlessinger, K.; Levy, D.E.; Stark, G.R. Novel roles of unphosphorylated STAT3 in oncogenesis and transcriptional regulation. *Cancer Res.* **2005**, *65*, 939–947.

187. Park, H.J.; Li, J.; Hannah, R.; Biddie, S.; Leal-Cervantes, A.I.; Kirschner, K.; Flores Santa Cruz, D.; Sexl, V.; Göttgens, B.; Green, A.R. Cytokine-induced megakaryocytic differentiation is regulated by genome-wide loss of a uSTAT transcriptional program. *EMBO J.* **2016**, *35*, 580–594. [CrossRef]

188. Nagy, Z.S.; Rui, H.; Stepkowski, S.M.; Karras, J.; Kirken, R.A. A preferential role for STAT5, not constitutively active STAT3, in promoting survival of a human lymphoid tumor. *J. Immunol.* **2006**, *177*, 5032–5040. [CrossRef]

189. Xu, D.; Qu, C.K. Protein tyrosine phosphatases in the jak/stat pathway. *Front. Biosci.* **2008**, *13*, 4925–4932. [CrossRef]

190. Krebs, D.L.; Hilton, D.J. Socs: Physiological suppressors of cytokine signaling. *J. Cell Sci.* **2000**, *113*, 2813–2819.

191. Shuai, K. Modulation of stat signaling by stat-interacting proteins. *Oncogene* **2000**, *19*, 2638–2644. [CrossRef] [PubMed]

192. Turkson, J.; Ryan, D.; Kim, J.S.; Zhang, Y.; Chen, Z.; Haura, E.; Laudano, A.; Sebti, S.; Hamilton, A.D.; Jove, R. Phosphotyrosyl peptides block Stat3-mediated DNA binding activity, gene regulation, and cell transformation. *J. Biol. Chem.* **2001**, *276*, 45443–45455. [CrossRef] [PubMed]

193. Turkson, J.; Kim, J.S.; Zhang, S.; Yuan, J.; Huang, M.; Glenn, M.; Haura, E.; Sebti, S.; Hamilton, A.D.; Jove, R. Novel peptidomimetic inhibitors of signal transducer and activator of transcription 3 dimerization and biological activity. *Mol. Cancer Ther.* **2004**, *3*, 261–269. [PubMed]

194. Yue, P.; Turkson, J. Targeting STAT3 in cancer: How successful are we? *Expert Opin. Investig. Drugs* **2009**, *18*, 45–56. [CrossRef]

195. Sen, M.; Grandis, J.R. Nucleic acid-based approaches to STAT inhibition. *JAKSTAT* **2012**, *1*, 285–291. [CrossRef] [PubMed]

196. Nelson, E.A.; Sharma, S.V.; Settleman, J.; Frank, D.A. A chemical biology approach to developing STAT inhibitors: Molecular strategies for accelerating clinical translation. *Oncotarget* **2011**, *2*, 518–524. [CrossRef] [PubMed]

197. Walker, S.R.; Frank, D.A. Screening approaches to generating STAT inhibitors: Allowing the hits to identify the targets. *JAKSTAT* **2012**, *1*, 292–299. [CrossRef]

198. Druker, B.J.; Talpaz, M.; Resta, D.J.; Peng, B.; Buchdunger, E.; Ford, J.M.; Lydon, N.B.; Kantarjian, H.; Capdeville, R.; Ohno-Jones, S.; et al. Efficacy and safety of a specific inhibitor of the BCR-ABL tyrosine kinase in chronic myeloid leukemia. *N. Engl. J. Med.* **2001**, *344*, 1031–1037. [CrossRef]

199. Druker, B.J.; Sawyers, C.L.; Kantarjian, H.; Resta, D.J.; Reese, S.F.; Ford, J.M.; Capdeville, R.; Talpaz, M. Activity of a specific inhibitor of the BCR-ABL tyrosine kinase in the blast crisis of chronic myeloid leukemia and acute lymphoblastic leukemia with the Philadelphia chromosome. *N. Engl. J. Med.* **2001**, *344*, 1038–1042. [CrossRef]

200. Talpaz, M.; Shah, N.P.; Kantarjian, H.; Donato, N.; Nicoll, J.; Paquette, R.; Cortes, J.; O'Brien, S.; Nicaise, C.; Bleickardt, E.; et al. Dasatinib in imatinib-resistant Philadelphia chromosome-positive leukemias. *N. Engl. J. Med.* **2006**, *354*, 2531–2541. [CrossRef]

201. Golas, J.M.; Arndt, K.; Etienne, C.; Lucas, J.; Nardin, D.; Gibbons, J.; Frost, P.; Ye, F.; Boschelli, D.H.; Boschelli, F. SKI-606, a 4-anilino-3-quinolinecarbonitrile dual inhibitor of Src and Abl kinases, is a potent antiproliferative agent against chronic myelogenous leukemia cells in culture and causes regression of K562 xenografts in nude mice. *Cancer Res.* **2003**, *63*, 375–381. [PubMed]

202. Konig, H.; Holyoake, T.L.; Bhatia, R. Effective and selective inhibition of chronic myeloid leukemia primitive hematopoietic progenitors by the dual Src/Abl kinase inhibitor SKI-606. *Blood* **2008**, *111*, 2329–2338. [CrossRef] [PubMed]

203. Weisberg, E.; Manley, P.W.; Breitenstein, W.; Brüggen, J.; Cowan-Jacob, S.W.; Ray, A.; Huntly, B.; Fabbro, D.; Fendrich, G.; Hall-Meyers, E.; et al. Characterization of AMN107, a selective inhibitor of native and mutant Bcr-Abl. *Cancer Cell* **2005**, *7*, 129–141. [CrossRef] [PubMed]

204. O'Hare, T.; Shakespeare, W.C.; Zhu, X.; Eide, C.A.; Rivera, V.M.; Wang, F.; Adrian, L.T.; Zhou, T.; Huang, W.S.; Xu, Q.; et al. AP24534, a pan-BCR-ABL inhibitor for chronic myeloid leukemia, potently inhibits the T315I mutant and overcomes mutation-based resistance. *Cancer Cell* **2009**, *16*, 401–412. [CrossRef]

205. Shah, N.P.; Nicoll, J.M.; Nagar, B.; Gorre, M.E.; Paquette, R.L.; Kuriyan, J.; Sawyers, C.L. Multiple BCR-ABL kinase domain mutations confer polyclonal resistance to the tyrosine kinase inhibitor imatinib (STI571) in chronic phase and blast crisis chronic myeloid leukemia. *Cancer Cell* **2002**, *2*, 117–125. [CrossRef]

206. Daver, N.; Schlenk, R.F.; Russell, N.H.; Levis, M.J. Targeting FLT3 mutations in AML: Review of current knowledge and evidence. *Leukemia* **2019**, *33*, 299–312. [CrossRef]

207. Levis, M. Midostaurin approved for FLT3-mutated AML. *Blood* **2017**, *129*, 3403–3406. [CrossRef]

208. Crump, M.; Hedley, D.; Kamel-Reid, S.; Leber, B.; Wells, R.; Brandwein, J.; Buckstein, R.; Kassis, J.; Minden, M.; Matthews, J.; et al. A randomized phase I clinical and biologic study of two schedules of sorafenib in patients with myelodysplastic syndrome or acute myeloid leukemia: A NCIC (National Cancer Institute of Canada) Clinical Trials Group Study. *Leuk. Lymphoma* **2010**, *51*, 252–260. [CrossRef]

209. Borthakur, G.; Kantarjian, H.; Ravandi, F.; Zhang, W.; Konopleva, M.; Wright, J.J.; Faderl, S.; Verstovsek, S.; Mathews, S.; Andreeff, M.; et al. Phase I study of sorafenib in patients with refractory or relapsed acute leukemias. *Haematologica* **2011**, *96*, 62–68. [CrossRef]

210. Ravandi, F.; Cortes, J.E.; Jones, D.; Faderl, S.; Garcia-Manero, G.; Konopleva, M.Y.; O'Brien, S.; Estrov, Z.; Borthakur, G.; Thomas, D.; et al. Phase I/II study of combination therapy with sorafenib, idarubicin, and cytarabine in younger patients with acute myeloid leukemia. *J. Clin. Oncol.* **2010**, *28*, 1856–1862. [CrossRef]

211. Knapper, S.; Burnett, A.K.; Littlewood, T.; Kell, W.J.; Agrawal, S.; Chopra, R.; Clark, R.; Levis, M.J.; Small, D. A phase 2 trial of the FLT3 inhibitor lestaurtinib (CEP701) as first-line treatment for older patients with acute myeloid leukemia not considered fit for intensive chemotherapy. *Blood* **2006**, *108*, 3262–3270. [CrossRef] [PubMed]

212. Shiotsu, Y.; Kiyoi, H.; Ishikawa, Y.; Tanizaki, R.; Shimizu, M.; Umehara, H.; Ishii, K.; Mori, Y.; Ozeki, K.; Minami, Y.; et al. KW-2449, a novel multikinase inhibitor, suppresses the growth of leukemia cells with FLT3 mutations or T315I-mutated BCR/ABL translocation. *Blood* **2009**, *114*, 1607–1617. [CrossRef] [PubMed]

213. Pratz, K.W.; Cortes, J.; Roboz, G.J.; Rao, N.; Arowojolu, O.; Stine, A.; Shiotsu, Y.; Shudo, A.; Akinaga, S.; Small, D.; et al. A pharmacodynamic study of the FLT3 inhibitor KW-2449 yields insight into the basis for clinical response. *Blood* **2009**, *113*, 3938–3946. [CrossRef] [PubMed]

214. O'Farrell, A.M.; Foran, J.M.; Fiedler, W.; Serve, H.; Paquette, R.L.; Cooper, M.A.; Yuen, H.A.; Louie, S.G.; Kim, H.; Nicholas, S.; et al. An innovative phase I clinical study demonstrates inhibition of FLT3 phosphorylation by SU11248 in acute myeloid leukemia patients. *Clin. Cancer Res.* **2003**, *9*, 5465–5476. [PubMed]

215. Fiedler, W.; Kayser, S.; Kebenko, M.; Janning, M.; Krauter, J.; Schittenhelm, M.; Götze, K.; Weber, D.; Göhring, G.; Teleanu, V.; et al. A phase I/II study of sunitinib and intensive chemotherapy in patients over 60 years of age with acute myeloid leukaemia and activating FLT3 mutations. *Br. J. Haematol.* **2015**, *169*, 694–700. [CrossRef] [PubMed]

216. Cheng, Y.; Paz, K. Tandutinib, an oral, small-molecule inhibitor of FLT3 for the treatment of AML and other cancer indications. *IDrugs* **2008**, *11*, 46–56. [PubMed]

217. Zhou, F.; Ge, Z.; Chen, B. Quizartinib (AC220): A promising option for acute myeloid leukemia. *Drug Des. Dev. Ther.* **2019**, *13*, 1117–1125. [CrossRef]

218. Sutamtewagul, G.; Vigil, C.E. Clinical use of FLT3 inhibitors in acute myeloid leukemia. *Onco Targets Ther.* **2018**, *11*, 7041–7052. [CrossRef]

219. Zhao, J.; Song, Y.; Liu, D. Gilteritinib: A novel FLT3 inhibitor for acute myeloid leukemia. *Biomark Res.* **2019**, *7*, 19. [CrossRef]

220. Zhang, H.; Savage, S.; Schultz, A.R.; Bottomly, D.; White, L.; Segerdell, E.; Wilmot, B.; McWeeney, S.K.; Eide, C.A.; Nechiporuk, T.; et al. Clinical resistance to crenolanib in acute myeloid leukemia due to diverse molecular mechanisms. *Nat. Commun.* **2019**, *10*, 244. [CrossRef]

221. Ghiaur, G.; Levis, M. Mechanisms of Resistance to FLT3 Inhibitors and the Role of the Bone Marrow Microenvironment. *Hematol. Oncol. Clin. N. Am.* **2017**, *31*, 681–692. [CrossRef] [PubMed]

222. Baird, J.H.; Gotlib, J. Clinical Validation of KIT Inhibition in Advanced Systemic Mastocytosis. *Curr. Hematol. Malig. Rep.* **2018**, *13*, 407–416. [CrossRef] [PubMed]

223. Kampa-Schittenhelm, K.M.; Frey, J.; Haeusser, L.A.; Illing, B.; Pavlovsky, A.A.; Blumenstock, G.; Schittenhelm, M.M. Crenolanib is a type I tyrosine kinase inhibitor that inhibits mutant KIT D816 isoforms prevalent in systemic mastocytosis and core binding factor leukemia. *Oncotarget* **2017**, *8*, 82897–82909. [CrossRef] [PubMed]

224. Corbin, A.S.; Griswold, I.J.; La Rosée, P.; Yee, K.W.; Heinrich, M.C.; Reimer, C.L.; Druker, B.J.; Deininger, M.W. Sensitivity of oncogenic KIT mutants to the kinase inhibitors MLN518 and PD180970. *Blood* **2004**, *104*, 3754–3757. [CrossRef] [PubMed]

225. Mascarenhas, J.; Baer, M.R.; Kessler, C.; Hexner, E.; Tremblay, D.; Price, L.; Sandy, L.; Weinberg, R.; Pahl, H.; Silverman, L.R.; et al. Phase II trial of Lestaurtinib, a JAK2 inhibitor, in patients with myelofibrosis. *Leuk. Lymphoma* **2019**, *60*, 1343–1345. [CrossRef] [PubMed]

226. Hexner, E.; Roboz, G.; Hoffman, R.; Luger, S.; Mascarenhas, J.; Carroll, M.; Clementi, R.; Bensen-Kennedy, D.; Moliterno, A. Open-label study of oral CEP-701 (lestaurtinib) in patients with polycythaemia vera or essential thrombocythaemia with JAK2-V617F mutation. *Br. J. Haematol.* **2014**, *164*, 83–93. [CrossRef]

227. Vainchenker, W.; Leroy, E.; Gilles, L.; Marty, C.; Plo, I.; Constantinescu, S.N. JAK inhibitors for the treatment of myeloproliferative neoplasms and other disorders. *F1000 Res.* **2018**, *7*, 82. [CrossRef]

228. Deininger, M.; Radich, J.; Burn, T.C.; Huber, R.; Paranagama, D.; Verstovsek, S. The effect of long-term ruxolitinib treatment on JAK2p.V617F allele burden in patients with myelofibrosis. *Blood* **2015**, *126*, 1551–1554. [CrossRef]

229. Gallipoli, P.; Cook, A.; Rhodes, S.; Hopcroft, L.; Wheadon, H.; Whetton, A.D.; Jørgensen, H.G.; Bhatia, R.; Holyoake, T.L. JAK2/STAT5 inhibition by nilotinib with ruxolitinib contributes to the elimination of CML CD34+ cells in vitro and in vivo. *Blood* **2014**, *124*, 1492–1501. [CrossRef]

230. Van Den Neste, E.; André, M.; Gastinne, T.; Stamatoullas, A.; Haioun, C.; Belhabri, A.; Reman, O.; Casasnovas, O.; Ghesquieres, H.; Verhoef, G.; et al. A phase II study of the oral JAK1/JAK2 inhibitor ruxolitinib in advanced relapsed/refractory Hodgkin lymphoma. *Haematologica* **2018**, *103*, 840–848. [CrossRef]

231. Ghermezi, M.; Spektor, T.M.; Berenson, J.R. The role of JAK inhibitors in multiple myeloma. *Clin. Adv. Hematol. Oncol.* **2019**, *17*, 500–505. [PubMed]

232. Pemmaraju, N.; Kantarjian, H.; Kadia, T.; Cortes, J.; Borthakur, G.; Newberry, K.; Garcia-Manero, G.; Ravandi, F.; Jabbour, E.; Dellasala, S.; et al. A phase I/II study of the Janus kinase (JAK)1 and 2 inhibitor ruxolitinib in patients with relapsed or refractory acute myeloid leukemia. *Clin. Lymphoma Myeloma Leuk.* **2015**, *15*, 171–176. [CrossRef] [PubMed]

233. Diaz, A.E.; Mesa, R.A. Pacritinib and its use in the treatment of patients with myelofibrosis who have thrombocytopenia. *Future Oncol.* **2018**, *14*, 797–807. [CrossRef] [PubMed]

234. Younes, A.; Romaguera, J.; Fanale, M.; McLaughlin, P.; Hagemeister, F.; Copeland, A.; Neelapu, S.; Kwak, L.; Shah, J.; de Castro Faria, S.; et al. Phase I study of a novel oral Janus kinase 2 inhibitor, SB1518, in patients with relapsed lymphoma: Evidence of clinical and biologic activity in multiple lymphoma subtypes. *J. Clin. Oncol.* **2012**, *30*, 4161–4167. [CrossRef]

235. Pardanani, A.; Hood, J.; Lasho, T.; Levine, R.L.; Martin, M.B.; Noronha, G.; Finke, C.; Mak, C.C.; Mesa, R.; Zhu, H.; et al. TG101209, a small molecule JAK2-selective kinase inhibitor potently inhibits myeloproliferative disorder-associated JAK2V617F and MPLW515L/K mutations. *Leukemia* **2007**, *21*, 1658–1668. [CrossRef]

236. Ramakrishnan, V.; Kimlinger, T.; Haug, J.; Timm, M.; Wellik, L.; Halling, T.; Pardanani, A.; Tefferi, A.; Rajkumar, S.V.; Kumar, S. TG101209, a novel JAK2 inhibitor, has significant in vitro activity in multiple myeloma and displays preferential cytotoxicity for CD45+ myeloma cells. *Am. J. Hematol.* **2010**, *85*, 675–686. [CrossRef]

237. Cheng, Z.; Yi, Y.; Xie, S.; Yu, H.; Peng, H.; Zhang, G. The effect of the JAK2 inhibitor TG101209 against T cell acute lymphoblastic leukemia (T-ALL) is mediated by inhibition of JAK-STAT signaling and activation of the crosstalk between apoptosis and autophagy signaling. *Oncotarget* **2017**, *8*, 106753–106763. [CrossRef]

238. Meyer, S.C.; Keller, M.D.; Chiu, S.; Koppikar, P.; Guryanova, O.A.; Rapaport, F.; Xu, K.; Manova, K.; Pankov, D.; O'Reilly, R.J.; et al. CHZ868, a Type II JAK2 Inhibitor, Reverses Type I JAK Inhibitor Persistence and Demonstrates Efficacy in Myeloproliferative Neoplasms. *Cancer Cell* **2015**, *28*, 15–28. [CrossRef]

239. Wu, S.C.; Li, L.S.; Kopp, N.; Montero, J.; Chapuy, B.; Yoda, A.; Christie, A.L.; Liu, H.; Christodoulou, A.; van Bodegom, D.; et al. Activity of the Type II JAK2 Inhibitor CHZ868 in B Cell Acute Lymphoblastic Leukemia. *Cancer Cell* **2015**, *28*, 29–41. [CrossRef]

240. Berdeja, J.; Palandri, F.; Baer, M.R.; Quick, D.; Kiladjian, J.J.; Martinelli, G.; Verma, A.; Hamid, O.; Walgren, R.; Pitou, C.; et al. Phase 2 study of gandotinib (LY2784544) in patients with myeloproliferative neoplasms. *Leuk. Res.* **2018**, *71*, 82–88. [CrossRef]

241. Blair, H.A. Fedratinib: First Approval. *Drugs* **2019**, *79*, 1719–1725. [CrossRef] [PubMed]

242. Hao, Y.; Chapuy, B.; Monti, S.; Sun, H.H.; Rodig, S.J.; Shipp, M.A. Selective JAK2 inhibition specifically decreases Hodgkin lymphoma and mediastinal large B-cell lymphoma growth in vitro and in vivo. *Clin. Cancer Res.* **2014**, *20*, 2674–2683. [CrossRef] [PubMed]

243. Verstovsek, S.; Hoffman, R.; Mascarenhas, J.; Soria, J.C.; Bahleda, R.; McCoon, P.; Tang, W.; Cortes, J.; Kantarjian, H.; Ribrag, V. A phase I, open-label, multi-center study of the JAK2 inhibitor AZD1480 in patients with myelofibrosis. *Leuk. Res.* **2015**, *39*, 157–163. [CrossRef] [PubMed]

244. Suryani, S.; Bracken, L.S.; Harvey, R.C.; Sia, K.C.; Carol, H.; Chen, I.M.; Evans, K.; Dietrich, P.A.; Roberts, K.G.; Kurmasheva, R.T.; et al. Evaluation of the in vitro and in vivo efficacy of the JAK inhibitor AZD1480 against JAK-mutated acute lymphoblastic leukemia. *Mol. Cancer Ther.* **2015**, *14*, 364–374. [CrossRef]

245. Derenzini, E.; Lemoine, M.; Buglio, D.; Katayama, H.; Ji, Y.; Davis, R.E.; Sen, S.; Younes, A. The JAK inhibitor AZD1480 regulates proliferation and immunity in Hodgkin lymphoma. *Blood Cancer J.* **2011**, *1*, e46. [CrossRef]

246. Scuto, A.; Krejci, P.; Popplewell, L.; Wu, J.; Wang, Y.; Kujawski, M.; Kowolik, C.; Xin, H.; Chen, L.; Wang, Y.; et al. The novel JAK inhibitor AZD1480 blocks STAT3 and FGFR3 signaling, resulting in suppression of human myeloma cell growth and survival. *Leukemia* **2011**, *25*, 538–550. [CrossRef]

247. Burger, R.; Le Gouill, S.; Tai, Y.T.; Shringarpure, R.; Tassone, P.; Neri, P.; Podar, K.; Catley, L.; Hideshima, T.; Chauhan, D.; et al. Janus kinase inhibitor INCB20 has antiproliferative and apoptotic effects on human myeloma cells in vitro and in vivo. *Mol. Cancer Ther.* **2009**, *8*, 26–35. [CrossRef]

248. Li, J.; Favata, M.; Kelley, J.A.; Caulder, E.; Thomas, B.; Wen, X.; Sparks, R.B.; Arvanitis, A.; Rogers, J.D.; Combs, A.P.; et al. INCB16562, a JAK1/2 selective inhibitor, is efficacious against multiple myeloma cells and reverses the protective effects of cytokine and stromal cell support. *Neoplasia* **2010**, *12*, 28–38. [CrossRef]

249. Faderl, S.; Ferrajoli, A.; Harris, D.; Van, Q.; Kantarjian, H.M.; Estrov, Z. Atiprimod blocks phosphorylation of JAK-STAT and inhibits proliferation of acute myeloid leukemia (AML) cells. *Leuk. Res.* **2007**, *31*, 91–95. [CrossRef]

250. Quintás-Cardama, A.; Manshouri, T.; Estrov, Z.; Harris, D.; Zhang, Y.; Gaikwad, A.; Kantarjian, H.M.; Verstovsek, S. Preclinical characterization of atiprimod, a novel JAK2 AND JAK3 inhibitor. *Investig. New Drugs* **2011**, *29*, 818–826. [CrossRef]

251. Yang, J.; Ikezoe, T.; Nishioka, C.; Furihata, M.; Yokoyama, A. AZ960, a novel Jak2 inhibitor, induces growth arrest and apoptosis in adult T-cell leukemia cells. *Mol. Cancer Ther.* **2010**, *9*, 3386–3395. [CrossRef] [PubMed]

252. Mascarenhas, J.O.; Talpaz, M.; Gupta, V.; Foltz, L.M.; Savona, M.R.; Paquette, R.; Turner, A.R.; Coughlin, P.; Winton, E.; Burn, T.C.; et al. Primary analysis of a phase II open-label trial of INCB039110, a selective JAK1 inhibitor, in patients with myelofibrosis. *Haematologica* **2017**, *102*, 327–335. [CrossRef] [PubMed]

253. Phillips, T.J.; Forero-Torres, A.; Sher, T.; Diefenbach, C.S.; Johnston, P.; Talpaz, M.; Pulini, J.; Zhou, L.; Scherle, P.; Chen, X.; et al. Phase 1 study of the PI3Kδ inhibitor INCB040093 ± JAK1 inhibitor itacitinib in relapsed/refractory B-cell lymphoma. *Blood* **2018**, *132*, 293–306. [CrossRef] [PubMed]

254. Verstovsek, S.; Talpaz, M.; Ritchie, E.; Wadleigh, M.; Odenike, O.; Jamieson, C.; Stein, B.; Uno, T.; Mesa, R.A. A phase I, open-label, dose-escalation, multicenter study of the JAK2 inhibitor NS-018 in patients with myelofibrosis. *Leukemia* **2017**, *31*, 393–402. [CrossRef] [PubMed]

255. Honda, A.; Kuramoto, K.; Niwa, T.; Naito, H. NS-018 reduces myeloma cell proliferation and suppresses osteolysis through inhibition of the JAK2 and Src signaling pathways. *Blood Cancer J.* **2018**, *8*, 62. [CrossRef] [PubMed]

256. Baffert, F.; Régnier, C.H.; De Pover, A.; Pissot-Soldermann, C.; Tavares, G.A.; Blasco, F.; Brueggen, J.; Chène, P.; Drueckes, P.; Erdmann, D.; et al. Potent and selective inhibition of polycythemia by the quinoxaline JAK2 inhibitor NVP-BSK805. *Mol. Cancer Ther.* **2010**, *9*, 1945–1955. [CrossRef] [PubMed]

257. Ringel, F.; Kaeda, J.; Schwarz, M.; Oberender, C.; Grille, P.; Dörken, B.; Marque, F.; Manley, P.W.; Radimerski, T.; le Coutre, P. Effects of Jak2 type 1 inhibitors NVP-BSK805 and NVP-BVB808 on Jak2 mutation-positive and Bcr-Abl-positive cell lines. *Acta Haematol.* **2014**, *132*, 75–86. [CrossRef]

258. Jatiani, S.S.; Cosenza, S.C.; Reddy, M.V.; Ha, J.H.; Baker, S.J.; Samanta, A.K.; Olnes, M.J.; Pfannes, L.; Sloand, E.M.; Arlinghaus, R.B.; et al. A Non-ATP-Competitive Dual Inhibitor of JAK2 and BCR-ABL Kinases: Elucidation of a Novel Therapeutic Spectrum Based on Substrate Competitive Inhibition. *Genes Cancer.* **2010**, *1*, 331–345. [CrossRef]

259. Mesa, R.A.; Kiladjian, J.J.; Catalano, J.V.; Devos, T.; Egyed, M.; Hellmann, A.; McLornan, D.; Shimoda, K.; Winton, E.F.; Deng, W.; et al. SIMPLIFY-1: A Phase III Randomized Trial of Momelotinib Versus Ruxolitinib in Janus Kinase Inhibitor-Naïve Patients With Myelofibrosis. *J. Clin. Oncol.* **2017**, *35*, 3844–3850. [CrossRef]

260. Harrison, C.N.; Vannucchi, A.M.; Platzbecker, U.; Cervantes, F.; Gupta, V.; Lavie, D.; Passamonti, F.; Winton, E.F.; Dong, H.; Kawashima, J.; et al. Momelotinib versus best available therapy in patients with myelofibrosis previously treated with ruxolitinib (SIMPLIFY 2): A randomised, open-label, phase 3 trial. *Lancet Haematol.* **2018**, *5*, e73–e81. [CrossRef]

261. Monaghan, K.A.; Khong, T.; Burns, C.J.; Spencer, A. The novel JAK inhibitor CYT387 suppresses multiple signalling pathways, prevents proliferation and induces apoptosis in phenotypically diverse myeloma cells. *Leukemia* **2011**, *25*, 1891–1899. [CrossRef] [PubMed]

262. Lipka, D.B.; Hoffmann, L.S.; Heidel, F.; Markova, B.; Blum, M.C.; Breitenbuecher, F.; Kasper, S.; Kindler, T.; Levine, R.L.; Huber, C.; et al. LS104, a non-ATP-competitive small-molecule inhibitor of JAK2, is potently inducing apoptosis in JAK2V617F-positive cells. *Mol. Cancer Ther.* **2008**, *7*, 1176–1184. [CrossRef] [PubMed]

263. Kasper, S.; Breitenbuecher, F.; Hoehn, Y.; Heidel, F.; Lipka, D.B.; Markova, B.; Huber, C.; Kindler, T.; Fischer, T. The kinase inhibitor LS104 induces apoptosis, enhances cytotoxic effects of chemotherapeutic drugs and is targeting the receptor tyrosine kinase FLT3 in acute myeloid leukemia. *Leuk. Res.* **2008**, *32*, 1698–1708. [CrossRef] [PubMed]

264. Grunberger, T.; Demin, P.; Rounova, O.; Sharfe, N.; Cimpean, L.; Dadi, H.; Freywald, A.; Estrov, Z.; Roifman, C.M. Inhibition of acute lymphoblastic and myeloid leukemias by a novel kinase inhibitor. *Blood* **2003**, *102*, 4153–4158. [CrossRef]

265. Verstovsek, S.; Manshouri, T.; Quintás-Cardama, A.; Harris, D.; Cortes, J.; Giles, F.J.; Kantarjian, H.; Priebe, W.; Estrov, Z. WP1066, a novel JAK2 inhibitor, suppresses proliferation and induces apoptosis in erythroid human cells carrying the JAK2 V617F mutation. *Clin. Cancer Res.* **2008**, *14*, 788–796. [CrossRef]

266. Ferrajoli, A.; Faderl, S.; Van, Q.; Koch, P.; Harris, D.; Liu, Z.; Hazan-Halevy, I.; Wang, Y.; Kantarjian, H.M.; Priebe, W.; et al. WP1066 disrupts Janus kinase-2 and induces caspase-dependent apoptosis in acute myelogenous leukemia cells. *Cancer Res.* **2007**, *67*, 11291–11299. [CrossRef]

267. Faderl, S.; Ferrajoli, A.; Harris, D.; Van, Q.; Priebe, W.; Estrov, Z. WP-1034, A Novel JAK-STAT Inhibitor, with Proapoptotic and Antileukemic Activity in Acute Myeloid Leukemia (AML). *Anticancer Res.* **2005**, *25*, 1841–1850.

268. Verstovsek, S.; Tam, C.S.; Wadleigh, M.; Sokol, L.; Smith, C.C.; Bui, L.A.; Song, C.; Clary, D.O.; Olszynski, P.; Cortes, J.; et al. Phase I evaluation of XL019, an oral, potent, and selective JAK2 inhibitor. *Leuk. Res.* **2014**, *38*, 316–322. [CrossRef]

269. Haridas, V.; Nishimura, G.; Xu, Z.X.; Connolly, F.; Hanausek, M.; Walaszek, Z.; Zoltaszek, R.; Gutterman, J.U. Avicin D: A protein reactive plant isoprenoid dephosphorylates Stat 3 by regulating both kinase and phosphatase activities. *PLoS ONE* **2009**, *4*, e5578. [CrossRef]

270. Zhang, C.; Li, B.; Gaikwad, A.S.; Haridas, V.; Xu, Z.; Gutterman, J.U.; Duvic, M. Avicin D selectively induces apoptosis and downregulates p-STAT-3, bcl-2, and survivin in cutaneous T-cell lymphoma cells. *J. Investig. Dermatol.* **2008**, *128*, 2728–2735. [CrossRef]

271. Pandey, M.K.; Sung, B.; Ahn, K.S.; Aggarwal, B.B. Butein suppresses constitutive and inducible signal transducer and activator of transcription (STAT) 3 activation and STAT3-regulated gene products through the induction of a protein tyrosine phosphatase SHP-1. *Mol. Pharmacol.* **2009**, *75*, 525–533. [CrossRef] [PubMed]

272. Bhutani, M.; Pathak, A.K.; Nair, A.S.; Kunnumakkara, A.B.; Guha, S.; Sethi, G.; Aggarwal, B.B. Capsaicin is a novel blocker of constitutive and interleukin-6-inducible STAT3 activation. *Clin. Cancer Res.* **2007**, *13*, 3024–3032. [CrossRef] [PubMed]

273. Kannaiyan, R.; Hay, H.S.; Rajendran, P.; Li, F.; Shanmugam, M.K.; Vali, S.; Abbasi, T.; Kapoor, S.; Sharma, A.; Kumar, A.P.; et al. Celastrol inhibits proliferation and induces chemosensitization through down-regulation of NF-κB and STAT3 regulated gene products in multiple myeloma cells. *Br. J. Pharmacol.* **2011**, *164*, 1506–1521. [CrossRef] [PubMed]

274. Ishdorj, G.; Johnston, J.B.; Gibson, S.B. Inhibition of constitutive activation of STAT3 by curcurbitacin-I (JSI-124) sensitized human B-leukemia cells to apoptosis. *Mol. Cancer Ther.* **2010**, *9*, 3302–3314. [CrossRef]

275. van Kester, M.S.; Out-Luiting, J.J.; von dem Borne, P.A.; Willemze, R.; Tensen, C.P.; Vermeer, M.H. Cucurbitacin I inhibits Stat3 and induces apoptosis in Sézary cells. *J. Investig. Dermatol.* **2008**, *128*, 1691–1695. [CrossRef]

276. Bi, L.; Yu, Z.; Wu, J.; Yu, K.; Hong, G.; Lu, Z.; Gao, S. Honokiol Inhibits Constitutive and Inducible STAT3 Signaling via PU.1-Induced SHP1 Expression in Acute Myeloid Leukemia Cells. *Tohoku J. Exp. Med.* **2015**, *237*, 163–172. [CrossRef]

277. Yao, Y.; Sun, Y.; Shi, M.; Xia, D.; Zhao, K.; Zeng, L.; Yao, R.; Zhang, Y.; Li, Z.; Niu, M.; et al. Piperlongumine induces apoptosis and reduces bortezomib resistance by inhibiting STAT3 in multiple myeloma cells. *Oncotarget* **2016**, *7*, 73497–73508. [CrossRef]

278. Kuttan, G.; Kumar, K.B.; Guruvayoorappan, C.; Kuttan, R. Antitumor, anti-invasion, and antimetastatic effects of curcumin. *Adv. Exp. Med. Biol.* **2007**, *595*, 173–184.

279. Blasius, R.; Reuter, S.; Henry, E.; Dicato, M.; Diederich, M. Curcumin regulates signal transducer and activator of transcription (STAT) expression in K562 cells. *Biochem. Pharmacol.* **2006**, *72*, 1547–1554. [CrossRef]

280. Rajasingh, J.; Raikwar, H.P.; Muthian, G.; Johnson, C.; Bright, J.J. Curcumin induces growth-arrest and apoptosis in association with the inhibition of constitutively active JAK-STAT pathway in T cell leukemia. *Biochem. Biophys. Res. Commun.* **2006**, *340*, 359–368. [CrossRef]

281. Uddin, S.; Hussain, A.R.; Manogaran, P.S.; Al-Hussein, K.; Platanias, L.C.; Gutierrez, M.I.; Bhatia, K.G. Curcumin suppresses growth and induces apoptosis in primary effusion lymphoma. *Oncogene* **2005**, *24*, 7022–7030. [CrossRef] [PubMed]

282. Mackenzie, G.G.; Queisser, N.; Wolfson, M.L.; Fraga, C.G.; Adamo, A.M.; Oteiza, P.I. Curcumin induces cell-arrest and apoptosis in association with the inhibition of constitutively active NF-kappaB and STAT3 pathways in Hodgkin's lymphoma cells. *Int. J. Cancer.* **2008**, *123*, 56–65. [CrossRef] [PubMed]

283. Bharti, A.C.; Donato, N.; Aggarwal, B.B. Curcumin (diferuloylmethane) inhibits constitutive and IL-6-inducible STAT3 phosphorylation in human multiple myeloma cells. *J. Immunol.* **2003**, *171*, 3863–3871. [CrossRef] [PubMed]

284. Lin, L.; Deangelis, S.; Foust, E.; Fuchs, J.; Li, C.; Li, P.K.; Schwartz, E.B.; Lesinski, G.B.; Benson, D.; Lü, J.; et al. A novel small molecule inhibits STAT3 phosphorylation and DNA binding activity and exhibits potent growth suppressive activity in human cancer cells. *Mol. Cancer.* **2010**, *9*, 217. [CrossRef]

285. Liby, K.; Voong, N.; Williams, C.R.; Risingsong, R.; Royce, D.B.; Honda, T.; Gribble, G.W.; Sporn, M.B.; Letterio, J.J. The synthetic triterpenoid CDDO-Imidazolide suppresses STAT phosphorylation and induces apoptosis in myeloma and lung cancer cells. *Clin. Cancer Res.* **2006**, *12*, 4288–4293. [CrossRef]

286. Nam, S.; Scuto, A.; Yang, F.; Chen, W.; Park, S.; Yoo, H.S.; Konig, H.; Bhatia, R.; Cheng, X.; Merz, K.H.; et al. Indirubin derivatives induce apoptosis of chronic myelogenous leukemia cells involving inhibition of Stat5 signaling. *Mol. Oncol.* **2012**, *6*, 276–283. [CrossRef]

287. Zhou, J.; Bi, C.; Janakakumara, J.V.; Liu, S.C.; Chng, W.J.; Tay, K.G.; Poon, L.F.; Xie, Z.; Palaniyandi, S.; Yu, H.; et al. Enhanced activation of STAT pathways and overexpression of survivin confer resistance to FLT3 inhibitors and could be therapeutic targets in AML. *Blood* **2009**, *113*, 4052–4062. [CrossRef]

288. Nam, S.; Buettner, R.; Turkson, J.; Kim, D.; Cheng, J.Q.; Muehlbeyer, S.; Hippe, F.; Vatter, S.; Merz, K.H.; Eisenbrand, G.; et al. Indirubin derivatives inhibit Stat3 signaling and induce apoptosis in human cancer cells. *Proc. Natl. Acad. Sci. USA* **2005**, *102*, 5998–6003. [CrossRef]

289. Pinz, S.; Unser, S.; Rascle, A. The natural chemopreventive agent sulforaphane inhibits STAT5 activity. *PLoS ONE* **2014**, *9*, e99391. [CrossRef]

290. Liu, S.; Walker, S.R.; Nelson, E.A.; Cerulli, R.; Xiang, M.; Toniolo, P.A.; Qi, J.; Stone, R.M.; Wadleigh, M.; Bradner, J.E.; et al. Targeting STAT5 in hematologic malignancies through inhibition of the bromodomain and extra-terminal (BET) bromodomain protein BRD2. *Mol. Cancer Ther.* **2014**, *13*, 1194–1205. [CrossRef]

291. Guerra, B.; Martín-Rodríguez, P.; Díaz-Chico, J.C.; McNaughton-Smith, G.; Jiménez-Alonso, S.; Hueso-Falcón, I.; Montero, J.C.; Blanco, R.; León, J.; Rodríguez-González, G.; et al. CM363, a novel naphthoquinone derivative which acts as multikinase modulator and overcomes imatinib resistance in chronic myelogenous leukemia. *Oncotarget* **2017**, *8*, 29679–29698. [CrossRef] [PubMed]

292. Juen, L.; Brachet-Botineau, M.; Parmenon, C.; Bourgeais, J.; Hérault, O.; Gouilleux, F.; Viaud-Massuard, M.C.; Prié, G. New inhibitor targeting signal transducer and activator of transcription 5 (STAT5) signaling in myeloid leukemias. *J. Med. Chem.* **2017**, *60*, 6119–6136. [CrossRef] [PubMed]

293. Brachet-Botineau, M.; Deynoux, M.; Vallet, N.; Polomski, M.; Juen, L.; Hérault, O.; Mazurier, F.; Viaud-Massuard, M.C.; Prié, G.; Gouilleux, F. A Novel Inhibitor of STAT5 Signaling Overcomes Chemotherapy Resistance in Myeloid Leukemia Cells. *Cancers* **2019**, *11*, 2043. [CrossRef] [PubMed]

294. Nelson, E.A.; Walker, S.R.; Kepich, A.; Gashin, L.B.; Hideshima, T.; Ikeda, H.; Chauhan, D.; Anderson, K.C.; Frank, D.A. Nifuroxazide inhibits survival of multiple myeloma cells by directly inhibiting STAT3. *Blood* **2008**, *112*, 5095–5102. [CrossRef]

295. Khanim, F.L.; Merrick, B.A.; Giles, H.V.; Jankute, M.; Jackson, J.B.; Giles, L.J.; Birtwistle, J.; Bunce, C.M.; Drayson, M.T. Redeployment-based drug screening identifies the anti-helminthic niclosamide as anti-myeloma therapy that also reduces free light chain production. *Blood Cancer J.* **2011**, *1*, e39. [CrossRef]

296. Sharma, A.; Jyotsana, N.; Lai, C.K.; Chaturvedi, A.; Gabdoulline, R.; Görlich, K.; Murphy, C.; Blanchard, J.E.; Ganser, A.; Brown, E.; et al. Pyrimethamine as a Potent and Selective Inhibitor of Acute Myeloid Leukemia Identified by High-throughput Drug Screening. *Curr. Cancer Drug Targets* **2016**, *16*, 818–828. [CrossRef]

297. Liu, H.; Qin, Y.; Zhai, D.; Zhang, Q.; Gu, J.; Tang, Y.; Yang, J.; Li, K.; Yang, L.; Chen, S.; et al. Antimalarial Drug Pyrimethamine Plays a Dual Role in Antitumor Proliferation and Metastasis through Targeting DHFR and TP. *Mol. Cancer Ther.* **2019**, *18*, 541–555. [CrossRef]

298. Nelson, E.A.; Walker, S.R.; Weisberg, E.; Bar-Natan, M.; Barrett, R.; Gashin, L.B.; Terrell, S.; Klitgaard, J.L.; Santo, L.; Addorio, M.R.; et al. The STAT5 inhibitor pimozide decreases survival of chronic myelogenous leukemia cells resistant to kinase inhibitors. *Blood* **2011**, *117*, 3421–3429. [CrossRef]

299. Bar-Natan, M.; Nelson, E.A.; Walker, S.R.; Kuang, Y.; Distel, R.J.; Frank, D.A. Dual inhibition of Jak2 and STAT5 enhances killing of myeloproliferative neoplasia cells. *Leukemia* **2012**, *26*, 1407–1410. [CrossRef]

300. Tabe, Y.; Konopleva, M.; Andreeff, M.; Ohsaka, A. Effects of PPARγ Ligands on Leukemia. *PPAR Res.* **2012**, *2012*. [CrossRef]

301. Rousselot, P.; Prost, S.; Guilhot, J.; Roy, L.; Etienne, G.; Legros, L.; Charbonnier, A.; Coiteux, V.; Cony-Makhoul, P.; Huguet, F.; et al. Pioglitazone together with imatinib in chronic myeloid leukemia: A proof of concept study. *Cancer* **2017**, *123*, 1791–1799. [CrossRef] [PubMed]

302. Liu, H.; Zang, C.; Fenner, M.H.; Liu, D.; Possinger, K.; Koeffler, H.P.; Elstner, E. Growth inhibition and apoptosis in human Philadelphia chromosome-positive lymphoblastic leukemia cell lines by treatment with the dual PPARα/γ ligand TZD18. *Blood* **2006**, *107*, 3683–3692. [CrossRef] [PubMed]

303. Wang, L.H.; Yang, X.Y.; Zhang, X.; Huang, J.; Hou, J.; Li, J.; Xiong, H.; Mihalic, K.; Zhu, H.; Xiao, W.; et al. Transcriptional inactivation of STAT3 by PPARgamma suppresses IL-6-responsive multiple myeloma cells. *Immunity* **2004**, *20*, 205–218. [CrossRef]

304. Ren, Z.; Cabell, L.A.; Schaefer, T.S.; McMurray, J.S. Identification of a high-affinity phosphopeptide inhibitor of Stat3. *Bioorg. Med. Chem. Lett.* **2003**, *13*, 633–636. [CrossRef]

305. Coleman, D.R., IV; Ren, Z.; Mandal, P.K.; Cameron, A.G.; Dyer, G.A.; Muranjan, S.; Campbell, M.; Chen, X.; McMurray, J.S. Investigation of the binding determinants of phosphopeptides targeted to the SRC homology 2 domain of the signal transducer and activator of transcription 3. Development of a high-affinity peptide inhibitor. *J. Med. Chem.* **2005**, *48*, 6661–6670. [CrossRef]

306. Siddiquee, K.A.; Gunning, P.T.; Glenn, M.; Katt, W.P.; Zhang, S.; Schrock, C.; Sebti, S.M.; Jove, R.; Hamilton, A.D.; Turkson, J. An oxazole-based small-molecule Stat3 inhibitor modulates Stat3 stability and processing and induces antitumor cell effects. *ACS Chem. Biol.* **2007**, *2*, 787–798. [CrossRef]

307. Becker, S.; Groner, B.; Müller, C.W. Three-dimensional structure of the Stat3beta homodimer bound to DNA. *Nature* **1998**, *394*, 145–151. [CrossRef]

308. Siddiquee, K.; Zhang, S.; Guida, W.C.; Blaskovich, M.A.; Greedy, B.; Lawrence, H.R.; Yip, M.L.; Jove, R.; McLaughlin, M.M.; Lawrence, N.J.; et al. Selective chemical probe inhibitor of Stat3, identified through structure-based virtual screening, induces antitumor activity. *Proc. Natl. Acad. Sci. USA* **2007**, *104*, 7391–7396. [CrossRef]

309. Song, H.; Wang, R.; Wang, S.; Lin, J. A low-molecular-weight compound discovered through virtual database screening inhibits Stat3 function in breast cancer cells. *Proc. Natl. Acad. Sci. USA* **2005**, *102*, 4700–4705. [CrossRef]

310. Xu, X.; Kasembeli, M.; Jiang, X.; Tweardy, B.J.; Tweardy, D.J. Chemical probes that competitively and selectively inhibit Stat3 activation. *PLoS ONE* **2009**, *4*, e4783. [CrossRef]

311. Matsuno, K.; Masuda, Y.; Uehara, Y.; Sato, H.; Muroya, A.; Takahashi, O.; Yokotagawa, T.; Furuya, T.; Okawara, T.; Otsuka, M.; et al. Identification of a New Series of STAT3 Inhibitors by Virtual Screening. *ACS Med. Chem. Lett.* **2010**, *1*, 371–375. [CrossRef] [PubMed]

312. Zhang, X.; Sun, Y.; Pireddu, R.; Yang, H.; Urlam, M.K.; Lawrence, H.R.; Guida, W.C.; Lawrence, N.J.; Sebti, S.M. A novel inhibitor of STAT3 homodimerization selectively suppresses STAT3 activity and malignant transformation. *Cancer Res.* **2013**, *73*, 1922–1933. [CrossRef] [PubMed]

313. Huang, Y.H.; Vakili, M.R.; Molavi, O.; Morrissey, Y.; Wu, C.; Paiva, I.; Soleimani, A.H.; Sanaee, F.; Lavasanifar, A.; Lai, R. Decoration of Anti-CD38 on Nanoparticles Carrying a STAT3 Inhibitor Can Improve the Therapeutic Efficacy Against Myeloma. *Cancers* **2019**, *11*, 248. [CrossRef] [PubMed]

314. Fletcher, S.; Singh, J.; Zhang, X.; Yue, P.; Page, B.D.; Sharmeen, S.; Shahani, V.M.; Zhao, W.; Schimmer, A.D.; Turkson, J.; et al. Disruption of transcriptionally active Stat3 dimers with non-phosphorylated, salicylic acid-based small molecules: Potent in vitro and tumor cell activities. *Chembiochem* **2009**, *10*, 1959–1964. [CrossRef]

315. Fletcher, S.; Page, B.D.; Zhang, X.; Yue, P.; Li, Z.H.; Sharmeen, S.; Singh, J.; Zhao, W.; Schimmer, A.D.; Trudel, S.; et al. Antagonism of the Stat3-Stat3 protein dimer with salicylic acid based small molecules. *Chem. Med. Chem.* **2011**, *6*, 1459–1470. [CrossRef]

316. Page, B.D.; Fletcher, S.; Yue, P.; Li, Z.; Zhang, X.; Sharmeen, S.; Datti, A.; Wrana, J.L.; Trudel, S.; Schimmer, A.D.; et al. Identification of a non-phosphorylated, cell permeable, small molecule ligand for the Stat3 SH2 domain. *Bioorg. Med. Chem. Lett.* **2011**, *21*, 5605–5609. [CrossRef]

317. Page, B.D.; Croucher, D.C.; Li, Z.H.; Haftchenary, S.; Jimenez-Zepeda, V.H.; Atkinson, J.; Spagnuolo, P.A.; Wong, Y.L.; Colaguori, R.; Lewis, A.M.; et al. Inhibiting aberrant signal transducer and activator of transcription protein activation with tetrapodal, small molecule Src homology 2 domain binders: Promising agents against multiple myeloma. *J. Med. Chem.* **2013**, *56*, 7190–7200. [CrossRef]

318. Eiring, A.M.; Page, B.D.G.; Kraft, I.L.; Mason, C.C.; Vellore, N.A.; Resetca, D.; Zabriskie, M.S.; Zhang, T.Y.; Khorashad, J.S.; Engar, A.J.; et al. Combined STAT3 and BCR-ABL1 inhibition induces synthetic lethality in therapy-resistant chronic myeloid leukemia. *Leukemia* **2015**, *29*, 586–597. [CrossRef]

319. Page, B.D.; Khoury, H.; Laister, R.C.; Fletcher, S.; Vellozo, M.; Manzoli, A.; Yue, P.; Turkson, J.; Minden, M.D.; Gunning, P.T. Small molecule STAT5-SH2 domain inhibitors exhibit potent antileukemia activity. *J. Med. Chem.* **2012**, *55*, 1047–1055. [CrossRef]

320. Cumaraswamy, A.A.; Lewis, A.M.; Geletu, M.; Todic, A.; Diaz, D.B.; Cheng, X.R.; Brown, C.E.; Laister, R.C.; Muench, D.; Kerman, K.; et al. Nanomolar-potency small molecule inhibitor of STAT5 protein. *ACS Med. Chem. Lett.* **2014**, *5*, 1202–1206. [CrossRef]

321. Liao, Z.; Gu, L.; Vergalli, J.; Mariani, S.A.; De Dominici, M.; Lokareddy, R.K.; Dagvadorj, A.; Purushottamachar, P.; McCue, P.A.; Trabulsi, E.; et al. Structure-Based Screen Identifies a Potent Small Molecule Inhibitor of Stat5a/b with Therapeutic Potential for Prostate Cancer and Chronic Myeloid Leukemia. *Mol. Cancer Ther.* **2015**, *14*, 1777–1793. [CrossRef] [PubMed]

322. Müller, J.; Sperl, B.; Reindl, W.; Kiessling, A.; Berg, T. Discovery of chromone-based inhibitors of the transcription factor STAT5. *Chembiochem* **2008**, *9*, 723–727. [CrossRef] [PubMed]

323. Elumalai, N.; Berg, A.; Natarajan, K.; Scharow, A.; Berg, T. Nanomolar inhibitors of the transcription factor STAT5b with high selectivity over STAT5a. *Angew. Chem. Int. Ed. Engl.* **2015**, *54*, 4758–4763. [CrossRef] [PubMed]

324. Elumalai, N.; Berg, A.; Rubner, S.; Blechschmidt, L.; Song, C.; Natarajan, K.; Matysik, J.; Berg, T. Rational development of Stafib-2: A selective, nanomolar inhibitor of the transcription factor STAT5b. *Sci. Rep.* **2017**, *7*, 819. [CrossRef]

325. Jerez, A.; Clemente, M.J.; Makishima, H.; Koskela, H.; Leblanc, F.; Peng Ng, K.; Olson, T.; Przychodzen, B.; Afable, M.; Gomez-Segui, I.; et al. STAT3 mutations unify the pathogenesis of chronic lymphoproliferative disorders of NK cells and T-cell large granular lymphocyte leukemia. *Blood* **2012**, *120*, 3048–3057. [CrossRef]

326. Kaneko, N.; Kita, A.; Yamanaka, K.; Mori, M. Combination of YM155, a survivin suppressant with a STAT3 inhibitor: A new strategy to treat diffuse large B-cell lymphoma. *Leuk. Res.* **2013**, *37*, 1156–1161. [CrossRef]

327. Mencalha, A.L.; Du Rocher, B.; Salles, D.; Binato, R.; Abdelhay, E. LLL-3, a STAT3 inhibitor, represses BCR-ABL-positive cell proliferation, activates apoptosis and improves the effects of Imatinib mesylate. *Cancer Chemother. Pharmacol.* **2010**, *65*, 1039–1046. [CrossRef]

328. Lin, L.; Benson, D.M., Jr.; DeAngelis, S.; Bakan, C.E.; Li, P.K.; Li, C.; Lin, J. A small molecule, LLL12 inhibits constitutive STAT3 and IL-6-induced STAT3 signaling and exhibits potent growth suppressive activity in human multiple myeloma cells. *Int. J. Cancer* **2012**, *130*, 1459–1469. [CrossRef]

329. Redell, M.S.; Ruiz, M.J.; Alonzo, T.A.; Gerbing, R.B.; Tweardy, D.J. Stat3 signaling in acute myeloid leukemia: Ligand-dependent and -independent activation and induction of apoptosis by a novel small-molecule Stat3 inhibitor. *Blood* **2011**, *117*, 5701–5709. [CrossRef]

330. Schust, J.; Sperl, B.; Hollis, A.; Mayer, T.U.; Berg, T. Stattic: A small-molecule inhibitor of STAT3 activation and dimerization. *Chem. Biol.* **2006**, *13*, 1235–1242. [CrossRef]

331. Huang, Y.H.; Molavi, O.; Alshareef, A.; Haque, M.; Wang, Q.; Chu, M.P.; Venner, C.P.; Sandhu, I.; Peters, A.C.; Lavasanifar, A.; et al. Constitutive Activation of STAT3 in Myeloma Cells Cultured in a Three-Dimensional, Reconstructed Bone Marrow Model. *Cancers* **2018**, *10*, 206. [CrossRef] [PubMed]

332. Liu, J.; Liang, L.; Li, D.; Nong, L.; Zheng, Y.; Huang, S.; Zhang, B.; Li, T. JAK3/STAT3 oncogenic pathway and PRDM1 expression stratify clinicopathologic features of extranodal NK/T-cell lymphoma, nasal type. *Oncol. Rep.* **2019**, *41*, 3219–3232. [CrossRef] [PubMed]

333. Beebe, J.D.; Liu, J.Y.; Zhang, J.T. Two decades of research in discovery of anticancer drugs targeting STAT3, how close are we? *Pharmacol. Ther.* **2018**, *191*, 74–91. [CrossRef] [PubMed]

334. Hayakawa, F.; Sugimoto, K.; Harada, Y.; Hashimoto, N.; Ohi, N.; Kurahashi, S.; Naoe, T. A novel STAT inhibitor, OPB-31121, has a significant antitumor effect on leukemia with STAT-addictive oncokinases. *Blood Cancer J.* **2013**, *3*, e166. [CrossRef]

335. Brambilla, L.; Genini, D.; Laurini, E.; Merulla, J.; Perez, L.; Fermeglia, M.; Carbone, G.M.; Pricl, S.; Catapano, C.V. Hitting the right spot: Mechanism of action of OPB-31121, a novel and potent inhibitor of the Signal Transducer and Activator of Transcription 3 (STAT3). *Mol. Oncol.* **2015**, *9*, 1194–1206. [CrossRef]

336. Ogura, M.; Uchida, T.; Terui, Y.; Hayakawa, F.; Kobayashi, Y.; Taniwaki, M.; Takamatsu, Y.; Naoe, T.; Tobinai, K.; Munakata, W.; et al. Phase I study of OPB-51602, an oral inhibitor of signal transducer and activator of transcription 3, in patients with relapsed/refractory hematological malignancies. *Cancer Sci.* **2015**, *106*, 896–901. [CrossRef]

337. Bai, L.; Zhou, H.; Xu, R.; Zhao, Y.; Chinnaswamy, K.; McEachern, D.; Chen, J.; Yang, C.Y.; Liu, Z.; Wang, M.; et al. A Potent and Selective Small-Molecule Degrader of STAT3 Achieves Complete Tumor Regression In Vivo. *Cancer Cell* **2019**, *36*, 498–511. [CrossRef]

338. Yco, L.P.; Mocz, G.; Opoku-Ansah, J.; Bachmann, A.S. Withaferin A Inhibits STAT3 and Induces Tumor Cell Death in Neuroblastoma and Multiple Myeloma. *Biochem. Insights* **2014**, *7*, 1–13. [CrossRef]

339. Wang, Y.; Wu, L.; Cai, H.; Lei, H.; Ma, C.M.; Yang, L.; Xu, H.; Zhu, Q.; Yao, Z.; Wu, Y. Sinomenine derivative YL064: A novel STAT3 inhibitor with promising anti-myeloma activity. *Cell Death Dis.* **2018**, *9*, 1093. [CrossRef]

340. Turkson, J.; Zhang, S.; Palmer, J.; Kay, H.; Stanko, J.; Mora, L.B.; Sebti, S.; Yu, H.; Jove, R. Inhibition of constitutive signal transducer and activator of transcription 3 activation by novel platinum complexes with potent antitumor activity. *Mol. Cancer Ther.* **2004**, *3*, 1533–1542.

341. Turkson, J.; Zhang, S.; Mora, L.B.; Burns, A.; Sebti, S.; Jove, R. A novel platinum compound inhibits constitutive Stat3 signaling and induces cell cycle arrest and apoptosis of malignant cells. *J. Biol. Chem.* **2005**, *280*, 32979–32988. [CrossRef] [PubMed]

342. Nagel-Wolfrum, K.; Buerger, C.; Wittig, I.; Butz, K.; Hoppe-Seyler, F.; Groner, B. The interaction of specific peptide aptamers with the DNA binding domain and the dimerization domain of the transcription factor Stat3 inhibits transactivation and induces apoptosis in tumor cells. *Mol. Cancer Res.* **2004**, *2*, 170–182. [PubMed]

343. Wang, X.; Zeng, J.; Shi, M.; Zhao, S.; Bai, W.; Cao, W.; Tu, Z.; Huang, Z.; Feng, W. Targeted blockage of signal transducer and activator of transcription 5 signaling pathway with decoy oligodeoxynucleotides suppresses leukemic k562 cell growth. *DNA Cell Biol.* **2011**, *30*, 71–78. [CrossRef] [PubMed]

344. Zhao, X.; Zhang, Z.; Moreira, D.; Su, Y.L.; Won, H.; Adamus, T.; Dong, Z.; Liang, Y.; Yin, H.H.; Swiderski, P.; et al. B Cell Lymphoma Immunotherapy Using TLR9-Targeted Oligonucleotide STAT3 Inhibitors. *Mol. Ther.* **2018**, *26*, 695–707. [CrossRef]

345. Jing, N.; Li, Y.; Xiong, W.; Sha, W.; Jing, L.; Tweardy, D.J. G-quartet oligonucleotides: A new class of signal transducer and activator of transcription 3 inhibitors that suppresses growth of prostate and breast tumors through induction of apoptosis. *Cancer Res.* **2004**, *64*, 6603–6609. [CrossRef]

346. Hillion, J.; Belton, A.M.; Shah, S.N.; Turkson, J.; Jing, N.; Tweardy, D.J.; di Cello, F.; Huso, D.L.; Resar, L.M. Nanoparticle delivery of inhibitory signal transducer and activator of transcription 3 G-quartet oligonucleotides blocks tumor growth in HMGA1 transgenic model of T-cell leukemia. *Leuk. Lymphoma* **2014**, *55*, 1194–1197. [CrossRef]

347. Shastri, A.; Choudhary, G.; Teixeira, M.; Gordon-Mitchell, S.; Ramachandra, N.; Bernard, L.; Bhattacharyya, S.; Lopez, R.; Pradhan, K.; Giricz, O.; et al. Antisense STAT3 inhibitor decreases viability of myelodysplastic and leukemic stem cells. *J. Clin. Investig.* **2018**, *128*, 5479–5488. [CrossRef]

348. Hong, D.; Kurzrock, R.; Kim, Y.; Woessner, R.; Younes, A.; Nemunaitis, J.; Fowler, N.; Zhou, T.; Schmidt, J.; Jo, M.; et al. AZD9150, a next-generation antisense oligonucleotide inhibitor of STAT3 with early evidence of clinical activity in lymphoma and lung cancer. *Sci. Transl. Med.* **2015**, *7*, 314. [CrossRef]

349. Reilley, M.J.; McCoon, P.; Cook, C.; Lyne, P.; Kurzrock, R.; Kim, Y.; Woessner, R.; Younes, A.; Nemunaitis, J.; Fowler, N.; et al. STAT3 antisense oligonucleotide AZD9150 in a subset of patients with heavily pretreated lymphoma: Results of a phase 1b trial. *J. Immunother. Cancer* **2018**, *6*, 119. [CrossRef]

350. Zhang, Q.; Hossain, D.M.; Nechaev, S.; Kozlowska, A.; Zhang, W.; Liu, Y.; Kowolik, C.M.; Swiderski, P.; Rossi, J.J.; Forman, S.; et al. TLR9-mediated siRNA delivery for targeting of normal and malignant human hematopoietic cells in vivo. *Blood* **2013**, *121*, 1304–1315. [CrossRef]

351. Von Manstein, V.; Yang, C.M.; Richter, D.; Delis, N.; Vafaizadeh, V.; Groner, B. Resistance of cancer cells to targeted therapies through the activation of compensating signaling loops. *Curr. Signal. Transduct. Ther.* **2013**, *8*, 193–202. [CrossRef] [PubMed]

Review

Direct Targeting Options for STAT3 and STAT5 in Cancer

Anna Orlova [1], Christina Wagner [1], Elvin D. de Araujo [2,3], Dávid Bajusz [4], Heidi A. Neubauer [1], Marco Herling [5], Patrick T. Gunning [2,3], György M. Keserű [4] and Richard Moriggl [1,*]

[1] Institute of Animal Breeding and Genetics, University of Veterinary Medicine, 1210 Vienna, Austria; anna.orlova@vetmeduni.ac.at (A.O.); Christina-Maria.Wagner@vetmeduni.ac.at (C.W.); Heidi.Neubauer@vetmeduni.ac.at (H.A.N.)

[2] Department of Chemical and Physical Sciences, University of Toronto Mississauga, Mississauga, ON L5L 1C6, Canada; e.dearaujo@mail.utoronto.ca (E.D.d.A.); patrick.gunning@utoronto.ca (P.T.G.)

[3] Centre for Medicinal Chemistry, University of Toronto Mississauga, Mississauga, ON L5L 1C6, Canada

[4] Medicinal Chemistry Research Group, Research Centre for Natural Sciences, H-1117 Budapest, Hungary; bajusz.david@ttk.mta.hu (D.B.); keseru.gyorgy@ttk.mta.hu (G.M.K.)

[5] Department I of Internal Medicine, Center for Integrated Oncology (CIO), Excellence Cluster for Cellular Stress Response and Aging-Associated Diseases (CECAD), and Center for Molecular Medicine Cologne (CMMC), Cologne University, 50937 Cologne, Germany; marco.herling@uk-koeln.de

* Correspondence: richard.moriggl@vetmeduni.ac.at

Received: 17 October 2019; Accepted: 29 November 2019; Published: 3 December 2019

Abstract: Signal transducer and activator of transcription (STAT)3 and STAT5 are important transcription factors that are able to mediate or even drive cancer progression through hyperactivation or gain-of-function mutations. Mutated STAT3 is mainly associated with large granular lymphocytic T-cell leukemia, whereas mutated STAT5B is associated with T-cell prolymphocytic leukemia, T-cell acute lymphoblastic leukemia and $\gamma\delta$ T-cell-derived lymphomas. Hyperactive STAT3 and STAT5 are also implicated in various hematopoietic and solid malignancies, such as chronic and acute myeloid leukemia, melanoma or prostate cancer. Classical understanding of STAT functions is linked to their phosphorylated parallel dimer conformation, in which they induce gene transcription. However, the functions of STAT proteins are not limited to their phosphorylated dimerization form. In this review, we discuss the functions and the roles of unphosphorylated STAT3/5 in the context of chromatin remodeling, as well as the impact of STAT5 oligomerization on differential gene expression in hematopoietic neoplasms. The central involvement of STAT3/5 in cancer has made these molecules attractive targets for small-molecule drug development, but currently there are no direct STAT3/5 inhibitors of clinical grade available. We summarize the development of inhibitors against the SH2 domains of STAT3/5 and discuss their applicability as cancer therapeutics.

Keywords: STAT3; STAT5; cancer; small-molecule inhibitors

---

1. Introduction

The Janus kinase/Signal transducer and activator of transcription (JAK-STAT) pathway is one of the core cancer pathways that integrates signals from cytokines, hormones and growth factors to induce or repress gene expression in cells [1]. The pathway consists of four JAK kinases (JAK1, JAK2, JAK3 and TYK2) and seven STAT transcription factors (STAT1, STAT2, STAT3, STAT4, STAT5A, STAT5B and STAT6). Despite conserved structure and a common mechanism of action, STAT family members show distinct and even opposite functions in tumor biology.

STAT1 is generally not associated with promoting tumor growth and mostly mediates tumor-suppressive and pro-apoptotic functions [2,3]. Consistently, STAT1$^{-/-}$ mice are more prone to tumor development and *STAT1* deletion in leukemic cells decreases MHC class I expression [4,5]. Surprisingly, in a *v-abl*-driven model of STAT1$^{-/-}$ leukemic cells initially harboring low MHC class I, enhanced MHC class I expression was gained during the disease progression, thereby reducing tumor recognition by NK cells [6]. As an exception, STAT1 was shown to be an oncogenic driver in T-cell acute lymphoblastic leukemia (T-ALL) and ALK$^+$ anaplastic large cell lymphoma (ALCL), and STAT1 is associated with the JAK2 exon 12 mutation in the progression of myeloproliferative neoplasms (MPNs) [7–9]. Reports on direct involvements of STAT2, STAT4 and STAT6 in cancerous processes are scarce and are not discussed here further [10]. Associations of STAT3 and STAT5 with cancer progression are well established and heavily studied, and hence, are the main focus of this review.

2. Role of STAT3 and STAT5 in Cancer

STAT3 and STAT5 proteins are of particular interest in cancer research, as their hyperactivation was reported in processes ranging from inflammation and autoimmunity to infection and cancer. Within the latter, these proteins have been implicated in tumor initiation, as well as in metastasis or in conferring drug resistance mechanisms [10–12]. Constitutive hyperactivation of STAT3 and STAT5, as a result of their gain-of-function mutations or via enhanced signaling from upstream drivers, promotes tumor cell growth and survival [3].

STAT3 is the best-studied family member of the JAK-STAT pathway in cancer, and is a known oncogene in various types of solid malignancies, like melanoma or lung cancer [13–15]. Furthermore, both STAT3 and STAT5 are reported to play a major role in the progression and pathogenesis of prostate cancer [16–18]. Hyperactivation of STAT3 and STAT5 is commonly associated with an upstream oncogenic driver, such as hyperactive mutated tyrosine kinases, for example, JAK2^{V617F} or FLT3-ITD, or fusion proteins such as BCR-ABL, TEL-JAK2 or TEL-ABL1 [3,19–21]. The role of STAT5 in the transformation process induced by BCR-ABL p210 fusion protein has been particularly well-studied. It was shown that the absence of STAT5 diminishes the ability of cells to transform even upon harboring potent oncogenes such as BCR-ABL. Inducible deletion of STAT5 arrests and kills chronic myeloid leukemia (CML) cell lines, defining STAT5 as a therapeutic cancer target [22]. Hyperactivation of STAT3 and STAT5 can also occur via direct mutation in these genes, which is also associated with cancer progression in patients [1]. Interestingly, mutated STAT3 is mainly associated with large granular lymphocytic T-cell leukemia (T-LGLL), whereas mutated STAT5B is found in patients with T-cell prolymphocytic leukemia (T-PLL), T-ALL, γδ T-cell-derived lymphoma and monomorphic epitheliotropic intestinal T-cell lymphoma (MEITL) [23–25].

Inhibitors of kinases upstream of STAT3/5 are available, but patients often relapse by developing drug resistance through persistent signaling or enhanced upregulation of STAT3 or STAT5 expression [21,26]. Targeting STAT3 and STAT5 directly or in combination with tyrosine kinase inhibitors might be an attractive way to overcome these resistance mechanisms [27].

3. Non-Canonical Functions of STAT3 and STAT5

STAT proteins share similar structural architecture including five domains: an N-terminal domain, a coiled-coil domain, a DNA-binding domain, an SH2 domain and a C-terminal transactivation domain [11] (Figure 1a). Upon phosphorylation, STAT5 undergoes a conformational change and forms parallel dimers via the SH2 domain (Figure 1b). This conformation allows the dimers to be recognized by importins and facilitates transport into the nucleus, where they bind GAS consensus sequences to induce target gene transcription. Such dramatic conformational rearrangements are possible due to the flexible linkers that are connecting the core fragment of STAT5 with its N- and C-terminus [28–30]. The complex approach of molecular dynamic simulations and bioinformatic analyses identified three distinct interaction surfaces within the dimer unique to STAT5, which include intramolecular interactions between the SH2 domain and the phosphotyrosine motif [28,31].

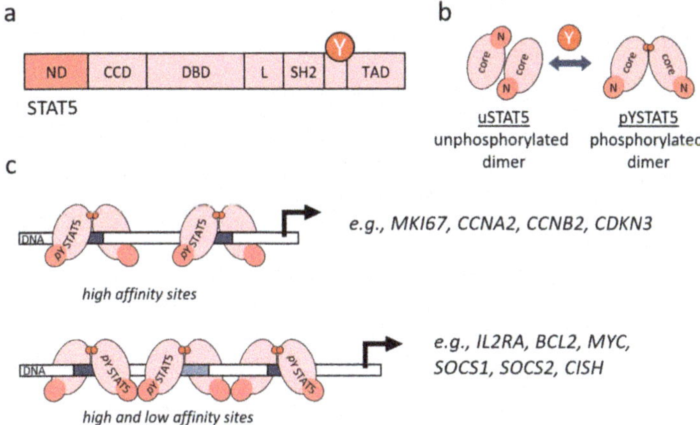

Figure 1. Functions of STAT5. (**a**) Domain structure of STAT5. Position of the critical activating tyrosine phosphorylation site is depicted with Y. ND—N-domain; CCD—coiled-coil domain; DBD—DNA-binding domain; L—linker; SH2—SH2 domain; TAD—transactivation domain. (**b**) Conformational changes of STAT5 from an unphosphorylated antiparallel dimer (uSTAT5) to a phosphorylated parallel dimer (pYSTAT5). (**c**) Dimer and oligomer conformations of STAT5 result in binding to GAS sites on DNA with different affinities, resulting in expression of different genes. Examples of dimer versus tetramer target genes are incorporated from [32,33].

Interestingly, the SH2 domain of the hyperactivated gain-of-function variant STAT5B[N642H] was also crystallized and compared to wild type human STAT5B. The crystal structure revealed two conformations of the mutant SH2 domain: one conformation is more closed, potentially allowing longer tyrosine phosphorylation lifetimes by blocking phosphatase attack. The other conformation, preserved in the crystal structure, revealed a more open conformation, which could facilitate phospho-peptide or protein interactions leading to hyperactivation [34].

Interestingly, the functions of STAT proteins are not limited to their state as phosphorylated dimers. Unphosphorylated STAT dimers (uSTAT), as well as tetramer/oligomer conformations, are involved in the functionality of some STATs [32,35]. Two phosphorylated dimers can form tetramers via their N-terminal oligomerization domains. This interaction stabilizes DNA binding and allows attraction of the tetramers to low-affinity sites, thereby providing altered binding selectivity and fine-tuning of transcriptional responses [36] (Figure 1c).

STAT5 is particularly known for forming oligomers, a function shared with STAT1 and STAT4. In contrast, STAT3 shows weaker tetramer formation upon activation. STAT2 and STAT6 were not reported to form oligomers in cellulo [32,37–39]. However, yeast hybridization assays showed that recombinant N-domains of all STATs are able to self-dimerize [40]. Interestingly, this interaction is clearly homotypic, which might facilitate individual functions of different STATs. In addition, recombinant STAT2 and STAT6 N-domains can form oligomers with the same affinity as other STAT members, but this was not observed in cellulo [40].

The N-terminus also plays an important role in the function of STAT3: it mediates dimerization of uSTAT3, whereas for phosphorylated STAT3 (pYSTAT3) it can enable tetramerization [41,42]. Interestingly, oligomerization of STAT3 is not commonly observed in cellulo. Still, expression of certain genes like A2M (α2-macroglobulin) was shown to depend on an N-domain interaction of STAT3 [42].

Hu et al. showed that in STAT3[−/−] MEFs, exogenous expression of an N-terminally truncated STAT3 protein leads to a decrease in expression of a small subset of genes, compared to expression of wild type STAT3 [43]. Not surprisingly, mutated STAT3[Y705F], which is unable to become phosphorylated and to form parallel dimers, significantly affected global gene expression. However, certain genes (e.g., *MRAS, MET*) were still activated even by uSTAT3 [44]. Interestingly, STAT3^{Y705F} was able to weakly bind to selected promoters and to induce gene expression, thereby promoting an anti-viral, anti-proliferative effect in response to interferon stimulation. It would be interesting to determine if this activity is connected to the heterodimerization of STAT3 with STAT1 or if it is fully mediated through uSTAT3 [45]. uSTAT3 was also reported to be involved in interactions and complex formation with unphosphorylated nuclear factor kappa-light-chain-enhancer of activated B cells (NF-κB), which resulted in activation of a subset of NF-κB-dependent genes. Of note, the authors showed an additional subset of uSTAT3-dependent, but NF-κB-independent genes [46].

Timofeeva et al. showed that uSTAT3 plays a repressive role in the apoptosis of cancer cells and that inhibition of STAT3 N-domain functions can abolish this repressive effect. Treatment with the STAT3 N-domain peptidomimetic, ST3-H2A2, hindered STAT3 binding to the regulatory domains of various genes, including the gene for the proapoptotic C/EBP-homologous protein (CHOP). This, in turn, led to a decrease in the heterochromatin mark H3K9me3 in the promoter region of this gene [47].

uSTAT5 is mostly localized in the cytoplasm in the form of anti-parallel dimers. In mammalian cells, cytoplasmic uSTAT5 was shown to associate with and stabilize the Golgi apparatus [48]. On the other hand, a smaller proportion of uSTAT5 was also found in the nucleus where it colocalizes with the transcriptional repressor CTCF, thereby diminishing megakaryocyte differentiation via competition for DNA binding with the transcription factor ERG [35]. Additionally, there is evidence that uSTAT5 can migrate into the nucleus and bind heterochromatin protein 1α (HP1α) to promote the formation of heterochromatin. This leads to repression of various genes, among which many are found to be involved in cancer development [49].

One function of the STAT5 N-terminal domain involves docking of STAT5 to receptors. For example, it was shown that STAT5 interacts via the N-terminal domain with the glucocorticoid receptor [50]. The N-domain of STAT5 has an O-GlcNAc modification site on threonine 92. For STAT5A, it was shown that the presence of this glucose-derived modification within the N-domain is essential for full activation, suggesting cross-talk between the N- and C-domains and involvement of the N-domain in the regulation of metabolic functions in cells [51].

Another function of the STAT5 N-domain is mediating STAT5 tetramerization. Tetramerization was shown to be essential for proper T-cell development and was also found to be associated with enhanced activity of STAT5 as an oncogene [32]. STAT5 tetramers can bind to different motifs compared with STAT5 dimers, to induce or repress gene expression (Figure 1c) [52]. It was shown that deletion of the N-domain of STAT5 results in an absence of *c-MYC, BCL-2* and *cyclin D2* expression upon stimulation, indicating that oligomers can induce a different subset of genes than dimers [32]. Extensive studies by Lin et al., using a mutated STAT5 N-domain that is unable to form oligomers in vivo, defined the oligomer-dependent subset of genes as well as the importance of oligomerization in NK cell maturation [33,52].

The STAT5 N-domain was shown to be essential for leukemogenic transformation. Deletion of the N-domain or mutation of the O-GlcNAc-modified residue (T92A) abolished the initiation of the leukemic disease driven by gain-of-function STAT5A [32,51,53]. This suggests an important function of the STAT5 N-domain in oncogenic transformation and it proposes that the N-domain can serve as a novel targeting interface of STAT5.

4. Role of STAT3/5 in Chromatin Landscape

Over the last years, it was shown that STAT3 and STAT5 transcription factors can influence gene expression not only directly by binding to gene promoters but also through recruiting various chromatin remodelers and influencing gene expression and chromatin states on the global level. STAT3/5 can change the chromatin landscape in a cell by recruiting various chromatin-remodeling or DNA-modifying enzymes to the DNA, such as histone acetyltransferases (HATs), histone deacetylases (HDACs), as well as ten-eleven translocation methylcytosine dioxygenase 1/2 (TET1/2) or DNA (cytosine-5)-methyltransferase 1 (DNMT1) [54–57] (Figure 2a,b). These interactions influence eu- or hetero-chromatin formation or DNA methylation, thereby activating or repressing transcription. The STATs themselves can also be post-translationally modified by these enzymes: for example, methylation of STAT3 by the enhancer of zeste homolog 2/polycomb repressive complex 2 (EZH2/PRC2), or acetylation of STAT3 and STAT5 by CREB-binding protein (CBP)/p300 [1].

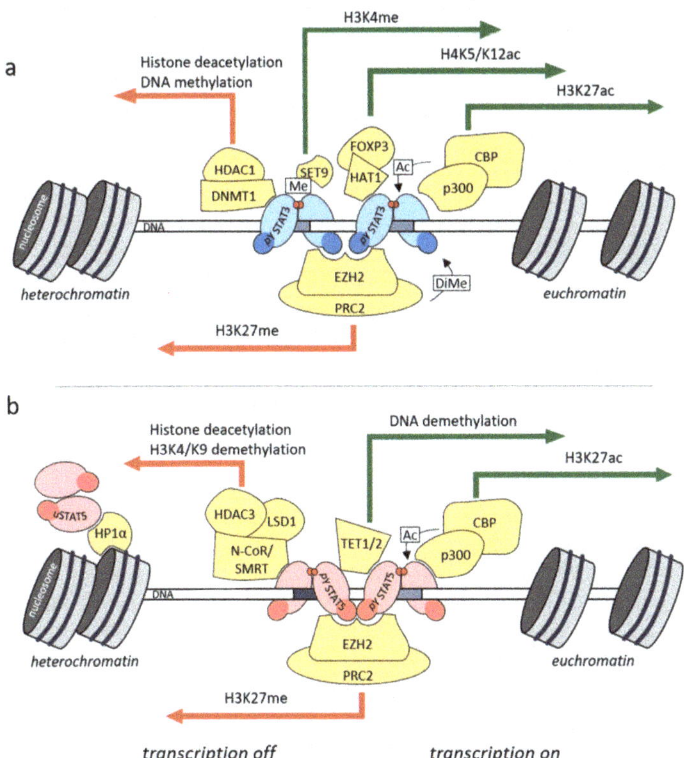

Figure 2. Role of STAT3/5 in regulating the chromatin landscape. (**a**) pYSTAT3 interacts with and recruits chromatin remodelers, and thereby promotes changes in chromatin compaction. (**b**) uSTAT5 and pYSTAT5 interact with different chromatin remodelers, thereby promoting chromatin changes associated with euchromatin (green arrows) or heterochromatin (red arrows) formation.

In the case of STAT3, acetylation or methylation at different residues leads to different activity and functionality of the protein (Figure 2a) [12]. For example, acetylation of STAT3 by CBP/p300 on lysine 685 in the C-terminal domain increases the DNA binding ability of STAT3 [58]. On the other hand, methylation of phosphorylated, promoter-bound STAT3 on lysine 140 by SET9 reduces its transcriptional activity on a subset of target genes [59], whereas dimethylation on lysine 49 by EZH2 is required for the expression of IL-6-dependent genes [60]. It was also shown that STAT3 binds to the promoter of the tyrosine phosphatase *SHP-1* and recruits DNMT1 and HDAC1 to silence its transcription in cutaneous T-cell lymphoma (CTCL) and in ALK$^+$ ALCL cell lines [57]. In regulatory T-cells, STAT3 was found to cooperate with FoxP3 and HAT1 to induce expression of IL-10 [61].

STAT5 is also known to recruit chromatin remodelers (Figure 2b). One example of such an interaction is co-activation of STAT5 by HAT nuclear receptor coactivator 1 (NCoA-1). NCoA-1 and STAT5A transiently co-transfected in HEK293T cells were shown to interact with each other by co-immunoprecipitation. This interaction required amino acids 751 to 753 in the STAT5 transactivation domain, which is conserved in both STAT5A and STAT5B [54]. Furthermore, in Ba/F3 cells, STAT5A was shown to interact with HDAC3 and lysine-specific demethylase 1 (LSD1), thereby activating or repressing gene expression [55]. The interaction between STAT5A and LSD1/HDAC3 is mediated by the STAT5A DNA-binding domain, linker and its SH2 domain and HDAC3 can additionally interact with the coiled-coil domain of STAT5A [55]. STAT5 influences not only the acetylation but also the methylation status of its surroundings, for example by recruiting TET1/2 to the *Foxp3* locus. This causes demethylation of the locus and plays an important role in regulatory T-cell differentiation [56].

Mandal et al. showed that during B-cell maturation and B-cell receptor rearrangement, gene regions required for immunoglobulin κ light chain expression are silenced by histone methyltransferase EZH2-STAT5 tetramer interactions (Figure 2b) [62]. This underlines the importance of the STAT5 N-terminal domain and tetramerization during B-cell development or as a mechanism for acute B-cell leukemia initiation or progression.

5. Direct STAT Targeting of the SH2 Domain

Treating cancers with hyperactivated or mutated STAT3 and STAT5 is currently achieved by targeting upstream kinases. However, despite tyrosine kinase inhibitors being significantly superior to classical chemotherapy, their application often causes severe side-effects and development of resistance [63]. Therefore, development of more specific and effective inhibitors that also target downstream components of hyperactivated pathways is desirable to overcome limitations of current strategies. In the following section, we focus specifically on small-molecule inhibitors of the STAT3/5 SH2 domains.

5.1. STAT3 Inhibitors

The first two small-molecule STAT3 inhibitors were fragment-sized compounds discovered by random biochemical and virtual screening. One compound, Stattic, was identified in a high-throughput screen of a diverse chemical library (Figure 3a) [64]. Another anthraquinone-based compound, STA-21 (Figure 3b), was discovered by structure-based virtual screening against the published X-ray structure of STAT3β [65,66]. Later, multiple analogs of STA-21 were reported, such as LLL-3 and LLL-12 (Figure 3c,d) [67,68]. Recently, several different chemotypes have emerged, often consisting of ring systems connected by amide-containing linkers. An important step forward was the identification of the salicylic acid moiety that is an efficient bioisostere of the phosphate group required for STAT-STAT dimer formation. Salicylic acid analogs were described as potent STAT3 inhibitors as exemplified by the inhibitor S31-201 (Figure 3e), as well as its optimized successors, SF-1-066, SF-1-121 and S31-1757 (Figure 3f,g) [69–71].

Virtual screening was useful to find new chemotypes of STAT3 inhibitors. Matsuno et al. and Xu et al. described further double-digit micromolar STAT3 inhibitors, STX-0119 and Cpd30-12 (in cellular assays; Figure 3h,i) [72,73]. Furthermore, a number of natural products (or analogs thereof) and antioxidants have been proposed and identified as potential inhibitors of STATs, most particularly STAT3, displaying even single-digit micromolar inhibitory activities [74–78].

Another noteworthy study reported an in silico fragment-based drug discovery approach, resulting in a single-digit micromolar STAT3 inhibitor, LY5 (in both cell-free and cell-based assays; Figure 3j), targeted towards the SH2 domain [79]. Recently, Zhang et al. identified benzothiazole as a novel scaffold among STAT3 inhibitors, resulting from a virtual screening of more than 200,000 compounds by a multistep protocol. In this study, four compounds were confirmed experimentally, with the benzothiazole-based compound 9 (Figure 3k) displaying a single-digit micromolar IC_{50} value against the IL-6/STAT3 signaling pathway [80]. Two further analogs were discovered by a similarity-based hit expansion. When the number of known STAT3 inhibitors reached a sufficient level to compile a training set for a three-dimensional (3D) pharmacophore-based virtual screening study, Leung et al. screened a small in-house dataset and tested five compounds, out of which one (Cpd1, Figure 3l) was experimentally confirmed in multiple STAT3-related endpoints [81].

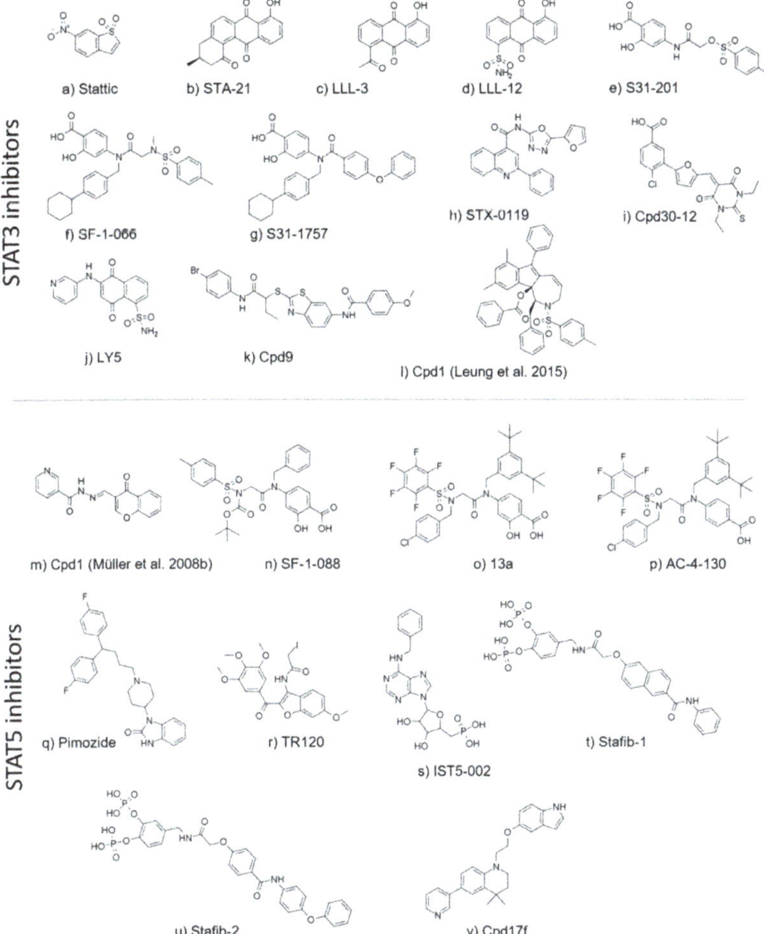

Figure 3. Chemical structures of STAT3 and STAT5 inhibitors (in their order of citation in the main text). Compounds with identical names (Cpd1) are further specified by citations.

The compounds OPB-31121, OPB-51602 and OPB-111077, which are substances from Otsuka Pharmaceuticals, were designed to inhibit STAT3 phosphorylation in cancer cell lines and xenograft models by targeting the STAT3 SH2 domain. Available studies with OPB-51602 indicate induction of STAT3 aggregation in autophagosomes. Additionally, there has been controversy on the activity of OPB-51602 against mitochondrial STAT3. Two of these STAT3 SH2 inhibitors, OPB-31121 and OPB-51602, were already tested in phase I/II clinical trials for solid tumors, non-Hodgkin lymphoma, myeloma and other hematopoietic malignancies [81]. In these studies, the compounds were promising based on a relatively long half-life, suggesting the possibility of a reduced dosing regimen to limit toxic side effects. However, the clinical trials for both lead structure drugs were terminated due to minimal antitumor activity, toxicity issues and poor pharmacokinetics. Another STAT3 inhibitor, OPB-111077, has completed a phase I trial for solid cancers (NCT02250170), and phase I/II trials are recruiting for acute myeloid leukemia (AML) (NCT03197714) and other refractory tumors (NCT03158324) [82,83]. Efficacies of the above-mentioned STAT3 SH2 inhibitors in vitro and in vivo are summarized in Table 1.

Table 1. Small-molecule compounds targeting the STAT3 SH2 domain.

Cpd	Protein-Based		Cell-Based			In Vivo Application Tested	Refs.
	Assay	IC$_{50}$ or Ki [μM]	Cell Line	Readout	IC$_{50}$ [μM]a		
Stattic	pY binding	5.1	HepG2, MDA-MB-231	viability	3.8		[64,84,85]
			RAW264.7	pYSTAT3	s (20)	Osteoclasto genesis in C57/BL6 mice (10 mg/kg)	
			M-SCC-17B, OSC-19, Cal33, UM-SCC-22	viability	2.2–3.5	Head and neck cancer xenograft (50 mg/kg)	
STA-21			Caov-3	reporter assay	s (20)	Psoriatic disease in mouse model and phase I in clinical trial (NCT0104794)	[66,86]
			MDA-MB-435s, MDA-MB-468, MDA-MB-231	DNA binding, viability	s (20)		
LLL-3			U373, MDA-MB-231	DNA binding	s (20)	Xenograft glioblastoma (50 mg/kg)	[68]
			MDA-MB-231	reporter assay	s (20)		
LLL-12			MDA-MB-231, SK-BR-3, PANC-1, HPAC, U87, U373, A549	viability, pYSTAT3, reporter assay	0.16–3.09	Glioblastoma, breast cancer xenograft (2.5, 5 mg/kg) Lung cancer xenograft (20, 10 mg/kg)	[67,87]
S31-201		80	NIH 3T3/v-Src	DNA binding	86	Xenograft breast cancer (5 mg/kg)	[69,71]
			MDA-MB-468, MDA-MB-231	pYSTAT3	s (100)		
			DU145, MDA468, OCI-AML-2	viability	28–112		
SF-1-066	FP	20	NIH 3T3/v-Src	DNA binding	35		[69]
			DU145, MDA468, OCI-AML-2	viability	17–37		
S31-1757	pY binding	13.5	HEK293	CoIP	s (50)		[88]
			MDA-MB-468, A549	pYSTAT3, reporter assay	s (50)		
STX-0119			HeLa	reporter assay	74	Xenograft lymphoma (160 mg/kg)	[73]
			HEK293	FRET-based dimerization	s (50)		
Cpd30-12	pY binding	114	HepG2, MEF/GFP-Stat3α, MDA-MB-468, MDA-MB-231, MBA-MD-435, MCF7	pYSTAT3, nuclear translocation, apoptosis	60		[72]
LY5	FP	2.5	U2OS, RH30, RD2, MDA-MB-231	viability, pYSTAT3	0.52–1.39	Xenograft breast cancer (5 mg/kg)	[79,89]
			UW426, UW288-1, DAOY	pYSTAT3	s (0.5)		
Cpd9			HepG2/STAT3	reporter assay, pYSTAT3	3.57		[80]
			MDA-MB-468	viability	8.83		
Cpd1	FP	~10	HeLa	reporter assay, DNA binding	~10		[81]

[a] s (20): significant effect at 20 μM (or other concentration, as indicated); cpd—compound; FP—fluorescence polarization.

5.2. STAT5 Inhibitors

The first landmark papers on STAT5 SH2 domain inhibitors were published in 2008 by the Berg group. These were driven by the development of a robust high-throughput screening assay for STAT5B inhibitors [90,91]. This facilitated the discovery of chromone-based STAT5B inhibitors as well as a nicotinoyl hydrazine derivative, Cpd1 (Figure 3m), as a selective STAT5B inhibitor (in comparison to STAT1 and STAT3) [91].

Later, four salicylic acid-based compounds (including the single-digit micromolar inhibitor SF-1-088, Figure 3n) were identified by the Gunning group [92]. These hits were optimized to generate the STAT5B inhibitor, 13a, through a structure-guided approach (Figure 3o). Studies with 13a have shown it can effectively reduce pYSTAT5B levels in cellulo, and follow up thermal stability studies showed an inhibitor-induced reduction in STAT5 stability and the potential to block de novo phosphorylation [28,93,94]. Additional studies extended this work to demonstrate AML cell viability inhibition with AC-4-130, a further improved lead compound (Figure 3p) [95].

It is worth noting that in a 2011 study, Nelson et al. identified the neuroleptic drug pimozide as a STAT5 inhibitor (Figure 3q). However, recent studies have linked pimozide with proteolysis upstream of STAT5, rather than direct binding [96]. Nonetheless, the discovery later prompted Rondanin et al. to synthesize and screen a series of pimozide derivatives, two of which have surpassed the cytotoxic potency of pimozide as evaluated against imatinib-resistant BCR-ABL-expressing leukemia cells [97]. In 2016, the same group synthesized and tested 22 iodoacetamide-containing heterocycles for STAT5 inhibition, many of which (including TR120, Figure 3r) were confirmed experimentally in an in vitro cytotoxicity assay, although direct binding to STAT5 was not examined [98].

In 2015, Liao et al. conducted a large scale structure-based virtual screening campaign for STAT5A/B inhibitors [99]. Using a STAT3 based homology model of STAT5, the authors docked ~30 million compounds to the dimerization interface on the SH2 domain [65]. The top 30 hits were evaluated in various cell lines, with IST5-002 (Figure 3s) identified as a lead compound that inhibited STAT5A and STAT5B in the low micromolar range in cellulo. However, the first STAT5A/B gene product-selective inhibitor was developed in the same year, by the Berg group. Stafib-1 (Figure 3t), a catechol type bisphosphate-containing nanomolar STAT5B inhibitor (as evaluated in a fluorescence polarization assay), was identified by structure-based virtual screening against the SH2 domain of a STAT5B model derived from the structure of unphosphorylated STAT5A [30,100]. This compound was recently optimized into the single-digit nanomolar inhibitor Stafib-2 (Figure 3u) using a structure-based approach [101]. Stafib-2, similar to its predecessor, has shown high selectivity for STAT5B compared to other STAT proteins, including STAT5A. The hydrophilicity of the phosphate motifs was reduced by generating a pro-drug precursor, which employs pivaloyloxymethyl esters to conceal the negatively charged phosphate groups and increase cell penetrance. Recently, Natarajan et al. reported Stafia-1, a selective inhibitor for STAT5A, discovered by docking-based screening [102]. Tested applications for STAT5 inhibitors in vivo and in vitro are summarized in Table 2.

Berg et al. also identified nucleotide scaffolds with potentially selective inhibition properties against STAT5B. Although, the IC_{50} values in a fluorescence polarization assay are in the high micro-molar range (STAT5B, IC_{50}[ATP] = 97.4 ± 0.9 μM, IC_{50}[GTP] = 95.5 ± 3.2 μM; STAT5A, IC_{50}[ATP] = 443 ± 34 μM, IC_{50}[GTP] = 311 ± 46 μM), these concentrations are still well below intracellular ATP concentrations (>2 mM). It would be of high interest to validate these findings in cellulo. These results suggest STAT5B may also play a role in directly coupling gene expression to cellular metabolism, and also highlight a potential new targeting strategy using nucleotide-based inhibitor scaffolds [104]. Recently, a novel mechanism of selectively blocking STAT5B activity was proposed through protein-mediated Mannich reactions. The ligand-efficient 4-amino-furazan-3-carboxylic acid molecule (K_d = 420 μM, STAT5B fluorescent polarization assay) was identified from a library of 17,000 compounds. This phosphate-mimetic reacts with 1*H*-tetrazoles in the presence of formaldehyde, and this reaction only proceeds in acidic conditions (pH 5.0). However, the addition of MBP-tagged STAT5B-SH2 domain peptide catalyzed the reaction even at physiological conditions.

Although tetrazoles were not active against STAT5B (>10 mM), the ligation products showed substantial activity, which the authors attribute to super-additive binding interactions. Intracellular physiological concentrations of formaldehyde also emphasize the viability of this molecule as a STAT inhibitor and the utility of bio-catalytic Mannich reactions [105].

Table 2. Small-molecule compounds targeting the STAT5 SH2 domain.

Cpd	Protein-Based		Cell-Based			In Vivo Application Tested	Refs.
	Assay	IC_{50} or K_i $[\mu M]^a$	Cell Line	Readout	IC_{50} $[\mu M]^a$		
Cpd1	FP	47	K562, Daudi	DNA binding, pYSTAT5	s (100)		[91]
SF-1-088	FP	8.3	K562,MV4-11	viability, pYSTAT5	80–77		[92]
13a			K562	pYSTAT5, viability	s (15)		[28,93,94]
			MV4-11	viability	3.5		
AC-4-130	binding (thermal shift)	s (100)	MV4-11, MOLM-13	viability, reporter assay	1.7–1.9	AML xenograft (25 mg/kg)	[95]
			AML patient samples	viability	1.6–4.9		
Pimozide			KU812, K562	pYSTAT5, viability	s (5)	Approved by FDA as antipsychotic drug	[96]
TR120			K562	viability	0.12		[98]
				apoptosis	0.45		
IST5-002			K562, DU145, PC-3, COS-7	pYSTAT5, reporter assay, DNA binding	s (5)	Prostate cancer xenograft (25, 50, 100 mg/kg)	[99]
				viability	3.5		
Stafib-1	FP	0.044	K562	pYSTAT5	s (3)		[100]
Stafib-2	FP	0.009	K562	pYSTAT5	1.5		[101]
Cpd17f			K562, KU812, KG1a, MV4-11	viability, pYSTAT5	2.6–22.7		[103]

[a] s (100): significant effect at 100 μM (or other concentration, as indicated); cpd—compound; FP—fluorescence polarization.

A study by Juen et al. reported the optimization of the STAT5 inhibitor, Cpd17f (Figure 3v) [103]. Interestingly, the initial hit was originally intended for PPARα/γ inhibition, while the inhibition of STAT5 phosphorylation was an off-target effect [106]. Since the compound did not inhibit PPARs, but showed a considerable inhibition of STAT5 phosphorylation, 18 analogs were synthesized and tested in a cell-based assay. As shown by Brachet-Bottineau et al., in this special issue, Cpd17f inhibits STAT5B protein expression through non-transcriptional mechanisms. However, as with the majority of the aforementioned inhibitors, further confirmatory assays demonstrating the mechanism of action are required. Although multiple studies using molecular modeling or computational dynamics have been insightful in STAT5 inhibitor design, protein-inhibitor complexes characterized with atomic-level resolution, such as through X-ray crystallization, would provide a clearer understanding of target engagement in efficacy and selectivity studies.

Apart from these efforts targeting the SH2 domain of STAT3 or STAT5 molecules, studies have also reported describing approaches to block STAT3/5 DNA binding or the use of antisense RNA interference, discussed further in this special issue with separate overview articles.

6. Conclusions

STAT3 and STAT5 are the key nodes in transcriptional activation downstream of cytokine or kinase action in multiple cancers. This makes them attractive, but challenging targets for drug development. Recent findings expand on a previously rather simplistic understanding of STAT3/5 function as parallel phosphorylated dimers. It has become evident that higher-order conformations of pYSTATs, as well as uSTAT, are involved in chromatin landscape shaping, thereby acting beyond classical transcription factors. This knowledge provides a deeper understanding of their role in cancer biology, which together with various targeting efforts, will open novel therapeutic options.

Funding: C.W., H.A.N. and R.M. are supported by the Austrian Science Fund (FWF) [SFB-F04707, SFB-F06105, and under the frame of ERA PerMed (I 4218-B) and ERA-NET (I 4157-B)]. R.M., H.A.N. and A.O. were also generously supported by a private donation from Liechtenstein. D.B. and G.M.K. are supported by OTKA K 116904 (National Research, Development and Innovation Office, Hungary). P.T.G. is supported by research grants from NSERC (RGPIN-2014-05767), CIHR (MOP-130424, MOP-137036), Canada Research Chair (950-232042), Canadian Cancer Society (703963), Canadian Breast Cancer Foundation (705456) and infrastructure grants from CFI (33536) and the Ontario Research Fund (34876). M.H. was also supported by the EU Transcan-2 consortium 'ERANET-PLL' and by the ERA PerMed consortium 'JAKSTAT-TARGET'.

Acknowledgments: Open Access Funding by the Austrian Science Fund (FWF).

Conflicts of Interest: Authors declare no conflict of interest.

References

1. Wingelhofer, B.; Neubauer, H.A.; Valent, P.; Han, X.; Constantinescu, S.N.; Gunning, P.T.; Müller, M.; Moriggl, R. Implications of STAT3 and STAT5 signaling on gene regulation and chromatin remodeling in hematopoietic cancer. *Leukemia* **2018**, *32*, 1713–1726. [CrossRef] [PubMed]

2. Koromilas, A.E.; Sexl, V. The tumor suppressor function of STAT1 in breast cancer. *JAK-STAT* **2013**, *2*, e23353. [CrossRef] [PubMed]

3. Yu, H.; Jove, R. The stats of cancer—New molecular targets come of age. *Nat. Rev. Cancer* **2004**, *4*, 97–105. [CrossRef]

4. Meissl, K.; Macho-Maschler, S.; Müller, M.; Strobl, B. The good and the bad faces of STAT1 in solid tumours. *Cytokine* **2017**, *89*, 12–20. [CrossRef]

5. Klover, P.J.; Muller, W.J.; Robinson, G.W.; Pfeiffer, R.M.; Yamaji, D.; Hennighausen, L. Loss of STAT1 from Mouse Mammary Epithelium Results in an Increased Neu-Induced Tumor Burden. *Neoplasia* **2010**, *12*, 899–905. [CrossRef]

6. Kovacic, B.; Stoiber, D.; Moriggl, R.; Weisz, E.; Ott, R.G.; Kreibich, R.; Levy, D.E.; Beug, H.; Freissmuth, M.; Sexl, V. STAT1 acts as a tumor promoter for leukemia development. *Cancer Cell* **2006**, *10*, 77–87. [CrossRef]

7. Godfrey, A.L.; Chen, E.; Massie, C.E.; Silber, Y.; Pagano, F.; Bellosillo, B.; Guglielmelli, P.; Harrison, C.N.; Reilly, J.T.; Stegelmann, F.; et al. STAT1 activation in association with *JAK2* exon 12 mutations. *Haematologica* **2016**, *101*, e15. [CrossRef]

8. Prutsch, N.; Gurnhofer, E.; Suske, T.; Liang, H.C.; Schlederer, M.; Roos, S.; Wu, L.C.; Simonitsch-Klupp, I.; Alvarez-Hernandez, A.; Kornauth, C.; et al. Dependency on the TYK2/STAT1/MCL1 axis in anaplastic large cell lymphoma. *Leukemia* **2019**, *33*, 696–709. [CrossRef]

9. Sanda, T.; Tyner, J.W.; Gutierrez, A.; Ngo, V.N.; Glover, J.; Chang, B.H.; Yost, A.; Ma, W.; Fleischman, A.G.; Zhou, W.; et al. TYK2–STAT1–BCL2 Pathway Dependence in T-cell Acute Lymphoblastic Leukemia. *Cancer Discov.* **2013**, *3*, 564. [CrossRef]

10. Furth, P.A. STAT signaling in different breast cancer sub-types. *Mol. Cell. Endocrinol.* **2014**, *382*, 612–615. [CrossRef]

11. Levy, D.E.; Darnell, J.E. STATs: Transcriptional control and biological impact: Signalling. *Nat. Rev. Mol. Cell Biol.* **2002**, *3*, 651–662. [CrossRef] [PubMed]

12. Avalle, L.; Camporeale, A.; Camperi, A.; Poli, V. STAT3 in cancer: A double edged sword. *Cytokine* **2017**, *98*, 42–50. [CrossRef]

13. Bromberg, J.F.; Wrzeszczynska, M.H.; Devgan, G.; Zhao, Y.; Pestell, R.G.; Albanese, C.; Darnell, J.E., Jr. Stat3 as an Oncogene. *Cell* **1999**, *98*, 295–303. [CrossRef]

14. Corvinus, F.M.; Orth, C.; Moriggl, R.; Tsareva, S.A.; Wagner, S.; Pfitzner, E.B.; Baus, D.; Kaufman, R.; Huber, L.A.; Zatloukal, K.; et al. Persistent STAT3 Activation in Colon Cancer Is Associated with Enhanced Cell Proliferation and Tumor Growth. *Neoplasia* **2005**, *7*, 545–555. [CrossRef]

15. Yu, H.; Lee, H.; Herrmann, A.; Buettner, R.; Jove, R. Revisiting STAT3 signalling in cancer: new and unexpected biological functions. *Nat. Rev. Cancer* **2014**, *14*, 736. [CrossRef]

16. Mohanty, S.K.; Yagiz, K.; Pradhan, D.; Luthringer, D.J.; Amin, M.B.; Alkan, S.; Cinar, B. STAT3 and STAT5A are potential therapeutic targets in castration-resistant prostate cancer. *Oncotarget* **2017**, *8*. [CrossRef]

17. Gu, L.; Dagvadorj, A.; Lutz, J.; Leiby, B.; Bonuccelli, G.; Lisanti, M.P.; Addya, S.; Fortina, P.; Dasgupta, A.; Hyslop, T.; et al. Transcription Factor Stat3 Stimulates Metastatic Behavior of Human Prostate Cancer Cells in Vivo, whereas Stat5b Has a Preferential Role in the Promotion of Prostate Cancer Cell Viability and Tumor Growth. *Am. J. Pathol.* **2010**, *176*, 1959–1972. [CrossRef]

18. Boutillon, F.; Pigat, N.; Sackmann Sala, L.; Reyes-Gomez, E.; Moriggl, R.; Guidotti, J.-E.; Goffin, V. STAT5a/b Deficiency Delays, but does not Prevent, Prolactin-Driven Prostate Tumorigenesis in Mice. *Cancers* **2019**, *11*, 929. [CrossRef]

19. Cahu, X.; Constantinescu, S.N. Oncogenic Drivers in Myeloproliferative Neoplasms: From JAK2 to Calreticulin Mutations. *Curr. Hematol. Malig. Rep.* **2015**, *10*, 335–343. [CrossRef]

20. Staerk, J.; Constantinescu, S.N. The JAK-STAT pathway and hematopoietic stem cells from the JAK2 V617F perspective. *JAK-STAT* **2012**, *1*, 184–190. [CrossRef]

21. Warsch, W.; Walz, C.; Sexl, V. JAK of all trades: JAK2-STAT5 as novel therapeutic targets in BCR-ABL1+ chronic myeloid leukemia. *Blood* **2014**, *122*, 2167–2175. [CrossRef] [PubMed]

22. Hoelbl, A.; Kovacic, B.; Kerenyi, M.A.; Simma, O.; Warsch, W.; Cui, Y.; Beug, H.; Hennighausen, L.; Moriggl, R.; Sexl, V. Clarifying the role of Stat5 in lymphoid development and Abelson-induced transformation. *Blood* **2006**, *107*, 4898. [CrossRef]

23. Pham, H.T.T.; Maurer, B.; Prchal-Murphy, M.; Grausenburger, R.; Grundschober, E.; Javaheri, T.; Nivarthi, H.; Boersma, A.; Kolbe, T.; Elabd, M.; et al. STAT5BN642H is a driver mutation for T cell neoplasia. *J. Clin. Invest.* **2018**, *128*, 387–401. [CrossRef] [PubMed]

24. Waldmann, T.A.; Chen, J. Disorders of the JAK/STAT Pathway in T Cell Lymphoma Pathogenesis: Implications for Immunotherapy. *Annu. Rev. Immunol.* **2017**, *35*, 533–550. [CrossRef]

25. Schrader, A.; Crispatzu, G.; Oberbeck, S.; Mayer, P.; Pützer, S.; von Jan, J.; Vasyutina, E.; Warner, K.; Weit, N.; Pflug, N.; et al. Actionable perturbations of damage responses by TCL1/ATM and epigenetic lesions form the basis of T-PLL. *Nat. Commun.* **2018**, *9*, 697. [CrossRef]

26. Koppikar, P.; Bhagwat, N.; Kilpivaara, O.; Manshouri, T.; Adli, M.; Hricik, T.; Liu, F.; Saunders, L.M.; Mullally, A.; Abdel-Wahab, O.; et al. Heterodimeric JAK–STAT activation as a mechanism of persistence to JAK2 inhibitor therapy. *Nature* **2012**, *489*, 155–159. [CrossRef]

27. Berger, A.; Sexl, V.; Valent, P.; Moriggl, R. Inhibition of STAT5: a therapeutic option in BCR-ABL1-driven leukemia. *Oncotarget* **2014**, *5*, 9564–9576. [CrossRef]

28. Cumaraswamy, A.A.; Lewis, A.M.; Geletu, M.; Todic, A.; Diaz, D.B.; Cheng, X.R.; Brown, C.E.; Laister, R.C.; Muench, D.; Kerman, K.; et al. Nanomolar-Potency Small Molecule Inhibitor of STAT5 Protein. *ACS Med. Chem. Lett.* **2014**, *5*, 1202–1206. [CrossRef]

29. Mertens, C.; Zhong, M.; Krishnaraj, R.; Zou, W.; Chen, X.; Darnell, J.E. Dephosphorylation of phosphotyrosine on STAT1 dimers requires extensive spatial reorientation of the monomers facilitated by the N-terminal domain. *Genes Dev.* **2006**, *20*, 3372–3381.

30. Neculai, D.; Neculai, A.M.; Verrier, S.; Straub, K.; Klumpp, K.; Pfitzner, E.; Becker, S. Structure of the Unphosphorylated STAT5a Dimer. *J. Biol. Chem.* **2005**, *280*, 40782–40787. [CrossRef]

31. Fahrenkamp, D.; Li, J.; Ernst, S.; Schmitz-Van de Leur, H.; Chatain, N.; Küster, A.; Koschmieder, S.; Lüscher, B.; Rossetti, G.; Müller-Newen, G. Intramolecular hydrophobic interactions are critical mediators of STAT5 dimerization. *Sci. Rep.* **2016**, *6*, 35454. [CrossRef] [PubMed]

32. Moriggl, R.; Sexl, V.; Kenner, L.; Duntsch, C.; Stangl, K.; Gingras, S.; Hoffmeyer, A.; Bauer, A.; Piekorz, R.; Wang, D.; et al. Stat5 tetramer formation is associated with leukemogenesis. *Cancer Cell* **2005**, *7*, 87–99. [CrossRef]

33. Lin, J.-X.; Du, N.; Li, P.; Kazemian, M.; Gebregiorgis, T.; Spolski, R.; Leonard, W.J. Critical functions for STAT5 tetramers in the maturation and survival of natural killer cells. *Nat. Commun.* **2017**, *8*, 1320. [CrossRef]

34. de Araujo, E.D.; Erdogan, F.; Neubauer, H.A.; Meneksedag-Erol, D.; Manaswiyoungkul, P.; Eram, M.S.; Seo, H.-S.; Qadree, A.K.; Israelian, J.; Orlova, A.; et al. Structural and functional consequences of the STAT5BN642H driver mutation. *Nat. Commun.* **2019**, *10*, 2517. [CrossRef] [PubMed]

35. Park, H.J.; Li, J.; Hannah, R.; Biddie, S.; Leal-Cervantes, A.I.; Kirschner, K.; Flores Santa Cruz, D.; Sexl, V.; Go ttgens, B.; Green, A.R. Cytokine-induced megakaryocytic differentiation is regulated by genome-wide loss of a uSTAT transcriptional program. *EMBO J.* **2016**, *35*, 580–594. [CrossRef] [PubMed]

36. Zhao, Y.; Zeng, C.; Tarasova, N.I.; Chasovskikh, S.; Dritschilo, A.; Timofeeva, O.A. A new role for STAT3 as a regulator of chromatin topology. *Transcription* **2013**, *4*, 227–231. [CrossRef]

37. Kornfeld, J.-W.; Grebien, F.; Kerenyi, M.A.; Friedbichler, K.; Kovacic, B.; Zankl, B.; Hoelbl, A.; Nivarti, H.; Beug, H.; Sexl, V.; et al. The different functions of Stat5 and chromatin alteration through Stat5 proteins. *Front. Biosci. J. Virtual Libr.* **2008**, *13*, 6237–6254. [CrossRef]

38. Li, G.; Wang, Z.; Zhang, Y.; Kang, Z.; Haviernikova, E.; Cui, Y.; Hennighausen, L.; Moriggl, R.; Wang, D.; Tse, W.; et al. STAT5 requires the N-domain to maintain hematopoietic stem cell repopulating function and appropriate lymphoid-myeloid lineage output. *Exp. Hematol.* **2007**, *35*, 1684–1694. [CrossRef]

39. Murphy, T.; Yee, K.W.L. Cytarabine and daunorubicin for the treatment of acute myeloid leukemia. *Expert Opin. Pharmacother.* **2017**, *18*, 1765–1780. [CrossRef]

40. Ota, N.; Brett, T.J.; Murphy, T.L.; Fremont, D.H.; Murphy, K.M. N-domain–dependent nonphosphorylated STAT4 dimers required for cytokine-driven activation. *Nat. Immunol.* **2004**, *5*, 208–215. [CrossRef]

41. Vogt, M.; Domoszlai, T.; Kleshchanok, D.; Lehmann, S.; Schmitt, A.; Poli, V.; Richtering, W.; Muller-Newen, G. The role of the N-terminal domain in dimerization and nucleocytoplasmic shuttling of latent STAT3. *J. Cell Sci.* **2011**, *124*, 900–909. [CrossRef] [PubMed]

42. Zhang, X.; Darnell, J.E. Functional Importance of Stat3 Tetramerization in Activation of the α2-Macroglobulin Gene. *J. Biol. Chem.* **2001**, *276*, 33576–33581. [CrossRef]

43. Hu, T.; Yeh, J.E.; Pinello, L.; Jacob, J.; Chakravarthy, S.; Yuan, G.-C.; Chopra, R.; Frank, D.A. Impact of the N-Terminal Domain of STAT3 in STAT3-Dependent Transcriptional Activity. *Mol. Cell. Biol.* **2015**, *35*, 3284. [CrossRef]

44. Yang, J.; Chatterjee-Kishore, M.; Staugaitis, S.M.; Nguyen, H.; Schlessinger, K.; Levy, D.E.; Stark, G.R. Novel Roles of Unphosphorylated STAT3 in Oncogenesis and Transcriptional Regulation. *Cancer Res.* **2005**, *65*, 939. [PubMed]

45. Pfeffer, S.R.; Fan, M.; Du, Z.; Yang, C.H.; Pfeffer, L.M. Unphosphorylated STAT3 regulates the antiproliferative, antiviral, and gene-inducing actions of type I interferons. *Biochem. Biophys. Res. Commun.* **2017**, *490*, 739–745. [CrossRef]

46. Yang, J.; Liao, X.; Agarwal, M.K.; Barnes, L.; Auron, P.E.; Stark, G.R. Unphosphorylated STAT3 accumulates in response to IL-6 and activates transcription by binding to NFκB. *Genes Dev.* **2007**, *21*, 1396–1408.

47. Timofeeva, O.A.; Tarasova, N.I.; Zhang, X.; Chasovskikh, S.; Cheema, A.K.; Wang, H.; Brown, M.L.; Dritschilo, A. STAT3 suppresses transcription of proapoptotic genes in cancer cells with the involvement of its N-terminal domain. *Proc. Natl. Acad. Sci. USA* **2013**, *110*, 1267–1272. [CrossRef]

48. Lee, J.E.; Yang, Y.-M.; Liang, F.-X.; Gough, D.J.; Levy, D.E.; Sehgal, P.B. Nongenomic STAT5-dependent effects on Golgi apparatus and endoplasmic reticulum structure and function. *Am. J. Physiol. Cell Physiol.* **2011**, *302*, C804–C820. [CrossRef]

49. Hu, X.; Dutta, P.; Tsurumi, A.; Li, J.; Wang, J.; Land, H.; Li, W.X. Unphosphorylated STAT5A stabilizes heterochromatin and suppresses tumor growth. *Proc. Natl. Acad. Sci. USA* **2013**, *110*, 10213–10218. [CrossRef]

50. Engblom, D.; Kornfeld, J.-W.; Schwake, L.; Tronche, F.; Reimann, A.; Beug, H.; Hennighausen, L.; Moriggl, R.; Schutz, G. Direct glucocorticoid receptor-Stat5 interaction in hepatocytes controls body size and maturation-related gene expression. *Genes Dev.* **2007**, *21*, 1157–1162.

51. Freund, P.; Kerenyi, M.A.; Hager, M.; Wagner, T.; Wingelhofer, B.; Pham, H.T.T.; Elabd, M.; Han, X.; Valent, P.; Gouilleux, F.; et al. O-GlcNAcylation of STAT5 controls tyrosine phosphorylation and oncogenic transcription in STAT5-dependent malignancies. *Leukemia* **2017**, *31*, 2132–2142. [CrossRef]

52. Lin, J.-X.; Li, P.; Liu, D.; Jin, H.T.; He, J.; Rasheed, M.A.U.; Rochman, Y.; Wang, L.; Cui, K.; Liu, C.; et al. Critical Role of STAT5 Transcription Factor Tetramerization for Cytokine Responses and Normal Immune Function. *Immunity* **2012**, *36*, 586–599. [CrossRef]

53. Rauth, M.; Freund, P.; Orlova, A.; Grünert, S.; Tasic, N.; Han, X.; Ruan, H.-B.; Neubauer, A.H.; Moriggl, R. Cell Metabolism Control Through O-GlcNAcylation of STAT5: A Full or Empty Fuel Tank Makes a Big Difference for Cancer Cell Growth and Survival. *Int. J. Mol. Sci.* **2019**, *20*, 1028. [CrossRef] [PubMed]

54. Litterst, C.M.; Kliem, S.; Marilley, D.; Pfitzner, E. NCoA-1/SRC-1 Is an Essential Coactivator of STAT5 That Binds to the FDL Motif in the α-Helical Region of the STAT5 Transactivation Domain. *J. Biol. Chem.* **2003**, *278*, 45340–45351. [CrossRef]

55. Nanou, A.; Toumpeki, C.; Lavigne, M.D.; Lazou, V.; Demmers, J.; Paparountas, T.; Thanos, D.; Katsantoni, E. The dual role of LSD1 and HDAC3 in STAT5-dependent transcription is determined by protein interactions, binding affinities, motifs and genomic positions. *Nucleic Acids Res.* **2016**, *45*, 142–154. [CrossRef]

56. Yang, R.; Qu, C.; Zhou, Y.; Konkel, J.E.; Shi, S.; Liu, Y.; Chen, C.; Liu, S.; Liu, D.; Chen, Y.; et al. Hydrogen Sulfide Promotes Tet1- and Tet2-Mediated Foxp3 Demethylation to Drive Regulatory T Cell Differentiation and Maintain Immune Homeostasis. *Immunity* **2015**, *43*, 251–263. [CrossRef]

57. Zhang, Q.; Wang, H.Y.; Marzec, M.; Raghunath, P.N.; Nagasawa, T.; Wasik, M.A. STAT3- and DNA methyltransferase 1-mediated epigenetic silencing of SHP-1 tyrosine phosphatase tumor suppressor gene in malignant T lymphocytes. *Proc. Natl. Acad. Sci. USA* **2005**, *102*, 6948. [CrossRef]

58. Wang, R.; Cherukuri, P.; Luo, J. Activation of Stat3 Sequence-specific DNA Binding and Transcription by p300/CREB-binding Protein-mediated Acetylation. *J. Biol. Chem.* **2005**, *280*, 11528–11534. [CrossRef]

59. Yang, J.; Huang, J.; Dasgupta, M.; Sears, N.; Miyagi, M.; Wang, B.; Chance, M.R.; Chen, X.; Du, Y.; Wang, Y.; et al. Reversible methylation of promoter-bound STAT3 by histone-modifying enzymes. *Proc. Natl. Acad. Sci. USA* **2010**, *107*, 21499. [CrossRef]

60. Dasgupta, M.; Dermawan, J.K.T.; Willard, B.; Stark, G.R. STAT3-driven transcription depends upon the dimethylation of K49 by EZH2. *Proc. Natl. Acad. Sci. USA* **2015**, *112*, 3985. [CrossRef]

61. Hossain, D.M.S.; Panda, A.K.; Manna, A.; Mohanty, S.; Bhattacharjee, P.; Bhattacharyya, S.; Saha, T.; Chakraborty, S.; Kar, R.K.; Das, T.; et al. FoxP3 acts as a cotranscription factor with STAT3 in tumor-induced regulatory T cells. *Immunity* **2013**, *39*, 1057–1069. [CrossRef] [PubMed]

62. Mandal, M.; Powers, S.E.; Maienschein-Cline, M.; Bartom, E.T.; Hamel, K.M.; Kee, B.L.; Dinner, A.R.; Clark, M.R. Epigenetic repression of the Igk locus by STAT5-mediated recruitment of the histone methyltransferase Ezh2. *Nature Immunol.* **2011**, *12*, 1212–1220. [CrossRef] [PubMed]

63. Boumahdi, S.; de Sauvage, F.J. The great escape: tumour cell plasticity in resistance to targeted therapy. *Nat. Rev. Drug Discov.* **2019**. [CrossRef] [PubMed]

64. Schust, J.; Sperl, B.; Hollis, A.; Mayer, T.U.; Berg, T. Stattic: A Small-Molecule Inhibitor of STAT3 Activation and Dimerization. *Chem. Biol.* **2006**, *13*, 1235–1242.

65. Becker, S.; Groner, B.; Müller, C.W. Three-dimensional structure of the Stat3β homodimer bound to DNA. *Nature* **1998**, *394*, 145–151. [CrossRef]

66. Song, H.; Wang, R.; Wang, S.; Lin, J. A low-molecular-weight compound discovered through virtual database screening inhibits Stat3 function in breast cancer cells. *Proc. Natl. Acad. Sci. USA* **2005**, *102*, 4700. [CrossRef]

67. Lin, L.; Hutzen, B.; Li, P.-K.; Ball, S.; Zuo, M.; DeAngelis, S.; Foust, E.; Sobo, M.; Friedman, L.; Bhasin, D.; et al. A novel small molecule, LLL12, inhibits STAT3 phosphorylation and activities and exhibits potent growth-suppressive activity in human cancer cells. *Neoplasia* **2010**, *12*, 39–50. [CrossRef]

68. Fuh, B.; Sobo, M.; Cen, L.; Josiah, D.; Hutzen, B.; Cisek, K.; Bhasin, D.; Regan, N.; Lin, L.; Chan, C.; et al. LLL-3 inhibits STAT3 activity, suppresses glioblastoma cell growth and prolongs survival in a mouse glioblastoma model. *Br. J. Cancer* **2009**, *100*, 106–112. [CrossRef]

69. Fletcher, S.; Singh, J.; Zhang, X.; Yue, P.; Page, B.D.G.; Sharmeen, S.; Shahani, V.M.; Zhao, W.; Schimmer, A.D.; Turkson, J.; et al. Disruption of transcriptionally active Stat3 dimers with non-phosphorylated, salicylic acid-based small molecules: potent in vitro and tumor cell activities. *Chembiochem Eur. J. Chem. Biol.* **2009**, *10*, 1959–1964. [CrossRef]

70. Urlam, M.K.; Pireddu, R.; Ge, Y.; Zhang, X.; Sun, Y.; Lawrence, H.R.; Guida, W.C.; Sebti, S.M.; Lawrence, N.J. Development of new N-Arylbenzamides as STAT3 Dimerization Inhibitors. *MedChemComm* **2013**, *4*, 932–941. [CrossRef]

71. Siddiquee, K.; Zhang, S.; Guida, W.C.; Blaskovich, M.A.; Greedy, B.; Lawrence, H.R.; Yip, M.L.R.; Jove, R.; McLaughlin, M.M.; Lawrence, N.J.; et al. Selective chemical probe inhibitor of Stat3, identified through structure-based virtual screening, induces antitumor activity. *Proc. Natl. Acad. Sci. USA* **2007**, *104*, 7391. [CrossRef] [PubMed]

72. Xu, X.; Kasembeli, M.M.; Jiang, X.; Tweardy, B.J.; Tweardy, D.J. Chemical Probes that Competitively and Selectively Inhibit Stat3 Activation. *PLoS ONE* **2009**, *4*, e4783. [CrossRef] [PubMed]

73. Matsuno, K.; Masuda, Y.; Uehara, Y.; Sato, H.; Muroya, A.; Takahashi, O.; Yokotagawa, T.; Furuya, T.; Okawara, T.; Otsuka, M.; et al. Identification of a New Series of STAT3 Inhibitors by Virtual Screening. *ACS Med. Chem. Lett.* **2010**, *1*, 371–375. [CrossRef] [PubMed]

74. Shin, D.-S.; Kim, H.-N.; Shin, K.D.; Yoon, Y.J.; Kim, S.-J.; Han, D.C.; Kwon, B.-M. Cryptotanshinone Inhibits Constitutive Signal Transducer and Activator of Transcription 3 Function through Blocking the Dimerization in DU145 Prostate Cancer Cells. *Cancer Res.* **2008**, *69*. [CrossRef] [PubMed]

75. Lin, L.; Hutzen, B.; Ball, S.; Foust, E.; Sobo, M.; Deangelis, S.; Pandit, B.; Friedman, L.; Li, C.; Li, P.-K.; et al. New curcumin analogues exhibit enhanced growth-suppressive activity and inhibit AKT and signal transducer and activator of transcription 3 phosphorylation in breast and prostate cancer cells. *Cancer Sci.* **2009**, *100*, 1719–1727. [CrossRef]

76. Amani, H.; Ajami, M.; Nasseri Maleki, S.; Pazoki-Toroudi, H.; Daglia, M.; Tsetegho Sokeng, A.J.; Di Lorenzo, A.; Nabavi, S.F.; Devi, K.P.; Nabavi, S.M. Targeting signal transducers and activators of transcription (STAT) in human cancer by dietary polyphenolic antioxidants. *Biochimie* **2017**, *142*, 63–79. [CrossRef]

77. Li, M.; Yue, G.G.-L.; Song, L.-H.; Huang, M.-B.; Lee, J.K.-M.; Tsui, S.K.-W.; Fung, K.-P.; Tan, N.-H.; Lau, C.B.-S. Natural small molecule bigelovin suppresses orthotopic colorectal tumor growth and inhibits colorectal cancer metastasis via IL6/STAT3 pathway. *Biochem. Pharmacol.* **2018**, *150*, 191–201. [CrossRef]

78. Verdura, S.; Cuyàs, E.; Llorach-Parés, L.; Pérez-Sánchez, A.; Micol, V.; Nonell-Canals, A.; Joven, J.; Valiente, M.; Sánchez-Martínez, M.; Bosch-Barrera, J.; et al. Silibinin is a direct inhibitor of STAT3. *Food Chem. Toxicol.* **2018**, *116*, 161–172. [CrossRef]

79. Yu, W.; Xiao, H.; Lin, J.; Li, C. Discovery of Novel STAT3 Small Molecule Inhibitors via in Silico Site-Directed Fragment-Based Drug Design. *J. Med. Chem.* **2013**, *56*, 4402–4412. [CrossRef]

80. Zhang, M.; Zhu, W.; Li, Y. Discovery of novel inhibitors of signal transducer and activator of transcription 3 (STAT3) signaling pathway by virtual screening. *Eur. J. Med. Chem.* **2013**, *62*, 301–310. [CrossRef]

81. Leung, K.; Liu, L.; Lin, S.; Lu, L.; Zhong, H.; Susanti, D.; Rao, W.; Wang, M.; Ian, W.; Chan, D.S.; et al. Discovery of a small-molecule inhibitor of STAT3 by ligand-based pharmacophore screening. *Methods* **2015**, *71*, 38–43. [CrossRef]

82. Tolcher, A.; Flaherty, K.; Shapiro, G.I.; Berlin, J.; Witzig, T.; Habermann, T.; Bullock, A.; Rock, E.; Elekes, A.; Lin, C.; et al. A First-in-Human Phase I Study of OPB-111077, a Small-Molecule STAT3 and Oxidative Phosphorylation Inhibitor, in Patients with Advanced Cancers. *Oncologist* **2018**, *23*, 658–e72. [CrossRef] [PubMed]

83. Yoo, C.; Kang, J.; Lim, H.Y.; Kim, J.H.; Lee, M.-A.; Lee, K.-H.; Kim, T.-Y.; Ryoo, B.-Y. Phase I Dose-Finding Study of OPB-111077, a Novel STAT3 Inhibitor, in Patients with Advanced Hepatocellular Carcinoma. *Cancer Res. Treat.* **2018**, *51*, 510–518. [CrossRef] [PubMed]

84. Li, C.; Xu, L.; Jian, L.; Yu, R.; Zhao, J.; Sun, L.; Du, G.; Liu, X. Stattic inhibits RANKL-mediated osteoclastogenesis by suppressing activation of STAT3 and NF-κB pathways. *Int. Immunopharmacol.* **2018**, *58*, 136–144. [CrossRef] [PubMed]

85. Adachi, M.; Cui, C.; Dodge, C.T.; Bhayani, M.K.; Lai, S.Y. Targeting STAT3 inhibits growth and enhances radiosensitivity in head and neck squamous cell carcinoma. *Oral Oncol.* **2012**, *48*, 1220–1226. [CrossRef]

86. Miyoshi, K.; Takaishi, M.; Nakajima, K.; Ikeda, M.; Kanda, T.; Tarutani, M.; Iiyama, T.; Asao, N.; DiGiovanni, J.; Sano, S. Stat3 as a Therapeutic Target for the Treatment of Psoriasis: A Clinical Feasibility Study with STA-21, a Stat3 Inhibitor. *J. Investig. Dermatol.* **2011**, *131*, 108–117. [CrossRef]

87. Nie, Y.; Li, Y.; Hu, S. A novel small inhibitor, LLL12, targets STAT3 in non-small cell lung cancer in vitro and in vivo. *Oncol. Lett.* **2018**, *16*, 5349–5354. [CrossRef]

88. Zhang, X.; Sun, Y.; Pireddu, R.; Yang, H.; Urlam, M.K.; Lawrence, H.R.; Guida, W.C.; Lawrence, N.J.; Sebti, S.M. A novel inhibitor of STAT3 homodimerization selectively suppresses STAT3 activity and malignant transformation. *Cancer Res.* **2013**, *73*, 1922–1933. [CrossRef]

89. Xiao, H.; Bid, H.K.; Jou, D.; Wu, X.; Yu, W.; Li, C.; Houghton, P.J.; Lin, J. A Novel Small Molecular STAT3 Inhibitor, LY5, Inhibits Cell Viability, Cell Migration, and Angiogenesis in Medulloblastoma Cells. *J. Biol. Chem.* **2015**, *290*, 3418–3429. [CrossRef]

90. Müller, J.; Schust, J.; Berg, T. A high-throughput assay for signal transducer and activator of transcription 5b based on fluorescence polarization. *Anal. Biochem.* **2008**, *375*, 249–254.

91. Müller, J.; Sperl, B.; Reindl, W.; Kiessling, A.; Berg, T. Discovery of Chromone-Based Inhibitors of the Transcription Factor STAT5. *ChemBioChem* **2008**, *9*, 723–727. [CrossRef] [PubMed]
92. Page, B.D.G.; Khoury, H.; Laister, R.C.; Fletcher, S.; Vellozo, M.; Manzoli, A.; Yue, P.; Turkson, J.; Minden, M.D.; Gunning, P.T. Small Molecule STAT5-SH2 Domain Inhibitors Exhibit Potent Antileukemia Activity. *J. Med. Chem.* **2012**, *55*, 1047–1055. [CrossRef] [PubMed]
93. de Araujo, E.D.; Manaswiyoungkul, P.; Israelian, J.; Park, J.; Yuen, K.; Farhangi, S.; Berger-Becvar, A.; Abu-Jazar, L.; Gunning, P.T. High-throughput thermofluor-based assays for inhibitor screening of STAT SH2 domains. *J. Pharm. Biomed. Anal.* **2017**, *143*, 159–167. [CrossRef] [PubMed]
94. de Araujo, E.D.; Manaswiyoungkul, P.; Erdogan, F.; Qadree, A.K.; Sina, D.; Tin, G.; Toutah, K.; Yuen, K.; Gunning, P.T. A functional in vitro assay for screening inhibitors of STAT5B phosphorylation. *J. Pharm. Biomed. Anal.* **2019**, *162*, 60–65. [CrossRef]
95. Wingelhofer, B.; Maurer, B.; Heyes, E.C.; Cumaraswamy, A.C.; Berger-Becvar, A.; de Araujo, E.D.; Orlova, A.; Freund, P.; Ruge, F.; Park, J.; et al. Pharmacologic inhibition of STAT5 in acute myeloid leukemia. *Leukemia* **2018**, *32*, 1135–1146. [CrossRef]
96. Nelson, E.A.; Walker, S.R.; Weisberg, E.; Bar-natan, M.; Barrett, R.; Gashin, L.B.; Terrell, S.; Klitgaard, J.L.; Santo, L.; Addorio, M.R.; et al. The STAT5 inhibitor pimozide decreases survival of chronic myelogenous leukemia cells resistant to kinase inhibitors. *Blood* **2017**, *117*, 3421–3430. [CrossRef]
97. Rondanin, R.; Simoni, D.; Romagnoli, R.; Baruchello, R.; Marchetti, P.; Costantini, C.; Fochi, S.; Padroni, G.; Grimaudo, S.; Maria, R.; et al. Inhibition of activated STAT5 in Bcr/Abl expressing leukemia cells with new pimozide derivatives. *Bioorg. Med. Chem. Lett.* **2014**, *24*, 4568–4574.
98. Romagnoli, R.; Baraldi, P.G.; Prencipe, F.; Lopez-Cara, C.; Rondanin, R.; Simoni, D.; Hamel, E.; Grimaudo, S.; Pipitone, R.M.; Meli, M.; et al. Novel iodoacetamido benzoheterocyclic derivatives with potent antileukemic activity are inhibitors of STAT5 phosphorylation. *Eur. J. Med. Chem.* **2016**, *108*, 39–52. [CrossRef]
99. Liao, Z.; Gu, L.; Vergalli, J.; Mariani, S.A.; De Dominici, M.; Lokareddy, R.K.; Dagvadorj, A.; Purushottamachar, P.; McCue, P.A.; Trabulsi, E.; et al. Structure-based screen identifies a potent small-molecule inhibitor of Stat5a/b with therapeutic potential for prostate cancer and chronic myeloid leukemia. *Mol. Cancer Ther.* **2015**, *14*, 1777–1794. [CrossRef]
100. Elumalai, N.; Berg, A.; Natarajan, K.; Scharow, A.; Berg, T. Nanomolar Inhibitors of the Transcription Factor STAT5b with High Selectivity over STAT5a. *Angew. Chem. Int. Ed.* **2015**, *54*, 4758–4763. [CrossRef]
101. Elumalai, N.; Berg, A.; Rubner, S.; Blechschmidt, L.; Song, C.; Natarajan, K.; Matysik, J.; Berg, T. Rational development of Stafib-2: a selective, nanomolar inhibitor of the transcription factor STAT5b. *Sci. Rep.* **2017**, *7*, 819. [CrossRef] [PubMed]
102. Natarajan, K.; Müller-Klieser, D.; Rubner, S.; Berg, T. Stafia-1: a STAT5a-selective inhibitor developed via docking-based screening of in silico O-phosphorylated fragments. *Chem. Eur. J.* **2019**, in press. [CrossRef] [PubMed]
103. Juen, L.; Brachet-Botineau, M.; Parmenon, C.; Bourgeais, J.; Hérault, O.; Gouilleux, F.; Viaud-Massuard, M.-C.; Prié, G. New Inhibitor Targeting Signal Transducer and Activator of Transcription 5 (STAT5) Signaling in Myeloid Leukemias. *J. Med. Chem.* **2017**, *60*, 6119–6136. [CrossRef]
104. Berg, A.; Sperl, B.; Berg, T. ATP Inhibits the Transcription Factor STAT5b. *ChemBioChem* **2019**, *20*, 2227–2231. [CrossRef]
105. Wong, E.L.; Nawrotzky, E.; Arkona, C.; Kim, B.G.; Beligny, S.; Wang, X.; Wagner, S.; Lisurek, M.; Carstanjen, D.; Rademann, J. The transcription factor STAT5 catalyzes Mannich ligation reactions yielding inhibitors of leukemic cell proliferation. *Nat. Commun.* **2019**, *10*, 66. [CrossRef]
106. Parmenon, C.; Guillard, J.; Caignard, D.-H.; Hennuyer, N.; Staels, B.; Audinot-Bouchez, V.; Boutin, J.-A.; Dacquet, C.; Ktorza, A.; Viaud-Massuard, M.-C. 4,4-Dimethyl-1,2,3,4-tetrahydroquinoline-based PPARα/γ agonists. Part I: Synthesis and pharmacological evaluation. *Bioorg. Med. Chem. Lett.* **2008**, *18*, 1617–1622.

Review

Targeting STAT3 and STAT5 in Tumor-Associated Immune Cells to Improve Immunotherapy

Grégory Verdeil [1], Toby Lawrence [2,3], Anne-Marie Schmitt-Verhulst [2] and Nathalie Auphan-Anezin [2,*]

[1] Laboratory Regulation of immune dysfunction in cancer, Department of Oncology, University of Lausanne, CH-1066 Epalinges, Switzerland; gregory.verdeil@unil.ch

[2] Aix Marseille University, Centre National de la Recherche Scientifique (CNRS) UMR7280, Institut National de la Santé et de la Recherche Médicale (INSERM) U1104, Centre Immunologie Marseille-Luminy (CIML), Parc Scientifique de Luminy, Case 906, 13288 Marseille CEDEX 09, France; lawrence@ciml.univ-mrs.fr (T.L.); amsverhulst@gmail.com (A.-M.S.-V.)

[3] Centre for Inflammation Biology and Cancer Immunology, School of Immunology & Microbial Sciences, Faculty of Life Sciences and Medicine, King's College London, London SE1 1UL, UK

* Correspondence: auphan@ciml.univ-mrs.fr; Tel.: +33-4-9126-9189

Received: 24 October 2019; Accepted: 18 November 2019; Published: 21 November 2019

Abstract: Oncogene-induced STAT3-activation is central to tumor progression by promoting cancer cell expression of pro-angiogenic and immunosuppressive factors. STAT3 is also activated in infiltrating immune cells including tumor-associated macrophages (TAM) amplifying immune suppression. Consequently, STAT3 is considered as a target for cancer therapy. However, its interplay with other STAT-family members or transcription factors such as NF-κB has to be considered in light of their concerted regulation of immune-related genes. Here, we discuss new attempts at re-educating immune suppressive tumor-associated macrophages towards a CD8 T cell supporting profile, with an emphasis on the role of STAT transcription factors on TAM functional programs. Recent clinical trials using JAK/STAT inhibitors highlighted the negative effects of these molecules on the maintenance and function of effector/memory T cells. Concerted regulation of STAT3 and STAT5 activation in CD8 T effector and memory cells has been shown to impact their tumor-specific responses including intra-tumor accumulation, long-term survival, cytotoxic activity and resistance toward tumor-derived immune suppression. Interestingly, as an escape mechanism, melanoma cells were reported to impede STAT5 nuclear translocation in both CD8 T cells and NK cells. Ours and others results will be discussed in the perspective of new developments in engineered T cell-based adoptive therapies to treat cancer patients.

Keywords: inflammation; tumor-associated macrophages; adoptive T cell therapy; immune suppression; STAT transcription factors

---

1. Introduction

Inflammation is now considered as a hallmark of cancer [1] and the inflammatory context in many cancers is strongly linked to poor prognosis and resistance to therapy. Activating mutations in oncogenic RAS/BRAF/MEK pathways trigger a tumor-intrinsic inflammatory network with the concerted regulation of master transcription factors (TFs) including STAT3 [2], NF-κB and AP-1 [3,4] which in turn trigger the expression of cytokines, including IL-6, IL-1, IL-10, TNF and VEGF [5]. The presence of cytokines is a major regulator of immune cell differentiation/function and is a crucial factor to consider for immunotherapy protocols. Many of these cytokines signal through the stimulation of STAT TFs. In this review, we will comment on the role of STAT TFs (i) for the recruitment and function of tumor-associated macrophages (TAM) and (ii) for the regulation of T cell functions. As STAT

TFs are key players in the regulation of functions of these immune cells, their manipulation can have a beneficial or detrimental effect on the anti-tumor response depending on the targeted cell type.

2. Tumor-Induced Inflammation Drives Accumulation of Tumor-Infiltrating Myeloid Cells

Cytokine receptor-induced signaling sustains and amplifies the activation of STAT3, NF-κB and AP-1 in a positive amplification loop, fueling tumor-associated inflammation. As such, a core inflammation-related gene set regulated by STAT3, NF-κB and AP-1 has recently been proposed as an "inflammatory index" in breast cancer cell lines and patient samples [4]. Importantly, in correlation with this inflammatory index, this study reported a concerted regulation of (i) non-inflammatory genes related to angiogenesis, metastasis, and cell proliferation; (ii) tumor genome instability; and (iii) heterogeneity of the tumor microenvironment (TME), including the recruited immune cells. Remarkably, these co-regulated characteristics were found across several cancer types driven by distinct oncogenes [4].

Tumor-derived cytokines have been linked to the accumulation of immune-suppressive myeloid cells including both myeloid-derived suppressor cells (MDSCs; IMCs) and tumor-associated macrophages (TAM). In both human and mouse melanomas, IMCs have an important role in malignant progression and evasion from anti-tumor immunity that is linked to the suppression of T cell responses [6–9]. While increased TAM accumulation has been ascribed to a poor prognosis in established tumors [10], this notion should be refined given the extreme phenotypic and functional heterogeneity of these cells during tumor growth.

Cancer-related inflammation (i) promotes the recruitment of monocyte-derived cells into the tumor bed and (ii) acts systemically as shown by the dysregulated transcriptomic signature of circulating monocytes in breast cancer patients as compared to healthy controls [11]. Numerous therapeutic attempts (reviewed in [12]) to block cytokine-induced monocyte recruitment are under clinical trials using either blocking mAbs (anti-CCR2; anti-CSF1R) or small inhibitors for downstream cytokine receptor signaling (receptor tyrosine kinase inhibitor, Pexidartinib; CSF1R inhibitor, PLX3397).

Once recruited to the tumor, monocyte progenitors (CD11b$^+$ SiglecF$^-$ Ly-6G$^-$Ly6C$^+$ F4/80$^-$ CD169$^-$ MHC-II$^-$) undergo a multistep differentiation program [13,14], passing through an immature stage (Ly6C$^+$ F4/80$^-$ CD169^{int} MHC-II^{hi}) before reaching a mature state (Ly6C$^-$ F4/80$^+$ CD169^{hi} MHC-II$^{hi\ or\ low}$). Recently, the heterogeneity of TAM in several cancer types has been emphasized by single-cell RNA-sequencing paired with mass cytometry, including lung [15], kidney [16], breast and endometrial cancers [11], as well as in mouse sarcoma [17]. In the case of renal cell carcinoma, 17 TAM subsets have been characterized [16], most of them expressing CD169 while being discriminated by expression of CD163, CD204 and CD206. Of note, when compared to steady-state tissue-resident macrophages or monocytes, TAM exhibited peculiar gene expression profiles [11,16], highlighting a tumor-induced dysregulation.

There is compelling evidence for high levels of phenotypic plasticity in macrophages, exhibiting differential functional programs depending on their surrounding microenvironment. In response to microbial stimuli pro-inflammatory macrophages—often referred to as of "M1-type" - express cytokines supporting T cell activation. However, in malignancies, alternatively activated macrophages - referred as of "M2-type"-secrete cytokines that sustain tumor growth and exert immune suppressive functions. Therefore, differential signal transduction pathways downstream of cell surface sensors define gene expression programs underlying anti- or pro- tumor functions of macrophages [18,19]. As such, IFNγ stimulates STAT1/STAT2 and IRF1/8, which can further promote pro-inflammatory macrophages. While GM-CSF/STAT5 activates a pro-inflammatory signature in monocytes [20], it rather induces a unique reparative program in macrophages after sterile renal injury [21]. Tumor-induced inflammation involving the IL-6/gp130/STAT3 [22] and ERK5/STAT3 [23] axes was shown to drive a pro-tumor transcriptomic program. Additionally, we recently showed that ovarian cancer cells reprogram macrophages towards an IL-4/AKT/STAT6-mediated tumor-promoting phenotype through increased cholesterol efflux from the TAM membrane [24]. Moreover, IL-6 synergizes with IL-4 in the programming of human monocyte-derived macrophages through the concerted activation of STAT6 and STAT3 DNA-binding activities [25]. Given the spectrum of STAT3-regulated genes in TAM encoding pro-tumor and immune suppressive mediators (Table 1), STAT3-modulators are currently being developed to dampen the cancer supportive functions of TAM, as reported in this issue by Rébé and Ghiringhelli [26]. As such, the specific targeting of a highly immune suppressive TAM (CD11b+ CD163+) subset by liposome-encapsulated STAT3-inhibitors showed some success in reprogramming TAM towards a pro-inflammatory profile [27].

Table 1. STAT3–regulated genes in tumor-associated macrophages (TAM).

Target Genes	STAT3 Input	Cancer Type	TAM Phenotype
		Cytokines/Cytokine receptors	
Il10 *	positive	m-melanoma	CD11b+ [28]
		m-PDAC	CD68+, IL-10Ra+ [29]
Il23a *	positive	m-melanoma	CD11b+ CD11c− [30]
Il10ra, Il4ra	positive	m-PDAC	CD68+, IL-10Ra+ [29]
Tgfb1 *	positive	m-melanoma	CD11b+ CD11c− [30]
Il12a	negative	m-melanoma	CD11b+ [28]; CD11b+ CD11c− [30]
Ifng	negative	m-melanoma	CD11b+ [28]
		Chemokines/Chemokine Receptors	
Ccl5, Cxcl9-10-11	Negative	m-melanoma	CD11b+ [28]
Cxcl2, Cxcl12	positive	m-melanoma	BMDM+Tumor conditioned media [31] MDSC: CD11b+ GR1+ CD11c− [22]
		Scavenger receptors/Endocytosis	
Mrc1 (CD206)	positive	m-breast	CD11b+ Ly-6C^{lo} F4/80^{hi} CD24^{lo} MHC-II^{lo} [32]
		h-gastric	CD163+ CD209a+ [33]
CD163	positive	h-SCC	ERK5+ CD163+ [23]
		m-PDAC	CD68+, IL-10Ra+ [29]
Cd209a	positive	m-PDAC	CD68+, IL-10Ra+ [29]
		Immune suppression	
Arg1	positive	m-PDAC; h-PDAC	CD68+, IL-10Ra+ [29]; blood CD14+ [34]
Cox2	positive	m-melanoma	BMDM + Tumor conditioned media [31]
Ido1	positive	m-liver metastasis	liver-MDSC: CD11b+ Ly-6C$^{int/hi}$ Ly-6G^{+} [35]
		h- & m-glioma	h-CD68+; m-CD11b+ CD115+ [36]
Pdl1 (CD274)	positive	h-breast	CD163+ [37]
		m-liver metastasis	liver-MDSC: CD11b+ Ly-6C$^{int/hi}$ Ly-6G^{+} [35]
		Extra-cellular Matrix/Angiogenesis	
Mmp2	positive	m-melanoma	BMDM + Tumor conditioned media [31]
Vegf	positive	m-melanoma	CD11b+ [28]; MDSC: CD11b+ GR1+ CD11c- [22]
Cathepsin (B, L)	positive	m-PDAC	CD68+, IL-10Ra+ [29]
		Cell cycle/TFs	
Ccnd1	positive	m-melanoma	BMDM + Tumor conditioned media [31]
ATF6, sXBP1	positive	m-PDAC	CD68+, IL-10Ra+ [29]

* Immune suppressive cytokines.

3. Re-Educating TAM to Restore Anti-Tumor T Cell Functions.

Extensive studies are being conducted to reprogram pro-tumoral TAM towards an inflammatory T cell supporting profile (recently reviewed in [38]). Mature TAM exert both trophic functions—through the promotion of angiogenesis and tissue remodeling—and immune regulatory functions. Here we will focus on the immune aspects of TAM functions, even though these two activities might be closely inter-connected. Indeed, prolonged interactions between stromal TAM (mainly CD163^{+} CD206^{+}) and CD8 tumor-infiltrating lymphocytes (TILs) observed by dynamic imaging microscopy, are limiting CD8 T cell motility and their consecutive access to both human lung squamous cell carcinomas and mouse MMTV-PyMT tumors [39].

Recent parallel analyses of TAM and TILs from cancer patients have greatly expanded our knowledge on the reciprocal regulation of these cell lineages. Paired CyTOF-based analyses of CD8 TILs and TAM in human renal cell carcinomas [16] showed some correlation between exhausted CD8 TILs, CD4 regulatory T cells and a few peculiar TAM subpopulations (either CD169^{-} CD163^{-} CD68^{hi}

CD38[hi] CD204[+]; or CD169[+] CD163[+] CD68[hi] CD38[hi] CD204[+] CD206[+]). These molecular data had been further correlated with clinical features, with the result that patients with the former TIL/TAM subsets showed increased cancer progression.

We recently reported [13] that a subset of mature CD163[+] TAM present in mouse melanomas, exhibit transcripts related to T cell immune suppression. Interestingly, a high proportion of CD163[+] macrophages expressing phospho-STAT3 have been observed in human skin tumors [23]. Targeted-depletion of this minor CD163[+] TAM subset enhanced melanoma-infiltration by CD8 T cells and promoted CD8 T cell-mediated tumor regression in mice [13]; this was also accompanied by the recruitment of fresh monocytes and immature macrophages with an immune stimulatory phenotype (Figure 1).

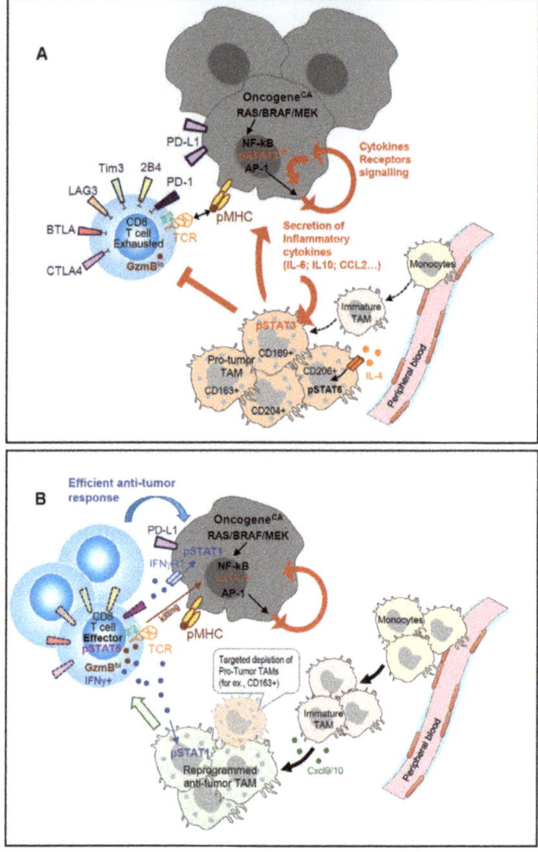

Figure 1. (**A**). Concerted regulation of tumor-induced inflammation promotes accumulation of immune-suppressive TAM. Oncogene activation in tumors induced secretion of inflammatory cytokines that activate STAT3 and promote the accumulation of immune-suppressive TAM; IL-4-derived signals through STAT6 sustain the pro-tumor TAM function that further sustains tumor growth in a positive feedback loop. (**B**). Targeted depletion of specific TAM subsets displaying strong T cell suppressive activity favors the recruitment of monocyte-derived immature TAM that secrete CXCL9/10 and display T cell activating capacities. Reinvigorated CD8 TILs with activated pSTAT5 secrete IFNγ that further maintains TAM anti-tumor functions and sensitizes tumors to T cell killing.

Cancer treatments have recently benefited from immune checkpoint therapies (ICT) targeting inhibitory signaling pathways in T lymphocytes (PD-1/PD-L1 and CTLA-4). Despite remarkable success in a subset of patients (objective responses ~20–30% for monotherapy and ~30–40% for combined therapy), primary resistance is observed in more than 50% of the patients and a subset of initial responders can later develop acquired resistance [40]. Analysis of pre-treatment tumor biopsy samples has revealed that patients with a pre-existing local CD8 T cell infiltrate (T cell inflamed) were more likely to show a clinical response to anti- PD-1/PD-L1 [41]. Furthermore, PD-L1-expression on dendritic cells and macrophages but not on cancer cells has been shown to shape the response to the PD-L1/PD-1 blockade [42,43].

Given the aforementioned reciprocal regulation of TAM and TILs, one should expect that ICT-based therapies could conjointly unleash tumor-specific T cells and reprogram TAM functions. Several groups exploring immune cells dynamics in vivo before and after ICT recently addressed this issue. In carcinoma-bearing mice, immune-PET monitoring showed that ICT induced an important infiltration of CD8 T cells within the tumor in responders as compared to non-responders; the responders displayed CD11b$^+$ TAM with a "M1-type" transcriptomic signature [44]. scRNA-seq and longitudinal CyTOF analyses also revealed ICT-induced dynamic changes in both TAM and TILs present in induced mouse sarcomas [17]. This study identified an ICT-induced decrease of both CD4 regulatory T cells and exhausted CD8 T cells. Meanwhile, conventional CD4 T cells and tumor-specific CD8 T cells displayed an activated profile as did recruited NK cells, with all these effector cells exhibiting an enhanced NF-κB and IFNγ-driven gene signatures. ICT consequently conferred a remarkable dynamic remodeling on intra-tumor monocytes and macrophages. Distinct TAM populations expressing IFNγ-induced iNOS, while down-regulating IL-4-responsive CD206 proteins, were enriched upon ICT Pathway analyses, which showed ICT-driven enrichment of IFNγ and NF-κB signaling as well as glucose metabolism. Concurrently, alternative "M2-type" TAM clusters with a MerTK^{hi} CD206$^+$ CCL2$^+$ CD274$^+$ IL4Rα$^+$ phenotype were reduced by approximately one-third. Kinetic CyTOF analyses further highlighted an important crossroads between days 7 and 9—a time window where the TME drives the TAM behavior towards a pro-inflammatory/T cell supporting or an anti-inflammatory/immune suppressive program. The TME agent mediating monocyte/macrophage reprogramming was ascribed to be IFNγ secreted by ICT-unleashed tumor-specific T cells. Importantly, T cell produced IFNγ not only engages its receptor signaling on surrounding TAM but also on tumor cells themselves. As such, STAT1-activation was reported to sensitize hepatocarcinomas to T cell mediated killing [45] (see Figure 1). Therefore, the therapeutic use of STAT/JAK-modulators should be carefully evaluated for their specific targeting of STAT3 activity while preserving IFNγ-induced STAT1/2 signaling, which appears to be beneficial for anti-tumor immune responses.

In the large fraction of patients who are non-responsive to ICT other therapeutic approaches must be proposed to reinvigorate anti-tumor responses. Providing an infusion of autologous tumor-specific CD8 T cells manipulated to express an effector (IFNγ-proficient) program could be a way to fight cancer and induce TAM reprogramming. This will be addressed in the next section.

4. STAT5 versus STAT3 in Adaptive CD8 T Cell Responses to Cancer

Cytokines are major regulators of T cell differentiation/function and are crucial factors to consider for immunotherapy protocols. Many of these cytokines signal through the stimulation of STAT3 and/or STAT5. This makes these two TFs key players in the regulation of functions of T cells. Both STAT3 and STAT5 activation can have a beneficial or detrimental effect on the anti-tumor response depending on the T cell type targeted. In this part, we will focus on the T cell-intrinsic role of these TFs and on the possibility to manipulate them for improving CD8 T cell anti-tumor responses.

4.1. STAT5 in the Adaptive T Cell Responses

The STAT5 pathway is predominantly activated in T cells by members of the γc cytokine family (IL-2, -7, -9 and -15) (reviewed in [46]). These cytokines signal through receptors containing the γc (CD132) subunit and lead to the activation of the JAK3 tyrosine kinase. In addition, activation of JAK1 occurs through its interaction with the IL-2Rβ (CD122) (for IL-2 and IL-15), but also, with the IL-4Rα, IL-7Rα and IL-9Rα units. For IL-2, high-affinity binding to its receptor requires the association of an additional subunit (IL-2Rα or CD25) to the IL-2Rβ and the γc components. IL-15 signaling is further peculiar in that it requires the association "in-trans" of the IL-15Rα component with the IL-2Rβ and the γc components [47]. The specificity of signaling by these cytokines is thus partially explained by the recruited STAT proteins, but also by the differential expression of the relevant receptors. STAT5-activating cytokines have a general role in the maintenance and expansion of T cell subsets. IL-2 has a central role in the expansion phase of T cells following their primary antigenic stimulation [47,48]. The roles of IL-7 or IL-15 are rather associated with the maintenance of naïve or memory T cells [47,49,50].

In adaptive immunity, except for a population of CD4 T cells with a regulatory function (Treg), IL-2Rα (CD25) is not expressed at a significant level at the surface of naïve CD4 and CD8 T cells. Although the two other chains of the IL-2R are expressed at low basal levels on naïve T cells, IL-2 fails to activate these T cells, consistent with the report that only Tregs respond promptly to in vivo IL-2 exposure [51]. Thus, antigen stimulation of the TCR/CD3 complex is a pre-requisite for the secondary signaling by the IL-2R. TCR/CD3 initiated signals, which do not involve STAT activation, lead to transient expression of genes, including *il2ra*. IL-2R mediated signaling through STAT5 activation further increases CD25 expression at the cell surface and leads to the stabilization of a panel of genes initially induced by the TCR [52,53]. We and others reported that TCR-initiated signaling is influenced by the avidity of TCR-peptide/MHC interaction, which impacts strength and duration of TCR engagement [54–56]. Strong TCR/CD3 stimulation causes production of IL-2, leading to an autocrine effect on IL-2R signaling, which is not induced by low avidity Ags. In this latter case, exogenous IL-2 or expression of a constitutive-active STAT5 protein [52,53] can serve as a substitute for the lack of the IL-2/IL-2R amplification loop. Interestingly, a study evaluating the IL-2R/JAK-regulated phospho-proteome in CD8 CTL revealed a dominance of proteins that control mRNA stability and components of the protein translational machinery leading to an accumulation of cytokines and effector molecules, as well as proteins that support metabolic processes essential for cell "fitness" and important oxygen-sensing components [57]. Of note, active STAT5 does not substitute for all IL-2 induced signaling events that also involve MAP kinase- and phospho-inositol 3-kinase-dependent pathways [58]. Altogether, for CD8 T cells, depending on their state of activation and on the dose of cytokine provided, IL-2 may amplify effector function and proliferation and induce terminal differentiation or activation-induced cell death [47,48].

The synergistic action of TCR-induced TFs and STAT5 in CD8 T cells mirrors the cooperation between STAT5 and (i) TCR-induced GATA3 in CD4 Th-2 cells to control the accessibility of *Il4* gene locus [59]; (ii) Tbet in Th-1/Tc-1 for the regulation of the *Ifng* locus [60,61]; and (iii) BCL6 in B lymphocytes for the generation of memory B cells [62]. Additionally, STAT5 activation was shown to promote GM-CSF [63] and IL-9 [64], producing T cells and to be a prerequisite for Foxp3-expressing Tregs [65,66]. By contrast, STAT5 is a negative regulator of Th-17 [67] and T-Fh [68] by competing with STAT3 and BCL6, respectively. Altogether, STAT5 appears to control secondary decisions in adaptive immunity (see Table 2).

Table 2. Concerted gene regulation by STAT3 and STAT5 in helper and cytotoxic lymphocytes.

STAT3 and STAT5-Regulated Genes in T and NK cells				
Target Genes	**STAT3 Input**	**Cell Subsets**	**STAT5 Input**	**Cell Subsets**
Cytokines/Cytokine receptors				
Il17	positive	hCD4 Th-17 [69]		
Il10	positive	mCD4 Th-2 [70]		
Il2ra (CD25)	positive	mCD4 Th-21 [71]	positive	mCD4 Th-1 [71], mCD8 [60], mNK [72]
Il7ra (CD127)			negative	mCD8 [60]
Il6st			negative	mCD8 [60]
Tgfb2r			negative	mCD8 [60]
Homing				
CCR7			negative	mCD8 [60]
Sell (CD62L)			negative	mCD8 [60]
TFs				
Prdm1 (Blimp)	positive	mCD4 Th-2 [70]		
Eomes			positive	mCD8 [60]
Tbx21 (Tbet)			positive	mCD8 [60]
Cytotoxicity				
Gzmb			positive	mCD8 [60], mNK [73]
Prf1			positive	mCD8 [60], mNK [73]
Ifng			positive	mCD8 [60], mNK [73]
Fasl			positive	mCD8 [60]
Cell cycle/Cell survival				
Myc	positive	hT lymphoma [74]	positive	mCD8 [60]
Pim1	positive	hT lymphoma [74]	positive	mCD8 [60,75]
Bcl2	positive	hT lymphoma [74]	positive	mCD8 [49,60]; NK [73]

Competitive gene regulation by STAT3 and STAT5 in T cells *			
Target Genes	**STAT3 Input**	**STAT5 Input**	**T Cell Subsets**
Il17	positive	competitor	mCD4 Th-17 [67]
Il9	competitor	positive	mCD4 Th-9 [64,76]
Bcl6	positive	competitor	mCD4 Tfh; mCD4 Th-1 [68,77]
Socs3	positive	positive	mCD4 Th-17/Treg balance [78]

* Mechanisms underlying competitive gene regulation by STAT3 and STAT5 are reviewed in [79].

Role of STAT5 in the Generation of Memory CD8 T Cells and Maintenance of Effector Functions

Several STAT5A mutants have been shown to confer cytokine-independent growth to Ba/F3 cells [72]. While a STAT5A S710F single-mutant displayed a strong constitutive activity sufficient to complement STAT5A^{null} bone marrow progenitors, a STAT5A double-mutant H298R/S710F was not able to complement STAT5A^{null} cells [80]. Further studies highlighted that the H298R mutation in the DNA-binding domain renders this protein inactive unless it can form heterodimers with endogenous wild-type STAT5 (see Figure 2). Therefore, the constitutive activity of STAT5 H298R/S710F is modest and limited by the bioavailability of wild-type STAT5 to pair with. By contrast, the single STAT5 S710F mutant exerts a strong constitutive activity in both CD8 T cells [81] and Ba/F3 cells [82].

Retroviral expression of STAT5A H298R/S710F (hereafter referred to as STAT5ca) in in vitro activated CD8 T cells led to the generation and maintenance of long-lived CD8 T effector cells upon their adoptive transfer [83]. Transcriptomic analyses of STAT5ca-expressing CD8 T cells highlighted a role for STAT5ca in the stabilization of a broad Tc-1 gene expression program initiated by TCR stimulation [60] (see Table 2, Figure 2). This observation is in agreement with the reported chromatin interactions of STAT5 in super-enhancers to activate IL-2 highly inducible genes [71].

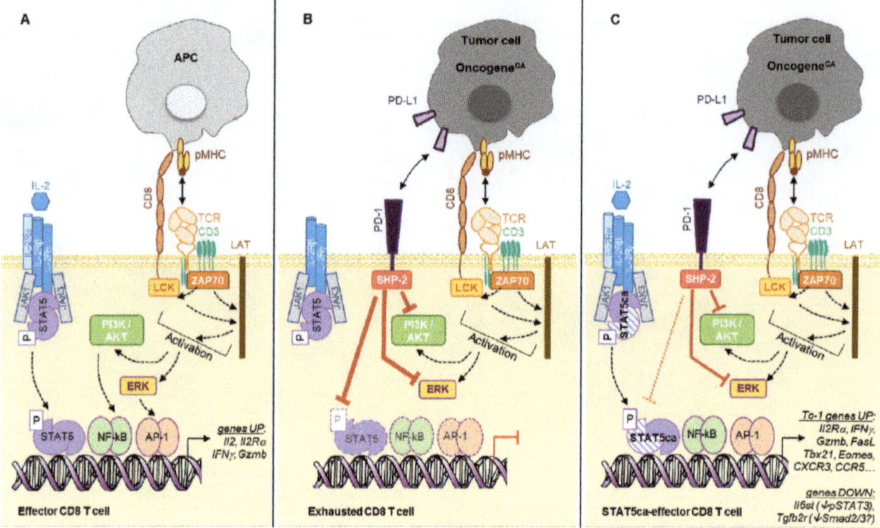

Figure 2. (**A**). Ag presented by APC (Ag-presenting cells) triggers a T cell activation cascade leading to gene transcription including *Il2* and *IL2Rα* genes. Binding of IL-2 to its receptor further amplifies the TCR-initiated gene transcription program. (**B**). Ag expressed on tumor cells mediates chronic TCR engagement on CD8 TILs leading to their "exhaustion", which is characterized by expression of multiple inhibitory receptors (as shown in Figure 1). For simplicity, we represent PD-1 only that recruits the phosphatase SHP-2 mediating inhibition of ERK and PI3K/AKT pathways as well as dephosphorylation of STAT5. (**C**). Expression of STAT5ca (H298R/S710F, here represented by dashed symbols as compared to the wild type (WT) protein) in CD8 T cells not only recapitulates the IL-2-mediated TCR-initiated gene transcription, but also stabilizes this functional program. This leads to a sustained Tc-1 program reminiscent of effector memory cells. Of note, while being PD-1^hi due to the chronic TCR engagement by their cognate Ag, STAT5ca-expressing T cells remain functional, as the S710F substitution reduces the SHP-2-mediated dephosphorylation. Additionally, STAT5ca represses the expression of *il6st* and *Tgfb2r* genes, rendering these cells insensitive to IL-6/STAT3 and TGFβ1/Smad signaling.

Of note, the in vivo maintenance of STAT5ca-expressing CD8 T cells remains under the control of γc-cytokines (IL-7, IL-15) and TCR tickling by self MHC class I [81]; these properties again point towards a moderate and controlled activity of this double-mutant. Accordingly, Kaech's group also reported that STAT5ca promoted memory CD8 T cells [49] that did not display any sign of transformation. However, Moriggl and colleagues recently demonstrated that high expression of S710F gain-of-function mutated STAT5A induced PTLC-nos (Peripheral T cell leukemia and lymphoma—not otherwise specified) cells when expressed during T cell development in transgenic mice [84].

Mice expressing a constitutively active STAT5Bca (H298R/S715F) transgene in the lymphoid lineage have been shown to present a selective expansion of memory-like CD8 T cells. Their analysis further suggested that moderate STAT5B activation underlies both IL-7/IL-15-dependent homeostatic proliferation of naive and memory CD8 T cells and IL-2-dependent development of CD4 CD25+ Tregs [85]. When expressed in the B cell lineage in mouse models, STAT5Bca (H298R/S715F) induces B cell acute lymphoblastic leukemia thanks to cooperative molecular events targeting JAK1 activity, tumor-suppressor genes, and pre-BCR signaling [86]. Indeed, mutated STAT5Bca was shown to antagonize preBCR-initiated TFs (NF-κB, IKAROS) for binding to B cell specific super enhancers [87]. Finally, mice which expressed a transgene, i.e., a human gain-of-function mutation of STAT5B (hSTAT5B N642H) identified in leukemic patients, developed lymphomas from multiple T cell subsets [88]. The recent crystal structure of hSTAT5B N642H highlighted important conformational changes in correlation to its resistance to dephosphorylation [89]. Overall, strong STAT5B hyperactivity appeared to trigger B or T lymphomas when express during lymphoid cell development and to directly influence disease aggressiveness and therapeutic resistance [87,89].

4.2. Role of STAT3 in T Cells

Several cytokines/cytokine receptors trigger STAT3 signaling in T cells, including IL-6-type cytokines (IL-6 and Oncostatin M), IL-10, IL-17, IL-21 and IL-27. Among the STAT3-activating cytokines, IL-21 is peculiar: While its receptor contains the γc (CD132) subunit, signals downstream of IL-21R converge on STAT3 and to a lesser extent on STAT1, rather than on STAT5 activation [46]. Intriguingly, IL-21 was shown to promote in vitro proliferation of CD8 T naïve and memory cells in synergy with STAT5-inducing cytokines IL-7/IL-15 [90]. Indeed, Leonard's group reported that IL-2 vs. IL-21 cytokine stimulation of pre-activated T cells induces STAT5 and STAT3 binding, respectively, to incompletely overlapping super enhancers [71]. Understanding the fine-tuning of cytokine-modulated STAT3/5 activities will require further effort as reviewed in [46]: This tight regulation relies on the T cell context, including cell lineage, differentiation state, and cytokine milieu, triggering synergistic/antagonistic STAT-mediated transcriptional regulation. Thus, the amplitude of STAT3 activation differs from one T cell subset to another in correlation with differential expression of the corresponding T cytokine receptor [91]. Interestingly, chronic TCR engagement was recently shown to enhance the sensitivity to IL-6/STAT3 signals [92], indicating again a cross-talk between TCR-induced and STAT TFs.

The function of STAT3 in various T cell subtypes is summarized in Table 2, but further details are described in this issue by both Rébé and Ghiringhelli [26] and Logotheti and Putzer [93].

The overall effect of STAT3 stimulation in T cells is generally correlated with a poor cytotoxic/increased regulatory response and as such, is associated with a defective anti-tumor immune response [94]. In the next part, we will focus on the role of STAT3 in the maintenance and formation of memory CD8 T cells.

4.2.1. Role of STAT3 in the Maintenance and Memory Formation of CD8 T Cells

Several studies have highlighted the role of IL-21 and STAT3 activation in T cell memory regulation. Using LCMV chronic infection in mice, Kaech's group showed that IL-10 and/or IL-21-induced STAT3 was required for the differentiation of memory CD8 T cells with a preserved replicative potential [95]. Indeed, in the absence of STAT3, memory precursor cells were replaced by T effector-like cells that failed to undergo homeostatic proliferation or to protect against secondary infection. Interestingly, STAT3 appeared to act both by induction of "pro-memory" TFs, such as Eomes and BCL6, and by dampening inflammation-driven TFs, including STAT4 and Tbet.

A similar role for STAT3 in human T cells [96] was reported in a cohort of patients suffering from autosomal-dominant hyper-IgE syndrome, which is caused by dominant-negative STAT3 mutations. These patients have increased numbers of naïve T cells but fewer central memory CD4$^+$ and CD8 memory T cells, leading to an increased susceptibility to EBV infections. This was associated with a defect of these naïve T cells to express the TFs important for memory formation.

Accordingly, IL-21 repressed the cytolytic effector program in tumor-specific CD8 T cells while preserving their replicative potential [97].

4.2.2. Role of STAT3 in the CD8 T Cell-Mediated Anti-Tumor Responses

STAT3-activating cytokines such as IL-6 and IL-10 are found in abundance within the tumor microenvironment. It was clearly shown that STAT3 inactivation in tumor-specific CD8 T cells increased their potential for tumor elimination after adoptive transfer [98,99]. Altogether, these studies suggest that a general inhibition of STAT3 in T cells would be a beneficial treatment for increasing tumor control. However, contradicting data exist in the literature [100]. In mice bearing transplanted tumors, a combination of anti-OX40 mAb and TGFβ/Smad inhibitors improved the CD8 T cell mediated tumor rejection. Use of OX40-Cre STAT3$^{fl/fl}$ mice further demonstrated that in absence of STAT3 in OX40-expressing T cells, the response to the treatment was lost [100]. These authors did not prove formally that the deletion of STAT3 in CD8 T effector cells only was mediating the same effect; but their data suggest that depending on the T cell subset or the immunotherapy used, inhibiting STAT3 could become detrimental and lead to poor survival of T cells and loss of the benefit of the treatment.

The reported discrepancies for the function of STAT3 in CD8 T cell mediated anti-tumor responses may rely on distinct roles of this TF depending on the T cell differentiation state (effector versus memory), the combined immunotherapy regimen and the STAT3-inducing cytokines in the TME. All these parameters may contribute to differential STAT3-activation thresholds triggering distinct T cell responses, a hypothesis that needs to be tested in refined analyses.

5. Adoptive CD8 T Cell Therapy and CAR-T Cell Generation for Cancer Immunotherapy

5.1. Increasing the Frequency of Tumor-Specific CD8 T Cells

As mentioned earlier in Section 3, beside tremendous achievements for patients with various types of cancers, immunotherapies aimed at blocking inhibiting receptors still failed for more than 50% of cancer patients. This could be the consequence of the paucity of tumor-specific CD8 T cells, an obstacle that could be circumvented by infusions of in vitro amplified tumor-specific T cells. Indeed, Adoptive Cell Transfer (ACT) of amplified TILs and Chimeric Antigenic Receptor (CAR) treatments are based on the use of the patient's T cells [101]. Historically, T cells used for ACT were obtained from tumor pieces/biopsies, amplified in vitro and transferred back into the patient when a sufficient number was reached (up to 10^{11} cells) [102]. The progress in gene engineering removed many hurdles to generate cancer specific T cells. CAR-T cells are gene-engineered products obtained from the expression in T cells of a construct encoding a domain of a single-chain variable antibody fragment recognizing a tumor antigen, fused to intracellular signaling motifs inducing T cell activation upon recognition of the target cells [103]. T cells from the blood are used for this therapy, allowing an increased potential to generate these cells on a large scale. CD19-directed CAR-T cells have shown impressive results in the treatment of patients with B-cell lymphoma [104]. However, evidence for the use of CAR-T cells against solid tumors is still sparse. The main limitation for application being the identification of Ags expressed exclusively on solid tumors and not on the non-transformed cell counterparts. Defining tumor-specific Ags for solid cancers is still a challenge [105,106]. Such an attempt was reported using mesothelin-specific CAR-T cells to treat patients bearing pancreatic ductal adenocarcinoma [107]. Of note, antibody-targeted chimeric T cells display receptors (i) of higher avidity for tumor-Ags than regular TCRs; (ii) which are not MHC restricted and thereby, are not impacted by tumor editing in the Ag processing/presentation pathway, (iii) interacting only with Ags expressed at the tumor cell surface.

Despite some clinical success using ACT or CAR-T cells, much effort is being made to improve T cell survival and resistance to immunosuppression in the frame of these therapies. Given the ability of IL-2 to expand CD8 T cells and to maintain their cytotoxic function, infusion of recombinant IL-2 in cancer patients was the first reported effective immunotherapy for human cancer [108]. However, IL-2 concomitantly acts on Tregs that suppress anti-tumor responses, and systemic IL-2 injections had severe adverse effects including vascular leak syndrome and pulmonary edema due to IL-2Rα expression on endothelial cells [109]. This has led to the development of (IL-2/anti-IL-2 mAb) complexes [110]: depending on the mAb the binding of the resulting complex is favored on IL-2Rβ/γc-expressing cells such as the memory CD8 T and NK cells, but prevented on IL-2Rα$^+$ Tregs and endothelial cells. Such approaches enhance the cytokine half-life in vivo and were shown to induce a massive (>100-fold) expansion of CD8 T cells in vivo [110,111] thereby boosting anti-tumor immune responses in mice and humans [112].

IL-21 is also considered to expand less differentiated CD8 T memory stem cells with both proliferative and multipotent potential [113] to be used in ACT.

The manipulation of cytokine-induced STAT-signaling in tumor-specific CD8 T lymphocytes may be another solution to potentiate adoptive T cell cancer therapies.

5.2. On CD8 T cell Intrinsic Modifications For adoptive T Cell Therapies

5.2.1. Promoting Effector Memory CD8 T Cells Through STAT5

Another hurdle for ACT is the persistence of the infused T cells, their capacity to home in tumor-invaded peripheral organs and to resist the local immune suppression. As such, CD19-directed CAR-T cells had limited capacities to control the growth of CD19$^+$ murine melanomas in relation with their scarcity in the tumor [114]. To infiltrate non hematologic solid tumors, T cells must acquire an effector phenotype: this includes the simultaneous loss of CD62L (L-selectin) and CCR7 expression, and acquisition of tissue-specific homing molecules such as adhesion molecules and chemokines/chemokine receptors [115]. Given the multiplicity of the molecular partners involved in T cell egress and migration, it is illusory to manipulate these targets, one by one or in combination. However, manipulating TFs known to be activated during T effector cell differentiation may have broader effects. We brought this proof of concept by expression of STAT5ca during the in vitro differentiation of CD8 T cells, which modified their in vivo migration upon ACT with increased infiltration of both non-lymphoid tissues in healthy mice and melanomas in tumor-bearing mice [60]. At the transcriptomic and protein levels, coincident down-regulation of CD62L and up-regulation of CCR2, CCR5 and CXCR3 was induced by STAT5ca as compared to standard CD8 T effector cells. Altogether, STAT5ca-CD8 T cells acquired a long-lived effector memory phenotype in correlation with a concerted STAT5-mediated regulation of *Tbx21* and *Eomes* genes (Table 2 and [60]). Coupled to their enhanced effector functions—Ag-driven cytotoxity and IFNγ production—these engineered STAT5ca-CD8 T cells mediated efficient melanoma rejection as compared with conventional CD8 T effector cells [83]. Importantly, this STAT5ca-induced reprogramming applies to both polyclonal and monoclonal (TCR transgenic) CD8 T cells, after primary or secondary stimulation [83]: these characteristics are of particular importance for translation to cancer patients' CD8 T cells. Finally, STAT5ca-expression in tumor-specific T cells had better therapeutic potential than the potentiation of standard CD8 T effector cells with combined infusions of IL-2 complexes, the effect of which disappeared at the end of treatment [83].

5.2.2. Promoting Central Memory CD8 T Cells Through STAT3

Recent studies have highlighted the importance of a memory-like CD8 T cell population to unleash an anti-tumor response by anti-PD-1 treatment [116,117]. In these studies, a small subset of CD8 TILs expressing PD-1 and the transcription factor TCF1 responded to anti-PD-1 immunotherapy by differentiating into highly cytotoxic TILs that mediated long-term tumor control. These studies highlight the importance of a long lasting, less differentiated population to mediate a more efficient control of the cancer.

This point is also valid for ACT, as transfer of expanded T cells that are too differentiated could lead to a failure of the treatment related to a lack of survival of the cells [118]. For that purpose, IL-21, a STAT3 stimulating cytokine, is efficient in driving the differentiation of central memory CAR or TCR engineered CD8 T cells in vitro [119]. Additionally, a recent study showed that CD19-directed CAR-T cells initially activated in vitro with optimal stimulation (anti-CD3/CD28) progressively acquired an "exhausted" phenotype upon transfer in tumor bearing mice. Thus, when retrieved from the tumor 12 days after their transfer, they presented a characteristic high surface expression of inhibitory receptors, diminished production of IFNγ and TNFα, and low cytolytic activity as well as the expression of the high-mobility group (HMG)-box TFs - TOX and TOX2. Expression of the latter TFs was recently described as resulting in commitment of CD8 T cells to an exhausted transcriptional and epigenetic developmental program distinct from that of functional CD8 T effector and T memory cells [120]. To understand further the factors associated with an efficient CAR response, Fraietta et al. [121] performed a transcriptomic profiling of CAR-T cells from complete-responding patients with chronic lymphocytic leukemia compared to non-responders. The gene signature was enriched in memory-related genes, including IL-6/STAT3 signatures, whereas T cells from non-responders upregulated programs involved in effector differentiation, glycolysis, exhaustion and apoptosis. Moreover, highly functional CAR-T cells from patients produced the STAT3-activating cytokine IL-6 and its concentration in serum was correlated with CAR-T cell expansion. These results showed STAT3 signaling is beneficial for an efficient CAR-T cell response [121]. This finding is further validated by the use of a new-generation CD19-CAR that encodes a truncated cytoplasmic domain from IL-2Rβ and a STAT3-binding tyrosine-X-X-glutamine motif. These CAR-T cells showed antigen-dependent activation of the JAK kinase and of the STAT3 and STAT5 TFs signaling pathways, which promoted their proliferation and prevented their terminal differentiation in vitro. The CAR-T cells demonstrated superior in vivo persistence and anti-tumor effects in models of hematologic malignancies as compared with CAR-T cells expressing a CD28 or 4-1BB co-stimulatory domain alone [114]. It should be noted that while a subtle increased STAT3 activation by these newly engineered IL-2Rβ-CAR-T cells was therapeutically beneficial, STAT3 gain-of-function mutants induced multi-organ autoimmunity and large granular lymphocytic leukemia [122].

In conclusion, these studies strengthen the interest of regulating STAT3/STAT5 activities in CD8 T cells to improve adoptive cell therapy: given their tightly balanced contribution to T cell memory fates, increased STAT3 or STAT5 activity is expected to improve the therapeutic success of hematological and non-hematological cancers, respectively. This hypothesis has to be validated in relevant preclinical mouse models.

5.2.3. On the STAT5 versus STAT3 Balance in T Cells

Several groups have provided data showing that the competitive binding of STAT3 and STAT5 dimers to DNA dictates the transcriptional regulation of certain target genes in T cells and thereby either trigger autoimmune disorders or promote tumor rejection (some examples are reported in Table 2). These data highlight the need for both in silico analyses [123] and appropriate cell/animal models to evaluate the outcome of competition between STAT TFs on T cell fates.

6. STAT5 and Resistance to Immunosuppression in the Tumor Microenvironment

Chronic stimulation of tumor-specific T cells promotes an exhausted phenotype, sharing strong [124] but not-completely overlapping similarities [125] with virus-induced exhausted CD8 T cells. Both the diversity and the high expression of inhibitory receptors by CD8 TILs (reviewed in [126]) can contribute to ICT resistance. However, combining blocking antibodies towards several inhibitory receptors to improve anti-tumor responses may further increase the adverse effects among which autoimmune attacks [127]. Therefore, manipulating CD8 T cells to resist local tumor-derived immune suppression may be of clinical benefit.

6.1. Immunosuppression by PD-1

PD-1 is related to the CD28 superfamily and is expressed on activated T cells, B cells, monocytes and macrophages. Engagement of PD-1 by its ligands (PD-L1 or PD-L2) induces inhibitory signals in T cells [128] through the induced phosphorylation of ITIM/ITSIM motives in its cytoplasmic domain, which leads to recruitment of the SHP-2/SHP-1 phosphatases. This dampens both TCR and CD28 signaling leading to abrogation of the PI3K/AKT and ERK pathways, as shown notably in primary CD4 T cells [129]. Interestingly, SHP-2 recruited by PD-1 was shown to counteract phospho-STAT5 signaling in human Tregs from HCV-infected patients' livers [130]. It was also demonstrated that PD-1 intrinsically regulates a subpopulation of high PD-1 expressing innate cells (ILC2) involved in helminth immunity (γc dependent proliferation) by inhibiting STAT5 phosphorylation through SHP-1/SHP-2 phosphatases [131]. Similarly, for T cells, the inhibitory effect of PD-1 is observed on populations expressing high levels of surface PD-1 such as exhausted T cells in contrast to acutely activated T cells [132]. However, in a melanoma model, in spite of upregulation of the PD-1/PD-L1 immunosuppressive pathway in the tumor microenvironment, STAT5ca-expressing CD8 T effector cells were found to be resistant to inhibition by PD-1/PD-L1 engagement as measured by their efficient IFNγ production, in contrast to their control counterpart [132]. Thus, the previously reported increased resistance to dephosphorylation of STAT5ca [72] may also contribute to its resistance to the phosphatases recruited to checkpoint inhibitors (Figure 2) [133,134].

6.2. Other Immune Suppressive Pathways

The diversity of immune suppressive mechanisms associated with inflammatory melanoma have previously been reported [133], with several involving tumor-derived proteins.

TGFβ1, an immunosuppressive cytokine produced by a number of cancer cells, was shown to inhibit both Th-1/Tc-1 cytokine production and Tc-1/NK cytolytic effector functions through the repression of *Ifng, Gzmb,* and *FasL* gene transcription [135,136]. Interestingly, STAT5ca-expressing T cells maintained efficient induction of both IFNγ and Gzmb in the presence of TGFβ1 [132]. The insensitivity to TGFβ1 may be linked to the STAT5ca-mediated negative regulation of *Tgfbr2* expression as well as its positive effects on *Ifng* and *Gzmb* gene transcription [60]; thus, STAT5ca-gene regulation counterbalances the transcriptional repression exerted by TGFβ1-induced Smad2/3. Indeed, two STAT5-activating cytokines, IL-2 [136] and IL-15 [137], have been shown to counteract TGFβ1-mediated immune suppression.

Of note, while TGFβ1 and IL-6 contribute to *Maf* up-regulation in exhausted CD8 T cells [125], STAT5ca-expressing T cells displayed reduced levels of *Il6st, Tgfbr2,* and *Maf* transcripts [60] and maintained a functional antitumor program.

Another tumor-secreted protein, the Dickkopf-related protein 2 (DKK2), was shown to be overexpressed in human colorectal cancers and melanomas, and to promote tumor progression by suppressing the activation of both CD8 T and NK cells [138]. Mechanistically, tumor cells secrete DKK2 that binds its receptor LRP5 inducing interactions between the intracellular domain of LRP5 and STAT5; while STAT5-phosphorylation was not precluded, STAT5 nuclear localization was impeded in cytotoxic cells. Conversely, DKK2 blockade enhanced cytotoxic immune cell activation.

Altogether, maintenance of active STAT5 signaling in cytotoxic cells appears to correlate with efficient anti-tumor immunity [73,83].

7. Conclusions

The frequency of tumor-specific CD8 T cells can now be increased in cancer patients through sophisticated personalized medical approaches including in vitro derived and expanded TILs or engineered CAR-T cells. Nevertheless the ability of adoptively transferred T cells to survive in cancer-invaded hosts and to efficiently penetrate into the tumor are dampened by the solid tumor landscape including cancer cells, tumor stroma and immune suppressive myeloid cells. Therefore, enhancing tumor T cell infiltration is one way to improve cancer immunotherapy. This could imply (i) manipulation of the T cells themselves through their transduction with chemokine receptors promoting non-hematopoietic tissue colonization [139,140] or manipulation of TFs that regulate expression of these homing receptors [60]; (ii) depletion of specific TAM subsets that exert trophic and/or immune suppressive functions either by ablation [13] or inactivation using STAT3 inhibitors [27]. The enhanced killing of cancer cells together with the recruitment of fresh monocytes will favor the engagement of the host immune system thanks to the presentation of tumor-derived Ags by professional APCs such as dendritic cells and/or immature TAM. This would result in spreading of tumor-derived epitopes ensuring a polyclonal anti-tumor response that could ultimately prevent antigen escape. Importantly, this favorable loop was observed when using STAT5ca-expressing CD8 T cells in ACT as their efficient capacity to infiltrate and kill inflammatory melanomas is accompanied by a concomitant infiltration of host T cells [83]. Compared to systemic treatment, such as IL-2 infusions, the targeted manipulation of STAT5 activity in tumor-specific CD8 T lymphocytes circumvents the negative stimulation of Tregs, which are also recruited into tumor beds. We here propose to transiently express (i.e., through mRNA transfection for example) the STAT5A H298R/S710F mutant in human CD8 T cells specific for Ags expressed on solid malignancies to be used in ACT: Both their efficient infiltration into tumors and IFNγ production resistant to tumor-derived suppression should help to restore endogenous T cell responses and to reprogram TAM. Although the IFNγ/STAT1 signature induced by T cells has recently been reported as a good prognostic factor for ICT response [141], a close control of this cytokine-induced signaling must be maintained since excess IFNγ can induce death of anti-tumor CD8 T cells at a certain stage of their differentiation/activation [142–144].

Funding: This research was funded by grants to T.L. from "L'Agence Nationale de la Recherche" (ANR-09-MIEN-029-01, ANR-10-BLAN-1302-01); "Ligue Nationale Contre le Cancer" and "Institut National contre le Cancer (INCa-PLBIO13-244), grants to G.V. from the Swiss National Foundation (310030_182680/1), the Pr. Max Cloëtta Foundation, the Swiss Cancer League (KFS-3679-08-2015) as well as grant to N.A.A. from "Association pour la Recherche sur le Cancer"(ARC-PJA20141201950).

Acknowledgments: The authors thank the CIML imaging, cytometry and animal facilities personnel for assistance. The authors also thank Lee Leserman for editing the manuscript.

Conflicts of Interest: The authors declare no conflict of interest.

References

1. Hanahan, D.; Weinberg, R.A. Hallmarks of cancer: The next generation. *Cell* **2011**, *144*, 646–674. [CrossRef] [PubMed]

2. Yu, H.; Pardoll, D.; Jove, R. STATs in cancer inflammation and immunity: A leading role for STAT3. *Nat. Rev. Cancer* **2009**, *9*, 798–809. [CrossRef] [PubMed]

3. Fleming, J.D.; Giresi, P.G.; Lindahl-Allen, M.; Krall, E.B.; Lieb, J.D.; Struhl, K. STAT3 acts through pre-existing nucleosome-depleted regions bound by FOS during an epigenetic switch linking inflammation to cancer. *Epigenetics Chromatin* **2015**, *8*, 7. [CrossRef] [PubMed]

4. Ji, Z.; He, L.; Regev, A.; Struhl, K. Inflammatory regulatory network mediated by the joint action of NF-kB, STAT3, and AP-1 factors is involved in many human cancers. *Proc Natl. Acad. Sci. USA* **2019**, *116*, 9453–9462. [CrossRef]

5. Taniguchi, K.; Karin, M. IL-6 and related cytokines as the critical lynchpins between inflammation and cancer. *Semin. Immunol.* **2014**, *26*, 54–74. [CrossRef]

6. Gabrilovich, D.I.; Ostrand-Rosenberg, S.; Bronte, V. Coordinated regulation of myeloid cells by tumours. *Nat. Rev. Immunol.* **2012**, *12*, 253–268. [CrossRef]

7. Riesenberg, S.; Groetchen, A.; Siddaway, R.; Bald, T.; Reinhardt, J.; Smorra, D.; Kohlmeyer, J.; Renn, M.; Phung, B.; Aymans, P.; et al. MITF and c-Jun antagonism interconnects melanoma dedifferentiation with pro-inflammatory cytokine responsiveness and myeloid cell recruitment. *Nat. Commun.* **2015**, *6*, 8755. [CrossRef]

8. Soudja, S.M.; Wehbe, M.; Mas, A.; Chasson, L.; De Tenbossche, C.P.; Huijbers, I.; Van den Eynde, B.; Schmitt-Verhulst, A.M. Tumor-initiated inflammation overrides protective adaptive immunity in an induced melanoma model in mice. *Cancer Res.* **2010**, *70*, 3515–3525. [CrossRef]

9. Wehbe, M.; Soudja, S.M.; Mas, A.; Chasson, L.; Guinamard, R.; Powis de Tenbossche, C.; Verdeil, G.; Van den Eynde, B.; Schmitt-Verhulst, A.M. Epithelial-Mesenchymal-Transition-like and TGFβ pathways associated with autochthonous inflammatory melanoma development in mice. *PLoS ONE* **2012**, *7*, e49419. [CrossRef]

10. Noy, R.; Pollard, J.W. Tumor-associated macrophages: From mechanisms to therapy. *Immunity* **2014**, *41*, 49–61. [CrossRef]

11. Cassetta, L.; Fragkogianni, S.; Sims, A.H.; Swierczak, A.; Forrester, L.M.; Zhang, H.; Soong, D.Y.H.; Cotechini, T.; Anur, P.; Lin, E.Y.; et al. Human Tumor-Associated Macrophage and Monocyte Transcriptional Landscapes Reveal Cancer-Specific Reprogramming, Biomarkers, and Therapeutic Targets. *Cancer Cell* **2019**, *35*, 588–602. [CrossRef]

12. Bercovici, N.; Guerin, M.V.; Trautmann, A.; Donnadieu, E. The Remarkable Plasticity of Macrophages: A Chance to Fight Cancer. *Front. Immunol.* **2019**, *10*, 1563. [CrossRef] [PubMed]

13. Etzerodt, A.; Tsalkitzi, K.; Maniecki, M.; Damsky, W.; Delfini, M.; Baudoin, E.; Moulin, M.; Bosenberg, M.; Graversen, J.H.; Auphan-Anezin, N.; et al. Specific targeting of CD163(+) TAMs mobilizes inflammatory monocytes and promotes T cell-mediated tumor regression. *J. Exp. Med.* **2019**, *216*, 2394–2411. [CrossRef]

14. Movahedi, K.; Laoui, D.; Gysemans, C.; Baeten, M.; Stange, G.; Van den Bossche, J.; Mack, M.; Pipeleers, D.; In't Veld, P.; De Baetselier, P.; et al. Different tumor microenvironments contain functionally distinct subsets of macrophages derived from Ly6C(high) monocytes. *Cancer Res.* **2010**, *70*, 5728–5739. [CrossRef]

15. Lavin, Y.; Kobayashi, S.; Leader, A.; Amir, E.D.; Elefant, N.; Bigenwald, C.; Remark, R.; Sweeney, R.; Becker, C.D.; Levine, J.H.; et al. Innate Immune Landscape in Early Lung Adenocarcinoma by Paired Single-Cell Analyses. *Cell* **2017**, *169*, 750–765. [CrossRef]

16. Chevrier, S.; Levine, J.H.; Zanotelli, V.R.T.; Silina, K.; Schulz, D.; Bacac, M.; Ries, C.H.; Ailles, L.; Jewett, M.A.S.; Moch, H.; et al. An Immune Atlas of Clear Cell Renal Cell Carcinoma. *Cell* **2017**, *169*, 736–749. [CrossRef]

17. Gubin, M.M.; Esaulova, E.; Ward, J.P.; Malkova, O.N.; Runci, D.; Wong, P.; Noguchi, T.; Arthur, C.D.; Meng, W.; Alspach, E.; et al. High-Dimensional Analysis Delineates Myeloid and Lymphoid Compartment Remodeling during Successful Immune-Checkpoint Cancer Therapy. *Cell* **2018**, *175*, 1014–1030. [CrossRef]

18. Lawrence, T.; Natoli, G. Transcriptional regulation of macrophage polarization: Enabling diversity with identity. *Nat. Rev. Immunol.* **2011**, *11*, 750–761. [CrossRef]

19. Martinez, F.O.; Gordon, S. The M1 and M2 paradigm of macrophage activation: Time for reassessment. *F1000Prime Rep.* **2014**, *6*, 13. [CrossRef]

20. Croxford, A.L.; Lanzinger, M.; Hartmann, F.J.; Schreiner, B.; Mair, F.; Pelczar, P.; Clausen, B.E.; Jung, S.; Greter, M.; Becher, B. The Cytokine GM-CSF Drives the Inflammatory Signature of CCR2+ Monocytes and Licenses Autoimmunity. *Immunity* **2015**, *43*, 502–514. [CrossRef]

21. Huen, S.C.; Huynh, L.; Marlier, A.; Lee, Y.; Moeckel, G.W.; Cantley, L.G. GM-CSF Promotes Macrophage Alternative Activation after Renal Ischemia/Reperfusion Injury. *J. Am. Soc. Nephrol.* **2015**, *26*, 1334–1345. [CrossRef]

22. Kujawski, M.; Kortylewski, M.; Lee, H.; Herrmann, A.; Kay, H.; Yu, H. Stat3 mediates myeloid cell-dependent tumor angiogenesis in mice. *J. Clin. Investig.* **2008**, *118*, 3367–3377. [CrossRef]

23. Giurisato, E.; Xu, Q.; Lonardi, S.; Telfer, B.; Russo, I.; Pearson, A.; Finegan, K.G.; Wang, W.; Wang, J.; Gray, N.S.; et al. Myeloid ERK5 deficiency suppresses tumor growth by blocking protumor macrophage polarization via STAT3 inhibition. *Proc. Natl. Acad. Sci. USA* **2018**, *115*, E2801–E2810. [CrossRef] [PubMed]

24. Goossens, P.; Rodriguez-Vita, J.; Etzerodt, A.; Masse, M.; Rastoin, O.; Gouirand, V.; Ulas, T.; Papantonopoulou, O.; Van Eck, M.; Auphan-Anezin, N.; et al. Membrane Cholesterol Efflux Drives Tumor-Associated Macrophage Reprogramming and Tumor Progression. *Cell Metab.* **2019**, *29*, 1376–1389. [CrossRef] [PubMed]

25. Gupta, S.; Jain, A.; Syed, S.N.; Snodgrass, R.G.; Pfluger-Muller, B.; Leisegang, M.S.; Weigert, A.; Brandes, R.P.; Ebersberger, I.; Brune, B.; et al. IL-6 augments IL-4-induced polarization of primary human macrophages through synergy of STAT3, STAT6 and BATF transcription factors. *Oncoimmunology* **2018**, *7*, e1494110. [CrossRef] [PubMed]

26. Rebe, C.; Ghiringhelli, F. STAT3, a Master Regulator of Anti-Tumor Immune Response. *Cancers* **2019**, *11*, 1280. [CrossRef]

27. Andersen, M.N.; Etzerodt, A.; Graversen, J.H.; Holthof, L.C.; Moestrup, S.K.; Hokland, M.; Moller, H.J. STAT3 inhibition specifically in human monocytes and macrophages by CD163-targeted corosolic acid-containing liposomes. *Cancer Immunol. Immunother.* **2019**, *68*, 489–502. [CrossRef] [PubMed]

28. Lee, H.; Deng, J.; Xin, H.; Liu, Y.; Pardoll, D.; Yu, H. A requirement of STAT3 DNA binding precludes Th-1 immunostimulatory gene expression by NF-kappaB in tumors. *Cancer Res.* **2011**, *71*, 3772–3780. [CrossRef] [PubMed]

29. Yan, D.; Wang, H.W.; Bowman, R.L.; Joyce, J.A. STAT3 and STAT6 Signaling Pathways Synergize to Promote Cathepsin Secretion from Macrophages via IRE1alpha Activation. *Cell Rep.* **2016**, *16*, 2914–2927. [CrossRef]

30. Kortylewski, M.; Xin, H.; Kujawski, M.; Lee, H.; Liu, Y.; Harris, T.; Drake, C.; Pardoll, D.; Yu, H. Regulation of the IL-23 and IL-12 balance by Stat3 signaling in the tumor microenvironment. *Cancer Cell* **2009**, *15*, 114–123. [CrossRef]

31. Deng, J.; Liu, Y.; Lee, H.; Herrmann, A.; Zhang, W.; Zhang, C.; Shen, S.; Priceman, S.J.; Kujawski, M.; Pal, S.K.; et al. S1PR1-STAT3 signaling is crucial for myeloid cell colonization at future metastatic sites. *Cancer Cell* **2012**, *21*, 642–654. [CrossRef] [PubMed]

32. Jones, L.M.; Broz, M.L.; Ranger, J.J.; Ozcelik, J.; Ahn, R.; Zuo, D.; Ursini-Siegel, J.; Hallett, M.T.; Krummel, M.; Muller, W.J. STAT3 Establishes an Immunosuppressive Microenvironment during the Early Stages of Breast Carcinogenesis to Promote Tumor Growth and Metastasis. *Cancer Res.* **2016**, *76*, 1416–1428. [CrossRef] [PubMed]

33. Cheng, Z.; Zhang, D.; Gong, B.; Wang, P.; Liu, F. CD163 as a novel target gene of STAT3 is a potential therapeutic target for gastric cancer. *Oncotarget* **2017**, *8*, 87244–87262. [CrossRef] [PubMed]

34. Trovato, R.; Fiore, A.; Sartori, S.; Cane, S.; Giugno, R.; Cascione, L.; Paiella, S.; Salvia, R.; De Sanctis, F.; Poffe, O.; et al. Immunosuppression by monocytic myeloid-derived suppressor cells in patients with pancreatic ductal carcinoma is orchestrated by STAT3. *J. Immunother. Cancer* **2019**, *7*, 255. [CrossRef] [PubMed]

35. Thorn, M.; Guha, P.; Cunetta, M.; Espat, N.J.; Miller, G.; Junghans, R.P.; Katz, S.C. Tumor-associated GM-CSF overexpression induces immunoinhibitory molecules via STAT3 in myeloid-suppressor cells infiltrating liver metastases. *Cancer Gene.* **2016**, *23*, 188–198. [CrossRef] [PubMed]

36. Lamano, J.B.; Lamano, J.B.; Li, Y.D.; DiDomenico, J.D.; Choy, W.; Veliceasa, D.; Oyon, D.E.; Fakurnejad, S.; Ampie, L.; Kesavabhotla, K.; et al. Glioblastoma-Derived IL6 Induces Immunosuppressive Peripheral Myeloid Cell PD-L1 and Promotes Tumor Growth. *Clin. Cancer Res.* **2019**, *25*, 3643–3657. [CrossRef]

37. Zerdes, I.; Wallerius, M.; Sifakis, E.G.; Wallmann, T.; Betts, S.; Bartish, M.; Tsesmetzis, N.; Tobin, N.P.; Coucoravas, C.; Bergh, J.; et al. STAT3 Activity Promotes Programmed-Death Ligand 1 Expression and Suppresses Immune Responses in Breast Cancer. *Cancers* **2019**, *11*, 1479. [CrossRef] [PubMed]

38. Kowal, J.; Kornete, M.; Joyce, J.A. Re-education of macrophages as a therapeutic strategy in cancer. *Immunotherapy* **2019**, *11*, 677–689. [CrossRef]

39. Peranzoni, E.; Lemoine, J.; Vimeux, L.; Feuillet, V.; Barrin, S.; Kantari-Mimoun, C.; Bercovici, N.; Guerin, M.; Biton, J.; Ouakrim, H.; et al. Macrophages impede CD8 T cells from reaching tumor cells and limit the efficacy of anti-PD-1 treatment. *Proc Natl. Acad. Sci. USA* **2018**, *115*, E4041–E4050. [CrossRef]

40. O'Donnell, J.S.; Long, G.V.; Scolyer, R.A.; Teng, M.W.; Smyth, M.J. Resistance to PD1/PDL1 checkpoint inhibition. *Cancer Treat. Rev.* **2017**, *52*, 71–81. [CrossRef]

41. Tumeh, P.C.; Harview, C.L.; Yearley, J.H.; Shintaku, I.P.; Taylor, E.J.; Robert, L.; Chmielowski, B.; Spasic, M.; Henry, G.; Ciobanu, V.; et al. PD-1 blockade induces responses by inhibiting adaptive immune resistance. *Nature* **2014**, *515*, 568–571. [CrossRef] [PubMed]

42. Lin, H.; Wei, S.; Hurt, E.M.; Green, M.D.; Zhao, L.; Vatan, L.; Szeliga, W.; Herbst, R.; Harms, P.W.; Fecher, L.A.; et al. Host expression of PD-L1 determines efficacy of PD-L1 pathway blockade-mediated tumor regression. *J. Clin. Investig.* **2018**, *128*, 805–815. [CrossRef] [PubMed]

43. Tang, H.; Liang, Y.; Anders, R.A.; Taube, J.M.; Qiu, X.; Mulgaonkar, A.; Liu, X.; Harrington, S.M.; Guo, J.; Xin, Y.; et al. PD-L1 on host cells is essential for PD-L1 blockade-mediated tumor regression. *J. Clin. Investig.* **2018**, *128*, 580–588. [CrossRef] [PubMed]

44. Rashidian, M.; LaFleur, M.W.; Verschoor, V.L.; Dongre, A.; Zhang, Y.; Nguyen, T.H.; Kolifrath, S.; Aref, A.R.; Lau, C.J.; Paweletz, C.P.; et al. Immuno-PET identifies the myeloid compartment as a key contributor to the outcome of the antitumor response under PD-1 blockade. *Proc. Natl. Acad. Sci. USA* **2019**, *116*, 16971–16980. [CrossRef]

45. Wei, Y.; Zhao, Q.; Gao, Z.; Lao, X.M.; Lin, W.M.; Chen, D.P.; Mu, M.; Huang, C.X.; Liu, Z.Y.; Li, B.; et al. The local immune landscape determines tumor PD-L1 heterogeneity and sensitivity to therapy. *J. Clin. Investig.* **2019**, *129*, 3347–3360. [CrossRef]

46. Leonard, W.J.; Lin, J.X.; O'Shea, J.J. The gammac Family of Cytokines: Basic Biology to Therapeutic Ramifications. *Immunity* **2019**, *50*, 832–850. [CrossRef]

47. Raeber, M.E.; Zurbuchen, Y.; Impellizzieri, D.; Boyman, O. The role of cytokines in T-cell memory in health and disease. *Immunol Rev.* **2018**, *283*, 176–193. [CrossRef]

48. Spolski, R.; Li, P.; Leonard, W.J. Biology and regulation of IL-2: From molecular mechanisms to human therapy. *Nat. Rev. Immunol.* **2018**, *18*, 648–659. [CrossRef]

49. Hand, T.W.; Cui, W.; Jung, Y.W.; Sefik, E.; Joshi, N.S.; Chandele, A.; Liu, Y.; Kaech, S.M. Differential effects of STAT5 and PI3K/AKT signaling on effector and memory CD8 T-cell survival. *Proc. Natl. Acad. Sci. USA* **2010**, *107*, 16601–16606. [CrossRef]

50. Tripathi, P.; Kurtulus, S.; Wojciechowski, S.; Sholl, A.; Hoebe, K.; Morris, S.C.; Finkelman, F.D.; Grimes, H.L.; Hildeman, D.A. STAT5 is critical to maintain effector CD8+ T cell responses. *J. Immunol.* **2010**, *185*, 2116–2124. [CrossRef]

51. O'Gorman, W.E.; Dooms, H.; Thorne, S.H.; Kuswanto, W.F.; Simonds, E.F.; Krutzik, P.O.; Nolan, G.P.; Abbas, A.K. The initial phase of an immune response functions to activate regulatory T cells. *J. Immunol.* **2009**, *183*, 332–339. [CrossRef]

52. Verdeil, G.; Chaix, J.; Schmitt-Verhulst, A.M.; Auphan-Anezin, N. Temporal cross-talk between TCR and STAT signals for CD8 T cell effector differentiation. *Eur. J. Immunol.* **2006**, *36*, 3090–3100. [CrossRef]

53. Verdeil, G.; Puthier, D.; Nguyen, C.; Schmitt-Verhulst, A.M.; Auphan-Anezin, N. STAT5-mediated signals sustain a TCR-initiated gene expression program toward differentiation of CD8 T cell effectors. *J. Immunol.* **2006**, *176*, 4834–4842. [CrossRef]

54. Altan-Bonnet, G.; Germain, R.N. Modeling T cell antigen discrimination based on feedback control of digital ERK responses. *PLoS Biol.* **2005**, *3*, e356. [CrossRef]

55. Auphan-Anezin, N.; Mazza, C.; Guimezanes, A.; Barrett-Wilt, G.A.; Montero-Julian, F.; Roussel, A.; Hunt, D.F.; Malissen, B.; Schmitt-Verhulst, A.M. Distinct orientation of the alloreactive monoclonal CD8 T cell activation program by three different peptide/MHC complexes. *Eur. J. Immunol.* **2006**, *36*, 1856–1866. [CrossRef]

56. Chau, L.A.; Bluestone, J.A.; Madrenas, J. Dissociation of intracellular signaling pathways in response to partial agonist ligands of the T cell receptor. *J. Exp. Med.* **1998**, *187*, 1699–1709. [CrossRef]

57. Rollings, C.M.; Sinclair, L.V.; Brady, H.J.M.; Cantrell, D.A.; Ross, S.H. Interleukin-2 shapes the cytotoxic T cell proteome and immune environment-sensing programs. *Sci. Signal* **2018**, *11*, eaap8112. [CrossRef]

58. Ross, S.H.; Rollings, C.; Anderson, K.E.; Hawkins, P.T.; Stephens, L.R.; Cantrell, D.A. Phosphoproteomic Analyses of Interleukin 2 Signaling Reveal Integrated JAK Kinase-Dependent and -Independent Networks in CD8(+) T Cells. *Immunity* **2016**, *45*, 685–700. [CrossRef]

59. Cote-Sierra, J.; Foucras, G.; Guo, L.; Chiodetti, L.; Young, H.A.; Hu-Li, J.; Zhu, J.; Paul, W.E. Interleukin 2 plays a central role in Th2 differentiation. *Proc. Natl. Acad. Sci. USA* **2004**, *101*, 3880–3885. [CrossRef]

60. Grange, M.; Verdeil, G.; Arnoux, F.; Griffon, A.; Spicuglia, S.; Maurizio, J.; Buferne, M.; Schmitt-Verhulst, A.M.; Auphan-Anezin, N. Active STAT5 regulates T-bet and eomesodermin expression in CD8 T cells and imprints a T-bet-dependent Tc1 program with repressed IL-6/TGF-beta1 signaling. *J. Immunol.* **2013**, *191*, 3712–3724. [CrossRef]

61. Shi, M.; Lin, T.H.; Appell, K.C.; Berg, L.J. Janus-kinase-3-dependent signals induce chromatin remodeling at the Ifng locus during T helper 1 cell differentiation. *Immunity* **2008**, *28*, 763–773. [CrossRef]

62. Scheeren, F.A.; Naspetti, M.; Diehl, S.; Schotte, R.; Nagasawa, M.; Wijnands, E.; Gimeno, R.; Vyth-Dreese, F.A.; Blom, B.; Spits, H. STAT5 regulates the self-renewal capacity and differentiation of human memory B cells and controls Bcl-6 expression. *Nat. Immunol.* **2005**, *6*, 303–313. [CrossRef]

63. Sheng, W.; Yang, F.; Zhou, Y.; Yang, H.; Low, P.Y.; Kemeny, D.M.; Tan, P.; Moh, A.; Kaplan, M.H.; Zhang, Y.; et al. STAT5 programs a distinct subset of GM-CSF-producing T helper cells that is essential for autoimmune neuroinflammation. *Cell Res.* **2014**, *24*, 1387–1402. [CrossRef]

64. Olson, M.R.; Verdan, F.F.; Hufford, M.M.; Dent, A.L.; Kaplan, M.H. STAT3 Impairs STAT5 Activation in the Development of IL-9-Secreting T Cells. *J. Immunol.* **2016**, *196*, 3297–3304. [CrossRef]

65. Burchill, M.A.; Yang, J.; Vogtenhuber, C.; Blazar, B.R.; Farrar, M.A. IL-2 receptor beta-dependent STAT5 activation is required for the development of Foxp3+ regulatory T cells. *J. Immunol.* **2007**, *178*, 280–290. [CrossRef]

66. Yao, Z.; Kanno, Y.; Kerenyi, M.; Stephens, G.; Durant, L.; Watford, W.T.; Laurence, A.; Robinson, G.W.; Shevach, E.M.; Moriggl, R.; et al. Nonredundant roles for Stat5a/b in directly regulating Foxp3. *Blood* **2007**, *109*, 4368–4375. [CrossRef]

67. Yang, X.P.; Ghoreschi, K.; Steward-Tharp, S.M.; Rodriguez-Canales, J.; Zhu, J.; Grainger, J.R.; Hirahara, K.; Sun, H.W.; Wei, L.; Vahedi, G.; et al. Opposing regulation of the locus encoding IL-17 through direct, reciprocal actions of STAT3 and STAT5. *Nat. Immunol.* **2011**, *12*, 247–254. [CrossRef]

68. Nurieva, R.I.; Podd, A.; Chen, Y.; Alekseev, A.M.; Yu, M.; Qi, X.; Huang, H.; Wen, R.; Wang, J.; Li, H.S.; et al. STAT5 protein negatively regulates T follicular helper (Tfh) cell generation and function. *J. Biol. Chem.* **2012**, *287*, 11234–11239. [CrossRef]

69. Tripathi, S.K.; Chen, Z.; Larjo, A.; Kanduri, K.; Nousiainen, K.; Aijo, T.; Ricano-Ponce, I.; Hrdlickova, B.; Tuomela, S.; Laajala, E.; et al. Genome-wide Analysis of STAT3-Mediated Transcription during Early Human Th17 Cell Differentiation. *Cell Rep.* **2017**, *19*, 1888–1901. [CrossRef]

70. Poholek, A.C.; Jankovic, D.; Villarino, A.V.; Petermann, F.; Hettinga, A.; Shouval, D.S.; Snapper, S.B.; Kaech, S.M.; Brooks, S.R.; Vahedi, G.; et al. IL-10 induces a STAT3-dependent autoregulatory loop in TH2 cells that promotes Blimp-1 restriction of cell expansion via antagonism of STAT5 target genes. *Sci. Immunol.* **2016**, *1*, eaaf8612.

71. Li, P.; Mitra, S.; Spolski, R.; Oh, J.; Liao, W.; Tang, Z.; Mo, F.; Li, X.; West, E.E.; Gromer, D.; et al. STAT5-mediated chromatin interactions in superenhancers activate IL-2 highly inducible genes: Functional dissection of the Il2ra gene locus. *Proc. Natl. Acad. Sci. USA* **2017**, *114*, 12111–12119. [CrossRef]

72. Onishi, M.; Nosaka, T.; Misawa, K.; Mui, A.L.; Gorman, D.; McMahon, M.; Miyajima, A.; Kitamura, T. Identification and characterization of a constitutively active STAT5 mutant that promotes cell proliferation. *Mol. Cell. Biol.* **1998**, *18*, 3871–3879. [CrossRef]

73. Gotthardt, D.; Putz, E.M.; Grundschober, E.; Prchal-Murphy, M.; Straka, E.; Kudweis, P.; Heller, G.; Bago-Horvath, Z.; Witalisz-Siepracka, A.; Cumaraswamy, A.A.; et al. STAT5 Is a Key Regulator in NK Cells and Acts as a Molecular Switch from Tumor Surveillance to Tumor Promotion. *Cancer Discov.* **2016**, *6*, 414–429. [CrossRef]

74. Cayrol, F.; Praditsuktavorn, P.; Fernando, T.M.; Kwiatkowski, N.; Marullo, R.; Calvo-Vidal, M.N.; Phillip, J.; Pera, B.; Yang, S.N.; Takpradit, K.; et al. THZ1 targeting CDK7 suppresses STAT transcriptional activity and sensitizes T-cell lymphomas to BCL2 inhibitors. *Nat. Commun.* **2017**, *8*, 14290. [CrossRef]

75. Nivarthi, H.; Prchal-Murphy, M.; Swoboda, A.; Hager, M.; Schlederer, M.; Kenner, L.; Tuckermann, J.; Sexl, V.; Moriggl, R.; Ermakova, O. Stat5 gene dosage in T cells modulates CD8+ T-cell homeostasis and attenuates contact hypersensitivity response in mice. *Allergy* **2015**, *70*, 67–79. [CrossRef]

76. Liao, W.; Spolski, R.; Li, P.; Du, N.; West, E.E.; Ren, M.; Mitra, S.; Leonard, W.J. Opposing actions of IL-2 and IL-21 on Th9 differentiation correlate with their differential regulation of BCL6 expression. *Proc. Natl. Acad. Sci. USA* **2014**, *111*, 3508–3513. [CrossRef]

77. Oestreich, K.J.; Mohn, S.E.; Weinmann, A.S. Molecular mechanisms that control the expression and activity of Bcl-6 in TH1 cells to regulate flexibility with a TFH-like gene profile. *Nat. Immunol.* **2012**, *13*, 405–411. [CrossRef]

78. Son, H.J.; Lee, S.H.; Lee, S.Y.; Kim, E.K.; Yang, E.J.; Kim, J.K.; Seo, H.B.; Park, S.H.; Cho, M.L. Oncostatin M Suppresses Activation of IL-17/Th17 via SOCS3 Regulation in CD4+ T Cells. *J. Immunol.* **2017**, *198*, 1484–1491. [CrossRef]

79. Wingelhofer, B.; Neubauer, H.A.; Valent, P.; Han, X.; Constantinescu, S.N.; Gunning, P.T.; Muller, M.; Moriggl, R. Implications of STAT3 and STAT5 signaling on gene regulation and chromatin remodeling in hematopoietic cancer. *Leukemia* **2018**, *32*, 1713–1726. [CrossRef]

80. Moriggl, R.; Sexl, V.; Kenner, L.; Duntsch, C.; Stangl, K.; Gingras, S.; Hoffmeyer, A.; Bauer, A.; Piekorz, R.; Wang, D.; et al. Stat5 tetramer formation is associated with leukemogenesis. *Cancer Cell* **2005**, *7*, 87–99. [CrossRef]

81. Grange, M.; Giordano, M.; Mas, A.; Roncagalli, R.; Firaguay, G.; Nunes, J.A.; Ghysdael, J.; Schmitt-Verhulst, A.M.; Auphan-Anezin, N. Control of CD8 T cell proliferation and terminal differentiation by active STAT5 and CDKN2A/CDKN2B. *Immunology* **2015**, *145*, 543–557. [CrossRef]

82. Freund, P.; Kerenyi, M.A.; Hager, M.; Wagner, T.; Wingelhofer, B.; Pham, H.T.T.; Elabd, M.; Han, X.; Valent, P.; Gouilleux, F.; et al. O-GlcNAcylation of STAT5 controls tyrosine phosphorylation and oncogenic transcription in STAT5-dependent malignancies. *Leukemia* **2017**, *31*, 2132–2142. [CrossRef]

83. Grange, M.; Buferne, M.; Verdeil, G.; Leserman, L.; Schmitt-Verhulst, A.M.; Auphan-Anezin, N. Activated STAT5 promotes long-lived cytotoxic CD8+ T cells that induce regression of autochthonous melanoma. *Cancer Res.* **2012**, *72*, 76–87. [CrossRef]

84. Maurer, B.; Nivarthi, H.; Wingelhofer, B.; Pham, H.T.T.; Schlederer, M.; Suske, T.; Grausenburger, R.; Schiefer, A.I.; Prchal-Murphy, M.; Chen, D.; et al. High activation of STAT5A drives peripheral T-cell lymphoma and leukemia. *Haematologica* **2019**. [CrossRef]

85. Burchill, M.A.; Goetz, C.A.; Prlic, M.; O'Neil, J.J.; Harmon, I.R.; Bensinger, S.J.; Turka, L.A.; Brennan, P.; Jameson, S.C.; Farrar, M.A. Distinct effects of STAT5 activation on CD4+ and CD8+ T cell homeostasis: Development of CD4+CD25+ regulatory T cells versus CD8+ memory T cells. *J. Immunol.* **2003**, *171*, 5853–5864. [CrossRef]

86. Heltemes-Harris, L.M.; Larson, J.D.; Starr, T.K.; Hubbard, G.K.; Sarver, A.L.; Largaespada, D.A.; Farrar, M.A. Sleeping Beauty transposon screen identifies signaling modules that cooperate with STAT5 activation to induce B-cell acute lymphoblastic leukemia. *Oncogene* **2016**, *35*, 3454–3464. [CrossRef]

87. Katerndahl, C.D.S.; Heltemes-Harris, L.M.; Willette, M.J.L.; Henzler, C.M.; Frietze, S.; Yang, R.; Schjerven, H.; Silverstein, K.A.T.; Ramsey, L.B.; Hubbard, G.; et al. Antagonism of B cell enhancer networks by STAT5 drives leukemia and poor patient survival. *Nat. Immunol.* **2017**, *18*, 694–704. [CrossRef]

88. Pham, H.T.T.; Maurer, B.; Prchal-Murphy, M.; Grausenburger, R.; Grundschober, E.; Javaheri, T.; Nivarthi, H.; Boersma, A.; Kolbe, T.; Elabd, M.; et al. STAT5BN642H is a driver mutation for T cell neoplasia. *J. Clin. Investig.* **2018**, *128*, 387–401. [CrossRef]

89. De Araujo, E.D.; Erdogan, F.; Neubauer, H.A.; Meneksedag-Erol, D.; Manaswiyoungkul, P.; Eram, M.S.; Seo, H.S.; Qadree, A.K.; Israelian, J.; Orlova, A.; et al. Structural and functional consequences of the STAT5B(N642H) driver mutation. *Nat. Commun.* **2019**, *10*, 2517. [CrossRef]

90. Zeng, R.; Spolski, R.; Finkelstein, S.E.; Oh, S.; Kovanen, P.E.; Hinrichs, C.S.; Pise-Masison, C.A.; Radonovich, M.F.; Brady, J.N.; Restifo, N.P.; et al. Synergy of IL-21 and IL-15 in regulating CD8+ T cell expansion and function. *J. Exp. Med.* **2005**, *201*, 139–148. [CrossRef]

91. Jones, L.L.; Alli, R.; Li, B.; Geiger, T.L. Differential T Cell Cytokine Receptivity and Not Signal Quality Distinguishes IL-6 and IL-10 Signaling during Th17 Differentiation. *J. Immunol.* **2016**, *196*, 2973–2985. [CrossRef]

92. Ashouri, J.F.; Hsu, L.Y.; Yu, S.; Rychkov, D.; Chen, Y.; Cheng, D.A.; Sirota, M.; Hansen, E.; Lattanza, L.; Zikherman, J.; et al. Reporters of TCR signaling identify arthritogenic T cells in murine and human autoimmune arthritis. *Proc. Natl. Acad. Sci. USA* **2019**, *116*, 18517–18527. [CrossRef]

93. Logotheti, S.; Putzer, B.M. STAT3 and STAT5 Targeting for Simultaneous Management of Melanoma and Autoimmune Diseases. *Cancers* **2019**, *11*, 1448. [CrossRef]

94. Wang, Y.; Shen, Y.; Wang, S.; Shen, Q.; Zhou, X. The role of STAT3 in leading the crosstalk between human cancers and the immune system. *Cancer Lett.* **2018**, *415*, 117–128. [CrossRef]

95. Cui, W.; Liu, Y.; Weinstein, J.S.; Craft, J.; Kaech, S.M. An interleukin-21-interleukin-10-STAT3 pathway is critical for functional maturation of memory CD8+ T cells. *Immunity* **2011**, *35*, 792–805. [CrossRef]

96. Siegel, A.M.; Heimall, J.; Freeman, A.F.; Hsu, A.P.; Brittain, E.; Brenchley, J.M.; Douek, D.C.; Fahle, G.H.; Cohen, J.I.; Holland, S.M.; et al. A critical role for STAT3 transcription factor signaling in the development and maintenance of human T cell memory. *Immunity* **2011**, *35*, 806–818. [CrossRef]

97. Hinrichs, C.S.; Spolski, R.; Paulos, C.M.; Gattinoni, L.; Kerstann, K.W.; Palmer, D.C.; Klebanoff, C.A.; Rosenberg, S.A.; Leonard, W.J.; Restifo, N.P. IL-2 and IL-21 confer opposing differentiation programs to CD8+ T cells for adoptive immunotherapy. *Blood* **2008**, *111*, 5326–5333. [CrossRef]

98. Kujawski, M.; Zhang, C.; Herrmann, A.; Reckamp, K.; Scuto, A.; Jensen, M.; Deng, J.; Forman, S.; Figlin, R.; Yu, H. Targeting STAT3 in adoptively transferred T cells promotes their in vivo expansion and antitumor effects. *Cancer Res.* **2010**, *70*, 9599–9610. [CrossRef]

99. Yue, C.; Shen, S.; Deng, J.; Priceman, S.J.; Li, W.; Huang, A.; Yu, H. STAT3 in CD8+ T Cells Inhibits Their Tumor Accumulation by Downregulating CXCR3/CXCL10 Axis. *Cancer Immunol. Res.* **2015**, *3*, 864–870. [CrossRef]

100. Triplett, T.A.; Tucker, C.G.; Triplett, K.C.; Alderman, Z.; Sun, L.; Ling, L.E.; Akporiaye, E.T.; Weinberg, A.D. STAT3 Signaling Is Required for Optimal Regression of Large Established Tumors in Mice Treated with Anti-OX40 and TGFbeta Receptor Blockade. *Cancer Immunol. Res.* **2015**, *3*, 526–535. [CrossRef]

101. Rosenberg, S.A.; Restifo, N.P. Adoptive cell transfer as personalized immunotherapy for human cancer. *Science* **2015**, *348*, 62–68. [CrossRef]

102. Dudley, M.E.; Wunderlich, J.R.; Yang, J.C.; Sherry, R.M.; Topalian, S.L.; Restifo, N.P.; Royal, R.E.; Kammula, U.; White, D.E.; Mavroukakis, S.A.; et al. Adoptive cell transfer therapy following non-myeloablative but lymphodepleting chemotherapy for the treatment of patients with refractory metastatic melanoma. *J. Clin. Oncol.* **2005**, *23*, 2346–2357. [CrossRef]

103. Sadelain, M.; Riviere, I.; Riddell, S. Therapeutic T cell engineering. *Nature* **2017**, *545*, 423–431. [CrossRef]

104. Schuster, S.J.; Svoboda, J.; Chong, E.A.; Nasta, S.D.; Mato, A.R.; Anak, O.; Brogdon, J.L.; Pruteanu-Malinici, I.; Bhoj, V.; Landsburg, D.; et al. Chimeric Antigen Receptor T Cells in Refractory B-Cell Lymphomas. *N. Engl. J. Med.* **2017**, *377*, 2545–2554. [CrossRef]

105. Majzner, R.G.; Mackall, C.L. Clinical lessons learned from the first leg of the CAR T cell journey. *Nat. Med.* **2019**, *25*, 1341–1355. [CrossRef]

106. Yamamoto, T.N.; Kishton, R.J.; Restifo, N.P. Developing neoantigen-targeted T cell-based treatments for solid tumors. *Nat. Med.* **2019**, *25*, 1488–1499. [CrossRef]

107. Beatty, G.L.; O'Hara, M.H.; Lacey, S.F.; Torigian, D.A.; Nazimuddin, F.; Chen, F.; Kulikovskaya, I.M.; Soulen, M.C.; McGarvey, M.; Nelson, A.M.; et al. Activity of Mesothelin-Specific Chimeric Antigen Receptor T Cells Against Pancreatic Carcinoma Metastases in a Phase 1 Trial. *Gastroenterology* **2018**, *155*, 29–32. [CrossRef]

108. Rosenberg, S.A. IL-2: The first effective immunotherapy for human cancer. *J. Immunol.* **2014**, *192*, 5451–5458. [CrossRef]

109. Krieg, C.; Letourneau, S.; Pantaleo, G.; Boyman, O. Improved IL-2 immunotherapy by selective stimulation of IL-2 receptors on lymphocytes and endothelial cells. *Proc. Natl. Acad. Sci. USA* **2010**, *107*, 11906–11911. [CrossRef]

110. Boyman, O.; Kovar, M.; Rubinstein, M.P.; Surh, C.D.; Sprent, J. Selective stimulation of T cell subsets with antibody-cytokine immune complexes. *Science* **2006**, *311*, 1924–1927. [CrossRef]

111. Verdeil, G.; Marquardt, K.; Surh, C.D.; Sherman, L.A. Adjuvants targeting innate and adaptive immunity synergize to enhance tumor immunotherapy. *Proc. Natl. Acad. Sci. USA* **2008**, *105*, 16683–16688. [CrossRef]

112. Boyman, O.; Arenas-Ramirez, N. Development of a novel class of interleukin-2 immunotherapies for metastatic cancer. *Swiss Med. Wkly.* **2019**, *149*, w14697. [CrossRef]

113. Sabatino, M.; Hu, J.; Sommariva, M.; Gautam, S.; Fellowes, V.; Hocker, J.D.; Dougherty, S.; Qin, H.; Klebanoff, C.A.; Fry, T.J.; et al. Generation of clinical-grade CD19-specific CAR-modified CD8+ memory stem cells for the treatment of human B-cell malignancies. *Blood* **2016**, *128*, 519–528. [CrossRef]

114. Kagoya, Y.; Tanaka, S.; Guo, T.; Anczurowski, M.; Wang, C.H.; Saso, K.; Butler, M.O.; Minden, M.D.; Hirano, N. A novel chimeric antigen receptor containing a JAK-STAT signaling domain mediates superior antitumor effects. *Nat. Med.* **2018**, *24*, 352–359. [CrossRef]

115. Sackstein, R.; Schatton, T.; Barthel, S.R. T-lymphocyte homing: An underappreciated yet critical hurdle for successful cancer immunotherapy. *Lab. Investig.* **2017**, *97*, 669–697. [CrossRef]

116. Miller, B.C.; Sen, D.R.; Al Abosy, R.; Bi, K.; Virkud, Y.V.; LaFleur, M.W.; Yates, K.B.; Lako, A.; Felt, K.; Naik, G.S.; et al. Subsets of exhausted CD8(+) T cells differentially mediate tumor control and respond to checkpoint blockade. *Nat. Immunol.* **2019**, *20*, 326–336. [CrossRef]

117. Siddiqui, I.; Schaeuble, K.; Chennupati, V.; Fuertes Marraco, S.A.; Calderon-Copete, S.; Pais Ferreira, D.; Carmona, S.J.; Scarpellino, L.; Gfeller, D.; Pradervand, S.; et al. Intratumoral Tcf1(+)PD-1(+)CD8(+) T Cells with Stem-like Properties Promote Tumor Control in Response to Vaccination and Checkpoint Blockade Immunotherapy. *Immunity* **2019**, *50*, 195–211. [CrossRef]

118. Robbins, P.F.; Dudley, M.E.; Wunderlich, J.; El-Gamil, M.; Li, Y.F.; Zhou, J.; Huang, J.; Powell, D.J., Jr.; Rosenberg, S.A. Cutting edge: Persistence of transferred lymphocyte clonotypes correlates with cancer regression in patients receiving cell transfer therapy. *J. Immunol.* **2004**, *173*, 7125–7130. [CrossRef]

119. Yang, S.; Ji, Y.; Gattinoni, L.; Zhang, L.; Yu, Z.; Restifo, N.P.; Rosenberg, S.A.; Morgan, R.A. Modulating the differentiation status of ex vivo-cultured anti-tumor T cells using cytokine cocktails. *Cancer Immunol. Immunother.* **2013**, *62*, 727–736. [CrossRef]

120. Khan, O.; Giles, J.R.; McDonald, S.; Manne, S.; Ngiow, S.F.; Patel, K.P.; Werner, M.T.; Huang, A.C.; Alexander, K.A.; Wu, J.E.; et al. TOX transcriptionally and epigenetically programs CD8(+) T cell exhaustion. *Nature* **2019**, *571*, 211–218. [CrossRef]

121. Fraietta, J.A.; Lacey, S.F.; Orlando, E.J.; Pruteanu-Malinici, I.; Gohil, M.; Lundh, S.; Boesteanu, A.C.; Wang, Y.; O'Connor, R.S.; Hwang, W.T.; et al. Determinants of response and resistance to CD19 chimeric antigen receptor (CAR) T cell therapy of chronic lymphocytic leukemia. *Nat. Med.* **2018**, *24*, 563–571. [CrossRef]

122. Milner, J.D.; Vogel, T.P.; Forbes, L.; Ma, C.A.; Stray-Pedersen, A.; Niemela, J.E.; Lyons, J.J.; Engelhardt, K.R.; Zhang, Y.; Topcagic, N.; et al. Early-onset lymphoproliferation and autoimmunity caused by germline STAT3 gain-of-function mutations. *Blood* **2015**, *125*, 591–599. [CrossRef]

123. Sadreev, I.I.; Chen, M.Z.Q.; Umezawa, Y.; Biktashev, V.N.; Kemper, C.; Salakhieva, D.V.; Welsh, G.I.; Kotov, N.V. The competitive nature of signal transducer and activator of transcription complex formation drives phenotype switching of T cells. *Immunology* **2018**, *153*, 488–501. [CrossRef]

124. McLane, L.M.; Abdel-Hakeem, M.S.; Wherry, E.J. CD8 T Cell Exhaustion During Chronic Viral Infection and Cancer. *Annu. Rev. Immunol.* **2019**, *37*, 457–495. [CrossRef]

125. Giordano, M.; Henin, C.; Maurizio, J.; Imbratta, C.; Bourdely, P.; Buferne, M.; Baitsch, L.; Vanhille, L.; Sieweke, M.H.; Speiser, D.E.; et al. Molecular profiling of CD8 T cells in autochthonous melanoma identifies Maf as driver of exhaustion. *EMBO J.* **2015**, *34*, 2042–2058. [CrossRef]

126. Speiser, D.E.; Ho, P.C.; Verdeil, G. Regulatory circuits of T cell function in cancer. *Nat. Rev. Immunol.* **2016**, *16*, 599–611. [CrossRef]

127. Ladak, K.; Bass, A.R. Checkpoint inhibitor-associated autoimmunity. *Best Pract. Res. Clin. Rheumatol.* **2018**, *32*, 781–802. [CrossRef]

128. Keir, M.E.; Butte, M.J.; Freeman, G.J.; Sharpe, A.H. PD-1 and its ligands in tolerance and immunity. *Annu. Rev. Immunol.* **2008**, *26*, 677–704. [CrossRef]

129. Celis-Gutierrez, J.; Blattmann, P.; Zhai, Y.; Jarmuzynski, N.; Ruminski, K.; Gregoire, C.; Ounoughene, Y.; Fiore, F.; Aebersold, R.; Roncagalli, R.; et al. Quantitative Interactomics in Primary T Cells Provides a Rationale for Concomitant PD-1 and BTLA Coinhibitor Blockade in Cancer Immunotherapy. *Cell Rep.* **2019**, *27*, 3315–3330. [CrossRef]

130. Franceschini, D.; Paroli, M.; Francavilla, V.; Videtta, M.; Morrone, S.; Labbadia, G.; Cerino, A.; Mondelli, M.U.; Barnaba, V. PD-L1 negatively regulates CD4+CD25+Foxp3+ Tregs by limiting STAT-5 phosphorylation in patients chronically infected with HCV. *J. Clin. Investig.* **2009**, *119*, 551–564. [CrossRef]

131. Taylor, D.K.; Walsh, P.T.; LaRosa, D.F.; Zhang, J.; Burchill, M.A.; Farrar, M.A.; Turka, L.A. Constitutive activation of STAT5 supersedes the requirement for cytokine and TCR engagement of CD4+ T cells in steady-state homeostasis. *J. Immunol.* **2006**, *177*, 2216–2223. [CrossRef] [PubMed]

132. Buferne, M.; Chasson, L.; Grange, M.; Mas, A.; Arnoux, F.; Bertuzzi, M.; Naquet, P.; Leserman, L.; Schmitt-Verhulst, A.M.; Auphan-Anezin, N. IFNgamma producing CD8+ T cells modified to resist major immune checkpoints induce regression of MHC class I-deficient melanomas. *Oncoimmunology* **2015**, *4*, e974959. [CrossRef] [PubMed]

133. Auphan-Anezin, N.; Verdeil, G.; Grange, M.; Soudja, S.M.; Wehbe, M.; Buferne, M.; Mas, A.; Schmitt-Verhulst, A.M. Immunosuppression in inflammatory melanoma: Can it be resisted by adoptively transferred T cells? *Pigment Cell Melanoma Res.* **2013**, *26*, 167–175. [CrossRef] [PubMed]

134. Verdeil, G.; Fuertes Marraco, S.A.; Murray, T.; Speiser, D.E. From T cell "exhaustion" to anti-cancer immunity. *Biochim. Biophys. Acta* **2016**, *1865*, 49–57. [CrossRef] [PubMed]

135. Sun, C.; Fu, B.; Gao, Y.; Liao, X.; Sun, R.; Tian, Z.; Wei, H. TGF-beta1 down-regulation of NKG2D/DAP10 and 2B4/SAP expression on human NK cells contributes to HBV persistence. *PLoS Pathog.* **2012**, *8*, e1002594. [CrossRef]

136. Thomas, D.A.; Massague, J. TGF-beta directly targets cytotoxic T cell functions during tumor evasion of immune surveillance. *Cancer Cell* **2005**, *8*, 369–380. [CrossRef]

137. Benahmed, M.; Meresse, B.; Arnulf, B.; Barbe, U.; Mention, J.J.; Verkarre, V.; Allez, M.; Cellier, C.; Hermine, O.; Cerf-Bensussan, N. Inhibition of TGF-beta signaling by IL-15: A new role for IL-15 in the loss of immune homeostasis in celiac disease. *Gastroenterology* **2007**, *132*, 994–1008. [CrossRef]

138. Xiao, Q.; Wu, J.; Wang, W.J.; Chen, S.; Zheng, Y.; Yu, X.; Meeth, K.; Sahraei, M.; Bothwell, A.L.M.; Chen, L.; et al. DKK2 imparts tumor immunity evasion through beta-catenin-independent suppression of cytotoxic immune-cell activation. *Nat Med.* **2018**, *24*, 262–270. [CrossRef]

139. Mardiana, S.; Lai, J.; House, I.G.; Beavis, P.A.; Darcy, P.K. Switching on the green light for chimeric antigen receptor T-cell therapy. *Clin. Transl. Immunol.* **2019**, *8*, e1046. [CrossRef]

140. Zhang, J.; Endres, S.; Kobold, S. Enhancing tumor T cell infiltration to enable cancer immunotherapy. *Immunotherapy* **2019**, *11*, 201–213. [CrossRef]

141. Ayers, M.; Lunceford, J.; Nebozhyn, M.; Murphy, E.; Loboda, A.; Kaufman, D.R.; Albright, A.; Cheng, J.D.; Kang, S.P.; Shankaran, V.; et al. IFN-gamma-related mRNA profile predicts clinical response to PD-1 blockade. *J. Clin. Investig.* **2017**, *127*, 2930–2940. [CrossRef] [PubMed]

142. Horton, B.L.; Williams, J.B.; Cabanov, A.; Spranger, S.; Gajewski, T.F. Intratumoral CD8(+) T-cell Apoptosis Is a Major Component of T-cell Dysfunction and Impedes Antitumor Immunity. *Cancer Immunol. Res.* **2018**, *6*, 14–24. [CrossRef] [PubMed]

143. Pai, C.S.; Huang, J.T.; Lu, X.; Simons, D.M.; Park, C.; Chang, A.; Tamaki, W.; Liu, E.; Roybal, K.T.; Seagal, J.; et al. Clonal Deletion of Tumor-Specific T Cells by Interferon-gamma Confers Therapeutic Resistance to Combination Immune Checkpoint Blockade. *Immunity* **2019**, *50*, 477–492. [CrossRef] [PubMed]
144. Zhu, J.; Powis de Tenbossche, C.G.; Cane, S.; Colau, D.; van Baren, N.; Lurquin, C.; Schmitt-Verhulst, A.M.; Liljestrom, P.; Uyttenhove, C.; Van den Eynde, B.J. Resistance to cancer immunotherapy mediated by apoptosis of tumor-infiltrating lymphocytes. *Nat. Commun.* **2017**, *8*, 1404. [CrossRef]

Review

STAT3, a Master Regulator of Anti-Tumor Immune Response

Cédric Rébé * and François Ghiringhelli *

Platform of Transfer in Cancer Biology, Centre Georges François Leclerc, INSERM LNC UMR1231, University of Bourgogne Franche-Comté, F-21000 Dijon, France
* Correspondence: crebe@cgfl.fr (C.R.); fghiringhelli@cgfl.fr (F.G.); Tel.: +33-3-8073-7790 (C.R. & F.G.)

Received: 18 July 2019; Accepted: 29 August 2019; Published: 30 August 2019

Abstract: Immune cells in the tumor microenvironment regulate cancer growth. Thus cancer progression is dependent on the activation or repression of transcription programs involved in the proliferation/activation of lymphoid and myeloid cells. One of the main transcription factors involved in many of these pathways is the signal transducer and activator of transcription 3 (STAT3). In this review we will focus on the role of STAT3 and its regulation, e.g., by phosphorylation or acetylation in immune cells and how it might impact immune cell function and tumor progression. Moreover, we will review the ability of STAT3 to regulate checkpoint inhibitors.

Keywords: STAT3; cancer; CD4$^+$ T cells; CD8$^+$ T cells; myeloid cells; immune check point

---

1. Introduction

STAT3 is part of the Signal Transducer and Activator of Transcription (STAT) family which includes seven members encoded by distinct genes. STAT3 has evolutionary conserved amino acid sequences between *H. sapiens* and *S. harrisii* (99.09%) [1]. In resting cells, STAT3 remains in an inactive form in the cytoplasm. Once activated, mainly through phosphorylation, STAT3 translocates to the nucleus to play its transcription activity for specific target genes [2]. STAT3 phosphorylation on tyrosine (Y705) is mainly regulated by members of Janus-activated kinases (JAK), whereas its phosphorylation on serine (S727) is commonly regulated by mitogen-activated protein kinases, CDK5 and protein kinase C [3]. Finally, histone acetyltransferase-mediated reversible acetylation of STAT3 on a single lysine residue (K685) is a third mechanism of STAT3 activation through STAT3 dimer stabilization [4]. However, the phosphorylation on S727 is responsible for a mitochondrial relocalization of STAT3 where it exerts non-transcriptional roles. This mitochondrial localization enables STAT3 to increase cell respiration (through electron transport chain complex activation) and Ras transformation [5]. Non-nuclear STAT3 can also regulate glycolysis, thus enhancing lactate production leading to the protection of cells from apoptosis and senescence and can also regulate calcium homeostasis, energy production and apoptosis at the endoplasmic reticulum level [6].

Regulation of STAT protein activation is controlled by negative regulators, e.g., PIAS (protein inhibitor of activated STAT) and SOCS (suppressors of cytokine signaling) proteins as well as protein tyrosine phosphatases. PIAS are nuclear factors that negatively regulate STAT transcriptional activity through many mechanisms, especially by interacting and thus blocking the DNA binding activity [7]. SOCS proteins directly or indirectly interact with tyrosine kinase SH2 domains to prevent JAK from activating STAT3 [8]. Protein tyrosine phosphatases (such as CD45, SHP-1 and SHP-2) remove phosphates from activated STATs, which represent a third level of STAT modulation [9–11]. Lastly, STAT3 has also been shown to go through ubiquitination-dependent proteosomal degradation [12]. Moreover, because of their homologies, STATs can form homodimer and heterodimers. Specificity depends on the activator signal and leads to the transcription of different target genes. For example, STAT3 can heterodimerized with STAT1, under IL-6 treatment [13].

It is now well-established that STAT3 signaling is a major intrinsic pathway driving apoptosis, inflammation, cellular transformation, survival, proliferation, invasion, angiogenesis and metastasis in cancer [14–17]. Moreover, STAT3 in cancer cells affects stromal cells function, establishing crosstalk between cancer cells and its microenvironment. For example STAT3 can dampen STAT1-mediated upregulation of MHC class I, allowing immune escape [1]. The other way for STAT3 to drive tumor immune escape is to regulate the function of stromal cells and more particularly immune cells.

In general, all seven STAT family members have prominent roles in T-cell function or T-cell differentiation, survival or expansion. STAT4 is essential for Th1 and STAT6 is important for Th2 differentiation. Similarly, all STAT proteins have all seven prominent roles in myeloid cells and they all influence each other's expression and activity status on complex and not understood chromatin regulation. All that makes the interpretation of complex immune cell scenarios triggered by multiple action of cytokines, growth factors, hormones and chemokines a tricky business to correctly relate functions to this or that STAT family member. Importantly, T-cell expansion by common γ-chain cytokines and many T-cell effector functions such as $CD8^+$ T-cell, $\gamma\delta$ T-cell generations and cytokine release function and mounting a killing or efficient cytokine signaling response against foreign or mutated antigen is a STAT5-mediated affair together with proper recognition and signaling through the T-cell receptor (TCR), where again interplays are not carefully understood or worked out [18,19]. Furthermore, STAT5 is also essential to generate Treg cells, where both *Foxp3* and *Cd25* are direct STAT5 target genes [20]. STAT5 has also essential functions in erythropoiesis or macrophage or dendritic cell (DC) polarization, but due to space constrains and focus on fine-tuning and twisting immune responses in health or disease we will here illuminate STAT3 function in immune cells. We illuminate many important immune modulatory interplays of STAT3 signaling in distinct T-cell and myeloid cell compartments. We describe current knowledge on the impact of STAT3 activation in immune cells on the balance between immunosurveillance and immunoescape. We will describe how STAT3 affects both myeloid and lymphoid cells usually in a way to inhibit anti-tumor immune response and to promote tumor growth.

2. STAT3 and T-Cells

T lymphocytes or T-cells play a central role in host adaptive immune response to cancer [21]. Tumor-infiltrating $CD4^+$ and $CD8^+$ T-cells are associated with varying clinical outcomes and survival in many types of cancer such as colorectal, [22] breast [23] and lung cancers [24]. Cytokines can shape T-cells immune response and tune $CD4^+$ T-cells differentiation and $CD8^+$ T-cells activation [25]. Among T-cells, different subsets have been described (regulatory T-cells, cytotoxic T-cells, T helper cells) with distinct functions that could be regulated by STAT3 (Table 1).

2.1. Th1/Th2

$CD4^+$ T helper (Th) cells assist other hematopoietic cells in immune processes, including activation of cytotoxic T lymphocytes (CTLs), natural killer (NK) cells and macrophages. Th cells are activated when stimulated with a peptidic sequence of the antigen they specifically recognize. These peptides are presented to Th cells by antigen presenting cells (APCs), such as DCs through MHC class II molecules. Once activated, they rapidly proliferate and secrete cytokines that will inhibit or assist the active immune response [26].

Table 1. Impact of STAT3 in T-cell subsets.

Immune Cell Family	STAT3 Role in	STAT3 Modulators	Effects	Reference
CD4+ T-cells	Th1	IL-27	↗ proliferation	[27]
	Th2	IL-21	IL-10 secretion	[28]
	Th17	KO STAT3	Inhibition of Differentiation	[29,30]
		IL-6, IL-21, IL-23	↗ RORγt and RORα → Differentiation (↗ IL-17)	[31–35]
		PPARγ ligand, Platelet factor 4	↗ SOCS3 → Inhibition of Differentiation	[36]
		SHP1	↘ pSTAT3 → Inhibition of Differentiation	[37]
		LOXL3	↘ STAT3 deacetylation → Differentiation	[38]
		Metformin, Resveratrol	SIRT1 activation → STAT3 acetylation → Th17 Differentiation	[39]
		IL-6 (+ TGF-β)	↗ STAT3 activation and ↘ Gfi-1 → ↗ CD39 and CD73 expression	[40]
		miR29a-3p, miR-21-5p	↘ STAT3 → Th17 Differentiation	[41]
	Treg	KO STAT3	↘ number of Treg	[42–44]
		IL-2	↗ STAT3 + STAT5 → ↗ FOXP3 → ↗ inhibitory functions	
		S1PR1	↗ STAT3 → Treg migration in the tumor	[45]
		IL-6, IL-27	↗ STAT3 → ↘ FOXP3 → ↘ Treg differentiation	[34,46]
		CDK5	↗ pSTAT3(S727) → ↗FOXP3	[47]
		GATA-3	↗ miR125a-5 → ↘ IL-6R + STAT3 → ↘ Treg conversion	[48]
		Wogonin	↘ pSTAT3(Y705) ↗ pSTAT3(S727) → ↘ Treg differentiation	[49]
		WP1066	↘ pSTAT3 → ↘ FOXP3+ Treg	[50]
		compound9# (fluorinated β-amino-ketone)	↘ FOXP3+ Treg	[51]
		anti- sense oligos + radiations	↘ FOXP3+ Treg	[52]
	Tfh	KO STAT3 or STAT3 siRNA	↘ BCL6 ↗ GATA-3, IL-4 (Th2)	[53]
		STAT DNA Binding domain mutation	↘ STAT3 activity → ↘ IL-21	[54]
		TGF-β	↗ pSTAT3 → ↘ GATA-3, IL-4 (Th2)	[53,55]
		Intratumoral Tfh-like cells	→ ↗ IL-21, IFN-γ → ↗ M2b	[56]
	Th9	KO STAT3	↗ pSTAT5 → ↗ IL-9	[57]
		Murine IL-10	↗ pSTAT3 → ↘ IL-9	[58]
		Human IL-21	↗ pSTAT3 → ↘ T-BET → ↗ IL-9	[59]
CD8+ T-cells		KO STAT3	↗ IFN-γ → ↗ CXCL10 production by myeloid cells ↗ CXCR3 (CXCL10 receptor) ↗ CD8+ proliferation and tumor invasion	[60] [61]
		CD28 stimulation	↗ pSTAT3 → ↗ NKG2D	[62]

Expression levels of Th1 cell genes coding for Interferon-γ (IFNG), TAP1, Granzym B (GZMB) are significantly higher in colorectal tumors than in normal tissue. A high expression of Th1 cytotoxic genes was associated with significantly improved disease-free survival whereas a low expression of those genes lead to disease recurrence [63]. T-BET, the master transcriptional regulator for Th1 differentiation is induced by TCRs and IL-12 stimulation [64]. In contrast, GATA-3 is the master assessor for Th2 differentiation, after stimulation with IL-4 [65]. IL-27, a member of the IL-6/IL-12 family produced by macrophages and DCs, favors Th1 differentiation by up-regulating T-BET, down-regulating GATA-3 and suppressing proinflammatory cytokine production such as IL-2, IL-4, and IL-13 [27,66]. In this context, only the IL-27-dependent Th1 proliferation was mediated by STAT3 [27]. In contrast, in a different cytokine context, in patients harboring STAT3 dominant negative mutations or STAT1 or IL-21R loss of function mutations, it was shown that IL-21/IL-21R, STAT3 and STAT1 signaling are required for in vitro differentiation of IL-10-secreting cells, related to Th2 [28].

2.2. Th17

Th17 cells are CD4$^+$ T-cells induced by TCR triggering together with IL-6, transforming growth factor (TGF)-β, and IL-23, an IL-12 family member, stimulation [67]. Th17 cells have emerged as key drivers of a wide range of autoimmune disorders, including inflammatory bowel disease, psoriasis, and ankylosing spondylitis [68]. Th17 cell expansion was observed in human cancers such as ovarian, melanoma, breast or colon cancers [69]. In colorectal cancer, patients with low expression of Th17 genes seem to have a prolonged disease-free survival [63]. However, a positive role of Th17 was proposed in melanoma and ovarian cancer. Therefore, the role of Th17 cells in cancer immunity remains controversial [70].

Th17 are characterized by the expression of the transcription factors RORγt and RORα and the production of IL-17A [71]. In addition, STAT3 is also essential for Th17 cell differentiation, since STAT3 ablation in mice CD4$^+$ T-cells, abrogates Th17 differentiation [29,30]. Moreover, in patients harboring STAT3 dominant negative mutations, IL-21R loss of function mutations or STAT1 gain of function mutations, it was shown that IL-21/IL-21R/STAT3 signaling is required for in vitro production of IL-17A/F by Th17 cells whereas STAT1 overexpression inhibits it [28]. STAT3 can associate with Trim28 and RORγt to drive the transcription of target cytokines such as IL-17A/F [72]. Moreover, many in vitro studies have shown that STAT3 can be activated by several pro-inflammatory cytokines including IL-6, IL-21 and IL-23 leading to the regulation of RORγt and RORα expression, along with the development and the stabilization of Th17 cells [31–35]. We found that the n-3 fatty acid docosahexaenoic acid (DHA) was able to induce the expression of SOCS3 in a PPARγ-dependent manner. SOCS3 then inhibits pSTAT3 and Th17 differentiation [36]. Another regulator of STAT3 activation in Th17 is the tyrosine phosphatase SHP1, which dampens IL-6- and IL-21-driven Th17 development and limits colitis in mice [37]. Similarly, the Platelet Factor 4 (PF4) can also up-regulate SOCS3 expression leading to the inhibition of STAT3, Th17 differentiation and tumor growth [73].

The STAT3 acetylation profile is also involved in Th17 polarization. First, lysyl oxidase-like 3 (LOXL3) is able to deacetylate STAT3 and to inhibit its transcriptional activity. *Loxl3* deficiency leads to constitutive STAT3 K685 acetylation causing reduced Th17 differentiation associated with resistance to DSS-induced colitis in mice [38]. Second, we showed that metformin and resveratrol, two SIRT1 activators, entail STAT3 acetylation leading to Th17 differentiation impairment and limit tumor growth in mice. The capacity of metformin to promote acetylation of STAT3 and to decrease Th17 differentiation was also shown in patients [39].

We have shown that in vitro Th17 cells differentiated with IL-6 and TGF-β and in vivo tumor-infiltrating Th17 cells express CD39 and CD73 ectonucleotidases. This ectonucleotidase catalytic machinery entails the degradation of extracellular ATP into adenosine, an immunosuppressive molecule which suppresses effector T-cells. The expression of ectonucleotidases is dependent on IL-6-driven STAT3 activation and TGF-β-mediated downregulation of the zinc finger protein Growth Factor Independent-1 (Gfi-1), both required for the transcriptional regulation of ectonucleotidase expression during Th17 cell differentiation. CD39 expression at the surface of Th17 cells fosters tumor growth, suggesting that the immunosuppressive functions of Th17 cells in cancer are dictated by ectonucleotidase expression [40]. It has been reported that naive T-cells can be differentiated into Th17 cells with IL-1β, IL-6, IL-23 and without TGF-β. Unlike Th17 cells generated with TGF-β and IL-6, these Th17 cells were highly pathogenic in vivo [74] and didn't express ectonucleotidases [40]. These observations propose STAT3 and Gfi-1 as key determinants in the immunoregulatory function of Th17 cells, at least in part through the regulation of ectonucleotidase expression [40].

The mechanistic role of STAT3 in Th17 positive effects on anti-tumoral response is less documented. TAM-derived exosomes can deliver miR-29a-3p and miR-21-5p to CD4$^+$ T-cells leading to the inhibition of STAT3 and consequently to Th17 polarization in favor to Treg, which is beneficial for epithelial ovarian cancer progression [41].

2.3. Treg

Suppressor regulatory T-cells (Treg) maintain peripheral immune tolerance [75,76]. Co-stimulation of naive CD4$^+$ T-cells with TCR and TGF-β, triggers the generation of CD4$^+$/CD25$^+$ Treg cells. This leads to the expression of FOXP3, Tregs master transcription factor [77]. These T-cells accumulate in tumors and in cancer patients peripheral blood [78]. An increase in Treg frequency is generally considered as a marker of poor prognosis in cancer, probably because Treg mediate suppression of anti-tumor immunity [79–81].

The role of STAT3 in *Foxp3* expression regulation in Tregs appears to be context-dependent. In vitro, IL-2 induces the binding of STAT3 and STAT5 to a highly conserved STAT-binding site located within the first intron of the *Foxp3* gene, leading to FOXP3 expression up-regulation in purified CD4$^+$CD25$^+$ T-cells but not in CD4$^+$CD25$^-$ cells [42].

In tumor-infiltrating Tregs, both STAT3 and STAT5 bind to a STAT consensus site in the *Foxp3* promoter to enhance FOXP3 expression which seems to be important in maintaining Tregs inhibitory functions [42–44]. S1PR1 (Sphingosine-1 Phosphate Receptor 1) signaling has been shown to restrain Treg number and functions. An increase in S1PR1 in CD4$^+$ T-cells promotes STAT3 activation and JAK/STAT3-dependent Treg tumor migration, whereas STAT3 ablation in T-cells diminishes tumor-associated Treg accumulation and tumor growth [45]. Treatment of metastatic cancer or chronic myelogenous leukemia after allogeneic hematopoietic stem cell transplantation in patients with low-dose IL-2, leads to an increase of peripheral blood CD4$^+$CD25$^+$ cells and to FOXP3 expression in CD3$^+$ T-cells [42]. STAT3 and FOXP3 co-operatively control a subset of genes, responsible for Treg cell ability to suppress Th17 cell-mediated inflammation [82]. In contrast, IL-21 which activates STAT3 but not STAT5 has no effect on Treg viability, activation or function, suggesting that in this context IL-21-mediated STAT3 activation is not sufficient [83]. The regulation of *Foxp3* expression by STAT3 was strengthened by other studies. CDK5 increases *Foxp3* gene expression through phosphorylation of STAT3 at serine 727 [47]. GATA-3 controls the expression of miR-125a-5p, which in turn inhibits the expression of IL-6R and STAT3 and dampens Treg conversion [48]. In contrast, differentiation of naive T-cells into Tregs in vitro, is impaired when STAT3 is activated (by IL-6 or IL-27) [34,46]. In a different context, STAT3 binds to a silencer element within the *Foxp3* locus [84] and could also inhibit STAT5 interacting with the *Foxp3* promoter [85], to prevent FOXP3 expression.

One possible explanation for these ambivalent actions of STAT3 on Treg differentiation could be the phosphorylation site. Indeed, wogonin, a natural flavonoid from *Scutellaria baicalensis,* inhibits Treg induction by down-regulating ERK and STAT3-Y705 phosphorylation and promoting NF-κB and STAT3-S727 activation [49]. The modulation of STAT3 activity by molecular compounds could lead to inhibition of Tregs activity. Thus, WP1066 (an inhibitor of STAT3 signaling) enhances T-cell cytotoxicity against melanoma through inhibition of FOXP3$^+$ Tregs [50]. Compound9#, a fluorinated β-amino-ketone molecule, also inhibits Treg induction both in vitro and in vivo, via blockage of JAK2 signaling [51]. Finally, STAT3 inhibition with anti-sense oligonucleotides in association with radiation is a potent therapeutic target against Tregs [52]. All these studies suggest that STAT3 is required for immunosuppressive functions of Tregs.

2.4. T Follicular Helper Cells

Tfh differentiation is complex because it requires interaction with other cells such as B cells or DC. In mice, IL-6, IL-21 and IL-27 are essential for Tfh formation while in humans, Tfh generation relies on TGF-β, IL-12, IL-23. In both mammalian species of rodents or humans, Tfh cells express BCL6, ASCL2, IL-21, PD-1, and ICOS and produce IL-21 and CXCL13 [86,87]. While the role of Tfh is ambiguous in lymphoid tumors, many studies report that accumulation of Tfh in tertiary structures within the tumor is of good prognosis for breast, colon and non-small cell lung cancer patients [88,89]. Even if the protective effects of Tfh cells seem to be dependent on IL-21 and CXCL13-mediated recruitment of leucocytes, little is known about the accurate mechanism of Tfh anti-tumoral effects.

Generation of Tfh cells in patients with impaired STAT3 DNA-binding function is compromised due to the inability of IL-12 to induce IL-21 production without affecting its capacity to induce ICOS, BCL6 or CXCR5 expression [54]. When siRNA specific for STAT3 was used, TGF-β, IL-12, IL-23 failed to induce BCL6 expression in vitro [53]. However, the requirement of STAT3 seems to depend on the differentiation status of Tfh cells: It is required for Tfh generation but once these cells are generated it is no longer required [90]. In the same context, murine STAT3-deficient CD4$^+$ T-cells, Tfh cells expressed less or more BCL6 and IL-21 according to the immune environment [91,92]. Moreover, in conditions where STAT3 is necessary for Tfh differentiation, two studies showed that it cooperates with the IkZF transcription factors Aiolos and Ikaros. Moreover, the kinase activity of ROCK2 (Rho-associated coiled-coil kinase 2, an actin cytoskeleton assembly regulator) is required to induce STAT3 phosphorylation, nuclear relocalization and DNA binding to regulate *Bcl6* expression [93,94]. Finally, the importance of STAT3 in Tfh differentiation was strengthened by its capacity to block the expression of the Th2-associated genes *Gata3* and *Il4* [53,55].

A new protumorigenic IL-21$^+$ Tfh-like cell subset with a CXCR5$^-$PD-1$^-$ BTLA$^-$CD69^{hi} phenotype was identified in hepatocellular carcinoma (HCC). STAT1 and STAT3 activation are critical for these Tfh-like cell induction which operate via IL-21-IFN-γ pathways to induce plasma cells and create conditions for M2b macrophage polarization and tumor growth [56].

The importance of STAT3 in Tfh differentiation and its pro- or anti-tumoral role is not clear and could be dependent on differentiation stage, localization and environment.

2.5. Th9

Th9 cells have been characterized as a proinflammatory CD4 T-cell subset that can be generated through TGF-β and IL-4 stimulation. These cells are characterized by IL-9 secretion. Th9 harbor potent IL-9-dependent anti-cancer properties in most solid tumors and especially in melanoma while they can promote the development of many hematological human tumors, including Hodgkin's lymphoma and other B cell lymphoma. Th9 cells activate both innate and adaptive immune responses, thereby favoring anti-cancer immunity and tumor elimination [95].

In this CD4 T-cell subset, STAT3 was shown to dampen IL-9 production through STAT5 inhibition in Th9 cells [57]. In vitro, Th9 long term ability to secrete IL-9 is inhibited by pSTAT3 through an IL-10 receptor signaling [58]. In contrast, in humans Th9, pSTAT3 (mainly driven by IL-21 self-induction) inhibits pSTAT1-mediated T-BET induction, through SOCS3 induction. Since T-BET is an inhibitor of IL-9 transcription, this sustains IL-9 production. Patient-derived Th9 cells with dysfunctional STAT3, lose their capacity to produce IL-9, because of SOCS3 expression down-regulation, which leads to an increase pSTAT1 and T-BET expression. In the same study, the loss of function mutations observed in patients were recapitulated with deletion studies in mice, revealing that absence of STAT3 culminates into increased IL-9 production [59].

2.6. CD8$^+$

CD8$^+$ T lymphocytes are central players in cancer immune response, through their capacity to kill malignant cells. Upon recognition by the TCR of specific antigenic peptides presented on the surface of target cells by human leukocyte antigen class I (HLA-I)/beta-2-microglobulin (β2m) complexes, the CTL effector functions are activated. These functions are mediated either directly, through exocytosis of cytotoxic granules containing perforin and granzym into the target cells, resulting in cancer cell destruction, or indirectly, through secretion of cytokines, including IFN-γ and TNF [96].

The stimulation of the human and murine CD8$^+$ T-cells CD28, stabilizes the tyrosine kinase Lck activity and pSTAT3-mediated transcription of NKG2D. NKG2D expressing CD8$^+$ T-cells exert cytolytic activity against target tumor cells in vitro and significantly improve the antitumor therapeutic effects in vivo [62]. Even if IL-10 is considered as an immune suppressor it can also increase expansion and cytotoxic activity of CD8$^+$ cells. Tumor-resident CD8$^+$ T-cells express high levels of IL-10R, leading to high levels of activated pSTAT3 and pSTAT1 in response to IL-10 [97]. In contrast, circulating CD8$^+$ T-cells from peripheral blood of HCC patients present high amounts of pSTAT3 which is correlated with high amount of IL-4, IL-6 and IL-10 and low quantity of IFN-γ which may result in abnormal immune surveillance against tumor cells [98]. In murine CD8$^+$ T-cells, STAT3 has been shown to inhibit both IFN-γ-mediated CXCL10 production by myeloid cells and CD8$^+$ T-cells CXCR3 expression (the receptor of CXCL10), blocking the migration of these cells to the tumor site [60].

In an adoptive transfer therapeutic strategy in mice, ablating STAT3 in CD8$^+$ T-cells prior to transfer, allows efficient tumor infiltration and robust CD8$^+$ T-cell proliferation, resulting in increased tumor antigen-specific T-cell activity and tumor growth inhibition [61].

Altogether STAT3 seems to inhibit CD8$^+$ T-cell expansion and cytolytic activity except in some particular conditions.

3. STAT3 and Myeloid Cells

APCs dictate immune system response, since these cells have been shown to capture antigens in the periphery, migrate to the lymphoid organs, and present processed peptides to T-cells in a way that may lead either to priming or tolerance induction [99]. Among myeloid cells, different subsets have been described (Macrophages, DCs, Myeloid-Derived Suppressor Cells (MDSCs)) with distinct functions that could be regulated by STAT3 (Table 2).

Table 2. Impact of STAT3 in myeloid cells.

Immune Cell Family	STAT3 Role in	STAT3 Modulators	Effects	Reference
yeloid cells	Macrophages	KO STAT3	Inflammatory macrophages ↗ CTL activation ↗ TLR9 pathway (IFN-γ, TNF-α, IL-12) ↗ IL-23 and ↘ IL-12	[100] [101] [102]
		STAT3 oxerexpression cancer cells (PAI-1, BMP6, IL-6)	↗ CD163 (M2 marker) ↗ pSTAT3 → polarization of M1 into M2 (CD163)	[103] [103–105]
		Tumor exosomes	↗ pSTAT3 → ↗ IL-6, IL-10, CCL2	[106]
		Corosolic acid, oleanic acid	↘ STAT3 activity → M1 polarization (↗ TNF-α, IFN-γ, IL-12, IL-2)	[107–109]
		ERK5	↗ pSTAT3 → pro-tumor macrophages	[110]
		M-CSF	↗ pSTAT3 → ↗ DC-SIGN	[111]
		KO SOCS3	↗ pSTAT3 → anti-tumor macrophages	[112]
		Tumor cells	↗ pSTAT3 → ↗ IL-10, VEGF, βFGF → angiogenesis	[113–115]
	Dendritic Cells	KO STAT3 or siRNA or shRNA	↗ DCs APC function and cytokine production (↗ IL-12, TNF-α and ↘ IL-10)	[102,116–121]
		Constitutive STAT3 activation	↘ MHC class II, ID-2	[117,122–124]
		Tumor-derived exosomes and IL-6	↘ PKCβII → DCs activation and differentiation	[125]
	MDSCs	KO STAT3 or siRNA or anti-sense oligonucleotides	↘ C/EBPβ ↘ immunosuppressive functions → MDSCs differentiation into DCs ↘ NOX2 → ↘ ROS → ↘ immunosuppression	[126] [117,127–130]
		Tumor-derived exosomes (HSP70)	↗ pSTAT3 → ↗ immunosuppressive functions (with an IL-6 amplification loop)	[131,132] [133]
		CCL2	↗ pSTAT3 → PMN-MDSCs-mediated T-cell suppression	
		IL-6	Early-stage MDSCs accumulation	
		G-CSF	↗ pSTAT3 → ↘ IRF8	[126,134]
		GM-CSF	↗ pSTAT3 → ↗ IDO → ↗ immunosuppression	[135]
		siRNA or STATTIC	↘ pSTAT3 → ↘ arginase-1 → ↘ immunosuppression	[136]
		JSI-124 (STAT3 inhibitor)	↘ pSTAT3 → ↘ IDO → ↘ immunosuppression	[137]
		CD45	↘ pSTAT3 → M-MDSCs differentiation into TAMs	[138]
		Embelin, PM01183, alisertib, STATTIC or BBI608	↘ pSTAT3	[139–142]
		Tumor cells	↗ pSTAT3 → ↗VEGF, βFGF → ↗ angiogenesis	[115]

3.1. Macrophages

Tumor-associated macrophages (TAMs) are subdivided into two subsets, M1 and M2 macrophages, based on their capacity to express or produce Nitric Oxide Synthase/IL-12/TNF-α or arginase-1/IL-10/TGF-β, respectively. M1 has potent microbicidal properties and promotes Th1 responses, whereas M2 supports Th2-associated effector functions [143,144]. M2 macrophages includes M2a, M2b or M2c subtypes. Tumor-derived signals, such as Macrophage-Colony Stimulating Factor (M-CSF/CSF-1), Monocyte Chemoattractant Protein-1 (MCP-1), or Chemokine (C-C motif) Ligand-2 (CCL2) entails the accumulation of M2 at the tumor site. M2 macrophages participate in tumor growth by releasing proangiogenic cytokines and growth factors, e.g. Epidermal Growth Factor (EGF), Vascular Endothelial Growth Factor (VEGF), Platelet-Derived Growth Factor (PDGF), Colony Stimulating Factor-1 (CSF-1), and basic Fibroblast Growth Factor (βFGF). They also generate arginase-1, IL-10 and TGF-β. These molecular messengers will inhibit the antitumor function of T-cells and NK cells. This will favor tumor tolerance and impairment of antitumor immunotherapies efficacy [145–147].

IL-6 inhibition of M-CSF-induced colony formation observed in animals was abolished in mice mutated for the gp130-STAT1/3 signaling, suggesting that the IL-6/STAT3 pathway could regulate macrophage homeostasis [148]. Moreover, breast cancer-derived exosomes are capable of inducing IL-6 secretion and a pro-tumoral phenotype (IL-6, IL-10, CCL2 production) in macrophages, partially via gp130/STAT3 signaling [106]. More particularly, STAT3 directly induces the expression of the M2 marker CD163, both in macrophages and tumor cells [103]. Prostate-cancer cells induce a change in macrophage phenotype from M1 into M2, through STAT3 activation [104]. This can be induced by Plasminogen Activating Inhibitor (PAI)-1 secreted by cancer cells [105].

The inhibition of STAT3 signaling in either macrophages or bone marrow-derived DCs is of great importance in cancer immunotherapy, because it allows these APCs to restore the responsiveness of tolerant T-cells from tumor-bearing mice. STAT3 signaling is a negative regulator in peritoneal elicited macrophages, as its targeted disruption gives a constitutively activated phenotype and an increased ability to produce inflammatory mediators in response to LPS. This may be the consequence of an increased STAT1 activity (leading to high production of inflammatory factors) or a lack of IL-10 production [100]. Moreover, macrophages derived from conditional STAT3 knockout mice are better than wild-type macrophages to prime cognate CTL responses and to cross-present tumor-derived antigen to CTLs in vitro. This leads to a more important proliferation of CTLs and an increased production of IFN-γ and TNF-α. Similarly, removing STAT3 in hematopoietic cells, leads to rapid activation of innate immunity by CpG (a TLR9 ligand), with enhanced activation of macrophages, neutrophils and NK cells and production of IFN-γ, TNF-α, IL-12 to eradicate B16 melanoma tumors [102]. Targeting STAT3 signaling therefore represents an attractive strategy to increase CTL responses in the tumor-bearing host [101]. Immunosuppressive activities of TAMs correlate with over-activated STAT3 signaling, whereas disruption of TAMs STAT3 activity can enhance rat immune response to breast cancer [149]. In glioblastoma patients, tumor-infiltrating macrophages were shown to be predominantly STAT3-positive M2 macrophages, which are associated with a poor prognosis [150]. The same team proposed that corosolic acid and oleanic acid can prevent tumor formation through their capacity to suppress macrophages M2 polarization and tumor cell proliferation by inhibiting STAT3 activation [107,108]. CD163-targeted corosolic acid-containing liposomes were also shown to reprogram M2 macrophages into M1 (increased expression of TNF-α, IFN-γ, IL-12, IL-2 and decreased expression of IL-10) [109]. Since ERK5 mediates Y705 phosphorylation of STAT3 in myeloid cells, blocking ERK5 might constitute a treatment strategy to reprogram macrophages toward an antitumor state by inhibiting STAT3-induced gene expression [110]. In intrahepatic cholangiocarcinoma (ICC), patients with high counts of CD163[+] M2 macrophages showed poor disease-free survival. Tumor cell supernatant from HuCCT1 ICC cell lines induces the production of IL-10 and VEGF-A by macrophages through activation of STAT3 and polarization towards the M2 phenotype [113]. Similarly, renal cell carcinoma-derived BMP (Bone Morphogenetic Protein)-6 mediates IL-10 expression in macrophages through Smad5 and STAT3, and M2 polarization [114]. These observations were confirmed by the fact that macrophages isolated from mouse tumors displayed activated STAT3 and induced angiogenesis in an in vitro tube formation assay via STAT3 induction of angiogenic factors, including VEGF and βFGF [115]. STAT3 signaling within the tumor microenvironment induces the production of IL-23, a procarcinogenic cytokine, via direct transcriptional activation of the IL-23/p19 gene in TAMs, while inhibiting the production of IL-12, a central anticarcinogenic cytokine, thereby shifting the balance of tumor immunity towards carcinogenesis [44]. The M-CSF-inducible DC-SIGN (Dendritic cell-specific ICAM-3-grabbing nonintegrin or CD209) expression along monocyte-to-macrophage differentiation is dependent on JNK and STAT3 activation. DC-SIGN contributes to the release of IL-10 that would maintain STAT3 activation in tumor cells, thus implying that DC-SIGN favors the maintenance of an activated STAT3 context in the tumor stroma. This will compromise the ability of TAMs and DCs to generate an effective antitumor response and to maintain an immunosuppressive environment [111]. This effect is potentiated by STAT3-activating cytokines IL-6 and IL-10 produced by STAT3-activated tumor cells. In the same way, IL-6-derived from gastric cancer cells induces normal macrophages differentiation to M2 macrophages with higher IL-10 and TGF-β expression, and lower IL-12 expression, via STAT3 phosphorylation. IL-6-induced M2 macrophages exert a pro-tumor function by promoting GC cell proliferation and migration [151].

However, an anti-tumoral role of STAT3 in macrophages has also been proposed, based on studies that investigated the importance of STAT3 in macrophages through an indirect manner, using SOCS3 conditional knockout mice in macrophages. Hyperactivation of STAT3 in myeloid cells simultaneously exerted anti-inflammatory as well as anti-tumor effects [112]. Thus, Lipoxin A4 induces STAT3 phosphorylation and mediates differentiation of monocytic-like cells into M2 subtypes with anti-tumorigenic activities [152]. The discrepancies between these studies could be explained by the fact that SOCS3 and lipoxin signaling should regulate other pathways such as NF-κB.

3.2. Dendritic Cells (DCs)

DCs are another differentiated stage of monocytes and are the key APCs of the immune system. DCs play a main role as immune sentinels in the initiation of T-cell response against microbial pathogens, tumors and inflammation [153,154].

The first evidence that STAT3 is important for DCs fate was made in mice lacking STAT3 expression in hematopoietic progenitors. These animals present a profound deficiency in DCs which are unresponsive to Flt3L stimulation [116]. However, the same mice bearing a tumor, present enhanced DC, T-cell, NK cell and neutrophil functions and a decreased tumor progression [117]. DCs derived from LysM^{cre}/STAT3$^{flox/flox}$ mice display higher cytokine production in response to TLR stimulation and activate more efficiently T-cells. Intratumoral administration of these DCs significantly inhibits MC38 tumor growth [119]. Moreover, ablating STAT3 in myeloid cells increases CpG-induced DCs maturation, T-cell activation, tumor antigen-specific T-cells generation and long-lasting antitumor immunity in B16 melanoma tumor model [102]. Similarly, CpG was used to administer STAT3 siRNA specifically to myeloid and B cells. Ablation of STAT3 in these cells increases DC engagement and adoptively transferred CD8$^+$ T-cells effector functions in vivo, with an upregulation of effector molecules such as perforin, granzym B, and IFN-γ [118]. Mice without STAT3 in myeloid compartment of tumor stroma, including DCs and macrophages, present reduced numbers of tumor-infiltrating CD4$^+$CD25$^+$/FOXP3$^+$/LAG3$^+$ Tregs, along with an increase in CD8$^+$ effector T-cells [117]. Constitutive STAT3 activation in tumor-residing DCs reduces expression of MHC class II and costimulatory molecules, impairs the antigen-presenting function of DC and contributes to the expansion of tumor-infiltrating FOXP3$^+$ T-cells and attenuates CD4$^+$ Th cell responses [117,122,123]. This can be partly explained by the fact that STAT3 inhibits the ID2 (inhibitor of differentiation 2) expression which promotes tumor immunity [124]. Finally, IL-6 is a suppressor of bone marrow-derived DC activation/maturation and a regulator of DC differentiation in vivo, through STAT3 phosphorylation. Then, DC differentiation/maturation is controlled by IL-6-gp130-STAT3, suggesting that this amplification loop may represent a target for controlling T-cell-mediated immune responses [155]. Similarly, mammary tumor-derived exosomes inhibit the differentiation of murine myeloid precursor cells into DCs in vitro. This was correlated with an increased IL-6 level and phosphorylated STAT3 and was blocked in bone marrow cells derived from IL-6 knockout mice. This suggests that tumor cells can dampen DCs differentiation through an autocrine activation of STAT3 by IL-6 [125].

In humans, tumor-derived factors suppress DC generation through STAT3-mediated PKCβII reduced expression [156]. STAT3-depleted DCs with adenoviral STAT3 short hairpin RNA (shRNA) or siRNA presented an altered cytokines production profile under TLR stimulation (such as more IL-12 and TNF-α and less IL-10), and induced tumor Ag-specific T-cells and IFN-γ-producing γδ T lymphocytes more efficiently than control DCs [119–121]. The impact on IL-12 can be explained by a competition of STAT3 with CDK9 on a binding site on the IL-12p35 promoter [157]. The effects of STAT3 on IL-10 expression can be explained by the fact that HDAC6 forms a complex with STAT3, recruited to a specific DNA sequence element in the *Il10* gene promoter [158].

3.3. Myeloid-Derived Suppressor Cells (MDSCs)

MDSCs have been identified in humans and mice as a population of immature myeloid cells with the ability to suppress T-cell activation [159]. In tumor-bearing mice, these cells have been shown to markedly expand in lymphoid organs and blood [160]. In addition, an increased MDSCs frequency was detected in the blood of patients with different types of cancers [161,162]. In mice and humans, MDSCs from tumor bearers induce antigen-specific MHC class I restricted tolerance of CD8$^+$ T-cells and are one of the major suppressors of antitumor immunity [163]. In humans the phenotype of MDSCs is a matter of debate. However, two major subsets of MDSCs have been identified: monocytic MDSCs (M-MDSCs), similar to monocytes and polymorphonuclear MDSCs (PMN-MDSCs) sharing phenotypic and morphologic features with neutrophils [164].

STAT3 is probably one of the main transcription factors that regulate MDSCs functions. MDSCs from tumor-bearing mice present high levels of phosphorylated STAT3, compared with immature myeloid cells from naive mice [128]. Moreover, ablation of STAT3 expression through the use of conditional knockout mice or selective STAT3 inhibitors (JSI-124) markedly reduce the expansion of MDSCs, promote accumulation of DCs and increase T-cell responses in tumor-bearing mice [117,127,128]. In mice, when STAT3 was specifically deleted in myeloid cells, the ability of MDSCs to inhibit CD4$^+$ and CD8$^+$ T-cell-dependent apoptosis by cell-cell contact and to induce Treg polarization through TGF-β, IL-10 and NOX2 secretion was decreased [131]. When STAT3 was specifically turned off in myeloid cells expressing TLR9 from prostate cancer patients, using a CpG-STAT3 siRNA or CpG-STAT3 antisense oligonucleotides, it abrogated the immunosuppressive effects of patient-derived MDSCs on effector CD8$^+$ T-cells [129,130]. In contrast, deletion of SOCS3 in myeloid cells leads to an increased activation of the STAT3 and to differentiation into Gr-1$^+$CD11b$^+$Ly6G$^+$ MDSCs, enhancing tumor growth [165]. STAT3 can also be regulated in MDSCs by the epigenetic-associated protein, p66a, which may physically interact with STAT3 to suppress its activity through posttranslational modification [166]. We have shown that tumor-derived exosome (TDE)-associated HSP70 led to STAT3 activation in MDSCs. This activation depends on TLR2/MyD88 and autocrine production of IL-6. TDEs from human tumor cell lines activate human MDSCs suppressive functions but not their expansion in an HSP70/TLR2-dependent manner, showing that the mechanism described in mice is also relevant in cancer patients [133]. In tumor cell supernatants, tumor soluble factors induce activation of ERK, which results in MDSCs expansion, while TDEs trigger STAT3 activation without promoting MDSCs expansion [133]. How can these discrepancies be explained? It is well known that STAT3 signaling in myeloid cells, entails the expression of Bcl-xL, c-myc, cyclin D1 or survivin, which favors cell proliferation and inhibits cell apoptosis and differentiation to mature cell types [15,167].

STAT3 controls the G-CSF-and G-MCSF responsive induction of C/EPBβ (CCAAT-enhancer-binding protein β) expression and the interferon regulatory factor 8 (IRF8) downregulation in myeloid cells [126,134]. The transcription factor C/EBPβ was reported to play a crucial role in controlling the differentiation of myeloid precursors to functional MDSCs [126] whereas IRF8 attenuated MDSC accumulation, phenotype and function [134]. These data suggest also a link between CSF and STAT3 pathway in the regulation of MDSC biology.

Finally, recent studies highlighted the importance of signaling pathways downstream STAT3 in MDSCs differentiation. MDSCs isolated from mouse tumors present activated STAT3. STAT3 favors the production of angiogenic factors, including VEGF and βFGF to induce angiogenesis in an in vitro tube formation assay [115]. STAT3 as well as STAT5 control the cytokine-induced expression of the cytoplasmic NADPH oxidase NOX2 [168], being e.g., crucial for DNA damage response in AML cells, leading together with mitochondrial ROS production, which is a predominant STAT3 affair to the production of ROS, responsible for MDSCs-induced immune suppression in murine colon, lung, mammary carcinoma, thymoma, sarcoma models and in patients with head and neck cancers [132]. STAT3 also favors immunosuppressive functions of MDSCs by inducing the expression and the activity of IDO (Indoleamine 2,3-dioxygenase 1) in breast [137] and liver [135] cancers or arginase-1 in head and neck squamous cell carcinoma [136]. STAT3-inducible up-regulation of the myeloid-related protein S100A9 enhances MDSCs production in cancer. Mice lacking this protein mount potent antitumor immune responses and reject implanted tumors, an effect reversed by administration of wild-type MDSCs [169].

However, STAT3 activity can be endogenously controlled in intra-tumoral MDSCs. Thus, tumor hypoxia led to CD45 protein tyrosine phosphatases activation, resulting in STAT3 activity downregulation and in M-MDSC differentiation into tumor-promoting TAMs [138].

Moreover, many inhibitors were proposed such as the XIAP inhibitor embelin [139], PM01183 a novel synthetic agent derived from trabectedin [140], alisertib (MLN8237), a small-molecule inhibitor of Auror-A kinase [141], STATTIC or BBI608 [142] to inhibit STAT3 in MDSCs with a potential clinical application to favor anti-tumor response.

4. STAT3 and Check Points Inhibitors

The emergence of immune 'checkpoint inhibitors', blocking negative regulators of T-cell immunity opened new clinical applications of cancer immunotherapies. The more widely studied are cytotoxic T lymphocyte antigen 4 (CTLA-4) and programmed death receptor 1 (PD-1). CTLA-4 is mainly expressed on T helper and Treg cells and binds to its ligands B7–1 (CD80) and B7–2 (CD86) on APCs [170]. PD-1 expression is induced both on activated CD8$^+$ T-cells, Tfh and Treg present in tumor microenvironment, and on activated B cells and NK cells. PD-1 has two ligands: PD-L1 (B7-H1) and PD-L2 (B7-DC). PD-1 signaling contributes to T-cell exhaustion [170]. Other checkpoints can also be implicated in tumor immune escape such as Lymphocyte activation gene 3 protein (LAG-3) and T-cell immunoglobulin and mucin domain-containing 3 (TIM-3), IDO1, B- and T-lymphocyte attenuator (BTLA), V-domain immunoglobulin suppressor of T-cell activation (VISTA) and the A2A adenosine receptor (A2A-R) [170,171].

STAT3 can bind the promoter of murine *Pdcd1* gene coding for PD-1 in T-cells [172]. STAT3 has been demonstrated to bind to the *Pdcd1l1* promoter (coding for PD-L1) to regulate its transcriptional expression in cancer cells. It requires either mutated oncogene chimeric nucleophosmin/anaplastic lymphoma kinase (ALK) in T-cell lymphoma [173], Latent membrane protein-1 (LMP1) of Epstein–Barr virus in nasopharyngeal carcinoma [174], HDAC6 in osteosarcoma [175], NF-κB in melanoma [176] or HIF-1α in pulmonary adenocarcinoma [177]. In contrast, a study shows that STAT3 is necessary for docetaxel-mediated inhibition of PD-1 expression [178].

In CRC patients, FGFR2 expression is correlated to PD-L1 expression. FGFR2 stimulation leads to STAT3 activation, PD-L1 expression and colon cancer cell death [179]. The EGFR signaling pathway can also regulate PD-L1 expression through the IL-6/JAK/STAT3 pathway in non-small cell lung cancer (NSCLC) cells [180,181]. The expression of PD-L1 is associated with a poor prognosis and inhibition of PD-L1 expression (e.g., through STAT3 or its partner inhibition) is associated to a decrease in cell proliferation and/or an increase in tumor cell death in most of these studies. However, the effects on T-cell anti-tumor response were not tested here.

PD-L1/L2 can be expressed in non-tumoral cells. In liver MDSCs, GM-CSF is responsible for STAT3 activation which in turn induces PD-L1 and IDO1 expression [135]. GBM cells secrete IL-6 which in turn activates STAT3 in infiltrated myeloid cells leading to STAT3 activation and PD-L1 expression. An anti-IL-6 therapy decreases PD-L1 expression and tumor size in a CD8[+] T-cell-dependent manner [182]. In HCC, a similar regulation was observed. In this setting, the HCC-associated fibroblasts produce IL-6 which in turn increases PD-L1 expression in neutrophils [183]. Similarly, in prostate carcinoma, DC generated in the presence of stromal myofibroblasts factors expressed significantly elevated levels of PD-L1 in a STAT3 and IL-6-dependent manner [184]. In chronic lymphocytic leukemia (CLL), STAT3 is required to PD-L1 expression and IL-10 production which in turn seems to be responsible for PD-1 expression in CD4[+] and CD8[+] T-cells [185]. In adult T-cell lymphoma, IL-27B production by Lymphoma cells and IL-27p28 production by macrophages lead to STAT3 activation and PD-L1/L2 expression in macrophages [186]. TGF-β is another cytokine secreted by inflammatory or tumor cells that can increase PD-L1 expression in DC in a STAT3-dependent manner [187]. The use of IFN-α in clinical oncology for many types of cancers is reconsidered, as IFN-α induces the expression of PD-L1 molecule, in the majority of the specific immune cell populations, particularly in DCs [188]. However, it should be noted that interferons are highly liver toxic and they act on all cell types in the body, where many unwanted side effects from neurotoxic problems to fever symptoms were reported in therapy, making their window of opportunity in fragile patients delicate.

Little is known about the ability of STAT3 to regulate other checkpoints. One study shows that STAT3/IRF1 are required for PD-L2 expression in melanoma cells [189]. In a non-cancerous context, it has been shown that CTLA-4 as well as PD-1, LAG-3 and TIM-3 expression is induced in HIV-infected DCs in a STAT3-dependent manner [190].

In contrast, the immunosuppressive effect of these checkpoints can be dependent on STAT3 activation in target-cells. For example, IDO1 may promote EMT (Epithelial-Mesenchymal Transition) by activation of the IL-6/STAT3/PD-L1 signaling pathway [191]. TIM-3[+] endothelial cells modulate T-cell response to lymphoma surrogate antigens by suppressing activation of CD4[+] T lymphocytes through the activation of the IL-6-STAT3 pathway, inhibiting Th1 polarization and providing protective immunity [192]. Finally, CTLA-4 critically shapes the characteristics of IL-17 producing CD8[+] cells (Tc17 a pro-tumoral population) in a STAT3 activation-dependent manner and inhibition of CTLA-4-induced STAT3 activity reverses Tc17 signature to Tc1-like cells with enhanced cytotoxic potential [193].

5. Conclusions

STAT3 regulates genes involved in biological functions of cancer and immune cells, rendering this pathway an interesting therapeutic target. STAT3 could be inhibited directly by peptides or natural compounds or indirectly by blocking upstream signaling pathways such as IL-6 and JAK2 pathways (For review see [194,195]). However, the complexity of STAT3 biology and its broad effects may render its clinical development complex. This review underlines the ambivalent effects of STAT3 on the antitumoral immune response, with both positive or negative effects, depending on the context or cell types involved. Additional translational studies on patients treated with STAT3 inhibitors are awaited to understand their effects on the complexity of tumor biology.

Funding: This research was funded by Ligue Nationale Contre le Cancer.

Acknowledgments: We thank Isabel Gregoire for carefully reading the manuscript.

Conflicts of Interest: The authors declare no conflict of interest.

References

1. Kosack, L.; Wingelhofer, B.; Popa, A.; Orlova, A.; Agerer, B.; Vilagos, B.; Majek, P.; Parapatics, K.; Lercher, A.; Ringler, A.; et al. The ERBB-STAT3 Axis Drives Tasmanian Devil Facial Tumor Disease. *Cancer Cell* **2019**, *35*, 125–139. [CrossRef] [PubMed]

2.   Yoshimura, A.; Naka, T.; Kubo, M. SOCS proteins, cytokine signalling and immune regulation. *Nat. Rev. Immunol.* **2007**, *7*, 454–465. [CrossRef] [PubMed]

3.   Hirano, T.; Ishihara, K.; Hibi, M. Roles of STAT3 in mediating the cell growth, differentiation and survival signals relayed through the IL-6 family of cytokine receptors. *Oncogene* **2000**, *19*, 2548–2556. [CrossRef] [PubMed]

4.   Yuan, Z.L.; Guan, Y.J.; Chatterjee, D.; Chin, Y.E. Stat3 dimerization regulated by reversible acetylation of a single lysine residue. *Science* **2005**, *307*, 269–273. [CrossRef] [PubMed]

5.   Meier, J.A.; Larner, A.C. Toward a new STATe: The role of STATs in mitochondrial function. *Semin. Immunol.* **2014**, *26*, 20–28. [CrossRef]

6.   Avalle, L.; Poli, V. Nucleus, Mitochondrion, or Reticulum? STAT3 a La Carte. *Int. J. Mol. Sci* **2018**, *19*, 2820. [CrossRef] [PubMed]

7.   Shuai, K.; Liu, B. Regulation of gene-activation pathways by PIAS proteins in the immune system. *Nat. Rev. Immunol.* **2005**, *5*, 593–605. [CrossRef]

8.   Alexander, W.S.; Hilton, D.J. The role of suppressors of cytokine signaling (SOCS) proteins in regulation of the immune response. *Annu Rev. Immunol.* **2004**, *22*, 503–529. [CrossRef]

9.   Migone, T.S.; Cacalano, N.A.; Taylor, N.; Yi, T.; Waldmann, T.A.; Johnston, J.A. Recruitment of SH2-containing protein tyrosine phosphatase SHP-1 to the interleukin 2 receptor; loss of SHP-1 expression in human T-lymphotropic virus type I-transformed T cells. *Proc. Natl. Acad. Sci. USA* **1998**, *95*, 3845–3850. [CrossRef]

10.   Schaper, F.; Gendo, C.; Eck, M.; Schmitz, J.; Grimm, C.; Anhuf, D.; Kerr, I.M.; Heinrich, P.C. Activation of the protein tyrosine phosphatase SHP2 via the interleukin-6 signal transducing receptor protein gp130 requires tyrosine kinase Jak1 and limits acute-phase protein expression. *Biochem. J.* **1998**, *335*, 557–565. [CrossRef]

11.   Irie-Sasaki, J.; Sasaki, T.; Matsumoto, W.; Opavsky, A.; Cheng, M.; Welstead, G.; Griffiths, E.; Krawczyk, C.; Richardson, C.D.; Aitken, K.; et al. CD45 is a JAK phosphatase and negatively regulates cytokine receptor signalling. *Nature* **2001**, *409*, 349–354. [CrossRef] [PubMed]

12.   Daino, H.; Matsumura, I.; Takada, K.; Odajima, J.; Tanaka, H.; Ueda, S.; Shibayama, H.; Ikeda, H.; Hibi, M.; Machii, T.; et al. Induction of apoptosis by extracellular ubiquitin in human hematopoietic cells: Possible involvement of STAT3 degradation by proteasome pathway in interleukin 6-dependent hematopoietic cells. *Blood* **2000**, *95*, 2577–2585. [PubMed]

13.   Delgoffe, G.M.; Vignali, D.A. STAT heterodimers in immunity: A mixed message or a unique signal? *JAKSTAT* **2013**, *2*, e23060. [CrossRef] [PubMed]

14.   Bromberg, J.; Darnell, J.E., Jr. The role of STATs in transcriptional control and their impact on cellular function. *Oncogene* **2000**, *19*, 2468–2473. [CrossRef] [PubMed]

15.   Yu, H.; Pardoll, D.; Jove, R. STATs in cancer inflammation and immunity: A leading role for STAT3. *Nat. Rev. Cancer* **2009**, *9*, 798–809. [CrossRef] [PubMed]

16.   Rebe, C.; Vegran, F.; Berger, H.; Ghiringhelli, F. STAT3 activation: A key factor in tumor immunoescape. *JAKSTAT* **2013**, *2*, e23010.

17.   Zhang, H.F.; Lai, R. STAT3 in Cancer-Friend or Foe? *Cancers* **2014**, *6*, 1408–1440. [CrossRef] [PubMed]

18.   Burchill, M.A.; Goetz, C.A.; Prlic, M.; O'Neil, J.J.; Harmon, I.R.; Bensinger, S.J.; Turka, L.A.; Brennan, P.; Jameson, S.C.; Farrar, M.A. Distinct effects of STAT5 activation on CD4+ and CD8+ T cell homeostasis: Development of CD4+CD25+ regulatory T cells versus CD8+ memory T cells. *J. Immunol.* **2003**, *171*, 5853–5864. [CrossRef]

19.   Agnello, D.; Lankford, C.S.; Bream, J.; Morinobu, A.; Gadina, M.; O'Shea, J.J.; Frucht, D.M. Cytokines and transcription factors that regulate T helper cell differentiation: New players and new insights. *J. Clin. Immunol.* **2003**, *23*, 147–161. [CrossRef]

20.   Yao, Z.; Kanno, Y.; Kerenyi, M.; Stephens, G.; Durant, L.; Watford, W.T.; Laurence, A.; Robinson, G.W.; Shevach, E.M.; Moriggl, R.; et al. Nonredundant roles for Stat5a/b in directly regulating Foxp3. *Blood* **2007**, *109*, 4368–4375. [CrossRef]

21.   Shiku, H. Importance of CD4+ helper T-cells in antitumor immunity. *Int. J. Hematol* **2003**, *77*, 435–438. [CrossRef] [PubMed]

22.   Galon, J.; Costes, A.; Sanchez-Cabo, F.; Kirilovsky, A.; Mlecnik, B.; Lagorce-Pages, C.; Tosolini, M.; Camus, M.; Berger, A.; Wind, P.; et al. Type, density, and location of immune cells within human colorectal tumors predict clinical outcome. *Science* **2006**, *313*, 1960–1964. [CrossRef] [PubMed]

23. Marrogi, A.J.; Munshi, A.; Merogi, A.J.; Ohadike, Y.; El-Habashi, A.; Marrogi, O.L.; Freeman, S.M. Study of tumor infiltrating lymphocytes and transforming growth factor-beta as prognostic factors in breast carcinoma. *Int. J. Cancer* **1997**, *74*, 492–501. [CrossRef]

24. Hiraoka, K.; Miyamoto, M.; Cho, Y.; Suzuoki, M.; Oshikiri, T.; Nakakubo, Y.; Itoh, T.; Ohbuchi, T.; Kondo, S.; Katoh, H. Concurrent infiltration by CD8+ T cells and CD4+ T cells is a favourable prognostic factor in non-small-cell lung carcinoma. *Br. J. Cancer* **2006**, *94*, 275–280. [CrossRef] [PubMed]

25. Tuting, T.; Storkus, W.J.; Lotze, M.T. Gene-based strategies for the immunotherapy of cancer. *J. Mol. Med.* **1997**, *75*, 478–491. [CrossRef]

26. Knutson, K.L.; Disis, M.L. Tumor antigen-specific T helper cells in cancer immunity and immunotherapy. *Cancer Immunol. Immunother.* **2005**, *54*, 721–728. [CrossRef]

27. Owaki, T.; Asakawa, M.; Morishima, N.; Mizoguchi, I.; Fukai, F.; Takeda, K.; Mizuguchi, J.; Yoshimoto, T. STAT3 is indispensable to IL-27-mediated cell proliferation but not to IL-27-induced Th1 differentiation and suppression of proinflammatory cytokine production. *J. Immunol.* **2008**, *180*, 2903–2911. [CrossRef]

28. Ma, C.S.; Wong, N.; Rao, G.; Nguyen, A.; Avery, D.T.; Payne, K.; Torpy, J.; O'Young, P.; Deenick, E.; Bustamante, J.; et al. Unique and shared signaling pathways cooperate to regulate the differentiation of human CD4+ T cells into distinct effector subsets. *J. Exp. Med.* **2016**, *213*, 1589–1608. [CrossRef]

29. Harris, T.J.; Grosso, J.F.; Yen, H.R.; Xin, H.; Kortylewski, M.; Albesiano, E.; Hipkiss, E.L.; Getnet, D.; Goldberg, M.V.; Maris, C.H.; et al. Cutting edge: An in vivo requirement for STAT3 signaling in TH17 development and TH17-dependent autoimmunity. *J. Immunol.* **2007**, *179*, 4313–4317. [CrossRef]

30. Zhang, Y.; Ma, C.A.; Lawrence, M.G.; Break, T.J.; O'Connell, M.P.; Lyons, J.J.; Lopez, D.B.; Barber, J.S.; Zhao, Y.; Barber, D.L.; et al. PD-L1 up-regulation restrains Th17 cell differentiation in STAT3 loss- and STAT1 gain-of-function patients. *J. Exp. Med.* **2017**, *214*, 2523–2533. [CrossRef]

31. Nishihara, M.; Ogura, H.; Ueda, N.; Tsuruoka, M.; Kitabayashi, C.; Tsuji, F.; Aono, H.; Ishihara, K.; Huseby, E.; Betz, U.A.; et al. IL-6-gp130-STAT3 in T cells directs the development of IL-17+ Th with a minimum effect on that of Treg in the steady state. *Int. Immunol.* **2007**, *19*, 695–702. [CrossRef] [PubMed]

32. Zhou, L.; Ivanov, I.I.; Spolski, R.; Min, R.; Shenderov, K.; Egawa, T.; Levy, D.E.; Leonard, W.J.; Littman, D.R. IL-6 programs T(H)-17 cell differentiation by promoting sequential engagement of the IL-21 and IL-23 pathways. *Nat. Immunol.* **2007**, *8*, 967–974. [CrossRef] [PubMed]

33. Yang, X.O.; Panopoulos, A.D.; Nurieva, R.; Chang, S.H.; Wang, D.; Watowich, S.S.; Dong, C. STAT3 regulates cytokine-mediated generation of inflammatory helper T cells. *J. Biol Chem* **2007**, *282*, 9358–9363. [CrossRef] [PubMed]

34. Laurence, A.; Tato, C.M.; Davidson, T.S.; Kanno, Y.; Chen, Z.; Yao, Z.; Blank, R.B.; Meylan, F.; Siegel, R.; Hennighausen, L.; et al. Interleukin-2 signaling via STAT5 constrains T helper 17 cell generation. *Immunity* **2007**, *26*, 371–381. [CrossRef] [PubMed]

35. Yang, X.O.; Pappu, B.P.; Nurieva, R.; Akimzhanov, A.; Kang, H.S.; Chung, Y.; Ma, L.; Shah, B.; Panopoulos, A.D.; Schluns, K.S.; et al. T helper 17 lineage differentiation is programmed by orphan nuclear receptors ROR alpha and ROR gamma. *Immunity* **2008**, *28*, 29–39. [CrossRef]

36. Berger, H.; Vegran, F.; Chikh, M.; Gilardi, F.; Ladoire, S.; Bugaut, H.; Mignot, G.; Chalmin, F.; Bruchard, M.; Derangere, V.; et al. SOCS3 transactivation by PPARgamma prevents IL-17-driven cancer growth. *Cancer Res.* **2013**, *73*, 3578–3590. [CrossRef] [PubMed]

37. Mauldin, I.S.; Tung, K.S.; Lorenz, U.M. The tyrosine phosphatase SHP-1 dampens murine Th17 development. *Blood* **2012**, *119*, 4419–4429. [CrossRef] [PubMed]

38. Ma, L.; Huang, C.; Wang, X.J.; Xin, D.E.; Wang, L.S.; Zou, Q.C.; Zhang, Y.S.; Tan, M.D.; Wang, Y.M.; Zhao, T.C.; et al. Lysyl Oxidase 3 Is a Dual-Specificity Enzyme Involved in STAT3 Deacetylation and Deacetylimination Modulation. *Mol. Cell* **2017**, *65*, 296–309. [CrossRef] [PubMed]

39. Limagne, E.; Thibaudin, M.; Euvrard, R.; Berger, H.; Chalons, P.; Vegan, F.; Humblin, E.; Boidot, R.; Rebe, C.; Derangere, V.; et al. Sirtuin-1 Activation Controls Tumor Growth by Impeding Th17 Differentiation via STAT3 Deacetylation. *Cell Rep.* **2017**, *19*, 746–759. [CrossRef]

40. Chalmin, F.; Mignot, G.; Bruchard, M.; Chevriaux, A.; Vegran, F.; Hichami, A.; Ladoire, S.; Derangere, V.; Vincent, J.; Masson, D.; et al. Stat3 and Gfi-1 transcription factors control Th17 cell immunosuppressive activity via the regulation of ectonucleotidase expression. *Immunity* **2012**, *36*, 362–373. [CrossRef]

41. Zhou, J.; Li, X.; Wu, X.; Zhang, T.; Zhu, Q.; Wang, X.; Wang, H.; Wang, K.; Lin, Y.; Wang, X. Exosomes Released from Tumor-Associated Macrophages Transfer miRNAs That Induce a Treg/Th17 Cell Imbalance in Epithelial Ovarian Cancer. *Cancer Immunol. Res.* **2018**, *6*, 1578–1592. [CrossRef] [PubMed]

42. Zorn, E.; Nelson, E.A.; Mohseni, M.; Porcheray, F.; Kim, H.; Litsa, D.; Bellucci, R.; Raderschall, E.; Canning, C.; Soiffer, R.J.; et al. IL-2 regulates FOXP3 expression in human CD4+CD25+ regulatory T cells through a STAT-dependent mechanism and induces the expansion of these cells in vivo. *Blood* **2006**, *108*, 1571–1579. [CrossRef] [PubMed]

43. Wan, Y.Y.; Flavell, R.A. Regulatory T-cell functions are subverted and converted owing to attenuated Foxp3 expression. *Nature* **2007**, *445*, 766–770. [CrossRef] [PubMed]

44. Kortylewski, M.; Xin, H.; Kujawski, M.; Lee, H.; Liu, Y.; Harris, T.; Drake, C.; Pardoll, D.; Yu, H. Regulation of the IL-23 and IL-12 balance by Stat3 signaling in the tumor microenvironment. *Cancer Cell* **2009**, *15*, 114–123. [CrossRef] [PubMed]

45. Priceman, S.J.; Shen, S.; Wang, L.; Deng, J.; Yue, C.; Kujawski, M.; Yu, H. S1PR1 is crucial for accumulation of regulatory T cells in tumors via STAT3. *Cell Rep.* **2014**, *6*, 992–999. [CrossRef] [PubMed]

46. Huber, M.; Steinwald, V.; Guralnik, A.; Brustle, A.; Kleemann, P.; Rosenplanter, C.; Decker, T.; Lohoff, M. IL-27 inhibits the development of regulatory T cells via STAT3. *Int. Immunol.* **2008**, *20*, 223–234. [CrossRef] [PubMed]

47. Lam, E.; Choi, S.H.; Pareek, T.K.; Kim, B.G.; Letterio, J.J. Cyclin-dependent kinase 5 represses Foxp3 gene expression and Treg development through specific phosphorylation of Stat3 at Serine 727. *Mol. Immunol.* **2015**, *67*, 317–324. [CrossRef] [PubMed]

48. Li, D.; Kong, C.; Tsun, A.; Chen, C.; Song, H.; Shi, G.; Pan, W.; Dai, D.; Shen, N.; Li, B. MiR-125a-5p Decreases the Sensitivity of Treg cells Toward IL-6-Mediated Conversion by Inhibiting IL-6R and STAT3 Expression. *Sci Rep.* **2015**, *5*, 14615. [CrossRef] [PubMed]

49. Xiao, W.; Yin, M.; Wu, K.; Lu, G.; Deng, B.; Zhang, Y.; Qian, L.; Jia, X.; Ding, Y.; Gong, W. High-dose wogonin exacerbates DSS-induced colitis by up-regulating effector T cell function and inhibiting Treg cell. *J. Cell Mol. Med.* **2017**, *21*, 286–298. [CrossRef] [PubMed]

50. Kong, L.Y.; Wei, J.; Sharma, A.K.; Barr, J.; Abou-Ghazal, M.K.; Fokt, I.; Weinberg, J.; Rao, G.; Grimm, E.; Priebe, W.; et al. A novel phosphorylated STAT3 inhibitor enhances T cell cytotoxicity against melanoma through inhibition of regulatory T cells. *Cancer Immunol. Immunother.* **2009**, *58*, 1023–1032. [CrossRef] [PubMed]

51. He, W.; Zhu, Y.; Mu, R.; Xu, J.; Zhang, X.; Wang, C.; Li, Q.; Huang, Z.; Zhang, J.; Pan, Y.; et al. A Jak2-selective inhibitor potently reverses the immune suppression by modulating the tumor microenvironment for cancer immunotherapy. *Biochem. Pharmacol.* **2017**, *145*, 132–146. [CrossRef] [PubMed]

52. Oweida, A.J.; Darragh, L.; Phan, A.; Binder, D.; Bhatia, S.; Mueller, A.; Van Court, B.; Milner, D.; Raben, D.; Woessner, R.; et al. The role of regulatory T cells in the response to radiation therapy in head and neck cancer. *J. Natl. Cancer Inst.* **2019**. [CrossRef] [PubMed]

53. Schmitt, N.; Liu, Y.; Bentebibel, S.E.; Munagala, I.; Bourdery, L.; Venuprasad, K.; Banchereau, J.; Ueno, H. The cytokine TGF-beta co-opts signaling via STAT3-STAT4 to promote the differentiation of human TFH cells. *Nat. Immunol.* **2014**, *15*, 856–865. [CrossRef] [PubMed]

54. Ma, C.S.; Avery, D.T.; Chan, A.; Batten, M.; Bustamante, J.; Boisson-Dupuis, S.; Arkwright, P.D.; Kreins, A.Y.; Averbuch, D.; Engelhard, D.; et al. Functional STAT3 deficiency compromises the generation of human T follicular helper cells. *Blood* **2012**, *119*, 3997–4008. [CrossRef] [PubMed]

55. Hercor, M.; Anciaux, M.; Denanglaire, S.; Debuisson, D.; Leo, O.; Andris, F. Antigen-presenting cell-derived IL-6 restricts the expression of GATA3 and IL-4 by follicular helper T cells. *J. Leukoc. Biol.* **2017**, *101*, 5–14. [CrossRef] [PubMed]

56. Chen, M.M.; Xiao, X.; Lao, X.M.; Wei, Y.; Liu, R.X.; Zeng, Q.H.; Wang, J.C.; Ouyang, F.Z.; Chen, D.P.; Chan, K.W.; et al. Polarization of Tissue-Resident TFH-Like Cells in Human Hepatoma Bridges Innate Monocyte Inflammation and M2b Macrophage Polarization. *Cancer Discov.* **2016**, *6*, 1182–1195. [CrossRef]

57. Olson, M.R.; Verdan, F.F.; Hufford, M.M.; Dent, A.L.; Kaplan, M.H. STAT3 Impairs STAT5 Activation in the Development of IL-9-Secreting T Cells. *J. Immunol.* **2016**, *196*, 3297–3304. [CrossRef]

58. Ulrich, B.J.; Verdan, F.F.; McKenzie, A.N.; Kaplan, M.H.; Olson, M.R. STAT3 Activation Impairs the Stability of Th9 Cells. *J. Immunol.* **2017**, *198*, 2302–2309. [CrossRef]

59. Zhang, Y.; Siegel, A.M.; Sun, G.; Dimaggio, T.; Freeman, A.F.; Milner, J.D. Human TH9 differentiation is dependent on signal transducer and activator of transcription (STAT) 3 to restrain STAT1-mediated inhibition. *J. Allergy Clin. Immunol.* **2019**, *143*, 1108–1118. [CrossRef]

60. Yue, C.; Shen, S.; Deng, J.; Priceman, S.J.; Li, W.; Huang, A.; Yu, H. STAT3 in CD8+ T Cells Inhibits Their Tumor Accumulation by Downregulating CXCR3/CXCL10 Axis. *Cancer Immunol. Res.* **2015**, *3*, 864–870. [CrossRef]

61. Kujawski, M.; Zhang, C.; Herrmann, A.; Reckamp, K.; Scuto, A.; Jensen, M.; Deng, J.; Forman, S.; Figlin, R.; Yu, H. Targeting STAT3 in adoptively transferred T cells promotes their in vivo expansion and antitumor effects. *Cancer Res.* **2010**, *70*, 9599–9610. [CrossRef] [PubMed]

62. Hu, J.; Batth, I.S.; Xia, X.; Li, S. Regulation of NKG2D(+)CD8(+) T-cell-mediated antitumor immune surveillance: Identification of a novel CD28 activation-mediated, STAT3 phosphorylation-dependent mechanism. *OncoImmunology* **2016**, *5*, e1252012. [CrossRef] [PubMed]

63. Tosolini, M.; Kirilovsky, A.; Mlecnik, B.; Fredriksen, T.; Mauger, S.; Bindea, G.; Berger, A.; Bruneval, P.; Fridman, W.H.; Pages, F.; et al. Clinical impact of different classes of infiltrating T cytotoxic and helper cells (Th1, th2, treg, th17) in patients with colorectal cancer. *Cancer Res.* **2011**, *71*, 1263–1271. [CrossRef] [PubMed]

64. Szabo, S.J.; Kim, S.T.; Costa, G.L.; Zhang, X.; Fathman, C.G.; Glimcher, L.H. A novel transcription factor, T-bet, directs Th1 lineage commitment. *Cell* **2000**, *100*, 655–669. [CrossRef]

65. Zheng, W.; Flavell, R.A. The transcription factor GATA-3 is necessary and sufficient for Th2 cytokine gene expression in CD4 T cells. *Cell* **1997**, *89*, 587–596. [CrossRef]

66. Yoshimoto, T.; Yasuda, K.; Mizuguchi, J.; Nakanishi, K. IL-27 suppresses Th2 cell development and Th2 cytokines production from polarized Th2 cells: A novel therapeutic way for Th2-mediated allergic inflammation. *J. Immunol.* **2007**, *179*, 4415–4423. [CrossRef] [PubMed]

67. Chen, Z.; O'Shea, J.J. Th17 cells: A new fate for differentiating helper T cells. *Immunol. Res.* **2008**, *41*, 87–102. [CrossRef]

68. Van den Berg, W.B.; Miossec, P. IL-17 as a future therapeutic target for rheumatoid arthritis. *Nat. Rev. Rheumatol.* **2009**, *5*, 549–553. [CrossRef]

69. Su, X.; Ye, J.; Hsueh, E.C.; Zhang, Y.; Hoft, D.F.; Peng, G. Tumor microenvironments direct the recruitment and expansion of human Th17 cells. *J. Immunol.* **2010**, *184*, 1630–1641. [CrossRef]

70. Hemdan, N.Y. Anti-cancer versus cancer-promoting effects of the interleukin-17-producing T helper cells. *Immunol. Lett* **2013**, *149*, 123–133. [CrossRef]

71. Ivanov, I.I.; McKenzie, B.S.; Zhou, L.; Tadokoro, C.E.; Lepelley, A.; Lafaille, J.J.; Cua, D.J.; Littman, D.R. The orphan nuclear receptor RORgammat directs the differentiation program of proinflammatory IL-17+ T helper cells. *Cell* **2006**, *126*, 1121–1133. [CrossRef] [PubMed]

72. Jiang, Y.; Liu, Y.; Lu, H.; Sun, S.C.; Jin, W.; Wang, X.; Dong, C. Epigenetic activation during T helper 17 cell differentiation is mediated by Tripartite motif containing 28. *Nat. Commun.* **2018**, *9*, 1424. [CrossRef]

73. Fang, S.; Liu, B.; Sun, Q.; Zhao, J.; Qi, H.; Li, Q. Platelet factor 4 inhibits IL-17/Stat3 pathway via upregulation of SOCS3 expression in melanoma. *Inflammation* **2014**, *37*, 1744–1750. [CrossRef] [PubMed]

74. Ghoreschi, K.; Laurence, A.; Yang, X.P.; Tato, C.M.; McGeachy, M.J.; Konkel, J.E.; Ramos, H.L.; Wei, L.; Davidson, T.S.; Bouladoux, N.; et al. Generation of pathogenic T(H)17 cells in the absence of TGF-beta signalling. *Nature* **2010**, *467*, 967–971. [CrossRef] [PubMed]

75. Sakaguchi, S. Regulatory T cells: Key controllers of Immunol.ogic self-tolerance. *Cell* **2000**, *101*, 455–458. [CrossRef]

76. Shevach, E.M. CD4+ CD25+ suppressor T cells: More questions than answers. *Nat. Rev. Immunol.* **2002**, *2*, 389–400. [CrossRef] [PubMed]

77. Chen, W.; Jin, W.; Hardegen, N.; Lei, K.J.; Li, L.; Marinos, N.; McGrady, G.; Wahl, S.M. Conversion of peripheral CD4+CD25- naive T cells to CD4+CD25+ regulatory T cells by TGF-beta induction of transcription factor Foxp3. *J. Exp. Med.* **2003**, *198*, 1875–1886. [CrossRef] [PubMed]

78. Nishikawa, H.; Sakaguchi, S. Regulatory T cells in tumor immunity. *Int. J. Cancer* **2010**, *127*, 759–767. [CrossRef] [PubMed]

79. Curiel, T.J.; Coukos, G.; Zou, L.; Alvarez, X.; Cheng, P.; Mottram, P.; Evdemon-Hogan, M.; Conejo-Garcia, J.R.; Zhang, L.; Burow, M.; et al. Specific recruitment of regulatory T cells in ovarian carcinoma fosters immune privilege and predicts reduced survival. *Nat. Med.* **2004**, *10*, 942–949. [CrossRef]

80. Bates, G.J.; Fox, S.B.; Han, C.; Leek, R.D.; Garcia, J.F.; Harris, A.L.; Banham, A.H. Quantification of regulatory T cells enables the identification of high-risk breast cancer patients and those at risk of late relapse. *J. Clin. Oncol.* **2006**, *24*, 5373–5380. [CrossRef]

81. Perrone, G.; Ruffini, P.A.; Catalano, V.; Spino, C.; Santini, D.; Muretto, P.; Spoto, C.; Zingaretti, C.; Sisti, V.; Alessandroni, P.; et al. Intratumoural FOXP3-positive regulatory T cells are associated with adverse prognosis in radically resected gastric cancer. *Eur. J. Cancer* **2008**, *44*, 1875–1882. [CrossRef] [PubMed]

82. Chaudhry, A.; Rudra, D.; Treuting, P.; Samstein, R.M.; Liang, Y.; Kas, A.; Rudensky, A.Y. CD4+ regulatory T cells control TH17 responses in a Stat3-dependent manner. *Science* **2009**, *326*, 986–991. [CrossRef] [PubMed]

83. Wuest, T.Y.; Willette-Brown, J.; Durum, S.K.; Hurwitz, A.A. The influence of IL-2 family cytokines on activation and function of naturally occurring regulatory T cells. *J. Leukoc. Biol.* **2008**, *84*, 973–980. [CrossRef] [PubMed]

84. Xu, L.; Kitani, A.; Stuelten, C.; McGrady, G.; Fuss, I.; Strober, W. Positive and negative transcriptional regulation of the Foxp3 gene is mediated by access and binding of the Smad3 protein to enhancer I. *Immunity* **2010**, *33*, 313–325. [CrossRef] [PubMed]

85. Laurence, A.; Amarnath, S.; Mariotti, J.; Kim, Y.C.; Foley, J.; Eckhaus, M.; O'Shea, J.J.; Fowler, D.H. STAT3 Transcription Factor Promotes Instability of nTreg Cells and Limits Generation of iTreg Cells during Acute Murine Graft-versus-Host Disease. *Immunity* **2012**, *37*, 209–222. [CrossRef] [PubMed]

86. Eivazi, S.; Bagheri, S.; Hashemzadeh, M.S.; Ghalavand, M.; Qamsari, E.S.; Dorostkar, R.; Yasemi, M. Development of T follicular helper cells and their role in disease and immune system. *Biomed. Pharmacother.* **2016**, *84*, 1668–1678. [CrossRef] [PubMed]

87. Qin, L.; Waseem, T.C.; Sahoo, A.; Bieerkehazhi, S.; Zhou, H.; Galkina, E.V.; Nurieva, R. Insights Into the Molecular Mechanisms of T Follicular Helper-Mediated Immunity and Pathology. *Front. Immunol.* **2018**, *9*, 1884. [CrossRef] [PubMed]

88. Gu-Trantien, C.; Loi, S.; Garaud, S.; Equeter, C.; Libin, M.; de Wind, A.; Ravoet, M.; Le Buanec, H.; Sibille, C.; Manfouo-Foutsop, G.; et al. CD4(+) follicular helper T cell infiltration predicts breast cancer survival. *J. Clin. Invest.* **2013**, *123*, 2873–2892. [CrossRef] [PubMed]

89. Bindea, G.; Mlecnik, B.; Tosolini, M.; Kirilovsky, A.; Waldner, M.; Obenauf, A.C.; Angell, H.; Fredriksen, T.; Lafontaine, L.; Berger, A.; et al. Spatiotemporal dynamics of intratumoral immune cells reveal the immune landscape in human cancer. *Immunity* **2013**, *39*, 782–795. [CrossRef]

90. Alshekaili, J.; Chand, R.; Lee, C.E.; Corley, S.; Kwong, K.; Papa, I.; Fulcher, D.A.; Randall, K.L.; Leiding, J.W.; Ma, C.S.; et al. STAT3 regulates cytotoxicity of human CD57+ CD4+ T cells in blood and lymphoid follicles. *Sci. Rep.* **2018**, *8*, 3529. [CrossRef]

91. Ray, J.P.; Marshall, H.D.; Laidlaw, B.J.; Staron, M.M.; Kaech, S.M.; Craft, J. Transcription factor STAT3 and type I interferons are corepressive insulators for differentiation of follicular helper and T helper 1 cells. *Immunity* **2014**, *40*, 367–377. [CrossRef] [PubMed]

92. Wu, H.; Xu, L.L.; Teuscher, P.; Liu, H.; Kaplan, M.H.; Dent, A.L. An Inhibitory Role for the Transcription Factor Stat3 in Controlling IL-4 and Bcl6 Expression in Follicular Helper T Cells. *J. Immunol.* **2015**, *195*, 2080–2089. [CrossRef] [PubMed]

93. Read, K.A.; Powell, M.D.; Baker, C.E.; Sreekumar, B.K.; Ringel-Scaia, V.M.; Bachus, H.; Martin, R.E.; Cooley, I.D.; Allen, I.C.; Ballesteros-Tato, A.; et al. Integrated STAT3 and Ikaros Zinc Finger Transcription Factor Activities Regulate Bcl-6 Expression in CD4(+) Th Cells. *J. Immunol.* **2017**, *199*, 2377–2387. [CrossRef] [PubMed]

94. Chen, W.; Nyuydzefe, M.S.; Weiss, J.M.; Zhang, J.; Waksal, S.D.; Zanin-Zhorov, A. ROCK2, but not ROCK1 interacts with phosphorylated STAT3 and co-occupies TH17/TFH gene promoters in TH17-activated human T cells. *Sci. Rep.* **2018**, *8*, 16636. [CrossRef] [PubMed]

95. Rivera Vargas, T.; Humblin, E.; Vegran, F.; Ghiringhelli, F.; Apetoh, L. TH9 cells in anti-tumor immunity. *Semin. Immunopathol.* **2017**, *39*, 39–46. [CrossRef]

96. Durgeau, A.; Virk, Y.; Corgnac, S.; Mami-Chouaib, F. Recent Advances in Targeting CD8 T-Cell Immunity for More Effective Cancer Immunotherapy. *Front. Immunol.* **2018**, *9*, 14. [CrossRef] [PubMed]

97. Emmerich, J.; Mumm, J.B.; Chan, I.H.; LaFace, D.; Truong, H.; McClanahan, T.; Gorman, D.M.; Oft, M. IL-10 directly activates and expands tumor-resident CD8(+) T cells without de novo infiltration from secondary lymphoid organs. *Cancer Res.* **2012**, *72*, 3570–3581. [CrossRef] [PubMed]

98. Wang, X.; Xin, W.; Zhang, H.; Zhang, F.; Gao, M.; Yuan, L.; Xu, X.; Hu, X.; Zhao, M. Aberrant expression of p-STAT3 in peripheral blood CD4+ and CD8+ T cells related to hepatocellular carcinoma development. *Mol. Med. Rep.* **2014**, *10*, 2649–2656. [CrossRef] [PubMed]

99. Marigo, I.; Dolcetti, L.; Serafini, P.; Zanovello, P.; Bronte, V. Tumor-induced tolerance and immune suppression by myeloid derived suppressor cells. *Immunol. Rev.* **2008**, *222*, 162–179. [CrossRef]

100. Cheng, F.; Wang, H.W.; Cuenca, A.; Huang, M.; Ghansah, T.; Brayer, J.; Kerr, W.G.; Takeda, K.; Akira, S.; Schoenberger, S.P.; et al. A critical role for Stat3 signaling in immune tolerance. *Immunity* **2003**, *19*, 425–436. [CrossRef]

101. Brayer, J.; Cheng, F.; Wang, H.; Horna, P.; Vicente-Suarez, I.; Pinilla-Ibarz, J.; Sotomayor, E.M. Enhanced CD8 T cell cross-presentation by macrophages with targeted disruption of STAT3. *Immunol. Lett.* **2010**, *131*, 126–130. [CrossRef] [PubMed]

102. Kortylewski, M.; Kujawski, M.; Herrmann, A.; Yang, C.; Wang, L.; Liu, Y.; Salcedo, R.; Yu, H. Toll-like receptor 9 activation of signal transducer and activator of transcription 3 constrains its agonist-based immunotherapy. *Cancer Res.* **2009**, *69*, 2497–2505. [CrossRef] [PubMed]

103. Cheng, Z.; Zhang, D.; Gong, B.; Wang, P.; Liu, F. CD163 as a novel target gene of STAT3 is a potential therapeutic target for gastric cancer. *Oncotarget* **2017**, *8*, 87244–87262. [CrossRef] [PubMed]

104. Solis-Martinez, R.; Cancino-Marentes, M.; Hernandez-Flores, G.; Ortiz-Lazareno, P.; Mandujano-Alvarez, G.; Cruz-Galvez, C.; Sierra-Diaz, E.; Rodriguez-Padilla, C.; Jave-Suarez, L.F.; Aguilar-Lemarroy, A.; et al. Regulation of immunophenotype modulation of monocytes-macrophages from M1 into M2 by prostate cancer cell-culture supernatant via transcription factor STAT3. *Immunol. Lett.* **2018**, *196*, 140–148. [CrossRef] [PubMed]

105. Kubala, M.H.; Punj, V.; Placencio-Hickok, V.R.; Fang, H.; Fernandez, G.E.; Sposto, R.; DeClerck, Y.A. Plasminogen Activator Inhibitor-1 Promotes the Recruitment and Polarization of Macrophages in Cancer. *Cell Rep.* **2018**, *25*, 2177–2191. [CrossRef]

106. Ham, S.; Lima, L.G.; Chai, E.P.Z.; Muller, A.; Lobb, R.J.; Krumeich, S.; Wen, S.W.; Wiegmans, A.P.; Moller, A. Breast Cancer-Derived Exosomes Alter Macrophage Polarization via gp130/STAT3 Signaling. *Front. Immunol.* **2018**, *9*, 871. [CrossRef]

107. Fujiwara, Y.; Komohara, Y.; Ikeda, T.; Takeya, M. Corosolic acid inhibits glioblastoma cell proliferation by suppressing the activation of signal transducer and activator of transcription-3 and nuclear factor-kappa B in tumor cells and tumor-associated macrophages. *Cancer Sci.* **2011**, *102*, 206–211. [CrossRef]

108. Fujiwara, Y.; Komohara, Y.; Kudo, R.; Tsurushima, K.; Ohnishi, K.; Ikeda, T.; Takeya, M. Oleanolic acid inhibits macrophage differentiation into the M2 phenotype and glioblastoma cell proliferation by suppressing the activation of STAT3. *Oncol. Rep.* **2011**, *26*, 1533–1537. [PubMed]

109. Andersen, M.N.; Etzerodt, A.; Graversen, J.H.; Holthof, L.C.; Moestrup, S.K.; Hokland, M.; Moller, H.J. STAT3 inhibition specifically in human monocytes and macrophages by CD163-targeted corosolic acid-containing liposomes. *Cancer Immunol. Immunother.* **2019**, *68*, 489–502. [CrossRef]

110. Giurisato, E.; Xu, Q.; Lonardi, S.; Telfer, B.; Russo, I.; Pearson, A.; Finegan, K.G.; Wang, W.; Wang, J.; Gray, N.S.; et al. Myeloid ERK5 deficiency suppresses tumor growth by blocking protumor macrophage polarization via STAT3 inhibition. *Proc. Natl. Acad. Sci. USA* **2018**, *115*, E2801–E2810. [CrossRef]

111. Dominguez-Soto, A.; Sierra-Filardi, E.; Puig-Kroger, A.; Perez-Maceda, B.; Gomez-Aguado, F.; Corcuera, M.T.; Sanchez-Mateos, P.; Corbi, A.L. Dendritic cell-specific ICAM-3-grabbing nonintegrin expression on M2-polarized and tumor-associated macrophages is macrophage-CSF dependent and enhanced by tumor-derived IL-6 and IL-10. *J. Immunol.* **2011**, *186*, 2192–2200. [CrossRef] [PubMed]

112. Hiwatashi, K.; Tamiya, T.; Hasegawa, E.; Fukaya, T.; Hashimoto, M.; Kakoi, K.; Kashiwagi, I.; Kimura, A.; Inoue, N.; Morita, R.; et al. Suppression of SOCS3 in macrophages prevents cancer metastasis by modifying macrophage phase and MCP2/CCL8 induction. *Cancer Lett.* **2011**, *308*, 172–180. [CrossRef] [PubMed]

113. Hasita, H.; Komohara, Y.; Okabe, H.; Masuda, T.; Ohnishi, K.; Lei, X.F.; Beppu, T.; Baba, H.; Takeya, M. Significance of alternatively activated macrophages in patients with intrahepatic cholangiocarcinoma. *Cancer Sci.* **2010**, *101*, 1913–1919. [CrossRef] [PubMed]

114. Lee, J.H.; Lee, G.T.; Woo, S.H.; Ha, Y.S.; Kwon, S.J.; Kim, W.J.; Kim, I.Y. BMP-6 in renal cell carcinoma promotes tumor proliferation through IL-10-dependent M2 polarization of tumor-associated macrophages. *Cancer Res.* **2013**, *73*, 3604–3614. [CrossRef] [PubMed]

115. Kujawski, M.; Kortylewski, M.; Lee, H.; Herrmann, A.; Kay, H.; Yu, H. Stat3 mediates myeloid cell-dependent tumor angiogenesis in mice. *J. Clin. Invest.* **2008**, *118*, 3367–3377. [CrossRef] [PubMed]

116. Laouar, Y.; Welte, T.; Fu, X.Y.; Flavell, R.A. STAT3 is required for Flt3L-dependent dendritic cell differentiation. *Immunity* **2003**, *19*, 903–912. [CrossRef]

117. Kortylewski, M.; Kujawski, M.; Wang, T.; Wei, S.; Zhang, S.; Pilon-Thomas, S.; Niu, G.; Kay, H.; Mule, J.; Kerr, W.G.; et al. Inhibiting Stat3 signaling in the hematopoietic system elicits multicomponent antitumor immunity. *Nat. Med.* **2005**, *11*, 1314–1321. [CrossRef]

118. Herrmann, A.; Kortylewski, M.; Kujawski, M.; Zhang, C.; Reckamp, K.; Armstrong, B.; Wang, L.; Kowolik, C.; Deng, J.; Figlin, R.; et al. Targeting Stat3 in the myeloid compartment drastically improves the in vivo antitumor functions of adoptively transferred T cells. *Cancer Res.* **2010**, *70*, 7455–7464. [CrossRef]

119. Iwata-Kajihara, T.; Sumimoto, H.; Kawamura, N.; Ueda, R.; Takahashi, T.; Mizuguchi, H.; Miyagishi, M.; Takeda, K.; Kawakami, Y. Enhanced cancer immunotherapy using STAT3-depleted dendritic cells with high Th1-inducing ability and resistance to cancer cell-derived inhibitory factors. *J. Immunol.* **2011**, *187*, 27–36. [CrossRef]

120. Sanseverino, I.; Purificato, C.; Varano, B.; Conti, L.; Gessani, S.; Gauzzi, M.C. STAT3-silenced human dendritic cells have an enhanced ability to prime IFNgamma production by both alphabeta and gammadelta T lymphocytes. *Immunobiology* **2014**, *219*, 503–511. [CrossRef]

121. Hoentjen, F.; Sartor, R.B.; Ozaki, M.; Jobin, C. STAT3 regulates NF-kappaB recruitment to the IL-12p40 promoter in dendritic cells. *Blood* **2005**, *105*, 689–696. [CrossRef] [PubMed]

122. Kitamura, H.; Kamon, H.; Sawa, S.; Park, S.J.; Katunuma, N.; Ishihara, K.; Murakami, M.; Hirano, T. IL-6-STAT3 controls intracellular MHC class II alphabeta dimer level through cathepsin S activity in dendritic cells. *Immunity* **2005**, *23*, 491–502. [CrossRef] [PubMed]

123. Ohno, Y.; Kitamura, H.; Takahashi, N.; Ohtake, J.; Kaneumi, S.; Sumida, K.; Homma, S.; Kawamura, H.; Minagawa, N.; Shibasaki, S.; et al. IL-6 down-regulates HLA class II expression and IL-12 production of human dendritic cells to impair activation of antigen-specific CD4(+) T cells. *Cancer Immunol. Immunother.* **2016**, *65*, 193–204. [CrossRef] [PubMed]

124. Li, H.S.; Liu, C.; Xiao, Y.; Chu, F.; Liang, X.; Peng, W.; Hu, J.; Neelapu, S.S.; Sun, S.C.; Hwu, P.; et al. Bypassing STAT3-mediated inhibition of the transcriptional regulator ID2 improves the antitumor efficacy of dendritic cells. *Sci. Signal.* **2016**, *9*, ra94. [CrossRef] [PubMed]

125. Yu, S.; Liu, C.; Su, K.; Wang, J.; Liu, Y.; Zhang, L.; Li, C.; Cong, Y.; Kimberly, R.; Grizzle, W.E.; et al. Tumor exosomes inhibit differentiation of bone marrow dendritic cells. *J. Immunol.* **2007**, *178*, 6867–6875. [CrossRef]

126. Zhang, H.; Nguyen-Jackson, H.; Panopoulos, A.D.; Li, H.S.; Murray, P.J.; Watowich, S.S. STAT3 controls myeloid progenitor growth during emergency granulopoiesis. *Blood* **2010**, *116*, 2462–2471. [CrossRef] [PubMed]

127. Nefedova, Y.; Huang, M.; Kusmartsev, S.; Bhattacharya, R.; Cheng, P.; Salup, R.; Jove, R.; Gabrilovich, D. Hyperactivation of STAT3 is involved in abnormal differentiation of dendritic cells in cancer. *J. Immunol.* **2004**, *172*, 464–474. [CrossRef] [PubMed]

128. Nefedova, Y.; Nagaraj, S.; Rosenbauer, A.; Muro-Cacho, C.; Sebti, S.M.; Gabrilovich, D.I. Regulation of dendritic cell differentiation and antitumor immune response in cancer by pharmacologic-selective inhibition of the janus-activated kinase 2/signal transducers and activators of transcription 3 pathway. *Cancer Res.* **2005**, *65*, 9525–9535. [CrossRef]

129. Hossain, D.M.; Pal, S.K.; Moreira, D.; Duttagupta, P.; Zhang, Q.; Won, H.; Jones, J.; D'Apuzzo, M.; Forman, S.; Kortylewski, M. TLR9-Targeted STAT3 Silencing Abrogates Immunosuppressive Activity of Myeloid-Derived Suppressor Cells from Prostate Cancer Patients. *Clin. Cancer Res.* **2015**, *21*, 3771–3782. [CrossRef]

130. Moreira, D.; Adamus, T.; Zhao, X.; Su, Y.L.; Zhang, Z.; White, S.V.; Swiderski, P.; Lu, X.; DePinho, R.A.; Pal, S.K.; et al. STAT3 Inhibition Combined with CpG Immunostimulation Activates Antitumor Immunity to Eradicate Genetically Distinct Castration-Resistant Prostate Cancers. *Clin. Cancer Res.* **2018**, *24*, 5948–5962. [CrossRef]

131. Zhou, J.; Qu, Z.; Sun, F.; Han, L.; Li, L.; Yan, S.; Stabile, L.P.; Chen, L.F.; Siegfried, J.M.; Xiao, G. Myeloid STAT3 Promotes Lung Tumorigenesis by Transforming Tumor Immunosurveillance into Tumor-Promoting Inflammation. *Cancer Immunol. Res.* **2017**, *5*, 257–268. [CrossRef] [PubMed]

132. Corzo, C.A.; Cotter, M.J.; Cheng, P.; Cheng, F.; Kusmartsev, S.; Sotomayor, E.; Padhya, T.; McCaffrey, T.V.; McCaffrey, J.C.; Gabrilovich, D.I. Mechanism regulating reactive oxygen species in tumor-induced myeloid-derived suppressor cells. *J. Immunol.* **2009**, *182*, 5693–5701. [CrossRef] [PubMed]

133. Chalmin, F.; Ladoire, S.; Mignot, G.; Vincent, J.; Bruchard, M.; Remy-Martin, J.P.; Boireau, W.; Rouleau, A.; Simon, B.; Lanneau, D.; et al. Membrane-associated Hsp72 from tumor-derived exosomes mediates STAT3-dependent immunosuppressive function of mouse and human myeloid-derived suppressor cells. *J. Clin. Invest.* **2010**, *120*, 457–471. [CrossRef] [PubMed]

134. Waight, J.D.; Netherby, C.; Hensen, M.L.; Miller, A.; Hu, Q.; Liu, S.; Bogner, P.N.; Farren, M.R.; Lee, K.P.; Liu, K.; et al. Myeloid-derived suppressor cell development is regulated by a STAT/IRF-8 axis. *J. Clin. Invest.* **2013**, *123*, 4464–4478. [CrossRef] [PubMed]

135. Thorn, M.; Guha, P.; Cunetta, M.; Espat, N.J.; Miller, G.; Junghans, R.P.; Katz, S.C. Tumor-associated GM-CSF overexpression induces immunoinhibitory molecules via STAT3 in myeloid-suppressor cells infiltrating liver metastases. *Cancer Gene Ther.* **2016**, *23*, 188–198. [CrossRef] [PubMed]

136. Vasquez-Dunddel, D.; Pan, F.; Zeng, Q.; Gorbounov, M.; Albesiano, E.; Fu, J.; Blosser, R.L.; Tam, A.J.; Bruno, T.; Zhang, H.; et al. STAT3 regulates arginase-I in myeloid-derived suppressor cells from cancer patients. *J. Clin. Invest.* **2013**, *123*, 1580–1589. [CrossRef]

137. Yu, J.; Du, W.; Yan, F.; Wang, Y.; Li, H.; Cao, S.; Yu, W.; Shen, C.; Liu, J.; Ren, X. Myeloid-derived suppressor cells suppress antitumor immune responses through IDO expression and correlate with lymph node metastasis in patients with breast cancer. *J. Immunol.* **2013**, *190*, 3783–3797. [CrossRef]

138. Kumar, V.; Cheng, P.; Condamine, T.; Mony, S.; Languino, L.R.; McCaffrey, J.C.; Hockstein, N.; Guarino, M.; Masters, G.; Penman, E.; et al. CD45 Phosphatase Inhibits STAT3 Transcription Factor Activity in Myeloid Cells and Promotes Tumor-Associated Macrophage Differentiation. *Immunity* **2016**, *44*, 303–315. [CrossRef]

139. Wu, T.; Wang, C.; Wang, W.; Hui, Y.; Zhang, R.; Qiao, L.; Dai, Y. Embelin impairs the accumulation and activation of MDSCs in colitis-associated tumorigenesis. *OncoImmunology* **2018**, *7*, e1498437. [CrossRef]

140. Kuroda, H.; Mabuchi, S.; Kozasa, K.; Yokoi, E.; Matsumoto, Y.; Komura, N.; Kawano, M.; Hashimoto, K.; Sawada, K.; Kimura, T. PM01183 inhibits myeloid-derived suppressor cells in vitro and in vivo. *Immunotherapy* **2017**, *9*, 805–817. [CrossRef]

141. Yin, T.; Zhao, Z.B.; Guo, J.; Wang, T.; Yang, J.B.; Wang, C.; Long, J.; Ma, S.; Huang, Q.; Zhang, K.; et al. Aurora-A inhibition eliminates myeloid cell-mediated immunosuppression and enhances the efficacy of anti-PD-L1 therapy in breast cancer. *Cancer Res.* **2019**. [CrossRef] [PubMed]

142. Guha, P.; Gardell, J.; Darpolor, J.; Cunetta, M.; Lima, M.; Miller, G.; Espat, N.J.; Junghans, R.P.; Katz, S.C. STAT3 inhibition induces Bax-dependent apoptosis in liver tumor myeloid-derived suppressor cells. *Oncogene* **2019**, *38*, 533–548. [CrossRef] [PubMed]

143. Mills, C.D.; Kincaid, K.; Alt, J.M.; Heilman, M.J.; Hill, A.M. M-1/M-2 macrophages and the Th1/Th2 paradigm. *J. Immunol.* **2000**, *164*, 6166–6173. [CrossRef] [PubMed]

144. Martinez, F.O.; Sica, A.; Mantovani, A.; Locati, M. Macrophage activation and polarization. *Front. Biosci.* **2008**, *13*, 453–461. [CrossRef] [PubMed]

145. Pollard, J.W. Tumour-educated macrophages promote tumour progression and metastasis. *Nat. Rev. Cancer* **2004**, *4*, 71–78. [CrossRef] [PubMed]

146. Sica, A.; Schioppa, T.; Mantovani, A.; Allavena, P. Tumour-associated macrophages are a distinct M2 polarised population promoting tumour progression: Potential targets of anti-cancer therapy. *Eur. J. Cancer* **2006**, *42*, 717–727. [CrossRef]

147. Lewis, C.E.; Pollard, J.W. Distinct role of macrophages in different tumor microenvironments. *Cancer Res.* **2006**, *66*, 605–612. [CrossRef]

148. Jenkins, B.J.; Grail, D.; Inglese, M.; Quilici, C.; Bozinovski, S.; Wong, P.; Ernst, M. Imbalanced gp130-dependent signaling in macrophages alters macrophage colony-stimulating factor responsiveness via regulation of c-fms expression. *Mol. Cell Biol.* **2004**, *24*, 1453–1463. [CrossRef]

149. Sun, Z.; Yao, Z.; Liu, S.; Tang, H.; Yan, X. An oligonucleotide decoy for Stat3 activates the immune response of macrophages to breast cancer. *Immunobiology* **2006**, *211*, 199–209. [CrossRef]

150. Komohara, Y.; Ohnishi, K.; Kuratsu, J.; Takeya, M. Possible involvement of the M2 anti-inflammatory macrophage phenotype in growth of human gliomas. *J. Pathol.* **2008**, *216*, 15–24. [CrossRef]

151. Fu, X.L.; Duan, W.; Su, C.Y.; Mao, F.Y.; Lv, Y.P.; Teng, Y.S.; Yu, P.W.; Zhuang, Y.; Zhao, Y.L. Interleukin 6 induces M2 macrophage differentiation by STAT3 activation that correlates with gastric cancer progression. *Cancer Immunol. Immunother.* **2017**, *66*, 1597–1608. [CrossRef] [PubMed]

152. Li, Y.; Cai, L.; Wang, H.; Wu, P.; Gu, W.; Chen, Y.; Hao, H.; Tang, K.; Yi, P.; Liu, M.; et al. Pleiotropic regulation of macrophage polarization and tumorigenesis by formyl peptide receptor-2. *Oncogene* **2011**, *30*, 3887–3899. [CrossRef] [PubMed]

153. Crowley, M.; Inaba, K.; Steinman, R.M. Dendritic cells are the principal cells in mouse spleen bearing immunogenic fragments of foreign proteins. *J. Exp. Med.* **1990**, *172*, 383–386. [CrossRef] [PubMed]

154. Steinman, R.M. The dendritic cell system and its role in immunogenicity. *Annu. Rev. Immunol.* **1991**, *9*, 271–296. [CrossRef]

155. Park, S.J.; Nakagawa, T.; Kitamura, H.; Atsumi, T.; Kamon, H.; Sawa, S.; Kamimura, D.; Ueda, N.; Iwakura, Y.; Ishihara, K.; et al. IL-6 regulates in vivo dendritic cell differentiation through STAT3 activation. *J. Immunol.* **2004**, *173*, 3844–3854. [CrossRef]

156. Farren, M.R.; Carlson, L.M.; Netherby, C.S.; Lindner, I.; Li, P.K.; Gabrilovich, D.I.; Abrams, S.I.; Lee, K.P. Tumor-induced STAT3 signaling in myeloid cells impairs dendritic cell generation by decreasing PKCbetaII abundance. *Sci. Signal.* **2014**, *7*, ra16. [CrossRef]

157. Wagner, A.H.; Conzelmann, M.; Fitzer, F.; Giese, T.; Gulow, K.; Falk, C.S.; Kramer, O.H.; Dietrich, S.; Hecker, M.; Luft, T. JAK1/STAT3 activation directly inhibits IL-12 production in dendritic cells by preventing CDK9/P-TEFb recruitment to the p35 promoter. *Biochem. Pharmacol.* **2015**, *96*, 52–64. [CrossRef]

158. Cheng, F.; Lienlaf, M.; Wang, H.W.; Perez-Villarroel, P.; Lee, C.; Woan, K.; Rock-Klotz, J.; Sahakian, E.; Woods, D.; Pinilla-Ibarz, J.; et al. A novel role for histone deacetylase 6 in the regulation of the tolerogenic STAT3/IL-10 pathway in APCs. *J. Immunol.* **2014**, *193*, 2850–2862. [CrossRef]

159. Gabrilovich, D.I.; Bronte, V.; Chen, S.H.; Colombo, M.P.; Ochoa, A.; Ostrand-Rosenberg, S.; Schreiber, H. The terminology issue for myeloid-derived suppressor cells. *Cancer Res.* **2007**, *67*, 425. [CrossRef]

160. Tu, S.; Bhagat, G.; Cui, G.; Takaishi, S.; Kurt-Jones, E.A.; Rickman, B.; Betz, K.S.; Penz-Oesterreicher, M.; Bjorkdahl, O.; Fox, J.G.; et al. Overexpression of interleukin-1beta induces gastric inflammation and cancer and mobilizes myeloid-derived suppressor cells in mice. *Cancer Cell* **2008**, *14*, 408–419. [CrossRef]

161. Almand, B.; Clark, J.I.; Nikitina, E.; van Beynen, J.; English, N.R.; Knight, S.C.; Carbone, D.P.; Gabrilovich, D.I. Increased production of immature myeloid cells in cancer patients: A mechanism of immunosuppression in cancer. *J. Immunol.* **2001**, *166*, 678–689. [CrossRef] [PubMed]

162. Diaz-Montero, C.M.; Salem, M.L.; Nishimura, M.I.; Garrett-Mayer, E.; Cole, D.J.; Montero, A.J. Increased circulating myeloid-derived suppressor cells correlate with clinical cancer stage, metastatic tumor burden, and doxorubicin-cyclophosphamide chemotherapy. *Cancer Immunol. Immunother.* **2009**, *58*, 49–59. [CrossRef] [PubMed]

163. Nagaraj, S.; Gupta, K.; Pisarev, V.; Kinarsky, L.; Sherman, S.; Kang, L.; Herber, D.L.; Schneck, J.; Gabrilovich, D.I. Altered recognition of antigen is a mechanism of CD8+ T cell tolerance in cancer. *Nat. Med.* **2007**, *13*, 828–835. [CrossRef] [PubMed]

164. Tcyganov, E.; Mastio, J.; Chen, E.; Gabrilovich, D.I. Plasticity of myeloid-derived suppressor cells in cancer. *Curr. Opin. Immunol.* **2018**, *51*, 76–82. [CrossRef] [PubMed]

165. Yu, H.; Liu, Y.; McFarland, B.C.; Deshane, J.S.; Hurst, D.R.; Ponnazhagan, S.; Benveniste, E.N.; Qin, H. SOCS3 Deficiency in Myeloid Cells Promotes Tumor Development: Involvement of STAT3 Activation and Myeloid-Derived Suppressor Cells. *Cancer Immunol. Res.* **2015**, *3*, 727–740. [CrossRef]

166. Xin, J.; Zhang, Z.; Su, X.; Wang, L.; Zhang, Y.; Yang, R. Epigenetic Component p66a Modulates Myeloid-Derived Suppressor Cells by Modifying STAT3. *J. Immunol.* **2017**, *198*, 2712–2720. [CrossRef]

167. Gabrilovich, D.I.; Nagaraj, S. Myeloid-derived suppressor cells as regulators of the immune system. *Nat. Rev. Immunol.* **2009**, *9*, 162–174. [CrossRef]

168. Mi, T.; Wang, Z.; Bunting, K.D. The Cooperative Relationship between STAT5 and Reactive Oxygen Species in Leukemia: Mechanism and Therapeutic Potential. *Cancers* **2018**, *10*, 359. [CrossRef]

169. Cheng, P.; Corzo, C.A.; Luetteke, N.; Yu, B.; Nagaraj, S.; Bui, M.M.; Ortiz, M.; Nacken, W.; Sorg, C.; Vogl, T.; et al. Inhibition of dendritic cell differentiation and accumulation of myeloid-derived suppressor cells in cancer is regulated by S100A9 protein. *J. Exp. Med.* **2008**, *205*, 2235–2249. [CrossRef]

170. Kyi, C.; Postow, M.A. Immune checkpoint inhibitor combinations in solid tumors: Opportunities and challenges. *Immunotherapy* **2016**, *8*, 821–837. [CrossRef]

171. Giuroiu, I.; Weber, J. Novel Checkpoints and Cosignaling Molecules in Cancer Immunotherapy. *Cancer J.* **2017**, *23*, 23–31. [CrossRef] [PubMed]

172. Austin, J.W.; Lu, P.; Majumder, P.; Ahmed, R.; Boss, J.M. STAT3, STAT4, NFATc1, and CTCF regulate PD-1 through multiple novel regulatory regions in murine T cells. *J. Immunol.* **2014**, *192*, 4876–4886. [CrossRef] [PubMed]

173. Marzec, M.; Zhang, Q.; Goradia, A.; Raghunath, P.N.; Liu, X.; Paessler, M.; Wang, H.Y.; Wysocka, M.; Cheng, M.; Ruggeri, B.A.; et al. Oncogenic kinase NPM/ALK induces through STAT3 expression of immunosuppressive protein CD274 (PD-L1, B7-H1). *Proc. Natl. Acad. Sci. USA* **2008**, *105*, 20852–20857. [CrossRef] [PubMed]

174. Fang, W.; Zhang, J.; Hong, S.; Zhan, J.; Chen, N.; Qin, T.; Tang, Y.; Zhang, Y.; Kang, S.; Zhou, T.; et al. EBV-driven LMP1 and IFN-gamma up-regulate PD-L1 in nasopharyngeal carcinoma: Implications for oncotargeted therapy. *Oncotarget* **2014**, *5*, 12189–12202. [CrossRef] [PubMed]

175. Keremu, A.; Aimaiti, A.; Liang, Z.; Zou, X. Role of the HDAC6/STAT3 pathway in regulating PD-L1 expression in osteosarcoma cell lines. *Cancer Chemother. Pharmacol.* **2019**, *83*, 255–264. [CrossRef] [PubMed]

176. Gowrishankar, K.; Gunatilake, D.; Gallagher, S.J.; Tiffen, J.; Rizos, H.; Hersey, P. Inducible but not constitutive expression of PD-L1 in human melanoma cells is dependent on activation of NF-kappaB. *PLoS ONE* **2015**, *10*, e0123410. [CrossRef]

177. Koh, J.; Jang, J.Y.; Keam, B.; Kim, S.; Kim, M.Y.; Go, H.; Kim, T.M.; Kim, D.W.; Kim, C.W.; Jeon, Y.K.; et al. EML4-ALK enhances programmed cell death-ligand 1 expression in pulmonary adenocarcinoma via hypoxia-inducible factor (HIF)-1alpha and STAT3. *OncoImmunology* **2016**, *5*, e1108514. [CrossRef] [PubMed]

178. Zhang, C.; Li, F.; Li, J.; Xu, Y.; Wang, L.; Zhang, Y. Docetaxel Down-Regulates PD-1 Expression via STAT3 in T Lymphocytes. *Clin. Lung. Cancer* **2018**, *19*, e675–e683. [CrossRef]

179. Li, P.; Huang, T.; Zou, Q.; Liu, D.; Wang, Y.; Tan, X.; Wei, Y.; Qiu, H. FGFR2 Promotes Expression of PD-L1 in Colorectal Cancer via the JAK/STAT3 Signaling Pathway. *J. Immunol.* **2019**. [CrossRef] [PubMed]

180. Zhang, N.; Zeng, Y.; Du, W.; Zhu, J.; Shen, D.; Liu, Z.; Huang, J.A. The EGFR pathway is involved in the regulation of PD-L1 expression via the IL-6/JAK/STAT3 signaling pathway in EGFR-mutated non-small cell lung cancer. *Int. J. Oncol.* **2016**, *49*, 1360–1368. [CrossRef]

181. Abdelhamed, S.; Ogura, K.; Yokoyama, S.; Saiki, I.; Hayakawa, Y. AKT-STAT3 Pathway as a Downstream Target of EGFR Signaling to Regulate PD-L1 Expression on NSCLC cells. *J. Cancer* **2016**, *7*, 1579–1586. [CrossRef] [PubMed]

182. Lamano, J.B.; Lamano, J.B.; Li, Y.D.; DiDomenico, J.D.; Choy, W.; Veliceasa, D.; Oyon, D.E.; Fakurnejad, S.; Ampie, L.; Kesavabhotla, K.; et al. Glioblastoma-Derived IL-6 Induces Immunosuppressive Peripheral Myeloid Cell PD-L1 and Promotes Tumor Growth. *Clin. Cancer Res.* **2019**. [CrossRef] [PubMed]

183. Cheng, Y.; Li, H.; Deng, Y.; Tai, Y.; Zeng, K.; Zhang, Y.; Liu, W.; Zhang, Q.; Yang, Y. Cancer-associated fibroblasts induce PDL1+ neutrophils through the IL6-STAT3 pathway that foster immune suppression in hepatocellular carcinoma. *Cell Death Dis.* **2018**, *9*, 422. [CrossRef] [PubMed]

184. Spary, L.K.; Salimu, J.; Webber, J.P.; Clayton, A.; Mason, M.D.; Tabi, Z. Tumor stroma-derived factors skew monocyte to dendritic cell differentiation toward a suppressive CD14(+) PD-L1(+) phenotype in prostate cancer. *OncoImmunology* **2014**, *3*, e955331. [CrossRef] [PubMed]

185. Kondo, K.; Shaim, H.; Thompson, P.A.; Burger, J.A.; Keating, M.; Estrov, Z.; Harris, D.; Kim, E.; Ferrajoli, A.; Daher, M.; et al. Ibrutinib modulates the immunosuppressive CLL microenvironment through STAT3-mediated suppression of regulatory B-cell function and inhibition of the PD-1/PD-L1 pathway. *Leukemia* **2018**, *32*, 960–970. [CrossRef] [PubMed]

186. Horlad, H.; Ma, C.; Yano, H.; Pan, C.; Ohnishi, K.; Fujiwara, Y.; Endo, S.; Kikukawa, Y.; Okuno, Y.; Matsuoka, M.; et al. An IL-27/Stat3 axis induces expression of programmed cell death 1 ligands (PD-L1/2) on infiltrating macrophages in lymphoma. *Cancer Sci.* **2016**, *107*, 1696–1704. [CrossRef]

187. Song, S.; Yuan, P.; Wu, H.; Chen, J.; Fu, J.; Li, P.; Lu, J.; Wei, W. Dendritic cells with an increased PD-L1 by TGF-beta induce T cell anergy for the cytotoxicity of hepatocellular carcinoma cells. *Int. Immunopharmacol.* **2014**, *20*, 117–123. [CrossRef]

188. Bazhin, A.V.; von Ahn, K.; Fritz, J.; Werner, J.; Karakhanova, S. Interferon-alpha Up-Regulates the Expression of PD-L1 Molecules on Immune Cells Through STAT3 and p38 Signaling. *Front. Immunol.* **2018**, *9*, 2129. [CrossRef]

189. Garcia-Diaz, A.; Shin, D.S.; Moreno, B.H.; Saco, J.; Escuin-Ordinas, H.; Rodriguez, G.A.; Zaretsky, J.M.; Sun, L.; Hugo, W.; Wang, X.; et al. Interferon Receptor Signaling Pathways Regulating PD-L1 and PD-L2 Expression. *Cell Rep.* **2017**, *19*, 1189–1201. [CrossRef]

190. Che, K.F.; Shankar, E.M.; Muthu, S.; Zandi, S.; Sigvardsson, M.; Hinkula, J.; Messmer, D.; Larsson, M. p38 Mitogen-activated protein kinase/signal transducer and activator of transcription-3 pathway signaling regulates expression of inhibitory molecules in T cells activated by HIV-1-exposed dendritic cells. *Mol. Med.* **2012**, *18*, 1169–1182. [CrossRef]

191. Zhang, W.; Zhang, J.; Zhang, Z.; Guo, Y.; Wu, Y.; Wang, R.; Wang, L.; Mao, S.; Yao, X. Overexpression of Indoleamine 2,3-Dioxygenase 1 Promotes Epithelial-Mesenchymal Transition by Activation of the IL-6/STAT3/PD-L1 Pathway in Bladder Cancer. *Transl. Oncol.* **2019**, *12*, 485–492. [CrossRef] [PubMed]

192. Huang, X.; Bai, X.; Cao, Y.; Wu, J.; Huang, M.; Tang, D.; Tao, S.; Zhu, T.; Liu, Y.; Yang, Y.; et al. Lymphoma endothelium preferentially expresses Tim-3 and facilitates the progression of lymphoma by mediating immune evasion. *J. Exp. Med.* **2010**, *207*, 505–520. [CrossRef] [PubMed]
193. Arra, A.; Lingel, H.; Kuropka, B.; Pick, J.; Schnoeder, T.; Fischer, T.; Freund, C.; Pierau, M.; Brunner-Weinzierl, M.C. The differentiation and plasticity of Tc17 cells are regulated by CTLA-4-mediated effects on STATs. *OncoImmunology* **2017**, *6*, e1273300. [CrossRef] [PubMed]
194. Mankan, A.K.; Greten, F.R. Inhibiting signal transducer and activator of transcription 3: Rationality and rationale design of inhibitors. *Expert Opin. Investig. Drugs* **2011**, *20*, 1263–1275. [CrossRef] [PubMed]
195. Shouksmith, A.E.; Gunning, P.T. Targeting Signal Transducer and Activator of Transcription (STAT) 3 with Small Molecules. In *Small-Molecule Transcription Factor Inhibitors in Oncology*; Khondaker, M.R.D.E.T., Ed.; Royal Society of Chemistry: London, UK, 2019; pp. 147–168.

Review

Structural Implications of STAT3 and STAT5 SH2 Domain Mutations

Elvin D. de Araujo [1,2], Anna Orlova [3], Heidi A. Neubauer [3], Dávid Bajusz [4], Hyuk-Soo Seo [5,6], Sirano Dhe-Paganon [5,6], György M. Keserű [4], Richard Moriggl [3] and Patrick T. Gunning [1,2,*]

[1] Centre for Medicinal Chemistry, University of Toronto at Mississauga, Mississauga, ON L5L 1C6, Canada; e.dearaujo@mail.utoronto.ca

[2] Department of Chemical & Physical Sciences, University of Toronto at Mississauga, Mississauga, ON L5L 1C6, Canada

[3] Institute of Animal Breeding and Genetics, University of Veterinary Medicine, A-1210 Vienna, Austria; Anna.Orlova@vetmeduni.ac.at (A.O.); Heidi.Neubauer@vetmeduni.ac.at (H.A.N.); Richard.Moriggl@vetmeduni.ac.at (R.M.)

[4] Medicinal Chemistry Research Group, Research Center for Natural Sciences, 1117 Budapest, Hungary; bajusz.david@ttk.mta.hu (D.B.); keseru.gyorgy@ttk.mta.hu (G.M.K.)

[5] Department of Cancer Biology, Dana-Farber Cancer Institute, Department of Biological Chemistry & Molecular Pharmacology, Harvard Medical School, Boston, MA 02115, USA; hux@crystal.harvard.edu (H.-S.S.); dhepag@crystal.harvard.edu (S.D.-P.)

[6] Department of Biological Chemistry, Department of Biological Chemistry & Molecular Pharmacology, Harvard Medical School, Boston, MA 02115, USA

[*] Correspondence: patrick.gunning@utoronto.ca; Tel.: +1-905-569-4588

Received: 30 September 2019; Accepted: 5 November 2019; Published: 8 November 2019

Abstract: Src Homology 2 (SH2) domains arose within metazoan signaling pathways and are involved in protein regulation of multiple pleiotropic cascades. In signal transducer and activator of transcription (STAT) proteins, SH2 domain interactions are critical for molecular activation and nuclear accumulation of phosphorylated STAT dimers to drive transcription. Sequencing analysis of patient samples has revealed the SH2 domain as a hotspot in the mutational landscape of STAT proteins although the functional impact for the vast majority of these mutations remains poorly characterized. Despite several well resolved structures for SH2 domain-containing proteins, structural data regarding the distinctive STAT-type SH2 domain is limited. Here, we review the unique features of STAT-type SH2 domains in the context of all currently reported STAT3 and STAT5 SH2 domain clinical mutations. The genetic volatility of specific regions in the SH2 domain can result in either activating or deactivating mutations at the same site in the domain, underscoring the delicate evolutionary balance of wild type STAT structural motifs in maintaining precise levels of cellular activity. Understanding the molecular and biophysical impact of these disease-associated mutations can uncover convergent mechanisms of action for mutations localized within the STAT SH2 domain to facilitate the development of targeted therapeutic interventions.

Keywords: STAT3; STAT5; SH2 domain; mutations; cancer; autosomal-dominant hyper IgE syndrome; inflammatory hepatocellular adenomas; T-cell large granular lymphocytic leukemia; T-cell prolymphocytic leukemia; growth hormone insensitivity syndrome

---

1. Introduction

Several key cellular pathways converge on the multidomain signal transducer and activator of transcription (STAT) proteins highlighting their importance in the development and progression of oncogenic and malignant diseases. Conventional STAT activation is initiated by cytokine or growth-factor interactions with extracellular receptors, stimulating SH2 domain-mediated recruitment of tyrosine kinases and STAT isoforms to the receptor cytoplasmic domains [1,2]. Nuclear translocation and accumulation of the resulting phosphorylated STAT dimers facilitates transcription of a wide array of gene products involved in proliferation and cellular survival including C-MYC [3], BCL-XL [3], MCL-1 [4], FOXP3 [5], BCL-2 [6], HIF [7], D-type cyclins [8], IGF-1 [9], and self-regulation of STAT3/STAT5 [10]. Normal STAT function is dependent on the SH2 domain which arbitrates homo- or hetero- STAT dimerization as well as multiple protein–protein interactions. As such, structurally altered SH2 domains exhibit considerable effects on STAT activity, leading to either hyperactivated or refractory STAT mutants. These critical roles in governing the transcriptional capacity, coupled with the relatively shallow binding surfaces elsewhere on the protein, resulted in the STAT SH2 domain dominating therapeutic interest for small molecule inhibitor development and intervention [11–15]. However, currently there are no clinical drug candidates directly targeting the STAT protein family. This is partially due to the limited structural data available on the STAT SH2 domains or their mutated disease-associated counterparts, and further compounded by observations that STAT SH2 domains are distinct from those found in other well characterized systems such as Src kinase. Here, we summarize structural features of STAT-type SH2 domains in the context of STAT3/STAT5 disease-associated mutations, and discuss their effects on protein activity, as well as potential new druggable pockets within the STAT SH2 domain.

2. Structure of STAT SH2 Domains

SH2 domains are modular units that arose within multicellular life, approximately 600 million years ago, and are therefore heavily tied to metazoan signal transduction [16]. There are 121 human SH2 domains that are classified into different groups based on their structural or phylogenetic characteristics [16]. Broadly, they have been classified into either STAT- or Src-type SH2 domains based on the presence of either an α-helix (STAT-type) or β-sheet (Src-type) at the C-terminus [17]. Alternatively, phylogenetic analysis has categorized SH2 domain-containing proteins into 38 different sub-families [16]. Functional activity-based screens have also been employed to stratify SH2 domain-containing proteins into four categories based on the identity of the fifth residue in the βD strand, which has been shown to be a critical determinant in phospho-peptide selectivity [18,19]. Despite different methods for classification, all SH2 domains contain conserved structural motifs that are canonical to the core function of phospho-Tyr (pY) peptide binding. These features represent an evolutionary compromise to preserving critical structural motifs while maintaining highly specific peptide recognition capacity.

The structure of an SH2 domain consists of a central anti-parallel β-sheet (with the three β-strands conventionally labeled βB-βD) interposed between two α-helices (αA and αB), often referred to as the αββββα motif [16]. The structure and nomenclature for the motifs of STAT SH2 domains is shown in Figure 1a,b. The β-sheet partitions the SH2 domain into two subpockets, referred to as the pY (phosphate-binding) and pY+3 (specificity) pocket [16]. The pY pocket is formed by the αA helix, the BC loop (region connecting βB-βC strands) and one face of the central β-sheet. The pY+3 pocket is created by the opposite face of the β-sheet as well as residues from the αB helix and CD and BC* loops (regions connecting βC-βD strands and αB-αC helices, respectively). Both the pY and pY+3 pockets are common targets for drug design due to well defined features and conserved residues. Within the pY+3 pocket, there are additional clefts that also have drug targeting potential. This includes the C-terminal region of the pY+3 pocket, also known as the evolutionary active region (EAR) [17]. The EAR contains an additional α-helix (αB') in STAT-type SH2 domains, as opposed to the Src-type which harbors a β-sheet (βE and βF although each strand is not always observed). Additionally, there is a cluster of non-polar residues (referred to as the hydrophobic system [20]) at the base of the pY+3 pocket that assists in stabilizing the conformation of the β-sheet and maintaining the integrity of the overall SH2 domain. The αB, αB', and BC* loop also participate in SH2-mediated STAT dimerization forming important cross-domain interactions. Therefore, residues in the pY+3 pocket can have a dual effect on STAT dimerization capacity and phospho-peptide binding. Conventional phospho-peptide binding occurs perpendicular to the β-sheet with the peptide adopting a binding mode as illustrated with STAT1 in Figure 1c. The phospho-Tyr interacts with conserved amino acids in the pY pocket, while the C-terminal residues stretch across the SH2 domain into the pY+3 pocket.

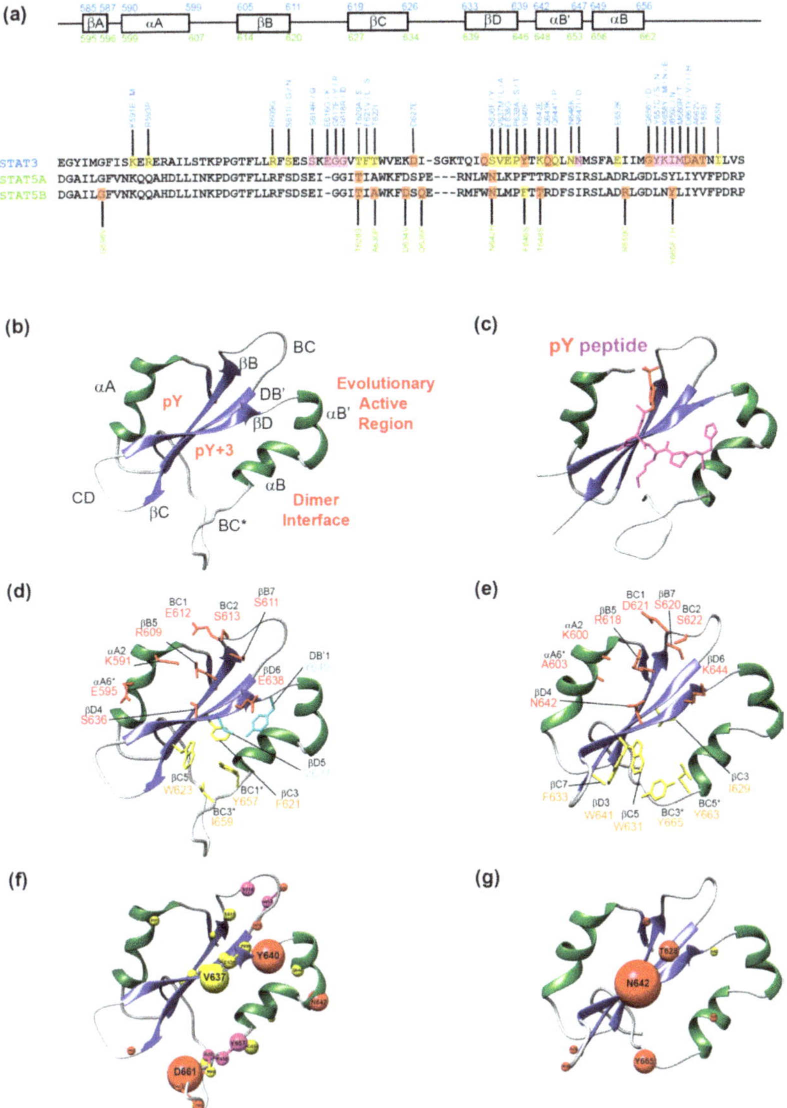

Figure 1. (**a**) Secondary structural motifs in STAT3 (blue) and STAT5 (green) with mutations annotated; (**b**) Structure of STAT3 SH2 domain; (**c**) Structure of pY-peptide-STAT1 SH2 domain. The pY residue is depicted in red with the C-terminal residues in violet; (**d**) Structure of STAT3 SH2 domain with Sheinerman residues (red), hydrophobic system residues (yellow) and the selectivity filter (cyan);

(**e**) Structure of STAT5B SH2 domain with the same color scheme as above; (**f**) STAT3 SH2 domain with all mutations highlighted in spheres. The volume of each sphere is proportional to frequency of cases identified. Red spheres indicate an activating mutation, yellow spheres indicate a destabilizing mutation and magenta spheres represent sites where both activating and refractory mutations are observed; (**g**) STAT5B SH2 domain with all mutations highlighted in spheres. The color scheme is the same as in (**f**). Protein structures were visualized using Chimera [21] with PDB codes: 4E68 [22] (STAT3), 6MBW (STAT5B) [23], 1YVL (STAT1) [24].

These residues are critical for maintaining proper binding interactions to facilitate protein dimerization, and specific mutations here can alter normal STAT function. An issue of particular relevance for drug discovery is protein flexibility, and indeed, STAT SH2 domains exhibit a particularly flexible behavior even in sub-microsecond timescales [23,25]. Notably, the accessible volume of the pY pocket varies dramatically. Also, crystal structures do not necessarily preserve even the main, targetable pockets in an accessible state. This further underlines the importance of accounting for protein dynamics in STAT-directed drug discovery efforts.

3. Disease-Associated Mutations in STAT3 and STAT5B SH2 Domains

Sequencing analysis of patient samples has identified multiple point mutations within the SH2 domain of STAT3 (Table 1) and STAT5B (Table 2) leading to variable effects on physiological activity. In mice, homozygous disruption of STAT3 is embryonically lethal [26], and correspondingly germline homozygous loss-of-function (LOF) mutations have not been identified in humans. Heterozygous loss of STAT3 can be tolerated to different extents and contributes to immunological deficiencies, most commonly autosomal-dominant Hyper IgE syndrome (AD-HIES) as a result of a reduced STAT3-mediated Th17 T-cell response [27–29]. Classical STAT3 function is implicated in Th17 T-cell lineage commitment, through upregulation of RORγt, promoting the release of IL-17 and IL-22. This stimulates transcription of genes associated with Th17 development. Loss of STAT3 function strongly diminishes Th17 T-cell expansion, thereby reducing the immunologic response leading to recurrent staphylococcal infections and exceedingly high levels of IgE that contribute to clinical presentations of eczema and eosinophilia.

Table 1. Disease-associated mutations in the STAT3 SH2 domain.

Mutation	Position	Location	Residue Relevance	Nucleotide Substitution	Cases	Pathology	Type	Ref
K591E	αA2	pY	Sheinerman	1771A>G	1	AD-HIES	Germline	[30]
K591M	αA2	pY	Sheinerman	1772A>T	1	AD-HIES	Germline	[31]
R593P	αA4	pY	-	1778G>C	1	AD-HIES	Germline	[32]
R609G	βB5	pY	Sheinerman& Signature	1825A>G	1	AD-HIES	Germline	[33]
S611G	βB7	pY	Sheinerman	1831A>G	2	AD-HIES	Germline	[34,35]
S611N	βB7	pY	& Signature Sheinerman	1832G>A	2	AD-HIES	Germline	[36,37]
S611I	βB7	pY	Sheinerman & Signature	1832G>T	1	AD-HIES	Germline	[38]
S614G	BC3	pY	Sheinerman	18040A>G	1	AD-HIES	Germline	[34]
S614R	BC3	pY	Sheinerman	1842C>G	1	T-LGLL	Somatic	[39–43]
					2	NK-LGLL		
					1	ALK⁻ALCL		
					1	HSTL		

Table 1. *Cont.*

Mutation	Position	Location	Residue Relevance	Nucleotide Substitution	Cases	Pathology	Type	Ref
E616G	BC5	pY	BC loop	1847A>G	1	DLBCL, NOS	Somatic	[44]
E616K	BC5	pY	BC loop	1846G>A	1	NKTL	Somatic	[45]
G617E	BC6	pY	BC loop	1850G>A	1	AD-HIES	Germline	[46]
G617V	BC6	pY	BC loop	1850G>T	1	AD-HIES	Germline	[34]
G617R	BC6	pY	BC loop	1849G>A	1	DLBCL, NOS	Somatic	[44]
G618R	BC6	pY	BC loop	1852G>C	1	ALK⁻ALCL	Somatic	[47,48]
					1	T-LGLL		
G618D	BC6	pY	BC loop	1853G>A	2	AD-HIES	Germline	[35,37]
T620A	βC2	pY	-	1858A>G	2	AD-HIES	Germline	[34,49]
T620S	βC2	pY	-	1859C>G	1	AD-HIES	Germline	[33]
F621V	βC3	pY+3	Hydro. Sys.	1861T>G	3	AD-HIES	Germline	[36,37,50]
F621L	βC3	pY+3	Hydro. Sys.	1863C>G	2	AD-HIES	Germline	[51,52]
F621S	βC3	pY+3	Hydro. Sys.	1862T>C	1	AD-HIES	Germline	[53]
T622I	βC4	pY	-	1865C>T	4	AD-HIES	Germline	[30,34,36]
D627E	CD1		-	1881C>A	1	ATLL	Somatic	[54]
S636F	βD4	pY	Sheinerman	1907C>T,	1	AD-HIES	Germline	[49]
S636Y	βD4	pY	Sheinerman	1907C>A	1	AD-HIES	Germline	[30]
V637M	βD5	pY+3	Sel. Filter	1909G>A	43	AD-HIES	Germline	[30,32–37,49,52,55,56]
V637L	βD5	pY+3	Sel. Filter	1909G>T	1	AD-HIES	Germline	[36]
V637A	βD5	pY+3	Sel. Filter	1910T>C	1	AD-HIES	Germline	[30]
E638G	βD6	pY	Sheinerman	1913A>G	4	AD-HIES	Germline	[49,51,57,58]
P639A	βD7	pY+3	-	1915C>G	1	AD-HIES	Germline	[37]
P639S	βD7	pY+3	-	1915C>T*	2	AD-HIES	Germline	[30,34]
P639T	βD7	pY+3	-	1915C>A	1	AD-HIES	Germline	[35]
Y640F	DB′1	pY+3	-	1919A>T	56	T-LGLL	Somatic	[43,44,47,48,57,59–69]
					2	IHT		
					3	CLPD-NKs		
					2	NK-LGLL		
					1	DLBCL, NOS		
					1	ANKL		
					1	NKTL		
					1	Sezary		
					3	HSTL		
K642E	αB′1	-	Dimer Inter	1924A>G	1	AD-HIES	Germline	[34]
Q643K	αB′2	-	Dimer Inter	1927C>A	1	T-LGLL	Somatic	[68]
Q644P	αB′3	-	Dimer Inter	1929A>C	3	AD-HIES	Germline	[52,70]
Q644del	αB′3	-	Dimer Inter	1930del CAG	2	AD-HIES	Germline	[35,36]
N646K	αB′5	pY+3	Dimer Inter	1938C>G	2	EOAD	Germline	[71]
N647D	αB′6	pY+3	Dimer Inter	1939A>G	8	AD-HIES	Germline	[36,72]
N647I	αB′6	pY+3	Dimer Inter	1940A>T	3	CLPD-NK	Somatic	[43,57,65,67]
					6	T-LGLL		
					1	HSTL		
E652K	αB3	pY+3	Dimer Inter	1954G>A	1	AD-HIES	Germline	[36]
G656D	αB7	pY+3	Dimer Inter	1967G>A	1	T-LGLL	Somatic	[73]
G656_Y657insF	αB7	pY+3	Hydro. Sys.	1968C>T; 1969_1970insTTT	1	IHCA	Somatic	[66]
Y657C	BC1*	pY+3	Hydro. Sys.	1970A>G	5	AD-HIES	Germline	[30,34,36,58]
Y657S	BC1*	pY+3	Hydro. Sys.	1970A>C	1	AD-HIES	Germline	[74]
Y657N	BC1*	pY+3	Hydro. Sys.	1969T>A	1	AD-HIES	Germline	[52]
Y657ins	BC1*	pY+3	Hydro. Sys.	-	1	TCL	Somatic	[47]
Y657dup	BC1*	pY+3	Hydro. Sys.	-	3	T-LGLL	Somatic	[48,65,72]

Table 1. *Cont.*

Mutation	Position	Location	Residue Relevance	Nucleotide Substitution	Cases	Pathology	Type	Ref
Y657_M660dup	BC1*	pY+3	Hydro. Sys.	1969T_1980G dup	1	IHCA	Somatic	[66]
K658M	BC2*	pY+3	Dimer Inter	1973A>T	2	T-LGLL	Somatic	[67,72]
K658N	BC2*	pY+3	Dimer Inter	1974G>T	1	T-LGLL EOAD	Somatic/Germline	[65,71]
K658Y	BC2*	pY+3	Dimer Inter	1972A>T; 1974G>T	1	IHCA	Somatic	[66]
K658E	BC2*	pY+3	Dimer Inter	1972A>G	1	AD-HIES	Germline	[75]
I659N	BC3*	pY+3	Hydro. Sys.	1976A>T	1	AD-HIES	Germline	[58]
I659L	BC3*	pY+3	Hydro. Sys.	1975A>C	2	T-LGLL	Somatic	[57,72]
M660R	BC4*	pY+3	Dimer Inter	1979T>G	1	AD-HIES	Germline	[55]
M660T	BC4*	pY+3	Dimer Inter	1978T>A	1	AD-HIES	Germline	[76]
D661I	BC5*	-	Dimer Inter	1981G>A; 1982A>T	1	CLPD-NK	Somatic	[67]
D661Y	BC5*	-	Dimer Inter	1981G>T	1	NK-LGL	Somatic	[41,47,48,60,63,65,67,68,72,77]
					56	T-LGL		
					1	HSTL		
					10	NKTL		
D661ins	BC5*	-	Dimer Inter	-	1	T-LGLL	Somatic	[47]
D661V	BC5*	-	Dimer Inter	1981A>T	10	T-LGLL	Somatic	[65,67]
D661H	BC5*	-	Dimer Inter	1981G>C	1	T-LGLL	Somatic	[65]
A662V	BC6*	-	Dimer Inter	1985C>T	1	ALK⁻ALCL	Somatic	[78]
T663I	BC7*	-	Dimer Inter	1988, 1989>TT	2	DLBCL/B1	Germline	[44,75]
I665N	BC9*	-	Dimer Inter	1998T>A	2	AD-HIES	Germline	[34]
V667L	BC11*	-	Dimer Inter	1999C>G	1	NKTL	Somatic	[45]
S668F	BC12*	-	Dimer Inter	2003C>T	3	AD-HIES	Germline	[30,34,37]
S668Y	BC12*	-	Dimer Inter	2003C>A	1	AD-HIES	Germline	[34]

The final search date for mutations from medical case reports and literature was 30 August, 2019. Abbreviations: AD-HIES, autosomal-dominant Hyper IgE syndrome; ALK⁻ALCL, anaplastic lymphoma kinase negative anaplastic large cell lymphoma; ANKL, aggressive natural killer cell leukemia; ATLL, adult T-cell leukemia lymphoma; EOAD, early onset autoimmune disease; CLPD-NKs, Chronic lymphoproliferative disorders of natural killer cells; DLBCL, NOS, Diffuse large B-cell lymphoma, not-otherwise-specified; NKTL, extranodal NK/T-cell lymphoma; HSTL, Hepatosplenic T-cell lymphoma; IHAC, inflammatory hepatocellular adenomas; IHT, inflammatory hepatocellular tumors; NK-LGLL, Natural killer cell large granular lymphocytic leukemia; T-LGLL, T-cell large granular lymphocytic leukemia; TCL, γδ-T-cell lymphoma.

Table 2. Disease-associated mutations in the STAT5B SH2 domain.

Mutation	Position	Location	Residue Relevance	Nucleotide Substitution	Cases	Pathology	Type	Ref.
G596V	βA2	-	-	1787G>T	1	APL	Somatic	[79]
T628S	βC2	pY	-	1883C>G	11	T-PLL	Somatic	[43,80,81]
					1	MEITL		
					3	HSTL		
					1	Eosinophilia		
A630P	βC4	pY	-	1888G>C	1	GHI	Germline *	[82]
D634V	βC8	pY	-	1901A>T	1	T-PLL	Somatic	[83]
Q636P	CD2	pY+3	-	1907A>C	1	MEITL	Somatic	[81]
N642H	βD4	pY	Sheinerman	1924A>C	39	MEITL	Somatic	[41,62,69,77,80,81,83–102]
					33	T-PLL		
					29	Eosinophilia		
					28	T-ALL		
					7	HSTL		
					11	LGLL		
					3	PCTL		
					3	Sézary		
					1	AAA		
					1	CNL		
					1	PTCL, NOS		
					1	AML		
F646S	DB'1	pY+3	-	1937T>C	1	GHI	Germline *	[103]
T648S	DB'3	pY+3	-	1942A>T	1	T-ALL	Somatic	[101]
R659C	αB'4	Dimer Inter	-	1975C>T	1	T-PLL	Somatic	[80]
Y665F	BC3*	Dimer Inter	Hydro. Sys.	1994A>T	6	T-PLL	Somatic	[43,80,101]
					3	HSTL		
					2	T-ALL		
					2	NKTL		
					5	LGLL		
Y665H	BC3*	Dimer Inter	Hydro. Sys.	1993T>C	2	T-PLL	Somatic	[80]

* Patients were homozygous for the point mutation. The final search date for mutations from medical case reports and literature was 30 August, 2019. Abbreviations: AAA, acquired aplastic anemia; AML, acute myeloid leukemia; APL, acute promyelocytic leukemia; CNL, chronic neutrophilic leukemia; GHI, growth hormone insensitivity; HSTL, hepatosplenic T-cell lymphoma; MEITL, monomorphic epitheliotropic intestinal T cell lymphoma; PCTL, primary cutaneous γδ T-cell lymphoma; PTCL-NOS, peripheral T-cell lymphoma not-other-specified; T-ALL, T-cell acute lymphoblastic leukemia; T-LGLL, T-cell large granular lymphocytic leukemia; T-PLL, T-cell prolymphocytic leukemia.

Comparatively, homozygous loss of both STAT5 gene products in mice is lethal late in embryonic development, due to the defective erythropoiesis and with loss of STAT5B manifesting with a myriad of physiological effects including sexual dimorphic body growth [104–106]. Clinical cases of patients with STAT5B LOF mutations on both alleles exhibit features similar to growth hormone insensitivity syndrome (GHIS). However, heterozygous human carriers of STAT5B LOF mutations generally do not present with any immunological deficiencies or growth complications, although there have been three germline dominant-negative heterozygous STAT5B mutations recently reported that result in postnatal growth impairment among other physiological symptoms [107]. Although these growth deficiencies are likely multifactorial in etiology, the growth hormone (GH)-growth hormone receptor (GHR) interactions that recruit JAK kinase and stimulate phosphorylation of STAT5B remain intact, suggesting a breakdown in corresponding STAT5B activity. STAT5B LOF patients often carry additional immunological burdens including reduced populations of several T-cell subtypes, suggesting multiple roles for STAT5 in T-cell differentiation [108].

Gain-of-function (GOF) germline mutations for STAT3 and STAT5 are rare and clinically diverse. STAT3 GOF mutations present with autoimmune responses, likely due to Th17 clonal expansion which also suppresses regulatory T-cell (Treg) formation. Clinical presentations of STAT3 GOF mutations also show parallels with STAT5 LOF mutations [28]. This is partially an effect of compensatory upregulation of SOCS3 (suppressor of cytokine signaling-3) which strongly inhibits hyperactivated STAT3, but by extension also dampens STAT5 and STAT1 activity. This potently reduces basal levels of STAT5 leading to growth immunodeficiencies.

Contrastingly, multiple de novo somatic GOF mutations arise in STAT3 and STAT5B leading to cancer pathogenesis, and such mutations have been implicated in both solid and liquid tumors. Hyperactivated STAT3 is identified in patients with diverse phenotypes and multiple somatic mutations have been associated with T-LGLL (T-cell large granular lymphocytic leukemia, 40–70% of all cases) and NK-LGLL (natural killer cell large granular lymphocytic leukemia, 30% of all cases), as well as different hepatocellular adenomas [40]. STAT5 is upregulated, directly and indirectly, through multiple mechanisms in several hematological malignancies leading to neoplastic transformation. For instance, in ~30% of acute myeloid leukemia (AML) cases, STAT5B is activated by mutated FLT3 (Fms-Like Tyrosine Kinase 3) [109]. Similarly, ~99% of all chronic myeloid leukemia (CML) cases, which result from the appearance of the Philadelphia chromosome, result in STAT5B hyperactivation [110]. B-cell cancers, such as B-ALL (B-cell acute lymphoid leukemia) and B-CLL (B-cell chronic lymphoid leukemia) are driven by upregulation of IL-7 and IL-22 which also simulate STAT5B [111–113]. Acute and chronic T-cell cancers have lower incidence, but a larger diversity, and are heavily implicated by STAT5B GOF driver mutations [84,114,115]. STAT GOF mutations generally stabilize protein structure thereby enhancing transcriptional output, leading to apoptosis-evading and survival phenotypes. LOF mutants generally distort secondary structure and contribute to either loss of activity or increased STAT degradation. Herein, we highlight the structural significance and mechanism of action of malignant STAT3/5B point mutations in the SH2 domain. Notably, STAT5A mutations are less frequently identified which is likely due to the differential roles of the protein isoforms. Although STAT5B has been characterized as a driver in several malignancies, STAT5A has been associated with tumor suppression [116,117]. STAT5A and STAT5B have similar SH2 domains (~93% amino acid similarity) with significant changes in the βD strand which likely contributes to varying peptide selectivity.

3.1. Mutations in the pY Pocket

As previously described, the pY pocket is formed by the αA helix, BC loop, and one face of the central β-sheet, and it harbors an overall positive electrostatic potential to stabilize binding with the electronegative phospho-Tyr side-chain (Figure 1). This region of the SH2 domain is characterized by strongly conserved residues that facilitate interactions with the phosphorylated peptide. It includes the SH2 domain signature sequence as well as a group of 8 phospho-Tyr interacting amino acids that have been collectively referred to as Sheinerman residues [20]. The SH2 domain signature sequence, FLXRXS (where X is a hydrophobic amino acid), corresponds to FLLRFS in all STAT proteins and is located on the βB strand. The eight Sheinerman residues correspond to the positions: αA2, αA6, βB5, βB7, BC1, BC2, βD4 and βD6 [20]. Critically, the βB5 residue is located within the SH2 domain signature sequence as an invariant Arg residue, which is conserved in 118 of 121 SH2 domain-containing proteins [16]. This indispensable Arg residue is the principal binding partner for phospho-Tyr with the side-chain guanidinium group participating in a bidentate ionic interaction with the phosphate. The side-chains of the αA2 (Arg/Lys in 118/121 SH2 domains), βB7 (Ser in 106/121 SH2 domains) and βD4 (His in 80/121 SH2 domains) residues also participate in direct interactions with the phospho-Tyr [16]. Notably, the αA2, βB7, and βD4 residues correspond to Lys591, Glu638, and Ser636 in STAT3 and Lys600, Ser620, and Asn642 in STAT5B. These interactions are highly important for phospho-peptide binding and are reported to contribute to >50% of Gibbs free energy of the protein–peptide interaction [16]. The high fidelity of these interactions ostensibly suggests that a mutation at these sites will have a dramatic effect on the activity or binding capacity of STAT3/5.

3.1.1. Mutations in Sheinerman Residues

Mutations identified at the αA2 site in STAT3 (Lys591Glu [30] and Lys591Met [31]) contribute to AD-HIES, a STAT3-deficient malignancy, due to removal of the positively-polarized Lys side-chain that directly coordinates the phospho-Tyr. Similarly, mutations at other critical residues in the pY pocket present with analogous clinical outcomes. Mutation of the invariant βB5 position (Arg609Gly [33]) in STAT3 leads to AD-HIES, presenting with reduced expression profiles of Th17 T-cells and high serum levels of IgE (11,300 IU/mL). Replacement of βD4 (Ser636Phe [49]) leads to strongly elevated IgE levels (17,407 IU/mL) with presentation of eczema, abscesses, and pneumonia, also characteristic of AD-HIES pathologies. Interestingly, a patient with a semi-conservative substitution (Ser636Tyr [30]) that retains H-bonding capacity at the βD4 position still presented with a reduced percentage of Th17 cells (0.31%) compared to healthy patients (>1%), highlighting the substantive role of changes in sterics at the phosphate binding positions. Examining the prevalence of mutations within the remaining Sheinerman residues shows infrequent mutations, such as STAT3 βB7 (in which Ser611Gly [34]/Asn [36] have been reported), which also lead to ablation of activity and AD-HIES. In cellulo studies with Ser611Asn indicate a reduced capacity for activation by phosphorylation consistent with altered phospho-Tyr binding [118]. Similarly, substitution at the Sheinerman βD6 position is also capable of triggering the AD-HIES phenotype. Notably, this βD6 (Glu638Gly [49,51]) mutation likely causes gross conformational changes in the β-sheet due to removal of a complete side-chain. Although mutations in the pY pocket of STAT3 tend to abolish peptide binding, specific mutations in the BC loop lead to either hyperactivation and LGL leukemias or protein dysfunction and AD-HIES. Since the BC loop is directly involved in multiple domain interactions including the pY and pY+3 pockets, mutations identified in this region will be discussed in detail in Section 3.3.

In STAT5, mutations at the conserved Sheinerman residues have not been identified apart from the βD4 position. In the majority of SH2 domain-containing proteins, the βD4 residue is a His, directly coordinating the phosphate, and mutagenesis of this residue abolishes peptide binding capacity [16]. As seen with STAT3 (Ser636Tyr), modification in sterics at this position greatly modulates phospho-Tyr peptide binding. In STAT5B, the βD4 residue is an Asn642, and the absence of a conserved His residue may represent an evolutionary response to tune down the basal activity of STAT5. Notably, this residue is most frequently mutated in STAT5B (Asn642His) and has been reported in multiple cancer phenotypes (>150 cases [41,62,69,77,80,81,83–102]), most commonly T-cell-prolymphocytic leukemia (T-PLL), monomorphic epitheliotropic intestinal T-cell lymphoma (MEITL), and T-cell acute lymphoblastic leukemia (T-ALL). Asn642His is an extremely aggressive oncodriver of T-cell neoplasia and previous studies have shown multiple T-cell subset organ infiltration and transformation in transgenic mice [23,119]. Recently, the crystal structure for the STAT5B Asn642His mutation was reported which suggested different SH2 domain conformations with either a neatly packed βD strand forming a tight central β-sheet or a more dissociated βD strand that provides greater access to the SH2 domain [23]. Additionally, different biophysical studies [23,41] have confirmed the substantial increase (~5–7 fold) in the affinity of pY containing peptides for mutated STAT5B (Asn642His) compared to wild type. This provides a molecular basis for the lower threshold of mutant STAT5B towards cytokine activation and the aggressive phenotype observed in patients. The Asn642His mutation is also predicted to lead to a more stable dimer interface and reduced dephosphorylation kinetics, prolonging the lifetime of the activation state [23].

3.1.2. Mutations Outside Sheinerman Residues

The mutational landscape of STAT3 and STAT5B within the pY pocket also extends to less conserved residues, predominantly on the βC strand. Mutations at βC4 have been identified in both STAT3 (Thr620Ser [33]/Ala [34,49]) and STAT5B (Thr628Ser). In STAT5B, this conservative mutation is commonly observed in T-PLL and T-ALL [86]. In vitro studies with STAT5B Asn642 variants have shown that bulkier substituents in the pY pocket reduce phospho-Tyr affinity [23]. Therefore, the loss of a single methylene group from Thr628 can better accommodate the cognate peptide yielding increased transcriptional activity. However, contrary effects are observed with Thr620Ser in STAT3, where this mutation, as well as Thr620Ala, both lead to reduced protein activity and AD-HIES. The contrasting effects of the same mutation at identical positions in STAT3 and STAT5B underscores the unique aspects of each structural motif, and more broadly, the potential for isoform specific drug targeting. Generally, STAT3 is less robust to molecular modifications with slight changes capable of strongly diminishing activity. This is unusual considering the melting temperature of isolated STAT3 (52.5 ± 0.7 °C) is higher than STAT5B (44.5 ± 0.3 °C) [120]. The total protein stability is likely the result of both structural and complex protein dynamics and requires further investigation. One destabilizing mutation has been observed in the pY pocket of STAT5B at the βC4 position (Ala630Pro [121]) which disrupts the β-sheet, reducing protein solubility and leading to misfolding and a clinical presentation of severe growth deficiency. Thus, destabilizing mutations are observed in STAT5B, but with a reduced frequency.

3.2. Mutations in the pY+3 Pocket

Adjacent to the pY pocket of the SH2 domain is the pY+3, or specificity pocket which interacts with C-terminal residues of the phospho-Tyr peptide. The βD strand is critical in facilitating these interactions, particularly the βD5 residue which controls accessibility to this pocket. In Src-type SH2 domains, additional interactions with the cognate phospho-peptide occur between residues in the evolutionary active region (EAR). This includes the βE and βF strands as well as the loops in between these structural elements. In STAT3 and STAT5, the EAR motif is comprised of an α-helix (αB'), and the corresponding interactions occur within the DB' region (loop in between βD strand and αB' helix), αB', and αB helices. There is also a clustering of predominantly aromatic, hydrophobic residues at the base of the pY+3 pocket which has been referred to as the hydrophobic system (βC3, βC5, βD3, BC1*, and BC3*). In STAT3, this includes residues Phe621 (βC3), Trp623 (βC5), Tyr657 (BC1*), and Ile659 (BC3*) and in STAT5B, Ile629 (βC3), Trp631 (βC5), Phe633 (βC7), Trp641 (βD3), Leu663 (BC1*) and Tyr665 (BC3*). In STAT3, the hydrophobicity of this pocket is reduced by the presence of Gln635 and Lys626 at the βD3 and βC7 positions, respectively. The increased polarity leads to a reduced STAT3 preference for phospho-peptides with non-polar residues in the C-terminal positions compared to STAT5B. The SH2 dimerization interface is in close proximity to the pY+3 pocket and is formed by the BC* loop, αB' helix and one face of the αB helix. As such, alterations in the pY+3 pocket directly affect the dimerization interface.

3.2.1. Mutations in the Hydrophobic System and βD Strand

The pY+3 pocket is a hotspot for STAT SH2 domain mutations. Generally, mutations that increase the polarity of this pocket result in protein destabilization and LOF. This was demonstrated in Epstein–Barr virus (EBV)-transformed B-cells expressing AD-HIES-associated STAT3 mutations. In these assays, the half-life of wild type STAT3 (25 ± 2 h) was shown to be substantially reduced by polar mutations at βD5 (Val637Met = 5.3 ± 4.5 h) and BC1* (Tyr657Cys = 5.5 ± 4.1 h) [122]. Conversely, STAT3 mutant Tyr640Phe (DB'1), which is a commonly identified mutation in solid and liquid tumors, leads to constitutive activation across several cell lines (hepatic epithelial cells, lung carcinomas, fibroblasts, etc.) through enhanced stability of STAT3 dimerization, nuclear accumulation and increased transcriptional activity following IFNγ stimulation [44,47,48,59–62,65–69]. In wild type STAT3, Tyr640 points directly into the hydrophobic system. Increasing hydrophobicity, through removal of the hydroxyl group tightens the packing of the pocket and enhances the activation potential. The STAT3 Tyr640Phe mutation has been identified in over >110 cancer cases in the COSMIC database and is most frequently observed in patients with T-LGLL. An analogous mutation is observed at the BC3* site in STAT5B where the second most frequent mutation (Tyr665Phe) results in STAT5B hyperactivation and has also been observed in patients with T-LGLL.

Within the hydrophobic system, the presence of aromaticity for π-π stacking interactions from key residue side-chains is strongly favored over non-aromatic Van der Waals interactions. This is especially seen at the βC3 position with patients harboring Phe621Val [36,37,50]/Leu [51,52]/Ser [53] mutations resulting in STAT3-deficient AD-HIES. In cellulo mutagenesis studies with the Phe621Val mutation have elucidated impaired STAT3 phosphorylation, resulting in defective DNA binding capacity. This strong requirement for aromaticity is also observed in STAT5B, where Tyr665His leads to a hyperactive mutant, since it retains the aromaticity of the imidazole ring to interact with Trp631 (βC5), despite the increase in side-chain polarity.

In addition to the hydrophobic system, the βD strand is critical for controlling pocket accessibility. This is primarily governed by the residue at the βD5 position in SH2 domains and corresponds to Val637 in STAT3. This residue serves as a selectivity filter, interacting with C-terminal amino acids of the phospho-Tyr peptide. In Src kinase, the βD5 residue corresponds to a Tyr and the aromatic ring is sandwiched between the Glu (pY+1 residue) and Ile (pY+3 residue) of the phospho-peptide [18]. In other SH2 domain-containing kinases, the βD5 residue interacts with all three C-terminal amino acids of the pY-peptide. Given the critical nature of this site, it is not unexpected that mutations strongly impair STAT3 activity. STAT3 Val637Met has been identified in a number of patient samples (>40 cases) and is associated with AD-HIES due to impaired response to cytokine activation and transcriptional activity. The importance of Val637Met [30,33,34,36,49,56] is further underscored by insensitivity to 100-fold increases in IL-6 to simulate phosphorylation. This inability to recognize specific phospho-peptides contributes to the defective STAT pathway observed in AD-HIES. As a crucial selectivity filter, even semi-conservative mutations Val637Leu [36] and Val637Ala [30] have also been shown to be disruptive to STAT3 activity and result in AD-HIES. Although STAT3 Val637Met may be a result of reduced protein stability [122], circular dichroism spectra for STAT3 Val637Ala suggest that this substitution does not cause large structural perturbations. Alternatively, this substitution likely reduces phospho-peptide binding in the pY+3 pocket [123]. The βD7 position also assists in the orientation of the βD5 residue, and mutation of the rigid Pro639 to 639Ala [37]/Ser [30,34]/Thr [35] also results in AD-HIES.

3.2.2. Mutations in the Dimerization Interface

In comparison to pY and pY+3 pockets of the SH2 domain, the dimerization interface represents a delicate balance in carefully regulating STAT activity. This region is littered with disease-causing mutations and slight changes to sterics or electronics at the αB, αB′, or BC* loop propagate their effects exponentially and lead to highly contrasting effects. There are multiple examples of such mutations throughout the dimerization interface. For instance, at the αB′6 site, patients with STAT3 Asn647Asp [36,72] exhibit symptoms of AD-HIES, but the Asn647Ile [57,65,67] mutation results in STAT3 hyperactivation manifesting as chronic lymphoproliferative disorder of NK cells (CPLD-NK) and T-LGLL. Comparable to trends observed at the central pY+3 pocket, hydrophobic or aromatic substitutions at the interface stabilize STAT3 dimer formation and subsequent phosphorylation, while changes in polarity, or in this case, an electrostatic reversal, effectively abolish STAT3 activity. Analogous effects are also observed at the BC1*–BC6* positions. At BC1* and BC2*, Tyr657Ser [74]/Asn [52]/Cys [30,34,36,58] and Lys658Glu [75] lead to AD-HIES, whereas Tyr657ins [47], Tyr657dup [48,72] and Lys658Met [67,72]/Asn [65]/Tyr [66] elicit several types of T-cell cancers [124]. At the BC3* position, Ile659Leu [57,72] has been characterized in T-LGLL, whereas the recently identified Ile659Asn [58] mutation distorts STAT3 activity leading to AD-HIES. Only destabilizing (AD-HIES causing) mutations have been identified at BC4* (Met660Arg [55] and Met660Thr [76]) in STAT3. The BC5* position was found to be genetically volatile with mutations occurring as Asp661His [65]/Val [78]/Tyr [41,47,48,60,63,65,67,68,72,77]/Ile [67], with all mutations resulting in STAT3 activation and enhanced response to cytokines. Finally, the BC6* (Ala662) and BC8* (Asn664) positions are critical SH2 domain interface determinants, where mutagenesis experiments have created artificial disulfide linked STAT3-Ala662Cys-Asn664Cys dimers that are constitutively active in cellulo and induce malignant transformation [66]. This further reinforces the role of the BC* loop in maintaining the dimer interface to control STAT activity. Individually, these Cys-mutations are likely destabilizing, but their pairing allows for covalent tethering of the STAT3 monomers and active dimer formation.

Within STAT3, the disordered BC* loop tends to be the only site amenable to insertion/deletion mutations. Since different mutations in this BC* loop lead to either hyper- or refractory activity, likely the substitutions have some degree of compensatory effect that allow them to persist compared to other regions of the SH2 domain. Furthermore, the remaining motifs of the SH2 domain are highly structured and less likely to tolerate insertions or deletions.

The volatility shown by each of the residues in the BC* loop to trigger such extreme changes in STAT3 behavior underscores both the importance of this region to STAT activity, but also the necessity to understand the underlying molecular dynamics that result in such variability. This is particularly true for the BC5* residue which has no prescribed role in other SH2 domains. However, as seen above, its malignant capacity is revealed in >100 identified cases featuring a mosaic of mutations, according to the COSMIC database. Specifically, the STAT3 Asp661Tyr mutation represents one of the most frequently occurring mutation in the SH2 domain of STAT3 along with Tyr640Phe (>100 cases). Although the increases in hydrophobicity and aromaticity have been speculated as critical determinants for the aggressive nature of this mutation, the site-specific mechanism of activation remains unclear. Notably, both of these frequently cancer-associated mutation sites in STAT3 (Tyr640 and Asp661) are by default Phe and Ile respectively in STAT5B. This substitution to the STAT3 cancer-associated genotype in STAT5B suggests that the protein may be more optimized for protein dimerization. This is an interesting observation and further suggests a delicate evolutionary balance in STAT5B by potentially improving interactions at the dimerization interface while reducing activity through the lack of an efficient phosphate-coordinating βD4 residue (Asn642). The functional significance of these changes in STAT5B, compared to STAT3, has not been biophysically characterized and may also suggest that additional mechanisms are relevant to the disease-associated phenotype including changes to protein stability or transcriptional regulation.

3.3. Mutations in the Additional Regions of the SH2 Domain

Mutational hotspots in regions outside the pY and pY+3 pockets may highlight additional areas that are important for protein regulation and exploitable for drug targeting and understanding disease progression. There is a tight clustering of mutations in the BC loop of STAT3 on the periphery of the pY pocket. Similar to the dimerization interface, mutations at these residues can either enhance or reduce STAT activity. This region is in close proximity to the pY pocket, pY+3 pocket, dimerization interface, and STAT linker domain and likely serves as an important allosteric communication bridge for interdomain signaling. As such, interactions at this region of the BC loop require a complex balance of flexibility and rigidity. For instance, at the STAT3 BC3 position, a mutation at Ser614Arg [39–42] leads to hyperactivation and LGLL, whereas a Ser614Gly [34] results in LOF and AD-HIES. Increasing the positive electrostatic potential at this region generally leads to STAT hyperactivation and draws the BC loop closer into the pY pocket. Mutations found at neighboring sites BC5, BC6, and B7 delineate similar trends, where Glu616Lys [45], Glu616Gly [44], Gly617Arg [44], and Gly618Arg [47,48] are found in diffuse large B-cell lymphoma and NK-malignancies. Corresponding electronegative or bulky substitutions are associated with AD-HIES and dysfunctional STAT3 (Gly617Glu [46], Gly618Asp [35,37], and Gly617Val [34]).

There are additional mutations that are located across STAT5 and identified in single patient cases. Given the general robust nature of STAT5 to mutations, it is difficult to assess the oncogenic driving capacity of a single mutation, or whether the disease is multifactorial in etiology. For instance, a mutation in the short βA strand was identified within STAT5B as Gly596Val [79]. However, this mutation was identified in a chimeric protein of STAT5B and retinoic acid receptor-α (STAT5B-RARα), which is associated with acute promyelocytic leukemia (APL) and is also resistant to all-trans retinoic acid therapy. All-trans retinoic acid (ATRA) is capable of inducing remission in almost all APL cases, with several exceptions [125]. Since this is a rare subtype of APL, where the fusion protein was identified in a small minority of cases (<10), it is difficult to judge the importance of the mutation to the progression of the disease, although the residue is conserved across species. Notably, a STAT3-RARα fusion was also recently discovered in an APL patient with a similar ATRA-resistance profile [126]. Surprisingly, no STAT5A fusions have been discovered. It would be interesting to define if the dimerization mechanism of these RARα fusions is mediated through a partially intact SH2 domain or other dimerization domain of STAT. Other mutations in STAT5B are located in non-hotspots including Asp634Val (βC8), Gln636Pro (CD2), and Arg659Cys (αB4) and were identified in patients with T-PLL and MEITL. Similar to STAT3, these mutations are in close proximity to the dimerization interface and increase the hydrophobicity of the region, which can facilitate hyperactivation. It is interesting to note that T-PLL represents the disease with most hyperactive STAT3/5B and JAK3 mutations among all subgroups of largely untargeted orphan T-cell neoplasias, and future targeting efforts in this pathway will likely benefit these patients.

4. Conclusions

Disease associated mutations are more frequently identified in STAT3 compared to STAT5B, suggesting that STAT5B is more robust to the alterations in structural motifs, or that STAT3 has a more pronounced role in normal physiological functioning. However, it is clear that even slight alterations to electronics or sterics in the SH2 domain can dramatically alter STAT3 activity. In STAT3, the majority of mutations identified in the pY pocket impair protein function, with the most substantial effects observed upon mutation of conserved Sheinerman residues (Figure 1f). In STAT5B, pY mutations are generally activating with the Asn642His substitution occurring most frequently in aggressive T-cell cancers (Figure 1g). The pY+3 specificity pockets are characterized by multiple mutations with variable effects. Broadly, mutations that improve hydrophobicity or introduce aromaticity lead to hyperactivation, while increases in pY+3 pocket polarity or removal of aromatic substituents diminish STAT function. This trend is also observed at the SH2 domain dimerization interface which is a hot-spot for mutations, and different substitutions at a single position can result in severe loss- or gain-of-function. Finally, the BC loop may be a critical region for allosteric communication pathways throughout the protein and has been evolutionarily tuned for the precise interactions. As such, marginally reducing electronegativity or increasing electropositivity leads to hyper- and hypo-activation, respectively.

Currently, additional structural studies and molecular dynamics simulations are required for a better understanding of the molecular mechanisms of STAT3/5B mutations at different sites within the SH2 domain. Considering the conformational flexibility of the main binding sites with state-of-the-art computational methods, for example thermodynamic integration, should be more thoroughly exploited in further work. These can be used to propose alternative treatments or highlight therapeutic approaches. For instance, the relative instability of the wild type STAT3 protein is shown to be amplified by AD-HIES-causing mutations. The use of small molecules that can trigger stimulation of protein chaperones to rescue dysfunctional STAT3 mutants has been shown to be effective in cellulo [122]. Alternatively, hyperactivated STAT3 is only marginally more stable than the wild-type protein which may be exploited by degradation enhancing therapeutic strategies such as the use of PROTACs and hydrophobic tagging. These efforts can be extended to STAT5 as well as examining the Asn642His site, as this hot-spot mutation mimics SH2 domain superbinders and is excessively aggressive due to its prime role in the pY pocket. Collectively, these structural studies offer a predictive approach for understanding the molecular foundations of additional mutations identified in the SH2 domain, based on their location and alterations to pocket electronics and sterics.

Funding: P.T.G is supported by research grants from NSERC (RGPIN-2014-05767), CIHR (MOP-130424, MOP-137036), Canada Research Chair (950-232042), Canadian Cancer Society (703963), Canadian Breast Cancer Foundation (705456) and infrastructure grants from CFI (33536) and the Ontario Research Fund (34876). R.M. is supported by the Austrian Science Fund (FWF) (SFB-F04707, SFB-F06105, under the frame of ERA-NET (I 4157-B)). H.A.N. is supported by the FWF, under the frame of ERA PerMed (I 4218-B). R.M., A.O., and H.A.N. were also generously supported by a private donation from Liechtenstein. D.B. and G.M.K. are supported by OTKA K 116904 (National Research, Development and Innovation Office, Hungary).

Acknowledgments: The authors also thank Open Access Funding by the Austrian Science Fund (FWF).

Conflicts of Interest: The authors declare no competing interests.

References

1. Wingelhofer, B.; Neubauer, H.A.; Valent, P.; Han, X.; Constantinescu, S.N.; Gunning, P.T.; Müller, M.; Moriggl, R. Implications of STAT3 and STAT5 signaling on gene regulation and chromatin remodeling in hematopoietic cancer. *Leuk.* **2018**, *32*, 1713–1726. [CrossRef]

2. O'Shea, J.J.; Schwartz, D.M.; Villarino, A.V.; Gadina, M.; McInnes, I.B.; Laurence, A. The JAK-STAT pathway: Impact on human disease and therapeutic intervention. *Annu. Rev. Med.* **2015**, *66*, 311–328. [CrossRef] [PubMed]

3. Bromberg, J.F.; Wrzeszczynska, M.H.; Devgan, G.; Zhao, Y.; Pestell, R.G.; Albanese, C.; E Darnell, J. Stat3 as an Oncogene. *Cell* **1999**, *98*, 295–303. [CrossRef]

4. Tsutsui, M.; Yasuda, H.; Suto, H.; Imai, H.; Isobe, Y.; Sasaki, M.; Kojima, Y.; Oshimi, K.; Sugimoto, K. Frequent STAT3 activation is associated with Mcl-1 expression in nasal NK-cell lymphoma. *Int. J. Lab. Hematol.* **2010**, *32*, 419–426. [CrossRef] [PubMed]

5. Passerini, L.; Allan, S.E.; Battaglia, M.; Di Nunzio, S.; Alstad, A.N.; Levings, M.K.; Roncarolo, M.G.; Bacchetta, R. STAT5-signaling cytokines regulate the expression of FOXP3 in CD4+CD25+ regulatory T cells and CD4+ CD25- effector T cells. *Int. Immunol.* **2008**, *20*, 421–431. [CrossRef] [PubMed]

6. Mirmohammadsadegh, A.; Hassan, M.; Bardenheuer, W.; Marini, A.; Gustrau, A.; Nambiar, S.; Tannapfel, A.; Bojar, H.; Ruzicka, T.; Hengge, U.R. STAT5 Phosphorylation in Malignant Melanoma Is Important for Survival and Is Mediated Through SRC and JAK1 Kinases. *J. Investig. Dermatol.* **2006**, *126*, 2272–2280. [CrossRef] [PubMed]

7. Xu, Q.; Briggs, J.; Park, S.; Niu, G.; Kortylewski, M.; Zhang, S.; Gritsko, T.; Turkson, J.; Kay, H.; Semenza, G.L.; et al. Targeting Stat3 blocks both HIF-1 and VEGF expression induced by multiple oncogenic growth signaling pathways. *Oncogene* **2005**, *24*, 5552–5560. [CrossRef]

8. Matsumura, I.; Kitamura, T.; Wakao, H.; Tanaka, H.; Hashimoto, K.; Albanese, C.; Downward, J.; Pestell, R.G.; Kanakura, Y. Transcriptional regulation of the cyclin D1 promoter by STAT5: Its involvement in cytokine-dependent growth of hematopoietic cells. *EMBO J.* **1999**, *18*, 1367–1377. [CrossRef]

9. Baik, M.; Yu, J.H.; Hennighausen, L. Growth hormone-STAT5 regulation of growth, hepatocellular carcinoma, and liver metabolism. *Ann. New York Acad. Sci.* **2011**, *1229*, 29–37. [CrossRef]

10. Nichane, M.; Ren, X.; Bellefroid, E.J. Self-regulation of Stat3 activity coordinates cell-cycle progression and neural crest specification. *EMBO J.* **2010**, *29*, 55–67. [CrossRef]

11. Ali, A.M.; Gómez-Biagi, R.F.; Rosa, D.A.; Lai, P.-S.; Heaton, W.L.; Park, J.S.; Eiring, A.M.; Vellore, N.A.; De Araujo, E.D.; Ball, D.P.; et al. Disarming an Electrophilic Warhead: Retaining Potency in Tyrosine Kinase Inhibitor (TKI)-Resistant CML Lines While Circumventing Pharmacokinetic Liabilities. *ChemMedChem* **2016**, *11*, 850–861. [CrossRef] [PubMed]

12. De Araujo, E.D.; Manaswiyoungkul, P.; Erdogan, F.; Qadree, A.K.; Sina, D.; Tin, G.; Toutah, K.; Yuen, K.; Gunning, P.T.; Erodgan, F. A functional in vitro assay for screening inhibitors of STAT5B phosphorylation. *J. Pharm. Biomed. Anal.* **2019**, *162*, 60–65. [CrossRef] [PubMed]

13. Wingelhofer, B.; Maurer, B.; Heyes, E.C.; Cumaraswamy, A.A.; Berger-Becvar, A.; De Araujo, E.D.; Orlova, A.; Freund, P.; Ruge, F.; Park, J.; et al. Pharmacologic inhibition of STAT5 in acute myeloid leukemia. *Leuk.* **2018**, *32*, 1135–1146. [CrossRef] [PubMed]

14. Elumalai, N.; Berg, A.; Rubner, S.; Blechschmidt, L.; Song, C.; Natarajan, K.; Matysik, J.; Berg, T. Rational development of Stafib-2: A selective, nanomolar inhibitor of the transcription factor STAT5b. *Sci. Rep.* **2017**, *7*, 819. [CrossRef]

15. Haftchenary, S.; Luchman, H.A.; Jouk, A.O.; Veloso, A.J.; Page, B.D.G.; Cheng, X.R.; Dawson, S.S.; Grinshtein, N.; Shahani, V.M.; Kerman, K.; et al. Potent Targeting of the STAT3 Protein in Brain Cancer Stem Cells: A Promising Route for Treating Glioblastoma. *ACS Med. Chem. Lett.* **2013**, *4*, 1102–1107. [CrossRef]

16. Liu, B.A.; Engelmann, B.W.; Nash, P.D. The language of SH2 domain interactions defines phosphotyrosine-mediated signal transduction. *FEBS Lett.* **2012**, *586*, 2597–2605. [CrossRef]

17. Gao, Q.; Hua, J.; Kimura, R.; Headd, J.J.; Fu, X.-Y.; Chin, Y.E. Identification of the Linker-SH2 Domain of STAT as the Origin of the SH2 Domain Using Two-dimensional Structural Alignment. *Mol. Cell. Proteom.* **2004**, *3*, 704–714. [CrossRef]

18. Zhou, S. SH2 domains recognize specific phosphopeptide sequences. *Cell* **1993**, *72*, 767–778. [CrossRef]

19. Songyang, Z.; Cantley, L.C. Recognition and specificity in protein tyrosine kinase-mediated signalling. *Trends Biochem. Sci.* **1995**, *20*, 470–475. [CrossRef]

20. Gianti, E.; Zauhar, R.J. An SH2 domain model of STAT5 in complex with phospho-peptides define "STAT5 Binding Signatures. " *J. Comput. Mol. Des.* **2015**, *29*, 451–470. [CrossRef]

21. Pettersen, E.F.; Goddard, T.D.; Huang, C.C.; Couch, G.S.; Greenblatt, D.M.; Meng, E.C.; Ferrin, T.E. UCSF Chimera?A visualization system for exploratory research and analysis. *J. Comput. Chem.* **2004**, *25*, 1605–1612. [CrossRef] [PubMed]

22. Nkansah, E.; Shah, R.; Collie, G.W.; Parkinson, G.N.; Palmer, J.; Rahman, K.M.; Bui, T.T.; Drake, A.F.; Husby, J.; Neidle, S.; et al. Observation of unphosphorylated STAT3 core protein binding to targetdsDNA by PEMSA and X-ray crystallography. *FEBS Lett.* **2013**, *587*, 833–839. [CrossRef] [PubMed]

23. De Araujo, E.D.; Erdogan, F.; Neubauer, H.A.; Meneksedag-Erol, D.; Manaswiyoungkul, P.; Eram, M.S.; Seo, H.-S.; Qadree, A.K.; Israelian, J.; Orlova, A.; et al. Structural and functional consequences of the STAT5BN642H driver mutation. *Nat. Commun.* **2019**, *10*, 2517. [CrossRef] [PubMed]

24. Mao, X.; Ren, Z.; Parker, G.N.; Sondermann, H.; Pastorello, M.A.; Wang, W.; McMurray, J.S.; Demeler, B.; Darnell, J.E.; Chen, X. Structural Bases of Unphosphorylated STAT1 Association and Receptor Binding. *Mol. Cell* **2005**, *17*, 761–771. [CrossRef]

25. Langenfeld, F.; Guarracino, Y.; Arock, M.; Trouve, A.; Tchertanov, L. How Intrinsic Molecular Dynamics Control Intramolecular Communication in Signal Transducers and Activators of Transcription Factor STAT5. *PLoS ONE* **2015**, *10*, e0145142. [CrossRef]

26. Takeda, K.; Noguchi, K.; Shi, W.; Tanaka, T.; Matsumoto, M.; Yoshida, N.; Kishimoto, T.; Akira, S. Targeted disruption of the mouse Stat3 gene leads to early embryonic lethality. *Proc. Natl. Acad. Sci. USA* **1997**, *94*, 3801–3804. [CrossRef]

27. Purvis, H.A.; Anderson, A.E.; Young, D.A.; Isaacs, J.D.; Hilkens, C.M.U. A Negative Feedback Loop Mediated by STAT3 Limits Human Th17 Responses. *J. Immunol.* **2014**, *193*, 1142–1150. [CrossRef]

28. Haddad, E. STAT3: Too much may be worse than not enough! *Blood* **2015**, *125*, 583–584. [CrossRef]

29. Laurence, A.D.J.; Uhlig, H.H. When half a glass of STAT3 is just not enough. *Blood* **2016**, *128*, 3020–3021. [CrossRef]

30. Woellner, C.; Gertz, E.M.; Schäffer, A.A.; Lagos, M.; Perro, M.; Glocker, E.-O.; Pietrogrande, M.C.; Cossu, F.; Franco, J.L.; Matamoros, N.; et al. Mutations in STAT3 and diagnostic guidelines for hyper-IgE syndrome. *J. Allergy Clin. Immunol.* **2010**, *125*, 424–432. [CrossRef]

31. Robinson, W.S.; Arnold, S.R.; Michael, C.F.; Vickery, J.D.; Schoumacher, R.A.; Pivnick, E.K.; Ward, J.C.; Nagabhushanam, V.; Lew, D.B. Case report of a young child with disseminated histoplasmosis and review of hyper immunoglobulin e syndrome (HIES). *Clin. Mol. Allergy* **2011**, *9*, 14. [CrossRef] [PubMed]

32. Ma, C.S.; Avery, D.T.; Chan, A.; Batten, M.; Bustamante, J.; Boisson-Dupuis, S.; Arkwright, P.D.; Kreins, A.Y.; Averbuch, D.; Engelhard, D.; et al. Functional STAT3 deficiency compromises the generation of human T follicular helper cells. *Blood* **2012**, *119*, 3997–4008. [CrossRef] [PubMed]

33. Zhang, L.-Y.; Tian, W.; Shu, L.; Jiang, L.-P.; Zhan, Y.-Z.; Zhao, X.-D.; Cui, Y.-X.; Tang, X.-M.; Wu, D.-Q.; Yang, X.-Q.; et al. Clinical Features, STAT3 Gene Mutations and Th17 Cell Analysis in Nine Children with Hyper-IgE Syndrome in Mainland China. *Scand. J. Immunol.* **2013**, *78*, 258–265. [CrossRef] [PubMed]

34. Chandesris, M.O.; Melki, I.; Natividad, A.; Puel, A.; Fieschi, C.; Yun, L.; Thumerelle, C.; Oksenhendler, E.; Boutboul, D.; Thomas, C.; et al. Autosomal dominant STAT3 deficiency and hyper-IgE syndrome: Molecular, cellular, and clinical features from a French national survey. *Medicine (United States)* **2012**, *91*. [CrossRef] [PubMed]

35. Freeman, A.F.; Renner, E.D.; Henderson, C.; Langenbeck, A.; Olivier, K.N.; Hsu, A.P.; Hagl, B.; Boos, A.; Davis, J.; Marciano, B.E.; et al. Lung Parenchyma Surgery in Autosomal Dominant Hyper-IgE Syndrome. *J. Clin. Immunol.* **2013**, *33*, 896–902. [CrossRef] [PubMed]

36. Holland, S.M.; DeLeo, F.R.; Elloumi, H.Z.; Hsu, A.P.; Uzel, G.; Brodsky, N.; Freeman, A.F.; Demidowich, A.; Davis, J.; Turner, M.L.; et al. STAT3Mutations in the Hyper-IgE Syndrome. *New Engl. J. Med.* **2007**, *357*, 1608–1619. [CrossRef] [PubMed]

37. Heimall, J.; Davis, J.; Shaw, P.A.; Hsu, A.P.; Gu, W.; Welch, P.; Holland, S.M.; Freeman, A.F. Paucity of genotype-phenotype correlations in STAT3 mutation positive Hyper IgE Syndrome (HIES). *Clin. Immunol.* **2011**, *139*, 75–84. [CrossRef]

38. Moreira Varanese, I.; Seminario, A.; Diaz Balve, D.; Uriarte, I.; Belardinelli, G.; Gomez Racio, A.; di Giovanni, D.; Bezrodnik, L. New STAT3 mutation in a patient with Hyper IgE Syndrome (HIGES) without criteria in HIES STAT3 score. *Transl. Biomed.* **2010**, *1*. [CrossRef]

39. Qiu, Z.-Y.; Fan, L.; Wang, L.; Qiao, C.; Wu, Y.-J.; Zhou, J.-F.; Xu, W.; Li, J.-Y. STAT3 mutations are frequent in T-cell large granular lymphocytic leukemia with pure red cell aplasia. *J. Hematol. Oncol.* **2013**, *6*, 82. [CrossRef]

40. Yan, Y.; Olson, T.L.; Nyland, S.B.; Feith, D.J.; Loughran, T.P. Emergence of a STAT3 mutated NK clone in LGL leukemia. *Leuk. Res. Rep.* **2015**, *4*, 4–7. [CrossRef]

41. Kűcűk, C.; Jiang, B.; Hu, X.; Zhang, W.; Chan, J.K.C.; Xiao, W.; Lack, N.; Alkan, C.; Williams, J.C.; Avery, K.N.; et al. Activating mutations of STAT5B and STAT3 in lymphomas derived from γδ-T or NK cells. *Nat. Commun.* **2015**, *6*, 6025. [CrossRef] [PubMed]

42. Blombery, P.; Thompson, E.R.; Jones, K.; Arnau, G.M.; Lade, S.; Markham, J.F.; Li, J.; Deva, A.; Johnstone, R.W.; Khot, A.; et al. Whole exome sequencing reveals activating JAK1 and STAT3 mutations in breast implant-associated anaplastic large cell lymphoma anaplastic large cell lymphoma. *Haematol.* **2016**, *101*, e387–e390. [CrossRef] [PubMed]

43. McKinney, M.; Moffitt, A.B.; Gaulard, P.; Travert, M.; De Leval, L.; Nicolae, A.; Raffeld, M.; Jaffe, E.S.; Pittaluga, S.; Xi, L.; et al. The Genetic Basis of Hepatosplenic T-cell Lymphoma. *Cancer Discov.* **2017**, *7*, 369–379. [CrossRef] [PubMed]

44. Ohgami, R.S.; Ma, L.; Monabati, A.; Zehnder, J.L.; Arber, D.A. STAT3 mutations are present in aggressive B-cell lymphomas including a subset of diffuse large B-cell lymphomas with CD30 expression. *Haematol.* **2014**, *99*, e105–e107. [CrossRef] [PubMed]

45. Song, T.L.; Nairismägi, M.L.; Laurensia, Y.; Lim, J.Q.; Tan, J.; Li, Z.M.; Kizhakeyil, A.; Wijaya, G.C.; Huang, D.C.; Nagarajan, S.; et al. Oncogenic activation of the STAT3 pathway drives PD-L1 expression in natural killer/T-cell lymphoma. *Blood* **2018**, *132*, 1146–1158. [CrossRef] [PubMed]

46. Schimke, L.F.; Sawalle-Belohradsky, J.; Roesler, J.; Wollenberg, A.; Rack, A.; Borte, M.; Rieber, N.; Cremer, R.; Maaß, E.; Dopfer, R.; et al. Diagnostic approach to the hyper-IgE syndromes: Immunologic and clinical key findings to differentiate hyper-IgE syndromes from atopic dermatitis. *J. Allergy Clin. Immunol.* **2010**, *126*, 611–617. [CrossRef] [PubMed]

47. Couronné, L.; Scourzic, L.; Pilati, C.; Della Valle, V.; Duffourd, Y.; Solary, E.; Vainchenker, W.; Merlio, J.-P.; Beylot-Barry, M.; Damm, F.; et al. STAT3 mutations identified in human hematologic neoplasms induce myeloid malignancies in a mouse bone marrow transplantation model. *Haematol.* **2013**, *98*, 1748–1752. [CrossRef]

48. Kristensen, T.; Larsen, M.; Rewes, A.; Frederiksen, H.; Thomassen, M.; Møller, M.B. Clinical Relevance of Sensitive and Quantitative STAT3 Mutation Analysis Using Next-Generation Sequencing in T-Cell Large Granular Lymphocytic Leukemia. *J. Mol. Diagn.* **2014**, *16*, 382–392. [CrossRef]

49. Renner, E.D.; Rylaarsdam, S.; Añover-Sombke, S.; Rack, A.L.; Reichenbach, J.; Carey, J.C.; Zhu, Q.; Jansson, A.F.; Barboza, J.; Schimke, L.F.; et al. Novel signal transducer and activator of transcription 3 (STAT3) mutations, reduced T(H)17 cell numbers, and variably defective STAT3 phosphorylation in hyper-IgE syndrome. *J. Allergy Clin. Immunol.* **2008**, *122*, 181–187. [CrossRef]

50. Odio, C.D.; Milligan, K.L.; Mc Gowan, K.; Spergel, A.K.R.; Bishop, R.; Boris, L.; Urban, A.; Welch, P.; Heller, T.; Kleiner, D.; et al. Endemic mycoses in patients with STAT3-mutated hyper-IgE (Job) syndrome. *J. Allergy Clin. Immunol.* **2015**, *136*, 1411–1413. [CrossRef]

51. Kumánovics, A.; Wittwer, C.T.; Pryor, R.J.; Augustine, N.H.; Leppert, M.F.; Carey, J.C.; Ochs, H.D.; Wedgwood, R.J.; Faville, R.J.; Quie, P.G.; et al. Rapid Molecular Analysis of the STAT3 Gene in Job Syndrome of Hyper-IgE and Recurrent Infectious Diseases. *J. Mol. Diagn.* **2010**, *12*, 213–219. [CrossRef] [PubMed]

52. Ives, M.L.; Ma, C.S.; Palendira, U.; Chan, A.; Bustamante, J.; Boisson-Dupuis, S.; Arkwright, P.D.; Engelhard, D.; Averbuch, D.; Magdorf, K.; et al. Signal transducer and activator of transcription 3 (STAT3) mutations underlying autosomal dominant hyper-IgE syndrome impair human CD8 + T-cell memory formation and function. *J. Allergy Clin. Immunol.* **2013**, *132*, 400–411. [CrossRef] [PubMed]

53. Mogensen, T.H.; Jakobsen, M.A.; Larsen, C.S. Identification of a novel STAT3 mutation in a patient with hyper-IgE syndrome. *Scand. J. Infect. Dis.* **2013**, *45*, 235–238. [CrossRef] [PubMed]

54. Bellon, M.; Lu, L.; Nicot, C. Constitutive activation of Pim1 kinase is a therapeutic target for adult T-cell leukemia. *Blood* **2016**, *127*, 2439–2450. [CrossRef]

55. Giacomelli, M.; Tamassia, N.; Moratto, D.; Bertolini, P.; Bertulli, C.; Plebani, A.; Cassatella, M.; Ricci, G.; Bazzoni, F.; Badolato, R. SH2-domain mutations in STAT3 in hyper-IgE syndrome patients result in impairment of IL-10 function. *Eur. J. Immunol.* **2011**, *41*, 3075–3084. [CrossRef]

56. Jiao, H.; Tóth, B.; Erdős, M.; Fransson, I.; Rakoczi, E.; Balogh, I.; Magyarics, Z.; Dérfalvi, B.; Csorba, G.; Szaflarska, A.; et al. Novel and recurrent STAT3 mutations in hyper-IgE syndrome patients from different ethnic groups. *Mol. Immunol.* **2008**, *46*, 202–206. [CrossRef]

57. Shi, M.; He, R.; Feldman, A.L.; Viswanatha, D.S.; Jevremovic, D.; Chen, D.; Morice, W.G. STAT3 mutation and its clinical and histopathologic correlation in T-cell large granular lymphocytic leukemia. *Hum. Pathol.* **2018**, *73*, 74–81. [CrossRef]

58. Goncalves, M.R.; Junior, P.R. Association between STAT3 mutations and phenotypic features in Hyper-IgE Syndrome. *J. Allergy Clin. Immunol.* **2019**, *143*, AB111. [CrossRef]

59. Calderaro, J.; Nault, J.C.; Balabaud, C.; Couchy, G.; Saint-Paul, M.C.; Azoulay, D.; Mehdaoui, D.; Luciani, A.; Zafrani, E.S.; Bioulac-Sage, P.; et al. Inflammatory hepatocellular adenomas developed in the setting of chronic liver disease and cirrhosis. *Mod. Pathol.* **2016**, *29*, 43–50. [CrossRef]

60. Dufva, O.; Kankainen, M.; Kelkka, T.; Sekiguchi, N.; Awad, S.A.; Eldfors, S.; Yadav, B.; Kuusanmäki, H.; Malani, D.; Andersson, E.I.; et al. Aggressive natural killer-cell leukemia mutational landscape and drug profiling highlight JAK-STAT signaling as therapeutic target. *Nat. Commun.* **2018**, *9*, 1567. [CrossRef]

61. Kataoka, K.; Nagata, Y.; Kitanaka, A.; Shiraishi, Y.; Shimamura, T.; Yasunaga, J.-I.; Totoki, Y.; Chiba, K.; Sato-Otsubo, A.; Nagae, G.; et al. Integrated molecular analysis of adult T cell leukemia/lymphoma. *Nat. Genet.* **2015**, *47*, 1304–1315. [CrossRef] [PubMed]

62. Kiel, M.J.; Sahasrabuddhe, A.A.; Rolland, D.C.M.; Velusamy, T.; Chung, F.; Schaller, M.; Bailey, N.G.; Betz, B.L.; Miranda, R.N.; Porcu, P.; et al. Genomic analyses reveal recurrent mutations in epigenetic modifiers and the JAK–STAT pathway in Sézary syndrome. *Nat. Commun.* **2015**, *6*, 8470. [CrossRef] [PubMed]

63. Ishida, F.; Matsuda, K.; Sekiguchi, N.; Makishima, H.; Taira, C.; Momose, K.; Nishina, S.; Senoo, N.; Sakai, H.; Ito, T.; et al. STAT3 gene mutations and their association with pure red cell aplasia in large granular lymphocyte leukemia. *Cancer Sci.* **2014**, *105*, 342–346. [CrossRef] [PubMed]

64. Bergmann, A.K.; Fataccioli, V.; Castellano, G.; Martin-Garcia, N.; Pelletier, L.; Ammerpohl, O.; Bergmann, J.; Bhat, J.; Pau, E.C.; Martín-Subero, J.I.; et al. DNA methylation profiling of hepatosplenic t-cell lymphoma. *Haematologica* **2019**, *104*, e104–e107. [CrossRef] [PubMed]

65. Koskela, H.L.M.; Eldfors, S.; Ellonen, P.; van Adrichem, A.J.; Kuusanmäki, H.; Andersson, E.I.; Lagström, S.; Clemente, M.J.; Olson, T.; Jalkanen, S.E.; et al. Somatic STAT3 mutations in large granular lymphocytic leukemia. *N. Engl. J. Med.* **2012**, *366*, 1905–1913. [CrossRef] [PubMed]

66. Pilati, C.; Amessou, M.; Bihl, M.P.; Balabaud, C.; Van Nhieu, J.T.; Paradis, V.; Nault, J.C.; Izard, T.; Bioulac-Sage, P.; Couchy, G.; et al. Somatic mutations activating STAT3 in human inflammatory hepatocellular adenomas. *J. Exp. Med.* **2011**, *208*, 1359–1366. [CrossRef]

67. Jerez, A.; Clemente, M.J.; Makishima, H.; Koskela, H.; Leblanc, F.; Ng, K.P.; Olson, T.; Przychodzen, B.; Afable, M.; Gomez-Segui, I.; et al. STAT3 mutations unify the pathogenesis of chronic lymphoproliferative disorders of NK cells and T-cell large granular lymphocyte leukemia. *Blood* **2012**, *120*, 3048–3057. [CrossRef]

68. Ohgami, R.S.; Ma, L.; Merker, J.D.; Martinez, B.; Zehnder, J.L.; Arber, D.A. STAT3 mutations are frequent in CD30+ T-cell lymphomas and T-cell large granular lymphocytic leukemia. *Leukemia* **2013**, *27*, 2244–2247. [CrossRef]

69. Nicolae, A.; Xi, L.; Pittaluga, S.; Abdullaev, Z.; Pack, S.D.; Chen, J.; Waldmann, T.A.; Jaffe, E.S.; Raffeld, M. Frequent STAT5B mutations in γδ hepatosplenic T-cell lymphomas. *Leuk.* **2014**, *28*, 2244–2248. [CrossRef]

70. Ma, C.S.; Chew, G.Y.; Simpson, N.; Priyadarshi, A.; Wong, M.; Grimbacher, B.; Fulcher, D.A.; Tangye, S.G.; Cook, M.C. Deficiency of Th17 cells in hyper IgE syndrome due to mutations in STAT3. *J. Exp. Med.* **2008**, *205*, 1551–1557. [CrossRef]

71. Flanagan, S.E.; Haapaniemi, E.; Russell, M.A.; Caswell, R.; Allen, H.L.; De Franco, E.; McDonald, T.J.; Rajala, H.; Ramelius, A.; Barton, J.; et al. Activating germline mutations in STAT3 cause early-onset multi-organ autoimmune disease. *Nat. Genet.* **2014**, *46*, 812–814. [CrossRef] [PubMed]

72. Fasan, A.; Kern, W.; Grossmann, V.; Haferlach, C.; Haferlach, T.; Schnittger, S. STAT3 mutations are highly specific for large granular lymphocytic leukemia. *Leukemia* **2013**, *27*, 1598–1600. [CrossRef] [PubMed]

73. Kim, M.S.; Lee, S.H.; Yoo, N.J.; Lee, S.H. STAT3 exon 21 mutation is rare in common human cancers. *Acta Oncol.* **2013**, *52*, 1221–1222. [CrossRef] [PubMed]

74. Liu, J.Y.; Li, Q.; Chen, T.T.; Guo, X.; Ge, J.; Yuan, L.X. Destructive pulmonary staphylococcal infection in a boy with hyper-IgE syndrome: A novel mutation in the signal transducer and activator of transcription 3 (STAT3) gene (p.Y657S). *Eur. J. Pediatr.* **2011**, *170*, 661–666. [CrossRef]
75. Milner, J.D.; Vogel, T.P.; Forbes, L.; Ma, C.A.; Stray-Pedersen, A.; Niemela, J.E.; Lyons, J.J.; Engelhardt, K.R.; Zhang, Y.; Topcagic, N.; et al. Early-onset lymphoproliferation and autoimmunity caused by germline STAT3 gain-of-function mutations. *Blood* **2015**, *125*, 591–599. [CrossRef]
76. Saikia, B.; Goel, S.; Suri, D.; Minz, R.W.; Rawat, A.; Singh, S. Novel Mutation in SH2 Domain of STAT3 (p.M660T) in Hyper-IgE Syndrome with Sterno-Clavicular and Paravertebral Abscesses. *Indian J. Pediatr.* **2017**, *203*, 244–495. [CrossRef]
77. Teramo, A.; Barila, G.; Calabretto, G.; Ercolin, C.; Lamy, T.; Moignet, A.; Roussel, M.; Pastoret, C.; Leoncin, M.; Gattazzo, C.; et al. STAT3 mutation impacts biological and clinical features of T-LGL leukemia. *Oncotarget* **2017**, *8*, 61876–61889. [CrossRef]
78. Crescenzo, R.; Abate, F.; Lasorsa, E.; Tabbo', F.; Gaudiano, M.; Chiesa, N.; Di Giacomo, F.; Spaccarotella, E.; Barbarossa, L.; Ercole, E.; et al. Convergent mutations and kinase fusions lead to oncogenic STAT3 activation in anaplastic large cell lymphoma. *Cancer Cell* **2015**, *27*, 516–532. [CrossRef]
79. Iwanaga, E.; Nakamura, M.; Nanri, T.; Kawakita, T.; Horikawa, K.; Mitsuya, H.; Asou, N. Acute promyelocytic leukemia harboring a STAT5B-RARA fusion gene and a G596V missense mutation in the STAT5B SH2 domain of the STAT5B-RARA. *Eur. J. Haematol.* **2009**, *83*, 499–501. [CrossRef]
80. Kiel, M.J.; Velusamy, T.; Rolland, D.; Sahasrabuddhe, A.A.; Chung, F.; Bailey, N.G.; Schrader, A.; Li, B.; Li, J.Z.; Ozel, A.B.; et al. Integrated genomic sequencing reveals mutational landscape of T-cell prolymphocytic leukemia. *Blood* **2014**, *124*, 1460–1472. [CrossRef]
81. Roberti, A.; Dobay, M.P.; Bisig, B.; Vallois, D.; Boéchat, C.; Lanitis, E.; Bouchindhomme, B.; Parrens, M.-C.; Bossard, C.; Quintanilla-Martinez, L.; et al. Type II enteropathy-associated T-cell lymphoma features a unique genomic profile with highly recurrent SETD2 alterations. *Nat. Commun.* **2016**, *7*, 12602–12613. [CrossRef] [PubMed]
82. Chia, D.J.; Subbian, E.; Buck, T.M.; Hwa, V.; Rosenfeld, R.G.; Skach, W.R.; Shinde, U.; Rotwein, P. Aberrant folding of a mutant Stat5b causes growth hormone insensitivity and proteasomal dysfunction. *J. Biol. Chem.* **2006**, *281*, 6552–6558. [CrossRef] [PubMed]
83. Schrader, A.; Crispatzu, G.; Oberbeck, S.; Mayer, P.; Pützer, S.; Von Jan, J.; Vasyutina, E.; Warner, K.; Weit, N.; Pflug, N.; et al. Actionable perturbations of damage responses by TCL1/ATM and epigenetic lesions form the basis of T-PLL. *Nat. Commun.* **2018**, *9*, 697. [CrossRef] [PubMed]
84. Rajala, H.L.M.; Eldfors, S.; Kuusanmäki, H.; Van Adrichem, A.J.; Olson, T.; Lagström, S.; Andersson, E.I.; Jerez, A.; Clemente, M.J.; Yan, Y.; et al. Discovery of somatic STAT5b mutations in large granular lymphocytic leukemia. *Blood* **2013**, *121*, 4541–4550. [CrossRef] [PubMed]
85. Babushok, D.V.; Perdigones, N.; Perin, J.C.; Olson, T.S.; Ye, W.; Roth, J.J.; Lind, C.; Cattier, C.; Li, Y.; Hartung, H.; et al. Emergence of clonal hematopoiesis in the majority of patients with acquired aplastic anemia. *Cancer Genet.* **2015**, *208*, 115–128. [CrossRef]
86. Lopez, C.; Bergmann, A.K.; Paul, U.; Penas, E.M.M.; Nagel, I.; Betts, M.; Johansson, P.; Ritgen, M.; Baumann, T.; Aymerich, M.; et al. Genes encoding members of the JAK-STAT pathway or epigenetic regulators are recurrently mutated in T-cell prolymphocytic leukaemia. *Br. J. Haematol.* **2016**, *173*, 265–273. [CrossRef]
87. Andersson, E.I.; Tanahashi, T.; Sekiguchi, N.; Gasparini, V.R.; Bortoluzzi, S.; Kawakami, T.; Matsuda, K.; Mitsui, T.; Eldfors, S.; Bortoluzzi, S.; et al. High incidence of activating STAT5B mutations in CD4-positive T-cell large granular lymphocyte leukemia. *Blood* **2016**, *128*, 2465–2468. [CrossRef]
88. Andersson, E.I.; Pützer, S.; Yadav, B.; Dufva, O.; Khan, S.; He, L.; Sellner, L.; Schrader, A.; Crispatzu, G.; Oleś, M.; et al. Discovery of novel drug sensitivities in T-PLL by high-throughput ex vivo drug testing and mutation profiling. *Leukemia* **2018**, *32*, 774–787. [CrossRef]
89. Bandapalli, O.R.; Schuessele, S.; Kunz, J.B.; Rausch, T.; Stütz, A.M.; Tal, N.; Geron, I.; Gershman, N.; Izraeli, S.; Eilers, J.; et al. The activating STAT5B N642H mutation is a common abnormality in pediatric T-cell acute lymphoblastic leukemia and confers a higher risk of relapse. *Haematol.* **2014**, *99*, e188–e192. [CrossRef]

90. Atak, Z.K.; Gianfelici, V.; Hulselmans, G.; De Keersmaecker, K.; Devasia, A.G.; Geerdens, E.; Mentens, N.; Chiaretti, S.; Durinck, K.; Uyttebroeck, A.; et al. Comprehensive Analysis of Transcriptome Variation Uncovers Known and Novel Driver Events in T-Cell Acute Lymphoblastic Leukemia. *PLoS Genet.* **2013**, *9*, e1003997.

91. Nairismägi, M.-L.; Tan, J.; Lim, J.Q.; Nagarajan, S.; Ng, C.C.Y.; Rajasegaran, V.; Huang, D.; Lim, W.K.; Laurensia, Y.; Wijaya, G.C.; et al. JAK-STAT and G-protein-coupled receptor signaling pathways are frequently altered in epitheliotropic intestinal T-cell lymphoma. *Leuk.* **2016**, *30*, 1311–1319. [CrossRef] [PubMed]

92. Lavallée, V.-P.; Krošl, J.; Lemieux, S.; Boucher, G.; Gendron, P.; Pabst, C.; Boivin, I.; Marinier, A.; Guidos, C.J.; Meloche, S.; et al. Chemo-genomic interrogation of CEBPA mutated AML reveals recurrent CSF3R mutations and subgroup sensitivity to JAK inhibitors. *Blood* **2016**, *127*, 3054–3061. [CrossRef] [PubMed]

93. Luo, Q.; Shen, J.; Yang, Y.; Tang, H.; Shi, M.; Liu, J.; Liu, Z.; Shi, X.; Yi, Y. CSF3R T618I, ASXL1 G942 fs and STAT5B N642H trimutation co-contribute to a rare chronic neutrophilic leukaemia manifested by rapidly progressive leucocytosis, severe infections, persistent fever and deep venous thrombosis. *Br. J. Haematol.* **2018**, *180*, 892–894. [CrossRef] [PubMed]

94. Prasad, A.; Rabionet, R.; Espinet, B.; Zapata, L.; Puiggros, A.; Melero, C.; Puig, A.; Sarria-Trujillo, Y.; Ossowski, S.; García-Muret, M.P.; et al. Identification of Gene Mutations and Fusion Genes in Patients with Sézary Syndrome. *J. Investig. Dermatol.* **2016**, *136*, 1490–1499. [CrossRef]

95. Ma, X.; Wen, L.; Wu, L.; Wang, Q.; Yao, H.; Wang, Q.; Ma, L.; Chen, S. Rare occurrence of a STAT5B N642H mutation in adult T-cell acute lymphoblastic leukemia. *Cancer Genet.* **2015**, *208*, 52–53. [CrossRef]

96. Cross, N.C.P.; Hoade, Y.; Tapper, W.J.; Carreno-Tarragona, G.; Fanelli, T.; Jawhar, M.; Naumann, N.; Pieniak, I.; Lübke, J.; Ali, S.; et al. Recurrent activating STAT5B N642H mutation in myeloid neoplasms with eosinophilia. *Leukemia* **2019**, *33*, 415–425. [CrossRef]

97. Ma, C.A.; Xi, L.; Cauff, B.; DeZure, A.; Freeman, A.F.; Hambleton, S.; Kleiner, G.; Leahy, T.R.; O'Sullivan, M.; Makiya, M.; et al. Somatic STAT5b gain-of-function mutations in early onset nonclonal eosinophilia, urticaria, dermatitis, and diarrhea. *Blood* **2017**, *129*, 650–653. [CrossRef]

98. Baer, C.; Muehlbacher, V.; Kern, W.; Haferlach, C.; Haferlach, T. Molecular genetic characterization of myeloid/lymphoid neoplasms associated with eosinophilia and rearrangement of PDGFRA, PDGFRB, FGFR1 or PCM1-JAK2. *Haematol.* **2018**, *103*, e348–e350. [CrossRef]

99. Simpson, H.M.; Khan, R.Z.; Song, C.; Sharma, D.; Sadashivaiah, K.; Furusawa, A.; Liu, X.; Nagaraj, S.; Sengamalay, N.; Sadzewicz, L.; et al. Concurrent Mutations in ATM and Genes Associated with Common γ Chain Signaling in Peripheral T Cell Lymphoma. *PLoS ONE* **2015**, *10*, e0141906. [CrossRef]

100. Jiang, L.; Gu, Z.-H.; Yan, Z.-X.; Zhao, X.; Xie, Y.-Y.; Zhang, Z.-G.; Pan, C.-M.; Hu, Y.; Cai, C.-P.; Dong, Y.; et al. Exome sequencing identifies somatic mutations of DDX3X in natural killer/T-cell lymphoma. *Nat. Genet.* **2015**, *47*, 1061–1066. [CrossRef]

101. Kontro, M.; Kuusanmäki, H.; Eldfors, S.; Burmeister, T.; Andersson, E.I.; Bruserud, O.; Brümmendorf, T.H.; Edgren, H.; Gjertsen, B.T.; Itälä-Remes, M.; et al. Novel activating STAT5B mutations as putative drivers of T-cell acute lymphoblastic leukemia. *Leukemia* **2014**, *28*, 1738–1742. [CrossRef] [PubMed]

102. Menon, M.P.; Nicolae, A.; Meeker, H.; Raffeld, M.; Xi, L.; Jegalian, A.G.; Miller, D.C.; Pittaluga, S.; Jaffe, E.S. Primary CNS T-cell Lymphomas: A Clinical, Morphologic, Immunophenotypic, and Molecular Analysis. *Am. J. Surg. Pathol.* **2015**, *39*, 1719–1729. [CrossRef] [PubMed]

103. Varco-Merth, B.; Feigerlová, E.; Shinde, U.; Rosenfeld, R.G.; Hwa, V.; Rotwein, P. Severe growth deficiency is associated with STAT5b mutations that disrupt protein folding and activity. *Mol. Endocrinol.* **2013**, *27*, 150–161. [CrossRef] [PubMed]

104. Hwa, V. STAT5B deficiency: Impacts on human growth and immunity. *Growth Horm. IGF Res.* **2016**, *28*, 16–20. [CrossRef]

105. Cui, Y.; Riedlinger, G.; Miyoshi, K.; Tang, W.; Li, C.; Deng, C.-X.; Robinson, G.W.; Hennighausen, L. Inactivation of Stat5 in Mouse Mammary Epithelium during Pregnancy Reveals Distinct Functions in Cell Proliferation, Survival, and Differentiation. *Mol. Cell. Boil.* **2004**, *24*, 8037–8047. [CrossRef]

106. Hoelbl, A.; Kovacic, B.; Kerenyi, M.A.; Simma, O.; Warsch, W.; Cui, Y.; Beug, H.; Hennighausen, L.; Moriggl, R.; Sexl, V. Clarifying the role of Stat5 in lymphoid development and Abelson-induced transformation. *Blood* **2006**, *107*, 4898–4906. [CrossRef]

107. Klammt, J.; Neumann, D.; Gevers, E.F.; Andrew, S.F.; Schwartz, I.D.; Rockstroh, D.; Colombo, R.; Sanchez, M.A.; Vokurkova, D.; Kowalczyk, J.; et al. Dominant-negative STAT5B mutations cause growth hormone insensitivity with short stature and mild immune dysregulation. *Nat. Commun.* **2018**, *9*, 2105. [CrossRef]

108. Kanai, T.; Jenks, J.; Nadeau, K.C. The STAT5b Pathway Defect and Autoimmunity. *Front. Immunol.* **2012**, *3*, 3. [CrossRef]

109. Sallmyr, A.; Fan, J.; Datta, K.; Kim, K.-T.; Grosu, D.; Shapiro, P.; Small, D.; Rassool, F. Internal tandem duplication of FLT3 (FLT3/ITD) induces increased ROS production, DNA damage, and misrepair: Implications for poor prognosis in AML. *Blood* **2008**, *111*, 3173–3182. [CrossRef]

110. Warsch, W.; Kollmann, K.; Eckelhart, E.; Fajmann, S.; Cerny-Reiterer, S.; Hölbl, A.; Gleixner, K.V.; Dworzak, M.; Mayerhofer, M.; Hoermann, G.; et al. High STAT5 levels mediate imatinib resistance and indicate disease progression in chronic myeloid leukemia. *Blood* **2011**, *117*, 3409–3420. [CrossRef]

111. Ribeiro, D.; Melão, A.; Van Boxtel, R.; Santos, C.I.; Silva, A.; Silva, M.C.; Cardoso, B.A.; Coffer, P.J.; Barata, J.T. STAT5 is essential for IL-7–mediated viability, growth, and proliferation of T-cell acute lymphoblastic leukemia cells. *Blood Adv.* **2018**, *2*, 2199–2213. [CrossRef] [PubMed]

112. Wang, Z.; Bunting, K.D. STAT5 activation in B-cell acute lymphoblastic leukemia: Damned if you do, damned if you don't. *Cancer Cell Microenviron.* **2016**, *3*, 1186.

113. Chiorazzi, N. Cell proliferation and death: Forgotten features of chronic lymphocytic leukemia B cells. *Best Pr. Res. Clin. Haematol.* **2007**, *20*, 399–413. [CrossRef] [PubMed]

114. Rajala, H.L.M.; Mustjoki, S. STAT5b in LGL leukemia—A novel therapeutic target? *Oncotarget* **2013**, *4*, 808–809. [CrossRef]

115. Mitchell, T.J.; Whittaker, S.J.; John, S. Dysregulated expression of COOH-terminally truncated Stat5 and loss of IL2-inducible Stat5-dependent gene expression in Sezary Syndrome. *Cancer Res.* **2003**, *63*, 9048–9054.

116. Zhang, Q.; Wang, H.Y.; Liu, X.; A Wasik, M. STAT5A is epigenetically silenced by the tyrosine kinase NPM1-ALK and acts as a tumor suppressor by reciprocally inhibiting NPM1-ALK expression. *Nat. Med.* **2007**, *13*, 1341–1348. [CrossRef]

117. Kollmann, S.; Grundschober, E.; Maurer, B.; Warsch, W.; Grausenburger, R.; Edlinger, L.; Huuhtanen, J.; Lagger, S.; Hennighausen, L.; Valent, P.; et al. Twins with different personalities: STAT5B-but not STAT5A-has a key role in BCR/ABL-induced leukemia. *Leuk.* **2019**, *33*, 1583–1597. [CrossRef]

118. He, J.; Shi, J.; Xu, X.; Zhang, W.; Wang, Y.; Chen, X.; Du, Y.; Zhu, N.; Zhang, J.; Wang, Q.; et al. STAT3 mutations correlated with hyper-IgE syndrome lead to blockage of IL-6/STAT3 signalling pathway. *J. Biosci.* **2012**, *37*, 243–257. [CrossRef]

119. Pham, H.T.T.; Maurer, B.; Prchal-Murphy, M.; Grausenburger, R.; Grundschober, E.; Javaheri, T.; Nivarthi, H.; Boersma, A.; Kolbe, T.; Elabd, M.; et al. STAT5B N642H is a driver mutation for T cell neoplasia. *J. Clin. Investig.* **2018**, *128*, 387–401. [CrossRef]

120. De Araujo, E.D.; Manaswiyoungkul, P.; Israelian, J.; Park, J.; Yuen, K.; Farhangi, S.; Berger-Becvar, A.; Abu-Jazar, L.; Gunning, P.T. High-throughput thermofluor-based assays for inhibitor screening of STAT SH2 domains. *J. Pharm. Biomed. Anal.* **2017**, *143*, 159–167. [CrossRef]

121. Kofoed, E.M.; Hwa, V.; Woods, K.A.; Buckway, C.K.; Tsubaki, J.; Pratt, K.L.; Bezrodnik, L.; Tepper, A.; Little, B.; Jasper, H.; et al. Growth Hormone Insensitivity Associated with aSTAT5bMutation. *New Engl. J. Med.* **2003**, *349*, 1139–1147. [CrossRef] [PubMed]

122. Bocchini, C.E.; Nahmod, K.; Katsonis, P.; Kim, S.; Kasembeli, M.M.; Freeman, A.; Lichtarge, O.; Makedonas, G.; Tweardy, D.J. Protein stabilization improves STAT3 function in autosomal dominant hyper-IgE syndrome. *Blood* **2016**, *128*, 3061–3072. [CrossRef] [PubMed]

123. Brambilla, L.; Genini, D.; Laurini, E.; Merulla, J.; Perez, L.; Fermeglia, M.; Carbone, G.M.; Pricl, S.; Catapano, C.V. Hitting the right spot: Mechanism of action of OPB-31121, a novel and potent inhibitor of the Signal Transducer and Activator of Transcription 3 (STAT3). *Mol. Oncol.* **2015**, *9*, 1194–1206. [CrossRef] [PubMed]

124. Chandrasekaran, P.; Zimmerman, O.; Paulson, M.; Sampaio, E.P.; Freeman, A.F.; Sowerwine, K.J.; Hurt, D.; Alcántara-Montiel, J.C.; Hsu, A.P.; Holland, S.M. Distinct mutations at the same positions of STAT3 cause either loss or gain of function. *J. Allergy Clin. Immunol.* **2016**, *138*, 1222–1224. [CrossRef] [PubMed]

125. Avvisati, G.; Tallman, M.S. All-trans retinoic acid in acute promyelocytic leukaemia. *Best Pr. Res. Clin. Haematol.* **2003**, *16*, 419–432. [CrossRef]

126. Yao, L.; Wen, L.; Wang, N.; Liu, T.; Xu, Y.; Ruan, C.; Wu, D.; Chen, S. Identification of novel recurrent STAT3-RARA fusions in acute promyelocytic leukemia lacking t(15;17)(q22;q12)/PML-RARA. *Blood* **2018**, *131*, 935–939. [CrossRef] [PubMed]

Article

STAT5 is Expressed in CD34$^+$/CD38$^-$ Stem Cells and Serves as a Potential Molecular Target in Ph-Negative Myeloproliferative Neoplasms

Emir Hadzijusufovic [1,2,3,*], Alexandra Keller [3], Daniela Berger [1,3], Georg Greiner [4], Bettina Wingelhofer [5], Nadine Witzeneder [4], Daniel Ivanov [1,3], Emmanuel Pecnard [6], Harini Nivarthi [7], Florian K. M. Schur [3], Yüksel Filik [1,3], Christoph Kornauth [3], Heidi A. Neubauer [5], Leonhard Müllauer [8], Gary Tin [9], Jisung Park [9], Elvin D. de Araujo [9], Patrick T. Gunning [9], Gregor Hoermann [4], Fabrice Gouilleux [6,10], Robert Kralovics [7], Richard Moriggl [5] and Peter Valent [1,2]

1. Ludwig Boltzmann Institute for Hematology and Oncology, Medical University of Vienna, 1090 Vienna, Austria; daniela.berger@meduniwien.ac.at (D.B.); daniel.ivanov@meduniwien.ac.at (D.I.); yueksel.filik@onc.lbg.ac.at (Y.F.); peter.valent@meduniwien.ac.at (P.V.)
2. Department/Hospital for Companion Animals and Horses, University Hospital for Small Animals, Internal Medicine Small Animals, University of Veterinary Medicine Vienna, 1210 Vienna, Austria
3. Department of Internal Medicine I, Division of Hematology & Hemostaseology, Medical University of Vienna, 1090 Vienna, Austria; alexandra.keller@meduniwien.ac.at (A.K.); florian.schur@ist.ac.at (F.K.M.S.); christoph.kornauth@meduniwien.ac.at (C.K.)
4. Department of Laboratory Medicine, Medical University of Vienna, 1090 Vienna, Austria; georg.greiner@meduniwien.ac.at (G.G.); nadine.witzeneder@meduniwien.ac.at (N.W.); gregor.hoermann@meduniwien.ac.at (G.H.)
5. Institute of Animal Breeding and Genetics, University of Veterinary Medicine Vienna, 1210 Vienna, Austria; bettina.wingelhofer@manchester.ac.uk (B.W.); heidi.neubauer@vetmeduni.ac.at (H.A.N.); richard.moriggl@vetmeduni.ac.at (R.M.)
6. INSERM, ERI-12, Faculté de Pharmacie, Université de Picardie Jules Verne, 80000 Amiens, France; emmanuel.pecnard@univ-tours.fr (E.P.); fabrice.gouilleux@univ-tours.fr (F.G.)
7. Research Center for Molecular Medicine (CeMM), 1090 Vienna, Austria; harini.nivarthi@meduniwien.ac.at (H.N.); robert.kralovics@meduniwien.ac.at (R.K.)
8. Department of Pathology, Medical University of Vienna, 1090 Vienna, Austria; leonhard.muellauer@meduniwien.ac.at
9. Department of Chemistry, University of Toronto, Toronto, ON M5S 1A1, Canada; gary.tin@mail.utoronto.ca (G.T.); ji.park@mail.utoronto.ca (J.P.); e.dearaujo@mail.utoronto.ca (E.D.d.A.); patrick.gunning@utoronto.ca (P.T.G.)
10. CNRS UMR 6239, GICC, Faculté de Médecine, Université François Rabelais, 37020 Tours, France
* Correspondence: emir.hadzijusufovic@meduniwien.ac.at; Tel.: +43-1-40400-49990

Received: 3 April 2020; Accepted: 14 April 2020; Published: 21 April 2020

Abstract: Janus kinase 2 (JAK2) and signal transducer and activator of transcription-5 (STAT5) play a key role in the pathogenesis of myeloproliferative neoplasms (MPN). In most patients, *JAK2* V617F or *CALR* mutations are found and lead to activation of various downstream signaling cascades and molecules, including STAT5. We examined the presence and distribution of phosphorylated (p) STAT5 in neoplastic cells in patients with MPN, including polycythemia vera (PV, $n = 10$), essential thrombocythemia (ET, $n = 15$) and primary myelofibrosis (PMF, $n = 9$), and in the *JAK2* V617F-positive cell lines HEL and SET-2. As assessed by immunohistochemistry, MPN cells displayed pSTAT5 in all patients examined. Phosphorylated STAT5 was also detected in putative CD34$^+$/CD38$^-$ MPN stem cells (MPN-SC) by flow cytometry. Immunostaining experiments and Western blotting demonstrated pSTAT5 expression in both the cytoplasmic and nuclear compartment of MPN cells. Confirming previous studies, we also found that JAK2-targeting drugs counteract the expression of pSTAT5 and growth in HEL and SET-2 cells. Growth-inhibition of MPN cells was also induced by the

STAT5-targeting drugs piceatannol, pimozide, AC-3-019 and AC-4-130. Together, we show that CD34$^+$/CD38$^-$ MPN-SC express pSTAT5 and that pSTAT5 is expressed in the nuclear and cytoplasmic compartment of MPN cells. Whether direct targeting of pSTAT5 in MPN-SC is efficacious in MPN patients remains unknown.

Keywords: MPN; STAT5; JAK2 V617F; neoplastic stem cells

1. Introduction

Classical myeloproliferative neoplasms (MPN) are incurable stem cell disorders characterized by the abnormal expansion of myeloid cells in the bone marrow (BM), elevated blood counts, extramedullary myelopoiesis, and a genetic instability with enhanced risk to transform to secondary acute myeloid leukemia (sAML) [1–6]. In most patients, a mutation in the *calreticulin* (*CALR*) gene or the *Janus kinase 2* (*JAK2*) point mutation V617F is found [3–6]. MPN-related morbidity and mortality are emerging health problems in the Western world. Notably, improved diagnostics and therapy, together with an enhanced life expectancy, have led to an increasing prevalence of MPN. For patients with advanced MPN or sAML, the only curative approach is allogeneic hematopoietic stem cell transplantation [7–9]. However, this therapy can only be offered to a subset of patients. In all other cases, disease management is based on symptom control and the use of growth-inhibitory drugs, including interferon-alpha, anagrelide, hydroxyurea or ruxolitinib [10–15]. However, these drugs have little if any curative potential and in many cases resistance develops during therapy [10–15]. Therefore, current research is seeking new molecular targets and is attempting to develop new targeted drugs for patients with MPN.

Based on the classification of the World Health Organization (WHO), three types of classical MPN have been defined: polycythemia vera (PV), essential thrombocythemia (ET) and primary myelofibrosis (PMF) [3]. Each of these neoplasms exhibits unique clinical, histopathological and molecular features [1–4,6]. However, the three MPN entities share molecular and pathologic characteristics, and in many cases an overlap or transition from one into another type of MPN is seen. In most patients, mutations in the *JAK2*, *CALR* or *thrombopoietin receptor* (*MPL*) gene are found [16–20]. Independent of the disease variant, JAK2 activation leads to a cascade of downstream signaling molecules and pathways in neoplastic cells [21–25]. One of the key downstream signaling molecules is the 'signal transducer and activator of transcription-5' (STAT5) protein [21–27].

Although STAT5 was initially characterized as a key transcription factor in various physiologic and pathologic processes, more recent data suggest that STAT5 also serves as a cytoplasmic signaling molecule that binds to and interacts with other signaling molecules in neoplastic cells to promote oncogenesis. Both the cytoplasmic and the nuclear fractions of STAT5 are considered to critically contribute to leukemogenesis in patients with myeloid neoplasms [28–30].

Similar to other myeloid neoplasms, MPN are considered to develop from transformed myeloid stem cells [31–35]. As only the neoplastic stem cells of an MPN (MPN-SC) can propagate the malignancy for unlimited time periods, they represent an important cellular target of therapy. However, little is known about the phenotype and target expression profiles of neoplastic stem cells in MPN. As in other myeloid neoplasms, MPN-SC are considered to reside within the CD34$^+$/CD38$^-$ population of the clone [32,33,35]. It has also been described that the immature CD34$^+$ cells in MPN patients express *JAK2* V617F [35].

The aims of the present study were to examine MPN cells for expression of phosphorylated (p) STAT5, to study the cellular distribution of pSTAT5 and to analyze the effects of pSTAT5-targeting drugs on MPN cells. Our data show that pSTAT5 is expressed in CD34$^+$/CD38$^-$ MPN stem cells and serves as a potential therapeutic target in MPN.

2. Results

2.1. Primary MPN Cells Express Nuclear and Cytoplasmic pSTAT5

As assessed by immunohistochemistry (IHC), primary MPN cells in the BM of patients with PV, ET and PMF expressed pSTAT5 in their nuclear and cytoplasmic compartment (Figure 1A and Table 1). The expression of pSTAT5 in normal BM cells (controls) was similar to that found in MPN BM sections examined by IHC. In all samples tested, megakaryocytes stained clearly positive for pSTAT5 (positive control), whereas erythroid cells stained negative for pSTAT5 (negative control). We were also able to confirm expression of cytoplasmic pSTAT5 in BM cells in patients with various MPN by multi-color flow cytometry (Figure 1B). In these experiments, all myeloid cells tested, including CD15[+] granulomonocytic cells, CD14[+] monocytes and CD34[+] stem and progenitor cells, were found to stain positive for pSTAT5 (Figure 1C). pSTAT5 was identified in BM cells in all three categories of MPN, regardless of expression of *JAK2* V617F and without major differences in staining intensities (Figure 1B, Table 1).

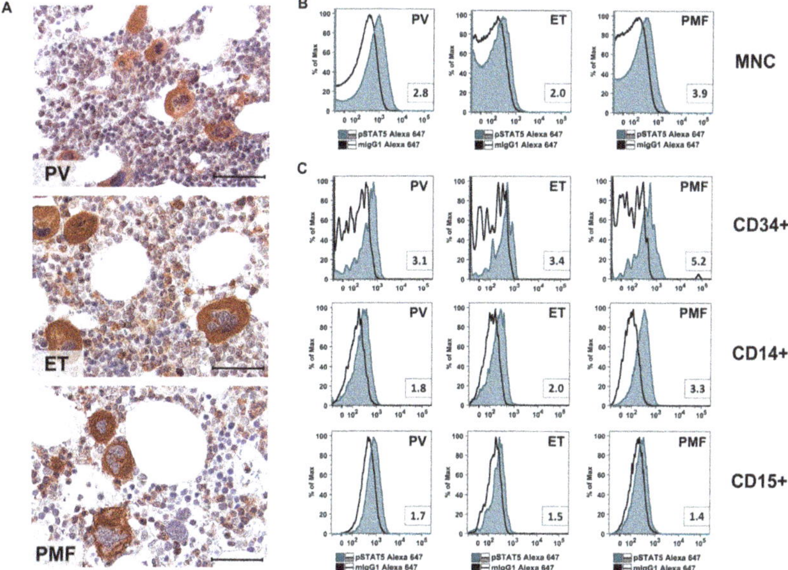

Figure 1. (**A**) Sections prepared from paraffin-embedded bone marrow (iliac crest) of patients with polycythemia vera (PV; patient #06), essential thrombocythemia (ET; patient #34) or primary myelofibrosis (PMF; patient #29) were stained with an anti-phosphorylated signal transducer and activator of transcription-5 (pSTAT5) antibody using immunohistochemistry. Examples of nuclear- and cytoplasmic staining are shown in Figure A1. Scale bar: 30 μm. Patient characteristics are shown in Table A1. (**B,C**) Bone marrow (BM) mononuclear cells (MNC) of patients with PV (patient #30), ET (patient #08) or PMF (patient #29) were stained with an anti-pSTAT5 Alexa-647 antibody. Intracellular expression levels of pSTAT5 were analyzed by flow cytometry in total MNC (**B**), or in cell subsets gated for CD34, CD14 or CD15 (**C**). The isotype-matched control antibody is also shown (open black histogram). Numbers in the small boxes represent the staining index defined as the ratio of the median fluorescence intensity (MFI) obtained with the anti-pSTAT5 antibody and MFI obtained with the isotype-matched control antibody (mIgG1).

Table 1. Immunohistochemical detection of pSTAT5 in bone marrow cells of MPN patients and controls.

Diagnosis	PV	PV	PV	PMF	PMF	PMF	ET	ET	ET	nBM	nBM	nBM	nBM
Patient #	18	06	30	20	29	16	34	02	23	46	48	50	51
Megakaryocytes	++	++	++	+	++	++	++	++	+	++	++	++	++
Myeloid prog.	+	+	+	+	+	+	+	+/−	+/−	+	+(+)	+	+
Neutrophil gran.	+	+	+/−	+	+	+	+	−	+/−	+/−	+	−	+/−
Eosinophil gran.	−	−	−	−	−	−	−	−	n.a.	−	−	n.a.	n.a.
Erythroid prog.	−	−	−	−	−	−	−	−	−	−	−	−	−

Score: ++, strong expression in most cells; +, clear expression in the majority of cases; +/-, expressed in subsets of cells or only weakly expressed; −, no expression (below detection limit). Abbreviations: CM, cutaneous mastocytosis; ET, essential thrombocythemia; gran., granulocytes; MPN, myeloproliferative neoplasm; n.a., not analyzed (no cells found by microscopy); nBM, normal bone marrow; NHL, Non-Hodgkin lymphoma; PMF, primary myelofibrosis; prog., progenitors; PV, polycythemia vera.

2.2. Primary CD34$^+$/CD38$^-$ MPN-SC Express pSTAT5

Expression of pSTAT5 in CD34$^+$/CD38$^-$ cells was examined in normal/reactive BM cells and in BM samples obtained from patients with MPN by multi-color flow cytometry. As visible in Figure 2, pSTAT5 was found to be expressed in normal CD34$^+$/CD38$^-$ stem cells as well as in CD34$^+$/CD38$^-$ MPN-SC. We also found that pSTAT5 levels were higher in *JAK2* V617F+ CD34$^+$/CD38$^-$ MPN-SC compared to normal stem cells ($p = 0.015$) (Figure 2A). In addition, we found that pSTAT5 is expressed at slightly higher levels in CD34$^+$/CD38$^-$ MPN-SC in *JAK2* V617F+ patients compared to *JAK2* V617F- patients, although the difference was not statistically significant ($p = 0.073$) (Figure 2B). However, no substantial differences in pSTAT5 expression in CD34$^+$/CD38$^-$ MPN cells were found when comparing various subsets of MPN (PV vs. ET vs. PMF) (Figure 2C).

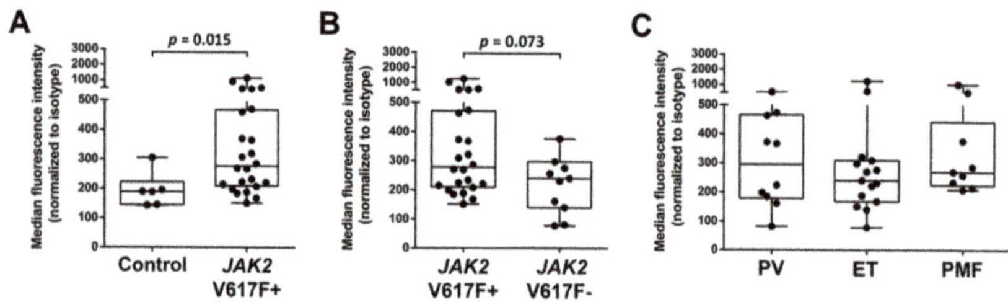

Figure 2. Bone marrow cells from patients with PV, ET or PMF were analyzed for intracellular expression of pSTAT5 in CD34$^+$/CD38$^-$/CD45^{dim} cells using an anti-pSTAT5 Alexa-647 antibody. (**A**) Expression of pSTAT5 in normal/reactive bone marrow (Control, $n = 6$) and bone marrow of MPN patients (MPN, $n = 24$). (**B**) Expression of pSTAT5 in CD34$^+$/CD38$^-$/CD45^{dim} bone marrow cells in *JAK2* V617F+ patients (V617F+, $n = 24$) and patients with wild type *JAK2*, a *CALR* mutation or an *MPL* mutation (V617F−, $n = 10$). (**C**) Expression of pSTAT5 in CD34$^+$/CD38$^-$/CD45^{dim} bone marrow cells in the three different MPN subgroups (PV, $n = 10$; ET, $n = 15$, PMF, $n = 9$). Boxes indicate the upper and lower quartiles, the median is defined by a horizontal line inside the boxes, and the whiskers show the highest and lowest values.

2.3. Detection of pSTAT5 in HEL and SET-2 Cells

We next examined the *JAK2* V617F+ cell lines HEL and SET-2 by immunocytochemistry (ICC). Confirming previous studies [36,37], we found that both cell lines express pSTAT5. Interestingly, both cell lines expressed pSTAT5 in their nuclear and cytoplasmic compartments (Figure 3A). Intracellular expression of pSTAT5 in these cell lines was also confirmed by flow cytometry (Figure 3B). Preincubation of HEL and SET-2 cells with JAK2-targeting drugs (ruxolitinib [38,39], R763 [40], TG101348 [41,42], AZD1480 [43,44]) or STAT5-targeting drugs (piceatannol [45], pimozide [46,47], AC-3-019 [48], AC-4-130 [48,49]) resulted in reduced pSTAT5 staining (Figure A2). The characteristics of the STAT5- and JAK2-targeting drugs used in these experiments together with their main targets are depicted in Table A2. We also compared pSTAT5 levels in purified nuclear and cytoplasmic fractions of HEL and SET-2 cells by Western blotting. Total STAT5 (STAT5A and STAT5B) was expressed more abundantly in the cytoplasmic extracts of HEL and SET-2 cells than in nuclear extracts (Figure 3C). Moreover, pSTAT5 was found at higher levels in the cytoplasm of HEL cells compared to nuclear fractions. By contrast, in SET-2 cells, pSTAT5 was found to be expressed more abundantly in the nuclear fractions compared to cytoplasmic fractions (Figure 3C).

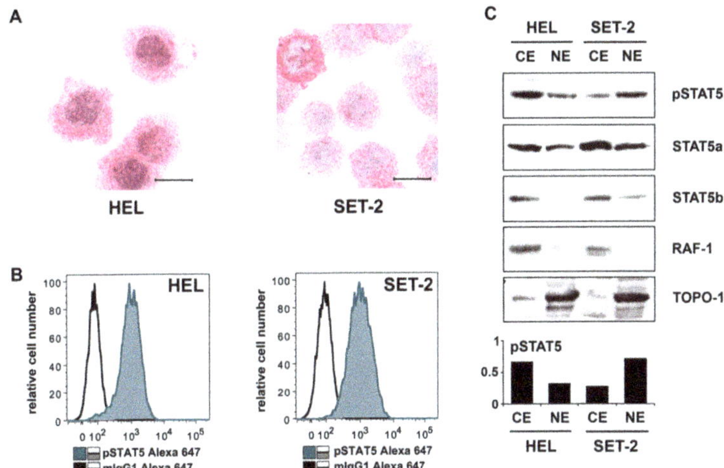

Figure 3. (**A**) Cytospin preparations of HEL cells and SET-2 cells were stained overnight with an anti-pSTAT5 antibody using immunocytochemistry. Scale bar: 20 µM. (**B**) HEL cells and SET-2 cells were stained with an anti-pSTAT5 Alexa-647 antibody for 30 min at room temperature and intracellular expression levels were analyzed by flow cytometry. (**C**) Nuclear (NE) and cytoplasmic (CE) fractions of HEL or SET-2 cells were analyzed for expression of phosphorylated STAT5 (pSTAT5), total STAT5a and total STAT5b by Western blotting. Antibodies against RAF-1 (cytoplasm) and TOPO-1 (nucleus) were used as fraction controls. The columns show the densitometry for pSTAT5. Uncropped blots are shown in Figure S1.

2.4. Effects of JAK2 V617F and CALR Mutants on Expression of Total STAT5 and pSTAT5 in Ba/F3 Cells

To explore the mechanism of expression and activation of STAT5 in MPN cells, we expressed *JAK2* V617F and mutated *CALR* in Ba/F3 cells containing human *MPL* (Ba/F3-MPL). As expected, expression of JAK2 V617F was followed by an increase in pSTAT5 levels in these cells (Figure A3). Moreover, we were able to show that drugs targeting STAT5 (piceatannol, pimozide) or JAK2 (AZD1480, TG101348, R763, ruxolitinib) counteract mutant-induced overexpression of pSTAT5 in our Ba/F3-MPL cells (Figure 4).

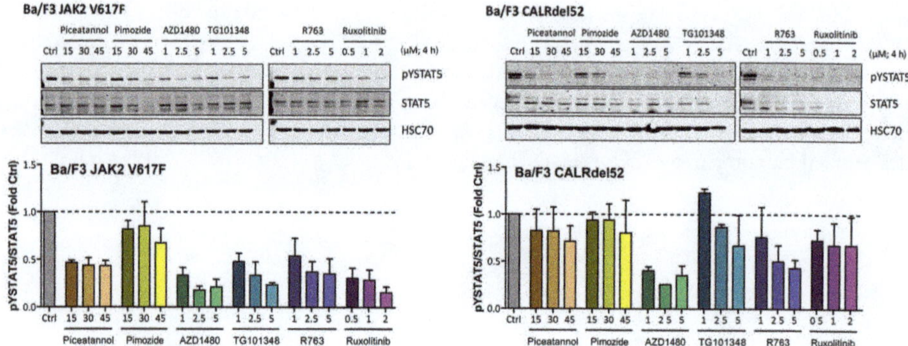

Figure 4. Ba/F3 cells modified to express JAK2 V617F or CALRdel52 were incubated with various JAK2 or STAT5 inhibitors (as indicated) for four hours at 37 °C. Thereafter, the expression of activated STAT5 (pSTAT5), total STAT5 (STAT5) and heat shock cognate 70 (HSC70), which was used as loading control, was analyzed using Western blotting. The results were quantified using densitometry and are expressed as the ratio of activated and total STAT5 normalized to the results obtained with control cells (Ctrl). Uncropped blots are shown in Figure S2.

2.5. Effects of Targeted Drugs on the Growth and Survival of HEL and SET-2 Cells

To confirm that pSTAT5 serves as a molecular target in MPN cells, we applied various JAK2-targeting drugs and STAT5-targeting drugs on HEL cells and SET-2 cells. As shown in Figure 5, the JAK2 blockers (ruxolitinib, R763, TG101348, AZD1480) and the STAT5 blockers (piceatannol, pimozide, AC-3-019, AC-4-130) were found to inhibit ^{3}[H]-thymidine uptake and thus proliferation with varying potency. For several drugs, these data confirmed previous studies [50–53]. In our study, the rank order of potency for HEL cells was AZD1480 > TG101348 > ruxolitinib > R763 > AC-4-130 > AC-3-019 > pimozide > piceatannol. The rank order of potency for SET-2 cells was R763 > ruxolitinib > AZD1480 > TG101348 > pimozide > AC-4-130 > AC-3-019 > piceatannol. Next, we examined the effects of various targeted drugs on the survival of MPN cells. In these experiments, we found that the JAK2 and STAT5 blockers examined induce apoptosis in HEL cells and SET-2 cells (Figure A4). In a separate set of experiments, we applied ruxolitinib and AC-4-130 in combination in HEL and SET-2 cells. However, no clear additive or synergistic effects of this drug combination was observed (Data not shown [54]). We also examined the effects of the JAK2 and STAT5 inhibitors on proliferation of two pSTAT5-low/negative solid cancer cell lines, A2780 and A375. As shown in Table A3, these cells were in general less sensitive to these drugs compared to HEL cells or SET2 cells.

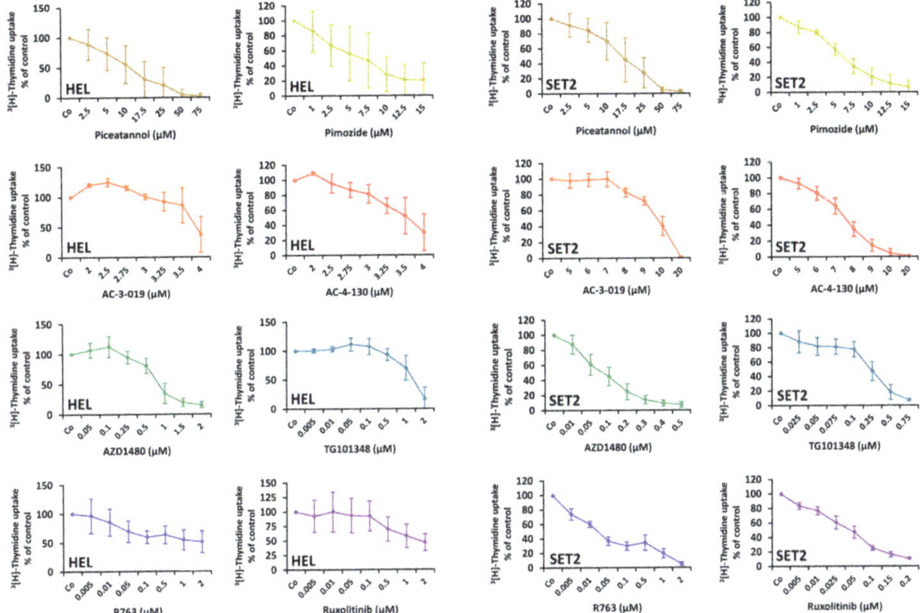

Figure 5. HEL and SET-2 cells were incubated with different concentrations of JAK2 or STAT5 targeting drugs for 48 h at 37 °C. Thereafter, ^{3}[H]-thymidine was added for 16 h and its incorporation was analyzed using a beta counter. The diagrams show the mean ± standard deviation (SD) of at least three independent experiments performed in triplicate.

2.6. Effects of Targeted Drugs on Primary Human MPN Cells

In a final step, we examined the effects of various JAK-2-targeting drugs and STAT5-targeting drugs on the growth and survival of primary mononuclear cells (MNC) obtained from the BM of patients with MPN. In these experiments, we found that the JAK-2 blockers and the STAT5 blockers used in this study are capable of inhibiting the proliferation of primary MPN cells (Figure 6A). Moreover, we found that these drugs decrease the relative numbers of primary CD34$^+$/CD38$^-$ MPN-SC in vitro (Figure 6B,C). Interestingly, ruxolitinib was found to decrease MPN-SC numbers after 48 h but not after 24 h. All drugs also decreased pSTAT5 levels in MPN-SC, although the downregulating effects of these drugs were rather weak (Figure 6D).

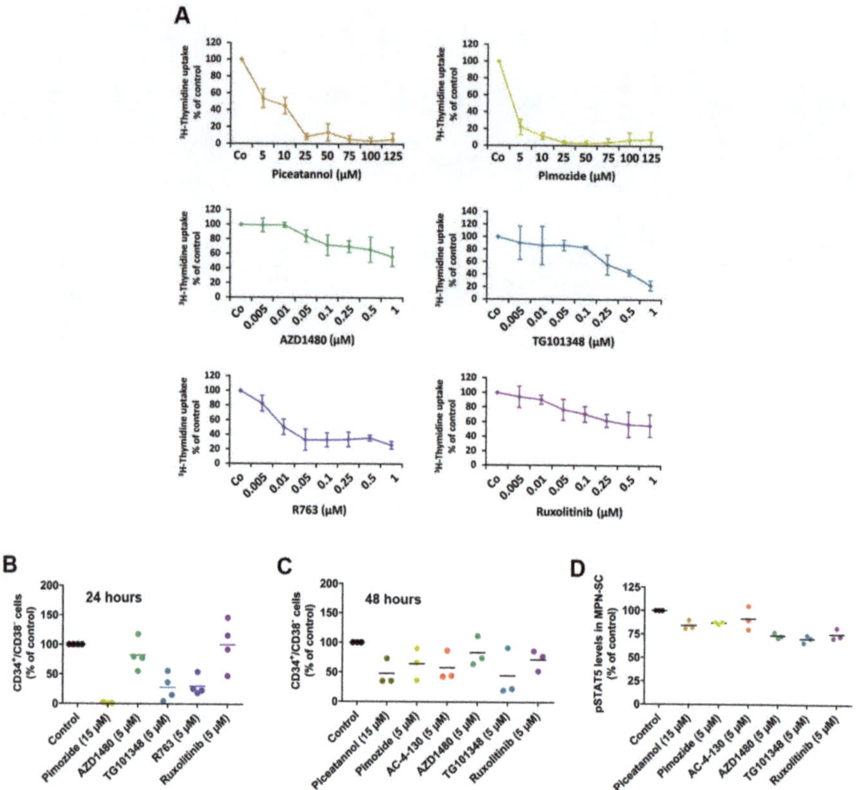

Figure 6. (**A**) Mononuclear cells isolated from the bone marrow of MPN patients were incubated with different concentrations of JAK2 or STAT5 targeting drugs for 48 h. Thereafter, ^{3}[H]-thymidine was added for 16 h and its incorporation was analyzed using a beta counter. The diagrams show the mean ± SD of three independent experiments for JAK2-targeting drugs (patients #07, #20 and #23) or five independent experiments for STAT5-targeting drugs (patients #20, #23, #37, #35 and #39). (**B,C**) Mononuclear cells isolated from the bone marrow of MPN patients were incubated with different concentrations of JAK2 or STAT5 targeting drugs for 24 h (**B**) or 48 h (**C**). Then, the numbers of putative MPN stem cells (MPN-SC), defined as CD34$^+$/CD38$^-$/CD45^{dim}, were assessed using counting beads. The graphs show MPN-SC as percent of control in four patients (#14, #29, #30 and #31) in (**B**) and in three patients (#31, #42 and #43) in (**C**). Horizontal lines indicate mean values. (**D**) pSTAT5 levels in CD34$^+$/CD38$^-$ MPN-SC determined by flow cytometry (same patients as in (**C**)) after an incubation with targeted drugs (as indicated) for 4 h. Values represent pSTAT5 levels relative to the control. The mean staining index is also shown (horizontal bars). Patient characteristics are shown in Table A1.

3. Discussion

Recent data suggest that STAT5 activation is a critical event triggering oncogenesis and growth of neoplastic cells in various hematologic malignancies [55,56]. It has also been described that leukemia-specific oncoproteins, such as BCR-ABL1, induce activation of STAT5 and thereby contribute to clonal expansion of neoplastic cells [57]. Several studies have shown that phosphorylated STAT5 (pSTAT5) is expressed in the cytoplasm and nuclei of neoplastic cells in patients with chronic myeloid leukemia (CML) and systemic mastocytosis (SM) [28,58]. Our study shows that neoplastic cells in PV and PMF also express nuclear and cytoplasmic pSTAT5 in a constitutive manner. In addition, we found that putative CD34$^+$/CD38$^-$ MPN-SC display pSTAT5. Finally, our data show that pharmacologic targeting of STAT5 reduces the growth of MPN cells and the numbers of MPN-SC. Together, these data suggest that the STAT5 pathway may contribute to oncogene-dependent growth of neoplastic cells in MPN. Whether inhibitors of pSTAT5, alone or in combination with other drugs, can exert clinically meaningful effects in MPN patients, remains unknown at present.

Recent data suggest that pSTAT5 is detectable in the cytoplasm of neoplastic cells in patients with AML, CML and SM, and that STAT5 acts as a pro-oncogenic driver in these myeloid neoplasms [28–30,58–60]. An interesting observation has been that the levels of cytoplasmic pSTAT5 often exceed the amounts of pSTAT5 found in the nuclear compartments in these cells [28,58]. In the current study, we found that pSTAT5 is located in both the cytoplasmic and nuclear fractions of primary MPN cells and in the MPN-related cell lines HEL and SET-2. In both cell lines, cytoplasmic and nuclear pSTAT5 were detected by ICC, flow cytometry and Western blotting (Figure 3). As assessed by Western blotting, the cytoplasmic and nuclear fractions of these cells displayed detectable levels of pSTAT5, and total STAT5 was found to be expressed more abundantly in the cytoplasmic fraction. Currently, it remains unknown whether nuclear pSTAT5 or cytoplasmic pSTAT5 plays a more important role in oncogenesis in MPN. Whereas nuclear pSTAT5 is considered to act as a pro-oncogenic transcription factors, cytoplasmic pSTAT5 may be involved in pro-oncogenic signaling involving the PI3 kinase pathway, similar to the situation in AML and SM [29,61].

MPN cells are considered to be organized in a 'stem cell hierarchy' similar to normal hematopoiesis [34]. In addition, MPN cells can undergo differentiation and terminal maturation in the same way as normal myeloid cells. As in normal hematopoiesis, only the most immature neoplastic stem cells (MPN-SC) have the capacity of self-renewal in MPN, a hypothesis that has major implications regarding drug therapy [62,63]. Notably, this model predicts that these cells can propagate the malignancy for unlimited time periods and that anti-neoplastic drugs have curative potential only when eliminating most or all of these neoplastic stem cells in a given neoplasm. So far, only little is known about the phenotype and function of MPN-SC [35,64]. Like in other myeloid neoplasms, these cells are considered to reside in a $CD34^+$ fraction of the clone [31]. In the present study, we were able to show that pSTAT5 is not only expressed in more mature clonal MPN cells, but also in immature $CD34^+/CD38^-$ (putative) MPN-SC (Figures 1 and 2). To the best of our knowledge, our study is the first to demonstrate that $CD34^+/CD38^-$ MPN-SC express pSTAT5. Regarding more mature clonal cells, monocytes, known to play a major role in MPN [65–67], were found to express high levels of pSTAT5, making these cells an additional potential target of STAT5 inhibition. As assessed by flow cytometry, we also found that the levels of pSTAT5 are higher in putative MPN-SC in patients with *JAK2* $V617F^+$ MPN compared to patients with *JAK2* $V617F^-$ MPN or normal $CD34^+/CD38^-$ stem cells. In contrast, no statistically significant difference was observed when comparing patients with *JAK2* $V617F^-$ MPN or normal $CD34^+/CD38^-$ stem cells, although some patients with *JAK2* $V617F^-$ MPN showed higher pSTAT5 levels than normal controls. Interestingly, when testing for pSTAT5 levels by IHC, no differences were found between the BM of MPN patients and normal controls, which seems to be in contrast to our flow cytometry results. However, our flow cytometry staining experiments were performed on stem cells, whereas IHC was performed on the bulk of MPN cells. Moreover, the IHC stain may be less capable of precisely quantifying differences in staining intensities compared to flow cytometry. This hypothesis is supported by the work of Teofili et al. [68], who did not detect differences in pSTAT5 levels between different subsets of MPN when using IHC in bulk cells. By contrast, using a flow cytometry approach, Abba et al. [69], were able to show differences in pSTAT5 levels when comparing $CD34^+$ MPN cells with normal $CD34^+$ cells.

In our study, no differences were seen when comparing pSTAT5 expression levels in $CD34^+/CD38^-$ MPN-SC among the three groups of MPN patients, and in each case, pSTAT5 was homogenously expressed in all MPN-SC in all patients. Collectively, these data suggest that neoplastic stem and progenitor cells in MPN patients express pSTAT5. As mentioned above, our results are also in line with the data published by Abba et al. [69], who showed that pSTAT5 levels are higher in $CD34^+$ MPN cells compared to normal CD34$^+$ cells. However, Abba et al. did not look into the more immature fraction of CD34$^+$/CD38$^-$ stem cells [69].

A number of previous and more recent studies have shown that STAT5 activation is a critical event triggering oncogenesis in MPN (stem) cells and that mutated forms of JAK2 and CALR can induce STAT5 activation [22–25]. In our study, we were able to show that putative SC in MPN express pSTAT5, but these cells only expressed pSTAT5 in excess over normal stem cells in patients with *JAK2*-mutated MPN. These data suggest that STAT5 may play a particular role in MPN-SC downstream of JAK2. However, more studies are required to define the exact role that STAT5 activation plays in the immature stem cell compartment in MPN.

Inhibitors of JAK2 and STAT5 were found to suppress expression of pSTAT5 in HEL and SET-2 cells (Figure A1). For several of these drugs, our data confirm the available literature [50–53]. An interesting observation was that whereas the JAK2-targeting drugs blocked STAT5 activation and MPN cell growth at relatively low concentrations, much higher concentrations of the STAT5 blockers were required to counteract proliferation in MPN cells (Figure 5). These observations suggest that targeting of STAT5 may be an interesting approach to block oncogenic signaling in MPN cells directly, but more potent and specific STAT5 inhibitors need to be developed to better inhibit MPN cell growth. Indeed, multiple attempts have been made recently to develop selective and potent STAT5 inhibitors. In the present study, we examined the effects of two such novel STAT5 inhibitors, AC-3-019 and AC-4-130 [49]. Similar to piceatannol and pimozide, these selective STAT5-SH2-domain-targeting drugs produced growth inhibition and apoptosis, albeit at relatively high concentrations (Figures 5 and A4). We also observed the effects of these inhibitors and of the other STAT5 or JAK2 blockers on MPN-SC. However, the effects on MPN-SC were weak, so these drugs are not expected to be able to eradicate the disease. One explanation for this result may be that other downstream signaling molecules and pathways are also involved in triggering oncogenesis and neoplastic stem cell growth in MPN. Therefore, we believe that more specific and more potent STAT5 blockers may be a reasonable approach to target MPN cells, and additional drugs or drug combinations, as proposed by Bar-Natan et al. [53] and others, may be required to elicit optimal anti-neoplastic or even curative effects.

4. Materials and Methods

4.1. Patients

Thirty-four patients with MPN, including 10 with PV, 15 with ET and 9 with PMF were examined for levels of pSTAT5 in MPN-SC. MPN variants were diagnosed according to WHO criteria [3]. Patient characteristics are shown in Table A1. Routine staging included physical examination, blood counts, morphologic examination of cells on BM smears, BM histology and immunohistochemistry, and analysis of BM and peripheral blood cells for expression of *JAK2* V617F, *CALR* and *MPL* mutations. Expression of *JAK2* V617F was determined by two PCR assays: the qualitative Ipsogena MutaScreen assay and the quantitative MutaQuant assay. Both assays were applied according to the recommendations of the manufacturer (Qiagen, Venlo, Netherlands). In *JAK2* V617F-negative patients, MPN cells were screened for *CALR* mutations using fluorescence- based PCR, followed by Sanger sequencing of codons 351–404 of *CALR* as described [16]. When no *JAK2* or *CALR* mutations were detected, the patients were analyzed for *MPL* mutations using an allelic discrimination assay as described [20]. In 15 patients (PV, $n = 4$; ET, $n = 5$; PMF, $n = 6$), BM mononuclear cells (MNC) were enriched using Ficoll. These MNC were used to assess drug effects on proliferation and/or survival. In addition, BM cells obtained from eight donors without MPN (control BM samples) were analyzed (Table A4). All investigations were approved by the local ethics committee of the Medical University of Vienna (ethic code: 224/2006; 1184/2014; 1063/2018). Informed consent was obtained from all patients.

4.2. Antibodies (Ab) and Other Reagents

The anti-pSTAT5 alpha (Tyr694) polyclonal Ab was purchased from Invitrogen (order number 71-6900; Carlsbad, CA, USA), anti-pSTAT5 (Y694) Alexa Fluor® 647 monoclonal Ab (mAb) 47 (order number: 612599) and an isotype-matched control antibody (mIgG1-Alexa Fluor® 647, order number: 557783) from BD Biosciences Pharmingen (San Jose, CA, USA), piceatannol (order number: P0453) and pimozide (order number: P1793) from Merck (Darmstadt, Germany), and the JAK2 blockers AZD1480 (order number: CT-A1480), TG101348 (order number: CT-TG101) and ruxolitinib (order number: CT-INCB) from ChemieTek (Indianapolis, IN, USA). The specificity of pSTAT5 (Y694) Alexa Fluor® 647 was confirmed by flow cytometry experiments performed with human cell lines with detectable pSTAT5 (KOPT-K1 and MYLA) or no detectable pSTAT5 (HUT78 and HH). In addition, we employed Ba/F3 cells with IL-3-inducible expression of pSTAT5. In these experiments, the specificity of the antibody was confirmed (Figure A5a,b). pSTAT5 expression in the cell lines tested was confirmed by Western Blotting (Figure A5c,d). Roswell Park Memorial Institute (RPMI) 1640 medium (order number: 10-041-CVR) was purchased from Corning (Manassas, VA, USA) and fetal calf serum (FCS; order number: 10270-106) from Gibco (Karlsbad, CA, USA). The JAK2 and Aurora kinase-targeting drug R763 was kindly provided by Yasumichi Hitoshi (Rigel Pharmaceuticals, San Francisco, CA, USA). The STAT5 inhibitors AC-3-019 and AC-4-130 were produced as described [48]. Stock solutions of drugs were prepared by dissolving in dimethyl sulfoxide (DMSO) (order number: D2650; Merck).

4.3. Cell Lines

A375 were purchased from LGC Standards (Wesel Germany); HEL and SET-2 cell lines were purchased from Deutsche Sammlung von Mikroorganismen und Zellkulturen (Braunschweig, Germany); KOPT-K1 cells were kindly provided by Takaomi Sanda (Cancer Science Institute of Singapore, Singapore); HH, HUT78 and MYLA were kindly provided by Marco Herling (University Hospital Cologne, Cologne, Germany). A2780 cells were purchased from Sigma (St. Louis, MO, USA). All cell lines were cultured in RPMI 1640 medium supplemented with 10% FCS.

4.4. Immunohistochemistry and Immunocytochemistry

Immunohistochemistry was performed on paraffin-embedded, formalin-fixed BM biopsy specimens using the indirect immunoperoxidase staining technique following established protocols [28]. Endogenous peroxidase was blocked by methanol/H_2O_2 and heat-induced epitope retrieval was performed (96 °C, 20 min, pH 9). A polyclonal anti-pSTAT5 antibody was applied at 1:100 for 20 h at 4 °C. Biotinylated goat anti-rabbit IgG (Vector Laboratories, Burlingham, CA, USA, order number: BA-1000) was applied as secondary antibody for 30 min at room temperature. Then, slides were exposed to an avidin/biotinylated peroxidase complex (Vectastain ABC Kit from Vector, order number: PK-6100) for 30 min. The chromogen 3-amino-9-ethylcarabzole (AEC) was then used. Finally, slides were counterstained in Mayer's hematoxylin (Morphisto, Frankfurt am Main, Germany, order number: 10231.01000). In each sample, the specificity of the anti-pSTAT5 stain was controlled by analyzing the internal negative control and positive control. In fact, erythroid cells always stained negative and megakaryocytes always stained positive as reported previously [28]. Immunocytochemistry was performed using HEL and SET-2 cells as described [28]. Cells were spun on cytospin slides and fixed with acetone for 8 min. Slides were pretreated in citrate buffer (pH 6.0) at 95 °C for 20 min and incubated with a polyclonal anti-pSTAT5 antibody (Invitrogen, Carlsbad, CA, USA order number: 71-6900) diluted 1:100, overnight at 4 °C. Slides were incubated with biotinylated goat anti-rabbit IgG (Biocare Medical, Walnut Creek, CA, USA) for 30 min at room temperature (RT) and then with streptavidin AP label (Biocare Medical, Walnut Creek, CA, USA) for 30 min. Neofuchsin (Nichirei, Tokyo, Japan) was used as the chromogen. All slides were counterstained in Mayer´s hematoxylin.

4.5. Flow Cytometry

Expression of cell surface antigens on primary neoplastic stem cells was analyzed by multicolor flow cytometry using antibodies against CD34, CD38 and CD45 as described [70]. Putative stem cells were defined as $CD34^+/CD38^-/CD45^{dim}$. The gating strategy is shown in Figure A6. For assessing the absolute numbers of stem cells by flow cytometry, Absolute Counting Beads (Thermo Fisher Scientific, Waltham, MA, USA, order number: C36950) were used according to the instructions by the manufacturer. For the flow cytometric detection of cytoplasmic pSTAT5, cells were first stained for cell surface antigens and then fixed in formaldehyde (2%). Cells were subsequently permeabilized by exposure to 50% methanol (-20 °C, 10 min), washed in phosphate-buffered saline containing 0.1% bovine serum albumin (order number: A4503; Merck) and stained with the Alexa 647-conjugated anti-pSTAT5 mAb 47 pY694 or an isotype-matched control antibody for 30 min at RT. Cells were then washed and analyzed on a FACSCanto (BD Biosciences). Staining reactions were expressed as median fluorescence intensity (MFI). pSTAT5 expression levels are shown as staining index (SI) defined as MFI produced by anti-pSTAT5 antibody:MFI of the isotype-matched control antibody.

4.6. Western Blot Analysis of Expression of pSTAT5 in Ba/F3 Cells

To confirm the selective effects of JAK2 or STAT5 targeting drugs, Western blot experiments were performed using the pSTAT5 antibody Tyr694 on Ba/F3 and Ba/F3-MPL cells engineered to express *JAK2* V617F and *CALR*del52, respectively. Ba/F3 cells were generated as described [71,72]. Cells were incubated with drugs targeting JAK2 (AZD1480, TG101348, ruxolitinib, R763; 1–5 µM) or STAT5 (piceatannol, pimozide; 10–45 µM) for 4 h. Then, pSTAT5 expression was analyzed by Western blotting essentially as described [49]. Nitrocellulose membranes (0.45 µm Amersham Protran; order number: 10600002; GE Healthcare, Buckinghamshire, UK) were incubated with the following antibodies at the dilution indicated: polyclonal rabbit anti-phospho-STAT5 (Y694) antibody (order number: 71-6900; 1:1000; Invitrogen, Camarillo, CA, USA), monoclonal mouse anti-STAT5 antibody 89/Stat5 (order number: 610191; 1:1000; BD Biosciences), mouse anti-HSC70 monoclonal antibody B-6 (order number: SC-7298; 1:10,000; Santa Cruz, St. Louis, MO, USA), IRDye® 680RD goat anti-rabbit IgG (order number: 925-68071; 1:10000; LI-COR, Lincoln, NE, USA), and IRDye® 800CW goat anti-mouse IgG (order number: 925-32210; 1:10000; LI-COR). pSTAT5 levels were quantified by densitometry and expressed as the pSTAT5/loading control ratio normalized to control (untreated) cells.

4.7. Isolation of the Cytoplasmic and Nuclear Fractions of HEL and SET-2 Cells

HEL and SET-2 cells were lysed in hypotonic buffer (20 mM HEPES, 10 mM KCl, 1 mM ethylenediaminetetraacetic acid, 0.2% NP40, 10% glycerol, 5 µg/mL aprotinin, 5 µg/mL leupeptin, 1 mM phenylmethylsulfonide fluoril, and 1 mM Na_2VO_4). Cell lysates were centrifuged (5 min, 800× g) to separate cytoplasmic and nuclear fractions [61]. Supernatants (cytoplasmic fraction) were frozen at -70 °C. Pelleted nuclei were resuspended in hypertonic buffer (hypotonic buffer plus 350 mM NaCl) and protein extracts were prepared by agitation (30 min, 4 °C). After debris was removed by centrifugation, nuclear extracts were frozen at -70 °C. Expression of pSTAT5 was determined by Western blotting and quantified by densitometry as reported [61]. Fractionation of subcellular compartments was controlled by applying anti-RAF-1 (cytoplasmic) and anti-topoisomerase-1 (nuclear) antibodies in parallel. All antibodies were from Santa Cruz.

4.8. Evaluation of Drug Effects on the Growth and Survival of MPN Cells

To further determine the functional role of STAT5 in MPN cells, we applied several targeting drugs: piceatannol, pimozide, AZD1480, TG101348, R763, ruxolitinib, AC-3-019 and AC-4-130 (Table A2). Primary human MPN cells (ET, $n = 2$; PMF, $n = 3$; PV, $n = 2$), HEL cells and SET-2 cells were incubated with increasing drug concentrations at 37 °C for 48 h. Thereafter, ^{3}[H]-thymidine was added, and its uptake was analyzed after 16 h using a beta-counter. For evaluation of apoptosis, HEL and SET-2 cells were incubated in control medium or in various drug concentrations for 24 h and 48 h at 37 °C. The percentage of apoptotic cells was quantified using Annexin V/Propidium Iodide staining as described [73]. In a subset of patients, we examined drug effects on putative CD34$^+$/CD38$^-$ MPN-SC.

5. Conclusions

STAT5 is a critical molecule in MPN cells that acts downstream of oncogenic JAK2 V617F and mutant CALR. We found that pSTAT5 is expressed abundantly in the nuclear and cytoplasmic compartment of MPN cells and that pSTAT5 is not only present in more mature clonal cells, but also in putative CD34$^+$/CD38$^-$ MPN-SC. Moreover, we show that STAT5 expression correlates with survival of MPN cells and that drugs targeting STAT5 can block growth and survival of these cells. Since MPN-SC display STAT5, and STAT5 is downstream of both JAK2 V617F and mutant CALR, targeting of STAT5 may be a promising approach to treat MPN.

Supplementary Materials: The following are available online at http://www.mdpi.com/2072-6694/12/4/1021/s1, Figure S1: Uncropped Blots of Figure 3C, Figure S2: Uncropped Blots of Figure 4, Figure S3: Uncropped Blots of Figure A3, Figure S4: Uncropped Blots of Figure A5.

Author Contributions: Conceptualization, E.H. and P.V.; methodology, validation and formal analysis, E.H., A.K., D.B., G.G., B.W., N.W., D.I., E.P., H.N., F.K.M.S., Y.F., C.K., H.A.N., L.M., G.H., F.G., R.K.; resources, G.T., J.P., E.D.d.A., P.T.G., L.M. R.K., R.M., P.V.; writing—original draft preparation, E.H. and P.V.; writing—review and editing, all authors; supervision, P.V.; project administration, E.H.; funding acquisition, E.H., R.K., H.A.N., P.T.G., R.M., P.V. All authors have read and agreed to the published version of the manuscript.

Funding: This research was funded by the National Bank of Austria, grant number 14835, by the Austrian Science Fund (FWF), SFB grant numbers F4701-B20, F4702-B20, F4704-B20, F4707-B20 and F6105-B20, and under the frame of ERA-NET (I 4157-B) and ERA PerMed (I 4218-B).

Acknowledgments: The authors would like to thank Sabine Cerny-Reiterer, Susanne Gamperl and Gabriele Stefanzl for their skillful technical assistance.

Conflicts of Interest: PTG is CSO of Janpix Inc., a Stat-targeting start-up company. The other authors declare no conflict of interest.

Appendix A

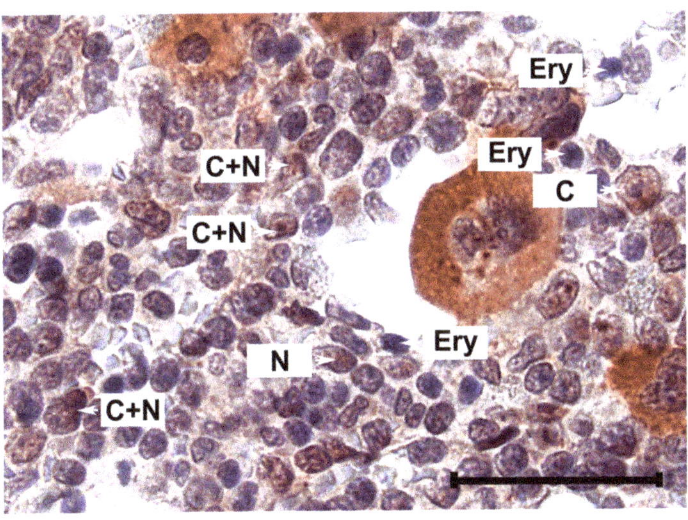

Figure A1. The picture shows an enlarged part of the picture of patient #06 (PV) from Figure 1A. The arrows point at examples of cytoplasmic 'C', nuclear 'N' and cytoplasmic/nuclear 'C+N' staining. 'Ery' marks examples of erythroid cells that were used as internal negative controls. Scale bar: 30 μm.

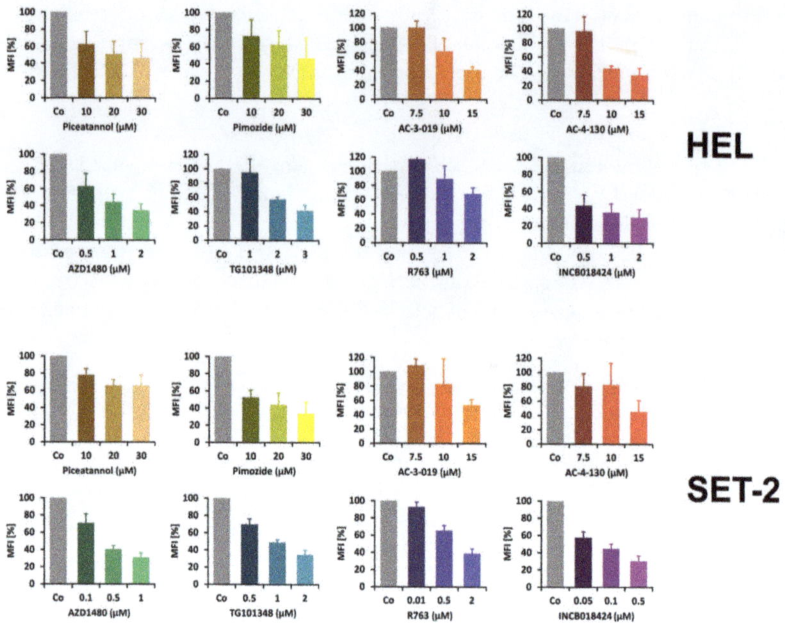

Figure A2. HEL and SET-2 cells were incubated with various concentrations of JAK2-targeting drugs (ruxolitinib, R763, TG101348, AZD1480) or STAT5-targeting drugs (piceatannol, pimozide, AC-3-019, AC-4-130) for 4 h. Thereafter, expression of pSTAT5 was analyzed by multicolor flow cytometry. The levels of pSTAT5 are depicted as median fluorescence index (MFI) values relative to the untreated control (Co).

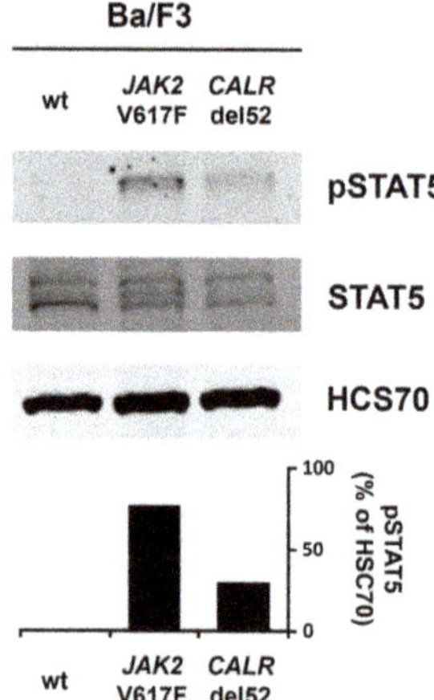

Figure A3. Wild type (wt) Ba/F3 cells or Ba/F3 cells transfected with either *JAK2* V617F or *CALRdel52* were incubated overnight in medium without IL-3. After, cell lysates were subjected to Western blotting using antibodies against pSTAT5, STAT5 and HSC70. The columns show the levels of pSTAT5 in percent of HSC70 as analyzed by densitometry. Uncropped blots are shown in Figure S3.

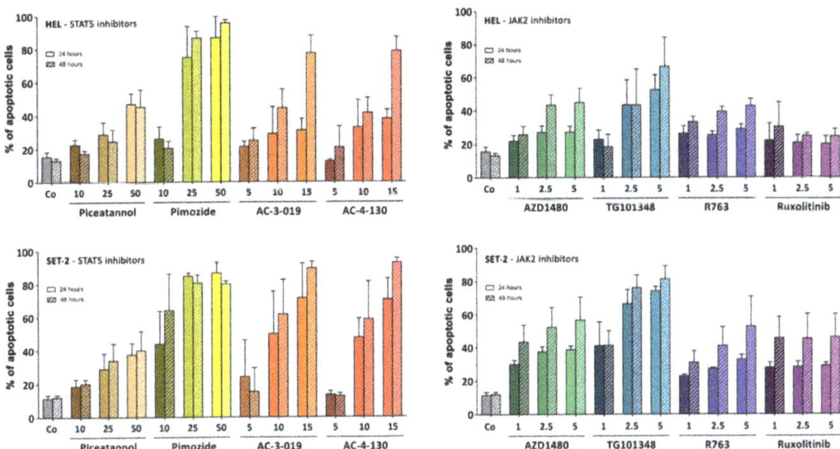

Figure A4. HEL and SET-2 cells (as indicated) were incubated with different concentrations of JAK2 or STAT5 targeting drugs for 24 h or 48 h at 37 °C. Thereafter, apoptosis induction was assessed using Annexin V/propidium iodide (PI) staining. The columns show the sum of cells positive for Annexin V-only, PI-only or Annexin V/PI for each condition and are depicted as mean ± SD of at least three independent experiments.

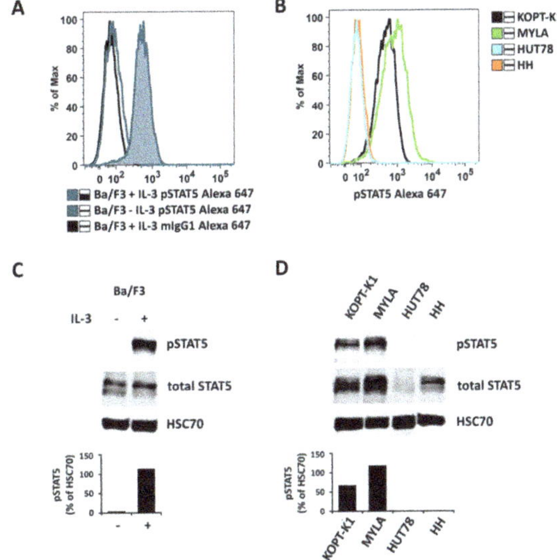

Figure A5. Ba/F3 cells were kept in medium without IL-3 overnight and split in the morning. One half was incubated with 10 ng/mL IL-3 for 40 min (Ba/F3 + IL-3), while the other half was kept in medium without IL-3 (Ba/F3 − IL-3). Thereafter, cells were analyzed for expression of pSTAT5 either by flow cytometry (**A**) or Western blotting (**C**). In (**A**), an Alexa-647-labeled anti-pSTAT5 antibody was used (with IL-3: blue histogram; without IL-3: open blue histogram). mIgG1 Alexa-647 was used as an isotype control (open black histogram) in Ba/F3 + IL-3 cells (**A**). Human lymphoma cell lines with detectable pSTAT5 levels (KOPT-K1 and MYLA) or no detectable pSTAT5 (HUT78 and HH) were also analyzed for pSTAT5 expression using flow cytometry (**B**) and Western blotting (**D**). In Western blot experiments, HSC70 served as a loading control. The columns show the levels of pSTAT5 in percent of HSC70 as analyzed by densitometry. Uncropped blots are shown in Figure S4.

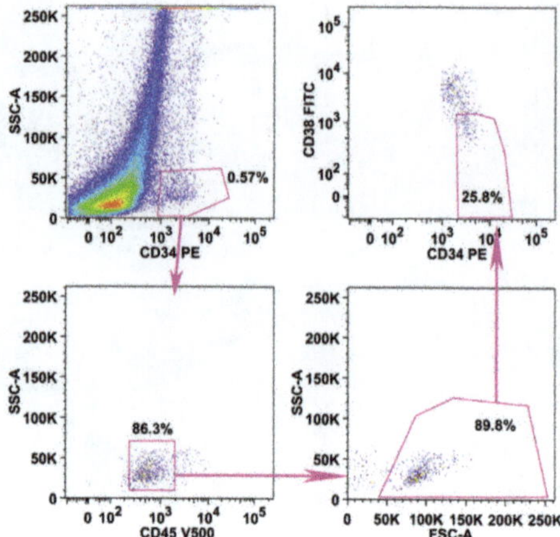

Figure A6. To identify immature CD34$^+$/CD38$^-$ stem cells (SC), we performed a sequential gating algorithm, indicated by pink arrows using FlowJo software. After excluding CD45-negative cells, immature cells were identified by their expression of CD34 (upper left panel) and were then gated as CD45 dim-positive blast cells (lower left panel). Thereafter, viable cells were selected by their light-scatter properties (lower right panel) and were then gated according to the lack (pink gate) of CD38 (upper right panel). The example shows bone marrow cells from patient #23.

Table A1. Patient characteristics.

#	Sex	Age at Sampling	Diagnosis	Mutated *JAK2, CALR* or *MPL*	Mutation	pSTAT5 FACS	pSTAT5 IHC	Proliferation	SC Assay	pSTAT5 Regulation
01	f	44	PMF	*JAK2*	V617F	yes	-	-	-	-
02	f	61	ET	none	–	yes	yes	-	-	-
03	f	53	ET	none	–	yes	-	-	-	-
04	f	71	ET	*JAK2*	V617F	yes	-	-	-	-
05	f	76	PV	*JAK2*	V617F	yes	-	-	-	-
06	f	76	PV	*JAK2*	V617F	yes	yes	-	-	-
07	f	53	PV	none	–	yes	-	yes	-	-
08	f	61	ET	*CALR*	ins5: c.1154_1155insTTGTC, p.Lys385Asnfs*47	yes	-	-	-	-
09	f	50	PMF	*JAK2*	V617F	yes	-	-	-	-
10	f	64	PMF	*JAK2*	V617F	yes	-	-	-	-
11	f	74	ET	*JAK2*	V617F	yes	-	-	-	-
12	f	52	PMF	*CALR*	ins5: c.1154_1155insTTGTC, p.Lys385Asnfs*47	yes	-	-	-	-
13	f	50	ET	*MPL*	W515L	yes	-	-	-	-
14	f	71	ET	*JAK2*	V617F	yes	-	-	yes	-
15	m	67	PV	none	–	yes	-	-	-	-
16	m	70	PMF	*CALR*	del52: c.1099_1150del, p. Leu367Thrfs*46	yes	yes	-	-	-
17	f	37	ET	none	–	yes	-	-	-	-
18	f	66	PV	*JAK2*	V617F	yes	yes	-	-	-
19	m	74	PV	*JAK2*	V617F	yes	-	-	-	-
20	m	53	PMF	*JAK2*	V617F	yes	yes	yes	-	-
21	m	71	PV	*JAK2*	V617F	yes	-	-	-	-
22	m	51	PMF	*JAK2*	V617F	yes	-	-	-	-
23	m	43	ET	*JAK2*	V617F	yes	yes	yes	-	-
24	m	59	ET	*CALR*	del52: c.1099_1150del, p. Leu367Thrfs*46	yes	-	-	-	-
25	f	84	ET	none	–	yes	-	-	-	-
26	m	37	ET	*JAK2*	V617F	yes	-	-	-	-
27	f	45	PMF	*JAK2*	V617F	yes	-	-	yes	-
28	f	31	ET	*JAK2*	V617F	yes	-	-	-	-
29	f	85	PMF	*JAK2*	V617F	yes	yes	-	yes	-
30	f	74	PV	*JAK2*	V617F	yes	yes	-	yes	-

Table A1. *Cont.*

#	Sex	Age at Sampling	Diagnosis	Mutated *JAK2, CALR or MPL*	Mutation	pSTAT5 FACS	pSTAT5 IHC	Proliferation	SC Assay	pSTAT5 Regulation
31	f	80	PV	*JAK2*	V617F	yes	-	-	yes	yes
32	f	72	ET	*JAK2*	V617F	yes	-	-	-	-
33	m	49	PV	*JAK2*	V617F	yes	-	-	-	-
34	m	64	ET	*JAK2*	V617F	yes	yes	-	-	-
35	f	71	PMF	*JAK2*	V617F	-	-	yes	-	-
36	f	68	ET	*JAK2*	V617F	-	-	yes	-	-
37	f	48	PMF	*JAK2*	V617F	-	-	yes	-	-
38	f	34	PV	*JAK2*	V617F	-	-	yes	-	-
39	m	88	PMF	*CALR*	del52: c.1099_1150del, p. Leu367Thrfs*46	-	-	-	yes	-
40	f	42	ET	*JAK2*	V617F	-	-	-	yes	-
41	f	38	ET	*JAK2*	V617F	-	-	-	yes	-
42	f	75	PV	*JAK2*	V617F	-	-	-	yes	yes
43	f	93	PMF	*JAK2*	V617F	-	-	-	yes	yes

CALR, calreticulin; ET, essential thrombocythemia; *JAK2*, *Janus kinase 2*; *MPL*, *MPL proto-oncogene, thrombopoietin receptor*; MPN, myeloproliferative neoplasm; PMF, primary myelofibrosis; Proliferation, effects of drugs on proliferation of MNC isolated from MPN patients; pSTAT5 FACS, sample analyzed for expression of pSTAT5 in CD34$^+$/CD38$^-$ putative stem cells; pSTAT5 IHC, sample analyzed by immunohistochemistry for expression of pSTAT5 in different bone marrow cells; pSTAT5 regulation, effects of JAK2 or STAT5 targeting drugs on pSTAT5 levels in CD34$^+$/CD38$^-$/CD45^{dim} cells; SC assay, effects of JAK2 or STAT5 targeting drugs on relative numbers of CD34$^+$/CD38$^-$/CD45^{dim} cells.

Table A2. Specifications of inhibitors used in this study.

Name [Ref]	Known Targets	Result of Target Inhibition (Drug Action)	Clinical Application	C_{max} in µM (at Dose/Day)	Concentration Range (µM) *	Supplier **
R763 [40]	JAK2, aurora kinase A, aurora kinase B, FLT3	G2/M phase arrest, endoreduplication	Phase 1 trial (human): 11.4–85.3 mg/day	0.09 (85.3 mg)	0.005–5	Rigel Pharmaceuticals
TG101348 [41,42]	JAK2, JAK1, JAK3, FLT3, RET, TYK2	proliferation inhibition, apoptosis induction	Phase 1 trial (human): 100–600 mg/day	2.8 (500 mg)	0.005–7.5	ChemieTek
AZD1480 [43,44]	JAK2, JAK1, JAK3, TYK2, FGFR3, STAT3	proliferation inhibition, apoptosis induction	Phase 1 trial (human): 5–80 mg/day	1.7 (30 mg)	0.005–7.5	ChemieTek
Ruxolitinib [38,39]	JAK2, JAK1, JAK3	proliferation inhibition, apoptosis induction	approved (human): PMF, PV (EU, USA)	2.3 (50 mg)	1–20	ChemieTek
Pimozide [46,47]	STAT5, D_2 dopamine receptor, 5-HT$_7$ receptor, Ca^{2+} channels	proliferation inhibition, apoptosis induction	approved (human): Tourette's Syndrome (EU, USA)	0.007 (2 mg)	0.5–25	Merck
Piceatannol [45]	STAT5, SYK, LCK	proliferation inhibition, apoptosis induction, histamine release blockade	n.a.	8.1 (80.7 mg/kg)	1–75	Merck
AC-3-019 [48]	STAT5	proliferation inhibition, apoptosis induction	n.a.	n.a.	0.05–15	P. Gunning
AC-4-130 [48,49]	STAT5	proliferation inhibition, apoptosis induction	n.a.	n.a.	0.25–15	P. Gunning

C_{max}, peak plasma concentration of a drug after administration; FGFR3, fibroblast growth factor receptor 3; FLT3, fms-like tyrosine kinase 3; JAK, janus kinase; LCK, leukocyte C-terminal SRC kinase; LYN, LCK/YES novel tyrosine kinase; n.a., not applicable; PMF, primary myelofibrosis; PV, polycythemia vera; Ref, Reference; STAT, signal transducer and activator of transcription; SYK, spleen tyrosine kinase; TYK2, tyrosine kinase 2; * Range of concentrations used in this study. ** ChemieTek (Indianapolis, IN, USA); LC Laboratories (Woburn, MA, USA); Merck, (Darmstadt, Germany); Rigel Pharmaceuticals (San Francisco, CA, USA); Patrick Gunning (Department of Chemistry, University of Toronto, Canada).

Table A3. IC_{50} concentrations of JAK2- and STAT5 targeting drugs in pSTAT5 high and low cell lines.

IC_{50} Obtained with Drugs in Various Cell Lines (SI pSTAT5)				
Inhibitor	HEL (13.4)	SET-2 (8.6)	A375 (1.93)	A2780 (1.97)
Piceatannol	10–17.5	10–17.5	>75	17.5–25
Pimozide	5–7.5	5–7.5	12.5–15	12.5–15
AC-3-019	3.5–4	9–10	10–12.5	12.5–15
AC-4-130	3.5–4	7–8	>15	>15
AZD1480	0.5–1	0.05–0.1	>2	0.5–1
TG1010348	1–2	0.1–0.25	>2	1–2
Ruxolitinib	1–2	0.0025–0.05	>2	>2

IC_{50} concentration: half maximal inhibitory concentration; SI: Staining Index (MFI produced by the anti-pSTAT5 antibody:MFI of the isotype-matched control antibody).

Table A4. Characteristics of control bone marrow donors.

#	Sex	Age at Sampling	Diagnosis	pSTAT5 FACS	pSTAT5 IHC
44	f	33	NHL	yes	-
45	m	37	nBM	yes	-
46	m	40	NHL	yes	yes
47	m	24	B-ALL (CR)	yes	-
48	f	47	CM	yes	yes
49	m	26	NHL	yes	-
50	f	33	NHL	-	yes
51	m	61	NHL	-	yes

B-ALL (CR), b-cell acute lymphoblastic leukemia in complete remission; CM, cutaneous mastocytosis; nBM, normal bone marrow; NHL, non-Hodgkin lymphoma; PMF, primary myelofibrosis; pSTAT5 FACS, sample analyzed for expression of pSTAT5 in CD34$^+$/CD38$^-$ putative stem cells; pSTAT5 IHC, sample analyzed by immunohistochemistry for expression of pSTAT5 in different bone marrow cells; PV, polycythemia vera.

References

1. Campbell, P.J.; Green, A.R. The myeloproliferative disorders. *N. Engl. J. Med.* **2006**, *355*, 2452–2466. [CrossRef] [PubMed]

2. Scott, L.M.; Tong, W.; Levine, R.L.; Scott, M.A.; Beer, P.A.; Stratton, M.R.; Futreal, P.A.; Erber, W.N.; McMullin, M.F.; Harrison, C.N.; et al. JAK2 exon 12 mutations in polycythemia vera and idiopathic erythrocytosis. *N. Engl. J. Med.* **2007**, *356*, 459–468. [CrossRef] [PubMed]

3. Arber, D.A.; Orazi, A.; Hasserjian, R.; Thiele, J.; Borowitz, M.J.; Le Beau, M.M.; Bloomfield, C.D.; Cazzola, M.; Vardiman, J.W. The 2016 revision to the World Health Organization classification of myeloid neoplasms and acute leukemia. *Blood* **2016**, *127*, 2391–2405. [CrossRef] [PubMed]

4. Skoda, R.C.; Duek, A.; Grisouard, J. Pathogenesis of myeloproliferative neoplasms. *Exp. Hematol.* **2015**, *43*, 599–608. [CrossRef] [PubMed]

5. Spivak, J.L. Myeloproliferative Neoplasms. *N. Engl. J. Med.* **2017**, *376*, 2168–2181. [CrossRef] [PubMed]

6. Tefferi, A. Primary myelofibrosis: 2012 update on diagnosis, risk stratification, and management. *Am. J. Hematol.* **2011**, *86*, 1017–1026. [CrossRef] [PubMed]

7. Cherington, C.; Slack, J.L.; Leis, J.; Adams, R.H.; Reeder, C.B.; Mikhael, J.R.; Camoriano, J.; Noel, P.; Fauble, V.; Betcher, J.; et al. Allogeneic stem cell transplantation for myeloproliferative neoplasm in blast phase. *Leuk. Res.* **2012**, *36*, 1147–1151. [CrossRef] [PubMed]

8. Deeg, H.J.; Bredeson, C.; Farnia, S.; Ballen, K.; Gupta, V.; Mesa, R.A.; Popat, U.; Hari, P.; Saber, W.; Seftel, M.; et al. Hematopoietic Cell Transplantation as Curative Therapy for Patients with Myelofibrosis: Long-Term Success in all Age Groups. *Biol. Blood Marrow Transpl.* **2015**, *21*, 1883–1887. [CrossRef]

9. Wolf, D.; Rudzki, J.; Gastl, G. Current treatment concepts of Philadelphia-negative MPN. *Curr. Cancer Drug Targets* **2011**, *11*, 44–55. [CrossRef]

10. Barosi, G.; Lupo, L.; Rosti, V. Management of myeloproliferative neoplasms: From academic guidelines to clinical practice. *Curr. Hematol. Malig. Rep.* **2012**, *7*, 50–56. [CrossRef]

11. Emadi, A.; Spivak, J.L. Anagrelide: 20 years later. *Expert Rev. Anticancer Ther.* **2009**, *9*, 37–50. [CrossRef] [PubMed]

12. Harrison, C.; Kiladjian, J.J.; Al-Ali, H.K.; Gisslinger, H.; Waltzman, R.; Stalbovskaya, V.; McQuitty, M.; Hunter, D.S.; Levy, R.; Knoops, L.; et al. JAK inhibition with ruxolitinib versus best available therapy for myelofibrosis. *N. Engl. J. Med.* **2012**, *366*, 787–798. [CrossRef] [PubMed]

13. Kiladjian, J.J.; Giraudier, S.; Cassinat, B. Interferon-alpha for the therapy of myeloproliferative neoplasms: Targeting the malignant clone. *Leukemia* **2016**, *30*, 776–781. [CrossRef] [PubMed]

14. Vannucchi, A.M.; Harrison, C.N. Emerging treatments for classical myeloproliferative neoplasms. *Blood* **2017**, *129*, 693–703. [CrossRef]

15. Verstovsek, S.; Mesa, R.A.; Gotlib, J.; Levy, R.S.; Gupta, V.; DiPersio, J.F.; Catalano, J.V.; Deininger, M.; Miller, C.; Silver, R.T.; et al. A double-blind, placebo-controlled trial of ruxolitinib for myelofibrosis. *N. Engl. J. Med.* **2012**, *366*, 799–807. [CrossRef] [PubMed]

16. Klampfl, T.; Gisslinger, H.; Harutyunyan, A.S.; Nivarthi, H.; Rumi, E.; Milosevic, J.D.; Them, N.C.; Berg, T.; Gisslinger, B.; Pietra, D.; et al. Somatic mutations of calreticulin in myeloproliferative neoplasms. *N. Engl. J. Med.* **2013**, *369*, 2379–2390. [CrossRef]

17. Kralovics, R.; Passamonti, F.; Buser, A.S.; Teo, S.S.; Tiedt, R.; Passweg, J.R.; Tichelli, A.; Cazzola, M.; Skoda, R.C. A gain-of-function mutation of JAK2 in myeloproliferative disorders. *N. Engl. J. Med.* **2005**, *352*, 1779–1790. [CrossRef]

18. Nangalia, J.; Massie, C.E.; Baxter, E.J.; Nice, F.L.; Gundem, G.; Wedge, D.C.; Avezov, E.; Li, J.; Kollmann, K.; Kent, D.G.; et al. Somatic CALR mutations in myeloproliferative neoplasms with nonmutated JAK2. *N. Engl. J. Med.* **2013**, *369*, 2391–2405. [CrossRef]

19. Pardanani, A.D.; Levine, R.L.; Lasho, T.; Pikman, Y.; Mesa, R.A.; Wadleigh, M.; Steensma, D.P.; Elliott, M.A.; Wolanskyj, A.P.; Hogan, W.J.; et al. MPL515 mutations in myeloproliferative and other myeloid disorders: A study of 1182 patients. *Blood* **2006**, *108*, 3472–3476. [CrossRef]

20. Pikman, Y.; Lee, B.H.; Mercher, T.; McDowell, E.; Ebert, B.L.; Gozo, M.; Cuker, A.; Wernig, G.; Moore, S.; Galinsky, I.; et al. MPLW515L is a novel somatic activating mutation in myelofibrosis with myeloid metaplasia. *PLoS Med.* **2006**, *3*, e270. [CrossRef]

21. Funakoshi-Tago, M.; Tago, K.; Abe, M.; Sonoda, Y.; Kasahara, T. STAT5 activation is critical for the transformation mediated by myeloproliferative disorder-associated JAK2 V617F mutant. *J. Biol. Chem.* **2010**, *285*, 5296–5307. [CrossRef] [PubMed]

22. Gibson, S.E.; Schade, A.E.; Szpurka, H.; Bak, B.; Maciejewski, J.P.; Hsi, E.D. Phospho-STAT5 expression pattern with the MPL W515L mutation is similar to that seen in chronic myeloproliferative disorders with JAK2 V617F. *Hum. Pathol.* **2008**, *39*, 1111–1114. [CrossRef] [PubMed]

23. Koschmieder, S.; Mughal, T.I.; Hasselbalch, H.C.; Barosi, G.; Valent, P.; Kiladjian, J.J.; Jeryczynski, G.; Gisslinger, H.; Jutzi, J.S.; Pahl, H.L.; et al. Myeloproliferative neoplasms and inflammation: Whether to target the malignant clone or the inflammatory process or both. *Leukemia* **2016**, *30*, 1018–1024. [CrossRef]

24. O'Sullivan, J.M.; Harrison, C.N. JAK-STAT signaling in the therapeutic landscape of myeloproliferative neoplasms. *Mol. Cell. Endocrinol.* **2017**, *451*, 71–79. [CrossRef] [PubMed]

25. Pasquier, F.; Cabagnols, X.; Secardin, L.; Plo, I.; Vainchenker, W. Myeloproliferative neoplasms: JAK2 signaling pathway as a central target for therapy. *Clin. Lymphoma Myeloma Leuk.* **2014**, *14*, S23–S35. [CrossRef] [PubMed]

26. Walz, C.; Ahmed, W.; Lazarides, K.; Betancur, M.; Patel, N.; Hennighausen, L.; Zaleskas, V.M.; Van Etten, R.A. Essential role for Stat5a/b in myeloproliferative neoplasms induced by BCR-ABL1 and JAK2(V617F) in mice. *Blood* **2012**, *119*, 3550–3560. [CrossRef]

27. Yan, D.; Hutchison, R.E.; Mohi, G. Critical requirement for Stat5 in a mouse model of polycythemia vera. *Blood* **2012**, *119*, 3539–3549. [CrossRef]

28. Baumgartner, C.; Cerny-Reiterer, S.; Sonneck, K.; Mayerhofer, M.; Gleixner, K.V.; Fritz, R.; Kerenyi, M.; Boudot, C.; Gouilleux, F.; Kornfeld, J.W.; et al. Expression of activated STAT5 in neoplastic mast cells in systemic mastocytosis: Subcellular distribution and role of the transforming oncoprotein KIT D816V. *Am. J. Pathol.* **2009**, *175*, 2416–2429. [CrossRef]

29. Harir, N.; Pecquet, C.; Kerenyi, M.; Sonneck, K.; Kovacic, B.; Nyga, R.; Brevet, M.; Dhennin, I.; Gouilleux-Gruart, V.; Beug, H.; et al. Constitutive activation of Stat5 promotes its cytoplasmic localization and association with PI3-kinase in myeloid leukemias. *Blood* **2007**, *109*, 1678–1686. [CrossRef]

30. Weber, A.; Borghouts, C.; Brendel, C.; Moriggl, R.; Delis, N.; Brill, B.; Vafaizadeh, V.; Groner, B. Stat5 Exerts Distinct, Vital Functions in the Cytoplasm and Nucleus of Bcr-Abl+ K562 and Jak2(V617F)+ HEL Leukemia Cells. *Cancers* **2015**, *7*, 503–537. [CrossRef]

31. Angona, A.; Alvarez-Larran, A.; Bellosillo, B.; Longaron, R.; Camacho, L.; Fernandez-Rodriguez, M.C.; Pairet, S.; Besses, C. Characterization of CD34+ hematopoietic progenitor cells in JAK2V617F and CALR-mutated myeloproliferative neoplasms. *Leuk. Res.* **2016**, *48*, 11–15. [CrossRef] [PubMed]

32. Chaligne, R.; James, C.; Tonetti, C.; Besancenot, R.; Le Couedic, J.P.; Fava, F.; Mazurier, F.; Godin, I.; Maloum, K.; Larbret, F.; et al. Evidence for MPL W515L/K mutations in hematopoietic stem cells in primitive myelofibrosis. *Blood* **2007**, *110*, 3735–3743. [CrossRef] [PubMed]

33. Jamieson, C.H.; Gotlib, J.; Durocher, J.A.; Chao, M.P.; Mariappan, M.R.; Lay, M.; Jones, C.; Zehnder, J.L.; Lilleberg, S.L.; Weissman, I.L. The JAK2 V617F mutation occurs in hematopoietic stem cells in polycythemia vera and predisposes toward erythroid differentiation. *Proc. Natl. Acad. Sci. USA* **2006**, *103*, 6224–6229. [CrossRef] [PubMed]

34. Mead, A.J.; Mullally, A. Myeloproliferative neoplasm stem cells. *Blood* **2017**, *129*, 1607–1616. [CrossRef] [PubMed]

35. Staerk, J.; Constantinescu, S.N. The JAK-STAT pathway and hematopoietic stem cells from the JAK2 V617F perspective. *JAK-STAT* **2012**, *1*, 184–190. [CrossRef]

36. Quentmeier, H.; Geffers, R.; Jost, E.; Macleod, R.A.; Nagel, S.; Rohrs, S.; Romani, J.; Scherr, M.; Zaborski, M.; Drexler, H.G. SOCS2: Inhibitor of JAK2V617F-mediated signal transduction. *Leukemia* **2008**, *22*, 2169–2175. [CrossRef]

37. Liu, R.Y.; Fan, C.; Garcia, R.; Jove, R.; Zuckerman, K.S. Constitutive activation of the JAK2/STAT5 signal transduction pathway correlates with growth factor independence of megakaryocytic leukemic cell lines. *Blood* **1999**, *93*, 2369–2379. [CrossRef]

38. Hornakova, T.; Springuel, L.; Devreux, J.; Dusa, A.; Constantinescu, S.N.; Knoops, L.; Renauld, J.C. Oncogenic JAK1 and JAK2-activating mutations resistant to ATP-competitive inhibitors. *Haematologica* **2011**, *96*, 845–853. [CrossRef]

39. Bartalucci, N.; Calabresi, L.; Balliu, M.; Martinelli, S.; Rossi, M.C.; Villeval, J.L.; Annunziato, F.; Guglielmelli, P.; Vannucchi, A.M. Inhibitors of the PI3K/mTOR pathway prevent STAT5 phosphorylation in JAK2V617F mutated cells through PP2A/CIP2A axis. *Oncotarget* **2017**, *8*, 96710–96724. [CrossRef]

40. Peter, B.; Bibi, S.; Eisenwort, G.; Wingelhofer, B.; Berger, D.; Stefanzl, G.; Blatt, K.; Herrmann, H.; Hadzijusufovic, E.; Hoermann, G.; et al. Drug-induced inhibition of phosphorylation of STAT5 overrides drug resistance in neoplastic mast cells. *Leukemia* **2018**, *32*, 1016–1022. [CrossRef]

41. Geron, I.; Abrahamsson, A.E.; Barroga, C.F.; Kavalerchik, E.; Gotlib, J.; Hood, J.D.; Durocher, J.; Mak, C.C.; Noronha, G.; Soll, R.M.; et al. Selective inhibition of JAK2-driven erythroid differentiation of polycythemia vera progenitors. *Cancer Cell* **2008**, *13*, 321–330. [CrossRef] [PubMed]

42. Wernig, G.; Kharas, M.G.; Okabe, R.; Moore, S.A.; Leeman, D.S.; Cullen, D.E.; Gozo, M.; McDowell, E.P.; Levine, R.L.; Doukas, J.; et al. Efficacy of TG101348, a selective JAK2 inhibitor, in treatment of a murine model of JAK2V617F-induced polycythemia vera. *Cancer Cell* **2008**, *13*, 311–320. [CrossRef] [PubMed]

43. Cook, A.M.; Li, L.; Ho, Y.; Lin, A.; Li, L.; Stein, A.; Forman, S.; Perrotti, D.; Jove, R.; Bhatia, R. Role of altered growth factor receptor-mediated JAK2 signaling in growth and maintenance of human acute myeloid leukemia stem cells. *Blood* **2014**, *123*, 2826–2837. [CrossRef] [PubMed]

44. Ioannidis, S.; Lamb, M.L.; Wang, T.; Almeida, L.; Block, M.H.; Davies, A.M.; Peng, B.; Su, M.; Zhang, H.J.; Hoffmann, E.; et al. Discovery of 5-chloro-N2-[(1S)-1-(5-fluoropyrimidin-2-yl)ethyl]-N4-(5-methyl-1H-pyrazol-3-yl)p yrimidine-2,4-diamine (AZD1480) as a novel inhibitor of the Jak/Stat pathway. *J. Med. Chem.* **2011**, *54*, 262–276. [CrossRef] [PubMed]

45. Su, L.; David, M. Distinct mechanisms of STAT phosphorylation via the interferon-alpha/beta receptor. Selective inhibition of STAT3 and STAT5 by piceatannol. *J. Biol. Chem.* **2000**, *275*, 12661–12666. [CrossRef]

46. Nelson, E.A.; Walker, S.R.; Weisberg, E.; Bar-Natan, M.; Barrett, R.; Gashin, L.B.; Terrell, S.; Klitgaard, J.L.; Santo, L.; Addorio, M.R.; et al. The STAT5 inhibitor pimozide decreases survival of chronic myelogenous leukemia cells resistant to kinase inhibitors. *Blood* **2011**, *117*, 3421–3429. [CrossRef]

47. Rondanin, R.; Simoni, D.; Maccesi, M.; Romagnoli, R.; Grimaudo, S.; Pipitone, R.M.; Meli, M.; Cascio, A.; Tolomeo, M. Effects of Pimozide Derivatives on pSTAT5 in K562 Cells. *ChemMedChem* **2017**, *12*, 1183–1190. [CrossRef]

48. Cumaraswamy, A.A.; Lewis, A.M.; Geletu, M.; Todic, A.; Diaz, D.B.; Cheng, X.R.; Brown, C.E.; Laister, R.C.; Muench, D.; Kerman, K.; et al. Nanomolar-Potency Small Molecule Inhibitor of STAT5 Protein. *ACS Med. Chem. Lett.* **2014**, *5*, 1202–1206. [CrossRef]

49. Wingelhofer, B.; Maurer, B.; Heyes, E.C.; Cumaraswamy, A.A.; Berger-Becvar, A.; de Araujo, E.D.; Orlova, A.; Freund, P.; Ruge, F.; Park, J.; et al. Pharmacologic inhibition of STAT5 in acute myeloid leukemia. *Leukemia* **2018**, *32*, 1135–1146. [CrossRef]

50. Szymanska, J.; Smolewski, P.; Majchrzak, A.; Cebula-Obrzut, B.; Chojnowski, K.; Trelinski, J. Pro-Apoptotic Activity of Ruxolitinib Alone and in Combination with Hydroxyurea, Busulphan, and PI3K/mTOR Inhibitors in JAK2-Positive Human Cell Lines. *Adv. Clin. Exp. Med.* **2015**, *24*, 195–202. [CrossRef]

51. Fiskus, W.; Verstovsek, S.; Manshouri, T.; Smith, J.E.; Peth, K.; Abhyankar, S.; McGuirk, J.; Bhalla, K.N. Dual PI3K/AKT/mTOR inhibitor BEZ235 synergistically enhances the activity of JAK2 inhibitor against cultured and primary human myeloproliferative neoplasm cells. *Mol. Cancer Ther.* **2013**, *12*, 577–588. [CrossRef] [PubMed]

52. Bogani, C.; Bartalucci, N.; Martinelli, S.; Tozzi, L.; Guglielmelli, P.; Bosi, A.; Vannucchi, A.M.; Associazione Italiana per la Ricerca sul Cancro, A.G.I.M.M. mTOR inhibitors alone and in combination with JAK2 inhibitors effectively inhibit cells of myeloproliferative neoplasms. *PLoS ONE* **2013**, *8*, e54826. [CrossRef] [PubMed]

53. Bar-Natan, M.; Nelson, E.A.; Walker, S.R.; Kuang, Y.; Distel, R.J.; Frank, D.A. Dual inhibition of Jak2 and STAT5 enhances killing of myeloproliferative neoplasia cells. *Leukemia* **2012**, *26*, 1407–1410. [CrossRef] [PubMed]

54. Emir Hadzijusufovic (Ludwig Boltzmann Institute for Hematology and Oncology, Medical University of Vienna, 1090 Vienna, Austria). *Personal observation*, 2018.

55. Freund, P.; Kerenyi, M.A.; Hager, M.; Wagner, T.; Wingelhofer, B.; Pham, H.T.T.; Elabd, M.; Han, X.; Valent, P.; Gouilleux, F.; et al. O-GlcNAcylation of STAT5 controls tyrosine phosphorylation and oncogenic transcription in STAT5-dependent malignancies. *Leukemia* **2017**, *31*, 2132–2142. [CrossRef]

56. Prestipino, A.; Emhardt, A.J.; Aumann, K.; O'Sullivan, D.; Gorantla, S.P.; Duquesne, S.; Melchinger, W.; Braun, L.; Vuckovic, S.; Boerries, M.; et al. Oncogenic JAK2(V617F) causes PD-L1 expression, mediating immune escape in myeloproliferative neoplasms. *Sci. Transl. Med.* **2018**, *10*. [CrossRef]

57. Reckel, S.; Hamelin, R.; Georgeon, S.; Armand, F.; Jolliet, Q.; Chiappe, D.; Moniatte, M.; Hantschel, O. Differential signaling networks of Bcr-Abl p210 and p190 kinases in leukemia cells defined by functional proteomics. *Leukemia* **2017**, *31*, 1502–1512. [CrossRef]

58. Chatain, N.; Ziegler, P.; Fahrenkamp, D.; Jost, E.; Moriggl, R.; Schmitz-Van de Leur, H.; Muller-Newen, G. Src family kinases mediate cytoplasmic retention of activated STAT5 in BCR-ABL-positive cells. *Oncogene* **2013**, *32*, 3587–3597. [CrossRef]

59. Bibi, S.; Arslanhan, M.D.; Langenfeld, F.; Jeanningros, S.; Cerny-Reiterer, S.; Hadzijusufovic, E.; Tchertanov, L.; Moriggl, R.; Valent, P.; Arock, M. Co-operating STAT5 and AKT signaling pathways in chronic myeloid leukemia and mastocytosis: Possible new targets of therapy. *Haematologica* **2014**, *99*, 417–429. [CrossRef]

60. Bunting, K.D.; Xie, X.Y.; Warshawsky, I.; Hsi, E.D. Cytoplasmic localization of phosphorylated STAT5 in human acute myeloid leukemia is inversely correlated with Flt3-ITD. *Blood* **2007**, *110*, 2775–2776. [CrossRef]

61. Harir, N.; Boudot, C.; Friedbichler, K.; Sonneck, K.; Kondo, R.; Martin-Lanneree, S.; Kenner, L.; Kerenyi, M.; Yahiaoui, S.; Gouilleux-Gruart, V.; et al. Oncogenic Kit controls neoplastic mast cell growth through a Stat5/PI3-kinase signaling cascade. *Blood* **2008**, *112*, 2463–2473. [CrossRef]

62. Schulenburg, A.; Blatt, K.; Cerny-Reiterer, S.; Sadovnik, I.; Herrmann, H.; Marian, B.; Grunt, T.W.; Zielinski, C.C.; Valent, P. Cancer stem cells in basic science and in translational oncology: Can we translate into clinical application? *J. Hematol. Oncol.* **2015**, *8*, 16. [CrossRef] [PubMed]

63. Adorno-Cruz, V.; Kibria, G.; Liu, X.; Doherty, M.; Junk, D.J.; Guan, D.; Hubert, C.; Venere, M.; Mulkearns-Hubert, E.; Sinyuk, M.; et al. Cancer stem cells: Targeting the roots of cancer, seeds of metastasis, and sources of therapy resistance. *Cancer Res.* **2015**, *75*, 924–929. [CrossRef] [PubMed]

64. Delhommeau, F.; Dupont, S.; Tonetti, C.; Masse, A.; Godin, I.; Le Couedic, J.P.; Debili, N.; Saulnier, P.; Casadevall, N.; Vainchenker, W.; et al. Evidence that the JAK2 G1849T (V617F) mutation occurs in a lymphomyeloid progenitor in polycythemia vera and idiopathic myelofibrosis. *Blood* **2007**, *109*, 71–77. [CrossRef] [PubMed]

65. Woods, B.; Chen, W.; Chiu, S.; Marinaccio, C.; Fu, C.; Gu, L.; Bulic, M.; Yang, Q.; Zouak, A.; Jia, S.; et al. Activation of JAK/STAT Signaling in Megakaryocytes Sustains Myeloproliferation In Vivo. *Clin. Cancer Res.* **2019**, *25*, 5901–5912. [CrossRef] [PubMed]

66. Vainchenker, W.; Constantinescu, S.N.; Plo, I. Recent advances in understanding myelofibrosis and essential thrombocythemia. *F1000Research* **2016**, *5*. [CrossRef]

67. Zhang, Y.; Lin, C.H.S.; Kaushansky, K.; Zhan, H. JAK2V617F Megakaryocytes Promote Hematopoietic Stem/Progenitor Cell Expansion in Mice Through Thrombopoietin/MPL Signaling. *Stem Cells* **2018**, *36*, 1676–1684. [CrossRef]

68. Teofili, L.; Martini, M.; Cenci, T.; Petrucci, G.; Torti, L.; Storti, S.; Guidi, F.; Leone, G.; Larocca, L.M. Different STAT-3 and STAT-5 phosphorylation discriminates among Ph-negative chronic myeloproliferative diseases and is independent of the V617F JAK-2 mutation. *Blood* **2007**, *110*, 354–359. [CrossRef]

69. Abba, C.; Campanelli, R.; Catarsi, P.; Villani, L.; Abbonante, V.; Sesta, M.A.; Barosi, G.; Rosti, V.; Massa, M. Constitutive STAT5 phosphorylation in CD34+ cells of patients with primary myelofibrosis: Correlation with driver mutation status and disease severity. *PLoS ONE* **2019**, *14*, e0220189. [CrossRef]

70. Herrmann, H.; Sadovnik, I.; Cerny-Reiterer, S.; Rulicke, T.; Stefanzl, G.; Willmann, M.; Hoermann, G.; Bilban, M.; Blatt, K.; Herndlhofer, S.; et al. Dipeptidylpeptidase IV (CD26) defines leukemic stem cells (LSC) in chronic myeloid leukemia. *Blood* **2014**, *123*, 3951–3962. [CrossRef]

71. Nivarthi, H.; Chen, D.; Cleary, C.; Kubesova, B.; Jager, R.; Bogner, E.; Marty, C.; Pecquet, C.; Vainchenker, W.; Constantinescu, S.N.; et al. Thrombopoietin receptor is required for the oncogenic function of CALR mutants. *Leukemia* **2016**, *30*, 1759–1763. [CrossRef]

72. Milosevic Feenstra, J.D.; Nivarthi, H.; Gisslinger, H.; Leroy, E.; Rumi, E.; Chachoua, I.; Bagienski, K.; Kubesova, B.; Pietra, D.; Gisslinger, B.; et al. Whole-exome sequencing identifies novel MPL and JAK2 mutations in triple-negative myeloproliferative neoplasms. *Blood* **2016**, *127*, 325–332. [CrossRef] [PubMed]

73. Wedeh, G.; Cerny-Reiterer, S.; Eisenwort, G.; Herrmann, H.; Blatt, K.; Hadzijusufovic, E.; Sadovnik, I.; Mullauer, L.; Schwaab, J.; Hoffmann, T.; et al. Identification of bromodomain-containing protein-4 as a novel marker and epigenetic target in mast cell leukemia. *Leukemia* **2015**, *29*, 2230–2237. [CrossRef] [PubMed]

Article

JAK/STAT-Activating Genomic Alterations Are a Hallmark of T-PLL

Linus Wahnschaffe [1,2,3,†], Till Braun [1,2,3,†], Sanna Timonen [4,5,6,†], Anil K. Giri [6],
Alexandra Schrader [1,2,3], Prerana Wagle [2], Henrikki Almusa [6], Patricia Johansson [7],
Dorine Bellanger [8,9], Cristina López [10,11], Claudia Haferlach [12], Marc-Henri Stern [8,9],
Jan Dürig [7,13], Reiner Siebert [10,11], Satu Mustjoki [4,5], Tero Aittokallio [6,14] and Marco Herling [1,2,3,*]

[1] Department I of Internal Medicine, Center for Integrated Oncology (CIO),
    Aachen-Bonn-Cologne-Duesseldorf, University of Cologne (UoC), 50937 Cologne, Germany;
    linus.wahnschaffe@uk-koeln.de (L.W.); till.braun@uk-koeln.de (T.B.); alexandra.schrader@uk-koeln.de (A.S.)
[2] Excellence Cluster for Cellular Stress Response and Aging-Associated Diseases (CECAD), University of
    Cologne, 50937 Cologne, Germany; prerana.wagle@uk-koeln.de
[3] Center for Molecular Medicine Cologne (CMMC), University of Cologne, 50937 Cologne, Germany
[4] Hematology Research Unit Helsinki, Helsinki University Hospital Comprehensive Cancer Center,
    FI-00029 Helsinki, Finland; sanna.timonen@helsinki.fi (S.T.); satu.mustjoki@helsinki.fi (S.M.)
[5] Translational Immunology Research program and Department of Clinical Chemistry and Hematology,
    University of Helsinki, FI-00014 Helsinki, Finland
[6] Institute for Molecular Medicine Finland (FIMM), University of Helsinki, FI-00014 Helsinki, Finland;
    anil.kumar@helsinki.fi (A.K.G.); henrikki.almusa@helsinki.fi (H.A.); tero.aittokallio@helsinki.fi (T.A.)
[7] Department of Hematology, University Hospital Essen, University of Duisburg-Essen, 45147 Essen,
    Germany; patricia.johansson@uk-essen.de (P.J.); jan.duerig@uk-essen.de (J.D.)
[8] Institut Curie, Centre de Recherche, 75248 Paris, France; dorine.bellanger@curie.fr (D.B.);
    marc-henri.stern@curie.fr (M.-H.S.)
[9] French National Institute of Health and Medical Research (INSERM) U830, 75248 Paris, France
[10] Institute for Human Genetics, Christian-Albrechts-University Kiel & University Hospital Schleswig Holstein,
    24105 Kiel, Germany; cristina.lopez@uni-ulm.de (C.L.); reiner.siebert@uni-ulm.de (R.S.)
[11] Institute of Human Genetics, Ulm University, Ulm University Medical Center, D-89081 Ulm, Germany
[12] MLL Munich Leukemia Laboratory, Max-Lebsche-Platz 31, 81377 Munich, Germany;
    claudia.haferlach@mll.com
[13] German Cancer Consortium (DKTK), 69120 Heidelberg, Germany
[14] Department of Mathematics and Statistics, University of Turku, FI-20014 Turku, Finland
*   Correspondence: marco.herling@uk-koeln.de; Tel.: +49-221-478-5969
†   Equal contribution.

Received: 17 October 2019; Accepted: 18 November 2019; Published: 21 November 2019

Abstract: T-cell prolymphocytic leukemia (T-PLL) is a rare and poor-prognostic mature T-cell leukemia. Recent studies detected genomic aberrations affecting *JAK* and *STAT* genes in T-PLL. Due to the limited number of primary patient samples available, genomic analyses of the JAK/STAT pathway have been performed in rather small cohorts. Therefore, we conducted—via a primary-data based pipeline—a meta-analysis that re-evaluated the genomic landscape of T-PLL. It included all available data sets with sequence information on *JAK* or *STAT* gene loci in 275 T-PLL. We eliminated overlapping cases and determined a cumulative rate of 62.1% of cases with mutated *JAK* or *STAT* genes. Most frequently, *JAK1* (6.3%), *JAK3* (36.4%), and *STAT5B* (18.8%) carried somatic single-nucleotide variants (SNVs), with missense mutations in the SH2 or pseudokinase domains as most prevalent. Importantly, these lesions were predominantly subclonal. We did not detect any strong association between mutations of a *JAK* or *STAT* gene with clinical characteristics. Irrespective of the presence of gain-of-function (GOF) SNVs, basal phosphorylation of STAT5B was elevated in all analyzed T-PLL. Fittingly, a significant proportion of genes encoding for potential negative regulators of STAT5B showed genomic losses (in 71.4% of T-PLL in total, in 68.4% of T-PLL without any *JAK* or *STAT* mutations). They included *DUSP4, CD45, TCPTP, SHP1, SOCS1, SOCS3,* and *HDAC9.*

Overall, considering such losses of negative regulators and the GOF mutations in *JAK* and *STAT* genes, a total of 89.8% of T-PLL revealed a genomic aberration potentially explaining enhanced STAT5B activity. In essence, we present a comprehensive meta-analysis on the highly prevalent genomic lesions that affect genes encoding JAK/STAT signaling components. This provides an overview of possible modes of activation of this pathway in a large cohort of T-PLL. In light of new advances in JAK/STAT inhibitor development, we also outline translational contexts for harnessing active JAK/STAT signaling, which has emerged as a 'secondary' hallmark of T-PLL.

Keywords: JAK; STAT; T-PLL; T-cell leukemia; meta-analysis; STAT5B signaling

---

1. Introduction

T-cell prolymphocytic leukemia (T-PLL) is an aggressive malignancy characterized by an expansion of mature T-lymphocytes [1]. Although with an incidence of <2.0/million/year in Western countries infrequently encountered, T-PLL is the most common mature T-cell leukemia [2]. Patients suffering from T-PLL typically present with exponentially rising white blood cell (WBC) counts accompanied by bone marrow (BM) infiltration and splenomegaly, and at lower frequencies by various other manifestations such as effusions or in skin [3]. Due to its chemotherapy-refractory behavior, T-PLL patients have a dismal prognosis with a median overall survival (OS) of <3 years [4–6]. Clinicians treating T-PLL face limited therapeutic options, mainly caused by a still incomplete mechanistic disease concept and, therefore, a rudimentary understanding of targetable vulnerabilities in T-PLL. The most potent single substance, the CD52-antibody alemtuzumab, leads to complete remissions in more than 80% of patients, however, nearly all eventually relapse (at median already within 12 months) [3,4,7].

Due to the limited number of primary patient samples available, genomic analyses studying the underlying disease mechanisms have been performed in rather small cohorts of T-PLL patients. In these studies, the most common molecular features revealed were rearrangements that involve chromosome 14 or X, resulting in juxtaposition of the *T-cell leukemia/lymphoma 1A (TCL1A)* or *Mature T-cell leukemia 1 (MTCP1)* proto-oncogenes to *TCRAD* gene enhancer elements [8]. The second most common lesions are genomic alterations of the tumor suppressor *ataxia telangiectasia mutated (ATM)*, found in >85% of cases [9]. In addition to these lesions, activating mutations targeting the Janus kinase (JAK)/signal transducer and activator of transcription (STAT) pathway components were identified in these series of genomic analyses [9–18].

The JAK/STAT pathway is a ubiquitous cytokine-mediated signaling cascade that regulates cell proliferation, differentiation, migration, and apoptosis [19]. After recruitment of JAKs (JAK1, JAK2, JAK3, TYK2) to cytokine receptors, they phosphorylate STAT proteins (STAT1, STAT2, STAT3, STAT4, STAT5A, STAT5B, STAT6), which then bind DNA and regulate target gene transcription [20]. Dysregulation of the JAK/STAT axis has been described as a key event in the pathogenesis of various hematologic malignancies [21]. Constitutively active JAK/STAT signaling induces T-cell tumors in mice [22,23]. Using different sequencing approaches, activating mutations of *JAK1*, *JAK3*, and *STAT5B* were identified as the most recurrent genomic aberrations affecting *JAK/STAT* genes in T-PLL [9–18]. However, prevalence of gene mutations, information on their allele frequencies, assessment of negative regulators of JAK/STAT signaling, and the phosphorylation status of the most recurrently affected JAK/STAT proteins vary considerably or are not reported in these studies [9–18].

Therapeutic approaches blocking JAK/STAT signaling have so far improved patient outcomes predominantly in autoimmune conditions and in graft-versus-host disease [24,25]. JAK inhibitors are currently tested for a number of new indications [26]. T-PLL cells have shown a notable in vitro sensitivity towards JAK inhibition, which was not directly linked to the *JAK/STAT* mutation status [15,16]. First reports present individual clinical activity of tofacitinib (pan JAK inhibitor) and ruxolitinib (JAK 1/2 inhibitor) in relapsed T-PLL [27,28].

Although many studies identified *JAK* and *STAT* genes to be commonly mutated in T-PLL, these analyses have been performed in rather small cohorts not providing a sufficient dataset to determine reliable mutation and variant allele frequencies (VAFs). In addition, the publication overlap of these studies was unresolved and a systematic assessment for other potential genomic causes (e.g., copy number alterations (CNAs)) has not been performed.

Here, we conducted a meta-analysis that was supplemented by new primary data, hence providing the largest cohort to date that evaluated the genomic aberrations affecting *JAK/STAT* signaling in T-PLL. In addition to summarizing information on the functional impact of the most recurrent lesions, we propose a model of potential mechanisms leading to constitutive JAK/STAT signaling in T-PLL cells.

2. Results

2.1. Characteristics and Overlaps of Included Studies

The meta-analysis considered all available publications that have analyzed variants of any *JAK* or *STAT* gene in cases of T-PLL, regardless of the sequencing approach used (Table S1). Redundantly sequenced cases were identified to eliminate overlaps between these 10 studies (Figure 1A). The most common sequencing approach was Sanger sequencing (Sanger seq., 7 studies), followed by targeted amplicon sequencing (TAS, 5 studies), whole exome sequencing (WES, 4 studies), and whole genome sequencing (WGS, 2 studies). *JAK3* (n = 272 T-PLL patients), *JAK1* (n = 246), and *STAT5B* (n = 209) were predominantly sequenced due to the bias by the targeted approaches. Germline controls were sequenced in 53 cases (19.3%). The number of analyzed patients varied from 3 to 71 patients across the 10 studies [9–18]. After subtracting all cases reported in more than one study, we identified 275 unique T-PLL cases as the core cohort.

Notably, data that were acquired by WGS, WES, and SNP array analysis were re-analyzed as primary data in cases, in which raw data were available (WES: 62.2%, n = 46/74 T-PLL patients; WGS: 50.0%, n = 4/8; SNP arrays: 100.0%, n = 71/71), while data gained through Sanger seq. or TAS were not re-evaluated. By re-analyzing the WES and SNP array data through a uniform pipeline, homogeneity of the resulting dataset was obtained. *JAK1*, *JAK2*, *JAK3*, *TYK2*, *STAT1*, *STAT2*, *STAT3*, *STAT4*, *STAT5A*, *STAT5B*, and *STAT6* are further referred to as 'any *JAK* or *STAT* gene'.

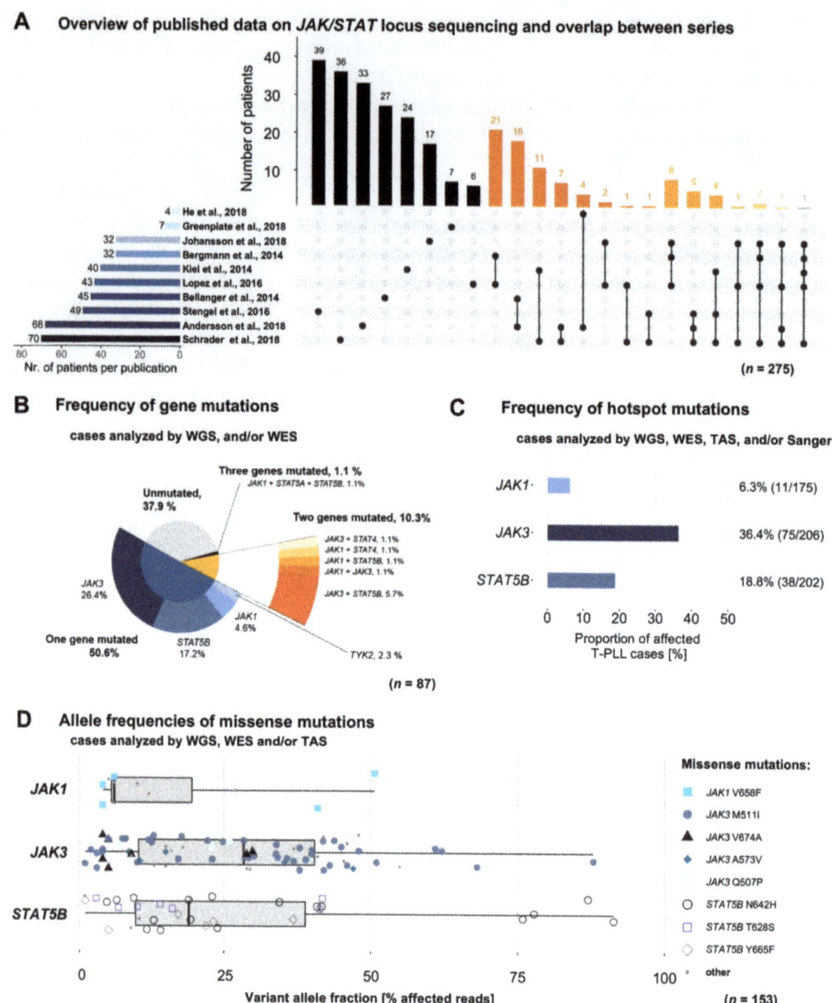

Figure 1. Meta-analyses of genomic profiling series in T-PLL underscore the high prevalence of mutations affecting *JAK* and *STAT* genes. (**A**) T-PLL patients (*n* = 275) sequenced for any *JAK* or *STAT* locus. Horizontal bar chart displays the total number of patients sequenced in each publication. Vertical bar chart indicates the size of intersections between sets of patients analyzed in one or more publications. Color-code of vertical bars indicates the number of studies reporting results of the same individual case (black: 1; dark-orange: 2; medium-orange: 3; light-orange: 4; light grey: 5). Dot connecting lines show the overlapping publications of each intersection. (**B**) Distribution of *JAK/STAT* mutations (*n* = 87 T-PLL analyzed by whole genome sequencing (WGS)/whole exome sequencing (WES)): in 62.1% of these cases (*n* = 54) at least one mutation in a gene of the *JAK/STAT* family was found to be mutated. (**C**) Relative frequencies of hotspot mutations of *JAK1*, *JAK3*, and *STAT5B* (as defined by the genomic region containing more than 90% of the respective mutations; *n* = 275 cases analyzed by WGS/WES/targeted amplicon sequencing (TAS)/Sanger sequencing). Every T-PLL case that was sequenced for the respective hotspot region of *JAK1* (Ex. 14–20), *JAK3* (Ex. 11–19), and/or *STAT5B* (Ex. 15–17), was included. (**D**) Box-whisker charted allele frequencies of missense mutations of *JAK1*, *JAK3*, and *STAT5B* (*n* = 153 T-PLL analyzed by WGS/WES and/or TAS) illustrate a high heterogeneity in variant allele fractions, with most lesions detected at lower frequencies (<50%).

2.2. Mutations in JAK and STAT Genes are Predominantly Found at Subclonal Levels

We included T-PLL cases with available data from WGS or WES in order to calculate the frequency of genomic variants in *JAK* and *STAT* genes. We identified a cumulative rate of 62.1% (n = 54/87, Figure 1B) cases with variants in any *JAK* or *STAT* gene. The vast majority of them were somatic single-nucleotide variants (SNVs) of missense type (96.0%, n = 72/75 of variants detected by WGS/WES). The remaining three lesions (4.0%, n = 3/75) were small in-frame deletions in *JAK1* or *JAK3*. In the following these SNVs and small in-frame deletions are referred to as 'mutations'. Somatic variants were called and known single nucleotide polymorphisms were filtered out in those tumor samples for which a matched germline sample was available. For T-PLL patients with a missing germline control (n = 38), we included only those variants which were detected in the WES with matched germline samples. *JAK3* (34.5%, n = 30/87), *STAT5B* (25.3%, n = 22/87), and *JAK1* (9.2%; n = 8/87) were the most recurrently mutated genes, while *TYK2* (2.3%, n = 2/87), *STAT4* (2.3%, n = 2/87), and *STAT5A* (1.1%, n = 1/87) were mutated at lower frequencies. Nine of 87 cases (10.3%) presented with mutations affecting two genes that encode for JAK or STAT proteins. Co-occuring ('double') *JAK1* and *JAK3* mutations were found in 1.1% of cases (n = 1/87), *JAK3* and *STAT4* in 1.1% (n = 1/87), and *JAK1* and *STAT4* mutations in 1.1% (n = 1/87); *JAK1* and *STAT5B* in 1.1% (n = 1/87), and *JAK3* and *STAT5B* in 5.8% (n = 5/87). Co-occuring *JAK1*, *STAT5A*, and *STAT5B* mutations ('triple') were detected in one case (1.1%).

Next, we calculated frequencies of missense mutations considering the hotspots of *JAK1* (Ex. 14–20), *JAK3* (Ex. 11–19), and *STAT5B* (Ex. 15–17). Hotspot regions were defined as genomic regions containing >90% of the respective mutations. By also including cases analyzed by TAS or Sanger seq. for the respective hotspot region, we arrived at a larger cohort for these analyses. Based on this, the hotspot of *JAK1* showed a relative mutation frequency of 6.3% (n = 11/175). For the hotspot of *JAK3* this was 36.4% (n = 75/206) and for the hotspot of *STAT5B* it was 18.8% (n = 38/202, Figure 1C).

As already described in smaller cohorts [9,16,18], most of these lesions presented as subclonal events with low VAFs (medians: *JAK1* 6.0%, *JAK3* 28.5%, *STAT5B* 19.0%, Figure 1D); however, we observed a high heterogeneity in VAFs. Notably, the N642H mutation of *STAT5B* was observed at a VAF of >50% in four cases and the M511I mutation of *JAK3* at VAFs >50% in five cases. Note that given the data provided in the reports, the percentage of treatment-naïve samples was 72.7%; sequential samples of individual T-PLL patients were not analyzed here.

2.3. Missense Mutations in JAK1, JAK3, and STAT5B Cluster Within the Conserved Pseudokinase and SH2 Domains

We further assessed the localization of the missense mutations of *JAK1*, *JAK3*, and *STAT5B* (Figure 2, upper protein schemes) as well as the frequencies of hotspot SNVs in these genes (lower inset). To assess the localization of the somatic SNVs of *JAK1*, *JAK3*, and *STAT5B*, we included all T-PLL cases which have been sequenced for the whole coding genome by WES or WGS. All T-PLL cases which have been sequenced for the respective lesion regardless of the sequencing approach were considered to determine the frequencies of hotspot mutations. The pseudokinase domain (JH2 domain) of *JAK1* was predominantly affected (62.5% of all *JAK1* mutations, n = 5/8) with V658F (3.7% of all analyzed cases, n = 9/246, regardless of the sequencing approach) as the most prominent lesion (Figure 2A). In *JAK3*, most somatic SNVs were found in the JH2-SH2-linker (69.7% of all *JAK3* missense mutations, n = 23/33 detected missense mutations), followed by the JH2 domain (27.3%, n = 9/33, Figure 2B). The M511I lesion was the most prevalent lesion in *JAK3* (27.2% of all analyzed cases, n = 74/272, regardless of the sequencing approach). Almost all missense mutations of *STAT5B* were found in the SH2 domain (96.2% of all *STAT5B* missense mutations, n = 25/26 detected missense mutations; Figure 2C). N642H (9.6% of all analyzed cases, n = 20/209, regardless of the sequencing approach), T628S (5.9%, n = 12/202), and Y665F (3.3%, n = 7/209) were the most recurrent mutations affecting *STAT5B*. Remarkably, we did not detect significant differences in the mutation frequencies with regard to the sequencing approach (see bars of the lower inset of Figure 2).

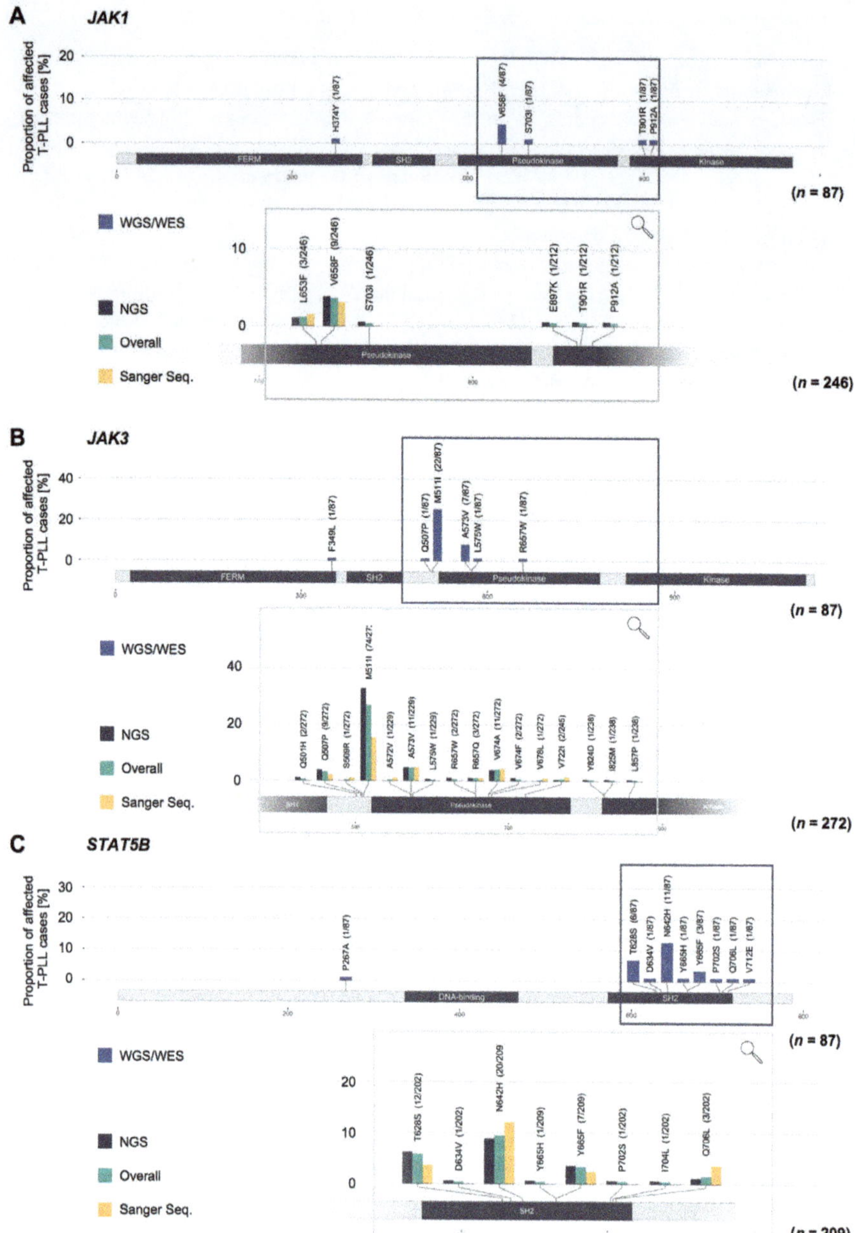

Figure 2. Missense mutations in *JAK1/JAK3/STAT5B* genes cluster within the conserved JH2 (pseudokinase) and SH2 domains. Upper protein scheme: gene-wide distribution of missense mutations of *JAK1* (**A**), *JAK3* (**B**), and *STAT5B* (**C**); *n* = 87 T-PLL analyzed by WES/WGS, representing the same cases as in Figure 1B. Lower inset for each sub-panel A–C: frequencies of hotspot mutations, stratified for applied sequencing methods (*n* = 275 cases analyzed with WES/WGS/TAS/Sanger seq). While *JAK1* and *JAK3* were predominately affected by lesions within the pseudokinase domains (JH2) and the SH2-JH2 linker regions, *STAT5B* missense mutations were presented mostly in the SH2 domain.

2.4. T-PLL Harboring any JAK or STAT Mutation Show Elevated TCL1A mRNA Level, but Present Comparable Clinical Outcomes

When analyzing associations of *JAK* and *STAT* mutations with immunophenotypic and cytogenetic data as well with the 18 most prevalent mutations found in T-PLL (as described by Schrader et al. [9]), we detected a decreased proportion of T-PLL cells with CD40L expression (p = 0.05, Fisher's exact test, Figure S1A) and an increased proportion of T-PLL cells harboring *NOTCH2* mutations (p = 0.03, Fisher's exact test, Figure S1B) in the *JAK3* mutated cohort (overview of tested associations is provided in Table S2). Notably, T-PLL with any *JAK* or *STAT* mutation showed elevated TCL1A mRNA levels as compared to patients without any *JAK* or *STAT* mutation (fold change (Fc) = 4.1, p = 0.01, Student's *t*-test, Figure 3A), further underlined by an increased proportion of cases harboring an inversion 14 (q11q32) in the *STAT5B* mutated cohort (p = 0.03, Fisher's exact test, Figure S1C). TCL1A mRNA levels were also elevated by just considering *JAK3* mutated cases (Fc = 6.5, p = 0.002, Student's *t*-test, Figure S1D) or *STAT5B* mutated cases (Fc = 3.6, p = 0.02, Student's *t*-test, Figure S1E) compared to T-PLL without either mutation, respectively. To assess the functional role of variants in *JAK* or *STAT* genes in T-PLL patients, we screened for significant associations between somatic *JAK/STAT* SNVs and clinical data, including outcome. For these analyses, we merged clinical information from nine studies [9–16,18]. As not all considered publications reported the same set of parameters, cohort sizes varied between the different analyses. When investigating an association of the *JAK/STAT* mutation status with overall survival, we did not observe differences in OS of T-PLL patients harboring any *JAK* or *STAT* mutation compared to patients without *JAK* or *STAT* mutation (p = 0.42, log rank test, Figure 3B). In contrast to studies of smaller cohorts [12,16], we could not observe a significantly shorter OS of *JAK3* mutated T-PLL patients (p = 0.26, log rank test, Figure S2A). Similarly, the *JAK3* M511 mutation was not associated with a shorter OS (p = 0.39, log rank test, Figure S2B). The OS of patients without any *JAK* or *STAT* mutations was comparable to those harboring a *STAT5B* (p = 0.55, log rank test, Figure S2C), *STAT5B* N642H (p = 0.60, log rank test, Figure S2D), or *JAK1* (p = 0.54, log rank test, Figure S2E) mutation. Interestingly, the *JAK1* L653F mutation status was associated with a significantly shorter OS (p = 0.003, log rank test, Figure S2F). In addition, cases with a *STAT5B* Y665F mutation also showed a shorter OS (p = 0.06, log rank test, Figure S2G). However, only two patients with a *JAK1* L653F mutation or five patients with a *STAT5B* Y665F mutation were included in the 'mutated' cohorts, so these results should be interpreted with caution. We did not detect any other association of clinical characteristics of T-PLL patients with the presence of *JAK* or *STAT* mutations (Table S3).

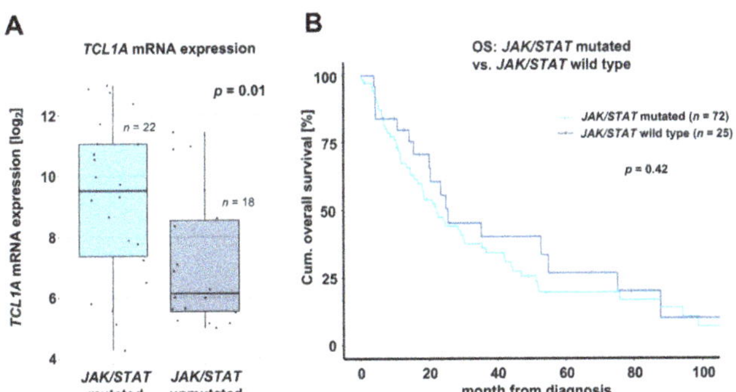

Figure 3. *JAK/STAT* mutation status shows association with elevated TCL1A mRNA expression, but not with patient outcomes. (**A**) TCL1A mRNA expression measured by array-based gene expression profiling (GEP) of *JAK/STAT* mutated cases compared to cases without any *JAK* or *STAT* mutation (fold change: 4.1; p = 0.01, Student's *t*-test). (**B**) Overall survival (OS) of *JAK/STAT* mutated cases (median OS: 21.0 months) compared to cases without any *JAK* or *STAT* mutation (median OS: 25.5 months, p = 0.42, log rank test). Significant associations between mutations affecting *JAK3* or *STAT5B* with immunophenotypic, mutational, and expression data are displayed in Figure S1. Associations of mutations affecting *JAK1*, *JAK3*, or *STAT5B* with OS are presented in Figure S2.

2.5. T-PLL Cells Show Basal STAT5B Phosphorylation, Regardless of Their JAK/STAT Mutation Status

Even though STAT5B was shown to be predominantly affected in T-PLL in previous series, its phosphorylation, and therefore activation state, was often not aligned with the genomic profiling data [9–11]. By analyzing primary samples of eight T-PLL cases, we observed noticeable basal STAT5B phosphorylation in every case (Figure 4A, no exposure to any culturing). We discovered no obvious association between STAT5B phosphorylation and the presence of any *JAK* or *STAT* mutation.

Samples without *JAK* or *STAT* mutation showed similar levels of high basal phospho-STAT5B to those that harbored such lesions, as already previously shown in a series of 6 T-PLL cases [9].

We further correlated the immunoblot data with alterations involving key regulators of STAT5B activity (either mutations or CNAs). For this, we pre-selected a set of 105 genes encoding for proteins known to regulate JAK/STAT signaling based on available literature (Table S4). We next identified of these 105 genes 7 which showed recurrent genomic lesions in the overall T-PLL cohort and which were reported to have a regulatory effect on STAT5B signaling in vitro or in vivo (Table S5). Overall, we identified recurrent mutations or genomic losses of seven negative regulators potentially leading to constitutive STAT5B activity (*DUSP4, CD45, TCPTP, SHP1, SOCS1, SOCS3,* and *HDAC9*) in 71.4% of analyzed T-PLL cases (*n* = 35/49 cases analyzed with WES/WGS and SNP array, Figure 4B). About 68.4% of T-PLL cases that did not reveal a mutation in any *JAK* or *STAT* gene showed such a CNA of a STAT5B signaling regulator. This resulted in a total of 89.8% of T-PLL cases that harbor a genomic lesion potentially explaining constitutive STAT5B signaling (either genomic losses of negative regulators, and/or mutations of *JAK* or *STAT* genes). Gene expression signatures (array-based) of cases that harbor a genomic loss or a mutation of a negative regulator of JAK/STAT signaling (and that lack a *JAK* or *STAT* mutation) were comparable to those of cases with a mutation in any *JAK* or *STAT* gene (principal component analysis in Figure S4A). Both cohorts showed similar activation (as per upregulated gene expression) of STAT5B target genes. All T-PLL with a mutation in a *JAK* or *STAT* gene and 90% of patients with a mutation or loss in a negative JAK/STAT regulator (and without any mutation in a *JAK* or *STAT* gene) showed overexpression of at least one STAT5 target gene. In contrast, the frequency of overexpressed STAT5 target genes was significantly lower in T-PLL without a genomic lesion potentially activating JAK/STAT signaling (60%, Figure S4B). In detail, *DUSP4, CD45, SOCS3, SHP1,* and *HDAC9* showed the highest prevalences of genomic losses with a higher frequency in T-PLL cases without any *JAK* or *STAT* mutation (Figure 4C). In addition, we identified genomic gains of *STAT2* (*n* = 1/49 T-PLL), *STAT3,* (*n* = 5/49 T-PLL), *STAT5A* (*n* = 5/49 T-PLL), and *STAT5B* (*n* = 5/49 T-PLL). These lesions were detected only in combination with mutations of a *JAK* or *STAT* gene or with genomic losses of a negative regulator of STAT5B signaling (Figure 4D).

Overall, we propose a model of constitutive JAK/STAT signaling induced by mutations of a predicted gain of function (GOF) in *JAK1, JAK3,* and *STAT5B* genes and by copy-number losses of negative regulators of STAT5B activity.

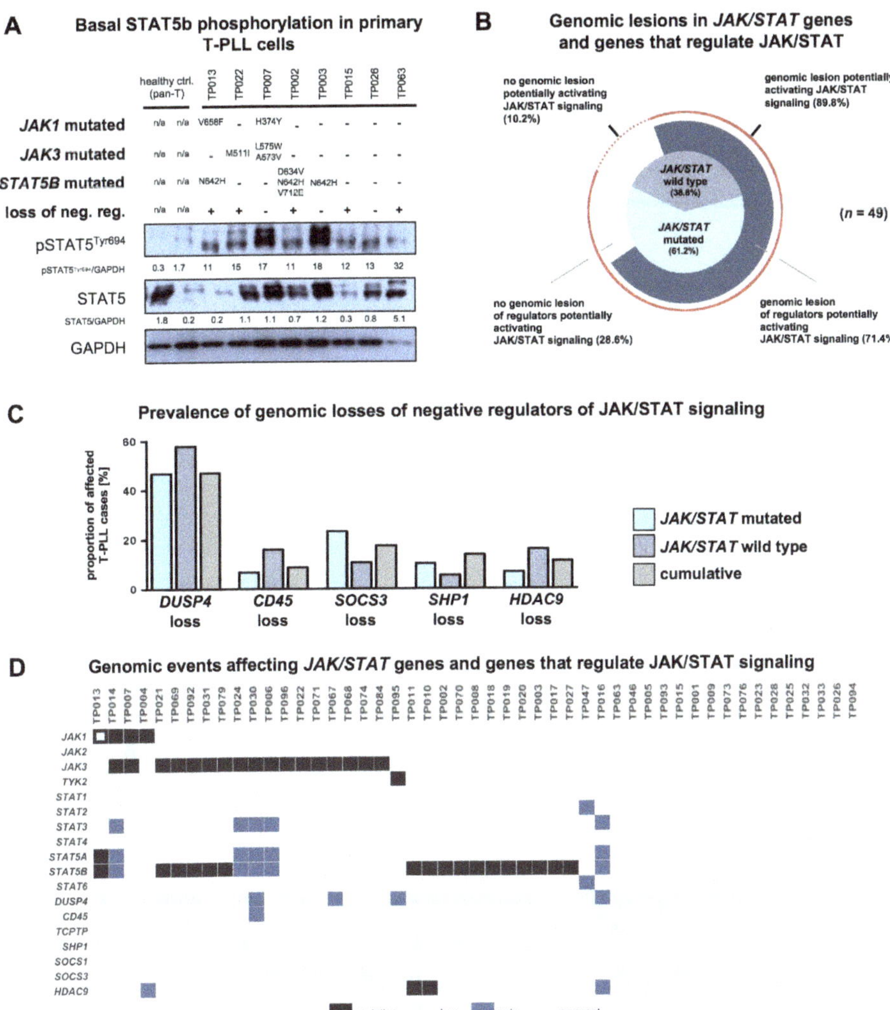

Figure 4. T-PLL cells show basal STAT5B phosphorylation, regardless of their *JAK/STAT* mutation status. (**A**) Elevated basal STAT5B phosphorylation levels of T-PLL cells. Controls: CD3$^+$ pan T-cells isolated from healthy individuals. 'Loss of neg. reg.': key regulators which negatively affect STAT5B activation (namely *DUSP4, CD45, TCPTP, SHP1, SOCS1, SOCS3,* and *HDAC*; see (D) for case-wise depiction) were considered based on available literature and based on data of their copy number alterations (CNA) in T-PLL. (**B**) Distribution of genomic lesions affecting any *JAK* or *STAT* gene and their regulators (*n* = 49 cases analyzed with WES/WGS and with SNP arrays). Inner pie chart: distribution of *JAK/STAT* mutations. Outer pie chart: Prevalence of genomic lesions of regulators activating JAK/STAT (either genomic losses of negative regulators or genomic gains of positive regulators). An overall proportion of 89.8% of T-PLL cases carried a genomic lesion potentially explaining constitutive STAT5B activation (mutation or CNA of JAK/STAT regulator). (**C**) Prevalence of the five most common genomic lesions affecting negative regulators of the JAK/STAT pathway (*n* = 49 cases analyzed with WES/WGS and SNP array): T-PLL cases without any mutation in a *JAK* or *STAT* gene showed a higher prevalence of genomic losses of *DUSP4, CD45,* and *HDAC9* as compared to *JAK/STAT* mutated cases. (**D**) Mapped genomic events involving *JAK* and *STAT* genes and their regulators (*n* = 49 cases analyzed with WES/WGS and SNP array). An overview of genomic lesions resulting in a suggested activation of the JAK/STAT pathway across all considered T-PLL cases is given in Figure S3.

3. Discussion

Here, we present a comprehensive meta-analysis of primary data from 275 T-PLL cases from 10 published studies that evaluated genomic lesions involving elements of JAK/STAT signaling. In this large cohort, we identified genomic aberrations potentially leading to constitutive JAK/STAT signaling in approximately 90% of cases, thus establishing this as a molecular hallmark of T-PLL. Our analysis provides an overview on the genomic causes of JAK/STAT signaling activation including *JAK/STAT* GOF mutations as well as CNAs of *JAK/STAT* genes or negative regulators of JAK/STAT signaling.

About 62% of T-PLL cases showed GOF mutations affecting any *JAK* or *STAT* gene, while previous publications reported a mutation frequency varying between 36% and 76% [10,18]. In accordance with these studies, we predominantly detected missense mutations in the SH2 or pseudokinase domains of *JAK1*, *JAK3*, and *STAT5B*, mostly occurring at a subclonal level. Mutations of genes encoding for components of the JAK/STAT pathway are frequently found in T-cell neoplasms (predominantly mutations of *JAK3*, *STAT3*, and *STAT5B*). Their reported incidences range from 10% (e.g., for Sézary syndrome [29]) to 70% (e.g., for intestinal T-cell lymphoma [30]). Compared to those, the cumulative mutation frequency of *JAK* and *STAT* genes in T-PLL documented in this meta-analysis is high [31].

Another important finding was the high basal phosphorylation of STAT5B in all analyzed T-PLL samples. In addition to a high proportion of T-PLL cases mutated in any *JAK* or *STAT* gene, we identified seven negative regulators to be commonly lost in T-PLL, potentially explaining cytokine-independent STAT5B activation. This affected 71.4% of all cases and 68.4% of those cases without a detectable *JAK/STAT* SNV. Of these negative regulators, *DUSP4*, *SOCS3*, and *CD45* were most frequently affected. While genomic losses of *DUSP4* and *SOCS3* were already reported in primary T-PLL [9,17,18], losses of *CD45* have not been described previously. Compared to CD4$^+$ T-cells isolated from healthy individuals, DUSP4 and CD45 mRNA was observed to be strongly downregulated in T-PLL cells [16]. Furthermore, the downregulation of CD45 transcripts was associated with an increased phosphorylation of STAT5, implicating a functional relevance of this negative regulator [17].

T-PLL cells usually show a complex karyotype and a high frequency of somatic CNAs [9,32]. In detail, the loci of the negative regulators *DUSP4* (chr.8p12) and of *SOCS3* (chr.17q25.3) are known to be highly affected by structural aberrations in T-PLL. Therefore, a secondary origin of the detected CNAs due to background imbalances has to be considered. T-PLL cases already harboring an activating mutation of a *JAK* or *STAT* gene or a genomic loss of a negative regulator showed in few cases (12.2%) also genomic gains of *JAK* or *STAT* genes, however, the functional impact of these genomic gains remains unclear.

Certainly, there is the possibility of other, yet-unknown regulators, which we did not include in our analysis. Furthermore, our analysis did not assess for epigenetic regulations or post-translational modifications of *JAK* and *STAT* genes or their regulators, as further causes of active JAK/STAT signaling. Moreover, constitutive T-cell-receptor or cytokine input of the malignant T-cell may represent other modes of (milieu-derived) JAK/STAT activation in T-PLL.

Another important result of the provided meta-analyses is the first description of elevated TCL1A in *JAK* and *STAT* mutated T-PLL. TCL1A overexpression is known to be associated with elevated levels of reactive oxygen species (ROS) [33]. In addition, an 8-oxoguanine DNA damage signature is characteristic for T-PLL, likely resulting from ROS-mediated DNA insults [9]. In this context, the initiating oncogene TCL1A is likely to contribute to the occurrence of new mutations, especially when the guardian of genome-integrity ATM is deleted or mutated, potentially explaining a higher prevalence of arising subclonal *JAK* and *STAT* mutations in T-PLL. Furthermore, a possible explanation for the association of reduced CD40L expression with the presence of *JAK3* mutations is a state of less T-cell receptor and cytokine dependent activation of T-PLL cells in the context of an acquired hyperactivating mutation of a signaling intermediate such as *JAK3*.

In line with our finding, that nearly every T-PLL case presents with a genomic aberration that potentially causes activated JAK/STAT signaling, we could not observe any differential association of the *JAK/STAT* mutation status with clinical characteristics, i.e., those indicating a more aggressive phenotype of T-PLL cells. Previous studies on small cohorts have shown a negative influence of the presence of *JAK3* mutations on OS [12,16]. However, as these studies used *JAK3* wild type T-PLL as a control group, they potentially included patients with mutations of other *JAK* or *STAT* genes, therefore not representing a proper control.

Although functional validations of the herein described genomic aberrations is outside the scope of this work, mechanistic analyses of many of the detected genomic aberrations and of their impact on tumorigenesis have been performed in various other entities. Such functional findings of mutations of *JAK1*, *JAK3*, and *STAT5B* are summarized in the following (an overview of literature is provided in Table S5).

JAK1, the third most frequently mutated *JAK* or *STAT* gene in this meta-analysis, showed a cumulative hotspot mutation rate of 6.3% in our analysis, in line with the mutation rates calculated in previous studies. V658F, the most prevalent lesion affecting *JAK1* in T-PLL, was initially described in the development of T-ALL [34]. The V658F lesion is homologous to the *JAK2* V617F mutation and was shown to lead to constitutive active JAK1 signaling in a cell line model [35]. Cells harboring the respective lesion presented enhanced basal JAK1, STAT5, and ERK phosphorylation and showed cytokine-independent growth [36]. S703I and L653F, less frequently detected *JAK1* mutations, were also proven to act as GOF mutations [37,38]. While the S703I mutation was shown to mediate proliferative effects in vitro and in vivo, the L653F mutation is predicted to soften the negative regulation of JAK1 in a biophysical model. The *JAK1* V617F, S703I, and L653F mutations potentially mediate cytokine-independent phosphorylation of STAT5B in T-PLL.

JAK3 mutations were the most recurrent genomic aberration affecting *JAK/STAT* genes in this meta-analysis with a hotspot mutation rate of 36.4% (literature: 21% [12]–71% [17]). The oncogenic potential of the most frequent *JAK3* mutation M511I as well as the less frequently occurring mutations A573V and V674A was shown in various systems: M511I, A573V, as well as V674A mutant cell lines demonstrated high phosphorylation of STAT5 and ERK, leading to cytokine independent growth [39,40]. When transplanting mice with bone marrow progenitor cells harboring the M511I or A573V mutation in *JAK3*, they develop a T-ALL-like disease, characterized by an expansion of immature CD8$^+$ T-cells. In contrast, mice transplanted with V674A mutant bone marrow progenitor cells presented severe lymphadenopathy and splenomegaly without a significant increase of the WBC count [41]. The mentioned *JAK3* mutations were also described in T-ALL [42]. The Q507P mutation, the third most frequent genomic aberration affecting *JAK3* in T-PLL, was not discovered in other entities and has so far not been investigated for its functional relevance.

In this analysis, 18.8% of T-PLL cases were mutated within the hotspot region of *STAT5B*, while the mutation rate in the literature varied from 7% [16] to 36% [10]. The most frequently found N642H mutation is a well described GOF mutation: *STAT5B* N642H transduced cell lines showed prolonged phosphorylation and dimerization of STAT5B, resulting in an enhanced transcriptional activity [43,44]. The N642H mutation leads to a sustained stability to the anti-parallel dimer as shown by crystallization [45]. Transgenic mice harboring the *STAT5B* N642H lesion rapidly developed aggressive CD8$^+$ T-cell neoplasms [46]. The *STAT5B* Y665F mutation, which had a notable influence on OS of T-PLL patients in our analysis, revealed highly phosphorylated STAT5B, an enhanced transcriptional activity and increased proliferation in a cell culture model, comparable to the *STAT5B* N642H mutation. The above-mentioned genomic aberrations, which have also been described in the mutational landscape of T-LGL [47], may play an important role in the activation of JAK/STAT signaling in T-PLL. The functional role of the second most detected *STAT5B* aberration, the T628S mutation, is much less established. This genetic aberration is previously described in hepatosplenic T-cell lymphoma and associated with an increased phosphorylation of STAT5 in a cell line model [48]. Notably, the three commonly affected *JAK/STAT* genes (*JAK1*, *JAK3*, *STAT5B*) are crucial for IL-2 signaling, the cytokine secreted at highest levels in T-PLL [49]. Targeted approaches in appropriate models have to address the chronological role of such lesions as most of them were detected at subclonal levels.

Given that nearly every T-PLL case harbors a genomic aberration potentially activating JAK/STAT signaling, JAK and STAT proteins are clinically relevant targets for T-PLL therapy [50]. Interestingly, in a large panel of blood cancers, primary T-PLL patient cells showed high sensitivity towards JAK-targeting compounds. Moreover, JAK inhibitors were among the top 25 most effective drugs in an *ex vivo* drug screen of 39 T-PLL patient samples exposed to 301 different substances [15,16]. Previous case reports also showed the initial promising results of a combination therapy of the JAK inhibitors ruxolitinib (JAK 1/2 inhibitor) and tofacitinib (pan JAK inhibitor) [27,28]. However, previous studies have also argued that the *JAK/STAT* mutation status alone does not predict ex-vivo sensitivity towards JAK inhibitors [16]. This is in line with the observation of multiple, potentially distinct mechanisms, apart from missense mutations of *JAK/STAT* genes, that are required for activating this pathway. In general, predicting sensitivity of kinase inhibitors often requires larger panels of biomarkers, some of which may be outside of the target pathways or direct drivers of the disease. Therefore, therapeutic targeting of JAK/STAT signaling has to be expanded on various levels, including the development of novel STAT3/5 inhibitors and the extension of pre-clinical and clinical studies to large cohorts of T-PLL and other T-cell lymphomas, all with well-characterized tumor material (e.g., JAK/STAT activation state). Selective STAT5 inhibitors which bind to the SH2 domain of *STAT5* are currently under development [51].

4. Materials and Methods

4.1. Data Acquisition

Available literature was systematically screened for all genomic profiling studies that have information on the mutational status of any *JAK* or *STAT* gene in T-PLL, regardless of the sequencing technique. We identified a total of 10 publications meeting these criteria (Table S6). Raw data of these publications was obtained from different sources, either publicly available or kindly provided by the corresponding authors. Full descriptions of the composed data set and detailed information on the sources of data are provided in Table S6. Considered T-PLL patients had provided written informed consent and all the studies were originally approved by their institutional review boards.

4.2. Data Merging

Considering the rarity of T-PLL and the strong collaboration network in T-PLL research, we assumed a pronounced exchange of patient samples between different centers and studies. To determine all redundantly analyzed patients in our cohort, we requested basic patient information to allow us to precisely identify potential overlaps while still ensuring anonymity of all the patients (Figure 1A). In cases where genomic regions were sequenced by different methods, we included the information provided by the widest type of analysis, ranking WGS > WES > TAS > Sanger sequencing. In total, 275 distinct T-PLL patients were included in our analyses.

4.3. Raw Data Analysis

Data acquired through whole-exome sequencing and SNP-array analysis was re-analyzed for those cases in which raw data was available, while data obtained through Sanger sequencing or targeted amplicon sequencing was not re-analyzed. Studies where raw data was not available were included by applying the already analyzed data. Detailed information on the data sources are listed in Table S6 and on the applied tools with references in Table S7. The below subsections describe the various genomic analyses performed for the meta-analysis.

4.3.1. Whole-Exome Sequencing (WES)

The sequenced raw exome reads were trimmed of B blocks from the end using Trimmomatic (RWTH Aachen, Aachen, Germany). The trimmed data were aligned to the human reference genome (GRCh build 37) using Burrows Wheeler Aligner (bwa-0.7.12, Wellcome Trust Sanger Institute, Cambridge, UK). After alignment the potential PCR duplicates were removed using Picard (Broad Institute of Harvard and MIT, Cambridge, MA, USA), and BAM files were generated, sorted, and indexed using SAMtools (Wellcome Trust Sanger Institute, Cambridge, UK). Exomic regions were re-aligned and the base quality scores were re-calibrated according to the Genome Analysis Toolkit Best Practices recommendations (GATK 4.1.3.0). VarScan2.2.3 (The Genome Institute, St. Louis, MO, USA) was used to call somatic mutations with the following parameters: strand-filter 1, min-coverage-normal 8, min-coverage-tumor 6, somatic-p-value 0.01, min-var-freq 0.05. The putative somatic mutations were annotated for functional consequences using SnpEff 4.03 (Institute of Environmental Health Sciences, Detroit, MI, USA). Variants present in 1000 genome or dbSNP 2.0 Build were filtered out from the analysis.

In order to call somatic single-nucleotide variants in 38 patients without matched germline samples, MuTect v2 (Broad Institute of Harvard and MIT) was employed with default parameters. Because we were mostly interested in identifying sequence variants that were likely to contribute to T-PLL pathogenesis, we focused on those variations that were identified in our analysis of 22 patients with matched healthy control samples. Only these filtered mutations have been reported for the patients without matched germline samples.

4.3.2. Somatic Copy-Number Alterations (sCNAs)

Genotyping of Affymetrix GenomeWide SNP6 array (Thermo Fisher Scientific, Waltham, MA, USA) was performed using Genotype Console Software 4 (GTC4, Thermo Fisher Scientific) with the default parameters using Birdseed v2 algorithms (Broad Institute of Harvard and MIT). In order to infer copy number variations in the T-PLL genome, we used the pooled controls as a reference (non-tumor hematopoietic cell DNA as 'germline' from T-PLL patients, n = 13) created using the GTC pipeline. The Canary algorithm was used to call CN state using the default settings in GTC. We performed segmentation analysis on the basis of identified SNPs/CNVs using inbuilt algorithm. Copy number values and segmentation were visually assessed in the Integrative Genomics Viewer (Broad Institute, Cambridge, MA, USA). Since the segmentation algorithm only reports significantly altered segments/regions, we mapped identified genomic regions to genes based on version 75 of the Ensembl annotation using BiomaRt (version 2.38 and the GenomicRanges R package, version 1.34.0).

4.3.3. Gene Expression Profiling (GEP)

The processed normalized expression profiles generated with GenomeStudio were downloaded from GEO. Data were merged and adjusted for batch effects using linear regression. A differential expression analysis was performed using Student's *t*-test on expression values of patients and healthy controls (CD3+ pan T-cells from 10 healthy donors). The differentially expressed probes were mapped to the respective genes by Illumina provided annotation files. Principal component analysis of the whole gene expression set was performed with R.

4.4. Clinical Data and Statistical Analyses

All publications were screened for accompanying clinical data and authors were contacted for further information. In total, we obtained and merged clinical data from nine publications [9–16,18]. For patients who were reported in multiple studies, we included the most detailed clinical data set. To assess possible correlations between mutational status and clinical phenotypes, we performed a screening approach testing for a broad range of parameters (Table S3), where the patients were stratified by overall mutational status, mutational status of specific genes, and by specific mutations. As a control, we used clinical data from T-PLL patients, who were unmutated in all *JAK/STAT* genes analyzed with WES/WGS.

All statistical analyses were performed using R software and R packages (Table S7). Analysis of overall survival, as measured from day of diagnosis until day of event or censoring, was performed and graphed with R survival and R survminer packages. Log rank statistics were calculated to test for differences in survival distributions. Continuous clinical parameters were tested for normal distribution with Shapiro–Wilk test and further assessed for their associations to *JAK/STAT* mutational status with either Student's *t*-test or Wilcoxon rank sum test, depending on the normality test. Fisher's exact test was used for testing differences in categorical data in patients with *JAK/STAT* mutations compared to unmutated patients. A $p < 0.05$ was considered to be statistically significant.

4.5. Immunoblots

We performed Western blots of whole-cell lysates of primary T-PLL cells and CD3+ pan T-cell isolated from age matched healthy controls by magnetic-bead based cell enrichment. Positive selection of CD3+ pan T-cells of healthy controls was performed according to the manufactures instruction (Biolegend). Following antibodies from Cell Signaling Technology were used in a 1:1000 dilution according to the manufacturer's instruction: phospho-STAT5TYR694 (clone C11C5, catalogue #9359), STAT5 (polyclonal, #9363), and GAPDH (clone 14C10, #2118). As a secondary antibody we made use of HRP-coupled anti-rabbit (Dianova) in a 1:5000 dilution. Immunoblots were done using Western Bright ECL (Advansta, Menlo Park, CA, USA) and luminescence intensities were evaluated by densitometry (ImageJ software, National Institute of Health, Bethesda, MD, USA). Uncropped images of the immunoblots are displayed in Figure S5.

4.6. Data Accessibility

All genomic data analyzed in this study was gathered from published series. For detailed sources of raw data and/or information on *JAK/STAT* mutations, see Table S6.

5. Conclusions

Overall, this meta-analysis provides an overview of the lesional landscape of *JAK/STAT* associated genes in the largest cohort of T-PLL cases to date and summarizes the potential functional impact of the most common genomic aberrations. Nearly every T-PLL case in our cohort had a genetic aberration, which can be intuitively implicated in constitutively active JAK/STAT signaling. Besides GOF missense mutations in *JAK/STAT* genes, genomic losses of negative regulators of the JAK-STAT signaling axis are frequently found in T-PLL, linking the high basal phosphorylation of STAT5B to genomic causes. We propose a model for these distinct mechanisms implicated in constitutive STAT5B signaling in T-PLL (Figure 5). Our data provide a rational framework for strategies to inhibit JAK/STAT signaling in T-PLL, even in patients whose leukemia does not carry mutations in a *JAK* or *STAT* gene. Development of novel STAT inhibitors and their application, including with synergistic partners, in model systems and in T-PLL patients, represent future tasks.

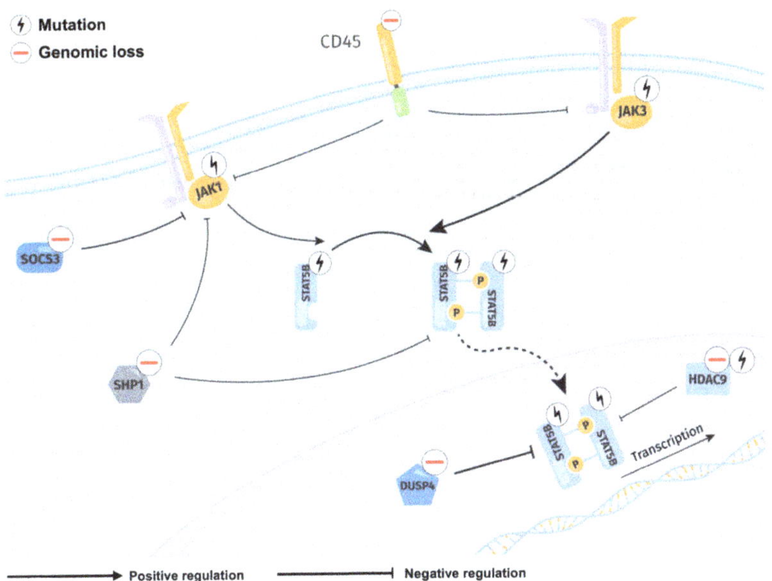

Figure 5. Proposed model of recurrent genomic lesions leading to an enhanced activation of STAT5B in T-PLL. Regulatory network summarizing detected genomic lesions (sCNAs, mutations) in *JAK/STAT* genes and their direct regulators. Mutations of *JAK* and *STAT* genes and genomic losses of negative STAT5B regulators being affected in more than 5% of T-PLL patients were included. Frequent missense mutations occur in the JH2 and SH2-JH2 linker of *JAK1* and *JAK3*, potentially leading to elevated phosphorylation and dimerization of STAT5B. Activation of JAK1 and JAK3 is potentially enhanced through genomic losses (*DUSP4, SOCS1, SOCS3, CD45, SHP1, HDAC9,* and *TCPTP*) and mutations (*HDAC9*) of negative regulators. STAT5B activation might be further increased through GOF mutations in its SH2 domain. Cytoplasmatic (*SHP1, TCPTP*) as well as nuclear (*DUSP4, HDAC9*) regulators are commonly affected by genomic losses in T-PLL, leading to intensified STAT5B signaling. Constitutive active STAT5B translocates into the nucleus and regulates transcription of many target genes relevant for T-cell development, differentiation, proliferation, migration, and apoptosis [52].

Supplementary Materials: The following materials are available online at http://www.mdpi.com/2072-6694/11/12/1833/s1, Figure S1: *JAK/STAT* mutation status shows association with the inversion 14 (q11q32) and elevated *TCL1A* mRNA expression; Figure S2: *JAK/STAT* mutation status shows no association with outcome in T-PLL patients; Figure S3: Overview of genomic lesions associated with an activation of the JAK/STAT pathway across all considered T-PLL cases; Figure S4: T-PLL cases harboring any *JAK* or *STAT* mutation show similar gene expression compared to T-PLL cases with a loss or mutation of a negative regulator potentially activating JAK/STAT signaling; Figure S5: Uncropped images of immunoblots; Table S1: Overview of considered publications; Table S2: Associations of *JAK* and *STAT* mutations with immunophenotypic, cytogenetic, and mutational data; Table S3: Screening of clinical parameters; Table S4: List of genes encoding for proteins which potentially regulate JAK/STAT signaling; Table S5: Overview of references; Table S6: Data sources; Table S7: Bioinformatic tools

Author Contributions: Conceptualization, L.W., T.B., A.S., and, M.H.; Formal analysis, L.W., S.T., A.K.G., H.A., and T.A.; Investigation, T.B.; Resources, D.B., P.J., C.L., C.H., M.-H.S., J.D., and R.S.; Data curation, L.W., S.T., and P.W.; Writing—original draft preparation, L.W., T.B., S.T., and A.K.G.; Writing—review and editing, A.S., R.S., S.M., T.A., and M.H.; Supervision, A.S., S.M., T.A., and M.H.; Project administration, L.W. and T.B.

Funding: This research was funded by the DFG Research Unit FOR1961 (Control-T; HE3553/4-2) and the DFG CRC SFB 1076 (Project B9), by the Köln Fortune program, by the Fritz Thyssen Foundation (10.15.2.034MN), by the Academy of Finland (grants 292611,295504, 310507, 326238), and by the Cancer Society of Finland. This work was also supported by the EU Transcan-2 consortium 'ERANET-PLL' and by the ERAPerMed consortium 'JAKSTAT-TARGET'. A.S. is funded by a scholarship through the José Carreras Leukemia Foundation (DJCLS 03 F/2016).

Acknowledgments: The authors gratefully thank all the authors of the original publications for providing either the raw or already processed data for the meta-analyses. The authors thank Michael Gruen for his technical support with Figure 5 and all the patients with their families for their invaluable contributions.

Conflicts of Interest: The authors declare no conflict of interest.

References

1. Staber, P.B.; Herling, M.; Bellido, M.; Jacobsen, E.D.; Davids, M.S.; Kadia, T.M.; Shustov, A.; Tournilhac, O.; Bachy, E.; Zaja, F.; et al. Consensus criteria for diagnosis, staging, and treatment response assessment of T-cell prolymphocytic leukemia. *Blood* **2019**, *134*, 1132–1143. [CrossRef] [PubMed]

2. Herling, M.; Khoury, J.D.; Washington, L.T.; Duvic, M.; Keating, M.J.; Jones, D. A systematic approach to diagnosis of mature T-cell leukemias reveals heterogeneity among WHO categories. *Blood* **2004**, *104*, 328–335. [CrossRef] [PubMed]

3. Dearden, C. How i treat prolymphocytic leukemia. *Blood* **2012**, *120*, 538–551. [CrossRef] [PubMed]

4. Hopfinger, G.; Busch, R.; Pflug, N.; Weit, N.; Westermann, A.; Fink, A.M.; Cramer, P.; Reinart, N.; Winkler, D.; Fingerle-Rowson, G.; et al. Sequential chemoimmunotherapy of fludarabine, mitoxantrone, and cyclophosphamide induction followed by alemtuzumab consolidation is effective in T-cell prolymphocytic leukemia. *Cancer* **2013**, *119*, 2258–2267. [CrossRef] [PubMed]

5. Herling, M.; Patel, K.A.; Teitell, M.A.; Konopleva, M.; Ravandi, F.; Kobayashi, R.; Jones, D. High TCL1 expression and intact T-cell receptor signaling define a hyperproliferative subset of T-cell prolymphocytic leukemia. *Blood* **2008**, *111*, 328–337. [CrossRef] [PubMed]

6. Ravandi, F.; O'Brien, S.; Jones, D.; Lerner, S.; Faderl, S.; Ferrajoli, A.; Wierda, W.; Garcia-Manero, G.; Thomas, D.; Koller, C.; et al. T-cell prolymphocytic leukemia: A single-institution experience. *Clin. Lymhoma Myeloma* **2005**, *6*, 234–239. [CrossRef]

7. Dearden, C.E.; Khot, A.; Else, M.; Hamblin, M.; Grand, E.; Roy, A.; Hewamana, S.; Matutes, E.; Catovsky, D. Alemtuzumab therapy in T-cell prolymphocytic leukemia: Comparing efficacy in a series treated intravenously and a study piloting the subcutaneous route. *Blood* **2011**, *118*, 5799–5802. [CrossRef]

8. Hoh, F.; Yang, Y.-S.; Guignard, L.; Padilla, A.; Stern, M.-H.; Lhoste, J.-M.; van Tilbeurgh, H. Crystal structure of p14TCL1, an oncogene product involved in T-cell prolymphocytic leukemia, reveals a novel β-barrel topology. *Structure* **1998**, *6*, 147–155. [CrossRef]

9. Schrader, A.; Crispatzu, G.; Oberbeck, S.; Mayer, P.; Pützer, S.; von Jan, J.; Vasyutina, E.; Warner, K.; Weit, N.; Pflug, N.; et al. Actionable perturbations of damage responses by TCL1/ATM and epigenetic lesions form the basis of T-PLL. *Nat. Commun.* **2018**, *9*, 697. [CrossRef]

10. Kiel, M.J.; Velusamy, T.; Rolland, D.; Sahasrabuddhe, A.A.; Chung, F.; Bailey, N.G.; Schrader, A.; Li, B.; Li, J.Z.; Ozel, A.B.; et al. Integrated genomic sequencing reveals mutational landscape of T-cell prolymphocytic leukemia. *Blood* **2014**, *124*, 1460–1472. [CrossRef]

11. Bellanger, D.; Jacquemin, V.; Chopin, M.; Pierron, G.; Bernard, O.A.; Ghysdael, J.; Stern, M.-H. Recurrent *JAK1* and *JAK3* somatic mutations in T-cell prolymphocytic leukemia. *Leukemia* **2013**, *28*, 417. [CrossRef] [PubMed]

12. Stengel, A.; Kern, W.; Zenger, M.; Perglerová, K.; Schnittger, S.; Haferlach, T.; Haferlach, C. Genetic characterization of T-PLL reveals two major biologic subgroups and *JAK3* mutations as prognostic marker. *Genes Chromosom. Cancer* **2016**, *55*, 82–94. [CrossRef] [PubMed]

13. López, C.; Bergmann, A.K.; Paul, U.; Murga Penas, E.M.; Nagel, I.; Betts, M.J.; Johansson, P.; Ritgen, M.; Baumann, T.; Aymerich, M.; et al. Genes encoding members of the JAK-STAT pathway or epigenetic regulators are recurrently mutated in T-cell prolymphocytic leukaemia. *Br. J. Haematol.* **2016**, *173*, 265–273. [CrossRef] [PubMed]

14. He, L.; Tang, J.; Andersson, E.I.; Timonen, S.; Koschmieder, S.; Wennerberg, K.; Mustjoki, S.; Aittokallio, T. Patient-customized drug combination prediction and testing for T-cell prolymphocytic leukemia patients. *Cancer Res.* **2018**, *78*, 2407–2418. [CrossRef]

15. Dietrich, S.; Oleś, M.; Lu, J.; Sellner, L.; Anders, S.; Velten, B.; Wu, B.; Hüllein, J.; da Silva Liberio, M.; Walther, T.; et al. Drug-perturbation-based stratification of blood cancer. *J. Clin. Investig.* **2018**, *128*, 427–445. [CrossRef]

16. Andersson, E.I.; Pützer, S.; Yadav, B.; Dufva, O.; Khan, S.; He, L.; Sellner, L.; Schrader, A.; Crispatzu, G.; Oleś, M.; et al. Discovery of novel drug sensitivities in T-PLL by high-throughput ex vivo drug testing and mutation profiling. *Leukemia* **2017**, *32*, 774. [CrossRef]

17. Greenplate, A.; Wang, K.; Tripathi, R.M.; Palma, N.; Ali, S.M.; Stephens, P.J.; Miller, V.A.; Shyr, Y.; Guo, Y.; Reddy, N.M.; et al. Genomic profiling of T-Cell neoplasms reveals frequent *JAK1* and *JAK3* mutations with clonal evasion from targeted therapies. *JCO Precis. Oncol.* **2018**. [CrossRef]

18. Johansson, P.; Klein-Hitpass, L.; Choidas, A.; Habenberger, P.; Mahboubi, B.; Kim, B.; Bergmann, A.; Scholtysik, R.; Brauser, M.; Lollies, A.; et al. SAMHD1 is recurrently mutated in T-cell prolymphocytic leukemia. *Blood Cancer J.* **2018**, *8*, 11. [CrossRef]

19. Rawlings, J.S.; Rosler, K.M.; Harrison, D.A. The JAK/STAT signaling pathway. *J. Cell Sci.* **2004**, *117*, 1281–1283. [CrossRef]

20. Villarino, A.V.; Kanno, Y.; O'Shea, J.J. Mechanisms and consequences of Jak-STAT signaling in the immune system. *Nat. Immunol.* **2017**, *18*, 374–384. [CrossRef]

21. Furqan, M.; Mukhi, N.; Lee, B.; Liu, D. Dysregulation of JAK-STAT pathway in hematological malignancies and JAK inhibitors for clinical application. *Biomark. Res.* **2013**, *1*, 5. [CrossRef] [PubMed]

22. Heinrich, T.; Rengstl, B.; Muik, A.; Petkova, M.; Schmid, F.; Wistinghausen, R.; Warner, K.; Crispatzu, G.; Hansmann, M.-L.; Herling, M.; et al. Mature T-cell lymphomagenesis induced by retroviral insertional activation of Janus kinase 1. *Mol. Ther.* **2013**, *21*, 1160–1168. [CrossRef] [PubMed]

23. Maurer, B.; Nivarthi, H.; Wingelhofer, B.; Pham, H.T.T.; Schlederer, M.; Suske, T.; Grausenburger, R.; Schiefer, A.-I.; Prchal-Murphy, M.; Chen, D.; et al. High activation of STAT5A drives peripheral T-cell lymphoma and leukemia. *Haematologica* **2019**. [CrossRef] [PubMed]

24. Taylor, P.C. Clinical efficacy of launched JAK inhibitors in rheumatoid arthritis. *Rheumatology* **2019**, *58* (Suppl. 1), i17–i26. [CrossRef]

25. Mannina, D.; Kröger, N. Janus Kinase inhibition for graft-versus-host disease: Current status and future prospects. *Drugs* **2019**, *79*, 1499–1509. [CrossRef]

26. Schwartz, D.M.; Bonelli, M.; Gadina, M.; O'Shea, J.J. Type I/II cytokines, JAKs, and new strategies for treating autoimmune diseases. *Nat. Rev. Rheumatol.* **2016**, *12*, 25–36. [CrossRef]

27. Gomez-Arteaga, A.; Margolskee, E.; Wei, M.T.; van Besien, K.; Inghirami, G.; Horwitz, S. Combined use of tofacitinib (pan-JAK inhibitor) and ruxolitinib (a JAK1/2 inhibitor) for refractory T-cell prolymphocytic leukemia (T-PLL) with a JAK3 mutation. *Leuk. Lymphoma* **2019**, *60*, 1626–1631. [CrossRef]

28. Wei, M.; Koshy, N.; van Besien, K.; Inghirami, G.; Horwitz, S.M. Refractory T-cell prolymphocytic leukemia with JAK3 mutation: In Vitro and clinical synergy of Tofacitinib and Ruxolitinib. *Blood* **2015**, *126*, 5486. [CrossRef]

29. Kiel, M.J.; Sahasrabuddhe, A.A.; Rolland, D.C.M.; Velusamy, T.; Chung, F.; Schaller, M.; Bailey, N.G.; Betz, B.L.; Miranda, R.N.; Porcu, P.; et al. Genomic analyses reveal recurrent mutations in epigenetic modifiers and the JAK–STAT pathway in Sézary syndrome. *Nat. Commun.* **2015**, *6*, 8470. [CrossRef]

30. Nicolae, A.; Xi, L.; Pham, T.H.; Pham, T.-A.; Navarro, W.; Meeker, H.G.; Pittaluga, S.; Jaffe, E.S.; Raffeld, M. Mutations in the JAK/STAT and RAS signaling pathways are common in intestinal T-cell lymphomas. *Leuemia* **2016**, *30*, 2245. [CrossRef]

31. Lone, W.; Alkhiniji, A.; Manikkam Umakanthan, J.; Iqbal, J. Molecular Insights into pathogenesis of peripheral T cell lymphoma: A review. *Curr. Hematol. Malig. Rep.* **2018**, *13*, 318–328. [CrossRef] [PubMed]

32. Maljaei, S.H.; Brito-Babapulle, V.; Hiorns, L.R.; Catovsky, D. Abnormalities of chromosomes 8, 11, 14, and X in T-prolymphocytic leukemia studied by fluorescence in situ hybridization. *Cancer Genet. Cytogenet.* **1998**, *103*, 110–116. [CrossRef]

33. Prinz, C.; Vasyutina, E.; Lohmann, G.; Schrader, A.; Romanski, S.; Hirschhäuser, C.; Mayer, P.; Frias, C.; Herling, C.D.; Hallek, M.; et al. Organometallic nucleosides induce non-classical leukemic cell death that is mitochondrial-ROS dependent and facilitated by TCL1-oncogene burden. *Mol. Cancer* **2015**, *14*, 114. [CrossRef] [PubMed]

34. Jeong, E.G.; Kim, M.S.; Nam, H.K.; Min, C.K.; Lee, S.; Chung, Y.J.; Yoo, N.J.; Lee, S.H. Somatic mutations of *JAK1* and *JAK3* in acute leukemias and solid cancers. *Clin. Cancer Res.* **2008**, *14*, 3716–3721. [CrossRef] [PubMed]

35. Gordon, G.M.; Lambert, Q.T.; Daniel, K.G.; Reuther, G.W. Transforming JAK1 mutations exhibit differential signalling, FERM domain requirements and growth responses to interferon-γ. *Biochem. J.* **2010**, *432*, 255–265. [CrossRef]

36. Hornakova, T.; Staerk, J.; Royer, Y.; Flex, E.; Tartaglia, M.; Constantinescu, S.N.; Knoops, L.; Renauld, J.-C. Acute lymphoblastic leukemia-associated JAK1 mutants activate the Janus kinase/STAT pathway via interleukin-9 receptor alpha homodimers. *J. Biol. Chem.* **2009**, *284*, 6773–6781. [CrossRef] [PubMed]

37. Flex, E.; Petrangeli, V.; Stella, L.; Chiaretti, S.; Hornakova, T.; Knoops, L.; Ariola, C.; Fodale, V.; Clappier, E.; Paoloni, F.; et al. Somatically acquired *JAK1* mutations in adult acute lymphoblastic leukemia. *J. Exp. Med.* **2008**, *205*, 751–758. [CrossRef]

38. Yang, S.; Luo, C.; Gu, Q.; Xu, Q.; Wang, G.; Sun, H.; Qian, Z.; Tan, Y.; Qin, Y.; Shen, Y.; et al. Activating JAK1 mutation may predict the sensitivity of JAK-STAT inhibition in hepatocellular carcinoma. *Oncotarget* **2016**, *7*, 5461–5469. [CrossRef]

39. Degryse, S.; de Bock, C.E.; Demeyer, S.; Govaerts, I.; Bornschein, S.; Verbeke, D.; Jacobs, K.; Binos, S.; Skerrett-Byrne, D.A.; Murray, H.C.; et al. Mutant JAK3 phosphoproteomic profiling predicts synergism between JAK3 inhibitors and MEK/BCL2 inhibitors for the treatment of T-cell acute lymphoblastic leukemia. *Leukemia* **2018**, *32*, 788. [CrossRef]

40. Martinez, G.S.; Ross, J.A.; Kirken, R.A. Transforming mutations of Jak3 (A573V and M511I) show differential sensitivity to selective JAK3 inhibitors. *Clin. Cancer Drugs* **2016**, *3*, 131–137. [CrossRef]

41. Degryse, S.; de Bock, C.E.; Cox, L.; Demeyer, S.; Gielen, O.; Mentens, N.; Jacobs, K.; Geerdens, E.; Gianfelici, V.; Hulselmans, G.; et al. JAK3 mutants transform hematopoietic cells through JAK1 activation, causing T-cell acute lymphoblastic leukemia in a mouse model. *Blood* **2014**, *124*, 3092–3100. [CrossRef] [PubMed]

42. Vicente, C.; Schwab, C.; Broux, M.; Geerdens, E.; Degryse, S.; Demeyer, S.; Lahortiga, I.; Elliott, A.; Chilton, L.; Starza, R.L.; et al. Targeted sequencing identifies associations between IL7R-JAK mutations and epigenetic modulators in T-cell acute lymphoblastic leukemia. *Haematologica* **2015**, *100*, 1301–1310. [CrossRef] [PubMed]

43. Rajala, H.L.M.; Eldfors, S.; Kuusanmäki, H.; van Adrichem, A.J.; Olson, T.; Lagström, S.; Andersson, E.I.; Jerez, A.; Clemente, M.J.; Yan, Y.; et al. Discovery of somatic STAT5b mutations in large granular lymphocytic leukemia. *Blood* **2013**, *121*, 4541–4550. [CrossRef] [PubMed]

44. Küçük, C.; Jiang, B.; Hu, X.; Zhang, W.; Chan, J.K.C.; Xiao, W.; Lack, N.; Alkan, C.; Williams, J.C.; Avery, K.N.; et al. Activating mutations of STAT5B and STAT3 in lymphomas derived from γδ-T or NK cells. *Nat. Commun.* **2015**, *6*, 6025. [CrossRef]

45. de Araujo, E.D.; Erdogan, F.; Neubauer, H.A.; Meneksedag-Erol, D.; Manaswiyoungkul, P.; Eram, M.S.; Seo, H.S.; Qadree, A.K.; Israelian, J.; Orlova, A.; et al. Structural and functional consequences of the STAT5BN642H driver mutation. *Nat. Commun.* **2019**, *10*, 2517. [CrossRef]

46. Pham, H.T.T.; Maurer, B.; Prchal-Murphy, M.; Grausenburger, R.; Grundschober, E.; Javaheri, T.; Nivarthi, H.; Boersma, A.; Kolbe, T.; Elabd, M.; et al. STAT5BN642H is a driver mutation for T cell neoplasia. *J. Clin. Investig.* **2018**, *128*, 387–401. [CrossRef]

47. Andersson, E.I.; Tanahashi, T.; Sekiguchi, N.; Gasparini, V.R.; Bortoluzzi, S.; Kawakami, T.; Matsuda, K.; Mitsui, T.; Eldfors, S.; Bortoluzzi, S.; et al. High incidence of activating STAT5B mutations in CD4-positive T-cell large granular lymphocyte leukemia. *Blood* **2016**, *128*, 2465–2468. [CrossRef]

48. McKinney, M.; Moffitt, A.B.; Gaulard, P.; Travert, M.; de Leval, L.; Raffeld, A.N.M.; Jaffe, E.S.; Pittaluga, S.; Xi, L.; Heavican, T.; et al. The genetic basis of hepatosplenic T-cell lymphoma. *Cancer Discov.* **2017**, *7*, 369–379. [CrossRef]

49. O'Shea, J.J.; Schwartz, D.M.; Villarino, A.V.; Gadina, M.; McInnes, I.B.; Laurence, A. The JAK-STAT pathway: impact on human disease and therapeutic intervention. *Annu. Rev. Med.* **2015**, *66*, 311–328. [CrossRef]

50. Schrader, A.; Braun, T.; Herling, M. The dawn of a new era in treating T-PLL. *Oncotarget* **2019**, *10*, 626–628. [CrossRef]

51. Wingelhofer, B.; Maurer, B.; Heyes, E.C.; Cumaraswamy, A.A.; Berger-Becvar, A.; de Araujo, E.D.; Orlova, A.; Freund, P.; Ruge, F.; Park, J.; et al. Pharmacologic inhibition of STAT5 in acute myeloid leukemia. *Leukemia* **2018**, *32*, 1135–1146. [CrossRef] [PubMed]

52. Kanai, T.; Seki, S.; Jenks, J.A.; Kohli, A.; Kawli, T.; Martin, D.P.; Snyder, M.; Bacchetta, R.; Nadeau, K.C. Identification of STAT5A and STAT5B target genes in human T cells. *PLoS ONE* **2014**, *9*, e86790. [CrossRef] [PubMed]

Article

STAT3 Mutation Is Associated with STAT3 Activation in CD30[+] ALK[−] ALCL

Emma I. Andersson [1,2,3,†], **Oscar Brück** [1,2,†], **Till Braun** [4], **Susanna Mannisto** [5], **Leena Saikko** [6], **Sonja Lagström** [7], **Pekka Ellonen** [7], **Sirpa Leppä** [5], **Marco Herling** [4], **Panu E. Kovanen** [6] and **Satu Mustjoki** [1,2,*]

[1] Hematology Research Unit Helsinki, Department of Hematology, University of Helsinki and Helsinki University Hospital Comprehensive Cancer Center, 00290 Helsinki, Finland; emma.andersson@hus.fi (E.I.A.); oscar.bruck@helsinki.fi (O.B.)

[2] Translational Immunology Research program and Department of Clinical Chemistry and Hematology, University of Helsinki, 00014 Helsinki, Finland

[3] HUSLAB, Laboratory of Genetics, Helsinki University Hospital, 00290 Helsinki, Finland

[4] Department I of Internal Medicine, CMMC, CECAD, CIO-ABCD, University of Cologne, 50931 Cologne, Germany; till.braun@uk-koeln.de (T.B.); marco.herling@uk-koeln.de (M.H.)

[5] Department of Oncology, Comprehensive Cancer Center, Helsinki University Hospital, 00290 Helsinki, Finland; susanna.mannisto@hus.fi (S.M.); sirpa.leppa@hus.fi (S.L.)

[6] Department of Pathology, HUSLAB, Helsinki University Central Hospital and University of Helsinki, 00290 Helsinki, Finland; leena.saikko@helsinki.fi (L.S.); panu.kovanen@helsinki.fi (P.E.K.)

[7] Institute for Molecular Medicine Finland (FIMM), University of Helsinki, 00014 Helsinki, Finland; sonja.lagstrom@kreftregisteret.no (S.L.); pekka.ellonen@helsinki.fi (P.E.)

*   Correspondence: satu.mustjoki@helsinki.fi; Tel.: +358-9-471-71898

†   These authors contributed equally to this work.

Received: 18 February 2020; Accepted: 13 March 2020; Published: 16 March 2020

Abstract: Peripheral T-cell lymphomas (PTCL) are a heterogeneous, and often aggressive group of non-Hodgkin lymphomas. Recent advances in the molecular and genetic characterization of PTCLs have helped to delineate differences and similarities between the various subtypes, and the JAK/STAT pathway has been found to play an important oncogenic role. Here, we aimed to characterize the JAK/STAT pathway in PTCL subtypes and investigate whether the activation of the pathway correlates with the frequency of *STAT* gene mutations. Patient samples from AITL ($n = 30$), ALCL ($n = 21$) and PTCL-NOS ($n = 12$) cases were sequenced for *STAT3*, *STAT5B*, *JAK1*, *JAK3*, and *RHOA* mutations using amplicon sequencing and stained immunohistochemically for pSTAT3, pMAPK, and pAKT. We discovered *STAT3* mutations in 13% of AITL, 13% of ALK[+] ALCL, 38% of ALK[−] ALCL and 17% of PTCL-NOS cases. However, no *STAT5B* mutations were found and *JAK* mutations were only present in ALK[−] ALCL (15%). Concurrent mutations were found in all subgroups except ALK[+] ALCL where *STAT3* mutations were always seen alone. High pY-STAT3 expression was observed especially in AITL and ALCL samples. When studying JAK-STAT pathway mutations, pY-STAT3 expression was highest in PTCLs harboring either *JAK1* or *STAT3* mutations and CD30[+] phenotype representing primarily ALK[−] ALCLs. Further investigation is needed to elucidate the molecular mechanisms of JAK-STAT pathway activation in PTCL.

Keywords: lymphoma; T-cells; STAT3; RHOA; NGS

---

1. Introduction

Peripheral T-cell lymphomas (PTCLs) form a heterogeneous, uncommon, and often aggressive group of non-Hodgkin's lymphomas (NHL) representing approximately 10–15% of all new NHL diagnoses [1]. The most prevalent PTLCs are PTCL not otherwise specified (NOS), angioimmunoblastic

T-cell lymphoma (AITL), anaplastic lymphoma kinase–negative (ALK⁻ ALCL), and anaplastic lymphoma kinase-positive anaplastic large cell lymphoma (ALK⁺ ALCL) [2,3].

Despite generally favorable response rates to chemotherapy, remissions are often not durable, and as such the natural history of PTCL is characterized by relapses and refractory disease. For this reason, upfront hematopoietic stem cell transplantation (HSCT) is often recommended in first remission for patients who are fit and have chemo-sensitive disease [4]. For patients who are not suitable for transplantation, chances of long-term disease control are very limited, with an average progression-free survival (PFS) of 5.5 months [5]. In this setting, optimal therapeutic approaches remain undefined representing an unmet clinical need.

Recent advances in the molecular and genetic characterization of PTCL have helped to delineate differences and similarities between the various subtypes [6]. Several recurrent mutations have been identified in small subsets of patients potentially enabling more accurate disease classification and guide treatment decisions. In AITL, the three most commonly identified genetic lesions occur in the Tet methylcytosine dioxygenase 2 gene (*TET2*), the Ras homolog gene family, member A (*RHOA*), and the isocitrate dehydrogenase 2 gene (*IDH2*). Besides AITL, *TET2* mutations can be seen with a high frequency also in PTCL-NOS with a T follicular helper cell phenotype [7]. RHOA, a small GTPase participating in T-cell activation and polarization, has recently been found to have a specific G17V mutation in 68% of AITL cases [8], predominantly in the background of *TET2* mutations. The overwhelming majority of *IDH2* mutations in PTCL affect the R172 residue [9]. However, *STAT3* mutations are relatively uncommon (5%) in AITL [10].

In ALK⁺ ALCL, the ALK chimeras activate STAT3, thus maintaining the neoplastic phenotype in ALK⁺ ALCL. In concordance with this, gene expression profiling has revealed a transcriptional gene signature including *ALK, TNFRSF8 (CD30), MUC1,* Th17-associated genes (*IL-17A, IL-17F,* and *ROR-γ*), a small group of immunoregulatory cytokines and receptors regulating STAT3 pathway [11].

A systematic characterization of the genetic alterations driving ALCL was recently undertaken using sequencing strategies. Activating mutations of *JAK1* and *STAT3* genes were found in 20% of ALK⁻ ALCLs, 38% of which displayed double lesions [12]. As the JAK/STAT pathway has a critical role in hematopoietic development, it is not surprising that it plays, when deregulated, an important oncogenic role in lymphoproliferative malignancies [13]. Many cancers and hematologic malignancies have been associated with the constitutive activation of the STAT family of proteins, which depends on JAK-mediated tyrosine phosphorylation for transcriptional activation [14]. In particular, activated JAK1/STAT3 and JAK2/STAT5 have been shown to facilitate T-cell transformation [15,16]. Activating mutations of *JAK1-3* and *STAT3-5* have also been found in a subset of NK/T-cell lymphomas, non-hepatosplenic gamma-delta T-cell lymphoma, large granular lymphocytic leukemia and T-cell prolymphocytic leukemia [10,17–22].

Here, we aimed to further characterize the JAK/STAT pathway in PTCL by investigating whether the activation of the pathway quantified with immunohistochemistry correlates with the frequency of *JAK, STAT,* and *RHOA* mutations. In addition, we associated JAK-STAT pathway activation with prognosis and essential clinical and pathological parameters. Finally, we identified novel therapeutic approaches for patients with high STAT3 phosphorylation using Reverse Phase Protein Array and drug sensitivity data from the Broad Institute Cancer Cell Line Encyclopedia (CCLE).

2. Results

2.1. STAT3 Mutations Are Frequent in PTCLs

As STAT3 is constitutively expressed in both AITLs and ALCLs, we sequenced for *STAT3* and *STAT5B* mutations. The entire *STAT3* gene and the hotspot SH2-domain of *STAT5B* were screened by targeted amplicon sequencing in 63 patients with T-cell lymphoma. We discovered that 13% of AITL and ALK⁺ ALCL cases harbored *STAT3* mutations, while 17% of PTCL-NOS cases were found to harbor *STAT3* mutations (Table 1). The highest frequency of *STAT3* mutations was found in the ALK⁻ ALCL subgroup (38%). The frequency of *STAT3* mutations in the entire patient cohort was 19%. While most mutations were found in the SH2-domain of *STAT3*, we observed one K290T mutation in the coiled coil domain and three patients harbored a P715L mutation in the transactivation domain (Figure 1 and Table S1). The most prevalent mutation type was the P715L mutation and the E616 deletion, which were identified in three patients each. No *STAT5B* SH2-domain mutations were identified in this cohort. *JAK1* mutations were found in two ALK⁻ ALCL patients (15%). Both mutations (G902R and G1097C) were seen in the protein tyrosine kinase domain of *JAK1*. No *JAK3* mutations were detected in any subgroup.

Table 1. *STAT3*, *JAK1/3*, and *RHOA* mutation frequencies in PTCLs.

PTCL Subtype	*n*	*STAT3* Mutation Frequency	*JAK1/3* Mutation Frequency	*RHOA* Mutation Frequency	Co-Occuring Mutations
AITL	30	13%	0%	70%	10%
NOS	12	17%	0%	17%	8%
ALK⁺ ALCL	8	13%	0%	0%	None
ALK⁻ ALCL	13	38%	15%	0%	8%
All	63	19%	3%	37%	8%

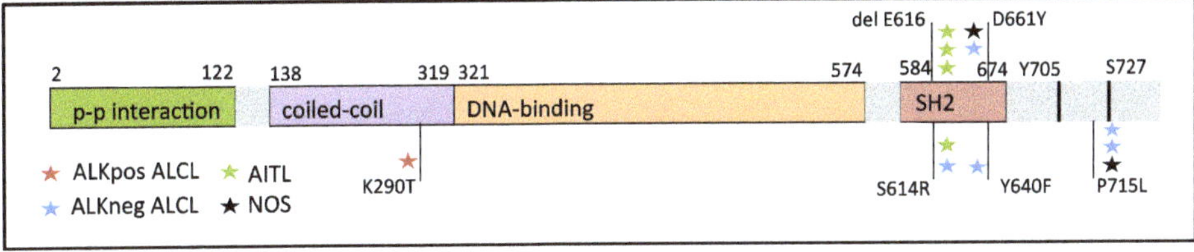

Figure 1. *STAT3* structure and mutation sites in ALCL, AITL, and PTCL-NOS patients of the study cohort.

As *RHOA* mutations are highly prevalent in AITL and have been found to cause activation of the MAPK/PI3K/AKT pathway and T cell receptor signaling, the hotspot mutation G17V was also screened [23,24]. The results showed that these mutations were frequent in AITL (70%) and PTCL-NOS (17%) but remained absent in ALCLs (Table 1). Some cases harbored both *STAT3* and *RHOA* mutations (10%). DNA samples from nodular lymphocyte predominant Hodgkin lymphoma with abundant follicular T helper cells (*n* = 13), were also available for *STAT3/STAT5B* and *RHOA* screening but no mutations were detected in these samples.

2.2. STAT3 Phosphorylation Levels Are Highest in ALK⁺ and ALK⁻ ALCL Cases

To characterize JAK-STAT pathway activation, we stained for pY-STAT3, pMAPK, and pAKT in AITL (*n* = 29), ALCL (*n* = 20) and PTCL-NOS (*n* = 12) patients and control lymph nodes (*n* = 4) using immunohistochemistry (Supplementary Figure S1). The differences between disease groups were significant notably for phosphorylation of pY-STAT3 and recapitulated well previous reports [25]. Further post-hoc analysis showed that pY-STAT3 was shown to be more frequently phosphorylated in AITL (mean = 43.7%) and ALCL (ALK⁻: mean = 62.3%, ALK⁺: mean = 79.7%) patients compared to control lymph nodes (mean 12.6%; Figure 2).

MAPK phosphorylation was significantly elevated in AITL patients (mean 10.5%) followed by ALK⁻ ALCL (mean 6.8%) and PTCL NOS (5.2%) when compared to control lymph node (Figure 2). No significant difference in pAKT levels between PTCL patients and control lymph nodes was observed.

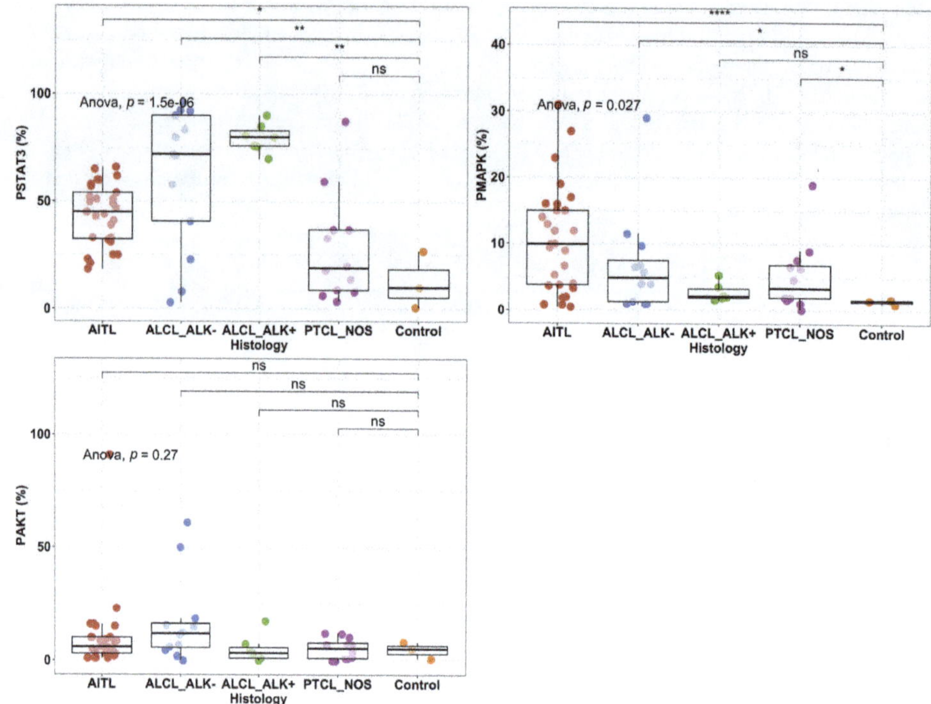

Figure 2. Phosphorylation percentage of pY-STAT3, MAPK, and AKT in PTCL subgroups and healthy controls (n = 2–3) analyzed by immunohistochemistry. Individual dots represent the proportion of positive cells for pSTAT3, pMAPK, and pAKT. Boxplots represent the median and interquartile range of protein phosphorylation levels. The antibodies used were specific for Tyr705 (pStat3), Thr202/Tyr204 (pMAPK), and Thr450 (pAKT). Comparison was analyzed with ANOVA and post-hoc tests between disease groups and healthy controls. *: $p < 0.05$; **: $p < 0.01$, ****: $p < 0.0001$; ns: not significant.

To confirm the findings and study STAT3 activation in other T cell malignancies, we measured STAT3 and pY-STAT3 levels by Western blotting in ALCL, acute T-lymphoblastic leukemia (T-ALL), NK/T-cell lymphoma (NKT), T-cell large granular lymphocyte leukemia (T-LGL), and PTCL cell lines (Figure 3). We confirmed high phosphorylation of pY-STAT3 (Tyr705) in ALCL but not in other T cell malignancies. Total STAT3 was most prominent in PTCL, ALCL and T-ALL cell lines. High JAK3 phosphorylation was observed in SupM2, Karpas299 and TLRB1 (ALCL) as well as NKL (NKT) cell lines and was not associated with pY-STAT3 phosphorylation.

To further investigate the association of protein expression to transcriptome data, we compared the *JAK1, JAK3, AKT*, and *MAPK* RNA level expression with *STAT3* RNA level expression and pSTAT3^{Tyr705} phosphorylation defined with *reverse phase protein arrays* (RPPA) from the CCLE database (Figure S2). Both *STAT3* expression (535 vs. 198 CPM $p = 0.003$, t-test) and phosphorylation (3.6 vs. −0.10 RPPA unit, $p < 0.001$) were observed to be superior in ALCL than non-ALCL malignant T-cells as shown in Figure 2. Higher *JAK3* expression was noted in ALCL over non-ALCL lines (364 vs. 53 CPM, $p < 0.001$), and correlated with *STAT3* expression ($r = 0.60$, $p = 0.01$). Moreover, higher *MAPK* expression was detected in ALCL over non-ALCL cell lines (127 vs. 68, $p = 0.05$), which correlated with *STAT3* expression ($r = 0.65$, $p = 0.006$). No correlation between *STAT3* and *JAK1* or *AKT* expression was observed.

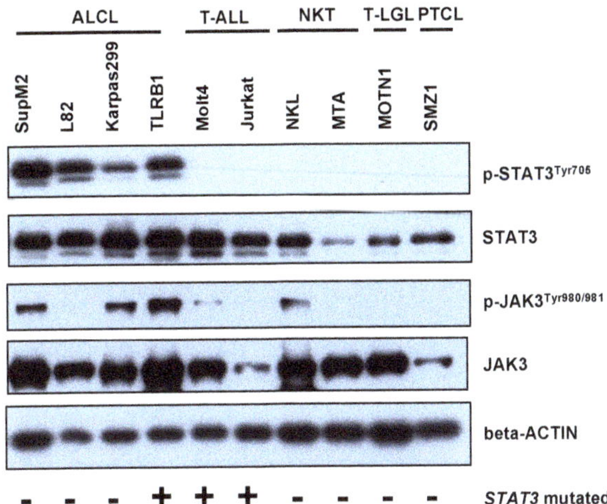

Figure 3. Western blot of p-STAT3 (Tyr705), STAT3, p-JAK3 (Tyr980/981), and JAK3 was characterized in T cell malignancy cell lines. Beta-ACTIN is used as a reference protein.

2.3. STAT3 Phosphorylation Is Associated with JAK1/STAT3 Mutation Status in CD30+ ALK− ALCLs

Next, we studied whether *STAT3* mutations would be associated with STAT3, MAPK, and AKT phosphorylation. No association between *STAT3* mutation status and pY-STAT3, pMAPK, or pAKT (%) expression was noted when studying all PTCL subtypes (Figure 4A). No association with pY-STAT3 phosphorylation was noted even after adjusting with tumor proportion (mean pSTAT3 77% vs. 66% in STAT3 mutated vs. non-mutated PTCLs). Most of the mutations were noted in relatively low variant allele frequencies (VAF 5–9%), which might in part explain the lack of correlation. Further evidence for this observation was found from cell line data, as the four ALCL cell lines in Figure 2 exhibit high pY-STAT3 phosphorylation but only one of the cell lines harbored a *JAK/STAT* mutation (Table S1). The TLBR1 cell line represents ALK− ALCL and harbors a known activating *STAT3* missense mutation S614R in the SH2 domain [26]. The other ALCL cell lines expressed ALK, which explains the constitutive STAT3 activation. Both T-ALL cell lines MOLT4 and Jurkat also harbor missense *STAT3/5B* mutations but conversely, the cell lines did not exhibit any pY-STAT3 phosphorylation. This could be explained by the mutations being localized elsewhere than in the activating SH2-domain.

Some PTCL cases harbored both *STAT3* and *RHOA* mutations (8%). Our results showed that the MAPK phosphorylation degree was higher in patients with *RHOA* mutations than patients without *RHOA* mutations (*RHOA*-positive mean = 12.4% vs. *RHOA*-negative mean 5.7%, $p = 0.005$, t-test; Figure 4B). Yet, *RHOA* mutation was not observed to be associated with STAT3 or AKT phosphorylation.

Phosphorylated STAT3 has been shown to induce CD30 expression in PTCL [27]. In addition, *STAT3* mutations are more frequent in CD30+ than CD30− T-cell lymphomas [28]. Therefore, we sought to investigate the interaction between mutations in the JAK-STAT pathway, CD30 phenotype and pY-STAT3 phosphorylation. Positive CD30 tumor phenotype was associated with higher STAT3 and AKT phosphorylation, but not with pMAPK (Figure S3A). Interestingly, pSTAT3 expression was highest in samples with CD30+ tumors and simultaneous mutation in *STAT3* or *JAK1* (Figure 4C). This phenotype was represented primarily by the ALK− ALCL subtype ($p < 0.001$, chi^2 test, Figure 4D), and their pSTAT3 expression (mean 68.7%) to stand above the mean of ALK− (62.3%) but below ALK+ ALCL (79.7%). We observed that 50% (6/12) of CD30+ ALK− ALCLs harbored a mutation in *JAK1* or *STAT3*. Higher *JAK1/STAT3* VAF trended for higher pY-STAT3 phosphorylation but the result remained non-significant possibly due to sample size (81.3% vs. 61.1% mean pSTAT3 expression in tumors with high ($n = 4$) vs. low ($n = 4$) *JAK1/STAT3* VAF $p = 0.22$, *t*-test). Unexpectedly, lowest STAT3 phosphorylation was observed in CD30⁻ patients with a *STAT3* gene mutation. Interestingly, all three in-frame E616 deletions were found in CD30⁻ AITL cases.

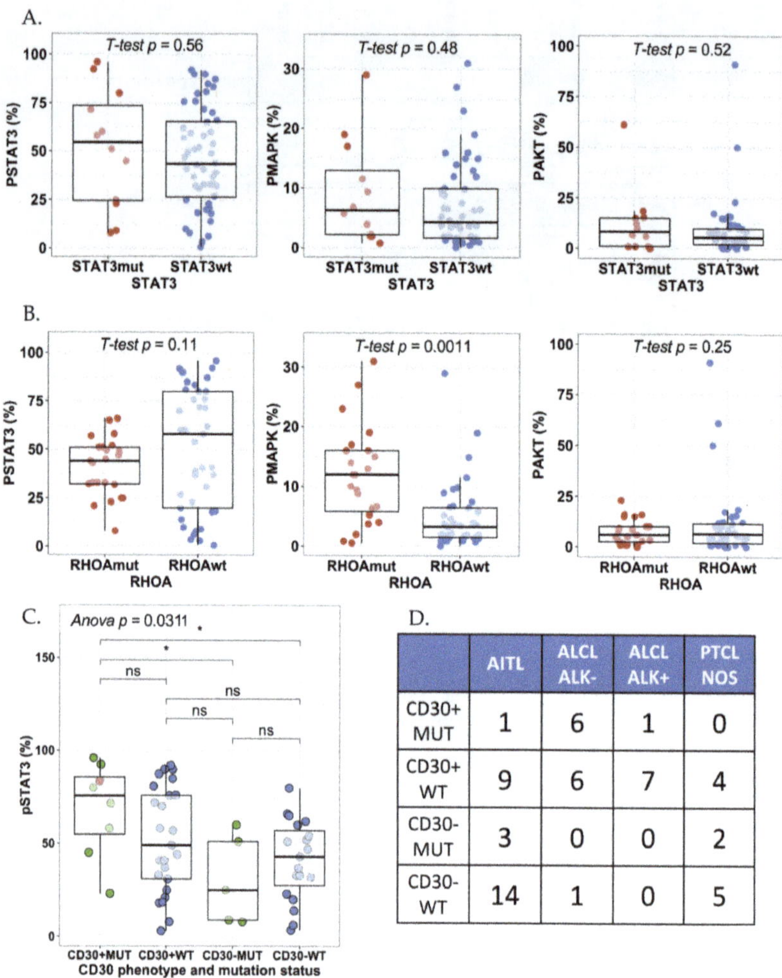

Figure 4. Comparison of the phosphorylation level of (**A**) STAT3, MAPK, and AKT by *STAT3* mutation status and by (**B**) *RHOA* mutation status (*t*-test). The phosphorylation level of pStat3 (Tyr705), pMAPK (Thr202/Tyr204) and pAKT (Thr450) was measured by immunohistochemistry. (**C**) Combinatory effect of CD30 phenotype and *JAK1* or *STAT3* mutation status on pSTAT3 expression was studied with one-way ANOVA and post-hoc correction. Green dots correspond to patients with *STAT3* mutations, the red dot correspond to a patient with *JAK1* mutation and blue dots correspond to samples without mutations in *JAK1* or *STAT3*. (**D**) Frequency table of patients by CD30 phenotype and *JAK1* or *STAT3* mutation status (rows) and different PTCL histologies (columns). *: $p < 0.05$; ns: not significant.

EBV status has been shown to be associated with STAT3 activation in B cell lymphomas [29]. No associations between STAT3, MAPK, and AKT phosphorylation and EBV⁺ tumors could be observed in our PTCL cohort (Figure S3B).

As fixation rate differs in samples of various origin and size, we investigated their association with STAT3, MAPK, and AKT phosphorylation. The sample cohort included 9 biopsy samples and 52 excised tumors. Biopsy samples were associated with higher pY-STAT3 but not with MAPK or AKT phosphorylation (Figure S4A). Moreover, correlation between pY-STAT3 and sample size remained non-significant (Figure S4B). Therefore, we hypothesize the difference to be due primarily to biological rather than technical reasons. Except for four samples, the sample cohort consisted of lymph node tissue samples. AKT phosphorylation was higher in lymph node samples, while no difference was observed between pY-STAT3 and MAPK phosphorylation levels (Figure S4B).

To further elucidate the spatial localization of STAT3 phosphorylation, 6 T-cell lymphoma samples (5 AITL, and 1 ALCL) and 2 NLPHL samples were stained for pY-STAT3, PD1, CD4, and CD8 using multiplex immunohistochemistry. Of the 5 AITL samples, two harbored *STAT3* mutations at 5% and 7% VAF, respectively. We found that the staining pattern or level of pY-STAT3 phosphorylation was not associated with *STAT3* mutation status (Figure 5).

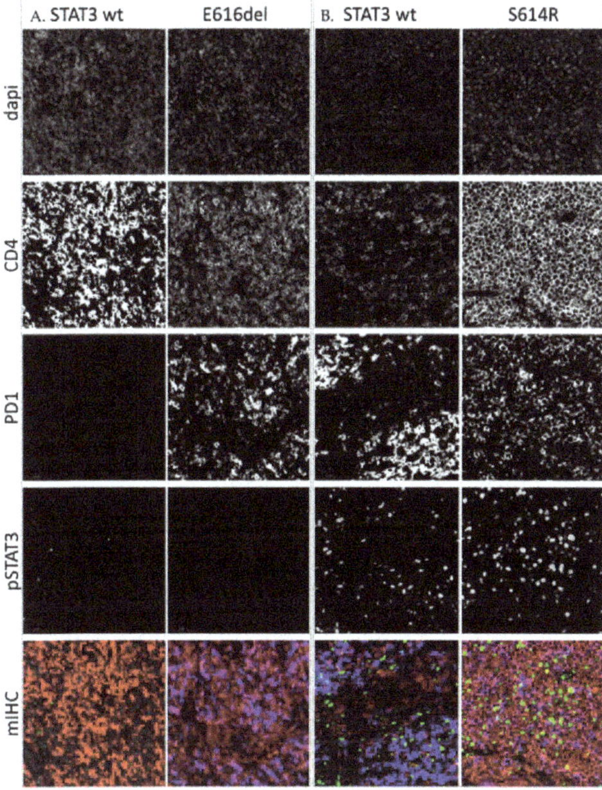

Figure 5. IHC stainings of four angioimmunoblastic T-cell lymphoma (AITL) lymph node samples. (**A**). Samples containing low amount of pSTAT3. (**B**). Samples containing high amounts of pSTAT3. Each slide has been digitalized with a 20× lens and similar exposure time per channel. Images have been further magnified 30× for visualization purposes. The first four rows of images represent single stainings of dapi counterstain, CD4, PD1, and pSTAT3 (Tyr705), respectively, and the last figure their composite image. Staining: red = CD4, blue = PD1, green = pSTAT3.

2.4. High pY-STAT3 Expression Is Associated with CD3⁻ CD5⁻ CD7⁻ CD30⁺ Immunophenotype Common to ALCL

To understand the association of JAK-STAT signaling pathway activity with clinical determinants, we combined pY-STAT3, pMAPK, and pAKT proportion defined by IHC with information on PTCL disease histology, stage, peripheral blood lactate dehydrogenase (LD) level at diagnosis, survival status, *STAT3* and *RHOA* mutation status (Figure 6). We noted that pY-STAT3, pMAPK, and pAKT expression did not correlate with each other. High pY-STAT3 expression was noted to be associated with ALCL histology as presented before (Figure 2). No association with survival status ($p = 0.17$, *t*-test), disease stage ($p = 0.40$, one-way ANOVA), LD level ($r = -0.13$, $p = 0.31$, Pearson correlation), *STAT3* ($p = 0.56$, *t*-test) or *RHOA* mutation status ($p = 0.11$, *t*-test) was noted.

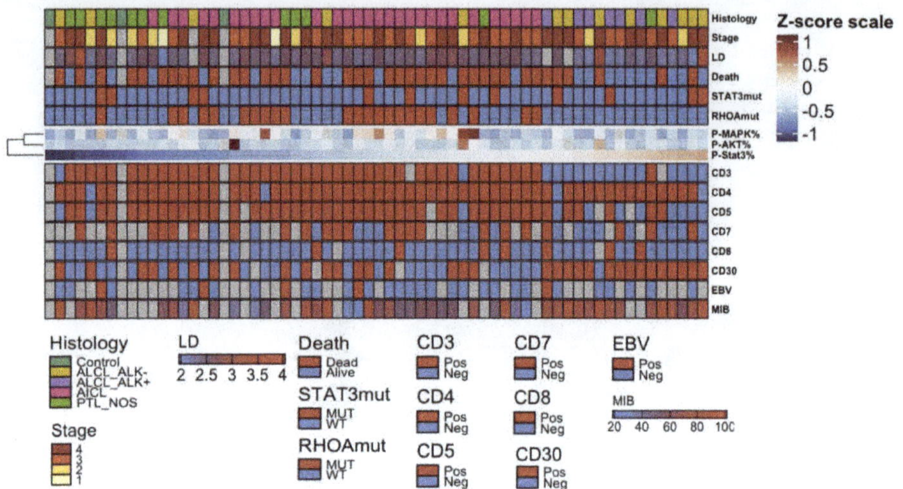

Figure 6. Heatmap visualizing the quantity of pY-STAT3, pMAPK, and pAKT and their association with clinicopathological parameters. The amount of phosphorylated STAT3, MAPK, and AKT have been median-centered and max-scaled, and organized columnwise by pY-STAT3 quantity and rowwise by hierarchical clustering using Spearman correlation distance and Ward linkage (ward.D2) method. Red color denotes higher and blue color lower proportion. Clinical parameters denoting disease histology, disease stage, log10-transformed lactate dehydrogenase (LD) level, survival status, and STAT3 and RHOA mutation status are added as annotations over the heatmap. CD3, CD4, CD5, CD7, CD8, and CD30 immunophenotype status classified as positive or negative, EBV seropositivity and the MIB proliferation index (%) are presented below the heatmap.

To correlate the expression of the JAK-STAT signaling pathway with immunohistochemical and serological factors of disease pathology, we investigated EBV seropositivity, and expression of immunophenotype markers as defined with IHC (Figure 6). Lower pY-STAT3 expression was associated with negative CD3 (37.7% and 75.4% mean pSTAT3 expression, $p < 0.001$, *t*-test), CD5 (42.8% and 68.4%, $p = 0.03$, *t*-test), CD7 (45.2% and 64.6%, $p = 0.01$) and positive CD30 phenotype (Figure S4A) designating ALCL immunophenotype as reported in Figure 2 [30]. No association with EBV seropositivity (Figure S4B), CD4 ($p = 0.86$, *t*-test) or CD8 ($p = 0.35$, *t*-test) immunophenotype nor MIB proliferation index ($r = 0.20$, $p = 0.19$, Pearson correlation) was noted.

2.5. Novel Potential Inhibitors of STAT3 Activation

In myeloproliferative neoplasms the JAK-STAT signaling pathway can be inhibited with the JAK1/JAK2 inhibitor ruxolitinib and in rheumatoid arthritis with less selective JAK inhibitors such as tofacitinib and baricitinib [31]. Several other pharmacological compounds affecting the JAK-STAT pathway are currently investigated in clinical trials [31]. We hypothesized that drugs investigated for other indications might induce differential sensitivity in cancer cell lines with active JAK-STAT signaling. By comparing cytotoxicity measured by the area under the curve (AUC) of 265 investigational and accepted compounds in pSTAT3 high ($n = 27$) and low ($n = 583$) cancer cell lines reported in the Sanger

GDSC and CCLE databases respectively, we identified the JAK-STAT inhibitor ruxolitinib to be more potent in pSTAT3 high cell lines (Figure 7). Interestingly, we discovered also the cell cycle checkpoint kinase Chk1/2 inhibitor AZD7762, the poly ADP ribose polymerase (PARP) inhibitor talazoparib and the nucleoside analog gemcitabine to exhibit most sensitivity in pSTAT3 cell lines.

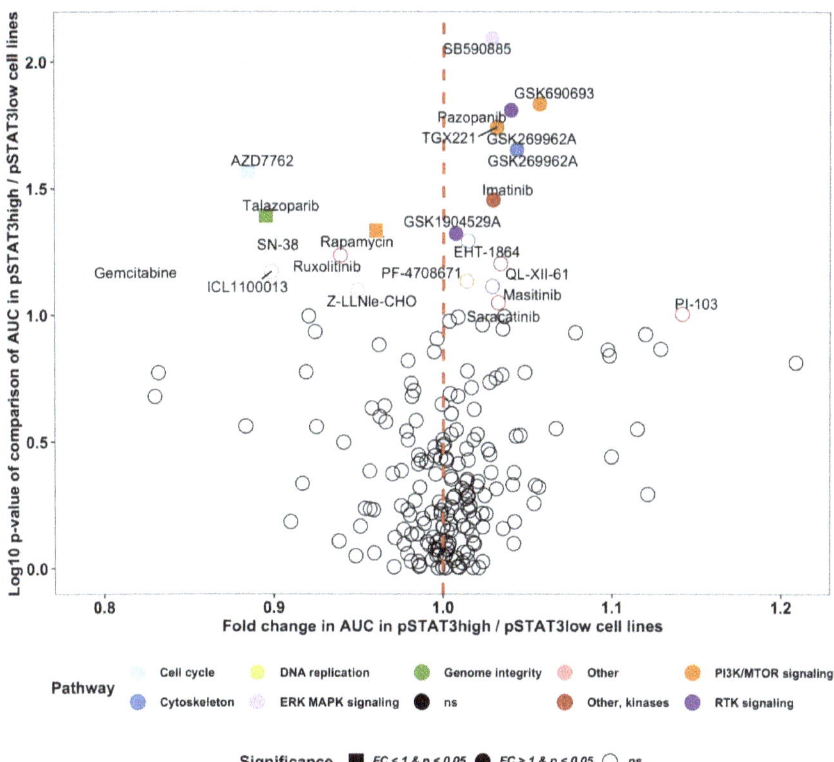

Figure 7. Differential drug sensitivity in pSTAT3 high ($n = 27$) and low ($n = 583$) cancer cell lines. The pSTAT3 phosphorylation level has been quantified with reverse phase protein array (RPPA, cut-off pSTAT3^{Tyr705} = 1). The drug sensitivity data represent the area under the curve (AUC) response of 265 different phamacological compounds, and has been retrieved from the Sanger GDSC dataset. The RPPA data are derived from the Broad Institute the Cancer Cell Line Encyclopedia (CCLE) database. Differential drug sensitivity has been computed with multiple T-tests, and the fold change defined as the mean AUC ratio in pSTAT3 high-to-low cell lines. Drugs with a fold change (FC) less than 1.0 are more sensitive in pSTAT3 high cell lines. The target pathway of the top pharmacological hits have been color-labeled.

3. Discussion

PTCL patients face unfavorable prognosis as chemotherapy results in poor 5-year overall survival rates [32]. Targeted treatment options only exist for ALK$^+$ ALCL, but are urgently needed for other PTCL entities. Using targeted sequencing, we demonstrated that *STAT3* alterations are prevalent in all PTCL subgroups. We discovered that 13% of AITL and ALK$^+$ ALCL cases and 17% of PTCL-NOS cases harbored *STAT3* mutations. The highest prevalence of *STAT3* mutations was seen in ALK$^-$ ALCL (38%). The mutations were found mostly in the SH2-domain of *STAT3* but also in the coiled coil (K290T) and transactivation domains (P715L). While we observed no association between overall *STAT3* mutation and pY-STAT3 phosphorylation, pY-STAT3 phosphorylation was most elevated in PTCL patients with combined CD30$^+$ tumor phenotype and mutations in either *STAT3* or *JAK1* representing primarily ALK$^-$ ALCLs and suggesting heterogenous mechanisms of STAT3 activation. Unexpectedly, PTCLs with CD30$^-$ phenotype and *STAT3* mutation were associated with lowest pY-STAT3 phosphorylation. As CD30 transcription is regulated by pSTAT3, the result might be explained by non-activating mutations or interference in the JAK-STAT pathway [27].

JAK/STAT mutations have been previously reported in PTCL-NOS, ALCL, and AITL [12,33,34], supporting the results of our study. The frequency of *JAK/STAT* mutations in different cohorts of PTCL has been reported to be around 20% which compares well to the mutation frequency seen in this study. In the whole cohort one of the most prevalent *STAT3* mutation was an in-frame deletion of E616, which was identified in three AITL patients. The E616 deletion has been shown to induce myeloid malignancy in a mouse bone marrow transplantation model [19]. The mutation has been observed previously in AITL, but also in adult T-cell lymphoma [10,35]. The *STAT3* mutation P715L observed in two ALK$^-$ ALCLs and one PTCL-NOS case has been observed in NKTL previously [36].

Some AITL and PTCL-NOS cases harbored concurrent *STAT3* and *RHOA* mutations (3/30 AITL and 1/12 PTCL-NOS). The incidence of *RHOA* mutations in our AITL cohort was 70%, which is somewhat higher than the incidence seen in previous AITL cohorts (68% and 53%) [8,23]. In PTCL-NOS the mutation incidence was 17% (2/12). Our results also showed that the MAPK phosphorylation degree correlated with *RHOA* mutations in our PTCL cohort.

Although this study included limited number of patients representing various PTCLs subtypes, our results suggest that pY-STAT3 is frequently constitutively phosphorylated in PTCL, especially in ALCLs and AITLs, but this was associated with *STAT3* mutations in only a fraction of PTCL samples. Previous studies have reported mechanisms underlying the constitutive activation of STAT3 independent of *STAT3* mutations. Gain-of–function mutations involving *JAK* genes have been implicated in activating STAT3 and contributing to the pathogenesis of hematologic malignancies and solid tumors [37,38]. Various cytokines (e.g., IL-6, IL-10, and IL-11) that are released into the microenvironment by tumor cells have also been shown to activate STAT3 [39]. Another mechanism of STAT3 activation is possible through the deregulation of suppressors of the JAK/STAT pathway such as SOCS3 [40] leading to constitutive STAT activation and oncogenesis. In addition, further studies are needed to combine any association with JAK-STAT pathway activation and expression of essential transcription factors such as GATA3 and Tbet, which define new subclasses of PTCL-NOS with prognostic consequences, where GATA3 signatures are associated with worse prognosis [41,42].

In our cohort high pSTAT3 expression correlated with the CD3$^-$ CD5$^-$ CD7$^-$ CD30$^+$ immunophenotype typical for ALCL. No association with neither immunophenotype markers nor essential clinical variables such as survival status, disease stage, MIB proliferation index or LD level was seen.

By comparing cytotoxicity measured by the area under the curve (AUC) of 265 investigational and accepted compounds in pSTAT3 high and low cancer cell lines, we identified novel potential inhibitors of STAT3 activation. The JAK-inhibitor ruxolitinib was among the top candidates. Ruxolitinib, a selective JAK1/2 inhibitor approved by the FDA for myeloproliferative neoplasms, is being studied in relapsed B-cell lymphoma and PTCL (NCT01431209).

Among compounds exhibiting most sensitivity in pSTAT3 high cell lines were the cell cycle checkpoint kinase Chk1/2 inhibitor AZD7762, the poly ADP ribose polymerase (PARP) inhibitor talazoparib, and the nucleoside analog gemcitabine. In phase II studies, gemcitabine monotherapy has shown activity against T-cell lymphomas. Zinzani et al. reported a 70% response rate in a phase II study of 44 pretreated patients with mycosis fungoides or cutaneous PTCL-unspecified [43]. In another phase II study, gemcitabine was given in combination with romidepsin to relapsed/refractory PTCL patients but the synergy observed in preclinical phase did not improve clinical outcomes [44]. Moreover, gemcitabine has been tested in combination with cisplatin and methylprednisolone as both upfront treatment and in relapsed/refractory PTCL, but did not significantly improve OS [45,46].

4. Materials and Methods

4.1. Patients and Cell Lines

Samples from 63 patients with PTCL were collected. Of these samples 30 were AITL, 12 PTCL-NOS and 21 ALCL (13 ALK$^-$ and 8 ALK$^+$). Four control samples from normal lymph nodes were also collected. The study was undertaken in compliance with the principles of the Helsinki Declaration and was approved by the ethics committee in the Helsinki University Hospital (Finland). As the samples studied were older diagnostic samples, the ethics committee required no written informed consent (TEO 5326/04/046/06, Valvira 9115/05.01.00.06/2011, HUS 302/E0/2006, HUS/1230/2017). The clinical and pathological characteristics of our study objects are summarized in Table S2.

4.2. Cell Lines and Cell Culture Conditions

The T-cell lines were cultured in RMPI-1640 medium including L-glutamine (Gibco, Grand Island, NY, USA) supplemented with 10% (L82, Karpas299, TLRB1, Molt4, Jurkat, and NKL cell line) or 20% (SupM2, MTA, MOTN1, and SMZ1 cell line) fetal bovine serum (FBS, Gibco) and Penicillin/Streptomycin (100 U/l, Gibco). IL2 (10ng/mL, MedChemExpress, Monmouth Junction, NJ, USA) was added for culturing of TLRB1 and NKL cells. Suspension cells were maintained at a density of 2.0–8.0 $\times$ 10^5. Culturing was done in an incubator at 37 °C and 5% CO$_2$ with 90% humidity.

4.3. Sample Preparation

According to standard clinical procedure, fresh tissue samples consisting mostly of lymph nodes (LN) were fixed in formalin and embedded in paraffin (FFPE) in the central pathology laboratory of Helsinki University Hospital (HUCH), Finland. Consequently, we cut 3.5 μm whole-tissue sections on Superfrost objective slides (Kindler O Gmbh, Freiburg, Germany).

4.4. DNA Extraction

DNA was extracted from FFPE tissue blocks. Ten 10 μm sections were cut with standard microtome (Leica SM2000 R Sliding Microtome, Wetzlar, Germany) using disposable blades. Excess paraffin was trimmed off and the sections were collected into sterile 1.5 mL microcentrifuge tubes. The sections were then deparaffinized with three pre-warmed (55 C) xylene washes followed by 95%, 75%, and 50% ethanol rinses. The tissue pellets were dried briefly at 37 °C to remove traces of ethanol. The pellets were then digested with 20 μL proteinase K (20 mg/mL proteinase K, Roche Diagnostics GmbH, (Mannheim, Germany) and 180 μL digestion buffer (10 mM Tris-HCl, pH 8.0 100 mM EDTA, pH 8.0 50 mM NaCl, and 0.5% SDS). Samples were incubated at 55 °C with mild agitation for 3 to 72 h. Fresh Proteinase K was added every 24 h followed by heat inactivation at 90 °C for 1 h after the tissue had been fully dissolved. For phenol-chloroform extraction, equal volume of phenol (Amresco, Solon, OH, USA) was added and vortexed. After spinning for 3 min at 14,000 rpm, the aqueous layer was transferred to a new tube. Then, an equal volume of phenol-chloroform-isoamyl alcohol (25:25:1) (Invitrogen, Carlsbad, CA, USA) was added, and the solution was vortexed and spun for 5 min at 14,000 rpm in a microcentrifuge. The aqueous layer was again transferred to a new tube and treated with RNase A at 100 μg/mL for 1 h at 37 °C. To remove any remaining RNase A phenol and phenol-chloroform-isoamyl alcohol steps were repeated. The aqueous layer was transferred to a new tube and the DNA was precipitated with 1/10 volume of 3 M sodium acetate and 1 volume of isopropanol. After thorough mixing, the solution was placed in a freezer for 30 min then spun at 14,000 rpm at 4 °C in a microcentrifuge for 10 min. The supernatant was discarded, and the pellet was washed with 1 mL 70% cold ethanol and spun at 14,000 rpm for 10 min at 4 °C. The supernatant was carefully discarded, and the pellet dried and finally suspended with 50 μL dH$_2$O.

4.5. Targeted Deep Amplicon Sequencing

Locus-specific primers were designed for *STAT3*, *STAT5B*, *JAK1*, *JAK3*, and *RHOA* mutation hotspots using Primer3 with user-defined parameters (http://primer3.wi.mit.edu/). After designing the locus-specific primer sequences (Table S3), sequence tails corresponding to the Illumina adapter sequences, were added to the 5′ end of the forward and reverse locus-specific primers, respectively. All oligonucleotides were ordered from Sigma-Aldrich (St. Louis, MO, USA). Deep targeted amplicon sequencing of known recurrent somatic mutations in the genes *STAT3*, *STAT5B*, *JAK1*, *JAK3*, and *RHOA* was performed with the Illumina MiSeq platform (San Diego, CA, USA)as previously described [47]. The data were analyzed with a bioinformatics pipeline, which is based on calling of variants with certain count/frequency of reads and filtering out false positives using the estimated error rate and quality data of amplicon reads.

4.6. Immunohistochemistry

4.6.1. Single Color Immunohistochemistry

Tissue sections were deparaffinized, rehydrated in graded ethanol series and then subjected to heat-induced epitope retrieval in Tris-HCl (10 mM Tris, pH 8.5, 25 min; for Phospho-MAPK and Phospho-AKT) or in EDTA buffer (10 mM EDTA, pH 8.0, 20 min; for Phospho-Stat3 and -5) in a Pre-Treatment module (DAKO/Agilent Technologies, Waltham, MA, USA). Samples were stained using LabVision autostainer (Thermo Fisher Scientific, Fremont, CA, USA). The endogenous peroxidase activity was blocked with Peroxidase-Blocking Solution (Dako/Agilent Tehnologies). The samples were treated with primary antibody for 1 h at RT. The antibodies used were rabbit monoclonal antibodies from Cell Signaling (Leiden, the Netherlands): clone Tyr705 for pY-STAT3 (1:300), clone Thr202/Tyr204 for Phospho-p44/42MAPK (1:300) and Thr450 for Phospho-AKT (1:100). BrightVision polymerisation technology was utilized to prepare polymeric HRP-linker antibody conjugates (ImmunoLogic, Duiven, the Netherlands) and 3,3-diaminobenzidine (DAB) was used as chromogen. Sections were counterstained with Mayer's hematoxylin (Dako/Agilent Tehnologies) and mounted with Eukitt (Honeywell Fluka, Frankfurt, Germany).

Protein expression was defined by visual examination. Five representative 400×-magnified high-power field-of-view (FOV) were selected, and protein expression (%) was estimated as the proportion of positive cells of 100 cells analyzed by FOV. Finally, the mean proportion of positive cells was calculated and used for statistical analyses.

4.6.2. Multiplex Immunohistochemistry

All phases were performed in room temperature (if not otherwise specified) and protein blocking as well as antibody incubations were performed in a humid chamber. The slides were washed with 0.1% Tween-20 (Thermo Fisher Scientific) diluted in 10 mM Tris-HCL buffered saline pH 7.4 (TBS) three times after peroxide block, each antibody staining, and fluorochrome reaction. Antibodies were tested with normal lymph node and pY-STAT3 was validated with tissue samples from a patient with large granular lymphocyte leukemia (Figure S5).

The slides were deparaffinized in xylene and rehydrated with graded series of ethanol and H2O. HIER was performed in 10 mM Tris-HCl - 1 mM EDTA buffer (pH 9) in +99 °C for 20 min (PT Module, Thermo Fisher Scientific). The peroxide activity was then blocked in 0.9% H2O2 solution for 15 min, and 10% normal goat serum was subsequently applied (TBS-NGS) for 15 min. We applied the primary antibody anti-PD1 (1:5000; clone PDCD1; LsBio, Seattle, WA, USA) diluted in TBS-NGS for 1 h 45 min, and then anti-mouse horseradish peroxidase (HRP) conjugated secondary antibodies (Immunologic) diluted 1:2 in washing buffer for 45 min. Then, we applied tyramide signal amplification (TSA) Alexa Fluor 488 (PerkinElmer, Boston, MA, USA) diluted 1:50 in TBS on the slides for 10 min. In order to multiplex antibodies, we denatured the primary antibody and quenched the enzymatic activity of HRP by repeating HIER as well as performing peroxide and protein block similarly as above. We applied the primary antibody against pY-STAT3 (Tyr705; 1:2500; clone D3A7; CellSignaling) overnight in +4 °C, anti-mouse HRP-conjugated secondary antibody diluted 1:5 in washing buffer for 45 min, and TSA Alexa Fluor 555 (PerkinElmer) for 10 min. We repeated HIER, peroxide block and protein block as above. Next, we applied anti-CD4 (1:150; clone EPR6855; Abcam, Cambridge, UK) diluted in TBS-NGS for 3 h and later AlexaFluor647-conjugated secondary antibody (1:300; Thermo Fisher Scientific) and Hoechst 33,342 (1 µg/mL; Sigma-Aldrich) counterstain diluted in washing buffer for 45 min. Finally, we applied ProLong Gold (Thermo Fisher Scientific) to mount the slides.

The tissue slides were scanned with 20x magnification with the Panoramic P250 Flash II whole slide scanner (3DHistech, Budapest, Hungary). We used DAPI, FITC, Cy3, and Cy5 filters and visually-optimized exposure times.

4.7. Western Blot

Immunoblots were performed on whole-cell protein lysates according to standard techniques. All primary antibodies (pJAK3$^{Tyr980/981}$, clone D44E3; JAK3, clone D7B12; pY-STAT3^{Tyr705}, clone D3A7; STAT3, clone 124H6; beta-ACTIN, clone 13E5) were ordered from Cell Signaling Technology and used at 1:1000 dilutions. Anti-mouse and anti-rabbit, both from Dianova, were used as secondary HRP-coupled antibodies in a 1:5000 dilution. Development of the immunoblots was performed by using Western Bright ECL (Advansta, Menlo Park, CA, USA). The original blots and densitometry readings of each band is available in the Supplementary Materials (Figure S6 and Table S4).

4.8. CCLE and DepMap Data

All CCLE pre-processed data were downloaded from the CCLE data portal (https://portals.broadinstitute.org/ccle/data). RPPA data dated 3 October2018, RNA-seq gene count data dated 29 September 2018, cell line annotation dated 26 December 2018, and mutation data dated 18 July 2018 were used. Gene counts were converted to CPM values using the edgeR and the TMM normalization method without log transformation. The cut-off for high vs. low STAT3 phosphorylation was set to pSTAT3^{Tyr705} = 1 (RPPA unit) based on the shape of the pSTAT3^{Tyr705} frequency histogram.

Area under the curve (AUC) drug sensitivity data annotated as "Sanger GDSC v17.3" and dated March 2018 were downloaded from the Cancer Dependency Map portal (https://depmap.org/portal/download/). Drugs not characterized in both pSTAT3 high and low cell line groups were removed from the analysis. Cell lines with no pSTAT3 RPPA data were removed from the analysis.

4.9. Statistical Analysis

Comparison of continuous variables between two groups was computed with unpaired *t*-test and between multiple groups with one-way ANOVA and post-hoc correction between disease groups and healthy controls.

Overall survival was defined as the time from diagnosis to death or last follow-up. Patients alive at the last follow-up date were censored. The log-rank test was used to compute Cox proportional-hazards.

Differential drug sensitivity was computed using t-test between pSTAT3 high ($n = 27$) and low ($n = 583$) cancer cell lines. The fold change in drug sensitivity was defined as the mean AUC ratio in pSTAT3 high-to-low cell lines. Drugs with a fold change (FC) less than 1.0 are more sensitive in pSTAT3 high cell lines.

R 3.5.1. [48] and Prism 6.0 was used for statistical analyses. R packages edgeR 3.24.0, forestplot 1.7.2, survminer 0.4.3, ggplot2 3.2.1, and ComplexHeatmap 2.3.2 were used.

5. Conclusions

In summary, this study discovered that pY-STAT3 phosphorylation was associated with both ALK$^+$ ALCLs and *JAK1/STAT3* mutated CD30$^+$ ALK$^-$ ALCLs suggesting a subgroup potentially benefitting of JAK/STAT targeted therapy.

Supplementary Materials: The following are available online at http://www.mdpi.com/2072-6694/12/3/702/s1, Figure S1: pMAPK, pAKT and pSTAT3 staining of PTCLs, Figure S2: *STAT3* expression correlation with *JAK1*, *JAK3*, *AKT* and *MAPK* expression in T-cell neoplasia cell lines from the CCLE dataset, Figure S3: Association of STAT3, MAPK and AKT phosphorylation with CD30 phenotype and EBV status, Figure S4: Association of STAT3, MAPK and AKT phosphorylation with sample type and organ origin, Figure S5: Multiplex immunohistochemistry staining of pSTAT3^{Tyr705} in CD8$^+$ LGL-leukemia as a positive control stain for pSTAT3 antibody, Figure S6: Full western blot of p-STAT3 (Tyr705), STAT3, p-JAK3 (Tyr980/981), and JAK3 in T cell malignancy cell lines, Table S1: Cell lines and *JAK/STAT* mutation status, Table S2: Clinical and pathological characteristics of PTCL patients, Table S3: Locus-specific primers for *STAT3* and *STAT5B* exons and *RHOA* G17V mutation hotspot, Table S4: Densitometry readings of Western blot stainings.

Author Contributions: conceptualization, S.M. (Satu Mustjoki), P.E.K., O.B. and E.I.A.; methodology, L.S., S.L. (Sonja Lagström) and P.E.; validation, E.I.A., T.B. and O.B.; formal analysis, E.I.A. and O.B.; data curation, S.M. (Susanna Mannisto); writing—original draft preparation, E.I.A. and O.B.; writing—review and editing, S.M. (Satu Mustjoki), P.E.K., S.L. (Sirpa Leppä) and M.H. All authors have read and agreed to the published version of the manuscript.

Funding: This research was funded by the Academy of Finland under the frame of ERA PerMed (JAKSTAT-TARGET consortium) and Terva program (Heal-Art consortium), European Research council (M-IMM project), the Finnish Cancer Societies, Finnish Cancer Institute, Biomedicum Helsinki Foundation, Sigrid Juselius Foundation, European Regional Development Fund, Signe and Ane Gyllenberg Foundation, Swedish Cultural Foundation, Blood Disease Foundation and the Finnish Cultural Foundation. M.H. and A.S. were supported by the DFG Research Unit FOR1961 (CONTROL-T; HE3553/4-2), by the Köln Fortune program, and by the Fritz Thyssen Foundation (10.15.2.034MN).

Acknowledgments: The personnel at the Hematology Research Unit Helsinki, HUSLAB Laboratory of Genetics (Soili Kytölä), and FIMM (Suvi Kyttänen and Tiina Hannunen) are acknowledged for their expert clinical and technical assistance. The authors thank Annabrita Schoonenberg from the Digital and Molecular Pathology Unit supported by Helsinki University and Biocenter Finland. Raphael Koch is acknowledged for providing cell lines. The authors thank Heidi Neubauer and Helena Sorger for valuable comments to the manuscript.

Conflicts of Interest: The authors declare no conflict of interest.

References

1. Project, T.N.-H.L.C. A Clinical Evaluation of the International Lymphoma Study Group Classification of Non-Hodgkin's Lymphoma. *Blood* **1997**, *89*, 3909–3918.

2. Ellin, F.; Landström, J.; Jerkeman, M.; Relander, T. Real-world data on prognostic factors and treatment in peripheral T-cell lymphomas: A study from the Swedish Lymphoma Registry. *Blood* **2014**, *124*, 1570–1577. [CrossRef]

3. International T-Cell Lymphoma Project; Vose, J.; Armitage, J.; Weisenburger, D. International Peripheral T-Cell and Natural Killer/T-Cell Lymphoma Study: Pathology Findings and Clinical Outcomes. *J. Clin. Oncol.* **2008**, *26*, 4124–4130. [CrossRef]

4. D'Amore, F.; Relander, T.; Lauritzsen, G.F.; Jantunen, E.; Hagberg, H.; Anderson, H.; Holte, H.; Osterborg, A.; Merup, M.; Brown, P.; et al. Up-Front Autologous Stem-Cell Transplantation in Peripheral T-Cell Lymphoma: NLG-T-01. *J. Clin. Oncol.* **2012**, *30*, 3093–3099. [CrossRef]

5. Mak, V.; Hamm, J.; Chhanabhai, M.; Shenkier, T.; Klasa, R.; Sehn, L.H.; Villa, D.; Gascoyne, R.D.; Connors, J.M.; Savage, K.J. Survival of Patients With Peripheral T-Cell Lymphoma After First Relapse or Progression: Spectrum of Disease and Rare Long-Term Survivors. *J. Clin. Oncol.* **2013**, *31*, 1970–1976. [CrossRef]

6. Iqbal, J.; Wilcox, R.; Naushad, H.; Rohr, J.; Heavican, T.B.; Wang, C.; Bouska, A.; Fu, K.; Chan, W.C.; Vose, J.M.; et al. Genomic signatures in T-cell lymphoma: How can these improve precision in diagnosis and inform prognosis? *Blood Rev.* **2016**, *30*, 89–100. [CrossRef]

7. Lemonnier, F.; Couronné, L.; Parrens, M.; Jaïs, J.-P.; Travert, M.; Lamant, L.; Tournillac, O.; Rousset, T.; Fabiani, B.; Cairns, R.A.; et al. Recurrent TET2 mutations in peripheral T-cell lymphomas correlate with TFH-like features and adverse clinical parameters. *Blood* **2012**, *120*, 1466–1469. [CrossRef]

8. Sakata-Yanagimoto, M.; Enami, T.; Yoshida, K.; Shiraishi, Y.; Ishii, R.; Miyake, Y.; Muto, H.; Tsuyama, N.; Sato-Otsubo, A.; Okuno, Y.; et al. Somatic RHOA mutation in angioimmunoblastic T cell lymphoma. *Nat. Genet.* **2014**, *46*, 171–175. [CrossRef]

9. Palomero, T.; Couronné, L.; Khiabanian, H.; Kim, M.-Y.; Ambesi-Impiombato, A.; Perez-Garcia, A.; Carpenter, Z.; Abate, F.; Allegretta, M.; Haydu, J.E.; et al. Recurrent mutations in epigenetic regulators, RHOA and FYN kinase in peripheral T cell lymphomas. *Nat. Genet.* **2014**, *46*, 166–170. [CrossRef]

10. Odejide, O.; Weigert, O.; Lane, A.A.; Toscano, D.; Lunning, M.A.; Kopp, N.; Kim, S.S.; Van Bodegom, D.; Bolla, S.; Schatz, J.H.; et al. A targeted mutational landscape of angioimmunoblastic T-cell lymphoma. *Blood* **2014**, *123*, 1293–1296. [CrossRef]

11. Iqbal, J.; Weisenburger, D.D.; Greiner, T.C.; Vose, J.M.; McKeithan, T.; Küçük, C.; Geng, H.; Deffenbacher, K.; Smith, L.; Dybkær, K.; et al. Molecular signatures to improve diagnosis in peripheral T-cell lymphoma and prognostication in angioimmunoblastic T-cell lymphoma. *Blood* **2010**, *115*, 1026–1036. [CrossRef]

12. Crescenzo, R.; Abate, F.; Lasorsa, E.; Tabbo', F.; Gaudiano, M.; Chiesa, N.; Di Giacomo, F.; Spaccarotella, E.; Barbarossa, L.; Ercole, E.; et al. Convergent mutations and kinase fusions lead to oncogenic STAT3 activation in anaplastic large cell lymphoma. *Cancer Cell* **2015**, *27*, 516–532. [CrossRef]

13. Pencik, J.; Pham, H.T.T.; Schmoellerl, J.; Javaheri, T.; Schlederer, M.; Culig, Z.; Merkel, O.; Moriggl, R.; Grebien, F.; Kenner, L. JAK-STAT signaling in cancer: From cytokines to non-coding genome. *Cytokine* **2016**, *87*, 26–36. [CrossRef]

14. Vainchenker, W.; Constantinescu, S.N. JAK/STAT signaling in hematological malignancies. *Oncogene* **2012**, *32*, 2601–2613. [CrossRef]

15. Migone, T.; Lin, J.; Cereseto, A.; Mulloy, J.; O'Shea, J.; Franchini, G.; Leonard, W. Constitutively activated Jak-STAT pathway in T cells transformed with HTLV-I. *Science* **1995**, *269*, 79–81. [CrossRef]

16. Heinrich, T.; Rengstl, B.; Muik, A.; Petkova, M.; Schmid, F.; Wistinghausen, R.; Warner, K.; Crispatzu, G.; Hansmann, M.-L.; Herling, M.; et al. Mature T-cell Lymphomagenesis Induced by Retroviral Insertional Activation of Janus Kinase 1. *Mol. Ther.* **2013**, *21*, 1160–1168. [CrossRef]

17. Nicolae, A.; Xi, L.; Pittaluga, S.; Abdullaev, Z.; Pack, S.D.; Chen, J.; Waldmann, T.A.; Jaffe, E.S.; Raffeld, M. Frequent STAT5B mutations in γδ hepatosplenic T-cell lymphomas. *Leukemia* **2014**, *28*, 2244–2248. [CrossRef] [PubMed]

18. Bamford, S.; Dawson, E.; Forbes, S.; Clements, J.; Pettett, R.; Dogan, A.; Flanagan, A.; Teague, J.; Futreal, P.A.; Stratton, M.R.; et al. The COSMIC (Catalogue of Somatic Mutations in Cancer) database and website. *Br. J. Cancer* **2004**, *91*, 355–358. [CrossRef]

19. Couronné, L.; Scourzic, L.; Pilati, C.; Della Valle, V.; Duffourd, Y.; Solary, E.; Vainchenker, W.; Merlio, J.-P.; Beylot-Barry, M.; Damm, F.; et al. STAT3 mutations identified in human hematologic neoplasms induce myeloid malignancies in a mouse bone marrow transplantation model. *Haematologica* **2013**, *98*, 1748–1752. [CrossRef]

20. Jerez, A.; Clemente, M.J.; Makishima, H.; Rajala, H.; Gómez-Seguí, I.; Olson, T.; McGraw, K.; Przychodzen, B.; Kulasekararaj, A.; Afable, M.; et al. STAT3 mutations indicate the presence of subclinical T-cell clones in a subset of aplastic anemia and myelodysplastic syndrome patients. *Blood* **2013**, *122*, 2453–2459. [CrossRef]

21. Morin, R.D.; Mendez-Lago, M.; Mungall, A.J.; Goya, R.; Mungall, K.L.; Corbett, R.D.; Johnson, N.A.; Severson, T.M.; Chiu, R.; Field, M.A.; et al. Frequent mutation of histone-modifying genes in non-Hodgkin lymphoma. *Nature* **2011**, *476*, 298–303. [CrossRef] [PubMed]

22. Schrader, A.; Crispatzu, G.; Oberbeck, S.; Mayer, P.; Pützer, S.; Von Jan, J.; Vasyutina, E.; Warner, K.; Weit, N.; Pflug, N.; et al. Actionable perturbations of damage responses by TCL1/ATM and epigenetic lesions form the basis of T-PLL. *Nat. Commun.* **2018**, *9*, 697. [CrossRef] [PubMed]

23. Yoo, H.Y.; Sung, M.; Lee, S.H.; Kim, S.; Lee, H.; Park, S.; Kim, S.C.; Lee, B.; Rho, K.; Lee, W.T.; et al. A recurrent inactivating mutation in RHOA GTPase in angioimmunoblastic T cell lymphoma. *Nat. Genet.* **2014**, *46*, 371–375. [CrossRef] [PubMed]

24. Fujisawa, M.; Sakata-Yanagimoto, M.; Nishizawa, S.; Komori, D.; Gershon, P.; Kiryu, M.; Tanzima, S.; Fukumoto, K.; Enami, T.; Muratani, M.; et al. Activation of RHOA–VAV1 signaling in angioimmunoblastic T-cell lymphoma. *Leukemia* **2017**, *32*, 694–702. [CrossRef]

25. Han, J.J.; O'Byrne, M.; Stenson, M.J.; Maurer, M.; Wellik, L.E.; Feldman, A.L.; McPhail, E.D.; Witzig, T.E.; Gupta, M. Prognostic and therapeutic significance of phosphorylated STAT3 and protein tyrosine phosphatase-6 in peripheral-T cell lymphoma. *Blood Cancer J.* **2018**, *8*, 110. [CrossRef]

26. Chen, J.; Zhang, Y.; Petrus, M.N.; Xiao, W.; Nicolae, A.; Raffeld, M.; Pittaluga, S.; Bamford, R.N.; Nakagawa, M.; Ouyang, S.T.; et al. Cytokine receptor signaling is required for the survival of ALK−anaplastic large cell lymphoma, even in the presence of JAK1/STAT3 mutations. *Proc. Natl. Acad. Sci. USA* **2017**, *114*, 3975–3980. [CrossRef]

27. Chiarle, R.; Voena, C.; Ambrogio, C.; Piva, R.; Inghirami, G. The anaplastic lymphoma kinase in the pathogenesis of cancer. *Nat. Rev. Cancer* **2008**, *8*, 11–23. [CrossRef]

28. Ohgami, R.S.; Ma, L.; Merker, J.D.; Martinez, B.; Zehnder, J.L.; A Arber, D. STAT3 mutations are frequent in CD30+ T-cell lymphomas and T-cell large granular lymphocytic leukemia. *Leukemia* **2013**, *27*, 2244–2247. [CrossRef]

29. Shair, K.H.Y.; Bendt, K.M.; Edwards, R.H.; Bedford, E.C.; Nielsen, J.N.; Raab-Traub, N. EBV Latent Membrane Protein 1 Activates Akt, NFκB, and Stat3 in B Cell Lymphomas. *PLOS Pathog.* **2007**, *3*, e166. [CrossRef]

30. Montes-Mojarro, I.; Steinhilber, J.; Bonzheim, I.; Quintanilla-Martinez, L.; Fend, F. The Pathological Spectrum of Systemic Anaplastic Large Cell Lymphoma (ALCL). *Cancers* **2018**, *10*, 107. [CrossRef]

31. Schwartz, D.M.; Kanno, Y.; Villarino, A.; Ward, M.; Gadina, M.; O'Shea, J.J. JAK inhibition as a therapeutic strategy for immune and inflammatory diseases. *Nat. Rev. Drug Discov.* **2018**, *17*, 78. [CrossRef] [PubMed]

32. Armitage, J.O. The aggressive peripheral T-cell lymphomas: 2017. *Am. J. Hematol.* **2017**, *92*, 706–715. [CrossRef] [PubMed]

33. Schatz, J.H.; Horwitz, S.; Teruya-Feldstein, J.; A Lunning, M.; Viale, A.; Huberman, K.; Socci, N.D.; Lailler, N.; Heguy, A.; Dolgalev, I.; et al. Targeted mutational profiling of peripheral T-cell lymphoma not otherwise specified highlights new mechanisms in a heterogeneous pathogenesis. *Leukemia* **2014**, *29*, 237–241. [CrossRef] [PubMed]

34. Manso, R.; Sánchez-Beato, M.; González-Rincón, J.; Gómez, S.; Rojo, F.; Mollejo, M.; García-Cosio, M.; Menárguez, J.; Piris, M.A.; Rodríguez-Pinilla, S.M. Mutations in the JAK/STAT pathway genes and activation of the pathway, a relevant finding in nodal Peripheral T-cell lymphoma. *Br. J. Haematol.* **2017**, *183*, 497–501. [CrossRef] [PubMed]

35. Kataoka, K.; Nagata, Y.; Kitanaka, A.; Shiraishi, Y.; Shimamura, T.; Yasunaga, J.-I.; Totoki, Y.; Chiba, K.; Sato-Otsubo, A.; Nagae, G.; et al. Integrated molecular analysis of adult T cell leukemia/lymphoma. *Nat. Genet.* **2015**, *47*, 1304–1315. [CrossRef]

36. Song, T.L.; Nairismägi, M.-L.; Laurensia, Y.; Lim, J.Q.; Tan, J.; Li, Z.-M.; Pang, W.-L.; Kizhakeyil, A.; Wijaya, G.-C.; Huang, D.; et al. Oncogenic activation of the STAT3 pathway drives PD-L1 expression in natural killer/T-cell lymphoma. *Blood* **2018**, *132*, 1146–1158. [CrossRef]

37. Kiel, M.J.; Velusamy, T.; Rolland, D.; Sahasrabuddhe, A.; Chung, F.; Bailey, N.G.; Schrader, A.; Li, B.; Li, J.Z.; Ozel, A.B.; et al. Integrated genomic sequencing reveals mutational landscape of T-cell prolymphocytic leukemia. *Blood* **2014**, *124*, 1460–1472. [CrossRef]

38. Gäbler, K.; Behrmann, I.; Haan, C. JAK2 mutants (e.g., JAK2V617F) and their importance as drug targets in myeloproliferative neoplasms. *JAK-STAT* **2013**, *2*, e25025. [CrossRef]

39. Fukuda, A.; Wang, S.; Morris, J.P.; Folias, A.E.; Liou, A.; Kim, G.E.; Akira, S.; Boucher, K.; Firpo, M.A.; Mulvihill, S.J.; et al. Stat3 and MMP7 Contribute to Pancreatic Ductal Adenocarcinoma Initiation and Progression. *Cancer Cell* **2011**, *19*, 441–455. [CrossRef]

40. Alexander, W.; Hilton, D.S. The Role of Suppressors of Cytokine Signaling (SOCS) Proteins in Regulation of the Immune Response. *Annu. Rev. Immunol.* **2004**, *22*, 503–529. [CrossRef]

41. Amador, C.; Greiner, T.; Heavican, T.B.; Smith, L.M.; Galvis, K.T.; Lone, W.G.; Bouska, A.; D'Amore, F.; Pedersen, M.B.; Pileri, S.A.; et al. Reproducing the Molecular Subclassification of Peripheral T-cell Lymphoma-NOS by Immunohistochemistry. *Blood* **2019**, *134*, 2159–2170. [CrossRef]

42. Heavican, T.B.; Bouska, A.; Yu, J.; Lone, W.; Amador, C.; Gong, Q.; Zhang, W.; Li, Y.; Dave, B.J.; Nairismägi, M.-L.; et al. Genetic drivers of oncogenic pathways in molecular subgroups of peripheral T-cell lymphoma. *Blood* **2019**, *133*, 1664–1676. [CrossRef] [PubMed]

43. Zinzani, P.L.; Baliva, G.; Magagnoli, M.; Bendandi, M.; Modugno, G.; Gherlinzoni, F.; Orcioni, G.F.; Ascani, S.; Simoni, R.; Pileri, S.A.; et al. Gemcitabine Treatment in Pretreated Cutaneous T-Cell Lymphoma: Experience in 44 Patients. *J. Clin. Oncol.* **2000**, *18*, 2603–2606. [CrossRef] [PubMed]

44. Pellegrini, C.; Dodero, A.; Chiappella, A.; Monaco, F.; Degl'Innocenti, D.; Salvi, F.; Vitolo, U.; Argnani, L.; Corradini, P.; Zinzani, P.L.; et al. A phase II study on the role of gemcitabine plus romidepsin (GEMRO regimen) in the treatment of relapsed/refractory peripheral T-cell lymphoma patients. *J. Hematol. Oncol.* **2016**, *9*, 1–7. [CrossRef] [PubMed]

45. Arkenau, H.-T.; Chong, G.; Cunningham, D.; Watkins, D.; Sirohi, B.; Chau, I.; Wotherspoon, A.; Norman, A.; Horwich, A.; Matutes, E. Gemcitabine, cisplatin and methylprednisolone for the treatment of patients with peripheral T-cell lymphoma: The Royal Marsden Hospital experience. *Haematology* **2007**, *92*, 271–272. [CrossRef] [PubMed]

46. Mahadevan, D.; Unger, J.M.; Spier, C.M.; Persky, D.O.; Young, F.; Leblanc, M.; Fisher, R.I.; Miller, T.P. Phase 2 trial of combined cisplatin, etoposide, gemcitabine, and methylprednisolone (PEGS) in peripheral T-cell non-Hodgkin lymphoma: Southwest Oncology Group Study S0350. *Cancer* **2012**, *119*, 371–379. [CrossRef] [PubMed]

47. Rajala, H.L.M.; Eldfors, S.; Kuusanmäki, H.; Van Adrichem, A.J.; Olson, T.; Lagström, S.; I Andersson, E.; Jerez, A.; Clemente, M.J.; Yan, Y.; et al. Discovery of somatic STAT5b mutations in large granular lymphocytic leukemia. *Blood* **2013**, *121*, 4541–4550. [CrossRef]

48. Team, R. *RStudio: Integrated Development for R*; RStudio: Boston, MA, USA, 2015.

Article

A Novel Inhibitor of STAT5 Signaling Overcomes Chemotherapy Resistance in Myeloid Leukemia Cells

Marie Brachet-Botineau [1], Margaux Deynoux [1], Nicolas Vallet [1,2], Marion Polomski [3], Ludovic Juen [3], Olivier Hérault [1,4], Frédéric Mazurier [1], Marie-Claude Viaud-Massuard [3], Gildas Prié [3] and Fabrice Gouilleux [1,*]

[1]   LNOx, GICC, CNRS ERL 7001, University of Tours, 37000 Tours, France; marie.brachet@univ-tours.fr (M.B.-B.); margaux.deynoux@etu.univ-tours.fr (M.D.); nls.vallet@gmail.com (N.V.); olivier.herault@univ-tours.fr (O.H.); frederic.mazurier@inserm.fr (F.M.)
[2]   Service d'Hématologie et Thérapie Cellulaire, CHRU de Tours, 37000 Tours, France
[3]   IMT, GICC, EA 7501, University of Tours, 37000 Tours, France; marion.polomski@etu.univ-tours.fr (M.P.); ludovic.juen@mcsaf.fr (L.J.); marie-claude.viaud-massuard@univ-tours.fr (M.-C.V.-M.); gildas.prie@univ-tours.fr (G.P.)
[4]   Service d'Hematologie Biologique, CHRU de Tours, 37000 Tours, France
*   Correspondence: fabrice.gouilleux@univ-tours.fr; Tel.: +33-(2)-47-36-62-91

Received: 8 August 2019; Accepted: 14 December 2019; Published: 17 December 2019

Abstract: Signal transducers and activators of transcription 5A and 5B (STAT5A and STAT5B) are crucial downstream effectors of tyrosine kinase oncogenes (TKO) such as BCR-ABL in chronic myeloid leukemia (CML) and FLT3-ITD in acute myeloid leukemia (AML). Both proteins have been shown to promote the resistance of CML cells to tyrosine kinase inhibitors (TKI) such as imatinib mesylate (IM). We recently synthesized and discovered a new inhibitor (17f) with promising antileukemic activity. 17f selectively inhibits STAT5 signaling in CML and AML cells by interfering with the phosphorylation and transcriptional activity of these proteins. In this study, the effects of 17f were evaluated on CML and AML cell lines that respectively acquired resistance to IM and cytarabine (Ara-C), a conventional therapeutic agent used in AML treatment. We showed that 17f strongly inhibits the growth and survival of resistant CML and AML cells when associated with IM or Ara-C. We also obtained evidence that 17f inhibits STAT5B but not STAT5A protein expression in resistant CML and AML cells. Furthermore, we demonstrated that 17f also targets oncogenic STAT5B N642H mutant in transformed hematopoietic cells.

Keywords: pharmacological inhibitor; STAT5 signaling; chemotherapy resistance; myeloid leukemia

---

1. Introduction

STAT5A and STAT5B are two closely related signal transducers and activators of transcription family members. Both proteins are crucial downstream effectors of tyrosine kinase oncogenes (TKO) such as Fms-like receptor tyrosine kinase 3 with internal tandem duplications (Flt3-ITD), BCR-ABL and JAK2^{V617F} which cause AML, CML and other myeloproliferative diseases (MPD), respectively [1]. STAT5 proteins are recognized as major drivers in the development and/or maintenance of CML as well as in the proliferation and survival of AML cells [2–4]. The development of tyrosine kinase inhibitors (TKI) targeting BCR-ABL such as imatinib mesylate (IM) has revolutionized the treatment of CML. Despite this success story, IM is not totally curative and approximately 50% of patients remain therapy-free after IM discontinuation. The inability of IM to completely eradicate leukemic stem cells (LSC) is probably responsible for the relapse of CML patients [5]. Moreover, the occurrence of *BCR-ABL* mutations in progressive or relapsed disease promotes IM resistance of CML cells [6]. Therefore, there is a need for complementary therapeutic strategies to cure CML. STAT5 fulfils all the criteria of a major drug target in CML [7]. High STAT5 expression levels have been shown not only to enhance IM resistance in CML cells but also to trigger *BCR-ABL* mutations by inducing the production of reactive oxygen species (ROS) responsible for DNA damage [8,9]. Moreover, STAT5 was shown to play a key role in the maintenance of chemoresistant CML stem cells [10]. Thus, targeting STAT5 would also benefit relapsed CML patients who became resistant to TKI. Several approaches have been used to target STAT5 in leukemia. Among them, cell-based screening with small molecule libraries of already approved drugs allowed the identification of the psychotropic drug pimozide as a potential STAT5 inhibitor in CML cells [11]. Pimozide decreased the tyrosine phosphorylation of STAT5 and induced growth arrest and apoptosis in CML cells. In addition, pimozide was shown to target the deubiquitinating (DUB) enzyme, USP1, in leukemic cells indicating that the effects of pimozide on STAT5 activity might be indirect [12]. Indirubin derivatives were also reported to inhibit STAT5 phosphorylation in CML cells but the mechanism of inhibition is most likely suppression of upstream tyrosine kinases [13]. More recently, a number of small inhibitors that bind to the Src homology domain 2 (SH_2) required for STAT5 activation and dimer formation, have been described [14]. These compounds exhibit potent and selective binding activity for STAT5 by effectively disrupting phosphopeptide interactions. Some of these inhibitors bind STAT5 proteins in a nanomolar range and inhibit the tyrosine phosphorylation of STAT5 and CML/AML cell growth in a micromolar range [15–17]. A final approach is to target STAT5 activity through the activation of peroxisome proliferator-activated receptor gamma (PPARγ) [18]. Indeed, the existence of cross-talk between PPARγ and STAT5 has been discussed. For instance, antidiabetic drugs such as glitazones, which are PPARγ agonists, were shown to have antileukemic activity [19,20]. Activation of PPARγ by pioglitazone not only decreases the phosphorylation of STAT5 in CML cells but also reduces expression of *STAT5* genes in quiescent and resistant CML stem cells [10]. Importantly, the combined use of pioglitazone and IM triggers apoptosis of these leukemic cells suggesting that besides phosphorylation, inhibition of STAT5 expression is of prime importance for resistant CML stem cell eradication. Based on these different data, we sought to identify new STAT5 inhibitors in a library of PPARα/γ ligands that were synthetized in our laboratory [21,22]. The synthesis of derivatives of a "hit" compound identified in the library screening allowed the discovery of a new inhibitor of STAT5 signaling in CML and AML cells [23]. This molecule (17f) selectively inhibits the phosphorylation and transcriptional activity of STAT5 and induces apoptosis of CML and AML cells. Herein, we showed that 17f associated with IM or Ara-C resensitizes CML and AML cells, respectively, that acquired resistance to these drugs. We demonstrated that 17f treatment reduces STAT5B protein levels in resistant CML and AML cells, suggesting that 17f overcomes chemotherapy resistance though the downregulation of this protein. We also found that 17f suppresses expression of oncogenic $STAT5^{N642H}$ mutant in transformed Ba/F3 cells.

2. Results

2.1. Effects of 17f Compound on Growth and Viability of IM-Sensitive and IM-Resistant BCR-ABL⁺ Cells

Initial experiments were carried out to determine the effects of 17f alone (see structure in Figure S1) on K562 cells that are sensitive (K562S) or resistant (K562R) to IM treatment. These in vitro models are depicted in Figure 1A. Sensitive and resistant cells were treated with various concentrations of 17f (ranging from 1 to 10 μM). Growth and viability were determined by trypan blue exclusion (Figure 1B) and MTT (3-(4,5-dimethylthiazol-2-yl)-2,5-diphenyltetrazolium bromide) (Figure 1C) assays. Addition of 17f clearly blocked the growth of K562S cells while K562R cells remain insensitive to 17f treatment at the same concentration. The EC_{50} value was found to be two times higher in K562R cells than in K562S cells (14.5 ± 4.8 μM vs. 6.9 ± 1.7 μM). We also observed that treatment with 5 μM 17f did not affect the growth and viability of K562R cells and used this suboptimal concentration in most experiments to evaluate the combined effects of 17f and IM.

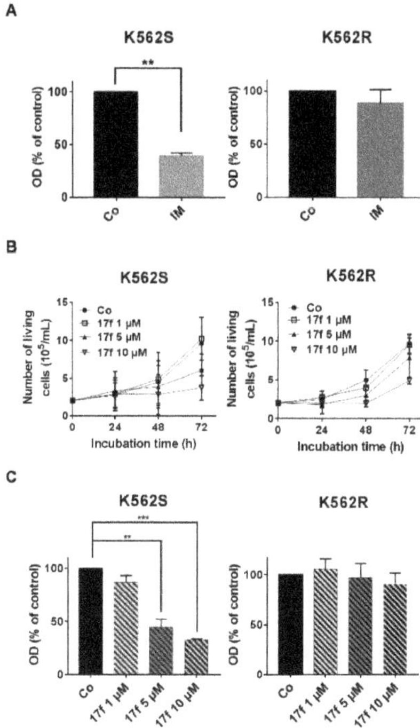

Figure 1. Effects of 17f molecule on K562S and K562R cell growth (**A**) Imatinib mesylate (IM)-sensitive K562 (K562S) and IM-resistant K562 cells (K562R) were treated with 1 µM IM or DMSO as control (Co) for 48 h. Cell viability was determined by MTT assays (data are presented as mean ± SD of three independent experiments ($n = 3$) in triplicates, *** $p < 0.001$; one sample t-test). (**B**) K562S and K562R cells were treated with 17f or DMSO as control (Co) for the indicated times. Growth kinetics were determined by trypan blue dye exclusion assays ($n = 3$ in triplicates, data are mean ± SD). (**C**) Cell viability was measured by MTT assays after treatment of K562S or K562R cells with increasing concentrations of 17f or DMSO as control (Co) during 48 h ($n = 3$ in triplicates, data are mean ± SD, ** $p < 0.01$, *** $p < 0.001$; one sample t-test).

2.2. 17f Induces Apoptosis and Cell Cycle Arrest in K562R Cells and Relieves the Resistance to IM

We then addressed whether 17f in combination with IM might directly abrogate the resistance of K562R cells to IM. K562S and K562R cells were treated with 17f in the presence of IM and cell growth and viability were determined by MTT assays (Figure 2A). As expected, IM strongly inhibited the growth of K562S cells. This inhibition was further enhanced by 17f in a dose-dependent fashion. Interestingly, we found that the addition of 1 µM 17f in the presence of IM was already enough to significantly reduce the growth and viability of K562R cells. Treatment with 5 µM and 10 µM of 17f further increased this inhibitory effect. To analyze the growth-suppressive properties of 17f in K562R cells, we determined the impact of this small molecule on apoptosis and the cell cycle. 17f induced apoptosis and changes in cell cycle phase distribution in a concentration-dependent manner (Figure 2B,C). 17f significantly increased the number of cells in the G_0 phase indicating that treatment with this compound induced quiescence of K562R cells.

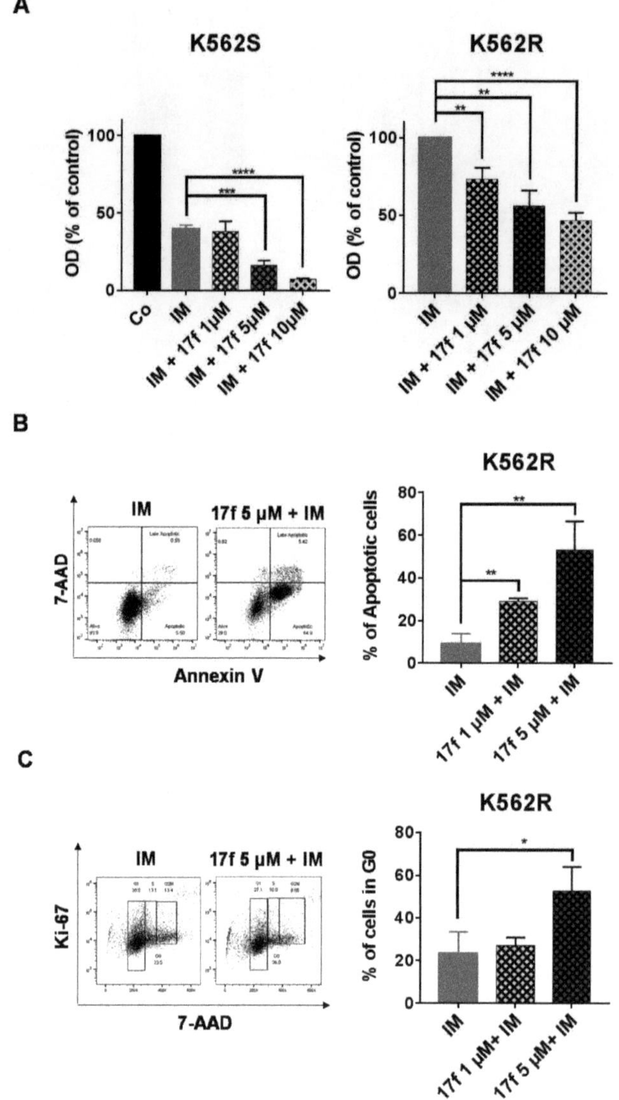

Figure 2. 17f overcomes the resistance of K562R cells to IM treatment. (**A**) K562S and K562R cells were treated with IM or not (Co) with or without 17f for 48 h. Cell viability was determined by MTT assays. (**B**) K562R cells cultured for 48 h with IM and 17f or IM vs. DMSO as control. Cells were stained with anti-annexin V coupled with FITC (fluorescein isothiocyanate) and with 7-amino-actinomycin D (7-AAD) to determine the percentages of apoptotic cells. One representative experiment is shown (left panel). (**C**) K562R cells treated for 48 h with IM and 17f or IM and DMSO as control were stained with 7-AAD and an Alexa Fluor 488-conjugated anti-Ki-67 antibody. Cell cycle phase distributions were then estimated by flow cytometry. The histogram presents the percentage of cells in the G_0 phase. One representative experiment is shown (left panel) ($n = 3$ in triplicates, data are mean ± SD, * $p < 0.05$, ** $p < 0.01$, *** $p < 0.001$, **** $p < 0.0001$).

2.3. 17f Inhibits STAT5-Dependent Transcriptional Activity in K562R Cells

We previously showed that 17f inhibits the transcriptional activity of STAT5 in CML cells. We then asked whether this small molecule also affects the activity of these proteins in IM-resistant K562R cells. We first determined the impact of this compound on the transcriptional activation of a reporter gene driven by a STAT5-specific promoter. K562S and K562R cells were transfected with a construct containing six tandem copies of the STAT5 response element in front of the minimal TK promoter fused to the luciferase reporter gene (6×(STAT5)-TK-luc). As control, cells were also transfected with a TK-luciferase vector without STAT5 response elements (TK-luc). Luciferase activity was determined 48 h post-transfection in K562S and K562R cells treated with DMSO as control, 17f (5 μM) and/or IM (1 μM). As expected, constitutive STAT5 activity induced by BCR-ABL increased luciferase activity in K562S cells transfected with the STAT5-dependent promoter construct compared to cells transfected with the control TK-luc vector (Figure 3A). This enhanced luciferase activity was strongly reduced after 17f or IM treatment. In sharp contrast, the luciferase activity remained elevated after treatment with IM in K562R cells transfected with the STAT5-dependent reporter construct, although this enzymatic activity was strongly decreased after the addition of 17f and IM. qRT-PCR experiments were then conducted to determine the effects of 17f on STAT5-dependent expression of target genes such as *PIM1* and *CISH* (Figure 3B). As expected, 17f or IM reduced expression of both genes in sensitive K562S cells while this effect was observed in resistant K562R cells after treatment with both compounds. Collectively, these data strongly suggest that 17f inhibits the transcriptional activity of STAT5 to bypass IM resistance in K562R cells.

2.4. 17f Inhibits STAT5B Protein Expression in IM-Resistant K562 Cells

We then determined the impact of 17f on BCR-ABL-induced tyrosine phosphorylation of STAT5 (P-Y$^{694/699}$-STAT5) by western blot and flow cytometry analysis (Figure 4A, Figures S2–S4). K562S and K562R cells were treated for 24 h instead of 48 h to analyze the early effects on STAT5 phosphorylation. IM strongly reduced P-Y-STAT5 levels in K562S, and the addition of 17f further enhanced this effect. P-Y-STAT5 levels were maintained in IM-treated K562R cells but were decreased after the addition of 17f. Interestingly, the level of STAT5 phosphorylation was strikingly enhanced in K562R cells after removal of IM and was weakly affected by the addition of 17f (Figure S5). To determine whether changes in P-Y-STAT5 levels reflect differences in protein abundance, immunoblots were performed with an anti-STAT5 antibody that recognizes STAT5A and STAT5B proteins (Figure 4B and Figure S2). As expected, IM inhibited the phosphorylation of STAT5 in sensitive and resistant cells. Interestingly, we observed that the association of 17f with IM reduces STAT5 expression in K562R cells but not in K562S cells (see Figure S3A,B for quantification). qRT-PCR experiments were then conducted to evaluate the impact of combination treatments on *STAT5A* and *STAT5B* gene expression in K562R cells. Results showed that *STAT5A/5B* mRNA levels were not affected by 17f when associated with IM (Figure 4B). In contrast, western blot analysis clearly evidenced that STAT5B protein expression was decreased after combination treatments suggesting that 17f sensitizes K562R cells to IM treatment by targeting STAT5B protein (Figure 4C).

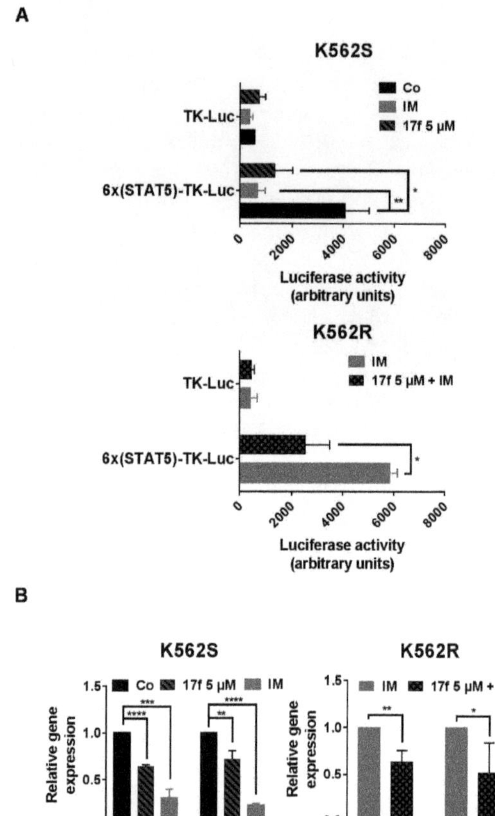

Figure 3. 17f associated with IM inhibit STAT5 activity in resistant K562R cells. (**A**) K562S or K562R cells transfected with a 6×(STAT5)-TK-luciferase reporter construct or a control TK-luciferase vector were treated or not (Co) with 17f (5 µM), IM (1 µM) or with the combination of 17f and IM for 48 h. Luciferase activities were then determined as described in Methods. Luciferase activity (arbitrary units) in the histogram represents the relative luminescence unit (rlu) values/mg of proteins ($n = 3$ in triplicates, data are mean ± SD, * $p < 0.05$, ** $p < 0.001$). (**B**) qRT-PCR analysis of *PIM1* and *CISH* expression in K562S and K562R treated or not (Co) with IM (1 µM),17f (5 µM) or with combined 17f and IM for 24 h. Results are presented as the fold change in *PIM1* and *CISH* gene expression in treated cells normalized to internal control genes (*GAPDH*, *ACTB* and *RPL13a*) and relative to control condition (normalized to 1) ($n = 3$ in triplicates, data are mean ± SD, * $p < 0.05$; one-sample *t*-test).

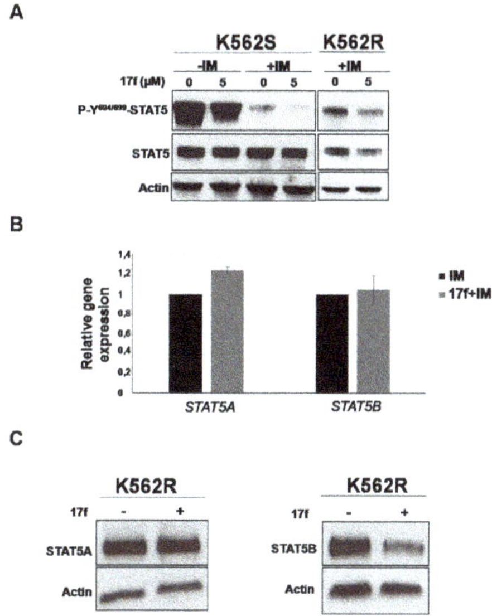

Figure 4. 17f associated with IM inhibits STAT5B protein expression in K562R cells (**A**) Protein extracts from K562S and K562R cells treated with 17f 5 μM or DMSO with or without IM for 24 h were analyzed by western blotting to detect P-Y$^{694/699}$-STAT5 and STAT5 protein expression ($n = 2$). Actin served as the loading control. (**B**) qRT-PCR analysis of *STAT5A* and *STAT5B* expression in K562R cultured with IM (1 μM) as control or treated with 17f (5μM) and IM for 24 h. Results are presented as the fold change in *STAT5A* and *STAT5B* gene expression in treated cells normalized to internal control genes (*GAPDH*, *ACTB* and *RPL13a*) and relative to control condition (normalized to 1) ($n = 3$ in triplicates, data are mean ± SD, * $p < 0.05$; one sample *t*-test). (**C**) Expression of STAT5A and STAT5B proteins in K562R cells treated or not with 17f (5 μM) was analyzed by western blot ($n = 2$). Actin served as the loading control.

2.5. Effects of 17f on Growth and Viability of Ara-C-Sensitive and Ara-C-Resistant FLT3-ITD Expressing Leukemic Cells

STAT5 is also phosphorylated by FLT3-ITD, a major TKO in AML cells. To exclude the possibility that 17f-mediated inhibition of STAT5 and cell growth is a peculiarity of IM-resistant BCR-ABL$^+$ cells, we used MV4-11 cells expressing FLT3-ITD that acquired resistance to Ara-C, a conventional therapeutic agent that affects DNA replication. Sensitive and resistant MV4-11 cell models are depicted in Figure 5A. We first evaluated the impact of 17f alone on MV4-11S and MV4-11R cell growth and showed that MV4-11R cells were more resistant to 17f treatment than MV4-11S cells (Figure 5B). Based on these data, IC$_{50}$ values were found to be three-fold higher in MV4-11R than in MV11-4S cells (10.79 ± 3.2 vs. 3.55 ± 0.47).

2.6. 17f Sensitizes MV4-11R Cells to Ara-C Treatment

We then analyzed the effects of 17f on MV4-11S and MV4-11R cell growth in the presence of Ara-C using trypan blue dye exclusion (Figure 6A) and MTT assays (Figure 6B). Addition of 17f significantly enhanced the growth inhibition and cytotoxic effect of Ara-C in MV4-11S cells. Importantly, 17f greatly reduced the growth of resistant MV4-11R cells cultured with Ara-C in a concentration-dependent fashion. This growth inhibition was already observed with 1 μM, a concentration that did not affect the growth of MV4-11R cells cultured in the absence of Ara-C. These data indicated that the addition of 17f overcomes the resistance of MV4-11R cells to Ara-C.

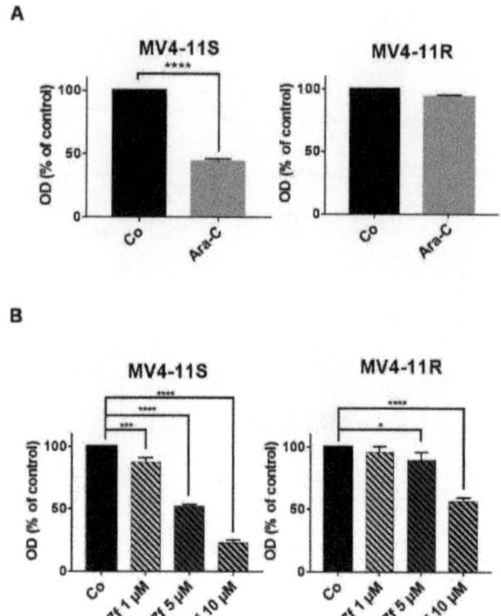

Figure 5. Effects of 17f on MV4-11S and MV4-11R cell growth (**A**) Ara-C-sensitive MV4-11 (MV4-11S) and Ara-C-resistant MV4-11 (MV4-11R) cells were treated with 1 µM Ara-C or DMSO as control (Co) for 48 h. Cell viability was then determined by MTT assays ($n = 3$ in triplicates, data are mean ± SD, **** $p < 0.0001$; one-sample t-test). (**B**) MV4-11S and MV4-11R cells were treated or not (Co) with increasing concentrations of 17f during 48 h. Cell viability was determined by MTT assays ($n = 3$ in triplicates, data are mean ± SD, ** $p < 0.01$, *** $p < 0.001$, **** $p < 0.0001$; one-sample t-test).

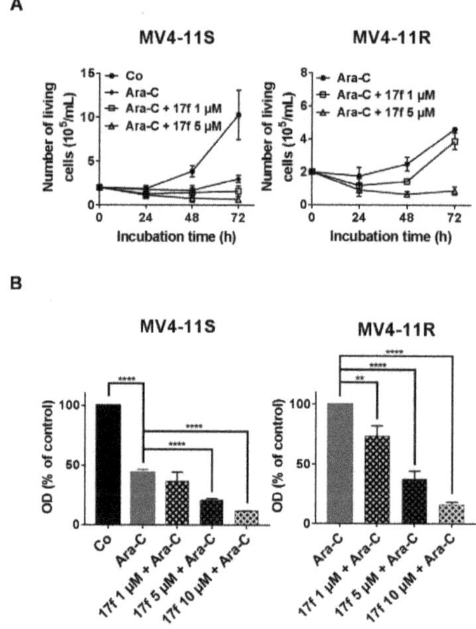

Figure 6. 17f relieves the resistance of MV4-11R cells to ARA-C treatment. (**A**) MV4-11S or MV4-11R cells were treated with Ara-C or not (Co) with or without 17f. Growth kinetics were determined by Trypan blue dye exclusion assays ($n = 3$ in triplicates, data are mean ± SD). (**B**) MV4-11S or MV4-11R cells were treated with Ara-C or not (Co) with or without 17f for 48 h. Cell viability was determined by MTT assays ($n = 3$ in triplicates, data are mean ± SD, * $p < 0.05$, ** $p < 0.01$, **** $p < 0.0001$; one-sample t-test).

2.7. 17f Triggers Apoptosis, Cell Cycle Arrest and Inhibition of STAT5B Expression in MV4-11R Cells

We then evaluated the effects of 17f on apoptosis and the cell cycle in MV4-11R cells. A significant increase in apoptotic cells was observed (Figure 7A) only after treatment with 5 µM 17f, while the addition of 1 µM was enough to enhance the number of cells in the G_0 phase of the cell cycle (Figure 7B). These results indicated that the growth-suppressive properties of 17f primarily affect the cell cycle in MV4-11R cells and apoptosis at higher concentrations. We then asked whether 17f interferes with STAT5 signaling in Ara-C-resistant AML cells and analyzed the impact of 17f on phosphorylation and expression of STAT5 in MV4-11R cells. In the absence of Ara-C, the level of STAT5 phosphorylation was slightly enhanced in MV4-11R cells (Figure S5C,D). The addition of 17f with or without Ara-C inhibited STAT5 expression in MV4-11R cells (Figure 7C and Figures S2B and S3C). Likewise, STAT5B expression was reduced after treatment with 17f alone or with Ara-C in resistant cells (Figure 7D).

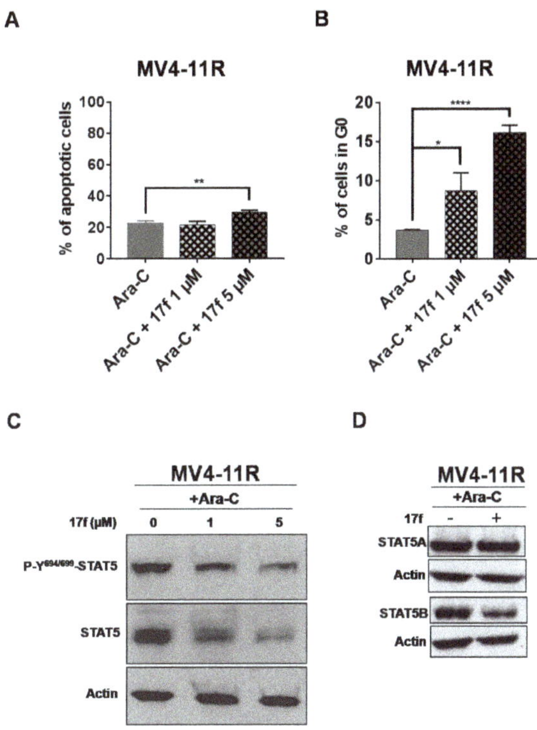

Figure 7. 17f promotes apoptosis, cell cycle arrest and inhibition of STAT5B protein expression in MV4-11R cells. (**A**) Flow cytometry histogram of MV4-11R cells cultured for 48 h with Ara-C and 17f or Ara-C and DMSO as control. Cells were stained with anti-annexin V coupled with FITC and with 7-AAD to determine the percentages of apoptotic cells ($n = 3$ in triplicates, data are mean ± SD, ** $p < 0.01$). (**B**) MV4-11R cells treated for 48 h with Ara-C and 17f or Ara-C and DMSO as control were stained with 7-AAD and an Alexa Fluor 488-conjugated anti-Ki67 antibody. Cell cycle phase distributions were then estimated by flow cytometry. The histogram presents the percentage of cells in the G_0 phase ($n = 3$ in triplicates, data are mean ± SD, * $p < 0.05$, *** $p < 0.001$). (**C**) Protein extracts from MV4-11R cells treated with Ara-C and 17f or Ara-C and DMSO for 24 h were analyzed by immunoblotting to detect P-Y$^{694/699}$-STAT5 and STAT5 protein expression ($n = 2$). Actin served as the loading control. (**D**) Expression of STAT5A and STAT5B proteins in MV4-11R cells treated or not with 17f (5 µM) was also analyzed by western blot ($n = 2$).

2.8. 17f Inhibits Expression of Oncogenic STAT5B[N642H] Mutant

Gain of function mutations of *STAT5B* have been described in hematopoietic malignancies. The recurrent hotspot mutation N642H has been identified in T cell leukemia and lymphomas and the STAT5B[N642H] mutant was shown to induce T cell neoplasia in transgenic mice [24–27]. We therefore tested the ability of 17f to inhibit STAT5B[N642H] expression and growth of hematopoietic cells transformed by this mutant. For this purpose, we used Ba/F3 cells expressing flag-tagged STAT5B[N642H] or flag-tagged wild-type STAT5B (wtSTAT5B) as control [27]. We found that Ba/F3-STAT5B[N642H] cells were more sensitive to 17f treatment than control Ba/F3-wtSTAT5B cells (Figure 8A). We then addressed whether STAT5[N642H] expression was impacted by 17f and showed that 17f reduces expression of this mutant in Ba/F3 cells but does not affect wtSTAT5B or endogenous STAT5A expression after 24 h treatment (Figure 8B).

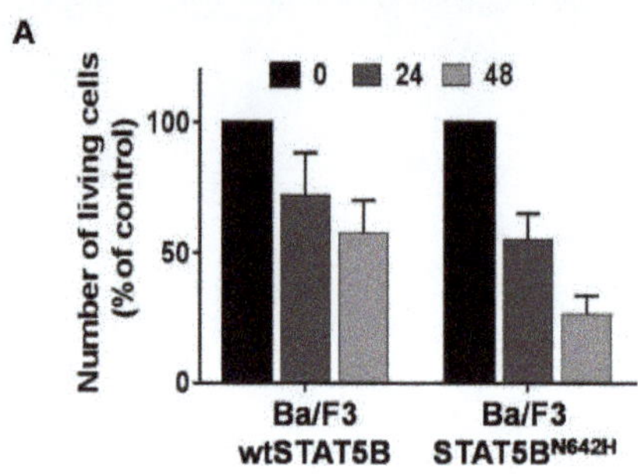

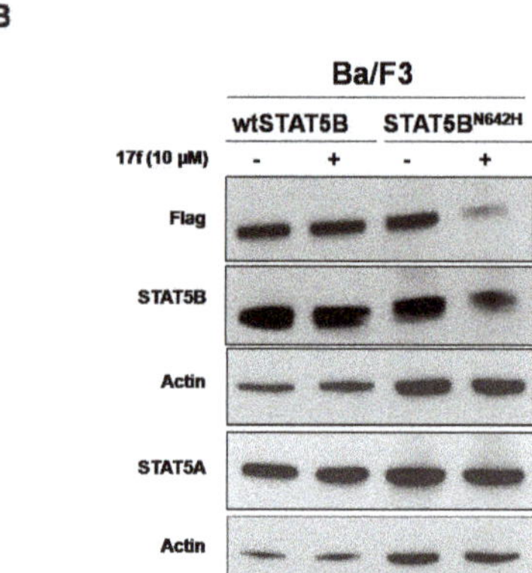

Figure 8. 17f inhibits STAT5B[N642H] activity and expression in Ba/F3 cells. (**A**) Cells were treated or not with 17f (10 µM). Growth were then determined by Trypan blue dye exclusion assays at the indicated times (*n* = 5 in triplicates, data are mean ± SD). (**B**) Protein extracts from MV4-11R cells treated with 17f for 24 h were analyzed by immunoblotting to detect flag-tagged wtSTAT5B, STAT5B[N642H] and endogenous STAT5A/STAT5B protein expression (*n* = 2). Actin served as the loading control.

3. Discussion

The development of pharmacological inhibitors targeting the JAK/STAT pathway has been the subject of intense investigation during the last decade. Among the STAT family members, STAT5 proteins are now recognized as important therapeutic targets in hematologic malignancies and also in certain solid tumors [28]. Distinct pharmacological compounds that directly or indirectly affect STAT5 activity and leukemia cell growth have been used or developed during these last years. We recently synthesized and discovered a new compound (17f) that inhibits STAT5 phosphorylation and transcriptional activity in various CML and AML cells, without detectable effects on other signal transduction molecules, such, as STAT3 and the protein kinases ERK1/2 and AKT [23]. We also demonstrated that 17f strongly reduces the growth of CML and AML cells with EC_{50} values below 10 µM close to EC_{50} values obtained with the STAT5 inhibitor pimozide (unpublished data) indicating that 17f as pimozide targets myeloid leukemia cells addicted to STAT5 signaling (see also Figure S1 for 17f and pimozide structures). In this study, we bring evidences that 17f also relieves the resistance of CML and AML cells to IM and Ara-C, respectively. Interestingly, we found that the concentrations of 17f required to restore the response to IM and Ara-C in resistant leukemic cells were much lower than EC_{50} values obtained for each resistant cell type. Indeed, inhibition of cell growth was already observed with 1 µM when combined with IM or Ara-C while EC_{50} values obtained for 17f compound alone were greater than 10 µM in these resistant cells. Depletion of IM or Ara-C in resistant cells might explain changes in the growth inhibitory effects of 17f. Indeed, we observed that the removal of IM strongly increases the phosphorylation of STAT5 in K562R cells. In these conditions, P-Y-STAT5 protein levels remain much higher in K562R cells after 17f treatment than in treated K562S cells, which are sensitive to lower concentrations of 17f. These data are in close agreement with a previously published study showing that high STAT5 levels mediate IM resistance in CML cells [8]. Although the removal of Ara-C results in a slight increase in STAT5 phosphorylation, the resistance of MV4-11R cells to this drug is not directly linked to overactivated STAT5. ERK1/2 and AKT kinases that also play a crucial role in cell survival, are involved in the resistance of MV4-11 cells to Ara-C [29]. It is then likely that Ara-C depletion may overexpress or overactivate these survival pathways in resistant MV4-11 cells. Whatever the resistance mechanism associated or not with STAT5 signaling, our data suggest that combination treatments with a STAT5 inhibitor might efficiently eliminate resistant CML and AML cells.

While 17f alone inhibited STAT5 phosphorylation in IM-depleted K562R cells, it decreased STAT5 expression in Ara-C-depleted MV4-11R cells. Importantly, combination treatments reduced expression of STAT5 in both resistant leukemic cells. The mechanisms involved in this downregulation remain unknown but are not associated with changes in *STAT5A* and *STAT5B* gene expression and specifically affect STAT5B protein. Importantly, we also demonstrated that 17f inhibits expression of STAT5B^{N642H} protein expression in transformed Ba/F3 cells. STAT5B^{N642H} is a driver mutation for T cell neoplasia and has been associated with aggressiveness, poor prognosis and an increased risk of relapse in T cell leukemia-lymphoma patients [24–27]. In addition to myeloid leukemia, 17f might be then employed to target lymphoproliferative disorders and lymphomas addicted to STAT5B^{N642H} signaling.

It is likely that 17f inhibits STAT5B expression via the ubiquitin/proteasome-dependent degradation of this protein. Indeed, STAT5 proteins were previously shown to be ubiquitinated and several ubiquitination sites have been identified in STAT5A and STAT5B protein sequences [30,31]. Cbl, a well-known E3 ubiquitin ligase was found to interact with STAT5 and to induce its ubiquitination [30]. Moreover, cytokine-mediated STAT5 phosphorylation was enhanced in hematopoietic stem cells from *c-cbl* knockout mice [32]. 17f alone or associated with IM or Ara-C might then promote ubiquitination and proteasomal degradation of STAT5B protein in resistant leukemic cells as well as in STAT5B^{N642H}-expressing cells. In a similar vein, pimozide was shown to target USP1, a ubiquitin specific protease involved in the deubiquitination of transcription factors such as ID-1. Pimozide-mediated inhibition of USP-1 promotes ID1 degradation and inhibition of leukemic cell growth [12]. It is therefore conceivable that 17f activity is connected to *a* proteasome regulatory network that controls STAT5B protein degradation. Alternatively, the combination of 17f and IM or Ara-C might also target chaperone molecules such as the heat shock proteins HSP90 or HSP70 proteins which were previously shown to regulate expression and/or stability of STAT5 [33,34]. The dual inhibition of BCR-ABL and HSP90 was shown to abrogate the growth of IM-resistant CML cells [35]. Furthermore, a key role of STAT5 has been demonstrated in the synergistic effects of FLT3 and HSP90 inhibitors in FLT3-ITD-expressing leukemic cells [36]. Importantly, HSP90 inhibitors not only target STAT5 but also overcome the resistance of AML cells to FLT3 inhibitors [37]. HSP70 was also found to induce STAT5 expression and drug resistance in AML and CML cells and inhibition of STAT5 activity was sufficient to resensitize resistant leukemic cells to chemotherapy [34,38].

If the downregulation of STAT5A and STAT5B expression can occur via ubiquitin/proteasome-dependent protein degradation, the selective effect of 17f on STAT5B still remains unclear. Nevertheless, using a bacterial two-hybrid screening approach, we previously identified the tumor suppressor hTid1 as a specific binding partner of STAT5B [39]. hTid1 belongs to the DnaJ chaperone protein family, which contains the J domain, a highly conserved domain that binds to Hsp70. The DnaJ-Hsp70 complexes are involved in protein folding and protein degradation and hTid1 was shown to promote the ubiquitination and degradation of various cellular proteins including transcription factors [40]. We demonstrated that overexpression of hTid1 specifically suppresses STAT5B protein expression and the transforming potential of a constitutively active STAT5B variant (STAT5B1*6) in hematopoietic cells. 17f might then target specific effectors of STAT5B protein stability/degradation, a hypothesis that has yet to be experimentally tested.

Besides these potential mechanisms, the capacity of 17f to restore the sensitivity of resistant CML or AML cells to IM or Ara-C suggests that inhibitors targeting STAT5 expression would also benefit AML or CML patients who have developed resistance to chemotherapy. Accordingly, PPARγ agonists were shown to inhibit *STAT5A* and *STAT5B* gene expression and to synergize with IM to eradicate resistant CML stem cells [10]. Our findings suggest that targeting STAT5B protein is a promising therapeutic strategy to eradicate leukemic cells that acquired resistance to chemotherapeutic agents. This is also supported by previous works showing that STAT5B but not STAT5A plays a key role in BCR-ABL-induced leukemogenesis and in the sensitivity of CML cells to TKI treatment [41,42]. Recent studies indicated that STAT5 proteins also exert important non canonical functions in normal and cancer cells. For instance, unphosphorylated STAT5 (uSTAT5: non phosphorylated on Y694/699 residues) were shown to be transcriptionally active in self-renewing hematopoietic stem cells and to promote leukemia/lymphoma cell survival [43,44]. Selective inhibitors that only block tyrosine phosphorylation and dimer formation might then be insufficient to fully abrogate STAT5 activity and resistance to chemotherapy. Herein, we showed that that inhibition of STAT5B expression elicited by 17f might unlock drug resistance in CML and AML cells. Using these promising data as a lead, we carried out a rational search for new derivatives of 17f with enhanced antileukemic activity. Modeling work was initiated to identify a pharmacophore that could help to optimize the development of 17f derivatives working in the nanomolar range. These new compounds could represent promising drugs to overcome chemotherapy resistance in leukemia or lymphomas.

4. Materials and Methods

4.1. Cell Cultures and Reagents

IM-sensitive (K562S) and IM-resistant (K562R) BCR-ABL$^+$ cells and MV4-11 cells were obtained from American Type Culture Collection (ATCC) and Deutsche Sammlung von Mikroorganismens und Zellkulturen (DSMZ), respectively, and maintained according to the supplier's recommendations. K562R and Ara-C-resistant MV4-11 (MV4-11R) cells were obtained after cultures of K562S and sensitive MV4-11 (MV4-11S) cells with increasing concentrations of IM and Ara-C (until 1 μM). Ba/F3-STAT5B^{N642H} and Ba/F3-wtSTAT5B cells were previously described in [27]. All cell lines were cultured in RPMI 1640, with 10% fetal bovine serum, 2 mM glutamine, 100 U/mL penicillin and 100 μg/mL streptomycin at 37 °C, 5% CO_2. Resistant cells were cultured with 1 μM IM or Ara-C. IM was purchased from Selleckchem (Houston, TX, USA) and Ara-C from Sandoz France (Levallois-Perret, France). Ba/F3-wtSTAT5B were cultured with IL-3. The synthesis of the 17f compound was previously described in [23].

4.2. Cell Proliferation Assays

Cell viability and proliferation were studied using a MTT cell proliferation assay (Sigma-Aldrich, St Louis, MO, USA). Briefly, 2×10^4 leukemic cells were cultured in 100 μL of RPMI medium in 96-well plates and treated with drugs for 48 h. Cells were then incubated with 10 μL of MTT working solution (5 g/L of methylthiazolyldiphenyl-tetrazolium bromide) for 4 h. Cells were lysed overnight at 37 °C with 100 μL of SDS 10%, HCl 0.003%. Optical density (OD) at 570 nm was then measured using a spectrophotometer CLARIOstar® (BMG Labtech, Offenburg, Germany). Living cells were also enumerated using the trypan blue dye exclusion method.

4.3. Apoptosis and Cell Cycle Analysis

Cells were washed with PBS, then stained (10^6 cells) in buffer containing FITC-annexin V and 7-amino-actinomycin D (7-AAD) (Beckmann Coulter, Fullerton, CA, USA) for 15 min at 4 °C and analyzed by flow cytometry (Becton Dickinson Accuri™ C6 flow cytometer). For cell cycle analysis [45], cells were first incubated with fixing solution (PFA 2%, Hepes 1%, saponin 0.03%) for 15 min and then in PFS permeabilization solution (PBS 1×, SVF 10%, saponin 0.03%, Hepes 1%). Cells were next stained for 30 min at room temperature with anti-Ki67-Alexa Fluor 488 monoclonal antibody or the corresponding isotype as control (Becton-Dickinson, Franklin Lakes, NJ, USA) before analysis by flow cytometry (Becton Dickinson Accuri™ C6 flow cytometer). The FlowJo® software (V10.1, BD Biosciences, Franklin Lakes, NJ, USA) was used to analyze data.

4.4. Plasmids, Transfection and Luciferase Reporter Assays

The 6×(STAT5)-TK-luc containing six tandem copies of the STAT5 binding site linked to the minimal TK-luciferase reporter gene and control TK-luc plasmids have been described elsewhere [46]. For transient transfection assays, cells were electroporated (270 V, 950 μF) with the different constructs (50 μg). Transfected cells were expanded for 24 h in medium and then treated for 48 h. Cell extracts were then prepared in luciferase buffer according to the manufacturer's protocol (One Glo luciferase assay kit, Promega, Madison, WI, USA). Luciferase activities were measured in a luminometer CLARIOstar® (BMG Labtech, Offenburg, Germany).

4.5. Western Blot

Cells were suspended in Laemmli's 2× buffer (Bio-Rad, Hercules, CA, USA), separated on SDS/PAGE and blotted onto nitrocellulose membrane. Blots were incubated with the following antibodies (Abs): P-Y$^{694/699}$-STAT5, Actin (Cell Signaling Technology, Danvers, MA, USA), STAT5 (BD Transduction Laboratories, Franklin Lakes, NJ, USA), STAT5A and STAT5B (Zymed/ThermoFisher Scientific, Waltham, MA, USA). Membranes were developed with the ECL chemiluminescence detection system (GE Healthcare, Little Chalfont Buckinghamshire, UK) using specific peroxidase (HRP) conjugated to rabbit or mouse IgG antibodies (Cell Signaling Technology).

4.6. P-Y$^{694/699}$-STAT5 Flow Cytometry Analysis

Cells were washed in PBS and incubated with a fixing solution PFA 4% for 15 min at room temperature. The first permeabilization solution PBS/Triton X-100 0.2% was then added and incubated for 30 min at 37 °C. After being washed with PBS/BSA 0.5%, cells were suspended with the second permeabilization solution PBS/MeOH 50% and incubated for 10 min on ice. Cells were then stained with anti P-Y$^{694/699}$-STAT5 antibodies or the corresponding isotype as control (BD Biosciences, NJ, USA) for 30 min at room temperature before analysis by flow cytometry (FACS Canto II, BD Biosciences).

4.7. qRT-PCR Analysis

RNA samples were reverse-transcribed using the SuperScript®VILO cDNA synthesis kit (Invitrogen, Carlsbad, CA, USA) as recommended by the supplier. The resulting cDNAs were used for quantitative real-time PCR (qRT-PCR). PCR primers (*PIM1: for* 5'-TTTCGAGCATGACGAAGAGA-3', *rev* 5'-GGGCCAAGCACCATCTAAT-3'; *CISH*: 5'- AGCCAAGACCTTCTCCTACCTT-3', *rev* 5'-TGGCATCTTCTGCAGGTGT-3'; *STAT5A*: *for* 5'-TCCCTATAACATGTACCCACA-3', *rev* 5'-ATGGTCTCATCCAGGTCGAA-3'; *STAT5B*: *for* 5'-TGAAGGCCACCATCATCAG-3', *rev* 5'-TGTTCAAGATCTCGCCACTG-3') were designed with the ProbeFinder software (Roche Applied Sciences, Basel, Switzerland) and used to amplify the RT-generated cDNAs. qRT-PCR analyses were performed on the Light Cycler 480 thermocycler II (Roche). *GAPDH* (glyceraldehyde-3-phosphate dehydrogenase), *ACTB* (actin beta) and *RPL13A* were used as reference genes for normalization of qRT-PCR experiments. Each reaction condition was performed in triplicate. Relative gene expression was analyzed using the $2^{-\Delta\Delta Ct}$ method [47].

5. Conclusions

In summary, this work shows for the first time that inhibition of STAT5B expression might be a promising targeting strategy to bypass the resistance of CML and AML cells to TKI or conventional chemotherapeutic agents. Investigations to elucidate the mechanisms involved in STAT5B downregulation induced by these combination therapies might help to design new inhibitors that specifically target cancer cells addicted to oncogenic STAT5B signaling.

Supplementary Materials: The following supplementary figures are available online at http://www.mdpi.com/2072-6694/11/12/2043/s1, Figure S1: Pimozide and 17f structures, Figure S2: Original western blot, Figure S3: Quantification of Western Blot data. Figure S4: Flow cytometry analysis of P-Y-STAT5 in K562S and K562R cells. Figure S5: Effects of 17f on P-Y-STAT5/STAT5 expression in IM-depleted K562R and Ara-C-depleted MV4-11R cells.

Author Contributions: Conceptualization, M.B.-B., G.P. and F.G.; Formal analysis, M.B.-B. and F.G.; Funding acquisition, G.P. and F.G.; Investigation, M.B.-B., M.D., N.V., M.P. and L.J.; Methodology, M.B.-B., M.D., N.V., M.P. and L.J.; Project administration, G.P. and F.G.; Resources, G.P. and F.G.; Supervision, G.P.; Validation, M.B.-B., M.D., N.V., M.P., F.M., G.P. and F.G.; Visualization, F.G.; Writing—original draft, M.B.-B., G.P. and F.G.; Writing—review & editing, M.D., N.V., L.J., F.M., M.-C.V.-M. and O.H.

Funding: This study was supported by FRM (grant number: DCM20181039564), CNRS, Ligue Contre le Cancer and University of Tours. MB-B was supported by the ARC foundation and FRM.

Acknowledgments: We would like to thank Emmanuel Pecnard, Farah Kouzi and Elodie Coste for technical assistance; Christina Maria Wagner and Heidi A. Neubauer for providing Ba/F3STAT5B^{N642H} and Ba/F3wtSTAT5B cells.

Conflicts of Interest: The authors declare no conflict of interest.

References

1. Bunting, K.D. STAT5 signaling in normal and pathologic hematopoiesis. *Front. Biosci.* **2007**, *12*, 2807–2820. [CrossRef] [PubMed]

2. Hoelbl, A.; Schuster, C.; Kovacic, B.; Zhu, B.; Wickre, M.; Hoelzl, M.A.; Fajmann, S.; Grebien, F.; Warsch, W.; Stengl, G.; et al. Stat5 is indispensable for the maintenance of Bcr/abl-positive leukemia. *Eur. Mol. Biol. Organ. Mol. Med.* **2010**, *2*, 98–110.

3. Ye, D.; Wolff, N.; Li, L.; Zhang, S.; Ilaria, R.L., Jr. STAT5 signaling is required for the efficient induction and maintenance of CML in mice. *Blood* **2006**, *107*, 4917–4925. [CrossRef] [PubMed]

4. Mizuki, M.; Fenski, R.; Halfter, H.; Matsumura, I.; Schmidt, R.; Müller, C.; Grüning, W.; Kratz-Albers, K.; Serve, S.; Steur, C.; et al. Flt3 mutations from patients with acute myeloid leukemia induce transformation of 32D cells mediated by the Ras and STAT5 pathways. *Blood* **2000**, *96*, 3907–3914. [CrossRef]

5. Chomel, J.C.; Bonnet, M.L.; Sorel, N.; Sloma, I.; Bennaceur-Griscelli, A.; Rea, D.; Legros, L.; Marfaing-Koka, A.; Bourhis, J.H.; Ame, S.; et al. Leukemic stem cell persistence in chronic myeloid leukemia patients in deep molecular response induced by tyrosine kinase inhibitors and the impact of therapy discontinuation. *Oncotarget* **2016**, *7*, 35293–35301. [CrossRef]

6. Gambacorti-Passerini, C.B.; Gunby, R.H.; Piazza, R.; Galietta, A.; Rostagno, R.; Scapozza, L. Molecular mechanisms of resistance to imatinib in Philadelphia-chromosome-positive leukaemias. *Lancet Oncol.* **2003**, *4*, 75–85. [CrossRef]

7. Cumaraswamy, A.A.; Todic, A.; Resetca, D.; Minden, M.D.; Gunning, P.T. Inhibitors of Stat5 protein signaling. *Med. Chem. Commun.* **2012**, *3*, 22–27. [CrossRef]
8. Warsch, W.; Kollmann, K.; Eckelhart, E.; Fajmann, S.; Cerny-Reiterer, S.; Hölbl, A.; Gleixner, K.V.; Dworzak, M.; Mayerhofer, M.; Hoermann, G.; et al. High STAT5 levels mediate imatinib resistance and indicate disease progression in chronic myeloid leukemia. *Blood* **2011**, *117*, 3409–3420. [CrossRef]
9. Warsch, W.; Grundschober, E.; Berger, A.; Gille, L.; Cerny-Reiterer, S.; Tigan, A.-S.; Hoelbl-Kovacic, A.; Valent, P.; Moriggl, R.; Sexl, V. STAT5 triggers BCR-ABL1 mutation by mediating ROS production in chronic myeloid leukemia. *Oncotarget* **2012**, *3*, 1669–1687. [CrossRef]
10. Prost, S.; Relouzat, F.; Spentchian, M.; Ouzegdouh, Y.; Saliba, J.; Massonnet, G.; Beressi, J.P.; Verhoeyen, E.; Raggueneau, V.; Maneglier, B.; et al. Erosion of the chronic myeloid leukemia stem cell pool by PPARγ agonists. *Nature* **2015**, *525*, 380–383. [CrossRef]
11. Nelson, E.A.; Walker, S.R.; Weisberg, E.; Bar-Natan, M.; Barrett, R.; Gashin, L.B.; Terrell, S.; Klitgaard, J.L.; Santo, L.; Addorio, M.R.; et al. The STAT5 inhibitor pimozide decreases survival of chronic myelogenous leukemia cells resistant to kinase inhibitors. *Blood* **2011**, *117*, 3421–3429. [CrossRef] [PubMed]
12. Mistry, H.; Hsieh, G.; Buhrlage, S.J.; Huang, M.; Park, E.; Cuny, G.D.; Galinsky, I.; Stone, R.M.; Gray, N.S.; D'Andrea, A.D.; et al. Small-molecule inhibitors of USP1 target ID1 degradation in leukemic cells. *Mol. Cancer Ther.* **2013**, *12*, 2651–2662. [CrossRef] [PubMed]
13. Nam, S.; Scuto, A.; Yang, F.; Chen, W.; Park, S.; Yoo, H.S.; Konig, H.; Bhatia, R.; Cheng, X.; Merz, K.H.; et al. Indirubin derivatives induce apoptosis of chronic myelogenous leukemia cells involving inhibition of Stat5 signaling. *Mol. Oncol.* **2012**, *6*, 276–283. [CrossRef] [PubMed]
14. Page, B.D.; Khoury, H.; Laister, R.C.; Fletcher, S.; Vellozo, M.; Manzoli, A.; Yue, P.; Turkson, J.; Minden, M.D.; Gunning, P.T. Small molecule STAT5-SH2 domain inhibitors exhibit potent antileukemia activity. *J. Med. Chem.* **2012**, *55*, 1047–1055. [CrossRef]
15. Cumaraswamy, A.A.; Lewis, A.M.; Geletu, M.; Todic, A.; Diaz, D.B.; Cheng, X.R.; Brown, C.E.; Laister, R.C.; Muench, D.; Kerman, K.; et al. Nanomolar-potency small molecule inhibitor of STAT5 protein. *Am. Chem. Soc. Med. Chem. Lett.* **2014**, *5*, 1202–1206. [CrossRef]
16. Elumalai, N.; Berg, A.; Rubner, S.; Blechschmidt, L.; Song, C.; Natarajan, K.; Matysik, J.; Berg, T. Rational development of Stafib-2: A selective, nanomolar inhibitor of the transcription factor STAT5b. *Sci. Rep.* **2017**, *7*, 819. [CrossRef]
17. Wingelhofer, B.; Maurer, B.; Heyes, E.C.; Cumaraswamy, A.A.; Berger-Becvar, A.; de Araujo, E.D.; Orlova, A.; Freund, P.; Ruge, F.; Park, J.; et al. Pharmacologic inhibition of STAT5 in acute myeloid leukemia. *Leukemia* **2018**, *32*, 1135–1146. [CrossRef]
18. Prost, S.; Le Dantec, M.; Augé, S.; Le Grand, R.; Derdouch, S.; Auregan, G.; Déglon, N.; Relouzat, F.; Aubertin, A.M.; Maillere, B.; et al. Human and simian immunodeficiency viruses deregulate early hematopoiesis through a Nef/PPARγ/STAT5 signaling pathway in macaques. *J. Clin. Investig.* **2008**, *118*, 1765–1775. [CrossRef]
19. Liu, H.; Zang, C.; Fenner, M.H.; Liu, D.; Possinger, K.; Koeffler, H.P.; Elstner, E. Growth inhibition and apoptosis in human Philadelphia chromosome-positive lymphoblastic leukemia cell lines by treatment with the dual PPARα/γ ligand TZD18. *Blood* **2006**, *107*, 3683–3692. [CrossRef]
20. Bertz, J.; Zang, C.; Liu, H.; Wächter, M.; Possinger, K.; Koeffler, H.P.; Elstner, E. Compound 48, a novel dual PPAR α/γ ligand, inhibits the growth of human CML cell lines and enhances the anticancer-effects of imatinib. *Leuk. Res.* **2009**, *33*, 686–692. [CrossRef]
21. Parmenon, C.; Guillard, J.; Caignard, D.-H.; Hennuyer, N.; Staels, B.; Audinot-Bouchez, V.; Boutin, J.A.; Dacquet, C.; Ktorza, A.; Viaud-Massuard, M.C. 4,4-Dimethyl-1,2,3,4-Tetrahydroquinoline-Based PPARα/γ agonists. Part I: Synthesis and pharmacological evaluation. *Bioorg. Med. Chem. Lett.* **2008**, *18*, 1617–1622. [CrossRef]
22. Parmenon, C.; Guillard, J.; Caignard, D.-H.; Hennuyer, N.; Staels, B.; Audinot-Bouchez, V.; Boutin, J.A.; Dacquet, C.; Ktorza, A.; Viaud-Massuard, M.C. 4,4-Dimethyl-1,2,3,4-Tetrahydroquinoline-Based PPARα/γ agonists. Part II: Synthesis and pharmacological evaluation of oxime and acidic head group structural variations. *Bioorg. Med. Chem. Lett.* **2009**, *19*, 2683–2687. [CrossRef]
23. Juen, L.; Brachet-Botineau, M.; Parmenon, C.; Bourgeais, J.; Hérault, O.; Gouilleux, F.; Viaud-Massuard, M.C.; Prié, G. New inhibitor targeting signal transducer and activator of transcription 5 (STAT5) signaling in myeloid leukemias. *J. Med. Chem.* **2017**, *60*, 6119–6136. [CrossRef]

24. Kontro, M.; Kuusanmäki, H.; Eldfors, S.; Burmeister, T.; Andersson, E.I.; Bruserud, O.; Brümmendorf, T.H.; Edgren, H.; Gjertsen, B.T.; Itälä-Remes, M.; et al. Novel activating STAT5B mutations as putative drivers of T-cell acute lymphoblastic leukemia. *Leukemia* **2014**, *28*, 1738–1742. [CrossRef]

25. Bandapalli, O.R.; Schuessele, S.; Kunz, J.B.; Rausch, T.; Stütz, A.M.; Tal, N.; Geron, I.; Gershman, N.; Izraeli, S.; Eilers, J.; et al. The activating STAT5B N642H mutation is a common abnormality in pediatric T-cell acute lymphoblastic leukemia and confers a higher risk of relapse. *Haematologica* **2014**, *99*, 188–192. [CrossRef]

26. Nicolae, A.; Xi, L.; Pittaluga, S.; Abdullaev, Z.; Pack, S.D.; Chen, J.; Waldmann, T.A.; Jaffe, E.S.; Raffeld, M. Frequent STAT5B mutations in γδ hepatosplenic T-cell lymphomas. *Leukemia* **2014**, *28*, 2244–2248. [CrossRef]

27. Pham, H.T.T.; Maurer, B.; Prchal-Murphy, M.; Grausenburger, R.; Grundschober, E.; Javaheri, T.; Nivarthi, H.; Boersma, A.; Kolbe, T.; Elabd, M.; et al. STAT5BN642H is a driver mutation for T cell neoplasia. *J. Clin. Investig.* **2018**, *128*, 387–401. [CrossRef]

28. Liao, Z.; Nevalainen, M.T. Targeting transcription factor Stat5a/b as a therapeutic strategy for prostate cancer. *Am. J. Transl. Res.* **2011**, *3*, 133–138.

29. Ko, Y.C.; Hu, C.Y.; Liu, Z.H.; Tien, H.F.; Ou, D.L.; Chien, H.F.; Lin, L.I. Cytarabine-Resistant FLT3-ITD Leukemia Cells are Associated with TP53 Mutation and Multiple Pathway Alterations—Possible Therapeutic Efficacy of Cabozantinib. *Int. J. Mol. Sci.* **2019**, *20*, 1230. [CrossRef]

30. Goh, E.L.; Zhu, T.; Leong, W.Y.; Lobie, P.E. c-Cbl is a negative regulator of GH-stimulated STAT5-mediated transcription. *Endocrinology* **2002**, *143*, 3590–3603. [CrossRef]

31. Chen, Y.; Dai, X.; Haas, A.L.; Wen, R.; Wang, D. Proteasome-dependent down-regulation of activated Stat5A in the nucleus. *Blood* **2006**, *108*, 566–574. [CrossRef]

32. Rathinam, C.; Thien, C.B.; Langdon, W.Y.; Gu, H.; Flavell, R.A. The E3 ubiquitin ligase c-Cbl restricts development and functions of hematopoietic stem cells. *Genes Dev.* **2008**, *22*, 992–997. [CrossRef]

33. Moulick, K.; Ahn, J.H.; Zong, H.; Rodina, A.; Cerchietti, L.; Gomes DaGama, E.M.; Caldas-Lopes, E.; Beebe, K.; Perna, F.; Hatzi, K.; et al. Affinity-based proteomics reveal cancer-specific networks coordinated by Hsp90. *Nat. Chem. Biol.* **2011**, *7*, 818–826. [CrossRef]

34. Guo, F.; Sigua, C.; Bali, P.; George, P.; Fiskus, W.; Scuto, A.; Annavarapu, S.; Mouttaki, A.; Sondarva, G.; Wei, S.; et al. Mechanistic role of heat shock protein 70 in Bcr-Abl-mediated resistance to apoptosis in human acute leukemia cells. *Blood* **2005**, *105*, 1246–1255. [CrossRef]

35. Wu, L.; Yu, J.; Chen, R.; Liu, Y.; Lou, L.; Wu, Y.; Huang, L.; Fan, Y.; Gao, P.; Huang, M.; et al. Dual inhibition of Bcr-Abl and Hsp90 by C086 potently inhibits the proliferation of imatinib-resistant CML cells. *Clin. Cancer Res.* **2015**, *21*, 833–843. [CrossRef]

36. Yao, Q.; Nishiuchi, R.; Kitamura, T.; Kersey, J.H. Human leukemias with mutated FLT3 kinase are synergistically sensitive to FLT3 and Hsp90 inhibitors: The key role of the STAT5 signal transduction pathway. *Leukemia* **2005**, *19*, 1605–1612. [CrossRef]

37. Katayama, K.; Noguchi, K.; Sugimoto, Y. Heat shock protein 90 inhibitors overcome the resistance to Fms-like tyrosine kinase 3 inhibitors in acute myeloid leukemia. *Oncotarget* **2018**, *9*, 34240–34258. [CrossRef]

38. Pocaly, M.; Lagarde, V.; Etienne, G.; Ribeil, J.A.; Claverol, S.; Bonneu, M.; Moreau-Gaudry, F.; Guyonnet-Duperat, V.; Hermine, O.; Melo, J.V.; et al. Overexpression of the heat-shock protein 70 is associated to imatinib resistance in chronic myeloid leukemia. *Leukemia* **2007**, *21*, 93–101. [CrossRef]

39. Dhennin-Duthille, I.; Nyga, R.; Yahiaoui, S.; Gouilleux-Gruart, V.; Régnier, A.; Lassoued, K.; Gouilleux, F. The tumor suppressor hTid1 inhibits STAT5b activity via functional interaction. *J. Biol. Chem.* **2011**, *286*, 5034–5042. [CrossRef]

40. Bae, M.K.; Jeong, J.W.; Kim, S.H.; Kim, S.Y.; Kang, H.J.; Kim, D.M.; Bae, S.K.; Yun, I.; Trentin, G.A.; Rozakis-Adcock, M.; et al. Tid-1 interacts with the von Hippel-Lindau protein and modulates angiogenesis by destabilization of HIF-1α. *Cancer Res.* **2005**, *65*, 2520–2525. [CrossRef]

41. Schaller-Schönitz, M.; Barzan, D.; Williamson, A.J.; Griffiths, J.R.; Dallmann, I.; Battmer, K.; Ganser, A.; Whetton, A.D.; Scherr, M.; Eder, M. BCR-ABL affects STAT5A and STAT5B differentially. *PLoS ONE* **2014**, *9*, e97243. [CrossRef]

42. Kollmann, S.; Grundschober, E.; Maurer, B.; Warsch, W.; Grausenburger, R.; Edlinger, L.; Huuhtanen, J.; Lagger, S.; Hennighausen, L.; Valent, P.; et al. Twins with different personalities: STAT5B-but not STAT5A-has a key role in BCR/ABL-induced leukemia. *Leukemia* **2019**, *33*, 1583–1597. [CrossRef]

43. Park, H.J.; Li, J.; Hannah, R.; Biddie, S.; Leal-Cervantes, A.I.; Kirschner, K.; Flores Santa Cruz, D.; Sexl, V.; Göttgens, B.; Green, A.R. Cytokine-induced megakaryocytic differentiation is regulated by genome-wide loss of a uSTAT transcriptional program. *Eur. Mol. Biol. Organ. J.* **2016**, *35*, 580–594. [CrossRef]
44. Nagy, Z.S.; Rui, H.; Stepkowski, S.M.; Karras, J.; Kirken, R.A. A preferential role for STAT5, not constitutively active STAT3, in promoting survival of a human lymphoid tumor. *J. Immunol.* **2006**, *177*, 5032–5040. [CrossRef]
45. Vignon, C.; Debeissat, C.; Georget, M.-T.; Bouscary, D.; Gyan, E.; Rosset, P.; Herault, O. Flow cytometric quantification of all phases of the cell cycle and apoptosis in a two-color fluorescence plot. *PLoS ONE* **2013**, *7*, e68425. [CrossRef]
46. Moriggl, R.; Gouilleux-Gruart, V.; Jähne, R.; Berchtold, S.; Gartmann, C.; Liu, X.; Hennighausen, L.; Sotiropoulos, A.; Groner, B.; Gouilleux, F. Deletion of the carboxyl-terminal transactivation domain of MGF-Stat5 results in sustained DNA binding and a dominant negative phenotype. *Mol. Cell. Biol.* **1996**, *16*, 5691–5700. [CrossRef]
47. Livak, K.J.; Schmittgen, T.D. Analysis of relative gene expression data using real-time quantitative PCR and the $2^{-\Delta\Delta CT}$ method. *Methods* **2001**, *25*, 402–408. [CrossRef]

Review

STAT3 and STAT5 Activation in Solid Cancers

Sebastian Igelmann [1,2], **Heidi A. Neubauer** [3] **and Gerardo Ferbeyre** [1,2,*]

[1]   Department of Biochemistry and Molecular Medicine, Université de Montréal, C.P. 6128, Succ. Centre-Ville, Montréal, QC H3C 3J7, Canada; sebastian.igelmann@umontreal.ca

[2]   CRCHUM, 900 Saint-Denis St, Montréal, QC H2X 0A9, Canada

[3]   Institute of Animal Breeding and Genetics, University of Veterinary Medicine Vienna, Vienna 1210, Austria; Heidi.Neubauer@vetmeduni.ac.at

*   Correspondence: g.ferbeyre@umontreal.ca

Received: 1 August 2019; Accepted: 18 September 2019; Published: 25 September 2019

Abstract: The Signal Transducer and Activator of Transcription (STAT)3 and 5 proteins are activated by many cytokine receptors to regulate specific gene expression and mitochondrial functions. Their role in cancer is largely context-dependent as they can both act as oncogenes and tumor suppressors. We review here the role of STAT3/5 activation in solid cancers and summarize their association with survival in cancer patients. The molecular mechanisms that underpin the oncogenic activity of STAT3/5 signaling include the regulation of genes that control cell cycle and cell death. However, recent advances also highlight the critical role of STAT3/5 target genes mediating inflammation and stemness. In addition, STAT3 mitochondrial functions are required for transformation. On the other hand, several tumor suppressor pathways act on or are activated by STAT3/5 signaling, including tyrosine phosphatases, the sumo ligase Protein Inhibitor of Activated STAT3 (PIAS3), the E3 ubiquitin ligase TATA Element Modulatory Factor/Androgen Receptor-Coactivator of 160 kDa (TMF/ARA160), the miRNAs miR-124 and miR-1181, the Protein of alternative reading frame 19 (p19ARF)/p53 pathway and the Suppressor of Cytokine Signaling 1 and 3 (SOCS1/3) proteins. Cancer mutations and epigenetic alterations may alter the balance between pro-oncogenic and tumor suppressor activities associated with STAT3/5 signaling, explaining their context-dependent association with tumor progression both in human cancers and animal models.

Keywords: solid cancers; cell cycle; apoptosis; inflammation; mitochondria; stemness; tumor suppression

---

1. Introduction

Activation of Signal Transducer and Activator of Transcription (STAT) proteins has been linked to many human cancers. STATs were initially discovered as latent cytosolic transcription factors that are phosphorylated by the Janus Kinase (JAK) family upon stimulation of membrane-associated cytokine and growth factor receptors. Phosphorylation triggers STAT dimerization and translocation to the nucleus to bind specific promoters and regulate transcription [1]. Here, we review the role of STAT family members STAT3 and STAT5 in solid human malignancies, as well as the mechanisms that may explain their association with either worse or better prognosis.

2. STAT3 and STAT5 in Solid Cancers

The discovery of cancer genes has been propelled by genetic analyses and more recently by next generation DNA sequencing technologies. Combined, these studies have identified 127 significantly mutated cancer genes that cover diverse signaling pathways [2]. Mutations acting as drivers in cancer are positively selected during tumor growth and constitute solid proof of the involvement of a particular gene as a driver in the disease. Mutations in STAT3 and STAT5 have been reported in patients with solid cancers, but unlike hyperactivation of the JAK/STAT pathway, STAT3/5 mutations in cancer are relatively infrequent and occur mostly in hematological malignancies.

An overview of reported STAT3/5 mutations in solid cancers is illustrated in Figure 1, based on data collected from the Catalogue of Somatic Mutations in Cancer (COSMIC) database. Mutations in *STAT3* are more prevalent than mutations in *STAT5A* or *STAT5B* genes. Noticeably, gastrointestinal cancers have the highest rates of STAT3/5 mutations compared with other solid cancers (Figure 1). Missense mutations tend to cluster within the SH2 domain, where gain-of-function mutations were previously characterized [3,4], as well as within the DNA binding domain and to an extent the N-terminal domain (Figure 1A). Interestingly, the *STAT3* Tyrosine 640 into Phenylalanine (Y640F) hotspot gain-of-function mutation reported in various lymphoid malignancies has also been detected in patients with liver cancer (Figure 1A). Nonsense and frameshift mutations are less frequent and more disperse, likely representing loss-of-function events (Figure 1B). Notably, a hotspot frameshift mutation at position Q368 within the DNA binding domain of STAT5B has been reported in 24 patients with various types of carcinoma; this frameshift generates a stop codon shortly after the mutation and is therefore likely to be loss-of-function, although characterization of this mutation has not been performed.

As opposed to mutation rates, STAT3/5 activation is very frequent in human cancers, perhaps reflecting increased cytokine signaling or mutations in cytokine receptors or negative regulators. STAT3/5 activation can be detected using antibodies that measure total levels or activation marks in STAT3/5 proteins (e.g. tyrosine phosphorylation). A better assessment of STAT3/5 activation can be obtained by measuring downstream signaling targets (i.e., mRNA levels of STAT3 [5] and STAT5 [6] target genes). A recent metanalysis of 63 different studies concluded that STAT3 protein overexpression was significantly associated with a worse 3-year overall survival (OS) (OR = 2.06, 95% CI = 1.57 to 2.71, $p < 0.00001$) and 5-year OS (OR = 2.00, 95% CI = 1.53 to 2.63, $p < 0.00001$) in patients with solid tumors [7]. Elevated STAT3 expression was associated with poor prognosis in gastric cancer, lung cancer, gliomas, hepatic cancer, osteosarcoma, prostate cancer and pancreatic cancer. However, high STAT3 protein expression levels predicted a better prognosis for breast cancer [7]. This study mixed data of both STAT3 and phospho-STAT3 (p-STAT3) expression limiting its ability to associate pathway activation to prognosis. Here, we summarize the data linking activation of STAT3/5 to overall survival in several major human solid cancers identifying the biomarkers used in each study (Table 1). Taken together, the results clearly show that *STAT3* and *STAT5* are important cancer genes despite their relatively low mutation frequency.

STAT3 activation is clearly a factor linked to bad prognosis in patients with lung cancer, liver cancer, renal cell carcinoma (RCC) and gliomas. In other tumors, the association is not significant. In solid tumors, STAT3 activation is more frequent than STAT5 activation although no explanation for this difference was proposed. In prostate cancer, both STAT3 and STAT5 have been associated with castration-resistant disease and proposed as therapeutic targets [8,9]. In colon cancer, the association between p-STAT3 and survival varies according to the study, but a high p-STAT3/p-STAT5 ratio indicates bad prognosis [10]. Also in breast cancer, p-STAT5 levels are clearly associated with better prognosis [11]. In liver cancer, STAT5 has ambivalent functions that were recently reviewed by Moriggl and colleagues [12]. Understanding mechanistically how STAT3/5 promote transformation and tumor suppression is important for the eventual design of new treatments. Also, survival data is highly influenced by the response of patients to their treatment and may not always reflect all mechanistic links between STAT3/5 activity and tumor biology. Of note, the effect of any gene is conditioned by the genetic context of gene action. Some genes can clearly exert a tumor suppressor effect in the initial stages of carcinogenesis that is lost when cancer mutations or epigenetic changes inactivate key effectors of these tumor suppressor pathways [13]. Human studies are usually limited to late stage tumors because it is easier to collect samples at that point. Studies in model systems, including primary cells, organoids and mouse models are thus required for a full understanding of how cancer genes work specifically at early stages in tumorigenesis.

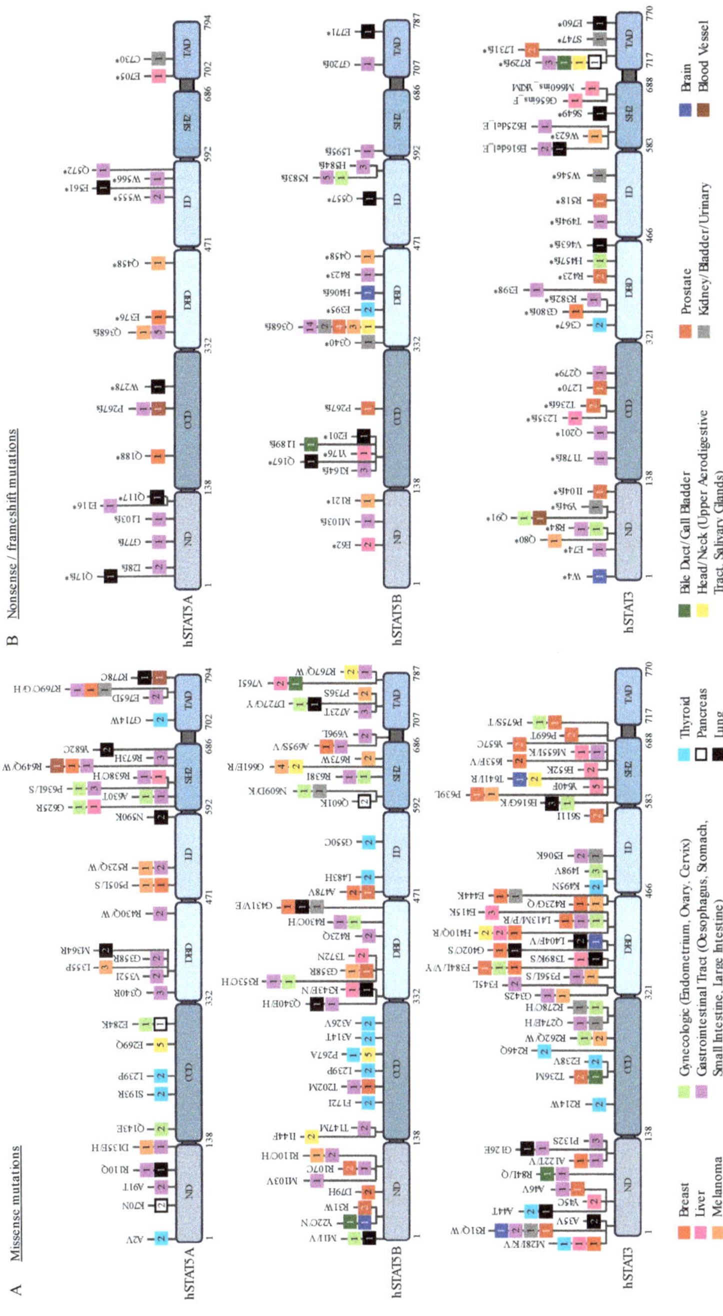

Figure 1. Map of somatic mutations detected in human Signal Transducer and Activator of Transcription (STAT)5A, STAT5B and STAT3 in patients with solid cancers. Individual missense mutations found in at least two patients (**A**), as well as all reported nonsense and frameshift mutations (**B**), are depicted. Numbers in each box represent the number of cases reported for each mutation. Data were mined from the Catalogue of Somatic Mutations In Cancer (COSMIC) database. ND, N-terminal domain; CCD, Coiled coil domain; DBD, DNA binding domain; LD, Linker domain; SH2, Src homology 2 domain; TAD, Transactivation domain.

Table 1. STAT3/5 activity and overall survival in major human solid tumors.

Tumor Type	Biomarker/Type of Study	Overall Survival	Ref
NSCLC	High p-STAT3/Meta-analysis of 9 studies	Log HR 0.67, 95% CI: 0.57–0.77, $p < 0.0001$	[14]
NSCLC	High p-STAT3/Cox regression multivariate analysis	HR 2.45, 95% CI: 1.084–5.556, $p = 0.031$	[15]
Lung cancer	High p-STAT3/Meta-analysis of 13 studies	HR 1.23, 95% CI: 1.04–1.46, $p = 0.02$	[16]
Pancreatic cancer	High p-STAT3/Log-rank test	No association $p > 0.05$	[17]
Liver cancer (HCC)	High p-STAT3/Meta-analysis of 8 studies	HR 1.69, 95% CI: 1.07–2.31, $p < 0.0001$ 3yr HR 1.67, 95% CI: 1.18–2.15, $p < 0.0001$ 5yr	[18]
Breast cancer	High p-STAT3/Meta-analysis of 12 studies	No association $p > 0.05$	[19]
Breast cancer (ER+)	High p-STAT3/Log-rank test	No association $p > 0.05$	[20]
GBM	High p-S727-STAT3/Cox regression multivariate analysis	HR 1.797, 95% CI: 1.028–3.142, $p = 0.040$	[21]
RCC	High p-S727-STAT3/Cox regression multivariate analysis	HR 3.32, 95% CI: 1.26–8.71, $p = 0.014$ 10yr	[22]
Colon cancer	High p-STAT3/p-STAT5 ratio/Cox regression multivariate analysis	HR 4.468, $p = 0.043$ 5yr	[10]
Colon cancer	High p-STAT3/Log-rank test	Worse overall survival, $p < 0.001$	[23]
Colon cancer	High p-STAT3/Cox regression multivariate analysis	HR 1.61, 95% CI: 1.11–2.34, $p = 0.015$	[24]
Breast cancer	Low p-STAT5/Cox regression multivariate analysis	HR 2.49, 95% CI: 1.23–5.05, $p = 0.012$ 5yr	[11]
Prostate cancer	High nuclear STAT5A/B/Cox regression multivariate analysis	HR 1.59, 95% CI: 1.04–2.44, $p = 0.034$	[9]

ER+, estrogen receptor-positive; HCC, hepatocellular carcinoma; GBM, glioblastoma multiforme; NSCLC, non-small-cell lung carcinoma; HR, hazard ratio; RCC, renal cell carcinoma.

3. Mechanisms of Transformation by STAT3/5 Proteins in Solid Cancers

STAT3 and STAT5 promote tumor progression by regulating the expression of cell cycle, survival and pro-inflammatory genes. In addition, they control mitochondrial functions, metabolism and stemness, as discussed below (Figure 2).

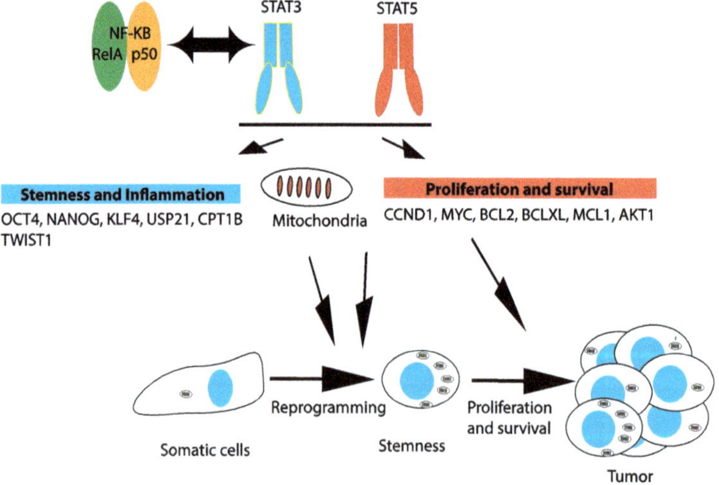

Figure 2. Mechanisms of tumorigenic activity of STAT3 and STAT5 signaling in solid tumors.

3.1. Cell Cycle and Apoptosis

As transcription factors, STAT3 and STAT5 regulate many genes required for cell cycle progression and cell survival. A major target of the transcriptional control of the mammalian cell cycle is cyclin

D. STAT3 regulates cyclin D expression in a complex with CD44 and the acetyltransferase p300. The latter acetylates STAT3 promoting its dimerization, nuclear translocation and binding to the cyclin D promoter [25]. Other cell cycle and survival genes regulated by STAT3 include *c-MYC (myc proto-oncogene)*, B-cell lymphoma 2 *(BCL2)*, *BCL2L1*/BCL-XL (B-cell lymphoma-extra large), *MCL1* (Myeloid Cell Leukemia Sequence 1) and *BIRC5*/survivin [26]. Recent studies combined ChIPSeq with whole transcriptome profiling in ABC DLBCL (activated B cell-like diffuse large B cell lymphoma) cell lines and revealed that STAT3 activates genes in the Phosphoinositide 3-Kinase (PI3K)/AKT/Mammalian Target of Rapamycin (mTOR) pathway, the Nuclear Factor Kappa-Light-Chain Enhancer of Activated B-Cells (NF-κB) pathway and the cell cycle regulation pathway, while repressing type I interferon signaling genes [27]. STAT5 also regulates the expression of cell cycle and cell survival genes [13] including *AKT1* [28], which encodes a pro-survival kinase.

3.2. Inflammation and Innate Immunity

Although the induction of cell proliferation and cell survival genes by STAT3/5 proteins contribute to their pro-cancer activity, in basal-like breast cancers the major genes associated with STAT3 activation control inflammation and the immune response [29]. Of note, inflammation is initially an adaptive response to pathological insults such as oncogenic stimuli, and it therefore exerts a tumor suppressive function. However, dysregulated inflammation in the long term provides a substrate for tumorigenesis [30]. STAT3 alone or in cooperation with NF-κB regulates the expression of many pro-inflammatory genes [31–33]. Starved tumor cells activate NF-κB and STAT3 via endoplasmic reticulum (ER) stress and secrete cytokines that stimulate tumor survival and clonogenic capacity [34]. The coactivation of these two transcription factors amplifies pro-inflammatory gene expression driving cancer-associated inflammation [35]. Of interest, the STAT3-NF-κB complex can repress the expression of DNA Damage Inducible Transcript 3 (DDIT3), an inhibitor of CCAAT Enhancer Binding Protein Beta (CEBPβ), another pro-inflammatory transcription factor [36].

Pharmacological agents that limit inflammation have been proposed for cancer prevention [37]. The use of metformin, a drug widely used to control diabetes, has been associated with a dramatic reduction in cancer incidence in many tissues [38]. Although the primary site of action of this drug is in mitochondria, a consequence of its effects is a potent reduction in the activation of NF-κB and STAT3, suggesting that the promising anticancer actions of metformin are related to its ability to curtail pro-inflammatory gene expression [39,40]. In contrast to STAT3, STAT5B inhibits NF-κB activity in the kidney fibroblast cell line COS by competing with coactivators of transcription [41], while it stimulates NF-κB in leukemia cells [42]. These results suggest the involvement of different regulatory mechanisms of STAT5 in hematopoietic cancers compared with solid cancers.

3.3. Mitochondria

In addition to their canonical roles in inflammation and immunity, STAT3 and STAT5 have been shown to localize to mitochondria. The mitochondrial localization of STAT3 is required for its ability to support malignant transformation in murine embryonic fibroblasts and breast cancer cells [43–46], and mito-STAT3 regulates mitochondrial metabolism and mitochondrial gene expression [45,47–51]. Several reports have suggested that STAT3 can be imported to mitochondria after phosphorylation on S727 [44,45] or upon acetylation [52,53]. Other studies have revealed that STAT3 mitochondrial translocation is mediated by interactions with Heat Shock Protein 22 (HSP22), Gene Associated with Retinoic and Interferon-Induced Mortality 19 (GRIM-19) or Translocase of Outer Mitochondrial Membrane 20 (TOM20) [54–56]. The mRNAs coding for some mitochondrial proteins are translated close to or in physical interaction with the import complex TOM [57,58]. The structural motifs mediating those interactions are located in the 3′ and 5′ UTRs of the mRNAs [59,60] and it will be interesting to investigate whether the mRNA of STAT3 also possesses RNA localization signals (zip codes) to localize in close proximity to mitochondria.

Whereas the role of mitochondrial STAT3 has been extensively studied, the role of STAT5 in mitochondria is less clear. The import of STAT5 to mitochondria is regulated by cytokines [43]. Once imported into the mitochondria, STAT5 binds the D-loop of mitochondrial DNA, although no increase in transcription of mitochondrial genes was observed [61]. Mito-STAT5 is also able to interact with the Pyruvate Dehydrogenase Complex (PDC) and was shown to regulate metabolism towards glycolysis, as observed in cells treated with cytokines [43,61]. In the same line, STAT3 was also shown to interact with the PDC in mitochondria [53].

3.4. Reprogramming and Stemness

The role of STAT3 in stem cell biology was initially recognized due to the requirement for the cytokine LIF to maintain pluripotency in cultures of mouse embryonic stem (ES) cells. STAT3 activation mediates the induction or repression of several genes in mouse ES cells including the pluripotency factors *Oct4*, *Klf4*, *Tfcp2l1* and polycomb proteins [62–64]. Many pluripotency factors, such as Homeobox Protein NANOG, are short-lived proteins. STAT3 controls protein stability by inducing the expression of the deubiquitinase Ubiquitin Specific Peptidase 21 (USP21), stabilizing NANOG in mouse ES cells. Induction of ES cell differentiation promotes the Extracellular Signal-Regulated Kinase (ERK)-dependent phosphorylation of USP21 and its dissociation from NANOG, leading to NANOG degradation [65]. STAT3 also plays a role in the reprogramming of somatic cells into induced pluripotent stem (IPS) cells [66] and it has been suggested that its effects depend on the demethylation of pluripotency factor promoters [67]. STAT3 also activates mitochondrial DNA transcription, promoting oxidative phosphorylation during maintenance and induction of pluripotency [68]. It is thus likely that the ability of STAT3 to stimulate stemness also plays a role in its oncogenic activity.

In many tumors, a subpopulation of cells possess a higher malignant capacity. These so-called tumor-initiating cells are suspected to regenerate the tumor after cancer chemotherapy and express many genes commonly expressed in ES cells [69]. It has been shown that STAT3 is required for the formation of tumor spheres and the viability of the cancer stem cell pool in many different tumors [39,40,70–83]. At least in breast cancer, a critical mechanism stimulated by STAT3 to regulate stemness involves genes in fatty acid oxidation [78,79] and the ability of STAT3 to adjust the levels of reactive oxygen species (ROS) produced in mitochondria [79]. In colorectal cancer cells, STAT3 forms a complex with the stem cell marker CD44 and the p300 acetyltransferase. Acetylation of STAT3 by this complex allows dimerization, nuclear translocation and binding to the promoters of genes required for stemness such as *c-MYC* and *TWIST1* [84].

The role of STAT5 in promoting cancer stemness does not affect many cell types and is mostly confined to hematopoietic cancers [85]. However, Nevalainen and colleagues reported that STAT5B induces stem cell properties in prostate cancer cells [86] in line with the increase in nuclear STAT5A/B observed in these tumors in correlation with bad prognosis [9]. Furthermore, transgenic mice with increased expression of prolactin in prostate epithelial cells displayed increases in the basal/stem cell compartment in association with activation of STAT5. This enrichment of stem cells was partially reversed by depletion of *Stat5a/b* [87]. The pro-stem cell oncogenic effect of STAT5 in the prostate contrasts with its effects in the mammary gland where STAT5 induces cell differentiation [88]. The ETS transcription factor Elf5 (E74-like factor 5) is a target of the prolactin-STAT5 axis and promotes mammary cell differentiation [89–91], supporting the tumor suppressive role of STAT5 in the mammary gland.

4. Tumor Suppressor Functions and Negative Regulation of STAT3/5 Signaling

The oncogenic activity of JAK/STAT signaling is controlled by several molecular barriers that limit the activation of this pathway. They include tyrosine phosphatases, E3 SUMO ligases of the Protein Inhibitor of Activated STAT3 (PIAS) family, E3 ubiquitin ligases and miRNAs. In addition, oncogenic STAT3/5 signaling can activate fail-safe tumor suppressors such as protein of alternative reading frame 19 (p19ARF), Suppressor of Cytokine Signaling 1 (SOCS1) and p53 that trigger apoptosis, ferroptosis and/or senescence in potentially malignant cells (Figure 3). Understanding these different responses to

STAT signaling in cancer is important to further distinguish tumors that would benefit from STAT3 or STAT5 inhibitors and those that would not.

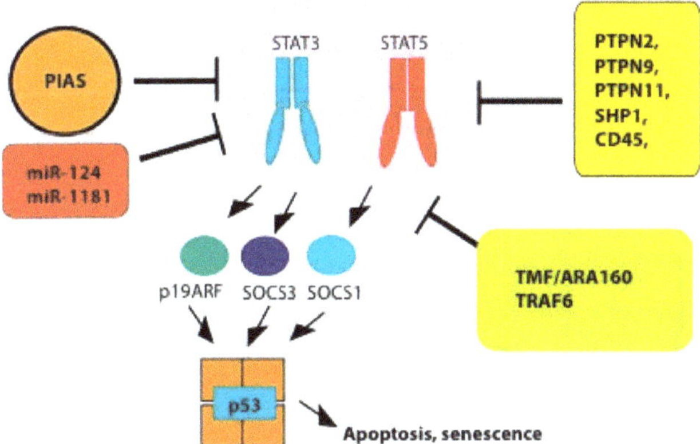

Figure 3. Tumor suppressor pathways acting on STAT3/5 activity (Protein Inhibitor of Activated STAT3 PIAS, miRNAs, E3 ligases, phosphatases) or activated by STAT3/5 transcriptional activity (Protein of alternative reading frame 19 (p19ARF) Suppressor of Cytokine Signaling 1 and 3 (SOCS1/3), p53). Abbreviations: (PTPN2 (Tyrosine-protein phosphatase non-receptor type 2), PTPN9/MEG2 (Tyrosine-protein phosphatase non-receptor type 9), PTPN11/SHP2 (Tyrosine-protein phosphatase non-receptor type 11), PTPN6/SHP1 (Tyrosine-protein phosphatase non-receptor type 6) and TNF receptor associated factor 6 (TRAF6)).

4.1. Tyrosine Phosphatases

Activation of STAT3 and STAT5 in tumors is often associated with tyrosine phosphorylation, a modification that can be reverted by several protein tyrosine phosphatases such as PTPN2 (Tyrosine-protein phosphatase non-receptor type 2), PTPN9/MEG2 (Tyrosine-protein phosphatase non-receptor type 9), PTPN11/SHP2 (Tyrosine-protein phosphatase non-receptor type 11) [92,93], CD45 [94] and PTPN6/SHP1 (Tyrosine-protein phosphatase non-receptor type 6) [95]. However, little is known about a possible role of these phosphatases in STAT3 activation in solid tumors. In liver cancers, SHP1 is downregulated in cells with mesenchymal features, and restoring its levels both reduced STAT3 phosphorylation and reversed the mesenchymal phenotype of liver cancer cells [95]. SHP1 and SHP2 also target STAT5 [96,97] but the significance of this regulation in solid tumors remains to be investigated.

4.2. PIAS

The Protein Inhibitor of Activated STAT3 (PIAS3) inhibits STAT3 transcriptional activity. In gliomas, PIAS3 expression is reduced [98]. Mechanistically, SMAD6 promotes PIAS3 degradation, promoting glioma cell growth and stem cell properties [76]. The PIAS proteins have SUMO E3 ligase activity acting on multiple proteins, and so their effects cannot be solely attributed to STAT3 inhibition [99]. Of interest, PIAS3 can bind NF-κB promoting its SUMOylation and inhibiting its activity [100,101], potentially targeting the expression of many pro-inflammatory genes required for tumor progression. Also, PIAS3 binds the N-terminus of p53 and prevents the interaction with its negative regulator MDM2, leading to p53 stabilization [102].

4.3. E3 Ligases

The Golgi resident and BC-box protein TATA Element Modulatory Factor/Androgen Receptor-Coactivator of 160 kDa (TMF/ARA160) was reported as an E3 ligase that catalyzes STAT3 ubiquitination leading to its proteasome-dependent degradation in myogenic C2C12 cells. The level of TMF/ARA160 was found to be significantly decreased in glioblastoma multiforme tumors, in benign meningioma and in malignant anaplastic meningioma, where STAT3 is known to play an oncogenic role [103]. TMF/ARA160 can also bind and ubiquitinate RELA/NF-κB leading to its proteasome-dependent degradation and a decrease in the expression of inflammatory genes [104]. Furthermore, the ubiquitin ligase TNF receptor associated factor 6 (TRAF6) binds and ubiquitinates STAT3 inhibiting the expression of STAT3 target genes [105]. During oncogene-induced senescence, STAT3 is degraded by the proteasome but the E3 ligase responsible has not been identified [106]. Recent results revealed that the long non-coding RNA (lncRNA) PVT1 (long non-coding RNA encoded by the human *PVT1* gene) binds STAT3 and protects it from ubiquitin-dependent degradation in gastric cancer [107]. PVT1 is upregulated in multiple cancers predicting poor prognosis for overall survival [108–110].

4.4. MiRNAs

The miRNA miR-124 regulates STAT3 signaling by targeting the mRNAs of interleukin-6 receptor (IL6R) [111] and STAT3 [112,113]. Suppression of this miRNA increases STAT3 phosphorylation and induces transformation in immortalized mouse hepatocytes. Of interest, systemic delivery of miR-124 prevented tumor growth in diethylnitrosamine (DEN)-treated mice, and miR-124 levels were found to be reduced in human hepatocellular carcinomas (HCC) [111]. In gliomas, miR-124 is poorly expressed but upregulation of its expression in glioma cancer stem cells inhibited the STAT3 pathway. In this model, STAT3 mediates immunosuppression, which was relieved upon systemic miR-124 delivery [114]. The circular RNA (circRNA) 100782 is upregulated in pancreatic cancer and its knockdown downregulates all miR-124 targets including IL6R and STAT3. This circRNA binds miR-124 suggesting that it may act as a miRNA sponge [115]. Furthermore, the miRNA miR-1181 also targets STAT3 and is downregulated in pancreatic cancer, predicting poorer overall survival. Overexpression of miR-1181 inhibited tumor formation and stem cell properties of pancreatic cancer cells [116].

4.5. The Suppressor of Cytokine Signaling SOCS

The members of the Suppressor of Cytokine Signaling (SOCS) family are major negative feedback regulators of JAK/STAT signaling and their expression is dysregulated in many human cancers [117–119]. These genes provide a barrier for cells with aberrant cytokine activation by inhibiting cytokine signaling [120]. In STAT3 driven cancers, SOCS3 seems to be the most important negative feedback regulator and mouse models of SOCS3 ablation show strong STAT3 activation [119,121–124]. On the other hand, in solid cancers where STAT5 plays a causal role such as liver and prostate cancer, in addition to SOCS3, SOCS1 is frequently inactivated and mouse models of SOCS1 ablation increase both liver and prostate tumorigenesis [125–132]. In addition to their role as JAK/STAT signaling barriers, SOCS1 and SOCS3 can bind p53 and activate tumor suppressor responses such as senescence and ferroptosis when their expression is induced by aberrant STAT5 signaling in primary cells [133–138]. In this way, SOCS1 and SOCS3 also act as fail-safe tumor suppressors in response to aberrant JAK/STAT signaling. So far, the SOCS1-p53-senescence axis has been demonstrated in primary fibroblasts and mammary epithelial cells [133,139–141]. This mechanism may explain the better prognosis of some solid cancers with high p-STAT5 [142–144] and the high frequency of SOCS1 inactivation in STAT5-driven cancers [125–132]. However, it is difficult to obtain evidence of a senescence tumor-suppression response by studying established tumors that have already circumvented this pathway. Senescence is particularly noticeable in premalignant lesions and benign tumors [40,106,145–150], and can be reactivated by cancer chemotherapy [151,152]. For this reason, evidence of STAT5-induced senescence in human cancers is not yet available and should be studied in samples from premalignant tumors or after chemotherapy.

The mechanisms that disable SOCS1 and SOCS3 in human cancers are often epigenetic, mediated either by miRNAs, promoter methylation or protein phosphorylation [127,128,130,131,137,153–162]. The SRC family of kinases (SFK) phosphorylate SOCS1 at Y80, interfering with p53-SOCS1 interactions. SFK inhibitors can reverse this effect and could be used to restore the SOCS1-p53 axis in tumors where these two proteins remain intact [162]. It is also possible to consider treatments that re-express SOCS1/3 in tumors. Indeed, in liver cancer *SOCS3* gene expression can be re-established by drugs that activate the Farnesoid X receptor (FXR) [163,164]. Gene therapy strategies are also under development to re-express SOCS1 or SOCS3 in tumors [165–167].

4.6. P19ARF-p53 Pathway

One of the first reports demonstrating that STAT3 can act as a tumor suppressor was shown in glioblastoma multiforme (GBM) [168] where a combination of low Phosphatase and tensin homolog (PTEN) expression and loss of STAT3 in astrocytes increased their tumorigenicity. This observation is in contrast to papers cited above on the requirement for STAT3 to maintain tumor stem cells in GBM [73,75,169]. This could be explained if STAT3 acts early in tumorigenesis as a tumor suppressor but gains oncogenic functions in the context of the cancer genome and epigenome. An interesting mechanism for the tumor suppressor role of STAT3 was recently described in the prostate where STAT3 induces the expression of p19ARF [170]. The latter is a tumor suppressor that activates p53 and inhibits ribosome biogenesis inducing cellular senescence and apoptosis [171–174]. Loss of STAT3 disrupts this STAT3-ARF-p53 axis and permits tumor progression [175]. STAT3 and other STATs can also induce p21 leading to cell cycle arrest or cellular senescence [176,177]. Further evidence for STAT3 as a tumor suppressor has been reported in lung [178], colon [179,180], thyroid [181], liver [182,183], skin [184], neck [185], nasopharynx, rectum [186], salivary gland [187] and breast cancers [188] but the mechanisms remain to be investigated.

5. Concluding Remarks

Context-dependent activities of STAT3 and STAT5 in solid human cancers justify detailed molecular studies that will clarify the specific molecular mechanisms of action of these two cancer genes. The cancer genome and transcriptome are shaped and selected to favor cancer cell survival and proliferation. Although restoring mutated genes is technologically difficult, reprograming the transcriptome to restore tumor suppression may be feasible. Drugs acting on STAT3/5 and their regulators may restore the control of cell proliferation in cancer cells.

Funding: G.F. is supported by the CIBC chair for breast cancer research at the CRCHUM. This work was funded by a grant from the Canadian Institute of Health and Research (CIHR-MOP229774) to G.F. H.A.N. is supported by the Austrian Science Fund (FWF), under the frame of ERA PerMed (I 4218-B). H.A.N. is also generously supported by a private donation from Liechtenstein.

Acknowledgments: We thank V. Bourdeau for comments.

Conflicts of Interest: The authors declare no conflict of interest.

References

1. Bromberg, J.; Darnell, J.E., Jr. The role of STATs in transcriptional control and their impact on cellular function. *Oncogene* **2000**, *19*, 2468–2473. [CrossRef] [PubMed]
2. Kandoth, C.; McLellan, M.D.; Vandin, F.; Ye, K.; Niu, B.; Lu, C.; Xie, M.; Zhang, Q.; McMichael, J.F.; Wyczalkowski, M.A.; et al. Mutational landscape and significance across 12 major cancer types. *Nature* **2013**, *502*, 333–339. [CrossRef] [PubMed]
3. de Araujo, E.D.; Erdogan, F.; Neubauer, H.A.; Meneksedag-Erol, D.; Manaswiyoungkul, P.; Eram, M.S.; Seo, H.S.; Qadree, A.K.; Israelian, J.; Orlova, A.; et al. Structural and functional consequences of the STAT5B(N642H) driver mutation. *Nat. Commun.* **2019**, *10*, 2517. [CrossRef] [PubMed]

4. Koskela, H.L.; Eldfors, S.; Ellonen, P.; van Adrichem, A.J.; Kuusanmaki, H.; Andersson, E.I.; Lagstrom, S.; Clemente, M.J.; Olson, T.; Jalkanen, S.E.; et al. Somatic STAT3 mutations in large granular lymphocytic leukemia. *N. Engl. J. Med.* **2012**, *366*, 1905–1913. [CrossRef] [PubMed]

5. Carpenter, R.L.; Lo, H.W. STAT3 Target Genes Relevant to Human Cancers. *Cancers (Basel)* **2014**, *6*, 897–925. [CrossRef]

6. Basham, B.; Sathe, M.; Grein, J.; McClanahan, T.; D'Andrea, A.; Lees, E.; Rascle, A. In vivo identification of novel STAT5 target genes. *Nucleic Acids Res.* **2008**, *36*, 3802–3818. [CrossRef]

7. Wu, P.; Wu, D.; Zhao, L.; Huang, L.; Shen, G.; Huang, J.; Chai, Y. Prognostic role of STAT3 in solid tumors: A systematic review and meta-analysis. *Oncotarget* **2016**, *7*, 19863–19883. [CrossRef]

8. Mohanty, S.K.; Yagiz, K.; Pradhan, D.; Luthringer, D.J.; Amin, M.B.; Alkan, S.; Cinar, B. STAT3 and STAT5A are potential therapeutic targets in castration-resistant prostate cancer. *Oncotarget* **2017**, *8*, 85997–86010. [CrossRef]

9. Mirtti, T.; Leiby, B.E.; Abdulghani, J.; Aaltonen, E.; Pavela, M.; Mamtani, A.; Alanen, K.; Egevad, L.; Granfors, T.; Josefsson, A.; et al. Nuclear Stat5a/b predicts early recurrence and prostate cancer-specific death in patients treated by radical prostatectomy. *Hum. Pathol.* **2013**, *44*, 310–319. [CrossRef]

10. Klupp, F.; Diers, J.; Kahlert, C.; Neumann, L.; Halama, N.; Franz, C.; Schmidt, T.; Lasitschka, F.; Warth, A.; Weitz, J.; et al. Expressional STAT3/STAT5 Ratio is an Independent Prognostic Marker in Colon Carcinoma. *Ann. Surg. Oncol.* **2015**, *22*, S1548–S1555. [CrossRef]

11. Peck, A.R.; Witkiewicz, A.K.; Liu, C.; Stringer, G.A.; Klimowicz, A.C.; Pequignot, E.; Freydin, B.; Tran, T.H.; Yang, N.; Rosenberg, A.L.; et al. Loss of nuclear localized and tyrosine phosphorylated Stat5 in breast cancer predicts poor clinical outcome and increased risk of antiestrogen therapy failure. *J. Clin. Oncol.* **2011**, *29*, 2448–2458. [CrossRef] [PubMed]

12. Kaltenecker, D.; Themanns, M.; Mueller, K.M.; Spirk, K.; Suske, T.; Merkel, O.; Kenner, L.; Luis, A.; Kozlov, A.; Haybaeck, J.; et al. Hepatic growth hormone-JAK2-STAT5 signalling: Metabolic function, non-alcoholic fatty liver disease and hepatocellular carcinoma progression. *Cytokine* **2018**. [CrossRef] [PubMed]

13. Ferbeyre, G.; Moriggl, R. The role of Stat5 transcription factors as tumor suppressors or oncogenes. *Biochim. Biophys. Acta* **2011**, *1815*, 104–114. [CrossRef] [PubMed]

14. Xu, Y.H.; Lu, S. A meta-analysis of STAT3 and phospho-STAT3 expression and survival of patients with non-small-cell lung cancer. *Eur. J. Surg. Oncol.* **2014**, *40*, 311–317. [CrossRef] [PubMed]

15. Sun, Z.G.; Zhang, M.; Yang, F.; Gao, W.; Wang, Z.; Zhu, L.M. Clinical and prognostic significance of signal transducer and activator of transcription 3 and mucin 1 in patients with non-small cell lung cancer following surgery. *Oncol. Lett.* **2018**, *15*, 4278–4288. [CrossRef] [PubMed]

16. Tong, M.; Wang, J.; Jiang, N.; Pan, H.; Li, D. Correlation between p-STAT3 overexpression and prognosis in lung cancer: A systematic review and meta-analysis. *PLoS ONE* **2017**, *12*, e0182282. [CrossRef] [PubMed]

17. Koperek, O.; Aumayr, K.; Schindl, M.; Werba, G.; Soleiman, A.; Schoppmann, S.; Sahora, K.; Birner, P. Phosphorylation of STAT3 correlates with HER2 status, but not with survival in pancreatic ductal adenocarcinoma. *APMIS* **2014**, *122*, 476–481. [CrossRef] [PubMed]

18. Liang, C.; Xu, Y.; Ge, H.; Li, G.; Wu, J. Clinicopathological significance and prognostic role of p-STAT3 in patients with hepatocellular carcinoma. *Onco-Targets Ther.* **2018**, *11*, 1203–1214. [CrossRef]

19. Liu, Y.; Huang, J.; Li, W.; Chen, Y.; Liu, X.; Wang, J. Meta-analysis of STAT3 and phospho-STAT3 expression and survival of patients with breast cancer. *Oncotarget* **2018**, *9*, 13060–13067. [CrossRef] [PubMed]

20. Sonnenblick, A.; Salgado, R.; Brohee, S.; Zahavi, T.; Peretz, T.; Van den Eynden, G.; Rouas, G.; Salmon, A.; Francis, P.A.; Di Leo, A.; et al. p-STAT3 in luminal breast cancer: Integrated RNA-protein pooled analysis and results from the BIG 2-98 phase III trial. *Int. J. Oncol.* **2018**, *52*, 424–432. [CrossRef]

21. Lin, G.S.; Chen, Y.P.; Lin, Z.X.; Wang, X.F.; Zheng, Z.Q.; Chen, L. STAT3 serine 727 phosphorylation influences clinical outcome in glioblastoma. *Int. J. Clin. Exp. Pathol.* **2014**, *7*, 3141–3149.

22. Vilardell, J.; Alcaraz, E.; Sarro, E.; Trilla, E.; Cuadros, T.; de Torres, I.; Plana, M.; Ramon, Y.C.S.; Pinna, L.A.; Ruzzene, M.; et al. Under-expression of CK2beta subunit in ccRCC represents a complementary biomarker of p-STAT3 Ser727 that correlates with patient survival. *Oncotarget* **2018**, *9*, 5736–5751. [CrossRef] [PubMed]

23. Kusaba, T.; Nakayama, T.; Yamazumi, K.; Yakata, Y.; Yoshizaki, A.; Inoue, K.; Nagayasu, T.; Sekine, I. Activation of STAT3 is a marker of poor prognosis in human colorectal cancer. *Oncol. Rep.* **2006**, *15*, 1445–1451. [CrossRef]

24. Morikawa, T.; Baba, Y.; Yamauchi, M.; Kuchiba, A.; Nosho, K.; Shima, K.; Tanaka, N.; Huttenhower, C.; Frank, D.A.; Fuchs, C.S.; et al. STAT3 expression, molecular features, inflammation patterns, and prognosis in a database of 724 colorectal cancers. *Clin. Cancer Res.* **2011**, *17*, 1452–1462. [CrossRef]

25. Lee, J.L.; Wang, M.J.; Chen, J.Y. Acetylation and activation of STAT3 mediated by nuclear translocation of CD44. *J. Cell Biol.* **2009**, *185*, 949–957. [CrossRef] [PubMed]

26. Banerjee, K.; Resat, H. Constitutive activation of STAT3 in breast cancer cells: A review. *Int. J. Cancer* **2016**, *138*, 2570–2578. [CrossRef] [PubMed]

27. Lu, L.; Zhu, F.; Zhang, M.; Li, Y.; Drennan, A.C.; Kimpara, S.; Rumball, I.; Selzer, C.; Cameron, H.; Kellicut, A.; et al. Gene regulation and suppression of type I interferon signaling by STAT3 in diffuse large B cell lymphoma. *Proc. Natl. Acad. Sci. USA* **2018**, *115*, E498–E505. [CrossRef] [PubMed]

28. Creamer, B.A.; Sakamoto, K.; Schmidt, J.W.; Triplett, A.A.; Moriggl, R.; Wagner, K.U. Stat5 promotes survival of mammary epithelial cells through transcriptional activation of a distinct promoter in Akt1. *Mol. Cell. Biol.* **2010**, *30*, 2957–2970. [CrossRef]

29. Tell, R.W.; Horvath, C.M. Bioinformatic analysis reveals a pattern of STAT3-associated gene expression specific to basal-like breast cancers in human tumors. *Proc. Natl. Acad. Sci. USA* **2014**, *111*, 12787–12792. [CrossRef]

30. Karki, R.; Kanneganti, T.D. Diverging inflammasome signals in tumorigenesis and potential targeting. *Nat. Rev. Cancer* **2019**, *19*, 197–214. [CrossRef]

31. Martincuks, A.; Andryka, K.; Kuster, A.; Schmitz-Van de Leur, H.; Komorowski, M.; Muller-Newen, G. Nuclear translocation of STAT3 and NF-kappaB are independent of each other but NF-kappaB supports expression and activation of STAT3. *Cell. Signal.* **2017**, *32*, 36–47. [CrossRef] [PubMed]

32. Goldstein, I.; Paakinaho, V.; Baek, S.; Sung, M.H.; Hager, G.L. Synergistic gene expression during the acute phase response is characterized by transcription factor assisted loading. *Nat. Commun.* **2017**, *8*, 1849. [CrossRef] [PubMed]

33. Grivennikov, S.I.; Karin, M. Dangerous liaisons: STAT3 and NF-kappaB collaboration and crosstalk in cancer. *Cytokine Growth Factor Rev.* **2010**, *21*, 11–19. [CrossRef] [PubMed]

34. Yoon, S.; Woo, S.U.; Kang, J.H.; Kim, K.; Shin, H.J.; Gwak, H.S.; Park, S.; Chwae, Y.J. NF-kappaB and STAT3 cooperatively induce IL6 in starved cancer cells. *Oncogene* **2012**, *31*, 3467–3481. [CrossRef] [PubMed]

35. Atsumi, T.; Singh, R.; Sabharwal, L.; Bando, H.; Meng, J.; Arima, Y.; Yamada, M.; Harada, M.; Jiang, J.J.; Kamimura, D.; et al. Inflammation amplifier, a new paradigm in cancer biology. *Cancer Res.* **2014**, *74*, 8–14. [CrossRef] [PubMed]

36. Canino, C.; Luo, Y.; Marcato, P.; Blandino, G.; Pass, H.I.; Cioce, M. A STAT3-NFkB/DDIT3/CEBPbeta axis modulates ALDH1A3 expression in chemoresistant cell subpopulations. *Oncotarget* **2015**, *6*, 12637–12653. [CrossRef] [PubMed]

37. Thi, H.T.H.; Hong, S. Inflammasome as a Therapeutic Target for Cancer Prevention and Treatment. *J. Cancer Prev.* **2017**, *22*, 62–73. [CrossRef] [PubMed]

38. Pollak, M.N. Investigating metformin for cancer prevention and treatment: The end of the beginning. *Cancer Discov.* **2012**, *2*, 778–790. [CrossRef]

39. Hirsch, H.A.; Iliopoulos, D.; Struhl, K. Metformin inhibits the inflammatory response associated with cellular transformation and cancer stem cell growth. *Proc. Natl. Acad. Sci. USA* **2013**, *110*, 972–977. [CrossRef]

40. Deschenes-Simard, X.; Parisotto, M.; Rowell, M.C.; Le Calve, B.; Igelmann, S.; Moineau-Vallee, K.; Saint-Germain, E.; Kalegari, P.; Bourdeau, V.; Kottakis, F.; et al. Circumventing senescence is associated with stem cell properties and metformin sensitivity. *Aging Cell* **2019**. [CrossRef]

41. Luo, G.; Yu-Lee, L. Stat5b inhibits NFkappaB-mediated signaling. *Mol. Endocrinol.* **2000**, *14*, 114–123. [CrossRef] [PubMed]

42. Kawashima, T.; Murata, K.; Akira, S.; Tonozuka, Y.; Minoshima, Y.; Feng, S.; Kumagai, H.; Tsuruga, H.; Ikeda, Y.; Asano, S.; et al. STAT5 induces macrophage differentiation of M1 leukemia cells through activation of IL-6 production mediated by NF-kappaB p65. *J. Immunol.* **2001**, *167*, 3652–3660. [CrossRef] [PubMed]

43. Chueh, F.Y.; Leong, K.F.; Yu, C.L. Mitochondrial translocation of signal transducer and activator of transcription 5 (STAT5) in leukemic T cells and cytokine-stimulated cells. *Biochem. Biophys. Res. Commun.* **2010**, *402*, 778–783. [CrossRef] [PubMed]

44. Gough, D.J.; Corlett, A.; Schlessinger, K.; Wegrzyn, J.; Larner, A.C.; Levy, D.E. Mitochondrial STAT3 supports Ras-dependent oncogenic transformation. *Science* **2009**, *324*, 1713–1716. [CrossRef] [PubMed]

45. Wegrzyn, J.; Potla, R.; Chwae, Y.J.; Sepuri, N.B.; Zhang, Q.; Koeck, T.; Derecka, M.; Szczepanek, K.; Szelag, M.; Gornicka, A.; et al. Function of mitochondrial Stat3 in cellular respiration. *Science* **2009**, *323*, 793–797. [CrossRef] [PubMed]

46. Zhang, Q.; Raje, V.; Yakovlev, V.A.; Yacoub, A.; Szczepanek, K.; Meier, J.; Derecka, M.; Chen, Q.; Hu, Y.; Sisler, J.; et al. Mitochondrial localized Stat3 promotes breast cancer growth via phosphorylation of serine 727. *J. Biol. Chem.* **2013**, *288*, 31280–31288. [CrossRef] [PubMed]

47. Macias, E.; Rao, D.; Carbajal, S.; Kiguchi, K.; DiGiovanni, J. Stat3 binds to mtDNA and regulates mitochondrial gene expression in keratinocytes. *J. Investig. Dermatol.* **2014**, *134*, 1971–1980. [CrossRef]

48. Rincon, M.; Pereira, F.V. A New Perspective: Mitochondrial Stat3 as a Regulator for Lymphocyte Function. *Int. J. Mol. Sci.* **2018**, *19*, 1656. [CrossRef]

49. Avalle, L.; Poli, V. Nucleus, Mitochondrion, or Reticulum? STAT3 a La Carte. *Int. J. Mol. Sci.* **2018**, *19*, 2820. [CrossRef]

50. Luo, D.; Fraga-Lauhirat, M.; Millings, J.; Ho, C.; Villarreal, E.M.; Fletchinger, T.C.; Bonfiglio, J.V.; Mata, L.; Nemesure, M.D.; Bartels, L.E.; et al. Phospho-valproic acid (MDC-1112) suppresses glioblastoma growth in preclinical models through the inhibition of STAT3 phosphorylation. *Carcinogenesis* **2019**. [CrossRef]

51. Sala, D.; Cunningham, T.J.; Stec, M.J.; Etxaniz, U.; Nicoletti, C.; Dall'Agnese, A.; Puri, P.L.; Duester, G.; Latella, L.; Sacco, A. The Stat3-Fam3a axis promotes muscle stem cell myogenic lineage progression by inducing mitochondrial respiration. *Nat. Commun.* **2019**, *10*, 1796. [CrossRef]

52. Bernier, M.; Paul, R.K.; Martin-Montalvo, A.; Scheibye-Knudsen, M.; Song, S.; He, H.J.; Armour, S.M.; Hubbard, B.P.; Bohr, V.A.; Wang, L.; et al. Negative regulation of STAT3 protein-mediated cellular respiration by SIRT1 protein. *J. Biol. Chem.* **2011**, *286*, 19270–19279. [CrossRef]

53. Xu, Y.S.; Liang, J.J.; Wang, Y.; Zhao, X.J.; Xu, L.; Xu, Y.Y.; Zou, Q.C.; Zhang, J.M.; Tu, C.E.; Cui, Y.G.; et al. STAT3 Undergoes Acetylation-dependent Mitochondrial Translocation to Regulate Pyruvate Metabolism. *Sci. Rep.* **2016**, *6*, 39517. [CrossRef]

54. Boengler, K.; Hilfiker-Kleiner, D.; Heusch, G.; Schulz, R. Inhibition of permeability transition pore opening by mitochondrial STAT3 and its role in myocardial ischemia/reperfusion. *Basic Res. Cardiol.* **2010**, *105*, 771–785. [CrossRef] [PubMed]

55. Tammineni, P.; Anugula, C.; Mohammed, F.; Anjaneyulu, M.; Larner, A.C.; Sepuri, N.B. The import of the transcription factor STAT3 into mitochondria depends on GRIM-19, a component of the electron transport chain. *J. Biol. Chem.* **2013**, *288*, 4723–4732. [CrossRef]

56. Qiu, H.; Lizano, P.; Laure, L.; Sui, X.; Rashed, E.; Park, J.Y.; Hong, C.; Gao, S.; Holle, E.; Morin, D.; et al. H11 kinase/heat shock protein 22 deletion impairs both nuclear and mitochondrial functions of STAT3 and accelerates the transition into heart failure on cardiac overload. *Circulation* **2011**, *124*, 406–415. [CrossRef] [PubMed]

57. Gadir, N.; Haim-Vilmovsky, L.; Kraut-Cohen, J.; Gerst, J.E. Localization of mRNAs coding for mitochondrial proteins in the yeast Saccharomyces cerevisiae. *RNA* **2011**, *17*, 1551–1565. [CrossRef]

58. Fox, T.D. Mitochondrial protein synthesis, import, and assembly. *Genetics* **2012**, *192*, 1203–1234. [CrossRef]

59. Zhang, Y.; Chen, Y.; Gucek, M.; Xu, H. The mitochondrial outer membrane protein MDI promotes local protein synthesis and mtDNA replication. *EMBO J.* **2016**, *35*, 1045–1057. [CrossRef]

60. Hansen, K.G.; Herrmann, J.M. Transport of Proteins into Mitochondria. *Protein J.* **2019**. [CrossRef]

61. Richard, A.J.; Hang, H.; Stephens, J.M. Pyruvate dehydrogenase complex (PDC) subunits moonlight as interaction partners of phosphorylated STAT5 in adipocytes and adipose tissue. *J. Biol. Chem.* **2017**, *292*, 19733–19742. [CrossRef] [PubMed]

62. Kidder, B.L.; Yang, J.; Palmer, S. Stat3 and c-Myc genome-wide promoter occupancy in embryonic stem cells. *PLoS ONE* **2008**, *3*, e3932. [CrossRef] [PubMed]

63. Hall, J.; Guo, G.; Wray, J.; Eyres, I.; Nichols, J.; Grotewold, L.; Morfopoulou, S.; Humphreys, P.; Mansfield, W.; Walker, R.; et al. Oct4 and LIF/Stat3 additively induce Kruppel factors to sustain embryonic stem cell self-renewal. *Cell Stem Cell* **2009**, *5*, 597–609. [CrossRef] [PubMed]

64. Ye, S.; Li, P.; Tong, C.; Ying, Q.L. Embryonic stem cell self-renewal pathways converge on the transcription factor Tfcp2l1. *EMBO J.* **2013**, *32*, 2548–2560. [CrossRef] [PubMed]

65. Jin, J.; Liu, J.; Chen, C.; Liu, Z.; Jiang, C.; Chu, H.; Pan, W.; Wang, X.; Zhang, L.; Li, B.; et al. The deubiquitinase USP21 maintains the stemness of mouse embryonic stem cells via stabilization of Nanog. *Nat. Commun.* **2016**, *7*, 13594. [CrossRef] [PubMed]

66. Yang, J.; van Oosten, A.L.; Theunissen, T.W.; Guo, G.; Silva, J.C.; Smith, A. Stat3 activation is limiting for reprogramming to ground state pluripotency. *Cell Stem Cell* **2010**, *7*, 319–328. [CrossRef]

67. Tang, Y.; Luo, Y.; Jiang, Z.; Ma, Y.; Lin, C.J.; Kim, C.; Carter, M.G.; Amano, T.; Park, J.; Kish, S.; et al. Jak/Stat3 signaling promotes somatic cell reprogramming by epigenetic regulation. *Stem Cells* **2012**, *30*, 2645–2656. [CrossRef]

68. Carbognin, E.; Betto, R.M.; Soriano, M.E.; Smith, A.G.; Martello, G. Stat3 promotes mitochondrial transcription and oxidative respiration during maintenance and induction of naive pluripotency. *EMBO J.* **2016**, *35*, 618–634. [CrossRef]

69. Ben-Porath, I.; Thomson, M.W.; Carey, V.J.; Ge, R.; Bell, G.W.; Regev, A.; Weinberg, R.A. An embryonic stem cell-like gene expression signature in poorly differentiated aggressive human tumors. *Nat. Genet.* **2008**, *40*, 499–507. [CrossRef]

70. Zhou, J.; Wulfkuhle, J.; Zhang, H.; Gu, P.; Yang, Y.; Deng, J.; Margolick, J.B.; Liotta, L.A.; Petricoin, E., 3rd; Zhang, Y. Activation of the PTEN/mTOR/STAT3 pathway in breast cancer stem-like cells is required for viability and maintenance. *Proc. Natl. Acad. Sci. USA* **2007**, *104*, 16158–16163. [CrossRef]

71. Marotta, L.L.; Almendro, V.; Marusyk, A.; Shipitsin, M.; Schemme, J.; Walker, S.R.; Bloushtain-Qimron, N.; Kim, J.J.; Choudhury, S.A.; Maruyama, R.; et al. The JAK2/STAT3 signaling pathway is required for growth of CD44(+)CD24(-) stem cell-like breast cancer cells in human tumors. *J. Clin. Investig.* **2011**, *121*, 2723–2735. [CrossRef] [PubMed]

72. Iliopoulos, D.; Hirsch, H.A.; Struhl, K. An epigenetic switch involving NF-kappaB, Lin28, Let-7 MicroRNA, and IL6 links inflammation to cell transformation. *Cell* **2009**, *139*, 693–706. [CrossRef] [PubMed]

73. Gong, A.H.; Wei, P.; Zhang, S.; Yao, J.; Yuan, Y.; Zhou, A.D.; Lang, F.F.; Heimberger, A.B.; Rao, G.; Huang, S. FoxM1 Drives a Feed-Forward STAT3-Activation Signaling Loop That Promotes the Self-Renewal and Tumorigenicity of Glioblastoma Stem-like Cells. *Cancer Res.* **2015**, *75*, 2337–2348. [CrossRef] [PubMed]

74. Peng, L.; Jiang, D. Resveratrol eliminates cancer stem cells of osteosarcoma by STAT3 pathway inhibition. *PLoS ONE* **2018**, *13*, e0205918. [CrossRef] [PubMed]

75. Man, J.; Yu, X.; Huang, H.; Zhou, W.; Xiang, C.; Huang, H.; Miele, L.; Liu, Z.; Bebek, G.; Bao, S.; et al. Hypoxic Induction of Vasorin Regulates Notch1 Turnover to Maintain Glioma Stem-like Cells. *Cell Stem Cell* **2018**, *22*, 104–118.e6. [CrossRef] [PubMed]

76. Jiao, J.; Zhang, R.; Li, Z.; Yin, Y.; Fang, X.; Ding, X.; Cai, Y.; Yang, S.; Mu, H.; Zong, D.; et al. Nuclear Smad6 promotes gliomagenesis by negatively regulating PIAS3-mediated STAT3 inhibition. *Nat. Commun.* **2018**, *9*, 2504. [CrossRef] [PubMed]

77. He, W.; Wu, J.; Shi, J.; Huo, Y.M.; Dai, W.; Geng, J.; Lu, P.; Yang, M.W.; Fang, Y.; Wang, W.; et al. IL22RA1/STAT3 Signaling Promotes Stemness and Tumorigenicity in Pancreatic Cancer. *Cancer Res.* **2018**, *78*, 3293–3305. [CrossRef]

78. Wang, T.; Fahrmann, J.F.; Lee, H.; Li, Y.J.; Tripathi, S.C.; Yue, C.; Zhang, C.; Lifshitz, V.; Song, J.; Yuan, Y.; et al. JAK/STAT3-Regulated Fatty Acid beta-Oxidation Is Critical for Breast Cancer Stem Cell Self-Renewal and Chemoresistance. *Cell Metab.* **2018**, *27*, 136–150. [CrossRef]

79. Kitajima, S.; Yoshida, A.; Kohno, S.; Li, F.; Suzuki, S.; Nagatani, N.; Nishimoto, Y.; Sasaki, N.; Muranaka, H.; Wan, Y.; et al. The RB-IL-6 axis controls self-renewal and endocrine therapy resistance by fine-tuning mitochondrial activity. *Oncogene* **2017**, *36*, 5145–5157. [CrossRef]

80. Chen, M.W.; Yang, S.T.; Chien, M.H.; Hua, K.T.; Wu, C.J.; Hsiao, S.M.; Lin, H.; Hsiao, M.; Su, J.L.; Wei, L.H. The STAT3-miRNA-92-Wnt Signaling Pathway Regulates Spheroid Formation and Malignant Progression in Ovarian Cancer. *Cancer Res.* **2017**, *77*, 1955–1967. [CrossRef]

81. Schroeder, A.; Herrmann, A.; Cherryholmes, G.; Kowolik, C.; Buettner, R.; Pal, S.; Yu, H.; Muller-Newen, G.; Jove, R. Loss of androgen receptor expression promotes a stem-like cell phenotype in prostate cancer through STAT3 signaling. *Cancer Res.* **2014**, *74*, 1227–1237. [CrossRef] [PubMed]

82. Tang, Y.; Kitisin, K.; Jogunoori, W.; Li, C.; Deng, C.X.; Mueller, S.C.; Ressom, H.W.; Rashid, A.; He, A.R.; Mendelson, J.S.; et al. Progenitor/stem cells give rise to liver cancer due to aberrant TGF-beta and IL-6 signaling. *Proc. Natl. Acad. Sci. USA* **2008**, *105*, 2445–2450. [CrossRef] [PubMed]

83. Thiagarajan, P.S.; Zheng, Q.; Bhagrath, M.; Mulkearns-Hubert, E.E.; Myers, M.G.; Lathia, J.D.; Reizes, O. STAT3 activation by leptin receptor is essential for TNBC stem cell maintenance. *Endocr. Relat. Cancer* **2017**, *24*, 415–426. [CrossRef] [PubMed]

84. Su, Y.J.; Lai, H.M.; Chang, Y.W.; Chen, G.Y.; Lee, J.L. Direct reprogramming of stem cell properties in colon cancer cells by CD44. *EMBO J.* **2011**, *30*, 3186–3199. [CrossRef] [PubMed]

85. Schepers, H.; van Gosliga, D.; Wierenga, A.T.; Eggen, B.J.; Schuringa, J.J.; Vellenga, E. STAT5 is required for long-term maintenance of normal and leukemic human stem/progenitor cells. *Blood* **2007**, *110*, 2880–2888. [CrossRef] [PubMed]

86. Talati, P.G.; Gu, L.; Ellsworth, E.M.; Girondo, M.A.; Trerotola, M.; Hoang, D.T.; Leiby, B.; Dagvadorj, A.; McCue, P.A.; Lallas, C.D.; et al. Jak2-Stat5a/b Signaling Induces Epithelial-to-Mesenchymal Transition and Stem-Like Cell Properties in Prostate Cancer. *Am. J. Pathol.* **2015**, *185*, 2505–2522. [CrossRef] [PubMed]

87. Boutillon, F.; Pigat, N.; Sala, L.S.; Reyes-Gomez, E.; Moriggl, R.; Guidotti, J.E.; Goffin, V. STAT5a/b Deficiency Delays, but does not Prevent, Prolactin-Driven Prostate Tumorigenesis in Mice. *Cancers (Basel)* **2019**, *11*, 929. [CrossRef] [PubMed]

88. Liu, X.; Robinson, G.W.; Wagner, K.U.; Garrett, L.; Wynshaw-Boris, A.; Hennighausen, L. Stat5a is mandatory for adult mammary gland development and lactogenesis. *Genes Dev.* **1997**, *11*, 179–186. [CrossRef] [PubMed]

89. Oakes, S.R.; Naylor, M.J.; Asselin-Labat, M.L.; Blazek, K.D.; Gardiner-Garden, M.; Hilton, H.N.; Kazlauskas, M.; Pritchard, M.A.; Chodosh, L.A.; Pfeffer, P.L.; et al. The ETS transcription factor Elf5 specifies mammary alveolar cell fate. *Genes Dev.* **2008**, *22*, 581–586. [CrossRef] [PubMed]

90. Metser, G.; Shin, H.Y.; Wang, C.; Yoo, K.H.; Oh, S.; Villarino, A.V.; O'Shea, J.J.; Kang, K.; Hennighausen, L. An autoregulatory enhancer controls mammary-specific STAT5 functions. *Nucleic Acids Res.* **2016**, *44*, 1052–1063. [CrossRef] [PubMed]

91. Stute, P.; Sielker, S.; Wood, C.E.; Register, T.C.; Lees, C.J.; Dewi, F.N.; Williams, J.K.; Wagner, J.D.; Stefenelli, U.; Cline, J.M. Life stage differences in mammary gland gene expression profile in non-human primates. *Breast Cancer Res. Treat.* **2012**, *133*, 617–634. [CrossRef] [PubMed]

92. Kim, M.; Morales, L.D.; Jang, I.S.; Cho, Y.Y.; Kim, D.J. Protein Tyrosine Phosphatases as Potential Regulators of STAT3 Signaling. *Int. J. Mol. Sci.* **2018**, *19*, 2708. [CrossRef] [PubMed]

93. Huang, Y.; Wang, J.; Cao, F.; Jiang, H.; Li, A.; Li, J.; Qiu, L.; Shen, H.; Chang, W.; Zhou, C.; et al. SHP2 associates with nuclear localization of STAT3: Significance in progression and prognosis of colorectal cancer. *Sci. Rep.* **2017**, *7*, 17597. [CrossRef] [PubMed]

94. Kumar, V.; Cheng, P.; Condamine, T.; Mony, S.; Languino, L.R.; McCaffrey, J.C.; Hockstein, N.; Guarino, M.; Masters, G.; Penman, E.; et al. CD45 Phosphatase Inhibits STAT3 Transcription Factor Activity in Myeloid Cells and Promotes Tumor-Associated Macrophage Differentiation. *Immunity* **2016**, *44*, 303–315. [CrossRef] [PubMed]

95. Fan, L.C.; Shiau, C.W.; Tai, W.T.; Hung, M.H.; Chu, P.Y.; Hsieh, F.S.; Lin, H.; Yu, H.C.; Chen, K.F. SHP-1 is a negative regulator of epithelial-mesenchymal transition in hepatocellular carcinoma. *Oncogene* **2015**, *34*, 5252–5263. [CrossRef] [PubMed]

96. Chen, Y.; Wen, R.; Yang, S.; Schuman, J.; Zhang, E.E.; Yi, T.; Feng, G.S.; Wang, D. Identification of Shp-2 as a Stat5A phosphatase. *J. Biol. Chem.* **2003**, *278*, 16520–16527. [CrossRef]

97. Xiao, W.; Ando, T.; Wang, H.Y.; Kawakami, Y.; Kawakami, T. Lyn- and PLC-beta3-dependent regulation of SHP-1 phosphorylation controls Stat5 activity and myelomonocytic leukemia-like disease. *Blood* **2010**, *116*, 6003–6013. [CrossRef]

98. Brantley, E.C.; Nabors, L.B.; Gillespie, G.Y.; Choi, Y.H.; Palmer, C.A.; Harrison, K.; Roarty, K.; Benveniste, E.N. Loss of protein inhibitors of activated STAT-3 expression in glioblastoma multiforme tumors: Implications for STAT-3 activation and gene expression. *Clin. Cancer Res.* **2008**, *14*, 4694–4704. [CrossRef] [PubMed]

99. Sundvall, M.; Korhonen, A.; Vaparanta, K.; Anckar, J.; Halkilahti, K.; Salah, Z.; Aqeilan, R.I.; Palvimo, J.J.; Sistonen, L.; Elenius, K. Protein inhibitor of activated STAT3 (PIAS3) protein promotes SUMOylation and nuclear sequestration of the intracellular domain of ErbB4 protein. *J. Biol. Chem.* **2012**, *287*, 23216–23226. [CrossRef]

100. Jang, H.D.; Yoon, K.; Shin, Y.J.; Kim, J.; Lee, S.Y. PIAS3 suppresses NF-kappaB-mediated transcription by interacting with the p65/RelA subunit. *J. Biol. Chem.* **2004**, *279*, 24873–24880. [CrossRef]

101. Liu, Y.; Bridges, R.; Wortham, A.; Kulesz-Martin, M. NF-kappaB repression by PIAS3 mediated RelA SUMOylation. *PLoS ONE* **2012**, *7*, e37636. [CrossRef]

102. Zhao, Z.; Wu, L.; Shi, H.; Wu, C. p53 Nterminal binding and stabilisation by PIAS3 inhibits MDM2induced p53 ubiquitination and regulates cell growth. *Mol. Med. Rep.* **2014**, *9*, 1903–1908. [CrossRef] [PubMed]

103. Perry, E.; Tsruya, R.; Levitsky, P.; Pomp, O.; Taller, M.; Weisberg, S.; Parris, W.; Kulkarni, S.; Malovani, H.; Pawson, T.; et al. TMF/ARA160 is a BC-box-containing protein that mediates the degradation of Stat3. *Oncogene* **2004**, *23*, 8908–8919. [CrossRef] [PubMed]

104. Abrham, G.; Volpe, M.; Shpungin, S.; Nir, U. TMF/ARA160 downregulates proangiogenic genes and attenuates the progression of PC3 xenografts. *Int. J. Cancer* **2009**, *125*, 43–53. [CrossRef] [PubMed]

105. Wei, J.; Yuan, Y.; Jin, C.; Chen, H.; Leng, L.; He, F.; Wang, J. The ubiquitin ligase TRAF6 negatively regulates the JAK-STAT signaling pathway by binding to STAT3 and mediating its ubiquitination. *PLoS ONE* **2012**, *7*, e49567. [CrossRef] [PubMed]

106. Deschenes-Simard, X.; Gaumont-Leclerc, M.F.; Bourdeau, V.; Lessard, F.; Moiseeva, O.; Forest, V.; Igelmann, S.; Mallette, F.A.; Saba-El-Leil, M.K.; Meloche, S.; et al. Tumor suppressor activity of the ERK/MAPK pathway by promoting selective protein degradation. *Genes Dev.* **2013**, *27*, 900–915. [CrossRef] [PubMed]

107. Zhao, J.; Du, P.; Cui, P.; Qin, Y.; Hu, C.; Wu, J.; Zhou, Z.; Zhang, W.; Qin, L.; Huang, G. LncRNA PVT1 promotes angiogenesis via activating the STAT3/VEGFA axis in gastric cancer. *Oncogene* **2018**. [CrossRef] [PubMed]

108. You, L.; Chang, D.; Du, H.Z.; Zhao, Y.P. Genome-wide screen identifies PVT1 as a regulator of Gemcitabine sensitivity in human pancreatic cancer cells. *Biochem. Biophys. Res. Commun.* **2011**, *407*, 1–6. [CrossRef] [PubMed]

109. Xie, Z.; Chen, X.; Li, J.; Guo, Y.; Li, H.; Pan, X.; Jiang, J.; Liu, H.; Wu, B. Salivary HOTAIR and PVT1 as novel biomarkers for early pancreatic cancer. *Oncotarget* **2016**, *7*, 25408–25419. [CrossRef]

110. Pan, X.; Li, B.; Fan, N.; Li, J.; Cai, F.; Zhao, G.; Zheng, G.; Gao, C. Long Noncoding RNA PVT1 as a Potent Predictor of Prognosis in Cancers: A Meta-Analysis. *Clin. Lab.* **2017**, *63*, 1657–1666. [CrossRef]

111. Hatziapostolou, M.; Polytarchou, C.; Aggelidou, E.; Drakaki, A.; Poultsides, G.A.; Jaeger, S.A.; Ogata, H.; Karin, M.; Struhl, K.; Hadzopoulou-Cladaras, M.; et al. An HNF4alpha-miRNA inflammatory feedback circuit regulates hepatocellular oncogenesis. *Cell* **2011**, *147*, 1233–1247. [CrossRef] [PubMed]

112. Cai, B.; Li, J.; Wang, J.; Luo, X.; Ai, J.; Liu, Y.; Wang, N.; Liang, H.; Zhang, M.; Chen, N.; et al. microRNA-124 regulates cardiomyocyte differentiation of bone marrow-derived mesenchymal stem cells via targeting STAT3 signaling. *Stem Cells* **2012**, *30*, 1746–1755. [CrossRef] [PubMed]

113. Zhang, J.; Lu, Y.; Yue, X.; Li, H.; Luo, X.; Wang, Y.; Wang, K.; Wan, J. MiR-124 suppresses growth of human colorectal cancer by inhibiting STAT3. *PLoS ONE* **2013**, *8*, e70300. [CrossRef] [PubMed]

114. Wei, J.; Wang, F.; Kong, L.Y.; Xu, S.; Doucette, T.; Ferguson, S.D.; Yang, Y.; McEnery, K.; Jethwa, K.; Gjyshi, O.; et al. miR-124 inhibits STAT3 signaling to enhance T cell-mediated immune clearance of glioma. *Cancer Res.* **2013**, *73*, 3913–3926. [CrossRef] [PubMed]

115. Chen, G.; Shi, Y.; Zhang, Y.; Sun, J. CircRNA_100782 regulates pancreatic carcinoma proliferation through the IL6-STAT3 pathway. *Onco-Targets Ther.* **2017**, *10*, 5783–5794. [CrossRef]

116. Jiang, J.; Li, Z.; Yu, C.; Chen, M.; Tian, S.; Sun, C. MiR-1181 inhibits stem cell-like phenotypes and suppresses SOX2 and STAT3 in human pancreatic cancer. *Cancer Lett.* **2015**, *356*, 962–970. [CrossRef] [PubMed]

117. Jiang, M.; Zhang, W.W.; Liu, P.; Yu, W.; Liu, T.; Yu, J. Dysregulation of SOCS-Mediated Negative Feedback of Cytokine Signaling in Carcinogenesis and Its Significance in Cancer Treatment. *Front. Immunol.* **2017**, *8*, 70. [CrossRef]

118. Ghafouri-Fard, S.; Oskooei, V.K.; Azari, I.; Taheri, M. Suppressor of cytokine signaling (SOCS) genes are downregulated in breast cancer. *World J. Surg. Oncol.* **2018**, *16*, 226. [CrossRef]

119. He, B.; You, L.; Uematsu, K.; Zang, K.; Xu, Z.; Lee, A.Y.; Costello, J.F.; McCormick, F.; Jablons, D.M. SOCS-3 is frequently silenced by hypermethylation and suppresses cell growth in human lung cancer. *Proc. Natl. Acad. Sci. USA* **2003**, *100*, 14133–14138. [CrossRef]

120. Inagaki-Ohara, K.; Kondo, T.; Ito, M.; Yoshimura, A. SOCS, inflammation, and cancer. *JAKSTAT* **2013**, *2*, e24053. [CrossRef]

121. Suzuki, A.; Hanada, T.; Mitsuyama, K.; Yoshida, T.; Kamizono, S.; Hoshino, T.; Kubo, M.; Yamashita, A.; Okabe, M.; Takeda, K.; et al. CIS3/SOCS3/SSI3 plays a negative regulatory role in STAT3 activation and intestinal inflammation. *J. Exp. Med.* **2001**, *193*, 471–481. [CrossRef] [PubMed]

122. Niwa, Y.; Kanda, H.; Shikauchi, Y.; Saiura, A.; Matsubara, K.; Kitagawa, T.; Yamamoto, J.; Kubo, T.; Yoshikawa, H. Methylation silencing of SOCS-3 promotes cell growth and migration by enhancing JAK/STAT and FAK signalings in human hepatocellular carcinoma. *Oncogene* **2005**, *24*, 6406–6417. [CrossRef] [PubMed]

123. Rigby, R.J.; Simmons, J.G.; Greenhalgh, C.J.; Alexander, W.S.; Lund, P.K. Suppressor of cytokine signaling 3 (SOCS3) limits damage-induced crypt hyper-proliferation and inflammation-associated tumorigenesis in the colon. *Oncogene* **2007**, *26*, 4833–4841. [CrossRef] [PubMed]

124. Lesina, M.; Kurkowski, M.U.; Ludes, K.; Rose-John, S.; Treiber, M.; Kloppel, G.; Yoshimura, A.; Reindl, W.; Sipos, B.; Akira, S.; et al. Stat3/Socs3 activation by IL-6 transsignaling promotes progression of pancreatic intraepithelial neoplasia and development of pancreatic cancer. *Cancer Cell* **2011**, *19*, 456–469. [CrossRef] [PubMed]

125. Gui, Y.; Yeganeh, M.; Ramanathan, S.; Leblanc, C.; Pomerleau, V.; Ferbeyre, G.; Saucier, C.; Ilangumaran, S. SOCS1 controls liver regeneration by regulating HGF signaling in hepatocytes. *J. Hepatol.* **2011**, *55*, 1300–1308. [CrossRef] [PubMed]

126. Yeganeh, M.; Gui, Y.; Kandhi, R.; Bobbala, D.; Tobelaim, W.S.; Saucier, C.; Yoshimura, A.; Ferbeyre, G.; Ramanathan, S.; Ilangumaran, S. Suppressor of cytokine signaling 1-dependent regulation of the expression and oncogenic functions of p21(CIP1/WAF1) in the liver. *Oncogene* **2016**, *35*, 4200–4211. [CrossRef] [PubMed]

127. Yoshikawa, H.; Matsubara, K.; Qian, G.S.; Jackson, P.; Groopman, J.D.; Manning, J.E.; Harris, C.C.; Herman, J.G. SOCS-1, a negative regulator of the JAK/STAT pathway, is silenced by methylation in human hepatocellular carcinoma and shows growth-suppression activity. *Nat. Genet.* **2001**, *28*, 29–35. [CrossRef] [PubMed]

128. Zhao, R.C.; Zhou, J.; He, J.Y.; Wei, Y.G.; Qin, Y.; Li, B. Aberrant promoter methylation of SOCS-1 gene may contribute to the pathogenesis of hepatocellular carcinoma: A meta-analysis. *J. BUON* **2016**, *21*, 142–151. [PubMed]

129. Suzuki, M.; Shigematsu, H.; Shivapurkar, N.; Reddy, J.; Miyajima, K.; Takahashi, T.; Gazdar, A.F.; Frenkel, E.P. Methylation of apoptosis related genes in the pathogenesis and prognosis of prostate cancer. *Cancer Lett.* **2006**, *242*, 222–230. [CrossRef]

130. Kobayashi, N.; Uemura, H.; Nagahama, K.; Okudela, K.; Furuya, M.; Ino, Y.; Ito, Y.; Hirano, H.; Inayama, Y.; Aoki, I.; et al. Identification of miR-30d as a novel prognostic maker of prostate cancer. *Oncotarget* **2012**, *3*, 1455–1471. [CrossRef]

131. Chevrier, M.; Bobbala, D.; Villalobos-Hernandez, A.; Khan, M.G.; Ramanathan, S.; Saucier, C.; Ferbeyre, G.; Geha, S.; Ilangumaran, S. Expression of SOCS1 and the downstream targets of its putative tumor suppressor functions in prostate cancer. *BMC Cancer* **2017**, *17*, 157. [CrossRef] [PubMed]

132. Villalobos-Hernandez, A.; Bobbala, D.; Kandhi, R.; Khan, M.G.; Mayhue, M.; Dubois, C.M.; Ferbeyre, G.; Saucier, C.; Ramanathan, S.; Ilangumaran, S. SOCS1 inhibits migration and invasion of prostate cancer cells, attenuates tumor growth and modulates the tumor stroma. *Prostate Cancer Prostatic Dis.* **2017**, *20*, 36–47. [CrossRef] [PubMed]

133. Calabrese, V.; Mallette, F.A.; Deschenes-Simard, X.; Ramanathan, S.; Gagnon, J.; Moores, A.; Ilangumaran, S.; Ferbeyre, G. SOCS1 links cytokine signaling to p53 and senescence. *Mol. Cell* **2009**, *36*, 754–767. [CrossRef] [PubMed]

134. Mallette, F.A.; Calabrese, V.; Ilangumaran, S.; Ferbeyre, G. SOCS1, a novel interaction partner of p53 controlling oncogene-induced senescence. *Aging* **2010**, *2*, 445–452. [CrossRef] [PubMed]

135. Saint-Germain, E.; Mignacca, L.; Vernier, M.; Bobbala, D.; Ilangumaran, S.; Ferbeyre, G. SOCS1 regulates senescence and ferroptosis by modulating the expression of p53 target genes. *Aging* **2017**, *9*, 2137–2162. [CrossRef] [PubMed]

136. Kong, X.; Feng, D.; Wang, H.; Hong, F.; Bertola, A.; Wang, F.S.; Gao, B. Interleukin-22 induces hepatic stellate cell senescence and restricts liver fibrosis in mice. *Hepatology* **2012**, *56*, 1150–1159. [CrossRef] [PubMed]

137. Bouamar, H.; Jiang, D.; Wang, L.; Lin, A.P.; Ortega, M.; Aguiar, R.C. MicroRNA 155 control of p53 activity is context dependent and mediated by Aicda and Socs1. *Mol. Cell. Biol.* **2015**, *35*, 1329–1340. [CrossRef] [PubMed]

138. Cui, X.; Shan, X.; Qian, J.; Ji, Q.; Wang, L.; Wang, X.; Li, M.; Ding, H.; Liu, Q.; Chen, L.; et al. The suppressor of cytokine signaling SOCS1 promotes apoptosis of intestinal epithelial cells via p53 signaling in Crohn's disease. *Exper. Mol. Pathol.* **2016**, *101*, 1–11. [CrossRef]

139. Mallette, F.A.; Gaumont-Leclerc, M.F.; Ferbeyre, G. The DNA damage signaling pathway is a critical mediator of oncogene-induced senescence. *Genes Dev.* **2007**, *21*, 43–48. [CrossRef]

140. Mallette, F.A.; Gaumont-Leclerc, M.F.; Huot, G.; Ferbeyre, G. Myc Down-regulation as a Mechanism to Activate the Rb Pathway in STAT5A-induced Senescence. *J. Biol. Chem.* **2007**, *282*, 34938–34944. [CrossRef]

141. Mallette, F.A.; Moiseeva, O.; Calabrese, V.; Mao, B.; Gaumont-Leclerc, M.F.; Ferbeyre, G. Transcriptome analysis and tumor suppressor requirements of STAT5-induced senescence. *Ann. N. Y. Acad. Sci.* **2010**, *1197*, 142–151. [CrossRef] [PubMed]

142. Barash, I. Stat5 in the mammary gland: Controlling normal development and cancer. *J. Cell. Physiol.* **2006**, *209*, 305–313. [CrossRef] [PubMed]

143. Nevalainen, M.T.; Xie, J.; Torhorst, J.; Bubendorf, L.; Haas, P.; Kononen, J.; Sauter, G.; Rui, H. Signal transducer and activator of transcription-5 activation and breast cancer prognosis. *J. Clin. Oncol.* **2004**, *22*, 2053–2060. [CrossRef]

144. Peck, A.R.; Witkiewicz, A.K.; Liu, C.; Klimowicz, A.C.; Stringer, G.A.; Pequignot, E.; Freydin, B.; Yang, N.; Ertel, A.; Tran, T.H.; et al. Low levels of Stat5a protein in breast cancer are associated with tumor progression and unfavorable clinical outcomes. *Breast Cancer Res.* **2012**, *14*, R130. [CrossRef] [PubMed]

145. Castro, P.; Giri, D.; Lamb, D.; Ittmann, M. Cellular senescence in the pathogenesis of benign prostatic hyperplasia. *Prostate* **2003**, *55*, 30–38. [CrossRef] [PubMed]

146. Gray-Schopfer, V.C.; Cheong, S.C.; Chong, H.; Chow, J.; Moss, T.; Abdel-Malek, Z.A.; Marais, R.; Wynford-Thomas, D.; Bennett, D.C. Cellular senescence in naevi and immortalisation in melanoma: A role for p16? *Br. J. Cancer* **2006**, *95*, 496–505. [CrossRef] [PubMed]

147. Maldonado, J.L.; Timmerman, L.; Fridlyand, J.; Bastian, B.C. Mechanisms of cell-cycle arrest in Spitz nevi with constitutive activation of the MAP-kinase pathway. *Am. J. Pathol.* **2004**, *164*, 1783–1787. [CrossRef]

148. Michaloglou, C.; Vredeveld, L.C.; Mooi, W.J.; Peeper, D.S. BRAF(E600) in benign and malignant human tumours. *Oncogene* **2008**, *27*, 877–895. [CrossRef]

149. Michaloglou, C.; Vredeveld, L.C.; Soengas, M.S.; Denoyelle, C.; Kuilman, T.; van der Horst, C.M.; Majoor, D.M.; Shay, J.W.; Mooi, W.J.; Peeper, D.S. BRAFE600-associated senescence-like cell cycle arrest of human naevi. *Nature* **2005**, *436*, 720–724. [CrossRef] [PubMed]

150. Vernier, M.; Bourdeau, V.; Gaumont-Leclerc, M.F.; Moiseeva, O.; Begin, V.; Saad, F.; Mes-Masson, A.M.; Ferbeyre, G. Regulation of E2Fs and senescence by PML nuclear bodies. *Genes Dev.* **2011**, *25*, 41–50. [CrossRef]

151. Dorr, J.R.; Yu, Y.; Milanovic, M.; Beuster, G.; Zasada, C.; Dabritz, J.H.; Lisec, J.; Lenze, D.; Gerhardt, A.; Schleicher, K.; et al. Synthetic lethal metabolic targeting of cellular senescence in cancer therapy. *Nature* **2013**, *501*, 421–425. [CrossRef] [PubMed]

152. Jing, H.; Kase, J.; Dorr, J.R.; Milanovic, M.; Lenze, D.; Grau, M.; Beuster, G.; Ji, S.; Reimann, M.; Lenz, P.; et al. Opposing roles of NF-kappaB in anti-cancer treatment outcome unveiled by cross-species investigations. *Genes Dev.* **2011**, *25*, 2137–2146. [CrossRef] [PubMed]

153. Zardo, G.; Tiirikainen, M.I.; Hong, C.; Misra, A.; Feuerstein, B.G.; Volik, S.; Collins, C.C.; Lamborn, K.R.; Bollen, A.; Pinkel, D.; et al. Integrated genomic and epigenomic analyses pinpoint biallelic gene inactivation in tumors. *Nat. Genet.* **2002**, *32*, 453–458. [CrossRef] [PubMed]

154. Liu, T.C.; Lin, S.F.; Chang, J.G.; Yang, M.Y.; Hung, S.Y.; Chang, C.S. Epigenetic alteration of the SOCS1 gene in chronic myeloid leukaemia. *Br. J. Haematol.* **2003**, *123*, 654–661. [CrossRef] [PubMed]

155. Chim, C.S.; Wong, K.Y.; Loong, F.; Srivastava, G. SOCS1 and SHP1 hypermethylation in mantle cell lymphoma and follicular lymphoma: Implications for epigenetic activation of the Jak/STAT pathway. *Leukemia* **2004**, *18*, 356–358. [CrossRef] [PubMed]

156. Ekmekci, C.G.; Gutierrez, M.I.; Siraj, A.K.; Ozbek, U.; Bhatia, K. Aberrant methylation of multiple tumor suppressor genes in acute myeloid leukemia. *Am. J. Hematol.* **2004**, *77*, 233–240. [CrossRef]

157. Sutherland, K.D.; Lindeman, G.J.; Choong, D.Y.; Wittlin, S.; Brentzell, L.; Phillips, W.; Campbell, I.G.; Visvader, J.E. Differential hypermethylation of SOCS genes in ovarian and breast carcinomas. *Oncogene* **2004**, *23*, 7726–7733. [CrossRef]

158. Hatirnaz, O.; Ure, U.; Ar, C.; Akyerli, C.; Soysal, T.; Ferhanoglu, B.; Ozcelik, T.; Ozbek, U. The SOCS-1 gene methylation in chronic myeloid leukemia patients. *Am. J. Hematol.* **2007**, *82*, 729–730. [CrossRef]

159. Jiang, S.; Zhang, H.W.; Lu, M.H.; He, X.H.; Li, Y.; Gu, H.; Liu, M.F.; Wang, E.D. MicroRNA-155 functions as an OncomiR in breast cancer by targeting the suppressor of cytokine signaling 1 gene. *Cancer Res.* **2010**, *70*, 3119–3127. [CrossRef]

160. Merkel, O.; Hamacher, F.; Griessl, R.; Grabner, L.; Schiefer, A.I.; Prutsch, N.; Baer, C.; Egger, G.; Schlederer, M.; Krenn, P.W.; et al. Oncogenic role of miR-155 in anaplastic large cell lymphoma lacking the t(2;5) translocation. *J. Pathol.* **2015**, *236*, 445–456. [CrossRef]

161. Zhao, X.D.; Zhang, W.; Liang, H.J.; Ji, W.Y. Overexpression of miR -155 promotes proliferation and invasion of human laryngeal squamous cell carcinoma via targeting SOCS1 and STAT3. *PLoS ONE* **2013**, *8*, e56395. [CrossRef]

162. Saint-Germain, E.; Mignacca, L.; Huot, G.; Acevedo, M.; Moineau-Vallee, K.; Calabrese, V.; Bourdeau, V.; Rowell, M.C.; Ilangumaran, S.; Lessard, F.; et al. Phosphorylation of SOCS1 inhibits the SOCS1-p53 tumor suppressor axis. *Cancer Res.* **2019**. [CrossRef] [PubMed]

163. Guo, F.; Xu, Z.; Zhang, Y.; Jiang, P.; Huang, G.; Chen, S.; Lyu, X.; Zheng, P.; Zhao, X.; Zeng, Y.; et al. FXR induces SOCS3 and suppresses hepatocellular carcinoma. *Oncotarget* **2015**, *6*, 34606–34616. [CrossRef]

164. Attia, Y.M.; Tawfiq, R.A.; Ali, A.A.; Elmazar, M.M. The FXR Agonist, Obeticholic Acid, Suppresses HCC Proliferation & Metastasis: Role of IL-6/STAT3 Signalling Pathway. *Sci. Rep.* **2017**, *7*, 12502. [CrossRef] [PubMed]

165. Sugase, T.; Takahashi, T.; Serada, S.; Fujimoto, M.; Hiramatsu, K.; Ohkawara, T.; Tanaka, K.; Miyazaki, Y.; Makino, T.; Kurokawa, Y.; et al. SOCS1 Gene Therapy Improves Radiosensitivity and Enhances Irradiation-Induced DNA Damage in Esophageal Squamous Cell Carcinoma. *Cancer Res.* **2017**, *77*, 6975–6986. [CrossRef] [PubMed]

166. Sugase, T.; Takahashi, T.; Serada, S.; Fujimoto, M.; Ohkawara, T.; Hiramatsu, K.; Nishida, T.; Hirota, S.; Saito, Y.; Tanaka, K.; et al. SOCS1 gene therapy has antitumor effects in imatinib-resistant gastrointestinal stromal tumor cells through FAK/PI3 K signaling. *Gastric Cancer* **2018**, *21*, 968–976. [CrossRef] [PubMed]

167. Yoneda, T.; Kunimura, N.; Kitagawa, K.; Fukui, Y.; Saito, H.; Narikiyo, K.; Ishiko, M.; Otsuki, N.; Nibu, K.I.; Fujisawa, M.; et al. Overexpression of SOCS3 mediated by adenovirus vector in mouse and human castration-resistant prostate cancer cells increases the sensitivity to NK cells in vitro and in vivo. *Cancer Gene Ther.* **2019**. [CrossRef] [PubMed]

168. de la Iglesia, N.; Konopka, G.; Puram, S.V.; Chan, J.A.; Bachoo, R.M.; You, M.J.; Levy, D.E.; Depinho, R.A.; Bonni, A. Identification of a PTEN-regulated STAT3 brain tumor suppressor pathway. *Genes Dev.* **2008**, *22*, 449–462. [CrossRef] [PubMed]

169. Sherry, M.M.; Reeves, A.; Wu, J.K.; Cochran, B.H. STAT3 is required for proliferation and maintenance of multipotency in glioblastoma stem cells. *Stem Cells* **2009**, *27*, 2383–2392. [CrossRef]

170. Pencik, J.; Schlederer, M.; Gruber, W.; Unger, C.; Walker, S.M.; Chalaris, A.; Marie, I.J.; Hassler, M.R.; Javaheri, T.; Aksoy, O.; et al. STAT3 regulated ARF expression suppresses prostate cancer metastasis. *Nat. Commun.* **2015**, *6*, 7736. [CrossRef]

171. Lessard, F.; Morin, F.; Ivanchuk, S.; Langlois, F.; Stefanovsky, V.; Rutka, J.; Moss, T. The ARF tumor suppressor controls ribosome biogenesis by regulating the RNA polymerase I transcription factor TTF-I. *Mol. Cell* **2010**, *38*, 539–550. [CrossRef] [PubMed]

172. Sherr, C.J. The INK4a/ARF network in tumour suppression. *Nat. Rev. Mol. Cell Biol.* **2001**, *2*, 731–737. [CrossRef] [PubMed]

173. Ko, A.; Han, S.Y.; Song, J. Regulatory Network of ARF in Cancer Development. *Mol. Cell* **2018**, *41*, 381–389. [CrossRef]

174. Ferbeyre, G.; de Stanchina, E.; Lin, A.W.; Querido, E.; McCurrach, M.E.; Hannon, G.J.; Lowe, S.W. Oncogenic ras and p53 cooperate to induce cellular senescence. *Mol. Cell. Biol.* **2002**, *22*, 3497–3508. [CrossRef] [PubMed]

175. Pencik, J.; Wiebringhaus, R.; Susani, M.; Culig, Z.; Kenner, L. IL-6/STAT3/ARF: The guardians of senescence, cancer progression and metastasis in prostate cancer. *Swiss Med. Wkly.* **2015**, *145*, w14215. [CrossRef] [PubMed]

176. Chin, Y.E.; Kitagawa, M.; Su, W.C.; You, Z.H.; Iwamoto, Y.; Fu, X.Y. Cell growth arrest and induction of cyclin-dependent kinase inhibitor p21 WAF1/CIP1 mediated by STAT1. *Science* **1996**, *272*, 719–722. [CrossRef] [PubMed]

177. Bellido, T.; O'Brien, C.A.; Roberson, P.K.; Manolagas, S.C. Transcriptional activation of the p21(WAF1,CIP1,SDI1) gene by interleukin-6 type cytokines. A prerequisite for their pro-differentiating and anti-apoptotic effects on human osteoblastic cells. *J. Biol. Chem.* **1998**, *273*, 21137–21144. [CrossRef] [PubMed]

178. Grabner, B.; Schramek, D.; Mueller, K.M.; Moll, H.P.; Svinka, J.; Hoffmann, T.; Bauer, E.; Blaas, L.; Hruschka, N.; Zboray, K.; et al. Disruption of STAT3 signalling promotes KRAS-induced lung tumorigenesis. *Nat. Commun.* **2015**, *6*, 6285. [CrossRef] [PubMed]

179. Musteanu, M.; Blaas, L.; Mair, M.; Schlederer, M.; Bilban, M.; Tauber, S.; Esterbauer, H.; Mueller, M.; Casanova, E.; Kenner, L.; et al. Stat3 is a negative regulator of intestinal tumor progression in Apc(Min) mice. *Gastroenterology* **2010**, *138*, 1003–1011. [CrossRef]
180. Lee, J.; Kim, J.C.; Lee, S.E.; Quinley, C.; Kim, H.; Herdman, S.; Corr, M.; Raz, E. Signal transducer and activator of transcription 3 (STAT3) protein suppresses adenoma-to-carcinoma transition in Apcmin/+ mice via regulation of Snail-1 (SNAI) protein stability. *J. Biol. Chem.* **2012**, *287*, 18182–18189. [CrossRef]
181. Couto, J.P.; Daly, L.; Almeida, A.; Knauf, J.A.; Fagin, J.A.; Sobrinho-Simoes, M.; Lima, J.; Maximo, V.; Soares, P.; Lyden, D.; et al. STAT3 negatively regulates thyroid tumorigenesis. *Proc. Natl. Acad. Sci. USA* **2012**, *109*, E2361–E2370. [CrossRef] [PubMed]
182. Wang, H.; Lafdil, F.; Wang, L.; Park, O.; Yin, S.; Niu, J.; Miller, A.M.; Sun, Z.; Gao, B. Hepatoprotective versus oncogenic functions of STAT3 in liver tumorigenesis. *Am. J. Pathol.* **2011**, *179*, 714–724. [CrossRef] [PubMed]
183. Schneller, D.; Machat, G.; Sousek, A.; Proell, V.; van Zijl, F.; Zulehner, G.; Huber, H.; Mair, M.; Muellner, M.K.; Nijman, S.M.; et al. p19(ARF)/p14(ARF) controls oncogenic functions of signal transducer and activator of transcription 3 in hepatocellular carcinoma. *Hepatology* **2011**, *54*, 164–172. [CrossRef] [PubMed]
184. Zhang, H.F.; Chen, Y.; Wu, C.; Wu, Z.Y.; Tweardy, D.J.; Alshareef, A.; Liao, L.D.; Xue, Y.J.; Wu, J.Y.; Chen, B.; et al. The Opposing Function of STAT3 as an Oncoprotein and Tumor Suppressor Is Dictated by the Expression Status of STAT3beta in Esophageal Squamous Cell Carcinoma. *Clin. Cancer Res.* **2016**, *22*, 691–703. [CrossRef] [PubMed]
185. Pectasides, E.; Egloff, A.M.; Sasaki, C.; Kountourakis, P.; Burtness, B.; Fountzilas, G.; Dafni, U.; Zaramboukas, T.; Rampias, T.; Rimm, D.; et al. Nuclear localization of signal transducer and activator of transcription 3 in head and neck squamous cell carcinoma is associated with a better prognosis. *Clin. Cancer Res.* **2010**, *16*, 2427–2434. [CrossRef]
186. Gordziel, C.; Bratsch, J.; Moriggl, R.; Knosel, T.; Friedrich, K. Both STAT1 and STAT3 are favourable prognostic determinants in colorectal carcinoma. *Br. J. Cancer* **2013**, *109*, 138–146. [CrossRef] [PubMed]
187. Ettl, T.; Stiegler, C.; Zeitler, K.; Agaimy, A.; Zenk, J.; Reichert, T.E.; Gosau, M.; Kuhnel, T.; Brockhoff, G.; Schwarz, S. EGFR, HER2, survivin, and loss of pSTAT3 characterize high-grade malignancy in salivary gland cancer with impact on prognosis. *Hum. Pathol.* **2012**, *43*, 921–931. [CrossRef] [PubMed]
188. Sato, T.; Neilson, L.M.; Peck, A.R.; Liu, C.; Tran, T.H.; Witkiewicz, A.; Hyslop, T.; Nevalainen, M.T.; Sauter, G.; Rui, H. Signal transducer and activator of transcription-3 and breast cancer prognosis. *Am. J. Cancer Res.* **2011**, *1*, 347–355. [PubMed]

Review

Balancing STAT Activity as a Therapeutic Strategy

Kelsey L. Polak [1,†], Noah M. Chernosky [1,†], Jacob M. Smigiel [1], Ilaria Tamagno [1] and Mark W. Jackson [1,2,*]

[1] Department of Pathology, Case Western Reserve University, School of Medicine, Cleveland, OH 44106, USA; kxp328@case.edu (K.L.P.); nmc71@case.edu (N.M.C.); jxs1094@case.edu (J.M.S.); ixt64@case.edu (I.T.)
[2] Case Comprehensive Cancer Center, Case Western Reserve University, School of Medicine, Cleveland, OH 44106, USA
* Correspondence: mark.w.jackson@case.edu
† These authors contributed equally to this work.

Received: 17 September 2019; Accepted: 31 October 2019; Published: 3 November 2019

Abstract: Driven by dysregulated IL-6 family member cytokine signaling in the tumor microenvironment (TME), aberrant signal transducer and activator of transcription (STAT3) and (STAT5) activation have been identified as key contributors to tumorigenesis. Following transformation, persistent STAT3 activation drives the emergence of mesenchymal/cancer-stem cell (CSC) properties, important determinants of metastatic potential and therapy failure. Moreover, STAT3 signaling within tumor-associated macrophages and neutrophils drives secretion of factors that facilitate metastasis and suppress immune cell function. Persistent STAT5 activation is responsible for cancer cell maintenance through suppression of apoptosis and tumor suppressor signaling. Furthermore, STAT5-mediated CD4+/CD25+ regulatory T cells (T_regs) have been implicated in suppression of immunosurveillance. We discuss these roles for STAT3 and STAT5, and weigh the attractiveness of different modes of targeting each cancer therapy. Moreover, we discuss how anti-tumorigenic STATs, including STAT1 and STAT2, may be leveraged to suppress the pro-tumorigenic functions of STAT3/STAT5 signaling.

Keywords: STAT3; STAT5; cancer progression; cancer-stem cell; cytokine; therapy resistance; metastasis; immunosuppression; tumor microenvironment; proliferation

---

1. Introduction

A complex milieu of both cellular and non-cellular components creates a heterogeneous tumor and microenvironment [1–8]. As a tumor becomes more heterogeneous, the risk of metastasis and therapy failure increases [9–12]. Cancer cells, pericytes, immune cells, fibroblasts, and endothelial cells are just some of the cellular components of the tumor [13,14]. A highly dysregulated network of cytokines and growth factors, emanating from cancer and stromal tumor microenvironment (TME) cells, contributes to the evolution of cancer cells, and often, suppression of immune cell function. Many of these cytokines and growth factors result in the phosphorylation and activation of signal transducers and activators of transcription (STAT) proteins, which can drive cell-specific changes in gene expression. STAT3 and STAT5 activity is often elevated in aggressive subtypes of cancer and serve as prognostic indicators [15–24]. Here, we discuss the impact of microenvironmental signals on STAT3/STAT5 activation during cancer development and progression. Elevated STAT3 activity promotes epithelial-mesenchymal transition (EMT) and a stem cell program in cancer cells, while also suppressing the function of immune cells within the tumor, all of which are important steps that underlie metastasis and therapy failure [25–30]. Likewise, persistent STAT5 activity induces (i) transformation, (ii) proliferation, and (iii) anti-apoptotic signals that contribute to hematological malignancies, while also suppressing anti-tumor immunity by expanding CD4+/CD25+ regulatory T cells (T_regs) [28,31–36]. We will discuss options for targeting STAT3/5 in cancer, either directly or indirectly through the inhibition of upstream kinases, receptors and/or ligands. In addition, we will discuss how STAT1/2 activity can counter the more aggressive phenotypes induced by STAT3/5. We propose that balancing STAT-activated cytokine signaling in the TME may serve as an effective therapeutic strategy.

2. Activating STAT3 and STAT5

The STAT family of transcription factors is comprised of seven different members, STAT1, -2, -3, -4, -5a, -5b, and -6. Hereafter, when discussing the overlapping functions of STAT5A and STAT5B, we will refer to them as "STAT5". STAT proteins transduce signals from the cell membrane to the nucleus, bypassing the need for second messengers [37–39]. STATs are phosphorylated by Janus Kinases (JAK1, 2, 3) or Tyrosine Kinase 2 (Tyk2), which are recruited to ligand-activated receptors, including cytokine, growth factor, or g-protein associated receptors. Most important of the STAT3/5 activators are the IL-6 family members, which include IL-6, IL-11, ciliary neurotrophic factor (CNTF), leukemia inhibitory factor (LIF), oncostatin-m (OSM), cardiotrophin 1 (CT-1), cardiotrophin-like cytokine (CLC), and IL-27. IL-6 family cytokines form a heterodimeric complex consisting of gp130 and a cytokine specific subunit (IL-6Rα, IL-11Rα, CNTFRα, gp130, LIFR, and OSMR), recruiting and activating the JAKs or Tyk2. Once phosphorylated, STATs dimerize and translocate to the nucleus, where they bind to short stretches of DNA and act as transcription factors to induce the expression of genes implicated in cell proliferation, survival, differentiation, motility, apoptosis, and metabolism (Figure 1) [37,40–43].

The crucial role STATs play in these normal physiological processes was first demonstrated in STAT-deficient mouse models. The generation of tissue-specific STAT3 knockout models have identified STAT3 as a key component in a wide variety of processes, including, but not limited to: T-cell proliferation, suppression of apoptosis, epidermal regeneration during wound healing, macrophage and neutrophil anti-inflammatory responses, and mammary gland involution [44–46]. Deletion of STAT5 in mice demonstrated high incidence of perinatal lethality, prevented the appropriate development of B-cells and T-cells, and inhibited the function of hematopoietic stem cells [47]. Under normal conditions, JAK/STAT activation is tightly regulated by protein inhibitors of activated STATs (PIAS), tyrosine phosphatases, and suppressors of cytokine signaling (SOCS) proteins that inhibit JAK catalytic activity [48,49]. In cancer however, STAT3 and STAT5 activity becomes dysregulated, resulting in elevated STAT3/5-driven responses in tumor, stromal, and immune cells.

Advances in sequencing technologies have allowed scientists to investigate the frequency of mutations in the JAK/STAT signaling pathway, resulting in the identification of mutations that constitutively activate STAT3, STAT5, and JAK2. The majority of STAT mutations occur in the SH2 and C-terminal domains of STAT3 and STAT5B and associate with leukemias and lymphomas [50,51]. STAT3 is the most frequently mutated member of the STAT family, with high incidence of mutation in T-cell large granular lymphocytic leukemias and NK lymphoproliferative disorders [52,53]. STAT5B mutations were similarly identified in these diseases, but with lower frequency. In addition, sequencing of STAT3 exons in diffuse large B-cell lymphoma patients identified a missense point mutation (M206K) in the coiled-coil domain that affected SH2 domain function and drove robust proliferation [54,55]. Furthermore, JAK2 V617F, a constitutively active JAK2 mutant, phosphorylates both STAT3 and STAT5 proteins and has high frequency in patients with hematopoietic stem cell diseases, such as myeloproliferative diseases, essential thrombocythemia, polycythemia vera (PV), and idiopathic myelofibrosis (IMF) [56]. In addition, JAK2 V617F mutations have transformation potential in in vivo bone marrow transplantation assays and induce persistent activation of STAT5 and a PV phenotype [57]. While persistent activation of both STAT3 and STAT5 has been implicated in the transformation process, the greater impact of dysregulated STAT3/5 appears to be their influence on the induction of aggressive cancer cell properties and immunosuppression. The impacts of both cell-intrinsic reprogramming and immune dysfunction are discussed below.

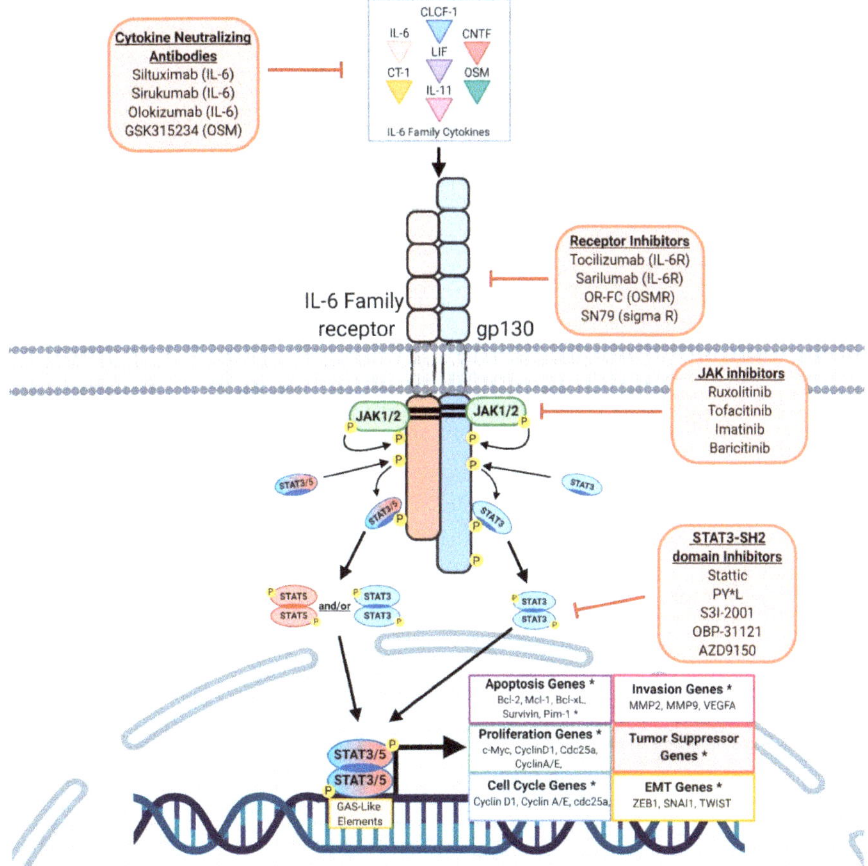

Figure 1. STAT3/5 Signaling Cascades and Therapy Targets. Schematic representation of canonical activation of signal transducer and activator of transcription-3 (STAT3) and STAT5 by IL-6 family member cytokines. IL-6 family cytokines drive receptor heterodimerization and subsequent Janus Kinase (JAK activation). JAKs phosphorylate tyrosine residues along the cytoplasmic domain of the receptor dimer, which recruits STAT proteins and facilitates their binding to interferon-gamma activation site-like (GAS-like) elements and regulation of large sets of genes. Asterisks (*) denote that additional info can be found in Table 1.

Table 1. Summary of the Roles of STAT3 and STAT5 in Cancer and Strategies for their Inhibition.

STAT-Family Protein	pYSTAT3/5-Activating Cytokines and Growth Factors	Normal Immune Function	Pro-Tumorigenic Genes Activated	Pro-Tumorigenic Effects	Therapies and Drugs in Development
STAT3	**Growth Factors:** EGFs, FGFs, HGFs, Leptin (REF), PDGFs **Lymphoid Cytokines:** IL-4, IL-13 **Myeloid Cytokines:** G-CSF, IL-10 **Ubiquitous Cytokines:** CNTF, LIF, IL-6, OSM	Development of Mature Neutrophils [58] Neutrophil Mobilization from Bone Marrow [59]	**Apoptosis:** BCL2, BCL-XL, MCL1, PIM1 **Cell Cycle:** Cyclin A, Cyclin D1, CDC25A **EMT:** ZEB1, SNAI1, TWIST, CDH2, Vimentin **Migration/Invasion:** MMP2, MMP9, VEGFA **Proliferation:** cMYC, NOTCH4 **Stem Cell:** SOX2, NANOG, OCT4 **Tumor Suppressor:** p53	Transformation Activation of Anti-Apoptotic Proteins EMT M2 Polarization N2 Polarization Enhanced Metastasis Therapy Resistance	Imatinib (JAK1/2) Ruxolitinib (JAK1/2) Tofacitinib (JAK1/2) *Siltuximab* (IL-6) *Tocilizumab* (IL-6R) *GSK315234* (OSM) *OS-FC* (OSMR) *PY*L* (SH2 Domain) *S3I-2001* (SH2 Domain) *SN79* (Sigma Receptor) IFNβ

Table 1. *Cont.*

STAT-Family Protein	pYSTAT3/5-Activating Cytokines and Growth Factors	Normal Immune Function	Pro-Tumorigenic Genes Activated	Pro-Tumorigenic Effects	Therapies and Drugs in Development
STAT3	**Growth Factors:** EGFs, FGFs, HGFs, Leptin (REF), PDGFs **Lymphoid Cytokines:** IL-4, IL-13 **Myeloid Cytokines:** G-CSF, IL-10 **Ubiquitous Cytokines:** CNTF, LIF, IL-6, OSM	Development of Mature Neutrophils [58] Neutrophil Mobilization from Bone Marrow [59]	**Apoptosis:** BCL2, BCL-XL, MCL1, PIM1 **Cell Cycle:** Cyclin A, Cyclin D1, CDC25A **EMT:** ZEB1, SNAI1, TWIST, CDH2, Vimentin **Migration/Invasion:** MMP2, MMP9, VEGFA **Proliferation:** cMYC, NOTCH4 **Stem Cell:** SOX2, NANOG, OCT4 **Tumor Suppressor:** p53	Transformation Activation of Anti-Apoptotic Proteins EMT M2 Polarization N2 Polarization Enhanced Metastasis Therapy Resistance	Imatinib (JAK1/2) Ruxolitinib (JAK1/2) Tofacitinib (JAK1/2) *Siltuximab* (IL-6) *Tocilizumab* (IL-6R) *GSK315234* (OSM) *OS-FC* (OSMR) *PY*L* (SH2 Domain) *S3I-2001* (SH2 Domain) *SN79* (Sigma Receptor) IFNβ
STAT5	**Growth Factors:** EGFs, FGFs, FLT3L, Leptin [60], PDGFs, Prolactin, SCF **Lymphoid Cytokines:** IL-2, IL-4, IL-7 [61], IL-9 [62], IL-13, IL-15, Thymic Stromal Lymphopoietin [63] **Myeloid Cytokines:** Erythropoietin [64], G-CSF, GM-CSF, IL-3, IL-5, IL-10, Thrombopoietin [65] **Ubiquitous Cytokines:** IL-21 [66], IL-31 [67], OSM	Differentiation, Survival, and Lineage Expansion of NK and NKT Cells Dendritic Cell Function [68] T Cell Differentiation and Expansion Effector Memory CD8+ T Cell-Mediated Cancer Cell Clearance [69] Myeloid Cell Differentiation and Survival [70]	**Apoptosis:** BCL-2, BCL-XL, MCL1, PIM1, Survivin **Cell Cycle:** Cyclin D1, Cyclin E **Immune Function:** PRDM1, BCL-6 **Migration/Invasion:** MMP2, MMP9, VEGFA **Proliferation:** AKT1, cMYC, PI3K **Tumor Suppressor:** p53	Transformation Activation of Anti-Apoptotic Proteins T_{reg} Population Expansion M2 Polarization [71] Megakaryopoeisis [72] Erythropoeisis [73]	Imatinib (JAK1/2) Ruxolitinib (JAK1/2) Tofacitinib (JAK1/2) *Siltuximab* (IL-6) *Tocilizumab* (IL-6R) *GSK315234* (OSM) *OS-FC* (OSMR) *AC-4-130* (SH2 Domain) [74] *BP-1-107* (SH2 Domain) [75] *BP-1-108* (SH2 Domain) [75] *Pomstafib-2* (SH2 Domain) [76] *SF-1-087* (SH2 Domain) [75] *SF-1-088* (SH2 Domain) [75] IFNβ

Therapies included in the far right column are listed by their respective name followed by their biological target in parentheses. Italicized font denotes therapies that are not currently approved by the FDA as cancer therapies.

3. IL-6 Family Cytokine Dysregulation in the TME

A number of IL-6 family members have long been recognized for their involvement in the pathogenesis of aggressive cancers [19,77–81]. For example, IL-6 levels serve as a prognostic biomarker and predictor of therapeutic response in pancreatic ductal adenocarcinoma (PDAC), bladder, gastric, lung adenocarcinoma, colorectal, cervical, liver, and breast cancers: all of which have a high incidence of metastasis and resistance [77,82–86]. This is due to the pro-tumor effects of IL-6 in tumor cells and stromal components. In tumor cells, IL-6 can drive EMT, therapy resistance, and invasive characteristics [87–89]. Concurrently, IL-6 can shift the anti-tumor immune responses towards immunosuppression via recruitment of myeloid-derived suppressor cells (MDSCs) and expansion of FoxP3+ T_{regs} [90–92]. Furthermore, IL-6 secretion by certain stromal components (cancer-associated fibroblasts, macrophages, and neutrophils) drives EMT in associated tumor populations, which is supported by immunohistochemical staining at the invasive edge of patient breast tumors, where levels of phosphorylated STAT3 and IL-6 is elevated [93,94].

Interestingly, IL-6 activation of STAT3 functions in a positive feed-forward loop to drive the secretion of new IL-6 into the TME, which then interacts with IL-6R/gp130 to further activate JAK/STAT signaling [93,95]. OSM also potently activates a feed-forward loop, resulting in the de novo production of additional OSM and OSMR by tumor and immune cells [96]. Interestingly, OSM was first identified to play a tumor suppressive role and inhibited the proliferation of melanoma cell models [97]. However, later studies correlated OSM-OSMR signaling with robust STAT3 and STAT5 activation and more aggressive tumor phenotypes [98–106]. Like IL-6, OSM feed-forward signaling is observed in aggressive cancers with limited-therapeutic options, including glioblastoma (GBM), non-small cell lung carcinoma (NSCLC), PDAC, and triple negative breast cancer (TNBC) [107–111]. Elevated signaling through the OSM/OSMR axis induces high levels of an 'inflammatory module', which includes IL-6, CCL2, IL-1, CXCL1, CXCL9, CXCL10, and CXCL11—all of which have been implicated in migration, invasion, therapy failure, and dedifferentiation to a cancer stem cell (CSC) program. Importantly, neutralization of OSM by treatment with an Fc-tagged soluble OSMR suppresses this inflammatory module suggesting that the OSM/OSMR feed-forward loop may be critical for the long-term maintenance of inflammatory signaling [112,113].

IL-11 similarly engages in an autocrine feed-forward mechanism to drive persistent JAK2/STAT3 activation, which can promote resistance to platinum-based therapies [114]. Furthermore, IL-11 and LIF both contribute to tumorigenesis by enhancing tumor cell survival through STAT3-mediated activation of anti-apoptotic proteins, Bcl-2 and Survivin, and the inhibition of tumor suppressor p53 [114–116]. IL-11 and LIF also contribute to cancer progression by driving EMT through STAT3 and Akt/mTOR signaling, thereby conferring an enhanced migratory capacity [81,117,118]. The impact of CT-1, CLC, CNTF, and IL-27 on cancer progression is currently understudied. While IL-6 family cytokines are important mediators of STAT3 activation, other cancer-associated receptors can also activate STAT3. STAT3-driven tumor progression can also be achieved by epidermal growth factor receptors (both wild-type EGFR and EGFRvIII), fibroblast growth factor receptors (FGFR), insulin-like growth factor receptor (IGFR), hepatocyte growth factor (HGFR, also known as MET), platelet-derived growth factor receptor (PDGFR), vascular endothelial growth factor receptor (VEGFR), v-src, and Bcr-Abl [119–127]. Interestingly, acetylation of STAT3 also induces tumor progression through enhanced pro-tumorigenic IL-17A secretion [128,129]. These examples demonstrate the abundance of mechanisms through which STAT3 can become activated.

4. STAT3 Activation of a Mesenchymal/CSC Program in Cancer Cells

The majority of cancers have constitutive activation of STAT3 [130–132]. STAT3 was first termed oncogenic when its persistent activation was discovered in v-src transformed mouse embryonic fibroblasts [133]. Subsequent studies demonstrated that expression of a constitutively active STAT3 isoform (STAT3-C) can drive transformation of pre-malignant human mammary epithelial cells (HMEC) and MCF-10A cells to malignant breast cancer [134]. In addition, Ras-induced transformation in bladder and breast carcinoma models exhibit mitochondrial accumulation of STAT3 and more robust cellular glycolysis, a characteristic of cancer cells [135]. In PDAC, STAT3 is required for both the development of pre-malignant pancreatic interepithelial neoplasias (PanINs), as well as the progression of PanINs to PDAC [136]. While STAT3 has been strongly implicated as a driver of oncogenesis, evidence suggests that persistent cytokine activation of STAT3 in pre-malignant lesions, engages a tumor suppressive senescence response. In non-transformed HMEC models, OSM engages senescence through a direct STAT3 interaction with mothers against decapentaplegic-3 (SMAD3). However, downstream constitutive expression of c-Myc could overcome OSM-induced senescence and drive EMT and invasion [137]. Beyond the genetic events that occur as a normal cell becomes transformed, cancer progression relies heavily on an evolving microenvironment, which impacts both the cancer cells and the immune system, ultimately influencing patient outcomes. In cancer cells, increased STAT3 activity induces EMT-driving transcription factors, such as ZEB1, SNAIL, and Twist, which initiate the repression of epithelial markers and expression of mesenchymal markers (N-cadherin, Vimentin) [138–141]. In addition, STAT3 induces the expression of matrix metalloproteinases (MMP), which can also contribute to the loss of cell-cell contacts [142–144].

While EMT is important in normal physiology to allow cells to migrate during development and wound healing, inappropriate EMT and the loss of cell-cell interactions contributes to aggressive cancer cell behaviors, including metastasis and resistance to therapy. In TNBC, the loss of E-cadherin or gain of Vimentin, N-Cadherin, or Snail expression, typically at the tumor's edge, correlates with poor clinical outcome [145–149]. Likewise, the presence and abundance of circulating tumor cells (CTC) that express mesenchymal markers have been referred to as the "silent predictors of metastasis" because they correlate with tumor cell dissemination [150]. Moreover, mesenchymal CTC can be used to track a patient's response to therapy, with increasing numbers correlating with increased risk of relapse [151–155].

Likewise, the emergence of CSC properties, which are often defined experimentally as using tumorsphere- or tumor-initiating assays, frequently occur concomitantly with a mesenchymal phenotype [156–163]. STAT3 has been identified as an essential driver of the de novo reprogramming to a CSC state through activation of Sox2, Nanog, and Oct4 [164–170]. Further evidence linking EMT and a CSC phenotype is provided by the observation that a subset of CTC, responsible for initiating metastasis termed the metastasis-initiating cells, express high levels of the STAT3-driven CSC marker CD44 [151–153,171–174]

The importance of mesenchymal/CSC reprogramming in metastasis and therapy failure continues to emerge. For example, single-cell analysis identified a predominantly mesenchymal/CSC program in early-stage TNBC micro-metastases, in contrast to late-stage metastases [154]. The findings are consistent with the idea that mesenchymal/CSC initiate metastatic outgrowth at a secondary site, followed by differentiation. Single cell RNA sequencing (scRNA-seq) confirmed that EMT in primary tumors proceeds through distinct, hybrid states, ranging from completely epithelial to completely mesenchymal [155]. The epithelial-mesenchymal hybrids, which harbor the greatest level of phenotypic plasticity, are more efficient at intravasating, surviving in circulation, extravasating to the lungs, and forming metastases [175]. Acute exposure to Adriamycin or Taxanes drives the adaptive emergence of therapy-resistant, CD44^{HIGH} CSC in both breast tumor explants and breast cancer cell lines [176]. scRNA-seq determined that chemo-resistant cells activate an EMT program, which was not evident before treatment [177]. In mouse models, the ability to undergo EMT is important for therapeutic resistance [178,179]. These findings suggest that potent mesenchymal/CSC programming has significant consequences that allow cancer cells to adapt to chemotherapy and survive, ultimately contributing to therapy failure.

Indeed, STAT3 has also been implicated in acquired therapy resistance in many cancers including NSCLC, GBM, PDAC, melanoma, and breast cancer [107,180–185]. Elevated STAT3 phosphorylation in ovarian cancer is associated with paclitaxel resistance and increased tumor cell invasion post-therapy, consistent with the elevated expression of mesenchymal/CSC genes and increased tumor initiating potential [176,186,187]. In Trastuzumab-resistant HER2+ breast cancer, STAT3 feed-forward loops generate a TME rich with STAT3-activating cytokines that promote and maintain the mesenchymal/CSC phenotypes. Inhibition of the STAT3 feed-forward activation diminished tumor growth and metastasis and resensitized cells to therapy [188]. In cancer cells driven by diverse receptor tyrosine kinases (RTKs) (EGFR, HER2, ALK, and MET), MEK inhibitors drive feedback activation of STAT3 through FGFR and JAKs, resulting in therapy failure [189]. Likewise, in NSCLC cells resistant to molecularly-targeted therapies (EGFR-TKI, ALK inhibitor Crizotinib, and MEK inhibitor Selumetinib), OSM/JAK1/STAT3–signaling protects cells from targeted drug-induced apoptosis [190]. JAK/STAT3 signaling also interacts with numerous other growth, proliferative, and survival pathways, such as PI3K/Akt, MAPK, NF-kB, Notch, Wnt/β-Catenin, and TGF-β, among others [191–196]. Tamoxifen-resistant breast cancers have elevated STAT3 and Notch4 expression associated with metastasis and tumorigenicity. Interestingly, inhibition of Notch4 was able to reduce the phosphorylation of STAT3 and suppress metastasis, suggesting STAT3 and Notch4 cooperate to promote therapy resistance [197]. In addition, a STAT3/NF-kB complex promotes cisplatin resistance in malignant mesothelioma, and targeted inhibition of this complex inhibits the growth of refractory tumors [198]. Such evidence suggests STAT3 may not be solely responsible for therapy failure. Instead, STAT3-mediated cross-talk with additional signaling effectors may coordinate resistance [140,199–201]. Nonetheless, the common theme appears to be a STAT3-activated, mesenchymal/CSC program.

5. A STAT3-Generated Pro-Metastatic Immune Microenvironment

An important and rapidly emerging area of research is the impact of the immune microenvironment on metastasis and therapy failure. Beyond its impact on tumor cells, STAT3 has an important role in restricting immune cell functions and producing immunosuppressive factors. STAT3-induced cytokines are important mediators of the crosstalk between tumor cells, tumor-associated macrophages (TAMs), and tumor-associated neutrophils (TANs), which are responsible for generating a pro-metastatic and pro-angiogenic TME [202]. In the TME, macrophages and neutrophils are exposed to a host of STAT3-induced cytokines, such as IL-4, IL-6, IL-10, IL-13, VEGFA, and TGF-β that drive their polarization to a pro-tumorigenic M2 (macrophages) or N2 (neutrophils) state [203,204]. Conversely, robust STAT3 signaling activity in M2-TAMs and N2-TANs correlates with the production of factors able to drive cancer cell EMT (OSM, IL-6, TGF-β) and angiogenesis (VEGF, TGF-β, PDGF, and FGF) [205–208].

Importantly, because of their localization at the invasive edge of tumors, the contribution of M2-TAMs and N2-TANs to metastasis is increasingly being recognized. For example, TAMs have been demonstrated to form simultaneous physical contacts with tumor cells and endothelial cells that result in the formation of invadopodia, which assist cancer cells in transendothelial migration and escape into the circulatory system. These sites of cancer cell intravasation are called "tumor microenvironment of metastasis (TMEM)" and have been validated as prognostic markers of metastasis. Chemotherapy increases TMEM in breast cancer patients, thereby potentially facilitating metastasis [209,210]. STAT3 inhibition in TAM populations re-sensitizes breast cancer cells to paclitaxel, further suggesting tumor and TAM crosstalk is an essential component in STAT3-driven therapy resistance [205,211]. TANs, on the other hand, were found to function as circulatory escorts of CTC and promote their proliferation, survival, and seeding of secondary sites [212–214]. Importantly, OSM and IL-6 were two of the 4 most frequent cytokines secreted by neutrophils found clustered with CTC. These cytokines interact with OSMR and IL-6R, which are expressed on cancer cells [212]. The neutrophil/CTC cross-talk promotes the cell cycle progression in CTC, thereby expanding their metastatic potential.

Following the classical events of the metastatic cascade, after a tumor cell intravasates into circulation, it extravasates to establish secondary sites. An emerging concept in the metastatic cascade is the role of myeloid cells- such as basophils, neutrophils, eosinophils, monocytes, and macrophages- in the establishment of a pre-metastatic niche [215]. Although the exact TME components required for colonization of secondary sites remain a mystery, the activation of a STAT3-sphingosine 1-phosphate receptor-1 (S1PR1) axis in tumor cells, and secretion of IL-6 and IL-10, was observed to persistently activate STAT3 at distant, pre-metastatic sites. At these pre-metastatic sites, persistent STAT3 phosphorylation was associated with myeloid cell migration from the primary tumor to the secondary site. Subsequent targeting of STAT3 in myeloid cells disrupted metastatic tumor outgrowth, suggesting STAT3 plays an integral role in priming distant metastatic sites for tumor cell outgrowth [216].

6. STAT5 in Cancer Cells

As previously mentioned, cells produce two different STAT5 proteins, STAT5A and STAT5B, which share greater than 90% sequence homology [217]. However, evidence suggests STAT5A and STAT5B play different functional roles in normal and cancer cell systems. Genetic deletion of STAT5 in pure mouse backgrounds are embryonic lethal, due to the essential roles of STAT5 in erythropoiesis and iron metabolism [218–220]. While the distinct roles of STAT5A and STAT5B remain understudied, data from mammary gland-specific knockouts suggest that STAT5A is required for lactogenesis in the mammary gland, while STAT5B is imperative for mammary gland differentiation and development [221]. The function of STAT5A and STAT5B appear to be cell-specific. They can have either synergistic or opposing effects, such as in memory B-cell differentiation, which may be due to (i) the formation of STAT5A/STAT5B homo- or hetero-dimers, and/or (ii) differences in nuclear shuttling mechanisms [222–224]. Furthermore, genetic tuning models depleting STAT5A and/or STAT5B have demonstrated a critical role for STAT5 in the accumulation and development of innate lymphoid cells, such as NK cells [225–227].

Given its role in lymphoid cell development and differentiation, it is not surprising that STAT5 activity contributes to hematologic malignancies. Deletion of STAT5 prevents transformation by the Abl oncogene, thereby preventing leukemia development [228]. Genetic and pharmacologic inhibition of STAT5 activity decreases expression of apoptosis inhibitors MCL1 and BCL2 and inhibits leukemogenesis of BCR-ABL1+ acute lymphoblastic leukemia (ALL), both in cell lines and newly diagnosed and relapsed/TKI-resistant ALL patients [229]. Likewise, a new, STAT5 inhibitor suppressed the proliferation of human acute myeloid leukemia (AML) cell lines and primary FLT3-ITD+ AML patient cells [74]. Combined inhibition of STAT3 and STAT5 by shRNAs also suppressed growth in chronic myeloid leukemia, suggesting that combinatorial suppression of STAT3 and STAT5 may be efficacious in treating hematological malignancies [230].

STAT5 activation has also been implicated in the progression of solid tumor malignancies. Deletion of STAT5 in the mammary gland, hepatocytes, and prostate cells delays the development of mammary, liver, and prostate cancer [32,231,232]. Like STAT3, experimental evidence implicates STAT5B as a driver of tumorigenesis, as it can drive EMT and increased invasiveness in hepatocellular carcinoma (HCC) [233]. Moreover, in mammary epithelial cells, thymocytes, and epithelial prostate cancer cells, persistent activation of STAT5 is sufficient to drive transformation [234–236]. Furthermore, the function of STAT5 in solid tumors extends beyond oncogenesis as evidence has emerged that STAT5 signaling can induce a metastatic cascade. For example, STAT5 inhibition in colorectal cancer induces G1 cell cycle arrest and reduces cancer cell migration, demonstrating the role of STAT5 in proliferation and metastasis [237]. Additional studies demonstrate that STAT5A/B signaling in prostate cancer and squamous cell carcinoma of the head and neck drive EMT programming, which results in enhanced cell migration, invasion, and formation of metastases [238,239].

An increasing number of reports demonstrate that STAT5 drives tumorigenesis and cancer progression through cooperation with other intracellular signaling cascades and activation of additional feed-forward loops. STAT5 activates transcription of AKT1 and PI3K, and, in turn, Akt1 phosphorylates STAT5 to induce cell survival [240,241]. Furthermore, STAT5-dependent Akt restores cyclin D expression, which promotes proliferation [241]. Once cells aberrantly proliferate, apoptosis suppressors Bcl-xL and Bcl-2 are activated via persistent STAT5 signaling, driving tumor cell survival [16,234,242–244]. BCL-XL expression is also enhanced via formation of a transcription factor complex comprised of phosphorylated STAT5 and nuclear EGFRvIII. This transcription factor complex binds to the BCL-XL promoter to induce its transcription and promote anti-apoptotic signaling [245]. Similar to STAT3, STAT5 translocates to tumor cell mitochondria, suggesting an interaction with the mitochondrial genome to promote aerobic glycolysis (the Warburg Effect), a defining characteristic of cancer cells [246]. Collectively, this data demonstrates that STAT5 mediates crosstalk between cancer cell survival, and proliferation, and metabolism signaling pathways.

Surprisingly, STAT5 in certain model systems has been demonstrated to function in a tumor suppressive manner. Human breast cancers infrequently (~7%) show signs of STAT5 activation (compared to 40% of STAT3 activation). This elevated STAT5 activity trends with more differentiated and lower grade tumors, suggesting that STAT5 does not induce the aggressive cancer cell program initiated by STAT3, at least in breast cancer [247]. STAT5 expression in these models stabilizes E-cadherin surface marker expression and reverses the undifferentiated mesenchymal phenotype [248]. In normal human fibroblasts, aberrant activation of STAT5A induces a senescence response concurrent with accumulation of p53 and DNA damaged foci. Furthermore, knockdown DNA-repair kinase, ATM, and tumor suppressor, retinoblastoma protein did not eliminate damaged foci, providing evidence for the persistence of DNA damage in pre-malignant lesions [249]. Interestingly in HCC models, liver specific STAT5 knockout results in tumor formation through the enhanced activation of TGFβ/STAT3 signaling [20]. Physiologically, STAT proteins have been identified as drivers of erythropoiesis. STAT5A/B double knockout mice in a mixed genetic background (Sv129 x C57Bl/6) results in mild hematopoietic phenotypes, due to compensatory activation and enhanced DNA-binding of STAT1/3 [218].

7. STAT5 in T_{reg}-Associated Immunosuppression

In addition to its impact on cancer cells, activated STAT5 dampens anti-tumor immune function. This immunosuppressive function is largely driven by CD4+/CD25+ T_{regs}, a subset of T cells that contribute to tumor progression and metastasis [250] and correlates with poor patient prognosis [251,252]. Sustained STAT5 phosphorylation in progenitor T cells induces the differentiation to a T_{reg} population that, in turn, significantly diminishes the function of cytotoxic and helper T cells [253–255]. Experimental depletion of T_{regs} from the TME results in enhanced infiltration of mature CD4+ and CD8+ T cells into the tumor, leading to tumor rejection [256,257]. Furthermore, STAT5-mediated T_{reg} expansion increases IL-10, IL-4, and IL-13, which skew TAMs to an M2 immunosuppressive phenotype. M2 macrophages are immunosuppressive because they release elevated levels of IL-10 and transforming growth factor-β (TGF-β) and restrict secretion of immune stimulatory cytokines via NF-kB repression [258,259]. STAT5-induced T_{regs} and M2 macrophage populations secrete VEGF-A, which, along with TGF-β, promotes angiogenesis [255,260,261].

Tumor clearance mechanisms are also suppressed by T_{reg} expansion through impediment of B cell development and maturation [262]. STAT5-activated T_{regs} result in a decrease in follicular helper T (T_{fh}) cell populations via Blimp-1, which severely hampers germinal center formation in lymph nodes [263]. This reduction in germinal centers diminishes the number of B cells that can be recruited and primed to aid in an anti-tumor immune response. Interestingly, the differentiation and self-renewal of memory B cells are respectively influenced by STAT5A-mediated repression and STAT5B-mediated induction of BCL-6. Immunosuppression driven by the expansion of T_{regs}, relies on STAT5-mediated alteration in T-cell metabolism. Mature effector T cells preferentially undergo glycolysis and require a de novo fatty acid synthesis reliant on acetyl-coA carboxylase 1 whereas T_{regs} undergo lipid-oxidation and readily synthesize fatty acids because of a structural reconfiguration of mitochondrial cristae [264–266]. Accumulation of intracellular lipids impairs autophagy, providing a mechanism for T_{reg}-mediated immunosuppression [267,268]. Just as STAT5 function in the mitochondria impacts tumor cell functions, as described earlier, mitochondrial STAT5 activation drives metabolic shifts in the immune compartment, inducing an expansion of T_{regs}.

However, while the suppression of T_{regs} increases tumor immunity, provided STAT5 remains functional, the overall impact of suppressing STAT5 signaling in the remaining immune cells of the TME remains open to debate. Mouse models deficient in STAT5 have depleted CD8+ T cell, NK cell, and T_{reg} populations, suggesting STAT5 plays an integral role in the proper development of multiple immune cell types [269]. STAT5 signaling contributes to the differentiation of naive CD4+ T cells into CD8+ T cells, T_h1, T_h2, T_h9, T_hGM, and T_{regs}, while inhibition of STAT5 is required for the generation of T_h17 and T_{fh} cells [270,271]. Importantly, STAT5 is heavily involved in the development, survival, and lytic function of NK cells. Knock-out or suppression of STAT5 in NK populations sparked the secretion of VEGF-A, a growth factor that supports tumor-associated angiogenesis in melanoma and leukemia models [269,272]. Taken together, these findings suggest that targeting STAT5 may threaten the integrity of anti-tumor immune functions and drive worse outcome in patients [269,272].

8. Targeting STAT Activity

Given the roles of STAT3 and STAT5 in tumor progression and immunosuppression discussed above (Figure 2), multiple methods of inhibiting their activity are being pursued. Successful inhibition of STAT3 would prevent the acquisition of, and potentially revert, a mesenchymal/CSC program, making cancer cells less invasive and more sensitive to therapy. Moreover, STAT3 inhibition would help activate anti-tumor immunity by reducing immunosuppressive factors and increasing the infiltration of immune cells into the TME. Likewise, suppressing STAT5 in cancer cells, particularly leukemia, halts proliferation and induces apoptosis, suggesting that STAT5 may be a valuable therapeutic target [74]. However, due to STAT5's controversial roles in tumor progression and immune cell maturation and differentiation, further studies are required to elucidate the effects of targeting STAT5 in cancer patients. [273]. Generally speaking, the majority of small molecule inhibitors are designed to have high affinity for the catalytic domain of an enzyme; the ATP binding site for example, which STAT proteins lack. Therefore, direct inhibition of STAT3 relies on disruption of binding motifs necessary for downstream signal transduction. A series of STAT3 phospho-ester SH-2 domain inhibitors have been developed, with the intent to inhibit dimerization of activated STAT3 (PY*L, S3I-2001, OPB-31121, etc.). While these approaches are promising, the compounds continue to be refined [274–277]. An alternative approach would involve targeting the upstream activator of STAT3, which could include neutralizing antibodies for specific IL-6 family cytokines, competitive antibodies hindering cytokine-receptor interactions, and JAK inhibitors.

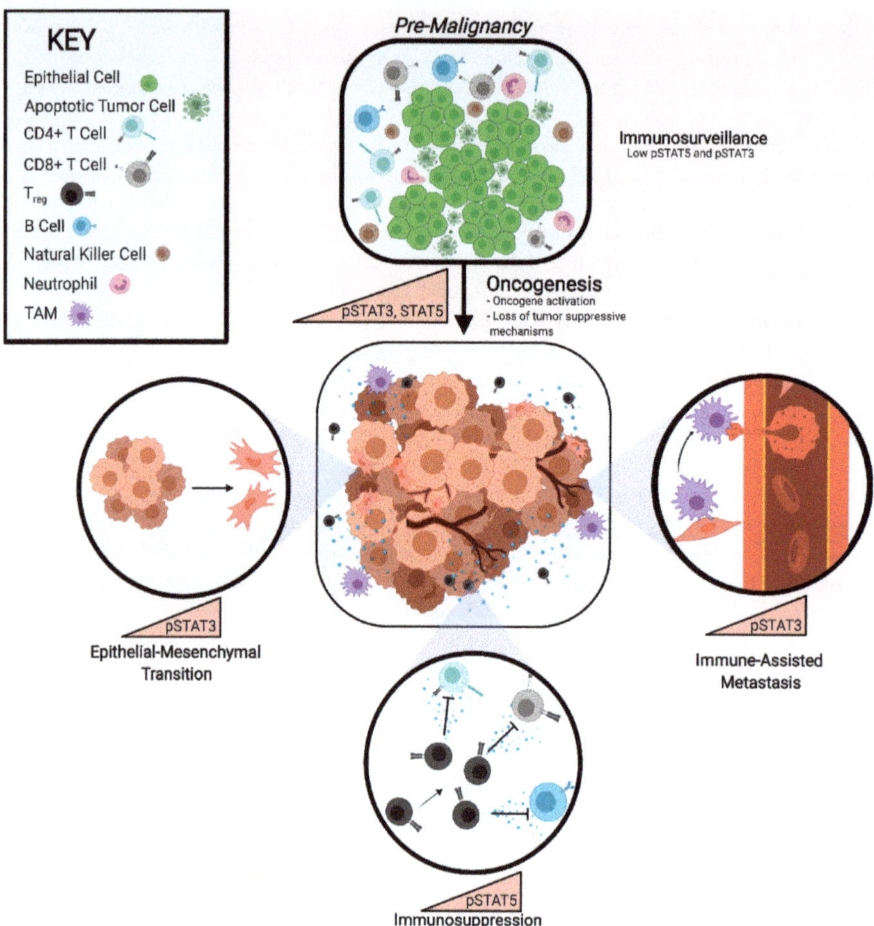

Figure 2. Biological Impact of STAT3 and STAT5 Activation. Pre-malignant cell populations, in which apoptotic signaling and immunosurveillance are functional, exhibit low levels of phosphorylated STAT3 (pSTAT3) and STAT5 (pSTAT5). Elevated activity of pSTAT3 and/or pSTAT5 accompanies tumorigenesis, leading to the inhibition of apoptotic pathways and repression of immune cell recognition of a burgeoning tumor. pSTAT3 is utilized by malignant cell populations to drive epithelial-mesenchymal transition (EMT) and by tumor-associated macrophages (TAMs) and tumor-associated neutrophils (TANs) to drive metastasis. pSTAT5 activity in progenitor T-cells drives expansion of a Treg population that then secretes factors that inhibit the function of CD4+ and CD8+ T-cells as well as B-cells.

One of the first IL-6 monoclonal antibodies generated was Tocilizumab, which inhibits IL-6 signaling by preventing IL-6 binding to both the soluble and transmembrane forms of IL-6R [278,279]. Pre-clinical studies demonstrated strong anti-tumor cell activity in multiple myeloma and therapeutic efficacy in Castelman's disease, rheumatoid arthritis, and cytokine release syndrome, however, Tocilizumab is not currently FDA-approved for cancer treatments [280–283]. Clinical trials using Tocilizumab in combination with other monoclonal antibodies, chemotherapies, and immunotherapies to treat cancers such as HER2+ breast cancer, B-cell Non-Hodgkin Lymphoma, metastatic NSCLC, and recurrent metastatic colorectal adenocarcinoma are ongoing [284]. In addition to Tocilizumab, Siltuximab is an IL-6 specific neutralizing antibody that has emerged as another promising therapy. Pre-clinical studies support Siltuximab use in KRAS-mutant lung cancer, particularly in tumors with elevated stromal production of IL-6 [285]. Similar to Tocilizumab, Siltuximab is undergoing clinical trials for patients with malignant solid tumors in cancers with elevated stromal presence, such as ovarian, pancreatic, colorectal, head and neck, and lung neoplasms [286]. Furthermore, phase II studies of combination therapy with Siltuximab have demonstrated anti-tumor effects for patients diagnosed with metastatic prostate cancer, suggesting clinical efficacy of targeting the IL-6-STAT signaling axis with combination therapies [287].

Our lab has focused on OSM, given its potent activity at inducing numerous inflammatory cytokines that promote mesenchymal/CSC reprogramming in cancer cells and generate a pro-tumor immune microenvironment [106,288–290]. Neither OSM nor OSMR protein are abundantly expressed in normal tissues in the absence of inflammation, in contrast to JAK1, JAK2, and STAT3 (key OSM effectors) [109]. This finding is supported by the observation that knocking out OSM or OSMR from mice results in only mild phenotypes [291]. Moreover, OSMR has characteristics unique from other IL-6 family co-receptors, resulting in distinct signaling and biological effects [43,292,293]. For example, OSMR strongly recruits SHC, resulting in the hyper-activation of the MAPK signaling cascade [293]. Other gp130 co-receptors fail to do this, and rely solely on the gp130-mediated SHP-2 recruitment for MAPK activation, which is less robust than the gp130/OSMR heterodimer [43]. Therefore, suppressing the OSM signaling axis may have benefits in aggressive subtypes of cancer. Anti-OSM antibody (GSK315234) and OSMR-fusion protein (OR-FC) were first examined in pre-clinical studies in chronic inflammatory disease models, such as rheumatoid arthritis and inflammatory heart disease [294,295]. Antibodies suppressing OSM signaling ameliorate these pathological conditions, suggesting that OSM is a driver of these chronic hyper-inflamed states. More recently, OSM signaling was identified as a driver of inflammatory bowel disease (IBD), especially in patients who fail to respond to anti-tumor necrosis factor-α (TNF-α) antibodies. Neutralization of OSM in IBD models suppresses a cadre of inflammatory cytokines and reduces colitis severity, further supporting the OSM feed-forward loop as a critical mediator in the long-term maintenance of inflammatory signaling, a state common in the TME [112].

Another recent study investigating the effects of an anti-OSM antibody in a murine model of lupus nephritis demonstrated that OSM-driven EMT and extracellular matrix secretion, leading to renal fibrosis, could be suppressed, concomitant with the suppression of JAK/STAT3 activation [296]. More recently, a clinical grade OSM neutralizing antibody was used to treat pre-clinical models of squamous cell carcinoma. Again, OSM neutralizing antibodies suppressed the STAT3 feed forward signaling, resulting in reduced invasion and metastasis [96,297]. Disrupting cell surface receptor activation of STAT3 by inhibiting OSMR signaling was recently described in a study of an OSMR/gp130 antagonist (SN79), which prevents STAT3 phosphorylation in astrogliosis [298]. While these studies demonstrate the potential of OSM and OSMR inhibition as a therapeutic strategy, a number of ligand/receptors activate STAT3 and STAT5, as described above. Therefore, additional studies will be needed to define whether single ligand-receptor inhibitors can sufficiently impact STAT3 activation, thus suppressing tumor growth.

Currently, JAK inhibitors are the most promising inhibitors of STAT-driven phenotypes. Commonly used JAK 1/2 inhibitor, Ruxolitinib, has been shown to robustly block both STAT3 and STAT5 activation.

Yet, while JAK inhibitors can reduce STAT3 activation, there is conflicting data on the impact of JAK inhibitors. Some studies find that JAK inhibition suppresses tumor growth [93], while other studies find that they ultimately enhance metastasis, likely because they also suppress the positive influence of JAK activity on other STAT proteins (including STAT5 and STAT1/2, as discussed below) [299]. In addition, side effects of JAK inhibitors may be more pronounced, as JAKs activate other pathways, such as MAPK and PI3K in normal cells as well as cancer [43].

9. Balancing Opposing STAT-Activated Cytokine Signaling as a Therapeutic Strategy

Though STAT proteins are grouped together based on their structural similarities and common functions as transcription factors, individual STAT proteins can have diverse functions. We have focused extensively on STAT3 and STAT5 and their described roles in cancer progression, however this is not the universal effect among all STATs. For example, Interferon-β (IFNβ) induces the phosphorylation and activation of STAT1/2, which form a complex with IRF9 to create the transcription factor complex ISGF3. IFNβ/P-ISGF3 signaling induces interferon-stimulated genes (ISG), mesenchymal-epithelial transition (MET), and the differentiation of CSC into a less aggressive epithelial, non-CSC state with reduced migratory potential and reduced tumorsphere forming capabilities [111,300,301]. Importantly, the IFNβ and OSM/STAT3 signaling pathways strongly oppose one another. OSM represses transcription of IFNβ, thereby eliminating autocrine and paracrine IFNβ-mediated activation of P-ISGF3 and repressing ISG expression in both cancer cells and immune cells [301].

In addition to the impact of IFNβ on cancer cells, increasing rationale supports developing methods for the delivery of P-ISGF3 activators (or Type I IFN-agonists more generally) directly to the TME. First, favorable responses to frontline chemotherapy correlate with robust IFN signaling in both mouse and human studies [111,302–304]. Elevated IFN signaling in the tumor correlates with immunologically "hot" tumors harboring elevated numbers of tumor infiltrating lymphocytes (TILs), activated immune surveillance, increased tumor antigen cross presentation, and diminished numbers of immunosuppressive cells including MDSCs and T_{regs} [111,302,305–307]. Second, loss of Type I IFN signaling correlates with metastasis and decreased survival. Restoration of Type I IFN signaling significantly decreases metastasis and improves survival outcome [303,304]. Third, in contrast to STAT3 activators (which promote a pro-tumorigenic M2 state), addition of type I IFN inhibits macrophage polarization to an M2 state [308,309]. Type I IFNs also induce the differentiation of neutrophils into anti-tumor N1s [310]. Fourth, administration of IFNβ prior to surgical resection significantly improves response rates to immunotherapies such as anti-PD1/anti-PDL-1 [311]. Therefore, we propose balancing pro-tumor STAT3 activation with anti-tumor STAT1/STAT2 activation as a novel therapeutic approach. This STAT3/STAT1 balancing would (i) reprogram mesenchymal/CSC to a non-CSC state, making them more susceptible to chemotherapy and (ii) enhance anti-tumor immunity, thereby facilitating immune cell-mediated tumor cell killing. Yet, while IFN treatment is currently approved to treat hematological malignancies and some solid tumors (melanoma), the high doses of IFNs needed to inhibit cancer cell proliferation or induce cell death result in side-effects that limit its effectiveness [311–313]. We suggest tumor targeting antibodies (or nanobodies) linked to IFNβ. Generation of an oncogenic cytokine or receptor antibody conjugated to an IFNβ first demonstrated success in limiting resistance to EGFR inhibitors in breast cancer [314]. The designed therapy sought to re-activate innate and adaptive immune components, while simultaneously targeting the oncogenic receptor EGFR [303]. IFNβ-conjugated antibodies show immense promise. They would limit toxicity by using tumor-associated receptors to target IFNs to the TME and suppress STAT3 activation, while simultaneously, activating STAT1/2 in both tumor cells and immune cells [315].

10. Conclusions

As discussed throughout this review, STAT3 and STAT5 have emerged as essential components involved in regulating tumor progression. Cytokines and cytokine receptors of the IL-6 family are some of the most widely recognized STAT activators and are abundantly expressed on cancer cells as well as tumor-infiltrating immune cells. The role of STAT3 in promoting molecular programs in cancer cells that induce tumor metastasis and therapy resistance mechanisms continue to emerge, as does the impact of STAT3 as a suppressor of immune cell function in the TME. These findings suggest that specifically suppressing STAT3 activation would be beneficial to patients. Targeting STAT5 in hematological malignancies is gaining traction, and the clinical successes of JAK and tyrosine kinase inhibitors in disrupting STAT3 and STAT5 activation, provide strong support for the development of direct STAT3 and STAT5 inhibitors, summarized in Table 1. Specific suppression of STAT5 in immune-suppressive T_{regs} could also prove beneficial, but STAT5 is essential for many other immune cells as well. Thus, systemic suppression of STAT5 activity could undermine tumor immunity and promote tumor progression, as recently reported for STAT5 knock-out from NK cells [272]. However, reducing the aberrant activation of STAT5 without complete ablation may have therapeutic efficacy upon combination with other vulnerabilities. We conclude that targeting STAT3, either directly by disrupting STAT3-homodimer formation or indirectly by suppressing the activation of receptors responsible for persistent STAT3 phosphorylation, would reverse the cellular programs driving metastasis and therapy failure. Furthermore, by activating STAT1/2 within TME cells, the programs that prevent metastasis and enhance therapeutic efficacy could be re-engaged. This STAT balancing would improve outcomes for patients, particularly those with aggressive cancers that may currently have limited therapeutic options.

Funding: This research is supported by T32 (CA059366) to Kelsey Polak.

Acknowledgments: Both Figures 1 and 2 were created with BioRender.

Conflicts of Interest: The authors declare no conflict of interest.

References

1. Lin, E.Y.; Nguyen, A.V.; Russell, R.G.; Pollard, J.W. Colony-stimulating factor 1 promotes progression of mammary tumors to malignancy. *J. Exp. Med.* **2001**, *193*, 727–740. [CrossRef]

2. Vaupel, P. Tumor microenvironmental physiology and its implications for radiation oncology. *Semin. Radiat. Oncol.* **2004**, *14*, 198–206. [CrossRef]

3. Jimenez-Sanchez, A.; Memon, D.; Pourpe, S.; Veeraraghavan, H.; Li, Y.; Vargas, H.A.; Gill, M.B.; Park, K.J.; Zivanovic, O.; Konner, J.; et al. Heterogeneous Tumor-Immune Microenvironments among Differentially Growing Metastases in an Ovarian Cancer Patient. *Cell* **2017**, *170*, 927–938.e920. [CrossRef]

4. Johnson, B.E.; Mazor, T.; Hong, C.; Barnes, M.; Aihara, K.; McLean, C.Y.; Fouse, S.D.; Yamamoto, S.; Ueda, H.; Tatsuno, K.; et al. Mutational analysis reveals the origin and therapy-driven evolution of recurrent glioma. *Science* **2014**, *343*, 189–193. [CrossRef]

5. De Bruin, E.C.; McGranahan, N.; Mitter, R.; Salm, M.; Wedge, D.C.; Yates, L.; Jamal-Hanjani, M.; Shafi, S.; Murugaesu, N.; Rowan, A.J.; et al. Spatial and temporal diversity in genomic instability processes defines lung cancer evolution. *Science* **2014**, *346*, 251–256. [CrossRef]

6. McGranahan, N.; Favero, F.; de Bruin, E.C.; Birkbak, N.J.; Szallasi, Z.; Swanton, C. Clonal status of actionable driver events and the timing of mutational processes in cancer evolution. *Sci. Transl. Med.* **2015**, *7*, 283ra254. [CrossRef]

7. Natrajan, R.; Sailem, H.; Mardakheh, F.K.; Arias Garcia, M.; Tape, C.J.; Dowsett, M.; Bakal, C.; Yuan, Y. Microenvironmental Heterogeneity Parallels Breast Cancer Progression: A Histology-Genomic Integration Analysis. *PLoS Med.* **2016**, *13*, e1001961. [CrossRef]

8. Azizi, E.; Carr, A.J.; Plitas, G.; Cornish, A.E.; Konopacki, C.; Prabhakaran, S.; Nainys, J.; Wu, K.; Kiseliovas, V.; Setty, M. Single-Cell Map of Diverse Immune Phenotypes in the Breast Tumor Microenvironment. *Cell* **2018**, *174*, 1293–1308.e1236. [CrossRef]

9. Rybinski, B.; Yun, K. Addressing intra-tumoral heterogeneity and therapy resistance. *Oncotarget* **2016**, *7*, 72322–72342. [CrossRef]

10. Mroz, E.A.; Tward, A.D.; Pickering, C.R.; Myers, J.N.; Ferris, R.L.; Rocco, J.W. High intratumor genetic heterogeneity is related to worse outcome in patients with head and neck squamous cell carcinoma. *Cancer* **2013**, *119*, 3034–3042. [CrossRef]

11. Caswell, D.R.; Swanton, C. The role of tumour heterogeneity and clonal cooperativity in metastasis, immune evasion and clinical outcome. *BMC Med.* **2017**, *15*, 133. [CrossRef]

12. Jamal-Hanjani, M.; Quezada, S.A.; Larkin, J.; Swanton, C. Translational implications of tumor heterogeneity. *Clin. Cancer Res.* **2015**, *21*, 1258–1266. [CrossRef]

13. Plava, J.; Cihova, M.; Burikova, M.; Matuskova, M.; Kucerova, L.; Miklikova, S. Recent advances in understanding tumor stroma-mediated chemoresistance in breast cancer. *Mol. Cancer* **2019**, *18*, 67. [CrossRef]

14. Marusyk, A.; Almendro, V.; Polyak, K. Intra-tumour heterogeneity: A looking glass for cancer? *Nat. Rev. Cancer* **2012**, *12*, 323. [CrossRef]

15. Wagner, K.U.; Rui, H. Jak2/Stat5 signaling in mammogenesis, breast cancer initiation and progression. *J. Mammary Gland. Biol. Neoplasia* **2008**, *13*, 93–103. [CrossRef]

16. Ahonen, T.J.; Xie, J.; LeBaron, M.J.; Zhu, J.; Nurmi, M.; Alanen, K.; Rui, H.; Nevalainen, M.T. Inhibition of transcription factor Stat5 induces cell death of human prostate cancer cells. *J. Biol. Chem.* **2003**, *278*, 27287–27292. [CrossRef]

17. Tan, S.H.; Dagvadorj, A.; Shen, F.; Gu, L.; Liao, Z.; Abdulghani, J.; Zhang, Y.; Gelmann, E.P.; Zellweger, T.; Culig, Z.; et al. Transcription factor Stat5 synergizes with androgen receptor in prostate cancer cells. *Cancer Res.* **2008**, *68*, 236–248. [CrossRef]

18. Leong, P.L.; Xi, S.; Drenning, S.D.; Dyer, K.F.; Wentzel, A.L.; Lerner, E.C.; Smithgall, T.E.; Grandis, J.R. Differential function of STAT5 isoforms in head and neck cancer growth control. *Oncogene* **2002**, *21*, 2846–2853. [CrossRef]

19. Grivennikov, S.; Karin, E.; Terzic, J.; Mucida, D.; Yu, G.Y.; Vallabhapurapu, S.; Scheller, J.; Rose-John, S.; Cheroutre, H.; Eckmann, L.; et al. IL-6 and Stat3 are required for survival of intestinal epithelial cells and development of colitis-associated cancer. *Cancer Cell* **2009**, *15*, 103–113. [CrossRef]

20. Hosui, A.; Kimura, A.; Yamaji, D.; Zhu, B.M.; Na, R.; Hennighausen, L. Loss of STAT5 causes liver fibrosis and cancer development through increased TGF-{beta} and STAT3 activation. *J. Exp. Med.* **2009**, *206*, 819–831. [CrossRef]

21. Lesina, M.; Kurkowski, M.U.; Ludes, K.; Rose-John, S.; Treiber, M.; Klöppel, G.; Yoshimura, A.; Reindl, W.; Sipos, B.; Akira, S.; et al. Stat3/Socs3 activation by IL-6 transsignaling promotes progression of pancreatic intraepithelial neoplasia and development of pancreatic cancer. *Cancer Cell* **2011**, *19*, 456–469. [CrossRef] [PubMed]

22. Burke, W.M.; Jin, X.; Lin, H.J.; Huang, M.; Liu, R.; Reynolds, R.K.; Lin, J. Inhibition of constitutively active Stat3 suppresses growth of human ovarian and breast cancer cells. *Oncogene* **2001**, *20*, 7925–7934. [CrossRef] [PubMed]

23. Wu, P.; Wu, D.; Zhao, L.; Huang, L.; Shen, G.; Huang, J.; Chai, Y. Prognostic role of STAT3 in solid tumors: A systematic review and meta-analysis. *Oncotarget* **2016**, *7*, 19863–19883. [CrossRef] [PubMed]

24. Mao, Y.L.; Li, Z.W.; Lou, C.J.; Pang, D.; Zhang, Y.Q. Phospho-STAT5 expression is associated with poor prognosis of human colonic adenocarcinoma. *Pathol. Oncol. Res.* **2011**, *17*, 333–339. [CrossRef] [PubMed]

25. Ma, J.H.; Qi, J.; Lin, S.Q.; Zhang, C.Y.; Liu, F.Y.; Xie, W.D.; Li, X. STAT3 targets ERR-alpha to promote epithelial-mesenchymal transition, migration and invasion in triple negative breast cancer cells. *Mol. Cancer Res.* **2019**. [CrossRef] [PubMed]

26. Xie, T.X.; Huang, F.J.; Aldape, K.D.; Kang, S.H.; Liu, M.; Gershenwald, J.E.; Xie, K.; Sawaya, R.; Huang, S. Activation of stat3 in human melanoma promotes brain metastasis. *Cancer Res.* **2006**, *66*, 3188–3196. [CrossRef] [PubMed]

27. Abdulghani, J.; Gu, L.; Dagvadorj, A.; Lutz, J.; Leiby, B.; Bonuccelli, G.; Lisanti, M.P.; Zellweger, T.; Alanen, K.; Mirtti, T.; et al. Stat3 promotes metastatic progression of prostate cancer. *Am. J. Pathol.* **2008**, *172*, 1717–1728. [CrossRef]

28. Klupp, F.; Diers, J.; Kahlert, C.; Neumann, L.; Halama, N.; Franz, C.; Schmidt, T.; Lasitschka, F.; Warth, A.; Weitz, J.; et al. Expressional STAT3/STAT5 Ratio is an Independent Prognostic Marker in Colon Carcinoma. *Ann. Surg. Oncol.* **2015**, *22* (Suppl. 3), S1548–S1555. [CrossRef]

29. Real, P.J.; Sierra, A.; De Juan, A.; Segovia, J.C.; Lopez-Vega, J.M.; Fernandez-Luna, J.L. Resistance to chemotherapy via Stat3-dependent overexpression of Bcl-2 in metastatic breast cancer cells. *Oncogene* **2002**, *21*, 7611–7618. [CrossRef]

30. Fan, Q.M.; Jing, Y.Y.; Yu, G.F.; Kou, X.R.; Ye, F.; Gao, L.; Li, R.; Zhao, Q.D.; Yang, Y.; Lu, Z.H.; et al. Tumor-associated macrophages promote cancer stem cell-like properties via transforming growth factor-beta1-induced epithelial-mesenchymal transition in hepatocellular carcinoma. *Cancer Lett.* **2014**, *352*, 160–168. [CrossRef]
31. Baskiewicz-Masiuk, M.; Machalinski, B. The role of the STAT5 proteins in the proliferation and apoptosis of the CML and AML cells. *Eur. J. Haematol.* **2004**, *72*, 420–429. [CrossRef] [PubMed]
32. Morcinek, J.C.; Weisser, C.; Geissinger, E.; Schartl, M.; Wellbrock, C. Activation of STAT5 triggers proliferation and contributes to anti-apoptotic signalling mediated by the oncogenic Xmrk kinase. *Oncogene* **2002**, *21*, 1668–1678. [CrossRef] [PubMed]
33. Mahmud, S.A.; Manlove, L.S.; Farrar, M.A. Interleukin-2 and STAT5 in regulatory T cell development and function. *JAKSTAT* **2013**, *2*, e23154. [CrossRef] [PubMed]
34. Rani, A.; Murphy, J.J. STAT5 in Cancer and Immunity. *J. Interferon Cytokine Res.* **2016**, *36*, 226–237. [CrossRef] [PubMed]
35. Han, S.; Toker, A.; Liu, Z.Q.; Ohashi, P.S. Turning the Tide Against Regulatory T Cells. *Front. Oncol.* **2019**, *9*, 279. [CrossRef]
36. Shou, J.; Zhang, Z.; Lai, Y.; Chen, Z.; Huang, J. Worse outcome in breast cancer with higher tumor-infiltrating FOXP3+ Tregs: A systematic review and meta-analysis. *BMC Cancer* **2016**, *16*, 687. [CrossRef]
37. Lim, C.P.; Cao, X. Structure, function, and regulation of STAT proteins. *Mol. Biosyst.* **2006**, *2*, 536–550. [CrossRef]
38. Bromberg, J. Stat proteins and oncogenesis. *J. Clin. Investig.* **2002**, *109*, 1139–1142. [CrossRef]
39. Clevenger, C.V. Roles and Regulation of Stat Family Transcription Factors in Human Breast Cancer. *Am. J. Pathol.* **2004**, *165*, 1449–1460. [CrossRef]
40. Decker, T.; Kovarik, P.; Meinke, A. GAS elements: A few nucleotides with a major impact on cytokine-induced gene expression. *J. Interferon Cytokine Res.* **1997**, *17*, 121–134. [CrossRef]
41. Rawlings, J.S.; Rosler, K.M.; Harrison, D.A. The JAK/STAT signaling pathway. *J. Cell Sci.* **2004**, *117*, 1281–1283. [CrossRef] [PubMed]
42. Murakami, M.; Kamimura, D.; Hirano, T. Pleiotropy and Specificity: Insights from the Interleukin 6 Family of Cytokines. *Immunity* **2019**, *50*, 812–831. [CrossRef] [PubMed]
43. Heinrich, P.C.; Behrmann, I.; Haan, S.; Hermanns, H.M.; Muller-Newen, G.; Schaper, F. Principles of interleukin (IL)-6-type cytokine signalling and its regulation. *Biochem. J.* **2003**, *374*, 1–20. [CrossRef] [PubMed]
44. Raz, R.; Lee, C.K.; Cannizzaro, L.A.; d'Eustachio, P.; Levy, D.E. Essential role of STAT3 for embryonic stem cell pluripotency. *Proc. Natl. Acad. Sci. USA* **1999**, *96*, 2846–2851. [CrossRef] [PubMed]
45. Kirito, K.; Osawa, M.; Morita, H.; Shimizu, R.; Yamamoto, M.; Oda, A.; Fujita, H.; Tanaka, M.; Nakajima, K.; Miura, Y.; et al. A functional role of Stat3 in in vivo megakaryopoiesis. *Blood* **2002**, *99*, 3220–3227. [CrossRef] [PubMed]
46. Akira, S. Roles of STAT3 defined by tissue-specific gene targeting. *Oncogene* **2000**, *19*, 2607–2611. [CrossRef]
47. Yao, Z.; Cui, Y.; Watford, W.T.; Bream, J.H.; Yamaoka, K.; Hissong, B.D.; Li, D.; Durum, S.K.; Jiang, Q.; Bhandoola, A.; et al. Stat5a/b are essential for normal lymphoid development and differentiation. *Proc. Natl. Acad. Sci. USA* **2006**, *103*, 1000–1005. [CrossRef]
48. Wu, M.; Song, D.; Li, H.; Yang, Y.; Ma, X.; Deng, S.; Ren, C.; Shu, X. Negative regulators of STAT3 signaling pathway in cancers. *Cancer Manag. Res.* **2019**, *11*, 4957–4969. [CrossRef]
49. Kershaw, N.J.; Murphy, J.M.; Lucet, I.S.; Nicola, N.A.; Babon, J.J. Regulation of Janus kinases by SOCS proteins. *Biochem. Soc. Trans.* **2013**, *41*, 1042–1047. [CrossRef]
50. Wingelhofer, B.; Neubauer, H.A.; Valent, P.; Han, X.; Constantinescu, S.N.; Gunning, P.T.; Muller, M.; Moriggl, R. Implications of STAT3 and STAT5 signaling on gene regulation and chromatin remodeling in hematopoietic cancer. *Leukemia* **2018**, *32*, 1713–1726. [CrossRef]
51. Orlova, A.; Wingelhofer, B.; Neubauer, H.A.; Maurer, B.; Berger-Becvar, A.; Keseru, G.M.; Gunning, P.T.; Valent, P.; Moriggl, R. Emerging therapeutic targets in myeloproliferative neoplasms and peripheral T-cell leukemia and lymphomas. *Expert Opin. Targets* **2018**, *22*, 45–57. [CrossRef] [PubMed]
52. Forbes, S.A.; Beare, D.; Boutselakis, H.; Bamford, S.; Bindal, N.; Tate, J.; Cole, C.G.; Ward, S.; Dawson, E.; Ponting, L.; et al. COSMIC: Somatic cancer genetics at high-resolution. *Nucleic Acids Res.* **2017**, *45*, D777–D783. [CrossRef] [PubMed]

53. Shahmarvand, N.; Nagy, A.; Shahryari, J.; Ohgami, R.S. Mutations in the signal transducer and activator of transcription family of genes in cancer. *Cancer Sci.* **2018**, *109*, 926–933. [CrossRef] [PubMed]

54. Zhang, T.; Kee, W.H.; Seow, K.T.; Fung, W.; Cao, X. The coiled-coil domain of Stat3 is essential for its SH2 domain-mediated receptor binding and subsequent activation induced by epidermal growth factor and interleukin-6. *Mol. Cell Biol.* **2000**, *20*, 7132–7139. [CrossRef] [PubMed]

55. Hu, G.; Witzig, T.E.; Gupta, M. A novel missense (M206K) STAT3 mutation in diffuse large B cell lymphoma deregulates STAT3 signaling. *PLoS ONE* **2013**, *8*, e67851. [CrossRef]

56. Levine, R.L.; Wadleigh, M.; Cools, J.; Ebert, B.L.; Wernig, G.; Huntly, B.J.; Boggon, T.J.; Wlodarska, I.; Clark, J.J.; Moore, S.; et al. Activating mutation in the tyrosine kinase JAK2 in polycythemia vera, essential thrombocythemia, and myeloid metaplasia with myelofibrosis. *Cancer Cell* **2005**, *7*, 387–397. [CrossRef]

57. James, C.; Ugo, V.; Le Couedic, J.P.; Staerk, J.; Delhommeau, F.; Lacout, C.; Garcon, L.; Raslova, H.; Berger, R.; Bennaceur-Griscelli, A.; et al. A unique clonal JAK2 mutation leading to constitutive signalling causes polycythaemia vera. *Nature* **2005**, *434*, 1144–1148. [CrossRef]

58. Panopoulos, A.D.; Zhang, L.; Snow, J.W.; Jones, D.M.; Smith, A.M.; El Kasmi, K.C.; Liu, F.; Goldsmith, M.A.; Link, D.C.; Murray, P.J.; et al. STAT3 governs distinct pathways in emergency granulopoiesis and mature neutrophils. *Blood* **2006**, *108*, 3682–3690. [CrossRef]

59. Kang, S.; Tanaka, T.; Narazaki, M.; Kishimoto, T. Targeting Interleukin-6 Signaling in Clinic. *Immunity* **2019**, *50*, 1007–1023. [CrossRef]

60. Mutze, J.; Roth, J.; Gerstberger, R.; Hubschle, T. Nuclear translocation of the transcription factor STAT5 in the rat brain after systemic leptin administration. *Neurosci. Lett.* **2007**, *417*, 286–291. [CrossRef]

61. Ribeiro, D.; Melao, A.; van Boxtel, R.; Santos, C.I.; Silva, A.; Silva, M.C.; Cardoso, B.A.; Coffer, P.J.; Barata, J.T. STAT5 is essential for IL-7-mediated viability, growth, and proliferation of T-cell acute lymphoblastic leukemia cells. *Blood Adv.* **2018**, *2*, 2199–2213. [CrossRef] [PubMed]

62. Demoulin, J.B.; Uyttenhove, C.; Lejeune, D.; Mui, A.; Groner, B.; Renauld, J.C. STAT5 activation is required for interleukin-9-dependent growth and transformation of lymphoid cells. *Cancer Res.* **2000**, *60*, 3971–3977. [PubMed]

63. Rochman, Y.; Kashyap, M.; Robinson, G.W.; Sakamoto, K.; Gomez-Rodriguez, J.; Wagner, K.U.; Leonard, W.J. Thymic stromal lymphopoietin-mediated STAT5 phosphorylation via kinases JAK1 and JAK2 reveals a key difference from IL-7-induced signaling. *Proc. Natl. Acad. Sci. USA* **2010**, *107*, 19455–19460. [CrossRef] [PubMed]

64. Okutani, Y.; Kitanaka, A.; Tanaka, T.; Kamano, H.; Ohnishi, H.; Kubota, Y.; Ishida, T.; Takahara, J. Src directly tyrosine-phosphorylates STAT5 on its activation site and is involved in erythropoietin-induced signaling pathway. *Oncogene* **2001**, *20*, 6643–6650. [CrossRef] [PubMed]

65. Kirito, K.; Watanabe, T.; Sawada, K.; Endo, H.; Ozawa, K.; Komatsu, N. Thrombopoietin regulates Bcl-xL gene expression through Stat5 and phosphatidylinositol 3-kinase activation pathways. *J. Biol. Chem.* **2002**, *277*, 8329–8337. [CrossRef]

66. Scheeren, F.A.; Diehl, S.A.; Smit, L.A.; Beaumont, T.; Naspetti, M.; Bende, R.J.; Blom, B.; Karube, K.; Ohshima, K.; van Noesel, C.J.; et al. IL-21 is expressed in Hodgkin lymphoma and activates STAT5: Evidence that activated STAT5 is required for Hodgkin lymphomagenesis. *Blood* **2008**, *111*, 4706–4715. [CrossRef]

67. Maier, E.; Mittermeir, M.; Ess, S.; Neuper, T.; Schmiedlechner, A.; Duschl, A.; Horejs-Hoeck, J. Prerequisites for Functional Interleukin 31 Signaling and Its Feedback Regulation by Suppressor of Cytokine Signaling 3 (SOCS3). *J. Biol. Chem.* **2015**, *290*, 24747–24759. [CrossRef]

68. Toniolo, P.A.; Liu, S.; Yeh, J.E.; Moraes-Vieira, P.M.; Walker, S.R.; Vafaizadeh, V.; Barbuto, J.A.; Frank, D.A. Inhibiting STAT5 by the BET bromodomain inhibitor JQ1 disrupts human dendritic cell maturation. *J. Immunol.* **2015**, *194*, 3180–3190. [CrossRef]

69. Tripathi, P.; Kurtulus, S.; Wojciechowski, S.; Sholl, A.; Hoebe, K.; Morris, S.C.; Finkelman, F.D.; Grimes, H.L.; Hildeman, D.A. STAT5 is critical to maintain effector CD8+ T cell responses. *J. Immunol.* **2010**, *185*, 2116–2124. [CrossRef]

70. Kieslinger, M.; Woldman, I.; Moriggl, R.; Hofmann, J.; Marine, J.C.; Ihle, J.N.; Beug, H.; Decker, T. Antiapoptotic activity of Stat5 required during terminal stages of myeloid differentiation. *Genes Dev.* **2000**, *14*, 232–244.

71. Kuroda, E.; Ho, V.; Ruschmann, J.; Antignano, F.; Hamilton, M.; Rauh, M.J.; Antov, A.; Flavell, R.A.; Sly, L.M.; Krystal, G. SHIP represses the generation of IL-3-induced M2 macrophages by inhibiting IL-4 production from basophils. *J. Immunol.* **2009**, *183*, 3652–3660. [CrossRef] [PubMed]

72. Park, H.J.; Li, J.; Hannah, R.; Biddie, S.; Leal-Cervantes, A.I.; Kirschner, K.; Flores Santa Cruz, D.; Sexl, V.; Gottgens, B.; Green, A.R. Cytokine-induced megakaryocytic differentiation is regulated by genome-wide loss of a uSTAT transcriptional program. *EMBO J.* **2016**, *35*, 580–594. [CrossRef] [PubMed]

73. Kim, A.R.; Ulirsch, J.C.; Wilmes, S.; Unal, E.; Moraga, I.; Karakukcu, M.; Yuan, D.; Kazerounian, S.; Abdulhay, N.J.; King, D.S.; et al. Functional Selectivity in Cytokine Signaling Revealed Through a Pathogenic EPO Mutation. *Cell* **2017**, *168*, 1053–1064 e1015. [CrossRef] [PubMed]

74. Wingelhofer, B.; Maurer, B.; Heyes, E.C.; Cumaraswamy, A.A.; Berger-Becvar, A.; de Araujo, E.D.; Orlova, A.; Freund, P.; Ruge, F.; Park, J.; et al. Pharmacologic inhibition of STAT5 in acute myeloid leukemia. *Leukemia* **2018**, *32*, 1135–1146. [CrossRef]

75. Page, B.D.; Khoury, H.; Laister, R.C.; Fletcher, S.; Vellozo, M.; Manzoli, A.; Yue, P.; Turkson, J.; Minden, M.D.; Gunning, P.T. Small molecule STAT5-SH2 domain inhibitors exhibit potent antileukemia activity. *J. Med. Chem.* **2012**, *55*, 1047–1055. [CrossRef]

76. Elumalai, N.; Berg, A.; Rubner, S.; Blechschmidt, L.; Song, C.; Natarajan, K.; Matysik, J.; Berg, T. Rational development of Stafib-2: A selective, nanomolar inhibitor of the transcription factor STAT5b. *Sci. Rep.* **2017**, *7*, 819. [CrossRef]

77. Unver, N.; McAllister, F. IL-6 family cytokines: Key inflammatory mediators as biomarkers and potential therapeutic targets. *Cytokine Growth Factor Rev.* **2018**, *41*, 10–17. [CrossRef]

78. Sullivan, N.J.; Sasser, A.K.; Axel, A.E.; Vesuna, F.; Raman, V.; Ramirez, N.; Oberyszyn, T.M.; Hall, B.M. Interleukin-6 induces an epithelial-mesenchymal transition phenotype in human breast cancer cells. *Oncogene* **2009**, *28*, 2940–2947. [CrossRef]

79. Putoczki, T.L.; Ernst, M. IL-11 signaling as a therapeutic target for cancer. *Immunotherapy* **2015**, *7*, 441–453. [CrossRef]

80. Lu, J.; Ksendzovsky, A.; Yang, C.; Mehta, G.U.; Yong, R.L.; Weil, R.J.; Park, D.M.; Mushlin, H.M.; Fang, X.; Balgley, B.M.; et al. CNTF receptor subunit alpha as a marker for glioma tumor-initiating cells and tumor grade: Laboratory investigation. *J. Neurosurg.* **2012**, *117*, 1022–1031. [CrossRef]

81. Li, X.; Yang, Q.; Yu, H.; Wu, L.; Zhao, Y.; Zhang, C.; Yue, X.; Liu, Z.; Wu, H.; Haffty, B.G.; et al. LIF promotes tumorigenesis and metastasis of breast cancer through the AKT-mTOR pathway. *Oncotarget* **2014**, *5*, 788–801. [CrossRef] [PubMed]

82. Liao, W.C.; Lin, J.T.; Wu, C.Y.; Huang, S.P.; Lin, M.T.; Wu, A.S.; Huang, Y.J.; Wu, M.S. Serum interleukin-6 level but not genotype predicts survival after resection in stages II and III gastric carcinoma. *Clin. Cancer Res.* **2008**, *14*, 428–434. [CrossRef] [PubMed]

83. Jiang, X.P.; Yang, D.C.; Elliott, R.L.; Head, J.F. Reduction in serum IL-6 after vacination of breast cancer patients with tumour-associated antigens is related to estrogen receptor status. *Cytokine* **2000**, *12*, 458–465. [CrossRef] [PubMed]

84. Chung, Y.-C.; Chang, Y.-F. Serum interleukin-6 levels reflect the disease status of colorectal cancer. *J. Surg. Oncol.* **2003**, *83*, 222–226. [CrossRef] [PubMed]

85. Tawara, K.; Oxford, J.T.; Jorcyk, C.L. Clinical significance of interleukin (IL)-6 in cancer metastasis to bone: Potential of anti-IL-6 therapies. *Cancer Manag. Res.* **2011**, *3*, 177–189. [CrossRef] [PubMed]

86. Komoda, H.; Tanaka, Y.; Honda, M.; Matsuo, Y.; Hazama, K.; Takao, T. Interleukin-6 levels in colorectal cancer tissues. *World J. Surg.* **1998**, *22*, 895–898. [CrossRef]

87. Ebbing, E.A.; van der Zalm, A.P.; Steins, A.; Creemers, A.; Hermsen, S.; Rentenaar, R.; Klein, M.; Waasdorp, C.; Hooijer, G.K.J.; Meijer, S.L.; et al. Stromal-derived interleukin 6 drives epithelial-to-mesenchymal transition and therapy resistance in esophageal adenocarcinoma. *Proc. Natl. Acad. Sci. USA* **2019**, *116*, 2237–2242. [CrossRef]

88. Yadav, A.; Kumar, B.; Datta, J.; Teknos, T.N.; Kumar, P. IL-6 promotes head and neck tumor metastasis by inducing epithelial-mesenchymal transition via the JAK-STAT3-SNAIL signaling pathway. *Mol. Cancer Res.* **2011**, *9*, 1658–1667. [CrossRef]

89. Gyamfi, J.; Lee, Y.H.; Eom, M.; Choi, J. Interleukin-6/STAT3 signalling regulates adipocyte induced epithelial-mesenchymal transition in breast cancer cells. *Sci. Rep.* **2018**, *8*, 8859. [CrossRef]

90. Jiang, M.; Chen, J.; Zhang, W.; Zhang, R.; Ye, Y.; Liu, P.; Yu, W.; Wei, F.; Ren, X.; Yu, J. Interleukin-6 Trans-Signaling Pathway Promotes Immunosuppressive Myeloid-Derived Suppressor Cells via Suppression of Suppressor of Cytokine Signaling 3 in Breast Cancer. *Front. Immunol.* **2017**, *8*, 1840. [CrossRef]

91. Liu, Q.; Yu, S.; Li, A.; Xu, H.; Han, X.; Wu, K. Targeting interlukin-6 to relieve immunosuppression in tumor microenvironment. *Tumour Biol.* **2017**, *39*, 1010428317712445. [CrossRef] [PubMed]

92. Kato, T.; Noma, K.; Ohara, T.; Kashima, H.; Katsura, Y.; Sato, H.; Komoto, S.; Katsube, R.; Ninomiya, T.; Tazawa, H.; et al. Cancer-Associated Fibroblasts Affect Intratumoral CD8(+) and FoxP3(+) T Cells Via IL6 in the Tumor Microenvironment. *Clin. Cancer Res.* **2018**, *24*, 4820–4833. [CrossRef] [PubMed]

93. Chang, Q.; Bournazou, E.; Sansone, P.; Berishaj, M.; Gao, S.P.; Daly, L.; Wels, J.; Theilen, T.; Granitto, S.; Zhang, X.; et al. The IL-6/JAK/Stat3 feed-forward loop drives tumorigenesis and metastasis. *Neoplasia* **2013**, *15*, 848–862. [CrossRef] [PubMed]

94. Wu, X.; Tao, P.; Zhou, Q.; Li, J.; Yu, Z.; Wang, X.; Li, J.; Li, C.; Yan, M.; Zhu, Z.; et al. IL-6 secreted by cancer-associated fibroblasts promotes epithelial-mesenchymal transition and metastasis of gastric cancer via JAK2/STAT3 signaling pathway. *Oncotarget* **2017**, *8*, 20741–20750. [CrossRef] [PubMed]

95. Chang, Q.; Daly, L.; Bromberg, J. The IL-6 feed-forward loop: A driver of tumorigenesis. *Semin. Immunol.* **2014**, *26*, 48–53. [CrossRef] [PubMed]

96. Smigiel, J.; Parvani, J.G.; Tamagno, I.; Polak, K.; Jackson, M.W. Breaking the oncostatin M feed-forward loop to suppress metastasis and therapy failure. *J. Pathol.* **2018**, *245*, 6–8. [CrossRef]

97. Zarling, J.M.; Shoyab, M.; Marquardt, H.; Hanson, M.B.; Lioubin, M.N.; Todaro, G.J. Oncostatin M: A growth regulator produced by differentiated histiocytic lymphoma cells. *Proc. Natl. Acad. Sci. USA* **1986**, *83*, 9739–9743. [CrossRef]

98. Auguste, P.; Guillet, C.; Fourcin, M.; Olivier, C.; Veziers, J.; Pouplard-Barthelaix, A.; Gascan, H. Signaling of type II oncostatin M receptor. *J. Biol. Chem.* **1997**, *272*, 15760–15764. [CrossRef]

99. Mirmohammadsadegh, A.; Hassan, M.; Bardenheuer, W.; Marini, A.; Gustrau, A.; Nambiar, S.; Tannapfel, A.; Bojar, H.; Ruzicka, T.; Hengge, U.R. STAT5 phosphorylation in malignant melanoma is important for survival and is mediated through SRC and JAK1 kinases. *J. Investig. Derm.* **2006**, *126*, 2272–2280. [CrossRef]

100. Lacreusette, A.; Nguyen, J.M.; Pandolfino, M.C.; Khammari, A.; Dreno, B.; Jacques, Y.; Godard, A.; Blanchard, F. Loss of oncostatin M receptor β in metastatic melanoma cells. *Oncogene* **2006**, *26*, 881–892. [CrossRef]

101. Hassel, J.C.; Winnemoller, D.; Schartl, M.; Wellbrock, C. STAT5 contributes to antiapoptosis in melanoma. *Melanoma Res.* **2008**, *18*, 378–385. [CrossRef] [PubMed]

102. West, N.R.; Murray, J.I.; Watson, P.H. Oncostatin-M promotes phenotypic changes associated with mesenchymal and stem cell-like differentiation in breast cancer. *Oncogene* **2014**, *33*, 1485–1494. [CrossRef] [PubMed]

103. Lapeire, L.; Hendrix, A.; Lambein, K.; Van Bockstal, M.; Braems, G.; Van Den Broecke, R.; Limame, R.; Mestdagh, P.; Vandesompele, J.; Vanhove, C.; et al. Cancer-associated adipose tissue promotes breast cancer progression by paracrine oncostatin M and Jak/STAT3 signaling. *Cancer Res.* **2014**, *74*, 6806–6819. [CrossRef] [PubMed]

104. Kim, M.S.; Louwagie, J.; Carvalho, B.; Terhaar Sive Droste, J.S.; Park, H.L.; Chae, Y.K.; Yamashita, K.; Liu, J.; Ostrow, K.L.; Ling, S.; et al. Promoter DNA methylation of oncostatin m receptor-beta as a novel diagnostic and therapeutic marker in colon cancer. *PLoS ONE* **2009**, *4*, e6555. [CrossRef] [PubMed]

105. Queen, M.M.; Ryan, R.E.; Holzer, R.G.; Keller-Peck, C.R.; Jorcyk, C.L. Breast cancer cells stimulate neutrophils to produce oncostatin M: Potential implications for tumor progression. *Cancer Res.* **2005**, *65*, 8896–8904. [CrossRef]

106. Junk, D.J.; Bryson, B.L.; Smigiel, J.M.; Parameswaran, N.; Bartel, C.A.; Jackson, M.W. Oncostatin M promotes cancer cell plasticity through cooperative STAT3-SMAD3 signaling. *Oncogene* **2017**, *36*, 4001–4013. [CrossRef]

107. Jahani-Asl, A.; Yin, H.; Soleimani, V.D.; Haque, T.; Luchman, H.A.; Chang, N.C.; Sincennes, M.-C.; Puram, S.V.; Scott, A.M.; Lorimer, I.A.J.; et al. Control of glioblastoma tumorigenesis by feed-forward cytokine signaling. *Nat. Neurosci.* **2016**, *19*, 798–806. [CrossRef]

108. Shien, K.; Papadimitrakopoulou, V.A.; Ruder, D.; Behrens, C.; Shen, L.; Kalhor, N.; Song, J.; Lee, J.J.; Wang, J.; Tang, X.; et al. JAK1/STAT3 Activation through a Proinflammatory Cytokine Pathway Leads to Resistance to Molecularly Targeted Therapy in Non-Small Cell Lung Cancer. *Mol. Cancer* **2017**, *16*, 2234–2245. [CrossRef]

109. Smigiel, J.M.; Parameswaran, N.; Jackson, M.W. Potent EMT and CSC Phenotypes Are Induced By Oncostatin-M in Pancreatic Cancer. *Mol. Cancer Res.* **2017**, *15*, 478–488. [CrossRef]

110. Tawara, K.; Bolin, C.; Koncinsky, J.; Kadaba, S.; Covert, H.; Sutherland, C.; Bond, L.; Kronz, J.; Garbow, J.R.; Jorcyk, C.L. OSM potentiates preintravasation events, increases CTC counts, and promotes breast cancer metastasis to the lung. *Breast Cancer Res.* **2018**, *20*, 53. [CrossRef]

111. Doherty, M.R.; Cheon, H.; Junk, D.J.; Vinayak, S.; Varadan, V.; Telli, M.L.; Ford, J.M.; Stark, G.R.; Jackson, M.W. Interferon-beta represses cancer stem cell properties in triple-negative breast cancer. *Proc. Natl. Acad. Sci. USA* **2017**, *114*, 13792–13797. [CrossRef] [PubMed]

112. West, N.R.; Hegazy, A.N.; Owens, B.M.J.; Bullers, S.J.; Linggi, B.; Buonocore, S.; Coccia, M.; Gortz, D.; This, S.; Stockenhuber, K.; et al. Oncostatin M drives intestinal inflammation and predicts response to tumor necrosis factor-neutralizing therapy in patients with inflammatory bowel disease. *Nat. Med.* **2017**, *23*, 579–589. [CrossRef] [PubMed]

113. Diveu, C.; Venereau, E.; Froger, J.; Ravon, E.; Grimaud, L.; Rousseau, F.; Chevalier, S.; Gascan, H. Molecular and functional characterization of a soluble form of oncostatin M/interleukin-31 shared receptor. *J. Biol. Chem.* **2006**, *281*, 36673–36682. [CrossRef] [PubMed]

114. Tao, L.; Huang, G.; Wang, R.; Pan, Y.; He, Z.; Chu, X.; Song, H.; Chen, L. Cancer-associated fibroblasts treated with cisplatin facilitates chemoresistance of lung adenocarcinoma through IL-11/IL-11R/STAT3 signaling pathway. *Sci. Rep.* **2016**, *6*, 38408. [CrossRef]

115. Putoczki, T.L.; Thiem, S.; Loving, A.; Busuttil, R.A.; Wilson, N.J.; Ziegler, P.K.; Nguyen, P.M.; Preaudet, A.; Farid, R.; Edwards, K.M.; et al. Interleukin-11 is the dominant IL-6 family cytokine during gastrointestinal tumorigenesis and can be targeted therapeutically. *Cancer Cell* **2013**, *24*, 257–271. [CrossRef]

116. Yu, H.; Yue, X.; Zhao, Y.; Li, X.; Wu, L.; Zhang, C.; Liu, Z.; Lin, K.; Xu-Monette, Z.Y.; Young, K.H.; et al. LIF negatively regulates tumour-suppressor p53 through Stat3/ID1/MDM2 in colorectal cancers. *Nat. Commun.* **2014**, *5*, 5218. [CrossRef]

117. Yuan, J.H.; Yang, F.; Wang, F.; Ma, J.Z.; Guo, Y.J.; Tao, Q.F.; Liu, F.; Pan, W.; Wang, T.T.; Zhou, C.C.; et al. A long noncoding RNA activated by TGF-beta promotes the invasion-metastasis cascade in hepatocellular carcinoma. *Cancer Cell* **2014**, *25*, 666–681. [CrossRef]

118. Yue, X.; Zhao, Y.; Zhang, C.; Li, J.; Liu, Z.; Liu, J.; Hu, W. Leukemia inhibitory factor promotes EMT through STAT3-dependent miR-21 induction. *Oncotarget* **2016**, *7*, 3777–3790. [CrossRef]

119. Bohrer, L.R.; Chuntova, P.; Bade, L.K.; Beadnell, T.C.; Leon, R.P.; Brady, N.J.; Ryu, Y.; Goldberg, J.E.; Schmechel, S.C.; Koopmeiners, J.S.; et al. Activation of the FGFR-STAT3 pathway in breast cancer cells induces a hyaluronan-rich microenvironment that licenses tumor formation. *Cancer Res.* **2014**, *74*, 374–386. [CrossRef]

120. Quintanal-Villalonga, A.; Ojeda-Marquez, L.; Marrugal, A.; Yague, P.; Ponce-Aix, S.; Salinas, A.; Carnero, A.; Ferrer, I.; Molina-Pinelo, S.; Paz-Ares, L. The FGFR4-388arg Variant Promotes Lung Cancer Progression by N-Cadherin Induction. *Sci. Rep.* **2018**, *8*, 2394. [CrossRef]

121. Yao, C.; Su, L.; Shan, J.; Zhu, C.; Liu, L.; Liu, C.; Xu, Y.; Yang, Z.; Bian, X.; Shao, J.; et al. IGF/STAT3/NANOG/Slug Signaling Axis Simultaneously Controls Epithelial-Mesenchymal Transition and Stemness Maintenance in Colorectal Cancer. *Stem Cells* **2016**, *34*, 820–831. [CrossRef] [PubMed]

122. Ding, X.; Ji, J.; Jiang, J.; Cai, Q.; Wang, C.; Shi, M.; Yu, Y.; Zhu, Z.; Zhang, J. HGF-mediated crosstalk between cancer-associated fibroblasts and MET-unamplified gastric cancer cells activates coordinated tumorigenesis and metastasis. *Cell Death Dis.* **2018**, *9*, 867. [CrossRef] [PubMed]

123. Cheng, C.C.; Liao, P.N.; Ho, A.S.; Lim, K.H.; Chang, J.; Su, Y.W.; Chen, C.G.; Chiang, Y.W.; Yang, B.L.; Lin, H.C.; et al. STAT3 exacerbates survival of cancer stem-like tumorspheres in EGFR-positive colorectal cancers: RNAseq analysis and therapeutic screening. *J. Biomed. Sci.* **2018**, *25*, 60. [CrossRef] [PubMed]

124. Zhao, D.; Pan, C.; Sun, J.; Gilbert, C.; Drews-Elger, K.; Azzam, D.J.; Picon-Ruiz, M.; Kim, M.; Ullmer, W.; El-Ashry, D.; et al. VEGF drives cancer-initiating stem cells through VEGFR-2/Stat3 signaling to upregulate Myc and Sox2. *Oncogene* **2015**, *34*, 3107–3119. [CrossRef]

125. Ando, T.; Kudo, Y.; Iizuka, S.; Tsunematsu, T.; Umehara, H.; Shrestha, M.; Matsuo, T.; Kubo, T.; Shimose, S.; Arihiro, K.; et al. Ameloblastin induces tumor suppressive phenotype and enhances chemosensitivity to doxorubicin via Src-Stat3 inactivation in osteosarcoma. *Sci. Rep.* **2017**, *7*, 40187. [CrossRef]

126. Ma, L.; Xu, Z.; Wang, J.; Zhu, Z.; Lin, G.; Jiang, L.; Lu, X.; Zou, C. Matrine inhibits BCR/ABL mediated ERK/MAPK pathway in human leukemia cells. *Oncotarget* **2017**, *8*, 108880–108889. [CrossRef]

127. Puram, S.V.; Yeung, C.M.; Jahani-Asl, A.; Lin, C.; de la Iglesia, N.; Konopka, G.; Jackson-Grusby, L.; Bonni, A. STAT3-iNOS Signaling Mediates EGFRvIII-Induced Glial Proliferation and Transformation. *J. Neurosci.* **2012**, *32*, 7806–7818. [CrossRef]

128. Yuan, Z.L.; Guan, Y.J.; Chatterjee, D.; Chin, Y.E. Stat3 dimerization regulated by reversible acetylation of a single lysine residue. *Science* **2005**, *307*, 269–273. [CrossRef]

129. Limagne, E.; Thibaudin, M.; Euvrard, R.; Berger, H.; Chalons, P.; Vegan, F.; Humblin, E.; Boidot, R.; Rebe, C.; Derangere, V.; et al. Sirtuin-1 Activation Controls Tumor Growth by Impeding Th17 Differentiation via STAT3 Deacetylation. *Cell Rep.* **2017**, *19*, 746–759. [CrossRef]

130. Johnston, P.A.; Grandis, J.R. STAT3 signaling: Anticancer strategies and challenges. *Mol. Interv.* **2011**, *11*, 18–26. [CrossRef]

131. Bromberg, J.F.; Wrzeszczynska, M.H.; Devgan, G.; Zhao, Y.; Pestell, R.G.; Albanese, C.; Darnell, J.E. Stat3 as an Oncogene. *Cell* **1999**, *98*, 295–303. [CrossRef]

132. Avalle, L.; Pensa, S.; Regis, G.; Novelli, F.; Poli, V. STAT1 and STAT3 in tumorigenesis: A matter of balance. *JAKSTAT* **2012**, *1*, 65–72. [CrossRef] [PubMed]

133. Bromberg, J.F.; Horvath, C.M.; Besser, D.; Lathem, W.W.; Darnell, J.E., Jr. Stat3 activation is required for cellular transformation by v-src. *Mol. Cell Biol.* **1998**, *18*, 2553–2558. [CrossRef] [PubMed]

134. Dechow, T.N.; Pedranzini, L.; Leitch, A.; Leslie, K.; Gerald, W.L.; Linkov, I.; Bromberg, J.F. Requirement of matrix metalloproteinase-9 for the transformation of human mammary epithelial cells by Stat3-C. *Proc. Natl. Acad. Sci. USA* **2004**, *101*, 10602–10607. [CrossRef]

135. Gough, D.J.; Corlett, A.; Schlessinger, K.; Wegrzyn, J.; Larner, A.C.; Levy, D.E. Mitochondrial STAT3 supports Ras-dependent oncogenic transformation. *Science* **2009**, *324*, 1713–1716. [CrossRef]

136. Fukuda, A.; Wang, S.C.; Morris, J.P.t.; Folias, A.E.; Liou, A.; Kim, G.E.; Akira, S.; Boucher, K.M.; Firpo, M.A.; Mulvihill, S.J.; et al. Stat3 and MMP7 contribute to pancreatic ductal adenocarcinoma initiation and progression. *Cancer Cell* **2011**, *19*, 441–455. [CrossRef]

137. Bryson, B.L.; Junk, D.J.; Cipriano, R.; Jackson, M.W. STAT3-mediated SMAD3 activation underlies Oncostatin M-induced Senescence. *Cell Cycle* **2017**, *16*, 319–334. [CrossRef]

138. Rokavec, M.; Oner, M.G.; Li, H.; Jackstadt, R.; Jiang, L.; Lodygin, D.; Kaller, M.; Horst, D.; Ziegler, P.K.; Schwitalla, S.; et al. IL-6R/STAT3/miR-34a feedback loop promotes EMT-mediated colorectal cancer invasion and metastasis. *J. Clin. Investig.* **2014**, *124*, 1853–1867. [CrossRef]

139. Cho, K.H.; Jeong, K.J.; Shin, S.C.; Kang, J.; Park, C.G.; Lee, H.Y. STAT3 mediates TGF-beta1-induced TWIST1 expression and prostate cancer invasion. *Cancer Lett.* **2013**, *336*, 167–173. [CrossRef]

140. Xiong, H.; Hong, J.; Du, W.; Lin, Y.-w.; Ren, L.-l.; Wang, Y.-c.; Su, W.-y.; Wang, J.-l.; Cui, Y.; Wang, Z.-h.; et al. Roles of STAT3 and ZEB1 proteins in E-cadherin down-regulation and human colorectal cancer epithelial-mesenchymal transition. *J. Biol. Chem.* **2012**, *287*, 5819–5832. [CrossRef]

141. Singh, A.; Settleman, J. EMT, cancer stem cells and drug resistance: An emerging axis of evil in the war on cancer. *Oncogene* **2010**, *29*, 4741–4751. [CrossRef] [PubMed]

142. Xie, T.X.; Wei, D.; Liu, M.; Gao, A.C.; Ali-Osman, F.; Sawaya, R.; Huang, S. Stat3 activation regulates the expression of matrix metalloproteinase-2 and tumor invasion and metastasis. *Oncogene* **2004**, *23*, 3550–3560. [CrossRef] [PubMed]

143. Jia, Z.H.; Jia, Y.; Guo, F.J.; Chen, J.; Zhang, X.W.; Cui, M.H. Phosphorylation of STAT3 at Tyr705 regulates MMP-9 production in epithelial ovarian cancer. *PLoS ONE* **2017**, *12*, e0183622. [CrossRef] [PubMed]

144. Lin, C.Y.; Tsai, P.H.; Kandaswami, C.C.; Lee, P.P.; Huang, C.J.; Hwang, J.J.; Lee, M.T. Matrix metalloproteinase-9 cooperates with transcription factor Snail to induce epithelial-mesenchymal transition. *Cancer Sci.* **2011**, *102*, 815–827. [CrossRef] [PubMed]

145. Doyle, S.; Evans, A.J.; Rakha, E.A.; Green, A.R.; Ellis, I.O. Influence of E-cadherin expression on the mammographic appearance of invasive nonlobular breast carcinoma detected at screening. *Radiology* **2009**, *253*, 51–55. [CrossRef]

146. Rakha, E.A.; Abd El Rehim, D.; Pinder, S.E.; Lewis, S.A.; Ellis, I.O. E-cadherin expression in invasive non-lobular carcinoma of the breast and its prognostic significance. *Histopathology* **2005**, *46*, 685–693. [CrossRef]

147. Rakha, E.A.; Patel, A.; Powe, D.G.; Benhasouna, A.; Green, A.R.; Lambros, M.B.; Reis-Filho, J.S.; Ellis, I.O. Clinical and biological significance of E-cadherin protein expression in invasive lobular carcinoma of the breast. *Am. J. Surg. Pathol.* **2010**, *34*, 1472–1479. [CrossRef]

148. Liu, T.; Zhang, X.; Shang, M.; Zhang, Y.; Xia, B.; Niu, M.; Liu, Y.; Pang, D. Dysregulated expression of Slug, vimentin, and E-cadherin correlates with poor clinical outcome in patients with basal-like breast cancer. *J. Surg. Oncol.* **2013**, *107*, 188–194. [CrossRef]

149. Markiewicz, A.; Welnicka-Jaskiewicz, M.; Seroczynska, B.; Skokowski, J.; Majewska, H.; Szade, J.; Zaczek, A.J. Epithelial-mesenchymal transition markers in lymph node metastases and primary breast tumors—Relation to dissemination and proliferation. *Am. J. Transl. Res.* **2014**, *6*, 793–808.

150. Zhou, L.; Dicker, D.T.; Matthew, E.; El-Deiry, W.S.; Alpaugh, R.K. Circulating tumor cells: Silent predictors of metastasis. *F1000Research* **2017**, *6*. [CrossRef]

151. Yu, M.; Bardia, A.; Wittner, B.S.; Stott, S.L.; Smas, M.E.; Ting, D.T.; Isakoff, S.J.; Ciciliano, J.C.; Wells, M.N.; Shah, A.M.; et al. Circulating breast tumor cells exhibit dynamic changes in epithelial and mesenchymal composition. *Science* **2013**, *339*, 580–584. [CrossRef] [PubMed]

152. Wu, S.; Liu, S.; Liu, Z.; Huang, J.; Pu, X.; Li, J.; Yang, D.; Deng, H.; Yang, N.; Xu, J. Classification of circulating tumor cells by epithelial-mesenchymal transition markers. *PLoS ONE* **2015**, *10*, e0123976. [CrossRef] [PubMed]

153. Satelli, A.; Mitra, A.; Brownlee, Z.; Xia, X.; Bellister, S.; Overman, M.J.; Kopetz, S.; Ellis, L.M.; Meng, Q.H.; Li, S. Epithelial-mesenchymal transitioned circulating tumor cells capture for detecting tumor progression. *Clin. Cancer Res.* **2015**, *21*, 899–906. [CrossRef] [PubMed]

154. Lawson, D.A.; Bhakta, N.R.; Kessenbrock, K.; Prummel, K.D.; Yu, Y.; Takai, K.; Zhou, A.; Eyob, H.; Balakrishnan, S.; Wang, C.Y.; et al. Single-cell analysis reveals a stem-cell program in human metastatic breast cancer cells. *Nature* **2015**, *526*, 131–135. [CrossRef] [PubMed]

155. Doherty, M.R.; Smigiel, J.M.; Junk, D.J.; Jackson, M.W. Cancer Stem Cell Plasticity Drives Therapeutic Resistance. *Cancers* **2016**, *8*, 8. [CrossRef]

156. Reya, T.; Morrison, S.J.; Clarke, M.F.; Weissman, I.L. Stem cells, cancer, and cancer stem cells. *Nature* **2001**, *414*, 105–111. [CrossRef]

157. Kim, C.F.; Jackson, E.L.; Woolfenden, A.E.; Lawrence, S.; Babar, I.; Vogel, S.; Crowley, D.; Bronson, R.T.; Jacks, T. Identification of bronchioalveolar stem cells in normal lung and lung cancer. *Cell* **2005**, *121*, 823–835. [CrossRef]

158. O'Brien, C.A.; Pollett, A.; Gallinger, S.; Dick, J.E. A human colon cancer cell capable of initiating tumour growth in immunodeficient mice. *Nature* **2007**, *445*, 106–110. [CrossRef]

159. Collins, A.T.; Berry, P.A.; Hyde, C.; Stower, M.J.; Maitland, N.J. Prospective identification of tumorigenic prostate cancer stem cells. *Cancer Res.* **2005**, *65*, 10946–10951. [CrossRef]

160. Hayashida, T.; Jinno, H.; Kitagawa, Y.; Kitajima, M. Cooperation of cancer stem cell properties and epithelial-mesenchymal transition in the establishment of breast cancer metastasis. *J. Oncol.* **2011**, *2011*, 591427. [CrossRef]

161. Choi, S.; Yu, J.; Park, A.; Dubon, M.J.; Do, J.; Kim, Y.; Nam, D.; Noh, J.; Park, K.S. BMP-4 enhances epithelial mesenchymal transition and cancer stem cell properties of breast cancer cells via Notch signaling. *Sci. Rep.* **2019**, *9*, 11724. [CrossRef]

162. Hung, J.J.; Kao, Y.S.; Huang, C.H.; Hsu, W.H. Overexpression of Aiolos promotes epithelial-mesenchymal transition and cancer stem cell-like properties in lung cancer cells. *Sci. Rep.* **2019**, *9*, 2991. [CrossRef] [PubMed]

163. Reiman, J.M.; Knutson, K.L.; Radisky, D.C. Immune promotion of epithelial-mesenchymal transition and generation of breast cancer stem cells. *Cancer Res.* **2010**, *70*, 3005–3008. [CrossRef] [PubMed]

164. Kashyap, V.; Rezende, N.C.; Scotland, K.B.; Shaffer, S.M.; Persson, J.L.; Gudas, L.J.; Mongan, N.P. Regulation of stem cell pluripotency and differentiation involves a mutual regulatory circuit of the NANOG, OCT4, and SOX2 pluripotency transcription factors with polycomb repressive complexes and stem cell microRNAs. *Stem Cells Dev.* **2009**, *18*, 1093–1108. [CrossRef] [PubMed]

165. Loh, Y.H.; Wu, Q.; Chew, J.L.; Vega, V.B.; Zhang, W.; Chen, X.; Bourque, G.; George, J.; Leong, B.; Liu, J.; et al. The Oct4 and Nanog transcription network regulates pluripotency in mouse embryonic stem cells. *Nat. Genet.* **2006**, *38*, 431–440. [CrossRef]

166. Zhou, J.; Wulfkuhle, J.; Zhang, H.; Gu, P.; Yang, Y.; Deng, J.; Margolick, J.B.; Liotta, L.A.; Petricoin, E., 3rd; Zhang, Y. Activation of the PTEN/mTOR/STAT3 pathway in breast cancer stem-like cells is required for viability and maintenance. *Proc. Natl. Acad. Sci. USA* **2007**, *104*, 16158–16163. [CrossRef]

167. Kim, S.Y.; Kang, J.W.; Song, X.; Kim, B.K.; Yoo, Y.D.; Kwon, Y.T.; Lee, Y.J. Role of the IL-6-JAK1-STAT3-Oct-4 pathway in the conversion of non-stem cancer cells into cancer stem-like cells. *Cell Signal* **2013**, *25*, 961–969. [CrossRef]

168. Bareiss, P.M.; Paczulla, A.; Wang, H.; Schairer, R.; Wiehr, S.; Kohlhofer, U.; Rothfuss, O.C.; Fischer, A.; Perner, S.; Staebler, A.; et al. SOX2 expression associates with stem cell state in human ovarian carcinoma. *Cancer Res.* **2013**, *73*, 5544–5555. [CrossRef]

169. Jiang, J.; Li, Z.; Yu, C.; Chen, M.; Tian, S.; Sun, C. MiR-1181 inhibits stem cell-like phenotypes and suppresses SOX2 and STAT3 in human pancreatic cancer. *Cancer Lett.* **2015**, *356*, 962–970. [CrossRef]

170. Yin, X.; Zhang, B.H.; Zheng, S.S.; Gao, D.M.; Qiu, S.J.; Wu, W.Z.; Ren, Z.G. Coexpression of gene Oct4 and Nanog initiates stem cell characteristics in hepatocellular carcinoma and promotes epithelial-mesenchymal transition through activation of Stat3/Snail signaling. *J. Hematol. Oncol.* **2015**, *8*, 23. [CrossRef]

171. Van den Hoogen, C.; van der Horst, G.; Cheung, H.; Buijs, J.T.; Lippitt, J.M.; Guzman-Ramirez, N.; Hamdy, F.C.; Eaton, C.L.; Thalmann, G.N.; Cecchini, M.G.; et al. High aldehyde dehydrogenase activity identifies tumor-initiating and metastasis-initiating cells in human prostate cancer. *Cancer Res.* **2010**, *70*, 5163–5173. [CrossRef] [PubMed]

172. Liu, C.; Kelnar, K.; Liu, B.; Chen, X.; Calhoun-Davis, T.; Li, H.; Patrawala, L.; Yan, H.; Jeter, C.; Honorio, S.; et al. The microRNA miR-34a inhibits prostate cancer stem cells and metastasis by directly repressing CD44. *Nat. Med.* **2011**, *17*, 211–215. [CrossRef] [PubMed]

173. Pascual, G.; Avgustinova, A.; Mejetta, S.; Martin, M.; Castellanos, A.; Attolini, C.S.; Berenguer, A.; Prats, N.; Toll, A.; Hueto, J.A.; et al. Targeting metastasis-initiating cells through the fatty acid receptor CD36. *Nature* **2017**, *541*, 41–45. [CrossRef] [PubMed]

174. Mimeault, M.; Batra, S.K. Hypoxia-inducing factors as master regulators of stemness properties and altered metabolism of cancer- and metastasis-initiating cells. *J. Cell Mol. Med.* **2013**, *17*, 30–54. [CrossRef]

175. Pastushenko, I.; Brisebarre, A.; Sifrim, A.; Fioramonti, M.; Revenco, T.; Boumahdi, S.; Van Keymeulen, A.; Brown, D.; Moers, V.; Lemaire, S.; et al. Identification of the tumour transition states occurring during EMT. *Nature* **2018**, *556*, 463–468. [CrossRef]

176. Goldman, A.; Majumder, B.; Dhawan, A.; Ravi, S.; Goldman, D.; Kohandel, M.; Majumder, P.K.; Sengupta, S. Temporally sequenced anticancer drugs overcome adaptive resistance by targeting a vulnerable chemotherapy-induced phenotypic transition. *Nat. Commun.* **2015**, *6*, 6139. [CrossRef]

177. Kim, C.; Gao, R.; Sei, E.; Brandt, R.; Hartman, J.; Hatschek, T.; Crosetto, N.; Foukakis, T.; Navin, N.E. Chemoresistance Evolution in Triple-Negative Breast Cancer Delineated by Single-Cell Sequencing. *Cell* **2018**, *173*, 879–893 e813. [CrossRef]

178. Zheng, X.; Carstens, J.L.; Kim, J.; Scheible, M.; Kaye, J.; Sugimoto, H.; Wu, C.C.; LeBleu, V.S.; Kalluri, R. Epithelial-to-mesenchymal transition is dispensable for metastasis but induces chemoresistance in pancreatic cancer. *Nature* **2015**, *527*, 525–530. [CrossRef]

179. Fischer, K.R.; Durrans, A.; Lee, S.; Sheng, J.; Li, F.; Wong, S.T.; Choi, H.; El Rayes, T.; Ryu, S.; Troeger, J.; et al. Epithelial-to-mesenchymal transition is not required for lung metastasis but contributes to chemoresistance. *Nature* **2015**, *527*, 472–476. [CrossRef]

180. Zhu, X.; Shen, H.; Yin, X.; Long, L.; Chen, X.; Feng, F.; Liu, Y.; Zhao, P.; Xu, Y.; Li, M.; et al. IL-6R/STAT3/miR-204 feedback loop contributes to cisplatin resistance of epithelial ovarian cancer cells. *Oncotarget* **2017**, *8*, 39154–39166. [CrossRef]

181. Ganguly, D.; Fan, M.; Yang, C.H.; Zbytek, B.; Finkelstein, D.; Roussel, M.F.; Pfeffer, L.M. The critical role that STAT3 plays in glioma-initiating cells: STAT3 addiction in glioma. *Oncotarget* **2018**, *9*, 22095–22112. [CrossRef] [PubMed]

182. Liu, F.; Cao, J.; Wu, J.; Sullivan, K.; Shen, J.; Ryu, B.; Xu, Z.; Wei, W.; Cui, R. Stat3-targeted therapies overcome the acquired resistance to vemurafenib in melanomas. *J. Investig. Derm.* **2013**, *133*, 2041–2049. [CrossRef] [PubMed]

183. Nagathihalli, N.S.; Castellanos, J.A.; Lamichhane, P.; Messaggio, F.; Shi, C.; Dai, X.; Rai, P.; Chen, X.; VanSaun, M.N.; Merchant, N.B. Inverse Correlation of STAT3 and MEK Signaling Mediates Resistance to RAS Pathway Inhibition in Pancreatic Cancer. *Cancer Res.* **2018**, *78*, 6235–6246. [CrossRef] [PubMed]

184. Zhang, X.; Ren, D.; Wu, X.; Lin, X.; Ye, L.; Lin, C.; Wu, S.; Zhu, J.; Peng, X.; Song, L. miR-1266 Contributes to Pancreatic Cancer Progression and Chemoresistance by the STAT3 and NF-kappaB Signaling Pathways. *Mol. Nucleic Acids* **2018**, *11*, 142–158. [CrossRef]

185. Kettner, N.M.; Vijayaraghavan, S.; Durak, M.G.; Bui, T.; Kohansal, M.; Ha, M.J.; Liu, B.; Rao, X.; Wang, J.; Yi, M.; et al. Combined Inhibition of STAT3 and DNA Repair in Palbociclib-Resistant ER-Positive Breast Cancer. *Clin. Cancer Res.* **2019**, *25*, 3996–4013. [CrossRef]

186. Wang, L.; Zhang, F.; Cui, J.Y.; Chen, L.; Chen, Y.T.; Liu, B.W. CAFs enhance paclitaxel resistance by inducing EMT through the IL6/JAK2/STAT3 pathway. *Oncol. Rep.* **2018**, *39*, 2081–2090. [CrossRef]

187. Gupta, P.B.; Fillmore, C.M.; Jiang, G.; Shapira, S.D.; Tao, K.; Kuperwasser, C.; Lander, E.S. Stochastic state transitions give rise to phenotypic equilibrium in populations of cancer cells. *Cell* **2011**, *146*, 633–644. [CrossRef]

188. Korkaya, H.; Kim, G.I.; Davis, A.; Malik, F.; Henry, N.L.; Ithimakin, S.; Quraishi, A.A.; Tawakkol, N.; D'Angelo, R.; Paulson, A.K.; et al. Activation of an IL6 inflammatory loop mediates trastuzumab resistance in HER2+ breast cancer by expanding the cancer stem cell population. *Mol. Cell* **2012**, *47*, 570–584. [CrossRef]

189. Lee, H.J.; Zhuang, G.; Cao, Y.; Du, P.; Kim, H.J.; Settleman, J. Drug resistance via feedback activation of Stat3 in oncogene-addicted cancer cells. *Cancer Cell* **2014**, *26*, 207–221. [CrossRef]

190. Simon, G.R.; Somaiah, N. A tabulated summary of targeted and biologic therapies for non-small-cell lung cancer. *Clin. Lung Cancer* **2014**, *15*, 21–51. [CrossRef]

191. Abell, K.; Watson, C.J. The Jak/Stat pathway: A novel way to regulate PI3K activity. *Cell Cycle* **2005**, *4*, 897–900. [CrossRef] [PubMed]

192. Avila, M.A.; Kagan, P.; Sultan, M.; Tachlytski, I.; Safran, M.; Ben-Ari, Z. Both MAPK and STAT3 signal transduction pathways are necessary for IL-6-dependent hepatic stellate cells activation. *PLoS ONE* **2017**, *12*. [CrossRef]

193. Fan, Y.; Mao, R.; Yang, J. NF-kappaB and STAT3 signaling pathways collaboratively link inflammation to cancer. *Protein Cell* **2013**, *4*, 176–185. [CrossRef] [PubMed]

194. Kamakura, S.; Oishi, K.; Yoshimatsu, T.; Nakafuku, M.; Masuyama, N.; Gotoh, Y. Hes binding to STAT3 mediates crosstalk between Notch and JAK-STAT signalling. *Nat. Cell Biol.* **2004**, *6*, 547–554. [CrossRef] [PubMed]

195. Fragoso, M.A.; Patel, A.K.; Nakamura, R.E.; Yi, H.; Surapaneni, K.; Hackam, A.S. The Wnt/beta-catenin pathway cross-talks with STAT3 signaling to regulate survival of retinal pigment epithelium cells. *PLoS ONE* **2012**, *7*, e46892. [CrossRef]

196. Tang, L.Y.; Heller, M.; Meng, Z.; Yu, L.R.; Tang, Y.; Zhou, M.; Zhang, Y.E. Transforming Growth Factor-beta (TGF-beta) Directly Activates the JAK1-STAT3 Axis to Induce Hepatic Fibrosis in Coordination with the SMAD Pathway. *J. Biol. Chem.* **2017**, *292*, 4302–4312. [CrossRef]

197. Bui, Q.T.; Im, J.H.; Jeong, S.B.; Kim, Y.M.; Lim, S.C.; Kim, B.; Kang, K.W. Essential role of Notch4/STAT3 signaling in epithelial-mesenchymal transition of tamoxifen-resistant human breast cancer. *Cancer Lett.* **2017**, *390*, 115–125. [CrossRef]

198. Canino, C.; Luo, Y.; Marcato, P.; Blandino, G.; Pass, H.I.; Cioce, M. A STAT3-NFkB/DDIT3/CEBPbeta axis modulates ALDH1A3 expression in chemoresistant cell subpopulations. *Oncotarget* **2015**, *6*, 12637–12653. [CrossRef]

199. Alt-Holland, A.; Sowalsky, A.G.; Szwec-Levin, Y.; Shamis, Y.; Hatch, H.; Feig, L.A.; Garlick, J.A. Suppression of E-cadherin function drives the early stages of Ras-induced squamous cell carcinoma through upregulation of FAK and Src. *J. Investig. Derm.* **2011**, *131*, 2306–2315. [CrossRef]

200. Onder, T.T.; Gupta, P.B.; Mani, S.A.; Yang, J.; Lander, E.S.; Weinberg, R.A. Loss of E-cadherin promotes metastasis via multiple downstream transcriptional pathways. *Cancer Res.* **2008**, *68*, 3645–3654. [CrossRef]

201. D'Amico, S.; Shi, J.; Martin, B.L.; Crawford, H.C.; Petrenko, O.; Reich, N.C. STAT3 is a master regulator of epithelial identity and KRAS-driven tumorigenesis. *Genes Dev.* **2018**, *32*, 1175–1187. [CrossRef] [PubMed]

202. Aras, S.; Zaidi, M.R. TAMeless traitors: Macrophages in cancer progression and metastasis. *Br. J. Cancer* **2017**, *117*, 1583–1591. [CrossRef] [PubMed]

203. Qian, B.Z.; Li, J.; Zhang, H.; Kitamura, T.; Zhang, J.; Campion, L.R.; Kaiser, E.A.; Snyder, L.A.; Pollard, J.W. CCL2 recruits inflammatory monocytes to facilitate breast-tumour metastasis. *Nature* **2011**, *475*, 222–225. [CrossRef]

204. Wu, L.; Saxena, S.; Awaji, M.; Singh, R.K. Tumor-Associated Neutrophils in Cancer: Going Pro. *Cancers* **2019**, *11*, 564. [CrossRef]

205. Yang, C.; He, L.; He, P.; Liu, Y.; Wang, W.; He, Y.; Du, Y.; Gao, F. Increased drug resistance in breast cancer by tumor-associated macrophages through IL-10/STAT3/bcl-2 signaling pathway. *Med. Oncol.* **2015**, *32*, 352. [CrossRef]

206. Webb, N.J.; Myers, C.R.; Watson, C.J.; Bottomley, M.J.; Brenchley, P.E. Activated human neutrophils express vascular endothelial growth factor (VEGF). *Cytokine* **1998**, *10*, 254–257. [CrossRef] [PubMed]

207. Li, S.; Cong, X.; Gao, H.; Lan, X.; Li, Z.; Wang, W.; Song, S.; Wang, Y.; Li, C.; Zhang, H.; et al. Tumor-associated neutrophils induce EMT by IL-17a to promote migration and invasion in gastric cancer cells. *J. Exp. Clin. Cancer Res.* **2019**, *38*, 6. [CrossRef] [PubMed]

208. Liu, C.Y.; Xu, J.Y.; Shi, X.Y.; Huang, W.; Ruan, T.Y.; Xie, P.; Ding, J.L. M2-polarized tumor-associated macrophages promoted epithelial-mesenchymal transition in pancreatic cancer cells, partially through TLR4/IL-10 signaling pathway. *Lab. Investig.* **2013**, *93*, 844–854. [CrossRef]

209. Pignatelli, J.; Bravo-Cordero, J.J.; Roh-Johnson, M.; Gandhi, S.J.; Wang, Y.; Chen, X.; Eddy, R.J.; Xue, A.; Singer, R.H.; Hodgson, L.; et al. Macrophage-dependent tumor cell transendothelial migration is mediated by Notch1/MenaINV-initiated invadopodium formation. *Sci. Rep.* **2016**, *6*. [CrossRef]

210. Pignatelli, J.; Goswami, S.; Jones, J.G.; Rohan, T.E.; Pieri, E.; Chen, X.; Adler, E.; Cox, D.; Maleki, S.; Bresnick, A.; et al. Invasive breast carcinoma cells from patients exhibit MenaINV- and macrophage-dependent transendothelial migration. *Sci. Signal* **2014**, *7*, ra112. [CrossRef]

211. Yang, J.; Liao, D.; Chen, C.; Liu, Y.; Chuang, T.H.; Xiang, R.; Markowitz, D.; Reisfeld, R.A.; Luo, Y. Tumor-associated macrophages regulate murine breast cancer stem cells through a novel paracrine EGFR/Stat3/Sox-2 signaling pathway. *Stem Cells* **2013**, *31*, 248–258. [CrossRef] [PubMed]

212. Szczerba, B.M.; Castro-Giner, F.; Vetter, M.; Krol, I.; Gkountela, S.; Landin, J.; Scheidmann, M.C.; Donato, C.; Scherrer, R.; Singer, J.; et al. Neutrophils escort circulating tumour cells to enable cell cycle progression. *Nature* **2019**, *566*, 553–557. [CrossRef] [PubMed]

213. Spicer, J.D.; McDonald, B.; Cools-Lartigue, J.J.; Chow, S.C.; Giannias, B.; Kubes, P.; Ferri, L.E. Neutrophils Promote Liver Metastasis via Mac-1–Mediated Interactions with Circulating Tumor Cells. *Cancer Res.* **2012**, *72*, 3919–3927. [CrossRef] [PubMed]

214. Wculek, S.K.; Malanchi, I. Neutrophils support lung colonization of metastasis-initiating breast cancer cells. *Nature* **2015**, *528*, 413–417. [CrossRef]

215. Psaila, B.; Lyden, D. The metastatic niche: Adapting the foreign soil. *Nat. Rev. Cancer* **2009**, *9*, 285–293. [CrossRef]

216. Deng, J.; Liu, Y.; Lee, H.; Herrmann, A.; Zhang, W.; Zhang, C.; Shen, S.; Priceman, S.J.; Kujawski, M.; Pal, S.K.; et al. S1PR1-STAT3 signaling is crucial for myeloid cell colonization at future metastatic sites. *Cancer Cell* **2012**, *21*, 642–654. [CrossRef]

217. Kanai, T.; Seki, S.; Jenks, J.A.; Kohli, A.; Kawli, T.; Martin, D.P.; Snyder, M.; Bacchetta, R.; Nadeau, K.C. Identification of STAT5A and STAT5B target genes in human T cells. *PLoS ONE* **2014**, *9*, e86790. [CrossRef]

218. Grebien, F.; Kerenyi, M.A.; Kovacic, B.; Kolbe, T.; Becker, V.; Dolznig, H.; Pfeffer, K.; Klingmuller, U.; Muller, M.; Beug, H.; et al. Stat5 activation enables erythropoiesis in the absence of EpoR and Jak2. *Blood* **2008**, *111*, 4511–4522. [CrossRef]

219. Zhu, B.M.; McLaughlin, S.K.; Na, R.; Liu, J.; Cui, Y.; Martin, C.; Kimura, A.; Robinson, G.W.; Andrews, N.C.; Hennighausen, L. Hematopoietic-specific Stat5-null mice display microcytic hypochromic anemia associated with reduced transferrin receptor gene expression. *Blood* **2008**, *112*, 2071–2080. [CrossRef]

220. Kerenyi, M.A.; Grebien, F.; Gehart, H.; Schifrer, M.; Artaker, M.; Kovacic, B.; Beug, H.; Moriggl, R.; Mullner, E.W. Stat5 regulates cellular iron uptake of erythroid cells via IRP-2 and TfR-1. *Blood* **2008**, *112*, 3878–3888. [CrossRef]

221. Barash, I. Stat5 in the mammary gland: Controlling normal development and cancer. *J. Cell Physiol.* **2006**, *209*, 305–313. [CrossRef] [PubMed]

222. Cella, N.; Groner, B.; Hynes, N.E. Characterization of Stat5a and Stat5b homodimers and heterodimers and their association with the glucocortiocoid receptor in mammary cells. *Mol. Cell Biol.* **1998**, *18*, 1783–1792. [CrossRef] [PubMed]

223. Zeng, R.; Aoki, Y.; Yoshida, M.; Arai, K.; Watanabe, S. Stat5B shuttles between cytoplasm and nucleus in a cytokine-dependent and -independent manner. *J. Immunol.* **2002**, *168*, 4567–4575. [CrossRef] [PubMed]

224. Meinke, A.; Barahmand-Pour, F.; Wohrl, S.; Stoiber, D.; Decker, T. Activation of different Stat5 isoforms contributes to cell-type-restricted signaling in response to interferons. *Mol. Cell Biol.* **1996**, *16*, 6937–6944. [CrossRef] [PubMed]

225. Villarino, A.V.; Sciumè, G.; Davis, F.P.; Iwata, S.; Zitti, B.; Robinson, G.W.; Hennighausen, L.; Kanno, Y.; O'Shea, J.J. Subset- and tissue-defined STAT5 thresholds control homeostasis and function of innate lymphoid cells. *J. Exp. Med.* **2017**, *214*, 2999–3014. [CrossRef]

226. Maurer, B.; Farlik, M.; Sexl, V. It is a differentiation game: STAT5 in a new role. *Cell Death Differ.* **2017**, *24*, 953–954. [CrossRef]

227. Heltemes-Harris, L.M.; Farrar, M.A. The role of STAT5 in lymphocyte development and transformation. *Curr. Opin. Immunol.* **2012**, *24*, 146–152. [CrossRef]

228. Hoelbl, A.; Kovacic, B.; Kerenyi, M.A.; Simma, O.; Warsch, W.; Cui, Y.; Beug, H.; Hennighausen, L.; Moriggl, R.; Sexl, V. Clarifying the role of Stat5 in lymphoid development and Abelson-induced transformation. *Blood* **2006**, *107*, 4898–4906. [CrossRef]

229. Minieri, V.; De Dominici, M.; Porazzi, P.; Mariani, S.A.; Spinelli, O.; Rambaldi, A.; Peterson, L.F.; Porcu, P.; Nevalainen, M.T.; Calabretta, B. Targeting STAT5 or STAT5-Regulated Pathways Suppresses Leukemogenesis of Ph+ Acute Lymphoblastic Leukemia. *Cancer Res.* **2018**, *78*, 5793–5807. [CrossRef]

230. Gleixner, K.V.; Schneeweiss, M.; Eisenwort, G.; Berger, D.; Herrmann, H.; Blatt, K.; Greiner, G.; Byrgazov, K.; Hoermann, G.; Konopleva, M.; et al. Combined targeting of STAT3 and STAT5: A novel approach to overcome drug resistance in chronic myeloid leukemia. *Haematologica* **2017**, *102*, 1519–1529. [CrossRef]

231. Kaltenecker, D.; Themanns, M.; Mueller, K.M.; Spirk, K.; Golob-Schwarzl, N.; Friedbichler, K.; Kenner, L.; Haybaeck, J.; Moriggl, R. STAT5 deficiency in hepatocytes reduces diethylnitrosamine-induced liver tumorigenesis in mice. *Cytokine* **2018**. [CrossRef] [PubMed]

232. Boutillon, F.; Pigat, N.; Sala, L.S.; Reyes-Gomez, E.; Moriggl, R.; Guidotti, J.E.; Goffin, V. STAT5a/b Deficiency Delays, but does not Prevent, Prolactin-Driven Prostate Tumorigenesis in Mice. *Cancers* **2019**, *11*, 929. [CrossRef] [PubMed]

233. Lee, T.K.; Man, K.; Poon, R.T.P.; Lo, C.M.; Yuen, A.P.; Ng, I.O.; Ng, K.T.; Leonard, W.; Fan, S.T. Signal Transducers and Activators of Transcription 5b Activation Enhances Hepatocellular Carcinoma Aggressiveness through Induction of Epithelial-Mesenchymal Transition. *Cancer Res.* **2006**, *66*, 9948–9956. [CrossRef] [PubMed]

234. Johnston, A.N.; Bu, W.; Hein, S.; Garcia, S.; Camacho, L.; Xue, L.; Qin, L.; Nagi, C.; Hilsenbeck, S.G.; Kapali, J.; et al. Hyperprolactinemia-inducing antipsychotics increase breast cancer risk by activating JAK-STAT5 in precancerous lesions. *Breast Cancer Res.* **2018**, *20*, 42. [CrossRef]

235. Vafaizadeh, V.; Klemmt, P.; Brendel, C.; Weber, K.; Doebele, C.; Britt, K.; Grez, M.; Fehse, B.; Desrivieres, S.; Groner, B. Mammary epithelial reconstitution with gene-modified stem cells assigns roles to Stat5 in luminal alveolar cell fate decisions, differentiation, involution, and mammary tumor formation. *Stem Cells* **2010**, *28*, 928–938. [CrossRef]

236. Bessette, K.; Lang, M.L.; Fava, R.A.; Grundy, M.; Heinen, J.; Horne, L.; Spolski, R.; Al-Shami, A.; Morse, H.C., 3rd; Leonard, W.J.; et al. A Stat5b transgene is capable of inducing CD8+ lymphoblastic lymphoma in the absence of normal TCR/MHC signaling. *Blood* **2008**, *111*, 344–350. [CrossRef]

237. Xiong, H.; Su, W.Y.; Liang, Q.C.; Zhang, Z.G.; Chen, H.M.; Du, W.; Chen, Y.X.; Fang, J.Y. Inhibition of STAT5 induces G1 cell cycle arrest and reduces tumor cell invasion in human colorectal cancer cells. *Lab. Investig.* **2009**, *89*, 717–725. [CrossRef]

238. Koppikar, P.; Lui, V.W.; Man, D.; Xi, S.; Chai, R.L.; Nelson, E.; Tobey, A.B.; Grandis, J.R. Constitutive activation of signal transducer and activator of transcription 5 contributes to tumor growth, epithelial-mesenchymal transition, and resistance to epidermal growth factor receptor targeting. *Clin. Cancer Res.* **2008**, *14*, 7682–7690. [CrossRef]

239. Talati, P.G.; Gu, L.; Ellsworth, E.M.; Girondo, M.A.; Trerotola, M.; Hoang, D.T.; Leiby, B.; Dagvadorj, A.; McCue, P.A.; Lallas, C.D.; et al. Jak2-Stat5a/b Signaling Induces Epithelial-to-Mesenchymal Transition and Stem-Like Cell Properties in Prostate Cancer. *Am. J. Pathol.* **2015**, *185*, 2505–2522. [CrossRef]

240. Creamer, B.A.; Sakamoto, K.; Schmidt, J.W.; Triplett, A.A.; Moriggl, R.; Wagner, K.U. Stat5 promotes survival of mammary epithelial cells through transcriptional activation of a distinct promoter in Akt1. *Mol. Cell Biol.* **2010**, *30*, 2957–2970. [CrossRef]

241. Schmidt, J.W.; Wehde, B.L.; Sakamoto, K.; Triplett, A.A.; Anderson, S.M.; Tsichlis, P.N.; Leone, G.; Wagner, K.U. Stat5 regulates the phosphatidylinositol 3-kinase/Akt1 pathway during mammary gland development and tumorigenesis. *Mol. Cell Biol.* **2014**, *34*, 1363–1377. [CrossRef] [PubMed]

242. Tsuruyama, T.; Hiratsuka, T.; Jin, G.; Imai, Y.; Takeuchi, H.; Maruyama, Y.; Kanaya, K.; Ozeki, M.; Takakuwa, T.; Haga, H.; et al. Murine leukemia retrovirus integration induces the formation of transcription factor complexes on palindromic sequences in the signal transducer and activator of transcription factor 5a gene during the development of pre-B lymphomagenesis. *Am. J. Pathol.* **2011**, *178*, 1374–1386. [CrossRef] [PubMed]

243. Chen, Y.; Zhou, Q.; Zhang, L.; Zhong, Y.; Fan, G.; Zhang, Z.; Wang, R.; Jin, M.; Qiu, Y.; Kong, D. Stellettin B induces apoptosis in human chronic myeloid leukemia cells via targeting PI3K and Stat5. *Oncotarget* **2017**, *8*, 28906–28921. [CrossRef] [PubMed]

244. Dagvadorj, A.; Collins, S.; Jomain, J.B.; Abdulghani, J.; Karras, J.; Zellweger, T.; Li, H.; Nurmi, M.; Alanen, K.; Mirtti, T.; et al. Autocrine prolactin promotes prostate cancer cell growth via Janus kinase-2-signal transducer and activator of transcription-5a/b signaling pathway. *Endocrinology* **2007**, *148*, 3089–3101. [CrossRef] [PubMed]

245. Latha, K.; Li, M.; Chumbalkar, V.; Gururaj, A.; Hwang, Y.; Dakeng, S.; Sawaya, R.; Aldape, K.; Cavenee, W.K.; Bogler, O.; et al. Nuclear EGFRvIII-STAT5b complex contributes to glioblastoma cell survival by direct activation of the Bcl-XL promoter. *Int. J. Cancer* **2013**, *132*, 509–520. [CrossRef]

246. Chueh, F.-Y.; Leong, K.-F.; Yu, C.-L. Mitochondrial translocation of signal transducer and activator of transcription 5 (STAT5) in leukemic T cells and cytokine-stimulated cells. *Biochem. Biophys. Res. Commun.* **2010**, *402*, 778–783. [CrossRef]

247. Walker, S.R.; Xiang, M.; Frank, D.A. Distinct roles of STAT3 and STAT5 in the pathogenesis and targeted therapy of breast cancer. *Mol. Cell Endocrinol.* **2014**, *382*, 616–621. [CrossRef]

248. Sultan, A.S.; Xie, J.; LeBaron, M.J.; Ealley, E.L.; Nevalainen, M.T.; Rui, H. Stat5 promotes homotypic adhesion and inhibits invasive characteristics of human breast cancer cells. *Oncogene* **2005**, *24*, 746–760. [CrossRef]

249. Mallette, F.A.; Gaumont-Leclerc, M.F.; Ferbeyre, G. The DNA damage signaling pathway is a critical mediator of oncogene-induced senescence. *Genes Dev.* **2007**, *21*, 43–48. [CrossRef]

250. Nishikawa, H.; Sakaguchi, S. Regulatory T cells in cancer immunotherapy. *Curr. Opin. Immunol.* **2014**, *27*, 1–7. [CrossRef]

251. Knol, A.C.; Nguyen, J.M.; Quereux, G.; Brocard, A.; Khammari, A.; Dreno, B. Prognostic value of tumor-infiltrating Foxp3+ T-cell subpopulations in metastatic melanoma. *Exp. Derm.* **2011**, *20*, 430–434. [CrossRef] [PubMed]

252. Sasada, T.; Kimura, M.; Yoshida, Y.; Kanai, M.; Takabayashi, A. CD4+CD25+ regulatory T cells in patients with gastrointestinal malignancies: Possible involvement of regulatory T cells in disease progression. *Cancer* **2003**, *98*, 1089–1099. [CrossRef] [PubMed]

253. Zhao, P.; Li, J.; Tian, Y.; Mao, J.; Liu, X.; Feng, S.; Li, J.; Bian, Q.; Ji, H.; Zhang, L. Restoring Th17/Treg balance via modulation of STAT3 and STAT5 activation contributes to the amelioration of chronic obstructive pulmonary disease by Bufei Yishen formula. *J. Ethnopharmacol.* **2018**, *217*, 152–162. [CrossRef] [PubMed]

254. Burchill, M.A.; Yang, J.; Vogtenhuber, C.; Blazar, B.R.; Farrar, M.A. IL-2 receptor beta-dependent STAT5 activation is required for the development of Foxp3+ regulatory T cells. *J. Immunol.* **2007**, *178*, 280–290. [CrossRef]

255. O'Shea, J.J.; Paul, W.E. Mechanisms underlying lineage commitment and plasticity of helper CD4+ T cells. *Science* **2010**, *327*, 1098–1102. [CrossRef]

256. Li, X.; Kostareli, E.; Suffner, J.; Garbi, N.; Hammerling, G.J. Efficient Treg depletion induces T-cell infiltration and rejection of large tumors. *Eur. J. Immunol.* **2010**, *40*, 3325–3335. [CrossRef]

257. Fisher, S.A.; Aston, W.J.; Chee, J.; Khong, A.; Cleaver, A.L.; Solin, J.N.; Ma, S.; Lesterhuis, W.J.; Dick, I.; Holt, R.A.; et al. Transient Treg depletion enhances therapeutic anti-cancer vaccination. *Immun. Inflamm. Dis.* **2017**, *5*, 16–28. [CrossRef]

258. Schmidt, A.; Zhang, X.M.; Joshi, R.N.; Iqbal, S.; Wahlund, C.; Gabrielsson, S.; Harris, R.A.; Tegner, J. Human macrophages induce CD4(+)Foxp3(+) regulatory T cells via binding and re-release of TGF-beta. *Immunol. Cell Biol.* **2016**, *94*, 747–762. [CrossRef]

259. Goswami, K.K.; Ghosh, T.; Ghosh, S.; Sarkar, M.; Bose, A.; Baral, R. Tumor promoting role of anti-tumor macrophages in tumor microenvironment. *Cell Immunol.* **2017**, *316*, 1–10. [CrossRef]

260. Müller-Hermelink, N.; Braumüller, H.; Pichler, B.; Wieder, T.; Mailhammer, R.; Schaak, K.; Ghoreschi, K.; Yazdi, A.; Haubner, R.; Sander, C.A.; et al. TNFR1 Signaling and IFN-γ Signaling Determine whether T Cells Induce Tumor Dormancy or Promote Multistage Carcinogenesis. *Cancer Cell* **2008**, *13*, 507–518. [CrossRef]

261. Edwards, J.P.; Thornton, A.M.; Shevach, E.M. Release of active TGF-beta1 from the latent TGF-beta1/GARP complex on T regulatory cells is mediated by integrin beta8. *J. Immunol.* **2014**, *193*, 2843–2849. [CrossRef] [PubMed]

262. Wang, Y.; Schafer, C.C.; Hough, K.P.; Tousif, S.; Duncan, S.R.; Kearney, J.F.; Ponnazhagan, S.; Hsu, H.C.; Deshane, J.S. Myeloid-Derived Suppressor Cells Impair B Cell Responses in Lung Cancer through IL-7 and STAT5. *J. Immunol.* **2018**, *201*, 278–295. [CrossRef] [PubMed]

263. Nurieva, R.I.; Podd, A.; Chen, Y.; Alekseev, A.M.; Yu, M.; Qi, X.; Huang, H.; Wen, R.; Wang, J.; Li, H.S.; et al. STAT5 protein negatively regulates T follicular helper (Tfh) cell generation and function. *J. Biol. Chem.* **2012**, *287*, 11234–11239. [CrossRef] [PubMed]

264. Buck, M.D.; O'Sullivan, D.; Klein Geltink, R.I.; Curtis, J.D.; Chang, C.H.; Sanin, D.E.; Qiu, J.; Kretz, O.; Braas, D.; van der Windt, G.J.; et al. Mitochondrial Dynamics Controls T Cell Fate through Metabolic Programming. *Cell* **2016**, *166*, 63–76. [CrossRef]

265. Berod, L.; Friedrich, C.; Nandan, A.; Freitag, J.; Hagemann, S.; Harmrolfs, K.; Sandouk, A.; Hesse, C.; Castro, C.N.; Bahre, H.; et al. De novo fatty acid synthesis controls the fate between regulatory T and T helper 17 cells. *Nat. Med.* **2014**, *20*, 1327–1333. [CrossRef]

266. Michalek, R.D.; Gerriets, V.A.; Jacobs, S.R.; Macintyre, A.N.; MacIver, N.J.; Mason, E.F.; Sullivan, S.A.; Nichols, A.G.; Rathmell, J.C. Cutting edge: Distinct glycolytic and lipid oxidative metabolic programs are essential for effector and regulatory CD4+ T cell subsets. *J. Immunol.* **2011**, *186*, 3299–3303. [CrossRef]

267. Galluzzi, L.; Baehrecke, E.H.; Ballabio, A.; Boya, P.; Bravo-San Pedro, J.M.; Cecconi, F.; Choi, A.M.; Chu, C.T.; Codogno, P.; Colombo, M.I.; et al. Molecular definitions of autophagy and related processes. *EMBO J.* **2017**, *36*, 1811–1836. [CrossRef]

268. Singh, R.; Kaushik, S.; Wang, Y.; Xiang, Y.; Novak, I.; Komatsu, M.; Tanaka, K.; Cuervo, A.M.; Czaja, M.J. Autophagy regulates lipid metabolism. *Nature* **2009**, *458*, 1131–1135. [CrossRef]

269. Lin, J.X.; Li, P.; Liu, D.; Jin, H.T.; He, J.; Ata Ur Rasheed, M.; Rochman, Y.; Wang, L.; Cui, K.; Liu, C.; et al. Critical Role of STAT5 transcription factor tetramerization for cytokine responses and normal immune function. *Immunity* **2012**, *36*, 586–599. [CrossRef]

270. Owen, D.L.; Farrar, M.A. STAT5 and CD4 (+) T Cell Immunity. *F1000Research* **2017**, *6*, 32. [CrossRef]

271. Grange, M.; Buferne, M.; Verdeil, G.; Leserman, L.; Schmitt-Verhulst, A.M.; Auphan-Anezin, N. Activated STAT5 promotes long-lived cytotoxic CD8+ T cells that induce regression of autochthonous melanoma. *Cancer Res.* **2012**, *72*, 76–87. [CrossRef] [PubMed]

272. Gotthardt, D.; Putz, E.M.; Grundschober, E.; Prchal-Murphy, M.; Straka, E.; Kudweis, P.; Heller, G.; Bago-Horvath, Z.; Witalisz-Siepracka, A.; Cumaraswamy, A.A.; et al. STAT5 Is a Key Regulator in NK Cells and Acts as a Molecular Switch from Tumor Surveillance to Tumor Promotion. *Cancer Discov.* **2016**, *6*, 414–429. [CrossRef] [PubMed]

273. Wolf, D.; Sopper, S.; Pircher, A.; Gastl, G.; Wolf, A.M. Treg(s) in Cancer: Friends or Foe? *J. Cell Physiol.* **2015**, *230*, 2598–2605. [CrossRef] [PubMed]

274. Turkson, J.; Kim, J.S.; Zhang, S.; Yuan, J.; Huang, M.; Glenn, M.; Haura, E.; Sebti, S.; Hamilton, A.D.; Jove, R. Novel peptidomimetic inhibitors of signal transducer and activator of transcription 3 dimerization and biological activity. *Mol. Cancer* **2004**, *3*, 261–269.

275. Fontaine, F.; Overman, J.; Francois, M. Pharmacological manipulation of transcription factor protein-protein interactions: Opportunities and obstacles. *Cell Regen.* **2015**, *4*, 2. [CrossRef]

276. Siddiquee, K.; Zhang, S.; Guida, W.C.; Blaskovich, M.A.; Greedy, B.; Lawrence, H.R.; Yip, M.L.; Jove, R.; McLaughlin, M.M.; Lawrence, N.J.; et al. Selective chemical probe inhibitor of Stat3, identified through structure-based virtual screening, induces antitumor activity. *Proc. Natl. Acad. Sci. USA* **2007**, *104*, 7391–7396. [CrossRef]

277. Siddiquee, K.A.; Gunning, P.T.; Glenn, M.; Katt, W.P.; Zhang, S.; Schrock, C.; Sebti, S.M.; Jove, R.; Hamilton, A.D.; Turkson, J. An oxazole-based small-molecule Stat3 inhibitor modulates Stat3 stability and processing and induces antitumor cell effects. *ACS Chem. Biol.* **2007**, *2*, 787–798. [CrossRef]

278. Sato, K.; Tsuchiya, M.; Saldanha, J.; Koishihara, Y.; Ohsugi, Y.; Kishimoto, T.; Bendig, M.M. Reshaping a human antibody to inhibit the interleukin 6-dependent tumor cell growth. *Cancer Res.* **1993**, *53*, 851–856.

279. Mihara, M.; Kasutani, K.; Okazaki, M.; Nakamura, A.; Kawai, S.; Sugimoto, M.; Matsumoto, Y.; Ohsugi, Y. Tocilizumab inhibits signal transduction mediated by both mIL-6R and sIL-6R, but not by the receptors of other members of IL-6 cytokine family. *Int. Immunopharmacol.* **2005**, *5*, 1731–1740. [CrossRef]

280. Nishimoto, N.; Sasai, M.; Shima, Y.; Nakagawa, M.; Matsumoto, T.; Shirai, T.; Kishimoto, T.; Yoshizaki, K. Improvement in Castleman's disease by humanized anti-interleukin-6 receptor antibody therapy. *Blood* **2000**, *95*, 56–61. [CrossRef]

281. Nishimoto, N.; Yoshizaki, K.; Miyasaka, N.; Yamamoto, K.; Kawai, S.; Takeuchi, T.; Hashimoto, J.; Azuma, J.; Kishimoto, T. Treatment of rheumatoid arthritis with humanized anti-interleukin-6 receptor antibody: A multicenter, double-blind, placebo-controlled trial. *Arthritis Rheum.* **2004**, *50*, 1761–1769. [CrossRef] [PubMed]

282. Dhillon, S. Intravenous Tocilizumab: A Review of Its Use in Adults with Rheumatoid Arthritis. *BioDrugs* **2013**, *28*, 75–106. [CrossRef] [PubMed]

283. Le, R.Q.; Li, L.; Yuan, W.; Shord, S.S.; Nie, L.; Habtemariam, B.A.; Przepiorka, D.; Farrell, A.T.; Pazdur, R. FDA Approval Summary: Tocilizumab for Treatment of Chimeric Antigen Receptor T Cell-Induced Severe or Life-Threatening Cytokine Release Syndrome. *Oncologist* **2018**, *23*, 943–947. [CrossRef] [PubMed]

284. Dambacher, J.; Beigel, F.; Seiderer, J.; Haller, D.; Goke, B.; Auernhammer, C.J.; Brand, S. Interleukin 31 mediates MAP kinase and STAT1/3 activation in intestinal epithelial cells and its expression is upregulated in inflammatory bowel disease. *Gut* **2007**, *56*, 1257–1265. [CrossRef] [PubMed]

285. Song, L.; Smith, M.A.; Doshi, P.; Sasser, K.; Fulp, W.; Altiok, S.; Haura, E.B. Antitumor efficacy of the anti-interleukin-6 (IL-6) antibody siltuximab in mouse xenograft models of lung cancer. *J. Thorac. Oncol.* **2014**, *9*, 974–982. [CrossRef] [PubMed]

286. Chen, R.; Chen, B. Siltuximab (CNTO 328): A promising option for human malignancies. *Drug Des. Devel.* **2015**, *9*, 3455–3458. [CrossRef] [PubMed]

287. Fizazi, K.; De Bono, J.S.; Flechon, A.; Heidenreich, A.; Voog, E.; Davis, N.B.; Qi, M.; Bandekar, R.; Vermeulen, J.T.; Cornfeld, M.; et al. Randomised phase II study of siltuximab (CNTO 328), an anti-IL-6 monoclonal antibody, in combination with mitoxantrone/prednisone versus mitoxantrone/prednisone alone in metastatic castration-resistant prostate cancer. *Eur. J. Cancer* **2012**, *48*, 85–93. [CrossRef]

288. Takata, F.; Dohgu, S.; Matsumoto, J.; Machida, T.; Sakaguchi, S.; Kimura, I.; Yamauchi, A.; Kataoka, Y. Oncostatin M-induced blood-brain barrier impairment is due to prolonged activation of STAT3 signaling in vitro. *J. Cell Biochem.* **2018**, *119*, 9055–9063. [CrossRef]

289. Jorcyk, C.L.; Holzer, R.G.; Ryan, R.E. Oncostatin M induces cell detachment and enhances the metastatic capacity of T-47D human breast carcinoma cells. *Cytokine* **2006**, *33*, 323–336. [CrossRef]

290. Sterbova, S.; Karlsson, T.; Persson, E. Oncostatin M induces tumorigenic properties in non-transformed human prostate epithelial cells, in part through activation of signal transducer and activator of transcription 3 (STAT3). *Biochem. Biophys. Res. Commun.* **2018**, *498*, 769–774. [CrossRef]

291. Richards, C.D. The enigmatic cytokine oncostatin m and roles in disease. *ISRN Inflamm.* **2013**, *2013*, 512103. [CrossRef] [PubMed]

292. Hintzen, C.; Evers, C.; Lippok, B.E.; Volkmer, R.; Heinrich, P.C.; Radtke, S.; Hermanns, H.M. Box 2 region of the oncostatin M receptor determines specificity for recruitment of Janus kinases and STAT5 activation. *J. Biol. Chem.* **2008**, *283*, 19465–19477. [CrossRef]

293. Hermanns, H.M.; Radtke, S.; Schaper, F.; Heinrich, P.C.; Behrmann, I. Non-redundant signal transduction of interleukin-6-type cytokines. The adapter protein Shc is specifically recruited to rhe oncostatin M receptor. *J. Biol. Chem.* **2000**, *275*, 40742–40748. [CrossRef] [PubMed]

294. Choy, E.H.; Bendit, M.; McAleer, D.; Liu, F.; Feeney, M.; Brett, S.; Zamuner, S.; Campanile, A.; Toso, J. Safety, tolerability, pharmacokinetics and pharmacodynamics of an anti- oncostatin M monoclonal antibody in rheumatoid arthritis: Results from phase II randomized, placebo-controlled trials. *Arthritis Res.* **2013**, *15*, R132. [CrossRef] [PubMed]

295. Poling, J.; Gajawada, P.; Richter, M.; Lorchner, H.; Polyakova, V.; Kostin, S.; Shin, J.; Boettger, T.; Walther, T.; Rees, W.; et al. Therapeutic targeting of the oncostatin M receptor-beta prevents inflammatory heart failure. *Basic Res. Cardiol.* **2014**, *109*, 396. [CrossRef] [PubMed]

296. Liu, Q.; Du, Y.; Li, K.; Zhang, W.; Feng, X.; Hao, J.; Li, H.; Liu, S. Anti-OSM Antibody Inhibits Tubulointerstitial Lesion in a Murine Model of Lupus Nephritis. *Mediat. Inflamm.* **2017**, *2017*, 3038514. [CrossRef] [PubMed]

297. Kucia-Tran, J.A.; Tulkki, V.; Scarpini, C.G.; Smith, S.; Wallberg, M.; Paez-Ribes, M.; Araujo, A.M.; Botthoff, J.; Feeney, M.; Hughes, K.; et al. Anti-oncostatin M antibody inhibits the pro-malignant effects of oncostatin M receptor overexpression in squamous cell carcinoma. *J. Pathol.* **2018**, *244*, 283–295. [CrossRef] [PubMed]

298. Robson, M.J.; Turner, R.C.; Naser, Z.J.; McCurdy, C.R.; O'Callaghan, J.P.; Huber, J.D.; Matsumoto, R.R. SN79, a sigma receptor antagonist, attenuates methamphetamine-induced astrogliosis through a blockade of OSMR/gp130 signaling and STAT3 phosphorylation. *Exp. Neurol.* **2014**, *254*, 180–189. [CrossRef]

299. Bottos, A.; Gotthardt, D.; Gill, J.W.; Gattelli, A.; Frei, A.; Tzankov, A.; Sexl, V.; Wodnar-Filipowicz, A.; Hynes, N.E. Decreased NK-cell tumour immunosurveillance consequent to JAK inhibition enhances metastasis in breast cancer models. *Nat. Commun.* **2016**, *7*, 12258. [CrossRef]

300. Doherty, M.R.; Jackson, M.W. Using interferon-beta to combat cancer stem cell properties in triple negative breast cancer. *Oncoscience* **2018**, *5*, 169–171. [CrossRef]

301. Doherty, M.R.; Parvani, J.G.; Tamagno, I.; Junk, D.J.; Bryson, B.L.; Cheon, H.J.; Stark, G.R.; Jackson, M.W. The opposing effects of interferon-beta and oncostatin-M as regulators of cancer stem cell plasticity in triple-negative breast cancer. *Breast Cancer Res.* **2019**, *21*, 54. [CrossRef] [PubMed]

302. Sistigu, A.; Yamazaki, T.; Vacchelli, E.; Chaba, K.; Enot, D.P.; Adam, J.; Vitale, I.; Goubar, A.; Baracco, E.E.; Remedios, C.; et al. Cancer cell-autonomous contribution of type I interferon signaling to the efficacy of chemotherapy. *Nat. Med.* **2014**, *20*, 1301–1309. [CrossRef] [PubMed]

303. Yang, X.; Zhang, X.; Fu, M.L.; Weichselbaum, R.R.; Gajewski, T.F.; Guo, Y.; Fu, Y.X. Targeting the tumor microenvironment with interferon-beta bridges innate and adaptive immune responses. *Cancer Cell* **2014**, *25*, 37–48. [CrossRef] [PubMed]

304. Ling, X.; Marini, F.; Konopleva, M.; Schober, W.; Shi, Y.; Burks, J.; Clise-Dwyer, K.; Wang, R.Y.; Zhang, W.; Yuan, X.; et al. Mesenchymal Stem Cells Overexpressing IFN-beta Inhibit Breast Cancer Growth and Metastases through Stat3 Signaling in a Syngeneic Tumor Model. *Cancer Microenviron.* **2010**, *3*, 83–95. [CrossRef]

305. Burstein, M.D.; Tsimelzon, A.; Poage, G.M.; Covington, K.R.; Contreras, A.; Fuqua, S.A.; Savage, M.I.; Osborne, C.K.; Hilsenbeck, S.G.; Chang, J.C.; et al. Comprehensive genomic analysis identifies novel subtypes and targets of triple-negative breast cancer. *Clin. Cancer Res.* **2015**, *21*, 1688–1698. [CrossRef]

306. Bidwell, B.N.; Slaney, C.Y.; Withana, N.P.; Forster, S.; Cao, Y.; Loi, S.; Andrews, D.; Mikeska, T.; Mangan, N.E.; Samarajiwa, S.A.; et al. Silencing of Irf7 pathways in breast cancer cells promotes bone metastasis through immune escape. *Nat. Med.* **2012**, *18*, 1224–1231. [CrossRef]

307. Liu, S.; Lachapelle, J.; Leung, S.; Gao, D.; Foulkes, W.D.; Nielsen, T.O. CD8+ lymphocyte infiltration is an independent favorable prognostic indicator in basal-like breast cancer. *Breast Cancer Res.* **2012**, *14*, R48. [CrossRef]

308. Xie, C.; Liu, C.; Wu, B.; Lin, Y.; Ma, T.; Xiong, H.; Wang, Q.; Li, Z.; Ma, C.; Tu, Z. Effects of IRF1 and IFN-beta interaction on the M1 polarization of macrophages and its antitumor function. *Int. J. Mol. Med.* **2016**, *38*, 148–160. [CrossRef]

309. Moreira-Teixeira, L.; Sousa, J.; McNab, F.W.; Torrado, E.; Cardoso, F.; Machado, H.; Castro, F.; Cardoso, V.; Gaifem, J.; Wu, X.; et al. Type I IFN Inhibits Alternative Macrophage Activation during Mycobacterium tuberculosis Infection and Leads to Enhanced Protection in the Absence of IFN-gamma Signaling. *J. Immunol.* **2016**, *197*, 4714–4726. [CrossRef]

310. Pylaeva, E.; Lang, S.; Jablonska, J. The Essential Role of Type I Interferons in Differentiation and Activation of Tumor-Associated Neutrophils. *Front. Immunol.* **2016**, *7*, 629. [CrossRef]

311. Brockwell, N.K.; Owen, K.L.; Zanker, D.; Spurling, A.; Rautela, J.; Duivenvoorden, H.M.; Baschuk, N.; Caramia, F.; Loi, S.; Darcy, P.K.; et al. Neoadjuvant Interferons: Critical for Effective PD-1-Based Immunotherapy in TNBC. *Cancer Immunol. Res.* **2017**, *5*, 871–884. [CrossRef] [PubMed]

312. Parker, B.S.; Rautela, J.; Hertzog, P.J. Antitumour actions of interferons: Implications for cancer therapy. *Nat. Rev. Cancer* **2016**, *16*, 131–144. [CrossRef] [PubMed]

313. Essers, M.A.; Offner, S.; Blanco-Bose, W.E.; Waibler, Z.; Kalinke, U.; Duchosal, M.A.; Trumpp, A. IFNalpha activates dormant haematopoietic stem cells in vivo. *Nature* **2009**, *458*, 904–908. [CrossRef] [PubMed]

314. Ma, H.; Jin, S.; Yang, W.; Zhou, G.; Zhao, M.; Fang, S.; Zhang, Z.; Hu, J. Interferon-alpha enhances the antitumour activity of EGFR-targeted therapies by upregulating RIG-I in head and neck squamous cell carcinoma. *Br. J. Cancer* **2018**, *118*, 509–521. [CrossRef] [PubMed]

315. Cheon, H.; Holvey-Bates, E.G.; Schoggins, J.W.; Forster, S.; Hertzog, P.; Imanaka, N.; Rice, C.M.; Jackson, M.W.; Junk, D.J.; Stark, G.R. IFNbeta-dependent increases in STAT1, STAT2, and IRF9 mediate resistance to viruses and DNA damage. *EMBO J.* **2013**, *32*, 2751–2763. [CrossRef]

 *cancers*

Review

STAT3 and STAT5 Targeting for Simultaneous Management of Melanoma and Autoimmune Diseases

Stella Logotheti [1] and Brigitte M. Pützer [1,2,*

[1] Institute of Experimental Gene Therapy and Cancer Research, Rostock University Medical Center, 18057 Rostock, Germany; Styliani.Logotheti@med.uni-rostock.de

[2] Department Life, Light & Matter, University of Rostock, 18059 Rostock, Germany

* Correspondence: brigitte.puetzer@med.uni-rostock.de; Tel.: +49-381-4495068

Received: 8 August 2019; Accepted: 23 September 2019; Published: 27 September 2019

Abstract: Melanoma is a skin cancer which can become metastatic, drug-refractory, and lethal if managed late or inappropriately. An increasing number of melanoma patients exhibits autoimmune diseases, either as pre-existing conditions or as sequelae of immune-based anti-melanoma therapies, which complicate patient management and raise the need for more personalized treatments. STAT3 and/or STAT5 cascades are commonly activated during melanoma progression and mediate the metastatic effects of key oncogenic factors. Deactivation of these cascades enhances antitumor-immune responses, is efficient against metastatic melanoma in the preclinical setting and emerges as a promising targeting strategy, especially for patients resistant to immunotherapies. In the light of the recent realization that cancer and autoimmune diseases share common mechanisms of immune dysregulation, we suggest that the systemic delivery of STAT3 or STAT5 inhibitors could simultaneously target both, melanoma and associated autoimmune diseases, thereby decreasing the overall disease burden and improving quality of life of this patient subpopulation. Herein, we review the recent advances of STAT3 and STAT5 targeting in melanoma, explore which autoimmune diseases are causatively linked to STAT3 and/or STAT5 signaling, and propose that these patients may particularly benefit from treatment with STAT3/STAT5 inhibitors.

Keywords: melanoma; autoimmune disease; inflammation; STAT3; STAT5; immunotherapy; tumor–immune cell interactions

---

1. Introduction

Within the last decade, the cancer field has witnessed a rapid paradigm shift from traditional chemotherapy to immunotherapy. A major advantage of immunotherapeutics is that, instead of directly targeting and killing the tumor as cytotoxic drugs do, they stimulate a person's own immune system to recognize and destroy cancer cells. In this respect, they ally with the immune cells not only to shrink the primary tumor but also to establish durable responses against circulating cancer cells that might lurk beyond the primary site. Cancer immunotherapy approaches include (a) checkpoint inhibitors, which act by releasing the brakes that prevent T-cells from killing cancer cells, (b) monoclonal antibodies, that are designed to attach to cancer cell-specific antigens, (c) cancer vaccines, boosting the immune system's response to tumors, or (d) cell-based therapies, such as chimeric antigen receptor (CAR) T-cell therapy, where T cells taken from a patient's tumor are expanded and/or genetically engineered ex vivo and then administered back to the patient, a process termed adoptive transfer [1]. The field of successful application of immunotherapy includes, but is not limited to, metastatic melanoma [2], a type of cancer that arises from melanocytes and represents the deadliest form of skin cancer, with increasing prevalence. Once it becomes metastatic, the prognosis is very unfavorable and, thus, early diagnosis is crucial for effective management [3].

Since 2011, several next-generation immune-based formulations, such as the checkpoint inhibitors ipilimumab, pembrolizumab, and nivolumab, received approval by the Food and Drug Administration (FDA) for the indication of metastatic melanoma [2] and led to significant clinical improvements, either as monotherapies or in combination regimens [4]. However, despite their remarkable success, only up to 20% to 30% of patients have benefited from these treatments, while the rest are either non-responders (primary resistance) or partial responders (acquired resistance). Unresponsiveness is attributed to factors such as CD8+ T cell density in the tumor microenvironment, monocyte frequency, tumor heterogeneity, and neoantigen load, as well as the composition of patient's gut microbiota [4].

Other challenges regarding immunotherapy include immune-related toxicities. In particular, 85% of melanoma patients under ipilimumab treatment have experienced immune-related adverse events of any grade, with over one-third discontinuing therapy or requiring additional systemic treatment to manage side effects [5]. Immunotherapy faces limitations in patients with both an overactive (autoimmune disease patients) or a suppressed (organ transplant recipients) immune system. On the one hand, melanoma patients with pre-existing autoimmune diseases who receive ipilimumab treatment present frequent disease flares and exacerbations, requiring additional immunosuppression or therapy discontinuation [6,7]. On the other hand, solid organ transplant recipients are at increased risk of developing metastatic melanoma, and when they do, they exhibit a higher probability for graft rejection upon immune checkpoint inhibitor treatment [8,9]. In general, a dysregulated immune system poses as the Sword of Damocles in the decision of clinicians to prescribe immunotherapy. In this regard, there is an increasing need to develop drugs for the treatment of melanoma that are not only safer for such patients but are also able to manage this cancer along with a co-existing immune-related disease.

Herein, we hypothesize that therapeutic management of various clinical disorders simultaneously can be achieved by targeting their major common pathways. As a representative case, we consider the Janus kinase/signal transducers and activators of transcription (JAK/STAT) pathway, which is activated in a wide range of autoimmune diseases (AD), as well as in many different cancer types. It mediates immune responses to several insults from resisting infection to maintaining immune tolerance, enforcing barrier functions, and guarding against cancer [10]. In this review, we summarize ADs that are associated with malignant melanoma, we explore how STAT3 and STAT5 signaling contributes to their pathogenesis, and we evaluate STAT3/STAT5 inhibition as a feasible strategy to target these diseases in order to achieve their simultaneous management with a single drug.

2. Associations among Melanoma, Inflammation and Autoimmune Diseases

A cancer patient can frequently experience disorders, such as inflammation and autoimmune diseases, that occur as frequent pre-existing, predisposing, or intercurrent conditions [11–13]. Intriguingly, these diseases seem to be etiologically interrelated with one another. On one side, cancer has been linked with chronic inflammation. Moreover, there are interconnections between autoimmune diseases and cancer, since certain autoimmune disorders predispose to neoplasias [14]. Patients suffering from dermatomyositis, inflammatory bowel disease, systemic lupus erythematosus, rheumatoid arthritis, psoriasis or Sjögren syndrome may have increased risks for malignancies, whereas the type(s) of cancer they tend to develop depend on the type of the autoimmune disease [15]. For example, rheumatoid arthritis patients are at a greater risk of developing clonal expansions of large granular lymphocytes [16]. Although an autoimmune disorder does not necessarily lead to cancer, it is, however, a phenotypic manifestation of a deregulated immune system, which is generally a favorable background for cancer development [17]. Vice versa, cancer immunotherapy triggers autoimmunity towards several anatomical sites and can lead to conditions that range from relatively minor, such as skin depigmentation, to severe colitis, pancreatitis, lung or liver toxicity [18]. Furthermore, a persistent inflammation can be a fertile ground for the development of an AD [19]. Overall, an inflammation can potentially progress to a neoplasia or an autoimmune disorder, and this process can be facilitated by deregulation of the innate immune system [11,14] (Figure 1).

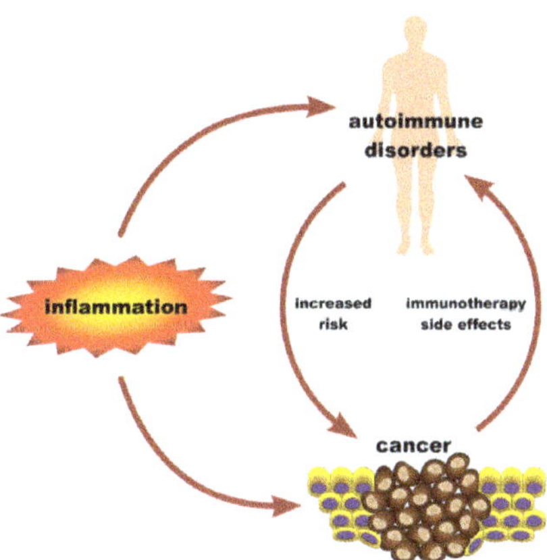

Figure 1. Associations between inflammation, cancer, and autoimmune disease. An inflammation can eventually lead to cancer or autoimmune disease. Some autoimmune disorders may predispose to malignancies. Cancer immunotherapy is associated to autoimmunity.

This interplay among ADs and cancer is evident in the case of melanoma patients. First of all, there is a long-standing inverse association between melanoma and vitiligo, a condition where the immune system produces autoantibodies against immunogenic, melanocytic-specific molecules (melanocytic-differentiation antigens, MDAs) that propel melanin production. These autoantibodies attack melanocytes, ultimately producing white skin patches. Thus, vitiligo patients show a lower melanoma risk because the antibodies against melanocytes confer natural cancer protection [20]. Vice versa, some melanoma patients develop a similar antibody-based condition, called melanoma-associated hypopigmentation or vitiligo-like depigmentation, which is usually considered as a predictor of better outcomes [21], although there are also reports demonstrating that hypopigmentation can be associated with disease progression [22–24]. Given that malignant melanoma co-evolves with immune cell phenotypes [25], this dynamic interaction between autoimmunity-based hypopigmentation and skin cancer could, in the long run, become a double-edged sword. Specifically, these tumors exhibit a high intratumoral heterogeneity and plasticity [25]. Therefore, over the course of the disease, tight immunosurveillance against MDA-expressing cells could serve as a microenvironmental cue that adds evolutionary pressure for immunoselection of low-MDA expressing cell variants that are poorly recognized by the autoantibodies. Such MDA-negative cells possess enhanced invasive capabilities. In this way, anti-MDA responses provoked by the melanoma tumor, can, in turn, promote clonal expansion of low-MDA-expressing cell variants with activated prometastatic programs that can migrate to distant sites, giving rise to secondary tumors [24].

Other autoimmune diseases are positively correlated with melanoma progression [2,26]. A retrospective meta-analysis assessed that the prevalence of pre-existing AD in melanoma patients increased by 1.7-fold within a decade. Prevalence rates were higher in metastatic melanoma patients compared to primary, non-metastatic skin cancer patients or the general population, suggesting that a pre-existing AD could possibly favor metastatic progression. The most common ADs among metastatic melanoma patients were myositis, peripheral neuropathy, type 1 diabetes mellitus, rheumatoid arthritis, psoriasis, autoimmune pancreatitis, autoimmune aplastic anemia, relapsing polychondritis, Hashimoto's encephalopathy, and inflammatory bowel disease. The authors suggested that disturbances in shared molecular or immune pathways could underlie increased susceptibility of AD patients to melanoma [2]. Another clinical study showed that only a small percentage of melanoma patients have a pre-existing AD, but in these patients, progression is faster and prognosis is worse. This subpopulation has a significantly shorter median overall survival and disease-free survival after first metastasis versus cases with a primary tumor, while poorer prognosis was independent of the side effects of AD treatment. Those patients with an antibody-mediated AD had a worse prognosis in relation to those with a T cell-related AD. [27]. Together, these data unveil a potential link between malignant melanoma and immune dysregulation, implying that a pre-existing autoimmune disease could be a predisposing factor for melanoma development and progression. If that is indeed the case, further experimental investigations are needed.

The relationship between inflammation and cancer is well established [28] and inflammation contributes to melanoma initiation and progression [25]. Upregulation of key cancer-related inflammation genes, such as the transcription factors nuclear factor 'kappa-light-chain-enhancer' of activated B-cells (NF-κB) and STAT3, as well as several inflammatory cytokines and angiogenic factors, facilitate melanoma development. Indeed, inoculation of B16-F10 melanoma cells in chronic skin inflammation mouse models led to the formation of larger volume tumors in comparison with the non-inflamed controls, followed by increased periostin expression and M2 macrophage recruitment in the tumor microenvironment (TME) [29]. Similarly, in a mouse model of melanoma, repetitive UV exposure promoted metastasis via neutrophilic skin inflammation. In particular, UV irradiation-induced neutrophil recruitment and activation, which was initiated by the release of high mobility group box 1 (HMGB1) from UV-damaged epidermal keratinocytes. The UV-induced neutrophilic inflammatory response stimulated angiogenesis and increased the number of lung metastases through modulation of melanoma–endothelial cell interactions [30]. The association between inflammation and melanoma is also supported by the fact that anti-inflammatory drugs contribute to melanoma prevention or treatment. Population-based case-control studies showed that continuous use of low-dose non-steroid anti-inflammatory drugs (NSAIDs), such as diclofenac, ibuprofen, and naproxen, reduced the incidence of skin cancer, even though in a sex-dependent manner [31]. Other studies showed that combining celecoxib, a cyclooxygenase COX-2 selective NSAID, with chemotherapeutics can be an effective strategy for melanoma treatment [32].

3. A Basis for the Development of Multipotent Drugs against Melanoma, Inflammation, and Autoimmune Disorders

It is becoming increasingly evident that common molecular mechanisms underlie inflammatory-driven [12], as well as inflammatory/autoimmune-driven neoplasias [17]. The emerging similarities and interconnections among these diseases imply that despite their diverse clinical manifestations, outcome, and response to therapy, a limited number of mutual molecular events might drive their pathogenesis. In this regard, the mechanistic commonalities between cancer, inflammation, and ADs have created a basis for drug repurposing. For instance, anti-inflammatory drugs are used in combination with anticancer regimens [12,33]. Additionally, immunomodulatory drugs used for the treatment of ADs show efficacy against neoplastic diseases [34,35] and vice-versa, anticancer therapeutics can be applied to treat non-cancer immune-mediated diseases [36]. Recent network-based studies demonstrate that disorders with distinct clinical manifestations share a common genetic background [37]. Therefore, identifying shared signaling networks and cellular pathophenotypes at the core of these diseases is anticipated to further shape drug development in the future. We use the term 'multipotent drug' for any targeted inhibitor with the potential to act against two or more pathologies, even if these are different diseases, such as autoimmunity and cancer as an example. Moreover, given that they are mechanistically interrelated [11,12], we further postulate that the effectiveness of such a multipotent drug may depend on its interference with 'root' pathways that pose as common denominators in the pathology of these diseases. Hence, identification of these common traits is a prerequisite for the development of multipotent drugs, which could simultaneously manage cancer in conjunction with coexisting, pre-existing, or predisposing disorders (Figure 2).

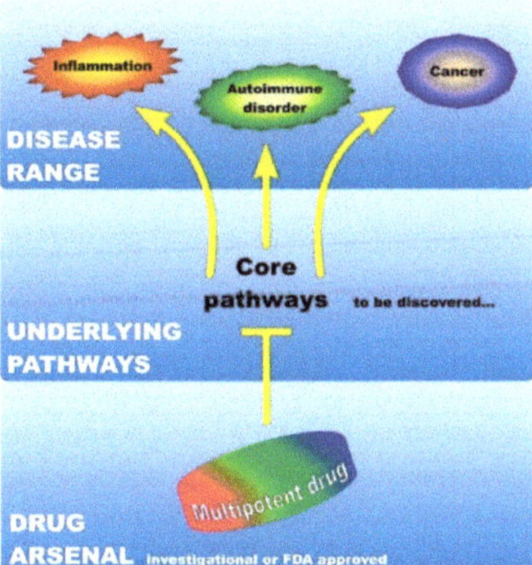

Figure 2. Proposed rationale for treating melanoma and associated inflammation or autoimmunity with multipotent drugs. Investigational or Food and Drug Administration (FDA)-approved drugs could exert anti-inflammatory, immunomodulatory, and anticancer effects. This, to an extent, might be due to their interference with signaling pathways commonly activated in inflammation, autoimmune disease, or cancer. These core pathways underlying all-three pathological conditions can be further characterized to facilitate simultaneous management of associated pathologies due to distinct diseases.

To this end, considering that deregulation of immunity is the main culprit in inflammation, autoimmune disease, and melanoma, we hypothesized that pathways mediating immune responses might be suitable therapeutic targets. Given that the JAK/STAT pathway is commonly activated in these diseases, we postulate that it is an emerging candidate 'core' pathway. Herein, we review the current knowledge of the effects of STAT3 and STAT5 signaling in melanoma, inflammation, and autoimmune disorders and estimate their value for concurrent management of these interconnected conditions.

4. STAT3 and STAT5 Signaling in Melanoma Initiation and Progression

STAT proteins, particularly STAT3 and, to a lesser extent, STAT5 govern fundamental oncogenic processes. STAT family members respond to cytokines, growth factors, and hormones, and transduce signals from the cell membrane to the nucleus to transactivate genes involved in cellular immunity, proliferation, apoptosis, and differentiation. The STAT family encompasses seven members, STAT1, STAT2, STAT3, STAT4, STAT5A, STAT5B, and STAT6, which show high homology in their functional domains. Their main difference lies in the C-terminal transactivation domain that varies among family members and modulates transcriptional activation of target genes. Under physiologic conditions, STAT signaling is stimulus-dependent and tightly regulated. The regulation of STAT activity via phosphorylation of a conserved C-terminal tyrosine residue has been previously described in detail [38,39]. Constitutive activation of STAT3 and STAT5 is involved in tumor formation and progression. However, not all STAT members are promoting cancer progression. STAT1, for instance, exerts proapoptotic and antiproliferative effects, and it seems that both, STAT3 and STAT5 can antagonize its function [40].

STAT3 is constitutively activated across many types of human cancers, including melanoma, as a consequence of aberrant autocrine or paracrine stimulation by cytokines and growth factors, such as interleukins (IL-6, IL-10, IL-12), interferons (IFNs), granulocyte-colony stimulating factor (G-CSF or CSF3), leptin, prolactin (PRL), growth hormone (HGH), epidermal growth factor (EGF), hepatocyte growth factor (HGF), basic fibroblast growth factor (FGF2), and virus proteins (e.g., v-Src, v-Fps, v-Sis). Permanent STAT3 activation can also result from receptors with intrinsic tyrosine kinase activities (such as erb-b2 receptor tyrosine kinase 2 (ERBB2), epidermal growth factor receptor (EGFR), and hepatocyte growth factor receptor HGFR), non-receptor tyrosine kinases (such as c-Src and c-abl), or G-protein coupled receptors [38,41]. An overactive STAT3 is associated with poor prognosis in melanoma patients and drives tumor initiation and malignant progression via induction of several cancer hallmarks, such as apoptosis inhibition, tumor angiogenesis, epithelial–mesenchymal transition (EMT), and stemness. In particular, upon c-Src activation, STAT3 promotes tumor cell survival and proliferation by transactivating *BCL2L1* (that encodes the Bcl-XL protein) and *MCL1*, two crucial anti-apoptotic genes that override cell death regulatory pathways. Additionally, STAT3 regulates the *MYC* gene, which mediates escape of melanoma tumor cells from both terminal differentiation and G0/G1 arrest. Beyond its function as a transcriptional activator, STAT3 can also suppress transcription of the well-established tumor suppressor *TP53* [38]. This suppression can provide an explanation on why melanomas that lack mutations in either *TP53* or its main negative regulator, such as *MDM2*, exhibit aggressive characteristics. STAT3 upregulates nodal factors of angiogenesis, mainly vascular endothelial growth factor (VEGF), hypoxia-inducible factor 1a (HIF-1α) and matrix metalloproteinase-2 (MMP-2) [38] and fosters brain metastasis in melanoma [42]. In addition, drivers of melanoma invasion and metastasis, such as ΔEx2/3p73 (named DNp73), a transactivation-deficient N-terminally truncated oncogenic isoform of the *TP73* gene, can trigger an IGF1R-AKT/STAT3 signaling cascade, which leads to the activation of EMT markers and acquisition of mesenchymal cell phenotypes [43,44]. Additionally, STAT3 is often mentioned in connection with ΔNp63 [45]. ΔNp63α indirectly drives STAT3 (Tyr705) phosphorylation in an autocrine loop by transactivating interleukin 6 (IL-6) and IL-8. Phosphorylated STAT3 stabilizes HIF-1α, resulting in an increased production of VEGF in ΔNp63α-overexpressing cells [44]. In melanoma, ΔEx2/3p73, but not ΔNp63α, mediates STAT3 (Tyr705) phosphorylation in an EPLIN/IGF-1R-dependent manner [43]. With respect to its ability to promote cancer through regulating cancer stem cell (CSC) activities [46], STAT3 upregulation mediates reprogramming of melanoma cells to melanoma stem cells by inducing expression of the Yamanaka factors, providing hints of STAT3 implication in cancer stemness [47]. Moreover, STAT3 increases chemoresistance of melanomas to selective inhibitors of BRAF, which are exploited in the treatment of unresectable or metastatic melanoma with a *BRAF-V600* gain-of-function mutation [48].

Similar to STAT3, STAT5 also exerts oncogenic functions. There are two distinct genes, *STAT5A* and *STAT5B*, that have arisen by gene duplication and encode protein products which share more than 90% peptide sequence similarity and differ primarily in the C-terminus-encoding region [49]. STAT5A is predominantly expressed in the mammary gland, while STAT5B is prevalent in muscle and liver. They respond to a number of factors, such as prolactin, Growth Hormone, erythropoietin, thrombopoietin, EGF, IL-2, IL-3, IL-6, IL-7, IL-9, and IL-15 [40]. STAT5 promotes malignant transformation in hematological malignancies, breast and prostate cancer, non-small cell lung carcinoma, and melanoma [50]. The initial correlation between STAT5 activation and aggressive characteristics of melanoma was first described in a Xiphophorus fish melanoma model [51]. Subsequently, Hassel and colleagues showed that constitutive activation of STAT5 correlated with the upregulation of its antiapoptotic target gene *BCL2L1*, an effect which was conserved in both, human melanoma cells and murine melanocytes [52]. In melanoma patients, STAT5B transcripts were significantly upregulated, whereas STAT5 was found phosphorylated in 62% of the metastatic cases versus normal human melanocytes and benign nevi. STAT5 activation is induced by the EGF/JAK1 axis in melanoma cell lines. Knockdown of *STAT5B* in these cells causes downregulation of BCL2, which triggers cell death and G1 arrest, thus underscoring that STAT5B acts as a survival factor in melanoma [50]. In addition to its antiapoptotic role in melanoma, STAT5 influences the sensitivity to anti-melanoma treatment, including interferon alpha (IFN-α) immunotherapy and BRAF inhibitors. For instance, in skin cancer patients under adjuvant IFN-α therapy, STAT5 expression emerged as an independent predictor of progression-free survival. Recurrence in patients with STAT5-expressing tumors was observed either during or after IFN-α therapy [53]. In accordance herewith, STAT5 has been shown to be overexpressed in IFNα-resistant melanoma cells [54]. Moreover, this transcription factor regulates the nicotinamide phosphoribosyltransferase (NAMPT), a key enzyme in the maintenance of cellular nicotinamide adenine dinucleotide (NAD+) levels, in response to activated B-Raf/extracellular signal-regulated kinases (BRAF/ERK) signaling. NAMPT induces melanoma cell proliferation accompanied by a phenotypic switch towards an invasive phenotype and resistance to BRAF inhibitors [55]. Consistently, a combination of the neuroleptic drug pimozide with indoleamine 2,3-dioxygenase sensitizes melanoma cells to apoptosis and inhibits cell migration via STAT5 suppression [56]. However, pimozide is not a direct STAT5 inhibitor, but rather targets upstream pathway activation for degradation.

5. STAT3 and STAT5 in the Crosstalk between Melanoma and Immune Cells

Melanomas display qualitative and quantitative changes in the density, composition, functional state, and organization of immune infiltrates, the so-called immune contexture, which render immune cells tolerant to tumors or exhaust their ability to attack tumor cells [57]. STAT3 influences the interplay between melanoma cells and components of the immune system, thus contributing to immune evasion. Tumor and immune cells expressing STAT3 develop sophisticated interactions to overall support an immunosuppressive tumor environment that propels metastatic progression. In detail, melanoma cells which overexpress STAT3 protein inhibit the expression of proinflammatory cytokines and chemokines, such as VEGF, IL-10, and IL-6. These, in turn, induce STAT3 activity in hematopoietic progenitor cells (HPCs) to promote the production of immature myeloid cells (IMCs) and plasmacytoid dendritic cells (pDCs). Through IL-10, IMCs block the maturation of dendritic cells into antigen-presenting cells. Thus, these immature dendritic cells are unable to stimulate the antitumor effects of CD8+ T cells and natural killer (NK) cells. In parallel, pDCs promote the accumulation of regulatory T cells (Treg) in the tumor microenvironment. IL-10 and TGF-β secreted by Treg cells further enhance the immunosuppressive microenvironment by restraining both, CD8+ effector T-cell function and DC maturation [38,58–60]. Direct STAT3 targeting in melanoma cell lines sufficed to modulate these crosstalks and enhanced antitumor activity through increasing interferon gamma (IFNγ) levels, mature dendritic cells, and CD8+ T cells. This resulted in reduced tumor growth and prolonged survival of tumor-bearing mice [61]. There is the first evidence that STAT3 activation in NK cells that trigger innate immune responses by the destruction of the tumor cells, can suppress their cytotoxicity against melanoma cells in vivo. The dual effect of STAT3 in melanoma, as well as tumor-attacking immune cells, is of great therapeutic interest since STAT3 inhibition not only blocks survival of cancer cells themselves but also boosts the cytotoxic behavior of infiltrating immune cells [62].

The effect of STAT5 on the melanoma–immune cell interplay is just beginning to be unveiled, mainly via studies in NK cells. The data demonstrate that STAT5 is indispensable for NK cell survival. Conditional knockdown of STAT5 in NK cells led to an almost entire loss of this cell population in transgenic mouse models. Subsequent intravenous injections of these mice with B16F10 melanoma cells increased tumor cell infiltration in the lungs when compared to the control groups [63]. In accordance with the results described above, Sathe et al. reported that an IL-15/JAK1/STAT5 pathway supports the generation of NK cells via transactivation of the antiapoptotic protein MCL1, and that *MCL1* conditional knockdown in NK cells in mice led to multiorgan metastases upon transplantation of melanoma tumors [64]. Other investigations revealed that forced expression of antiapoptotic *BCL2* rescues survival of STAT5-deficient NK cells [65]. Nevertheless, the presence of NK cells expressing high levels of BCL2, but lacking STAT5 was not sufficient to control tumor growth. As this study shows, lowering STAT5 levels in NK cells caused exacerbation of tumor growth in vivo, irrespective of the ability of NK cells to recognize and eradicate tumor cells [65]. They further demonstrated that upregulation of perforin, granzymes, and IFNγ by STAT5 is the key to the enhancement of NK cell cytotoxicity against melanoma [65]. In addition to NK cells, a STAT5-induced cytotoxic potential of immune cells towards skin cancer has also been reported for CD8+ T cells. In particular, genetic modification of CD8+ effector T cells with a constitutively active form of STAT5 gives them high efficiency for host colonization after adoptive transfer and transforms them into long-lived antigen-responsive cells. Upon transfer into melanoma-bearing hosts, these cells accumulate in the tumor site, become activated by tumor antigens, and express the cytolytic factor granzyme B, resulting in tumor regression [66]. It was recently shown that malignant melanoma cells induce inhibition of STAT5 signaling on cytotoxic NK and CD8+ T cells to achieve immune evasion and tumor progression. This effect is mediated by the melanoma-secreted dickkopf WNT signaling pathway inhibitor 2 (DKK2) [67].

In conclusion, STAT3 establishes a reciprocal relationship between melanoma cells and immune cells in favor of tumor immune evasion. STAT3-expressing cancer cells dysregulate immune cells, while at the same time, immune cells with high levels of STAT3 are unable to target melanoma cells. In contrast, STAT5 expression in tumor-attacking immune cells supports antitumor cytotoxicity (Figure 3). These sophisticated interactions between melanoma and immune cells via STAT proteins (and perhaps other factors) could provide a precocious basis towards interpreting the associations between melanoma and co-occurrent autoimmune conditions. Future studies are anticipated to shed more light on common mechanisms underlying melanoma and AD which could, at least in part, explain why these distinct pathologies of melanoma and autoimmunity appear, for example, in the same patients.

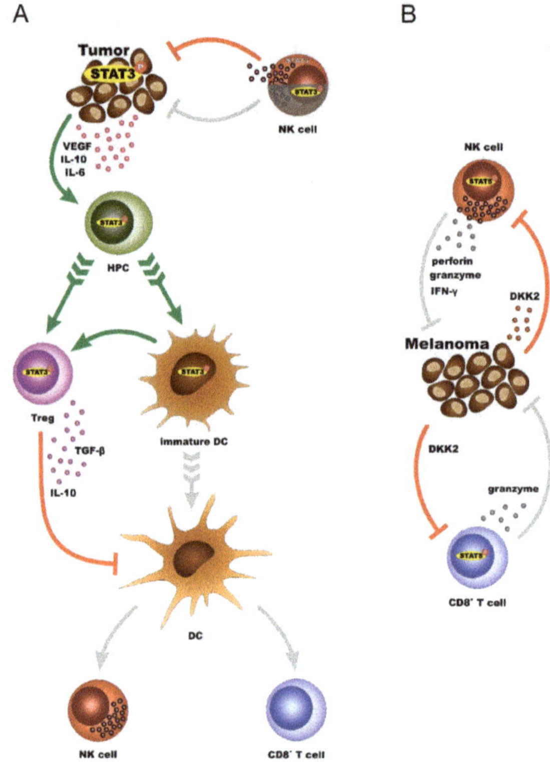

Figure 3. STAT3 and STAT5-mediated crosstalks between melanoma and immune cells create an immunosuppressive network that propels metastatic progression. (**A**) STAT3-expressing tumors secrete factors that upregulate STAT3 in hematopoietic progenitor cells (HPCs). This results in the accumulation of Tregs and immature dendritic cells in the tumor microenvironment, which further inhibit the maturation of these cells. Due to the lack of mature dendritic cells, the tumor-attacking CD8+ T cells and NK cells cannot be stimulated. NK cells with inactive STAT3 can repress tumors, while STAT3 activation in NK cells blocks their tumor-lysing properties. (**B**) STAT5-activated NK cells are effective against melanoma cells. This property is suppressed by secretion of DKK2 from melanoma cells (see main text for details).

6. STAT3 and STAT5 Pathways in Inflammation and Melanoma-Associated Autoimmune Diseases

During inflammation, lymphoid and myeloid cells are recruited to the site of the lesion. By secreting cytokines, immune cells and macrophages control the development of inflammation. Effective and timely management of an inflammation is crucial for maintaining the immune system homeostasis. Any factor that activates inflammation or triggers immune reactions could lead to an autoimmune disease, whenever immune regulation of the inflammatory response fails to manipulate the situation appropriately. Thus, the root cause of most autoimmune diseases is the failure of the immune response to orchestrate the dynamics or magnitude of an inflammatory response, limit it to a relevant body site, and terminate it at the proper time [19]. The balance between Th17 cells, a subset of proinflammatory T helper cells which are characterized by IL-17 production, and Tregs that maintain tolerance to self-antigens is a key determinant in the development of an AD, and the signals that induce Th17 differentiation from CD4+ T cells inhibit Treg differentiation [68]. STATs affect inflammatory disease since they control the development of hematopoietic cells that regulate inflammation and mediate the responses of target cells to inflammatory cytokines [69]. For instance, STAT members, such as STAT1 and STAT3, are highly interlinked with NF-κB [70,71]. Not only does STAT3 respond to NF-κB-induced IL-6, a critical mediator of the acute phase response in inflammation [72], but it also physically interacts with this transcription factor to co-regulate inflammatory gene targets [71]. However, as the damage response persists, chronic activation of STAT proteins leads to inflammatory, idiopathic, and autoimmune conditions [72]. In line with this, *STAT3* gain-of-function mutations can cause early-onset lymphoproliferation and autoimmunity [73] and have been associated with Th17 hyperactivation [74]. STAT5 signaling is also involved in autoimmunity. Patients with STAT5B deficiency have decreased numbers of Treg cells and exhibit immunological aberrations, whereas most of them suffer from severe eczema and AD [49]. Herein, we briefly summarize our current knowledge of how STAT3 and STAT5 are implicated in inflammatory and autoinflammatory conditions as well as in autoimmune diseases. The effect of STAT3 and STAT5 on autoimmune diseases with a demonstrated link to melanoma [2] is described in Section 6.1 to Section 6.8 and depicted in Table 1. STAT3/STAT5-related inflammatory and autoimmune diseases with a still understudied effect on melanoma incidence are also mentioned (Sections 6.9–6.12).

Table 1. Effects of STAT3/STAT5 in autoimmune diseases associated with melanoma (conditions are listed by decreasing prevalence rate in melanoma patients).

AD	STAT3 Input	STAT5 Input	STAT3 Inhibition	STAT5 Inhibition	Refs
Peripheral Neuropathy	activated JAK/STAT3 signaling pathway	n.d.	n.d.	n.d.	[75,76]
Type 1 Diabetes Mellitus	activated in T cells, premature differentiation of stem cells into cells of the pancreatic endocrine lineage	STAT5 phosphorylation in monocytes	therapeutic	n.d.	[77–79]
Rheumatoid Arthritis	activated in CD4+ T cells, increase of Th17/Treg ratio, induction of inflammatory and osteoclastogenic factors	decrease of Th17/Treg ratio	therapeutic	n.d.	[80–83]
Psoriasis	STAT3 activation in the skin	n.d.	therapeutic	n.d.	[84,85]
Autoimmune Pancreatitis	STAT3 loss-of-function mutation	n.d.	n.d.	n.d.	[86,87]

Table 1. *Cont.*

AD	STAT3 Input	STAT5 Input	STAT3 Inhibition	STAT5 Inhibition	Refs
Autoimmune Aplastic Anemia	STAT3 gain-of-function mutations in T cells	IL-2/STAT5B active in Tregs, STAT5B gain-of-function mutations in T cells	n.d.	n.d.	[88–90]
Hashimoto's Encephalopathy	n.d.	activation of neuroinflammation-promoting T helper cells	n.d.	n.d.	[91,92]
Inflammatory Bowel Disease	STAT3 activation in T-cells contributes to colitis, STAT3 activation in myeloid cells protects from colitis	proliferation of intestinal epithelial stem cells, regeneration of the crypt epithelium, regulation of Th17/Treg balance, protection of dendritic cells from apoptosis, mucosal wound healing, resistance to intestinal injury	therapeutic	n.d.	[72,93–97]

n.d.: not determined.

6.1. Peripheral Neuropathy

Peripheral neuropathy is an AD of paraneoplastic etiology, which may appear as melanoma comorbidity. It has been attributed to an immune response directed against antigens in the tumor that subsequently cross-react with the same or similar epitopes in the nervous system [75]. First evidence associates neuropathic pain with an activated JAK/STAT3 signaling pathway [76]. The exact role of STATs in this condition is yet to be determined.

6.2. Type 1 Diabetes Mellitus

Type 1 diabetes (T1D) is a chronic AD causing immune-mediated loss of pancreatic β cells. The key mediators of β cell destruction are the CD4+ and CD8+ T cells. Treg cells are also important for the maintenance of immune tolerance and prevention of autoimmunity. STAT3 is activated in T cells of T1D patients and favors their resistance to suppression [77]. It also influences diabetes by altering pancreatic cells development. This was shown in a study where an activating STAT3 mutation caused premature differentiation of stem cells into cells of the pancreatic endocrine lineage [78]. Similar to STAT3, STAT5 activation has been correlated with T1D, since persistent STAT5 phosphorylation characterizes monocytes of individuals with or at-risk for T1D and is suspected of altering the epigenetic regulation of inflammatory response genes [79].

6.3. Rheumatoid Arthritis

Rheumatoid arthritis (RA) is a chronic disorder manifested with continuous inflammation, swelling, destruction, and pain in multiple joints. Th17 cells are the pathogenic culprit in RA, while Treg cells suppress RA. STAT3 and STAT5 knockdown in RA demonstrated that STAT3 increases Th17 but decreases Treg proportions, whereas STAT5 has the opposite effects on these cell populations [80], providing insights that STAT5 counteracts STAT3 activity on the reciprocal balance of Th17 and Tregs in the context of RA. In addition, STAT3 induces factors that lead to inflammation and osteoclast genesis, while its conditional knockout in mice blocks joint inflammation and destruction and offers resistance to collagen-induced arthritis [81]. Consistently, STAT3 phosphorylation and subsequent upregulation of its downstream gene targets are discriminative for CD4+ T cells of RA patients [82]. It is possible that STAT5 counteracts STAT3 in the context of RA via exerting opposite effects in the Th17/Treg balance, which is crucial in the pathogenesis of this disease [83].

6.4. Psoriasis

Psoriasis is a chronic autoimmune disease characterized by patches of abnormal skin, which are typically red, dry, itchy, and scaly. Its most prevalent type is psoriasis vulgaris. It is considered a T cell-mediated AD modified by genetic susceptibility and environmental stimuli. Th17 lymphocytes, which differentiate from naïve T cells upon IL-6 induction, play a central role in its pathogenesis, and recent evidence suggests that inflammatory circuits are established among Th1 and Th17 cells and keratinocytes, promoting psoriasis. This leads to the hypothesis that STAT3 and STAT5, which regulate Th17, are possibly associated with psoriasis. Indeed, STAT3 activation has been observed in the skin of psoriatic patients [84]. Consistently, transgenic mice, in which keratinocytes express a constitutively active form of STAT3, develop psoriasis-like skin lesions. Topical administration of STAT3 inhibitor improved psoriatic lesions in both, these mice and in clinical patients [85]. The role of STAT5 in psoriasis needs to be assessed.

6.5. Autoimmune Pancreatitis

Autoimmune pancreatitis is a unique form of chronic pancreatitis initially reported in 1995. It exhibits characteristic histological features, frequent elevations of serum IgG4 antibodies, and a predictable response to steroid therapy. Potential long-term sequelae include pancreatic duct stones and malignancy [86]. A recent report describes a case of acute pancreatitis that presented in a patient with a loss-of-function STAT3 mutation [87]. The involvement of STATs in this newly-identified autoimmune condition remains to be elucidated.

6.6. Autoimmune Aplastic Anemia

Aplastic anemia (AA) is a non-malignant blood disorder, caused by immune destruction of hematopoietic and progenitor cells, which leads to a failure of bone marrow to sustain blood production. As a result, the patients develop pancytopenia, evidenced by low levels of all blood cell types. Clonal hematopoiesis with the formation of a genetically distinct subpopulation of blood cells has been suggested as a basis of this disease, where clonal cells have acquired the ability of increased proliferation and immune escape [88]. *STAT5B* [88] and *STAT3* [89] gain-of-function mutations have been found in T cells of AA patients, while an IL-2/STAT5B pathway characterized patients' Treg subpopulations [90]. Overall, these data link STATs with AA, although continued investigation into the mechanistic connections is necessary.

6.7. Hashimoto's Encephalopathy

Hashimoto's encephalopathy is a rare disease of autoimmune origin, characterized by high titers of antithyroid antibodies. It has been proposed that an abnormal function of the immune system causes neuronal inflammation, leading to impaired brain function. The condition co-exists with other ADs, such as T1D mellitus, systemic lupus erythematosus, and Sjögren syndrome [91]. Although the molecular mechanism of this rare disease remains elusive, a study showing that STAT5 programs a distinct subset of Granulocyte-macrophage colony-stimulating factor (GM-CSF)-producing T helper cells, which is essential for autoimmune neuroinflammation [92], could provide a potential link between STATs and this disorder. Future investigations could shed more light on possible associations of STATs with the etiopathology of encephalitis.

6.8. Inflammatory Bowel Disease

Inflammatory bowel disease (IBD) describes a group of chronic, relapsing, autoimmune diseases of complex etiology that become manifest at the small intestine and colon. It encompasses ulcerative colitis and Crohn's disease, which differ in the location of inflammation within the gastrointestinal tract. Several cytokines that activate STAT3 drive IBD pathogenesis. Data obtained from IBD patients showed that STAT3 is activated in their inflamed colons. Intriguingly, expression of STAT3 in different immune cell populations exerts opposing effects on IBD. In particular, phosphorylated protein in T-cells contributes to colitis, while its activation in myeloid cells (neutrophils and macrophages) and enterocytes is protective against colitis. In mouse models of IBD, several proinflammatory cytokines, such as IL-2, IL-6, IL-15, IL-21, IL-23, IL-17, and IL-18, have detrimental effects, whereas anti-inflammatory cytokines, including IFNα/β, IL-10, IL-11, and IL-22, are beneficial. STAT3 targeting with small molecule inhibitors ameliorates IBD in vivo and is under consideration as a new approach to IBD management, particularly in patients that are refractory to current therapies [72]. However, given the opposing functions of STAT3 on different immune cell types, targeting should be directed towards the specific cell population which promotes IBD, without jeopardizing those immune cells that protect from disease [93].

STAT5 has rather anti-inflammatory properties both in the intestinal epithelium and in adaptive immune cells and has been shown to oppose STAT3 functions in the context of IBD. A GH/STAT5 axis is protective against colitis in animal models. On one side, STAT5 is critical for the proliferation of intestinal epithelial stem cells and regeneration of the crypt epithelium. STAT5B-deficient mice are particularly susceptible to chemically induced colitis, a fact that was attributed to apoptosis of epithelial cells and damage of the mucosal barrier. In agreement, activated STAT5 promotes mucosal wound healing, increases intestinal epithelial stem cell proliferation, accelerates crypt regeneration, and confers resistance to intestinal injury [95,96]. Lack of active STAT5 predisposes to *Clostridium difficile* infection-induced colitis [97]. Otherwise, STAT5 regulates T cell differentiation, especially Treg cells. In addition, IL-2 signaling via STAT5 limits Th17 in favor of Treg development. This is opposed to STAT3 function, which induces Th17 differentiation and negatively controls Treg differentiation. The balance between Tregs and effector T cells is key in IBD suppression. Moreover, GM-CSF-induced STAT5 signaling protects dendritic cells from apoptosis, in contrast to the apoptotic effects of an IL-21/STAT3 axis in these cells [94].

6.9. Asthma

Asthma is a condition characterized by inflammation, which leads to hyperresponsiveness of the airways and remodeling of the airway wall. Airway inflammation and remodeling is accompanied by STAT3 activation and by increased levels of Th2- and Th17-type cytokines in the lung. STAT3 regulates immune cell recruitment, specifically Th2 cells, during allergic inflammation. Targeted STAT3 depletion in the airway epithelium prevents house dust mite-mediated allergic inflammation and airway hyperresponsiveness. Interestingly, other studies suggest that Th2 cells respond to STAT3 activation differentially, depending on their location in the lung [72]. In a similar manner, in animal models of asthma, the STAT5 pathway-regulates lymphocyte proliferation induced by ovalbumin [98], while a GM-CSF/STAT5 axis enhances survival of lung granulocytes [99]. Observations on a large cohort of hospitalized asthma patients revealed an increased cancer incidence, which, however, was highly cancer type-dependent. These patients presented an increased risk for stomach and colon cancer but decreased for endometrial cancer and melanoma [100]. Further studies which could also include the population of outpatient asthma patients are needed to investigate the potential link between asthma and several cancer types.

6.10. Cachexia and Polymyositis

Cachexia is an inflammation-driven metabolic syndrome, where impaired regulation of the balance between anabolism and catabolism results in the loss of adipose tissue and skeletal muscle, eventually leading to muscle atrophy. Mechanistic studies have shown that STAT3 phosphorylation in response to overexpression of its upstream cytokine IL-6 is associated with this condition [72]. In a similar manner, levels of phosphorylated STAT3 and cytokine IL-22 have been found elevated in muscle tissues of patients with polymyositis, an idiopathic inflammatory myopathy that causes muscle weakness [101]. Several known STAT3 inhibitors that directly target hyperphosphorylated STAT3 also reduced markers of cachexia in cell culture models and antagonized catabolic signaling in mice, supporting the concept that STAT3 is a valid target for cachexia treatment [72]. No study has correlated STAT5 and polymyositis thus far. An association between cachexia and melanoma has been suggested based on mice studies. In detail, B16F10 melanoma-bearing mice depicted skeletal muscle and epididymal fat mass reduction, muscle strength loss, and locomotor activity impairment, associated with elevated IL-6 levels in the plasma [102]. Further studies are needed to investigate if these effects are mediated by STATs.

6.11. Fibrosis

Tissue fibrosis is an inflammatory condition which is caused upon dysregulation of the tightly-controlled process of wound healing and is characterized by overproduction of the extracellular matrix (ECM) and excess matrix contraction. STAT3 drives wound healing following inflammation and contributes to fibrosis by inducing production of ECM or by transactivating profibrotic metalloproteases, such as MMP-9. The blockade of STAT3 phosphorylation with small molecule inhibitors could ameliorate the development of pulmonary fibrosis in mice exposed to bleomycin [72]. In contrast to the profibrotic function of STAT3, STAT5 seems to protect from fibrosis, perhaps through counteracting STAT3. In particular, loss of STAT5 causes liver fibrosis through STAT3 activation [103]. It was recently shown that the dorsal skin of mice contains multiple and diverse lineages of fibroblasts. One of those lineages is responsible for the fibrotic response to injury and cancer-stroma formation. Ablation of this specific lineage diminished connective tissue deposition in wounds and reduced melanoma growth in vivo [104]. This study provides an initial link between melanoma and fibrotic injury, although further investigation is required to confirm this finding in the cancer patient population and to unveil the underlying mechanisms.

6.12. Systemic Lupus Erythematosus

Systemic lupus erythematosus (SLE) is a chronic, systemic AD that distinguishes itself by autoreactive lymphocytes and proinflammatory cytokine production. It can often lead to renal injury and multiple organ damage due to the formation of autoantibodies against nuclear antigens, cell surface, and serum proteins [105]. Both STAT3 and STAT5 are implicated in SLE etiopathology and have been found activated in immune cells of SLE patients [106,107]. In T cells of SLE patients, STAT3 and STAT5 synergistically transactivate IL-10 through epigenetic remodeling [108]. In the development of SLE, dysregulation of the Th17/Treg balance may play an important role and could be responsible for an increased proinflammatory response, especially in the active form of the disease [109,110]. In this context, SLE has been recently associated with an active IL-17/STAT3 axis, accompanied by higher Th17 cell numbers [111]. STAT3 inhibitors in combination with immunosuppressive drugs, improve SLE via restoration of this balance [105], while agents inhibiting STAT3 phosphorylation have shown efficacy in the treatment of SLE [106]. STAT5 is also implicated in SLE, since activated STAT5 upregulates the antiapoptotic targets Bcl-2 and Ki-67 in CD4+ T cells, perhaps providing them with a survival and proliferative advantage over Treg cells [112]. A large patient population-based, retrospective study recently highlighted a statistically significant association between SLE and melanoma [113]. Dreyer et al. also described that SLE tends to be positively associated with melanoma, although the trend was not statistically significant [114]. However, another meta-analysis study reported a decreased risk of melanoma in SLE patients [115]. Further exploration of the association between malignant melanoma and SLE is anticipated to shed more light on this issue.

7. Perspectives, Challenges, and Limitations Towards Translating STAT3/STAT5 Targeting to Therapeutic Solutions for Patients with AD and Melanoma

Recent studies converge to the revolutionary idea of targeting a single nodal molecule to manage two inter-related pathological conditions simultaneously [116–118]. In line with this notion, we evaluated whether STAT targeting can be considered for the management of melanoma and its associated autoimmune comorbidities. An advantage of this approach is that there is an already established arsenal of STAT3 and STAT5 inhibitors, which are currently under investigation for the treatment of cancer or autoimmune diseases, as reviewed in detail by Loh and colleagues [40]. This arsenal is being further enriched by indirect epigenetic modifiers of STAT proteins [70,119,120]. STAT3 inhibitors have demonstrated their efficacy in animal tumor models [121] and are currently under ongoing clinical trials for melanoma treatment (ClinicalTrials.gov Identifiers: NCT01904123, NCT03195699). In a similar manner, these inhibitory molecules are also efficient in treating several autoimmune diseases in vivo [85,106] and are now tested in a clinical anti-T1D setting (ClinicalTrials.gov Identifier: NCT02641522), which has been reported as melanoma-associated comorbidity. Moreover, it has been reported that cancer patients who exhibit cachexia may benefit from STAT3 inactivation, as it preserves fat and tissue mass and rescues cachectic phenotypes, in addition to eliminating tumors [122]. Another approach for simultaneous disease management could be achieved through nutritherapeutic strategies. Several dietary compounds are potent inhibitors of STAT3 regulatory networks [123,124]. These nutriceuticals could have potential applications in patients with chronic ADs, in the sense that their long-term use as dietary supplements might not only mitigate AD symptoms, but also reduce the cancer risk. Hence, STAT3 inhibitors may hold therapeutic promise as multipurpose drugs for the treatment of melanoma in conjunction with associated ADs. Due to its roles in cancer and immunity, STAT5 is also a proposed therapeutic target [125]. Studies in AD animal models with transplanted melanoma tumors versus immunocompetent controls will not only provide proof-of-concept about the feasibility of using agents such as STAT inhibitors against different diseases but might also establish an experimental approach to facilitate insights into the complex interactions between melanoma and immune cells.

One challenge regarding the proposed approach remains, namely which melanoma–immune cell interactions would be suitable for modulation to achieve dual targeting of these diseases without reducing the beneficial effects of other immune cell populations. Following this direction, it is intriguing to hypothesize that a possible interplay among melanoma, Th17, and Treg cells could be manipulated towards this goal. In detail, it has been recognized that an unbalanced Th17/Treg response, on the one hand, mediates several melanoma-associated Ads, such as psoriasis, IBD, RA, and SLE, and on the other hand, contributes to carcinogenesis mainly via propelling chronic inflammation [126]. Interestingly, STAT3 and STAT5 emerge as fine-tuners of the Th17/Treg balance, since STAT3 is responsible for Treg cells and STAT5 mediates Th17 differentiation [126]. Given that Th17 [127] and Tregs [128] have been found dysregulated during melanoma progression, it is appealing to postulate that disturbance of the Th17/Treg homeostasis may favor melanoma cell proliferation in a STAT3/STAT5-dependent manner. Future studies could shed more light on this issue.

Targeting of the interplay between melanoma cells and NK cells poses as another challenge. While STAT3 inhibition reactivates NK cells against melanoma, STAT5 inhibition suppresses tumor cytotoxic activities of NK cells. A hypothetical explanation is that NK cells could be particularly sensitive to a still understudied equilibrium between STAT3 and STAT5. Such a scenario poses limitations to the straightforward inhibition of either STAT3 or STAT5 in NK cells and necessitates elucidating how the STAT3/STAT5 ratio is regulated and what is its contribution to the melanoma–NK cells interplay. In NK cells, STAT3 inhibition leads to STAT5 activation, and a delicate balance exists between STAT3-mediated suppression and STAT5-induced activation of NK-cell cytotoxicity [62]. It has been hypothesized that this inverse correlation between STAT3 and STAT5 could serve as a regulator of autoimmunity, given that both factors respond to the same cytokines. In particular, IL-15-induced STAT3 activation could counteract the IL-15-STAT5-mediated NK-cell cytotoxicity to prevent autoimmunity [62]. A detailed understanding of the mechanisms governing transactions between melanoma and immune cell populations could catalyze the development of anticancer strategies aiming at increasing the potential of killer cells while suppressing self-destruction.

Along with hopes, targeting of STAT factors has also raised skepticism, since its pleiotropic effects in non-tumor tissues are predicted to potentiate the risk of adverse events [129]. Indeed, STATs are points of convergence of several signal transduction pathways that affect diverse processes [130], and as such, their inhibition will have multiple effects, some of which may be manifested as side effects in nontumor tissues. However, this disadvantage can theoretically turn into an advantage in the event of multimorbid cancer patients, since in these patients systemic administration of inhibitors against pleiotropic molecules that are commonly activated in melanoma and associated comorbidities could, in the long run, simultaneously manage lesions. In this case, the side effect at a site outside the tumor could be proven beneficial if it actually counteracts a co-existing AD. Such directions could inspire next-generation therapeutics, especially in melanoma patients whose pre-existing conditions refrain them from getting benefit from current immunotherapy regimens.

8. Conclusions

Inflammation and autoimmune disorders emerge as comorbidities in melanoma, providing hints that deregulation in common processes may govern all-three diseases. Crosstalks are established among melanoma cells and several subpopulations of immune cells, reciprocally modifying their behavior. Molecular drivers of these crosstalks, such as STAT3 and STAT5 signaling pathways, could be targeted to manipulate interactions between melanoma cells and immune cells, to develop therapeutics that coordinately regulate skin cancer and inflammatory and/or autoimmune conditions. In this review, we (a) state that the high-risk patients with a non-competent immune system (i.e., those with a pre-existing AD or transplant organ recipients) constitute a special population which requires personalized management with novel therapeutics that are able to both kill tumor cells and suppress immune system responses, (b) evaluate if such an approach is feasible, based on the molecular commonalities between tumors, inflammation, and autoimmune disease, and (c) estimate STAT targeting as a representative example, although targeting of additional, probably still undiscovered molecules, may also exert a similar multipotent effect. Immunotherapeutics have taken center stage and emerge as the blockbusters of pharmacopoieia. Their favorable effects on cancer patients keep therapeutic hopes high. However, the use of cancer immunotherapy on patients with associated autoimmune disease could lead to life-threatening side effects. In contrast, drugs with an inherent ability to cause immune suppression which could theoretically counteract an overactive immune system might constitute a more suitable option in the context of personalized therapy. To this end, selected next-generation therapeutics, e.g., STAT inhibitors, may have the potential to shape the first line of treatment for these patient populations in the future.

Funding: This work has received financial support from the Deutsche Krebshilfe (German Cancer Aid, grant 70112353), DFG (grant PU 188/17-1), BMBF, and Rostock University Medical School.

Acknowledgments: The authors are especially grateful to Stephan Marquardt for support in image preparation.

Conflicts of Interest: The authors declare no conflict of interest.

References

1. Ventola, C.L. Cancer Immunotherapy, Part 1: Current Strategies and Agents. *Pharm. Ther.* **2017**, *42*, 375–383.
2. Ma, Q.; Shilkrut, M.; Zhao, Z.; Li, M.; Batty, N.; Barber, B. Autoimmune comorbidities in patients with metastatic melanoma: A retrospective analysis of us claims data. *BMC Cancer* **2018**, *18*, 145. [CrossRef] [PubMed]
3. Domingues, B.; Lopes, J.M.; Soares, P.; Pópulo, H. Melanoma treatment in review. *ImmunoTargets Ther.* **2018**, *7*, 35–49. [CrossRef] [PubMed]
4. Park, Y.J.; Kuen, D.S.; Chung, Y. Future prospects of immune checkpoint blockade in cancer: From response prediction to overcoming resistance. *Exp. Mol. Med.* **2018**, *50*, 109. [CrossRef] [PubMed]
5. Horvat, T.Z.; Adel, N.G.; Dang, T.O.; Momtaz, P.; Postow, M.A.; Callahan, M.K.; Carvajal, R.D.; Dickson, M.A.; D'Angelo, S.P.; Woo, K.M.; et al. Immune-Related Adverse Events, Need for Systemic Immunosuppression, and Effects on Survival and Time to Treatment Failure in Patients with Melanoma Treated with Ipilimumab at Memorial Sloan Kettering Cancer Center. *J. Clin. Oncol.* **2015**, *33*, 3193–3198. [CrossRef] [PubMed]
6. Johnson, D.B.; Sullivan, R.J.; Ott, P.A.; Carlino, M.S.; Khushalani, N.I.; Ye, F.; Guminski, A.; Puzanov, I.; Lawrence, D.P.; Buchbinder, E.I.; et al. Ipilimumab Therapy in Patients with Advanced Melanoma and Preexisting Autoimmune Disorders. *JAMA Oncol.* **2016**, *2*, 234–240. [CrossRef]
7. Menzies, A.M.; Johnson, D.B.; Ramanujam, S.; Atkinson, V.G.; Wong, A.N.M.; Park, J.J.; McQuade, J.L.; Shoushtari, A.N.; Tsai, K.K.; Eroglu, Z.; et al. Anti-PD-1 therapy in patients with advanced melanoma and preexisting autoimmune disorders or major toxicity with ipilimumab. *Ann. Oncol.* **2017**, *28*, 368–376. [CrossRef] [PubMed]
8. Tripathi, S.V.; Morris, C.R.; Alhamad, T.; Fields, R.C.; Linette, G.P.; Cornelius, L.A. Metastatic melanoma after solid organ transplantation: An interdisciplinary, institution-based review of management with systemic and targeted therapies. *J. Am. Acad. Dermatol.* **2018**, *78*, 184–185. [CrossRef]
9. Abdel-Wahab, N.; Safa, H.; Abudayyeh, A.; Johnson, D.H.; Trinh, V.A.; Zobniw, C.M.; Lin, H.; Wong, M.K.; Abdelrahim, M.; Gaber, A.O.; et al. Checkpoint inhibitor therapy for cancer in solid organ transplantation recipients: An institutional experience and a systematic review of the literature. *J. Immunother. Cancer* **2019**, *7*, 106. [CrossRef]
10. Villarino, A.V.; Kanno, Y.; Ferdinand, J.R.; O'Shea, J.J. Mechanisms of Jak/STAT signaling in immunity and disease. *J. Immunol.* **2015**, *194*, 21–27. [CrossRef]
11. Karin, M.; Lawrence, T.; Nizet, V. Innate immunity gone awry: Linking microbial infections to chronic inflammation and cancer. *Cell* **2006**, *124*, 823–835. [CrossRef] [PubMed]
12. Wogan, G.N.; Dedon, P.C.; Tannenbaum, S.R.; Fox, J.G. Infection, inflammation and colon carcinogenesis. *Oncotarget* **2012**, *3*, 737–738. [CrossRef] [PubMed]
13. Smith, A.W.; Reeve, B.B.; Bellizzi, K.M.; Harlan, L.C.; Klabunde, C.N.; Amsellem, M.; Bierman, A.S.; Hays, R.D. Cancer, comorbidities, and health-related quality of life of older adults. *Health Care Financ. Rev.* **2008**, *29*, 41–56. [PubMed]
14. Franks, A.L.; Slansky, J.E. Multiple associations between a broad spectrum of autoimmune diseases, chronic inflammatory diseases and cancer. *Anticancer Res.* **2012**, *32*, 1119–1136. [PubMed]
15. Beyaert, R.; Beaugerie, L.; Van Assche, G.; Brochez, L.; Renauld, J.C.; Viguier, M.; Cocquyt, V.; Jerusalem, G.; Machiels, J.P.; Prenen, H.; et al. Cancer risk in immune-mediated inflammatory diseases (IMID). *Mol. Cancer* **2013**, *12*, 98. [CrossRef] [PubMed]
16. Schwaneck, E.C.; Renner, R.; Junker, L.; Einsele, H.; Gadeholt, O.; Geissinger, E.; Kleinert, S.; Gernert, M.; Tony, H.P.; Schmalzing, M. Prevalence and Characteristics of Persistent Clonal T Cell Large Granular Lymphocyte Expansions in Rheumatoid Arthritis: A Comprehensive Analysis of 529 Patients. *Arthritis Rheumatol.* **2018**, *70*, 1914–1922. [CrossRef]
17. Pateras, I.S.; Havaki, S.; Nikitopoulou, X.; Vougas, K.; Townsend, P.A.; Panayiotidis, M.I.; Georgakilas, A.G.; Gorgoulis, V.G. The DNA damage response and immune signaling alliance: Is it good or bad? Nature decides when and where. *Pharmacol. Ther.* **2015**, *154*, 36–56. [CrossRef]
18. Amos, S.M.; Duong, C.P.; Westwood, J.A.; Ritchie, D.S.; Junghans, R.P.; Darcy, P.K.; Kershaw, M.H. Autoimmunity associated with immunotherapy of cancer. *Blood* **2011**, *118*, 499–509. [CrossRef]
19. Cohen, I.R. Activation of benign autoimmunity as both tumor and autoimmune disease immunotherapy: A comprehensive review. *J. Autoimmun.* **2014**, *54*, 112–117. [CrossRef]

20. Teulings, H.E.; Overkamp, M.; Ceylan, E.; Nieuweboer-Krobotova, L.; Bos, J.D.; Nijsten, T.; Wolkerstorfer, A.W.; Luiten, R.M.; van der Veen, J.P. Decreased risk of melanoma and nonmelanoma skin cancer in patients with vitiligo: A survey among 1307 patients and their partners. *Br. J. Dermatol.* **2013**, *168*, 162–171. [CrossRef]

21. Teulings, H.E.; Limpens, J.; Jansen, S.N.; Zwinderman, A.H.; Reitsma, J.B.; Spuls, P.I.; Luiten, R.M. Vitiligo-like depigmentation in patients with stage III-IV melanoma receiving immunotherapy and its association with survival: A systematic review and meta-analysis. *J. Clin. Oncol.* **2015**, *33*, 773–781. [CrossRef] [PubMed]

22. Cho, E.A.; Lee, M.A.; Kang, H.; Lee, S.D.; Kim, H.O.; Park, Y.M. Vitiligo-like Depigmentation Associated with Metastatic Melanoma of an Unknown Origin. *Ann. Dermatol.* **2009**, *21*, 178–181. [CrossRef]

23. Pinner, S.; Jordan, P.; Sharrock, K.; Bazley, L.; Collinson, L.; Marais, R.; Bonvin, E.; Goding, C.; Sahai, E. Intravital imaging reveals transient changes in pigment production and Brn2 expression during metastatic melanoma dissemination. *Cancer Res.* **2009**, *69*, 7969–7977. [CrossRef] [PubMed]

24. Fürst, K.; Steder, M.; Logotheti, S.; Angerilli, A.; Spitschak, A.; Marquardt, S.; Schumacher, T.; Engelmann, D.; Herchenröder, O.; Rupp, R.A.W.; et al. DNp73-induced degradation of tyrosinase links depigmentation with EMT-driven melanoma progression. *Cancer Lett.* **2019**, *442*, 299–309. [CrossRef] [PubMed]

25. Hölzel, M.; Tüting, T. Inflammation-Induced Plasticity in Melanoma Therapy and Metastasis. *Trends Immunol.* **2016**, *37*, 364–374. [CrossRef] [PubMed]

26. Wang, Y.; Deng, O.; Feng, Z.; Du, Z.; Xiong, X.; Lai, J.; Yang, X.; Xu, M.; Wang, H.; Taylor, D.; et al. RNF126 promotes homologous recombination via regulation of E2F1-mediated BRCA1 expression. *Oncogene* **2016**, *35*, 1363–1372. [CrossRef]

27. Bottoni, U.; Paolino, G.; Ambrifi, M.; Didona, D.; Albanesi, M.; Clerico, R.; Lido, P.; Brachini, A.; Corsetti, P.; Richetta, A.G.; et al. Association between autoimmune disease and cutaneous melanoma with regard to melanoma prognosis. *Clin. Exp. Dermatol.* **2015**, *40*, 254–259. [CrossRef]

28. Grivennikov, S.I.; Greten, F.R.; Karin, M. Immunity, inflammation, and cancer. *Cell* **2010**, *140*, 883–899. [CrossRef]

29. Ohno, F.; Nakahara, T.; Kido-Nakahara, M.; Ito, T.; Nunomura, S.; Izuhara, K.; Furue, M. Periostin Links Skin Inflammation to Melanoma Progression in Humans and Mice. *Int. J. Mol. Sci.* **2019**, *20*, 169. [CrossRef]

30. Bald, T.; Quast, T.; Landsberg, J.; Rogava, M.; Glodde, N.; Lopez-Ramos, D.; Kohlmeyer, J.; Riesenberg, S.; van den Boorn-Konijnenberg, D.; Hömig-Hölzel, C.; et al. Ultraviolet-radiation-induced inflammation promotes angiotropism and metastasis in melanoma. *Nature* **2014**, *507*, 109–113. [CrossRef]

31. Joosse, A.; Koomen, E.R.; Casparie, M.K.; Herings, R.M.; Guchelaar, H.J.; Nijsten, T. Non-steroidal anti-inflammatory drugs and melanoma risk: Large Dutch population-based case-control study. *J. Investig. Dermatol.* **2009**, *129*, 2620–2627. [CrossRef]

32. Gogas, H.; Polyzos, A.; Stavrinidis, I.; Frangia, K.; Tsoutsos, D.; Panagiotou, P.; Markopoulos, C.; Papadopoulos, O.; Pectasides, D.; Mantzourani, M.; et al. Temozolomide in combination with celecoxib in patients with advanced melanoma. A phase II study of the Hellenic Cooperative Oncology Group. *Ann. Oncol.* **2006**, *17*, 1835–1841. [CrossRef]

33. Logotheti, S.; Khoury, N.; Vlahopoulos, S.A.; Skourti, E.; Papaevangeliou, D.; Liloglou, T.; Gorgoulis, V.; Budunova, I.; Kyriakopoulos, A.M.; Zoumpourlis, V. N-bromotaurine surrogates for loss of antiproliferative response and enhances cisplatin efficacy in cancer cells with impaired glucocorticoid receptor. *Transl. Res.* **2016**, *173*, 58–73.e52. [CrossRef]

34. Ben-Ari, E.T. Dual purpose: Some cancer therapies used to treat autoimmune diseases. *J. Natl. Cancer Inst.* **2004**, *96*, 577–579. [CrossRef]

35. Chanan-Khan, A.A.; Swaika, A.; Paulus, A.; Kumar, S.K.; Mikhael, J.R.; Rajkumar, S.V.; Dispenzieri, A.; Lacy, M.Q. Pomalidomide: The new immunomodulatory agent for the treatment of multiple myeloma. *Blood Cancer J.* **2013**, *3*, e143. [CrossRef]

36. Lash, A.A. Anticancer drugs for noncancer diseases. *Medsurg Nurs.* **1996**, *5*, 177–180.

37. Kontou, P.I.; Pavlopoulou, A.; Dimou, N.L.; Pavlopoulos, G.A.; Bagos, P.G. Network analysis of genes and their association with diseases. *Gene* **2016**, *590*, 68–78. [CrossRef]

38. Kortylewski, M.; Jove, R.; Yu, H. Targeting STAT3 affects melanoma on multiple fronts. *Cancer Metastasis Rev.* **2005**, *24*, 315–327. [CrossRef]

39. Heppler, L.N.; Frank, D.A. Rare mutations provide unique insight into oncogenic potential of STAT transcription factors. *J. Clin. Investig.* **2018**, *128*, 113–115. [CrossRef]

40. Loh, C.Y.; Arya, A.; Naema, A.F.; Wong, W.F.; Sethi, G.; Looi, C.Y. Signal Transducer and Activator of Transcription (STATs) Proteins in Cancer and Inflammation: Functions and Therapeutic Implication. *Front. Oncol.* **2019**, *9*, 48. [CrossRef]

41. Lim, C.P.; Cao, X. Structure, function, and regulation of STAT proteins. *Mol. Biosyst.* **2006**, *2*, 536–550. [CrossRef]

42. Xie, T.X.; Huang, F.J.; Aldape, K.D.; Kang, S.H.; Liu, M.; Gershenwald, J.E.; Xie, K.; Sawaya, R.; Huang, S. Activation of stat3 in human melanoma promotes brain metastasis. *Cancer Res.* **2006**, *66*, 3188–3196. [CrossRef]

43. Steder, M.; Alla, V.; Meier, C.; Spitschak, A.; Pahnke, J.; Fürst, K.; Kowtharapu, B.S.; Engelmann, D.; Petigk, J.; Egberts, F.; et al. DNp73 exerts function in metastasis initiation by disconnecting the inhibitory role of EPLIN on IGF1R-AKT/STAT3 signaling. *Cancer Cell* **2013**, *24*, 512–527. [CrossRef]

44. Engelmann, D.; Pützer, B.M. Emerging from the shade of p53 mutants: N-terminally truncated variants of the p53 family in EMT signaling and cancer progression. *Sci. Signal.* **2014**, *7*, re9. [CrossRef]

45. Galoczova, M.; Coates, P.; Vojtesek, B. STAT3, stem cells, cancer stem cells and p63. *Cell. Mol. Biol. Lett.* **2018**, *23*, 12. [CrossRef]

46. Marquardt, S.; Solanki, M.; Spitschak, A.; Vera, J.; Pützer, B.M. Emerging functional markers for cancer stem cell-based therapies: Understanding signaling networks for targeting metastasis. *Semin. Cancer Biol.* **2018**, *53*, 90–109. [CrossRef]

47. Wang, Y.; Mou, Y.; Zhang, H.; Wang, X.; Li, R.; Cheng, Z.; Liu, X. Reprogramming Factors Remodel Melanoma Cell Phenotype by Changing Stat3 Expression. *Int. J. Med. Sci.* **2017**, *14*, 1402–1409. [CrossRef]

48. Liu, F.; Cao, J.; Wu, J.; Sullivan, K.; Shen, J.; Ryu, B.; Xu, Z.; Wei, W.; Cui, R. Stat3-targeted therapies overcome the acquired resistance to vemurafenib in melanomas. *J. Investig. Dermatol.* **2013**, *133*, 2041–2049. [CrossRef]

49. Kanai, T.; Jenks, J.; Nadeau, K.C. The STAT5b Pathway Defect and Autoimmunity. *Front. Immunol.* **2012**, *3*, 234. [CrossRef]

50. Mirmohammadsadegh, A.; Hassan, M.; Bardenheuer, W.; Marini, A.; Gustrau, A.; Nambiar, S.; Tannapfel, A.; Bojar, H.; Ruzicka, T.; Hengge, U.R. STAT5 phosphorylation in malignant melanoma is important for survival and is mediated through SRC and JAK1 kinases. *J. Investig. Dermatol.* **2006**, *126*, 2272–2280. [CrossRef]

51. Wellbrock, C.; Geissinger, E.; Gómez, A.; Fischer, P.; Friedrich, K.; Schartl, M. Signalling by the oncogenic receptor tyrosine kinase Xmrk leads to activation of STAT5 in Xiphophorus melanoma. *Oncogene* **1998**, *16*, 3047–3056. [CrossRef]

52. Hassel, J.C.; Winnemöller, D.; Schartl, M.; Wellbrock, C. STAT5 contributes to antiapoptosis in melanoma. *Melanoma Res.* **2008**, *18*, 378–385. [CrossRef]

53. Machiraju, D.; Moll, I.; Gebhardt, C.; Sucker, A.; Buder-Bakhaya, K.; Schadendorf, D.; Hassel, J.C. STAT5 expression correlates with recurrence and survival in melanoma patients treated with interferon-α. *Melanoma Res.* **2018**, *28*, 204–210. [CrossRef]

54. Wellbrock, C.; Weisser, C.; Hassel, J.C.; Fischer, P.; Becker, J.; Vetter, C.S.; Behrmann, I.; Kortylewski, M.; Heinrich, P.C.; Schartl, M. STAT5 contributes to interferon resistance of melanoma cells. *Curr. Biol.* **2005**, *15*, 1629–1639. [CrossRef]

55. Ohanna, M.; Cerezo, M.; Nottet, N.; Bille, K.; Didier, R.; Beranger, G.; Mograbi, B.; Rocchi, S.; Yvan-Charvet, L.; Ballotti, R.; et al. Pivotal role of NAMPT in the switch of melanoma cells toward an invasive and drug-resistant phenotype. *Genes Dev.* **2018**, *32*, 448–461. [CrossRef]

56. Jia, H.; Ren, W.; Feng, Y.; Wei, T.; Guo, M.; Guo, J.; Zhao, J.; Song, X.; Wang, M.; Zhao, T.; et al. The enhanced antitumour response of pimozide combined with the IDO inhibitor L-MT in melanoma. *Int. J. Oncol.* **2018**, *53*, 949–960. [CrossRef]

57. Fridman, W.H.; Zitvogel, L.; Sautès-Fridman, C.; Kroemer, G. The immune contexture in cancer prognosis and treatment. *Nat. Rev. Clin. Oncol.* **2017**. [CrossRef]

58. Lee, H.; Pal, S.K.; Reckamp, K.; Figlin, R.A.; Yu, H. STAT3: A target to enhance antitumor immune response. *Curr. Top. Microbiol. Immunol.* **2011**, *344*, 41–59. [CrossRef]

59. Wang, Y.; Shen, Y.; Wang, S.; Shen, Q.; Zhou, X. The role of STAT3 in leading the crosstalk between human cancers and the immune system. *Cancer Lett.* **2018**, *415*, 117–128. [CrossRef]

60. Yu, H.; Kortylewski, M.; Pardoll, D. Crosstalk between cancer and immune cells: Role of STAT3 in the tumour microenvironment. *Nat. Rev. Immunol.* **2007**, *7*, 41–51. [CrossRef]

61. Emeagi, P.U.; Maenhout, S.; Dang, N.; Heirman, C.; Thielemans, K.; Breckpot, K. Downregulation of Stat3 in melanoma: Reprogramming the immune microenvironment as an anticancer therapeutic strategy. *Gene Ther.* **2013**, *20*, 1085–1092. [CrossRef]
62. Gotthardt, D.; Sexl, V. STATs in NK-Cells: The Good, the Bad, and the Ugly. *Front. Immunol.* **2016**, *7*, 694. [CrossRef]
63. Eckelhart, E.; Warsch, W.; Zebedin, E.; Simma, O.; Stoiber, D.; Kolbe, T.; Rülicke, T.; Mueller, M.; Casanova, E.; Sexl, V. A novel Ncr1-Cre mouse reveals the essential role of STAT5 for NK-cell survival and development. *Blood* **2011**, *117*, 1565–1573. [CrossRef]
64. Sathe, P.; Delconte, R.B.; Souza-Fonseca-Guimaraes, F.; Seillet, C.; Chopin, M.; Vandenberg, C.J.; Rankin, L.C.; Mielke, L.A.; Vikstrom, I.; Kolesnik, T.B.; et al. Innate immunodeficiency following genetic ablation of Mcl1 in natural killer cells. *Nat. Commun.* **2014**, *5*, 4539. [CrossRef]
65. Gotthardt, D.; Putz, E.M.; Grundschober, E.; Prchal-Murphy, M.; Straka, E.; Kudweis, P.; Heller, G.; Bago-Horvath, Z.; Witalisz-Siepracka, A.; Cumaraswamy, A.A.; et al. STAT5 Is a Key Regulator in NK Cells and Acts as a Molecular Switch from Tumor Surveillance to Tumor Promotion. *Cancer Discov.* **2016**, *6*, 414–429. [CrossRef]
66. Grange, M.; Buferne, M.; Verdeil, G.; Leserman, L.; Schmitt-Verhulst, A.M.; Auphan-Anezin, N. Activated STAT5 promotes long-lived cytotoxic CD8+ T cells that induce regression of autochthonous melanoma. *Cancer Res.* **2012**, *72*, 76–87. [CrossRef]
67. Xiao, Q.; Wu, J.; Wang, W.J.; Chen, S.; Zheng, Y.; Yu, X.; Meeth, K.; Sahraei, M.; Bothwell, A.L.M.; Chen, L.; et al. DKK2 imparts tumor immunity evasion through β-catenin-independent suppression of cytotoxic immune-cell activation. *Nat. Med.* **2018**, *24*, 262–270. [CrossRef]
68. Lee, G.R. The Balance of Th17 versus Treg Cells in Autoimmunity. *Int. J. Mol. Sci.* **2018**, *19*, 730. [CrossRef]
69. Kaplan, M.H. STAT signaling in inflammation. *JAKSTAT* **2013**, *2*, e24198. [CrossRef]
70. Wieczorek, M.; Ginter, T.; Brand, P.; Heinzel, T.; Krämer, O.H. Acetylation modulates the STAT signaling code. *Cytokine Growth Factor Rev.* **2012**, *23*, 293–305. [CrossRef]
71. Yu, H.; Pardoll, D.; Jove, R. STATs in cancer inflammation and immunity: A leading role for STAT3. *Nat. Rev. Cancer* **2009**, *9*, 798–809. [CrossRef]
72. Kasembeli, M.M.; Bharadwaj, U.; Robinson, P.; Tweardy, D.J. Contribution of STAT3 to Inflammatory and Fibrotic Diseases and Prospects for its Targeting for Treatment. *Int. J. Mol. Sci.* **2018**, *19*, 2299. [CrossRef]
73. Milner, J.D.; Vogel, T.P.; Forbes, L.; Ma, C.A.; Stray-Pedersen, A.; Niemela, J.E.; Lyons, J.J.; Engelhardt, K.R.; Zhang, Y.; Topcagic, N.; et al. Early-onset lymphoproliferation and autoimmunity caused by germline STAT3 gain-of-function mutations. *Blood* **2015**, *125*, 591–599. [CrossRef]
74. Wienke, J.; Janssen, W.; Scholman, R.; Spits, H.; van Gijn, M.; Boes, M.; van Montfrans, J.; Moes, N.; de Roock, S. A novel human STAT3 mutation presents with autoimmunity involving Th17 hyperactivation. *Oncotarget* **2015**, *6*, 20037–20042. [CrossRef]
75. Kloos, L.; Sillevis Smitt, P.; Ang, C.W.; Kruit, W.; Stoter, G. Paraneoplastic ophthalmoplegia and subacute motor axonal neuropathy associated with anti-GQ1b antibodies in a patient with malignant melanoma. *J. Neurol. Neurosurg. Psychiatry* **2003**, *74*, 507–509. [CrossRef]
76. Tsuda, M.; Kohro, Y.; Yano, T.; Tsujikawa, T.; Kitano, J.; Tozaki-Saitoh, H.; Koyanagi, S.; Ohdo, S.; Ji, R.R.; Salter, M.W.; et al. JAK-STAT3 pathway regulates spinal astrocyte proliferation and neuropathic pain maintenance in rats. *Brain* **2011**, *134*, 1127–1139. [CrossRef]
77. Ihantola, E.L.; Viisanen, T.; Gazali, A.M.; Näntö-Salonen, K.; Juutilainen, A.; Moilanen, L.; Rintamäki, R.; Pihlajamäki, J.; Veijola, R.; Toppari, J.; et al. Effector T Cell Resistance to Suppression and STAT3 Signaling during the Development of Human Type 1 Diabetes. *J. Immunol.* **2018**, *201*, 1144–1153. [CrossRef]
78. Saarimäki-Vire, J.; Balboa, D.; Russell, M.A.; Saarikettu, J.; Kinnunen, M.; Keskitalo, S.; Malhi, A.; Valensisi, C.; Andrus, C.; Eurola, S.; et al. An Activating STAT3 Mutation Causes Neonatal Diabetes through Premature Induction of Pancreatic Differentiation. *Cell Rep.* **2017**, *19*, 281–294. [CrossRef]
79. Garrigan, E.; Belkin, N.S.; Alexander, J.J.; Han, Z.; Seydel, F.; Carter, J.; Atkinson, M.; Wasserfall, C.; Clare-Salzler, M.J.; Amick, M.A.; et al. Persistent STAT5 phosphorylation and epigenetic dysregulation of GM-CSF and PGS2/COX2 expression in Type 1 diabetic human monocytes. *PLoS ONE* **2013**, *8*, e76919. [CrossRef]

80. Ju, J.H.; Heo, Y.J.; Cho, M.L.; Jhun, J.Y.; Park, J.S.; Lee, S.Y.; Oh, H.J.; Moon, S.J.; Kwok, S.K.; Park, K.S.; et al. Modulation of STAT-3 in rheumatoid synovial T cells suppresses Th17 differentiation and increases the proportion of Treg cells. *Arthritis Rheum.* **2012**, *64*, 3543–3552. [CrossRef]

81. Oike, T.; Sato, Y.; Kobayashi, T.; Miyamoto, K.; Nakamura, S.; Kaneko, Y.; Kobayashi, S.; Harato, K.; Saya, H.; Matsumoto, M.; et al. Stat3 as a potential therapeutic target for rheumatoid arthritis. *Sci. Rep.* **2017**, *7*, 10965. [CrossRef]

82. Anderson, A.E.; Maney, N.J.; Nair, N.; Lendrem, D.W.; Skelton, A.J.; Diboll, J.; Brown, P.M.; Smith, G.R.; Carmody, R.J.; Barton, A.; et al. Expression of STAT3-regulated genes in circulating CD4+ T cells discriminates rheumatoid arthritis independently of clinical parameters in early arthritis. *Rheumatology* **2019**, *58*, 1250–1258. [CrossRef]

83. Lee, S.H.; Park, J.S.; Byun, J.K.; Jhun, J.; Jung, K.; Seo, H.B.; Moon, Y.M.; Kim, H.Y.; Park, S.H.; Cho, M.L. PTEN ameliorates autoimmune arthritis through down-regulating STAT3 activation with reciprocal balance of Th17 and Tregs. *Sci. Rep.* **2016**, *6*, 34617. [CrossRef]

84. Calautti, E.; Avalle, L.; Poli, V. Psoriasis: A STAT3-Centric View. *Int. J. Mol. Sci.* **2018**, *19*, 171. [CrossRef]

85. Miyoshi, K.; Takaishi, M.; Nakajima, K.; Ikeda, M.; Kanda, T.; Tarutani, M.; Iiyama, T.; Asao, N.; DiGiovanni, J.; Sano, S. Stat3 as a therapeutic target for the treatment of psoriasis: A clinical feasibility study with STA-21, a Stat3 inhibitor. *J. Investig. Dermatol.* **2011**, *131*, 108–117. [CrossRef]

86. Hart, P.A.; Kamisawa, T.; Brugge, W.R.; Chung, J.B.; Culver, E.L.; Czakó, L.; Frulloni, L.; Go, V.L.; Gress, T.M.; Kim, M.H.; et al. Long-term outcomes of autoimmune pancreatitis: A multicentre, international analysis. *Gut* **2013**, *62*, 1771–1776. [CrossRef]

87. Michaud, C.; Peppers, B.; Frith, J.; Tcheurekdjian, H.; Hostoffer, R. Idiopathic pancreatitis in a patient with a STAT3 mutation. *Allergy Rhinol.* **2016**, *7*, 42–44. [CrossRef]

88. Babushok, D.V.; Perdigones, N.; Perin, J.C.; Olson, T.S.; Ye, W.; Roth, J.J.; Lind, C.; Cattier, C.; Li, Y.; Hartung, H.; et al. Emergence of clonal hematopoiesis in the majority of patients with acquired aplastic anemia. *Cancer Genet.* **2015**, *208*, 115–128. [CrossRef]

89. Jerez, A.; Clemente, M.J.; Makishima, H.; Rajala, H.; Gómez-Seguí, I.; Olson, T.; McGraw, K.; Przychodzen, B.; Kulasekararaj, A.; Afable, M.; et al. STAT3 mutations indicate the presence of subclinical T-cell clones in a subset of aplastic anemia and myelodysplastic syndrome patients. *Blood* **2013**, *122*, 2453–2459. [CrossRef]

90. Kordasti, S.; Costantini, B.; Seidl, T.; Perez Abellan, P.; Martinez Llordella, M.; McLornan, D.; Diggins, K.E.; Kulasekararaj, A.; Benfatto, C.; Feng, X.; et al. Deep phenotyping of Tregs identifies an immune signature for idiopathic aplastic anemia and predicts response to treatment. *Blood* **2016**, *128*, 1193–1205. [CrossRef]

91. Mocellin, R.; Walterfang, M.; Velakoulis, D. Hashimoto's encephalopathy: Epidemiology, pathogenesis and management. *CNS Drugs* **2007**, *21*, 799–811. [CrossRef]

92. Sheng, W.; Yang, F.; Zhou, Y.; Yang, H.; Low, P.Y.; Kemeny, D.M.; Tan, P.; Moh, A.; Kaplan, M.H.; Zhang, Y.; et al. STAT5 programs a distinct subset of GM-CSF-producing T helper cells that is essential for autoimmune neuroinflammation. *Cell Res.* **2014**, *24*, 1387–1402. [CrossRef]

93. Sugimoto, K. Role of STAT3 in inflammatory bowel disease. *World J. Gastroenterol.* **2008**, *14*, 5110–5114. [CrossRef]

94. Zundler, S.; Neurath, M.F. Integrating Immunologic Signaling Networks: The JAK/STAT Pathway in Colitis and Colitis-Associated Cancer. *Vaccines* **2016**, *4*, 5. [CrossRef]

95. Gilbert, S.; Zhang, R.; Denson, L.; Moriggl, R.; Steinbrecher, K.; Shroyer, N.; Lin, J.; Han, X. Enterocyte STAT5 promotes mucosal wound healing via suppression of myosin light chain kinase-mediated loss of barrier function and inflammation. *EMBO Mol. Med.* **2012**, *4*, 109–124. [CrossRef]

96. Gilbert, S.; Nivarthi, H.; Mayhew, C.N.; Lo, Y.H.; Noah, T.K.; Vallance, J.; Rülicke, T.; Müller, M.; Jegga, A.G.; Tang, W.; et al. Activated STAT5 confers resistance to intestinal injury by increasing intestinal stem cell proliferation and regeneration. *Stem Cell Rep.* **2015**, *4*, 209–225. [CrossRef]

97. Liu, R.; Moriggl, R.; Zhang, D.; Li, H.; Karns, R.; Ruan, H.B.; Niu, H.; Mayhew, C.; Watson, C.; Bangar, H.; et al. Constitutive STAT5 activation regulates Paneth and Paneth-like cells to control. *Life Sci. Alliance* **2019**, *2*, e201900296. [CrossRef]

98. Li, G.; Liu, Z.; Ran, P.; Qiu, J.; Zhong, N. Activation of signal transducer and activator of transcription 5 (STAT5) in splenocyte proliferation of asthma mice induced by ovalbumin. *Cell Mol. Immunol.* **2004**, *1*, 471–474.

99. Turlej, R.K.; Fiévez, L.; Sandersen, C.F.; Dogné, S.; Kirschvink, N.; Lekeux, P.; Bureau, F. Enhanced survival of lung granulocytes in an animal model of asthma: Evidence for a role of GM-CSF activated STAT5 signalling pathway. *Thorax* **2001**, *56*, 696–702. [CrossRef]

100. Ji, J.; Shu, X.; Li, X.; Sundquist, K.; Sundquist, J.; Hemminki, K. Cancer risk in hospitalised asthma patients. *Br. J. Cancer* **2009**, *100*, 829–833. [CrossRef]

101. Ciccia, F.; Rizzo, A.; Alessandro, R.; Guggino, G.; Maugeri, R.; Saieva, L.; Cannizzaro, A.; Giardina, A.; De Leo, G.; Gerardo Iacopino, D.; et al. Activated IL-22 pathway occurs in the muscle tissues of patients with polymyositis or dermatomyositis and is correlated with disease activity. *Rheumatology* **2014**, *53*, 1307–1312. [CrossRef]

102. Voltarelli, F.A.; Frajacomo, F.T.; Padilha, C.S.; Testa, M.T.J.; Cella, P.S.; Ribeiro, D.F.; de Oliveira, D.X.; Veronez, L.C.; Bisson, G.S.; Moura, F.A.; et al. Syngeneic B16F10 Melanoma Causes Cachexia and Impaired Skeletal Muscle Strength and Locomotor Activity in Mice. *Front. Physiol.* **2017**, *8*, 715. [CrossRef]

103. Hosui, A.; Kimura, A.; Yamaji, D.; Zhu, B.M.; Na, R.; Hennighausen, L. Loss of STAT5 causes liver fibrosis and cancer development through increased TGF-{beta} and STAT3 activation. *J. Exp. Med.* **2009**, *206*, 819–831. [CrossRef]

104. Rinkevich, Y.; Walmsley, G.G.; Hu, M.S.; Maan, Z.N.; Newman, A.M.; Drukker, M.; Januszyk, M.; Krampitz, G.W.; Gurtner, G.C.; Lorenz, H.P.; et al. Skin fibrosis. Identification and isolation of a dermal lineage with intrinsic fibrogenic potential. *Science* **2015**, *348*, aaa2151. [CrossRef]

105. Park, J.S.; Kim, S.M.; Hwang, S.H.; Choi, S.Y.; Kwon, J.Y.; Kwok, S.K.; Cho, M.L.; Park, S.H. Combinatory treatment using tacrolimus and a STAT3 inhibitor regulate Treg cells and plasma cells. *Int. J. Immunopathol. Pharmacol.* **2018**, *32*, 2058738418778724. [CrossRef]

106. Slight-Webb, S.; Guthridge, J.M.; Chakravarty, E.F.; Chen, H.; Lu, R.; Macwana, S.; Bean, K.; Maecker, H.T.; Utz, P.J.; James, J.A. Mycophenolate mofetil reduces STAT3 phosphorylation in systemic lupus erythematosus patients. *JCI Insight* **2019**, *4*, 124575. [CrossRef]

107. Meshaal, S.; El Refai, R.; El Saie, A.; El Hawary, R. Signal transducer and activator of transcription 5 is implicated in disease activity in adult and juvenile onset systemic lupus erythematosus. *Clin. Rheumatol.* **2016**, *35*, 1515–1520. [CrossRef]

108. Hedrich, C.M.; Rauen, T.; Apostolidis, S.A.; Grammatikos, A.P.; Rodriguez Rodriguez, N.; Ioannidis, C.; Kyttaris, V.C.; Crispin, J.C.; Tsokos, G.C. Stat3 promotes IL-10 expression in lupus T cells through trans-activation and chromatin remodeling. *Proc. Natl. Acad. Sci. USA* **2014**, *111*, 13457–13462. [CrossRef]

109. Yang, J.; Yang, X.; Zou, H.; Chu, Y.; Li, M. Recovery of the immune balance between Th17 and regulatory T cells as a treatment for systemic lupus erythematosus. *Rheumatology* **2011**, *50*, 1366–1372. [CrossRef]

110. Talaat, R.M.; Mohamed, S.F.; Bassyouni, I.H.; Raouf, A.A. Th1/Th2/Th17/Treg cytokine imbalance in systemic lupus erythematosus (SLE) patients: Correlation with disease activity. *Cytokine* **2015**, *72*, 146–153. [CrossRef]

111. Chen, S.Y.; Liu, M.F.; Kuo, P.Y.; Wang, C.R. Upregulated expression of STAT3/IL-17 in patients with systemic lupus erythematosus. *Clin. Rheumatol.* **2019**, *38*, 1361–1366. [CrossRef]

112. Goropevšek, A.; Gorenjak, M.; Gradišnik, S.; Dai, K.; Holc, I.; Hojs, R.; Krajnc, I.; Pahor, A.; Avčin, T. STAT5 phosphorylation in CD4 T cells from patients with SLE is related to changes in their subsets and follow-up disease severity. *J. Leukoc. Biol.* **2017**, *101*, 1405–1418. [CrossRef]

113. Grushchak, S.; Mathieu, R.J.; Orrell, K.A.; Hagstrom, E.L.; Laumann, A.E.; West, D.P.; Nardone, B. Malignant melanoma association with systemic lupus erythematosus in a large midwestern U.S. patient population: A retrospective study. *Int. J. Dermatol.* **2018**, *57*, e34–e36. [CrossRef]

114. Dreyer, L.; Faurschou, M.; Mogensen, M.; Jacobsen, S. High incidence of potentially virus-induced malignancies in systemic lupus erythematosus: A long-term followup study in a Danish cohort. *Arthritis Rheum.* **2011**, *63*, 3032–3037. [CrossRef]

115. Cao, L.; Tong, H.; Xu, G.; Liu, P.; Meng, H.; Wang, J.; Zhao, X.; Tang, Y.; Jin, J. Systemic lupus erythematous and malignancy risk: A meta-analysis. *PLoS ONE* **2015**, *10*, e0122964. [CrossRef]

116. Colonna, M. TREM1 Blockade: Killing Two Birds with One Stone. *Trends Immunol.* **2019**, *40*, 781–783. [CrossRef]

117. Asteriou, E.; Gkoutzourelas, A.; Mavropoulos, A.; Katsiari, C.; Sakkas, L.I.; Bogdanos, D.P. Curcumin for the Management of Periodontitis and Early ACPA-Positive Rheumatoid Arthritis: Killing Two Birds with One Stone. *Nutrients* **2018**, *10*, 908. [CrossRef]

118. Cai, Z.; Zhang, J.; Li, H. Two Birds with One Stone: Regular Use of PDE5 Inhibitors for Treating Male Patients with Erectile Dysfunction and Cardiovascular Diseases. *Cardiovasc. Drugs Ther.* **2019**, *33*, 119–128. [CrossRef]

119. Pietschmann, K.; Bolck, H.A.; Buchwald, M.; Spielberg, S.; Polzer, H.; Spiekermann, K.; Bug, G.; Heinzel, T.; Böhmer, F.D.; Krämer, O.H. Breakdown of the FLT3-ITD/STAT5 axis and synergistic apoptosis induction by the histone deacetylase inhibitor panobinostat and FLT3-specific inhibitors. *Mol. Cancer Ther.* **2012**, *11*, 2373–2383. [CrossRef]

120. Kosan, C.; Ginter, T.; Heinzel, T.; Krämer, O.H. STAT5 acetylation: Mechanisms and consequences for immunological control and leukemogenesis. *JAKSTAT* **2013**, *2*, e26102. [CrossRef]

121. Lesinski, G.B. The potential for targeting the STAT3 pathway as a novel therapy for melanoma. *Future Oncol.* **2013**, *9*, 925–927. [CrossRef]

122. Piccirillo, R.; Giavazzi, R. Inactivating STAT3: Bad for tumor, good for muscle. *Cell Cycle* **2015**, *14*, 939–940. [CrossRef]

123. Trécul, A.; Morceau, F.; Dicato, M.; Diederich, M. Dietary compounds as potent inhibitors of the signal transducers and activators of transcription (STAT) 3 regulatory network. *Genes Nutr.* **2012**, *7*, 111–125. [CrossRef]

124. Li, T.; Fu, X.; Tse, A.K.; Guo, H.; Lee, K.W.; Liu, B.; Su, T.; Wang, X.; Yu, Z. Inhibiting STAT3 signaling is involved in the anti-melanoma effects of a herbal formula comprising Sophorae Flos and Lonicerae Japonicae Flos. *Sci. Rep.* **2017**, *7*, 3097. [CrossRef]

125. Rani, A.; Murphy, J.J. STAT5 in Cancer and Immunity. *J. Interferon Cytokine Res.* **2016**, *36*, 226–237. [CrossRef]

126. Knochelmann, H.M.; Dwyer, C.J.; Bailey, S.R.; Amaya, S.M.; Elston, D.M.; Mazza-McCrann, J.M.; Paulos, C.M. When worlds collide: Th17 and Treg cells in cancer and autoimmunity. *Cell. Mol. Immunol.* **2018**, *15*, 458–469. [CrossRef]

127. Chen, C.; Gao, F.H. Th17 Cells Paradoxical Roles in Melanoma and Potential Application in Immunotherapy. *Front. Immunol.* **2019**, *10*, 187. [CrossRef]

128. Jacobs, J.F.; Nierkens, S.; Figdor, C.G.; de Vries, I.J.; Adema, G.J. Regulatory T cells in melanoma: The final hurdle towards effective immunotherapy? *Lancet Oncol.* **2012**, *13*, e32–e42. [CrossRef]

129. Wong, A.L.A.; Hirpara, J.L.; Pervaiz, S.; Eu, J.Q.; Sethi, G.; Goh, B.C. Do STAT3 inhibitors have potential in the future for cancer therapy? *Expert Opin. Investig. Drugs* **2017**, *26*, 883–887. [CrossRef]

130. Harada, D.; Takigawa, N.; Kiura, K. The Role of STAT3 in Non-Small Cell Lung Cancer. *Cancers* **2014**, *6*, 708–722. [CrossRef]

Review

ADAM17 Activity and IL-6 Trans-Signaling in Inflammation and Cancer

Neele Schumacher and Stefan Rose-John *

Biochemical Institute, University of Kiel, 24118 Kiel, Germany; nschumacher@biochem.uni-kiel.de
* Correspondence: rosejohn@biochem.uni-kiel.de

Received: 30 September 2019; Accepted: 2 November 2019; Published: 5 November 2019

Abstract: All ligands of the epidermal growth factor receptor (EGF-R) are transmembrane proteins, which need to be proteolytically cleaved in order to be systemically active. The major protease responsible for this cleavage is the membrane metalloprotease ADAM17, which also has been implicated in cleavage of TNFα and interleukin-6 (IL-6) receptor. It has been recently shown that in the absence of ADAM17, the main protease for EGF-R ligand processing, colon cancer formation is largely abrogated. Intriguingly, colon cancer formation depends on EGF-R activity on myeloid cells rather than on intestinal epithelial cells. A major activity of EGF-R on myeloid cells is the stimulation of IL-6 synthesis. Subsequently, IL-6 together with the ADAM17 shed soluble IL-6 receptor acts on intestinal epithelial cells via IL-6 trans-signaling to induce colon cancer formation, which can be blocked by the inhibitor of IL-6 trans-signaling, sgp130Fc. Blockade of IL-6 trans-signaling therefore offers a new therapeutic window downstream of the EGF-R for the treatment of colon cancer and possibly of other EGF-R related neoplastic diseases.

Keywords: ADAM17; interleukin-6; trans-signaling; epidermal growth factor receptor (EGF-R); shedding; metalloprotease; tumor necrosis factor alpha (TNFα); inflammation associated cancer; colon cancer; lung cancer

---

1. Introduction

IL-6 is a four-helical cytokine with pleiotropic activities, which is synthesized by many cell types upon appropriate stimulation and which can act on many cell types during several disease states such as inflammation and cancer [1]. IL-6 exerts its function through binding to the alpha-receptor Interleukin-6 receptor (IL-6R). The complex of IL-6 and IL-6R binds to the beta-receptor glycoprotein 130 (gp130) and induces dimerization of gp130 [2]. This dimerized gp130 leads to an activation of the tyrosine kinase Janus kinase 1 (JAK1), which is constitutively associated with the cytoplasmic portion of gp130. JAK1 phosphorylates tyrosine residues within the cytoplasmic portion of gp130. This leads to the recruitment of the adapter protein and phosphatase SHP2 that initiates MAP kinase and PI3 kinase signaling [3]. Furthermore, phosphorylated tyrosine residues recruit the cytoplasmic transcription factors STAT1 and STAT3, which thereupon become phosphorylated, dimerize, and translocate into the nucleus, where they bind to DNA and stimulate the transcription of gp130 target genes [3]. One of the earliest gp130 target genes is the gene coding for SOCS3. The SOCS3 protein is recruited to the membrane proximal tyrosine residue within the gp130 cytoplasmic tail where it inhibits the activity of JAK1. Therefore, SOCS3 is a negative feedback inhibitor of gp130 signaling [3]. Recently, it was shown that gp130 activation leads to the phosphorylation and activation of the YAP pathway (Figure 1). This pathway was shown to be important in the development of colon cancer [4,5].

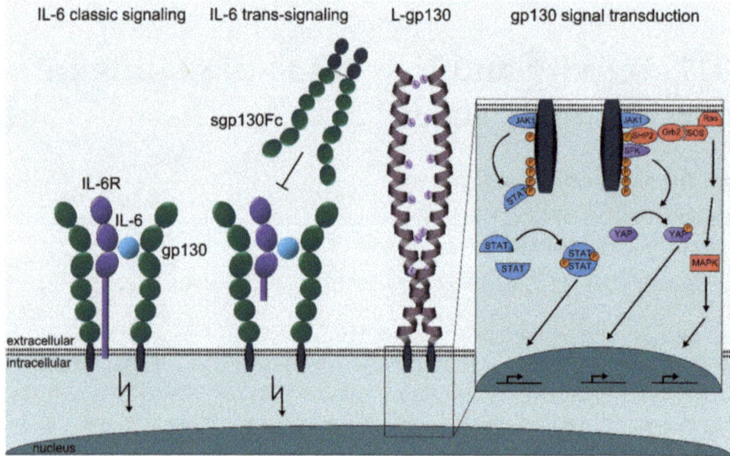

Figure 1. Modes of IL-6 signaling. IL-6 can signal via the membrane-bound IL-6R, which, however, is only expressed on some cell types. Alternatively, IL-6 can bind to the soluble IL-6R (sIL-6R) and the complex of IL-6/sIL-6R can stimulate virtually all cells in the body since gp130 is ubiquitously expressed. In an artificial model of cell-autonomous gp130 stimulation, the extracellular portion of gp130 has been exchanged for a leucine zipper, leading to constitutive gp130 signaling in the absence of any ligand [6]. The insert shows the main intracellular signaling pathways stimulated by gp130 activation.

While gp130 is expressed on all cells in the body, IL-6R is only expressed on some cells including hepatocytes, some leukocytes and some epithelial cells. Since IL-6 shows only measurable binding to the IL-6R but not to gp130, it follows that IL-6 can only act on cells, which express the IL-6R [2,7]. Interestingly, the membrane-bound IL-6R can be cleaved from the cell membrane by A disintegrin and metalloproteinases (ADAMs), resulting in a soluble form of the receptor [8–10]. The soluble IL-6R (sIL-6R) can still bind its ligand IL-6 and the complex of IL-6 and sIL-6R can bind to gp130 and induce dimerization of signaling. This mode of signaling has been called IL-6 trans-signaling [11]. Since gp130 is ubiquitously expressed, IL-6 trans-signaling can lead to the stimulation of virtually every cell in the body (Figure 1) [12]. However, there are few proteases described that can cleave membrane-bound IL-6R thus enabling IL-6 trans-signaling. Besides, the closely related ADAM proteases ADAM10 and ADAM17, meprin metalloproteases were reported to release the soluble form of the IL-6R [13–16].

Moreover, since on most cells gp130 is expressed at higher levels than IL-6R, stimulation by IL-6 leads to the stimulation of only some gp130 molecules whereas stimulation with the IL-6/sIL-6R complex leads to the activation and stimulation of all gp130 proteins resulting in a higher signal amplitude as a consequence of IL-6 trans-signaling as compared with IL-6 signaling via the membrane-bound IL-6R. IL-6 upon binding to the IL-6R is rapidly internalized in contrast to the IL-6/sIL-6R complex, which is internalized only with low efficacy [17]. This resulted in significantly longer stimulation mediated by the IL-6/sIL-6R complex as compared to IL-6 alone [17]. This property of IL-6 trans-signaling might be important for several cellular or tissue responses, which can only be induced by IL-6 trans-signaling but not by IL-6 classic signaling via the membrane-bound IL-6R [18].

Importantly, IL-6 trans-signaling can be specifically blocked by the extracellular portion of gp130 fused to the Fc-portion of a human IgG1 antibody (sgp130Fc) [1,19]. This blockade of IL-6 trans-signaling does not affect activities of IL-6 via the membrane-bound IL-6R. In transgenic mice overexpressing the sgp130Fc, IL-6 trans-signaling is completely blocked. These mice have been widely used to define the role of IL-6 classic signaling and IL-6 trans-signaling in many mouse models of human diseases (Figure 1) [1,12,20].

In healthy individuals, IL-6 levels are in the range between 1 and 5 pg/mL, but they can increase by many orders of magnitude during inflammatory states or upon infection. Plasma IL-6 levels of up to several µg/mL have been reported in patients with septic shock [21]. Interestingly, the level of sIL-6R in the blood of healthy individuals is in the range of 40–80 ng/mL, more than 1000 times higher than IL-6 [20]. A soluble form of gp130, which is generated by differential splicing [22] is found in the plasma of healthy individuals at approximately 400 ng/mL [20]. IL-6 binds to the IL-6R with an affinity of about 1 nM, whereas the complex of IL-6 and sIL-6R binds to gp130 with an affinity of 10 pM. Therefore, IL-6 released into the circulation will bind to sIL-6R and thereafter be trapped by sgp130. Thus, sIL-6R and sgp130 form a buffer for IL-6 [23]. The important function of the IL-6 buffer was underlined by studies on the coding single nucleotide polymorphism (SNP; rs2228145) in the human IL-6R that results in the alteration of Asp358 into Ala358. These amino acid residues are in close proximity of the proteolytic cleavage site of the IL-6R [16]. Interestingly, the IL-6R with Ala358 is a better protease substrate and is shed more efficiently leading to significantly increased plasma levels of sIL-6R [24]. Patients with the Ala358 variant of the IL-6R not only show about 2-fold elevated levels of sIL-6R [25], but they are also protected from inflammatory diseases such as congestive heart disease, abdominal aneurism, and rheumatoid arthritis [26–28]. Moreover, patients with the rs2228145 SNP showed a significant lower risk of Crohn's disease and ulcerative colitis as compared to control individuals [29]. Furthermore, growth rates of abdominal aortic aneurisms in patients were smaller with the Ala358 variant of the IL-6R [30]. These results can be explained by a higher capacity of the IL-6 buffer in the blood, which is caused by the higher circulating sIL-6R levels [20,23].

IL-6 is synthesized and secreted by many cell types and the cytokine also acts on many cell types. IL-6 transgenic animals and IL-6 knock-out animals have been generated to define the major activities of this cytokine [31–37]. In order to explore cell-specific activities of IL-6, we generated a constitutively active gp130 construct. To this end, we replaced the entire extracellular portion of gp130 by a leucine zipper of the c-jun oncogene. Such an arrangement had been used to generate constitutively active cytokine receptor complexes [38]. Transfection of various cell types with the leucine zipper dimerized gp130 molecule, which we called L-gp130, was shown to lead to phosphorylation of STAT3 and to long-term growth of IL-6 dependent cells. Moreover, mouse embryonic stem cells transfected with the L-gp130 cDNA construct remained undifferentiated even in the absence of exogenously supplied LIF, indicating that cell autonomous gp130 activation had been achieved [6]. Interestingly, these data also supported the view that sustained gp130 signaling could not be completely switched off by the induction of the negative feed-back inhibitor protein SOCS3 (Figure 1) [6]. Recently, we inserted the L-gp130 cDNA construct into the ROSA26 locus of the mouse. In these mice, by breeding to an appropriate cre transgenic mouse, the L-gp130 molecule can be switched on constitutively in a cell autonomous manner. Activating gp130 in B-cells or in the entire hematopoietic system resulted in B-cell lymphomas and plasma cell disorders with full penetrance [39]. These mice will be useful in the future to study many more cell-specific activities of IL-6 without confounding activities of other cells or tissues.

ADAM17 is a membrane-bound metalloprotease, which was originally identified as the protease responsible for cleavage of the membrane-bound cytokine TNFα [40,41]. Upon gene deletion of ADAM17, it turned out that this protease is essential for life and it is also responsible for cleavage of ligands of the EGF-R [42]. In the following years, it was found that ADAM17 is also responsible for cleavage of the membrane-bound IL-6R [43] and for around 80 additional transmembrane protein substrates [44,45]. Conditional deletion of ADAM17 [46] and analysis of hypomorphic ADAM17$^{ex/ex}$ mice [47] revealed important roles of the protease in immune regulation and homeostasis of many different tissues [44], including the central nervous system [45]. ADAM17 is expressed in most types of tissues but the cell surface activity of ADAM17 seems to be tightly regulated [44]. It turned out that intracellular trafficking of ADAM17 through the Golgi to the plasma membrane requires the presence of iRhoms, which are members of the intramembrane rhomboid proteases and which have lost protease activity in the course of evolution [48]. In the absence of iRhom1 or iRhom2, ADAM17 mRNA is expressed but the protein does not reach the cell surface [49,50]. Moreover, an additional protein, FRMD8 or iTAP, which binds to iRhom2 is involved in the activation of the ADAM17 protease via phosphorylation [51,52]. An additional mechanism of ADAM17 activation is the cell surface exposure of phosphatidylserine, which binds to a cationic phosphatidylserine-binding motif within the membrane-proximal domain of ADAM17 [53]. This mechanism explains the profound activation of ADAM17 by all inducers of apoptosis analyzed thus far [54].

Since ADAM17 orchestrates at least three major signaling pathways—namely TNFα, IL-6R, and EGF-R signaling—it is not surprising that the regulation of protease activity is under tight control. Moreover, the involvement of ADAM17 in fundamental pathophysiologic processes such as inflammation and cancer can be envisaged. Of note, the importance of ADAM17 has been largely overlooked by all transcriptomics based studies since the mRNA expression of ADAM17 does not show major changes between organs as well as in health and disease [44,45]. The role of ADAM17 and IL-6 trans-signaling in inflammation and cancer will therefore be reviewed in the following paragraphs (Figure 2).

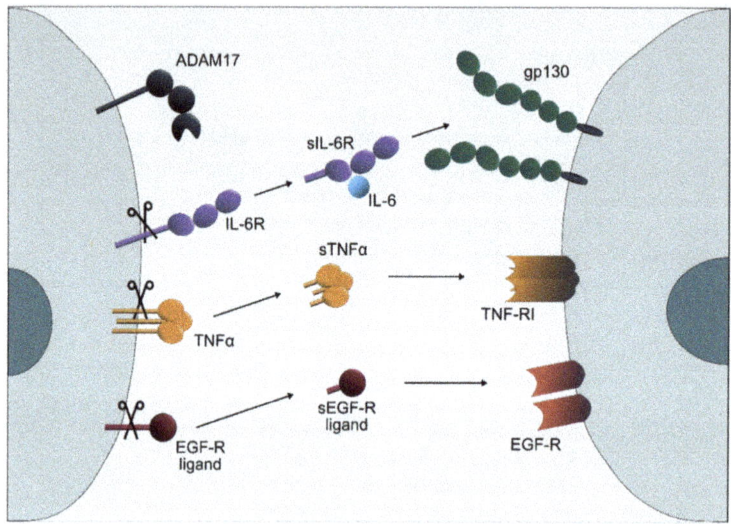

Figure 2. ADAM17 is a membrane-bound metalloprotease, which cleaves more than 80 substrates [44,45]. Major substrates of ADAM17 are IL-6R, TNFα, and ligands of the EGF-R. Cleavage of the IL-6R is a prerequisite of the pro-inflammatory IL-6 trans-signaling pathway via the sIL-6R. Cleavage of TNFα and the formation of soluble TNFα (sTNFα) leads to the stimulation of the pro-inflammatory TNFα receptor 1 (TNF-R$_I$). Cleavage of ligands of the EGF-R is needed for the systemic activity of the ligands and therefore for the full stimulation of the EGF-R [47].

2. Results

2.1. IL-6 Trans-Signaling and ADAM17 in Inflammation

IL-6 is essential for the induction, progression and resolution of local and systemic inflammatory responses that activate acute phase reaction, promote tissue damage and development of autoimmune reactions, and plays a pivotal role in transition from innate to adaptive immune responses [55].

IL-6 knock-out mice are protected from development of experimental autoimmune encephalomyelitis (EAE), the mouse model of human multiple sclerosis [56–58] and mouse models of rheumatoid arthritis [59,60]. These mice also fail to efficiently control bacterial and viral infections with vaccinia virus, *Listeria monocytogenes* and *Mycobacterium tuberculosis* [31,61]. Similar to IL-6-deficient mice, IL-6R knock-out mice display a deficit in acute phase response, compromised wound healing, and reduced cellular infiltration during inflammation [62]. While acute phase reactions and defense against bacteria is mediated via membrane-bound IL-6R [63] delayed macrophage invasion in mouse models of inflammation was dependent on the availability of sIL-6R [64–66].

The cellular origin of sIL-6R was longtime unknown. Since it is evident that increased sIL-6R levels correlate with infiltrated leukocytes in various inflammatory pathologies like arthritic joints [67] and acute inflammation [64] macrophages were considered as a main source of sIL-6R. Furthermore, IL-6 stimulates polarization and proliferation of M2 macrophages via induction of IL-4R expression as shown in mouse models of obesity [68,69].

T cell responses can also be governed by both forms of IL-6 activities. Whereas T_{reg} development during experimental airway inflammation is controlled via membrane-bound IL-6R [70], IL-6 trans-signaling orchestrates T cell recruitment in an experimental peritoneal inflammation model [71]. Recently, it was found that TGFß, together with IL-6, drives the initial differentiation from naïve T-cells to pathogenic IL-17-producing T-cells (T_H17 cells), which are key factors for induction of tissue damage in a variety of chronic inflammatory and autoimmune diseases [72,73]. Due to its complex role in immune modulation, it is essential to understand how IL-6 contributes to the respective pathology.

2.1.1. Rheumatoid Arthritis

The first successful biologics-based therapy of rheumatoid arthritis patients was developed in the early 1990s by Marc Feldman and Ravinder Maini at the Charing Cross Hospital, London, UK and made use of antibodies neutralizing TNFα [74]. These TNFα antibodies had originally been prepared for clinical trials in human sepsis, which, however, entirely failed [75,76]. The first clinical trial with TNFα antibodies, which only involved few rheumatoid arthritis patients, was highly successful and led finally to the development of several TNFα-blocking drugs, which by now changed the perception of rheumatoid arthritis from a vastly debilitating disease to a largely manageable condition [76,77].

A decade ago, blocking IL-6R signaling with anti-IL-6R antibodies was approved for treatment of rheumatoid arthritis [78]. Beforehand, mouse models of rheumatoid arthritis had shown that IL-6 signaling drives disease progression by stimulating synovial hyperplasia, preservation of joint inflammation, and damage of underlying cartilage and bone [79–81]. All these disease symptoms are largely regulated by STAT3. Interestingly, it turned out that monotherapy with anti-IL-6R antibodies was superior to treatment of patients with the TNFα antibody adalimumab [82,83].

Mice which carry a Y757F mutation in the cytoplasmic tail of gp130 are unable to initiate the SHP2–MAP kinase–PI3 kinase axis because Y757 of gp130 is the docking site of SHP2 [84,85]. These mice can still activate the STAT1/STAT3 pathway [86,87]. Furthermore, these mice do not show negative regulation by SOCS3 since also this protein requires phosphorylated Y757 for its negative feedback activity [88,89]. These so-called gp130$^{F/F}$ mice therefore show increased STAT1/STAT3, but no SHP2–MAP kinase–PI3 kinase signaling [86]. Besides enhanced autoantibody production against DNA, amplified cell infiltration into the joints and enhanced osteoclast activation, gp130$^{F/F}$ mice develop arthritis-like symptoms within one year of age, thus resembling human rheumatoid arthritis [87]. Furthermore, clonal deletion of activated T cells was altered and these chronically activated T cells showed persistent IL-6-induced STAT3 and JAK1 phosphorylation indicating that IL-6 signaling in T cells plays a critical role in disease progression of rheumatoid arthritis [87]. In contrast, a dominant-negative STAT3 mutant with reduced STAT3 activity exhibits diminished cell infiltration into the joints, pannus formation, and cartilage damage in experimental arthritis in mice [90].

IL-6 signaling initiates a range of degenerative and inflammatory processes during rheumatoid arthritis. In vitro experiments suggested a possible effect of IL-6 and its soluble receptor on synovial fibroblasts proliferation [91] and survival [92]. Furthermore, IL-6 trans-signaling promotes the generation of osteoclasts [93–95] and bone resorption by stimulating the expression of receptor activator of nuclear factor kappa κ B (NFκB) (RANK) ligand (RANKL), which stimulates RANK expressed on osteoclast precursors leading to differentiation to bone-resorbing osteoclasts [96]. Binding of RANKL to RANK activates downstream signaling pathways like NFκB, MAPK, JNK, and Akt, which consequently stimulate the expression of osteoclastic transcription factors in osteoclast precursors. The study of Feng et al. [97] showed that IL-6 trans-signaling attenuated RANKL-induced ERK and JNK activation suggesting that IL-6 trans-signaling differentially regulates RANKL-induced osteoclastogenesis.

Enhanced proliferation and activation of synovial fibroblasts and osteoclasts leads to irreversible destruction of cartilage and bone of the affected joint. The majority of these processes could be abrogated by blocking IL-6 trans signaling, which coordinated leukocyte infiltration and severity of joint damage [79,98,99].

Interestingly, EGF-R inhibition reduced the severity of experimental arthritis, most likely due to reduction of synovial fibroblasts proliferation, cytokine production, and reduction of osteoclastogenesis [100]. Furthermore, iRhom2-dependent trafficking of the IL-6R sheddase ADAM17 and its activity on myeloid cells was shown to be involved in development of experimental arthritis in mice [101]. Since ADAM17 mediates the release of TNFα, sIL-6R, and EGF-R ligands, its activity is likely to play an important role in the development of rheumatoid arthritis (see below).

2.1.2. Acute Inflammation

Rapid infiltration of neutrophils followed by replacement of mononuclear leukocytes is crucial for proper resolution of an acute inflammatory situation. IL-6 signaling via its soluble receptor regulates expression of CXC and CC chemokines that are needed for the switch of initial neutrophil influx to attraction of monocytes. In this context, IL-6 controls neutrophil infiltration by suppressing neutrophil-attracting chemokines and directs their apoptosis [64,102]. These effects were shown to be dependent on STAT3 activity [103]. The switch from neutrophil to monocytic cells could be abrogated by blocking IL-6 trans-signaling with the sgp130Fc protein in an acute peritonitis model [64,104] and in an air pouch model of acute inflammation in mice [65]. Furthermore, it was described that depletion of infiltrating neutrophils reduced the amount of sIL-6R in a model of local and acute inflammation. Thus, it was reasoned that in local and acute inflammation processes, neutrophils were a major source of sIL-6R [54]. This led us to investigate if proteolysis by ADAM17 is needed for the local release of sIL-6R and resolution of an acute inflammation. Contrary to our expectations, local ADAM17 activity was not critical for proper cell infiltration in acute inflammation in mice since the sIL-6R was infiltrating from the circulation in an ADAM17-independent process. These studies, however, confirmed the critical role of IL-6 trans-signaling in local and acute inflammation [105].

Initial recruitment of neutrophils and IL-6 secretion during acute peritonitis was described to be dependent on IFNγ. Defective IL-6 secretion into the site of inflammation in IFN$\gamma^{-/-}$ mice could be restored by treatment with recombinant IFNγ and IL-1ß, but not with IFNγ alone [102]. IL-6 trans-signaling was furthermore shown to be critical for transition between innate and acquired immune responses. T cell recruitment was impaired in IL-6$^{-/-}$ mice in a mouse model of peritonitis due to reduced chemokine expression which was mediated by STAT3 activation [71].

Unresolved inflammation leads inevitably to chronic fibrotic tissue damage. IL-6$^{-/-}$ mice were protected from the development of fibrosis in recurrent inflammatory activation of peritonitis. In this particular model, IL-6 was needed for survival and expansion of Th1 cells, which were shown to be crucial for STAT1 activation in stromal cells [106]. Recently, it was shown that IL-6-mediated STAT1 phosphorylation, which reflects the effector characteristics of CD4$^+$ T cells, is critically regulated by the tyrosine phosphatases PTPN2 and PTPN22. These two phosphatases were described to limit STAT1-mediated IL-6 signaling in both activated and memory CD4$^+$ T cells thus modulating the expression of genes associated with particular immune responses [107].

2.1.3. Sepsis

Mice with gp130-dependent STAT3 hyperactivation (gp130$^{F/F}$) were hypersensitive to LPS-induced endotoxic shock. Overall survival of these mice was reduced accompanied by increased IL-6 expression and neutrophil influx into the peritoneal cavity. Both, genetic ablation of IL-6 in gp130$^{F/F}$ or wildtype mice and specific inhibition of IL-6 trans-signaling led to an increased overall survival of LPS-induced endotoxic shock, suggesting that IL-6 signaling via the sIL-6R triggers the proinflammatory actions of IL-6 during LPS-induced endotoxic shock [108].

In a large animal study with 120 mice, the effect of total IL-6 blockade or selective inhibition of IL-6 trans-signaling on survival of sepsis induced by cecal ligation puncture was compared. Interestingly, inhibition of IL-6 trans-signaling via sgp130Fc was far superior to global IL-6 signaling blockade with an IL-6 neutralizing antibody. In this model, mice treated with sgp130Fc showed an up to 100% survival rate, an intact acute phase response, and less intestinal epithelial cell apoptosis, whereas no benefit was seen upon global IL-6 signaling blockade [109]. This enhanced therapeutic efficacy of inhibiting only the IL-6 trans-signaling pathway might be due to regenerative effects of IL-6 via the classic signaling pathway that was left intact when only IL-6 trans-signaling was blocked [12,110].

Deficiency of ADAM17 in inflammatory mouse models was shown to greatly diminish inflammatory responses [46,111,112]. When myeloid cells lacked ADAM17, mice showed slightly better survival rate after cecal ligation puncture. Blood and peritoneal bacteria levels, as well as inflammatory cytokine and chemokine plasma concentrations, were reduced in myeloid ADAM17-deficient mice, while increased neutrophil influx was observed [113]. This was possibly due to reduced proteolysis of the chemokine receptor CXCR2 or L-selectin [114,115].

2.1.4. Inflammatory Bowel Disease

IL-6 has been described as a T cell activation factor and shown to induce T cell proliferation and survival [116,117]. This ability has been linked to the pathology of several models of chronic inflammatory disease [87,118]. In several animal models of intestinal inflammation, the disease could be alleviated by specific blockade of IL-6 trans-signaling with sgp130Fc. Blockade of IL-6 trans-signaling led to beneficial T-cell apoptosis in the colon of mice with chronic intestinal inflammation [118], since an enhanced T-cell resistance against apoptosis may contribute to disease perpetuation in Crohn's disease [119]. In a murine model of intestinal inflammation using SAMP1/Yit mice, which is considered to be similar to inflammatory bowel disease in humans, treatment with sgp130Fc improved inflammation, as validated by less destruction of the epithelial layer and by less inflammatory infiltrates in the lamina propria of the small intestine [120]. In patients with inflammatory bowel disease, treatment with the IL-6R neutralizing antibody Tocilizumab suggested a clinical benefit [121], but this treatment has not been approved by the FDA. It is possible that, like in the murine cecal ligation puncture model of sepsis, specific blockade of IL-6 trans-signaling with sgp130Fc might be superior to global IL-6 blockade due to regenerative effects of IL-6 via the classic signaling pathway [12,110]. Of note, the sgp130Fc protein under the WHO name Olamkicept is currently tested in phase II clinical trials in patients with inflammatory bowel disease [20].

2.1.5. Lung Pathophysiology: Emphysema, Asthma, and Idiopathic Pulmonary Fibrosis

Proinflammatory actions of IL-6 are implicated in the pathogenesis of lung emphysema and patients with COPD [122]. Analysis in mice revealed that development of emphysema was reliant on IL-6. In contrast to wt mice, gp130$^{F/F}$ mice showed spontaneous development of lung emphysema at 6 months of age due to alveolar cell apoptosis, which was abrogated by genetic deletion of IL-6. Treatment of wt mice with cigarette smoke increased emphysema-associated lung volume increase, which was not seen in IL-6-deficient animals [123]. In both models, in the spontaneous development of lung emphysema seen in gp130$^{F/F}$ mice and in cigarette smoke treated wildtype mice, treatment with sgp130Fc abrogated alveolar cell apoptosis and disease pathogenesis. This study furthermore revealed a possible connection of IL-6 trans-signaling and the mTOR pathway. After blockade of the serine/threonine kinase mTOR with rapamycin, IL-6 mediated lung pathologies were significantly reduced [124].

IL-6 was claimed to be important for the maintenance of the balance of effector T cells and regulatory T cells in lung pathogenesis. In a mouse model of asthma, IL-6 trans-signaling rather supported Th2 cytokine production, whereas classic IL-6 signaling was needed for the development of

FoxP3[+] regulatory T cells in the lung [70]. A recent study revealed an IL-6 trans-signaling specific gene signature in patients with asthma, which was accompanied by an increase in submucosal T cell and macrophage infiltration. Stimulation of primary human bronchial epithelial with both IL-6 and sIL-6R induced genes associated with airway remodeling such as matrix metalloproteinases and reduced expression of genes associated with epithelial barrier function such as ß-catenin. The gene signature found in this study was hypothesized to be helpful in determining patient subsets with high submucosal-inflammation and poor asthma control [125].

Idiopathic pulmonary fibrosis (IPF) characterizes a heterogeneous, chronic, and irreversible disorder in which deposition of collagen results in the loss of lung function. Although it is unsure if pulmonary fibrosis is the result of an unresolved inflammatory process the role of inflammatory cytokines is discussed. A study of cytokine levels in serum of patients with IPF revealed that IL-6 and IL-8 were increased in individuals with exacerbated disease state and that these increased levels correlated with a worse outcome [126]. This deleterious impact of IL-6 on the development of lung fibrosis was analyzed in different mouse models. In adenosine deaminase-deficient mice, which develop pulmonary inflammation and remodeling as well as in bleomycin-induced lung injury, IL-6 was shown to drive cellular infiltration and collagen deposition in the lungs [127,128]. A role for IL-6 trans-signaling in IPF was suggested as elevated sIL-6R levels were found in the lungs of patients with IPF and of mice with bleomycin-induced pulmonary fibrosis. Neutralization of IL-6 trans-signaling with sgp130Fc led to reduction of pulmonary inflammation, diminished myofibroblast accumulation and improved respiratory functions [129]. The development of pulmonary fibrosis in mice and differentiation of lung fibroblasts to collagen-producing myofibroblasts involves STAT3 activation by TGFß or IL-6 trans-signaling [130]. It was suggested that increased STAT3 activity in fibroblasts from IPF patients led to reduced apoptosis and thus contributes to the persistence of these cells [131,132].

2.2. IL-6 Trans-Signaling and ADAM17 in Cancer

It is widely accepted that inflammatory cytokine signaling plays key roles at many stages of tumorigenesis and supports malignant cell proliferation and metastasis [133]. Intestinal inflammation in several different animal models was efficiently blocked by a blockade of IL-6 trans-signaling [118,120] using sgp130Fc. On the other hand, intestinal inflammation was exacerbated by the application of Hyper-IL-6, a fusion protein of IL-6 and sIL-6R, which is a strong stimulator of IL-6 trans-signaling [134]. Therefore, it was reasonable to assume that inflammation associated cancer might also be driven by IL-6 trans-signaling [135].

2.2.1. Colorectal Cancer

When mice were treated with a low dose of the mutagen azoxymethane (AOM) followed by consecutive cycles of orally administered dextran sulfate sodium (DSS), they developed small visible lesions around day 20, followed by the appearance of large tumors until day 80 [136]. When analyzing RAG-1 knockout mice, which are defective in B-cell and T-cell receptor recombination and therefore lack B- and T-lymphocytes and hence are severely immune-deficient, it became clear that these animals did not develop large tumors indicating that lymphocytes are involved in the regulation of tumor growth [136]. Colon samples from different time points after AOM/DSS treatment demonstrated massive IL-6 and sIL-6R accumulation from day 21 onwards. Treatment of the mice with a neutralizing antibody against IL-6R drastically reduced the tumor score [136]. Interestingly, also selective blockade of IL-6 trans-signaling by injection of the sgp130Fc protein effectively suppressed colon carcinogenesis, suggesting that IL-6 trans-signaling by inducing STAT3 phosphorylation and expression of the anti-apoptotic proteins bcl-xl and bcl-2 was responsible for growth of epithelial tumor cells [136]. In human colon tumor tissues, it was shown that the expression and activity of ADAM17 was significantly increased as compared with normal tissue, underlining an important role of ADAM17 in mediating the IL-6 trans-signaling response via the sIL-6R [137]. In a different intestinal tumor model, mice repeatedly treated only with 9 DSS but not with AOM developed intestinal tumors after 18 weeks. The incidence and number of tumors was drastically reduced when the mice were treated with the sgp130Fc protein. Interestingly, it was shown that lamina propria macrophages via induction of ADAM17 were important for the generation of sIL-6R and therefore for the induction of IL-6 trans-signaling [138].

When IL-6$^{-/-}$ mice were compared to wt mice in their susceptibility to inflammatory colon cancer in the AOM/DSS model, it turned out that IL-6$^{-/-}$ mice had less tumors but more inflammation in the intestine, arguing for a role of IL-6 in the intestinal regeneration response. [110]. Indeed, it was shown in the same study that deletion of STAT3 in the intestinal epithelium led to more severe DSS colitis with pronounced colonic ulcerations and body weight loss, indicating that the IL-6 response in the intestinal epithelial cells was important for regeneration [110]. These data were further corroborated by a parallel study in which the authors used transgenic mice expressing in intestinal epithelial cells a constitutively active form of gp130 dimerized by a leucine zipper [6], which led to cell-autonomous STAT3 activation. These mice were largely resistant to DSS-induced colitis [139]. The above data suggested a role of IL-6 in the regeneration of the intestine. Indeed, treatment of mice with recombinant IL-6 protected the animals from DSS-induced colitis whereas treatment with a neutralizing IL-6 antibody aggravated DSS-induced colitis [140,141].

Inflammatory hepatocellular adenomas often harbor mutations in the gp130 gene, which leads to ligand independent activation of gp130 [142]. Transgenic mice, which express constitutively active gp130 transgene developed aberrant proliferation of intestinal epithelial cells. Interestingly, it was observed that the YAP and Notch signaling pathway were activated by the constitutively active gp130 mutant [5]. Furthermore, it was reported by the same authors that loss of APC in mice and in human patients led to the activation of Src, YAP, Notch, and STAT3. The presence of the activated gp130 transgene in intestinal epithelial cells accelerated the development of colorectal cancer and depended on the activities of Src family and JAK kinases [4].

Mice with a complete deletion of the ADAM17 gene are not viable [42]. Although, ADAM17 floxed mice are available [46], it was not completely clear in which cell types we wanted to delete the ADAM17 in order to show a decisive role of the protease, since many cells including macrophages can be the source of the sIL-6R in vivo [2]. We therefore made use of ADAM17 hypomorphic ADAM17$^{ex/ex}$ mice, which we had engineered to only express about 5% of ADAM17 protein levels in all cells [47]. ADAM17 hypomorphic ADAM17$^{ex/ex}$ mice were shown to be hypersensitive to DSS-induced colitis due to a failure to induce EGF-R signaling, which was needed for intestinal regeneration. EGF-R activation could be compensated by treatment of the mice with recombinant ligands of the EGF-R, which compensated for the lack of cleavage of these membrane-bound proteins [47]. We therefore applied a colon cancer model to ADAM17$^{ex/ex}$ mice. In APC$^{min/+}$ mice, a heterozygous germ line truncation in the APC gene results in the formation of intestinal tumors after spontaneous loss of heterozygosity of the remaining APC wt allele [143,144]. Strikingly, in the absence of ADAM17 activity, almost no intestinal tumors were detected and the few remaining tumors were only of low-grade dysplasia whereas in the presence of ADAM17, also high-grade dysplasias and invasive carcinomas were detected [144]. Surprisingly, we failed to detect significantly elevated levels of phosphorylated EGF-R on intestine sections of wt and ADAM17$^{ex/ex}$ mice (Figure 3) [144].

Elevated activity of the EGF-R has been associated with an increased risk of colon carcinogenesis [144,145]. Recently, it was reported that in patients, EGF-R was expressed on myeloid cells within intestinal tumors and that in mice, EGF-R on myeloid cells but not on intestinal epithelial cells promoted colitis associated cancer in the AOM/DSS model and tumorigenesis in the APC$^{min/+}$ model and protects from colitis in an independent fashion [141]. Moreover, EGF-R activation in myeloid cells led to strongly increased synthesis and secretion of IL-6 [141].

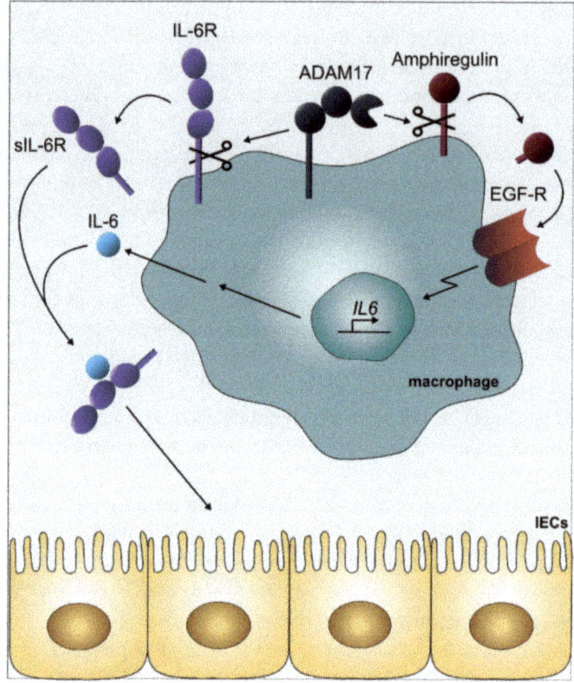

Figure 3. Involvement of ADAM17 in the formation of colon cancer. Under inflammatory conditions, ADAM17 cleaves the membrane-bound EGF-R ligand amphiregulin, leading to the stimulation of the EGF-R on macrophages. This EGF-R stimulation on macrophages leads to increased IL-6 synthesis via transcriptional stimulation. Moreover, activation of ADAM17 under inflammatory conditions leads to cleavage of the membrane-bound IL-6R on macrophages to generate soluble IL-6R (sIL-6R). IL-6 in complex with sIL-6R leads to stimulation of intestinal epithelial cells (IECs) via IL-6 trans-signaling resulting in colon cancer formation [144].

ADAM17 on myeloid cells can lead to the shedding of ligands of the EGF-R such as amphiregulin and to the shedding of the IL-6R [10]. We therefore speculated that IL-6 trans-signaling initiated by the EGF-R mediated IL-6 secretion and the ADAM17-mediated IL-6R shedding might contribute to colon cancer formation in the APC$^{min/+}$ model. We therefore crossed APC$^{min/+}$ mice with either IL-6$^{-/-}$ mice or mice, which were transgenic for sgp130Fc and in which IL-6 trans-signaling was blocked [19,65]. Indeed, tumorigenesis was strongly reduced to the same extent in IL-6$^{-/-}$ mice and sgp130Fc transgenic mice indicating that IL-6 trans-signaling strongly contributed to tumor formation in the APC$^{min/+}$ model [144]. Furthermore, in a chemical colon cancer model, in which mice were repeatedly treated with AOM but not with DSS, sgp130Fc transgenic mice were protected from colon cancer [144]. We concluded from these data that in colon cancer, IL-6 trans-signaling apparently acted downstream of the EGF-R and blockade of IL-6 trans-signaling might represent a novel therapeutic window for patients resistant to anti-EGF-R antibodies (Figure 3) [146].

2.2.2. Pancreatic Cancer

In the Kras^{G12D} pancreatic cancer model, strong non-cell-autonomous STAT3 activation in tumor cells was observed [147]. IL-6 mRNA levels were significantly increased in tumor tissue. The cellular source of IL-6 was infiltrating immune cells, mainly F4/80-positive macrophages. When the pancreatic Kras^{G12D} mice were crossed with IL-6$^{-/-}$ mice or sgp130Fc transgenic mice. IL-6$^{-/-}$ mice had fewer and predominantly low grade pancreatic intraepithelial neoplasias. Interestingly, blocking IL-6 trans-signaling had a similar effect as deletion of the IL-6 gene since sgp130Fc transgenic mice also showed strongly reduced tumor numbers. This decrease in tumor development was paralleled by a marked reduction of STAT3 phosphorylation [147]. It was speculated that the concept of non-cell-autonomous STAT3 activation of tumor cells brought about by IL-6 trans-signaling was a general phenomenon, which might also be applied to other neoplasias and cancer diseases [147].

2.2.3. Liver Cancer

During liver regeneration, quiescent hepatocytes regain their ability to enter the cell cycle. Importantly, IL-6 [148] and in particular IL-6 trans-signaling has been found to be required for liver regeneration [149]. Accordingly, in IL-6/sIL-6R double transgenic mice, permanent hepatocyte proliferation and formation of adenomas was observed, indicating a possible role of IL-6 trans-signaling for hepatocellular neoplasia [150,151]. Treatment of mice with diethylnitrosamine is used as one of the standard models for liver cancer [152]. In this model, it has been recognized that IL-6 deficient mice develop substantially less tumors as compared to wt mice [153]. When sgp130Fc transgenic mice were treated with diethylnitrosamine, it turned out that myeloid cell derived sIL-6R increased upon treatment and liver tumor development the animals was largely blocked [154]. We concluded from these experiments that IL-6 trans-signaling but not IL-6 classic signaling was involved in the development of hepatocellular carcinoma, at least in the diethylnitrosamine model [154].

2.2.4. Lung Cancer

As mentioned above, gp130 signal transduction is mediated by the SHP2–MAP kinase–PI3 kinase axis. In addition, STAT1 and STAT3 become recruited to the cytoplasmic tail of gp130, are phosphorylated by JAK1, and dimerize and travel to the nucleus to act as transcription factors [3]. Mice, which carry a Y757F mutation in the cytoplasmic tail of gp130 are unable to initiate the SHP2–MAP kinase–PI3 kinase axis because Y757 of gp130 is the docking site of SHP2. These mice can still activate the STAT1/STAT3 pathway [86]. Furthermore, these mice do not show negative regulation by SOCS3 since also this protein requires phosphorylated Y757 for its negative feedback activity [86]. These so-called gp130$^{F/F}$ mice therefore show increased STAT1/STAT3 but no SHP2–MAP kinase–PI3 kinase signaling [86].

The gp130$^{F/F}$ mouse model has been widely used to explore a range of gp130 mediated disease states. The more pronounced gp130/STAT1/STAT3 response seen in these mice has been used to analyze novel therapeutic strategies for inflammatory diseases. Gp130$^{F/F}$ mice are more sensitive than wt mice towards endotoxin. It was found that genetic ablation of IL-6, antibody-mediated inhibition of the IL-6R or blockade of IL-6 trans-signaling by the sgp130Fc protein completely protected gp130$^{F/F}$ mice from endotoxin hypersensitivity [108]. Gp130$^{F/F}$ mice spontaneously develop lung emphysema, which could be prevented by blocking IL-6 trans-signaling [124]. When floxed Kras^{G12D} mice were treated via intranasal inhalation with adenovirus, which encoded cre recombinase 6 weeks prior to analysis, massive lung cancer development was observed, which was much more pronounced in gp130$^{F/F}$ mice as compared to wt mice [155]. Ablation of IL-6 or STAT3 suppressed the extent of lung cancer in this model [155]. Furthermore, we could demonstrate that specific inhibition of IL-6 trans-signaling by the sgp130Fc protein significantly ameliorated lung cancer pathogenesis [155].

Since IL-6 trans-signaling depends on the proteolysis of the membrane-bound IL-6R by the metalloprotease ADAM17, we asked whether in the absence of ADAM17 activity, lung tumor formation in the Kras^{G12D} mouse model would be blocked. When ADAM17$^{ex/ex}$ mice were crossed with Kras^{G12D} mice, tumor formation in the lung was largely inhibited [156]. Interestingly, the cytoplasmic portion of ADAM17 was phosphorylated in Kras^{G12D} mice but not in wt mice [156]. Phosphorylation of the cytoplasmic portion of ADAM17 has been shown to lead to activation of the enzyme [157]. Interestingly, in Kras^{G12D} mice, the levels of sIL-6R were significantly elevated, whereas in compound ADAM17$^{ex/ex}$: Kras^{G12D} mice, sIL-6R levels were comparable to wt mice. No changes in soluble TNFα levels were observed [156]. Moreover, in a xenograft model with ADAM17-deleted A549 cells, tumor formation was abrogated as compared to unedited A549 cells [156]. Along the same line, pharmacologic inhibition of ADAM17 activity also led to decreased tumor formation in Kras^{G12D} mice together with decreased sIL-6R levels [156]. Moreover, in a nicotine-derived nitrosamine ketone model of tobacco-related lung cancer, genetic or pharmacologic blockade of ADAM17 led to a significant reduction of lung lesions and concomitant reduced sIL-6R levels [158]. Together, these data indicated that ADAM17 via IL-6 trans-signaling was a prerequisite of lung tumor formation in these animal models.

3. Conclusions and Outlook

ADAM17 activation orchestrates three major signal transduction pathways, namely IL-6 trans-signaling, stimulation of the TNFα signaling pathway via TNF-R$_I$ and the EGF-R pathway via cleavage of EGF-R ligands such as amphiregulin, TGFα, or Hb-EGF. Apparently, in many inflammatory diseases as well as in colon cancer and lung cancer, the IL-6 trans-signaling pathway plays a dominant role as compared to the TNFα signaling pathway (Table 1).

Table 1. Efficacy of IL-6 trans-signaling blockade by sgp130Fc in preclinical models of inflammation and inflammation associated cancer.

Disease Model	Outcome of the Study
Intestinal inflammation [118,138]	Suppression of colitis activity
Acute inflammation [54,64,65]	Blockade of inflammatory processes
Atherosclerosis [159]	Regression of advanced atherosclerosis
Rheumatoid arthritis [79,98,99]	Improvement of established arthritis
Sepsis [108,109]	Up to 100% survival in different sepsis models
Pancreatitis-lung failure [160]	100% survival of severe acute pancreatitis
Lung emphysema [124]	Improvement by blockade of alveolar cell apoptosis
Abdominal aortic aneurism [30]	Improved survival in two animal models
Colon cancer [110,136–138,144]	Blockade of tumor formation
Pancreatic cancer [147]	Inhibition of pancreatic neoplasia progression
Liver cancer [154]	Protection from tumor formation
Lung cancer [156,158]	Amelioration of lung cancer pathogenesis

The dependence of IL-6 trans-signaling via the generation of IL-6 and sIL-6R by the EGF-R on macrophages demonstrates the existence of a possible novel therapeutic window for the treatment of colon cancer and lung cancer, which might be important in view of the fact that EGF-R antibodies including cetuximab or pantuximab are only useful in patients in whom there are no activating KRAS mutations [161,162]. Even in such patients without KRAS mutations, who show an initial response, almost invariably resistance against EGF-R blockade is seen [163]. In all of these cases, blockade of the IL-6 trans-signaling pathway might offer an additional and alternative treatment option. This treatment option, however, will need to be verified in future human studies.

Author Contributions: N.S. and S.R.-J. wrote and edited the review article.

Funding: The work of S.R.-J. is supported by grants from the Deutsche Forschungsgemeinschaft, Bonn, Germany (SFB841 project C1; SFB877 project A1) and by the Cluster of Excellence 'Inflammation at Interfaces'.

Conflicts of Interest: S.R.-J. has acted as a consultant and speaker for AbbVie, Chugai, Genentech Roche, Pfizer, and Sanofi. He also declares that he is an inventor on patents owned by CONARIS Research Institute, which develops the sgp130Fc protein Olamkicept together with the company I-Mab. S.R.-J. has stock ownership in CONARIS. N.S. declares no competing interests.

Abbreviations

ADAM	a disintegrin and metalloprotease
AOM	azoxymethane
APC	adenomatous polyposis coli protein
COPD	chronic obstructive pulmonary disease
DSS	dextransulfate sodium
EAE	experimental autoimmune encephalomyelitis
EGF	epidermal growth factor
FDA	US food and drug administration
FoxP3	forkhead box protein 3
FRMD	FERM domain containing protein, identical to iTAP
Gp130	glycoprotein 130 kDa
IECs	intestinal epithelial cells
IFN	interferon
IL-6	Interleukin-6
IL-6R	Interleukin-6 receptor
iRhom	inactive rhomboid protease
iTAP	iRhom Tail-Associated Protein identical to FRMD8
JAK	Janus kinase
JNK	c-Jun N-terminal kinase
Kras	oncogene first identified in Kirsten RAt Sarcoma virus
MAP	mitogen-activated protein
mTOR	mechanistic target of rapamycin
NFκB	nuclear factor kappa-light-chain-enhancer
Notch	cell signaling receptor present in most animals
PTPN	tyrosine phosphatase
RANK	Receptor Activator of NFκB
RANKL	ligand of RANK
ROSA26	reverse oriented splice acceptor, Clone 26
SHP2	protein-tyrosine phosphatase shp2
SOCS	suppressor of cytokine signaling
Src	proto-oncogene tyrosine-protein kinase
STAT	signal transducer and activator of transcription
TNF	tumor necrosis factor
WHO	world health organization
YAP	proto-oncogene yes associated protein.

References

1. Jones, S.A.; Scheller, J.; Rose-John, S. Therapeutic strategies for the clinical blockade of IL-6/gp130 signaling. *J. Clin. Invest.* **2011**, *121*, 3375–3383. [CrossRef]
2. Rose-John, S. IL-6 Trans-Signaling via the Soluble IL-6 Receptor: Importance for the Pro-Inflammatory Activities of IL-6. *Int. J. Biol. Sci.* **2012**, *8*, 1237–1247. [CrossRef] [PubMed]
3. Schaper, F.; Rose-John, S. Interleukin-6: Biology, signaling and strategies of blockade. *Cytokine Growth Factor Rev.* **2015**, *26*, 475–487. [CrossRef] [PubMed]
4. Taniguchi, K.; Moroishi, T.; De Jong, P.R.; Krawczyk, M.; Grebbin, B.M.; Luo, H.; Xu, R.H.; Golob-Schwarzl, N.; Schweiger, C.; Wang, K.; et al. YAP–IL-6ST autoregulatory loop activated on APC loss controls colonic tumorigenesis. *Proc. Natl. Acad. Sci. USA* **2017**, *114*, 1643–1648. [CrossRef] [PubMed]
5. Taniguchi, K.; Wu, L.W.; Grivennikov, S.I.; De Jong, P.R.; Lian, I.; Yu, F.X.; Wang, K.; Ho, S.B.; Boland, B.S.; Chang, J.T.; et al. A gp130–Src–YAP module links inflammation to epithelial regeneration. *Nature* **2015**, *519*, 57–62. [CrossRef] [PubMed]

6. Stuhlmann-Laeisz, C.; Lang, S.; Chalaris, A.; Krzysztof, P.; Enge, S.; Eichler, J.; Klingmüller, U.; Samuel, M.; Ernst, M.; Rose-John, S.; et al. Forced Dimerization of gp130 Leads to Constitutive STAT3 Activation, Cytokine-independent Growth, and Blockade of Differentiation of Embryonic Stem Cells. *Mol. Biol. Cell* **2006**, *17*, 2986–2995. [CrossRef] [PubMed]
7. Rose-John, S. Interleukin-6 family cytokines. *Cold Spring Harb. Perspect. Biol.* **2018**, *10*. [CrossRef] [PubMed]
8. Mülberg, J.; Schooltink, H.; Stoyan, T.; Günther, M.; Graeve, L.; Buse, G.; Mackiewicz, A.; Heinrich, P.C.; Rose-John, S. The soluble interleukin-6 receptor is generated by shedding. *Eur. J. Immunol.* **1993**, *23*, 473–480. [CrossRef]
9. Yan, I.; Schwarz, J.; Lücke, K.; Schumacher, N.; Schumacher, V.; Schmidt, S.; Rabe, B.; Saftig, P.; Donners, M.; Rose-John, S.; et al. Adam17 controls IL-6 signaling by cleavage of the murine IL-6Rα from the cell surface of leukocytes during inflammatory responses. *J. Leukoc. Biol.* **2016**, *99*, 749–760. [CrossRef]
10. Schumacher, N.; Meyer, D.; Mauermann, A.; Von Der Heyde, J.; Wolf, J.; Schwarz, J.; Knittler, K.; Murphy, G.; Michalek, M.; Garbers, C.; et al. Shedding of Endogenous Interleukin-6 Receptor (IL-6R) Is Governed by A Disintegrin and Metalloproteinase (ADAM) Proteases while a Full-length IL-6R Isoform Localizes to Circulating Microvesicles. *J. Biol. Chem.* **2015**, *290*, 26059–26071. [CrossRef]
11. Rose-John, S.; Heinrich, P.C. Soluble receptors for cytokines and growth factors: Generation and biological function. *Biochem. J.* **1994**, *300*, 281–290. [CrossRef] [PubMed]
12. Scheller, J.; Chalaris, A.; Schmidt-Arras, D.; Rose-John, S. The pro-and anti-inflammatory properties of the cytokine interleukin-6. *Biochim. Biophys. Acta* **2011**, *1813*, 878–888. [CrossRef] [PubMed]
13. Arnold, P.; Boll, I.; Rothaug, M.; Schumacher, N.; Schmidt, F.; Wichert, R.; Schneppenheim, J.; Lokau, J.; Pickhinke, U.; Koudelka, T.; et al. Meprin Metalloproteases Generate Biologically Active Soluble Interleukin-6 Receptor to Induce Trans-Signaling. *Sci. Rep.* **2017**, *7*, 44053. [CrossRef] [PubMed]
14. Garbers, C.; Jänner, N.; Chalaris, A.; Moss, M.L.; Floss, D.M.; Meyer, D.; Koch-Nolte, F.; Rose-John, S.; Scheller, J. Species Specificity of ADAM10 and ADAM17 Proteins in Interleukin-6 (IL-6) Trans-signaling and Novel Role of ADAM10 in Inducible IL-6 Receptor Shedding. *J. Biol. Chem.* **2011**, *286*, 14804–14811. [CrossRef]
15. Riethmueller, S.; Ehlers, J.C.; Lokau, J.; Düsterhöft, S.; Knittler, K.; Dombrowsky, G.; Grötzinger, J.; Rabe, B.; Rose-John, S.; Garbers, C. Cleavage Site Localization Differentially Controls Interleukin-6 Receptor Proteolysis by ADAM10 and ADAM17. *Sci. Rep.* **2016**, *6*, 25550. [CrossRef]
16. Riethmueller, S.; Somasundaram, P.; Ehlers, J.C.; Hung, C.W.; Flynn, C.M.; Lokau, J.; Agthe, M.; Düsterhöft, S.; Zhu, Y.; Grötzinger, J.; et al. Proteolytic origin of the soluble human IL-6R in vivo and a decisive role of n-glycosylation. *PLoS Biol.* **2017**, *15*, e2000080. [CrossRef]
17. Peters, M.; Blinn, G.; Solem, F.; Fischer, M.; Zum Büschenfelde, K.H.M.; Rose-John, S. In vivo and in vitro activities of the gp130-stimulating designer cytokine Hyper-IL-6. *J. Immunol.* **1998**, *161*, 3575–3581.
18. Schmidt-Arras, D.; Rose-John, S. IL-6 pathway in the liver: From physiopathology to therapy. *J. Hepatol.* **2016**, *64*, 1403–1415. [CrossRef]
19. Jostock, T.; Müllberg, J.; Özbek, S.; Atreya, R.; Blinn, G.; Voltz, N.; Fischer, M.; Neurath, M.F.; Rose-John, S. Soluble gp130 is the natural inhibitor of soluble interleukin-6 receptor transsignaling responses. *Eur. J. Biochem.* **2001**, *268*, 160–167. [CrossRef]
20. Garbers, C.; Heink, S.; Korn, T.; Rose-John, S. Interleukin-6: Designing specific therapeutics for a complex cytokine. *Nat. Rev. Drug Discov.* **2018**, *17*, 395–412. [CrossRef]
21. Waage, A. The complex pattern of cytokines in serum from patients with meningococcal septic shock. Association between interleukin 6, interleukin 1, and fatal outcome. *J. Exp. Med.* **1989**, *169*, 333–338. [CrossRef] [PubMed]
22. Wolf, J.; Waetzig, G.H.; Chalaris, A.; Reinheimer, T.M.; Wege, H.; Rose-John, S.; Garbers, C. Different Soluble Forms of the Interleukin-6 Family Signal Transducer gp130 Fine-tune the Blockade of Interleukin-6 Trans-signaling. *J. Biol. Chem.* **2016**, *291*, 16186–16196. [CrossRef] [PubMed]
23. Scheller, J.; Rose-John, S. The interleukin 6 pathway and atherosclerosis. *Lancet* **2012**, *380*, 338. [CrossRef]
24. Garbers, C.; Monhasery, N.; Aparicio-Siegmund, S.; Lokau, J.; Baran, P.; Nowell, M.A.; Jones, S.A.; Rose-John, S.; Scheller, J. The interleukin-6 receptor Asp358Ala single nucleotide polymorphism rs2228145 confers increased proteolytic conversion rates by ADAM proteases. *Biochim. Biophys. Acta* **2014**, *1842*, 1485–1494. [CrossRef] [PubMed]

25. Aparicio-Siegmund, S.; Garbers, Y.; Flynn, C.M.; Waetzig, G.H.; Gouni-Berthold, I.; Krone, W.; Berthold, H.K.; Laudes, M.; Rose-John, S.; Garbers, C. The IL-6-neutralizing sIL-6R-sgp130 buffer system is disturbed in patients with type 2 diabetes. *Am. J. Physiol. Endocrinol. Metab.* **2019**, *317*, 411–420. [CrossRef] [PubMed]

26. Ferreira, R.C.; Freitag, D.F.; Cutler, A.J.; Howson, J.M.M.; Rainbow, D.B.; Smyth, D.J.; Kaptoge, S.; Clarke, P.; Boreham, C.; Coulson, R.M.; et al. Functional IL6R 358Ala Allele Impairs Classical IL-6 Receptor Signaling and Influences Risk of Diverse Inflammatory Diseases. *PLoS Genet.* **2013**, *9*, e1003444. [CrossRef]

27. IL6R Genetics Consortium Emerging Risk Factors Collaboration; Sarwar, N.; Butterworth, A.S.; Freitag, D.F.; Gregson, J.; Willeit, P.; Gorman, D.N.; Gao, P.; Saleheen, D.; Rendon, A.; et al. Interleukin-6 receptor pathways in coronary heart disease: A collaborative meta-analysis of 82 studies. *Lancet* **2012**, *379*, 1205–1213. [PubMed]

28. Interleukin-6 Receptor Mendelian Randomisation Analysis (IL6R MR) Consortium; Sarwar, N.; Butterworth, A.S.; Freitag, D.F.; Gregson, J.; Willeit, P.; Gorman, D.N.; Gao, P.; Saleheen, D.; Rendon, A.; et al. The interleukin-6 receptor as a target for prevention of coronary heart disease: A mendelian randomisation analysis. *Lancet* **2012**, *379*, 1214–1224.

29. Parisinos, C.A.; Serghiou, S.; Katsoulis, M.; George, M.J.; Patel, R.S.; Hemingway, H.; Hingorani, A.D. Variation in Interleukin 6 Receptor Gene Associates with Risk of Crohn's Disease and Ulcerative Colitis. *Gastroenterology* **2018**, *155*, 303–306. [CrossRef]

30. Paige, E.; Clement, M.; Lareyre, F.; Sweeting, M.; Raffort, J.; Grenier, C.; Finigan, A.; Harrison, J.; Peters, J.E.; Sun, B.B.; et al. Interleukin-6 receptor signaling and abdominal aortic aneurysm growth rates. *Circ. Genom. Precis. Med.* **2019**, *12*, e002413. [CrossRef]

31. Kopf, M.; Baumann, H.; Freer, G.; Freudenberg, M.; Lamers, M.; Kishimoto, T.; Zinkernagel, R.; Bluethmann, H.; Köhler, G. Impaired immune and acute-phase responses in interleukin-6-deficient mice. *Nature* **1994**, *368*, 339–342. [CrossRef] [PubMed]

32. Poli, V.; Balena, R.; Fattori, E.; Markatos, A.; Yamamoto, M.; Tanaka, H.; Ciliberto, G.; Rodan, G.; Costantini, F. Interleukin-6 deficient mice are protected from bone loss caused by estrogen depletion. *EMBO J.* **1994**, *13*, 1189–1196. [CrossRef] [PubMed]

33. Fattori, E.; Cappelletti, M.; Costa, P.; Sellitto, C.; Cantoni, L.; Carelli, M.; Faggioni, R.; Fantuzzi, G.; Ghezzi, P.; Poli, V. Defective inflammatory response in interleukin 6-deficient mice. *J. Exp. Med.* **1994**, *180*, 1243–1250. [CrossRef] [PubMed]

34. Suematsu, S.; Matsuda, T.; Aozasa, K.; Akira, S.; Nakano, N.; Ohno, S.; Miyazaki, J.; Yamamura, K.; Hirano, T.; Kishimoto, T. IgG1 plasmacytosis in interleukin 6 transgenic mice. *Proc. Natl. Acad. Sci. USA* **1989**, *86*, 7547–7551. [CrossRef] [PubMed]

35. Suematsu, S.; Matsusaka, T.; Matsuda, T.; Ohno, S.; Miyazaki, J.; Yamamura, K.; Hirano, T.; Kishimoto, T. Generation of plasmacytomas with the chromosomal translocation t (12;15) in interleukin 6 transgenic mice. *Proc. Natl. Acad. Sci. USA* **1992**, *89*, 232–235. [CrossRef]

36. Peters, M.; Jacobs, S.; Ehlers, M.; Vollmer, P.; Müllberg, J.; Wolf, E.; Brem, G.; Meyer zum Büschenfelde, K.H.; Rose-John, S. The function of the soluble interleukin 6 (IL-6) receptor in vivo: Sensitization of human soluble IL-6 receptor transgenic mice towards IL-6 and prolongation of the plasma half-life of IL-6. *J. Exp. Med.* **1996**, *183*, 1399–1406. [CrossRef]

37. Peters, M.; Schirmacher, P.; Goldschmitt, J.; Odenthal, M.; Peschel, C.; Fattori, E.; Ciliberto, G.; Dienes, H.P.; Meyer zum Büschenfelde, K.H.; Rose-John, S. Extramedullary Expansion of Hematopoietic Progenitor Cells in Interleukin (IL)-6–sIL-6R Double Transgenic Mice. *J. Exp. Med.* **1997**, *185*, 755–766. [CrossRef]

38. Herrman, J.M.; Patel, N.; Timans, J.C.; Kastelein, R.A. Functional Replacement of Cytokine Receptor Extracellular Domains by Leucine Zippers. *J. Biol. Chem.* **1996**, *271*, 30386–30391.

39. Scherger, A.K.; Al-Maarri, M.; Maurer, H.C.; Schick, M.; Maurer, S.; Ollinger, R.; Gonzalez-Menendez, I.; Martella, M.; Thaler, M.; Pechloff, K.; et al. Activated gp130 signaling selectively targets b cell differentiation to induce mature lymphoma and plasmacytoma. *JCI Insight* **2019**, *4*, e128435. [CrossRef]

40. Black, R.A.; Rauch, C.T.; Kozlosky, C.J.; Peschon, J.J.; Slack, J.L.; Wolfson, M.F.; Castner, B.J.; Stocking, K.L.; Reddy, P.; Srinivasan, S.; et al. A metalloproteinase disintegrin that releases tumour-necrosis factor-alpha from cells. *Nature* **1997**, *385*, 729–733. [CrossRef]

41. Moss, M.L.; Jin, S.L.; Milla, M.E.; Bickett, D.M.; Burkhart, W.; Carter, H.L.; Chen, W.J.; Clay, W.C.; Didsbury, J.R.; Hassler, D.; et al. Cloning of a disintegrin metalloproteinase that processes precursor tumour-necrosis factor-alpha. *Nature* **1997**, *385*, 733–736. [CrossRef] [PubMed]

42. Peschon, J.J. An Essential Role for Ectodomain Shedding in Mammalian Development. *Science* **1998**, *282*, 1281–1284. [CrossRef] [PubMed]

43. Matthews, V.; Schuster, B.; Schütze, S.; Bussmeyer, I.; Ludwig, A.; Hundhausen, C.; Sadowski, T.; Saftig, P.; Hartmann, D.; Kallen, K.J.; et al. Cellular Cholesterol Depletion Triggers Shedding of the Human Interleukin-6 Receptor by ADAM10 and ADAM17 (TACE). *J. Biol. Chem.* **2003**, *278*, 38829–38839. [CrossRef] [PubMed]

44. Scheller, J.; Chalaris, A.; Garbers, C.; Rose-John, S. ADAM17: A molecular switch to control inflammation and tissue regeneration. *Trends Immunol.* **2011**, *32*, 380–387. [CrossRef] [PubMed]

45. Zunke, F.; Rose-John, S. The shedding protease adam17: Physiology and pathophysiology. *Biochim. Biophys. Acta* **2017**, *1864*, 2059–2070. [CrossRef] [PubMed]

46. Horiuchi, K.; Kimura, T.; Miyamoto, T.; Takaishi, H.; Okada, Y.; Toyama, Y.; Blobel, C.P. Cutting Edge: TNF-α-Converting Enzyme (TACE/ADAM17) Inactivation in Mouse Myeloid Cells Prevents Lethality from Endotoxin Shock. *J. Immunol.* **2007**, *179*, 2686–2689. [CrossRef] [PubMed]

47. Chalaris, A.; Adam, N.; Sina, C.; Rosenstiel, P.; Lehmann-Koch, J.; Schirmacher, P.; Hartmann, D.; Cichy, J.; Gavrilova, O.; Schreiber, S.; et al. Critical role of the disintegrin metalloprotease ADAM17 for intestinal inflammation and regeneration in mice. *J. Exp. Med.* **2010**, *207*, 1617–1624. [CrossRef]

48. Dulloo, I.; Muliyil, S.; Freeman, M. The molecular, cellular and pathophysiological roles of iRhom pseudoproteases. *Open Biol.* **2019**, *9*, 190003. [CrossRef]

49. Adrain, C.; Freeman, M. New lives for old: Evolution of pseudoenzyme function illustrated by iRhoms. *Nat. Rev. Mol. Cell Biol.* **2012**, *13*, 489–498. [CrossRef]

50. McIlwain, D.R.; Lang, P.A.; Maretzky, T.; Hamada, K.; Ohishi, K.; Maney, S.K.; Berger, T.; Murthy, A.; Duncan, G.; Xu, H.C.; et al. iRhom2 regulation of TACE controls TNF-mediated protection against Listeria and responses to LPS. *Science* **2012**, *335*, 229–232. [CrossRef]

51. Künzel, U.; Grieve, A.G.; Meng, Y.; Sieber, B.; Cowley, S.A.; Freeman, M. FRMD8 promotes inflammatory and growth factor signalling by stabilising the iRhom/ADAM17 sheddase complex. *Elife* **2018**, *7*, e35012. [CrossRef] [PubMed]

52. Oikonomidi, I.; Burbridge, E.; Cavadas, M.; Sullivan, G.; Collis, B.; Naegele, H.; Clancy, D.; Brezinova, J.; Hu, T.; Bileck, A.; et al. Itap, a novel irhom interactor, controls tnf secretion by policing the stability of irhom/tace. *Elife* **2018**, *7*, e35032. [CrossRef] [PubMed]

53. Sommer, A.; Kordowski, F.; Büch, J.; Maretzky, T.; Evers, A.; Andrä, J.; Düsterhöft, S.; Michalek, M.; Lorenzen, I.; Somasundaram, P.; et al. Phosphatidylserine exposure is required for ADAM17 sheddase function. *Nat. Commun.* **2016**, *7*, 11523. [CrossRef] [PubMed]

54. Chalaris, A.; Rabe, B.; Paliga, K.; Lange, H.; Laskay, T.; Fielding, C.A.; Jones, S.A.; Rose-John, S.; Scheller, J. Apoptosis is a natural stimulus of IL6R shedding and contributes to the proinflammatory trans-signaling function of neutrophils. *Blood* **2007**, *110*, 1748–1755. [CrossRef]

55. Hunter, C.A.; Jones, S.A. IL-6 as a keystone cytokine in health and disease. *Nat. Immunol.* **2015**, *16*, 448–457. [CrossRef]

56. Samoilova, E.B.; Horton, J.L.; Hilliard, B.; Liu, T.S.; Chen, Y. IL-6-deficient mice are resistant to experimental autoimmune encephalomyelitis: Roles of IL-6 in the activation and differentiation of autoreactive T cells. *J. Immunol.* **1998**, *161*, 6480–6486.

57. Okuda, Y.; Sakoda, S.; Bernard, C.C.; Fujimura, H.; Saeki, Y.; Kishimoto, T.; Yanagihara, T. IL-6-deficient mice are resistant to the induction of experimental autoimmune encephalomyelitis provoked by myelin oligodendrocyte glycoprotein. *Int. Immunol.* **1998**, *10*, 703–708. [CrossRef]

58. Eugster, H.P.; Frei, K.; Kopf, M.; Lassmann, H.; Fontana, A.; Eugster, H. IL-6-deficient mice resist myelin oligodendrocyte glycoprotein-induced autoimmune encephalomyelitis. *Eur. J. Immunol.* **1998**, *28*, 2178–2187. [CrossRef]

59. Ohshima, S.; Saeki, Y.; Mima, T.; Sasai, M.; Nishioka, K.; Nomura, S.; Kopf, M.; Katada, Y.; Tanaka, T.; Suemura, M.; et al. Interleukin 6 plays a key role in the development of antigen-induced arthritis. *Proc. Natl. Acad. Sci. USA* **1998**, *95*, 8222–8226. [CrossRef]

60. Alonzi, T.; Fattori, E.; Lazzaro, D.; Costa, P.; Probert, L.; Kollias, G.; De Benedetti, F.; Poli, V.; Ciliberto, G. Interleukin 6 Is Required for the Development of Collagen-induced Arthritis. *J. Exp. Med.* **1998**, *187*, 461–468. [CrossRef]

61. Ladel, C.H.; Blum, C.; Dreher, A.; Reifenberg, K.; Kopf, M.; Kaufmann, S.H. Lethal tuberculosis in interleukin-6-deficient mutant mice. *Infect. Immun.* **1997**, *65*, 4843–4849. [PubMed]

62. McFarland-Mancini, M.M.; Funk, H.M.; Paluch, A.M.; Zhou, M.; Giridhar, P.V.; Mercer, C.A.; Kozma, S.C.; Drew, A.F. Differences in Wound Healing in Mice with Deficiency of IL-6 versus IL-6 Receptor. *J. Immunol.* **2010**, *184*, 7219–7228. [CrossRef] [PubMed]

63. Hoge, J.; Yan, I.; Jänner, N.; Schumacher, V.; Chalaris, A.; Steinmetz, O.M.; Engel, D.R.; Scheller, J.; Rose-John, S.; Mittrücker, H.W. IL-6 controls the innate immune response against listeria monocytogenes via classical IL-6 signaling. *J. Immunol.* **2013**, *190*, 703–711. [CrossRef] [PubMed]

64. Hurst, S.M.; Wilkinson, T.S.; McLoughlin, R.M.; Jones, S.; Horiuchi, S.; Yamamoto, N.; Rose-John, S.; Fuller, G.M.; Topley, N.; Jones, S.A. IL-6 and Its Soluble Receptor Orchestrate a Temporal Switch in the Pattern of Leukocyte Recruitment Seen during Acute Inflammation. *Immunity* **2001**, *14*, 705–714. [CrossRef]

65. Rabe, B.; Chalaris, A.; May, U.; Waetzig, G.H.; Seegert, D.; Williams, A.S.; Jones, S.A.; Rose-John, S.; Scheller, J. Transgenic blockade of interleukin 6 transsignaling abrogates inflammation. *Blood* **2008**, *111*, 1021–1028. [CrossRef]

66. Kraakman, M.J.; Kammoun, H.L.; Allen, T.L.; Deswaerte, V.; Henstridge, D.C.; Estevez, E.; Matthews, V.B.; Neill, B.; White, D.A.; Murphy, A.J.; et al. Blocking IL-6 trans-Signaling Prevents High-Fat Diet-Induced Adipose Tissue Macrophage Recruitment but Does Not Improve Insulin Resistance. *Cell Metab.* **2015**, *21*, 403–416. [CrossRef]

67. Desgeorges, A.; Gabay, C.; Silacci, P.; Novick, D.; Roux-Lombard, P.; Grau, G.; Dayer, J.M.; Vischer, T.; Guerne, P.A. Concentrations and origins of soluble interleukin 6 receptor-alpha in serum and synovial fluid. *J. Rheumatol.* **1997**, *24*, 1510–1516.

68. Braune, J.; Weyer, U.; Hobusch, C.; Mauer, J.; Brüning, J.C.; Bechmann, I.; Gericke, M. IL-6 Regulates M2 Polarization and Local Proliferation of Adipose Tissue Macrophages in Obesity. *J. Immunol.* **2017**, *198*, 2927–2934. [CrossRef]

69. Mauer, J.; Chaurasia, B.; Goldau, J.; Vogt, M.C.; Ruud, J.; Nguyen, K.D.; Theurich, S.; Hausen, A.C.; Schmitz, J.; Brönneke, H.S.; et al. Signaling by IL-6 promotes alternative activation of macrophages to limit endotoxemia and obesity-associated resistance to insulin. *Nat. Immunol.* **2014**, *15*, 423–430. [CrossRef]

70. Doganci, A.; Eigenbrod, T.; Krug, N.; De Sanctis, G.T.; Hausding, M.; Erpenbeck, V.J.; Haddad, E.B.; Schmitt, E.; Bopp, T.; Kallen, K.J.; et al. The IL-6R alpha chain controls lung CD4+ CD25+ Treg development and function during allergic airway inflammation in vivo. *J. Clin. Invest.* **2005**, *115*, 313–325. [CrossRef]

71. McLoughlin, R.M.; Jenkins, B.J.; Grail, D.; Williams, A.S.; Fielding, C.A.; Parker, C.R.; Ernst, M.; Topley, N.; Jones, S.A. IL-6 trans-signaling via STAT3 directs T cell infiltration in acute inflammation. *Proc. Natl. Acad. Sci. USA* **2005**, *102*, 9589–9594. [CrossRef] [PubMed]

72. Veldhoen, M.; Hocking, R.J.; Atkins, C.J.; Locksley, R.M.; Stockinger, B. Tgfbeta in the context of an inflammatory cytokine milieu supports de novo differentiation of IL-17-producing T cells. *Immunity* **2006**, *24*, 179–189. [CrossRef] [PubMed]

73. Bettelli, E.; Carrier, Y.; Gao, W.; Korn, T.; Strom, T.B.; Oukka, M.; Weiner, H.L.; Kuchroo, V.K. Reciprocal developmental pathways for the generation of pathogenic effector TH17 and regulatory T cells. *Nature* **2006**, *441*, 235–238. [CrossRef] [PubMed]

74. Elliott, M.J.; Maini, R.N.; Feldmann, M.; Kalden, J.R.; Antoni, C.; Smolen, J.S.; Leeb, B.; Breedveld, F.C.; Macfarlane, J.D.; Bijl, H. Randomised double-blind comparison of chimeric monoclonal antibody to tumour necrosis factor alpha (cA2) versus placebo in rheumatoid arthritis. *Lancet* **1994**, *344*, 1105–1110. [CrossRef]

75. Grau, G.E.; Maennel, D.N. TNF inhibition and sepsis—Sounding a cautionary note. *Nat. Med.* **1997**, *3*, 1193–1195. [CrossRef]

76. Feldmann, M. Development of anti-TNF therapy for rheumatoid arthritis. *Nat. Rev. Immunol.* **2002**, *2*, 364–371. [CrossRef]

77. Aggarwal, B.B.; Gupta, S.C.; Kim, J.H. Historical perspectives on tumor necrosis factor and its superfamily: 25 years later, a golden journey. *Blood* **2012**, *119*, 651–665. [CrossRef]

78. Ogata, A.; Hirano, T.; Hishitani, Y.; Tanaka, T. Safety and Efficacy of Tocilizumab for the Treatment of Rheumatoid Arthritis. *Clin. Med. Insights Arthritis Musculoskelet. Disord.* **2012**, *5*. [CrossRef]

79. Nowell, M.A.; Williams, A.S.; Carty, S.A.; Scheller, J.; Hayes, A.J.; Jones, G.W.; Richards, P.J.; Slinn, S.; Ernst, M.; Jenkins, B.J.; et al. Therapeutic targeting of IL-6 trans signaling counteracts STAT3 control of experimental inflammatory arthritis. *J. Immunol.* **2009**, *182*, 613–622. [CrossRef]

80. Ivashkiv, L.B.; Hu, X. The JAK/STAT pathway in rheumatoid arthritis: Pathogenic or protective? *Arthritis Rheum.* **2003**, *48*, 2092–2096. [CrossRef]

81. Walker, J.G.; Smith, M.D. The Jak-STAT pathway in rheumatoid arthritis. *J. Rheumatol.* **2005**, *32*, 1650–1653. [PubMed]

82. Burmester, G.R.; Lin, Y.; Patel, R.; van Adelsberg, J.; Mangan, E.K.; Graham, N.M.; van Hoogstraten, H.; Bauer, D.; Ignacio Vargas, J.; Lee, E.B. Efficacy and safety of sarilumab monotherapy versus adalimumab monotherapy for the treatment of patients with active rheumatoid arthritis (monarch): A randomised, double-blind, parallel-group phase iii trial. *Ann. Rheum. Dis* **2017**, *76*, 840–847. [CrossRef] [PubMed]

83. Gabay, C.; Emery, P.; Van Vollenhoven, R.; Dikranian, A.; Alten, R.; Pavelka, K.; Klearman, M.; Musselman, D.; Agarwal, S.; Green, J.; et al. Tocilizumab monotherapy versus adalimumab monotherapy for treatment of rheumatoid arthritis (ADACTA): A randomised, double-blind, controlled phase 4 trial. *Lancet* **2013**, *381*, 1541–1550. [CrossRef]

84. Sengupta, T.K.; Talbot, E.S.; Scherle, P.A.; Ivashkiv, L.B. Rapid inhibition of interleukin-6 signaling and Stat3 activation mediated by mitogen-activated protein kinases. *Proc. Natl. Acad. Sci. USA* **1998**, *95*, 11107–11112. [CrossRef]

85. Jain, N.; Zhang, T.; Fong, S.L.; Lim, C.P.; Cao, X. Repression of Stat3 activity by activation of mitogen-activated protein kinase (MAPK). *Oncogene* **1998**, *17*, 3157–3167. [CrossRef]

86. Tebbutt, N.C.; Giraud, A.S.; Inglese, M.; Jenkins, B.; Waring, P.; Clay, F.J.; Malki, S.; Alderman, B.M.; Grail, D.; Hollande, F.; et al. Reciprocal regulation of gastrointestinal homeostasis by SHP2 and STAT-mediated trefoil gene activation in gp130 mutant mice. *Nat. Med.* **2002**, *8*, 1089–1097. [CrossRef]

87. Atsumi, T.; Ishihara, K.; Kamimura, D.; Ikushima, H.; Ohtani, T.; Hirota, S.; Kobayashi, H.; Park, S.J.; Saeki, Y.; Kitamura, Y.; et al. A Point Mutation of Tyr-759 in Interleukin 6 Family Cytokine Receptor Subunit gp130 Causes Autoimmune Arthritis. *J. Exp. Med.* **2002**, *196*, 979–990. [CrossRef]

88. Nicholson, S.E.; De Souza, D.; Fabri, L.J.; Corbin, J.; Willson, T.A.; Zhang, J.G.; Silva, A.; Asimakis, M.; Farley, A.; Nash, A.D.; et al. Suppressor of cytokine signaling-3 preferentially binds to the shp-2-binding site on the shared cytokine receptor subunit gp130. *Proc. Natl. Acad. Sci. USA* **2000**, *97*, 6493–6498. [CrossRef]

89. Schmitz, J.; Weissenbach, M.; Haan, S.; Heinrich, P.C.; Schaper, F. SOCS3 Exerts Its Inhibitory Function on Interleukin-6 Signal Transduction through the SHP2 Recruitment Site of gp130. *J. Biol. Chem.* **2000**, *275*, 12848–12856. [CrossRef]

90. Shouda, T.; Yoshida, T.; Hanada, T.; Wakioka, T.; Oishi, M.; Miyoshi, K.; Komiya, S.; Kosai, K.I.; Hanakawa, Y.; Hashimoto, K.; et al. Induction of the cytokine signal regulator SOCS3/CIS3 as a therapeutic strategy for treating inflammatory arthritis. *J. Clin. Invest.* **2001**, *108*, 1781–1788. [CrossRef]

91. Mihara, M.; Moriya, Y.; Kishimoto, T.; Ohsugi, Y. Interleukin-6 (IL-6) induces the proliferation of synovial fibroblastic cells in the presence of soluble IL-6 receptor. *Rheumatology* **1995**, *34*, 321–325. [CrossRef] [PubMed]

92. Krause, A.; Scaletta, N.; Ji, J.D.; Ivashkiv, L.B. Rheumatoid arthritis synoviocyte survival is dependent on Stat3. *J. Immunol.* **2002**, *169*, 6610–6616. [CrossRef] [PubMed]

93. Tamura, T.; Udagawa, N.; Takahashi, N.; Miyaura, C.; Tanaka, S.; Yamada, Y.; Koishihara, Y.; Ohsugi, Y.; Kumaki, K.; Taga, T. Soluble interleukin-6 receptor triggers osteoclast formation by interleukin. *Proc. Natl. Acad. Sci. USA* **1993**, *90*, 11924–11928. [CrossRef] [PubMed]

94. Udagawa, N. Interleukin (IL)-6 induction of osteoclast differentiation depends on IL-6 receptors expressed on osteoblastic cells but not on osteoclast progenitors. *J. Exp. Med.* **1995**, *182*, 1461–1468. [CrossRef] [PubMed]

95. Kotake, S.; Sato, K.; Kim, K.J.; Takahashi, N.; Udagawa, N.; Nakamura, I.; Yamaguchi, A.; Kishimoto, T.; Suda, T.; Kashiwazaki, S. Interleukin-6 and soluble interleukin-6 receptors in the synovial fluids from rheumatoid arthritis patients are responsible for osteoclast-like cell formation. *J. Bone Miner. Res.* **1996**, *11*, 88–95. [CrossRef]

96. Palmqvist, P.; Persson, E.; Conaway, H.H.; Lerner, U.H. IL-6, leukemia inhibitory factor, and oncostatin M stimulate bone resorption and regulate the expression of receptor activator of NF-kappa B ligand, osteoprotegerin, and receptor activator of NF-kappa B in mouse calvariae. *J. Immunol.* **2002**, *169*, 3353–3362. [CrossRef]

97. Feng, W.; Liu, H.; Luo, T.; Liu, D.; Du, J.; Sun, J.; Wang, W.; Han, X.; Yang, K.; Guo, J.; et al. Combination of IL-6 and sIL-6R differentially regulate varying levels of RANKL-induced osteoclastogenesis through NF-kappab, ERK and JNK signaling pathways. *Sci. Rep.* **2017**, *7*, 41411. [CrossRef] [PubMed]

98. Nowell, M.A.; Richards, P.J.; Horiuchi, S.; Yamamoto, N.; Rose-John, S.; Topley, N.; Williams, A.S.; Jones, S.A. Soluble IL-6 receptor governs IL-6 activity in experimental arthritis: Blockade of arthritis severity by soluble glycoprotein130. *J. Immunol.* **2003**, *171*, 3202–3209. [CrossRef]

99. Richards, P.J.; Nowell, M.A.; Horiuchi, S.; McLoughlin, R.M.; Fielding, C.A.; Grau, S.; Yamamoto, N.; Ehrmann, M.; Williams, A.S.; Topley, N.; et al. Functional characterization of a soluble gp130 isoform and its therapeutic capacity in an experimental model of inflammatory arthritis. *Arthritis Rheum.* **2006**, *54*, 1662–1672. [CrossRef]

100. Swanson, C.D.; Akama-Garren, E.H.; Stein, E.A.; Petralia, J.D.; Ruiz, P.J.; Edalati, A.; Lindstrom, T.M.; Robinson, W.H. Inhibition of epidermal growth factor receptor tyrosine kinase ameliorates collagen-induced arthritis. *J. Immunol.* **2012**, *188*, 3513–3521. [CrossRef]

101. Issuree, P.D.A.; Maretzky, T.; McIlwain, D.R.; Monette, S.; Qing, X.; Lang, P.A.; Swendeman, S.L.; Park-Min, K.H.; Binder, N.; Kalliolias, G.D.; et al. iRHOM2 is a critical pathogenic mediator of inflammatory arthritis. *J. Clin. Invest.* **2013**, *123*, 928–932. [CrossRef] [PubMed]

102. McLoughlin, R.M.; Witowski, J.; Robson, R.L.; Wilkinson, T.S.; Hurst, S.M.; Williams, A.S.; Williams, J.D.; Rose-John, S.; Jones, S.A.; Topley, N. Interplay between IFN-gamma and IL-6 signaling governs neutrophil trafficking and apoptosis during acute inflammation. *J. Clin. Invest.* **2003**, *112*, 598–607. [CrossRef] [PubMed]

103. Fielding, C.A.; McLoughlin, R.M.; McLeod, L.; Colmont, C.S.; Najdovska, M.; Grail, D.; Ernst, M.; Jones, S.A.; Topley, N.; Jenkins, B.J. IL-6 regulates neutrophil trafficking during acute inflammation via STAT3. *J. Immunol.* **2008**, *181*, 2189–2195. [CrossRef] [PubMed]

104. Romano, M.; Sironi, M.; Toniatti, C.; Polentarutti, N.; Fruscella, P.; Ghezzi, P.; Faggioni, R.; Luini, W.; Van Hinsbergh, V.; Sozzani, S.; et al. Role of IL-6 and Its Soluble Receptor in Induction of Chemokines and Leukocyte Recruitment. *Immunity* **1997**, *6*, 315–325. [CrossRef]

105. Schumacher, N.; Schmidt, S.; Schwarz, J.; Dohr, D.; Lokau, J.; Scheller, J.; Garbers, C.; Chalaris, A.; Rose-John, S.; Rabe, B. Circulating Soluble IL-6R but Not ADAM17 Activation Drives Mononuclear Cell Migration in Tissue Inflammation. *J. Immunol.* **2016**, *197*, 3705–3715. [CrossRef] [PubMed]

106. Fielding, C.A.; Jones, G.W.; McLoughlin, R.M.; McLeod, L.; Hammond, V.J.; Uceda, J.; Williams, A.S.; Lambie, M.; Foster, T.L.; Liao, C.T.; et al. Interleukin-6 signaling drives fibrosis in unresolved inflammation. *Immunity* **2014**, *40*, 40–50. [CrossRef] [PubMed]

107. Twohig, J.P.; Cardus Figueras, A.; Andrews, R.; Wiede, F.; Cossins, B.C.; Derrac Soria, A.; Lewis, M.J.; Townsend, M.J.; Millrine, D.; Li, J.; et al. Activation of naive CD4$^+$ T cells re-tunes stat1 signaling to deliver unique cytokine responses in memory CD4$^+$ T cells. *Nat. Immunol.* **2019**, *20*, 458–470. [CrossRef] [PubMed]

108. Greenhill, C.J.; Rose-John, S.; Lissilaa, R.; Ferlin, W.; Ernst, M.; Hertzog, P.J.; Mansell, A.; Jenkins, B.J. IL-6 trans-signaling modulates TLR4-dependent inflammatory responses via STAT3. *J. Immunol.* **2011**, *186*, 1199–1208. [CrossRef]

109. Barkhausen, T.; Tschernig, T.; Rosenstiel, P.; van Griensven, M.; Vonberg, R.P.; Dorsch, M.; Mueller-Heine, A.; Chalaris, A.; Scheller, J.; Rose-John, S.; et al. Selective blockade of interleukin-6 trans-signaling improves survival in a murine polymicrobial sepsis model. *Crit. Care Med.* **2011**, *39*, 1407–1413. [CrossRef]

110. Grivennikov, S.; Karin, E.; Terzic, J.; Mucida, D.; Yu, G.Y.; Vallabhapurapu, S.; Scheller, J.; Rose-John, S.; Cheroutre, H.; Eckmann, L.; et al. IL-6 and Stat3 Are Required for Survival of Intestinal Epithelial Cells and Development of Colitis-Associated Cancer. *Cancer Cell* **2009**, *15*, 103–113. [CrossRef]

111. Arndt, P.G.; Strahan, B.; Wang, Y.; Long, C.; Horiuchi, K.; Walcheck, B. Leukocyte ADAM17 Regulates Acute Pulmonary Inflammation. *PLoS ONE* **2011**, *6*, e19938. [CrossRef] [PubMed]

112. Nicolaou, A.; Zhao, Z.; Northoff, B.H.; Sass, K.; Herbst, A.; Kohlmaier, A.; Chalaris, A.; Wolfrum, C.; Weber, C.; Steffens, S.; et al. Adam17 Deficiency Promotes Atherosclerosis by Enhanced TNFR2 Signaling in Mice. *Arterioscler. Thromb. Vasc. Biol.* **2017**, *37*, 247–257. [CrossRef] [PubMed]

113. Mishra, H.K.; Johnson, T.J.; Seelig, D.M.; Walcheck, B. Targeting ADAM17 in leukocytes increases neutrophil recruitment and reduces bacterial spread during polymicrobial sepsis. *J. Leukoc. Biol.* **2016**, *100*, 999–1004. [CrossRef] [PubMed]

114. Mishra, H.K.; Long, C.; Bahaie, N.S.; Walcheck, B. Regulation of cxcr2 expression and function by a disintegrin and metalloprotease-17 (ADAM17). *J. Leukoc. Biol.* **2015**, *97*, 447–454. [CrossRef]

115. Tang, J.; Zarbock, A.; Gomez, I.; Wilson, C.L.; Lefort, C.T.; Stadtmann, A.; Bell, B.; Huang, L.C.; Ley, K.; Raines, E.W. ADAM17-dependent shedding limits early neutrophil influx but does not alter early monocyte recruitment to inflammatory sites. *Blood* **2011**, *118*, 786–794. [CrossRef]

116. Teague, T.K.; Schaefer, B.C.; Hildeman, D.; Bender, J.; Mitchell, T.; Kappler, J.W.; Marrack, P. Activation-Induced Inhibition of Interleukin 6–Mediated T Cell Survival and Signal Transducer and Activator of Transcription 1 Signaling. *J. Exp. Med.* **2000**, *191*, 915–926. [CrossRef]

117. Takeda, K.; Kaisho, T.; Yoshida, N.; Takeda, J.; Kishimoto, T.; Akira, S. Stat3 activation is responsible for IL-6-dependent T cell proliferation through preventing apoptosis: Generation and characterization of T cell-specific Stat3-deficient mice. *J. Immunol.* **1998**, *161*, 4652–4660. [CrossRef]

118. Atreya, R.; Mudter, J.; Finotto, S.; Müllberg, J.; Jostock, T.; Wirtz, S.; Schütz, M.; Bartsch, B.; Holtmann, M.; Becker, C.; et al. Blockade of interleukin 6 trans signaling suppresses T-cell resistance against apoptosis in chronic intestinal inflammation: Evidence in Crohn disease and experimental colitis in vivo. *Nat. Med.* **2000**, *6*, 583–588. [CrossRef]

119. Ina, K.; Itoh, J.; Fukushima, K.; Kusugami, K.; Yamaguchi, T.; Kyokane, K.; Imada, A.; Binion, D.G.; Musso, A.; West, G.A.; et al. Resistance of Crohn's disease T cells to multiple apoptotic signals is associated with a Bcl-2/Bax mucosal imbalance. *J. Immunol.* **1999**, *163*, 1081–1090.

120. Mitsuyama, K.; Matsumoto, S.; Rose-John, S.; Suzuki, A.; Hara, T.; Tomiyasu, N.; Handa, K.; Tsuruta, O.; Funabashi, H.; Scheller, J.; et al. STAT3 activation via interleukin 6 trans-signalling contributes to ileitis in SAMP1/Yit mice. *Gut* **2006**, *55*, 1263–1269. [CrossRef]

121. Ito, H.; Takazoe, M.; Fukuda, Y.; Hibi, T.; Kusugami, K.; Andoh, A.; Matsumoto, T.; Yamamura, T.; Azuma, J.; Nishimoto, N.; et al. A pilot randomized trial of a human anti-interleukin-6 receptor monoclonal antibody in active Crohn's disease. *Gastroenterology* **2004**, *126*, 989–996. [CrossRef] [PubMed]

122. Papaioannou, A.I.; Mazioti, A.; Kiropoulos, T.; Tsilioni, I.; Koutsokera, A.; Tanou, K.; Nikoulis, D.J.; Georgoulias, P.; Zakynthinos, E.; Gourgoulianis, K.I.; et al. Systemic and airway inflammation and the presence of emphysema in patients with COPD. *Respir. Med.* **2010**, *104*, 275–282. [CrossRef] [PubMed]

123. Ruwanpura, S.M.; McLeod, L.; Miller, A.; Jones, J.; Bozinovski, S.; Vlahos, R.; Ernst, M.; Armes, J.; Bardin, P.G.; Anderson, G.P.; et al. Interleukin-6 promotes pulmonary emphysema associated with apoptosis in mice. *Am. J. Respir. Cell Mol. Biol.* **2011**, *45*, 720–730. [CrossRef] [PubMed]

124. Ruwanpura, S.M.; McLeod, L.; Dousha, L.F.; Seow, H.J.; Alhayyani, S.; Tate, M.D.; Deswaerte, V.; Brooks, G.D.; Bozinovski, S.; MacDonald, M.; et al. Therapeutic targeting of the IL-6 trans-signaling/mechanistic target of rapamycin complex 1 axis in pulmonary emphysema. *Am. J. Respir. Crit. Care Med.* **2016**, *194*, 1494–1505. [CrossRef]

125. Jevnikar, Z.; Östling, J.; Ax, E.; Calvén, J.; Thörn, K.; Israelsson, E.; Oberg, L.; Singhania, A.; Lau, L.C.; Ward, J.A.; et al. Epithelial IL-6 trans-signaling defines a new asthma phenotype with increased airway inflammation. *J. Allergy Clin. Immunol.* **2019**, *143*, 577–590. [CrossRef]

126. Papiris, S.A.; Tomos, I.P.; Karakatsani, A.; Spathis, A.; Korbila, I.; Analitis, A.; Kolilekas, L.; Kagouridis, K.; Loukides, S.; Karakitsos, P.; et al. High levels of IL-6 and IL-8 characterize early-on idiopathic pulmonary fibrosis acute exacerbations. *Cytokine* **2018**, *102*, 168–172. [CrossRef]

127. Pedroza, M.; Schneider, D.J.; Karmouty-Quintana, H.; Coote, J.; Shaw, S.; Corrigan, R.; Molina, J.G.; Alcorn, J.L.; Galas, D.; Gelinas, R.; et al. Interleukin-6 contributes to inflammation and remodeling in a model of adenosine mediated lung injury. *PLoS ONE* **2011**, *6*, e22667. [CrossRef]

128. Saito, F.; Tasaka, S.; Inoue, K.I.; Miyamoto, K.; Nakano, Y.; Ogawa, Y.; Yamada, W.; Shiraishi, Y.; Hasegawa, N.; Fujishima, S.; et al. Role of Interleukin-6 in Bleomycin-Induced Lung Inflammatory Changes in Mice. *Am. J. Respir. Cell Mol. Biol.* **2008**, *38*, 566–571. [CrossRef]

129. Le, T.T.T.; Karmouty-Quintana, H.; Melicoff, E.; Le, T.T.T.; Weng, T.; Chen, N.Y.; Pedroza, M.; Zhou, Y.; Davies, J.; Philip, K.; et al. Blockade of IL-6 Trans Signaling Attenuates Pulmonary Fibrosis. *J. Immunol.* **2014**, *193*, 3755–3768. [CrossRef]

130. Pedroza, M.; Le, T.T.; Lewis, K.; Karmouty-Quintana, H.; To, S.; George, A.T.; Blackburn, M.R.; Tweardy, D.J.; Agarwal, S.K. Stat-3 contributes to pulmonary fibrosis through epithelial injury and fibroblast-myofibroblast differentiation. *FASEB J.* **2016**, *30*, 129–140. [CrossRef]

131. Moodley, Y.P.; Scaffidi, A.K.; Misso, N.L.; Keerthisingam, C.; McAnulty, R.J.; Laurent, G.J.; Mutsaers, S.E.; Thompson, P.J.; Knight, D.A. Fibroblasts Isolated from Normal Lungs and Those with Idiopathic Pulmonary Fibrosis Differ in Interleukin-6/gp130-Mediated Cell Signaling and Proliferation. *Am. J. Pathol.* **2003**, *163*, 345–354. [CrossRef]

132. Pechkovsky, D.V.; Prele, C.M.; Wong, J.; Hogaboam, C.M.; McAnulty, R.J.; Laurent, G.J.; Zhang, S.S.M.; Selman, M.; Mutsaers, S.E.; Knight, D.A. STAT3-Mediated Signaling Dysregulates Lung Fibroblast-Myofibroblast Activation and Differentiation in UIP/IPF. *Am. J. Pathol.* **2012**, *180*, 1398–1412. [CrossRef] [PubMed]

133. Balkwill, F.; Mantovani, A.; Balkwill, F. Inflammation and cancer: Back to Virchow? *Lancet* **2001**, *357*, 539–545. [CrossRef]

134. Fischer, M.; Goldschmitt, J.; Peschel, C.; Brakenhoff, J.P.G.; Kallen, K.J.; Wollmer, A.; Grötzinger, J.; Rose-John, S. A bioactive designer cytokine for human hematopoietic progenitor cell expansion. *Nat. Biotechnol.* **1997**, *15*, 142–145. [CrossRef] [PubMed]

135. Mitsuyama, K.; Sata, M.; Rose-John, S. Interleukin-6 trans-signaling in inflammatory bowel disease. *Cytokine Growth Factor Rev.* **2006**, *17*, 451–461. [CrossRef] [PubMed]

136. Becker, C.; Fantini, M.; Schramm, C.; Lehr, H.; Wirtz, S.; Burg, J.; Strand, S.; Kiesslich, R.; Huber, S.; Galle, P.; et al. TGF-beta suppresses tumor progression in colon cancer by inhibition of IL-6 trans-signaling. *Immunity* **2004**, *21*, 491–501. [CrossRef] [PubMed]

137. Becker, C.; Fantini, M.C.; Wirtz, S.; Nikolaev, A.; Lehr, H.A.; Galle, P.R.; Rose-John, S.; Neurath, M.F. IL-6 signaling promotes tumor growth in colorectal cancer. *Cell Cycle* **2005**, *4*, 217–220. [CrossRef] [PubMed]

138. Matsumoto, S.; Hara, T.; Mitsuyama, K.; Yamamoto, M.; Tsuruta, O.; Sata, M.; Scheller, J.; Rose-John, S.; Kado, S.; Takada, T. Essential roles of IL-6 trans-signaling in colonic epithelial cells, induced by the IL-6/soluble-IL-6 receptor derived from lamina propria macrophages, on the development of colitis-associated premalignant cancer in a murine model. *J. Immunol.* **2010**, *184*, 1543–1551. [CrossRef]

139. Bollrath, J.; Phesse, T.J.; Von Burstin, V.A.; Putoczki, T.; Bennecke, M.; Bateman, T.; Nebelsiek, T.; Lundgren-May, T.; Canli, O.; Schwitalla, S.; et al. gp130-Mediated Stat3 Activation in Enterocytes Regulates Cell Survival and Cell-Cycle Progression during Colitis-Associated Tumorigenesis. *Cancer Cell* **2009**, *15*, 91–102. [CrossRef]

140. Spehlmann, M.E.; Manthey, C.F.; Dann, S.M.; Hanson, E.; Sandhu, S.S.; Liu, L.Y.; Abdelmalak, F.K.; Diamanti, M.A.; Retzlaff, K.; Scheller, J.; et al. Trp53 deficiency protects against acute intestinal inflammation. *J. Immunol.* **2013**, *191*, 837–847. [CrossRef]

141. Srivatsa, S.; Paul, M.C.; Cardone, C.; Holcmann, M.; Amberg, N.; Pathria, P.; Diamanti, M.A.; Linder, M.; Timelthaler, G.; Dienes, H.P.; et al. EGFR in Tumor-Associated Myeloid Cells Promotes Development of Colorectal Cancer in Mice and Associates with Outcomes of Patients. *Gastroenterology* **2017**, *153*, 178–190. [CrossRef] [PubMed]

142. Rebouissou, S.; Amessou, M.; Couchy, G.; Poussin, K.; Imbeaud, S.; Pilati, C.; Izard, T.; Balabaud, C.; Bioulac-Sage, P.; Zucman-Rossi, J. Frequent in-frame somatic deletions activate gp130 in inflammatory hepatocellular tumours. *Nature* **2009**, *457*, 200–204. [CrossRef] [PubMed]

143. Yamada, Y.; Mori, H. Multistep carcinogenesis of the colon in Apc$^{min/+}$ mouse. *Cancer Sci.* **2007**, *98*, 6–10. [CrossRef] [PubMed]

144. Schmidt, S.; Schumacher, N.; Schwarz, J.; Tangermann, S.; Kenner, L.; Schlederer, M.; Sibilia, M.; Linder, M.; Altendorf-Hofmann, A.; Knösel, T.; et al. ADAM17 is required for EGF-R–induced intestinal tumors via IL-6 trans-signaling. *J. Exp. Med.* **2018**, *215*, 1205–1225. [CrossRef] [PubMed]

145. Sibilia, M.; Kroismayr, R.; Lichtenberger, B.M.; Natarajan, A.; Hecking, M.; Holcmann, M. The epidermal growth factor receptor: From development to tumorigenesis. *Differentiation* **2007**, *75*, 770–787. [CrossRef]

146. Van Emburgh, B.O.; Sartore-Bianchi, A.; Di Nicolantonio, F.; Siena, S.; Bardelli, A. Acquired resistance to EGFR-targeted therapies in colorectal cancer. *Mol. Oncol.* **2014**, *8*, 1084–1094. [CrossRef]

147. Lesina, M.; Kurkowski, M.U.; Ludes, K.; Rose-John, S.; Treiber, M.; Klöppel, G.; Yoshimura, A.; Reindl, W.; Sipos, B.; Akira, S.; et al. Stat3/Socs3 Activation by IL-6 Transsignaling Promotes Progression of Pancreatic Intraepithelial Neoplasia and Development of Pancreatic Cancer. *Cancer Cell* **2011**, *19*, 456–469. [CrossRef]

148. Cressman, D.E.; Greenbaum, L.E.; DeAngelis, R.A.; Ciliberto, G.; Furth, E.E.; Poli, V.; Taub, R. Liver Failure and Defective Hepatocyte Regeneration in Interleukin-6-Deficient Mice. *Science* **1996**, *274*, 1379–1383. [CrossRef]

149. Peters, M.; Blinn, G.; Jostock, T.; Schirmacher, P.; Büschenfelde, K.M.Z.; Galle, P.R.; Rose–John, S. Combined interleukin 6 and soluble interleukin 6 receptor accelerates murine liver regeneration. *Gastroenterology* **2000**, *119*, 1663–1671. [CrossRef]

150. Schirmacher, P.; Peters, M.; Ciliberto, G.; Blessing, M.; Lotz, J.; Meyer zum Büschenfelde, K.H.; Rose-John, S. Hepatocellular Hyperplasia, Plasmacytoma Formation, and Extramedullary Hematopoiesis in Interleukin (IL)-6/Soluble IL-6 Receptor Double-Transgenic Mice. *Am. J. Pathol.* **1998**, *153*, 639–648. [CrossRef]

151. Maione, D.; Di Carlo, E.; Li, W.; Musiani, P.; Modesti, A.; Peters, M.; Rose-John, S.; Della Rocca, C.; Tripodi, M.; Lazzaro, D.; et al. Coexpression of IL-6 and soluble IL-6R causes nodular regenerative hyperplasia and adenomas of the liver. *EMBO J.* **1998**, *17*, 5588–5597. [CrossRef] [PubMed]
152. Maeda, S.; Kamata, H.; Luo, J.L.; Leffert, H.; Karin, M. Ikkbeta couples hepatocyte death to cytokine-driven compensatory proliferation that promotes chemical hepatocarcinogenesis. *Cell* **2005**, *121*, 977–990. [CrossRef] [PubMed]
153. Naugler, W.E.; Sakurai, T.; Kim, S.; Maeda, S.; Kim, K.; Elsharkawy, A.M.; Karin, M. Gender Disparity in Liver Cancer Due to Sex Differences in MyD88-Dependent IL-6 Production. *Science* **2007**, *317*, 121–124. [CrossRef] [PubMed]
154. Bergmann, J.; Muller, M.; Baumann, N.; Reichert, M.; Heneweer, C.; Bolik, J.; Lucke, K.; Gruber, S.; Carambia, A.; Boretius, S.; et al. IL-6 trans-signaling is essential for the development of hepatocellular carcinoma in mice. *Hepatology* **2017**, *65*, 89–103. [CrossRef] [PubMed]
155. Brooks, G.D.; McLeod, L.; Alhayyani, S.; Miller, A.; Russell, P.A.; Ferlin, W.; Rose-John, S.; Ruwanpura, S.; Jenkins, B. IL-6 trans-signaling promotes KRAS-driven lung carcinogenesis. *Cancer Res.* **2016**, *76*, 866–876. [CrossRef]
156. Saad, M.I.; Alhayyani, S.; McLeod, L.; Yu, L.; Alanazi, M.; Deswaerte, V.; Tang, K.; Jarde, T.; Smith, J.A.; Prodanovic, Z.; et al. ADAM17 selectively activates the IL-6 trans-signaling/erk mapk axis in kras-addicted lung cancer. *EMBO Mol. Med.* **2019**, *11*, e9976. [CrossRef]
157. Xu, P.; Derynck, R. Direct activation of tace-mediated ectodomain shedding by p38 map kinase regulates egf receptor-dependent cell proliferation. *Mol. Cell* **2009**, *37*, 551–566. [CrossRef]
158. Saad, M.I.; McLeod, L.; Yu, L.; Ebi, H.; Ruwanpura, S.; Sagi, I.; Rose-John, S.; Jenkins, B.J. The ADAM17 protease promotes tobacco smoke carcinogen-induced lung tumourigenesis. *Carcinogenesis* **2019**. [CrossRef]
159. Schuett, H.; Oestreich, R.; Waetzig, G.H.; Annema, W.; Luchtefeld, M.; Hillmer, A.; Bavendiek, U.; Von Felden, J.; Divchev, D.; Kempf, T.; et al. Transsignaling of Interleukin-6 Crucially Contributes to Atherosclerosis in Mice. *Arterioscler. Thromb. Vasc. Biol.* **2012**, *32*, 281–290. [CrossRef]
160. Zhang, H.; Neuhöfer, P.; Song, L.; Rabe, B.; Lesina, M.; Kurkowski, M.U.; Treiber, M.; Wartmann, T.; Regnér, S.; Thorlacius, H.; et al. IL-6 trans-signaling promotes pancreatitis-associated lung injury and lethality. *J. Clin. Invest.* **2013**, *123*, 1019–1031. [CrossRef]
161. Linardou, H.; Dahabreh, I.J.; Kanaloupiti, D.; Siannis, F.; Bafaloukos, D.; Kosmidis, P.; Papadimitriou, C.A.; Murray, S. Assessment of somatic k-RAS mutations as a mechanism associated with resistance to EGFR-targeted agents: A systematic review and meta-analysis of studies in advanced non-small-cell lung cancer and metastatic colorectal cancer. *Lancet Oncol.* **2008**, *9*, 962–972. [CrossRef]
162. Tobin, N.P.; Foukakis, T.; De Petris, L.; Bergh, J. The importance of molecular markers for diagnosis and selection of targeted treatments in patients with cancer. *J. Intern. Med.* **2015**, *278*, 545–570. [CrossRef] [PubMed]
163. Pietrantonio, F.; Vernieri, C.; Siravegna, G.; Mennitto, A.; Berenato, R.; Perrone, F.; Gloghini, A.; Tamborini, E.; Lonardi, S.; Morano, F.; et al. Heterogeneity of acquired resistance to anti-EGFR monoclonal antibodies in patients with metastatic colorectal cancer. *Clin. Cancer Res.* **2017**, *23*, 2414–2422. [CrossRef] [PubMed]

Review

Activation of STAT3 and STAT5 Signaling in Epithelial Ovarian Cancer Progression: Mechanism and Therapeutic Opportunity

Chin-Jui Wu [1,†], Vignesh Sundararajan [2,†], Bor-Ching Sheu [1], Ruby Yun-Ju Huang [3,4] and Lin-Hung Wei [1,*]

[1] Department of Obstetrics & Gynecology, National Taiwan University Hospital, College of Medicine, National Taiwan University, Taipei 10002, Taiwan; cjwu00@gmail.com (C.-J.W.); bcsheu@ntu.edu.tw (B.-C.S.)

[2] Cancer Science Institute of Singapore, National University of Singapore, Center for Translational Medicine, Singapore 117599, Singapore; csivsun@nus.edu.sg

[3] Department of Obstetrics and Gynaecology, National University of Singapore, Singapore 119077, Singapore; rubyhuang@ntu.edu.tw

[4] School of Medicine, College of Medicine, National Taiwan University, Taipei 10051, Taiwan

* Correspondence: weilh1966@gmail.com; Tel.: +886-2-2312-3456 (ext. 71570); Fax: +886-2-2311-4965

† These authors contributed equally to this work.

Received: 27 October 2019; Accepted: 17 December 2019; Published: 19 December 2019

Abstract: Epithelial ovarian cancer (EOC) is the most lethal of all gynecologic malignancies. Despite advances in surgical and chemotherapeutic options, most patients with advanced EOC have a relapse within three years of diagnosis. Unfortunately, recurrent disease is generally not curable. Recent advances in maintenance therapy with anti-angiogenic agents or Poly ADP-ribose polymerase (PARP) inhibitors provided a substantial benefit concerning progression-free survival among certain women with advanced EOC. However, effective treatment options remain limited in most recurrent cases. Therefore, validated novel molecular therapeutic targets remain urgently needed in the management of EOC. Signal transducer and activator of transcription-3 (STAT3) and STAT5 are aberrantly activated through tyrosine phosphorylation in a wide variety of cancer types, including EOC. Extrinsic tumor microenvironmental factors in EOC, such as inflammatory cytokines, growth factors, hormones, and oxidative stress, can activate STAT3 and STAT5 through different mechanisms. Persistently activated STAT3 and, to some extent, STAT5 increase EOC tumor cell proliferation, survival, self-renewal, angiogenesis, metastasis, and chemoresistance while suppressing anti-tumor immunity. By doing so, the STAT3 and STAT5 activation in EOC controls properties of both tumor cells and their microenvironment, driving multiple distinct functions during EOC progression. Clinically, increasing evidence indicates that the activation of the STAT3/STAT5 pathway has significant correlation with reduced survival of recurrent EOC, suggesting the importance of STAT3/STAT5 as potential therapeutic targets for cancer therapy. This review summarizes the distinct role of STAT3 and STAT5 activities in the progression of EOC and discusses the emerging therapies specifically targeting STAT3 and STAT5 signaling in this disease setting.

Keywords: ovarian cancer; STAT3; STAT5

---

1. Introduction

Epithelial ovarian cancer (EOC) is a heterogenous entity comprised of different histotypes [1] with unique molecular features and clinical characteristics that influence chemosensitivity and the probability of survival [2]. Cytoreductive surgery and platinum/taxane-combination chemotherapy remain the mainstay of primary treatment for advance-staged diseases, resulting in the initial remission in up to 80% of EOC patients. Nonetheless, approximately 75% of patients with advanced diseases develop recurrence within three years of diagnosis [3], which is generally not curable, owing to the development of chemoresistance. In this respect, EOC remains the most lethal of all gynecologic malignancies, and the relative survival rates at ten years for stage III and IV disease are 23% and 8%, respectively [4]. Molecular therapeutics targeting the angiogenesis and DNA damage repair pathways have provided significant steps forward in the management of certain EOC patients. However, there is still a clear unmet need for most patients with recurrence. Identifying novel molecular therapeutic targets relevant to the disease progression is thus highly anticipated.

Signal transducers and activators of transcription (STATs) belong to a family of cytoplasmic transcription factors that communicate signals from the cell membrane to the nucleus. The STAT family includes seven structurally and functionally related proteins: STAT1, STAT2, STAT3, STAT4, STAT5A, STAT5B, and STAT6 [5]. STATs have essential roles in fundamental processes, including sustaining proliferation, evading apoptosis, inducing angiogenesis, promoting invasion, and suppressing anti-tumor immunity [6,7]. Upon the binding of cytokines or growth factors to cognate receptors on the cell surface, STATs are tyrosine phosphorylated, particularly by Janus kinase (JAK), Abelson (Abl) kinase or SRC kinase families. Phosphorylated STAT (pY-STAT) dimer undergoes conformation change and shuttles into the nucleus, functioning as a transcription factor. Each STAT protein appears to have distinct physiologic functions in the development, differentiation, and immune response (For a comprehensive review, please see [7]). In particular, mice carrying homozygous deletion for *STAT5* (STAT5a$^{-/-}$5b$^{-/-}$) which later turned out to be hypermorphic *STAT5* deletion mice lacking the N-domains were infertile, with defects in the differentiation of functional corpora lutea, disrupting ovarian development [8].

STATs activation is rapid and transient under most physiological conditions. Notably, compelling evidence indicates that constitutive activation of STAT proteins, particularly STAT3 and STAT5, plays a critical role in oncogenic transformation. Clinically, aberrant activation of STAT3 and, to some extent, STAT5, is associated with both solid and hematopoietic cancers [9–12]. Accumulating evidence has indicated that downregulating STAT3/STAT5 mitigates the malignant behavior of cancer cells [13], highlighting the potential of STAT3/STAT5 as a therapeutic target. Collecting data has shown the role of STAT3 in the disease progression mechanism of EOC. Compared to normal or benign ovarian tumors, pY-STAT3/pY-STAT5 protein expression was significantly higher in the malignant EOC tissues, supporting its role in ovarian carcinogenesis [14,15]. The activation of the STAT3 pathway and the increase in pY-STAT3 (Tyr705) expression directly correlated with higher clinical stage, lower degree of differentiation, presence of lymph node metastasis, and more reduced survival in EOC [15–17]. Moreover, elevated pY-STAT3 expression in the omentum was associated with poor survival in patients with high-grade EOC. The activation and translocation of pY-STAT3 to the nucleus was observed in 29–58% of all EOC histotypes [13,16]. Specifically, nuclear pY-STAT3 expression was found to be associated with clear cell and serous carcinoma [17]. The activation of STAT3 pathway was, in particular, related to overall survival in ovarian clear cell carcinoma patients [16]. In recurrent diseases, levels of STAT3 activation were doubled, indicating that STAT3 activation could be directly associated with disease relapse [18]. Moreover, one study suggests STAT5 may be related to RELA (p65 subunit of NF-kB) and carboplatin resistance in EOC [19].

2. Regulation of STAT3/STAT5 Activation in EOC

Constitutive activation of STAT3/STAT5 has been identified in a wide range of human cancers. As a primary event during malignant transformation, somatic *STAT3* and *STAT5* driver mutations have been identified in hematopoietic neoplasms. For example, somatic mutations in the *STAT3* gene were found in 40% of granular lymphocytic leukemia and T-cell lymphoma patients, with recurrent mutations located on the gene segment encoding the SH2 domain, which mediates STAT3 dimerization and activation [20,21]. Also, a small percentage of granular lymphocytic leukemia patients harbored *STAT5B* mutations, resulting in increased transcriptional activity and phosphorylation [22]. However, genetic mutations that result in hyperactivated *STAT3/STAT5* have not been reported in EOC [23]. In EOC, constitutive upregulation of *STATs* in the absence of somatic mutations is primarily contributed through persistent Tyr phosphorylation signals. In general, *STAT3/STAT5* are activated in response to the binding of numerous cytokines, hormones, and growth factors to their receptors and by the activation of intracellular kinases, mostly in case of tyrosine phosphorylation by the four JAK family kinases. Typically, STAT3/STAT5 are activated by phosphorylation on critical residues (STAT3 Tyr residue 705 and Ser727 (ERK, JNK, and other stress kinases); STAT5A Tyr residue 694, Ser725 (CDK8) and Ser779 (PAK1/2) and STAT5B Tyr residue 699 and Ser730 (CDK8)) [9]. The JAK-STAT signaling in EOC can be further modulated by various molecular pathways, as summarized in Figure 1.

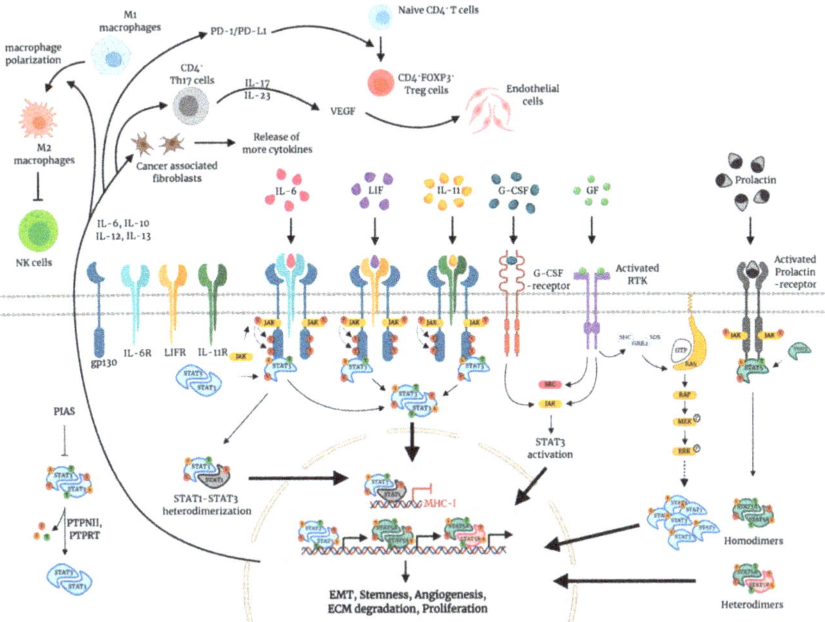

Figure 1. Signal transducers and activators of transcription (STAT)3 and STAT5 signaling in epithelial ovarian cancer (EOC) and tumor microenvironment. Distinct families of cytokines such as Interleukins (IL-6,IL-11) and leukemia inhibitory factor (LIF) bind to their homodimeric cognate receptors IL-6R, IL-11R and LIFR respectively, and share a signal-transducing receptor gp130. Janus kinase (JAK) phosphorylate gp130 to enable docking and phosphorylation of STAT3 at Tyrosine (symbol Y or Tyr) residue 705. Tyrosine phosphorylation of STAT3 can also be mediated by activation of other oncogenic proteins including growth factor (GF)-mediated receptor Tyrosine kinase (RTK) activation, granulocyte colony-stimulating factor (G-CSF)-mediated activation, SRC and RAS/MEK/ERK pathway. Phosphorylated STAT3 dynamically undergo dimerization and nuclear translocation to trigger STAT3-mediated transcription of target genes. Binding of Prolactin to its receptor facilitate JAK-mediated phosphorylation of STAT5A and STAT5B at Tyr residue 694 and Tyr residue 699, respectively, leading to homodimerization or heterodimerization before nuclear translocation for target gene activation. STAT3 and STAT5 signaling in cancer cells release more cytokines into tumor microenvironment that generate a plethora of immune-compromising functions (highlighted in the main text). Figure created with Biorender.com.

2.1. IL-6 Pathway

Upstream to the JAK-STAT signaling, the IL-6 family of cytokines is critical for signal transduction. The binding of IL-6 family cytokines to the ligand-binding subunit of gp130 initiates its homodimerization, activates JAK to bind to gp130, and then triggers downstream signaling cascades [24]. In EOC, The IL-6 family of cytokines is one of the major families of immunoregulatory cytokines. IL-6, LIF, and IL-11 secreted in the EOC tumor microenvironment function in concert to induce ovarian cancer cell JAK-STAT signaling [25–27].

2.2. TP53 Mutation

Notably, Tyr phosphorylation of JAK2 can be diminished by wild type but not mutant p53 in EOC cells [28]. Since there is a high frequency of somatic *TP53* mutation in high-grade serous carcinoma (HGSC), this suggests that STAT3 phosphorylation and its DNA binding activity can be modulated by the p53 status in HGSC.

2.3. Lipid Metabolism Pathway

STAT3 is also regulated by the cellular redox state controlling the transcription of genes related to the invasive phenotype. Pathways related to lipid metabolism known to affect the redox state thus are intriguing mechanisms since obesity is known to be a risk factor for poor EOC survival [29]. For example, leukotriene B_4, a lipoxygenase pathway metabolite, together with their cognate receptor leukotriene B_4 receptor 2 (BLT2) contribute to the generation of reactive oxygen species (ROS) in EOC cells, which results in the activation of JAK and the STAT3-MMP2 cascade [30]. Leptin is an adipokine that exerts its activity through the membrane receptor, the obesity receptor (OB-R). The overexpression of OB-R in EOC significantly correlates with poor progression-free survival [31]. It has been shown that Leptin/OB-R signaling may phosphorylate STAT3 through the activation of JAK in EOC cells [32]. Recently, CD97, a member of the EGF-TM7 family of G-protein coupled receptors, is known to be expressed in several malignancies, including EOC. The interaction between overexpressed CD97 and its ligand CD55 activates JAK2/STAT3 signaling and confers an invasive cell phenotype of EOC [33].

2.4. Receptor Tyrosine Kinases (RTKs)

Several RTKs have been suggested to modulate the JAK-STAT signaling pathway. Vascular endothelial growth factor (VEGF), a key mediator of angiogenesis in EOC, induces pY-STAT3 through the binding of VEGF receptor 2 (VEGFR2) in EOC cells [34]. The expressions of VEGF, VEGFR1, and VEGFR2 are significantly correlated with pY-STAT3/pY-STAT5 in EOC [14]. Epidermal growth factor (EGF) is known either to directly activate JAK2/STAT3 signaling pathway or to induce the secondary mediator-like response mediated by the IL-6 axis in an autocrine manner in EOC [35]. EGFR which is overexpressed in EOC is significantly correlated with pY-STAT3 [17]. Oncogenic RAS and RAF mutations are prevalent in EOC, and these driver mutations are highly associated with aberrant ERK signaling, resulting in uncontrolled cellular proliferation [36,37]. Mechanistically, activated RTKs stimulate RAS activation, which then activates RAF. RAF phosphorylates and activated MEK, which in turn activates ERK through phosphorylation. Aberrant phosphorylation of ERK mediates phosphorylation of S727-STAT3 (Figure 1) and contributes to cisplatin resistance in certain EOC cell lines [38].

2.5. Alternative Cytokine or Non-Receptor Tyrosine Kinases

Granulocyte-colony stimulation factor (G-CSF), a commonly used cytokine receptor to aid hematopoietic recovery following chemotherapy, activates the JAK/STAT pathway in EOC through its cognate receptor, the G-CSFR, which is predominantly expressed in HGSC [39]. Furthermore, non-receptor tyrosine kinase SRC family members can alternatively tyrosine phosphorylate STAT proteins, and SRC is overexpressed and activated in late staged EOC. SRC family tyrosine kinases are essential for STAT activation in EOC, especially during metastasis. It is known that activated STAT localizes not only to nuclei, but also to focal adhesions. SRC family kinases induce pY-STAT3 and contribute to strong interactions between the pY-STAT3 and the focal adhesion complex [13]. In particular, c-SRC, a member of SRC family kinases, is primarily involved in hypoxia-triggered intracellular signaling. Under hypoxia, activation of c-SRC induces nuclear pY-STAT3 and enhances the binding ability of STAT3 to Hypoxia-inducible factor 1-alpha (HIF-1α), which contributes to chemoresistance in EOC [40]. Moreover, recombinant human erythropoietin (rhEpo) has been shown to bind to an alternative Epo receptor, EphB4, to activate the SRC-STAT pathway, triggering tumor growth, and resulting in decreased survival of EOC [41]. This further supports the notion that the use of rhEpo to treat anemia in cancer patients can compromise their overall survival [42].

2.6. Other Gene Regulatory Mechanisms

A dysregulated transcriptional control of STAT3 by miRNAs was reported in EOC. Increased expression of miR551b-3p, which is resultant from the frequent q26.2 amplification in HGSC, interacted with the *STAT3* promoter by recruiting RNA-pol-II and TWIST1 to turn on *STAT3* transcription [43]. The upregulation of STAT3 is subsequently required for miR551b-induced growth and metastasis of EOC cells. Enhancer of zeste homolog 2 (EZH2), a member of the polycomb repressor complex 2, functions primarily to promote transcriptional silencing via histone 3 on lysine 27 trimethylation (H3K27me3) and plays an essential role in EOC progression [44,45]. Interestingly, upon Tyr372 phosphorylation of EZH2 by protein kinase A, pY372-EZH2 efficiently interacted with STAT3 protein. This non-canonical EZH2 interaction reduced cellular levels of STAT3 and altered STAT3 activation, leading to the downregulation of downstream target IL-6R in EOC [46]. Furthermore, tyrosine phosphorylation of EZH2 by JAK3 in lymphoid neoplasia was reported to promote dissociation of the polycomb repressive complex 2 (PRC2) complex leading to decreased global H3K27me3 levels while it switches EZH2 to a transcriptional activator [47], but studies in EOC are lacking. Moreover, EZH2 can also interact with the STAT5 N-terminal oligomerisation domain, which was shown to be essential for B-cell acute leukemic transformation silencing the kappa light chain expression [48]. Such studies postulate that STAT3/5 interaction with EZH2 could be a valuable target interface for future therapeutic intervention [49], but future studies with EOC model systems are needed.

3. The Function of STAT3 and STAT5 in EOC

3.1. Apoptosis

Under normal physiological conditions, apoptosis is a process that governs programmed cell death and is responsible for the elimination of cells in normal tissues, to maintain tissue homeostasis. This process is commonly observed in several self-renewing tissues, including the gut, bone marrow, and skin, to accommodate newly generated cells daily. Emphasizing the crucial role of apoptosis in normal cell turnover, apoptosis remains to be one of the critical cell processes that are highly dysregulated during cancer progression. Evasion of apoptosis by cancer cells often results in excessive tumor growth, metastatic spread, and even causing resistance to cancer treatment. Apoptosis primarily occurs through two pathways. In the extrinsic pathway, binding of external death ligands to death receptors triggers the cascade, resulting in caspase-mediated cell death. While, in the intrinsic pathway, internal stimuli such as DNA damage, cellular stress triggers the activation of proapoptotic factors such as B-cell lymphoma 2 (BCL-2) family members that resemble three functional groups: inhibitors of apoptosis (BCL-2, BCL-$_{XL}$, BCL-W, Mcl-1, BCL-B, and A1), promoters of apoptosis (BAX, BAK, and BOK); and regulatory proteins (BAD, BIK, BID, HrK, BIM, BMF, NOXA, and PUMA) [50]. Notably, tumor suppressor p53 is a transcription factor with primary pro-apoptotic function, and it remains one of the most commonly mutated genes in human cancers, underlining the significance of apoptosis deregulation in cancer [51].

Activated STAT3/5 proteins target specific inhibitors of apoptosis that mainly act on the intrinsic pathway to suspend cell death in cancer cells. Inhibition of STAT3 signaling leads to apoptosis of ovarian clear cell carcinoma and decreased BCL-2 expression [52]. G-CSF and Leptin can both activate STAT3 phosphorylation, and both can promote increased cellular BCL-2 levels, thereby protecting EOC cells against apoptosis [39,53]. RELA and STAT5 proteins transcriptionally activate the expression of *Bcl-x$_L$* through direct promoter binding in ovarian, non-small-cell lung carcinoma and transformed leukemia cells, as well as render chemoresistance in carboplatin resistant ovarian cancer cell lines [19,54,55]. Also, cell survival signaling diminishes the effectiveness of chemotherapy, which contributes towards the acquisition and development of chemoresistance in cancer cells. In paclitaxel-resistant ovarian cancer cells, blocking of STAT3 activity suppresses STAT3 downstream antiapoptotic regulatory genes *BCL2L1*, *MCL1*, and *BIRC5*, which increases paclitaxel sensitivity in paclitaxel-resistant ovarian cancer cells in vitro [56]. Similarly, abrogation of constitutive STAT3 activity shows significant reductions in the expression of the BCL-2, BCL-$_{XL}$, and Survivin protein, which circumvents cisplatin resistance in EOC [57].

MicroRNAs contribute to diverse physiological and pathological processes by involving in the regulation of several essential biological processes. An array of microRNAs induces apoptosis in cancer cells through JAK/STAT signaling in multiple cancer types. miR-17-5p, miR-133b, miR-134, miR-13, miR-147, miR-182, miR-204, miR-874 are among a few miRNAs that require either STAT3 or STAT5 to induce apoptosis. In EOC, miRNA-519a promotes apoptosis of SKOV3 cells by directly targeting the 3'UTR of STAT3, which results in a decrease of the mRNA and protein expression levels of STAT3, Mcl-1, and BCL-$_{XL}$ [58] (For a detailed list, see Table A1).

3.2. Proliferation

EOC is one of the few cancer types that is partially regulated by hormones. IL-6-induced STAT3 phosphorylation levels were found to be increased in cells treated with follicle-stimulating hormone (FSH), luteinizing hormone (LH), 17β-estradiol or estrogen and it facilitated cell proliferation in human ovarian surface epithelial (HOSE) and ovarian cancer (OVCA) cell lines [59]. In the case of ovarian cancer cell proliferation, Leptin, a hormone secreted by adipose tissue interlinks obesity and EOC progression [31,60]. At the molecular level, Leptin simultaneous increased pY-STAT3 (Tyr705) and nuclear localization of Estrogen Receptor (ER)α, and induced STAT3 binding to ERα that resulted in a significant increase in cell proliferation and migration [32,53,61]. Also, Leptin promoted ovarian cancer proliferation by induced phosphorylation of STAT5 in SKOV3 and A2780 cells, which mediated leptin-induced expression of miR-182 and miR-96 and subsequent inhibition of Forkhead box O3 [62]. Prolactin-mediated pY-STAT5 activation complexes with tumor suppressor *BRCA1* in the nucleus, and this complex hinders the transcription of cell-cycle inhibitor p21, leading to increased proliferation [63]. Treatment of metastatic ovarian carcinoma cell line CaOV-3 with Leptin receptor blockers: SHLA and Lan-2 resulted in a predominantly inhibitory effect on STAT3 phosphorylation and downregulated cell proliferation through blocking *Cyclin D1* and *E2F1* [64].

Expression of phosphorylated STAT3 coupled with Ki-67 expression, was found to be increased in primary human ovarian carcinoma, particularly in patients with high nuclear expression of pY-STAT3 exhibited poor prognosis [15]. EOC cells cultured under hypoxic conditions revealed higher levels of pY-STAT3 (Tyr705); however, with knockdown of STAT3 expression, the proliferation rate of cancer cells was significantly reduced [65]. Alternatively, the hyperbaric oxygenation method (a systemic increase of dissolved oxygen delivery in serum) substantially decreased pY-STAT3 (Tyr705) levels, along with decreased tumor volume in a murine xenograft model [66]. Recently, it was found that activated STAT3 may deploy specific microRNAs to promote ovarian cancer cell proliferation and to generate associated phenotypes. For instance, high STAT3 levels in SKOV3 cells increased the levels of oncogenic microRNA-216a, which in turn directly targeted tumor suppressor phosphatase and tensin homolog (PTEN) expression and rendered cisplatin resistance [67].

Interference of STAT3 activity has been employed as a strategy to hinder ovarian cancer cell proliferation. In vitro and in vivo studies with ovarian cancer cell lines have shown that siRNA and shRNA mediated STAT3 depletion, downregulated the expression of Cyclin D1, Survivin as well as reduced tumor weight [68,69]. Another study has shown that siRNA mediated STAT3 downregulation suppressed SKOV3 cell growth and arrested the cell cycle in the G1 phase [70]. Several studies have highlighted that plant-derived phytochemicals, such as Pterostilbene, Cryptotanshinone and Curcumin, suppressed STAT3 activation, thereby reduced cancer cell proliferation and have been proposed as possible adjuvants of conventional chemotherapy [71–73].

3.3. Angiogenesis

In general, tumor outgrowth exceeding 2 mm in diameter must gain access to an increased supply of oxygen and nutrients. These requirements are fulfilled through angiogenesis, a process that involves the formation of new blood vessels from the existing vasculature. Therefore, tumor angiogenesis is a hallmark of cancer that promotes tumor progression and metastasis. Tumor cells cause an angiogenic switch by secreting pro-angiogenic proteins and/or repressing the expression of anti-angiogenic factors. Most notably, vascular endothelial growth factor (VEGF) and its receptors are crucial in instigating angiogenesis in tumor cells, exert vascular permeability activity and stimulate cell migration in macrophage and endothelial cell populations. Activated STAT3 and STAT5 regulate the expression of VEGF and increased angiogenesis in a variety of cancer, including EOC [14,74,75]. Immunohistochemical staining of patient-derived ovarian epithelial carcinoma tissues identified significant overlap of expression between VEGF, pY-STAT3, and pY-STAT5 in ovarian carcinoma cells, compared to the benign and normal group [14]. In ovarian clear cell carcinoma, immunohistochemical analysis of primary tumors showed high nuclear expression of pY-STAT3 and HIF1α [76]. It is known that IL-6 signals via STAT3, not only directly induces the transcription of VEGF, but also activates expression of downstream gene HIF1α, where HIF1α is a paramount transcription factor controlling VEGF expression [77]. These studies indicate an IL-6/STAT3/HIF1α/VEGF autocrine activation loop in EOC, especially clear cell carcinoma histotype. (For a detailed list, see Table A2).

Targeting STAT3 activation in EOC directly or indirectly affected angiogenesis. Extract from cinnamon was a potent inhibitor of VEGF secretion, inhibited the expression and phosphorylation of STAT3 and AKT, suppressed HIF1α expression as well as significantly reduced tumor growth and blood vessel formation in mice models [78]. 3,3′-Diindolylmethane (DIM; an active metabolite found in cruciferous vegetables) treatment blocked IL-6-induced STAT3 phosphorylation, attenuated angiogenesis by suppressing HIF1α and VEGF expression in SKOV3 cells [79]. In the same study, oral administration of DIM in combination with intraperitoneal administration of cisplatin treatment reduced tumor volume by 65% due to downregulation of pY-STAT3, STAT3, and Mcl-1 levels with simultaneously increased cleavage of Caspase 3 and poly ADP-ribose polymerase (PARP) activity. It is important to note that clinical trials of EOC patients treated with the VEGF inhibitor aflibercept revealed more reduced survival rates in patients harboring high levels of circulating IL-6, indicating that IL-6/STAT3 activation in tumor cells may provide a survival gain during anti-VEGF treatment [80]. The fact that STAT3 is critically involved in the VEGF pathway and tumor angiogenesis indicates that blockade of STAT3 is a therapeutic target to heighten an effective antiangiogenic treatment in EOC.

3.4. Tumor Progression and Metastasis

Cancer metastasis is a multi-step process that occurs when a selected group of cells detach from the primary tumor, utilize blood or lymphatic system to gain access into circulation, and instigate tumor colonies in secondary organ sites. The metastatic pattern of EOC differs from that of most other epithelial malignant diseases. After the direct extension, EOC most frequently disseminates via the transcoelomic route, forming diffuse multifocal intraperitoneal nodes and malignant ascites [81]. Accumulation of pY-STAT3 expression coupled with loss of protein inhibitor of activated STAT3 (PIAS3) in fallopian tube secretory epithelial cells displayed common peritoneal metastatic nodules that eventually led to the progression of HGSC. *AKT2*, one of the most frequent amplicon alterations in HGSC, activates PKM2–STAT3/NF-κB axis, ultimately results in increased migratory and invasive potential of EOC cells as well as promoting lung metastasis in mouse models [82].

Several hypotheses have been put forward to comprehend the multistep process of metastasis. Yet, recapitulation of an embryonic cell differentiation program known as epithelial to mesenchymal transition (EMT) has gained recent advances to support its inevitable role. Briefly stating, EMT is a process when epithelial cells lose their signature characteristics like cell-cell adhesion to acquire a mesenchymal phenotype with migratory and invasive behavior [83]. Apart from its role in EOC metastasis, EMT can promote tumor initiation, stemness, and chemoresistance in EOC [84]. STAT proteins have been well documented to play a role in fostering EMT in several cancer entities [85–87]. In EOC, inhibition of EGFR or STAT3 activity reduced N-cadherin, Vimentin expression, decreased colony-forming ability, cell motility, and migration behavior [35,38]. Also, extracellular heat shock protein (HSP90) promotes the binding of STAT3 to the TWIST1 promoter and thereby increasing TWIST1 transcription, and these effects were diminished after HSP90 inhibitor treatment [88].

Another critical aspect of EMT that strongly associates with STAT proteins is their role in self-renewal. STAT3 and STAT5 have essential roles in regulating cancer stem cells (CSCs) of EOC [89]. Over 90% of ascites cells derived from EOC patients showed activated pY-STAT3 signaling (Tyr705), which increased the migratory potential of EOC cells, and it also increased widespread peritoneal metastasis [90,91]. Tumor spheroids isolated from the ascites of recurrent EOC patients are enriched with tumor cells overexpressing STAT3 compared with cells isolated from the ascites of chemotherapy-naïve patients [92]. EOC spheroids serve as the vehicle for ovarian cancer cell dissemination in the peritoneal cavity and represent a significant impediment in the efficacy of chemotherapy agents [93]. Mechanistically, STAT3 signaling regulates ovarian CSCs by targeting miR-92a/DKK1 and subsequently activating Wnt/β-catenin signaling [18]. Therefore, inhibition of STAT3 signaling effectively eliminates the formation of the metastatic niche, and it suppresses cancer cell persistence after chemotherapy.

Extracellular matrix (ECM) is a complex network of the microenvironment, stabilized through structural proteins such as Laminins, Collagens, and Fibronectins, that holds epithelial-derived cancer cells, endothelial and stromal cells in proximity. Degradation of ECM is a characteristic feature of disseminating cancer cells to determine whether metastatic tumors form or not. Activation of STAT3 in several cancers contributed towards degradation of ECM, primarily mediated through increased activity of matrix-degrading Matrix Metallo-Proteinases (MMPs). In EOC cells, ligand-mediated and stress hormone (norepinephrine) mediated activation of STAT3, via nuclear translocation, induced MMP-2 and MMP-9 expressions and siRNA mediated STAT3 silencing declined MMPs release, denoting the direct regulation of MMPs by STAT3 [30,94–96]. ECM component hyaluronan (HA) associated with CD44 to mediate nuclear Nanog-STAT3 interaction, that activated EMT, increased cell migration and invasion and also triggered the expression of MDR1, which rendered multi-drug resistance to EOC cells [97,98]. A BET bromodomain and extra-terminal inhibitor, i-BET151, reduced viability migration and invasion of EOC cells as well as decreased expression of MMP-2 and MMP-9 expression, highlighting the potential usage of this inhibitor as a treatment strategy [99]. siRNA mediated inhibition of IL-6R or anti-human IL-6R antibody (tocilizumab) reduced pY-STAT3 and MMP-9 expression levels suggesting that interference of STAT3 signaling in ovarian clear cell carcinoma could be an effective therapeutic strategy [100]. (For a detailed list, see Tables A3, A4 and A6).

Growing evidence emphasizes the role of nano-sized secretory vesicles known as "exosomes" released from the plasma membrane to facilitate intercellular communication during various physiological processes such as antigen presentation or exchange of membrane proteins. In particular, exosomes are highly implicated in mediating cancer metastasis, angiogenesis, and drug resistance [101,102]. EOC-derived exosomes changed the morphology of human peritoneal mesothelial cells to a mesenchymal phenotype, through CD44 internalization, and blockage of exosome release, which suppressed ovarian cancer invasion [103]. Besides, hypoxia is known to drive excessive exosome release from cancer cells in several tumor types [104]. Under hypoxic conditions, EOC cells showed activated STAT3 levels, increased the release of exosomes to promote proliferation, which occurs through altering proteins of the Rab family [105]. A microfluidic ChIP-based exosomes isolation method confirmed elevated pY-STAT3 levels in exosomes isolated from high-grade serous ovarian cancer cell lines and patients, implying that vesicles secreted from cancer patients have activated STAT3 signaling that could foster cancer metastasis [101].

4. Tumor Microenvironment

Emerging evidence reveals the presence of vibrant multicellular interactions between malignant and non-malignant somatic cells, which generate a complex milieu referred to as the "tumor microenvironment" [106,107]. The tumor microenvironment encompasses a multitude of distinct cell types, which primarily includes endothelial cells, infiltrating immune cells (tumor-associated neutrophils, T and B lymphocytes, natural killer cells, tumor-associated macrophages, mast cells) and cancer-associated fibroblasts. These non-malignant somatic cells are often engaged in delivering tumor-promoting factors and are involved in every aspect of tumorigenesis as well as in metastasis. Cellular communications between these cell types are governed by the copious release of cytokines, growth factors, inflammatory and matrix remodeling components from the tumor bulk (Figure 1). Although the immune cells of the tumor microenvironment are efficient enough to combat and evade the cancer cells, these cells are instead confined and manipulated by cancer cells to promote their growth and distribution, ultimately influencing the patient's clinical outcome [108]. Henceforth, understanding the biology of the tumor-host hostile environment becomes inevitable for improving treatment strategies. Several lines of evidence have highlighted the role of STAT3/5 activation in non-malignant somatic cells of the EOC tumor microenvironment and the crosstalk between tumor and host stromal cells.

4.1. Non-Immune Stromal Cells

In the tumor microenvironment, non-immune stromal cells comprise endothelial cells, pericytes, fibroblasts, and mesenchymal stem cells. STAT3 pathway facilitates crosstalk between tumor cells and endothelial cells that mediates pro-angiogenic signaling. Conditioned media obtained from EOC cells stimulated rapid, transient STAT3 phosphorylation and nuclear translocation in cord blood CD34$^+$ progenitor cells, as well as initiated early capillary-like structure formation in human microvascular endothelial cells [109,110]. CD133$^+$ ovarian cancer stem cells cultured on a cell culture matrix formed fluid-conducting tube networks activated NF-κB and STAT3 signal pathways, through autocrine chemokine (C-C motif) ligand 5 (CCL5) upregulation promoting differentiation into endothelial cells [111]. Implantation of HeyA8 and SKOV3 cells into the peritoneal cavity of female nude mice secreted significant levels of IL-6 and soluble IL-6 receptor (sIL-6R), activated STAT3 signaling and facilitated migration of endothelial cells [112].

Cancer-associated fibroblasts (CAFs) are a heterogeneous cell population that is usually identified by their high expression of mesenchymal markers including Vimentin, fibroblast-secreted protein-1 (FSP-1), α-smooth muscle actin (αSMA), and fibroblast activation protein (FAP). CAFs favor tumor growth through increased secretion of cytokines, metabolites, and ECM molecules, thus promoting angiogenesis and they interfere with antitumor immune response [113]. CAFs mediate EOC cell proliferation through NF-κB and JNK signaling activation. Moreover, they release VEGF to promote tumor angiogenesis [114]. CAFs induce EMT in EOC cell lines through IL-6/JAK2/STAT3 pathway, which subsequently renders resistance to paclitaxel treatment [115]. Also, CAFs release chemokine CCL5 to increase STAT3 and Akt phosphorylation that mediate cisplatin resistance in EOC cells [116,117].

In the EOC tumor microenvironment, the intra-abdominal fat deposition is regarded as a significant source of cytokines and has been shown to stimulate growth and promote EOC metastasis [118]. Conditioned media obtained from subcutaneous and visceral fat-derived adipose stromal cells enhanced growth and migration of EOC cells, through activation of JAK2/STAT3 signaling pathway [119]. Studies demonstrating the distinct regulation of STAT3 in other non-immune stromal cell populations are quite preliminary and need further in-depth investigations.

4.2. Immune Function

T lymphocytes play a crucial role in stimulating the adaptive immune response to target the expanding tumor mass. Depending on the nature of the tissue, T-cells can either be pro- or anti-tumorigenic. Tumor-infiltrating T-cells exist as distinct populations, and here we highlight a few main types of T-cell players that are regulated through STAT3/5 signaling.

CD4$^+$ Th17 cell population in EOC is identified to be pro-tumorigenic. EOC cells secrete cytokines, including IL-1β, IL-6, and IL-23, which are involved in the expansion of CD4$^+$ Th17 cell population (Figure 1), through activation of STAT3 [120,121]. Phosphorylation and acetylation profiles of STAT3 determine the differentiation and polarization of CD4$^+$ Th17 cells that form the majority of tumor infiltrated T cells [122,123]. An increase in the Th17 cell population sustains the secretion of more cytokines (IL-17, IL-23) that eventually stimulates the release of angiogenesis factors VEGF and TGFβ in fibroblasts and endothelial cells. One study indicated exosomes transfer miRNAs, including miR-29a-3p and miR-21-5p, to synergistically induce the Treg/Th17 cell imbalance through direct targeting of STAT3 in CD4$^+$ T cells [124]. Thus, active STAT3 in CD4$^+$ T cells generates an inflammatory environment around the budding tumor aids its growth by stimulating angiogenesis and abrogates antitumor response. (For a detailed list, see Table A5).

Another subpopulation of CD4$^+$ cells known as regulatory T (Treg) cells that are involved in the dampening of antitumor activity in the tumor microenvironment. Accumulation of Treg cells in tumors and ascites of patients with EOC showed reduced survival rate [125]. STAT3 and STAT5 are known to bind to the promoter and increase the transcription of FOXP3, which is essential for the conversion of naive CD4$^+$ T cells into CD4$^+$CD25$^+$ FOXP3$^+$ Treg cells [126,127]. In response to IL-6 and IL-12 stimulation, STAT3 also positively regulates immune check point proteins such as programmed cell death receptor 1 (PD-1), programmed death-ligand (PD-L1) expression through direct promoter binding in CD4$^+$ T cells. PD-1/PD-L1 axis promotes the differentiation of CD4$^+$ T cells into FOXP3$^+$ Treg cells (Figure 1), and EOC patients with high PD-1 expression showed poor prognosis [128]. Also, STAT3 induces the expression of IL-10 and TGF-β1, through direct promoter binding, which ultimately restrains the tumor-suppressive role of CD8$^+$ effector T-cell function and dendritic cells [129,130]. Another important finding revealed that STAT3 sequesters STAT1 through cytoplasmic heterodimerization and hinders STAT1 mediated transcription of major histocompatibility complex (MHC) class I genes [131]. Reduced surface expression of MHC class-I genes in cancer cells pose unfavorable presentation of cancer cells to CD8$^+$ effector cells and thereby diminishing tumor immunosurveillance.

CD8$^+$ T lymphocytes in the tumor microenvironment play a crucial role in combating cancer cells through antigen-specific blockage of immunosuppressive Treg cells. This blockage is regulated by STAT3 through the secretion of IFN-γ [132]. Using the Mx1-Cre-loxP mice model, Kortylewski et al. showed that STAT3$^{-/-}$ CD8$^+$ T cells can secrete increased IFN-γ levels, which displayed markedly enhanced tumor-suppressive functions of dendritic cells, natural killer cells and neutrophils [133]. Moreover, a study reported that when tumor supernatants (from EOC cell lines OVCAR3, CAOV3 and SKOV3) were co-cultured with CD8$^+$ T cells reduced STAT5 phosphorylation which diminished CD8$^+$ T cell proliferation [134]. EOC patients with increased levels of CD8$^+$ T cells infiltration and a high CD8$^+$/Treg ratio showed favorable prognosis [135,136].

4.3. Tumor-Associated Macrophages

Tumor-associated macrophages (TAMs) form a prominent component of the inflammatory tumor microenvironment, which primarily performs a pro-tumorigenic role, including the release of proangiogenic cytokines, matrix proteases and growth factors, suppression of adaptive immunity, self-renewal and chemotherapeutic resistance of cancer stem cells [137]. siRNA-mediated STAT3 inhibition in macrophages increased IL-6 and IL-10 secretion, induced Cyclin-D1 mediated cell proliferation in co-cultured SKOV3 cells [138]. Other studies have shown that the interaction of TAMs and ovarian cancer stem-like cells (OCSLCs) being involved in the occurrence, recurrence, and multidrug resistance of ovarian cancers [139]. The OCSLCs co-cultured with macrophages also induces SKOV3 cell stemness via IL-8/STAT3 signaling [140].

Macrophages exist as a heterogeneous population, that can be broadly classified into two main phenotypes: classically activated M1 macrophages and alternatively activated M2 macrophages. The polarization of M1 or M2 macrophages is highly dependent on the cytokines involved in their activation. M1 macrophages are activated by cytokines such as interferon-γ (IFNγ), IL-12 and they play a crucial role in recruiting T cells to the tumor microenvironment. M1 macrophages enhanced immune response in order to restrain tumor development. On the contrary, M2 macrophages are activated by Th2 cytokines (IL-4, IL-10, and IL-13). M2 macrophages interfered with the antitumor activity of T cells or NK cells. EOC patients with a high number of M2 CD68$^+$ and CD163$^+$ macrophages were associated with advanced stage and poor progression-free and overall survival [141]. Accordingly, ovarian cancer patients with high M1 to M2 ratio of TAMs were associated with extended survival rate, indicating the pro-tumorigenic role of M2 macrophages [142]. Moreover, ascites from EOC patients polarized macrophages toward the M2 phenotype through STAT3 activation, while non-EOC did not activate STAT3. Also, the expression of transmembrane protein B4-T7 in TAMs, but not in ovarian carcinoma cells suppressed the T-cell immune response, inversely correlated with patient survival [143]. These studies have demonstrated that STAT3 mediated polarization of macrophages is crucial in determining the pro-tumorigenic nature of tumor microenvironment. (For a detailed list, see Table A5).

5. Targeting of the STAT3/STAT5 Signaling Pathway in EOC

In principle, targeting the constitutive activation of STAT3/5 could be approached in various ways. These strategies include (1) inhibiting the upstream of STAT3 activation pathway (for example, ligands antagonist, JAK or SRC inhibitor) (2) blocking the SH2 domain that inhibit STAT dimerization (3) inhibiting the translocation of phosphorylated dimerized STAT3 into the nucleus (4) inhibit the binding of activated dimerized STAT to DNA [144,145]. Table 1 lists the names and properties of agents targeting JAK/STAT that are FDA approved or in clinical development.

The anti-human-IL-6 antibody Siltuximab substantially reduced nuclear pY-STAT3 expression in IL-6-producing intraperitoneal EOC xenografts [146]. Siltuximab significantly inhibited the tumor growth of IGROV1 intraperitoneal xenograft, accompanied by reductions in angiogenesis and macrophage infiltration. In phase 2 clinical trial, single-agent Siltuximab showed modest response rate (5.6%) with reductions in serum CCL2, CXCL12, and VEGF, in recurrent, platinum-resistant diseases [146]. In phase 1 dose-finding trial, the anti-IL-6R monoclonal antibody Tocilizumab combined with carboplatin/pegylated liposomal doxorubicin and interferon α2b in patients with advanced EOC was studied. A functional blockade of IL-6 signaling by Tocilizumab decreased pY-STAT3 by both myeloid cells and the different populations of T cells, leading to increased production of the tumor-immunity promoting cytokine secretion of IL-12 and IL-1β [147]. The clinical benefit was observed in 85% (17/20) in a patient with recurrent EOC. The revamping of immunity support for better tumor control by targeting IL-6/STAT3 signaling in EOC.

JAK inhibitors have been developed in recent years. The utilization of JAK inhibitors has been attempted to interfere with JAK-mediated STAT3/5 activation and evaluate therapeutic efficacy in various EOC models. Ruxolitinib, a potent oral JAK1, and JAK2 inhibitor has been approved by FDA to treat myelofibrosis in 2011. Ruxolitinib significantly inhibited pY-STAT3 in EOC cells. Single-agent Ruxolitinib suppresses EOC tumor growth in mice. More importantly, Ruxolitinib substantially enhances the anti-tumor activity of chemotherapy agents, including paclitaxel, cisplatin, carboplatin, doxorubicin, and topotecan in EOC cells. In an OVCAR-8 murine model, Ruxolitinib synergistically increased tumor control by paclitaxel [147]. A phase I/II clinical trial of EOC has been conducted with combination with or without paclitaxel and carboplatin since 2016 and is under recruiting now (ClinicalTrials.gov Identifier: NCT02713386). The FDA also approved other JAK inhibitors, tofacitinib (pan JAK with preferentially selectively JAK 3/1) in 2016 and baricitinib (selectively JAK 1/2) in 2018 for the treatment of rheumatoid arthritis [148]. However, there are, as yet, no studies or trials mentioning the relationship between these two and gynecologic cancers. Itacitinib, another JAK 1/2 inhibitor, has been tested in non-small cell lung cancer, lymphoma, and pancreatic cancer. A previous terminated trial showed an acceptable safety profile in combination with nab-paclitaxel and gemcitabine [149]. It has been tested in multiple cancers including endometrial cancer and breast cancer in the proceeding phase Ib/II clinical trial (ClinicalTrials.gov Identifier: NCT02646748). In the preclinical stage, paclitaxel and Momelotinib (ATP-competitive inhibitor of JAK1/2, previously named CYT387) inhibited JAK2/STAT3 activation, reduced tumor burden, abrogated cancer stem cell expressions and prolonged disease-free survival in a murine xenograft model [89,150]. MLS-2384 is a synthetic 6-bromoindirubin derivative with potent dual JAK/SRC inhibitory activity in EOC cells. In vitro, MLS-2384 suppresses the viability of A2780 cells, which is consistent with the inhibition of phosphorylation of JAK2, SRC, and STAT3 [151]. Although these JAK inhibitors are widely used in medicine, the side effect of JAK inhibitors is significant, ranging from immunosuppression to organ toxicity. AZD1480, a selective JAK2 inhibitor, reduced tumor growth, decreased peritoneal dissemination and diminished ascites production in a murine model for advanced EOC [152]. However, the previous phase I study of the solid tumor was terminated due to neuropsychiatric side effects [153].

Table 1. Agents targeting the Janus kinase/signal transducers and activators of transcription (JAK/STAT) that are U.S. Food and Drug Administration (FDA) approved or in clinical development.

Category	Drug	Target	Phase	Histology	Remark	References
Upstream ligand	Siltuximab (CNTO 328)	IL-6	Phase II	19 serous, 1 CC	• Modest RR as monotherapy • Reduce angiogenesis and macrophage infiltration • Clinial approval for Castleman's disease.	[146]
	Tociclizumab	IL-6R	Phase I	15 serous, 3 CC, 3 EM	• A comination with interferon-alpha showed clinical benefit and immune cell change, but not powered. • Proceed to phase II in EOC. • Clinical approval for RA, juvanile idiopathic arthritis, CAR T-cell-induced cytokine-release syndrome	[154]
Upstream kinase inhibitors	Ruxolitinib (INC424)	JAK 1, JAK 2	Phase I/II	Not specified	• In EOC, synergistically increase tumor control with multiple chemotherapy agents (see context) • Phase I/II study, combined with paclitaxel and carboplatin in advanced EOC • Clinical approval for MF, PCV	[147] NCT02713386.
	Tofacitinib (CP690550)	JAK3JAK1JAK2	Not tested in EOC	-	• Clinical approval for RA	
	Baricitinib (INCB28050)	JAK 1, JAK 2	Not tested in EOC	-	• Clinical approval for RA	
	Itacitinib (ABT494)	JAK 1, JAK 2	Not tested in EOC	-	• Phase Ib/II study, combined with pembrolizumab to test multiple cancers including endometrial cancer and breast cancer • EMA orphan approval for GVHD.	NCT02646748
	Momelotinib (CYT387)	JAK 1, JAK 2	Preclinical	Serous and CC cell lines	• Reduce tumor burden in combination with paclitaxel compared with paclitaxel only arm in xenograft mice. • No significant effect was noted on CYT378 only treatment compared with control. • Phase III trials and FDA Fast-track designation in MF patients	[150,155]
	MLS2384	c-SRC, JAK 2, TYK2	Preclinical	Not specified	• Suppress cell viability in diverse human cancer cell lines in vitro.	[151]

Abbreviations: AE; adverse effect, CAR; chimeric antigen receptor, CC; clear cell, DFS; disease-free survival, EM; endometrioid, EMA; European medicines agency, GVHD; graft-versus-host disease, IP; intraperitoneal, MF; myelofibrosis, PCV; polycythemia vera; RR; response rate; TYK 2; tyrosine kinase 2.EOC: epithelial ovarian cancer, RA; rheumatic arthritis, IL; interleukin.

Pilot studies focusing on analogs from natural compounds have been developed to inhibit STAT3 activation. The anticancer analogs from curcumin, diarylidenyl piperidone (DAP) derivatives such as HO-3867, HO-4200, HO-4318, inhibit STAT3 activity and sensitize drug-resistant ovarian carcinoma and clear cell carcinoma cells to paclitaxel or cisplatin [52,156–158]. Among them, HO-3867 is the most studied and effective in selectively targeting STAT3 and inhibiting EOC tumor growth. Specifically, HO-3867 effectively blocked ascites-mediated activation of STAT3 in EOC cells, inhibited invasion, and metastasis in a murine orthotopic EOC model [91]. Additionally, HO-3867 targets hypoxia-stimulated pY-STAT3 (Tyr705) via the ubiquitin–proteasome degradation pathway, leading to tumor growth suppression in xenograft mice and the downregulation of proteins involved in cell survival, proliferation, and angiogenesis [159]. Furthermore, HO-4200 and HO-4318 significantly inhibited fatty acid synthase and pY-STAT3 and decreased the expression of STAT3 target proteins in primary platinum-resistant EOC cells, resulting in decreased expression of Ki67 and VEGF in ex-vivo human tumor specimens [156].

Direct inhibition of STAT3 activity through the interference of SH2 domain dimerization, nucleus transportation, or DNA binding is another highly investigated therapeutic strategy. However, most of the molecules proposed are still in preliminary stage of development for EOC treatment. Small-molecules such as decoy oligo-deoxy-nucleotides (ODN) inhibit STAT3 activation by blocking the pY-STAT3 nuclear translocation for subsequent transcriptional target gene activation. STAT3-ODN has been examined in several in vitro EOC cell models and demonstrated significant inhibition of invasiveness and enhancement sensitivity to chemotherapeutic agents [160–162].

Lastly, proteomic analysis has identified RELA and STAT5 as two vital proteins associated with carboplatin resistance in HGCS patients. Small-molecule mediated inhibition of NF-kB (by BMS-345541) and STAT5 (by dasatinib) synergistically sensitized carboplatin-resistant EOC cells towards carboplatin treatment [19].

6. Conclusions

Research in past decades has facilitated our comprehension of the critical roles of aberrant STAT3/STAT5 activation in EOC cells as well as in the tumor microenvironment. The STAT3/STAT5 signaling dysregulates a plethora of cellular processes in EOC, which results in uncontrolled cancer cell proliferation, induction of angiogenesis, promotion of metastasis factors, and suppression of host immune response. In addition to extracellular signals that activate STAT, phosphorylated STATs positively regulate the expression of interleukins and growth factors, generating a vicious autocrine feedback loop that sustains the constitutive STAT3/STAT5 signaling cascade. Accordingly, multiple studies have provided ample evidence to show that interfering with STAT3/STAT5 signaling, through knockdown or inhibitor intervention, resulted in antitumor effects in both in vitro and in vivo animal models carrying human tumors. However, none of the existing candidate compounds showed anti-tumor efficacy in EOC patients. Given the importance of STAT3/STAT5 as a promising target, future research should explore inhibitors against upstream regulators for the development of clinically useful anticancer therapeutics. A histotype-specific approach to target ovarian clear cell carcinoma with the STAT3 pathway might be an avenue worth pursuing.

Other abbreviations: SHC; SRC homology and collagen family, GRB2; Growth factor receptor-bound protein 2, SOS; Son of sevenless, NK; Natural killer, PIAS; Protein inhibitor of activated STAT, PD-1; Programmed cell death protein 1, PD-L1; Programmed death-ligand 1, PTPN11; Tyrosine-protein phosphatase non-receptor type 11, PTPRT; Protein Tyrosine phosphatase receptor type T, MHC-I; Major histocompatibility complex class I, VEGF; Vascular endothelial growth factor, RTK; Receptor Tyrosine kinase.

Funding: This research was funded by Ministry of Science and Technology grant number 108-2811-B-002-512.

Conflicts of Interest: The authors declare no conflict of interest.

Appendix A

Table A1. STAT3 downstream for apoptosis.

Mediators	Effect	References
miR-17-5b	increase	[163]
miR-133b	increase	[164]
miR-134	increase	[165]
miR-135a	increase	[164]
miR-147	increase	[166]
miR-182	increase	[167]
miR-204	increase	[168]
miR-874	increase	[169]
Bcl-x$_L$	increase	[57,170,171]
Bcl-2	increase	[52,57,172]
cIAP-1	increase	[173]
Mcl-1L	increase	[172,174,175]
Fas	decrease	[171,176]
Survivin	increase	[57]

Table A2. STAT3 downstream for angiogenesis.

Mediators	Effect	References
VEGF	increase	[80,177–179]
HIF1	increase	[76,159,180,181]
CCL5	increase	[111,117]

Table A3. STAT3 downstream for ECM degradation.

Mediators	Effect	References
MMP-2	increase	[182,183]
MMP-7	increase	[184]
MMP-9	increase	[103]

Table A4. STAT3 downstream for epithelial-mesenchymal transition(EMT).

Mediators	Effect	References
Twist1	increase	[88,185–188]
Snail1	increase	[87,186,189]
Zeb	increase	[86]
miR-200 family	decrease	[86]
let7	decrease	[86]

Table A5. STAT3 downstream for immune invasion.

Mediators	Effect	References
IL-6	increase	[35,177,190–194]
IL-10	increase	[129,138,143]
IL-11	increase	[27]
IL-23	increase	[120,121]
LIF	increase	[26,195]

Table A6. STAT3 downstream for stemness.

Mediators	Effect	References
BMI-1	increase	[140,188]
Oct3/4	increase	[89,92,196,197]
Nanog	increase	[196,197]

References

1. Meinhold-Heerlein, I.; Fotopoulou, C.; Harter, P.; Kurzeder, C.; Mustea, A.; Wimberger, P.; Hauptmann, S.; Sehouli, J. The new WHO classification of ovarian, fallopian tube, and primary peritoneal cancer and its clinical implications. *Arch. Gynecol. Obstet.* **2016**, *293*, 695–700. [CrossRef] [PubMed]

2. Soslow, R.A. Histologic Subtypes of Ovarian Carcinoma. *Int. J. Gynecol. Pathol.* **2008**, *27*, 161–174. [CrossRef] [PubMed]

3. Bookman, M.A.; Brady, M.F.; McGuire, W.P.; Harper, P.G.; Alberts, D.S.; Friedlander, M.; Colombo, N.; Fowler, J.M.; Argenta, P.A.; De Geest, K.; et al. Evaluation of New Platinum-Based Treatment Regimens in Advanced-Stage Ovarian Cancer: A Phase III Trial of the Gynecologic Cancer InterGroup. *J. Clin. Oncol.* **2009**, *27*, 1419–1425. [CrossRef]

4. Baldwin, L.A.; Miller, R.W.; Tucker, T.; Goodrich, S.T.; Podzielinski, I.; Ueland, F.R.; Seamon, L.G.; Huang, B.; DeSimone, C.P.; Van Nagell, J.R. Ten-Year Relative Survival for Epithelial Ovarian Cancer. *Obstet. Gynecol.* **2012**, *120*, 612–618. [CrossRef] [PubMed]

5. Darnell, J.E. STATs and Gene Regulation. *Science* **1997**, *277*, 1630–1635. [CrossRef]

6. Levy, D.E.; Darnell, J.E., Jr. Signalling: STATs: Transcriptional control and biological impact. *Nat. Rev. Mol. Cell Biol.* **2002**, *3*, e651. [CrossRef] [PubMed]

7. Yu, H.; Pardoll, E.; Jove, R. STATs in cancer inflammation and immunity: A leading role for STAT3. *Nat. Rev. Cancer* **2009**, *9*, 798–809. [CrossRef]

8. Teglund, S.; McKay, C.; Schuetz, E.; Van Deursen, J.M.; Stravopodis, D.; Wang, D.; Brown, M.; Bodner, S.; Grosveld, G.; Ihle, J.N. Stat5a and Stat5b proteins have essential and nonessential, or redundant, roles in cytokine responses. *Cell* **1998**, *93*, 841–850. [CrossRef]

9. Yu, H.; Jove, R. The STATs of cancer-New molecular targets come of age. *Nat. Rev. Cancer* **2004**, *4*, 97–105. [CrossRef]

10. Burke, W.M.; Jin, X.; Lin, H.-J.; Huang, M.; Liu, R.; Reynolds, R.K.; Lin, J. Inhibition of constitutively active Stat3 suppresses growth of human ovarian and breast cancer cells. *Oncogene* **2001**, *20*, 7925–7934. [CrossRef]

11. Huang, M.; Page, C.; Reynolds, R.; Lin, J. Constitutive Activation of Stat 3 Oncogene Product in Human Ovarian Carcinoma Cells. *Gynecol. Oncol.* **2000**, *79*, 67–73. [CrossRef] [PubMed]

12. Savarese, T.M.; Campbell, C.L.; McQuain, C.; Mitchell, K.; Guardiani, R.; Quesenberry, P.J.; Nelson, B.E. Coexpression of oncostatin m and its receptors and evidence for stat3 activation in human ovarian carcinomas. *Cytokine* **2002**, *17*, e324. [CrossRef]

13. Silver, D.L.; Naora, H.; Liu, J.; Cheng, W.; Montell, D.J. Activated signal transducer and activator of transcription (STAT) 3: Localization in focal adhesions and function in ovarian cancer cell motility. *Cancer Res.* **2004**, *64*, e3550. [CrossRef] [PubMed]

14. Chen, H.; Ye, D.; Xie, X.; Chen, B.; Lü, W. VEGF, VEGFRs expressions and activated STATs in ovarian epithelial carcinoma. *Gynecol. Oncol.* **2004**, *94*, 630–635. [CrossRef] [PubMed]

15. Min, H.; Wei-Hong, Z. Constitutive activation of signal transducer and activator of transcription 3 in epithelial ovarian carcinoma. *J. Obstet. Gynaecol. Res.* **2009**, *35*, 918–925. [CrossRef]

16. Yoshikawa, T.; Miyamoto, M.; Aoyama, T.; Soyama, H.; Goto, T.; Hirata, J.; Suzuki, A.; Nagaoka, I.; Tsuda, H.; Furuya, K.; et al. JAK2/STAT3 pathway as a therapeutic target in ovarian cancers. *Oncol. Lett.* **2018**, *15*, 5772–5780. [CrossRef]

17. Rosen, D.G.; Yang, G.; Bast, R.C.; Amin, H.M.; Lai, R.; Liu, J.; Mercado-Uribe, I. The role of constitutively active signal transducer and activator of transcription 3 in ovarian tumorigenesis and prognosis. *Cancer* **2006**, *107*, 2730–2740. [CrossRef]

18. Chen, M.-W.; Yang, S.-T.; Chien, M.-H.; Wu, C.-J.; Hua, K.-T.; Hsiao, S.; Lin, H.; Su, J.-L.; Wei, L.-H. The STAT3-miRNA-92-Wnt Signaling Pathway Regulates Spheroid Formation and Malignant Progression in Ovarian Cancer. *Cancer Res.* **2017**, *77*, 1955–1967. [CrossRef]

19. Jinawath, N.; Vasoontara, C.; Jinawath, A.; Fang, X.; Zhao, K.; Yap, K.-L.; Guo, T.; Lee, C.S.; Wang, W.; Balgley, B.M.; et al. Oncoproteomic Analysis Reveals Co-Upregulation of RELA and STAT5 in Carboplatin Resistant Ovarian Carcinoma. *PLoS ONE* **2010**, *5*, e11198. [CrossRef]

20. Koskela, H.L.; Eldfors, S.; Ellonen, P.; Van Adrichem, A.J.; Kuusanmäki, H.; Andersson, E.I.; Lagström, S.; Clemente, M.J.; Olson, T.; Jalkanen, S.E.; et al. Somatic STAT3 mutations in large granular lymphocytic leukemia. *N. Engl. J. Med.* **2012**, *366*, 1905–1913. [CrossRef]

21. Ohgami, R.; Ma, L.; Merker, J.; Martinez, B.; Zehnder, J.; Arber, D. STAT3 mutations are frequent in CD30$^+$ T-cell lymphomas and T-cell large granular lymphocytic leukemia. *Leukemia* **2013**, *27*, e2244. [CrossRef] [PubMed]

22. Rajala, H.L.M.; Eldfors, S.; Kuusanmäki, H.; Van Adrichem, A.J.; Olson, T.; Lagström, S.; Andersson, E.I.; Jerez, A.; Clemente, M.J.; Yan, Y.; et al. Discovery of somatic STAT5b mutations in large granular lymphocytic leukemia. *Blood* **2013**, *121*, 4541–4550. [CrossRef] [PubMed]

23. Shahmarvand, N.; Nagy, A.; Shahryari, J.; Ohgami, R.S. Mutations in the signal transducer and activator of transcription family of genes in cancer. *Cancer Sci.* **2018**, *109*, 926–933. [CrossRef] [PubMed]

24. Murakami, M.; Hibi, M.; Nakagawa, N.; Nakagawa, T.; Yasukawa, K.; Yamanishi, K.; Taga, T.; Kishimoto, T. IL-6-induced homodimerization of gp130 and associated activation of a tyrosine kinase. *Science* **1993**, *260*, 1808–1810. [CrossRef]

25. Kumar, J.; Ward, A.C. Role of the interleukin 6 receptor family in epithelial ovarian cancer and its clinical implications. *Biochim. Biophys. Acta (BBA) Bioenerg.* **2014**, *1845*, 117–125. [CrossRef]

26. McLean, K.; Tan, L.; Bolland, D.E.; Coffman, L.G.; Peterson, L.F.; Talpaz, M.; Neamati, N.; Buckanovich, R.J. Leukemia inhibitory factor functions in parallel with interleukin-6 to promote ovarian cancer growth. *Oncogene* **2019**, *38*, e1576. [CrossRef]

27. Zhou, W.; Sun, W.; Yung, M.M.H.; Dai, S.; Cai, Y.; Chen, C.-W.; Meng, Y.; Lee, J.B.; Braisted, J.C.; Xu, Y.; et al. Autocrine activation of JAK2 by IL-11 promotes platinum drug resistance. *Oncogene* **2018**, *37*, 3981–3997. [CrossRef]

28. Reid, T.; Jin, X.; Song, H.; Tang, H.-J.; Reynolds, R.K.; Lin, J. Modulation of Janus kinase 2 by p53 in ovarian cancer cells. *Biochem. Biophys. Res. Commun.* **2004**, *321*, 441–447. [CrossRef]

29. Pavelka, J.C.; Brown, R.S.; Karlan, B.Y.; Cass, I.; Leuchter, R.S.; Lagasse, L.D.; Li, A.J. Effect of obesity on survival in epithelial ovarian cancer. *Cancer* **2006**, *107*, 1520–1524. [CrossRef]

30. Seo, J.-M.; Park, S.; Kim, J.-H. Leukotriene B4 Receptor-2 Promotes Invasiveness and Metastasis of Ovarian Cancer Cells through Signal Transducer and Activator of Transcription 3 (STAT3)-dependent Up-regulation of Matrix Metalloproteinase 2*. *J. Boil. Chem.* **2012**, *287*, 13840–13849. [CrossRef]

31. Uddin, S.; Bu, R.; Ahmed, M.; Abubaker, J.; Al-Dayel, F.; Bavi, P.; Al-Kuraya, K.S. Overexpression of leptin receptor predicts an unfavorable outcome in Middle Eastern ovarian cancer. *Mol. Cancer* **2009**, *8*, e74. [CrossRef] [PubMed]

32. Choi, J.-H.; Lee, K.-T.; Leung, P.C. Estrogen receptor alpha pathway is involved in leptin-induced ovarian cancer cell growth. *Carcinogenesis* **2011**, *32*, e589. [CrossRef] [PubMed]

33. Bin Park, G.; Kim, D. MicroRNA-503-5p Inhibits the CD97-Mediated JAK2/STAT3 Pathway in Metastatic or Paclitaxel-Resistant Ovarian Cancer Cells. *Neoplasia* **2019**, *21*, 206–215. [CrossRef] [PubMed]

34. Lu, W.; Chen, H.; Yel, F.; Wang, F.; Xie, X. VEGF induces phosphorylation of STAT3 through binding VEGFR2 in ovarian carcinoma cells in vitro. *Eur. J. Gynaecol. Oncol.* **2006**, *27*, 363–369. [PubMed]

35. Colomiere, M.; Ward, A.; Riley, C.; Trenerry, M.; Cameron-Smith, D.; Findlay, J.; Ackland, L.; Ahmed, N. Cross talk of signals between EGFR and IL-6R through JAK2/STAT3 mediate epithelial–mesenchymal transition in ovarian carcinomas. *Br. J. Cancer* **2009**, *100*, e134. [CrossRef] [PubMed]

36. Garnett, M.J.; Marais, R. Guilty as charged B-RAF is a human oncogene. *Cancer Cell* **2004**, *6*, 313–319. [CrossRef]

37. Dobrzycka, B.; Terlikowski, S.J.; Kowalczuk, O.; Niklińska, W.; Chyczewski, L.; Kulikowski, M. Mutations in the KRAS gene in ovarian tumors. *Folia Histochem. Cytobiol.* **2009**, *47*, 221–224. [CrossRef]

38. Yue, P.; Zhang, X.; Paladino, D.; Sengupta, B.; Ahmad, S.; Holloway, R.W.; Ingersoll, S.B.; Turkson, J. Hyperactive EGF receptor, Jaks and Stat3 signaling promote enhanced colony-forming ability, motility and migration of cisplatin-resistant ovarian cancer cells. *Oncogene* **2011**, *31*, 2309–2322. [CrossRef]

39. Kumar, J.; Fraser, F.; Riley, C.; Ahmed, N.; McCulloch, D.; Ward, A. Granulocyte colony-stimulating factor receptor signalling via Janus kinase 2/signal transducer and activator of transcription 3 in ovarian cancer. *Br. J. Cancer* **2014**, *110*, e133. [CrossRef]

40. Guo, Q.; Lu, L.; Liao, Y.; Wang, X.; Zhang, Y.; Liu, Y.; Huang, S.; Sun, H.; Li, Z.; Zhao, L. Influence of c-Src on hypoxic resistance to paclitaxel in human ovarian cancer cells and reversal of FV-429. *Cell Death Dis.* **2018**, *8*, e3178. [CrossRef]

41. Pradeep, S.; Huang, J.; Mora, E.M.; Nick, A.M.; Cho, M.S.; Wu, S.Y.; Noh, K.; Pecot, C.V.; Rupaimoole, R.; Stein, M.A.; et al. Erythropoietin Stimulates Tumor Growth via EphB4. *Cancer Cell* **2015**, *28*, 610–622. [CrossRef] [PubMed]

42. Aapro, M.; Jelkmann, W.; Constantinescu, S.N.; Leyland-Jones, B. Effects of erythropoietin receptors and erythropoiesis-stimulating agents on disease progression in cancer. *Br. J. Cancer* **2012**, *106*, 1249–1258. [CrossRef] [PubMed]

43. Chaluvally-Raghavan, P.; Jeong, K.J.; Pradeep, S.; Silva, A.M.; Yu, S.; Liu, W.; Moss, T.; Rodriguez-Aguayo, C.; Zhang, N.; Ram, P.; et al. Direct Upregulation of STAT3 by MicroRNA-551b-3p Deregulates Growth and Metastasis of Ovarian Cancer. *Cell Rep.* **2016**, *15*, 1493–1504. [CrossRef] [PubMed]

44. Hu, S.; Yu, L.; Li, Z.; Shen, Y.; Wang, J.; Cai, J.; Xiao, L.; Wang, Z. Overexpression of EZH2 contributes to acquired cisplatin resistance in ovarian cancer cells in vitro and in vivo. *Cancer Boil. Ther.* **2010**, *10*, 788–795. [CrossRef]

45. Li, H.; Cai, Q.; Godwin, A.K.; Zhang, R. Enhancer of zeste homolog 2 promotes the proliferation and invasion of epithelial ovarian cancer cells. *Mol. Cancer Res.* **2010**, *8*, 1610–1618. [CrossRef]

46. Özeş, A.R.; Pulliam, N.; Ertosun, M.G.; Yılmaz, Ö.; Tang, J.; Çopuroğlu, E.; Matei, D.; Ozes, O.N.; Nephew, K.P. Protein kinase A-mediated phosphorylation regulates STAT3 activation and oncogenic EZH2 activity. *Oncogene* **2018**, *37*, 3589–3600. [CrossRef]

47. Yan, J.; Li, B.; Lin, B.; Lee, P.T.; Chung, T.-H.; Tan, J.; Bi, C.; Lee, X.T.; Selvarajan, V.; Ng, S.-B.; et al. EZH2 phosphorylation by JAK3 mediates a switch to noncanonical function in natural killer/T-cell lymphoma. *Blood* **2016**, *128*, 948–958. [CrossRef]

48. Mandal, M.; Powers, S.E.; Maienschein-Cline, M.; Bartom, E.T.; Hamel, K.M.; Kee, B.L.; Dinner, A.R.; Clark, M.R. Epigenetic repression of the Igk locus by STAT5-mediated recruitment of the histone methyltransferase Ezh2. *Nat. Immunol.* **2011**, *12*, 1212–1220. [CrossRef]

49. Orlova, A.; Wagner, C.; De Araujo, E.D.; Bajusz, D.; Neubauer, H.A.; Herling, M.; Gunning, P.T.; Keserű, G.M.; Moriggl, R. Direct Targeting Options for STAT3 and STAT5 in Cancer. *Cancers* **2019**, *11*, 1930. [CrossRef]

50. Cory, S.; Adams, J.M. The Bcl2 family: Regulators of the cellular life-or-death switch. *Nat. Rev. Cancer* **2002**, *2*, 647–656. [CrossRef]

51. Muller, P.A.; Vousden, K.H. p53 mutations in cancer. *Nat. Cell Biol.* **2013**, *15*, e2. [CrossRef] [PubMed]

52. Bixel, K.; Saini, U.; Bid, H.K.; Fowler, J.; Riley, M.; Wanner, R.; Dorayappan, K.D.P.; Rajendran, S.; Konishi, I.; Matsumura, N.; et al. Targeting STAT3 by HO3867 induces apoptosis in ovarian clear cell carcinoma. *Int. J. Cancer* **2017**, *141*, 1856–1866. [CrossRef] [PubMed]

53. Kumar, J.; Fang, H.; McCulloch, D.R.; Crowley, T.; Ward, A.C. Leptin receptor signaling via Janus kinase 2/Signal transducer and activator of transcription 3 impacts on ovarian cancer cell phenotypes. *Oncotarget* **2017**, *8*, 93530–93540. [CrossRef] [PubMed]

54. De Groot, R.P.; Raaijmakers, J.A.; Lammers, J.-W.J.; Koenderman, L. STAT5-Dependent CyclinD1 and Bcl-xL Expression in Bcr-Abl-Transformed Cells. *Mol. Cell Boil. Res. Commun.* **2000**, *3*, 299–305. [CrossRef]

55. Sánchez-Ceja, S.; Reyes-Maldonado, E.; Vázquez-Manríquez, M.; López-Luna, J.; Belmont, A.; Gutiérrez-Castellanos, S. Differential expression of STAT5 and Bcl-xL, and high expression of Neu and STAT3 in non-small-cell lung carcinoma. *Lung Cancer* **2006**, *54*, 163–168. [CrossRef]

56. Guo, Y.; Nemeth, J.; O'Brien, C.; Susa, M.; Liu, X.; Zhang, Z.; Choy, E.; Mankin, H.; Hornicek, F.; Duan, Z. Effects of Siltuximab on the IL-6-Induced Signaling Pathway in Ovarian Cancer. *Clin. Cancer Res.* **2010**, *16*, 5759–5769. [CrossRef]

57. Ji, T.; Gong, D.; Han, Z.; Wei, X.; Yan, Y.; Ye, F.; Ding, W.; Wang, J.; Xia, X.; Li, F.; et al. Abrogation of constitutive Stat3 activity circumvents cisplatin resistant ovarian cancer. *Cancer Lett.* **2013**, *341*, 231–239. [CrossRef]

58. Tian, F.; Jia, L.; Chu, Z.; Han, H.; Zhang, Y.; Cai, J. MicroRNA-519a inhibits the proliferation and promotes the apoptosis of ovarian cancer cells through targeting signal transducer and activator of transcription 3. *Exp. Ther. Med.* **2017**, *15*, 1819–1824. [CrossRef]

59. Syed, V. Reproductive Hormone-Induced, STAT3-Mediated Interleukin 6 Action in Normal and Malignant Human Ovarian Surface Epithelial Cells. *J. Natl. Cancer Inst.* **2002**, *94*, 617–629. [CrossRef]

60. Nagle, C.M.; Australian Ovarian Cancer Study Group; Dixon, S.C.; Jensen, A.; Kjaer, S.K.; Modugno, F.; DeFazio, A.; Fereday, S.; Hung, J.; Johnatty, S.E.; et al. Obesity and survival among women with ovarian cancer: Results from the Ovarian Cancer Association Consortium. *Br. J. Cancer* **2015**, *113*, 817–826. [CrossRef]

61. Chin, Y.-T.; Wang, L.-M.; Hsieh, M.-T.; Shih, Y.-J.; Nana, A.W.; Changou, C.A.; Yang, Y.-C.S.H.; Chiu, H.-C.; Fu, E.; Davis, P.J.; et al. Leptin OB3 peptide suppresses leptin-induced signaling and progression in ovarian cancer cells. *J. Biomed. Sci.* **2017**, *24*, e51. [CrossRef] [PubMed]

62. Xu, X.; Dong, Z.; Li, Y.; Yang, Y.; Yuan, Z.; Qu, X.; Kong, B. The upregulation of signal transducer and activator of transcription 5-dependent microRNA-182 and microRNA-96 promotes ovarian cancer cell proliferation by targeting forkhead box O3 upon leptin stimulation. *Int. J. Biochem. Cell Boil.* **2013**, *45*, 536–545. [CrossRef] [PubMed]

63. Chen, K.-H.E.; Walker, A.M. Prolactin inhibits a major tumor-suppressive function of wild type BRCA1. *Cancer Lett.* **2016**, *375*, 293–302. [CrossRef] [PubMed]

64. Fiedor, E.; Gregoraszczuk, E. Łucja the molecular mechanism of action of superactive human leptin antagonist (SHLA) and quadruple leptin mutein Lan-2 on human ovarian epithelial cell lines. *Cancer Chemother. Pharmacol.* **2016**, *78*, 611–622. [CrossRef]

65. Selvendiran, K.; Bratasz, A.; Kuppusamy, M.L.; Tazi, M.F.; Rivera, B.K.; Kuppusamy, P. Hypoxia induces chemoresistance in ovarian cancer cells by activation of signal transducer and activator of transcription 3. *Int. J. Cancer* **2009**, *125*, 2198–2204. [CrossRef]

66. Selvendiran, K.; Kuppusamy, M.L.; Ahmed, S.; Bratasz, A.; Meenakshisundaram, G.; Rivera, B.K.; Khan, M.; Kuppusamy, P. Oxygenation inhibits ovarian tumor growth by downregulating STAT3 and cyclin-D1 expressions. *Cancer Boil. Ther.* **2010**, *10*, 386–390. [CrossRef]

67. Jin, P.; Liu, Y.; Wang, R. STAT3 regulated miR-216a promotes ovarian cancer proliferation and cisplatin resistance. *Biosci. Rep.* **2018**, *38*, BSR20180547. [CrossRef]

68. Jiang, Q.; Dai, L.; Cheng, L.; Chen, X.; Li, Y.; Zhang, S.; Su, X.; Zhao, X.; Wei, Y.; Deng, H. Efficient inhibition of intraperitoneal ovarian cancer growth in nude mice by liposomal delivery of short hairpin RNA against STAT3. *J. Obstet. Gynaecol. Res.* **2013**, *39*, 701–709. [CrossRef]

69. Pan, Y.; Zhang, L.; Zhang, X.; Liu, R. Synergistic effects of eukaryotic co-expression plasmid-based STAT3-specific siRNA and LKB1 on ovarian cancer in vitro and in vivo. *Oncol. Rep.* **2015**, *33*, e774. [CrossRef]

70. Zhao, S.-H.; Zhao, F.; Zheng, J.-Y.; Gao, L.-F.; Zhao, X.-J.; Cui, M.-H. Knockdown of stat3 expression by RNAi inhibits in vitro growth of human ovarian cancer. *Radiol. Oncol.* **2011**, *45*, 196–203. [CrossRef]

71. Saydmohammed, M.; Joseph, D.; Syed, V. Curcumin suppresses constitutive activation of STAT-3 by up-regulating protein inhibitor of activated STAT-3 (PIAS-3) in ovarian and endometrial cancer cells. *J. Cell. Biochem.* **2010**, *110*, 447. [CrossRef] [PubMed]

72. Wen, W.; Lowe, G.; Roberts, C.M.; Finlay, J.; Han, E.S.; Glackin, C.A.; Dellinger, T.H. Pterostilbene Suppresses Ovarian Cancer Growth via Induction of Apoptosis and Blockade of Cell Cycle Progression Involving Inhibition of the STAT3 Pathway. *Int. J. Mol. Sci.* **2018**, *19*, 1983. [CrossRef] [PubMed]

73. Yang, Y.; Cao, Y.; Chen, L.; Liu, F.; Qi, Z.; Cheng, X.; Wang, Z. Cryptotanshinone suppresses cell proliferation and glucose metabolism via STAT3/SIRT3 signaling pathway in ovarian cancer cells. *Cancer Med.* **2018**, *7*, 4610–4618. [CrossRef] [PubMed]

74. Gong, W.; Wang, L.; Yao, J.C.; Ajani, J.A.; Wei, D.; Aldape, K.D.; Xie, K.; Sawaya, R.; Huang, S. Expression of Activated Signal Transducer and Activator of Transcription 3 Predicts Expression of Vascular Endothelial Growth Factor in and Angiogenic Phenotype of Human Gastric Cancer. *Clin. Cancer Res.* **2005**, *11*, 1386–1393. [CrossRef] [PubMed]

75. Moser, C.; Ruemmele, P.; Gehmert, S.; Schenk, H.; Kreutz, M.P.; Mycielska, M.E.; Hackl, C.; Kroemer, A.; ASchnitzbauer, A.; Stoeltzing, O.; et al. STAT5b as Molecular Target in Pancreatic Cancer-Inhibition of Tumor Growth, Angiogenesis, and Metastases. *Neoplasia* **2012**, *14*, 915–925. [CrossRef]

76. Anglesio, M.S.; George, J.; Kulbe, H.; Friedlander, M.; Rischin, D.; Lemech, C.; Power, J.; Coward, J.; Cowin, P.A.; House, C.M.; et al. IL6-STAT3-HIF Signaling and Therapeutic Response to the Angiogenesis Inhibitor Sunitinib in Ovarian Clear Cell Cancer. *Clin. Cancer Res.* **2011**, *17*, 2538–2548. [CrossRef]

77. Wei, L.-H.; Kuo, M.-L.; Chen, C.-A.; Chou, C.-H.; Lai, K.-B.; Lee, C.-N.; Hsieh, C.-Y. Interleukin-6 promotes cervical tumor growth by VEGF-dependent angiogenesis via a STAT3 pathway. *Oncogene* **2003**, *22*, 1517–1527. [CrossRef]

78. Zhang, K.; Han, E.S.; Dellinger, T.H.; Lu, J.; Nam, S.; Anderson, R.A.; Yim, J.H.; Wen, W. Cinnamon extract reduces VEGF expression via suppressing HIF-1α gene expression and inhibits tumor growth in mice. *Mol. Carcinog.* **2017**, *56*, 436–446. [CrossRef]

79. Kandala, P.K.; Srivastava, S.K. Regulation of Janus-activated kinase-2 (JAK2) by diindolylmethane in ovarian cancer in vitro and in vivo. *Drug Discov. Ther.* **2012**, *6*, 94–101. [CrossRef]

80. Eichten, A.; Su, J.; Adler, A.P.; Zhang, L.; Ioffe, E.; Parveen, A.A.; Yancopoulos, G.D.; Rudge, J.S.; Lowy, I.; Lin, H.C.; et al. Resistance to Anti-VEGF Therapy Mediated by Autocrine IL6/STAT3 Signaling and Overcome by IL6 Blockade. *Cancer Res.* **2016**, *76*, 2327–2339. [CrossRef]

81. Güth, U.; Huang, D.J.; Bauer, G.; Stieger, M.; Wight, E.; Singer, G. Metastatic patterns at autopsy in patients with ovarian carcinoma. *Cancer* **2007**, *110*, 1272–1280. [CrossRef] [PubMed]

82. Zheng, B.; Geng, L.; Zeng, L.; Liu, F.; Huang, Q. AKT2 contributes to increase ovarian cancer cell migration and invasion through the AKT2-PKM2-STAT3/NF-κB axis. *Cell. Signal.* **2018**, *45*, 122–131. [CrossRef] [PubMed]

83. Brabletz, T.; Kalluri, R.; Nieto, A.M.; Weinberg, R.A. EMT in cancer. *Nat. Rev. Cancer* **2018**, *18*, e128. [CrossRef] [PubMed]

84. Loret, N.; Denys, H.; Tummers, P.; Berx, G. The Role of Epithelial-to-Mesenchymal Plasticity in Ovarian Cancer Progression and Therapy Resistance. *Cancers* **2019**, *11*, 838. [CrossRef] [PubMed]

85. Balanis, N.; Wendt, M.K.; Schiemann, B.J.; Wang, Z.; Schiemann, W.P.; Carlin, C.R. Epithelial to Mesenchymal Transition Promotes Breast Cancer Progression via a Fibronectin-dependent STAT3 Signaling Pathway*. *J. Boil. Chem.* **2013**, *288*, 17954–17967. [CrossRef]

86. Guo, L.; Chen, C.; Shi, M.; Wang, F.; Chen, X.; Diao, D.; Hu, M.; Yu, M.; Qian, L.; Guo, N. Stat3-coordinated Lin-28–let-7–HMGA2 and miR-200–ZEB1 circuits initiate and maintain oncostatin M-driven epithelial–mesenchymal transition. *Oncogene* **2013**, *32*, 5272–5282. [CrossRef]

87. Kim, M.-J.; Lim, J.; Yang, Y.; Lee, M.-S.; Lim, J.-S. N-myc downstream-regulated gene 2 (NDRG2) suppresses the epithelial–mesenchymal transition (EMT) in breast cancer cells via STAT3/Snail signaling. *Cancer Lett.* **2014**, *354*, 33–42. [CrossRef]

88. Chong, K.Y.; Kang, M.; Garofalo, F.; Ueno, D.; Liang, H.; Cady, S.; Madarikan, O.; Pitruzzello, N.; Tsai, C.-H.; Hartwich, T.M.; et al. Inhibition of Heat Shock Protein 90 suppresses TWIST1 Transcription. *Mol. Pharmacol.* **2019**, *96*, 168–179. [CrossRef]

89. Abubaker, K.; Luwor, R.B.; Escalona, R.; McNally, O.; Quinn, M.A.; Thompson, E.W.; Findlay, J.K.; Ahmed, N. Targeted Disruption of the JAK2/STAT3 Pathway in Combination with Systemic Administration of Paclitaxel Inhibits the Priming of Ovarian Cancer Stem Cells Leading to a Reduced Tumor Burden. *Front. Oncol.* **2014**, *4*, e75. [CrossRef]

90. Kim, S.; Gwak, H.; Kim, H.S.; Kim, B.; Dhanasekaran, D.N.; Song, Y.S. Malignant ascites enhances migratory and invasive properties of ovarian cancer cells with membrane bound IL-6R in vitro. *Oncotarget* **2016**, *7*, 83148–83159. [CrossRef]

91. Saini, U.; Naidu, S.; ElNaggar, A.; Bid, H.; Wallbillich, J.; Bixel, K.; Bolyard, C.; Suarez, A.; Kaur, B.; Kuppusamy, P.; et al. Elevated STAT3 expression in ovarian cancer ascites promotes invasion and metastasis: A potential therapeutic target. *Oncogene* **2017**, *36*, e168. [CrossRef] [PubMed]

92. Latifi, A.; Luwor, R.B.; Bilandzic, M.; Nazaretian, S.; Stenvers, K.; Pyman, J.; Zhu, H.; Thompson, E.W.; Quinn, M.A.; Findlay, J.K.; et al. Isolation and Characterization of Tumor Cells from the Ascites of Ovarian Cancer Patients: Molecular Phenotype of Chemoresistant Ovarian Tumors. *PLoS ONE* **2012**, *7*, e46858. [CrossRef] [PubMed]

93. Shield, K.; Ackland, M.L.; Ahmed, N.; Rice, G.E. Multicellular spheroids in ovarian cancer metastases: Biology and pathology. *Gynecol. Oncol.* **2009**, *113*, 143–148. [CrossRef] [PubMed]

94. Jia, Z.-H.; Jia, Y.; Guo, F.-J.; Chen, J.; Zhang, X.-W.; Cui, M.-H. Phosphorylation of STAT3 at Tyr705 regulates MMP-9 production in epithelial ovarian cancer. *PLoS ONE* **2017**, *12*, e0183622. [CrossRef] [PubMed]

95. Landen, C.N.; Lin, Y.G.; Armaiz-Pena, G.N.; Das, P.D.; Arevalo, J.M.; Kamat, A.A.; Han, L.Y.; Jennings, N.B.; Spannuth, W.A.; Thaker, P.H.; et al. Neuroendocrine Modulation of Signal Transducer and Activator of Transcription-3 in Ovarian Cancer. *Cancer Res.* **2007**, *67*, 10389–10396. [CrossRef]

96. Zou, M.; Zhang, X.; Xu, C. IL6-induced metastasis modulators p-STAT3, MMP-2 and MMP-9 are targets of 3,3'-diindolylmethane in ovarian cancer cells. *Cell Oncol.* **2016**, *39*, e47. [CrossRef]

97. Bourguignon, L.Y.W.; Peyrollier, K.; Xia, W.; Gilad, E. Hyaluronan-CD44 interaction activates stem cell marker Nanog, Stat-3-mediated MDR1 gene expression, and ankyrin-regulated multidrug efflux in breast and ovarian tumor cells. *J. Boil. Chem.* **2008**, *283*, 17635–17651. [CrossRef]

98. Liu, S.; Sun, J.; Cai, B.; Xi, X.; Yang, L.; Zhang, Z.; Feng, Y.; Sun, Y. NANOG regulates epithelial-mesenchymal transition and chemoresistance through activation of the STAT3 pathway in epithelial ovarian cancer. *Tumor Boil.* **2016**, *37*, 9671–9680. [CrossRef]

99. Liu, A.; Fan, D.; Wang, Y. The BET bromodomain inhibitor i-BET151 impairs ovarian cancer metastasis and improves antitumor immunity. *Cell Tissue Res.* **2018**, *374*, 577–585. [CrossRef]

100. Yanaihara, N.; Hirata, Y.; Yamaguchi, N.; Noguchi, Y.; Saito, M.; Nagata, C.; Takakura, S.; Yamada, K.; Okamoto, A. Antitumor effects of interleukin-6 (IL-6)/interleukin-6 receptor (IL-6R) signaling pathway inhibition in clear cell carcinoma of the ovary. *Mol. Carcinog.* **2015**, *55*, 832–841. [CrossRef]

101. Dorayappan, K.D.P.; Wallbillich, J.J.; Cohn, D.E.; Selvendiran, K. The biological significance and clinical applications of exosomes in ovarian cancer. *Gynecol. Oncol.* **2016**, *142*, 199–205. [CrossRef] [PubMed]

102. Shen, J.; Zhu, X.; Fei, J.; Shi, P.; Yu, S.; Zhou, J. Advances of exosome in the development of ovarian cancer and its diagnostic and therapeutic prospect. *OncoTargets Ther.* **2018**, *11*, 2831–2841. [CrossRef] [PubMed]

103. Nakamura, K.; Sawada, K.; Kinose, Y.; Yoshimura, A.; Toda, A.; Nakatsuka, E.; Hashimoto, K.; Mabuchi, S.; Morishige, K.; Kurachi, H.; et al. Exosomes Promote Ovarian Cancer Cell Invasion through Transfer of CD44 to Peritoneal Mesothelial Cells. *Mol. Cancer Res.* **2017**, *15*, 78–92. [CrossRef] [PubMed]

104. Syn, N.; Wang, L.; Sethi, G.; Thiery, J.-P.; Goh, B.-C. Exosome-Mediated Metastasis: From Epithelial–Mesenchymal Transition to Escape from Immunosurveillance. *Trends Pharmacol. Sci.* **2016**, *37*, 606–617. [CrossRef] [PubMed]

105. Dorayappan, K.D.P.; Wanner, R.; Wallbillich, J.J.; Saini, U.; Zingarelli, R.; Suarez, A.A.; Cohn, D.E.; Selvendiran, K. Hypoxia-induced exosomes contribute to a more aggressive and chemoresistant ovarian cancer phenotype: A novel mechanism linking STAT3/Rab proteins. *Oncogene* **2018**, *37*, 3806–3821. [CrossRef] [PubMed]

106. Balkwill, F.R.; Capasso, M.; Hagemann, T. The tumor microenvironment at a glance. *J. Cell Sci.* **2012**, *125*, 5591–5596. [CrossRef] [PubMed]

107. Hanahan, D.; Coussens, L.M. Accessories to the Crime: Functions of Cells Recruited to the Tumor Microenvironment. *Cancer Cell* **2012**, *21*, 309–322. [CrossRef]

108. Galon, J.; Fridman, W.-H.; Pagès, F. The Adaptive Immunologic Microenvironment in Colorectal Cancer: A Novel Perspective. *Cancer Res.* **2007**, *67*, 1883–1886. [CrossRef]

109. Gest, C.; Mirshahi, P.; Li, H.; Pritchard, L.-L.; Joimel, U.; Blot, E.; Chidiac, J.; Poletto, B.; Vannier, J.-P.; Varin, R.; et al. Ovarian cancer: Stat3, RhoA and IGF-IR as therapeutic targets. *Cancer Lett.* **2012**, *317*, 207–217. [CrossRef]

110. Ye, F.; Chen, H.-Z.; Xie, X.; Ye, D.-F.; Lu, W.-G.; Ding, Z.-M. Vascular endothelial growth factor (VEGF) and ovarian carcinoma cell supernatant activate signal transducers and activators of transcription (STATs) via VEGF receptor-2 (KDR) in human hemopoietic progenitor cells. *Gynecol. Oncol.* **2004**, *94*, e125. [CrossRef]

111. Tang, S.; Xiang, T.; Huang, S.; Zhou, J.; Wang, Z.; Xie, R.; Long, H.; Zhu, B. Ovarian cancer stem-like cells differentiate into endothelial cells and participate in tumor angiogenesis through autocrine CCL5 signaling. *Cancer Lett.* **2016**, *376*, 137–147. [CrossRef] [PubMed]

112. Nilsson, M.B.; Langley, R.R.; Fidler, I.J. Interleukin-6, secreted by human ovarian carcinoma cells, is a potent proangiogenic cytokine. *Cancer Res.* **2005**, *65*, 10794–10800. [CrossRef] [PubMed]

113. Kalluri, R. The biology and function of fibroblasts in cancer. *Nat. Rev. Cancer* **2016**, *16*, 582–598. [CrossRef] [PubMed]

114. Sun, W.; Fu, S. Role of cancer-associated fibroblasts in tumor structure, composition and the microenvironment in ovarian cancer. *Oncol. Lett.* **2019**, *18*, 2173–2178. [CrossRef] [PubMed]

115. Wang, L.; Zhang, F.; Cui, J.-Y.; Chen, L.; Chen, Y.-T.; Liu, B.-W. CAFs enhance paclitaxel resistance by inducing EMT through the IL-6/JAK2/STAT3 pathway. *Oncol. Rep.* **2018**, *39*, 2081–2090. [CrossRef] [PubMed]

116. Yan, H.; Guo, B.-Y.; Zhang, S. Cancer-associated fibroblasts attenuate Cisplatin-induced apoptosis in ovarian cancer cells by promoting STAT3 signaling. *Biochem. Biophys. Res. Commun.* **2016**, *470*, 947–954. [CrossRef]

117. Zhou, B.; Sun, C.; Li, N.; Shan, W.; Lu, H.; Guo, L.; Guo, E.; Xia, M.; Weng, D.; Meng, L.; et al. Cisplatin-induced CCL5 secretion from CAFs promotes cisplatin-resistance in ovarian cancer via regulation of the STAT3 and PI3K/Akt signaling pathways. *Int. J. Oncol.* **2016**, *48*, 2087–2097. [CrossRef]

118. Nieman, K.M.; Kenny, H.A.; Penicka, C.V.; Ladányi, A.; Buell-Gutbrod, R.; Zillhardt, M.R.; Romero, I.L.; Carey, M.S.; Mills, G.B.; Hotamisligil, G.S.; et al. Adipocytes promote ovarian cancer metastasis and provide energy for rapid tumor growth. *Nat. Med.* **2011**, *17*, 1498–1503. [CrossRef]

119. Kim, B.; Kim, H.S.; Kim, S.; Haegeman, G.; Tsang, B.K.; Dhanasekaran, D.N.; Song, Y.S. Adipose Stromal Cells from Visceral and Subcutaneous Fat Facilitate Migration of Ovarian Cancer Cells via IL-6/JAK2/STAT3 Pathway. *Cancer Res. Treat.* **2017**, *49*, 338–349. [CrossRef]

120. Miyahara, Y.; Odunsi, K.; Chen, W.; Peng, G.; Matsuzaki, J.; Wang, R.-F. Generation and regulation of human CD4$^+$ IL-17-producing T cells in ovarian cancer. *Proc. Natl. Acad. Sci. USA* **2008**, *105*, 15505–15510. [CrossRef]

121. Yang, X.O.; Panopoulos, A.D.; Nurieva, R.; Chang, S.H.; Wang, D.; Watowich, S.S.; Dong, C. STAT3 Regulates Cytokine-mediated Generation of Inflammatory Helper T Cells. *J. Boil. Chem.* **2007**, *282*, 9358–9363. [CrossRef] [PubMed]

122. Harris, T.J.; Grosso, J.F.; Yen, H.-R.; Xin, H.; Kortylewski, M.; Albesiano, E.; Hipkiss, E.L.; Getnet, D.; Goldberg, M.V.; Maris, C.H.; et al. Cutting Edge: An In Vivo Requirement for STAT3 Signaling in TH17 Development and TH17-Dependent Autoimmunity. *J. Immunol.* **2007**, *179*, 4313–4317. [CrossRef] [PubMed]

123. Ma, L.; Huang, C.; Wang, X.-J.; Xin, D.E.; Wang, L.-S.; Zou, Q.C.; Zhang, Y.-N.S.; Tan, M.-D.; Wang, Y.-M.; Zhao, T.C.; et al. Lysyl Oxidase 3 Is a Dual-Specificity Enzyme Involved in STAT3 Deacetylation and Deacetylimination Modulation. *Mol. Cell* **2017**, *65*, 296–309. [CrossRef] [PubMed]

124. Zhou, J.; Li, X.; Wu, X.; Zhang, T.; Zhu, Q.; Wang, X.; Wang, H.; Wang, K.; Lin, Y.; Wang, X. Exosomes Released from Tumor-Associated Macrophages Transfer miRNAs That Induce a Treg/Th17 Cell Imbalance in Epithelial Ovarian Cancer. *Cancer Immunol. Res.* **2018**, *6*, 1578–1592. [CrossRef] [PubMed]

125. Curiel, T.J.; Coukos, G.; Zou, L.; Alvarez, X.; Cheng, P.; Mottram, P.; Evdemon-Hogan, M.; Conejo-Garcia, J.R.; Zhang, L.; Burow, M.; et al. Specific recruitment of regulatory T cells in ovarian carcinoma fosters immune privilege and predicts reduced survival. *Nat. Med.* **2004**, *10*, 942–949. [CrossRef]

126. Chen, W.; Jin, W.; Hardegen, N.; Lei, K.; Li, L.; Marinos, N.; McGrady, G.; Wahl, S.M. Conversion of Peripheral CD4$^+$CD25$^-$ Naive T Cells to CD4$^+$CD25$^+$ Regulatory T Cells by TGF-β Induction of Transcription Factor Foxp3. *J. Exp. Med.* **2003**, *198*, 1875–1886. [CrossRef]

127. Zorn, E.; Nelson, E.A.; Mohseni, M.; Porcheray, F.; Kim, H.; Litsa, D.; Bellucci, R.; Raderschall, E.; Canning, C.; Soiffer, R.J.; et al. IL-2 regulates FOXP3 expression in human CD4$^+$CD25$^+$ regulatory T cells through a STAT-dependent mechanism and induces the expansion of these cells in vivo. *Blood* **2006**, *108*, 1571–1579. [CrossRef]

128. Hamanishi, J.; Mandai, M.; Iwasaki, M.; Okazaki, T.; Tanaka, Y.; Yamaguchi, K.; Higuchi, T.; Yagi, H.; Takakura, K.; Minato, N.; et al. Programmed cell death 1 ligand 1 and tumor-infiltrating CD8$^+$ T lymphocytes are prognostic factors of human ovarian cancer. *Proc. Natl. Acad. Sci. USA* **2007**, *104*, 3360–3365. [CrossRef]

129. Dercamp, C.; Chemin, K.; Caux, C.; Trinchieri, G.; Vicari, A.P. Distinct and Overlapping Roles of Interleukin-10 and CD25$^+$ Regulatory T Cells in the Inhibition of Antitumor CD8 T-Cell Responses. *Cancer Res.* **2005**, *65*, 8479–8486. [CrossRef]

130. Kinjyo, I.; Inoue, H.; Hamano, S.; Fukuyama, S.; Yoshimura, T.; Koga, K.; Takaki, H.; Himeno, K.; Takaesu, G.; Kobayashi, T.; et al. Loss of SOCS3 in T helper cells resulted in reduced immune responses and hyperproduction of interleukin 10 and transforming growth factor–β1. *J. Exp. Med.* **2006**, *203*, 1021–1031. [CrossRef]

131. Kosack, L.; Wingelhofer, B.; Popa, A.; Orlova, A.; Agerer, B.; Vilagos, B.; Majek, P.; Parapatics, K.; Lercher, A.; Ringler, A.; et al. The ERBB-STAT3 Axis Drives Tasmanian Devil Facial Tumor Disease. *Cancer Cell* **2019**, *35*, 125–139.e9. [CrossRef] [PubMed]

132. Nishikawa, H.; Kato, T.; Tawara, I.; Ikeda, H.; Kuribayashi, K.; Allen, P.M.; Schreiber, R.D.; Old, L.J.; Shiku, H. IFN-γ Controls the Generation/Activation of CD4⁺CD25⁺ Regulatory T Cells in Antitumor Immune Response. *J. Immunol.* **2005**, *175*, 4433–4440. [CrossRef] [PubMed]

133. Kortylewski, M.; Kujawski, M.; Wang, T.; Wei, S.; Zhang, S.; Pilon-Thomas, S.; Niu, G.; Kay, H.; Mulé, J.; Kerr, W.G.; et al. Inhibiting Stat3 signaling in the hematopoietic system elicits multicomponent antitumor immunity. *Nat. Med.* **2005**, *11*, 1314–1321. [CrossRef] [PubMed]

134. Wang, H.; Xie, X.; Lu, W.-G.; Ye, D.-F.; Chen, H.-Z.; Li, X.; Cheng, Q. Ovarian carcinoma cells inhibit T cell proliferation: Suppression of IL-2 receptor beta and gamma expression and their JAK-STAT signaling pathway. *Life Sci.* **2004**, *74*, 1739–1749. [CrossRef]

135. Preston, C.C.; Maurer, M.J.; Oberg, A.L.; Visscher, D.W.; Kalli, K.R.; Hartmann, L.C.; Goode, E.L.; Knutson, K.L. The Ratios of CD8⁺ T Cells to CD4⁺CD25⁺ FOXP3⁺ and FOXP3- T Cells Correlate with Poor Clinical Outcome in Human Serous Ovarian Cancer. *PLoS ONE* **2013**, *8*, e80063. [CrossRef]

136. Sato, E.; Olson, S.H.; Ahn, J.; Bundy, B.; Nishikawa, H.; Qian, F.; Jungbluth, A.A.; Frosina, D.; Gnjatic, S.; Ambrosone, C.; et al. Intraepithelial CD8⁺ tumor-infiltrating lymphocytes and a high CD8⁺/regulatory T cell ratio are associated with favorable prognosis in ovarian cancer. *Proc. Natl. Acad. Sci. USA* **2005**, *102*, 18538–18543. [CrossRef]

137. Chanmee, T.; Ontong, P.; Konno, K.; Itano, N. Tumor-Associated Macrophages as Major Players in the Tumor Microenvironment. *Cancers* **2014**, *6*, 1670–1690. [CrossRef]

138. Takaishi, K.; Komohara, Y.; Tashiro, H.; Ohtake, H.; Nakagawa, T.; Katabuchi, H.; Takeya, M. Involvement of M2-polarized macrophages in the ascites from advanced epithelial ovarian carcinoma in tumor progression via Stat3 activation. *Cancer Sci.* **2010**, *101*, 2128–2136. [CrossRef]

139. Yin, M.; Li, X.; Tan, S.; Zhou, H.J.; Ji, W.; Bellone, S.; Xu, X.; Zhang, H.; Santin, A.D.; Lou, G.; et al. Tumor-associated macrophages drive spheroid formation during early transcoelomic metastasis of ovarian cancer. *J. Clin. Investig.* **2016**, *126*, 4157–4173. [CrossRef]

140. Ning, Y.; Cui, Y.; Li, X.; Cao, X.; Chen, A.; Xu, C.; Cao, J.; Luo, X. Co-culture of ovarian cancer stem-like cells with macrophages induced SKOV3 cells stemness via IL-8/STAT3 signaling. *Biomed. Pharmacother.* **2018**, *103*, 262–271. [CrossRef]

141. Lan, C.; Huang, X.; Lin, S.; Huang, H.; Cai, Q.; Wan, T.; Lu, J.; Liu, J. Expression of M2-Polarized Macrophages is Associated with Poor Prognosis for Advanced Epithelial Ovarian Cancer. *Technol. Cancer Res. Treat.* **2012**, *12*, 259–267. [CrossRef] [PubMed]

142. Zhang, M.; He, Y.; Sun, X.; Li, Q.; Wang, W.; Zhao, A.; Di, W. A high M1/M2 ratio of tumor-associated macrophages is associated with extended survival in ovarian cancer patients. *J. Ovarian Res.* **2014**, *7*, e19. [CrossRef] [PubMed]

143. Kryczek, I.; Zou, L.; Rodriguez, P.; Zhu, G.; Wei, S.; Mottram, P.; Brumlik, M.; Cheng, P.; Curiel, T.; Myers, L.; et al. B7-H4 expression identifies a novel suppressive macrophage population in human ovarian carcinoma. *J. Exp. Med.* **2006**, *203*, 871–881. [CrossRef] [PubMed]

144. Bharadwaj, U.; Kasembeli, M.M.; Tweardy, D.J. *STAT3 Inhibitors in Cancer: A Comprehensive Update*; Humana Press: Cham, Switzerland, 2016; pp. 95–161. ISBN 978-3-319-42949-6.

145. Wong, A.L.; Hirpara, J.L.; Pervaiz, S.; Eu, J.-Q.; Sethi, G.; Goh, B.-C. Do STAT3 inhibitors have potential in the future for cancer therapy? *Expert Opin. Investig. Drugs* **2017**, *26*, 1–5. [CrossRef] [PubMed]

146. Coward, J.; Kulbe, H.; Chakravarty, P.; Leader, D.; Vassileva, V.; Leinster, D.A.; Thompson, R.; Schioppa, T.; Nemeth, J.; Vermeulen, J.; et al. Interleukin-6 as a therapeutic target in human ovarian cancer. *Clin. Cancer Res.* **2011**, *17*, 6083–6096. [CrossRef] [PubMed]

147. Han, E.S.; Wen, W.; Dellinger, T.H.; Wu, J.; Lu, S.A.; Jove, R.; Yim, J.H. Ruxolitinib synergistically enhances the anti-tumor activity of paclitaxel in human ovarian cancer. *Oncotarget* **2018**, *9*, 24304–24319. [CrossRef] [PubMed]

148. Schwartz, D.M.; Kanno, Y.; Villarino, A.; Ward, M.; Gadina, M.; O'Shea, J.J. JAK inhibition as a therapeutic strategy for immune and inflammatory diseases. *Nat. Rev. Drug Discov.* **2017**, *16*, 843–862. [CrossRef] [PubMed]

149. Beatty, G.L.; Shahda, S.; Beck, T.; Uppal, N.; Cohen, S.J.; Donehower, R.; Gabayan, A.; Assad, A.; Switzky, J.; Zhen, H.; et al. A Phase Ib/II Study of the JAK1 Inhibitor, Itacitinib, plus nab-Paclitaxel and Gemcitabine in Advanced Solid Tumors. *Oncologist* **2019**, *24*, 14-e10. [CrossRef] [PubMed]

150. Chan, E.; Luwor, R.; Burns, C.; Kannourakis, G.; Findlay, J.K.; Ahmed, N. Momelotinib decreased cancer stem cell associated tumor burden and prolonged disease-free remission period in a mouse model of human ovarian cancer. *Oncotarget* **2018**, *9*, 16599–16618. [CrossRef]

151. Liu, L.; Gaboriaud, N.; Vougogianopoulou, K.; Tian, Y.; Wu, J.; Wen, W.; Skaltsounis, L.; Jove, R. MLS-2384, a new 6-bromoindirubin derivative with dual JAK/Src kinase inhibitory activity, suppresses growth of diverse cancer cells. *Cancer Biol. Ther.* **2014**, *15*, 178–184. [CrossRef]

152. Wen, W.; Liang, W.; Wu, J.; Kowolik, C.M.; Buettner, R.; Scuto, A.; Hsieh, M.-Y.; Hong, H.; Brown, C.E.; Forman, S.J.; et al. Targeting JAK1/STAT3 signaling suppresses tumor progression and metastasis in a peritoneal model of human ovarian cancer. *Mol. Cancer Ther.* **2014**, *13*, 3037–3048. [CrossRef] [PubMed]

153. Plimack, E.R.; Lorusso, P.M.; McCoon, P.; Tang, W.; Krebs, A.D.; Curt, G.; Eckhardt, S.G. AZD1480: A Phase I Study of a Novel JAK2 Inhibitor in Solid Tumors. *Oncologist* **2013**, *18*, 819–820. [CrossRef] [PubMed]

154. Goedemans, R.; Dijkgraaf, E.M.; Santegoets, S.J.A.M.; Reyners, A.K.L.; Wouters, M.C.A.; Kenter, G.G.; Van Erkel, A.R.; Van Poelgeest, M.I.E.; Nijman, H.W.; Van Der Hoeven, J.J.M.; et al. A phase I trial combining carboplatin/doxorubicin with tocilizumab, an anti-IL-6R monoclonal antibody, and interferon-α2b in patients with recurrent epithelial ovarian cancer. *Ann. Oncol.* **2015**, *26*, 2141–2149.

155. Abubaker, K.; Luwor, R.B.; Zhu, H.; McNally, O.; Quinn, M.A.; Burns, C.J.; Thompson, E.W.; Findlay, J.K.; Ahmed, N. Inhibition of the JAK2/STAT3 pathway in ovarian cancer results in the loss of cancer stem cell-like characteristics and a reduced tumor burden. *BMC Cancer* **2014**, *14*, e317. [CrossRef] [PubMed]

156. Elnaggar, A.C.; Saini, U.; Naidu, S.; Wanner, R.; Sudhakar, M.; Fowler, J.; Nagane, M.; Kuppusamy, P.; Cohn, D.E.; Selvendiran, K. Anticancer potential of diarylidenyl piperidone derivatives, HO-4200 and H-4318, in cisplatin resistant primary ovarian cancer. *Cancer Boil. Ther.* **2016**, *17*, 1107–1115. [CrossRef]

157. Selvendiran, K.; Ahmed, S.; Dayton, A.; Kuppusamy, M.L.; Rivera, B.K.; Kálai, T.; Hideg, K.; Kuppusamy, P. HO-3867, a curcumin analog, sensitizes cisplatin-resistant ovarian carcinoma, leading to therapeutic synergy through STAT3 inhibition. *Cancer Boil. Ther.* **2011**, *12*, 837–845. [CrossRef]

158. Selvendiran, K.; Tong, L.; Bratasz, A.; Kuppusamy, M.L.; Ahmed, S.; Ravi, Y.; Trigg, N.J.; Rivera, B.K.; Kálai, T.; Hideg, K.; et al. Anticancer efficacy of a difluorodiarylidenyl piperidone (HO-3867) in human ovarian cancer cells and tumor xenografts. *Mol. Cancer Ther.* **2010**, *9*, 1169–1179. [CrossRef]

159. McCann, G.A.; Naidu, S.; Rath, K.S.; Bid, H.K.; Tierney, B.J.; Suarez, A.; Varadharaj, S.; Zhang, J.; Hideg, K.; Houghton, P.; et al. Targeting constitutively-activated STAT3 in hypoxic ovarian cancer, using a novel STAT3 inhibitor. *Oncoscience* **2014**, *1*, 216–228. [CrossRef]

160. Liu, M.; Wang, F.; Wen, Z.; Shi, M.; Zhang, H. Blockage of STAT3 Signaling Pathway with a Decoy Oligodeoxynucleotide Inhibits Growth of Human Ovarian Cancer Cells. *Cancer Investig.* **2013**, *32*, 8–12. [CrossRef]

161. Ma, Y.; Zhang, X.; Xu, X.; Shen, L.; Yao, Y.; Yang, Z.; Liu, P. STAT3 Decoy Oligodeoxynucleotides-Loaded Solid Lipid Nanoparticles Induce Cell Death and Inhibit Invasion in Ovarian Cancer Cells. *PLoS ONE* **2015**, *10*, e0124924. [CrossRef]

162. Zhang, X.; Liu, P.; Zhang, B.; Wang, A.; Yang, M. Role of STAT3 decoy oligodeoxynucleotides on cell invasion and chemosensitivity in human epithelial ovarian cancer cells. *Cancer Genet. Cytogenet.* **2010**, *197*, 46–53. [CrossRef] [PubMed]

163. Liao, X.-H.; Xiang, Y.; Yu, C.-X.; Li, J.-P.; Li, H.; Nie, Q.; Hu, P.; Zhou, J.; Zhang, T.-C. STAT3 is required for MiR-17-5p-mediated sensitization to chemotherapy-induced apoptosis in breast cancer cells. *Oncotarget* **2014**, *5*, e15763. [CrossRef] [PubMed]

164. Zhou, W.; Bi, X.; Gao, G.; Sun, L. miRNA-133b and miRNA-135a induce apoptosis via the JAK2/STAT3 signaling pathway in human renal carcinoma cells. *Biomed. Pharmacother.* **2016**, *84*, 722–729. [CrossRef] [PubMed]

165. Zhang, Y.; Kim, J.; Mueller, A.C.; Dey, B.; Yang, Y.; Lee, D.-H.; Hachmann, J.; Finderle, S.; Park, D.M.; Christensen, J.; et al. Multiple receptor tyrosine kinases converge on microRNA-134 to control KRAS, STAT5B, and glioblastoma. *Cell Death Differ.* **2014**, *21*, 720–734. [CrossRef]

166. Li, Z.-Y.; Yang, L.; Liu, X.-J.; Wang, X.-Z.; Pan, Y.-X.; Luo, J.-M. The Long Noncoding RNA MEG3 and its Target miR-147 Regulate JAK/STAT Pathway in Advanced Chronic Myeloid Leukemia. *EBioMedicine* **2018**, *34*, 61–75. [CrossRef]

167. Gheysarzadeh, A.; Yazdanparast, R. STAT5 Reactivation by Catechin Modulates H2O2-Induced Apoptosis Through miR-182/FOXO1 Pathway in SK-N-MC Cells. *Cell Biochem. Biophys.* **2015**, *71*, 649–656. [CrossRef]

168. Wang, X.; Qiu, W.; Zhang, G.; Xu, S.; Gao, Q.; Yang, Z. MicroRNA-204 targets JAK2 in breast cancer and induces cell apoptosis through the STAT3/BCl-2/survivin pathway. *Int. J. Clin. Exp. Pathol.* **2015**, *8*, 5017–5025.

169. Zhao, B.; Dong, A.-S. MiR-874 inhibits cell growth and induces apoptosis by targeting STAT3 in human colorectal cancer cells. *Eur. Rev. Med. Pharmacol. Sci.* **2016**, *20*, 269–277.

170. Zaanan, A.; Okamoto, K.; Kawakami, H.; Khazaie, K.; Huang, S.; Sinicrope, F.A. The Mutant KRAS Gene Up-regulates BCL-XL Protein via STAT3 to Confer Apoptosis Resistance That Is Reversed by BIM Protein Induction and BCL-XL Antagonism*. *J. Boil. Chem.* **2015**, *290*, 23838–23849. [CrossRef]

171. Catlett-Falcone, R.; Landowski, T.H.; Oshiro, M.M.; Turkson, J.; Levitzki, A.; Savino, R.; Ciliberto, G.; Moscinski, L.; Fernández-Luna, J.L.; Nuñez, G.; et al. Constitutive activation of Stat3 signaling confers resistance to apoptosis in human U266 myeloma cells. *Immunity* **1999**, *10*, 105–115. [CrossRef]

172. Bhattacharya, S.; Ray, R.M.; Johnson, L.R. STAT3-mediated transcription of Bcl-2, Mcl-1 and c-IAP2 prevents apoptosis in polyamine-depleted cells. *Biochem. J.* **2005**, *392*, 335–344. [CrossRef] [PubMed]

173. Lanuti, P.; Bertagnolo, V.; Pierdomenico, L.; Bascelli, A.; Santavenere, E.; Alinari, L.; Capitani, S.; Miscia, S.; Marchisio, M. Enhancement of TRAIL cytotoxicity by AG-490 in human ALL cells is characterized by downregulation of cIAP-1 and cIAP-2 through inhibition of Jak2/Stat3. *Cell Res.* **2009**, *19*, 1079–1089. [CrossRef] [PubMed]

174. Fukumoto, T.; Iwasaki, T.; Okada, T.; Hashimoto, T.; Moon, Y.; Sakaguchi, M.; Fukami, Y.; Nishigori, C.; Oka, M. High expression of Mcl-1L via the MEK-ERK-phospho-STAT3 (Ser727) pathway protects melanocytes and melanoma from UVB-induced apoptosis. *Genes Cells* **2016**, *21*, 185–199. [CrossRef] [PubMed]

175. Becker, T.; Boyd, S.; Mijatov, B.; Gowrishankar, K.; Snoyman, S.; Pupo, G.; Scolyer, R.; Mann, G.; Kefford, R.; Zhang, X.; et al. Mutant B-RAF-Mcl-1 survival signaling depends on the STAT3 transcription factor. *Oncogene* **2014**, *33*, 1158–1166. [CrossRef] [PubMed]

176. Haga, S.; Terui, K.; Zhang, H.Q.; Enosawa, S.; Ogawa, W.; Inoue, H.; Okuyama, T.; Takeda, K.; Akira, S.; Ogino, T.; et al. Stat3 protects against Fas-induced liver injury by redox-dependent and -independent mechanisms. *J. Clin. Investig.* **2003**, *112*, 989–998. [CrossRef]

177. Nagasaki, T.; Hara, M.; Nakanishi, H.; Takahashi, H.; Takeyama, H. Interleukin-6 released by colon cancer-associated fibroblasts is critical for tumour angiogenesis: Anti-interleukin-6 receptor antibody suppressed angiogenesis and inhibited tumour–stroma interaction. *Br. J. Cancer* **2014**, *110*, 469–478. [CrossRef]

178. Wei, D.; Le, X.; Zheng, L.; Wang, L.; Frey, J.A.; Gao, A.C.; Peng, Z.; Huang, S.; Xiong, H.Q.; Abbruzzese, J.L.; et al. Stat3 activation regulates the expression of vascular endothelial growth factor and human pancreatic cancer angiogenesis and metastasis. *Oncogene* **2003**, *22*, 319–329. [CrossRef]

179. Zou, M.; Xu, C.; Li, H.; Zhang, X.; Fan, W. 3,3′-Diindolylmethane suppresses ovarian cancer cell viability and metastasis and enhances chemotherapy sensitivity via STAT3 and Akt signaling in vitro and in vivo. *Arch. Biochem. Biophys.* **2018**, *651*. [CrossRef]

180. Ganji, P.N.; Park, W.; Wen, J.; Mahaseth, H.; Landry, J.; Farris, A.B.; Willingham, F.; Sullivan, P.S.; Proia, D.A.; El-Hariry, I.; et al. Antiangiogenic effects of ganetespib in colorectal cancer mediated through inhibition of HIF-1α and STAT-3. *Angiogenesis* **2013**, *16*, 903–917. [CrossRef]

181. Wincewicz, A.; Koda, M.; Sulkowska, M.; Kanczuga-Koda, L.; Sulkowski, S. Comparison of STAT3 with HIF-1alpha, Ob and ObR expressions in human endometrioid adenocarcinomas. *Tissue Cell* **2008**, *40*, 405–410. [CrossRef]

182. Xie, T.-X.; Huang, F.-J.; Aldape, K.D.; Kang, S.-H.; Liu, M.; Gershenwald, J.E.; Xie, K.; Sawaya, R.; Huang, S. Activation of Stat3 in Human Melanoma Promotes Brain Metastasis. *Cancer Res.* **2006**, *66*, 3188–3196. [CrossRef] [PubMed]

183. Xie, T.-X.; Wei, D.; Liu, M.; Gao, A.C.; Ali-Osman, F.; Sawaya, R.; Huang, S. Stat3 activation regulates the expression of matrix metalloproteinase-2 and tumor invasion and metastasis. *Oncogene* **2004**, *23*, 3550–3560. [CrossRef] [PubMed]

184. Fukuda, A.; Wang, S.C.; Morris, J.P.; Folias, A.E.; Liou, A.; Kim, G.E.; Akira, S.; Boucher, K.M.; Firpo, M.A.; Mulvihill, S.J.; et al. Stat3 and MMP7 contribute to pancreatic ductal adenocarcinoma initiation and progression. *Cancer Cell* **2011**, *19*, 441–455. [CrossRef] [PubMed]

185. Zhang, C.; Guo, F.; Xu, G.; Ma, J.; Shao, F. STAT3 cooperates with Twist to mediate epithelial-mesenchymal transition in human hepatocellular carcinoma cells. *Oncol. Rep.* **2015**, *33*, 1872–1882. [CrossRef] [PubMed]

186. Zhao, C.; Wang, Q.; Wang, B.; Sun, Q.; He, Z.; Hong, J.; Kuehn, F.; Liu, E.; Zhang, Z. IGF-1 induces the epithelial-mesenchymal transition via Stat5 in hepatocellular carcinoma. *Oncotarget* **2014**, *5*, 111922–111930. [CrossRef] [PubMed]

187. Siddiqui, I.; Erreni, M.; Kamal, M.A.; Porta, C.; Marchesi, F.; Pesce, S.; Pasqualini, F.; Schiarea, S.; Chiabrando, C.; Mantovani, A.; et al. Differential role of Interleukin-1 and Interleukin-6 in K-Ras-driven pancreatic carcinoma undergoing mesenchymal transition. *OncoImmunology* **2017**, *7*, e1388485. [CrossRef] [PubMed]

188. Talati, P.G.; Gu, L.; Ellsworth, E.M.; Girondo, M.A.; Trerotola, M.; Hoang, D.T.; Leiby, B.; Dagvadorj, A.; McCue, P.A.; Lallas, C.D.; et al. Jak2-Stat5a/b Signaling Induces Epithelial-to-Mesenchymal Transition and Stem-Like Cell Properties in Prostate Cancer. *Am. J. Pathol.* **2015**, *185*, 2505–2522. [CrossRef]

189. Sultan, A.S.; Brim, H.; Sherif, Z.A. Co-overexpression of Janus kinase 2 and signal transducer and activator of transcription 5a promotes differentiation of mammary cancer cells through reversal of epithelial–mesenchymal transition. *Cancer Sci.* **2008**, *99*, 272–279. [CrossRef]

190. Chang, Q.; Bournazou, E.; Sansone, P.; Berishaj, M.; Gao, S.P.; Daly, L.; Wels, J.; Theilen, T.; Granitto, S.; Zhang, X.; et al. The IL-6/JAK/Stat3 Feed-Forward Loop Drives Tumorigenesis and Metastasis. *Neoplasia* **2013**, *15*, 848–862. [CrossRef]

191. Hadjidaniel, M.D.; Muthugounder, S.; Hung, L.T.; Sheard, M.A.; Shirinbak, S.; Chan, R.Y.; Nakata, R.; Borriello, L.; Malvar, J.; Kennedy, R.J.; et al. Tumor-associated macrophages promote neuroblastoma via STAT3 phosphorylation and up-regulation of c-MYC. *Oncotarget* **2014**, *5*, 91516–91529. [CrossRef]

192. Hamada, S.; Masamune, A.; Yoshida, N.; Takikawa, T.; Shimosegawa, T. IL-6/STAT3 Plays a Regulatory Role in the Interaction Between Pancreatic Stellate Cells and Cancer Cells. *Dig. Dis. Sci.* **2016**, *61*, 1561–1571. [CrossRef] [PubMed]

193. Song, L.; Turkson, J.; Karras, J.G.; Jove, R.; Haura, E.B. Activation of Stat3 by receptor tyrosine kinases and cytokines regulates survival in human non-small cell carcinoma cells. *Oncogene* **2003**, *22*, 4150–4165. [CrossRef] [PubMed]

194. Wu, X.; Tao, P.; Zhou, Q.; Li, J.; Yu, Z.; Wang, X.; Li, J.; Li, C.; Yan, M.; Zhu, Z.; et al. IL-6 secreted by cancer-associated fibroblasts promotes epithelial-mesenchymal transition and metastasis of gastric cancer via JAK2/STAT3 signaling pathway. *Oncotarget* **2014**, *5*, e20741. [CrossRef] [PubMed]

195. Chen, H.; Aksoy, I.; Gonnot, F.; Osteil, P.; Aubry, M.; Hamela, C.; Rognard, C.; Hochard, A.; Voisin, S.; Fontaine, E.; et al. Reinforcement of STAT3 activity reprogrammes human embryonic stem cells to naive-like pluripotency. *Nat. Commun.* **2015**, *6*, e7095. [CrossRef]

196. Wang, D.; Xiang, T.; Zhao, Z.; Lin, K.; Yin, P.; Jiang, L.; Liang, Z.; Zhu, B. Autocrine interleukin-23 promotes self-renewal of CD133$^+$ ovarian cancer stem-like cells. *Oncotarget* **2016**, *7*, e76006. [CrossRef]

197. Li, X.; Wang, H.; Ding, J.; Nie, S.; Wang, L.; Zhang, L.; Ren, S. Celastrol strongly inhibits proliferation, migration and cancer stem cell properties through suppression of Pin1 in ovarian cancer cells. *Eur. J. Pharmacol.* **2019**, *842*, 146–156. [CrossRef]

Article

STAT5a/b Deficiency Delays, but does not Prevent, Prolactin-Driven Prostate Tumorigenesis in Mice

Florence Boutillon [1,2], Natascha Pigat [1,2], Lucila Sackmann Sala [1,2], Edouard Reyes-Gomez [3,4,5], Richard Moriggl [6,7], Jacques-Emmanuel Guidotti [1,2,†] and Vincent Goffin [1,2,*,†]

1   Institut Necker Enfants Malades, Inserm U1151, 75014 Paris, France
2   Faculté de Médecine, Université Paris Descartes, 75014 Paris, France
3   Unité d'Histologie et d'Anatomie Pathologique, Ecole Nationale Vétérinaire d'Alfort, 94704 Maisons-Alfort, France
4   Laboratoire d'Anatomo-Cytopathologie, BioPôle Alfort, Ecole Nationale Vétérinaire d'Alfort, 94704 Maisons-Alfort, France
5   U955—IMRB, Inserm, Ecole Nationale Vétérinaire d'Alfort, UPEC, 94704 Maisons-Alfort, France
6   Institute of Animal Breeding and Genetics, University of Veterinary Medicine Vienna, 1210 Vienna, Austria
7   Medical University of Vienna, 1090 Vienna, Austria
*   Correspondence: Vincent.goffin@inserm.fr
†   These authors contributed equally to this paper.

Received: 11 June 2019; Accepted: 1 July 2019; Published: 2 July 2019

Abstract: The canonical prolactin (PRL) Signal Transducer and Activator of Transcription (STAT) 5 pathway has been suggested to contribute to human prostate tumorigenesis via an autocrine/paracrine mechanism. The probasin (Pb)-PRL transgenic mouse models this mechanism by overexpressing PRL specifically in the prostate epithelium leading to strong STAT5 activation in luminal cells. These mice exhibit hypertrophic prostates harboring various pre-neoplastic lesions that aggravate with age and accumulation of castration-resistant stem/progenitor cells. As STAT5 signaling is largely predominant over other classical PRL-triggered pathways in Pb-PRL prostates, we reasoned that Pb-Cre recombinase-driven genetic deletion of a floxed *Stat5a/b* locus should prevent prostate tumorigenesis in so-called Pb-PRL$^{\Delta STAT5}$ mice. Anterior and dorsal prostate lobes displayed the highest *Stat5a/b* deletion efficiency with no overt compensatory activation of other PRLR signaling cascade at 6 months of age; hence the development of tumor hallmarks was markedly reduced. *Stat5a/b* deletion also reversed the accumulation of stem/progenitor cells, indicating that STAT5 signaling regulates prostate epithelial cell hierarchy. Interestingly, ERK1/2 and AKT, but not STAT3 and androgen signaling, emerged as escape mechanisms leading to delayed tumor development in aged Pb-PRL$^{\Delta STAT5}$ mice. Unexpectedly, we found that Pb-PRL prostates spontaneously exhibited age-dependent decline of STAT5 signaling, also to the benefit of AKT and ERK1/2 signaling. As a consequence, both Pb-PRL and Pb-PRL$^{\Delta STAT5}$ mice ultimately displayed similar pathological prostate phenotypes at 18 months of age. This preclinical study provides insight on STAT5-dependent mechanisms of PRL-induced prostate tumorigenesis and alternative pathways bypassing STAT5 signaling down-regulation upon prostate neoplasia progression.

Keywords: STAT5; AKT; ERK1/2; prolactin; androgens; prostate cancer; knockout; escape mechanisms; stem/progenitor cells; cell hierarchy

---

1. Introduction

Studies of human prostate cancer specimens support a role for prolactin (PRL) signaling in disease progression and recurrence [1]. Indeed, PRL is expressed in more than half of prostate tumors (including local, locally advanced, and hormone refractory) and over 60% metastases [2,3], and its level of expression in primary tumors is positively associated with high Gleason score (i.e., high disease severity) [2]. Since circulating PRL levels are not correlated to prostate cancer risk [4,5], this suggests that the contribution of PRL signaling to prostate tumorigenesis mainly occurs through an autocrine/paracrine mechanism.

The canonical signaling pathways activated by the PRL receptor (PRLR) involve Janus kinase 2 (JAK2)/Signal Transducer and Activator of Transcription (STAT) pathway, the extracellular regulated kinase (ERK) 1/2 pathway and the phosphoinositide 3-kinase (PI3K)–AKT pathway [6,7]. In human and rodent prostate, the PRLR preferentially signals via the STAT5 pathway; in fact, activation of the other pathways is not detected [2,8,9]. STAT5 involves two highly homologous proteins referred to as STAT5A (94 kDa) and STAT5B (92 kDa) that are encoded by two distinct genes [10]. In PRL signaling STAT5A and STAT5B are activated by JAK2-mediated phosphorylation of a conserved C-terminal tyrosine upon which they form a parallel dimer that allows more efficient translocation into the nucleus where they activate the transcription of target genes involving many protooncogenes. In various preclinical models of prostate cancer, STAT5A/B (hereafter referred to as STAT5 unless specifically discriminated for gene product) has been shown to be critical for cell survival and proliferation through both androgen-dependent and androgen-independent mechanisms [9,11–14]. Recently, STAT5 was shown to promote epithelial-to-mesenchymal transition (EMT) and stem-like features in human prostate cancer cells [15], supporting the earlier finding that this cascade promotes metastatic properties of prostate cancer cells [16]. Consistent with these observations, inhibition of STAT5 signaling using a dominant-negative STAT5B mutant inhibited the in vitro growth and invasiveness of cell lines derived from the TRAMP (transgenic adenocarcinoma of the mouse prostate) mouse model [17]. Additionally, pharmacological inhibition of the JAK2/STAT5 cascade using the JAK2 inhibitor AZD1480 blocked the growth of primary androgen-dependent as well as the growth of recurrent castrate-resistant prostate cancer (CRPC) xenografts [18]. However, so far no transgenic mouse model studies exist for complete STAT5 deletion to ultimately test its role in prostate cancer initiation and progression, which we investigated here.

In clinical specimens of human prostate cancer, the *STAT5a/b* gene locus was shown to undergo amplification during prostate cancer progression towards metastatic CRPC [19]. Accordingly, the *STAT5a* gene locus was found to be amplified in up to 20% of metastatic CRPC with the neuroendocrine phenotype [20]. Furthermore, STAT5 was shown to be overexpressed in prostate cancer compared to healthy prostate samples, to be positively correlated with Gleason score and to predict recurrence after prostatectomy [2,19,21,22]. These findings corroborated the results obtained in preclinical models and further supported the relevance to develop therapeutic strategies aimed to inhibit STAT5 signaling in prostate cancer. Small molecule inhibitors were designed to block the docking of the SH2 domain of STAT5 to the critical tyrosine of the receptor-JAK2 complex and these have shown potency to inhibit STAT5 tyrosine phosphorylation, nuclear translocation and transcriptional activity [23]. The lead compound (IST5) induced massive apoptosis of prostate cancer cell lines and explant culture of patient-derived prostate cancer [23]. More recently this inhibitor was shown to sensitize prostate cancer to radiation by inhibiting STAT5-mediated DNA repair via the homologous recombination mechanism [24]. Additional observations supporting the role of PRL/STAT5 signaling in prostate tumorigenesis and cancer progression can be found in various review articles published within the last decade [1,14,25–27].

In contrast to the human prostate [28], expression of the *Prl* gene is not detected in the mouse prostate. Hence, to decipher the molecular and cellular mechanisms linking the autocrine PRL/STAT5 loop to prostate tumorogenesis, we use here a prostate-specific PRL transgenic mouse model. This model involves expression of rat PRL under the control of the prostate-specific, androgen-regulated probasin (Pb) minimal promoter [29]. At six months of age, Pb-PRL mice exhibit dramatically hypertrophied prostates (all lobes) harboring various pre-neoplastic lesions including prostate-intraepithelial neoplasia (PINs), increased stroma density, and inflammation [1,30,31]. They also displayed distended ducts filled with abundant secretions. All these phenotypes aggravate with age. While our seminal report indicated occasional occurrence of invasive cancer in 20 month-old mice [30], these findings were not consistently found in subsequent studies, suggesting possible interference with the genetic background and/or health/microbiota/immune cell status of transgenics that could drift particularly if animal facilities were relocated, as was the case.

According to earlier findings mentioned above, STAT5 was the only typical PRLR signaling pathway that was found to be activated in Pb-PRL prostates [30]. STAT5 was highly activated in dorsal and lateral lobe, while the ventral lobe displayed much lower level of activation presumably due to PRLR down-regulation specifically in that lobe [31]. Of interest, we discovered that Pb-PRL prostates displayed altered cell hierarchy. First, the basal/stem cell compartment was found to be markedly amplified in the prostate epithelium [30]. Second, these prostates showed amplification of another primitive cell population that had never been described before as it is very rare in the healthy prostate. This newly-identified epithelial cell pool, called LSC^{med} according to its cell sorting profile (Lin−/Sca-1+/CD49f^{med}), combines luminal (Cytokeratin [CK] 8 positive) and stem (Stem-cell antigen-1 positive) phenotypic features and exhibits stem/progenitor properties in functional assays [31,32]. Notably, while the basal/stem cell compartment has been long described to be castrate-resistant in rats [33], we recently reported that the LSC^{med} cell compartment survived castration even better than basal/stem cells in mice [32].

The amplification of castrate-resistant, stem/progenitor cells in Pb-PRL prostates further supports a role for PRL signaling in the progression of prostate tumors including escape to androgen deprivation. Of interest, careful analysis of immunostaining data suggested that the emergence of these two primitive cell populations in Pb-PRL tumors may be zonally correlated to elevated STAT5 activation [31]. However, as neither basal/stem nor LSC^{med} cell population exhibit detectable levels of PRLR expression [32], the role of STAT5 signaling in their amplification remains elusive and may involve paracrine mechanisms. In order to elucidate the actual contribution of STAT5 signaling in the various hallmarks of PRL-induced prostate tumorigenesis, we took advantage of previously developed floxed *Stat5a/b* mice [34] to abolish STAT5 expression in the epithelial cells of Pb-PRL mice. As reported below, in the anterior and dorsal prostate lobes in which *Stat5a/b* locus deletion was highly efficient, STAT5 deficiency delayed the occurrence of PRL-induced pathological phenotypes in young mice, but could not prevent prostate tumorigenesis in older animals due to the emergence of alternative signaling pathways.

2. Results

2.1. STAT5 Deletion has no Detectable Effect on the Prostate Tissue

The analyses of STAT5 expression in the prostate were performed on 6 month-old animals. As shown by RT-qPCR (Figure S1A) *Stat5a* expression was largely prevalent (~5 fold) over *Stat5b* expression in control mice (STAT5$^{f/f}$), but both genes displayed similar lobe-specific expression patterns with the highest expression in anterior and lateral lobes. While the lobe differences were less marked at the protein level, the same profile was observed by immunoblot using the G2 antibody that cross-reacts with both STAT5 isoforms (Figure S1B,C). In ΔSTAT5 mice the fold-reduction in STAT5 expression was the highest in lateral lobe and the lowest in ventral lobe (Figure S1B,C). The residual expression of STAT5 detected at the mRNA and protein levels in tissue extracts of the different lobes of ΔSTAT5 mouse prostates accounted for the unaltered STAT5 expression in non-epithelial cells and possibly to incomplete *Stat5* deletion in the epithelium especially in the ventral lobe (mosaicism). To address this further, we monitored STAT5 expression by immunohistochemistry (IHC) using the G2 antibody (Figure S1D). In ΔSTAT5 prostates a homogenous reduction of STAT5 immunostaining was observed in the epithelium of all but the ventral lobes compared to STAT5$^{f/f}$ prostates (Figure S1D). Together these analyses indicated down-regulation of STAT5A/B expression in the epithelium of the lateral, dorsal and anterior lobes of ΔSTAT5 mouse prostates.

Prostates harvested from 6, 12, and 18 months old ΔSTAT5 animals failed to display any macroscopic alterations (e.g., organ atrophy or hypertrophy). Accordingly, the weights of the different lobes (normalized to mouse weight) were unaffected by STAT5 deletion (Figure S2A). Histopathological analyses of prostate sections from these animals were performed blinded by a trained veterinarian pathologist (E.R.-G.). This analysis failed to identify detectable alterations in ΔSTAT5 compared to STAT5$^{F/F}$ prostates regarding prostate architecture and histology (Figure S2B). These findings are consistent with an older study that failed to observe histological defects in dorsal and lateral lobes of *Stat5a*-null mice [35]. In fact, systemic STAT5A deficiency induced local disorganization within acinar epithelium of the ventral lobe only, which could not be confirmed in our study as Pb-Cre4-driven *Stat5* gene ablation was inefficient in that lobe.

Taken together, these results indicate that Pb-Cre4 recombinase-driven *Stat5* deletion has no major impact on prostate tissue integrity.

2.2. Lobe-Specific Pattern of STAT5 Deletion in Pb-PRL^{ΔSTAT5} Mice

We next generated Pb-PRL mice harboring *Stat5a/b* gene deletion in the epithelium (hereafter called Pb-PRL^{ΔSTAT5}). The analyses of *Stat5a/b* expression were performed on 6 month-old animals. As observed in STAT5$^{f/f}$ mice, *Stat5a* mRNA was predominant over *Stat5b* in Pb-PRL$^{STAT5f/f}$ prostates, with highest levels in anterior and lateral lobes (Figure 1A). In Pb-PRL^{ΔSTAT5} the reduction in *Stat5a/b* mRNA expression was significant in all but the ventral lobes, and more marked in anterior and dorsal (>3-fold) than lateral (<2 fold) lobes (Figure 1A). This pattern was globally confirmed at the protein level using immunoblot and IHC analyses. As shown in Figure 1B (immunoblot) and 1C (quantification), STAT5 protein was efficiently deleted in anterior and dorsal lobes of Pb-PRL^{ΔSTAT5} mice, and the deletion was homogeneous in the epithelium (Figure 1E). In contrast, no significant reduction in STAT5 protein levels were observed in ventral prostate, while deletion in the lateral lobe was intermediate (Figure 1B,C and Figure S3A).

As earlier reported, STAT5 is massively activated by transgenic PRL overexpressed in Pb-PRL prostates. Therefore tyrosine phosphorylated (p) STAT5 was also monitored by immunoblot (Figure 1B,D) and IHC (Figure 1F,G and Figure S3B) using a validated anti-pSTAT5 antibody [30,31]. Thanks to the contrasted nuclear staining obtained in IHC using the latter antibody (Figure 1F), we could quantify the actual level of activated (nuclear) STAT5 in the luminal epithelium of the four lobes. As expected from above-mentioned studies, the level of STAT5 tyrosine phosphorylation in Pb-PRL$^{STAT5f/f}$ mice was the highest in dorsal lobe and the lowest in ventral lobe (Figure 1G). Also, the pattern of nuclear STAT5 was zonal in the anterior and ventral prostates and more uniformly spread in dorsal and lateral prostates (Figures 1F and S3B). In Pb-PRL^{ΔSTAT5} mice, the significant reduction in STAT5 phosphorylation in anterior and dorsal lobes was assessed by immunoblot (Figure 1B,D) confirming our IHC results of STAT5 down-regulation in epithelial cells (Figure 1F,G). In agreement with the low efficiency of STAT5 deletion in the ventral lobe (Figure 1A,C), no significant reduction of pSTAT5 could be detected in that lobe (Figure 1D,G). Finally, in the lateral lobe, despite of the reduction of STAT5 expression (Figure 1C) there was no significant reduction of STAT5 tyrosine phosphorylation, as determined by blot (Figure 1B) or by IHC (Figure 1G). This is consistent with the fact that the degree of STAT5 phosphorylation in that the lobe is intrinsically moderate in Pb-PRL$^{STAT5f/f}$ mice (Figure 1G), which limits the impact of mild decreased STAT5 expression.

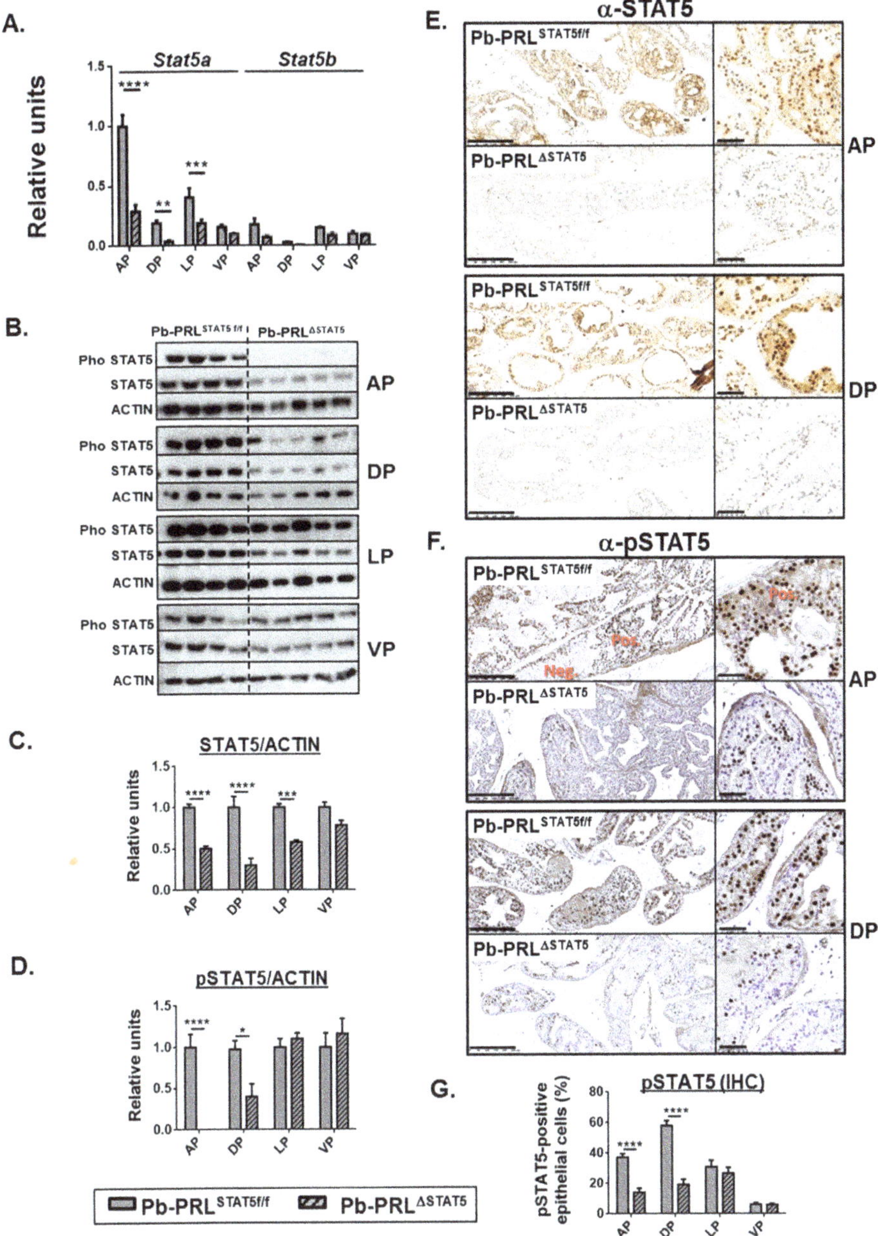

Figure 1. Lobe-specific pattern of STAT5 deletion in 6 month-old Pb-PRL^ΔSTAT5 mice. (**A**). Lobe-specific expression of *Stat5a* and *Stat5b* mRNA in Pb-PRL^STAT5f/f and Pb-PRL^ΔSTAT5 mice as determined by RT-qPCR is shown. (**B–D**). Lobe-specific expression and phosphorylation of STAT5 protein in Pb-PRL^STAT5f/f and Pb-PRL^ΔSTAT5 mice as determined by immunoblot is shown. Quantification of STAT5/ACTIN (C) and pSTAT5/ACTIN (D) was performed by densitometry and is shown as fold change versus Pb-PRL^STAT5f/f mice for each lobe. (**E**,**F**). Immunohistochemical analysis of STAT5 protein expression (**E**) and phosphorylation (**F**) in anterior (AP) and dorsal (DP) prostates of Pb-PRL^STAT5f/f and Pb-PRL^ΔSTAT5 mice, as indicated. In panel F, negatively (neg) next to positively (pos) immunostained

areas are highlighted. See Figure S3 for the other lobes. (**G**). The percentage of pSTAT5-positive epithelial cells as determined using Calopix software is shown for each lobe. *Statistics*: Stars (* $p < 0.05$, ** $p < 0.01$, *** $p < 0.001$, **** $p < 0.0001$; idem in all figures below) denote significant differences in a repeated-measures two-way ANOVA with Sidak's multiple comparisons. Size bars: 250 μm in large images and 50 μm in insets.

To confirm that the down-regulation of STAT5 phosphorylation resulted in lower STAT5 signaling, we monitored the levels of expression of Suppressor of Cytokine Signaling (SOCS)/Cytokine-Inducible SH2 Containing Protein (CISH) genes as typical negative regulators and direct targets of JAK/STAT signaling. Based on former reports, SOCS1/2/3 and CISH can be efficiently induced by PRLR/JAK/STAT signaling [7]. SOCS2 was predominantly expressed in the mouse prostate, while SOCS1 was hardly detected (see Figure 6D below for relative expression levels). SOCS2, SOCS3 and CISH were all up-regulated in Pb-PRL$^{STAT5f/f}$ dorsal prostates and significantly down-regulated in Pb-PRL$^{\Delta STAT5}$ confirming lower STAT5 signaling in the latter (Figure S3C). In addition to SOCS genes, we earlier reported that PRLR expression (all isoforms) was down-regulated in Pb-PRL prostates compared to wild-type (WT) littermates [31]; this was confirmed in this study using the dorsal prostate of Pb-PRL$^{STAT5f/f}$ mice (Figure S3D). Notably, normal levels of PRLR expression were restored in Pb-PRL$^{\Delta STAT5}$ prostates suggesting that PRLR expression is negatively regulated by STAT5 signaling (Figure S3D).

In summary, Pb-Cre4-driven *Stat5a/b* deletion was highly efficient to down-regulate STAT5 expression and signaling in the anterior and dorsal prostates but not in the ventral prostate; lateral prostate *Stat5a/b* deletion was intermediate. As described below, these lobe-specific effects were used to delineate the actual involvement of epithelial STAT5 pathway in the various tumor-related phenotypes analyzed.

2.3. STAT5 Deletion Reduces Hallmarks of PRL-Induced Prostate Tumorigenesis at 6 Month of Age

To evaluate the role of STAT5 pathway in PRL-driven prostate phenotypes, we compared the prostates of Pb-PRL$^{STAT5f/f}$ and Pb-PRL$^{\Delta STAT5}$ mice regarding the major hallmarks of early tumorigenesis previously characterized in Pb-PRL prostates [30–32,36].

2.3.1. Prostate Growth

According to former reports involving Pb-PRL mice [30], this study confirmed that prostate hypertrophy was detectable from 3 months of age in Pb-PRL$^{STAT5f/f}$ mice compared to control mice. At 6 months of age prostate weight was more than doubled compared to controls (Figure 2A). All lobes significantly contributed to prostate hypertrophy (Figures 2A and S4A). This phenotype was primarily due to cell hyperplasia, as reflected by the higher number of cells obtained in cell sorting (see below Section 2.4) and was further magnified by the increase in prostatic secretions, especially in the anterior prostate. Prostate weight of Pb-PRL$^{\Delta STAT5}$ was significantly reduced compared to Pb-PRL$^{STAT5f/f}$ mice, but remained higher than in control STAT5$^{f/f}$ mice (Figure 2A, top panel). In fact, anterior and dorsal lobes of Pb-PRL$^{\Delta STAT5}$ mice exhibited significant weight loss compared to their counterparts in Pb-PRL$^{STAT5f/f}$ mice (Figure 2A, bottom panels), but this was not the case for ventral and lateral prostates (Figure S4A). These results are in agreement with the lobe-specific pattern of STAT5 signaling down-regulation described above.

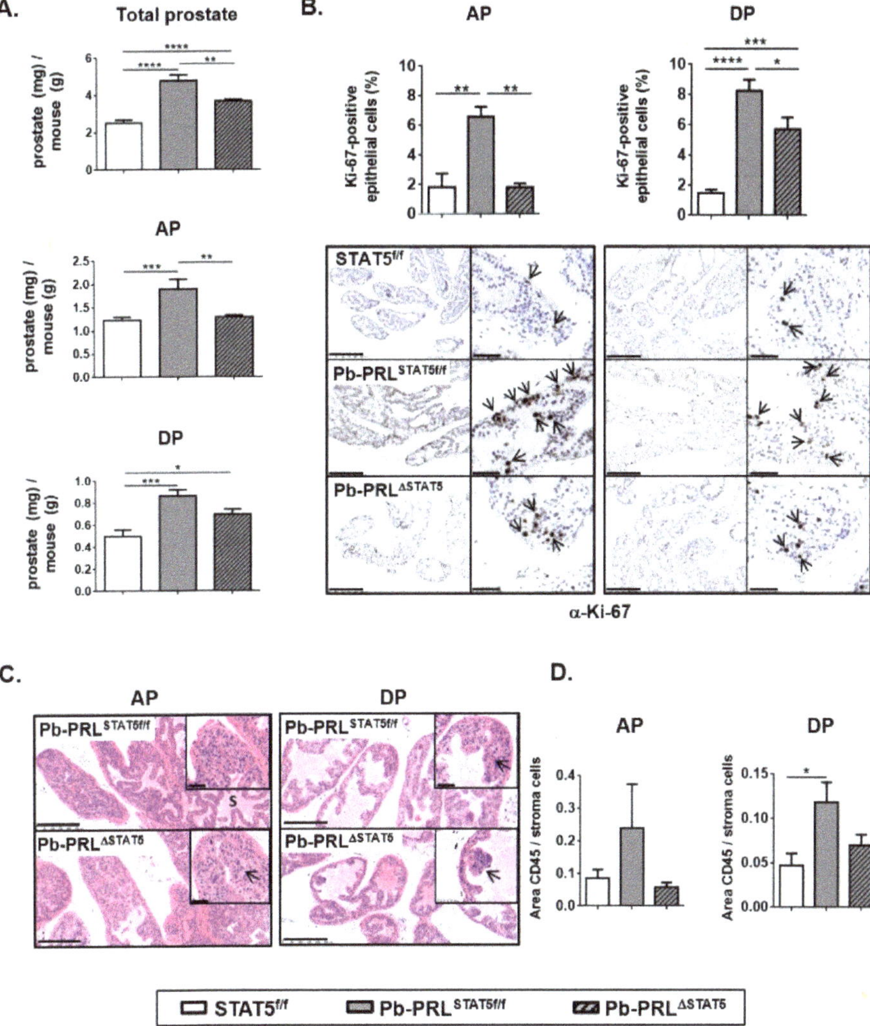

Figure 2. STAT5 deletion reduces hallmarks of early prostate tumorigenesis in 6 month-old Pb-PRL$^{\Delta STAT5}$ mice. (**A**). The weight of total prostate and of anterior (AP) and dorsal (DP) prostate lobes of STAT5$^{f/f}$, Pb-PRL$^{STAT5f/f}$ and Pb-PRL$^{\Delta STAT5}$ mice is expressed as the ratio of prostate tissue weight normalized to the weight of corresponding animal (see Figure S4A for other lobes). (**B**). Cell proliferation in anterior and dorsal lobes was assessed by immunostaining using anti-Ki-67 antibody (hematoxylin counterstaining). Arrows in insets show representative Ki-67 nuclear immunostaining in epithelial cells. The proliferation index (ratio of Ki-67-positive versus total epithelial cells) was quantified using Calopix software (see Figure S4 for other lobes). (**C**). Histological analysis (hematoxylin counterstaining) of anterior and dorsal lobes of Pb-PRL$^{STAT5f/f}$ and Pb-PRL$^{\Delta STAT5}$ mice showing similar hyperplasia in both genotypes (see Figure S4 for other lobes). (**D**). Inflammation was identified using CD45 immunostaining. The degree of inflammation was quantified using Calopix software and is represented as the ratio of CD45+ area versus stroma area. *Statistics*: Stars denote significant differences in a repeated-measures one-way ANOVA with Tukey's multiple comparisons. Size bars: 250 μm in large images and 50 μm in insets.

2.3.2. Cell Proliferation

According to their hypertrophic phenotype, all prostate lobes of Pb-PRL mice were previously shown to exhibit increased cell proliferation index (Ki-67 IHC) compared to control mice [36]. Using a computer-assisted image analysis methodology for the quantification of Ki-67-positive cells, this was confirmed in this study involving Pb-PRL$^{STAT5f/f}$ compared to STAT5$^{f/f}$ mice. In agreement with lobe weight reduction and STAT5 signaling down-regulation (see above), a significant reduction in cell proliferation index was observed in the anterior, and to a lesser extent, in the dorsal lobes of Pb-PRL$^{\Delta STAT5}$ mice (Figure 2B), while no change was observed in ventral prostate (Figure S4B).

2.3.3. Histopathology and Inflammation

At 6 months of age, Pb-PRL displays relatively mild histopathological phenotypes mainly including low grade PINs (early tumorigenesis step) [37,38]. In this study, we failed to find obvious differences between the histopathological features of Pb-PRL$^{STAT5f/f}$ compared to Pb-PRL$^{\Delta STAT5}$ prostates (Figure 2C and Figure S4C). The main phenotypical hallmark was in fact the reduction of the secretory phenotype in the latter, in agreement with the role of PRL signaling in prostate secretory function [39].

Inflammatory cell infiltrates are scarce in WT prostates. In contrast, according to previous reports [38,40], Pb-PRL$^{STAT5f/f}$ mouse prostates displayed several foci of CD45+ inflammatory cells especially in the ventral lobe (Figure 2D and Figure S4D). Pb-PRL$^{\Delta STAT5}$ mouse prostates displayed a less marked inflammatory phenotype in the anterior and dorsal lobes (Figure 2D), suggesting a role for luminal STAT5 in this phenotype. In contrast, there was a trend for an increased number of inflammatory cell foci in the two other lobes compared to Pb-PRL$^{STAT5f/f}$ prostates (Figure S4D).

Taken together, our analyses support a protective role of *Stat5* ablation against the early steps of PRL-triggered prostate tumorigenesis, weakening hyperplasia, lowering cell proliferation, and reducing chronic inflammation.

2.4. STAT5 Deletion Alters Epithelial Cell Hierarchy

The composition of the prostate epithelium can be determined using cell sorting strategies based on cell surface expression of Stem cell antigen-1 (Sca-1) and Integrin alpha 6 (CD49f). Using these validated markers, luminal and basal (also called LSC^{high}) cells can be discriminated [41]. We recently discovered in the mouse prostate a third epithelial cell population that we called LSC^{med} [31]. The basal and LSC^{med} cell populations include stem/progenitor cells and exhibit tumor-initiating capacities when transformed [32,42]. Of interest, in pre-tumoral Pb-PRL prostates both basal and LSC^{med} cell populations are amplified at the expense of the luminal compartment [30–32].

Due to the lobe-specific efficacy of *Stat5* deletion (see Figure 1), we investigated the distribution of these three epithelial cell populations individually in each lobe of the three genotypes of interest using cell sorting. Data in Figure 3A were obtained by pooling 3–6 animals to ensure enough material for analysis.

In STAT5$^{f/f}$ prostates luminal cells were predominant in all lobes (60% to 90%) and the analysis revealed very low prevalence of basal/LSC^{med} cells in ventral prostate compared to other lobes (Figure 3A,B). In Pb-PRL$^{STAT5f/f}$ prostate, basal and LSC^{med} cells were dramatically amplified in all lobes at the expense of luminal cells that dropped to <20% of the hyperplastic epithelium (Figure 3B). As shown in Figure 3C, the contents in basal and LSC^{med} cells in the various lobes were inversely related (the more basal, the less LSC^{med}). Plotting these data against the level of STAT5 phosphorylation in the epithelium of the various lobes of Pb-PRL$^{STAT5f/f}$ mouse prostates, as determined by IHC (Figure 1G) revealed that higher levels of STAT5 activation were associated with higher content in basal cells (Figure 3C).

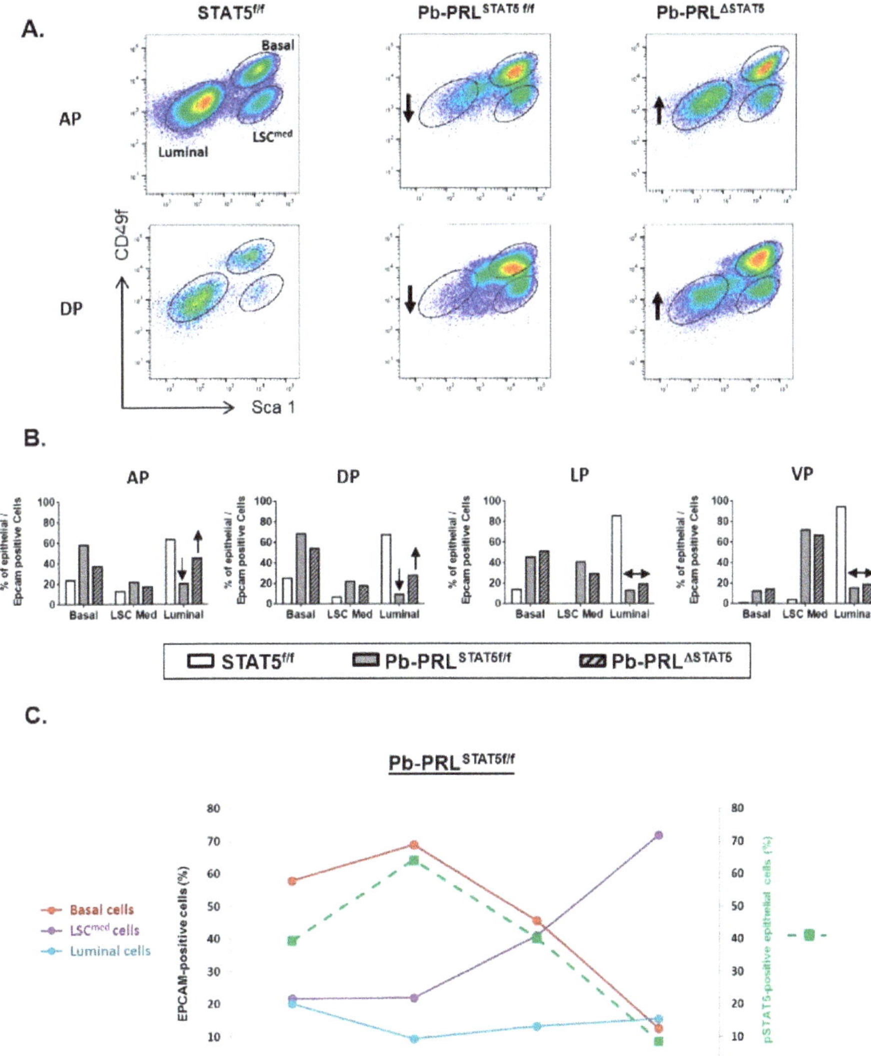

Figure 3. STAT5 deletion alters epithelial cell hierarchy in 6 month-old Pb-PRL$^{\Delta STAT5}$ mice. (**A**). Representative FACS profiles of anterior (AP) and dorsal (DP) prostate lobes from STAT5$^{f/f}$, Pb-PRL$^{STAT5f/f}$ and Pb-PRL$^{\Delta STAT5}$ mice (3–6 animals per genotype). Graphs depict epithelial cells only (gated as Lin-/Epithelial cell adhesion molecule (EPCAM)+), with large gates for percentage analysis. Each FACS profile shows gated epithelial populations: basal/stem (LSC^{high}), LSC^{med} and luminal cells. (**B**). Quantification of the three cell populations in the four prostate lobes from the three genotypes is shown. In panels A and B, the arrows indicate the loss of luminal cells in Pb-PRL$^{STAT5f/f}$ versus STAT5$^{f/f}$ mice, and their partial rescue in anterior and dorsal prostate of Pb-PRL$^{\Delta STAT5}$ mice (no similar effect was observed in the two other lobes). (**C**). For each prostate lobe of Pb-PRL$^{STAT5f/f}$ mice, the percentage of the three epithelial cell populations was plotted versus the level of STAT5 phosphorylation, as determined in Figure 1G.

Analysis of Pb-PRL$^{\Delta STAT5}$ mouse prostates showed that this enrichment in basal/ LSC^{med} cells was partly reversed upon STAT5 signaling down-regulation. Compared to Pb-PRL$^{STAT5f/f}$ mice the content in luminal cells was partly rescued in the anterior and dorsal lobes of Pb-PRL$^{\Delta STAT5}$ mice at the expense of basal/LSC^{med} cells (Figure 3A,B). No similar effect was observed in lateral and ventral lobes of Pb-PRL$^{\Delta STAT5}$ mice (Figure 3B) in agreement with the unaltered levels of epithelial pSTAT5 compared to Pb-PRL$^{STAT5f/f}$ mice (Figure 1G).

Taken together, these observations revealed the central role of epithelial STAT5 in the control of epithelial cell hierarchy by local PRL in the mouse prostate. In particular, up-regulation of STAT5 activation led to the accumulation of stem/progenitor cells previously shown to exhibit tumor-initiating capacities when transformed.

2.5. STAT5 Deletion does not Promote Alternative PRLR Signaling Cascades

In order to investigate whether *Stat5* gene deletion on the Pb-PRL background could promote compensatory PRLR-triggered signaling pathways that may be responsible for the effects reported above in Pb-PRL$^{\Delta STAT5}$ animals, we analyzed the phosphorylation status of STAT5, STAT3, ERK1/2, and AKT in prostates of 6 month old mice of the three genotypes. The immunoblots are shown in Figure 4A (anterior), B (dorsal) and C (ventral) and quantification for each lobe is shown in Figure 4D.

Virtually no basal phosphorylation could be detected for STAT5 in any lobe of STAT5$^{f/f}$ prostates, which contrasted with STAT3, ERK1/2, and AKT that all showed some background activation. Based on densitometric quantification, STAT5 was strongly activated in dorsal and anterior prostates of Pb-PRL$^{STAT5f/f}$ mice, reaching >30-fold induction compared to control mice. Mild activation (<2 fold) of STAT3 (anterior lobe) and AKT (dorsal lobe) was also observed. In ventral prostate, STAT5 was also activated, but to a much lower degree (~5 fold) than in the other lobes, in agreement with IHC data (Figure S3B). In contrast to the other lobes, the degree of activation of STAT3 in ventral lobe (~7 fold) was in the same range as STAT5 activation. Finally, ERK1/2 was not significantly triggered by PRL in any lobe as shown by unchanged pho/total ERK1/2 ratio (Figure 4D). However, due to slightly enhanced expression of ERK1/2 protein in the anterior lobe of Pb-PRL$^{STAT5f/f}$ prostates compared to STAT5$^{f/f}$ mice (Figure 4A), the tissue content in pERK1/2 in that lobe was significantly increased ($p < 0.01$), and the same trend (not significant) was observed in the other lobes (Figure 4B,C).

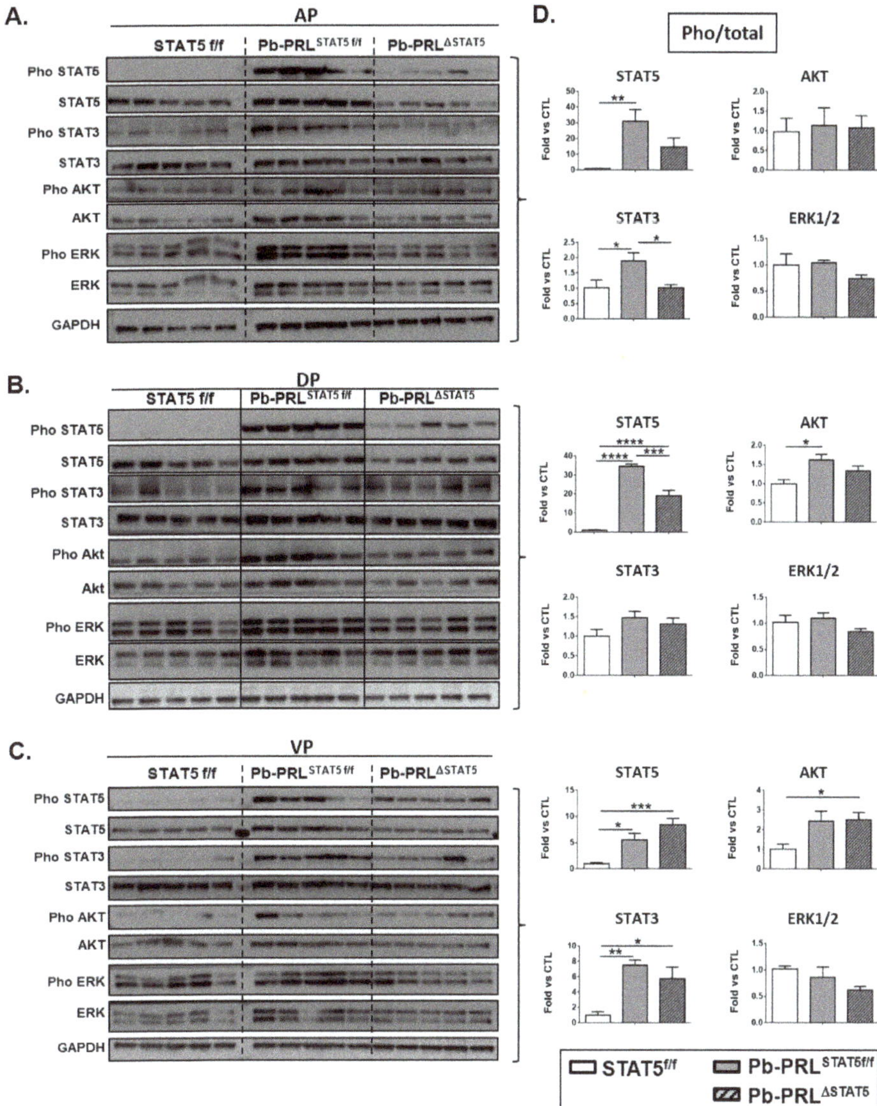

Figure 4. STAT5 deletion does not promote alternative signaling in 6 month-old Pb-PRL$^{\Delta STAT5}$ mice. (**A–C**). Canonical PRLR signaling pathways were analyzed by immunoblot in anterior (**A**), dorsal (**B**) and ventral (**C**) prostate lobes from STAT5$^{f/f}$, Pb-PRL$^{STAT5f/f}$ and Pb-PRL$^{\Delta STAT5}$ mice (each lane corresponds to a different mouse). (**D**). The activation of each pathway determined by the ratio of phosphorylated versus total protein (densitometry) is shown as fold-induction versus STAT5$^{f/f}$ samples. *Statistics*: Stars denote significant differences in a repeated-measures one-way ANOVA with Tukey's multiple comparisons.

The down-regulation of STAT5 in the anterior (Figure 4A) and dorsal (Figure 4B) lobes of Pb-PRL$^{\Delta STAT5}$ mice did not result in compensatory activation of alternative PRLR signaling pathways. In fact, the pathways that were mildly activated in these lobes in Pb-PRL$^{STAT5f/f}$ prostates, i.e., STAT3 in the anterior lobe and AKT in the dorsal lobe, were back to control levels in Pb-PRL$^{\Delta STAT5}$ prostates

(Figure 4D). While this effect suggests that activation of these pathways could be STAT5-dependent, it may also reflect the down-regulation of transgenic PRL protein that was observed in these lobes (Figure S4E). Since the level of PRL mRNA was unaltered (Figure S4F), this suggests that STAT5 signaling may contribute to PRL protein stability. Finally, in agreement with the poor efficiency of *Stat5* deletion in the ventral lobe of Pb-PRL$^{\Delta STAT5}$ mice, PRL expression (Figure S4E) and PRLR signaling pathways (Figure 4C,D) were unchanged in this lobe compared to Pb-PRL$^{STAT5f/f}$ mice.

Taken together, these data indicate that at 6 months of age, STAT5 deletion on the Pb-PRL background does not promote activation of alternative PRLR signaling pathways.

2.6. STAT5 Deletion does not Prevent Prostate Tumor Progression in Aged Pb-PRL Mice

In order to address the long-term effects of STAT5 signaling deletion on PRL-induced tumor progression, we analyzed Pb-PRL$^{STAT5f/f}$ and Pb-PRL$^{\Delta STAT5}$ mice at 12 and 18 months of age. In the latter, the ventral, lateral, and dorsal lobes were often undistinguishable (hereafter called half prostate) due to massive tissue hypertrophy so that lobe-specific interpretation could not be performed for all investigations.

The prostate weight of Pb-PRL$^{STAT5f/f}$ was significantly increased at 12 and even more at 18 months of age compared to STAT5$^{f/f}$ mouse prostates (Figure 5A for total prostate and Figure S5A for individual lobes). This weight gain was largely due to the dramatically increased secretory phenotype of Pb-PRL$^{STAT5f/f}$ prostates, especially in the anterior lobe that was extremely swollen/hypertrophic at dissection. This phenotype was confirmed on H&E sections by the presence of abundant and dense eosinophilic secretions in hypertrophied ducts ("S" on Figure 5B and Figure S5B). Compared to Pb-PRL$^{STAT5f/f}$ mice, prostate weight was significantly reduced in Pb-PRL$^{\Delta STAT5}$ mice at 12 months of age (Figure 5A). This effect was correlated to a less marked secretory phenotype and a reduction in cell proliferation that was significant in anterior and dorsal lobes only (Figure 5C). In 18 month-old mice, there was still a trend for reduced prostate weight (Figure 5A) and reduced proliferation index in Pb-PRL$^{\Delta STAT5}$ mice (Figure 5C).

At the histological level, prostates of 12 month old Pb-PRL$^{STAT5f/f}$ mice (Figure 5 and Figure S5B) exhibited various hallmarks of tumorigenesis that further aggravated with age including PINs, cribriform lesions ("c"), increased stromal density (stars) and inflammation (arrows), especially in ventral prostate. While these phenotypes tended to be less pronounced in age-matched Pb-PRL$^{\Delta STAT5}$ mice, they ultimately progressed in the latter so that at 18 months of age the mice of both genotypes became almost undistinguishable irrespective of the lobe (Figure 5B). The inflammatory phenotype was highly heterogeneous at 12 months of age precluding any clear genotype-related difference in any lobe (Figure 5D). There was nevertheless a global trend for lower inflammation in Pb-PRL$^{\Delta STAT5}$ mice, which persisted at 18 months of age (Figure 5D).

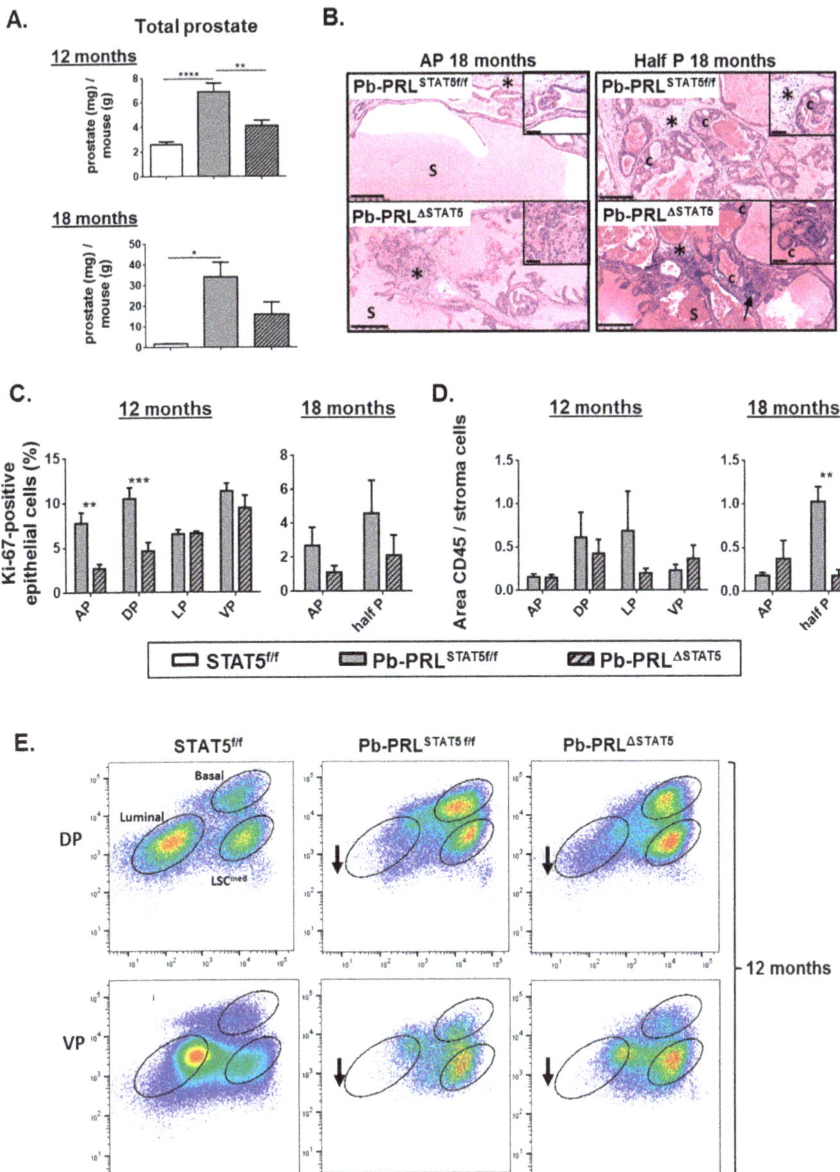

Figure 5. STAT5 deletion does not prevent prostate tumor progression in aged Pb-PRL mice. (**A**). The prostate weight in 12 and 18 month-old STAT5$^{f/f}$, Pb-PRL$^{STAT5f/f}$ and Pb-PRL$^{\Delta STAT5}$ mice is expressed as the ratio of total prostate weight normalized to the weight of corresponding animal (see Figure S5A for data per lobe). (**B**). Histological analysis (hematoxylin counterstaining) of prostate tumors of 18 month-old Pb-PRL$^{STAT5f/f}$ and Pb-PRL$^{\Delta STAT5}$ mice showing similar abnormal histology including PINs, cribriform lesions ("c"), increased stromal density (stars), inflammation (arrows), and dense eosinophilic secretions (S). Sections from the anterior lobe (AP) and from fused dorsal/lateral/ventral lobes are shown. See Figure S5B for 12 month-old animals. (**C**). The proliferation index in prostates from 12 (all lobes) and 18 month old (anterior lobe and total prostate) mice is shown (see Figure 2 for

details). (**D**). Inflammation was identified using CD45 immunostaining. The degree of inflammation was quantified using Calopix software and is represented as the ratio of CD45+ area versus stroma area. (**E**). Representative FACS profiles of dorsal (DP) and ventral (VP) prostate lobes from 12 month-old STAT5$^{f/f}$, Pb-PRL$^{STAT5f/f}$ and Pb-PRL$^{\Delta STAT5}$ mice (3–4 animals per genotype; see Figure 3A for details). Arrows show that the loss of luminal cells observed in Pb-PRL$^{STAT5f/f}$ versus STAT5$^{f/f}$ mice was not rescued in Pb-PRL$^{\Delta STAT5}$ mice. *Statistics:* Stars denote significant differences in a repeated-measures one-way ANOVA with Tukey's multiple comparisons (A) or two-way ANOVA with Sidak's multiple comparisons (C). Size bars: 250 μm in large images and 50 μm in insets.

According to the low efficiency of *Stat5* deletion in the ventral lobe, 12 month-old Pb-PRL$^{\Delta STAT5}$ and Pb-PRL$^{STAT5f/f}$ mice displayed very similar cell sorting profiles for this lobe, with accumulation of basal and LSC^{med} cells and low luminal cell content (Figure 5E). Importantly, the dorsal prostates of these animals also displayed very similar profiles enriched in stem/progenitor cells (Figure 5E), indicating that the partial rescue of the luminal cells observed in younger animals upon *Stat5* deletion (Figure 3A,B) was no longer maintained in aged mice.

Taken together these data indicate that the deletion of STAT5 delayed, but could not prevent the appearance of histopathological hallmarks of PRL-driven prostate tumorigenesis.

2.7. Spontaneous STAT5 Signaling Shutdown and Emergence of AKT and ERK1/2 Signaling in Aged Pb-PRL Mice

To understand further why the protective effect of STAT5 signaling inhibition observed at 6 months of age was progressively lost in old Pb-PRL$^{\Delta STAT5}$ mice, we compared the canonical PRLR signaling pathways in the dorsal prostates of the three genotypes at 6, 12, and 18 months of age (Figure 6A).

We discovered that in Pb-PRL$^{STAT5f/f}$ mice (lanes 1–3, 7–9 and 13–15 on Figure 6A), the degree of STAT5 activation (pSTAT5/STAT5 ratio) as well as the tissue content in activated STAT5 (pSTAT5/ACTIN) markedly declined between 6 and 12 months of age (Figure 6A,B). We investigated this phenotype further by monitoring the mRNA expression of key players of the PRLR/STAT5 pathway in Pb-PRL$^{STAT5f/f}$ prostate (Figure 6D). There was neither alteration of transgenic *Prl* and *Prlr* expression, nor up-regulation of short PRLR isoforms that have been shown to act as dominant-negative on STAT5 signaling [43]. The lower level of SOCS2 expression at 12- versus 6-months of age assessed that STAT5 signaling down-regulation in the latter was not due to unexpected SOCS overexpression; rather, it was associated with lower STAT5 signaling activity. SOCS7 [44] and nuclear receptor co-repressor 2 (NCOR2, also known as silencing mediator for retinoid or thyroid-hormone receptors, SMRT) [45] are two other negative regulators of STAT5 signaling. None of them were affected at the transcriptional level (Figure 6D and S6A). Otherwise, there was a significant age-related decrease in *Stat5a* mRNA expression (Figure 6D) that was also observed at the protein level in prostate lysates (STAT5/ACTIN ratio), although it did not reach significance, presumably due to unaltered *Stat5* expression in non-epithelial cells (Figure 6B). Comparison of pSTAT5 immunostaining in the epithelium of dorsal prostate of 6- versus 12-month old Pb-PRL$^{STAT5f/f}$ mice revealed that the reduced tissue content in activated STAT5 in older mice was associated with both lower number of pSTAT5-positive cells and, in the latter, reduced intensity of pSTAT5 staining (Figure 6E). Indeed, pSTAT5-positive cells in 12 month old Pb-PRL$^{STAT5f/f}$ mice exhibited faint nuclear staining similar to that observed in the few pSTAT5-positive epithelial cells of healthy prostates (Figure 6E, panels a and c), i.e., markedly lower compared to younger Pb-PRL$^{STAT5f/f}$ animals (Figure 6E, panel b). As a consequence, the level of STAT5 signaling in old Pb-PRL$^{STAT5f/f}$ mice became undistinguishable from that of age-matched Pb-PRL$^{\Delta STAT5}$ mice (see quantifications reported in Figure 6E). Accordingly, most of the targets that were regulated by STAT5 signaling in young animals were not significantly affected by *Stat5* deletion in older mice (Figure S6A).

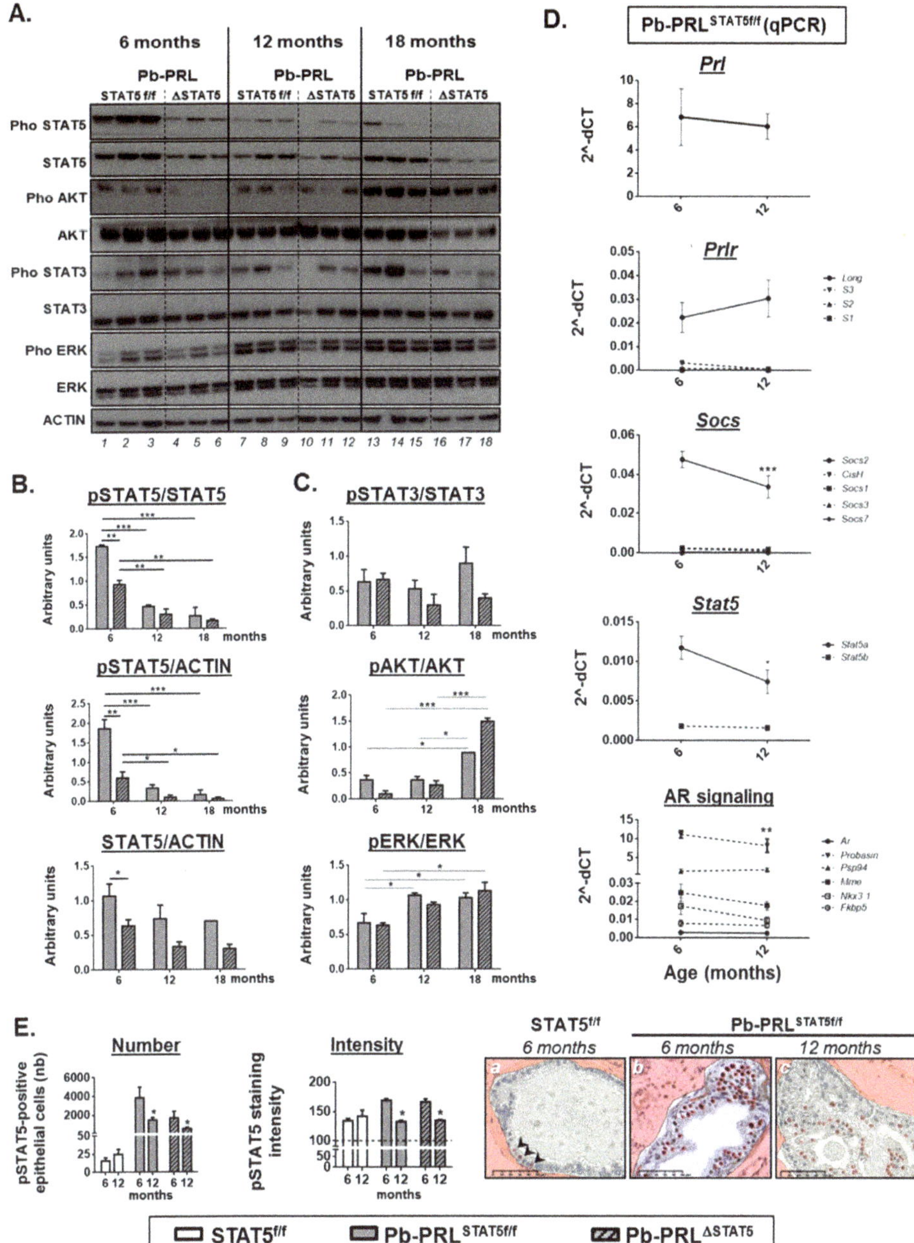

Figure 6. Spontaneous STAT5 signaling shutdown and emergence of AKT and ERK1/2 signaling in aged Pb-PRL mice. (**A**). Canonical PRLR signaling pathways activated in dorsal prostates of Pb-PRL$^{STAT5f/f}$ (lanes 1–3, 7–9, 13–15) and Pb-PRL$^{\Delta STAT5}$ (lanes 4–6, 10–12, 16–18) mice were compared at 6, 12 and 18 months of age as indicated. (**B**). Quantifications of the STAT5 pathway regarding STAT5 activation (pSTAT5/STAT5 ratio), tissue content in activated (pSTAT5/ACTIN) and total (STAT5/ACTIN) STAT5 as determined in panel A. (**C**). The activation of STAT3, AKT and ERK1/2 pathways as shown in panel A was quantified by densitometry as the ratio of phosphorylated versus total protein. (**D**). Age-dependent expression of various actors of the PRLR and AR pathways as determined by RT-qPCR in 6 and

12 month-old Pb-PRL$^{STAT5f/f}$ mice. The dotted line represents low expressed genes of the PRLR pathway (upper panels) and target genes of the AR pathway (bottom panel). See Figure S6 for genotype-dependent expression in 12 month-old mice. (**E**). Comparison of the number of pSTAT5-positive cells and of the intensity of pSTAT5 immunostaining (the horizontal dotted line is the background threshold) in 6 versus 12 month-old STAT5$^{f/f}$, Pb-PRL$^{STAT5f/f}$ and Pb-PRL$^{\Delta STAT5}$ mice. Representative images of low number (arrows)/low intensity (STAT5$^{f/f}$ mice), high number/high intensity (6 month-old Pb-PRL$^{STAT5f/f}$), and high number/low intensity (12 month-old Pb-PRL$^{STAT5f/f}$) are shown (positive cells are red circled). *Statistics*: Stars denote significant differences in a repeated-measures two-way ANOVA with Sidak's multiple comparisons. Size bars: 50 μm.

Concomitant to the age-dependent decline of STAT5 signaling in Pb-PRL$^{STAT5f/f}$ tumors, we observed a significant increase in ERK1/2 and AKT phosphorylation at 12 and 18 months of age, respectively (Figure 6A,C). Of interest, this phenomenon was also observed in Pb-PRL$^{\Delta STAT5}$ mice, indicating it occurred irrespective of *Stat5* status. There was no clear age-related alteration of STAT3 activation as it remained low in Pb-PRL$^{\Delta STAT5}$ mice and was more variable, but did not significantly increase in PRL$^{STAT5f/f}$ mice (Figure 6A,C).

Finally, we addressed potential contribution of androgen receptor (AR) signaling to the mechanisms favoring STAT5-independent tumor progression in aged mice of the Pb-PRL background. To that end we investigated the mRNA levels of *Ar* and of several AR target genes [32] in the dorsal prostate of 6 and 12 month old mice. Irrespective of the age, AR signaling was globally lower in Pb-PRL$^{STAT5f/f}$ mice compared to STAT5$^{f/f}$ mice based on the down-regulation of several genes of the AR signature (*Mme, Probasin, Nkx3.1*; Figure S6B). Additionally, there was also no evidence for age-related increase of AR signaling in Pb-PRL$^{STAT5f/f}$ mice (Figure 6D, bottom panel). Finally, *Stat5* deletion failed to rescue normal AR signaling levels in Pb-PRL$^{\Delta STAT5}$ mice, especially at 12 months of age (Figure S6B). These data clearly indicate that AR signaling up-regulation is not a compensatory mechanism of STAT5 signaling down-regulation in Pb-PRL mice.

Taken together, our analyses show that STAT5 signaling spontaneously decreases in PRL-driven prostate tumors with aging and our results suggest that alternative compensatory pathways such as AKT and ERK1/2 signaling, but not STAT3 and AR signaling, promote prostate tumor progression upon hyper PRL signaling in the absence of STAT5.

3. Discussion

Reports have provided strong evidence regarding the role of PRL/STAT5 pathway in the progression of human prostate cancer via an autocrine/paracrine mechanism (for reviews, [1,14,25–27]. Indeed, this pathway was shown to promote survival, proliferation, EMT, stemness, invasiveness, and DNA repair of/in prostate cancer cells, thereby contributing to prostate cancer progression and resistance to treatment. Several underlying molecular mechanisms have been described, non-exhaustively including up-regulation of PRL expression, STAT5 gene amplification and cooperation with androgen signaling pathway. While some of these biological effects are also observed in Pb-PRL mice (e.g., cell survival, proliferation, and stemness), the latter mice do not develop prostate cancer despite strong STAT5 action [30,31]. This suggested that hyper-activation of STAT5 signaling per se may not act as an oncogene in the prostate, but it could tweak prostate cancer progression in a more proto-oncogenic role driving e.g., enhanced Myc, Bcl2 or D type cyclin family member expressions. However, the promotion of prostate cancer progression with other oncogenic driver pathways identified in the prostate cancer genomic landscape could be complex and STAT5 signaling could be more important in other mutational context [46]. We here present the first genetically-modified mouse model allowing delineation of the actual role of STAT5 signaling in PRL-driven mouse prostate tumorigenesis.

The lobe-specific pattern of *Stat5* gene deficiency—high in dorsal and anterior lobes, low in ventral lobe of Pb-PRL$^{\Delta STAT5}$ mice—complicated the phenotype/genotype analyses, but at the same time provided the opportunity to assess the functions that are directly controlled by STAT5 signaling by comparing the different lobes within the same prostate sample. These analyses highlighted that epithelial cell survival/proliferation, immune cell infiltration, *Prlr* mRNA expression, and PRL protein stability are tightly dependent on epithelial STAT5 as their regulation (up or down) in Pb-PRL$^{STAT5f/f}$ mice (compared to control mice) was partly or totally reversed in the dorsal and anterior lobes, but not in the ventral lobe of Pb-PRL$^{\Delta STAT5}$ mice. While the STAT5-dependency of prostate epithelial cell survival/proliferation confirms many earlier reports (for a review, see Reference [26]), the molecular mechanisms mediating the epithelial STAT5-dependence of immune cell infiltration remains unknown. While high cell proliferation rate can in some instances be accompanied by increased apoptosis leading to the release of damage-associated molecular pattern (DAMPs) promoting inflammation. However, this is unlikely as Dillner and colleagues provided molecular and immunostaining evidence for reduced apoptosis in Pb-PRL prostates compared to healthy controls [47]. Alternatively, the increased secretory pattern of Pb-PRL prostates may feed the microenvironment with STAT5-induced cytokines/chemokines acting as immune cell chemoattractant. The locally high concentration of PRL may also contribute to inflammation as this hormone has been suggested to mediate inflammation in various pathological contexts, including association with prostate cancer inflammation [48,49]. Finally, the down-regulation of *Prlr* expression by PRL signaling confirmed our previous observation [31] and the reversibility of this phenotype upon *Stat5* ablation (Pb-PRL$^{\Delta STAT5}$ mice) suggested that this negative feedback effect is mediated by STAT5 signaling.

One of the striking phenotypes harbored by Pb-PRL mice is the amplification of the basal and LSC^{med} cell compartments. This was assessed both by cell sorting and by immunochemistry using cell population-specific phenotypic markers (p63/CK5 for basal and CK4 for LSC^{med} cells) [30–32]. The stem/progenitor cells contained in these two cell compartments are highly relevant to cancer progression as they are castrate-resistant and they exhibit tumor-initiating properties when transformed (e.g., by AKT/ERG overexpression or loss of key tumor suppressors in prostate cancer such as PTEN or p53) [32,42,50]. In xenograft experiments, the number of tumor-initiating cells present in the engrafted cell pool is directly correlated to the tumor incidence observed in host mice. In CWR22Rv xenografts, adenovirus-mediated overexpression of a dominant-negative STAT5A/B variant was shown to reduce tumor incidence [11], indicating a reduction in stem-like properties following STAT5 inhibition. This is in agreement with the capacity of STAT5 signaling to increase stem-like phenotypic features and functional properties of human prostate cancer cells in vitro [15]. Thus, the capacity of PRL/STAT5 signaling to induce the amplification of stem/progenitor cells may directly contribute to cancer progression by feeding the prostate tissue with cells able to resist treatments and to regrow a tumor (cancer recurrence).

In this study, the lobe-specific analyses of pSTAT5 by immunohistochemistry on the one hand, and of the three epithelial cell populations (luminal, basal, LSC^{med}) by cell sorting on the other hand, revealed that the prevalence of basal cells was positively correlated to the level of STAT5 activation in the luminal epithelium. The outcome of these analyses is schematized in Figure 7. This finding is in agreement with our former quantitative immunostaining investigations showing that basal cell clusters within the prostates epithelium of Pb-PRL mice are frequently surrounded by luminal cells displaying elevated levels of pSTAT5 [31]. Furthermore, the enrichment in basal/stem cells was partly reversed upon *Stat5* ablation in anterior and dorsal lobes of Pb-PRL$^{\Delta STAT5}$ mice. Another finding of these analyses is that the lobe-specific prevalence of basal/stem cells was inversely correlated to that of LSC^{med} cells. This supports that the latter emanate from the former in the prostate cell hierarchy, as proposed in our detailed study reporting the discovery of LSC^{med} progenitors [31]. Altogether, these observations demonstrate that the prostate cell hierarchy is tightly controlled by epithelial STAT5, which appears to limit luminal differentiation and to favor amplification of the basal/stem compartment (Figure 7). This effect is opposite to what happens in the mammary gland where STAT5A is mandatory for luminal cell differentiation [51]. The mechanism underlying the regulation of prostate cell hierarchy by STAT5 signaling is yet to be elucidated. Transcriptomic profiling of basal and LSC^{med} cells showed that they express only minimal levels of *Prlr* (Ref. [32] and our unpublished qPCR data) therefore their direct regulation by local PRL is unlikely. Paracrine regulation of basal cells by secreted factors downstream of PRL/STAT5 signaling in luminal cells is a possibility that is currently under investigation. As basal and LSC^{med} cells also express detectable levels of *Stat5* (Ref. [32] and our unpublished qPCR data), a cell-autonomous effect may also contribute to the partial reversion of basal cell enrichment in Pb-PRL$^{\Delta STAT5}$ mice. Indeed, although the androgen-regulated *Pb* promoter driving Cre recombinase expression is preferentially active in luminal cells, some reports suggest that it might also be active in basal cells [52].

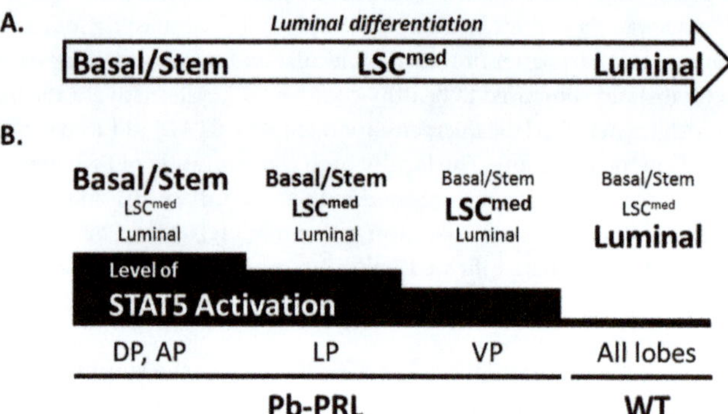

Figure 7. STAT5 signaling prevents luminal differentiation of the prostate epithelium. (**A**). Schematic representation of luminal differentiation placing LSC^{med} cells in-between basal and luminal cells. (**B**). The relative content in the three epithelial cell populations (basal/stem, LSC^{med}, luminal) in the various prostate lobes of 6-month old Pb-PRL mice or in WT mice (any lobe), as determined in Figure 3 is symbolized by the size of each population name. Below, the level of STAT5 activation in each lobe as determined in Figure 1 is also schematically represented by the black stair-bar. Based on these measurements, the highest the level of STAT5 activation, the lowest the level of epithelium differentiation was observed (see text for discussion).

While we earlier identified STAT5 as the main PRLR signaling pathway activated in Pb-PRL mouse prostates [30,31], the fact that pre-neoplastic prostate lesions observed in these animals never (or if at all, only exceptionally) evolve towards malignancy has always seemed in contradiction with the strong experimental and clinical evidence supporting the promoting role of this pathway in human prostate cancer (see Introduction). Human prostate specific PRL signaling could be more important than that of rodent prostates to drive tumorigenesis. Additionally, human prostate cancer is predominant in elder men and the various genetic and environmental risk factors (e.g., smoking, alcohol abuse, etc.) could insufficiently be recapitulated in inbred *C57Bl/6J* transgenics as used here as a study subject. Irrespective, the present work provides a possible explanation for the absence of prostate cancer development in Pb-PRL mice as we show that the prostates of these animals progressively lack STAT5 signaling with age. To address whether continuously elevated STAT5 signaling is actually required for the transformation of pre-neoplastic prostate lesions, it would be interesting to generate mice with enforced expression of a constitutively-active STAT5 variant in the prostate [53]. In conclusion, we show, for the first time, that aging can promote alternative compensatory pathway activations to bypass STAT5 function, but inflammation, cell proliferation and survival as well as prostate epithelium differentiation are significantly affected by PRL-STAT5 signaling.

The mechanism driving age-related decline of STAT5 signaling in the prostates of ageing Pb-PRL$^{STAT5f/f}$ mice remains unclear. No parallel up-regulation of the classical negative regulators of STAT5 signaling (SOCS proteins, short PRLR isoforms) could explain this observation. Although we noticed a decrease in STAT5 expression (mRNA and protein) in the dorsal prostates of 12 month old Pb-PRL$^{STAT5f/f}$ mice, this alone was insufficient to account for the pronounced loss of STAT5 activation (Figure 6B), suggesting that other mechanisms may be involved. In the breast cancer context, it has been shown that increased stiffness/density of the extracellular matrix around the tumor as a result of altered collagen deposition during cancer progression shifted the balance of PRLR signaling from JAK2/STAT5 to other pathways including ERK1/2 and AKT pathways through a Focal adhesion kinase-dependent mechanism [54,55]. Of interest, the pre-neoplastic lesions of Pb-PRL mouse prostates also exhibit strong remodeling of the extracellular matrix compared to controls, including >10 fold-increased expression of various Collagens [47]. This phenotype aggravated with age, concomitant to the decline in STAT5 signaling and up-regulation of AKT and ERK1/2 signaling, therefore it is possible that a similar mechanism as described in breast cancer contributes to the signaling shift observed in the prostates of ageing Pb-PRL$^{STAT5f/f}$ mice. No such up-regulation of AKT or ERK1/2 signaling was observed when STAT5 signaling was genetically impaired in younger Pb-PRL$^{\Delta STAT5}$ animals, further supporting that the progressive up-regulation of alternative signaling pathways in Pb-PRL$^{STAT5f/f}$ mice involves age-related mechanisms that require time to establish. These alternative signaling pathways allowed pre-neoplastic lesions to develop further in a STAT5-independent manner, as reflected by the elevated levels of cell proliferation and inflammation, and by the appearance of various histopathological phenotypes including cribriform lesions in old mice. In fact, these phenotypes were observed irrespective of *Stat5* status. Thus, although *Stat5* deficiency delayed the occurrence of various pre-neoplastic hallmarks, its protective effect was progressively diluted with age to become almost undetectable in 18 month-old animals. Of note, neither androgen nor STAT3 signaling could be identified as a compensatory pathway of age- or genetically-induced down-regulation of STAT5 signaling. There was even a trend for lower STAT3 signaling in *Stat5*-deficient prostates, suggesting some level of cooperation between both pathways.

Preventing the progression of established prostate cancers is where the therapeutic challenge stands. In agreement, there is currently strong research to develop STAT5 inhibitors that could be used for the treatment of various malignancies, including prostate cancer and myeloid leukemia [23,24,56]. Whatever the mechanisms driving the intrinsic decline of prostatic STAT5 signaling in Pb-PRL mice, the various genotype- and age-related phenomena observed in our study may shed light on possible outcomes of long-term pharmacological inhibition of STAT5 in prostate cancer context. On the good side, as low STAT5 signaling parallels luminal differentiation, pharmacological STAT5 inhibitors are not expected to result in tumor enrichment in stem/progenitor cells, but they can also act on the stromal cells such as immune cells. This effect may distinguish anti-STAT5 treatments from current therapeutic strategies (castration, chemotherapy) which generally kill luminal cells and let stem/progenitor cells unaffected, which is believed to favor tumor recurrence based on the tumor-initiating capacities of the latter. In addition, STAT5 inhibition may also decrease autocrine PRL stability as suggested from our results, which may further contribute to drug efficacy as PRL is presumably the major upstream factor triggering STAT5 pathway in the prostate. On the negative side, our study identified potential escape mechanisms to STAT5-targeted therapeutic strategies. In young Pb-PRL$^{\Delta STAT5}$ mice, which model early stages of STAT5 inhibition, no activation of alternative PRLR signaling pathway could be identified, and cell proliferation was actually reduced. With time, which is modeled by old Pb-PRL$^{\Delta STAT5}$ mice, we observed up-regulation of AKT and ERK1/2 signaling. One might speculate that in a context of pharmacological STAT5 inhibition these pathways may contribute to STAT5-independent prostate cancer progression. It is currently unknown whether AKT and ERK1/2 are triggered by PRL only or whether other paracrine growth factors will contribute. With the aim to design appropriate treatment combination limiting escape mechanisms to STAT5 inhibition, it may be relevant to address this mechanism in future studies.

4. Materials and Methods

4.1. Transgenic Mouse Strains

The Pb-PRL mouse colony carrying the rat prolactin transgene driven by the short *probasin* promoter was established on the *C57BL/6J* background (>20 backcrosses) in the Paris laboratory as previously described [30]. Mice carrying floxed *Stat5a/b* alleles (STAT5$^{f/f}$ mice) allowing targeted deletion of the whole *Stat5* locus (encompassing both *Stat5a* and *Stat5b* genes) were originally developed by L. Hennighausen [34]. Animals were obtained from the in-house colony of R. Moriggl (Vienna, Austria) established on a pure *C57Bl/6* genetic background. Pb-Cre mice (C57Bl/6 background) expressing the Cre recombinase under the control of the *short probasin* promoter [57] were purchased from the Frederick National Laboratory for Cancer Research (Frederick, MD, USA). Figure S7 displays the breeding scheme used to generate mice carrying or lacking prostate epithelium-specific deletion of *Stat5a/b* genes on the WT (STAT5$^{f/f}$ and ΔSTAT5 mice) and Pb-PRL (Pb-PRL$^{STAT5f/f}$ and Pb-PRL^{ΔSTAT5} mice) backgrounds. Note that only males were used to transmit the *Pb-Cre* allele.

Colonies were housed in controlled conditions, on a 12/12-hour light/dark cycle with normal food and water provided *ad libitum*. Mice were analyzed at 6, 12 and 18 months of age, and prostate were harvested immediately after sacrifice by cervical dislocation. To isolate the prostate, dissection of the urinary tract was performed and left lobes were separately dissected, weighed then rapidly snap frozen while the right-sided half prostate was fixed in paraformaldehyde (PFA) without being further dissected, so that tissue organization was preserved for histological analysis [36]. For all genotypes, we analyzed the four prostate lobes (anterior, dorsal, lateral, ventral) at all ages. Note that at 18 months of age, dorsal, lateral, and ventral lobes were often fused due to tissue hypertrophy.

Animal experiments were approved by the local ethical committee—Comité d'Ethique en matière d'Expérimentation Animale Paris Descartes (CEEA 34) for animal experimentation (authorization CEEA34.VG.095.12) and performed according to the European guidelines for animal experimentation.

4.2. Prostate Subpopulation Sorting by FACS

The procedures for sorting the three epithelial cell populations based on their Lin/Sca-1/CD49f antigenic profile was previously described [31,32,41]. Cell sorting was performed on a BD FACSAria III (BD Biosciences, San Jose, CA, USA). Sorted cells were collected in DMEM medium, supplemented with 50% FBS, glutamine, and penicillin-streptomycin, or in RA1 Lysis Buffer (Macherey-Nagel, Düren, Germany) to perform RNA extraction as earlier described [32].

4.3. Quantitative PCR

Total RNA was isolated from separate prostate lobes using the NucleoSpin® RNA XS (Macherey Nagel, Hoerd, France) according to manufacturer's instructions. RNA integrity was assessed on Agilent BioAnalyzer (all RNAs scored 7–10). RNA (250 ng) was reverse transcribed using SuperScript™II Reverse transcriptase with the SuperScript™II First-Strand Synthesis System for RT-PCR kit (Invitrogen, CA, USA). The cDNAs were then subjected to real-time PCR amplification using gene-specific primers (0.5 μM final concentration) purchased from IDT DNA (Integrated DNA Technologies, BVBA, Leuven, Belgium; HPLC purification; referred to as Mm.PT in Table S3) or Eurogentec (Liège, Belgium; Oligold quality, sequence given in Table S1). PPIA (Peptidyl Prolyl Isomerase A) that encodes Cyclophilin A that was used as the housekeeping gene in each reaction. Real-time PCR was performed using a Qtower 2.0 (Analytik Jena, Germany). The qPCR reaction contained 2 μL cDNA sample (12.5 ng) and 8 μL mastermix with 1× GoTaq® qPCR Master Mix (Promega, Charbonnières-les-Bains, France) and 0.5 μM primer. The Qtower 2.0 Instrument was used with the following program: Enzyme activation: 95 °C for 2 min; amplification (40 cycles): 95 °C for 15 s, 60 °C for 60 s. Results were generated with the Qtower 2.0 software and were analyzed by the comparative cycle threshold method and presented as fold change in gene expression relative to internal calibrators as mentioned in figures.

4.4. Western Blotting

Freshly dissected prostate lobes were snap frozen in liquid nitrogen and stored at −80 °C until processing as previously described [31]. Protein samples (20–50 μg) were resolved in 4–12% gradient SDS-PAGE in NuPAGE Bis-Tris Precast Gels (Life Technologies, Saint-Aubin, Fance) or 10% SDS-PAGE gels. Proteins were then transferred onto nitrocellulose membranes (BioRad, Marnes-la-coquette,

France). Membranes were cut horizontally according to the size of the proteins of interest and stained with primary antibodies as described in Table S1. For band detection, Horseradish Peroxidase (HRP)-coupled secondary anti-rabbit (7074, Cell Signaling, Saint-Quentin-en-Yvelines, France) or anti-mouse (NA931, GE Healthcare Europe, Freiburg, Germany) antibodies were added before Enhanced Chemiluminescent (ECL) substrate (Immobilon Western Chemiluminescent HRP Substrate, Millipore or LumiGLO Reagent, Cell Signaling). Representative parts of blots are shown in the Figures, whole blot membranes are displayed in Figure S8.

4.5. Prostate Histopathology

Histopathological diagnosis of prostate sections from all mice was performed in blind by an independent veterinary pathologist (E.R.-G.) trained for mouse prostate analysis [36] following the recommendations of the Mouse Models of Human Cancer Consortium Prostate Pathology Committee and the reference classification of prostate intraepithelial neoplasia (PIN) lesions in genetically-modified animals [58,59]. This qualitative analysis was complemented by quantitative IHC analyses, as described below.

4.6. Immunohistochemistry (IHC)

All prostate samples were fixed in 4% PFA, paraffin embedded, and underwent heat-induced antigen retrieval in citrate buffer at pH 6. IHC and IF were performed as previously described [31] using antibodies listed in Table S2. Vector Elite ABC HRP kit with DAB substrate (Vector Laboratories, Burlingame, CA, USA) was used for detection of IHC slides, with hematoxylin as counterstain.

4.7. Image Acquisition and Histology Quantifications

Prostate tissue sections (H&E or IHC) were digitally scanned using a NanoZoomer-2.0 RT scanner (Hamamatsu, Photonics, France) coupled to NDP.view2 software analysis beta version U12388-01 (Hamamatsu, Photonics, France). For quantification of immunostaining, computer-assisted analysis of digital (scanned) images was performed using Calopix software (http://www.tribvn.com/). To quantify nuclear immunostaining (pSTAT5, Ki-67) in the epithelium, first the Random Forest Tree for tissue recognition was used to delineate and only include the glandular areas in the analysis. Then, the "morphometry" software was applied to each prostate lobe to discriminate staining-positive (DAB+) versus staining-negative (DAB−) cells. A value of marking intensity (optical density, <100 considered as background) was also provided by the algorithm for both classes of cells. The index of proliferation (Ki-67+) and of STAT5 activation was calculated as the ratio of the number of positive versus total (positive + negative) cells counted. The whole procedure was validated by comparing the results obtained by Calopix-assisted versus manual counting of digital images as previously reported [36,38,60]. For the quantification of CD45 staining, results are expressed as the ratio between the surface of CD45+ cell foci versus the total surface of stroma [36,38,60].

4.8. Statistics

The specific statistical tests performed are described in the legends for all results reported. In summary, Analysis of Variance (ANOVA) tests were used to evaluate differences among three or more groups. Depending on the number of factors tested, One or Two-way ANOVAs were used. When prostate lobe samples issued from the same mice were compared in the analysis, a repeated-measures test was carried out. Post hoc multiple comparisons were performed with Sidak's or Tukey's multiple comparisons. One, two, three or four symbols illustrating significance represent p values <0.05; <0.01; <0.001 and <0.0001, respectively. A value of $p < 0.05$ was used as significance cutoff for all tests. Error bars represent S.E.M. All analyses were performed using GraphPad Prism version 6.00 for Windows (GraphPad Software, San Diego, CA, USA).

5. Conclusions

This work confirmed STAT5 signaling as a major driver of autocrine PRL-mediated prostate tumorigenesis as genetically-induced STAT5 down-regulation delayed the onset of various PRL-induced tumor hallmarks. The age-related decline in STAT5 signaling observed in Pb-PRL mice harboring intact *Stat5* genes was unexpected, and may explain why malignancies failed to develop despite alternative signaling pathways were switched on and presumably contributed to the progression of pre-neoplastic lesions. *Stat5* genotype-related differences of prostate phenotypes observed in young animals were progressively lost in ageing mice. Together, these findings highlight an important role of PRL/STAT5 in the development of prostate tumors.

Supplementary Materials: The following are available online at http://www.mdpi.com/2072-6694/11/7/929/s1, Figure S1. Lobe-specific pattern of STAT5 deletion in 6 month-old STAT5 mice, Figure S2. STAT5 deletion has no detectable effect on the prostate tissue, Figure S3. Lobe-specific pattern of STAT5 deletion of 6 month-old Pb-PRL STAT5 mice (complementary to Figure 1), Figure S4. Lobe-specific effects of STAT5 deletion of 6 month-old Pb-PRL STAT5 mice (complementary to Figure 4), Figure S5. STAT5 deletion does not prevent prostate tumor progression in aging Pb-PRL mice (complementary to Figure 5), Figure S6. Characterization of PRLR and AR signaling pathways in dorsal prostate of 12 month-old STAT5f/f, Pb-PRLSTAT5f/f and Pb-PRL STAT5 mice (complementary to Figure 6), Figure S7. Breeding scheme, Figure S8. Whole Western blot membranes corresponding to representative images shown in main and supplemental Figures, as indicated. Table S1: Antibody references and conditions of use, Table S2: Primers used for qPCR.

Author Contributions: Conceptualization, L.S.S., J.-E.G. and V.G.; Formal analysis, F.B. and N.P.; Funding acquisition, J.-E.G. and V.G.; Investigation, F.B., N.P. and L.S.S.; Methodology, F.B., N.P., L.S.S., J.-E.G. and V.G.; Project administration, J.-E.G.; Resources, R.M. and V.G.; Validation, E.R.-G., J.-E.G. and V.G.; Visualization, F.B. and N.P.; Writing—original draft, F.B., N.P. and V.G.; Writing—review & editing, L.S.S., R.M., J.-E.G. and V.G.

Funding: This research was funded by Cancéropôle Ile de France and Institut National du Cancer (INCa_6672) and annual institutional funding from Inserm, CNRS and the University Paris Descartes (all attributed to VG). RM was supported by the Austrian Science Fund (FWF), grants SFB-F4707 and SFB-F06105.

Acknowledgments: We thank Lothar Hennighausen for sharing STAT5-floxed mice originally developed in his laboratory. We are grateful to the personnel of the histology (Noémie Gadessaud, Sara Fontoura, Sophie Berissi), cytometry (Corinne Cordier, Jérome Mégret), and animal (Nadia Elganfoud and all zootechnicians) core facilities at the Structure Fédérative de Recherche Necker. We thank the Austrian Science Fund (FWF) for supporting the fees of this publication.

Conflicts of Interest: The authors declare no conflict of interest.

References

1. Goffin, V.; Hoang, D.T.; Bogorad, R.L.; Nevalainen, M.T. Prolactin regulation of the prostate gland: A female player in a male game. *Nat. Rev. Urol.* **2011**, *8*, 597–607. [CrossRef] [PubMed]

2. Li, H.; Ahonen, T.J.; Alanen, K.; Xie, J.; LeBaron, M.J.; Pretlow, T.G.; Ealley, E.L.; Zhang, Y.; Nurmi, M.; Singh, B.; et al. Activation of signal transducer and activator of transcription 5 in human prostate cancer is associated with high histological grade. *Cancer Res.* **2004**, *64*, 4774–4782. [CrossRef] [PubMed]

3. Dagvadorj, A.; collins, S.; Jomain, J.B.; Abdulghani, J.; Karras, J.; Zellweger, T.; Li, H.; Nurmi, M.; Alanen, K.; Mirtti, T.; et al. Autocrine prolactin promotes prostate cancer cell growth via Janus kinase-2-signal transducer and activator of transcription-5a/b signaling pathway. *Endocrinology* **2007**, *148*, 3089–3101. [CrossRef]

4. Stattin, P.; Rinaldi, S.; Stenman, U.H.; Riboli, E.; Hallmans, G.; Bergh, A.; Kaaks, R. Plasma prolactin and prostate cancer risk: A prospective study. *Int. J. Cancer* **2001**, *92*, 463–465. [CrossRef]

5. Berinder, K.; Akre, O.; Granath, F.; Hulting, A.L. Cancer risk in hyperprolactinemia patients: A population-based cohort study. *Eur. J. Endocrinol.* **2011**, *165*, 209–215. [CrossRef] [PubMed]

6. Aksamitiene, E.; Achanta, S.; Kolch, W.; Kholodenko, B.N.; Hoek, J.B.; Kiyatkin, A. Prolactin-stimulated activation of ERK1/2 mitogen-activated protein kinases is controlled by PI3-kinase/Rac/PAK signaling pathway in breast cancer cells. *Cell. Signal.* **2011**, *23*, 1794–1805. [CrossRef]

7. Bole-Feysot, C.; Goffin, V.; Edery, M.; Binart, N.; Kelly, P.A. Prolactin and its receptor: Actions, signal transduction pathways and phenotypes observed in prolactin receptor knockout mice. *Endocr. Rev.* **1998**, *19*, 225–268. [CrossRef]

8. Ahonen, T.J.; Harkonen, P.L.; Rui, H.; Nevalainen, M.T. PRL signal transduction in the epithelial compartment of rat prostate maintained as long-term organ cultures in vitro. *Endocrinology* **2002**, *143*, 228–238. [CrossRef]

9. Ahonen, T.J.; Xie, J.; LeBaron, M.J.; Zhu, J.; Nurmi, M.; Alanen, K.; Rui, H.; Nevalainen, M.T. Inhibition of transcription factor Stat5 induces cell death of human prostate cancer cells. *J. Biol. Chem.* **2003**, *278*, 27287–27292. [CrossRef]

10. Levy, D.E.; Darnell, J.E., Jr. Stats: Transcriptional control and biological impact. *Nat. Rev. Mol. Cell Biol.* **2002**, *3*, 651–662. [CrossRef]

11. Dagvadorj, A.; Kirken, R.A.; Leiby, B.; Karras, J.; Nevalainen, M.T. Transcription factor signal transducer and activator of transcription 5 promotes growth of human prostate cancer cells in vivo. *Clin. Cancer Res.* **2008**, *14*, 1317–1324. [CrossRef] [PubMed]

12. Thomas, C.; Zoubeidi, A.; Kuruma, H.; Fazli, L.; Lamoureux, F.; Beraldi, E.; Monia, B.P.; MacLeod, A.R.; Thuroff, J.W.; Gleave, M.E. Transcription factor Stat5 knockdown enhances androgen receptor degradation and delays castration-resistant prostate cancer progression in vivo. *Mol. Cancer Ther.* **2011**, *10*, 347–359. [CrossRef] [PubMed]

13. Tan, S.H.; Dagvadorj, A.; Shen, F.; Gu, L.; Liao, Z.; Abdulghani, J.; Zhang, Y.; Gelmann, E.P.; Zellweger, T.; Culig, Z.; et al. Transcription factor Stat5 synergizes with androgen receptor in prostate cancer cells. *Cancer Res.* **2008**, *68*, 236–248. [CrossRef] [PubMed]

14. Hoang, D.T.; Iczkowski, K.A.; Kilari, D.; See, W.; Nevalainen, M.T. Androgen receptor-dependent and -independent mechanisms driving prostate cancer progression: Opportunities for therapeutic targeting from multiple angles. *Oncotarget* **2017**, *8*, 3724–3745. [CrossRef] [PubMed]

15. Talati, P.G.; Gu, L.; Ellsworth, E.M.; Girondo, M.A.; Trerotola, M.; Hoang, D.T.; Leiby, B.; Dagvadorj, A.; McCue, P.A.; Lallas, C.D.; et al. Jak2-Stat5a/b Signaling Induces Epithelial-to-Mesenchymal Transition and Stem-Like Cell Properties in Prostate Cancer. *Am. J. Pathol.* **2015**, *185*, 2505–2522. [CrossRef] [PubMed]

16. Gu, L.; Vogiatzi, P.; Puhr, M.; Dagvadorj, A.; Lutz, J.; Ryder, A.; Addya, S.; Fortina, P.; Cooper, C.; Leiby, B.; et al. Stat5 promotes metastatic behavior of human prostate cancer cells in vitro and in vivo. *Endocr. Relat. Cancer* **2010**, *17*, 481–493. [CrossRef]

17. Kazansky, A.V.; Spencer, D.M.; Greenberg, N.M. Activation of signal transducer and activator of transcription 5 is required for progression of autochthonous prostate cancer: Evidence from the transgenic adenocarcinoma of the mouse prostate system. *Cancer Res.* **2003**, *63*, 8757–8762. [PubMed]

18. Gu, L.; Liao, Z.; Hoang, D.T.; Dagvadorj, A.; Gupta, S.; Blackmon, S.; Ellsworth, E.; Talati, P.; Leiby, B.; Zinda, M.; et al. Pharmacologic inhibition of Jak2-Stat5 signaling By Jak2 inhibitor AZD1480 potently suppresses growth of both primary and castrate-resistant prostate cancer. *Clin. Cancer Res.* **2013**, *19*, 5658–5674. [CrossRef] [PubMed]

19. Haddad, B.R.; Gu, L.; Mirtti, T.; Dagvadorj, A.; Vogiatzi, P.; Hoang, D.T.; Bajaj, R.; Leiby, B.; Ellsworth, E.; Blackmon, S.; et al. STAT5A/B gene locus undergoes amplification during human prostate cancer progression. *Am. J. Pathol.* **2013**, *182*, 2264–2275. [CrossRef]

20. Mohanty, S.K.; Yagiz, K.; Pradhan, D.; Luthringer, D.J.; Amin, M.B.; Alkan, S.; Cinar, B. STAT3 and STAT5A are potential therapeutic targets in castration-resistant prostate cancer. *Oncotarget* **2017**, *8*, 85997–86010. [CrossRef]

21. Mirtti, T.; Leiby, B.E.; Abdulghani, J.; Aaltonen, E.; Pavela, M.; Mamtani, A.; Alanen, K.; Egevad, L.; Granfors, T.; Josefsson, A.; et al. Nuclear Stat5a/b predicts early recurrence and prostate cancer-specific death in patients treated by radical prostatectomy. *Hum. Pathol.* **2013**, *44*, 310–319. [CrossRef] [PubMed]

22. Li, H.Z.; Zhang, Y.; Glass, A.; Zellweger, T.; Gehan, E.; Bubendorf, L.; Gelmann, E.P.; Nevalainen, M.T. Activation of signal transducer and activator of transcription-5 in prostate cancer predicts early recurrence. *Clin. Cancer Res.* **2005**, *11*, 5863–5868. [CrossRef] [PubMed]

23. Liao, Z.; Gu, L.; Vergalli, J.; Mariani, S.A.; De Dominici, M.; Lokareddy, R.K.; Dagvadorj, A.; Purushottamachar, P.; McCue, P.A.; Trabulsi, E.; et al. Structure-Based Screen Identifies a Potent Small Molecule Inhibitor of Stat5a/b with Therapeutic Potential for Prostate Cancer and Chronic Myeloid Leukemia. *Mol. Cancer Ther.* **2015**, *14*, 1777–1793. [CrossRef] [PubMed]

24. Maranto, C.; Udhane, V.; Hoang, D.T.; Gu, L.; Alexeev, V.; Malas, K.; Cardenas, K.; Brody, J.R.; Rodeck, U.; Bergom, C.; et al. STAT5A/B Blockade Sensitizes Prostate Cancer to Radiation through Inhibition of RAD51 and DNA Repair. *Clin. Cancer Res.* **2018**, *24*, 1917–1931. [CrossRef] [PubMed]

25. Liao, Z.; Nevalainen, M.T. Targeting transcription factor Stat5a/b as a therapeutic strategy for prostate cancer. *Am. J. Transl. Res.* **2011**, *3*, 133–138. [PubMed]

26. Tan, S.H.; Nevalainen, M.T. Signal transducer and activator of transcription 5A/B in prostate and breast cancers. *Endocr. Relat. Cancer* **2008**, *15*, 367–390. [CrossRef] [PubMed]

27. Goffin, V. Prolactin receptor targeting in breast and prostate cancers: New insights into an old challenge. *Pharmacol. Ther.* **2017**, *179*, 111–126. [CrossRef]

28. Nevalainen, M.T.; Valve, E.M.; Ingleton, P.M.; Nurmi, M.; Martikainen, P.M.; Harkonen, P.L. Prolactin and prolactin receptors are expressed and functioning in human prostate. *J. Clin. Investig.* **1997**, *99*, 618–627. [CrossRef]

29. Kindblom, J.; Dillner, K.; Sahlin, L.; Robertson, F.; Ormandy, C.; Tornell, J.; Wennbo, H. Prostate hyperplasia in a transgenic mouse with prostate-specific expression of prolactin. *Endocrinology* **2003**, *144*, 2269–2278. [CrossRef]

30. Rouet, V.; Bogorad, R.L.; Kayser, C.; Kessal, K.; Genestie, C.; Bardier, A.; Grattan, D.R.; Kelder, B.; Kopchick, J.J.; Kelly, P.A.; et al. Local prolactin is a target to prevent expansion of basal/stem cells in prostate tumors. *Proc. Natl. Acad. Sci. USA* **2010**, *107*, 15199–15204. [CrossRef]

31. Sackmann-Sala, L.; Chiche, A.; Mosquera-Garrote, N.; Boutillon, F.; Cordier, C.; Pourmir, I.; Pascual-Mathey, L.; Kessal, K.; Pigat, N.; Camparo, P.; et al. Prolactin-Induced Prostate Tumorigenesis Links Sustained Stat5 Signaling with the Amplification of Basal/Stem Cells and Emergence of Putative Luminal Progenitors. *Am. J. Pathol.* **2014**, *184*, 3105–3119. [CrossRef] [PubMed]

32. Sackmann Sala, L.; Boutillon, F.; Menara, G.; De Goyon-Pelard, A.; Leprevost, M.; Codzamanian, J.; Lister, N.; Pencik, J.; Clark, A.; Cagnard, N.; et al. A rare castration-resistant progenitor cell population is highly enriched in Pten-null prostate tumors. *J. Pathol.* **2017**, *243*, 54–64. [CrossRef] [PubMed]

33. English, H.F.; Santen, R.J.; Isaacs, J.T. Response of glandular versus basal rat ventral prostatic epithelial cells to androgen withdrawal and replacement. *Prostate* **1987**, *11*, 229–242. [CrossRef] [PubMed]

34. Cui, Y.; Riedlinger, G.; Miyoshi, K.; Tang, W.; Li, C.; Deng, C.X.; Robinson, G.W.; Hennighausen, L. Inactivation of Stat5 in mouse mammary epithelium during pregnancy reveals distinct functions in cell proliferation, survival, and differentiation. *Mol. Cell Biol.* **2004**, *24*, 8037–8047. [CrossRef] [PubMed]

35. Nevalainen, M.T.; Ahonen, T.J.; Yamashita, H.; Chandrashekar, V.; Bartke, A.; Grimley, P.M.; Robinson, G.W.; Hennighausen, L.; Rui, H. Epithelial defect in prostates of Stat5a-null mice. *Lab. Invest.* **2000**, *80*, 993–1006. [CrossRef] [PubMed]

36. Bernichtein, S.; Pigat, N.; Barry Delongchamps, N.; Boutillon, F.; Verkarre, V.; Camparo, P.; Reyes-Gomez, E.; Mejean, A.; Oudard, S.M.; Lepicard, E.M.; et al. Vitamin D3 prevents calcium-induced progression of early-stage prostate tumors by counteracting TRPC6 and calcium sensing receptor upregulation. *Cancer Res.* **2017**, *77*, 355–365. [CrossRef] [PubMed]

37. Capiod, T.; Barry Delongchamps, N.; Pigat, N.; Souberbielle, J.C.; Goffin, V. Do dietary calcium and vitamin D matter in men with prostate cancer? *Nat. Rev. Urol.* **2018**, *15*, 453–461. [CrossRef]

38. Bernichtein, S.; Pigat, N.; Camparo, P.; Latil, A.; Viltard, M.; Friedlander, G.; Goffin, V. Anti-inflammatory properties of Lipidosterolic extract of Serenoa repens (Permixon(R)) in a mouse model of prostate hyperplasia. *Prostate* **2015**, *75*, 706–722. [CrossRef]

39. Costello, L.C.; Franklin, R.B. Effect of prolactin on the prostate. *Prostate* **1994**, *24*, 162–166. [CrossRef]

40. Pigat, N.; Reyes-Gomez, E.; Boutillon, F.; Palea, S.; Barry Delongchamps, N.; Koch, E.; Goffin, V. Combined Sabal and Urtica Extracts (WS((R)) 1541) Exert Anti-proliferative and Anti-inflammatory Effects in a Mouse Model of Benign Prostate Hyperplasia. *Front. Pharmacol.* **2019**, *10*, 311. [CrossRef]

41. Lukacs, R.U.; Goldstein, A.S.; Lawson, D.A.; Cheng, D.; Witte, O.N. Isolation, cultivation and characterization of adult murine prostate stem cells. *Nat. Protoc.* **2010**, *5*, 702–713. [CrossRef] [PubMed]

42. Lawson, D.A.; Zong, Y.; Memarzadeh, S.; Xin, L.; Huang, J.; Witte, O.N. Basal epithelial stem cells are efficient targets for prostate cancer initiation. *Proc. Natl. Acad. Sci. USA* **2010**, *107*, 2610–2615. [CrossRef] [PubMed]

43. Berlanga, J.J.; Garcia-Ruiz, J.P.; Perrot-Applanat, M.; Kelly, P.A.; Edery, M. The short form of the prolactin receptor silences prolactin induction of the b-casein gene promoter. *Mol. Endocrinol.* **1997**, *11*, 1449–1457. [CrossRef]

44. Martens, N.; Uzan, G.; Wery, M.; Hooghe, R.; Hooghe-Peters, E.L.; Gertler, A. Suppressor of cytokine signaling 7 inhibits prolactin, growth hormone, and leptin signaling by interacting with STAT5 or STAT3 and attenuating their nuclear translocation. *J Biol. Chem.* **2005**, *280*, 13817–13823. [CrossRef] [PubMed]

45. Nakajima, H.; Brindle, P.K.; Handa, M.; Ihle, J.N. Functional interaction of STAT5 and nuclear receptor co-repressor SMRT: Implications in negative regulation of STAT5-dependent transcription. *EMBO J.* **2001**, *20*, 6836–6844. [CrossRef] [PubMed]

46. Taylor, B.S.; Schultz, N.; Hieronymus, H.; Gopalan, A.; Xiao, Y.; Carver, B.S.; Arora, V.K.; Kaushik, P.; Cerami, E.; Reva, B.; et al. Integrative genomic profiling of human prostate cancer. *Cancer Cell* **2010**, *18*, 11–22. [CrossRef]

47. Dillner, K.; Kindblom, J.; Flores-Morales, A.; Shao, R.; Tornell, J.; Norstedt, G.; Wennbo, H. Gene expression analysis of prostate hyperplasia in mice overexpressing the prolactin gene specifically in the prostate. *Endocrinology* **2003**, *144*, 4955–4966. [CrossRef]

48. McPherson, S.J.; Wang, H.; Jones, M.E.; Pedersen, J.; Iismaa, T.P.; Wreford, N.; Simpson, E.R.; Risbridger, G.P. Elevated androgens and prolactin in aromatase-deficient mice cause enlargement, but not malignancy, of the prostate gland. *Endocrinology* **2001**, *142*, 2458–2467. [CrossRef]

49. Tang, M.W.; Reedquist, K.A.; Garcia, S.; Fernandez, B.M.; Codullo, V.; Vieira-Sousa, E.; Goffin, V.; Reuwer, A.Q.; Twickler, M.T.; Gerlag, D.M.; et al. The prolactin receptor is expressed in rheumatoid arthritis and psoriatic arthritis synovial tissue and contributes to macrophage activation. *Rheumatology* **2016**, *55*, 2248–2259. [CrossRef]

50. Goldstein, A.S.; Huang, J.; Guo, C.; Garraway, I.P.; Witte, O.N. Identification of a cell of origin for human prostate cancer. *Science* **2010**, *329*, 568–571. [CrossRef]

51. Liu, X.; Robinson, G.W.; Wagner, K.U.; Garrett, L.; Wynshaw-Boris, A.; Hennighausen, L. Stat5a is mandatory for adult mammary gland development and lactogenesis. *Genes Dev.* **1997**, *11*, 179–186. [CrossRef] [PubMed]

52. Parisotto, M.; Metzger, D. Genetically engineered mouse models of prostate cancer. *Mol. Oncol.* **2013**, *7*, 190–205. [CrossRef] [PubMed]

53. Farrar, M.A. Design and use of constitutively active STAT5 constructs. *Methods Enzymol.* **2010**, *485*, 583–596. [CrossRef] [PubMed]

54. Barcus, C.E.; Keely, P.J.; Eliceiri, K.W.; Schuler, L.A. Prolactin signaling through focal adhesion complexes is amplified by stiff extracellular matrices in breast cancer cells. *Oncotarget* **2016**, *7*, 48093–48106. [CrossRef] [PubMed]

55. Barcus, C.E.; Keely, P.J.; Eliceiri, K.W.; Schuler, L.A. Stiff collagen matrices increase tumorigenic prolactin signaling in breast cancer cells. *J. Biol. Chem.* **2013**, *288*, 12722–12732. [CrossRef] [PubMed]

56. Wingelhofer, B.; Maurer, B.; Heyes, E.C.; Cumaraswamy, A.A.; Berger-Becvar, A.; de Araujo, E.D.; Orlova, A.; Freund, P.; Ruge, F.; Park, J.; et al. Pharmacologic inhibition of STAT5 in acute myeloid leukemia. *Leukemia* **2018**, *32*, 1135–1146. [CrossRef] [PubMed]

57. Wu, X.; Wu, J.; Huang, J.; Powell, W.C.; Zhang, J.; Matusik, R.J.; Sangiorgi, F.O.; Maxson, R.E.; Sucov, H.M.; Roy-Burman, P. Generation of a prostate epithelial cell-specific Cre transgenic mouse model for tissue-specific gene ablation. *Mech. Dev.* **2001**, *101*, 61–69. [CrossRef]

58. Park, J.H.; Walls, J.E.; Galvez, J.J.; Kim, M.J.; Abate-Shen, C.; Shen, M.M.; Cardiff, R.D. Prostatic Intraepithelial neoplasia in genetically engineered mice. *Am. J. Pathol.* **2002**, *161*, 727–735. [CrossRef]

59. Shappell, S.B.; Thomas, G.V.; Roberts, R.L.; Herbert, R.; Ittmann, M.M.; Rubin, M.A.; Humphrey, P.A.; Sundberg, J.P.; Rozengurt, N.; Barrios, R.; et al. Prostate pathology of genetically engineered mice: Definitions and classification. The consensus report from the Bar Harbor meeting of the Mouse Models of Human Cancer Consortium Prostate Pathology Committee. *Cancer Res.* **2004**, *64*, 2270–2305. [CrossRef]

60. Bernichtein, S.; Pigat, N.; Capiod, T.; Boutillon, F.; Verkarre, V.; Camparo, P.; Viltard, M.; Mejean, A.; Oudard, S.; Souberbielle, J.C.; et al. High milk consumption does not affect prostate tumor progression in two mouse models of benign and neoplastic lesions. *PLoS ONE* **2015**, *10*, e0125423. [CrossRef]

Article

STAT3 Activity Promotes Programmed-Death Ligand 1 Expression and Suppresses Immune Responses in Breast Cancer

Ioannis Zerdes [1,*], Majken Wallerius [1], Emmanouil G. Sifakis [1], Tatjana Wallmann [1], Stina Betts [1], Margarita Bartish [1], Nikolaos Tsesmetzis [2], Nicholas P. Tobin [1], Christos Coucoravas [3], Jonas Bergh [1,4], George Z. Rassidakis [1,5], Charlotte Rolny [1] and Theodoros Foukakis [1,4,*]

[1] Department of Oncology-Pathology, Karolinska Institutet, 17164 Stockholm, Sweden; majken.wallerius@gmail.com (M.W.); emmanouil.sifakis@ki.se (E.G.S.); tatjana_wallmann@web.de (T.W.); stina.betts@stud.ki.se (S.B.); margarita.bartish@ki.se (M.B.); nick.tobin@ki.se (N.P.T.); jonas.bergh@ki.se (J.B.); georgios.rassidakis@ki.se (G.Z.R.); charlotte.rolny@ki.se (C.R.)

[2] Department of Women's and Children's Health, Karolinska Institutet, 17177 Stockholm, Sweden; nikolaos.tsesmetzis@ki.se

[3] Department of Medical Biochemistry and Biophysics, Karolinska Institutet, 17165 Stockholm, Sweden; christos.coucoravas@ki.se

[4] Breast Center, Theme Cancer, Karolinska University Hospital, 17176 Stockholm, Sweden

[5] Department of Pathology and Cytology, Karolinska University Hospital, 17176 Stockholm, Sweden

[*] Correspondence: ioannis.zerdes@ki.se (I.Z.); theodoros.foukakis@ki.se (T.F.); Tel.: +46-76-560-6628 (I.Z.); +46-85-177-9953 (T.F.)

Received: 14 August 2019; Accepted: 27 September 2019; Published: 1 October 2019

Abstract: Signal transducer and activator of transcription 3 (STAT3) is an oncogene and multifaceted transcription factor involved in multiple cellular functions. Its role in modifying anti-tumor immunity has been recently recognized. In this study, the biologic effects of STAT3 on immune checkpoint expression and anti-tumor responses were investigated in breast cancer (BC). A transcriptional signature of phosphorylated STAT3 was positively correlated with PD-L1 expression in two independent cohorts of early BC. Pharmacologic inhibition and gene silencing of *STAT3* led to decreased Programmed Death Ligand 1 (PD-L1) expression levels in vitro, and resulted as well in reduction of tumor growth and decreased metastatic dissemination in a mammary carcinoma mouse model. The hampering of tumor progression was correlated to an anti-tumoral macrophage phenotype and accumulation of natural-killer cells, but also in reduced accrual of cytotoxic lymphocytes. In human BC, pro-tumoral macrophages correlated to PD-L1 expression, proliferation status and higher grade of malignancy, indicating a subset of patients with immunosuppressive properties. In conclusion, this study provides evidence for STAT3-mediated regulation of PD-L1 and modulation of immune microenvironment in BC.

Keywords: breast cancer; PD-L1; STAT3; M2 macrophages; NK cells; STAT3 inhibitor XIII

---

1. Introduction

Programmed Death 1 (PD-1, *CD279*) and its ligand Programmed Death Ligand 1 (PD-L1, *CD274*) are transmembrane proteins with role in autoimmunity, infection and anti-tumor immune response. PD-L1 is mostly expressed in tumor cells but also in dendritic cells and macrophages, while its receptor PD-1 is predominantly expressed in activated T-cells [1]. Their engagement leads to T-cell inactivation and to impairment of effective immune response against the tumor [2]. Therefore, the PD-1/PD-L1 axis represents an important immune checkpoint, and its targeting with monoclonal antibodies has been proven to be an effective immunotherapeutic strategy, demonstrating durable clinical responses and improved survival in several tumor types [3].

Breast cancer has been considered as a relatively non-immunogenic tumor due its low mutational burden. However, the presence of tumor-infiltrating lymphocytes has demonstrated prognostic and predictive value—at least in the triple negative and HER2 positive subtypes [4,5]. Data indicating the efficacy of immune checkpoint blockade in breast cancer patients are mostly derived from phase I and II clinical trial results [6], and recently the first phase III trial results showed that the addition of the anti-PD-L1 antibody atezolizumab to nab-paclitaxel was associated with improved progression-free survival in patients with triple-negative metastatic breast cancer [7].

Various genetic, transcriptional and post-translational factors have been involved in the regulation of PD-L1 and these may be tumor type-specific [8]. Among them, signal transducer and activator of transcription 3 (STAT3) represents a crucial transcription factor for cell proliferation, survival and tumor development [9]. A direct link between STAT3 and PD-L1 expression has been previously described [10], and recent data have provided insight into this regulatory mechanism [11]. STAT3 mediates the expression of important regulators of cell cycle and apoptosis but it can also play an important role in tumor-immune cells interaction by impairing effective antitumor immunity [12]. STAT3-mediated release of various cytokines and chemokines can interact and influence components of the tumor microenvironment (TME) and especially immune cell accumulation including T-cells, Natural Killer (NK) cells as well as tumor-associated macrophages (TAMs) [13]. Of note, TAMs represent a heterogeneous subpopulation with either pro-tumoral or anti-tumoral properties. Their accumulation has been correlated with a worse prognosis and therapeutic resistance in most solid tumors, including breast cancer [14].

In the present study, the role of STAT3 in the regulation of PD-L1 expression and in the potential modifications of the immune microenvironment in breast cancer was investigated. Our findings provide evidence for STAT3-mediated regulation of PD-L1 *in vitro* and impact on accumulation of pro-tumoral macrophages and other immune cell subpopulations in an *in vivo* murine mammary tumor model. The interactions of PD-L1 with STAT3, pro-tumoral macrophages and tumor characteristics have been explored as well in a well-characterized cohort of breast cancer patients.

2. Results

2.1. Association Between PD-L1 and STAT3 Expression in Breast Cancer Cell Lines and Human Breast Cancer

The association between PD-L1 and STAT3 expression was first assessed in human breast cancer cell lines. PD-L1 expression pattern was evaluated in three different human breast cancer cell lines by western blot analysis and immunohistochemistry. MDA-MB-231 and BT549 demonstrated high levels of PD-L1 compared to MCF7 cells, which showed very low—almost undetectable—PD-L1 protein (Figure 1A,B). STAT3 showed a ubiquitous expression in all three cell lines, as visualized by western blot (Figure 1C), while MDA-MB-231 and BT549 cells showed a higher degree of STAT3 phosphorylation (at the Y705 residue) compared to MCF7, as visualized by western blot (Figure 1C) and immunohistochemistry (Figure 1D), notably following the pattern of PD-L1 expression (Figure 1A,B). Importantly, no amplification of the *PDL1* gene locus was detected in any of the breast cancer cell lines as assessed by Fluorescence In Situ Hybridization (FISH) analysis performed in sections of FFPE cell blocks (Figure 1E). The association between pSTAT3 and PD-L1 was further assessed in a human breast cancer cohort for which primary tumor gene expression data were available ($n = 619$) and PD-L1 protein levels were assessed ($n = 539$). The scores of a previously published metagene signature of pSTAT3 in breast cancer (*pSTAT3*-GS) were positively correlated with *PD-L1* transcript expression levels (Spearman's rho = 0.34; $p < 2.2e-16$) (Figure 1F). The positive association between pSTAT3-GS score and PD-L1 transcript was also confirmed when RNA-sequencing data derived from the Cancer Genome Atlas (TCGA) Provisional dataset ($n = 1081$) (Spearman's rho = 0.38, $p < 0.01$; Supplementary Figure S2A) were used. Additionally, in positive cases for PD-L1, total cell protein expression was evaluated by immunohistochemistry (Figure 1G), and *pSTAT3*-GS scores were significantly higher than those in the PD-L1 negative cases ($p = 0.0027$) (Figure 1H). Moreover, in a subset of patients ($n = 83$) pSTAT3 was assessed by IHC, and high pSTAT3 protein levels were positively associated with PD-L1—especially in immune cells (Figure S3, Table S4).

Additionally, the expression of PD-L1 transcript and pSTAT3-GS score were higher in triple-negative (TN) versus non-TN breast tumors in both cohorts (Figure 2A,B and Figure S2) whereas total STAT3 gene expression did not differ between these two groups (Figure 2C and Figure S2).

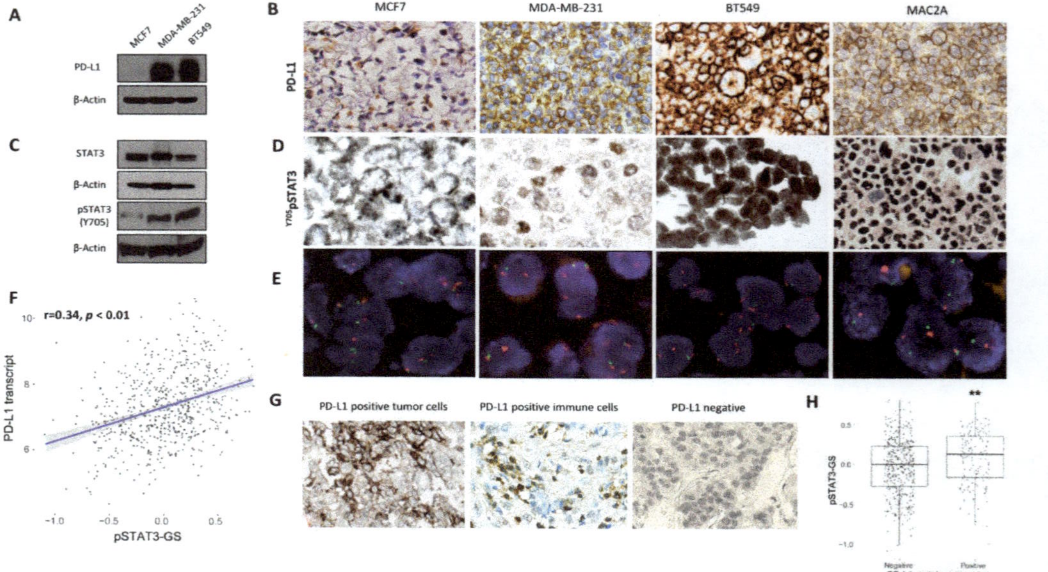

Figure 1. Expression of PD-L1 and STAT3 in breast cancer cell lines and in breast cancer patients. (**A**) Expression of PD-L1 protein in breast cancer cell lines in immunoblots. (**B**) Immunohistochemical expression of PD-L1 in breast cancer cell lines using FFPE cell blocks. (**C**) Protein expression of STAT3 and STAT3 phosphorylation at Tyr705 (Y705) residue in immunoblots in breast cancer cell lines. (**D**) Immunohistochemical expression pSTAT3 (Y705) in breast cancer cell lines using FFPE cell blocks. The anaplastic large cell lymphoma cell line Mac2A was used as a positive control. Original magnification: 400×. (**E**) Fluorescence in situ hybridization (FISH) analysis for PD-L1 probe performed on FFPE cell blocks. The validated probe (green signal) covers the gene locus at 9p24.1. A centromeric chromosome 9 probe (CEN9, red signal) was used as a control. No *PD-L1* gene amplification was demonstrated in Mac2A cell line. Original magnification: 630×. (**F**) Correlation of *PD-L1* transcript expression with a previously published pSTAT3-associated gene signature (pSTAT3-GS) reflecting the status of pSTAT3 expression in breast cancer patients with available gene expression profiling data ($n = 619$). (**G**) PD-L1 expression was evaluated in human breast cancer samples using immunohistochemistry in tissue microarrays. Representative patient cases with positive expression in tumor cells (left panel), immune cells (middle panel) and negative expression (right panel) are shown. Original magnification: 400×. (**H**) Correlation between samples with positive and negative PD-L1 protein expression on either tumor or immune cells (total cells) with pSTAT3-GS in breast cancer patients ($n = 539$, Wilcoxon-Mann-Whitney test, $p = 0.0027$, ** $p < 0.01$).

2.2. STAT3 Mediated Regulation of PD-L1 Expression In Vitro

Next, STAT3 regulation of PD-L1 expression was investigated using the STAT3 selective inhibitor C188-9 (XIII). Suppression of pSTAT3 activity decreased PD-L1 protein levels in MDA-MB-231 cell line, as detected by western blots (Figure 2D). Additionally, silencing *STAT3* gene expression using specific siRNA constructs led to decreased PD-L1 protein levels in transiently transfected BT549 cells (Figure 2E). Furthermore, PD-L1 expression in the SKBR3 breast cancer cell line stably transfected with a constitutively active STAT3 construct was examined. Constitutive activation of STAT3 resulted in increased protein and mRNA levels of PD-L1 (Figure 2F,G). Treatment of the PD-L1 negative BC cell line MCF7 with IL-6 resulted in increased protein levels of both PD-L1 and pSTAT3 (Figure 2H).

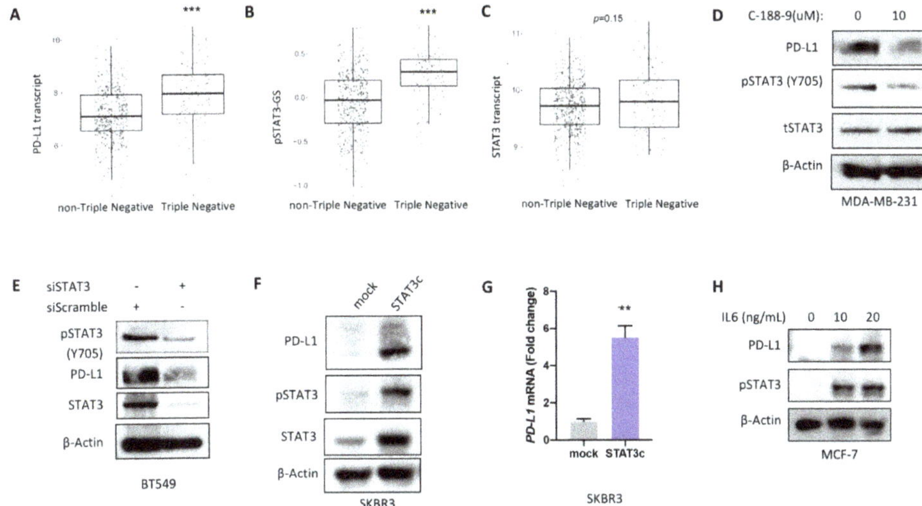

Figure 2. Expression patterns of STAT3 and PD-L1 in human breast cancer subtypes and regulation of PD-L1 by STAT3 in breast cancer cell lines. (**A**) Expression levels of *PD-L1* transcript, (**B**) pSTAT3-GS score and (**C**) STAT3 transcript in triple-negative versus non-triple negative breast cancer patients. (**D**) Inhibition of STAT3 activity by using STAT3 inhibitor C-188-9 (XIII) resulted in decreased levels of PD-L1 expression in immunoblot 48 h following treatment in MDA-MB-231 breast cancer cell line. (**E**) Knocking down *STAT3* using specific siRNA construct led to decreased levels of PD-L1 in the transiently transfected BT549 cell line. Stable transfection of SKBR3 breast cancer cell line with a STAT3 overexpressing plasmid (STAT3c) resulted in increased levels of PD-L1 (**F**) protein and (**G**) transcript expression. qPCR data are illustrated as the fold change relative to control and normalized to β-actin. They represent one out of three independent experiments and are depicted as the mean (±standard error of the mean, SEM; ** $p < 0.01$). (**H**) Increased protein levels of PD-L1 were noted in MCF7 cell line in response to treatment with IL-6.

2.3. Stat3 Silencing Downregulates PD-L1 in Mouse Cells, Restricts Tumor Growth and Metastatic Formation and Modifies Anti-Tumor Immune Response In Vivo

The impact of *Stat3* silencing on PD-L1 levels was explored in a murine model of breast cancer. More specifically, knocking down *Stat3* gene using a specific shRNA plasmid in mouse mammary carcinoma 4T1 cells (sh*Stat3* cells) resulted in decreased levels of *pd-l1* (Figure 3A–C) compared to cells transduced with corresponding empty vector (shCTR cells). When injected into the mammary fat pad of BALB/c mice, sh*Stat3* tumors displayed a decreased tumor volume (Day 25: 36.7%) and tumor weight (Day 25: 31.3%) compared to shCTR tumors (Figure 3E–G). In addition, suppression of *Stat3* expression resulted in decreased pulmonary metastatic index compared to control (Figure 3H,I).

Furthermore, it was investigated whether the reduction in tumor progression by silencing *Stat3* expression could affect the tumor immune profile. In fact, flow cytometry analysis (Figures S5 and S6) showed that silencing *Stat3* resulted in a significant increase in F4/80$^+$ TAMs (Figure 4A). However, these TAMs displayed increased levels of major histocompatibility complex class II (MHC II), indicating a more "M1"-like anti-tumoral phenotype compared to shCTR tumors (Figure 4B) [15,16]. Consistently with an anti-tumoral phenotype, NK cell accumulation was increased in sh*Stat3* tumors compared to controls (Figure 4C). Furthermore, these NK cells displayed higher expression of CD69 (Figure 4D,E) indicating that they were more activated in sh*Stat3* tumors compared to the controls. Silencing of *Stat3* resulted also in an increase of CD4$^+$ T cells (Figure 4F) while cytotoxic CD8$^+$ T cells accumulation was reduced, as compared to controls (Figure 4G). These CD4$^+$ T cells displayed increased expression of FoXP3$^+$ in combination with CD25 (Figure 4H,I), indicating an immunosuppressive feature.

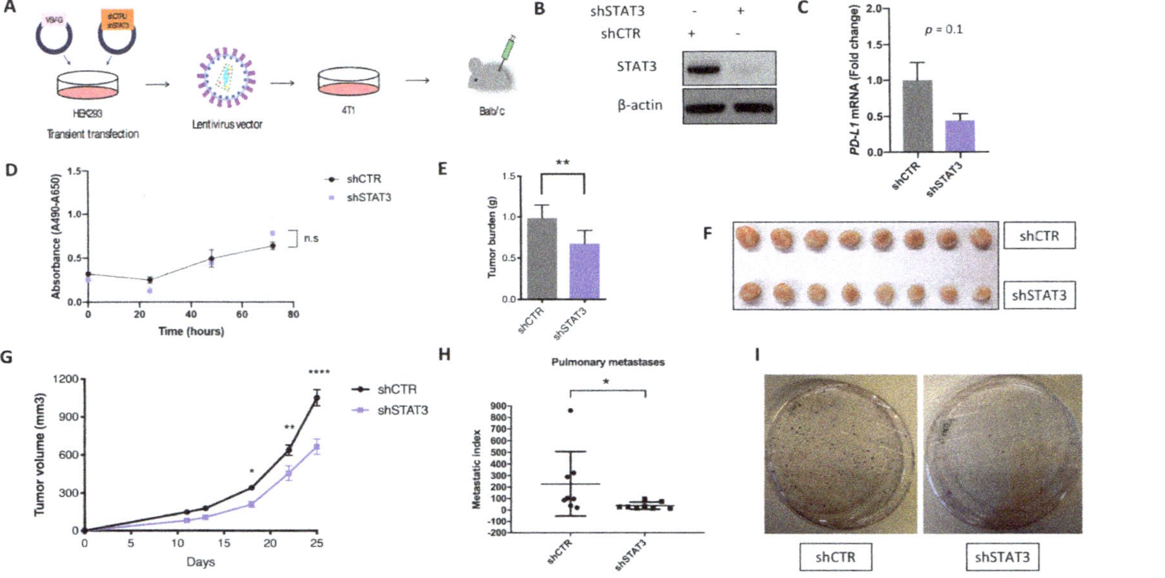

Figure 3. STAT3-mediated regulation of PD-L1 and effect on tumor growth and metastatic dissemination in vivo. (**A**) Schematic representation of the generation of the 4T1 breast cancer mouse model from transient transfection of HEK293 cells with shCTRL/sh*Sta3* plasmids and production of lentiviruses to transduction of 4T1 mouse breast cancer cell line, which was injected into the mammary fat pad of to Balb/c mice. (**B**) Downregulation of STAT3 protein levels in 4T1 cell line as assessed via immunoblotting upon sh*STAT3* plasmid transduction. (**C**) Decreased *pd-l1* transcript levels in sh*Stat3* 4T1 cell line as evaluated by qPCR. qPCR data are depicted as the fold change relative to control and are normalized to *18S* rRNA. Data represent one out of three independent experiments and presented as the mean ± SEM. (**D**) 4T1-shCTRL and 4T1-sh*STAT3* were cultured for three days under normal conditions and XTT proliferation assay was performed to assess cancer cell proliferation. The proliferation rate of *Stat3* silenced cells was not significantly changed compared to the control cells. Each time point denotes the mean of six replicates (±—standard deviation, SD). Two-way ANOVA test was used. (**E–G**) 4T1-shCTRL and 4T1-sh*Stat3* tumor cells were injected into the mammary fat pad of BALB/c mice (*n* = eight mice per group). Graphs display the tumor weight (**E**) and tumor volume (**G**). Student's t-test and two-way ANOVA with Sidak's multiple comparison test were used, respectively (* $p < 0.05$; ** $p < 0.01$, *** $p < 0.001$; **** $p < 0.0001$). shCTR and sh*Stat3* mice tumors are depicted in picture (**F**). (**H–I**) The graph depicts metastatic index, which is calculated by the number of well-formed colonies per tumor weight (**H**) (Wilcoxon-Mann-Whitney test, $p = 0.0148$). Representative photos from the colongenic assay are depicted in figure (**I**).

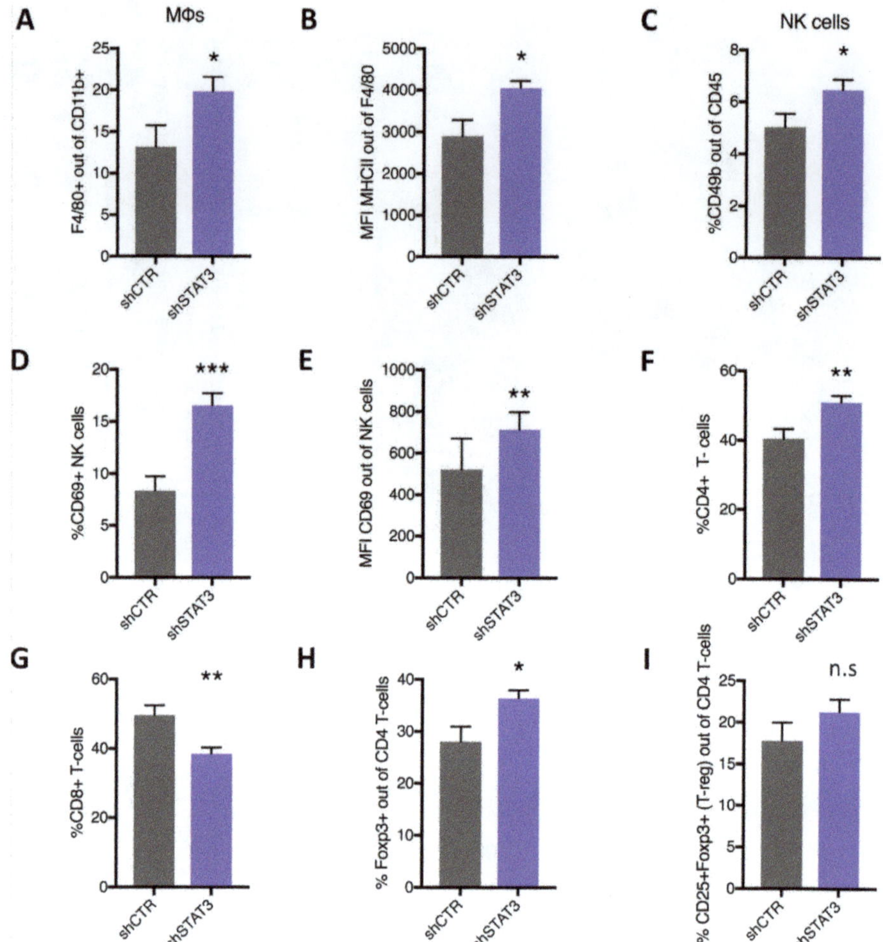

Figure 4. Effect of *Stat3* silencing on immunologic profile in mouse model. 4T1-shCTRL and 4T1-sh*Stat3* tumors were dissociated to a single cell level and analysed using flow cytometry for the percentage of (**A**). F4/80⁺ cells out of the gate of CD11b⁺. Graph (**B**) shows the mean fluorescence intensity (MFI) of MHCII out of F480+ cells in 4T1-shCTRL and 4T1-sh*Stat3* tumors. (**C**) Percentage of CD49b⁺ NK cells out of CD45 and percentage (**D**) and MFI (**E**) of CD69 out of NK cells in the same tumors. 4T1-shCTRL and 4T1-sh*Stat3* tumors were also analysed with flow cytometry for the percentages of CD4⁺ T-cells (**F**), CD8+ T-cells (**G**), FoxP3⁺ (**H**) and CD25⁺ FoxP3⁺ cells out of CD4⁺ T-cells (**I**). Student's t-test was performed for all comparisons depicted in the graphs and flow cytometry data are presented as mean ± SEM ($n = 8$; * $p < 0.05$; ** $p < 0.01$; *** $p < 0.001$); MHC: major histocompatibility complex.

2.4. CD163⁺ TAM Phenotype Is Associated with Higher PD-L1 Expression, Grade and Proliferation in Breast Cancer Patients

Apart from the mouse model, it was further explored whether PD-L1 expression correlated to a pro-tumoral "M2"-like (CD163⁺) TAM phenotype or an anti-tumoral CD11c⁺ macrophage/dendritic cell phenotype in breast cancer patient samples. Hence, 45 patient samples with PD-L1 positive ($n = 23$) or PD-L1 negative ($n = 22$) expression in tumor cells were stained for CD11c and CD163 (Figure 5A) and three different populations could be detected, i.e., CD163⁺CD11c⁻, CD163⁺CD11c⁺ and CD163⁻CD11c⁺.

Of note, both the percentage of CD163$^+$ and CD163$^+$CD11c$^+$ cells (Figure 5B,C) as well as their ratio to CD11c$^+$ antigen-presenting cells (CD163$^+$/CD11c$^+$; Figure 5D and CD163$^+$CD11c$^+$/CD11c$^+$; Figure 5E) were significantly higher in patients with PD-L1 positive expression in tumor cells. Similar results were observed when these subpopulations were associated with *PD-L1* transcript levels (Figure 5F–I). Conversely, higher percentage of CD11c$^+$ cells was observed in patients with PD-L1 protein negative tumors (Figure 5J) and were inversely correlated with *PD-L1* mRNA as well (data not shown). Accumulation of CD163$^+$CD11c$^-$, CD163$^+$CD11c$^+$ cells (Figure S4) and their respective ratio to CD11c$^+$ cells (Figure 6A,B) were observed in patients with grade 3 tumors and with high expression of the proliferation marker Ki67 (Figure 6C,D and Figure S4). In contrast, CD11c$^+$ were more prominent in patients with grade 1–2 tumors and with low expression of the proliferation marker Ki67 (Figure 6E,F). In corroborration, PD-L1 tumor IHC expression was significantly associated with high grade and Ki67 expression (Table S5).

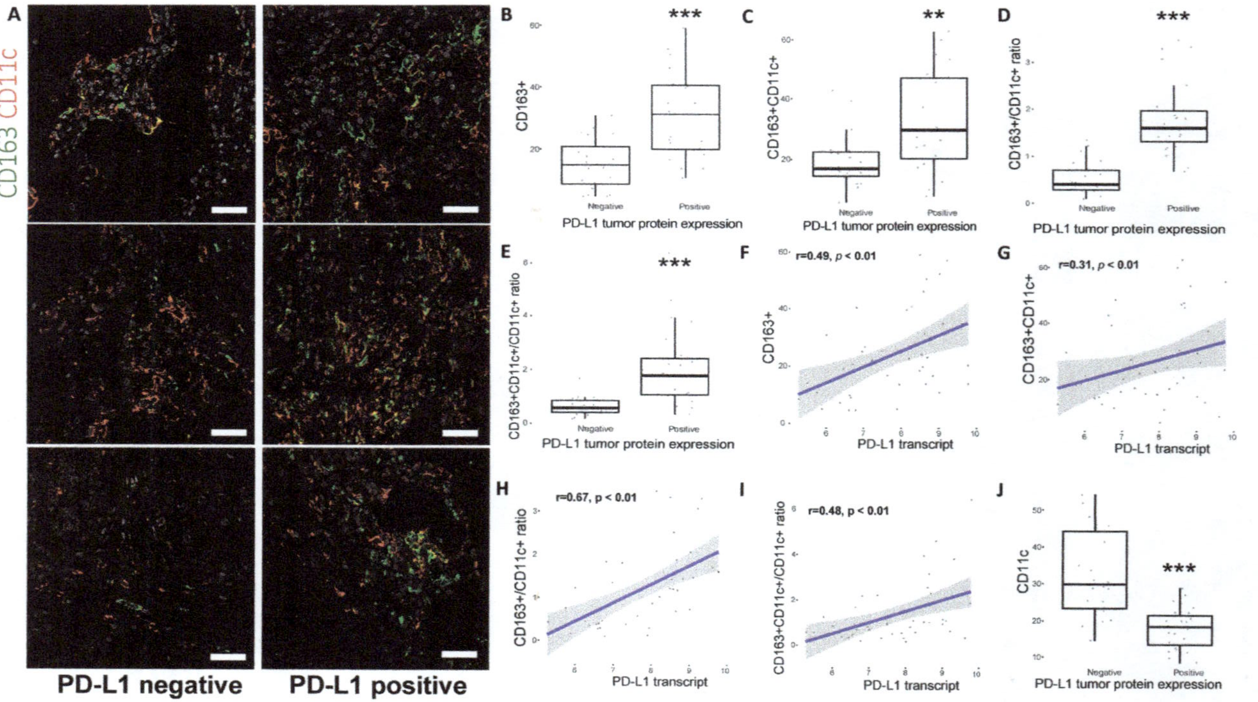

Figure 5. Pro-tumoral TAM phenotype correlates with PD-L1 expression levels in human breast cancer patients. (**A**) Double immunofluorescence staining with markers for antigen presenting cells such as M1-like TAMs (CD11c: red) and pro-tumoral M2-like TAMs (CD163: green) macrophages was performed in patients with PD-L1 positive ($n = 23$) and PD-L1 negative ($n = 22$) expression in tumor cells. Scale bar, 100 um. Correlations of PD-L1 protein expression in tumor cells with percentage of CD163$^+$ cells (**B**), percentage of CD163$^+$CD11c$^+$ cells (**C**), CD163$^+$/CD11c$^+$ ratio and CD163$^+$ CD11c+/CD11c+ ratio (M2-like subtype 2 versus M1-like phenotypes) (**D**–**E**). (**F**–**I**) The same percentages and ratios were also positively correlated with *PD-L1* transcript levels. An inverse correlation was noted between CD11c$^+$ cells with PD-L1 positive tumor protein expression (**J**). Wilcoxon-Mann-Whitney test and Spearman's rank correlation coefficient were used (* $p < 0.05$; ** $p < 0.01$; *** $p < 0.001$).

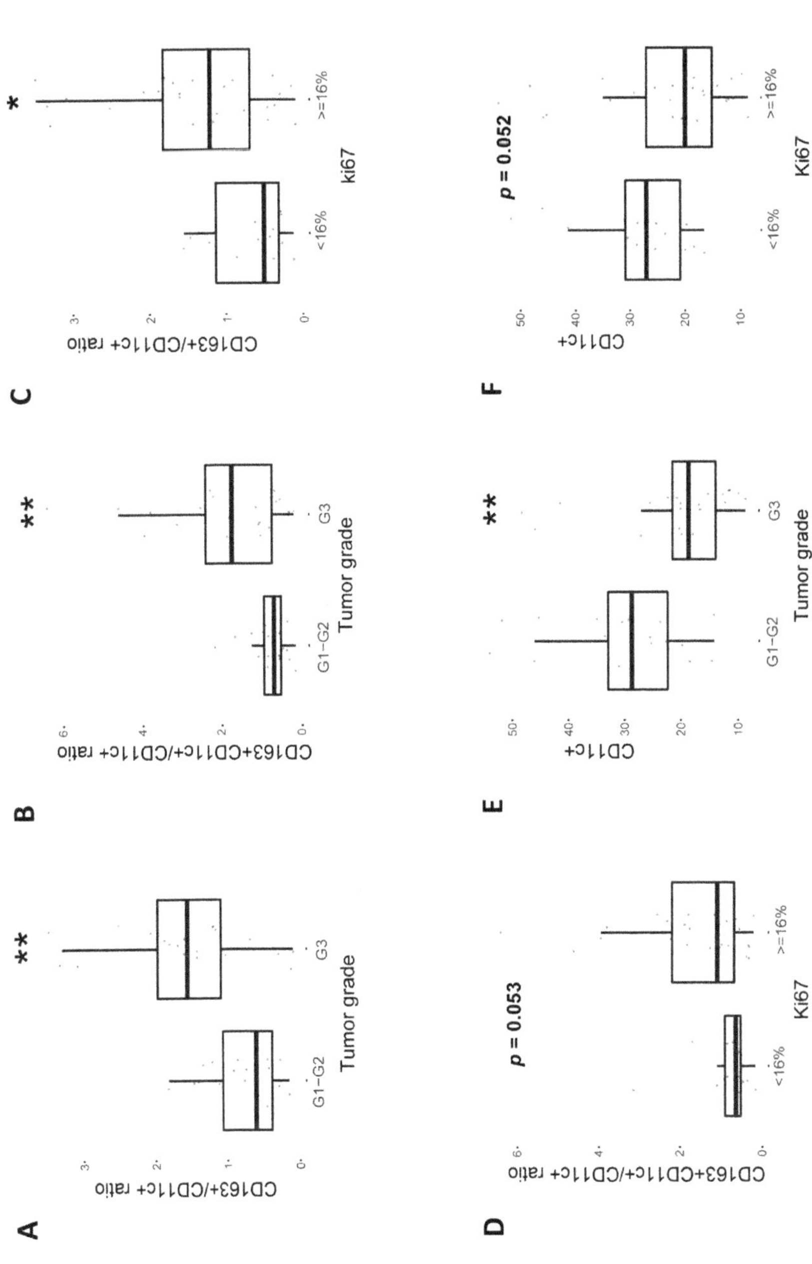

Figure 6. CD163$^+$ macrophage phenotype correlates with higher Ki67 and grade in human breast cancer patients. CD163$^+$/CD11c$^+$ and CD163$^+$CD11c$^+$/CD11c$^+$ ratios were positively correlated with high grade (**A,B**) and high Ki67 (**C,D**) in breast cancer patient patients ($n = 45$). CD11c$^+$ cell expression was associated with lower tumor grade and lower Ki67 (**E,F**). Wilcoxon-Mann-Whitney test was used (* $p < 0.05$; ** $p < 0.01$; *** $p < 0.001$).

3. Discussion

Immune checkpoint blockade has revolutionized cancer treatment, improving survival outcomes in cancer patients, but the underlying mechanisms of PD-L1 regulation are not yet fully understood [8]. Activated (phosphorylated) STAT3, which forms dimers and transports into the nucleus, represents a key transcription factor that critically controls proliferation, invasiveness, survival and metastasis [17]. It also modifies the immune response through various mechanisms, including regulation of PD-L1 expression [18]. Here, it was shown that STAT3 regulated PD-L1 expression in breast cancer cell lines and its silencing led to restriction of tumor growth and altered the immune profile in a murine breast cancer model. Moreover, PD-L1 expression was associated with a pro-tumoral TAM phenotype in breast cancer patients.

Even though a direct link between PD-L1 and STAT3 has been described in previous reports, in this study it was shown that STAT3 can influence immune response and PD-L1 expression in mouse models as well as in patients with early breast cancer. As indicated in a previous breast cancer cell line study, pSTAT1-pSTAT3 dimers bound on *PD-L1* gene promoter inducing its expression and therefore STAT3 inhibition led to partial downregulation of PD-L1 [19]. Similar studies suggesting a STAT3-mediated transcriptional regulation of PD-L1 have been performed in nucleophosmin—anaplastic lymphoma kinase (NPM-ALK) positive (+) anaplastic large cell lymphoma (ALCL) [10], in ALK (-) ALCL [11], in KRAS- and EGFR- mutant non-small cell lung cancer [20,21] and in head and neck squamous cell carcinoma [22]. By contrast, STAT3 was not directly bound on *PD-L1* gene promoter in melanoma cells [23]. In this study, it was confirmed that STAT3 can regulate PD-L1 expression in vitro.

STAT3 signaling contributes as well to a dynamic crosstalk between tumor cells and immune cells, including macrophages, CD8$^+$ T-cells, myeloid-derived suppressor cells, T-regs and NK cells [24]. In this study, it was demonstrated that *Stat3* gene silencing in 4T1 mouse cell line not only led to tumor growth restriction but also altered the immunologic profile in vivo. Importantly, a prominent anti-tumoral macrophage phenotype was denoted in sh*Stat3* tumors compared to shCTR. Decreased lung metastases were also observed in those tumors, further underscoring the role of skewing macrophage polarization in metastatic potential [25]. Similarly to our experiments, depletion of *Stat3* in epithelial cells of transgenic PyMT-MMTV mice resulted in decreased tumor growth and metastatic potential, and macrophage accumulation in *Stat3*-deficient mice [26]. However, in our experimental setting we could also show that TAMs had acquired an "M1"-like anti-tumoral phenotype that was correlated to accumulation of activated NK cells. Surprisingly, we also observed an accumulation of immunosupressive T-regs and less cytotoxic CD8$^+$ T-cells in sh*Stat3* tumors. This effect on T-cell activation could potentially be explained by the fact that PD-L1 is not only expressed on tumor cells, but also on immune cells—including macrophages. Indeed, approximately 50 % of TAMs express PD-L1 (data not shown) independent of *Stat3* expression in the 4T1 mammary tumor model system, but it is still unclear how this can influence the anti-tumor immunity. The expression of other co-inhibitory markers, transcriptional or post-transcriptional factors may also contribute to this restrained immune response. Increased PD-1 expression in TAMs was associated with impaired antitumor immune response in colorectal cancer patients [27], while in another study, PD-L1 expression in macrophages hindered their proliferation and activation [28]. On the other side, activated NK cells were elevated in sh*Stat3* tumors, indicating an anti-tumor immune activity, most likely responsible for the moderate effect on tumor growth. Of importance, the skewing of TAMs towards an anti-tumoral phenotype may instead dictate the metastatic dissemination, as these macrophages have previously been shown to hamper tumor cell extravasation into the blood vessels and form secondary tumors [25,29].

At the patient level, correlations between pSTAT3 and PD-L1 both at the mRNA and protein level were explored and demonstrated in a large breast cancer cohort. Of note, in a subset of patients PD-L1 expression in tumor cells was significantly associated with "M2"-like macrophage phenotype (driven by the expression of CD163) while both PD-L1 and "M2"-like macrophages were associated with higher Ki67 expression and tumor grade. Although it has long been described that TAM accumulation is correlated with tumor progression and a worse prognosis [30], recent reports featured their interaction with immune checkpoints in tumor cells and in TME components [31]. Specifically, the two distinct macrophage phenotypes, the anti-tumoral "M1"-like and the pro-tumoral "M2"-like can secrete cytokines and other factors, which in turn affect PD-L1 expression and anti-tumor immune response [32,33] and only few recent studies have shown correlation of PD-L1 levels in tumor cells with macrophages phenotypes. In a study of gastric adenocarcinoma "M2"-like macrophage infiltration was correlated with PD-L1 expression [34] while in a mouse model, anti-PD-1 therapy led to macrophage reprogramming from "M2"-like to "M1"-like phenotype and to a subsequent regression of osteosarcoma lung metastases [35].

The findings of this study have potential clinical implications. First, therapeutic strategies involving STAT3 inhibition could enhance the efficacy of anti-PD-L1/PD-1 monoclonal antibodies, which recently proved efficacy in patients with metastatic triple negative breast cancer [7]. The STAT3 SH2 domain binder inhibitor C188-9 inhibitor is currently tested in an ongoing phase I trial (NCT03195699) in patients with advanced-stage cancers, including breast cancer. Previous research findings indicated that combined targeting of IL-6/JAK/STAT pathway and PD-L1 resulted in restricted tumor growth in in vivo models [36,37], while in others, STAT3 targeting increased the efficacy of anti-PD1 mAb [38]. Currently, a phase II trial testing the combination of a pSTAT3 inhibitor (napabucasin) with the anti-PD-1 antibody nivolumab in microsatellite stable, refractory colorectal cancer (NCT03647839) is ongoing. Moreover, immune checkpoint blockade could be combined with agents (e.g., CSF-1 inhibitors) that can inhibit TAMs accumulation and/or re-programme macrophages towards a more anti-tumoral M1-like phenotype in selected breast cancer patients. Indeed, such therapeutic combinations are currently under investigation in clinical trials [31,39]. Nevertheless, the recognition of such immunosuppressive expression patterns may pave the way for the development of biomarkers and patient stratification. Hence, subgroups of patients who may gain the most benefit of conventional, targeted and/or immune therapy can be identified towards a more personalized cancer treatment approach, yet the importance of these findings still need to be prospectively investigated in breast cancer patients.

4. Materials and Methods

4.1. Cell Lines, Plasmids, and Reagents

The cell lines used in this study, along with their characteristics, are listed in Table S1. The human breast cancer cell lines BT549 and SKBR3, mouse mammary carcinoma cell line 4T1, as well as the control anaplastic large cell lymphoma (Mac2A) and Hodgkin lymphoma (HDLM2) cell lines were grown in complete RPMI-1640 medium (Gibco/Life Technologies, Waltham, MA, USA) supplemented with 10% fetal bovine serum (FBS) (Gibco/Life Technologies, Waltham, MA, USA), 1% L-glutamine (Gibco/Life Technologies, Waltham, MA, USA) and 1% penicillin/streptomycin (Gibco/Life Technologies, Waltham, MA, USA). The breast cancer cell lines MCF7 and MDA-MB-231 were grown in complete DMEM medium (Gibco/Life Technologies, Waltham, MA, USA) supplemented with 10% FBS, 1% L-glutamine, and 1% penicillin/streptomycin. Cells were cultured at 37 °C in 5% CO_2. Vector control and STAT3C-transfected SKBR3 cells were obtained from Sarah Walker and David Frank, Department of Medical Oncology, Dana-Farber Cancer Institute, and Department of Medicine, Harvard Medical School, Boston, MA, USA. Breast cancer cells were treated with STAT3 inhibitor C188-9/XIII (Calbiochem, St. Louis, MO, USA) and with recombinant human IL-6 (PeproTech, Rocky Hill, NJ, USA) at the indicated concentrations for 48 and 24 h respectively and whole cell lysates were prepared for western blot analysis.

4.2. Western Blot Analysis

Tumor cells were collected, washed in cold phosphate-buffered saline (PBS) (GE Healthcare Life Sciences, Chicago, IL, USA), lysed in lysis buffer and western blot analysis was performed as previously described [40]. The antibodies used in the present study are listed in Table S2.

4.3. RNA Extraction, cDNA Synthesis, and Real Time Quantitative Polymerase Chain Reaction (RT-qPCR)

Total RNA was extracted with the RNeasy® Plus Mini Kit (QIAGEN Inc., Hilden, Germany) and cDNA was synthesized using the Superscript First Strand Synthesis System (Invitrogen Life Technologies, Carlsbad, CA, USA) according to the manufacturer's protocol. The mRNA expression levels were quantified by RT-qPCR using the Power SYBR® Green PCR Master Mix (Applied Biosystems by Thermo Fisher Scientific, Waltham, MA, USA) in a one-step reaction and the mRNA expression levels were determined by the comparative CT ($\Delta\Delta$Ct) method. 18S rRNA and beta-actin were used as the endogenous control genes as indicated. The primer sequences used are listed in Table S3. The RT-qPCR program included Amplitaq Gold DNA polymerase activation at 95 °C (10 min) followed by 40 cycles of DNA denaturation (95 °C for 15 sec) and annealing/extension (60 °C for 30 sec). All reactions were performed using a Veriti 96-well thermal cycler (Applied Biosystems by Thermo Fisher Scientific, Waltham, MA, USA).

4.4. Tissue Microarrays, Immunohistochemical Methods and Scoring

Tissue microarrays (TMAs) were constructed using duplicate tumor cores from primary tumors and an automated tissue microarrayer (VTA-100, Veridiam, San Diego, CA, USA). Tissue sections from the TMAs were used for PD-L1 immunohistochemistry (IHC) using the Ventana autostainer system according to manufacturer's protocols. Positivity was defined by the presence of any single cell with membranous expression of PD-L1 either in tumor or in immune cells (total cells). In addition, whole tissue sections (4 μm) were prepared for a subset of patients in this cohort based on the expression of PD-L1 in tumor cells (positive and randomly selected negative cases) and stained using an anti-phosphorylated STAT3 (pSTAT3) antibody (Y705). At least 300 tumor cells were counted in five different high power fields in order to calculate the percentage of pSTAT3 positive cells. The median percentage of expression was used as a cut-off for dividing patient tumors in pSTAT3-high and pSTAT3-low expressing ones. Furthermore, cell pellets from cell lines were collected, fixed in formalin and embedded in paraffin to prepare cell blocks. Subsequently, IHC was performed using anti-pSTAT3 and anti-PD-L1 antibodies, as previously described [41]. The antibodies used are listed in Table S2. Moreover, Ki67 immunohistochemical staining and evaluation method have been previously described [42].

4.5. Fluorescence in Situ Hybridization

The *PDL1* gene locus was analyzed on formalin-fixed paraffin-embedded (FFPE) cell pellets by Fluorescence in situ hybridization using the probe and protocols recommended by the manufacturer (ZytoVision GmbH, Bremerhaven, Germany).

4.6. Transient Transfections and Gene Silencing

Cells were seeded at a density of 0.2–0.6×10^6 cells/ml 24 h before transfection. Silencing of breast cancer cell line BT549 with siRNA oligonucleotides specific for the gene sequence of *STAT3* was carried out with Lipofectamine 2000 (Thermofisher, Waltham, MA, USA) reagent according to the company's protocol. Approximately 3×10^6 cells were transfected with 300 nM specific siRNA or Control siRNA (All Stars Negative, QIAGEN, Hilden, Germany). Whole-cell lysates were prepared 48 h after transfection. The siRNA oligonucleotides or plasmids with which cells were transfected are listed in Table S3.

4.7. Proliferation Assay

Mouse 4T1-shCTR and 4T1-sh*Stat3* tumor cells were seeded at a density of 4×10^3 cells per well containing 100 uL in 96-well plates with complete RPMI-1640 medium for three days. For each cell line and each time point (0, 24, 48 and 72 h) six replicates were used. A mixture of XTT labeling with electron-coupling reagents was then made according to manufacturer's protocol (Cell Proliferation Kit

II (XTT), Roche, Basel, Switzerland). For each time point, measurement of cell viability was performed via microplate reader at a wavelength of 490 nm (reference wavelength: 650 nm).

4.8. Lentiviral Vectors

Tumor cells (4T1) were transduced with lentiviral vectors encoding shRNA for murine *Stat3* or corresponding control vector (listed in Table S3) as previously described [43].

4.9. 4T1 Breast Cancer Animal Model and Tumor Growth Assessment

For the in vivo tumor model, 2×10^5 4T1 cells (shCTR and sh*Stat3*) in a volume of 50 mL PBS were injected into the mammary fat pad of anesthetized four to six week-old female BALB/c mice, purchased from Charles River Laboratory. Mice were sacrificed three weeks after tumor cell injection, and tumors were weighed after dissection. Tumor size was measured externally using calipers, and tumor volumes were estimated using the following equation: $V = 4/3\pi \times (d/2)^2 \times D/2$, where d is the minor tumor axis and D is the major tumor axis. All ethical permits were obtained from the Swedish Board of Agriculture (N95/15).

4.10. Tumor Dissociation

Tumors were cut into smaller pieces using scalpels. Dissected tumors were minced in dissociation buffer (TrypLe and Stem Cell Pro Accutase (Life Technologies, Waltham, MA, USA); 1:1) and incubated at 37 °C for 30 mins in order single-cell suspensions to be obtained. Thereafter, suspensions were passed through a 19 G syringe needle, filtered and washed in PBS (10% FBS) to be used for flow cytometry.

4.11. Flow Cytometry

To prevent antibody nonspecific binding, single-cell suspensions of tumors were pre-incubated with anti-CD16/32 mAb (BioLegend, San Diego, CA, USA) on ice for 15 minutes before a 30-minute incubation on ice with specific antibodies. Cells were stained using antibodies for extracellular markers. T-regulatory (FoxP3$^+$) cells were stained as well according to manufacturer's instruction (BD Biosciences, San Jose, CA, USA). The viability of cells was verified using 7AAD or the Live/Dead fixable dead cell stain (Life Technologies, Waltham, MA, USA). Samples were acquired with a LSR II (BD Biosciences, San Jose, CA, USA) and analyzed using FlowJo software (Tree Star, Ashland, OR, USA) [15]. All antibodies used are listed in Table S2 and gating strategies are depicted in Supplementary Figures S5 and S6.

4.12. Lung Metastasis Colony Assay

Lungs from tumor-bearing mice were dissociated to single cells in an enzymatic buffer containing RPMI (Gibco/Life technologies, Waltham, MA, USA) 5% FBS (Gibco/Life technologies, Waltham, MA, USA) 0.2 mg/mL collagenase IV (Life technologies, Waltham, MA, USA) 0.2 mg/mL dispase (Life technologies, Waltham, MA, USA) and 0.1 mg/mL DNAse I (Sigma-Aldrich, St. Louis, MO, USA) as previously described [44,45]. Cell suspensions were plated in the presence of 60 uM 6-thioguanine (Sigma-Aldrich, St. Louis, MO, USA). Tumor cells resistant to 6-thioguanine were allowed to form colonies for approximately 10 days and then fixed with methanol, stained with crystal violet, and counted under a dissection microscope.

4.13. Immunofluorescence for Macrophage Markers in Human Breast Tumors

Paraffin-embedded breast cancer patient samples were cut in 4 μm thick whole-tissue sections and treated for antigen retrieval with sodium citrate buffer (Biocare Medical). All tumor sections were blocked in blocking buffer containing PBS (Sigma, St. Louis, MO, USA), 0.3% Triton X-100 (Sigma), 10 % fetal bovine serum (Gibco, Waltham, MA, USA) and 1% Bovine Serum Albumin (Sigma) and immunostained with the appropriate antibodies: anti-CD163 (Leica, Wetzlar, Germany) and anti-CD11c (Leica) and all secondary antibodies were conjugated with AlexaFluor 488 or AlexaFluor 546 fluorochromes (Life Technologies, Waltham, MA, USA). Cell nuclei were labeled with DAPI (Invitrogen Corp., Carlsbad, CA, USA). Isotype specific antibodies were used to ensure specificity of the antibodies. All antibodies used are listed in Table S2. Eight independent fields from each tumor section were analyzed by using LSM T-PMT Zeiss confocal microscope and quantified by ImageJ software (NIH, Bethesda, MD, USA).

4.14. Patient Cohorts

The patient cohort used in the study consists of women diagnosed with primary breast cancer between 1997 and 2005 in Stockholm health care region who were retrospectively selected using the Stockholm-Gotland Breast Cancer Registry. Data on clinical and pathological tumor characteristics, survival, loco-regional and systemic treatments, and follow-up have been collected and reported elsewhere [42]. The Cancer Genome Atlas (TCGA) provisional dataset [46], including 1081 patients with primary breast cancer, was used as a validation cohort. Available RNA-sequencing data were retrieved from cBioportal [47,48].

4.15. Gene Expression Profiling Using Microarrays

Gene expression profiling has been performed from all primary tumors in the cohort. Details regarding experimental methods have been previously described [42] and the gene expression microarray data can be accessed at the Gene Expression Omnibus database under accession number GSE48091. All ethical permits were obtained from Karolinska Institutet's (Stockholm, Sweden) ethics committee (Dnr 2006/1183-31/2, 2016/1505-32). According to the ethics board, no additional informed consent from the patients was required for these analyses.

4.16. Preprocessing and Normalization of the Microarray Gene Expression Data

The preprocessing and normalization of the microarray gene expression data were performed within R computing environment. Specifically, the raw data were background corrected, normalized and summarized to obtain a log-transformed expression value for each probe set using the RMA method [49] implemented in the aroma.affymetrix R package [50]. A nonspecific filter was employed and probe sets with the highest interquartile range were kept in the case of multiple mappings to the same Entrez Gene ID.

4.17. Phosphorylated STAT3-Associated Gene Signature

The phosphorylated STAT3-associated gene signature (pSTAT3-GS) was applied to the patient cohort as described in the original publication [51]. Specifically, the provided pSTAT3-GS's gene symbols were converted to Entrez Gene IDs using DAVID's Gene ID Conversion Tool [52] (DAVID Bioinformatics Resources version 6.8), and mapped to the microarray's probe sets. The (continuous) pSTAT3-GS scores, i.e., signed averages, were computed using the "sig.score" function from the genefu R package [53] (R package version 2.14.0). In total, 114 (out of the 123) pSTAT3-GS's genes were mapped and therefore used in the signature scores' calculation. For the TCGA dataset, in total 122 genes of the pSTAT3-GS were mapped.

4.18. Statistical Analyses

Statistical analyses and graphical representations regarding the in vitro and in vivo models were performed using GraphPad Prism software version 7.0 (GraphPad Software Inc., San Diego, CA, USA) while analyses concerning patient material were performed within R computing environment version 3.5.1. Wilcoxon–Mann–Whitney test, Student's t-test and two-way ANOVA were used as indicated. Spearman's rank correlation coefficient was used to evaluate the associations between continuous variables and Fisher's exact test for the associations between categorical variables. A *p*-value equal to or less than 0.05 was considered as statistically significant.

5. Conclusions

In this study, STAT3-mediated regulation of PD-L1 and modulation of immune microenvironment were shown in breast cancer. More research is warranted towards the further characterization of TME interactions and anti-tumor immunity, thus providing breast cancer patients better therapeutic options and prognostication factors.

Supplementary Materials: The following are available online at http://www.mdpi.com/2072-6694/11/10/1479/s1, Table S1. List of cancer cell lines used in the study; Table S2. List of antibodies used in the study; Table S3. List of primers, siRNA and shRNA target sequences used in the study; Table S4. Correlation of PD-L1 protein expression in tumor, immune and total cells with pSTAT3 protein expression in human breast cancer patients; Table S5. Correlation of PD-L1 protein expression in tumor cells with grade and Ki67 status in human breast cancer patients; Figure S1. Expression of PD-L1 and genetic alteration in control cell line; Figure S2. Correlation of PD-L1 transcript with pSTAT3-GS score and expression patterns in TCGA Provisional database. Figure S3. Expression of pSTAT3 protein in breast cancer patients; Figure S4. Correlations of CD163$^+$ cell and CD163$^+$ CD11c$^+$ cell percentage with proliferation status (Ki67) and tumor grade in human breast cancer patients. Figure S5. Gating strategies for macrophage panel MHC class II. Figure S6. Gating strategies for lymphocytic panel.

Author Contributions: Conceptualization: T.F., G.Z.R., C.R., I.Z.; formal analysis: I.Z., M.W., E.G.S., T.W., M.B.; funding acquisition: T.F., C.R., J.B.; investigation: I.Z., M.W., E.G.S., T.W., S.B., M.B., N.T.; methodology: I.Z., M.W., E.G.S., G.Z.R., T.F., C.R.; project administration: I.Z., E.G.S.; resources: T.F., C.R., G.Z.R., J.B.; validation: I.Z., M.W., E.G.S., T.W.; visualization: I.Z., M.W., E.G.S., T.W., S.B., M.B., N.T., C.C., G.Z.R.; writing—initial draft preparation: I.Z., E.G.S., G.Z.R., C.R., T.F.; writing—review and editing: I.Z., M.W., E.G.S., T.W., S.B., M.B., N.T., N.P.T., C.C., J.B., G.Z.R., C.R., T.F.

Funding: This study was supported by the Swedish Cancer Society (CAN 2015/213 to TF and CAN 2016/825 to CR); the Cancer Society in Stockholm (154132 to TF); the Swedish Breast Cancer Association (IZ, TF); The Iris, Stig och Gerry Castenbäcks Stiftelse for cancer research (NT), the King Gustav V Jubilee Foundation (184181 to NT) and the Swedish Research Council (VR 2018-02915 to CR).

Acknowledgments: We would like to thank Susanne Agartz and Jing Jing Ye for their technical support with patient material (TMAs), cell blocks and lysates derived from breast cancer cell lines. We would also like to thank John Lövrot for bioinformatics support and Ioannis Mantas, for his help with illustration work. We also thank Andreas Lundqvist (Karolinska Institute, Stockholm, Sweden) for providing us with reagents and Sarah Walker and David Frank (Department of Medical Oncology, Dana-Farber Cancer Institute, and Department of Medicine, Harvard Medical School, Boston, MA, USA) for providing us with SKBR3 breast cancer sub-lines.

Conflicts of Interest: Ioannis Zerdes, Majken Wallerius, Emmanouil G. Sifakis, Tatjana Wallmann, Stina Betts, Margarita Bartish, Nikolaos Tsesmetzis, Nicholas P. Tobin, Christos Coucoravas, George Z. Rassidakis and Charlotte Rolny have no conflicts of interest to disclose. Theodoros Foukakis: institutional grants from Roche and Pfizer and personal fees from Novartis, Pfizer, Roche and UpToDate; Jonas Bergh has received research funding from Merck paid to Karolinska Institutet and from Amgen, Bayer, Pfizer, Roche and Sanofi-Aventis paid to Karolinska University Hospital. No personal payments. Payment from UpToDate for a chapter in breast cancer prediction paid to Asklepios Medicine HB.

References

1. Keir, M.E.; Butte, M.J.; Freeman, G.J.; Sharpe, A.H. PD-1 and its ligands in tolerance and immunity. *Annu. Rev. Immunol.* **2008**, *26*, 677–704. [CrossRef] [PubMed]

2. Chen, L. Co-inhibitory molecules of the B7-CD28 family in the control of T-cell immunity. *Nat. Rev. Immunol.* **2004**, *4*, 336–347. [CrossRef] [PubMed]

3. Sharma, P.; Allison, J.P. The future of immune checkpoint therapy. *Science* **2015**, *348*, 56–61. [CrossRef] [PubMed]

4. Stanton, S.E.; Adams, S.; Disis, M.L. Variation in the Incidence and Magnitude of Tumor-Infiltrating Lymphocytes in Breast Cancer Subtypes: A Systematic Review. *JAMA Oncol.* **2016**, *2*, 1354–1360. [CrossRef] [PubMed]

5. Solinas, C.; Carbognin, L.; De Silva, P.; Criscitiello, C.; Lambertini, M. Tumor-infiltrating lymphocytes in breast cancer according to tumor subtype: Current state of the art. *Breast* **2017**, *35*, 142–150. [CrossRef] [PubMed]

6.   Solinas, C.; Gombos, A.; Latifyan, S.; Piccart-Gebhart, M.; Kok, M.; Buisseret, L. Targeting immune checkpoints in breast cancer: An update of early results. *ESMO Open* **2017**, *2*, e000255. [CrossRef] [PubMed]

7.   Schmid, P.; Adams, S.; Rugo, H.S.; Schneeweiss, A.; Barrios, C.H.; Iwata, H.; Dieras, V.; Hegg, R.; Im, S.A.; Shaw Wright, G.; et al. Atezolizumab and Nab-Paclitaxel in Advanced Triple-Negative Breast Cancer. *N. Engl. J. Med.* **2018**, *379*, 2108–2121. [CrossRef] [PubMed]

8.   Zerdes, I.; Matikas, A.; Bergh, J.; Rassidakis, G.Z.; Foukakis, T. Genetic, transcriptional and post-translational regulation of the programmed death protein ligand 1 in cancer: Biology and clinical correlations. *Oncogene* **2018**. [CrossRef] [PubMed]

9.   Bromberg, J.F.; Wrzeszczynska, M.H.; Devgan, G.; Zhao, Y.; Pestell, R.G.; Albanese, C.; Darnell, J.E., Jr. Stat3 as an oncogene. *Cell* **1999**, *98*, 295–303. [CrossRef]

10.   Marzec, M.; Zhang, Q.; Goradia, A.; Raghunath, P.N.; Liu, X.; Paessler, M.; Wang, H.Y.; Wysocka, M.; Cheng, M.; Ruggeri, B.A.; et al. Oncogenic kinase NPM/ALK induces through STAT3 expression of immunosuppressive protein CD274 (PD-L1, B7-H1). *Proc. Natl. Acad. Sci. USA* **2008**, *105*, 20852–20857. [CrossRef] [PubMed]

11.   Atsaves, V.; Tsesmetzis, N.; Chioureas, D.; Kis, L.; Leventaki, V.; Drakos, E.; Panaretakis, T.; Grander, D.; Medeiros, L.J.; Young, K.H.; et al. PD-L1 is commonly expressed and transcriptionally regulated by STAT3 and MYC in ALK-negative anaplastic large-cell lymphoma. *Leukemia* **2017**, *31*, 1633–1637. [CrossRef] [PubMed]

12.   Yu, H.; Pardoll, D.; Jove, R. STATs in cancer inflammation and immunity: A leading role for STAT3. *Nat. Rev. Cancer* **2009**, *9*, 798–809. [CrossRef] [PubMed]

13.   Huynh, J.; Chand, A.; Gough, D.; Ernst, M. Therapeutically exploiting STAT3 activity in cancer—Using tissue repair as a road map. *Nat. Rev. Cancer* **2019**, *19*, 82–96. [CrossRef] [PubMed]

14.   Ruffell, B.; Coussens, L.M. Macrophages and therapeutic resistance in cancer. *Cancer Cell* **2015**, *27*, 462–472. [CrossRef] [PubMed]

15.   Wallerius, M.; Wallmann, T.; Bartish, M.; Ostling, J.; Mezheyeuski, A.; Tobin, N.P.; Nygren, E.; Pangigadde, P.; Pellegrini, P.; Squadrito, M.L.; et al. Guidance Molecule SEMA3A Restricts Tumor Growth by Differentially Regulating the Proliferation of Tumor-Associated Macrophages. *Cancer Res.* **2016**, *76*, 3166–3178. [CrossRef] [PubMed]

16.   Movahedi, K.; Laoui, D.; Gysemans, C.; Baeten, M.; Stange, G.; Van den Bossche, J.; Mack, M.; Pipeleers, D.; In't Veld, P.; De Baetselier, P.; et al. Different tumor microenvironments contain functionally distinct subsets of macrophages derived from Ly6C(high) monocytes. *Cancer Res.* **2010**, *70*, 5728–5739. [CrossRef]

17.   Frank, D.A. Transcription factor STAT3 as a prognostic marker and therapeutic target in cancer. *J. Clin. Oncol.* **2013**, *31*, 4560–4561. [CrossRef]

18.   Johnson, D.E.; O'Keefe, R.A.; Grandis, J.R. Targeting the IL-6/JAK/STAT3 signalling axis in cancer. *Nat. Rev. Clin. Oncol.* **2018**, *15*, 234–248. [CrossRef]

19.   Sasidharan Nair, V.; Toor, S.M.; Ali, B.R.; Elkord, E. Dual inhibition of STAT1 and STAT3 activation downregulates expression of PD-L1 in human breast cancer cells. *Expert Opin. Ther. Targets* **2018**, *22*, 547–557. [CrossRef]

20.   Zhang, N.; Zeng, Y.; Du, W.; Zhu, J.; Shen, D.; Liu, Z.; Huang, J.A. The EGFR pathway is involved in the regulation of PD-L1 expression via the IL-6/JAK/STAT3 signaling pathway in EGFR-mutated non-small cell lung cancer. *Int. J. Oncol.* **2016**, *49*, 1360–1368. [CrossRef]

21.   Sumimoto, H.; Takano, A.; Teramoto, K.; Daigo, Y. RAS-Mitogen-Activated Protein Kinase Signal Is Required for Enhanced PD-L1 Expression in Human Lung Cancers. *PLoS ONE* **2016**, *11*, e0166626. [CrossRef] [PubMed]

22.   Bu, L.L.; Yu, G.T.; Wu, L.; Mao, L.; Deng, W.W.; Liu, J.F.; Kulkarni, A.B.; Zhang, W.F.; Zhang, L.; Sun, Z.J. STAT3 Induces Immunosuppression by Upregulating PD-1/PD-L1 in HNSCC. *J. Dent. Res.* **2017**, *96*, 1027–1034. [CrossRef] [PubMed]

23.   Garcia-Diaz, A.; Shin, D.S.; Moreno, B.H.; Saco, J.; Escuin-Ordinas, H.; Rodriguez, G.A.; Zaretsky, J.M.; Sun, L.; Hugo, W.; Wang, X.; et al. Interferon Receptor Signaling Pathways Regulating PD-L1 and PD-L2 Expression. *Cell Rep.* **2017**, *19*, 1189–1201. [CrossRef] [PubMed]

24.   Huynh, J.; Etemadi, N.; Hollande, F.; Ernst, M.; Buchert, M. The JAK/STAT3 axis: A comprehensive drug target for solid malignancies. *Semin. Cancer Biol.* **2017**, *45*, 13–22. [CrossRef] [PubMed]

25. Rolny, C.; Mazzone, M.; Tugues, S.; Laoui, D.; Johansson, I.; Coulon, C.; Squadrito, M.L.; Segura, I.; Li, X.; Knevels, E.; et al. HRG inhibits tumor growth and metastasis by inducing macrophage polarization and vessel normalization through downregulation of PlGF. *Cancer Cell* **2011**, *19*, 31–44. [CrossRef] [PubMed]

26. Jones, L.M.; Broz, M.L.; Ranger, J.J.; Ozcelik, J.; Ahn, R.; Zuo, D.; Ursini-Siegel, J.; Hallett, M.T.; Krummel, M.; Muller, W.J. STAT3 Establishes an Immunosuppressive Microenvironment during the Early Stages of Breast Carcinogenesis to Promote Tumor Growth and Metastasis. *Cancer Res.* **2016**, *76*, 1416–1428. [CrossRef] [PubMed]

27. Gordon, S.R.; Maute, R.L.; Dulken, B.W.; Hutter, G.; George, B.M.; McCracken, M.N.; Gupta, R.; Tsai, J.M.; Sinha, R.; Corey, D.; et al. PD-1 expression by tumour-associated macrophages inhibits phagocytosis and tumour immunity. *Nature* **2017**, *545*, 495–499. [CrossRef] [PubMed]

28. Hartley, G.P.; Chow, L.; Ammons, D.T.; Wheat, W.H.; Dow, S.W. Programmed Cell Death Ligand 1 (PD-L1) Signaling Regulates Macrophage Proliferation and Activation. *Cancer Immunol. Res.* **2018**, *6*, 1260–1273. [CrossRef] [PubMed]

29. Harney, A.S.; Arwert, E.N.; Entenberg, D.; Wang, Y.; Guo, P.; Qian, B.Z.; Oktay, M.H.; Pollard, J.W.; Jones, J.G.; Condeelis, J.S. Real-Time Imaging Reveals Local, Transient Vascular Permeability, and Tumor Cell Intravasation Stimulated by TIE2hi Macrophage-Derived VEGFA. *Cancer Discov.* **2015**, *5*, 932–943. [CrossRef] [PubMed]

30. Williams, C.B.; Yeh, E.S.; Soloff, A.C. Tumor-associated macrophages: Unwitting accomplices in breast cancer malignancy. *npj Breast Cancer* **2016**, *2*. [CrossRef] [PubMed]

31. Mantovani, A.; Marchesi, F.; Malesci, A.; Laghi, L.; Allavena, P. Tumour-associated macrophages as treatment targets in oncology. *Nat. Rev. Clin. Oncol.* **2017**, *14*, 399–416. [CrossRef] [PubMed]

32. De Palma, M.; Lewis, C.E. Macrophage regulation of tumor responses to anticancer therapies. *Cancer Cell* **2013**, *23*, 277–286. [CrossRef] [PubMed]

33. Galdiero, M.R.; Garlanda, C.; Jaillon, S.; Marone, G.; Mantovani, A. Tumor associated macrophages and neutrophils in tumor progression. *J. Cell. Physiol.* **2013**, *228*, 1404–1412. [CrossRef] [PubMed]

34. Harada, K.; Dong, X.; Estrella, J.S.; Correa, A.M.; Xu, Y.; Hofstetter, W.L.; Sudo, K.; Onodera, H.; Suzuki, K.; Suzuki, A.; et al. Tumor-associated macrophage infiltration is highly associated with PD-L1 expression in gastric adenocarcinoma. *Gastric Cancer* **2018**, *21*, 31–40. [CrossRef] [PubMed]

35. Dhupkar, P.; Gordon, N.; Stewart, J.; Kleinerman, E.S. Anti-PD-1 therapy redirects macrophages from an M2 to an M1 phenotype inducing regression of OS lung metastases. *Cancer Med.* **2018**, *7*, 2654–2664. [CrossRef] [PubMed]

36. Mace, T.A.; Shakya, R.; Pitarresi, J.R.; Swanson, B.; McQuinn, C.W.; Loftus, S.; Nordquist, E.; Cruz-Monserrate, Z.; Yu, L.; Young, G.; et al. IL-6 and PD-L1 antibody blockade combination therapy reduces tumour progression in murine models of pancreatic cancer. *Gut* **2018**, *67*, 320–332. [CrossRef]

37. Liu, H.; Shen, J.; Lu, K. IL-6 and PD-L1 blockade combination inhibits hepatocellular carcinoma cancer development in mouse model. *Biochem. Biophys. Res. Commun.* **2017**, *486*, 239–244. [CrossRef]

38. Lu, C.; Talukder, A.; Savage, N.M.; Singh, N.; Liu, K. JAK-STAT-mediated chronic inflammation impairs cytotoxic T lymphocyte activation to decrease anti-PD-1 immunotherapy efficacy in pancreatic cancer. *Oncoimmunology* **2017**, *6*, e1291106. [CrossRef]

39. Santoni, M.; Romagnoli, E.; Saladino, T.; Foghini, L.; Guarino, S.; Capponi, M.; Giannini, M.; Cognigni, P.D.; Ferrara, G.; Battelli, N. Triple negative breast cancer: Key role of Tumor-Associated Macrophages in regulating the activity of anti-PD-1/PD-L1 agents. *Biochim. Biophys. Acta Rev. Cancer* **2018**, *1869*, 78–84. [CrossRef]

40. Atsaves, V.; Lekakis, L.; Drakos, E.; Leventaki, V.; Ghaderi, M.; Baltatzis, G.E.; Chioureas, D.; Jones, D.; Feretzaki, M.; Liakou, C.; et al. The oncogenic JUNB/CD30 axis contributes to cell cycle deregulation in ALK+ anaplastic large cell lymphoma. *Br. J. Haematol.* **2014**, *167*, 514–523. [CrossRef]

41. Cotta, C.V.; Leventaki, V.; Atsaves, V.; Vidaki, A.; Schlette, E.; Jones, D.; Medeiros, L.J.; Rassidakis, G.Z. The helix-loop-helix protein Id2 is expressed differentially and induced by myc in T-cell lymphomas. *Cancer* **2008**, *112*, 552–561. [CrossRef] [PubMed]

42. Lundberg, A.; Lindstrom, L.S.; Harrell, J.C.; Falato, C.; Carlson, J.W.; Wright, P.K.; Foukakis, T.; Perou, C.M.; Czene, K.; Bergh, J.; et al. Gene Expression Signatures and Immunohistochemical Subtypes Add Prognostic Value to Each Other in Breast Cancer Cohorts. *Clin. Cancer Res.* **2017**, *23*, 7512–7520. [CrossRef] [PubMed]

43. Follenzi, A.; Naldini, L. HIV-based vectors. Preparation and use. *Methods Mol. Med.* **2002**, *69*, 259–274. [PubMed]

44. Witt, K.; Ligtenberg, M.A.; Conti, L.; Lanzardo, S.; Ruiu, R.; Wallmann, T.; Tufvesson-Stiller, H.; Chambers, B.J.; Rolny, C.; Lladser, A.; et al. Cripto-1 Plasmid DNA Vaccination Targets Metastasis and Cancer Stem Cells in Murine Mammary Carcinoma. *Cancer Immunol. Res.* **2018**, *6*, 1417–1425. [CrossRef] [PubMed]

45. Casazza, A.; Fu, X.; Johansson, I.; Capparuccia, L.; Andersson, F.; Giustacchini, A.; Squadrito, M.L.; Venneri, M.A.; Mazzone, M.; Larsson, E.; et al. Systemic and targeted delivery of semaphorin 3A inhibits tumor angiogenesis and progression in mouse tumor models. *Arterioscler. Thromb. Vasc. Biol.* **2011**, *31*, 741–749. [CrossRef] [PubMed]

46. Liu, J.; Lichtenberg, T.; Hoadley, K.A.; Poisson, L.M.; Lazar, A.J.; Cherniack, A.D.; Kovatich, A.J.; Benz, C.C.; Levine, D.A.; Lee, A.V.; et al. An Integrated TCGA Pan-Cancer Clinical Data Resource to Drive High-Quality Survival Outcome Analytics. *Cell.* **2018**, *173*, 400–416.e11. [CrossRef] [PubMed]

47. Gao, J.; Aksoy, B.A.; Dogrusoz, U.; Dresdner, G.; Gross, B.; Sumer, S.O.; Sun, Y.; Jacobsen, A.; Sinha, R.; Larsson, E.; et al. Integrative analysis of complex cancer genomics and clinical profiles using the cBioPortal. *Sci. Signal.* **2013**, *6*, pl1. [CrossRef] [PubMed]

48. Cerami, E.; Gao, J.; Dogrusoz, U.; Gross, B.E.; Sumer, S.O.; Aksoy, B.A.; Jacobsen, A.; Byrne, C.J.; Heuer, M.L.; Larsson, E.; et al. The cBio cancer genomics portal: An open platform for exploring multidimensional cancer genomics data. *Cancer Discov.* **2012**, *2*, 401–404. [CrossRef]

49. Irizarry, R.A.; Bolstad, B.M.; Collin, F.; Cope, L.M.; Hobbs, B.; Speed, T.P. Summaries of Affymetrix GeneChip probe level data. *Nucleic Acids Res.* **2003**, *31*, e15. [CrossRef]

50. Bengtsson, H.; Simpson, K.; Bullard, J.; Hansen, K. *Aroma. Affymetrix: A Generic Framework in R for Analyzing Small to Very Large Affymetrix Data Sets in Bounded Memory*; Tech Report; University of California, Berkeley: Berkeley, CA, USA, 2008.

51. Sonnenblick, A.; Brohee, S.; Fumagalli, D.; Vincent, D.; Venet, D.; Ignatiadis, M.; Salgado, R.; Van den Eynden, G.; Rothe, F.; Desmedt, C.; et al. Constitutive phosphorylated STAT3-associated gene signature is predictive for trastuzumab resistance in primary HER2-positive breast cancer. *BMC Med.* **2015**, *13*, 177. [CrossRef]

52. Huang, D.W.; Sherman, B.T.; Stephens, R.; Baseler, M.W.; Lane, H.C.; Lempicki, R.A. DAVID gene ID conversion tool. *Bioinformation* **2008**, *2*, 428–430. [CrossRef] [PubMed]

53. Gendoo, D.M.; Ratanasirigulchai, N.; Schröder, M.S.; Paré, L.; Parker, J.S.; Prat, A.; Haibe-Kains, B. Genefu: An R/Bioconductor package for computation of gene expression-based signatures in breast cancer. *Bioinformatics* **2015**, *32*, 1097–1099. [CrossRef] [PubMed]

Article

Therapeutic Targeting of Stat3 Using Lipopolyplex Nanoparticle-Formulated siRNA in a Syngeneic Orthotopic Mouse Glioma Model

Benedikt Linder [1,*], Ulrike Weirauch [2], Alexander Ewe [2], Anja Uhmann [3], Volker Seifert [4], Michel Mittelbronn [5,6,7,8,9], Patrick N. Harter [5,10], Achim Aigner [2,†] and Donat Kögel [1,†]

[1] Experimental Neurosurgery, Department of Neurosurgery, Neuroscience Center, Goethe University Hospital, 60528 Frankfurt am Main, Germany; koegel@em.uni-frankfurt.de

[2] Rudolf-Boehm-Institute for Pharmacology and Toxicology, Clinical Pharmacology, University of Leipzig, 04107 Leipzig, Germany; Ulrike.Weirauch@medizin.uni-leipzig.de (U.W.); Alexander.Ewe@medizin.uni-leipzig.de (A.E.); Achim.Aigner@medizin.uni-leipzig.de (A.A.)

[3] Institute of Human Genetics, Tumor Genetics Group, University of Göttingen, 37073 Göttingen, Germany; auhmann@gwdg.de

[4] Department of Neurosurgery, Goethe University Hospital, 60528 Frankfurt am Main, Germany; V.seifert@em.uni-frankfurt.de

[5] Institute of Neurology (Edinger-Institute), University Hospital Frankfurt, Goethe University, 60528 Frankfurt am Main, Germany; michel.mittelbronn@lns.etat.lu (M.M.); patrick.harter@kgu.de (P.N.H.)

[6] Luxembourg Centre for Systems Biomedicine (LCSB), University of Luxembourg, 4362 Esch-sur-Alzette, Luxembourg

[7] Laboratoire national de santé (LNS), 3555 Dudelange, Luxembourg

[8] Luxembourg Centre of Neuropathology (LCNP), 3555 Dudelange, Luxembourg

[9] NORLUX Neuro-Oncology Laboratory, Luxembourg Institute of Health (LIH), 1526 Luxembourg, Luxembourg

[10] German Cancer Consortium DKTK Partner site Frankfurt/Main, Frankfurt am Main, Germany and German Cancer Research Center DKFZ, 69120 Heidelberg, Germany

* Correspondence: linder@med.uni-frankfurt.de; Tel.: +49-69-6301-84051

† These authors contributed equally to this work.

Received: 21 December 2018; Accepted: 4 March 2019; Published: 8 March 2019

Abstract: Glioblastoma (GBM), WHO grade IV, is the most aggressive primary brain tumor in adults. The median survival time using standard therapy is only 12–15 months with a 5-year survival rate of around 5%. Thus, new and effective treatment modalities are of significant importance. Signal transducer and activator of transcription 3 (Stat3) is a key signaling protein driving major hallmarks of cancer and represents a promising target for the development of targeted glioblastoma therapies. Here we present data showing that the therapeutic application of siRNAs, formulated in nanoscale lipopolyplexes (LPP) based on polyethylenimine (PEI) and the phospholipid 1,2-dipalmitoyl-sn-glycero-3-phosphocholine (DPPC), represents a promising new approach to target Stat3 in glioma. We demonstrate that the LPP-mediated delivery of siRNA mediates efficient knockdown of Stat3, suppresses Stat3 activity and limits cell growth in murine (Tu2449) and human (U87, Mz18) glioma cells in vitro. In a therapeutic setting, intracranial application of the siRNA-containing LPP leads to knockdown of STAT3 target gene expression, decreased tumor growth and significantly prolonged survival in Tu2449 glioma-bearing mice compared to negative control-treated animals. This is a proof-of-concept study introducing PEI-based lipopolyplexes as an efficient strategy for therapeutically targeting oncoproteins with otherwise limited druggability.

Keywords: STAT3; siRNA/RNAi; polyethylenimine; PEI; lipopolyplex; siRNA delivery; glioma; glioblastoma

---

1. Introduction

Glioblastoma (grade IV glioma, WHO) is the most aggressive primary brain tumor in adults, with a median survival of ~15 months following standard radiochemotherapy with temozolomide (TMZ) and a 5-year survival rate below 5% [1]. Due to diffusely infiltrating tumor cells that cannot be surgically resected, most tumors will quickly recur and afflict the patients with even more aggressive disease [2]. GBM are further subdivided into distinct subtypes based on their genetic profile. Different groups reported different subtypes, but it is generally agreed that a proneural subtype with the best prognosis and a mesenchymal subtype with the worst prognosis exist [3,4]. Another subtype is the classical one [4], which is sometimes subdivided into a proneural and a neural subtype [3]. One key molecule that is frequently highly expressed and overactivated in GBM and associated with the most aggressive and treatment-resistant mesenchymal subtype [5,6] is the oncogenic transcription factor signal transducer and activator of transcription 3 (Stat3) that acts as a signaling hub promoting most hallmarks of cancer, including proliferation, cell survival, angiogenesis and immune evasion (reviewed in [7,8]). STAT3 activation is facilitated by phosphorylation of multiple upstream kinases [9]. Stat3 can be phosphorylated on two different sites (reviewed in [10]), Tyrosine705 (p^{Y705}Stat3) and Serine727 (p^{S727}Stat3), which can exert different functions in different cell types. In glioma it was shown that S727-phosphorylation is dependent on Y705-phoshorylation, that is necessary for maximal activation of Stat3 [11]. Activated Stat3 dimerizes and induces gene expression of a variety of genes, many of which are known to be important for hallmarks of cancers like migration/invasion (Mmp2, Mmp9; [12,13]) or EMT-like features and immune evasion/suppression (e.g., Snai1; [14,15]). Stat3 is activated in many other human cancers (reviewed in [16]) in addition to gliomas. In particular Stat3 expression increases with tumor grade in these tumors (Figure S1 and reviewed in [9,17]), and is negatively correlated with patient survival (reviewed in [9,17]). Additionally, Stat3 is known to be necessary for activation of microglia [18], the macrophage-like population of immune cells residing in the brain. In glioma, these cells usually acquire a pro-tumorigenic phenotype that promotes glioma cell migration and invasion [18].

Consequently, we and others previously demonstrated that targeting Stat3 by upstream signaling inhibitors of the Stat3 pathway decreases glioma cell proliferation and migration in vitro and prolongs overall survival of tumor-bearing mice in vivo [19–24].

One strategy to target aberrantly activated and/or overexpressed oncoproteins in tumor therapy is to ablate their expression by RNA interference (RNAi). This knockdown strategy relies on small interfering RNA (siRNA) that, due to high sequence specificity, generally exhibits fewer off-target effects in comparison to the application of pharmacological agents (reviewed in [25,26]). The second major advantage of RNAi is that it allows to selectively interfere with the activity of otherwise un-druggable or hard-to-drug targets like Stat3 and other oncogenic transcription factors. Using a pre-transplantational genetic knockdown approach, we could previously demonstrate that ablation of Stat3 function with lentiviral shRNA limits tumor growth of Tu2449 gliomas in an orthotopic, syngeneic mouse model [21], providing support for the concept of therapeutic intervention with Stat3 activity in vivo. Therapeutic intervention in already established glioma, however, requires the development of siRNA formulations for their direct application. We have previously established siRNA complexation with nanoscale polyethlyenimine (PEI)-based complexes for therapeutic siRNA delivery in vivo. This confers protection, cellular delivery and intracellular release of formulated small RNA molecules (siRNAs, miRNAs) in vitro and in vivo, and thus allows for their therapeutic application ([27–30]; see [31] for review). More recently, this was extended towards combinations with liposomes, leading to lipopolyplexes (LPP) that combine properties of both components [32].

For gene delivery, LPP comprising phospholipids had been previously shown to display strongly reduced surface charges, enhanced transfection efficiencies, decreased cytotoxicity and high colloidal stability as compared to their unmodified complex counterparts [33,34]. Likewise, we were able to demonstrate for siRNA delivery that DPPC-based LPP without co-lipids displayed very good transfection efficiency and further enhanced biocompatibility in cell culture. Intracellular siRNA release was not impaired by the liposomal component. Due to strongly reduced surface charges, storage stabilities under various conditions were markedly increased by protecting the LPP from aggregation [35]. Importantly, these favorable properties also translated into therapeutic in vivo efficacy upon systemic injection into tumor xenograft-bearing mice [32], thus providing the basis for now switching to another model and another mode of administration.

Here we investigated the therapeutic potential of intracranially applied LPP containing Stat3-specific siRNA in an orthotopic Tu2449 glioma model. We demonstrate that specific targeting of Stat3 using LPP significantly reduces tumor growth and improves overall survival. This proof-of-concept study thus supports the notion that highly therapy-resistant cancers such as malignant gliomas and hard-to-drug genes like Stat3 can be efficiently targeted using siRNA in vivo.

2. Results

2.1. LPP Nanoparticles Provide a Stable and Efficient siRNA Formulation

As shown previously, the complexation of siRNAs with the ~4–10 kDa low molecular weight polyethylenimine PEI F25-LMW [36] leads to the formation of positively charged nanoscale complexes ('polyplexes') in a size range of ~100–350 nm, dependent on the buffer conditions used for complexation [27]. This was confirmed here by NanoSight measurements, yielding PEI/siRNA complexes of ~130 nm diameter with a zeta potential of 27 mV (Figure 1a,c). The combination of these polyplexes with DPPC liposomes of about the same size (Figure 1a,b) resulted in the formation of slightly larger lipopolyplexes (Figure 1a,b,d; [37,38]). The liposomal contribution also affected the zeta potential of the LPP, which was almost neutral as opposed to the polyplexes (Figure 1a). Size determinations by NanoSight measurements were confirmed by dynamic light scattering (DLS) which, however, also indicated the tendency of PEI/siRNA complexes to aggregate, leading to larger particle sizes (Figure 1a, Supplementary Figure S2). Notably, this was inhibited by lipopolyplex (LPP) formation (Figure 1a; note PEI/siRNA complex size in DLS measurements). The colloidal stability of LPP provides an advantage over polyplexes and prompted us to prefer LPP for intracranial siRNA delivery. In an LDH release assay in Tu2449 cells, the LPP showed excellent biocompatibility in vitro (Figure 1e). The absence of appreciable cell damage upon transfection suggests the applicability of LPP also in a sensitive environment in vivo (CNS), and similar results in the case of the parent polyplexes also indicates that even an LPP decomposition with possible polyplex release would not lead to toxic effects. Since highly positive surface charges of nanoparticles may lead to enhanced cellular uptake, we tested next whether the considerably reduced zeta potentials of LPP would negatively affect biological efficacies. Notably, upon transfection of stable EGFP (Figure 1f) or luciferase (Luc3) reporter cell lines (Figure 1g) with siRNAs formulated in polyplexes or in LPP, no differences were observed. More specifically, the comparison of cells transfected with negative control siRNA (siLuc2) vs specific siRNA (siEGFP or siLuc3, respectively) revealed a profound 50–65% knockdown after 72 h, indicating biological efficacy of both, polyplexes and LPP. For the reasons stated above, LPP were selected for further studies.

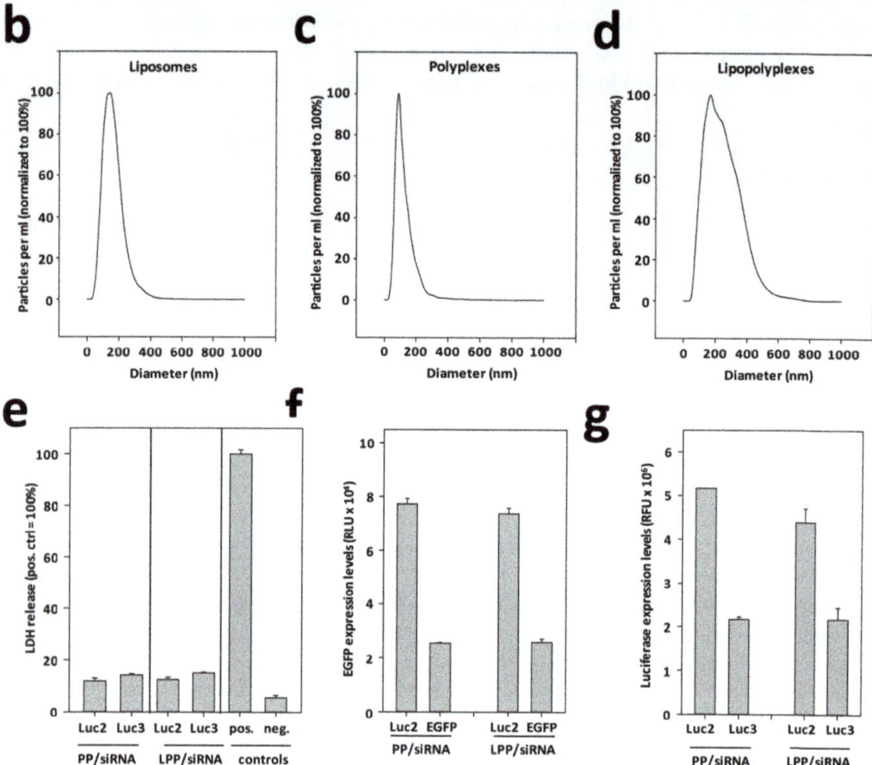

Figure 1. (**a**) Compilation of nanoparticle sizes and zeta potentials, as determined by Zetasizer and Nanosight. (**b–d**) Original diagrams of size measurements by Nanosight. (**e–g**) Biological properties of nanoparticles. (**e**) Determination of cytotoxicity by LDH release assay. (**f,g**) Determination of knockdown efficacies in EGFP/Luciferase (Luc3) Tu2449 reporter cells. Knockdown of (**f**) EGFP and (**g**) luciferase (Luc3) is shown. For (**a–d**) two independent samples were measured 10 times; (**e**) was performed twice in 4 biological replicates and (**f,g**) were performed twice in biological duplicate; shown are the summaries of all experiments.

2.2. Targeting STAT3 Reduces Cancer Cell Proliferation In Vitro

To test the therapeutic potential of LPP-formulated siRNA in vitro and in vivo, we selected Stat3 as a target because it is frequently overactivated in glioma and also hard-to-drug. Consistent with a number of previous reports, analysis of the TCGA data-set indicated that high expression of STAT3 is associated with an especially poor survival of GBM patients (Figure 2a). To validate these clinical observations in vitro, we knocked down STAT3 in two human GBM cell lines, U87 (Figure 2b) and Mz18 (Figure 2c). Using two different siRNAs we could significantly reduce *STAT3* mRNA expression in both cell lines, with siSTAT3-2 being more effective than siSTAT3-1. Consistently, STAT3 suppression was also achieved on the protein level in both cell lines (Figure 2d). Notably, we frequently observed a second band below the STAT3 signal in U87, but since both siRNAs target all three protein coding sequences of STAT3 (NM_213662.1, NM_003150.3 and NM_139276.2) this is likely an unspecific signal. To assess the antitumor effects of STAT3 depletion, we evaluated the growth kinetics of U87 (Figure 2e) and Mz18 cells (Figure 2f) after siSTAT3 treatment. Both cell lines showed significantly reduced proliferation 192 h after siSTAT3-treatment, with siSTAT3-2 again being more effective than siSTAT3-1. Of note, U87 cells were more sensitive to STAT3 depletion than Mz18 cells, indicating that this line may be particularly addicted to STAT3 activity, in line with findings described earlier [39]. Mz18 cells also express STAT3 and we could previously show that this line exhibits moderate levels of tyrosine-phosphorylated STAT3, which could be inhibited by upstream JAK2-inhibition [22]. We also tested the murine GBM cell line Tu2449, which we previously had used for in vivo experiments with pre-transplantational depletion of Stat3 with shRNA [21]. First, we sought out to test if siRNA-mediated Stat3-knockdown also inhibits proliferation and indeed we observed that siRNA delivery using conventional in vitro reagents like INTERFERin™ also achieved a reduction in proliferation (Figure 2g). Next, we applied siRNA complexed as polyplexes, in order to verify that the delivery method does not affect knockdown efficiency. Accordingly, LPP mediated siStat3 delivery strongly inhibited proliferation (Figure 2h) and was able to efficiently reduce Stat3 and phospho-Stat3 protein levels (Figure 2i), whereas polyplexes without liposomal content were accompanied by increased nonspecific toxicities although a knockdown could also be achieved (data not shown). Thus, in these experiments LPP were found to be superior over polyplexes.

Cell cycle analysis of Tu2449 cells showed a significant increase in G1 phase and concomitant decrease in G2 phase upon siStat3 transfection, suggesting that the observed antiproliferative effect is at least in part due to a G1 arrest upon Stat3 knockdown (Figure 3a). Decreased cell cycle progression was also confirmed in the human cell lines U87 and Mz18 (Supplementary Figure S3a,b). To further verify the dependency of Tu2449 cells on Stat3 in a more complex cell culture system, we generated Tu2449 tumor spheroids, which resemble an in vivo situation more closely with regard to gradient access to oxygen, nutrients, as well as therapeutics. siRNA-mediated knockdown of Stat3 lead to distinctly smaller spheroids than control treatment (Figure 3b,c), also demonstrating that LPP are efficient in transfecting cells in spheroids.

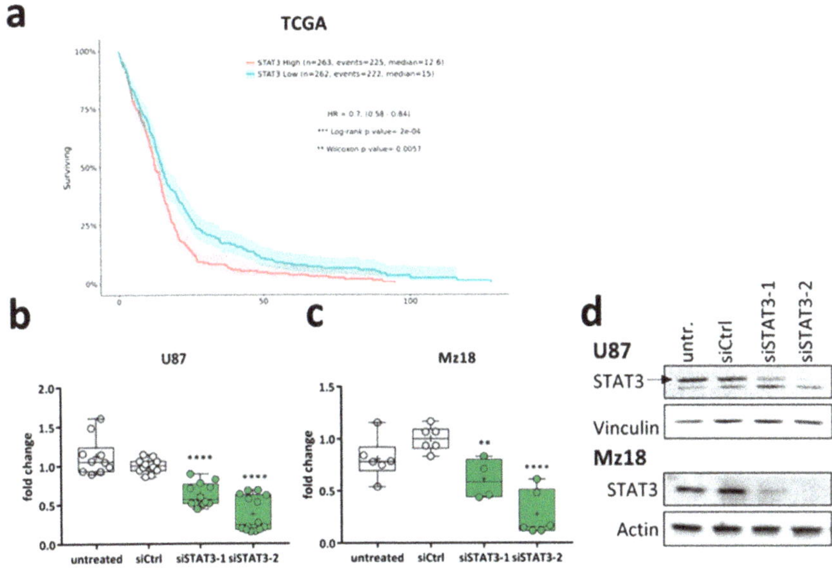

Figure 2. *Cont.*

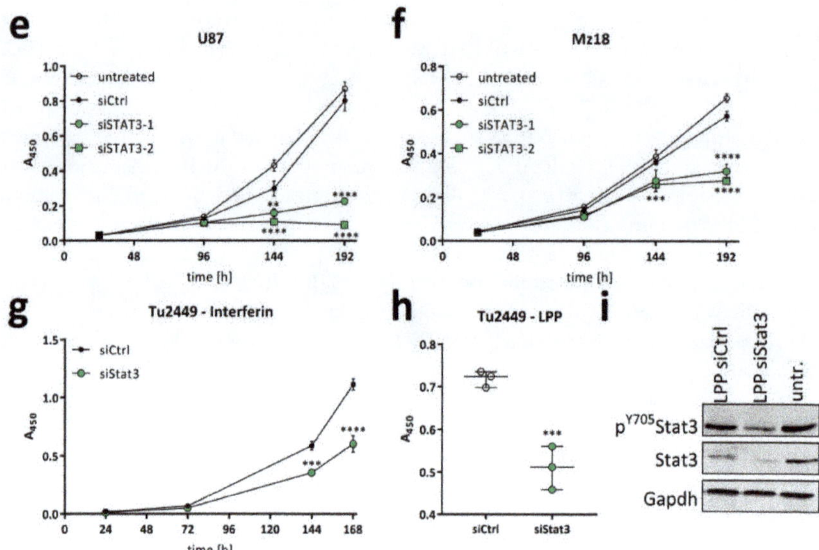

Figure 2. (**a**) Kaplan-Meier-Survival Plot from TCGA dataset GBM [40] showing that high STAT3 expression is associated with shorter survival; (**b**,**c**) qRT-PCR from (**b**) U87 and (**c**) Mz18 human glioma cell lines after transfection with control siRNA (siCtrl) or two siRNAs against STAT3 (siSTAT3-1 and siSTAT3-2). STAT3-expression was normalized to Actin as housekeeper and siCtrl-transfected cells as control sample using the ΔΔCt-method. The data are presented as box-plots (min-to-max) with all samples displayed as circles; the horizontal line in the box depicts the median value, the plus-symbol the mean. (**d**) Western Blot of U87 and Mz18 after transfection as in (**b**,**c**) after transfection of siCtrl, siSTAT3-1 or siSTAT3-2. (**e**–**h**) Proliferation (WST-1) assays of the human glioma cell lines (**e**) U87 and (**f**) Mz18, using INTERFERin and the two different siSTAT3 for comparison, and in the murine glioma cell line Tu2449 after transfection with (**g**) INTERFERin^{TM} or (**h**) LPP. The data in (**e**–**g**) are presented as mean +/− SEM; the data in (**h**) are presented as Box-Plots (min-to-max) with all samples displayed. (**i**) Western Blot of Tu2449 cells 96 h after transfection with 150 pmol LPP siCtrl or LPP siStat3. (**b**,**c**) shows the summary of at least three independent experiments performed in biological duplicates; (**d**) was performed twice; (**e**,**f**,**h**) were performed three (**g**) two times in biological triplicates; (**i**) was performed three times. **: $p < 0.01$; ***: $p < 0.001$ and ****: $p < 0.0001$ compared to siCtrl treatment.

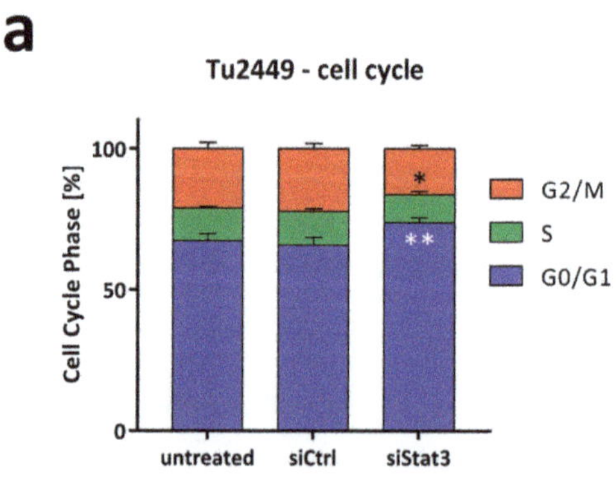

Figure 3. *Cont.*

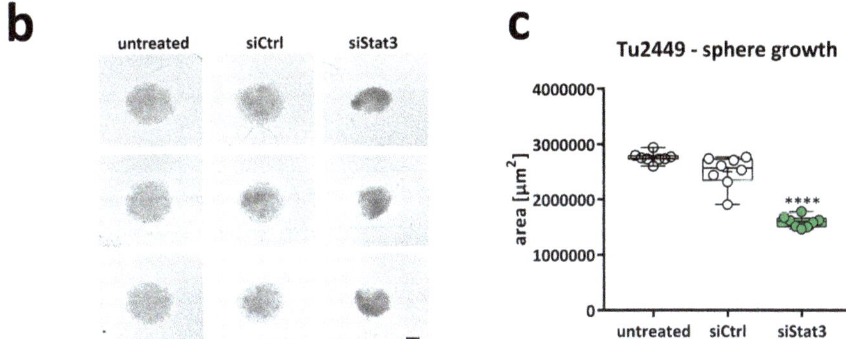

Figure 3. (**a**) Cell cycle analysis of Tu2449 cells after transfection of siCtrl or siStat3. A significant increase in the percentage of cells in the G1 phase suggests antiproliferative properties of Stat3-depletion. (**b**,**c**) Sphere growth assay of Tu2449 after transfection of siCtrl or siStat3. Three representative spheres for each condition are presented in (**b**) the quantification after 10 days of sphere growth is shown. Scale bar: 500 μm. (**c**) as box-plots (min-to-max) with all samples displayed as circles; the horizontal line in the box depicts the median value, the plus-symbol the mean. *: $p < 0.0.05$; **: $p < 0.01$ and ****: $p < 0.0001$ compared to siCtrl treatment. A representative experiment of at least three independent repetitions is shown.

In conclusion, we could successfully establish siRNA-mediated STAT3/Stat3 depletion and suppression of tumor cell proliferation by LPP siSTAT3 in vitro and verify Tu2449 as suitable cell line to explore a Stat3 targeting therapy.

2.3. LPP siStat3 Prolongs Overall Survival and Reduces Tumor Size in Glioma-Bearing Mice

Based on these findings, we pushed our system further towards therapeutic intervention, using a complex and pathophysiological relevant system. To this end, we employed our syngenic, orthotopic transplantation model [41] to assess LPP applicability and efficacy in vivo. The treatment scheme is outlined in Figure 4a. Briefly, after implantation of 10,000 Tu2449 cells into the striatum of 42 mice, the tumors were allowed to grow for one week. Hereafter, the mice were randomly divided in two groups of 21 animals each and the intracranial (i.e., intratumoral) treatment with LPPs was performed every third day. After three weeks of tumor growth, 8 mice were sacrificed for histological and molecular analyses. 1 tumor in the LPP siCtrl and LPP siStat3 cohort planned for histological assessment did not grow and were removed from the analysis. The remaining mice were monitored until they succumbed to the disease. Of those, 2 and 4 tumors of the LPP siCtrl and LPP siStat3 cohort did not grow well, respectively, and these animals were also removed from the analysis. The survival analysis (Figure 4b) showed that after LPP siStat3 none of the mice died during the treatment period, whereas six mice treated with LPP siCtrl had to be euthanized. The median survival was significantly improved after treatment with LPP siStat3 in comparison to LPP siCtrl from 26 to 33 days, respectively. Histological analyses (Figure 4c,d) confirmed that the tumors of mice treated with LPP siStat3 were significantly smaller compared to LPP siCtrl treated mice.

2.4. LPP siStat3 Slightly Reduces Stat3 Activation, But Does Not Affect Gross Tumor Proliferation Rates In Vivo

Next, we analyzed LPP siStat3 effects on the cellular and molecular level in vivo. When performing analyses of randomly chosen tumor areas (excluding the tumor border and necrotic areas), we found phospho-Stat3 (Tyr705) slightly reduced (Figure 5b). This, however, was not accompanied by alterations in proliferation rates, as determined by Ki67 staining for proliferating cells (Figure 5a). Decreased phospho-Stat3 levels were also confirmed in western blot analyses of whole tumor lysates (Figure 5c, upper panel), while Stat3 protein levels remained stable. Since Stat3 was proposed to modulate the host immune response including activation of microglia [18], we also analyzed the presence of T-cells and microglia in the tumor. Staining against CD4 and CD8a for T-cells and Iba1 for microglia showed a slight reduction of CD4-positive cells after LPP siStat3 treatment (Figure 6a) while no major differences between LPP siCtrl and LPP siStat3 treated animals were seen with regard to Cd8a and Iba1 (Figure 6b,c).

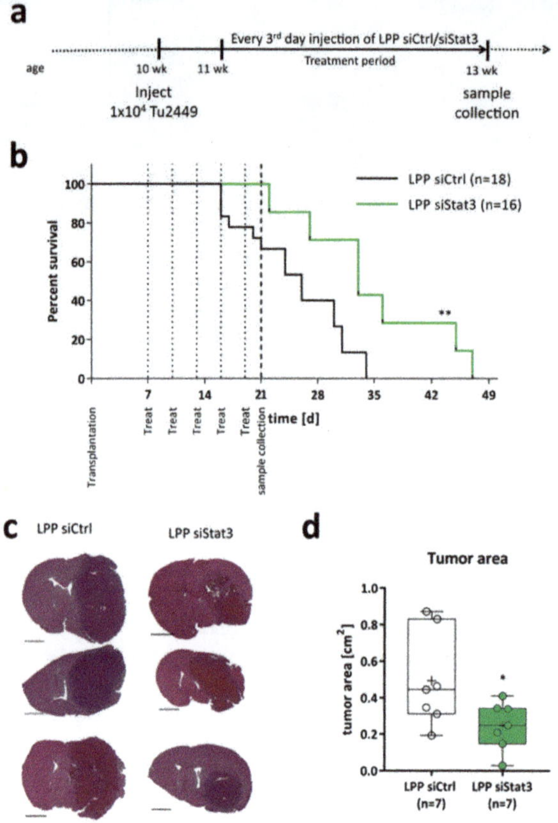

Figure 4. (**a**) Schematic presentation of the in vivo experiments. Briefly, 42 mice were injected with 10,000 Tu2449 cells into the right striatum. After 7 days the mice were randomly divided into two groups of 21 animals each and intracranial treatment was performed every third day. After 21 days 10 mice were euthanized for histological and molecular analysis. The remaining mice were monitored for long-term survival. Mice without established tumors were excluded from the analysis (3 (1 for histology; 2 for survival) for LPP siCtrl; 5 (1 for histology; 4 for survival) for LPP siStat3) (**b**) Kaplan-Meier-Survival Plot of B6C3F1 mice after implantation of Tu2449 and treatment with LPP siCtrl or LPP siStat3. The vertical dotted lines depict the treatment days, the vertical dashed line the day of sample collection. LPP siStat3 treatment significantly increases median survival from 26 days to 33 after LPP siCtrl and LPP siStat3 treatment respectively (**c**) Three representative brain sections with tumors 21 days after tumor cell implantation and treatment with LPP siCtrl (left side) or LPP siStat3 (right side); scale bar: 2 mm. (**d**) Box-Plots (min-to-max) with all samples displayed as circles; the horizontal line in the box depicts the median value, the plus-symbol the mean of tumor areas 21 days after tumor cell implantation after treatment with LPP siCtrl or LPP siStat3. *: $p < 0.05$; **: $p < 0.01$ compared to LPP siCtrl.

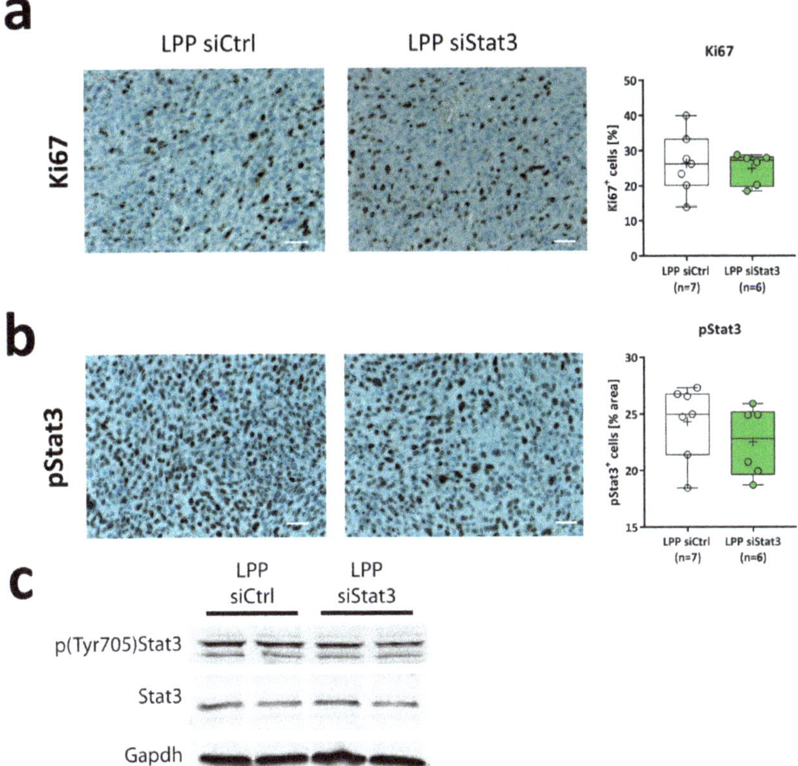

Figure 5. (**a**,**b**) Immunohistochemical analysis of (**a**) Ki67 and (**b**) tyrosine-phosphorylated Stat3 (pStat3) expressing tumor cells 21 days after implantation and treatment with LPP siCtrl and LPP siStat3. One representative picture is shown for each staining and treatment and the quantification is presented as box-plots (min-to-max) with all samples displayed as circles; the horizontal line in the box depicts the median value, the plus-symbol the mean of tumor areas 21 days after tumor cell implantation after treatment with LPP siCtrl or LPP siStat3. Scale bar: 200 μm. (**c**) Western Blot of whole-tumor lysates from two mice each after treatment with LPP siCtrl or LPP siStat3; Gapdh was used as the loading control.

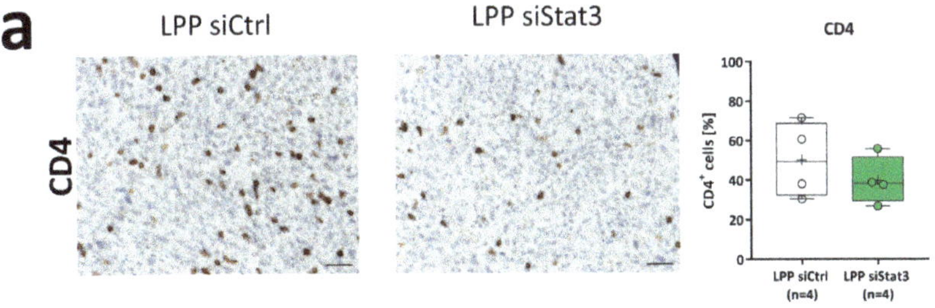

Figure 6. *Cont.*

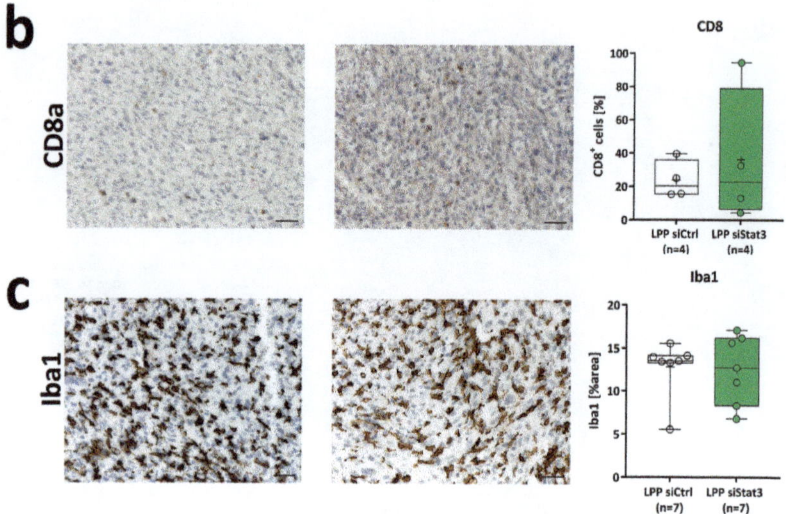

Figure 6. (**a–c**) Immunohistochemical analysis of (**a**) CD4, (**b**) CD8a and (**c**) Iba1 expressing tumor cells 21 days after implantation and treatment with LPP siCtrl and LPP siStat3. One representative picture is shown for each staining and treatment and the quantification is presented as box-plots (min-to-max) with all samples displayed as circles; the horizontal line in the box depicts the median value, the plus-symbol the mean value. scale bar: 200 μm.

2.5. LPP siStat3 Reduces Stat3 Expression in the Core Region of the Tumor

The fact that we observed significantly improved survival and reduced tumor sizes following treatment with LPP/siStat3, but could not observe a significant Stat3-repression in whole tumor samples led us to hypothesize that Stat3 depletion might be locally restricted to the area surrounding the injection site. To further investigate this hypothesis, we performed laser-capture microdissection to excise the central area of the tumor (Figure 7a) that equals the site of LPP injection. In case a sample showed necrotic areas in this region, these were first excised and discarded. Next, we isolated RNA from excised tumor core regions and performed qRT-PCR. We used 6 samples per cohort for the analyses, and obtained quantifiable data from 4 samples for *Stat3* and the housekeeping gene *Tbp*, but only 3 for the additional housekeeping *Hprt* (Figure 7b). These analyses showed that mRNA expression of *Stat3* and the known Stat3 target gene *Mmp9* was significantly decreased following LPP siStat3 treatment. Additionally, *Snail* showed a tendency for reduced expression. The expression of *Mmp2* remained unchanged. Due to the small sample sizes, this can be interpreted as a trend towards expressional suppression.

These results support our previously proposed hypothesis that LPP siStat3 can specifically target cancer cells to limit tumor growth in vivo. Despite the robust effects of LPP-formulated siStat3 on tumor growth, our results also indicate that only a small fraction of tumor cells can be reached using the current formulations and mode of application. These observations suggest that improvement of the bioavailability may allow further enhancement of therapeutic efficacies of this approach in vivo.

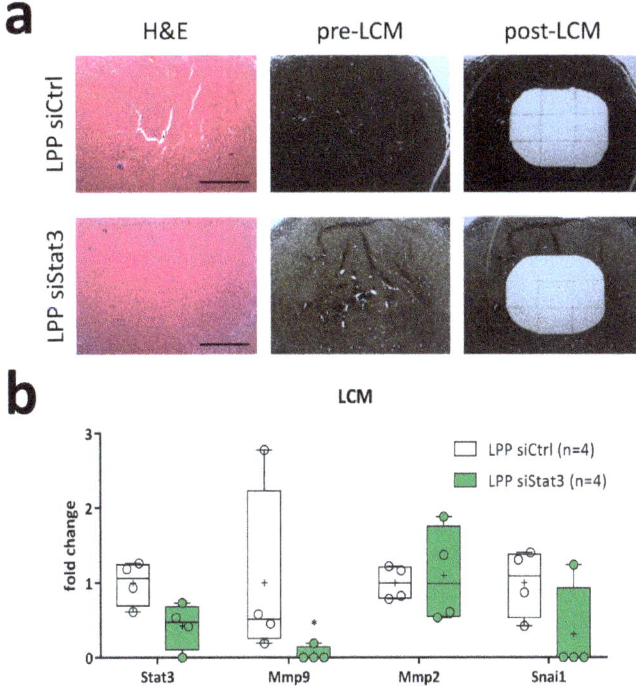

Figure 7. (**a**) Representative histology of tumor centers stained with H&E (left panel) and of unlabeled sections before (middle panel, pre-LCM) and after (right panel, post-LCM) microdissection of the central part of the tumor. Scale bar: 400 μm. (**b**) Taqman-based gene expression analysis normalized using the housekeepers *Hprt* and *Tbp* depicted as box-plots (min-to-max) with all samples displayed as circles; the horizontal line in the box depicts the median value, the plus-symbol the mean value. *: $p < 0.05$ compared to LPP siCtrl.

3. Discussion

Even after decades of research GBM still remains an incurable disease with one of the most dismal prognosis for cancer patients. The oncogenic transcription factor STAT3 represents a key signaling hub regulating many tumor-related processes including proliferation, migration, apoptosis-resistance, angiogenesis and immune evasion (reviewed in [12]). Furthermore, STAT3 expression is correlated with the highly aggressive mesenchymal subtype of glioma that is particularly resistant to conventional therapy [5,6] and is known to regulate stemness in glioma cells [6,23]. Previous findings from our and other groups had demonstrated that upstream pharmacological inhibition [22,42–44] and stable lentiviral depletion [17,21,23] of Stat3 provokes great antitumoral responses in vitro and improves survival of tumor-bearing mice in vivo.

To further advance the concept of targeting STAT3 in GBM, we decided to use a gene therapeutic approach aimed at interfering with STAT3 expression. To this end, we employed a novel approach by using siRNAs formulated in lipopolyplexes [32,37,45] that was investigated for its effects on cultured glioma cells and locally administered to syngeneic tumors of orthotopically transplanted mice that faithfully mimic all the hallmarks of human gliomas including high mitotic activity, focal necrosis surrounded by pseudopalisading cells and diffuse infiltration into the brain parenchyma [41]. Using this approach we could show that LPPs loaded with Stat3-siRNA inhibit proliferation of glioma cells in the absence of unspecific toxicity in vitro, significantly limit tumor growth in vivo and improve the overall survival of glioma-bearing mice. Our results also show that these antitumor effects were not associated with detectable differences in the numbers of infiltrating T-cells and microglia, suggesting that tumor cell-derived STAT3 represents the primary target of LPP-PEI-siRNA in vivo. The analyses of microglia were of particular interest because it was proposed that STAT3 is required for the activation of microglia [18,46] which usually acquire a pro-tumorigenic phenotype in glioma [18]. Since we did neither observe detectable differences in the number of microglia nor in their morphology, we concluded that Stat3 siRNA had no major influence on these brain-resident immune cells if supplied as LPP complexes, however.

Various approaches for siRNA delivery were also tested by other groups, e.g., by targeting Beclin1 using intranasal delivery of PEIs [47], by targeting Survivin upon stereotactic injection [28], by targeting Eg5 using siRNA delivered via an viral envelope [48], by using multifunctional surfactants as packaging reagents to target Hif-1α [49] or by dual targeting of EGFR and Akt2 using peptides [50]. This exemplifies that specific targeting of a molecular vulnerability of cancers can be exploited by various means. Accordingly, all these different approaches suffer, in principle, from the same drawbacks. These are, limited tissue distribution, securing the release in the target cells and avoidance of siRNA-degradation and/or capture by unintended cells (e.g., immune cells) or components of the extracellular matrix. Another recent approach using a compound drug consisting of an aptamer targeting PDGFbeta and an siRNA targeting Stat3 in glioma showed similar growth retardation in vitro and in vivo [19], although this study only used subcutaneous xenografts for the in vivo experiments. Interestingly, this study also indicated that a stable fraction of Stat3 protein remained present in the tumor cells that cannot be depleted using their approach, whereas reduction of Stat3 phosphorylation was very well possible. In line with these findings, we could observe depletion of phospho-Stat3 in vitro and in vivo, but did not observe depletion of total Stat3 following administration of LPP-siStat3 in vivo.

The major limitation of this study lies in the comparatively low perfusion range of siRNA achieved, providing an explanation why reduction of Stat3 expression (and its target genes) in the bulk tumor can be considered only as minor. Our data suggest that the LPP PEIs only reach a sub-fraction of the tumor cells, as seen by a clear tendency towards reduced *Stat3*-mRNA expression in the tumor core region. Due to the very low sample size (4 samples per group) a definitive answer cannot be inferred from this data. Based on the fact that (1) most samples after LPP siStat3 treatment have *Stat3* expression values below the mean of LPP siCtrl-treated animals and (2) the mRNA of the known Stat3 target gene *Mmp9* was significantly reduced after LPP siStat3 treatment, it can be deduced that LPP-mediated delivery of siRNA was successfully achieved. Additionally, based on the in vitro data we deduce that the siRNA can specifically and effectively deplete Stat3 expression, because we could successfully inhibit proliferation of Tu2449 cells and spheroids after LPP-mediated delivery of siStat3. Therefore we conclude that delivery to Tu2449 is possible even in more complex settings (i.e., spheroids). Hence, it is more likely that the limiting factor in our current approach is tissue dispersion of LPP complexes rather than siRNA specificity. Indeed, when exploring organotypic tissue slice cultures of intact tumor (xenograft) material with regard to nanoparticle tissue penetration using fluorophore-labeled siRNAs for microscopic evaluation of tissue penetrance, we did observe LPP to diffuse into the tissue only to a certain extent. LPP tissue penetration was found to be better than for polyplexes, reflecting lesser impairment of nanoparticle diffusion with reduced surface charge (zeta potential; [51]). However, in another in vivo system (non-tumorous mouse brain in an alpha-synuclein mouse model), even our polyplexes were found to distribute across the CNS down to the lumbar spinal cord after a single intracerebroventricular infusion, thus emphasizing a distribution sufficient for exerting biological effects [52].

The changes of Stat3-target genes like *Mmp9* and *Snail* also provide mechanistic hints on how Stat3-depletion increases survival. In particular, the repression of *Mmp9* expression might reduce the migratory capacity of tumor cells due to their inability to modify the ECM. This could be further potentiated by repression of *Snail*, which is a master regulator of EMT [15]. Furthermore, Snail has been shown to enhance an immunosuppressive phenotype in cancers [15]. LPP-siStat3-mediated

Snail depletion might therefore alleviate this phenotype and in turn make the tumors more accessible or amenable for immunotherapy. Future research should therefore be directed towards increasing intratumoral dispersion of siRNA in order to increase the knockdown efficiency. This could be achieved by further improving the existing complexation formulation for LPPs for example by reducing size and/or surface charges, or by enhancing the LPP concentration for being able to inject larger siRNA amounts in the maximum possible injection volume and/or development of alternative, increasingly sophisticated nanoparticle systems for delivery [26].

A different explanation for the limited Stat3-depletion could be changes in Stat3 turnover or stability. Accordingly, the half-life of Stat3 has been previously determined to be between 30 h in primary murine neurons to 90 h in hepatocytes [53] and around 50 h in Epstein-Barr-Virus-infected PBMCs [54] indicating pronounced variations between cell types. Considering that HeLa cells have a 50% turnover rate of ~60% of their entire proteome within 5 h [55] it might also be possible that the in vivo microenvironment stimulates the tumor cells to promote faster protein turnover, because of increased cell divisions. In this case, the treatment frequency every third day might be too low. However, the fact that we already observe strong increases in survival by 7 days (~25%) indicates that by continuous delivery of siRNA this issue can likely be resolved. One possibility to achieve continuous delivery could be to employ convection-enhanced delivery (CED) using implanted catheters. Accordingly, Chen et al. [56] could show that irinotecan-treatment using CED had the best survival responses in orthotopic, murine xenografts compared to conventional treatment approaches. In fact, the use of CED for treating glioma patients has also been tested in several clinical trials for the delivery of small molecules [57], antisense oligonucleotides [58] or siRNAs [59]. More recent reports also show the potential for multiplexed targeting using various siRNAs in established murine tumors [60]. Another study also reported the use of the so-called Cleveland Multiport Catheter (CMC) [61] which has already been tested in clinical trials (NCT02278510). The main advantage of the CMC is that it consists of four independent microcatheters and could therefore be employed to deliver multiple siRNAs and/or chemotherapeutics to more specifically target the vulnerabilities of a specific tumor. Despite these considerations, even the somewhat limited perfusion of Stat3-siRNA observed in our in vivo experiments with LPP therapy allowed to achieve impressive effects on tumor growth and overall survival, further supporting the suitability of STAT3 as an excellent target for GBM therapy. However, bearing in mind the enhanced colloidal stability of lipopolyplexes over their polyplex counterparts [35], the above mentioned approach of continuous delivery via direct infusion of the LPP into the tumor, using convection-enhanced delivery (CED) via an intratumoral catheter connected to a portable pump, also becomes feasible, especially because it is to be expected that siRNA stability or release kinetics would not be negatively affected by this approach.

4. Materials and Methods

4.1. Analysis of TCGA Data Sets

The publicly available TCGA datasets GBMLGG [62] and GBM [40] were accessed, analyzed and the plots were exported via the GlioVis portal (gliovis.bioinfo.cnio.es [63]). Comparison of Stat3 expression in different glioma grades was plotted based on the histology. The survival curves were derived from the GBM dataset of all GBM subtypes combined with a median cut-off, including confidence intervals.

4.2. Cells and Compounds

The murine GBM cell line Tu2449 [64] and double reporter cell line Tu2449-EGFP/Luc stably expressing EGFP and luciferase (Luc3) (kindly provided by Dr Alexander Wurm, Universitätsklinikum Leipzig, Leipzig, Germany), the human GBM cell line U-87 MG (U87, obtained from ATCC, Manassas, VA, USA) and Mz18 [65] were cultivated in DMEM Glutamax (high Glucose, Gibco/Thermo Fisher Scientific, Karlsruhe, Germany) with 10% heat-inactivated fetal calf serum and 1%

penicillin/streptomycin (all from Gibco). Cultures were maintained in a humidified incubator at 37 °C and 5% CO_2 and checked for mycoplasma contamination monthly by PCR Mycoplasma Test Kit II (AppliChem, Darmstadt, Germany) according to the manufacturer's instructions.

4.3. siRNAs + Sequences

Chemically synthesized siRNAs were purchased from MWG (Ebersberg, Germany), Dharmacon/GE Healthcare (Lafayette, CO, USA) or Eurogentec (Seraing, Belgium), with sequences as follows: siLuc2/siCtrl (human cell lines): 5′-CGU ACG CGG AAU ACU UCG AdTdT-3′ (sense), 5′-UCG AAG UAU UCC GCG UAC GdTdT-3′ (antisense); siLuc3: 5′-CUU ACG CUG AGU ACU UCG AdTdT-3′ (sense), 5′-UCG AAG UAC UCA GCG UAA GdTdT-3′ (antisense); siEGFP: 5′-GCA GCA CGA CUU CUU CAA G-dTdT-3′ (sense), 5′-CUU GAA GAA GUC GUG CUG CdTdT-3′ (antisense), human STAT3 (siSTAT3-1: CGU UAU AUA GGA ACC GUA AdTdT (sense) and 5′-UUA CGG UUC CUA UAU AAC GdTdT (antisense), siSTAT3-2: GCC UCU CUG CAG AAU UCA AdTdT (sense) and UUG AAU UCU GCA GAG AGG CdTdG (antisense) or murine Stat3 (siStat3: 5′-GGC AUA UCG AGC CAG CAA AdTdT-3′ (sense) and 5′-UUU GCU GGC UGC AUA UGC CdTdT-3′ (antisense)), or negative control siRNAs (mouse: 5′-AAU CCG CUG UCG GCU GGA AdTdT-3′ (sense) and 5′-UUC CAG CCG ACA GCG GAU UdTdT-3′ (antisense)).

4.4. Polyplex and Lipopolyplex Preparation

Liposomes from 1,2-dipalmitoyl-sn-glycero-3-phosphocholine (DPPC; Avanti Polar Lipids, Alabaster, AL, USA) were prepared by the hydration/extrusion method as described previously [32]. For polyplex formation, siRNAs were complexed in HN buffer (150 mM NaCl, 10 mM HEPES, pH 7.4) with branched 4–12 kDa PEI F25-LMW [36] at PEI/siRNA mass ratio = 7.5:1 (N/P ratio 57) as previously described [27]. Lipopolyplexes (LPP) for in vitro experiments were prepared by incubating equal volumes of PEI/siRNA complexes with DPPC liposomes at a PEI/lipid mass ratio of 2.6 [32]. For this, 25 µL polyplex containing 0.8 µg siRNA, 6 µg PEI F25 and 25 µL DPPC liposomes comprising 16 µg lipid were properly mixed by pipetting and vortexing, and incubated for at least 1 h at room temperature. For in vivo application, lipopolyplexes were prepared by complexing 0.5 µg siRNA (200 µM), 3.75 µg PEI F25 (4.2 mg/mL) in 3 µL and mixed with 10 µg DPPC (5 mg/mL; 2 µL) to a total volume of 5 µL per injection.

4.5. qRT-PCR

For preparation of total cellular RNA from cell lines, phenol/chloroform extraction using 250 µL TRI-Reagent (Sigma-Aldrich, Munich, Germany) according to the manufacturer's protocol was used. Employing the RevertAidTM H Minus First Strand cDNA Synthesis Kit (Fermentas, St. Leon-Roth, Germany), cDNA was transcribed from 800 ng RNA according to the manufacturer's protocol. Quantitative PCR (qPCR) was performed in a StepOnePlus Real-Time PCR System (Applied Biosystems, Darmstadt, Germany) using the PerfeCTa SYBR® Green FastMix ROX (Quantabio, Beverly, MA, USA). All procedures were conducted according to the manufacturers' protocols with 4 µL cDNA (diluted 1:10 with nuclease free water), 1 µL primers (5 µM) and 5 µL SYBR Green master mix. After pre-incubation for 15 s at 95 °C, 40 amplification cycles followed: 10 s at 95 °C, 10 s at 55 °C and 10 s at 72 °C. A melting curve for PCR product analysis was recorded by rapid cooling down from 95 °C to 65 °C followed by incubation at 65 °C for 15 s prior to heating to 95 °C. Actin-specific primer sets were always run in parallel for each sample, to normalize for equal mRNA/cDNA amounts. Target levels were determined by the $\Delta\Delta$Ct method [66]. Following primers were used: human STAT3 fwd: GAG GAC TGA GCA TCG AGC A and rev: CAT GTG ATC TGA CAC CCT GAA, human Actin fwd: CCA ACC GCG AGA AGA TGA and rev: CCA GAG GCG TAC AGG GAT AG.

Pooled samples from laser-capture-microdissected formalin-fixed paraffin embedded (FFPE) tissues were subjected to RNA-Isolation using the Arcturus® Paradise® PLUS FFPE RNA Isolation Kit (Thermo Fisher) and cDNA-synthesis using SuperScript IV Vilo (life technologies/Thermo

Fisher) per the manufacturer's instructions. Briefly, the RNA was eluted in 12 µL DEPC-treated H$_2$O and used entirely for the cDNA-synthesis without prior RNA quantification. The qPCR was performed using 20× Taqman Probes (Applied Biosystems, Darmstadt, Germany) and 2× Fast-Start Universal Probe Master Mix (Roche) and 2 µL of cDNA on a StepOne Plus System (Applied Biosystems) using the standard setting. The gene expression values were normalized to TATA-Box binding protein (Tbp) or Hypoxanthine-guanine phosphoribosyltransferase (Hprt) by calculating the mean Ct value of both housekeeping gene and using this value as reference for further normalization using the ΔΔCt method [66]. The following Probes were used: Hprt (Mm00446966_m1); Mmp2 (Mm00439506_m1); Mmp9 (Mm00442991_m1); Snail (Snail, Mm01249564_g1); Stat3 (Mm01219775_m1); Tbp (Mm01219775_m1).

4.6. Western Blotting

Preparation of protein lysates from Tu2449 cells [67] and mouse brain tumors [68] were prepared as described. SDS-PAGE and Western Blotting was performed as described [67]. Membranes were blocked in 5% BSA/TBS-Tween20 (TBS-T) for 1 h at room temperature. Antibodies were incubated at 4 °C overnight in 5% BSA/TBS-T, secondary goat anti-mouse or goat anti-rabbit (dilution 1:10,000, Li-Cor Biosciences, Bad Homburg, Germany) were incubated at room temperature for 1 h and detection was achieved using a LI-COR Odyssey reader (LI-COR Biosciences, Bad Homburg, Germany).

For protein analysis of human cell cultures, cells were seeded and transfected in 24-well plates as described above. After 72 h, medium was removed and cells were washed once with PBS. 80 µL RIPA lysis buffer (50 mM Tris (pH 7.4), 150 mM NaCl, 1% Triton X-100, 0.5% sodium deoxycholate, 0.1% SDS, 2.5 mM sodium pyrophosphate, 1 mM EDTA, 10 mM NaF, Protease Inhibitor Cocktail Set III (EDTA-free, Merck, Darmstadt, Germany) was added per well and plates were incubated on ice for 10 min. The suspension was transferred to Eppendorf tubes. After centrifugation (10,000 rpm, 4 °C, 10 min), the supernatant was transferred to a new Eppendorf tube.

Using the Bio-Rad DCTM Protein-Assay (Bio-Rad, Munich, Germany), protein concentration was determined according to manufacturer's protocol. 4× loading buffer was added (0.25 mM Tris-HCl, pH 6.8, 20% glycerol, 10% beta-mercaptoethanol, 8% SDS, 0.08% bromophenol blue) to yield a 1× concentration and 20 µg protein was loaded onto 10% polyacrylamide gels. Proteins were separated by SDS-PAGE and transferred to a 0.45 µm ImmobilonTM-P Transfer PVDF Membrane (Millipore, Burlington, MA, USA). Membranes were blocked with blocking buffer (5% (*w/v*) milk powder in TBST (10 mM Tris-HCl, pH 7.6, 150 mM NaCl, 0.1% Tween 20) for 30 min. After washing in TBST, membranes were incubated with primary antibodies diluted in 3% milk powder (*w/v*) in TBST: anti-human STAT3 (Thermo Fisher Scientific), anti-Actin (Cell signaling Technologies (CST (Frankfurt am Main, Germany)), or anti-Vinculin (Sigma-Aldrich) overnight at 4 °C. Then, blots were washed in TBST and incubated for 1 h with horseradish peroxidase-coupled goat anti-rabbit IgG (CST) or horseradish peroxidase-coupled goat anti-mouse IgG (Thermo Fisher Scientific) in 3% milk powder (*w/v*) in TBST before washing again. Bound antibodies were visualized by enhanced chemiluminescence (ECL kits: SignalFireTM (CST) or SuperSignal® West Femto (Thermo Fisher Scientific)).

4.7. Antibodies

The following antibodies and dilutions were used: Actin (CST, 13E5, 1:1000 in 3% milk); Cluster of Differentiation 4 (CD4); Affymetrix eBioscience/Thermo Fisher; clone 4SM95; 14-97664; 1:100 (IHC); CD8a (Synaptic Systems, Göttingen, Germany) 361003; 1:500 (IHC); Gapdh (Calbiochem, #CB1001, Darmstadt, Germany) 1:20,000; Iba1 (Wako, 019-19741); 1:1000 (IHC); Ki67 (SP6, Thermo Fisher, MA5-14520); 1:100 (IHC); Stat3 (Santa Cruz Biotechnology (Heidelberg, Germany)), c-20; sc-482; 1:200 (WB); Stat3 (phospho Tyr705; CST Technologies (CST (Frankfurt am Main, Germany)); D3A7; 9145S; 1:1000 for Western Blot (WB) and 1:500 for Immunohistochemistry (IHC)); STAT3 (Thermo Fisher, 9D8, 1:5000 in 3% milk); Vinculin (Sigma Aldrich, hVIN-1, 1:2000 in 3% milk).

4.8. Nanoparticle Characterization

Particle sizes and zeta potentials were determined by photon correlation spectroscopy (PCS) and phase analysis light scattering (PALS), using a Brookhaven ZetaPALS system (Brookhaven Instruments, Holtsville, NY, USA). The data were analyzed using the manufacturer's software and applying a viscosity and refractive index of pure water at 25 °C. For size determination, the complexes were analyzed in five runs with a run duration of 1 min. Results are expressed as intensity weighted mean diameter from different experiments. Zeta potentials were measured in ten runs, with each run containing ten cycles, and applying the Smoluchowski model. Additionally, hydrodynamic diameters of the nanoparticles were determined by nanoparticle tracking analysis (NTA), using a NanoSight LM 10 HS apparatus (Malvern) equipped with a 640 nm sCMOS camera and a temperature controlled sample chamber, as described previously [45,69].

4.9. Cell Transfection

EGFP and luciferase knockdown experiments were performed in 24-well plates. The day before, Tu2449-EGFP/Luc cells were seeded at a density of 30,000 cells per well in 0.5 mL fully supplemented medium. Polyplexes and lipopolyplexes comprising 0.8 µg siRNA per well were prepared as described above, containing siRNAs against the targets EGFP or luciferase, or the negative control siRNA (see above Section 4.3).

For transfection using INTERFERin^{TM} (Polyplus, Illkirch, France), 2×10^2 cells were seeded in a 96-well plate (proliferation assay; U87 or Mz18 cells) or 2×10^5 cells in a 24-well plate (cytotoxicity assay; Tu2449 cells), respectively. Cells were incubated under standard conditions unless stated otherwise. 10 nM siRNA were transfected according to the manufacturer's protocol and incubated for the indicated periods of time.

4.10. Determination of Knockdown Efficacies

Luciferase activities were determined in the double reporter cell line Tu2449-EGFP/Luc 96 h post transfection using the Beetle-Juice Kit (PJK, Kleinblittersdorf, Germany). The medium was aspirated and 300 µL lysis buffer (Promega, Mannheim, Germany) was added, prior to incubation for 30 min at RT. In a test tube, 25 µL luciferin substrate was mixed with 10 µL cell lysate and luminescence was immediately measured in a luminometer (Berthold, Bad Wildbad, Germany).

The EGFP knockdown was quantitated after 96 h by flow cytometry. Cells were trypsinized and centrifuged for 3 min at 3000 rpm followed by a washing step with 1% BSA in PBS (1 mL), and finally resuspended in 0.5 mL of the same buffer. The cells were measured in an Attune® Acoustic Focusing Cytometer (Life Technologies, Darmstadt, Germany) with appropriate instrument settings. 20,000 events were gated and analyzed using the Attune® software (V2.1.0). The data are presented as mean fluorescent intensities.

4.11. Cell Proliferation and Viability Assays

For proliferation assays, U87 or Mz18 cells were seeded and transfected in 96-well plates as described above. The number of viable cells was determined at the time points indicated using a colorimetric assay. Briefly, 50 µL of a 1:10 dilution of Cell proliferation Reagent WST-1 (Roche Molecular Biochemicals, Mannheim, Germany) in serum-free medium was added to the cells after aspirating the media. Cells were incubated for 1 h at 37 °C and absorbance at 450 nm was measured in an ELISA reader.

Acute cell damage upon transfection was analyzed in Tu2449 cells by the lactate dehydrogenase (LDH) release assay, using the Cytotoxicity Detection Kit (Roche, Mannheim, Germany) according to the manufacturer's protocol. Briefly, conditioned medium from knockdown experiments was collected after 24 h. Conditioned medium from untreated cells served as negative control, and for the determination of maximum LDH release (100% value), cells were lysed by adding Triton X-100 to final concentration of 2% into the medium. In a 96 well plate, 50 μL sample medium was mixed with 50 μL reagent mix and incubated for 30 min in the dark. The reaction was stopped with 50 μL 1 M acetic acid and the absorption at 490 nm and 620 nm as reference filter was measured in an ELISA reader. Fresh fully supplemented medium and reagent mix served as background and was subtracted from all values. Cytotoxicity values are shown in percent of the maximum value.

4.12. Cell Cycle Analysis

Cell cycle distribution was assessed using flow cytometry. Cells were seeded in 24-well plates and transfected as described above. 48 h after transfection, cells were trypsinized, transferred to Eppendorf tubes and centrifuged (1500 rpm, room temperature, 5 min). The supernatant was aspirated and the cell pellet was resuspended in ice cold 70% ethanol and incubated for at least 1 h at 4 °C for fixation. After centrifugation (1500 rpm, room temperature, 5 min) and discarding the supernatant, cells were resuspended in an RNase A solution (50 μg/mL RNase A in PBS) and incubated for 30 min at 37 °C. Propidium iodide solution was added to yield a 50 μg/mL concentration. Cell cycle distribution was analyzed in the BL2 channel using an Attune™ Flow Cytometer.

4.13. Spheroid Assay

To generate spheroids, 3×10^3 Tu2449 cells were seeded in a 96-well U-shaped ultra-low attachment plate and incubated under standard conditions. Using SAINT-sRNA transfection reagent (Synvolux, Leiden, The Netherlands), 90 nM siRNA were transfected according to the manufacturer's protocol. After 10 days, spheroid size was documented and measured using a Celigo Imaging Cytometer (Nexcelom Bioscience, Lawrence, MA, USA) and the corresponding software.

4.14. Animal Experiments

Intracranial implantation of tumor cells into the right striatum of B6C3F1 mice (Envigo, Huntingdon, UK; average weight 25 g) was performed as described [21]. Briefly, 10,000 viable Tu2449 in 1 μL were injected with a flow rate of 0.5 μL/min at a depth of 3 mm 1.5 mm posterior and 2 mm lateral to the bregma. 42 mice were implanted with tumor cells and one week after implantation the animals were randomly divided into two groups; one received LPP siCtrl treatment, the other LPP siStat3 treatment. Application of LPPs (equivalent to 0.5 μg siRNA in 5 μL which, due to limitations regarding the total lipopolyplex volume, was the maximum possible amount in this experimental setting) was performed using the same drill hole in the skull as for tumor cell implantation every third day one week after tumor cell implantation. 21 days after tumor implantation 10 mice per group were sacrificed; of those, 8 mice were used for histological analysis and two for molecular analyses. Those tumors were divided and one half was immediately frozen on dry ice and used for protein isolation, the other half was stored in RNAlater (Qiagen, Hilden, Germany) and used for RNA isolation. All intracranial injections were performed on anesthetized animals and treatment was done unblinded. Histoligical analysis showed that in the LPP siCtrl and LPP siStat3 group three and five tumors could not be established. Of those, one and two tumors were planned for histological assessment after 21 days. These mice were removed from the survival-analysis. Animals at the collection time point or those that showed neurological symptoms related to tumor burden were euthanized by cervical dislocation after they received a lethal injection of anesthetic. Hereafter the brains were immediately removed, rinsed in PBS and fixed for 2–7 days in 4% PFA (Chemcruz/Santa Cruz). Prior to paraffin embedding the fixed brains were cut coronally in 2–3 mm thick section and the resulting sections were aligned horizontally. Embedding in paraffin was performed according to standard paraffin embedding procedures known from routine pathology. Sildes were first stained with hematoxylin and eosin. Immunohistochemistry was performed on an automated staining system Bond III (Leica Microsystems, Wetzlar, Germany) using standard protocols. All animal experiments were approved by the local administrative council (Regierungspräsidium Darmstadt, FK/1011).

4.15. Determination of Tumor Size

The area of tumors was determined by measuring the longest distance (a) between tumor borders and the respective orthogonal line (b) from all tumor sections visible from H&E stained tumor sections. Using the formulas for ellipses the area (A) was calculated:

$$A = \pi \times 0.5 \times a \times b$$

If a brain contained multiple tumor sections those values were summed to account for the depth of the tumor. If multiple sections from one mouse were available the area, indicative for the tumor center, was chosen.

4.16. Laser-Capture Microdissection

Laser capture microdissection (LCM) was performed of 10 μm thick, serial sections mounted on UV-irradiated MMI membrane slides (MMI, Eching, Germany) and collected on MMI Isolation caps with diffuser caps (MMI). Prior to microdissection, the sections were dried at 37 °C for 1 h and rehydrated (2 times 10' xylene, 99%, 96%, 70% EtOH, 30'' each) and the first and last section of the series were stained with hematoxylin and eosin, dehydrated and mounted using Pertex (Leica, Wetzlar, Germany). LCM was performed using a laser capture microscope (Oylmpus IX71 equipped with MMI CellCamera und MMI CellCut) operated by the software MMI CellTools (v. 4.0).

4.17. Statistics

All statistical analysis were performed using GraphPad Prism 7 (GraphPad Software, La Jolla, CA, USA). Kaplan-Meier-Survival curves were analyzed using log-rank test, all other experiments using two-tailed T-Test, with assuming equal SD and without corrections for multiple comparisons. Cell cycle measurement was analyzed by two-way-ANOVA of matched samples with Dunnett's multiple comparison test against siCtrl treated cells. Unless stated otherwise, all samples are presented as Box-Plots (min-to-max), where the horizontal line depicts the median value and the "+" symbol deptics the calculated mean.

5. Conclusions

In conclusion, we provide first proof-of-concept evidence that intratumoral targeting of oncogenes like Stat3 using lipopolyplexsosomal siRNA for the treatment of brain tumors is feasible and leads to improved overall survival in a murine, syngeneic, orthotopic transplantation model. Future development of nanoparticle systems and the modes of their application will hopefully enhance the restricted tissue distribution achieved in this study, thereby further improving the tumor targeting capacity of siRNA. These efforts may lead to the development of new and highly specific treatments selectively targeting the expression of oncogenes such as STAT3, an avenue recently opened in a clinical trial focusing on another gene target in brain tumors (NCT03020017).

Supplementary Materials: The following are available online at http://www.mdpi.com/2072-6694/11/3/333/s1, Figure S1: STAT3-RNA expression from the TCGA dataset GBM-LGG plotted using the GlioVis portal, Figure S2: Nanoparticle sizes as determined by Zetasizer (dynamic light scattering), Figure S3: siStat3 hinders cell cycle progression in human GBM cell lines.

Author Contributions: Conceptualization, B.L., A.A. and D.K.; Formal analysis, B.L. and U.W.; Funding acquisition, V.S., A.A. and D.K.; Investigation, B.L., U.W., A.E., A.U., M.M. and P.N.H.; Project administration, A.A. and D.K.; Resources, A.U., V.S., M.M., P.N.H., A.A. and D.K.; Supervision, V.S., A.A. and D.K.; Visualization, B.L. and U.W.; Writing—original draft, B.L., U.W., A.A. and D.K.; Writing—review & editing, B.L., A.A. and D.K.

Funding: This research was funded by grants of the Deutsche Forschungsgemeinschaft (DFG, KO 1898 9/1 and AI 24/13-1) to DK and AA.

Acknowledgments: We would like to thank Gabriele Köpf and Hildegard König for excellent technical assistance and Tatjana Starzetz and Maika Dunst (Institute of Neurology (Edinger-Institute), University Hospital Frankfurt, Goethe University, Frankfurt am Main, Germany) for sectioning and staining of FFPE tissue. We are also grateful to Hanibal Bohnenberger (Institute of Pathology, University Medical Centre, Göttingen, Germany) for access to and help with the LCM microscope. M.M. would like to thank the Luxembourg National Research Fund (FNR) for the support (FNR PEARL P16/BM/11192868 grant).

Conflicts of Interest: The authors declare no conflict of interest.

References

1. Louis, D.N.; Perry, A.; Reifenberger, G.; von Deimling, A.; Figarella-Branger, D.; Cavenee, W.K.; Ohgaki, H.; Wiestler, O.D.; Kleihues, P.; Ellison, D.W. The 2016 World Health Organization Classification of Tumors of the Central Nervous System: A summary. *Acta Neuropathol.* **2016**, *131*, 803–820. [CrossRef]
2. Becker, K.P.; Yu, J. Status quo–standard-of-care medical and radiation therapy for glioblastoma. *Cancer J.* **2012**, *18*, 12–19. [CrossRef]
3. Verhaak, R.G.; Hoadley, K.A.; Purdom, E.; Wang, V.; Qi, Y.; Wilkerson, M.D.; Miller, C.R.; Ding, L.; Golub, T.; Mesirov, J.P.; et al. Integrated genomic analysis identifies clinically relevant subtypes of glioblastoma characterized by abnormalities in PDGFRA, IDH1, EGFR, and NF1. *Cancer Cell* **2010**, *17*, 98–110. [CrossRef]
4. Olar, A.; Aldape, K.D. Using the molecular classification of glioblastoma to inform personalized treatment. *J. Pathol.* **2014**, *232*, 165–177. [CrossRef]
5. Carro, M.S.; Lim, W.K.; Alvarez, M.J.; Bollo, R.J.; Zhao, X.; Snyder, E.Y.; Sulman, E.P.; Anne, S.L.; Doetsch, F.; Colman, H.; et al. The transcriptional network for mesenchymal transformation of brain tumours. *Nature* **2010**, *463*, 318–325. [CrossRef]
6. Hossain, A.; Gumin, J.; Gao, F.; Figueroa, J.; Shinojima, N.; Takezaki, T.; Priebe, W.; Villarreal, D.; Kang, S.G.; Joyce, C.; et al. Mesenchymal Stem Cells Isolated From Human Gliomas Increase Proliferation and Maintain Stemness of Glioma Stem Cells Through the IL-6/gp130/STAT3 Pathway. *Stem. Cells* **2015**, *33*, 2400–2415. [CrossRef]
7. Yu, H.; Lee, H.; Herrmann, A.; Buettner, R.; Jove, R. Revisiting STAT3 signalling in cancer: New and unexpected biological functions. *Nat. Rev. Cancer* **2014**, *14*, 736–746. [CrossRef]
8. Galoczova, M.; Coates, P.; Vojtesek, B. STAT3, stem cells, cancer stem cells and p63. *Cell. Mol. Biol. Lett.* **2018**, *23*, 12. [CrossRef]
9. Kim, J.E.; Patel, M.; Ruzevick, J.; Jackson, C.M.; Lim, M. STAT3 Activation in Glioblastoma: Biochemical and Therapeutic Implications. *Cancers (Basel)* **2014**, *6*, 376–395. [CrossRef]
10. Brantley, E.C.; Benveniste, E.N. Signal transducer and activator of transcription-3: A molecular hub for signaling pathways in gliomas. *Mol. Cancer Res.* **2008**, *6*, 675–684. [CrossRef]
11. Ganguly, D.; Fan, M.; Yang, C.H.; Zbytek, B.; Finkelstein, D.; Roussel, M.F.; Pfeffer, L.M. The critical role that STAT3 plays in glioma-initiating cells: STAT3 addiction in glioma. *Oncotarget* **2018**, *9*, 22095–22112. [CrossRef]
12. Luwor, R.B.; Baradaran, B.; Taylor, L.E.; Iaria, J.; Nheu, T.V.; Amiry, N.; Hovens, C.M.; Wang, B.; Kaye, A.H.; Zhu, H.J. Targeting Stat3 and Smad7 to restore TGF-beta cytostatic regulation of tumor cells in vitro and in vivo. *Oncogene* **2013**, *32*, 2433–2441. [CrossRef]
13. Westermarck, J.; Kahari, V.M. Regulation of matrix metalloproteinase expression in tumor invasion. *FASEB J.* **1999**, *13*, 781–792. [CrossRef]
14. Wendt, M.K.; Balanis, N.; Carlin, C.R.; Schiemann, W.P. STAT3 and epithelial-mesenchymal transitions in carcinomas. *JAKSTAT* **2014**, *3*, e28975. [CrossRef]
15. Kudo-Saito, C.; Shirako, H.; Takeuchi, T.; Kawakami, Y. Cancer metastasis is accelerated through immunosuppression during Snail-induced EMT of cancer cells. *Cancer Cell* **2009**, *15*, 195–206. [CrossRef]
16. Avalle, L.; Camporeale, A.; Camperi, A.; Poli, V. STAT3 in cancer: A double edged sword. *Cytokine* **2017**, *98*, 42–50. [CrossRef]
17. Liu, Y.; Li, C.; Lin, J. STAT3 as a Therapeutic Target for Glioblastoma. *Anticancer Agents Med. Chem* **2010**, *10*, 512–519. [CrossRef]
18. Li, W.; Graeber, M.B. The molecular profile of microglia under the influence of glioma. *Neuro Oncol.* **2012**, *14*, 958–978. [CrossRef]
19. Esposito, C.L.; Nuzzo, S.; Catuogno, S.; Romano, S.; de Nigris, F.; de Franciscis, V. STAT3 Gene Silencing by Aptamer-siRNA Chimera as Selective Therapeutic for Glioblastoma. *Mol. Nucleic Acids* **2018**, *10*, 398–411. [CrossRef]

20. Kiprianova, I.; Remy, J.; Milosch, N.; Mohrenz, I.V.; Seifert, V.; Aigner, A.; Kogel, D. Sorafenib Sensitizes Glioma Cells to the BH3 Mimetic ABT-737 by Targeting MCL1 in a STAT3-Dependent Manner. *Neoplasia* **2015**, *17*, 564–573. [CrossRef]

21. Priester, M.; Copanaki, E.; Vafaizadeh, V.; Hensel, S.; Bernreuther, C.; Glatzel, M.; Seifert, V.; Groner, B.; Kogel, D.; Weissenberger, J. STAT3 silencing inhibits glioma single cell infiltration and tumor growth. *Neuro Oncol.* **2013**, *15*, 840–852. [CrossRef]

22. Senft, C.; Priester, M.; Polacin, M.; Schroder, K.; Seifert, V.; Kogel, D.; Weissenberger, J. Inhibition of the JAK-2/STAT3 signaling pathway impedes the migratory and invasive potential of human glioblastoma cells. *J. Neurooncol.* **2011**, *101*, 393–403. [CrossRef]

23. Sherry, M.M.; Reeves, A.; Wu, J.L.K.; Cochran, B.H. STAT3 Is Required for Proliferation and Maintenance of Multipotency in Glioblastoma Stem Cells. *Stem. Cells* **2009**, *27*, 2383–2392. [CrossRef]

24. Takeda, K.; Kaisho, T.; Yoshida, N.; Takeda, J.; Kishimoto, T.; Akira, S. Stat3 activation is responsible for IL-6-dependent T cell proliferation through preventing apoptosis: Generation and characterization of T cell-specific Stat3-deficient mice. *J. Immunol.* **1998**, *161*, 4652–4660. [CrossRef]

25. Aigner, A. Applications of RNA interference: Current state and prospects for siRNA-based strategies in vivo. *Appl. Microbiol. Biotechnol.* **2007**, *76*, 9–21. [CrossRef]

26. Aigner, A.; Kogel, D. Nanoparticle/siRNA-based therapy strategies in glioma: Which nanoparticles, which siRNAs? *Nanomedicine (London)* **2018**, *13*, 89–103. [CrossRef]

27. Hobel, S.; Koburger, I.; John, M.; Czubayko, F.; Hadwiger, P.; Vornlocher, H.P.; Aigner, A. Polyethylenimine/small interfering RNA-mediated knockdown of vascular endothelial growth factor in vivo exerts anti-tumor effects synergistically with Bevacizumab. *J. Gene Med.* **2010**, *12*, 287–300.

28. Hendruschk, S.; Wiedemuth, R.; Aigner, A.; Topfer, K.; Cartellieri, M.; Martin, D.; Kirsch, M.; Ikonomidou, C.; Schackert, G.; Temme, A. RNA interference targeting survivin exerts antitumoral effects in vitro and in established glioma xenografts in vivo. *Neuro Oncol.* **2011**, *13*, 1074–1089. [CrossRef]

29. Ibrahim, A.F.; Weirauch, U.; Thomas, M.; Grunweller, A.; Hartmann, R.K.; Aigner, A. MicroRNA replacement therapy for miR-145 and miR-33a is efficacious in a model of colon carcinoma. *Cancer Res.* **2011**, *71*, 5214–5224. [CrossRef]

30. Hampl, V.; Martin, C.; Aigner, A.; Hoebel, S.; Singer, S.; Frank, N.; Sarikas, A.; Ebert, O.; Prywes, R.; Gudermann, T.; et al. Depletion of the transcriptional coactivators megakaryoblastic leukaemia 1 and 2 abolishes hepatocellular carcinoma xenograft growth by inducing oncogene-induced senescence. *EMBO Mol. Med.* **2013**, *5*, 1367–1382. [CrossRef]

31. Hobel, S.; Aigner, A. Polyethylenimines for siRNA and miRNA delivery in vivo. *Wiley Interdiscip. Rev. Nanomed. Nanobiotechnol.* **2013**, *5*, 484–501. [CrossRef]

32. Ewe, A.; Panchal, O.; Pinnapireddy, S.R.; Bakowsky, U.; Przybylski, S.; Temme, A.; Aigner, A. Liposome-polyethylenimine complexes (DPPC-PEI lipopolyplexes) for therapeutic siRNA delivery in vivo. *Nanomedicine* **2017**, *13*, 209–218. [CrossRef]

33. Heyes, J.; Palmer, L.; Chan, K.; Giesbrecht, C.; Jeffs, L.; MacLachlan, I. Lipid encapsulation enables the effective systemic delivery of polyplex plasmid DNA. *Molcure* **2007**, *15*, 713–720. [CrossRef]

34. Ko, Y.T.; Kale, A.; Hartner, W.C.; Papahadjopoulos-Sternberg, B.; Torchilin, V.P. Self-assembling micelle-like nanoparticles based on phospholipid-polyethyleneimine conjugates for systemic gene delivery. *J. Control. Release* **2009**, *133*, 132–138. [CrossRef]

35. Ewe, A.; Schaper, A.; Barnert, S.; Schubert, R.; Temme, A.; Bakowsky, U.; Aigner, A. Storage stability of optimal liposome-polyethylenimine complexes (lipopolyplexes) for DNA or siRNA delivery. *Acta Biomater.* **2014**, *10*, 2663–2673. [CrossRef]

36. Werth, S.; Urban-Klein, B.; Dai, L.; Hobel, S.; Grzelinski, M.; Bakowsky, U.; Czubayko, F.; Aigner, A. A low molecular weight fraction of polyethylenimine (PEI) displays increased transfection efficiency of DNA and siRNA in fresh or lyophilized complexes. *J. Control. Release* **2006**, *112*, 257–270. [CrossRef]

37. Schafer, J.; Hobel, S.; Bakowsky, U.; Aigner, A. Liposome-polyethylenimine complexes for enhanced DNA and siRNA delivery. *Biomaterials* **2010**, *31*, 6892–6900. [CrossRef]

38. Ewe, A.; Aigner, A. Cationic Lipid-Coated Polyplexes (Lipopolyplexes) for DNA and Small RNA Delivery. *Methods Mol. Biol.* **2016**, *1445*, 187–200.

39. Dasgupta, A.; Raychaudhuri, B.; Haqqi, T.; Prayson, R.; Van Meir, E.G.; Vogelbaum, M.; Haque, S.J. Stat3 activation is required for the growth of U87 cell-derived tumours in mice. *Eur. J. Cancer* **2009**, *45*, 677–684. [CrossRef]

40. The Cancer Genome Atlas Research Network. Comprehensive genomic characterization defines human glioblastoma genes and core pathways. *Nature* **2008**, *455*, 1061–1068.

41. Smilowitz, H.M.; Weissenberger, J.; Weis, J.; Brown, J.D.; O'Neill, R.J.; Laissue, J.A. Orthotopic transplantation of v-src-expressing glioma cell lines into immunocompetent mice: Establishment of a new transplantable in vivo model for malignant glioma. *J. Neurosurg.* **2007**, *106*, 652–659. [CrossRef] [PubMed]

42. Stechishin, O.D.; Luchman, H.A.; Ruan, Y.; Blough, M.D.; Nguyen, S.A.; Kelly, J.J.; Cairncross, J.G.; Weiss, S. On-target JAK2/STAT3 inhibition slows disease progression in orthotopic xenografts of human glioblastoma brain tumor stem cells. *Neuro Oncol.* **2013**, *15*, 198–207. [CrossRef]

43. Doucette, T.A.; Kong, L.Y.; Yang, Y.; Ferguson, S.D.; Yang, J.; Wei, J.; Qiao, W.; Fuller, G.N.; Bhat, K.P.; Aldape, K.; et al. Signal transducer and activator of transcription 3 promotes angiogenesis and drives malignant progression in glioma. *Neuro Oncol.* **2012**, *14*, 1136–1145. [CrossRef] [PubMed]

44. Weissenberger, J.; Priester, M.; Bernreuther, C.; Rakel, S.; Glatzel, M.; Seifert, V.; Kogel, D. Dietary curcumin attenuates glioma growth in a syngeneic mouse model by inhibition of the JAK1,2/STAT3 signaling pathway. *Clin. Cancer Res.* **2010**, *16*, 5781–5795. [CrossRef] [PubMed]

45. Ewe, A.; Aigner, A. Nebulization of liposome–polyethylenimine complexes (lipopolyplexes) for DNA or siRNA delivery: Physicochemical properties and biological activity. *Eur. J. Lipid Sci. Technol.* **2014**, *116*, 1195–1204. [CrossRef]

46. Planas, A.M.; Soriano, M.A.; Berruezo, M.; Justicia, C.; Estrada, A.; Pitarch, S.; Ferrer, I. Induction of Stat3, a signal transducer and transcription factor, in reactive microglia following transient focal cerebral ischaemia. *Eur. J. Neurosci.* **1996**, *8*, 2612–2618. [CrossRef]

47. Rodriguez, M.; Lapierre, J.; Ojha, C.R.; Kaushik, A.; Batrakova, E.; Kashanchi, F.; Dever, S.M.; Nair, M.; El-Hage, N. Intranasal drug delivery of small interfering RNA targeting Beclin1 encapsulated with polyethylenimine (PEI) in mouse brain to achieve HIV attenuation. *Sci. Rep.* **2017**, *7*, 1862. [CrossRef]

48. Matsuda, M.; Yamamoto, T.; Matsumura, A.; Kaneda, Y. Highly efficient eradication of intracranial glioblastoma using Eg5 siRNA combined with HVJ envelope. *Gene* **2009**, *16*, 1465–1476. [CrossRef]

49. Gillespie, D.L.; Aguirre, M.T.; Ravichandran, S.; Leishman, L.L.; Berrondo, C.; Gamboa, J.T.; Wang, L.; King, R.; Wang, X.; Tan, M.; et al. RNA interference targeting hypoxia-inducible factor 1alpha via a novel multifunctional surfactant attenuates glioma growth in an intracranial mouse model. *J. Neurosurg.* **2015**, *122*, 331–341. [CrossRef]

50. Michiue, H.; Eguchi, A.; Scadeng, M.; Dowdy, S.F. Induction of in vivo synthetic lethal RNAi responses to treat glioblastoma. *Cancer Biol.* **2009**, *8*, 2306–2313. [CrossRef]

51. Merz, L.; Hobel, S.; Kallendrusch, S.; Ewe, A.; Bechmann, I.; Franke, H.; Merz, F.; Aigner, A. Tumor tissue slice cultures as a platform for analyzing tissue-penetration and biological activities of nanoparticles. *Eur. J. Pharm. Biopharm.* **2017**, *112*, 45–50. [CrossRef]

52. Helmschrodt, C.; Hobel, S.; Schoniger, S.; Bauer, A.; Bonicelli, J.; Gringmuth, M.; Fietz, S.A.; Aigner, A.; Richter, A.; Richter, F. Polyethylenimine Nanoparticle-Mediated siRNA Delivery to Reduce alpha-Synuclein Expression in a Model of Parkinson's Disease. *Mol. Nucleic Acids* **2017**, *9*, 57–68. [CrossRef] [PubMed]

53. Mathieson, T.; Franken, H.; Kosinski, J.; Kurzawa, N.; Zinn, N.; Sweetman, G.; Poeckel, D.; Ratnu, V.S.; Schramm, M.; Becher, I.; et al. Systematic analysis of protein turnover in primary cells. *Nat. Commun.* **2018**, *9*, 689. [CrossRef] [PubMed]

54. Bocchini, C.E.; Nahmod, K.; Katsonis, P.; Kim, S.; Kasembeli, M.M.; Freeman, A.; Lichtarge, O.; Makedonas, G.; Tweardy, D.J. Protein stabilization improves STAT3 function in autosomal dominant hyper-IgE syndrome. *Blood* **2016**, *128*, 3061–3072. [CrossRef] [PubMed]

55. Boisvert, F.M.; Ahmad, Y.; Gierlinski, M.; Charriere, F.; Lamont, D.; Scott, M.; Barton, G.; Lamond, A.I. A quantitative spatial proteomics analysis of proteome turnover in human cells. *Mol. Cell. Proteom.* **2012**, *11*, M111.011429. [CrossRef] [PubMed]

56. Chen, P.Y.; Ozawa, T.; Drummond, D.C.; Kalra, A.; Fitzgerald, J.B.; Kirpotin, D.B.; Wei, K.C.; Butowski, N.; Prados, M.D.; Berger, M.S.; et al. Comparing routes of delivery for nanoliposomal irinotecan shows superior anti-tumor activity of local administration in treating intracranial glioblastoma xenografts. *Neuro Oncol.* **2013**, *15*, 189–197. [CrossRef] [PubMed]

57. Bruce, J.N.; Fine, R.L.; Canoll, P.; Yun, J.; Kennedy, B.C.; Rosenfeld, S.S.; Sands, S.A.; Surapaneni, K.; Lai, R.; Yanes, C.L.; et al. Regression of recurrent malignant gliomas with convection-enhanced delivery of topotecan. *Neurosurgery* **2011**, *69*, 1272–1279, discussion 1279–1280. [CrossRef] [PubMed]

58. Jaschinski, F.; Rothhammer, T.; Jachimczak, P.; Seitz, C.; Schneider, A.; Schlingensiepen, K.H. The antisense oligonucleotide trabedersen (AP 12009) for the targeted inhibition of TGF-beta2. *Curr. Pharm. Biotechnol.* **2011**, *12*, 2203–2213. [CrossRef] [PubMed]

59. Voges, J.; Reszka, R.; Gossmann, A.; Dittmar, C.; Richter, R.; Garlip, G.; Kracht, L.; Coenen, H.H.; Sturm, V.; Wienhard, K.; et al. Imaging-guided convection-enhanced delivery and gene therapy of glioblastoma. *Ann. Neurol.* **2003**, *54*, 479–487. [CrossRef] [PubMed]

60. Yu, D.; Khan, O.F.; Suva, M.L.; Dong, B.; Panek, W.K.; Xiao, T.; Wu, M.; Han, Y.; Ahmed, A.U.; Balyasnikova, I.V.; et al. Multiplexed RNAi therapy against brain tumor-initiating cells via lipopolymeric nanoparticle infusion delays glioblastoma progression. *Proc. Natl. Acad. Sci. USA* **2017**, *114*, E6147–E6156. [CrossRef] [PubMed]

61. Vogelbaum, M.A.; Brewer, C.; Barnett, G.H.; Mohammadi, A.M.; Peereboom, D.M.; Ahluwalia, M.S.; Gao, S. First-in-human evaluation of the Cleveland Multiport Catheter for convection-enhanced delivery of topotecan in recurrent high-grade glioma: Results of pilot trial 1. *J. Neurosurg.* **2018**. [CrossRef] [PubMed]

62. Ceccarelli, M.; Barthel, F.P.; Malta, T.M.; Sabedot, T.S.; Salama, S.R.; Murray, B.A.; Morozova, O.; Newton, Y.; Radenbaugh, A.; Pagnotta, S.M.; et al. Molecular Profiling Reveals Biologically Discrete Subsets and Pathways of Progression in Diffuse Glioma. *Cell* **2016**, *164*, 550–563. [CrossRef] [PubMed]

63. Bowman, R.L.; Wang, Q.; Carro, A.; Verhaak, R.G.; Squatrito, M. GlioVis data portal for visualization and analysis of brain tumor expression datasets. *Neuro Oncol.* **2017**, *19*, 139–141. [CrossRef] [PubMed]

64. Pohl, U.; Wick, W.; Weissenberger, J.; Steinbach, J.P.; Dichgans, J.; Aguzzi, A.; Weller, M. Characterization of Tu-2449, a glioma cell line derived from a spontaneous tumor in GFAP-v-src-transgenic mice: Comparison with established murine glioma cell lines. *Int. J. Oncol.* **1999**, *15*, 829–834. [CrossRef] [PubMed]

65. Hetschko, H.; Voss, V.; Horn, S.; Seifert, V.; Prehn, J.H.; Kogel, D. Pharmacological inhibition of Bcl-2 family members reactivates TRAIL-induced apoptosis in malignant glioma. *J. Neurooncol.* **2008**, *86*, 265–272. [CrossRef] [PubMed]

66. Livak, K.J.; Schmittgen, T.D. Analysis of relative gene expression data using real-time quantitative PCR and the 2(-Delta Delta C(T)) Method. *Methods* **2001**, *25*, 402–408. [CrossRef] [PubMed]

67. Antonietti, P.; Linder, B.; Hehlgans, S.; Mildenberger, I.C.; Burger, M.C.; Fulda, S.; Steinbach, J.P.; Gessler, F.; Rodel, F.; Mittelbronn, M.; et al. Interference with the HSF1/HSP70/BAG3 Pathway Primes Glioma Cells to Matrix Detachment and BH3 Mimetic-Induced Apoptosis. *Mol. Cancer* **2017**, *16*, 156–168. [CrossRef]

68. Weissenberger, J.; Loeffler, S.; Kappeler, A.; Kopf, M.; Lukes, A.; Afanasieva, T.A.; Aguzzi, A.; Weis, J. IL-6 is required for glioma development in a mouse model. *Oncogene* **2004**, *23*, 3308–3316. [CrossRef]

69. Schulze, J.; Hendrikx, S.; Schulz-Siegmund, M.; Aigner, A. Microparticulate poly(vinyl alcohol) hydrogel formulations for embedding and controlled release of polyethylenimine (PEI)-based nanoparticles. *Acta Biomater.* **2016**, *45*, 210–222. [CrossRef]

Article

Hepatic Stress Response in HCV Infection Promotes STAT3-Mediated Inhibition of HNF4A-*miR-122* Feedback Loop in Liver Fibrosis and Cancer Progression

Yucel Aydin [1], Ramazan Kurt [2], Kyoungsub Song [1], Dong Lin [1], Hanadi Osman [1], Brady Youngquist [1], John W. Scott [1], Nathan J. Shores [2], Paul Thevenot [3], Ari Cohen [3] and Srikanta Dash [1,2,*]

[1] Department of Pathology and Laboratory Medicine, Tulane University Health Sciences Center, New Orleans, LA 70112, USA; yaydin@tulane.edu (Y.A.); ksong@tulane.edu (K.S.); dlin6@tulane.edu (D.L.); hosman1@tulane.edu (H.O.); byoungquist@tulane.edu (B.Y.); jscottmd@tulane.edu (J.W.S.)

[2] Section of Gastroenterology and Hepatology, Tulane University Health Sciences Center, New Orleans, LA 70112, USA; rkurt@tulane.edu (R.K.); shoresnj@gmail.com (N.J.S.)

[3] Liver Transplant Surgery Section, Ochsner Medical Center, New Orleans, LA 70121, USA; paul.thevenot@oschner.org (P.T.); acohen@oschner.org (A.C.)

* Correspondence: sdash@tulane.edu; Tel.: +1-504-988-2519; Fax: +1-504-988-7389

Received: 20 August 2019; Accepted: 16 September 2019; Published: 20 September 2019

Abstract: Hepatitis C virus (HCV) infection compromises the natural defense mechanisms of the liver leading to a progressive end stage disease such as cirrhosis and hepatocellular carcinoma (HCC). The hepatic stress response generated due to viral replication in the endoplasmic reticulum (ER) undergoes a stepwise transition from adaptive to pro-survival signaling to improve host cell survival and liver disease progression. The minute details of hepatic pro-survival unfolded protein response (UPR) signaling that contribute to HCC development in cirrhosis are unknown. This study shows that the UPR sensor, the protein kinase RNA-like ER kinase (PERK), mediates the pro-survival signaling through nuclear factor erythroid 2-related factor 2 (NRF2)-mediated signal transducer and activator of transcription 3 (STAT3) activation in a persistent HCV infection model of Huh-7.5 liver cells. The NRF2-mediated STAT3 activation in persistently infected HCV cell culture model resulted in the decreased expression of hepatocyte nuclear factor 4 alpha (HNF4A), a major liver-specific transcription factor. The stress-induced inhibition of HNF4A expression resulted in a significant reduction of liver-specific *microRNA-122* (*miR-122*) transcription. It was found that the reversal of hepatic adaptive pro-survival signaling and restoration of *miR-122* level was more efficient by interferon (IFN)-based antiviral treatment than direct-acting antivirals (DAAs). To test whether *miR-122* levels could be utilized as a biomarker of hepatic adaptive stress response in HCV infection, serum *miR-122* level was measured among healthy controls, and chronic HCV patients with or without cirrhosis. Our data show that serum *miR-122* expression level remained undetectable in most of the patients with cirrhosis (stage IV fibrosis), suggesting that the pro-survival UPR signaling increases the risk of HCC through STAT3-mediated suppression of *miR-122*. In conclusion, our data indicate that hepatic pro-survival UPR signaling suppresses the liver-specific HNF4A and its downstream target *miR-122* in cirrhosis. These results provide an explanation as to why cirrhosis is a risk factor for the development of HCC in chronic HCV infection.

Keywords: hepatitis C virus (HCV); cirrhosis; hepatocellular carcinoma (HCC); endoplasmic reticulum (ER) stress; oxidative stress (OS); unfolded protein response (UPR); *microRNA-122* (*miR-122*); nuclear factor erythroid 2-related factor 2 (NRF2); signal transducer and activator of transcription 3 (STAT3); hepatocyte nuclear factor 4 alpha (HNF4A)

1. Introduction

Hepatocellular Carcinoma (HCC) is the most common form of liver cancer. The incidence of HCC has nearly doubled over the past decade, making it the third-leading cause of cancer-related death worldwide [1]. In most cases, HCC develops on the background of cirrhosis that is mainly associated with viral hepatitis, overconsumption of alcohol, and non-alcoholic fatty liver disease. In the USA, chronic hepatitis C virus (HCV) infection is the leading cause of HCC and liver transplantation [2]. Globally, 30% of all HCC cases are related with persistent HCV infection representing a significant public health problem [3]. Recently approved direct-acting antivirals (DAAs) enable to cure the majority of patients with chronic HCV infection, including patients with cirrhosis [4]. HCV treatment reduces liver inflammation, fibrosis, and HCC thus drops liver-disease related mortality in the future [5,6]. Although the causal relationship between HCV infection and chronic liver disease has been well documented, the minute mechanisms of the virus-cell interactions involved in disease progression are not well understood. Since there is no effective treatment option of cirrhosis and late-stage HCC, it is hoped that understanding the mechanisms of HCV-induced liver cancer will allow developing better treatment. This knowledge will also result in a better understanding of the pathogenesis of other agents such as hepatitis B virus (HBV) and non-viral agents leading to cirrhosis and HCC.

The mechanisms of how chronic HCV infections are associated with the development of cirrhosis and HCC have been a challenge to study due to lack of an appropriate small animal model [7]. In a previous publication, it was shown that hepatocytes develop an integrated stress response (ISR) due to the combination of endoplasmic reticulum (ER) and oxidative stress occurring with persistent HCV replication in the cell culture model [8]. The increased cellular stress response activates nuclear factor erythroid 2-related factor 2 (NRF2) signaling as a primary cell survival pathway by inducing the transcription of genes with the antioxidant response [9]. Previously published data have shown that the stress-induced NRF2 signaling induces expression of LAMP2A and HSC70 leading to induction of chaperone-mediated autophagy (CMA). The CMA induction in HCV-infected culture improves cell survival through degradation of major tumor suppressors: p53 and retinoblastoma protein (pRB) [10–12]. A recent publication from this laboratory has shown that HCV-mediated NRF2 signaling impairs autophagy process at the level of autophagosome-lysosome fusion leading to activation of the oncogenic signaling through the tyrosine kinase receptors (EGFR) [13]. These data support that the activation of NRF2 signaling contributes to cell survival and progression of preneoplastic lesion to HCC malignancies [14]. Signal transducer and activator of transcription 3 (STAT3) pathway is also activated as a major cell survival mechanism in HCC but not in the surrounding non-tumor tissue or normal liver [15,16]. The mechanisms of STAT3 activation could be due to an elevated expression of cytokines such as interleukin 6 (IL6), interleukin 22 (IL22), oxidative stress or epigenetic regulation [17,18]. However, studies have shown that STAT3 activation is transient in non-transformed cells even in the presence of cytokines and STAT3 activating mutations are rarely present in HCC [17,18]. The HCC-specific mechanism of STAT3 activation has not been addressed thoroughly.

Due to these reasons, we investigated the contribution of NRF2-related pro-survival mechanism through STAT3 activation in HCV culture in response to oxidative stress. We investigated STAT3-mediated pro-survival mechanism via a circuit that involves downregulation of hepatocyte nuclear factor 4 alpha (HNF4A) through *miR-24* and *miR-619* [19,20]. STAT3-inducible *miR-24* and *miR-629* destabilize the *HNF4A* mRNA leading to the permanent suppression of protein expression. HNF4A is a principal transcription factor required for liver development, hepatocyte differentiation, and hepatic function [21]. Many recent publications claim that the expression level of HNF4A and its target genes are impaired in cirrhosis and diminished in HCC [22–27]. HNF4A has been shown to play a role in HCC development related to chronic inflammation processes of the liver through regulating the transcription of *miR-122* [28–30]. The importance of this circuit in HCC development is further supported by previous reports showing that *miR-122* knockout mice develop HCC [31–33]. Based on these evidences, we hypothesized that the NRF2-mediated activation of the STAT3-HNF4A inflammatory loop could lead to a long-term suppression of *miR-122* that increases the HCC risk among patients with cirrhosis.

Using a persistently HCV-infected Huh-7.5 liver cell culture model, it was found that the hepatic adaptive response through the protein kinase RNA-like ER kinase (PERK)-NRF2 axis activates the STAT3-HNF4A inflammatory loop as a cell survival mechanism. Data presented in this manuscript suggest that the NRF2-mediated STAT3 activation silence the expression of HNF4A through *miR-24* upregulation. The activation of the STAT3-HNF4A loop leads to suppression of *miR-122* in persistently infected HCV cell culture and chronic HCV patients with cirrhosis. Finally, the data in the study demonstrate that the serum *miR-122* levels are depleted during advanced liver disease such as cirrhosis, which may explain why HCC develops most of the time on the background of chronic liver disease with cirrhosis.

2. Results

2.1. Persistent HCV Replication Leads to ER Stress and Oxidative Stress

To understand the hepatic adaptive response to ER stress, highly permissive Huh-7.5 cells were infected with JFH-AM120 at a multiplicity of infection (MOI) of 0.1, and HCV replication was studied over 21 days with a regular passage of infected cells at three-day intervals. Infected cells collected at different time points were examined for HCV protein expression by Western blotting. The efficiency of replication and spread of HCV-green fluorescence protein (GFP) chimera virus were examined by fluorescence microscopy with nuclear DAPI staining. Quantifications of GFP positive cells by ImageJ software show that the level of HCV replication increased with time. The percentage of Huh-7.5 cells expressing HCV-GFP fusion was quantified by flow cytometry analysis at 0, 3, 6, 9, and 12 days. These results indicated that about 5.9% of cells were GFP positive on day 3, and the number increased to 85.4% and 93.5% on day 9 and day 12, respectively. All these results support our previously published data showing that the high-level replication and spread of JFH-AM120 chimera HCV clone in Huh-7.5 liver cells (Figure S1). When the accumulation of viral protein exceeds the folding capacity of the ER resident chaperones, it creates a stress response called UPR. The activation of UPR secondary to ER stress response is the major driver of liver disease progression in HCV infection. Sustained HCV infection in Huh-7.5 liver cells is expected to accumulate unfolded protein load and expansion of the ER membranes [34]. We measured misfolded protein stress in the ER membranes by live cell imaging using a transient transfection-based approach with a commercially available ER-red fluorescence protein (RFP) construct (ER-RFP, BacMam 2.0, ThermoFisher, Waltham, MA, USA). The staining was intense in the rough ER around the nucleus of infected cells and the RFP signal overlapped with HCV-GFP. The ER staining of the uninfected cells was diffuse, whereas it was more intense in HCV-infected cells, suggesting the evidence of ER membrane expansion and accumulation of ER-RFP protein in the cells replicating HCV. Interestingly, HCV infection induced an extensive colocalization of ER-Tracker RFP with NS5A-GFP. The yellow fluorescence signal due to colocalization was markedly increased after overlaying the images of NS5A-GFP and ER-tracker RFP fluorescence. Quantitative assessment of the colocalization of HCV-GFP with ER-tracker RFP was significantly higher in HCV-infected cells than uninfected Huh-7.5 cells. These data confirm that HCV-GFP fusion protein is colocalized with the RFP protein in the perinuclear region of the infected cells that is consistent with increased GRP78/94 protein localization in the ER [35–37]. These results are also consistent with previous reports confirming the activation of the stress sensor of the UPR by Western blot analysis [13]. During HCV replication, several reactive oxidants such as reactive oxygen species (ROS) or reactive nitrogen species (RNS) are generated as a by-product of biochemical reactions in mitochondria, peroxisomes, and ER membranes leading to oxidative stress. A fluorescent-based assay was used to quantify ROS levels between uninfected and persistently infected cultures with HCV-Renilla luciferase virus (JFH1-V3-Rluc) on day 9. In this assay, H2DCFDA is converted to the highly fluorescent 2′7′-dichlorofluorescein due to the presence of ROS. The fluorescence intensity is proportional to the amount of oxidant present in HCV-infected cells. A flow cytometric analysis based on the quantitative approach displays that the majority of cells infected with HCV show punctate fluorescence staining due to the presence of high oxidative stress response. The oxidative stress was significantly higher in HCV-infected culture than uninfected Huh-7.5 cells by three repeated analyses. Results shown in Figure S2 confirm previously published data indicating that HCV infection also induces ER stress and oxidative stress [38–41].

2.2. Oxidative Stress and ER Stress Activate NRF2 Signaling in HCV Infection

A transcription factor, NRF2, plays a critical role in the control of genes involved in cell survival pathway. It induces varieties of cytoprotective genes harboring a short cis-acting sequence called the antioxidant response element (ARE) in their promoters [42,43]. First, we demonstrated that robust expression of the HCV core protein in the infected culture could be detected on day 3 and the number of positive cells was increased from day 6 and over 90% cells became core positive on day 12. The intensity of cytoplasmic core staining was quantified by ImageJ software (NIH, Bethesda, MD, USA) and found to increase over time, indicating active viral replication and spread after HCV infection. The activation of NRF2 signaling was examined by measuring the nuclear translocation in persistently infected culture in a kinetic study by immunocytochemical staining. NRF2 nuclear accumulation was found in 100% of HCV-infected culture starting from day 9 that correlated well with HCV core expression (Figure S3). These results indicate that HCV infection results in NRF2 activation and nuclear translocation.

2.3. Persistent HCV Infection Activates the NRF2-STAT3-HNF4A Regulatory Axis as a Cytoprotective Response

Increased oxidative stress has been reported in liver disease caused by viral hepatitis and non-viral etiologies [43]. The NRF2 has a crucial role in enabling adaptation to oxidative stress and ER stress by transcribing more than 2000 cytoprotective genes involved in cell survival [44,45]. The induction of NRF2 genes requires a common NRF2-binding motif on the promoter called ARE [46,47]. Five ARE (TGnnnnGC) and six ARE-like (TGAnnnnGC or TGAnnnnnGC) binding sites were identified in STAT3 promoter region (Figure 1A). Western blot analysis using phosphorylated and unphosphorylated specific antibodies revealed that HCV infection induced NRF2 and STAT3 activation (Figure 1B). The expression of STAT3 was found to be increased more in HCV-infected culture as compared to uninfected Huh-7.5 cells. The expression level of β-Actin did not change, suggesting that the differences were not at the level of protein loading or variation in the protein content of the lysates used. The inflammatory circuit consisting of HNF4A, *miR-122*, IL6, STAT3, and *miR-24* is implicated in hepatocellular transformation and liver oncogenesis [19]. The first component of the circuit is the STAT3-mediated suppression of HNF4A through *miR-24*. The second component is HNF4A-mediated suppression of *miR-122* transcription. Western blot analysis was performed to verify whether HCV-induced STAT3 also modulates expression of HNF4A pathway. Results shown in Figure 1B demonstrate that HCV infection increased NRF2, and STAT3 expression but suppressed HNF4A expression. Western blot data were quantified using ImageJ software to compare relative expression of NRF2, STAT3 and HNF4A (Figure S4). To determine the statistical correlation coefficient between the expression levels of STAT3 and HNF4A, the fraction of variance denoted by R^2 values was calculated (Figure 1C,D). Analyses of the R^2 values are very close to 1.0, suggesting that the induction of STAT3 correctly predicts the decrease of HNF4A in HCV-infected cells. Immunolocalization of STAT3 and HNF4A was performed in uninfected and infected cell culture on day 9 by confocal microscopy. This analysis showed that STAT3 activation in HCV-infected culture was associated with negative HNF4A expression in the nucleus (Figure 1E,F). These results indicate an inverse relationship between STAT3 activation and the expression of liver-specific transcription factor, HNF4A. Real-time RT-PCR data showed that HCV infection induced *NRF2* and *STAT3* mRNA levels, whereas *HNF4A* mRNA levels were decreased (Figure 1G–I), suggesting that activation of NRF2 signaling decreases the HNF4A protein level by altering the stability or reduced transcription of HNF4A. These results are consistent with the previous report suggesting that HCV infection inhibits *HNF4A* expression by reducing its mRNA levels [48]. To test the consistency of this observation, results were verified using an HCV infection model of primary human hepatocytes (PHHs). It was found that the expression of total NRF2 and phosphorylated NRF2 (pNRF2) were induced in HCV-infected PHHs by Western blot analysis measured for 12 days (Figure 2A). It was observed that the expression of total STAT3 and phosphorylated STAT3 (pSTAT3) were increased in a time-dependent manner in the infected PHHs culture as compared to uninfected PHHs, indicating that HCV replication induces STAT3 pathway. The expression level of HNF4A was decreased after HCV infection and inversely correlated with expression of HCV NS3 by Western blot analysis. All the Western blot bands were quantified using ImageJ software to compare the relative expression of NRF2, STAT3 and HNF4A (Figure S5). Total RNA isolated from infected PHHs was used to quantify the mRNA levels of STAT3, NRF2, HNF4A and internal control β-Actin by real-time RT-PCR. First, Ct values of individual gene were normalized with *β-Actin*mRNA and then, the fold change was determined as compared to uninfected PHHs. The mRNA levels of NRF2 and STAT3 were found to be increased in a time-dependent manner in HCV-infected PHHs (Figure 2B,C), whereas *HNF4A* mRNA levels were decreased over time post infection (Figure 2D).

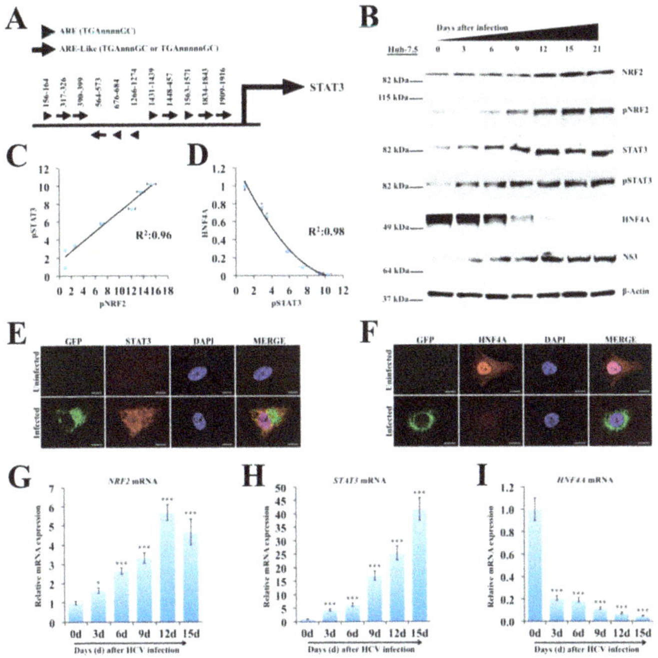

Figure 1. Persistent hepatitis C virus (HCV) replication induces nuclear factor erythroid 2-related factor 2 (NRF2)-mediated regulation of the signal transducer and activator of transcription 3 (STAT3)-hepatocyte nuclear factor 4 alpha (HNF4A) inflammatory loop. (**A**) STAT3-promoter region was examined for the presence of antioxidant response element (ARE) consensus sequences indicated by arrowheads, and ARE-like sequences shown by arrows. (**B**) Western blot analysis shows expression levels of total and phosphorylated NRF2, and STAT3 in infected Huh-7.5 liver cells over 21 days. The expression of HNF4A was inversely correlated with HCV NS3 protein expression. (**C**) Band intensities of phosphorylated STAT3 (pSTAT3) and phosphorylated NRF2 (pNRF2) were quantified using ImageJ software and R^2 values were determined by excel software. (**D**) Western blot results of HNF4A and pSTAT3 were quantified using ImageJ software and R^2 values were determined by excel software. (**E**) Colocalization of HCV-green fluorescence protein (GFP) with STAT3 by confocal microscopy. (**F**) Colocalization studies by confocal microscopy between HCV-GFP and nuclear expression of HNF4A in infected Huh-7.5 cells on day 9. Messenger RNA (mRNA) levels of individual genes in HCV-infected Huh-7.5 culture were measured by real-time RT-PCR analysis. (**G**) *NRF2* mRNA levels (**H**) *STAT3* mRNA levels. (**I**) *HNF4A* mRNA levels. The results are expressed as the mean ± standard deviation (SD) of three experiments. Error bars represent SD. *p*-values were calculated by ANOVA as compared to uninfected control. * *p*-value < 0.05, *** *p*-value < 0.001. Original magnification ×60. Scale bars = 10 μm.

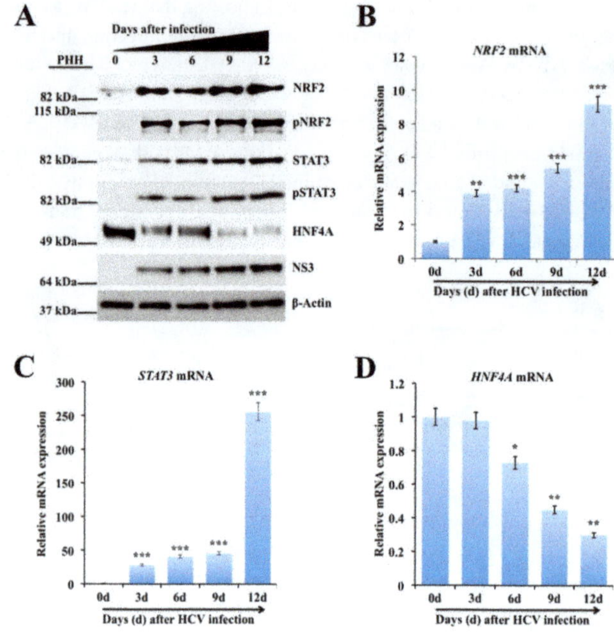

Figure 2. Hepatic stress response due to hepatitis C virus (HCV) replication in non-proliferative primary human hepatocytes (PHHs) activates nuclear factor erythroid 2-related factor 2 (NRF2)-mediated signal transducer and activator of transcription 3 (STAT3)-hepatocyte nuclear factor 4 alpha (HNF4A) expression. (**A**) Cell lysate was examined for the presence of HCV NS3 protein and for the activation of NRF2, STAT3 and HNF4A protein expression in infected PHH by Western blot. Messenger RNA (mRNA) levels of individual genes in HCV-infected PHHs were measured by real-time RT-PCR analysis. (**B**) Shown are *NRF2* mRNA levels (**C**) Shown are *STAT3* mRNA levels. (**D**) Shown are *HNF4A* mRNA levels. The results are expressed as the mean ± standard deviation (SD) of three experiments. Error bars represent SD. p-values were calculated by ANOVA as compared to day 0. * p-value < 0.05, ** p-value < 0.01, *** p-value < 0.001.

2.4. Silencing PERK and NRF2 Pathway Restore Expression of STAT3-HNF4A Inflammatory Loop

Persistent HCV replication induces sustained nuclear translocation of NRF2 as a cytoprotective mechanism in response to ER stress and oxidative stress. The impact of silencing NRF2 by a small interfering RNA (siRNA) treatment on the expression of STAT3 and HNF4A was examined in a persistently infected HCV cell culture model. The siRNA treatment was performed using day 9 infected HCV-GFP culture. As shown in Figure 3A, knockdown of NRF2 by siRNA decreased STAT3 expression and restored HNF4A expression indicating that the increased NRF2-mediated STAT3 expression is inversely associated with HNF4A downregulation. Uninfected Huh-7.5 cells treated with the NRF2 activator (sulforaphane) shows increased expression of phosphorylated NRF2 and STAT3 but suppressed HNF4A expression (Figure 3B). These data suggest that NRF2 mediated the regulation of the STAT3-HNF4A inflammatory loop in HCV infection. The relationship of NRF2 activation and ER stress was determined in infected Huh-7.5 cells after treatment with the ER stress inhibitor. As shown in Figure 3C, tauroursodeoxycholic acid (TUDCA) treatment decreased NRF2 expression, and STAT3 activation but restored HNF4A expression. To test the role of the PERK pathway on regulating the STAT3-HNF4A axis, the expression levels of STAT3 and HNF4A were measured after treatment with the PERK inhibitor by Western blotting. As shown in Figure 3D, inhibiting PERK blocked STAT3 induction and restored HNF4A expression. It is well known that STAT3 is phosphorylated by Janus kinases (JAK) leading to homo- or heterodimers, and nuclear translocation for transcriptional regulation of microRNAs and numerous genes involved in cell survival pathway. We examined whether treatment with the JAK inhibitor could also prevent STAT3 phosphorylation and restore HNF4A expression in HCV-infected culture. Day 9 infected Huh-7.5 cell culture was treated with increasing concentrations of the JAK inhibitor (pyridone 6, Calbiochem, San Diego, CA, USA) and after 72 hours, the expression of STAT3, pSTAT3 and HNF4A levels was examined by Western blotting (Figure 3E). Treatment with the JAK inhibitor did not block HCV replication but prevented STAT3 phosphorylation and restored expression of HNF4A. Furthermore, we found that infected Huh-7.5 cells treated with a highly specific STAT3 inhibitor (S3I-201, Selleck Chemicals, Houston, TX, USA) restored the expression level of HNF4A without altering level of HCV core expression (Figure 3F). Over expression of STAT3 by transient transfection downregulated the expression of HNF4A (Figure 3G). Huh-7.5 cells transfected with a control plasmid did not alter expression of HNF4A, suggesting specific regulation by STAT3 expression. The relative expression of NRF2, STAT3 and HNF4A under different siRNA or chemical treatments was compared by quantifying the Western blot results by ImageJ software (Figure S6).

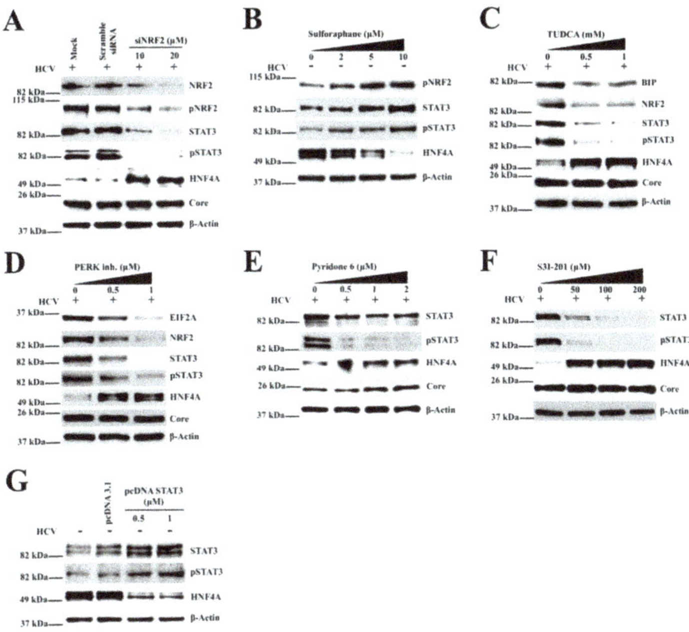

Figure 3. Shown is the nuclear factor erythroid 2-related factor 2 (NRF2)-dependent regulation of signal transducer and activator of transcription 3 (STAT3)-hepatocyte nuclear factor 4 alpha (HNF4A) expression in hepatitis C infection (HCV) infection. (**A**) The effect of NRF2 silencing on the expression of STAT3-HNF4A in HCV culture. Infected Huh-7.5 cells on day 9 were treated with combination of two small interfering RNAs (siRNAs) targeted to NRF2 and scrambled siRNA and the expression of STAT3 and HNF4A was measured by Western blot analysis after 72 hours. The effect of siRNA transfection did not alter HCV core expression or expression of β-Actin levels. (**B**) Shown is the effect of treatment with the NRF2 activator, sulforaphane, on the expression of STAT3 and HNF4A in uninfected Huh-7.5 liver cells. (**C**) Western blot analysis shows the effect of the endoplasmic reticulum (ER) stress inhibitor (TUDCA) on the expression levels of STAT3 and HNF4A in HCV-infected culture. (**D**) Shown are the expression levels of STAT3-HNF4A in HCV culture treated with the protein kinase RNA-like ER kinase (PERK) inhibitor by Western blot analysis. (**E**) Shown is the effect of the Janus kinase (JAK) inhibitor on the phosphorylation of STAT3 and HNF4A regulation. (**F**) Shown is the effect of the STAT3 inhibitor on the expression levels of HNF4A in infected cells by Western blot analysis. (**G**) The effect of STAT3 over expression by plasmid transfection on the levels of HNF4A in uninfected Huh-7.5 cells.

2.5. Persistent HCV Replication in Huh-7.5 Cells Leads to STAT3-HNF4A-Mediated Silencing of miR-122

STAT3-inducible *miR-24* controls the stability of *HNF4A* mRNA through binding to the 3′untranslated region (3′UTR). This regulation is linked to HNF4A-mediated induction of liver-specific *miR-122* transcription [49]. We tested the hypothesis whether NRF2-mediated cellular adaptive response to HCV infection regulates the STAT3-HNF4A inflammatory loop through modulating the expression of *miR-122* and *miR-24* (Figure 4A). We used a RT-PCR assay to measure their expression in persistently infected culture over 21 days. As expected, the *miR-24* expression progressively increased whereas the copy number of *miR-122* decreased in HCV culture over time (Figure 4B). We found that HCV-induced stress response decreased the expression of *miR-122* in a time-dependent manner. The appearance of control unrelated *miR-16* did not change under similar assay conditions. The impact of HCV replication on the regulation of microRNA loop was also verified in infected PHHs model. It was observed that *miR-122* level decreased in HCV-infected PHHs whereas *miR-24* level increased (Figure 4C). The level of *miR-16* did not change due to HCV infection in PHHs. We conducted as a next step a series of antiviral treatment using HCV-infected Huh-7.5 cells on day 9 to see whether HCV clearance could reverse the expression of *miR-122*. As shown in Figure 4D, IFN alpha (IFNα), sofosbuvir, ledipasvir and a combination of sofosbuvir and ledipasvir effectively cleared HCV-GFP expression without affecting viability of infected Huh-7.5 cells. Furthermore, it was found that inducing HCV clearance by treatment with either IFNα or DAAs (sofosbuvir, and ledipasvir) inhibited STAT3 protein levels and restored HNF4A expression (Figure 4E). The relative expression of STAT3, HNF4A and HCV core was determined quantifying the Western blot results by ImageJ software (Figure S7). The cross-regulatory effect of STAT3 and HNF4A expression on *miR-24* and *miR-122* levels was also verified after HCV clearance with different antivirals. It was found that restoration of *miR-122* level was more efficient by IFN-based antiviral treatment than by DAAs (Figure 4F,G). These results suggest that cellular adaptive response to HCV-induced stress activated the STAT3-HNF4A inflammatory loop that leads to decreased expression of *miR-122*.

It was also of interest to determine if the treatment of HCV-infected cells with small molecule inhibitors of PERK and ER stress could show a differential effect on the expression of microRNA that controls the expression of STAT3 and HNF4A. Infected Huh-7.5 cells on day 9 was treated with a PERK inhibitor or TUDCA for 72 hours and then the expression levels of *miR-24* and *miR-122* were examined by real-time RT-PCR. Data shown in Figure 5A indicate that this treatment did not inhibit HCV replication since no difference of HCV-GFP expression between untreated and treated groups was seen. The expression level of *miR-24* was higher in HCV-infected untreated culture due to STAT3 activation (Figure 5B). As expected, persistently infected HCV culture treated with the PERK inhibitor and TUDCA inhibited *miR-24* expression but restored expression of *miR-122* in HCV culture (Figure 5B,C). Taken together, these results suggested that HCV-induced PERK activation promoted the activation of the STAT3-HNF4A inflammatory loop to overcome the stress response associated with persistent HCV replication.

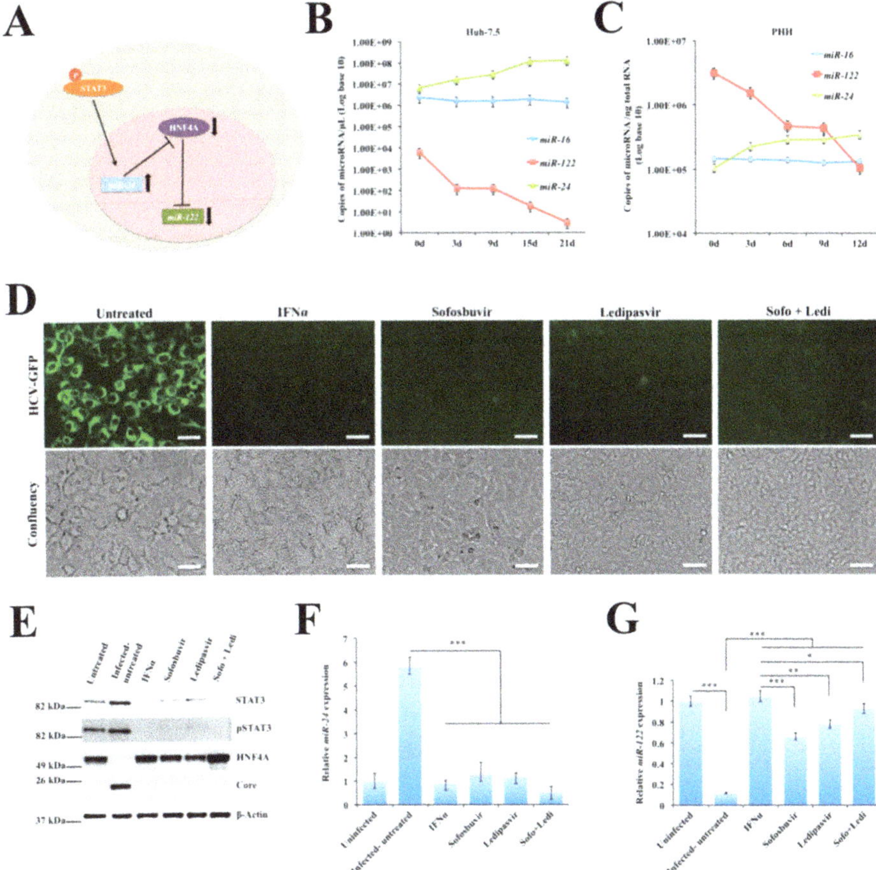

Figure 4. Relationship between signal transducer and activator of transcription 3 (STAT3)-hepatocyte nuclear factor 4 alpha (HNF4A) and the expression *miR-122* and *miR-24* in hepatitis C virus (HCV)-infected Huh-7.5 cells and primary human hepatocytes (PHHs). (**A**) Shown is the proposed hypothesis relating stress and STAT3-*miR-24*-HNF4A-*miR-122* feedback circuit in HCV infection. (**B**) Quantification of *miR-16*, *miR-24* and *miR-122* levels in persistently HCV-infected Huh-7.5 cells by real-time RT-PCR analysis over 21 days. (**C**) Quantification of *miR-122*, *miR-24* and *miR-122* levels in HCV-infected PHHs over 12 days. (**D**) Shown is the HCV-green fluorescence protein (GFP) expression in the infected culture with or without antiviral treatment. (**E**) Western blot shows the expression levels of STAT3 and HNF4A in the infected culture on day 9 with or without antiviral treatment. (**F**) The expression levels of *miR-24* in HCV-infected culture before and after treatment with interferon alpha (IFNα) or direct-acting antivirals (DAAs). (**G**) The expression levels of *miR-122* in HCV-infected culture treated with IFNα or DAA. The results are expressed as the mean ± standard deviation (SD) of three experiments. Error bars represent SD. *p*-values were calculated by ANOVA between different groups as compared to untreated HCV-infected group. * *p*-value < 0.05, ** *p*-value < 0.01, *** *p*-value < 0.001. Original magnification ×40, scale bars = 25 μm.

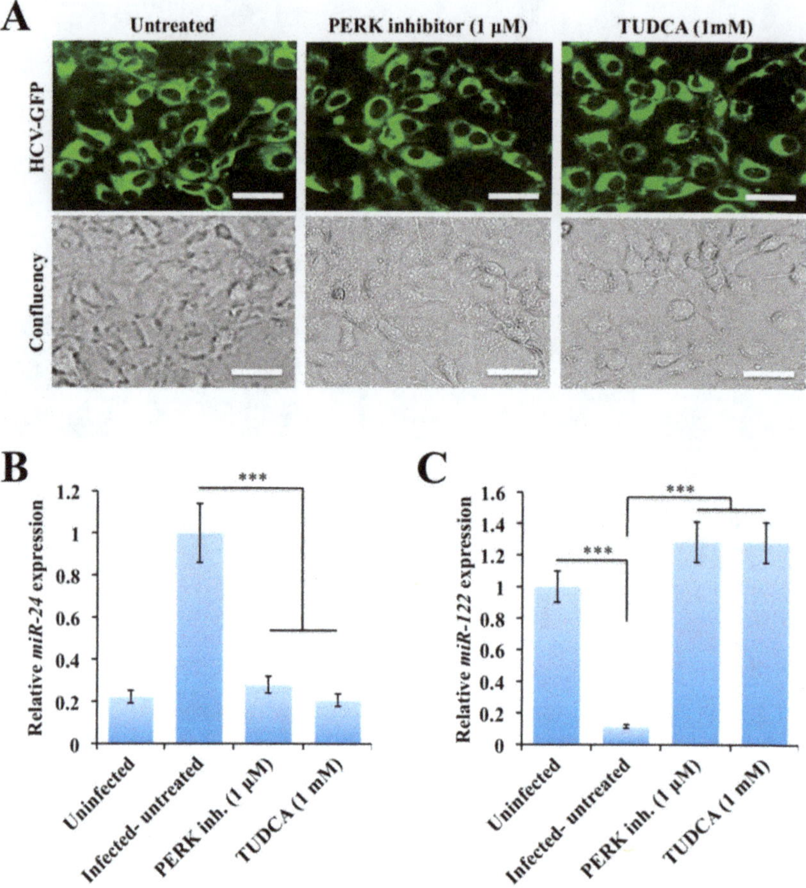

Figure 5. The expression of *miR-24* and *miR-122* in hepatitis C virus (HCV)-infected Huh-7.5 cells in the presence of the protein kinase RNA-like ER kinase (PERK) or endoplasmic reticulum (ER) stress inhibitors. Infected Huh-7.5 cells were treated with the PERK inhibitor or ER stress inhibitors for 72 hours. Cells were collected and the expression levels of *miR-24* and *miR-122* were measured by real-time RT-PCR. (**A**) Shown are expression levels of HCV-green fluorescence protein (GFP) chimera in day 9 HCV culture by fluorescence microscopy before and after treatment. (**B**) Shown are amounts of *miR-24* before and after treatment with the PERK inhibitor or tauroursodeoxycholic (TUDCA) in HCV-infected culture. (**C**) The levels of *miR-122* in uninfected and infected Huh-7.5 cells treated with either with the PERK inhibitor or TUDCA. The results are expressed as the mean ± standard deviation (SD) of three experiments. Error bars represent SD. *p*-values were calculated by ANOVA between different groups as compared to untreated HCV-infected group. *** *p*-value < 0.001. Original magnification ×40. Scale bars = 25 μm.

2.6. Persistent HCV Replication in Huh-7.5 Cells Leads to Decreased miR-122 Promoter Activity

An in vitro assessment was done to study the impact of the HCV-induced stress response on *miR-122* promoter activity in the presence and absence of a stress inducer and inhibitors. A firefly luciferase reporter construct p-(5.7/3.8k) that contained the 3726 to 5645 bp region (chr18: 54263641-54265560) in the pGL3-basic vector (Figure 6A) was used to measure the impact of HCV replication on *miR-122* promoter activity [50]. First, we measured the impact of HCV replication on the promoter activity in a transient transfection assay. The infected culture on day 9 was transfected with microRNA-promoter plasmid for 24 hours, and then firefly luciferase activity was measured. The firefly activity was normalized with protein concentration. It was found that HCV infection significantly suppressed *miR-122* promoter activity in Huh-7.5 cells. The promoter activity in the HCV culture was enhanced in the presence of the STAT3 inhibitor and JAK inhibitor. The PERK inhibitor as well as the ER stress inhibitor, TUDCA, also increased *miR-122* promoter activity (Figure 6B). Second, we tested whether uninfected Huh-7.5 cells treated with the ER stress or the NRF2 activator could modulate *miR-122* promoter activity. As shown in Figure 6C, inducing ER stress in uninfected Huh-7.5 cells by thapsigargin (TG) treatment suppressed the *miR-122* promoter activity significantly. The treatment with sulforaphane also suppressed the *miR-122* promoter activity in a concentration-dependent manner (Figure 6D). All these results support the conclusion that the ER stress and the NRF2 activators actively suppressed the *miR-122* promoter activity, but ER stress inhibitors restored the HCV-induced suppression of the promoter activity.

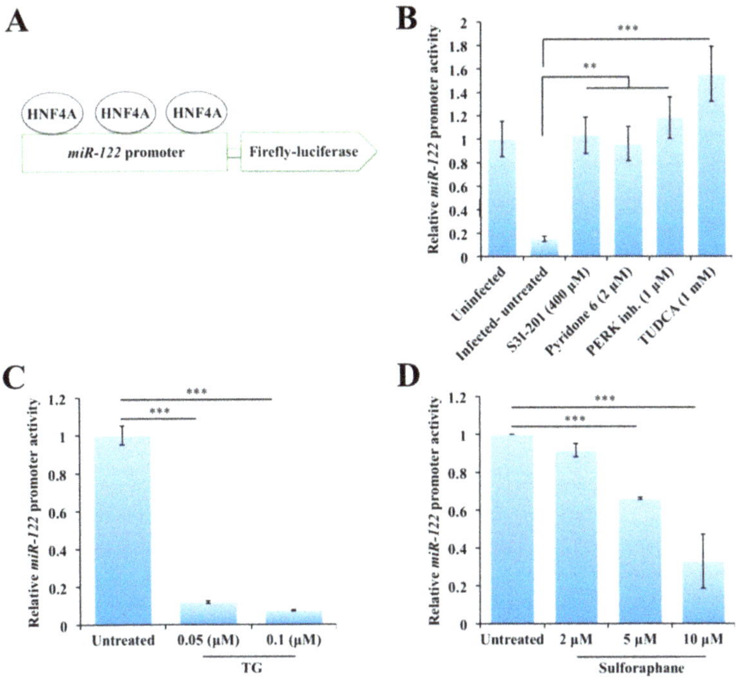

Figure 6. The impact of endoplasmic reticulum (ER) stress on the *miR-122* promoter activity. (**A**) There are three hepatocyte nuclear factor 4 alpha (HNF4A) binding sites present in *miR-122* promoter, which explain why HNF4A downregulation decreases *miR-122* under stress. (**B**) The effect of small molecule inhibitors targeted to signal transducer and activator of transcription 3 (STAT3), Janus kinase 1 (JAK1), protein kinase RNA-like ER kinase (PERK) and ER stress on the *miR-122* promoter regulation. Uninfected and hepatitis C virus (HCV)-infected Huh-7.5 cells were transfected with one microgram of *miR-122* promoter-construct with firefly luciferase. Following the transfection step, cells were treated with or without thapsigargin (TG), sulforaphane, and STAT3 inhibitors. After 24 hours, the luciferase activity was measured. Luciferase assays were performed three times. (**C**) ER stress activator regulation of *miR-122* promoter. (**D**) The impact of a nuclear factor erythroid 2-related factor 2 (NRF2) activator, sulforaphane, on *miR-122* promoter activity. The results are expressed as the mean ± standard deviation (SD) of three experiments. Error bars represent SD. *p*-values were calculated by ANOVA between different groups as compared to untreated HCV-infected group. ** *p*-value < 0.01, *** *p*-value < 0.001.

2.7. Decreased Expression of miR-122 Correlates with Patients with Cirrhosis

MiR-122 is the most abundant liver-specific microRNA that accounts for about 70% of the total microRNA population in the adult liver [49]. Based on the fact that the cellular stress adaptation to HCV infection leads to decreased expression of *miR-122* through STAT3-HNF4A, suggesting this liver-specific microRNA may be used as a biomarker for assessing the stress response among patients with chronic HCV infection. We next sought to test this hypothesis to see whether the expression levels of *miR-122* would be different among chronically infected patient with or without cirrhosis. A retrospective analysis of serum *miR-122* levels of patients with chronic hepatitis C patients with or without cirrhosis was performed. The analysis measured the levels of *miR-122* in sera from 18 healthy, 18 chronic HCV without cirrhosis and 18 chronic HCV patients with cirrhosis samples. Total RNA was isolated from 200 µL of serum samples, and *miR-122* levels were quantified by real-time reverse transcription. Tables 1 and 2 summarize the characteristics of the subjects included in this study. The viral titer was available in all patients with chronic hepatitis C. All cirrhotic patients were positive for HCV but viral titer was available for only 6 individuals (Table 2). It was found that the copy number of *miR-122* was higher in chronic HCV infection as compared to healthy control ($p < 0.01$). Interestingly, the *miR-122* levels were almost undetectable among most of the patients with stage IV fibrosis (Figure 7A). To rule out the possibility that decreased *miR-122* expression relates to the poor hepatic biosynthetic capability of cirrhotic liver, we measured expression of two additional microRNAs. A liver-specific *miR-221* level was measured in the same serum samples by real-time RT-PCR. These data showed that *miR-221* levels were detectable adequately (Figure 7B). The levels of STAT3-induced *miR-24* level were found to be increased in serum samples of patients with cirrhosis as compared to healthy individuals (Figure 7C). There was no difference in the serum *miR-16* level between the healthy control and chronic HCV infection without cirrhosis. The copy number of *miR-16* was comparable between normal, chronic HCV infection with or without cirrhosis (Figure 7D). There was no correlation between the age and serum *miR-122*, and *miR-16* levels, suggesting that the differences in the expression level are not related to the mean differences in the age of these patients. Serum *miR-122* levels were consistently found to be decreased since we found a significant difference between chronic HCV infection with or without cirrhosis ($p < 0.001$). Since the microRNA is required for HCV replication, we found that serum *miR-122* level was increased during chronic HCV infection. Difference in the *miR-122* level between normal healthy control and chronic HCV infection was also significant ($p < 0.05$) (Figure 8A). These data from serum testing suggest that *miR-122* expression is decreased in the advanced stage of liver disease. It was of interest to know if serum *miR-122* could be a potential diagnostic marker for cirrhosis through Receiver Operating Characteristic (ROC) plot analysis. The data showed that levels of serum *miR-122* are a potential marker for discriminating cirrhosis patients from chronic HCV and healthy control (Figure 8B). The ROC curve analysis revealed a strong separation between chronic HCV patients with cirrhosis versus without cirrhosis with an area under the curve (AUC) value 1.0, suggesting that *miR-122* is a handy marker for discriminating patients with chronic HCV from the cirrhosis group. The analysis revealed that serum levels of other *miR-16*, *miR-221* and *miR-24* were not reliable to accurately differentiate chronic HCV patients with or without cirrhosis. In summary, these results suggested that hepatic adaptive response to cellular stress associated with chronic HCV infection was stronger among the cirrhotic group that leads to depletion of serum *miR-122* level.

Table 1. Characteristics of the patients with Chronic Hepatitis C.

Patient Number	Age	Sex	HCV RNA IU/mL	HCV Genotype	Metavir Score
1	48	F	2,000,000	1a	3
2	46	F	29,000	1b	0
3	27	F	1,000,000	1a	1.2
4	59	F	553,000	1a	3
5	64	M	2,000,000	1a	1.2
6	34	F	7,000,000	1a	1,2
7	59	F	1,000,000	2	0
8	44	M	170,000	1a	1
9	50	M	3,000,000	1b	0
10	33	F	3,000,000	1b	1.2
11	53	M	904,000	1a	1
12	62	M	7062	1a	0
13	69	F	10,000,000	1b	2
14	47	F	2,000,000	2	0
15	50	F	10,000,000	1a	0
16	48	M	2,000,000	1b	0
17	40	F	200,000	1a	1
18	53	M	815,000	1a	1

HCV: hepatitis C virus; M: male; F: female.

Table 2. Characteristics of the patients with cirrhosis.

Patient Number	Age	Sex	HCV RNA IU/mL	HCV Genotype	Metavir Score
1	57	M	+	1a	4
2	52	M	+	1a	4
3	60	F	+	1a	4
4	64	F	+	N/A	4
5	62	M	2,700,000	1b	4
6	66	M	+	1a	4
7	54	M	+	1a	4
8	54	M	+	1b	4
9	62	F	+	N/A	4
10	58	F	+	N/A	4
11	64	M	+	2	4
12	66	F	+	N/A	4
13	57	F	+	N/A	4
14	56	M	5,940,000	N/A	4
15	60	M	7,200,000	1a	4
16	60	M	2,860,000	1b	4
17	56	M	4,310,000	1a	4
18	60	M	6,480,000	1a	4

HCV: hepatitis C virus; M: male; F: female; N/A: Not Available.

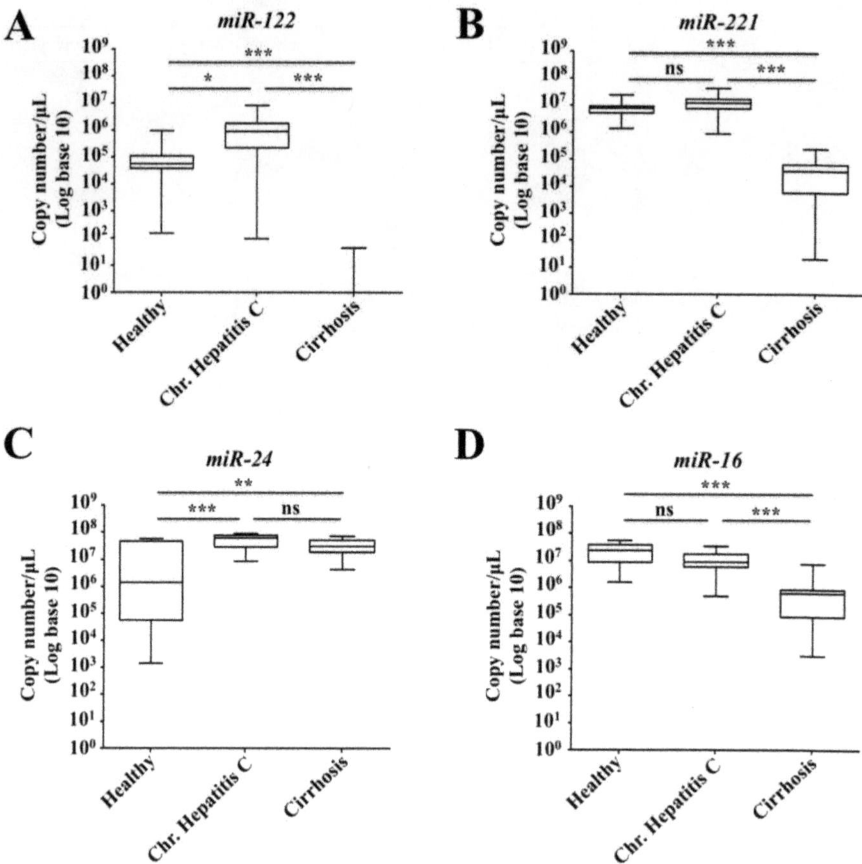

Figure 7. Serum levels of *miR-122*, *miR-221*, *miR-24* and *miR-16* were measured by real-time RT-PCR. (**A**) Shown are the serum *miR-122* levels among healthy individuals, chronic hepatitis C virus (HCV) infection and chronic HCV infection with cirrhosis. (**B**) Serum levels of another control liver-specific *miR-221* using the same set of serum samples. (**C**) Serum levels of signal transducer and activator of transcription 3 (STAT3)-regulated *miR-24* among healthy, chronic HCV infection and chronic HCV with cirrhosis. (**D**) Shown are the serum levels of *miR-16* in the same sets of serum samples. The results are expressed as the box plot with the median bar in triplicates. Whiskers represent minimum and maximum observed values. p-values were calculated by ANOVA between different groups. ns: not significant, * p-value < 0.05, ** p-value < 0.01, *** p-value < 0.001.

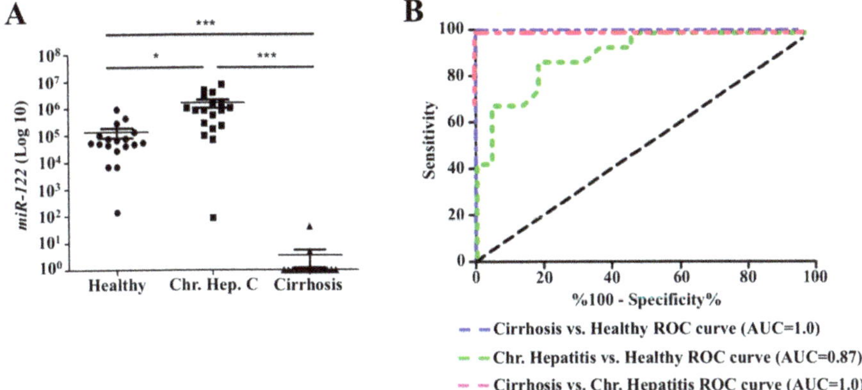

Figure 8. Serum *miR-122* levels discriminating patients with hepatitis C virus (HCV) infection and cirrhosis from healthy controls. (**A**) Copy number variation of serum *miR-122* levels between healthy, chronic HCV and chronic HCV with cirrhosis. (**B**) Receiver Operating Characteristic (ROC) plot. Data shown in panel A was used to generate ROC plot for determining diagnostic value of serum *miR-122* of cirrhosis and chronic HCV patients from healthy controls. The results are expressed as the mean ± standard error of mean (SEM). Error bars represent SEM. *p*-values were calculated by ANOVA between different groups. * *p*- value < 0.05, *** *p*-value < 0.001.

3. Discussion

Chronic liver disease associated with hepatitis virus infection (HBV and HCV) contributes to more than 70% of the HCC worldwide, whereas the remaining cases relate to non-viral etiologies [51]. Since most of HCC cases related to viral and non-viral etiology develop on the background of cirrhosis, suggesting that there may be a common pathway involved in the development of cirrhosis and HCC. However, approximately 20–35% of HCC develop without cirrhosis in the liver, suggesting that inflammation is not an absolute requirement for HCC development [52,53]. The molecular mechanism that drives the progression of cirrhosis and HCC during the chronic stage of liver disease is not understood. Whole genome sequencing found that many driver genes show altered expression in HCC related to chronic HCV infection [51]. Our hypothesis was that the ISR triggered in the infected cells as a compensatory mechanism to deal with varieties of stress responses such as oxidative stress, and ER stress during HCV replication in hepatocytes. The presence of long-lasting unresolved stress response also compromises liver function that leads to cirrhosis and HCC development. In support of this hypothesis, we showed that HCV infection induced ISR in response to ER stress and oxidative stress in cirrhosis that lead to decreased expression of type I IFN receptors [54]. Due to this reason, patients with cirrhosis frequently remain as non-responders to IFN/ribavirin-based antiviral therapy. Some of the pro-survival outcomes of the stress response are the inhibition of host protein synthesis due to increased eukaryotic translation initiation factor 2 subunit alpha (EIF2A) phosphorylation, degradation of misfolded proteins by ER-assisted protein degradation (ERAD) and autophagy. We showed that prolonged stress response activated PERK-induced NRF2 signaling as a cell survival mechanism [13]. The NRF2 signaling activates tumor-promoting autophagy (CMA) that degrades tumor suppressors and induces oncogenic signaling implicated in cell survival.

In this study, we found that NRF2 signaling activated transcription of STAT3, a member of STAT protein family, that is known to be induced by IL6 and participates in inflammation, tumorigenesis and autophagy [55]. STAT3 protein gets phosphorylated at tyrosine 705 by JAK2/tyrosine kinase 2 (TYK2), resulting from dissociation from the cytoplasmic tail of the cytokine receptors, homodimerizes and enters to the nucleus to activate gene transcriptions. Increased STAT3 activity is associated with HCC development and poor prognosis [56]. Activated STAT3 can inhibit cellular autophagy for promoting cell survival, supporting our hypothesis that prolonged stress response inhibits autophagy [57]. Other investigators and we have demonstrated that chronic inflammatory responses can activate STAT3 through IFN and IL6 production [22,58]. Although STAT3 activation occurs due to a variety of mechanisms, HCC-specific STAT3 activation mechanisms have not been well established. Our data show that cellular adaptive response to HCV-induced stress activates STAT3-mediated programing through HNF4A that leads to silencing of liver-specific *miR-122*. To support this hypothesis, we demonstrated that the excessive cellular stress associated with HCV infection activated STAT3 transcription through NRF2 signaling. Our data showed that persistent HCV infection promotes STAT3-inducible expression of *miR-24*. The increased *miR-24* transcription correlated with the decreased expression of *HNF4A* mRNA through binding to the 3'UTR in HCV infection. Our data support the data published earlier showing that an oncogenic circuit consisting of HNF4A, *miR-122*, IL6, STAT3, and *miR-24* is involved in the cell survival mechanisms under stress during HCV infection [19]. Cellular adaptive stress response associated with chronic HCV infection activated STAT3 and utilized this oncogenic circuit, which resulted in a decrease in the expression of HNF4A through *miR-24*. Nuclear HNF4A is a crucial transcription factor during embryonic liver development and hepatocyte differentiation [59]. Available literature shows that HNF4A is the major liver-specific transcription factor that modulates the expression of nearly 42% of the genes expressed in hepatocytes, involved in glucose, fatty acid metabolism, synthesis of blood coagulation factor VII, enzymes involved in hepatic detoxification, and hepatic differentiation [60–65]. Our data is supported by studies showing that mRNA and protein expression levels of HNF4A decreased severely patients with cirrhosis. Hemorrhage is a common cause of death in cirrhosis, especially variceal bleeding. Hepatocytes synthesize both clotting factors and endogenous anticoagulants. The levels of these proteins are reduced in cirrhosis. HNF4A induces the production of blood coagulation factor VII. Our study also provides an explanation why HNF4A targeted protein expression is decreased in cirrhosis due to severe hepatic stress response. Studies have shown that decreased expression of HNF4A is also correlate with bleeding disorder due to decreased expression of the blood coagulation factor VII [64,65].

The activation of this STAT3- HNF4A inflammatory loop led to decreased expression of *miR-124* and *miR-122*. These two microRNAs were regulated by HNF4A. This study was extended to see whether this hypothesis could be validated using prospectively collected serum samples of patients chronically infected with HCV. The measurement of serum *miR-122* levels was performed using serum samples from patients with chronic HCV infection with or without cirrhosis. It was observed that *miR-122* levels increased during the chronic stage of HCV infection, the levels significantly decreased in advanced liver disease, particularly with stage IV fibrosis called cirrhosis. Interestingly, we report here that the levels of *miR-122* have remained below the detection limit among all cirrhotic patients tested in our hand. Our study results are in agreement with a prior report showing that the serum *miR-122* level was found to decreased in liver injury in humans with advanced liver disease, including cirrhosis [66–69]. Decreased expression of *miR-122* has also been found in cirrhosis related to non-alcoholic steatohepatitis (NASH) [33]. The decreased production of *miR-122* in the cirrhotic patients was not related to the impaired biosynthetic capability of hepatocytes present in the cirrhotic liver because the expressions of a liver-specific *miR-221* and STAT3-specific *miR-24* were detectable. Our data show that unresolved stress response depletes the expression of *miR-122*, suggesting that the activation of the STAT3-HNF4A inflammatory loop also occurs in the chronic HCV infection in humans. Our findings provide two pieces of new information that may increase our understanding why STAT3 is transcriptionally activated in liver cancer. First, we show that cellular adaptive response to ER stress in HCV infection causes STAT3 activation at the level of mRNA transcription as well as phosphorylation. Second, we found that STAT3 activation is involved in silencing of *miR-122* in the cirrhosis. Recent studies have observed that epigenetic programing is involved in HCC development in cirrhosis after the viral cure [70–72]. We claimed that our results show a novel adaptive cell survival mechanism related to excessive ER stress that involved STAT3-mediated silencing of *miR-122* expression, the major liver-specific microRNA. The early silencing of *miR-122* and HNF4A increases the risk for HCC development in cirrhosis.

The *miR-122* plays a central role in liver development, differentiation, hepatic metabolic function and innate immunity in the liver [66]. However, the studies investigating the regulation of *miR-122* expression during chronic liver diseases related to viral and non-viral etiologies have produced conflicting results. A handful of publications showed that the levels of *miR-122* decreased during advanced liver disease related to hepatitis C, hepatitis B, and non-viral etiologies [73–75]. *MiR-122* expression in the liver is critical for maintaining innate immunity and IFN production. A report by Xu et al. [73] shows that *miR-122* supports innate immunity by removing the negative regulation of STAT3 signaling on IFN expression. They showed that *miR-122* targets three receptor tyrosine kinases that directly modulate STAT3 phosphorylation. Another recent study by Van Renne et al. [74] showed that the tumor suppressor protein, tyrosine phosphatase receptor delta (PTPRD) controls the STAT3 activation in HCV-induced HCC. They found that high expression of PTPRD suppressed STAT3 phosphorylation in healthy liver tissue but low expression of this protein in HCC resulted in the tumor-specific STAT3 activation. These reports are consistent with our results that HCV infection activates STAT3. On the other hand, the liver-specific *miR-122* expression is also decreased during chronic HBV infection. Chen et al. [75] demonstrated that miR-122 binds to the highly conserved region of HBV pre genomic RNA, causing RNA degradation and reduce viral replication. Data presented in this study show that HCV replication decreased the expression of miR-122 in a time-dependent manner. Data from recent publications confirm the downregulated expression of miR-122 during HCV-induced advanced liver disease [69,76,77]. A report by Luna et al. [76] showed that HCV RNA replication occurs independent of the miR-122 interaction. The team showed that the Argonaute (AGO) protein directly binds to the miR-122 binding sites in HCV RNA, specifically sequesters *miR-122*, to repress the liver target genes during chronic HCV infection. Their investigation showed that the *miR-122* expression decreased in the liver tissues of humans and chimpanzees infected with HCV. Another report by Hyrina et al. [77] showed that the plasma level of *miR-122* decreased during chronic HCV infection and were not restored after HCV cure. Their study also showed that serum *miR-122* level was correlated with liver transaminases. These results suggest that *miR-122* could be a liver-specific biomarker after HCV cure. A study by Sarasin-Filipowicz et al. [69] reported decreased expression of *miR-122* in the liver of HCV-infected patients who were resistant to IFN-based therapy. Our data showed that severe ER stress response during high-level HCV replication in cell culture models of Huh-7.5 liver cells and PHHs decreased the expression of *miR-122*. The expression levels of *miR-122* were depleted among patients with cirrhosis, but not with chronic HCV infections. The reason why the *miR-122* level decreased only in cirrhosis, but not during chronic liver disease is unclear. The host factor requirement for HCV replication has not been fully established. It is possible that HCV replication in the cirrhotic liver may depend on factors other than *miR-122*. The potential liver regeneration and increased inflammation may contribute to increased *miR-122* level in the blood during chronic HCV infection. However, data presented in this report do not support the earlier finding showing that *miR-122* is an essential host factor for HCV replication. The *miR-122* binding does not cause HCV RNA degradation instead it stabilizes the viral genome and promotes translation [78]. In summary, the results presented in this article are consistent with many earlier publications suggesting that decreased expression of *miR-122* in cirrhosis may be associated with loss of hepatic function and innate immunity, impaired liver regeneration, as well as increased risk of HCC development.

4. Materials and Methods

4.1. Cell Culture, Antibodies and Chemicals

The highly permissive transformed liver Huh-7.5 cell line was obtained from the laboratory of Charles M. Rice (Rockefeller University, New York). The Huh-7.5 cells were cultured in Dulbecco's Modified Eagle's Medium (DMEM) supplemented with two mM L-glutamine, sodium pyruvate, nonessential amino acids, 100 U/mL penicillin, 100 µg/mL streptomycin, and 10% fetal bovine serum. Huh-7.5 cell culture was infected with either JFH1-GFP chimera HCV or JFH-dV3-Rluc HCV using a protocol developed in our laboratory as previously described [79]. PHHs were obtained from XenoTech LLC (Kansas City, MO, USA) and cultured with hepatocyte culture media supplemented with 10% human serum (Invitrogen, Brown Deer, WI, USA). After 24 hours they were infected with cell culture grown HCV-GFP chimera virus with a MOI of 0.1. After 18 hours of infection, hepatocytes were replaced with fresh hepatocyte culture media (XenoTech, LLC, Kansas City, MO, USA) supplemented with 10% human serum (Invitrogen, Brown Deer, WI, USA). Uninfected or infected PHH were harvested every 3 days and cell pellets were used for RNA isolation and Western blot analysis. The success of HCV replication in the infected PHHs was assessed by Western blot analysis of HCV NS3 protein. Sofosbuvir was purchased from Acme Biosciences, Inc. (Palo Alto, CA, USA), and IFNα was purchased from EMD Merck (Billerica, MA, USA). Ledipasvir and S3I-201 were obtained from Selleck Chemicals (Houston, TX, USA). TG, TUDCA, the PERK inhibitor (GSK 2606414), and sulforaphane were obtained from Sigma-Aldrich (St. Louis, CA, USA). Pyridone 6 was obtained from Calbiochem (San Diego, CA, USA). The antibody to phospho-NRF2 was purchased from Abcam (Cambridge, MA, USA). Antibodies to GRP78 (BIP), EIF2A, STAT3, phospho-STAT3, HNF4A, and β-Actin were obtained from Cell Signaling (Beverly, MA, USA). Antibody to NS3 was purchased from Virogen Inc. (Boston, MA, USA). The Antibody to HCV core was purchased from Thermo Scientific (Waltham, MA, USA). The antibody to total NRF2 was purchased from Santa Cruz Biotechnology (Santa Cruz, CA, USA).

4.2. Quantitative Assessment of Misfolded Protein Burden in the ER

The accumulation of misfolded proteins in the infected culture was quantified using an ER-tracker reagent called CellLight ER-RFP BacMam 2.0 (Thermo Fisher Scientific). This construct expresses Red Fluorescence Protein (RFP) fused to the ER signal sequence of calreticulin and KDEL (ER-retention signal) to quantitate the extent of red fluorescence by fluorescence microscopy and flow cytometry. This reagent allows live multiplexing imaging of ER in the cells expressing HCV-GFP fusion protein. Briefly, HCV-infected Huh-7.5 cells, and uninfected Huh-7.5 cells were incubated with ER-tracker dye overnight at 37 °C, and the next day, cells were observed using a fluorescence microscope. For nuclear staining and imaging, the cells were directly incubated with Hoechst 33342 (Thermo Fisher Scientific) at 1 µg/mL for 5 minutes in 6-well plate. Then, the cells were washed with PBS and fixed for 10 minutes with 2% paraformaldehyde. The colocalization of GFP and RFP between infected and uninfected cells quantified by flow analysis.

4.3. Detection of Reactive Oxygen Species (ROS)

ROS in HCV culture measured using the cell-permeant H2DCFDA reagent (Thermo Fisher Scientific) according to the manufacturer's instructions. The 2′,7′-dichlorodihydrofluorescein diacetate (H2DCFDA) fluorescent probe reacts with ROS including hydrogen peroxide and hydroxyl radicals. The cell-permeant H2DCFDA diffuses into cells and is retained in the cytoplasm after cleavage by intracellular esterase. The ROS converts the non-fluorescent H2DCFDA to the highly fluorescent 2′,7′-dichlorofluorescein (DCF), which could be detected by fluorescent microscopy or quantified by flow cytometry.

4.4. Immunohistochemical Staining

Tissue culture cells were immobilized onto a glass slide by cytospin. Immunostaining of the cytospin slides of infected cells was performed using a standard protocol established in our laboratory [11]. We used a monoclonal antibody to HCV core protein (Thermo Scientific) and monoclonal antibody that detects phosphorylated NRF2 (Abcam).

4.5. SDS-PAGE and Western Blotting

Infected Huh-7.5 cells were harvested by the treatment of trypsin-ethylenediaminetetraacetic acid (EDTA). Cells were lysed in ice-cold lysis buffer (50 mM Tris HCl pH 8.0, 140 mM NaCl, 1.5 mM MgCl2, 0.5% NP-40 with complete protease inhibitor from Invitrogen) for 10 minutes in ice (about 1×10^6 cells/200 µL). Infected Huh-7.5 cells pelleted by low-speed centrifugation. The detergent compatible (DC) protein assay determined protein concentration. Samples were boiled for 10 minutes at 80 °C in the presence of 1× sodium dodecyl sulfate-polyacrylamide gel electrophoresis (SDS-PAGE)-loading buffer (250 mM Tris-HCl pH 6.8, 40% glycerol, 8% SDS, 0.57M β-mercaptoethanol, 0.12% bromophenol blue). Approximately 20 µg of protein was loaded onto 12% SDS-PAGE and transferred into a nitrocellulose membrane (Thermo Scientific). The membrane was blocked using a solution containing 5% of blotting-grade milk power (Bio-Rad, Hercules, CA, USA) for 2 hours then incubated with a primary antibody. After overnight incubation, the antigen-antibody complex was visualized with horseradish peroxidase (HRP)-conjugated goat anti-rabbit or anti-mouse Immunoglobulin G (IgG) and the ECL detection system (Amersham ECL, GE Healthcare Bio-Sciences, Pittsburgh, PA, USA).

4.6. siRNA Transfection

Persistently HCV-infected Huh-7.5 cells were cultured in 6-well plates (up to 60% confluence in DMEM supplemented with 10% FBS media) without antibiotics. The next day, culture media was replaced with fresh DMEM with 2% FBS and cells were then transfected with siRNA to either NRF2 (siRNA1: 5′-GAAUGGUCCUAAAACACCA-3′, siRNA2: 5′-UGACAGAAGUUGACAAUUA-3′) synthesized by Invitrogen [80,81] or scramble siRNA (Invitrogen) using Lipofectamine (Life Technology, Grand Island, NY, USA). The knockdown efficiency was analyzed by Western blot.

4.7. Quantification of mRNA Levels by RT-qPCR

Infected Huh-7.5 cells and PHHs were harvested at different time points post HCV infection, and total RNA was isolated using the RNeasy Mini kit (Qiagen, Germantown, MD, USA). The first strand of complementary DNA (cDNA) was synthesized from 1 µg of total RNA using an iScript Reverse Transcription Supermix (Bio-Rad). 100 ng cDNA was amplified using iTaq Universal SYBR Green Supermix (Bio-Rad) with gene-specific forward and reverse primers following instructions in the kit. The mRNA levels of NRF2, STAT3, HNF4A, and β-Actin (as internal control) were quantified using quantitative RT-PCR (RT-qPCR). Amplification, data acquisition, and analysis were carried out on the CFX96 real-time instrument using CFX manager software version 3.0 (Bio-Rad). The expression levels of mRNAs were normalized with the *β-Actin*mRNA level using the comparative threshold cycle method. The nucleotide sequences of oligonucleotide primers for *NRF2* mRNA (sense primer 5′-TACTCCCAGGTTGCCCACA-3′ and antisense primer 5′-CATCTACAAACGGGAATGTCTGC-3′) [82], for *STAT3* mRNA (sense primer 5′-TGGAAGAGGCGGCAGCAGATAGC-3′ and antisense primer 5′-CACGGCCCCCATTCCCACAT-3′) [83], for *HNF4A* mRNA (sense primer 5′-TGTCCCGACAGATCACCTC-3′ and antisense primer 5′-CACTCAACGAGAACCAGCAG-3′) [20], and for *β-Actin*mRNA (sense primer 5′-GAGCTACGAGCTGCCTGAC-3′ and antisense primer 5′-AGCACTGTGTTGGCGTACAG-3′) [84] were derived from previous published reports.

4.8. Quantification of Serum microRNA Levels by RT-qPCR

Total microRNA was isolated from 20 µL serum samples using miRNeasy kit (Qiagen) used for reverse transcription using miScript II RT kit (Qiagen) as described previously [85]. Individual microRNA was amplified using cDNA templates, a universal primer and a PCR kit using a recommended protocol. A standard curve was generated for determining microRNA copy number and Ct values using serially diluted synthetic microRNA. The copy number of each microRNA in the serum was calculated from the Ct values.

4.9. Statistical Analysis

Western blot, immunostaining, and immunofluorescence images were quantified using a computer image analysis software package (ImageJ version 1.52c, NIH, Bethesda, MD, USA). All measurements were made at least in triplicate ($n = 3$). To compare means within groups, we performed a one-factor analysis of variance (ANOVA) using the GraphPad Prism software (GraphPad Company, San Diego, CA, USA). Data were tested and found to be normally distributed. To determine the statistical correlation coefficient between protein expressions, fraction of variance denoted by R^2 values was determined using excel software. Sensitivity of the assays was plotted against the false positivity (1-specificity) using ROC curves using the GraphPad Prism software. Comparison of AUC was performed, which compares the AUC to the diagonal line of no information (AUC 0.5). We applied Dunnet's post hoc test to compare control samples with experimental samples when the overall p-value for the ANOVA analysis was significant ($p < 0.05$). When performing comparisons between multiple groups, each analyzed with ANOVA; we used the Bonferroni correction to determine a revised cutoff for statistical significance that gives a combined 5% type I error probability. * *p*-value < 0.05, ** *p*-value < 0.01, *** *p*-value < 0.001.

5. Conclusions

Hepatic adaptive response to HCV-induced stress reduces liver-specific miR-122 through activation of STAT3-HNF4A inflammatory feedback loop. Serum *miR-122* could be used as a biomarker to monitor the activation of this inflammatory loop in liver cirrhosis.

Supplementary Materials: The following are available online at http://www.mdpi.com/2072-6694/11/10/1407/s1, Figure S1: Persistent HCV replication model in permissive Huh-7.5 liver cells using HCV-GFP chimera virus, Figure S2: Live cell imaging of ER stress induced by HCV, Figure S3: Persistent replication of HCV-GFP chimera induces NRF2 activation, Figure S4: Quantification of Western blot of Figure 1, Figure S5: Quantification of Western blot of Figure 2A, Figure S6: Quantification of Western blot of Figure 3, Figure S7: Quantification of Western blot of Figure 4E.

Author Contributions: Y.A., R.K., K.S., D.L., H.O. and B.Y. performed various experiments included in this manuscript. Y.A. prepared figures for publication. J.W.S., N.J.S., P.T. and A.C. supplied serum samples and helped in data analysis. Y.A. and S.D. involved in analysis and wrote the paper.

Funding: This work was supported by NIH grants: CA089121, AI103106, 1P20GM11288-01, Louisiana Clinical and Translational Science (LACaTS) Center grant: U54 GM104940 and VA Merit Review Grant: 1I0IBX004516-01A1.

Acknowledgments: We acknowledge Professor Shi-Mei Zhuang, Key Laboratory of Gene Engineering of the Ministry of Education, State Key Laboratory of Biocontrol, School of Life Sciences Guangzhou, People's Republic of China who have supplied microRNA-122 promoter construct.

Conflicts of Interest: The authors declare that there is no conflict of interest associated with this manuscript.

References

1. Sayiner, M.; Golabi, P.; Younossi, Z.M. Disease Burden of Hepatocellular Carcinoma: A Global Perspective. *Dig. Dis. Sci.* **2019**, *64*, 910–917. [CrossRef] [PubMed]
2. Allison, R.D.; Tong, X.; Moorman, A.C.; Ly, K.N.; Rupp, L.; Xu, F.; Gordon, S.C.; Holmberg, S.D.; Chronic Hepatitis Cohort Study (CHeCS) Investigators. Increased incidence of cancer and cancer-related mortality among persons with chronic hepatitis C infection, 2006–2010. *J. Hepatol.* **2015**, *63*, 822–828. [CrossRef] [PubMed]
3. Thrift, A.P.; El-Serag, H.B.; Kanwal, F. Global epidemiology and burden of HCV infection and HCV-related disease. *Nat. Rev. Gastroenterol. Hepatol.* **2016**, *14*, 122. [CrossRef] [PubMed]
4. Pawlotsky, J.M.; Feld, J.J.; Zeuzem, S.; Hoofnagle, J.H. From non-A, non-B hepatitis to hepatitis C virus cure. *J. Hepatol.* **2015**, *62*, S87–S99. [CrossRef] [PubMed]
5. Lombardi, A.; Mondelli, M.U.; ESCMID Study Group for Viral Hepatitis (ESGVH). Hepatitis C: Is eradication possible? *Liver Int.* **2019**, *39*, 416–426. [CrossRef] [PubMed]

6. Thomas, D.L. Global control of hepatitis C: Where challenge meets opportunity. *Nat. Med.* **2013**, *19*, 850–858. [CrossRef] [PubMed]

7. Bartosch, B.; Thimme, R.; Blum, H.E.; Zoulim, F. Hepatitis C virus-induced hepatocarcinogenesis. *J. Hepatol.* **2009**, *51*, 810–820. [CrossRef]

8. Dash, S.; Aydin, Y.; Wu, T. Integrated stress response in hepatitis C promotes Nrf2-related chaperone-mediated autophagy: A novel mechanism for host-microbe survival and HCC development in liver cirrhosis. *Semin. Cell Dev. Biol.* **2019**. [CrossRef]

9. Raghunath, A.; Sundarraj, K.; Nagarajan, R.; Arfuso, F.; Bian, J.; Kumar, A.P.; Sethi, G.; Perumal, E. Antioxidant response elements: Discovery, classes, regulation and potential applications. *Redox Biol.* **2018**, *17*, 297–314. [CrossRef]

10. Chava, S.; Lee, C.; Aydin, Y.; Chandra, P.K.; Dash, A.; Chedid, M.; Thung, S.N.; Moroz, K.; Wu, T.; Nayak, N.C.; et al. Chaperone-mediated autophagy compensates for impaired macroautophagy in the cirrhotic liver to promote hepatocellular carcinoma. *Oncotarget* **2017**, *8*, 40019–40036. [CrossRef]

11. Aydin, Y.; Chatterjee, A.; Chandra, P.K.; Chava, S.; Chen, W.; Tandon, A.; Dash, A.; Chedid, M.; Moehlen, M.W.; Regenstein, F.; et al. Interferon-alpha-induced hepatitis C virus clearance restores p53 tumor suppressor more than direct-acting antivirals. *Hepatol. Commun.* **2017**, *1*, 256–269. [CrossRef] [PubMed]

12. Aydin, Y.; Chedid, M.; Chava, S.; Danielle Williams, D.; Liu, S.; Hagedorn, C.H.; Sumitran-Holgersson, S.; Reiss, K.; Moroz, K.; Lu, H.; et al. Activation of PERK-Nrf2 oncogenic signaling promotes Mdm2-mediated Rb degradation in persistently infected HCV culture. *Sci. Rep.* **2017**, *7*, 9223. [CrossRef] [PubMed]

13. Aydin, Y.; Stephens, C.M.; Chava, S.; Heidari, Z.; Panigrahi, R.; Williams, D.D.; Wiltz, K.; Bell, A.; Wilson, W.; Reiss, K.; et al. Chaperone-Mediated Autophagy Promotes Beclin1 Degradation in Persistently Infected Hepatitis C Virus Cell Culture. *Am. J. Pathol.* **2018**, *188*, 2339–2355. [CrossRef] [PubMed]

14. Petrelli, A.; Perra, A.; Cora, D.; Sulas, P.; Menegon, S.; Manca, C.; Migliore, C.; Kowalik, M.A.; Ledda-Columbano, G.M.; Giordano, S.; et al. MicroRNA/gene profiling unveils early molecular changes and nuclear factor erythroid related factor 2 (NRF2) activation in a rat model recapitulating human hepatocellular carcinoma (HCC). *Hepatology* **2014**, *59*, 228–241. [CrossRef] [PubMed]

15. Yang, S.F.; Wang, S.N.; Wu, C.F.; Yeh, Y.T.; Chai, C.Y.; Chunag, S.C.; Sheen, M.C.; Lee, K.T. Altered p-STAT3 (tyr705) expression is associated with histological grading and intratumour microvessel density in hepatocellular carcinoma. *J. Clin. Pathol.* **2007**, *60*, 642–648. [CrossRef] [PubMed]

16. He, G.; Yu, G.Y.; Temkin, V.; Ogata, H.; Kuntzen, C.; Sakurai, T.; Sieghart, W.; Peck-Radosavljevic, M.; Leffert, H.L.; Karin, M. Hepatocyte IKKbeta/NF-kappaB inhibits tumor promotion and progression by preventing oxidative stress-driven STAT3 activation. *Cancer Cell* **2010**, *17*, 286–297. [CrossRef] [PubMed]

17. He, G.; Karin, M. NF-kappaB and STAT3—Key players in liver inflammation and cancer. *Cell Res.* **2011**, *21*, 159–168. [CrossRef]

18. Svinka, J.; Mikulits, W.; Eferl, R. STAT3 in hepatocellular carcinoma: New perspectives. *Hepat. Oncol.* **2014**, *1*, 107–120. [CrossRef] [PubMed]

19. Hatziapostolou, M.; Polytarchou, C.; Aggelidou, E.; Drakaki, A.; Poultsides, G.A.; Jaeger, S.A.; Ogata, H.; Karin, M.; Struhl, K.; Hadzopoulou-Cladaras, M.; et al. An HNF4alpha-miRNA inflammatory feedback circuit regulates hepatocellular oncogenesis. *Cell* **2011**, *147*, 1233–1247. [CrossRef]

20. Aboulnasr, F.; Hazari, S.; Nayak, S.; Chandra, P.K.; Panigrahi, R.; Ferraris, P.; Chava, S.; Kurt, R.; Song, K.; Dash, A.; et al. IFN-lambda Inhibits MiR-122 Transcription through a Stat3-HNF4alpha Inflammatory Feedback Loop in an IFN-alpha Resistant HCV Cell Culture System. *PLoS ONE* **2015**, *10*, e0141655. [CrossRef]

21. Lau, H.H.; Ng, N.H.J.; Loo, L.S.W.; Jasmen, J.B.; Teo, A.K.K. The molecular functions of hepatocyte nuclear factors—In and beyond the liver. *J. Hepatol.* **2018**, *68*, 1033–1048. [CrossRef] [PubMed]

22. Yue, H.Y.; Yin, C.; Hou, J.L.; Zeng, X.; Chen, Y.X.; Zhong, W.; Hu, P.F.; Deng, X.; Tan, Y.X.; Zhang, J.P.; et al. Hepatocyte nuclear factor 4alpha attenuates hepatic fibrosis in rats. *Gut* **2010**, *59*, 236–246. [CrossRef] [PubMed]

23. Lazarevich, N.L.; Cheremnova, O.A.; Varga, E.V.; Ovchinnikov, D.A.; Kudrjavtseva, E.I.; Morozova, O.V.; Fleishman, D.I.; Engelhardt, N.V.; Duncan, S.A. Progression of HCC in mice is associated with a downregulation in the expression of hepatocyte nuclear factors. *Hepatology* **2004**, *39*, 1038–1047. [CrossRef] [PubMed]

24. Lazarevich, N.L.; Shavochkina, D.A.; Fleishman, D.I.; Kustova, I.F.; Morozova, O.V.; Chuchuev, E.S.; Patyutko, Y.I. Deregulation of hepatocyte nuclear factor 4 (HNF4) as a marker of epithelial tumors progression. *Exp. Oncol.* **2010**, *32*, 167–171. [PubMed]

25. Ning, B.F.; Ding, J.; Yin, C.; Zhong, W.; Wu, K.; Zeng, X.; Yang, W.; Chen, Y.X.; Zhang, J.P.; Zhang, X.; et al. Hepatocyte nuclear factor 4 alpha suppresses the development of hepatocellular carcinoma. *Cancer Res.* **2010**, *70*, 7640–7651. [CrossRef] [PubMed]

26. Tanaka, T.; Jiang, S.; Hotta, H.; Takano, K.; Iwanari, H.; Sumi, K.; Daigo, K.; Ohashi, R.; Sugai, M.; Ikegame, C.; et al. Dysregulated expression of P1 and P2 promoter-driven hepatocyte nuclear factor-4alpha in the pathogenesis of human cancer. *J. Pathol.* **2006**, *208*, 662–672. [CrossRef] [PubMed]

27. Yin, C.; Lin, Y.; Zhang, X.; Chen, Y.X.; Zeng, X.; Yue, H.Y.; Hou, J.L.; Deng, X.; Zhang, J.P.; Han, Z.G.; et al. Differentiation therapy of hepatocellular carcinoma in mice with recombinant adenovirus carrying hepatocyte nuclear factor-4alpha gene. *Hepatology* **2008**, *48*, 1528–1539. [CrossRef]

28. Li, C.; Deng, M.; Hu, J.; Li, X.; Chen, L.; Ju, Y.; Hao, J.; Meng, S. Chronic inflammation contributes to the development of hepatocellular carcinoma by decreasing miR-122 levels. *Oncotarget* **2016**, *7*, 17021–17034. [CrossRef]

29. Yang, Y.M.; Lee, C.G.; Koo, J.H.; Kim, T.H.; Lee, J.M.; An, J.; Kim, K.M.; Kim, S.G. Galpha12 overexpressed in hepatocellular carcinoma reduces microRNA-122 expression via HNF4alpha inactivation, which causes c-Met induction. *Oncotarget* **2015**, *6*, 19055–19069. [CrossRef]

30. Li, M.; Tang, Y.; Wu, L.; Mo, F.; Wang, X.; Li, H.; Qi, R.; Zhang, H.; Srivastava, A.; Ling, C. The hepatocyte-specific HNF4alpha/miR-122 pathway contributes to iron overload-mediated hepatic inflammation. *Blood* **2017**, *130*, 1041–1051. [CrossRef]

31. Hsu, S.H.; Wang, B.; Kota, J.; Yu, J.; Costinean, S.; Kutay, H.; Yu, L.; Bai, S.; La Perle, K.; Chivukula, R.R.; et al. Essential metabolic, anti-inflammatory, and anti-tumorigenic functions of miR-122 in liver. *J. Clin. Investig.* **2012**, *122*, 2871–2883. [CrossRef] [PubMed]

32. Tsai, W.C.; Hsu, S.D.; Hsu, C.S.; Lai, T.C.; Chen, S.J.; Shen, R.; Huang, Y.; Chen, H.C.; Lee, C.H.; Tsai, T.F.; et al. MicroRNA-122 plays a critical role in liver homeostasis and hepatocarcinogenesis. *J. Clin. Investig.* **2012**, *122*, 2884–2897. [CrossRef] [PubMed]

33. Takaki, Y.; Saito, Y.; Takasugi, A.; Toshimitsu, K.; Yamada, S.; Muramatsu, T.; Kimura, M.; Sugiyama, K.; Suzuki, H.; Arai, E.; et al. Silencing of microRNA-122 is an early event during hepatocarcinogenesis from non-alcoholic steatohepatitis. *Cancer Sci.* **2014**, *105*, 1254–1260. [CrossRef] [PubMed]

34. Shim, S.H.; Xia, C.; Zhong, G.; Babcock, H.P.; Vaughan, J.C.; Huang, B.; Wang, X.; Xu, C.; Bi, G.Q.; Zhuang, X. Super-resolution fluorescence imaging of organelles in live cells with photoswitchable membrane probes. *Proc. Natl. Acad. Sci. USA* **2012**, *109*, 13978–13983. [CrossRef] [PubMed]

35. Burban, A.; Sharanek, A.; Guguen-Guillouzo, C.; Guillouzo, A. Endoplasmic reticulum stress precedes oxidative stress in antibiotic-induced cholestasis and cytotoxicity in human hepatocytes. *Free Radic. Biol. Med.* **2018**, *115*, 166–178. [CrossRef]

36. Beriault, D.R.; Werstuck, G.H. Detection and quantification of endoplasmic reticulum stress in living cells using the fluorescent compound, Thioflavin T. *Biochim. Biophys. Acta* **2013**, *1833*, 2293–2301. [CrossRef] [PubMed]

37. Xing, Y.; Higuchi, K. Amyloid fibril proteins. *Mech. Ageing Dev.* **2002**, *123*, 1625–1636. [CrossRef]

38. Tardif, K.D.; Waris, G.; Siddiqui, A. Hepatitis C virus, ER stress, and oxidative stress. *Trends Microbiol.* **2005**, *13*, 159–163. [CrossRef]

39. Waris, G.; Turkson, J.; Hassanein, T.; Siddiqui, A. Hepatitis C virus (HCV) constitutively activates STAT-3 via oxidative stress: Role of STAT-3 in HCV replication. *J. Virol.* **2005**, *79*, 1569–1580. [CrossRef]

40. Gong, G.; Waris, G.; Tanveer, R.; Siddiqui, A. Human hepatitis C virus NS5A protein alters intracellular calcium levels, induces oxidative stress, and activates STAT-3 and NF-kappa B. *Proc. Natl. Acad. Sci. USA* **2001**, *98*, 9599–9604. [CrossRef]

41. Joyce, M.A.; Walters, K.A.; Lamb, S.E.; Yeh, M.M.; Zhu, L.F.; Kneteman, N.; Doyle, J.S.; Katze, M.G.; Tyrrell, D.L. HCV induces oxidative and ER stress, and sensitizes infected cells to apoptosis in SCID/Alb-uPA mice. *PLoS Pathog.* **2009**, *5*, e1000291. [CrossRef] [PubMed]

42. Burdette, D.; Olivarez, M.; Waris, G. Activation of transcription factor Nrf2 by hepatitis C virus induces the cell-survival pathway. *J. Gen. Virol.* **2010**, *91*, 681–690. [CrossRef]

43. Xu, D.; Xu, M.; Jeong, S.; Qian, Y.; Wu, H.; Xia, Q.; Kong, X. The Role of Nrf2 in Liver Disease: Novel Molecular Mechanisms and Therapeutic Approaches. *Front. Pharmacol.* **2018**, *9*, 1428. [CrossRef] [PubMed]

44. Ma, Q. Role of nrf2 in oxidative stress and toxicity. *Annu. Rev. Pharmacol. Toxicol.* **2013**, *53*, 401–426. [CrossRef] [PubMed]

45. Nguyen, T.; Nioi, P.; Pickett, C.B. The Nrf2-antioxidant response element signaling pathway and its activation by oxidative stress. *J. Biol. Chem.* **2009**, *284*, 13291–13295. [CrossRef]

46. Mitsuishi, Y.; Motohashi, H.; Yamamoto, M. The Keap1-Nrf2 system in cancers: Stress response and anabolic metabolism. *Front. Oncol.* **2012**, *2*, 200. [CrossRef] [PubMed]

47. Kansanen, E.; Kuosmanen, S.M.; Leinonen, H.; Levonen, A.L. The Keap1-Nrf2 pathway: Mechanisms of activation and dysregulation in cancer. *Redox Biol.* **2013**, *1*, 45–49. [CrossRef]

48. Vallianou, I.; Dafou, D.; Vassilaki, N.; Mavromara, P.; Hadzopoulou-Cladaras, M. Hepatitis C virus suppresses Hepatocyte Nuclear Factor 4 alpha, a key regulator of hepatocellular carcinoma. *Int. J. Biochem. Cell Biol.* **2016**, *78*, 315–326. [CrossRef]

49. Li, Z.Y.; Xi, Y.; Zhu, W.N.; Zeng, C.; Zhang, Z.Q.; Guo, Z.C.; Hao, D.L.; Liu, G.; Feng, L.; Chen, H.Z.; et al. Positive regulation of hepatic miR-122 expression by HNF4alpha. *J. Hepatol.* **2011**, *55*, 602–611. [CrossRef]

50. Zeng, C.; Wang, R.; Li, D.; Lin, X.J.; Wei, Q.K.; Yuan, Y.; Wang, Q.; Chen, W.; Zhuang, S.M. A novel GSK-3 beta-C/EBP alpha-miR-122-insulin-like growth factor 1 receptor regulatory circuitry in human hepatocellular carcinoma. *Hepatology* **2010**, *52*, 1702–1712. [CrossRef]

51. Kanda, T.; Goto, T.; Hirotsu, Y.; Moriyama, M.; Omata, M. Molecular Mechanisms Driving Progression of Liver Cirrhosis towards Hepatocellular Carcinoma in Chronic Hepatitis B and C Infections: A Review. *Int. J. Mol. Sci.* **2019**, *20*, 1358. [CrossRef] [PubMed]

52. Desai, A.; Sandhu, S.; Lai, J.P.; Sandhu, D.S. Hepatocellular carcinoma in non-cirrhotic liver: A comprehensive review. *World J. Hepatol.* **2019**, *11*, 1. [CrossRef] [PubMed]

53. Stine, J.G.; Wentworth, B.J.; Zimmet, A.; Rinella, M.E.; Loomba, R.; Caldwell, S.H.; Argo, C.K. Systematic review with meta-analysis: Risk of hepatocellular carcinoma in non-alcoholic steatohepatitis without cirrhosis compared to other liver diseases. *Aliment. Pharmacol. Ther.* **2018**, *48*, 696–703. [CrossRef] [PubMed]

54. Chandra, P.K.; Gunduz, F.; Hazari, S.; Kurt, R.; Panigrahi, R.; Poat, B.; Bruce, D.; Cohen, A.J.; Bohorquez, H.E.; Carmody, I.; et al. Impaired expression of type I and type II interferon receptors in HCV-associated chronic liver disease and liver cirrhosis. *PLoS ONE* **2014**, *9*, e108616. [CrossRef] [PubMed]

55. Levy, D.E.; Darnell, J.E., Jr. Stats: Transcriptional control and biological impact. *Nat. Rev. Mol. Cell Biol.* **2002**, *3*, 651–662. [CrossRef] [PubMed]

56. Calvisi, D.F.; Ladu, S.; Gorden, A.; Farina, M.; Conner, E.A.; Lee, J.S.; Factor, V.M.; Thorgeirsson, S.S. Ubiquitous activation of Ras and Jak/Stat pathways in human HCC. *Gastroenterology* **2006**, *130*, 1117–1128. [CrossRef]

57. Shen, S.; Niso-Santano, M.; Adjemian, S.; Takehara, T.; Malik, S.A.; Minoux, H.; Souquere, S.; Marino, G.; Lachkar, S.; Senovilla, L.; et al. Cytoplasmic STAT3 represses autophagy by inhibiting PKR activity. *Mol. Cell* **2012**, *48*, 667–680. [CrossRef]

58. Tacke, R.S.; Tosello-Trampont, A.; Nguyen, V.; Mullins, D.W.; Hahn, Y.S. Extracellular hepatitis C virus core protein activates STAT3 in human monocytes/macrophages/dendritic cells via an IL-6 autocrine pathway. *J. Biol. Chem.* **2011**, *286*, 10847–10855. [CrossRef]

59. Kyrmizi, I.; Hatzis, P.; Katrakili, N.; Tronche, F.; Gonzalez, F.J.; Talianidis, I. Plasticity and expanding complexity of the hepatic transcription factor network during liver development. *Genes Dev.* **2006**, *20*, 2293–2305. [CrossRef]

60. Odom, D.T.; Zizlsperger, N.; Gordon, D.B.; Bell, G.W.; Rinaldi, N.J.; Murray, H.L.; Volkert, T.L.; Schreiber, J.; Rolfe, P.A.; Gifford, D.K.; et al. Control of pancreas and liver gene expression by HNF transcription factors. *Science* **2004**, *303*, 1378–1381. [CrossRef]

61. Guzman-Lepe, J.; Cervantes-Alvarez, E.; Collin de l'Hortet, A.; Wang, Y.; Mars, W.M.; Oda, Y.; Bekki, Y.; Shimokawa, M.; Wang, H.; Yoshizumi, T.; et al. Liver-enriched transcription factor expression relates to chronic hepatic failure in humans. *Hepatol. Commun.* **2018**, *2*, 582–594. [CrossRef] [PubMed]

62. Liu, L.; Yannam, G.R.; Nishikawa, T.; Yamamoto, T.; Basma, H.; Ito, R.; Nagaya, M.; Dutta-Moscato, J.; Stolz, D.B.; Duan, F.; et al. The microenvironment in hepatocyte regeneration and function in rats with advanced cirrhosis. *Hepatology* **2012**, *55*, 1529–1539. [CrossRef] [PubMed]

63. Safdar, H.; Cheung, K.L.; Vos, H.L.; Gonzalez, F.J.; Reitsma, P.H.; Inoue, Y.; van Vlijmen, B.J. Modulation of mouse coagulation gene transcription following acute in vivo delivery of synthetic small interfering RNAs targeting HNF4alpha and C/EBPalpha. *PLoS ONE* **2012**, *7*, e38104. [CrossRef] [PubMed]

64. Zheng, X.W.; Kudaravalli, R.; Russell, T.T.; DiMichele, D.M.; Gibb, C.; Russell, J.E.; Margaritis, P.; Pollak, E.S. Mutation in the factor VII hepatocyte nuclear factor 4alpha-binding site contributes to factor VII deficiency. *Blood Coagul. Fibrinolysis* **2011**, *22*, 624–627. [CrossRef] [PubMed]

65. Inoue, Y.; Peters, L.L.; Yim, S.H.; Inoue, J.; Gonzalez, F.J. Role of hepatocyte nuclear factor 4alpha in control of blood coagulation factor gene expression. *J. Mol. Med.* **2006**, *84*, 334–344. [CrossRef] [PubMed]

66. Bandiera, S.; Pfeffer, S.; Baumert, T.F.; Zeisel, M.B. miR-122-a key factor and therapeutic target in liver disease. *J. Hepatol.* **2015**, *62*, 448–457. [CrossRef] [PubMed]

67. Trebicka, J.; Anadol, E.; Elfimova, N.; Strack, I.; Roggendorf, M.; Viazov, S.; Wedemeyer, I.; Drebber, U.; Rockstroh, J.; Sauerbruch, T.; et al. Hepatic and serum levels of miR-122 after chronic HCV-induced fibrosis. *J. Hepatol.* **2013**, *58*, 234–239. [CrossRef]

68. Cermelli, S.; Ruggieri, A.; Marrero, J.A.; Ioannou, G.N.; Beretta, L. Circulating microRNAs in patients with chronic hepatitis C and non-alcoholic fatty liver disease. *PLoS ONE* **2011**, *6*, e23937. [CrossRef]

69. Sarasin-Filipowicz, M.; Krol, J.; Markiewicz, I.; Heim, M.H.; Filipowicz, W. Decreased levels of microRNA miR-122 in individuals with hepatitis C responding poorly to interferon therapy. *Nat. Med.* **2009**, *15*, 31–33. [CrossRef]

70. Hamdane, N.; Juhling, F.; Crouchet, E.; El Saghire, H.; Thumann, C.; Oudot, M.A.; Bandiera, S.; Saviano, A.; Ponsolles, C.; Roca Suarez, A.A.; et al. HCV-Induced Epigenetic Changes Associated with Liver Cancer Risk Persist After Sustained Virologic Response. *Gastroenterology* **2019**, *156*, 2313–2329. [CrossRef]

71. Perez, S.; Kaspi, A.; Domovitz, T.; Davidovich, A.; Lavi-Itzkovitz, A.; Meirson, T.; Alison Holmes, J.; Dai, C.Y.; Huang, C.F.; Chung, R.T.; et al. Hepatitis C virus leaves an epigenetic signature post cure of infection by direct-acting antivirals. *PLoS Genet.* **2019**, *15*, e1008181. [CrossRef] [PubMed]

72. El-Araby, R.E.; Khalifa, M.A.; Zoheiry, M.M.; Zahran, M.Y.; Rady, M.I.; Ibrahim, R.A.; El-Talkawy, M.D.; Essawy, F.M. The interaction between microRNA-152 and DNA methyltransferase-1 as an epigenetic prognostic biomarker in HCV-induced liver cirrhosis and HCC patients. *Cancer Gene Ther.* **2019**. [CrossRef] [PubMed]

73. Chandra, P.K.; Bao, L.; Song, K.; Aboulnasr, F.M.; Baker, D.P.; Shores, N.; Wimley, W.C.; Liu, S.; Hagedorn, C.H.; Fuchs, S.Y.; et al. HCV infection selectively impairs type I but not type III IFN signaling. *Am. J. Pathol.* **2014**, *184*, 214–229. [CrossRef] [PubMed]

74. Florczyk, U.; Czauderna, S.; Stachurska, A.; Tertil, M.; Nowak, W.; Kozakowska, M.; Poellinger, L.; Jozkowicz, A.; Loboda, A.; Dulak, J. Opposite effects of HIF-1alpha and HIF-2alpha on the regulation of IL-8 expression in endothelial cells. *Free Radic. Biol. Med.* **2011**, *51*, 1882–1892. [CrossRef]

75. Son, Y.O.; Pratheeshkumar, P.; Roy, R.V.; Hitron, J.A.; Wang, L.; Zhang, Z.; Shi, X. Nrf2/p62 signaling in apoptosis resistance and its role in cadmium-induced carcinogenesis. *J. Biol. Chem.* **2014**, *289*, 28660–28675. [CrossRef] [PubMed]

76. Xu, H.; Xu, S.J.; Xie, S.J.; Zhang, Y.; Yang, J.H.; Zhang, W.Q.; Zheng, M.N.; Zhou, H.; Qu, L.H. MicroRNA-122 supports robust innate immunity in hepatocytes by targeting the RTKs/STAT3 signaling pathway. *Elife* **2019**, *8*. [CrossRef] [PubMed]

77. Van Renne, N.; Suarez, A.A.R.; Duong, F.H.T.; Gondeau, C.; Calabrese, D.; Fontaine, N.; Ababsa, A.; Bandiera, S.; Croonenborghs, T.; Pochet, N.; et al. miR-135a-5p-mediated downregulation of protein tyrosine phosphatase receptor delta is a candidate driver of HCV-associated hepatocarcinogenesis. *Gut* **2018**, *67*, 953–962. [CrossRef] [PubMed]

78. Chen, Y.N.; Shen, A.; Rider, P.J.; Yu, Y.; Wu, K.L.; Mu, Y.X.; Hao, Q.; Liu, Y.L.; Gong, H.; Zhu, Y.; et al. A liver-specific microRNA binds to a highly conserved RNA sequence of hepatitis B virus and negatively regulates viral gene expression and replication. *FASEB J.* **2011**, *25*, 4511–4521. [CrossRef] [PubMed]

79. Luna, J.M.; Scheel, T.K.H.; Danino, T.; Shaw, K.S.; Mele, A.; Fak, J.J.; Nishiuchi, E.; Takacs, C.N.; Catanese, M.T.; de Jong, Y.P.; et al. Hepatitis C Virus RNA Functionally Sequesters miR-122. *Cell* **2015**, *160*, 1099–1110. [CrossRef]

80. Hyrina, A.; Olmstead, A.D.; Steven, P.; Krajden, M.; Tam, E.; Jean, F. Treatment-Induced Viral Cure of Hepatitis C Virus-Infected Patients Involves a Dynamic Interplay among three Important Molecular Players

in Lipid Homeostasis: Circulating microRNA (miR)-24, miR-223, and Proprotein Convertase Subtilisin/Kexin Type 9. *Ebiomedicine* **2017**, *23*, 68–78. [CrossRef]

81. Jopling, C.L.; Yi, M.K.; Lancaster, A.M.; Lemon, S.M.; Sarnow, P. Modulation of hepatitis C virus RNA abundance by a liver-specific microRNA. *Science* **2005**, *309*, 1577–1581. [CrossRef] [PubMed]

82. Mylroie, H.; Dumont, O.; Bauer, A.; Thornton, C.C.; Mackey, J.; Calay, D.; Hamdulay, S.S.; Choo, J.R.; Boyle, J.J.; Samarel, A.M.; et al. PKC epsilon-CREB-Nrf2 signalling induces HO-1 in the vascular endothelium and enhances resistance to inflammation and apoptosis. *Cardiovasc. Res.* **2015**, *106*, 509–519. [CrossRef] [PubMed]

83. Gao, H.W.; Guo, R.F.; Speyer, C.L.; Reuben, J.; Neff, T.A.; Hoesel, L.M.; Riedemann, N.C.; McClintock, S.A.; Sarma, J.V.; Van Rooijen, N.; et al. Stat3 activation in acute lung injury. *J. Immunol.* **2004**, *172*, 7703–7712. [CrossRef] [PubMed]

84. Tawani, A.; Amanullah, A.; Mishra, A.; Kumar, A. Evidences for Piperine inhibiting cancer by targeting human G-quadruplex DNA sequences. *Sci. Rep.* **2016**, *6*, 39239. [CrossRef] [PubMed]

85. Mitchell, P.S.; Parkin, R.K.; Kroh, E.M.; Fritz, B.R.; Wyman, S.K.; Pogosova-Agadjanyan, E.L.; Peterson, A.; Noteboom, J.; O'Briant, K.C.; Allen, A.; et al. Circulating microRNAs as stable blood-based markers for cancer detection. *Proc. Natl. Acad. Sci. USA* **2008**, *105*, 10513–10518. [CrossRef] [PubMed]

Article

Decoration of Anti-CD38 on Nanoparticles Carrying a STAT3 Inhibitor Can Improve the Therapeutic Efficacy Against Myeloma

Yung-Hsing Huang [1], Mohammad Reza Vakili [2], Ommoleila Molavi [1,3], Yuen Morrissey [1], Chengsheng Wu [1,†], Igor Paiva [2], Amir Hasan Soleimani [2], Forugh Sanaee [2], Afsaneh Lavasanifar [2] and Raymond Lai [1,*]

[1] Department of Laboratory Medicine and Pathology, University of Alberta, Edmonton, AB T6G 2R3, Canada; yunghsin@ualberta.ca (Y.-H.H.); omolavi@ualberta.ca (O.M.); yuenw@ualberta.ca (Y.M.); chengshe@ualberta.ca (C.W.)

[2] Faculty of Pharmacy and Pharmaceutical Sciences, University of Alberta, Edmonton, AB T6G 2H7, Canada; vakili@ualberta.ca (M.R.V.); paiva@ualberta.ca (I.P.); asoleima@ualberta.ca (A.H.S.); fsanaee@ualberta.ca (F.S.); afsaneh@ualberta.ca (A.L.)

[3] Faculty of Pharmacy, Tabriz University of Medical Science, Tabriz P.O. Box 51664-14766, East Azerbaijan Province, Iran

[*] Correspondence: rlai@ualberta.ca; Tel.: +1-780-432-8454

[†] C.W. is now with the Department of Pathology, University of California San Diego, San Diego, CA 92093, USA.

Received: 10 January 2019; Accepted: 14 February 2019; Published: 20 February 2019

Abstract: STAT3 is an oncoprotein which has been shown to contribute to drug resistance in multiple myeloma (MM). Nonetheless, the clinical utility of STAT3 inhibitors in treating MM has been limited, partly related to some of their pharmacologic properties. To overcome these challenges, our group had previously packaged STAT3 inhibitors using a novel formulation of nanoparticles (NP) and found encouraging results. In this study, we aimed to further improve the pharmacologic properties of these NP by decorating them with monoclonal anti-CD38 antibodies. NP loaded with S3I-1757 (a STAT3 inhibitor), labeled as S3I-NP, were generated. S3I-NP decorated with anti-CD38 (labeled as CD38-S3I-NP) were found to have a similar nanoparticular size, drug encapsulation, and loading as S3I-NP. The release of S3I-1757 at 24 h was also similar between the two formulations. Using Cy5.5 labeling of the NP, we found that the decoration of anti-CD38 on these NP significantly increased the cellular uptake by two MM cell lines ($p < 0.001$). Accordingly, CD38-S3I-NP showed a significantly lower inhibitory concentration at 50% (IC50) compared to S3I-NP in two IL6-stimulated MM cell lines ($p < 0.001$). In a xenograft mouse model, CD38-S3I-NP significantly reduced the tumor size by 4-fold compared to S3I-NP on day 12 after drug administration ($p = 0.006$). The efficacy of CD38-S3I-NP in suppressing STAT3 phosphorylation in the xenografts was confirmed by using immunocytochemistry and Western blot analysis. In conclusion, our study suggests that the decoration of anti-CD38 on NP loaded with STAT3 inhibitors can further improve their therapeutic effects against MM.

Keywords: multiple myeloma; STAT3; S3I-1757; nanoparticle; CD38

---

1. Introduction

Recent studies have shown that nanoparticles (NP) can be an effective drug delivery system to treat cancers [1]. In addition to its usefulness in delivering hydrophobic drugs, NP can promote drug accumulation at tumor sites due to the fact that NP are too large to pass through the normal capillaries but small enough to leak through the poorly-formed vasculatures frequently present in malignant tumors [2]. To further increase the targeting ability and reduce drug toxicity, researchers have conjugated NP with various tumor-specific antibodies [3]. For example, different forms of NP conjugated with anti-human epidermal receptor-2 (HER2) antibodies have been generated to treat HER2-positive breast cancer [4–7]. Based on the results of a few studies, it appears that the decoration of NP with tumor-specific antibodies can indeed result in superior cellular uptake/cytotoxicity in vitro, as well as significantly improved tumor suppression in vivo, compared to their unconjugated counterparts [7–12]. Some studies also suggest that the conjugation of antibodies on the surface of NP is useful in overcoming drug resistance in cancers which overexpress drug efflux pumps (e.g., p-glycoprotein) [8,13]. The conjugation of antibodies on NP is versatile, and a variety of antibodies have been used to achieve specific experimental objectives, such as the use of antibodies targeting the vascular endothelial growth factor receptor (i.e., to block tumor angiogenesis) [12], matrix metalloproteinases (i.e., to block tumor invasion) [14], and transferrin receptor (i.e., to facilitate the crossing of NP through the blood-brain barrier) [15].

Multiple myeloma (MM) is a hematological disease which is characterized by a high frequency of relapses and resistance to chemotherapy. A signal transducer and activator of transcription 3 (STAT3), which is found to be active in more than 50% of MM, has been shown to contribute to the resistance to bortezomib, thalidomide, and dexamethasone in MM [16–19]. In view of its significance in cancer biology, STAT3 has been postulated to be an attractive anti-cancer target, and many STAT3 inhibitors (such as Stattic, S3I-201, and S3I-1757) have been developed [20–22]. However, the clinical utility of these compounds has been limited, which is likely related to some of their pharmacologic properties, such as their small size and hydrophobicity. Consequently, STAT3 inhibitors have been found to have relatively low therapeutic efficacy and high toxicity. In this regard, a clinical trial of OPB-31121 (an orally administered STAT3 inhibitor) in a cohort of patients with various types of solid cancer has reported that >80% of patients experienced significant nausea/diarrhea without therapeutic benefits [23]. To overcome these challenges, our group has recently generated NP to package STAT3 inhibitors. Specifically, we synthesized NP based on the poly(ethylene oxide)-*block*-poly(α-benzyl carboxylate-ε-caprolactone) (PEO-*b*-PBCL) backbone, and we used these NP to package S3I-1757 (denoted as S3I-NP); we found that S3I-NP exhibited significantly better anti-tumor effects in mice xenografted with a human melanoma cell line compared to free drugs [24].

In this study, we aimed to further improve the pharmacologic properties of S3I-NP by conjugating monoclonal antibodies against human CD38, a cell-surface marker highly expressed on MM cells, on the surface of S3I-NP (denoted as CD38-S3I-NP). We hypothesized that, due to the active targeting properties of anti-CD38 for MM cells, CD38-S3I-NP will demonstrate improved cellular uptake in vitro cytotoxicity and in vivo therapeutic efficacy compared to S3I-NP. Our results have provided the proof-of-principle that anti-CD38-conjugated NP loaded with STAT3 inhibitors are useful therapeutic agents for MM patients.

2. Results

2.1. Synthesis and Characterization of CD38-S3I-NP

To generate CD38-S3I-NP, we conjugated anti-CD38 monoclonal antibodies to the hydrophilic portion of PEO-*b*-PBCL. As illustrated in Figure 1A, anti-CD38 was first thiolated at the lysine residue located in the constant region of the heavy chain of the antibody. Thiolated anti-CD38 antibodies were then combined with maleimide-functionalized PEO-*b*-PBCL, such that antibodies were attached to the surface of the polymers. Lastly, anti-CD38-conjugated polymers were mixed with NP loaded with S3I-1757 (i.e., S3I-NP) to generate CD38-S3I-NP.

We then determined if the conjugation of anti-CD38 on the surface of S3I-NP significantly altered their physical properties. As summarized in Table 1, the average size of CD38-S3I-NP was 91.4 ± 9.4 nm, which is not significantly different from that of S3I-NP (97.4 ± 5.2 nm) (p = 0.39). Similarly, there is no significant difference in drug encapsulation efficiency between CD38-S3I-NP and S3I-NP (81.6 ± 7.2% versus 87.0 ± 9.2%, p = 0.47) as well as drug loading (14.7 ± 1.3% versus 15.7 ± 1.7%,

$p = 0.47$). The polydispersity index was significantly higher in CD38-S3I-NP compared to that of S3I-NP (0.367 ± 0.016 versus 0.273 ± 0.003, $p < 0.001$), suggesting that CD38-S3I-NP is less uniform in size compared to S3I-NP, possibly due to antibody aggregation. As shown in Figure 1B, significantly more S3I-1757 was found to be released from CD38-S3I-NP than that from S3I-NP after 1, 2, and 4 hours of incubation ($p < 0.001$, $p < 0.001$, and $p = 0.002$, respectively). Nevertheless, both formulations reached a comparable amount of S3I-1757 release (~68%, $p = 0.59$) at 24 h. Taken together, the physical properties between these two formulations are not substantially different.

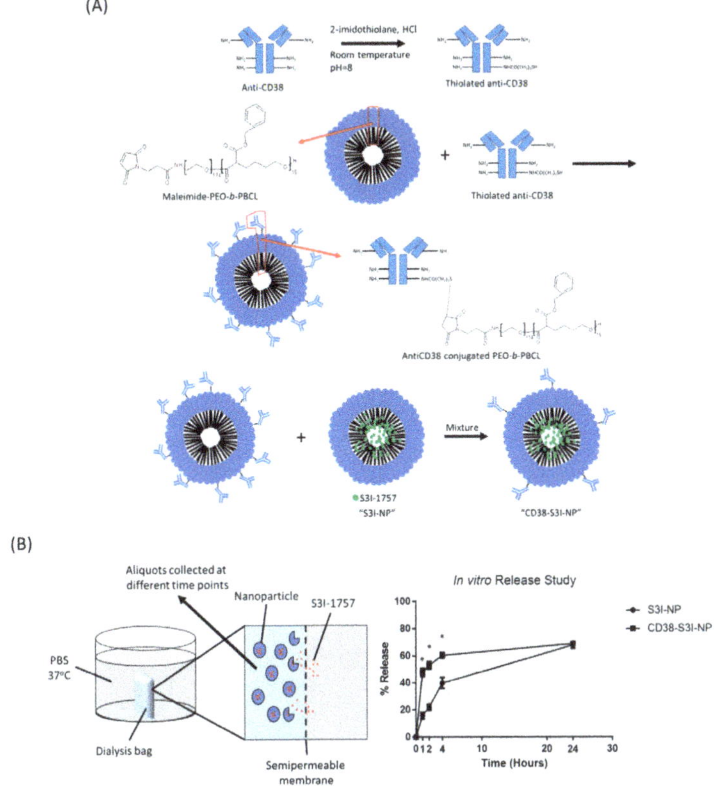

Figure 1. The synthesis of CD38-S3I-NP. (**A**) Chemical reactions of the anti-CD38 conjugation to PEO-*b*-PBCL, the building block of our nanoparticles (NP). The final product was mixed with S3I-NP to generate CD38-S3I-NP. (**B**) The release of S3I-1757 from S3I-NP or CD38-S3I-NP in vitro. The percentage of S3I-1757 released was calculated by the lost amount of S3I-1757 compared to the initial total amount of S3I-1757. The error bar represents the standard deviation from a triplicate experiment; * $p < 0.05$, via Student's *t*-test.

Table 1. Physical properties of S3I-NP and CD38-S3I-NP.

NP Formulation	Average Size (nm)	Polydispersity Index	Drug Encapsulation Efficiency (%)	Drug Loading (Weight %)
S3I-NP	97.4 ± 5.2	0.273 ± 0.003	$87.0 \pm 9.2\%$	$15.7 \pm 1.7\%$
CD38-S3I-NP	91.4 ± 9.4	$0.367 \pm 0.016*$	$81.6 \pm 7.2\%$	$14.7 \pm 1.3\%$

* $p < 0.05$, compared to S3I-NP.

2.2. Anti-CD38 Conjugation on NP Results in More Cellular Uptake by MM Cells

We then determined if the conjugation of anti-CD38 to NP can significantly improve the uptake of NP by MM cells. To facilitate the detection and quantification of NP in vitro, we synthesized Cy5.5 (a fluorophore)-conjugated NP with or without the coating of anti-CD38 (denoted as Cy5.5-CD38-NP and Cy5.5-NP, respectively). The NP used in these experiments was not loaded with the STAT3 inhibitor to avoid drug-induced cytotoxicity, which can potentially interfere with our assays. Two MM cell lines (U266 and RPMI8226) were used. SupM2, an ALK-positive anaplastic large cell lymphoma cell line, was used as a negative control. The CD38 expression in the two MM cell lines and the absence of CD38 expression in SupM2 are illustrated in Figure S1A,B. As shown in Figure 2, both MM cell lines incubated with Cy5.5-CD38-NP for 4 h exhibited a significantly higher level of intracellular Cy5.5 compared to cells incubated with Cy5.5-NP. Specifically, in U266 cells, Cy5.5-CD38-NP treatment yielded 43.2 ± 0.1% Cy5.5-positive cells, whereas Cy5.5-NP treatment resulted in only 0.4 ± 0.1% Cy5.5-positive cells ($p < 0.001$). Similarly, in RMMI8226 cells, Cy5.5-CD38-NP yielded significantly more Cy5.5-positive cells than Cy5.5-NP treatment (76.7 ± 1.1% versus 1.2 ± 0.1%) ($p < 0.001$). Compared to the background (i.e., no treatment), Cy5.5-CD38-NP only minimally increased the proportion of Cy5.5-positive cells in SupM2 cells (9.2 ± 0.3%).

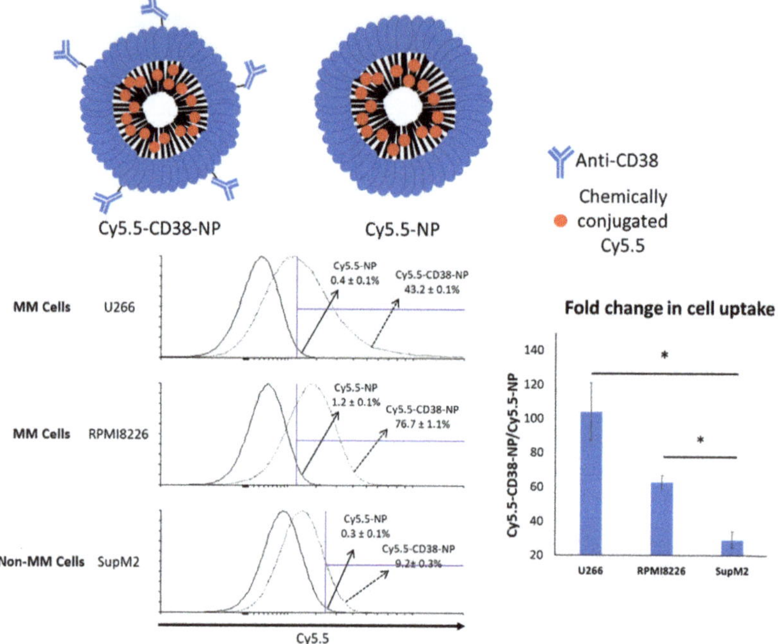

Figure 2. Flow cytometry analysis of the Cy5.5-positive cell population 4 h after treatment of Cy5.5-NP or Cy5.5-CD38-NP. Anti-CD38-conjugated NP exhibits improved cellular uptake of NP by multiple myeloma (MM) cells. Cy5.5 was chemically conjugated to the core of NP. The gated area was defined using the cells without NP treatment. The representative dot plot from a triplicate experiment is shown. The error values represent the standard deviation from the triplicate experiment. A non-MM cell line, SupM2, was included for comparison. The fold change in cell uptake was calculated by dividing the percentage of Cy5.5-positive cells with Cy5.5-CD38-NP treatment by that with Cy5.5-NP treatment. The error bar represents standard deviation from a triplicate experiment; * $p < 0.05$, via Student's *t*-test.

2.3. CD38-S3I-NP is More Cytotoxic to MM Cells than S3I-NP

We next assessed the cytotoxicity of CD38-S3I-NP compared to S3I-NP in two MM cell lines (U266 and RPMI8226) using MTS assay. In these experiments, we added exogenous IL6 (2 ng/mL) to the cell culture to enhance STAT3 activity in the cells. As shown in Figure 3A, in RPMI8226 cells, the addition of CD38-S3I-NP led to significantly lower cell viability compared to S3I-NP at 48 h (50 µM, $p = 0.001$). In U266 cells, significantly lower cell viability of CD38-S3I-NP was seen at 100 µM compared to S3I-NP ($p = 0.007$). In contrast, there was no significant difference in reducing cell viability of SupM2 between the two formulations, although these cells are known to be highly STAT3-active due to an endogenous tyrosine kinase, NPM-ALK [25]. As a comparison, we repeated the experiments using the combination of S3I-NP and free CD38 antibodies. As shown in Figure 3A, the results were similar to those of S3I-NP and significantly inferior to those CD38-S3I-NP in the two MM cell lines. The inhibitory concentration at 50% (IC_{50}) values generated from all of these experiments is summarized in Table 2. In the same experiments, we also confirmed that CD38-S3I-NP and S3I-NP were effective in suppressing STAT3 phosphorylation at residue Y705 (i.e., pSTAT3). Using Western blot analysis, we found that pSTAT3 induced by IL6 (2 ng/mL) in the two MM cell lines was substantially decreased by both formulations at 24 h (Figure 3B).

Table 2. IC_{50} values of U266, RPMI8226, and SupM2 cells treated with different NP.

Cell Line	Treatment	IC_{50} (µM)	
		24 h	**48 h**
U266	S3I-NP	136.7–163.6	115.4–148.6 **
	CD38-S3I-NP	127.7–151.3	106.3–114.0
	S3I-NP + Free Anti-CD38	143.5–172.4	128.6–139.4 **
RPMI8226	S3I-NP	110.9–124.5	88.2–98.1 **
	CD38-S3I-NP	100.4–109.8	64.0–73.6
	S3I-NP + Free Anti-CD38	108.9–142.2	87.2–99.5 **
SupM2	S3I-NP	88.8–105.6	110.4–144.8
	CD38-S3I-NP	86.0–98.9	83.16–119.4
	S3I-NP + Free Anti-CD38	90.1–110.0	106.9–134.9

** $p < 0.0001$, compared to the corresponding CD38-S3I-NP treatment.

2.4. CD38-S3I-NP is More Effective in Suppressing MM Tumor Growth In Vivo Compared to S3I-NP

We then elucidated if CD38-S3I-NP has therapeutic advantages over S3I-NP in a SCID mouse xenograft model. As detailed in Materials and Methods, we xenografted U266 cells stably expressing a luciferase expression construct (U266-luc) in SCID mice, such that the growth of tumors can be easily tracked ex vivo using bioluminescence imaging. When the tumors became detectable, S3I-NP or CD38-S3I-NP was injected intravenously. On day 15 after the injection of NP, 4/4 animals in the S3I-NP group reached the endpoints as defined in Materials and Methods, while 1/4 animals in the CD38-S3I-NP group did not. Statistical analysis reveals a trend for a longer survival for the CD38-S3I-NP group, although the difference between the two groups does not reach statistical significance ($p = 0.079$, Mantel-Cox test), most likely due to the small sample size. As a control, both mice treated with phosphate buffered saline (PBS) reached the endpoints on day 12.

Other than the time needed to reach the endpoints, we also directly assessed tumor growth in the two study groups. Specifically, we summed up the levels of detectable bioluminescence (radiance ranged between 4.00×10^5 and 1.00×10^7 p/sec/cm^2/scr) by using the IVIS Spectrum In Vivo Imaging System, as described in Materials and Methods. Images of a representative animal from each of the CD38-S3I-NP, S3I-NP, and PBS groups are shown in Figure 4A. Animals in the CD38-S3I-NP group had significantly lower tumor volume compared to the S3I-NP group at 240 and 288 h ($p = 0.018$ and $p = 0.006$, respectively). Since we had only two animals in the PBS group, the statistical significance cannot be determined for this group. Nonetheless, it is evident that the tumors grew substantially faster than those in the CD38-S3I-NP group (tumor volume at 288 h was 16.8 times higher).

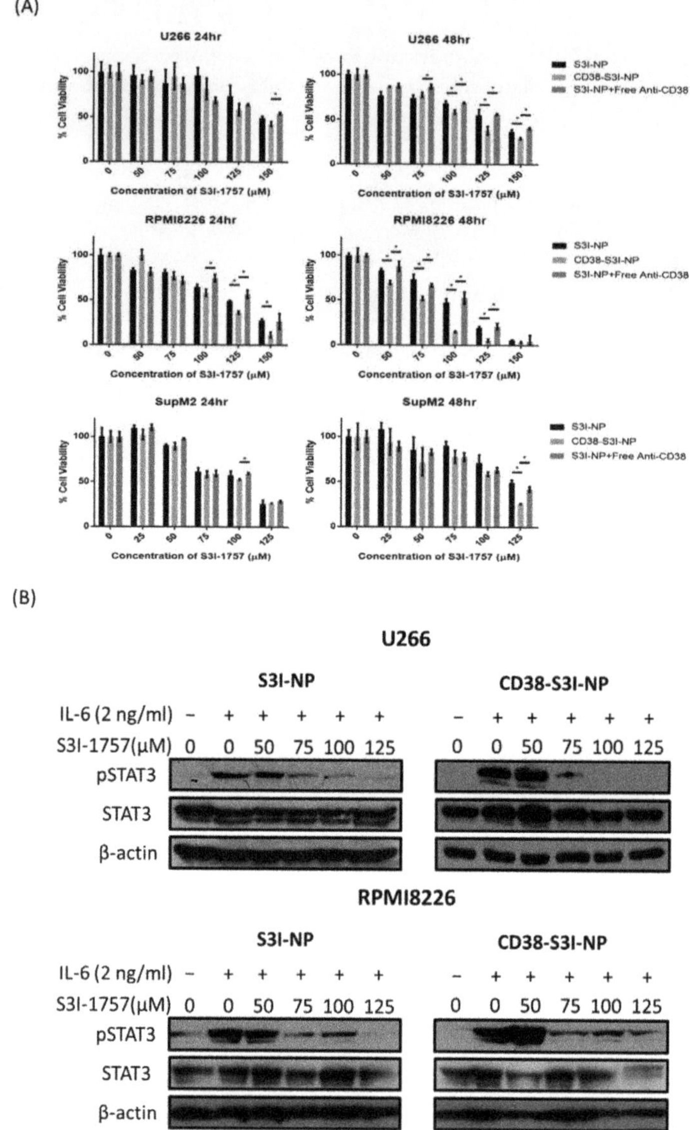

Figure 3. CD38-S3I-NP induces cytotoxicity and inhibits STAT3 activity in MM cells. (**A**) U266 and RPMI8226 cells were then treated with S3I-NP, CD38-S3I-NP, or S3I-NP with free CD38 antibody at a concentration which is equivalent to CD38-S3I-NP (1.4 mg/mL) with the presence of IL6 (2 ng/mL) for 24 and 48 h. Cell viability was measured using a 3-(4,5-dimethylthiazol-2-yl)-5-(3-carboxymethoxyphenyl)-2-(4-sulfophenyl)-2H-tetrazolium (MTS) cell viability assay in triplicate; * $p < 0.05$, via Student's *t*-test. (**B**) Western blot analysis of STAT3 and pSTAT3 levels in U266 and RPMI8226 cells treated with S3I-NP or CD38-S3I-NP with the presence of IL6 (2 ng/mL) for 24 h. β-actin was blotted as a loading control.

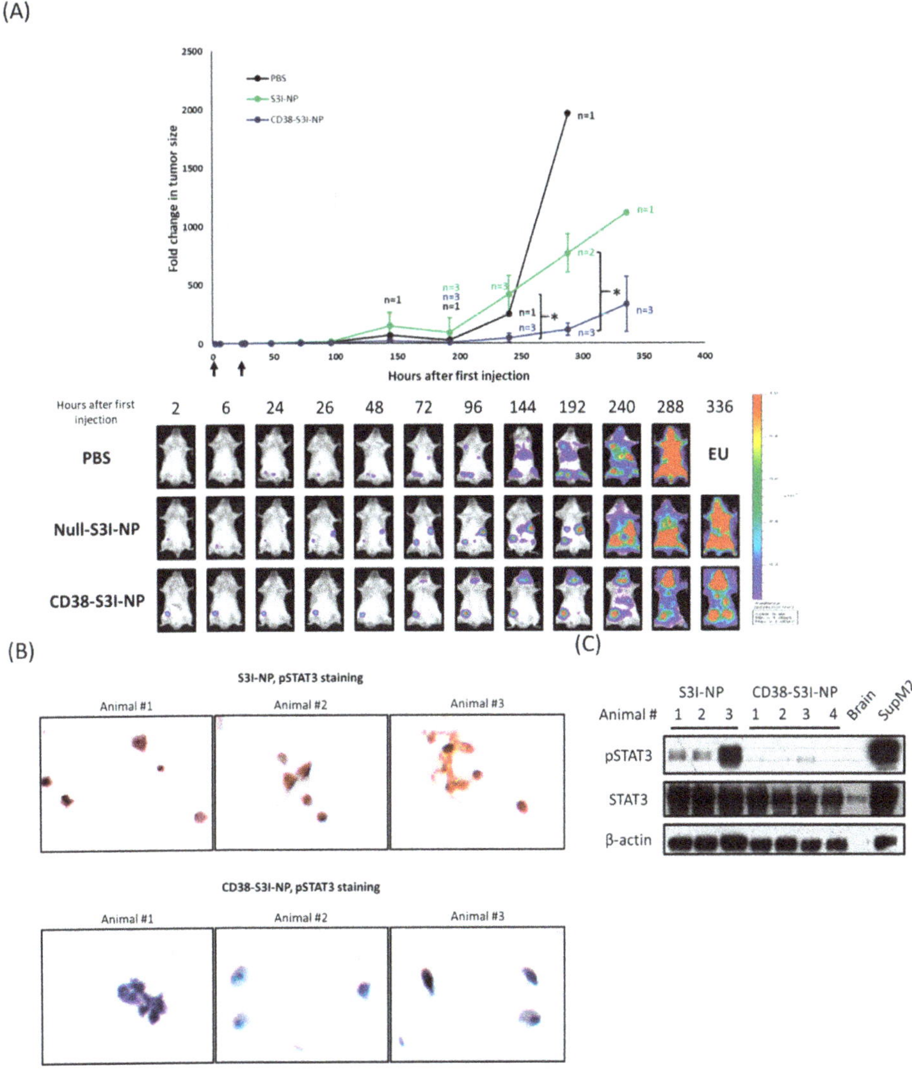

Figure 4. CD38-S3I-NP is more tumor suppressive than S3I-NP in MM xenograft. (**A**) SCID mice intravenously injected with PBS (black line, $n = 2$), 3 mg/kg S3I-NP (green line, $n = 4$) or CD38-S3I-NP (blue line, $n = 4$) every day for two days as indicated by arrows on the *x*-axis. Animal numbers other than the initial numbers at different time points are indicated. The tumor size was quantified by the bioluminescence intensity and normalized to the initial bioluminescence signal (i.e., 2 h post-injection). The representative bioluminescence images of animals treated with PBS, S3I-NP, or CD38-S3I-NP were shown. * $p < 0.05$, via Student's *t*-test; EU—euthanized. (**B**) The pSTAT3 levels in bone marrow mononuclear cells extracted from the SCID mice in (**A**) at the endpoint. Non-tumorous brain tissue from a SCID mouse was used as a negative control. SupM2 cells were used as a positive control for pSTAT3. β-actin was blotted as a loading control. (**C**) Immunocytochemical staining of pSTAT3 and in bone marrow mononuclear cells from (**B**). Each image represents bone marrow cells from one animal. The images were taken at a magnification of 400×.

We then assessed if the differences in tumor growth and survival between the CD38-S3I-NP and S3I-NP groups correlates with a difference in STAT3 down-regulation. To achieve this goal, we extracted bone marrow cells from specific bone fragments in which the involvement by MM was confirmed by the expression of bioluminescence. The expression of pSTAT3 was then detected using Western blot analysis and immunocytochemistry. The vast majority of cells extracted from the lesions (detailed in Materials and Methods) were confirmed to MM cells morphologically by using CD38 immunocytochemistry. pSTAT3 immunocytochemistry was then performed, and we found that MM cells from the CD38-S3I-NP group had no or barely detectable pSTAT3 signals, whereas MM cells from the S3I-NP group had relatively strong pSTAT3 signals in most of the cells examined (Figure 4B). This difference in pSTAT3 expression between the two groups was further confirmed by Western blot analysis (Figure 4C).

3. Discussion

It has been demonstrated that a number of novel NP drug delivery systems can significantly improve drug bioavailability in experimental models [26]. The mechanism for this improvement is believed to be attributed to the large size of the NP, which can pass through the leaky tumorous vasculature but not the normal blood vessels, resulting in the preferential accumulation of these NP in the tumors [26]. Many NP also allow the encapsulation of hydrophobic drugs which otherwise cannot be delivered to the cellular targets in their free form [27]. In this regard, our group had previously developed an NP which can effectively encapsulate and deliver cucurbitacin, a naturally occurring STAT3 inhibitor, making it more compatible for clinical use [28]. More recently, we used the same NP to encapsulate S3I-1757, a newly developed STAT3 inhibitor with higher potency and specificity, and we found that this NP (i.e., S3I-NP) demonstrated significant therapeutic efficacy against melanoma in a SCID mouse xenograft model [24].

To further improve the therapeutic potential of NP in treating cancer, various researchers have attempted to decorate NP with various targeting moieties such as monoclonal antibodies. For example, in a subcutaneous breast cancer xenograft mouse model, it was found that trastuzumab (an anti-HER2 antibody)-conjugated NP carrying doxorubicin was accumulated in the xenografts better than NP without antibodies, and this finding correlated with a >50% improvement in the reduction of tumor volume [7]. Similarly, in two other studies, cetuximab (an anti-EGFR antibody)-conjugated NP loaded with paclitaxel or gemcitabine was also found to show superior efficacy and tumor-targeting effects compared to NP without antibody conjugation [29,30]. Our group has previously found that anti-CD30 conjugated to a commercially available liposomal doxorubicin (Doxil®) was significantly more effective in treating ALK-positive anaplastic large cell lymphoma in a SCID mouse xenograft model [31]. In addition to the targeting function, antibodies conjugated on the surface of NP are also believed to prevent the uptake/removal of the drugs by the reticuloendothelial system [32].

The testing of STAT3 inhibitors in treating MM has been previously attempted. Previous studies have reported that novel STAT3 inhibitors such as atovaquone, SC99, and LLL12 can kill STAT3–active MM cells and significantly suppress subcutaneous tumors in SCID-mouse xenograft models [33–35]. However, these new anti-STAT3 agents appear to be structurally hydrophobic; without a carrier, their clinical uses will be limited as they are water-insoluble. Regarding CD38, we believed that this is a good target for MM because it is highly expressed in most cases of MM and its expression is relatively restricted in normal cells [36]. CD38 has been regarded as a therapeutic target for MM, and daratumumab is the first human anti-CD38 antibody approved by the Food and Drug Administration for treating MM [37,38]. To the best of our knowledge, our study is the second to report the development of an antibody-conjugated NP to treat MM. The first study was reported in early 2018, in which anti-CD38 conjugated NP carrying bortezomib exhibited a 2 to 3-fold increase in cell uptake by MM cells and a ~50% greater reduction of MM tumor growth compared to non-targeted NP [39]. These results are in line with our findings. Taken together, we believe that NP carrying a potent STAT3 inhibitor (such as S3I-1757) decorated by anti-CD38 is a reasonable approach to treat MM. Our data is in support of this concept.

The method we employed to conjugate anti-CD38 onto NP involved thiolation of the lysine residues of the constant region of immunoglobulins. The thiolated antibody then formed a highly stable thioester bond with the maleimide polymers. We believe that our conjugation method can provide two main advantages compared to that used in the other study of using anti-CD38 conjugated NP to treat MM, in which anti-CD38 was attached to NP via biotinylation [39]. First, the covalent thioester bond between the antibody and NP is substantially more stable than the non-covalent biotinylation bond, which likely results in a longer half-life of NP in vivo. Second, compared to biotinylation, the thiolation process of anti-CD38 is limited to its constant region, thus minimizing the risk of re-directing the antibodies in the wrong orientation (i.e., the hypervariable region of the immunoglobulin pointing inward). Thus, we believe that thiolation of the immunoglobulin can confer better stability and therapeutic efficacy to the NP.

Due to the NP barrier and the superior MM cell-targeting ability, we speculated that CD38-S3I-NP would possess a higher maximal tolerable dose and lower incidences of adverse effects compared to S3I-NP and free S3I-1757. Unlike OPB-31121, CD38-S3I-NP is not orally available since it can dissemble under conditions of adverse pH within the gastrointestinal tract. Thus, intravenous injection will be the best method to administer CD38-S3I-NP. Because STAT3 activity is linked to resistance to chemotherapy in MM, combined therapy of bortezomib, thalidomide, or dexamethasone with CD38-S3I-NP may improve response rates. One possible caveat of using CD38-S3I-NP in clinical settings is the fast clearance through the liver and kidneys. Therefore, approaches which prevent rapid NP clearance have to be developed to avoid toxicity to these organs.

4. Materials and Methods

4.1. Materials and Cell Culture

S3I-1757 (white powder with a molecular weight of 521.6 g/mol, soluble in DMSO) was obtained from Glixx Laboratories (Hopkinton, MA, USA). Methoxy-PEO (average molecular weight of 5000 g/mol), diisopropylamine (99%), benzyl chloroformate (tech 95%), sodium (in kerosin), butyllithium in hexane (2.5 M solution), palladium-coated charcoal, pyrene, and Cremophor® EL were purchased from Sigma Aldrich (St. Louis, MO, USA). α-benzyl carboxylate ε-caprolactone was prepared by Alberta Research Chemicals Incorporation (Edmonton, AB, Canada). Stannous octoate was purchased from MP Biomedicals Incorporation (Solon, OH, USA). All other chemicals were reagent grade. U266 and SupM2 cells were purchased from American Type Culture Collection (Manassas, VA, USA). RPMI8226 cells are a gift from Dr. Linda Pilarski (Department of Oncology, University of Alberta, Edmonton, AB, Canada). The cells were cultured in RPMI 1460 medium with L-glutamine (Life Technology, Carlsbad, CA, USA), 10% FBS (Life technology) and, 100U/mL penicillin/streptomycin (Sigma Aldrich). All cells were incubated at 37 °C supplied with 5% atmospheric CO_2.

4.2. Purification of Anti-CD38

The hybridoma cells (TBH-7) producing humanized anti-CD38 were cultured in RPMI1460 medium supplied with 10% fetal bovine serum. When the cells were confluent, they were transferred to RPMI1460 medium with 10% ultra-low IgG fetal bovine serum (ThermoFisher, Waltham, MA, USA). The cell supernatant was collected after 48 h. 150 mL of supernatant was concentrated to 5 mL using Amicon Ultra-15 Centrifugal Filter Unit (Millipore, Burlington, MA, USA). The concentrated supernatant was incubated in NAb™ Protein A/G Spin Column (ThermoFisher) for 10 min. The bound antibody was eluted out using the elution buffer (ThermoFisher). The concentration of purified anti-CD38 was measured by NanoDrop™ 1000 Spectrophotometer (ThermoFisher). The purified anti-CD38 was dialyzed in sterile PBS overnight prior to NP conjugation.

4.3. Preparation of NP

Poly-(ethylene oxide)-block-poly-(α-benzyl carboxylate ε-caprolactone) (PEO-b-PBCL) was synthesized using the method previously described [40]. In brief, α-benzyl carboxylate ε-caprolactone was mixed with methoxy-poly-(ethylene oxide) at 1:1.12 weight ratio with a trace amount of stannous octoate. The reaction mixture was incubated for 4 h at 140 °C in the vacuum oven and stopped by cooling the reaction at room temperature overnight. S3I-NP was prepared from PEO-b-PBCL block copolymers as previously described [24]. In brief, 20 mg of block copolymer was dissolved in tetrahydrofuran and was mixed with 2 mg S3I-1757 dissolved in DMSO. The mixture was incubated at room temperature with stirring overnight. The excess S3I-1757 was removed by centrifugation. For anti-CD38 conjugation, anti-CD38 was mixed with 2-imidothiolane at room temperature at pH 8.0 to synthesize thiolated anti-CD38. The maleimide PEO-b-PBCL was prepared by following a previously established protocol [41]. Micellized maleimide PEO-b-PBCL was mixed with thiolated anti-CD38. The anti-CD38 conjugated NP were mixed with S3I-NP in water to form CD38-S3I-NP through post-insertion method. The size and polydispersity index of S3I-NP and CD38-S3I-NP were measured by Zetasizer Nano®. The S3I-1757 concentrations in CD38-S3I-NP and S3I-NP were measured by high-performance liquid chromatography (HPLC) using a previously established protocol [24]. For the synthesis of Cy5.5-conjugated NP, a previously described protocol was followed [42].

4.4. In Vitro Release Assay

In vitro release assay was carried out as previously described [24]. In brief, 1 mL of CD38-S3I-NP or S3I-NP was put in a semi-permeable dialysis bag (molecular weight cutoff: 12,000–14,000 kDa). The bag was placed in sterile PBS and incubated in a shaking water bath at 37 °C. A 50 µL aliquot from the dialysis bag was collected at various time points for S3I-1757 concentration measurement by HPLC analysis. To maintain the total volume, 50 µL of PBS was added back to the dialysis bag after aliquot collection.

4.5. Cellular Uptake Assay

NP chemically conjugated with Cy5.5 (an amount equivalent to 0.2 ng Cy5.5) was added to 1.0×10^6 U266, RPMI8226 and SupM2 cells and cultured at 37 °C in dark for 4 h. Cells were washed with sterile PBS twice and subjected to flow cytometry (BD FACSCantoII, Franklin Lakes, NJ, USA) analysis using the Per-Cy5.5-APC fluorescence channel.

4.6. Cell Viability Assay

Cell proliferation was assessed by the CellTiter 96® AQueous one solution cell proliferation MTS assay (Promega, Madison, WI, USA). Approximately 2.5×10^4 U266 or RPMI8226 cells were seeded in each well of a 96-well plate and treated for 24 or 48 h, and 20 µL 3-(4,5-dimethylthiazol-2-yl)-5-(3-carboxymethoxyphenyl)-2-(4-sulfophenyl)-2H-tetrazolium (MTS) was added to each well. The absorbance of light at 490 nm was measured by a FLUOstar OPTIMA microplate reader (BMG Labtech, Cary, NC, USA). The IC_{50} values were calculated by Graph Prism 7 from the cell viability versus the logarithm of the concentration curve.

4.7. Western Blot Analysis

Total cell lysates were prepared and lysed with $1\times$ RIPA buffer ($10\times$ stock solution from Cell Signaling Technology, Danvers, MA, USA) with 0.05% protease inhibitor cocktail (Millipore) and 0.05% phosphatase inhibitor cocktail (Millipore). Protein concentrations were measured using a Pierce™ BCA Protein Assay Kit (ThermoFisher). Cell lysates treated with SDS were subject to SDS-PAGE and transferred onto a nitrocellulose membrane. The membrane was probed with anti-STAT3 (1:1000, CST, #9139), anti-pSTAT3 (Y705) (1:2000, CST, #9145), and anti-β-actin (1:1000, CST, #58169) diluted in 5% BSA in TBS-Tween20 (0.05%, *v/v*). These antibodies were probed with anti-mouse IgG conjugated with horseradish peroxidase (1:1000, Cell Signaling). The membrane was washed three times with TBS-T after secondary antibody treatment. The bands on the membrane were visualized with Pierce™ ECL Western blotting substrate (Thermo Scientific) and exposed to X-ray films (Fuji, Tokyo, Japan).

4.8. Preparation of U266-luc Cells by Lentiviral Transduction

Lentiviral particles carrying pLenti-Puro-luc were a kind gift from Dr. Kyle Potts and Dr. Mary Hitt (Department of Oncology, University of Alberta, Edmonton, AB, Canada). Also, 2×10^6 U266 cells were transduced with 2 mL of lentiviral particles and 6 µg/mL polybrene on a 6-well plate after 30 min of spin inoculation at $1000 \times g$. After 24 h, cells were washed and replenished with fresh lentiviral particles for another 24 h. Transduced cells were washed with PBS and resuspended in fresh growth medium for 2 days. The transduced cells were selected with 2 µg puromycin (ThermoFisher) in the medium. The luciferase activity of U266-luc cells was measured by bioluminescence imaging (Figure S2A). The growth rate and responsiveness to S3I-1757 of U266-luc were confirmed to be insignificantly different from parental U266 cells (Figure S2B,C).

4.9. In Vivo Studies Using MM Xenograft

The experimental protocols for all in vivo studies in this manuscript were approved by Animal Care and Use Committees, University of Alberta (#AUP00000282). Half of a million U266-luc was injected into SCID mice (Jackson, strain NOD.Cg-Prkdcscid Il2rgtm1Wjl/SzJ) intravenously. Twelve days after injection, 100 µL of S3I-NP or CD38-S3I-NP colloidal dispersions in dextrose 5 % was injected into the mice intravenously via the tail vein for two consecutive days at a dose of 3 mg/kg per day. The tumor size was measured by quantifying the total flux of bioluminescence signals detectable (radiance ranged between 4.00×10^5 and 1.00×10^7 p/sec/cm^2/scr) on the ventral side of each animal at various time points using Living Image Software (PerkinElmer, Waltham, MA, USA). At the endpoint defined by the approved protocol, the animals were euthanized and the bone marrow cells with MM bone lesion (visualized by bioluminescence) were removed from the femoral bone and split into two portions. One portion was stored at $-80\,°C$ as cell pellets for Western blot analysis, and another was stored in 4% paraformaldehyde at $4\,°C$ for immunocytochemistry.

4.10. Immunocytochemistry

Mouse bone marrow cells in 4% paraformaldehyde (Sigma Aldrich) were pelleted and resuspended in liquid histogel (Thermo Scientific) and transferred to a 15×15 mm^2 plastic mold (Leica, Wetzlar, Germany). Upon solidification, the cell-histogel was subjected to processing and embedding. The cell blocks were sectioned into 5-µm slides. Slides were rehydrated in xylene and decreasing concentrations of ethanol. The antigens were retrieved using $1\times$ citrate buffer (Sigma) by microwaving in a pressure cooker for 20 min. pSTAT3 (1:50, Santa Cruz, #sc-8059) and CD38 antibodies (1:100, Abcam, #ab108403) were diluted at 1:100 in antibody diluent (DAKO, Glostrup, Denmark). MACH2 mouse HRP polymer (Biocare Medical, Pacheco, CA, USA) was used as a secondary antibody. The chromogen and substrate were mixed and applied to each slide for 2 min for color development (DAKO).

4.11. Statistical Analysis

All numerical data in this study was presented as the mean from experiment replicates or independent experiments as described in the figure legends. For the comparison of IC$_{50}$ values in Table 2, two-way ANOVA and Tukey's multiple comparison test were employed. For survival analysis in the animal study, Statistical significance between groups was analyzed using the Mantel-Cox test ($\alpha = 0.05$). For the remaining comparisons, a Student's t-test with $\alpha = 0.05$ was used for the statistical analysis. The analysis was done using Microsoft Excel 365, except for Table 2, Figure 4A, and the overall survival analysis in the animal study, for which GraphPad Prism 7 was used for analysis.

5. Conclusions

In summary, this study has provided the proof-of-principle that the decoration of anti-CD38 on NP loaded with STAT3 inhibitors can further improve their therapeutic effects against MM. We also predict that the use of these NP can significantly lower the unwanted side effects of STAT3 inhibitors, as they are targeted to cancer cells, and preferentially released inside of these cells. These encouraging results provide sufficient basis for a Phase I clinical trial.

Supplementary Materials: The following are available online at http://www.mdpi.com/2072-6694/11/2/248/s1, Figure S1: MM cell lines express high levels of CD38, Figure S2: Characterization of U266-luc cells.

Author Contributions: Y.-H.H., A.L. and R.L. designed the experiments and wrote the manuscript. Y.-H.H., Y.M. and O.M. purified the CD38 monoclonal antibody. M.R.V. synthesized the antibody-conjugated polymers. Y.-H.H., C.W. and M.V. conducted the in vitro experiments. Y.-H.H., I.P. and A.H.S. conducted the animal experiments. F.S. optimized the protocols for characterization of the synthesized nanoparticles.

Funding: This study is supported by the funding of Alberta Innovates Health Solution (Collaborative Research and Innovation Opportunities—Project).

Acknowledgments: The authors would like to acknowledge the technical support from Kyle Potts and his supervisor Mary Hitt on the preparation of lentiviral particles carrying pLenti-Puro-luc plasmid.

Conflicts of Interest: The authors declare no conflict of interest.

References

1. Peer, D.; Karp, J.M.; Hong, S.; Farokhzad, O.C.; Margalit, R.; Langer, R. Nanocarriers as an emerging platform for cancer therapy. *Nat. Nanotechnol.* **2007**, *2*, 751–760. [CrossRef]

2. Fang, J.; Nakamura, H.; Maeda, H. The EPR effect: Unique features of tumor blood vessels for drug delivery, factors involved, and limitations and augmentation of the effect. *Adv. Drug Deliv. Rev.* **2011**, *63*, 136–151. [CrossRef]

3. Arruebo, M.; Valladares, M.; González-Fernández, Á. Antibody-conjugated nanoparticles for biomedical applications. *J. Nanomater.* **2009**, *2009*, 439389. [CrossRef]

4. Mi, Y.; Liu, X.; Zhao, J.; Ding, J.; Feng, S.S. Multimodality treatment of cancer with herceptin conjugated, thermomagnetic iron oxides and docetaxel loaded nanoparticles of biodegradable polymers. *Biomaterials* **2012**, *33*, 7519–7529. [CrossRef]

5. Parhi, P.; Sahoo, S.K. Trastuzumab guided nanotheranostics: A lipid based multifunctional nanoformulation for targeted drug delivery and imaging in breast cancer therapy. *J. Colloid Interface Sci.* **2015**, *451*, 198–211. [CrossRef]

6. Vivek, R.; Thangam, R.; NipunBabu, V.; Rejeeth, C.; Sivasubramanian, S.; Gunasekaran, P.; Muthuchelian, K.; Kannan, S. Multifunctional HER2-Antibody Conjugated Polymeric Nanocarrier-Based Drug Delivery System for Multi-Drug-Resistant Breast Cancer Therapy. *ACS Appl. Mater. Interfaces* **2014**, *6*, 6469–6480. [CrossRef]

7. Choi, W.I.; Lee, J.H.; Kim, J.Y.; Heo, S.U.; Jeong, Y.Y.; Kim, Y.H.; Tae, G. Targeted antitumor efficacy and imaging via multifunctional nano-carrier conjugated with anti-HER2 trastuzumab. *Nanomed. Nanotechnol. Biol. Med.* **2015**, *11*, 359–368. [CrossRef]

8. Punfa, W.; Yodkeeree, S.; Pitchakarn, P.; Ampasavate, C.; Limtrakul, P. Enhancement of cellular uptake and cytotoxicity of curcumin-loaded PLGA nanoparticles by conjugation with anti-P-glycoprotein in drug resistance cancer cells. *Acta Pharmacol. Sin.* **2012**, *33*, 823–831. [CrossRef]

9. Arya, G.; Vandana, M.; Acharya, S.; Sahoo, S.K. Enhanced antiproliferative activity of Herceptin (HER2)-conjugated gemcitabine-loaded chitosan nanoparticle in pancreatic cancer therapy. *Nanomed. Nanotechnol. Biol. Med.* **2011**, *7*, 859–870. [CrossRef]

10. Sreeranganathan, M.; Uthaman, S.; Sarmento, B.; Mohan, C.G.; Park, I.K.; Jayakumar, R. In vivo evaluation of cetuximab-conjugated poly(γ-glutamic acid)-docetaxel nanomedicines in EGFR-overexpressing gastric cancer xenografts. *Int. J. Nanomed.* **2017**, *12*, 7167–7182. [CrossRef]

11. Mukerjee, A.; Ranjan, A.P.; Vishwanatha, J.K. Targeted nanocurcumin therapy using annexin A2 antibody improves tumor accumulation and therapeutic efficacy against highly metastatic breast cancer. *J. Biomed. Nanotechnol.* **2016**, *12*, 1374–1392. [CrossRef]

12. Öztürk, K.; Esendağlı, G.; Gürbüz, M.U.; Tülü, M.; Çalış, S. Effective targeting of gemcitabine to pancreatic cancer through PEG-cored Flt-1 antibody-conjugated dendrimers. *Int. J. Pharm.* **2017**, *517*, 157–167. [CrossRef]

13. Punfa, W.; Suzuki, S.; Pitchakarn, P.; Yodkeeree, S.; Naiki, T.; Takahashi, S.; Limtrakul, P. Curcumin-loaded PLGA nanoparticles conjugated with anti-P-glycoprotein antibody to overcome multidrug resistance. *Asian Pac. J. Cancer Prev.* **2014**, *15*, 9249–9258. [CrossRef]

14. Hatakeyama, H.; Akita, H.; Ishida, E.; Hashimoto, K.; Kobayashi, H.; Aoki, T.; Yasuda, J.; Obata, K.; Kikuchi, H.; Ishida, T.; et al. Tumor targeting of doxorubicin by anti-MT1-MMP antibody-modified PEG liposomes. *Int. J. Pharm.* **2007**, *342*, 194–200. [CrossRef]

15. Sun, T.; Wu, H.; Li, Y.; Huang, Y.; Yao, L.; Chen, X.; Han, X.; Zhou, Y.; Du, Z. Targeting transferrin receptor delivery of temozolomide for a potential glioma stem cell-mediated therapy. *Oncotarget* **2017**, *8*, 74451–74465. [CrossRef]

16. Bharti, A.C.; Shishodia, S.; Reuben, J.M.; Weber, D.; Alexanian, R.; Raj-Vadhan, S.; Estrov, Z.; Talpaz, M.; Aggarwal, B.B. Nuclear factor–κB and STAT3 are constitutively active in CD138 + cells derived from multiple myeloma patients, and suppression of these transcription factors leads to apoptosis. *Blood* **2004**, *103*, 3175–3184. [CrossRef]

17. Liu, T.; Fei, Z.; Gangavarapu, K.J.; Agbenowu, S.; Bhushan, A.; Lai, J.C.; Daniels, C.K.; Cao, S. Interleukin-6 and JAK2/STAT3 signaling mediate the reversion of dexamethasone resistance after dexamethasone withdrawal in 7TD1 multiple myeloma cells. *Leuk. Res.* **2013**, *37*, 1322–1328. [CrossRef]

18. Zhang, X.D.; Baladandayuthapani, V.; Lin, H.; Mulligan, G.; Li, B.-Z.; Esseltine, D.L.W.; Qi, L.; Xu, J.; Hunziker, W.; Barlogie, B.; et al. Tight jnction protein 1 modulates proteasome capacity and proteasome inhibitor sensitivity in multiple myeloma via EGFR/JAK1/STAT3 signaling. *Cancer Cell* **2016**, *29*, 639–652. [CrossRef]

19. Wu, W.; Ma, D.; Wang, P.; Cao, L.; Lu, T.; Fang, Q.; Zhao, J.; Wang, J. Potential crosstalk of the interleukin-6-heme oxygenase-1-dependent mechanism involved in resistance to lenalidomide in multiple myeloma cells. *FEBS J.* **2016**, *283*, 834–849. [CrossRef]

20. Zhang, X.; Sun, Y.; Pireddu, R.; Yang, H.; Urlam, M.K.; Lawrence, H.R.; Guida, W.C.; Lawrence, N.J.; Sebti, S.M. A novel inhibitor of STAT3 homodimerization selectively suppresses STAT3 activity and malignant transformation. *Cancer Res.* **2013**, *73*, 1922–1933. [CrossRef]

21. Zhang, X.; Yue, P.; Fletcher, S.; Zhao, W.; Gunning, P.T.; Turkson, J. A novel small-molecule disrupts Stat3 SH2 domain–phosphotyrosine interactions and Stat3-dependent tumor processes. *Biochem. Pharmacol.* **2010**, *79*, 1398–1409. [CrossRef]

22. Schust, J.; Sperl, B.; Hollis, A.; Mayer, T.U.; Berg, T. Stattic: A small-molecule inhibitor of STAT3 activation and dimerization. *Chem. Biol.* **2006**, *13*, 1235–1242. [CrossRef]

23. Oh, D.-Y.; Lee, S.-H.; Han, S.-W.; Kim, M.-J.; Kim, T.-M.; Kim, T.-Y.; Heo, D.S.; Yuasa, M.; Yanagihara, Y.; Bang, Y.J. Phase I Study of OPB-31121, an Oral STAT3 Inhibitor, in Patients with Advanced Solid Tumors. *Cancer Res. Treat.* **2015**, *47*, 607–615. [CrossRef]

24. Soleimani, A.H.; Garg, S.M.; Paiva, I.M.; Vakili, M.R.; Alshareef, A.; Huang, Y.-H.; Molavi, O.; Lai, R.; Lavasanifar, A. Micellar nano-carriers for the delivery of STAT3 dimerization inhibitors to melanoma. *Drug Deliv. Transl. Res.* **2017**, 1–11. [CrossRef]

25. Zamo, A.; Chiarle, R.; Piva, R.; Howes, J.; Fan, Y.; Chilosi, M.; Levy, D.E.; Inghirami, G. Anaplastic lymphoma kinase (ALK) activates Stat3 and protects hematopoietic cells from cell death. *Oncogene* **2002**, *21*, 1038–1047. [CrossRef]

26. Nakamura, Y.; Mochida, A.; Choyke, P.L.; Kobayashi, H. Nanodrug Delivery: Is the Enhanced Permeability and Retention Effect Sufficient for Curing Cancer? *Bioconjug. Chem.* **2016**, *27*, 2225–2238. [CrossRef]

27. Martin, C.; Aibani, N.; Callan, J.F.; Callan, B. Recent advances in amphiphilic polymers for simultaneous delivery of hydrophobic and hydrophilic drugs. *Ther. Deliv.* **2016**, *7*, 15–31. [CrossRef]

28. Molavi, O.; Ma, Z.; Mahmud, A.; Alshamsan, A.; Samuel, J.; Lai, R.; Kwon, G.S.; Lavasanifar, A. Polymeric micelles for the solubilization and delivery of STAT3 inhibitor cucurbitacins in solid tumors. *Int. J. Pharm.* **2008**, *347*, 118–127. [CrossRef]

29. Wang, X.B.; Zhou, H.Y. Molecularly targeted gemcitabine-loaded nanoparticulate system towards the treatment of EGFR overexpressing lung cancer. *Biomed. Pharmacother.* **2015**, *70*, 123–128. [CrossRef]

30. Karra, N.; Nassar, T.; Ripin, A.N.; Schwob, O.; Borlak, J.; Benita, S. Antibody conjugated PLGA nanoparticles for targeted delivery of paclitaxel palmitate: Efficacy and biofate in a lung cancer mouse model. *Small* **2013**, *9*, 4221–4236. [CrossRef]

31. Molavi, O.; Xiong, X.B.; Douglas, D.; Kneteman, N.; Nagata, S.; Pastan, I.; Chu, Q.; Lavasanifar, A.; Lai, R. Anti-CD30 antibody conjugated liposomal doxorubicin with significantly improved therapeutic efficacy against anaplastic large cell lymphoma. *Biomaterials* **2013**, *34*, 8718–8725. [CrossRef]

32. Badkas, A.; Frank, E.; Zhou, Z.; Jafari, M.; Chandra, H.; Sriram, V.; Lee, J.Y.; Yadav, J.S. Modulation of in vitro phagocytic uptake and immunogenicity potential of modified Herceptin®-conjugated PLGA-PEG nanoparticles for drug delivery. *Colloids Surf. B Biointerfaces* **2018**, *162*, 271–278. [CrossRef]

33. Lin, L.; Benson, D.M.; Deangelis, S.; Bakan, C.E.; Li, P.K.; Li, C.; Lin, J. A small molecule, LLL12 inhibits constitutive STAT3 and IL-6-induced STAT3 signaling and exhibits potent growth suppressive activity in human multiple myeloma cells. *Int. J. Cancer* **2012**, *130*, 1459–1469. [CrossRef]

34. Zhang, Z.; Mao, H.; Du, X.; Zhu, J.; Xu, Y.; Wang, S. A novel small molecule agent displays potent anti-myeloma activity by inhibiting the JAK2-STAT3 signaling pathway. *Oncotarget* **2016**, *7*, 9296–9308. [CrossRef]

35. Xiang, M.; Kim, H.; Ho, V.T.; Walker, S.R.; Bar-Natan, M.; Anahtar, M.; Liu, S.; Toniolo, P.A.; Kroll, Y.; Jones, N.; et al. Gene expression-based discovery of atovaquone as a STAT3 inhibitor and anticancer agent. *Blood* **2016**, *128*, 1845–1853. [CrossRef]

36. Malavasi, F.; Funaro, A.; Alessio, M.; DeMonte, L.B.; Ausiello, C.M.; Dianzani, U.; Lanza, F.; Magrini, E.; Momo, M.; Roggero, S. CD38: A multi-lineage cell activation molecule with a split personality. *Int. J. Clin. Lab. Res.* **1992**, *22*, 73–80. [CrossRef]

37. De Weers, M.; Tai, Y.-T.; van derVeer, M.S.; Bakker, J.M.; Vink, T.; Jacobs, D.C.H.; Oomen, L.A.; Peipp, M.; Valerius, T.; Slootstra, J.W.; et al. Daratumumab, a Novel Therapeutic Human CD38 Monoclonal Antibody, Induces Killing of Multiple Myeloma and Other Hematological Tumors. *J. Immunol.* **2011**, *186*, 1840–1848. [CrossRef]

38. Beum, P.V.; Lindorfer, M.A.; Peek, E.M.; Stukenberg, P.T.; deWeers, M.; Beurskens, F.J.; Parren, P.W.; van de Winkel, J.G.; Taylor, R.P. Penetration of antibody-opsonized cells by the membrane attack complex of complement promotes Ca^{2+} influx and induces streamers. *Eur. J. Immunol.* **2011**, *41*, 2436–2446. [CrossRef]

39. Dela Puente, P.; Luderer, M.J.; Federico, C.; Jin, A.; Gilson, R.C.; Egbulefu, C.; Alhallak, K.; Shah, S.; Muz, B.; Sun, J.; et al. Enhancing proteasome-inhibitory activity and specificity of bortezomib by CD38 targeted nanoparticles in multiple myeloma. *J. Control. Release* **2018**, *270*, 158–176. [CrossRef]

40. Mahmud, A.; Xiong, X.B.; Lavasanifar, A. Novel self-associating poly(ethylene oxide)-block-poly(ε-caprolactone) block copolymers with functional side groups on the polyester block for drug delivery. *Macromolecules* **2006**, *39*, 9419–9428. [CrossRef]

41. Kogan, T. The synthesis of substituted methoxy-poly (ethyleneglycol) derivatives suitable for selective protein modification. *Synth. Commun.* **1992**, *22*, 2417–2424. [CrossRef]

42. Garg, S.M.; Paiva, I.M.; Vakili, M.R.; Soudy, R.; Agopsowicz, K.; Soleimani, A.H.; Hitt, M.; Kaur, K.; Lavasanifar, A. Traceable PEO-poly(ester) micelles for breast cancer targeting: The effect of core structure and targeting peptide on micellar tumor accumulation. *Biomaterials* **2017**, *144*, 17–29. [CrossRef] [PubMed]

Review

Companion Animals as Models for Inhibition of STAT3 and STAT5

Matthias Kieslinger *, Alexander Swoboda, Nina Kramer, Barbara Pratscher, Birgitt Wolfesberger and Iwan A. Burgener

Division of Small Animal Internal Medicine, Department of Companion Animals and Horses, University of Veterinary Medicine Vienna, 1210 Vienna, Austria; Alexander.Swoboda@vetmeduni.ac.at (A.S.); Nina.Kramer@vetmeduni.ac.at (N.K.); Barbara.Pratscher@vetmeduni.ac.at (B.P.); Birgitt.Wolfesberger@vetmeduni.ac.at (B.W.); Iwan.Burgener@vetmeduni.ac.at (I.A.B.)
* Correspondence: Matthias.Kieslinger@vetmeduni.ac.at; Tel.: +43-1-25077-5627

Received: 6 November 2019; Accepted: 13 December 2019; Published: 17 December 2019

Abstract: The use of transgenic mouse models has revolutionized the study of many human diseases. However, murine models are limited in their representation of spontaneously arising tumors and often lack key clinical signs and pathological changes. Thus, a closer representation of complex human diseases is of high therapeutic relevance. Given the high failure rate of drugs at the clinical trial phase (i.e., around 90%), there is a critical need for additional clinically relevant animal models. Companion animals like cats and dogs display chronic inflammatory or neoplastic diseases that closely resemble the human counterpart. Cat and dog patients can also be treated with clinically approved inhibitors or, if ethics and drug safety studies allow, pilot studies can be conducted using, e.g., inhibitors of the evolutionary conserved JAK-STAT pathway. The incidence by which different types of cancers occur in companion animals as well as mechanisms of disease are unique between humans and companion animals, where one can learn from each other. Taking advantage of this situation, existing inhibitors of known oncogenic STAT3/5 or JAK kinase signaling pathways can be studied in the context of rare human diseases, benefitting both, the development of drugs for human use and their application in veterinary medicine.

Keywords: cancer models; companion animals; STAT3; STAT5; comparative oncology

---

1. Introduction

Almost half of all households in the United States have at least one companion animal. This means that approximately 77 million dogs and 58 million cats share a common environment with their human owners and are largely exposed to the same health risk factors [1]. In the absence of significant cardiovascular disease, cancer is the number one cause of death for dogs, killing between 40% and 50% of individuals older than 10 years, and between 20% and 25% regardless of age [2–4]. Numbers are less detailed for cats, but the overall tumor incidence ranges between 30% and 35% [5]. The prevalence of cancer in companion animals has increased in the last decades, which may be the result of a real increase in cancer incidence, an increase in the population of companion animals at risk or the awareness and willingness of the animal owners to pursue diagnostic and treatment options [6]. While the full spectrum of tumor types seen in humans also occurs in cats and dogs, the rates for individual tumor types are often different. Canine osteosarcoma, soft tissue sarcoma and feline non-Hodgkin's lymphoma for example are significantly higher than in humans (Table 1), whereas other tumor entities like lung, prostate and colon tumors are rare in companion animals. Cancer in companion animals, particularly in dogs, resembles cancer in humans in many ways, including spontaneous disease occurring without an isogenic background or genetic engineering and shared environmental and societal status with owners. Further similarities in chronology of the disease adapted to lifespan, organization into various well-characterized breeds that show different incidences of tumor types, and shared environmental and societal status with owners makes them attractive objects for comparison.

Table 1. Incidence rates of various tumor types from human, dog and cat.

Tissue	Human	Dog	Cat
Mammary	127.5 [7]	250 [8]	13–25 [9]
Melanoma	22.2 [7]	19.8 [10]	ND
Testes	5.9 [7]	16.7 [11]	ND
Connective Tissue	3.5 [7]	40.1 [10]	17.0 [1]
Skin	98.85 [12]	103.3 [10]	34.7 [1]
Oral	11.3 [7]	20.4 [1]	11.6 [1]
NHL/Leukemia	33.7 [7]	76.3 [13]	41 [14]
Bone	1.0 [7]	27.2 [15]	3.1–4.9 [16]

Numbers represent cases per 100,000. ND = not determined.

The mouse has been an extremely useful tool to gain cellular and mechanistic understanding into the development of cancer [17]. The concept of oncogenes and tumor suppressors balancing cellular proliferation of multicellular organisms has been proven in vivo in genetically modified mice [18,19]. The use of mice deleted for a single gene has allowed us to determine the involvement of signaling pathways, genetic regulators like transcription factors, epigenetic factors, etc. in the development and sustained growth of cancer. Overall, experiments in mouse models have been invaluable in understanding the mechanistic basis of cancer biology. However, while offering critical insights into basic concepts, murine models underrepresent the heterogeneity and complex interplay between human immune and cancer cells [20,21].

The Janus kinase (JAK)—signal transducer and activator of transcription (STAT) signaling pathway—provides a fast and efficient way for relaying signals from the extracellular space to the nucleus and modifying gene expression [22]. Main targets of this pathway represent regulators of cell division and apoptosis as broadly discussed in several publications in this special issue [23–25]. As these are central aspects in the development of cancer, it is not surprising that members of this pathway are over-activated in many cancers [26,27]. The kinases of this pathway, JAK 1-3 and TYK2 are constantly activated in several different tumor types and are subject to pharmacological inhibition. Recently, also STAT proteins have come into focus of cancer researchers. Particularly STAT3 and STAT5 are activated in over 70% of all human cancer types and constitute a critical node in the signaling networks of tumor cells [28].

This review will highlight, why companion animals and particularly the dog represent an attractive link between murine models, addressing basic mechanistic aspects and human diseases specifically in the context of JAK-STAT signaling.

2. Preclinical Models

The vast majority of cancer models currently are represented by mice, and their fundamental importance for preclinical research is clearly established [20,29]. These animals come as various strains that have been inbred over many generations and thereby are genetically highly homogenous [30]. When used for experiments, they are matched for age, sex and size, receive the same sterile diets and are housed under specific pathogen-free (SPF) conditions. All of these factors are controlled for in order to generate standardized conditions that allow us to draw scientifically sound conclusions based on the variation of one single parameter [31]. Furthermore, this reduction of variables and thereby "noise" allows us to reduce the number of animals necessary to reach statistically significant results. While this concept has proven its scientific merits and is logical within itself, it is questionable, whether young, sex-matched and inbred mice on sterile nutrition are a good representation of the typically older, obese and genetically diverse human cancer patients [32].

The controlled environment also affects the outcome of cancer-related experiments. For example, it has become clear recently, that the microbiome influences the response to cancer treatment [33,34]. Accordingly, mice raised under SPF conditions in various research institutions show differences in the composition of their gut microbiome, affecting tumor growth rates [35]. Again, while differences in the gut microbiome are likely to exist in human cancer patients, the single standardized composition under which these experiments are carried out in laboratory mice are very likely no close representation of the much broader spectrum in humans. Additionally, rodents have adopted to most ecosystems metabolically, but in regard to drug metabolism, due to growth-hormone regulated p450 cytochrome components, pharmacodynamic and pharmacokinetic properties, canines are by far superior and therefore used as key models for FDA-drug approved testing [36].

Another important aspect is the genetic background and its modification particularly with respect to the immune system. Mice used for tumor experiments usually are highly inbred, reducing genetic variability [37]. Experiments regarding the consequences of a narrow genetic background have demonstrated potential phenotypic tilting, which may result in unrepresentative biased phenotypes [38]. Furthermore, the buffering of genetic variation, including disease-causing mutations, is impaired under these circumstances [39]. One effort to overcome limitations due to a narrow genetic background is to establish new reference populations derived from the crossing of several different mouse strains, as exemplified by the Collaborative Cross project [40]. By broadening and defining the genetic basis, it offers the perspective of enhancing genetic stability and reproducibility, thereby also representing a new and potentially better resource of murine models for human diseases.

Additionally, and probably of higher practical relevance is the fact that development and proliferation of tumor cells happens in a complex interplay with cells of the immune system [41]. However, cancer-studies in mice are often performed in the absence of a fully functional immune system, using immunocompromised mice as hosts for transplantations of human cell lines, patient-derived xenografts or human tumor-derived immune cell xenografts [42]. The immune system constitutes the major player in the counter-selection against tumor cells, thereby necessitating evasion or adaption strategies on the side of the tumor cells. Since this aspect is missing in such mouse models, the results from experiments thereof likely reflect only partial aspects of tumor biology and challenge their biological relevance.

One way to overcome these problems are humanized mice, which express human instead of murine components of the immune system like major histocompatibility complex, allowing the transplantation of tumor cells in an immunologically at least partially competent environment [43]. These models clearly represent a step forward, however, they are costly, technically complicated and still do not represent all components of a functional and homogenous immune system [44]. Moreover, human patients that have developed cancer are not living under special pathogen-free conditions and are very likely to have chronic viral infections, like the Epstein–Barr virus, cytomegalovirus or herpes simplex, which are present in up to 90% of the total population [45–47]. Such chronic infections exert a constant pressure and shape the immune system, which is not the case in mice under SPF conditions. Co-housing of "dirty" outbred immunocompetent mice could be a way for improvement, still mammalian viral species barriers exist, that make companion animals superior in these infectious aspects.

Finally, genetically engineered mice harboring the deletion of a tumor suppressor gene or the ectopic expression of an oncogene or combinations thereof can be used [48]. In this case, the tumors develop in the presence of a competent immune system and problems related to cross-species compatibility do not arise. This setting also enables the introduction of defined mutations that occur in human tumors. However, this comes at the price of costly and time-consuming development, often requiring years of work before availability. Furthermore, whereas the targeted insertion of defined mutations reflecting the human situation is a clear advantage, cancer development is a multi-step process, and the heterogeneity of these further steps is often different between engineered mice and human patients [49,50].

Overall, the mouse has been highly instructive in determining genes involved and their mechanistic contribution to the origin and development of cancer [51]. However, the limitations of this model, in particular the differences in environment and microbiome, life span, tumor etiology and genetic status may be the reason why certain aspects are not reflected closely in this model, resulting in only 11% of oncology drugs that work in mice being approved for human use [52]. Therefore, it is desirable from a translational perspective to add another layer that closer reflects human biology and cancer development.

3. Advantages and Disadvantages of Canine Tumor Models

Roughly 4,000,000 dogs and a similar number of cats are diagnosed with cancer each year in the U.S. [53]. Although exact epidemiological data are not available for companion animals, this translates into approximately 5300 cases per 100,000 dogs, which is around ten times higher than in humans with 500 cases per 100,000 persons. This large number of pets provides the opportunity to study spontaneous cancers that are highly similar to those occurring in humans, especially since most pet owners are highly motivated to seek out novel treatments for their companion animals.

There are several advantages of using companion animals as models for human cancer. Among them is the fact that tumors arise spontaneously, just as in humans, and that tumor initiation and progression are influenced by similar factors like age, nutrition, sex, reproductive status and environmental exposure [54]. The risk for developing cancer of the nasal cavity for example is increased up to 60% in animals that are kept by smokers in comparison to pets of non-smoking owners [55,56]. Although there are differences in the diet of humans and companion animals, many components such as meat, vegetables and carbohydrates are derived from the same sources and are consumed non-sterile as opposed to mice under SPF conditions [57]. Furthermore, studies have shown that the contact between owner and pet leads to a large overlap in the microbiome, the importance of which for human tumor development has come into focus recently [58].

Pets, and in particular dogs are large and relatively outbred in comparison to laboratory mice. In fact, the genetic variation across dog breeds or in mixed breeds is similar to the variation in humans on the basis of single nucleotide polymorphisms [59]. In individual pure breeds however, the level of genetic diversity is more restricted [60]. The canine genome has been sequenced with a coverage of 99%, revealing that the approximately 19,000 genes identified in the dog match to homologous or orthologous genes in humans [59]. Actually, for many gene families, particularly for those associated with cancer, the homologies are significantly closer than the relationship between human and mouse [61]. Accordingly, most oncogenes and tumor suppressors that are known from human cancers have been shown to contribute to canine cancers [62].

Dogs and cats of all breeds develop cancer, and the spectrum of cancers seen in companion animals is as diverse as that seen in human patients [63]. The dog is the species in which comparative oncology has shown the most growth, and where it is best characterized [64]. Interestingly, there are breed-specific differences as to the cancer subtypes, reflecting the underlying genetics of the various breeds (Table 2). Mast cell tumors and gliomas for example are over-represented in Boxers, Staffordshire, Weimaraner and Golden Retriever, osteosarcoma in Rottweilers, Greyhounds and Golden Retrievers, bladder cancer in Scottish Terriers, histiocytic sarcomas in Flat-Coated Retrievers and Bernese Mountain Dogs and melanoma and gastric carcinoma in Chow-Chows [65–67].

Table 2. Oncological disposition of various dog breeds.

Breeds	Most Frequent Tumor Types
Bernese Mountain Dog	Histiocytic sarcoma [68], Lymphoma [68,69], Osteosarcoma [68]
Boxer	Glioma [67,69], Mast cell tumor [10,67]
Flat-Coated Retriever	Soft tissue sarcoma, Histiocytic sarcoma, Hemangiosarcoma [70]
Golden Retriever	Mast cell tumor, Lymphoma, Oral Melanoma, Fibrosarcoma [67]
Magyar Viszla	Mast cell tumor, Hemangiosarcoma, Lymphoma [10,71]
Giant Schnauzer	Epidermal tumor, Hair follicle tumor, Melanocytic tumor [10]
Airedale Terrier	Melanoma [72], Lymphoma [67], Prostatic carcinoma [67]
Bullmastiff	Mast cell tumor, Lymphoma [67]
St. Bernard	Lymphoma [67], Osteosarcoma [73]
Irish Wolfhound	Osteosarcoma [67,74], Lymphoma [75]

The most frequent tumor types of dog breeds with high tumor incidence.

The global expression pattern of canine and human osteosarcoma for example shows a strong similarity, and cluster analysis of orthologous gene signatures does not segregate human and canine tumors [73]. Finally, many chemotherapy protocols used for the treatment of canine cancers have been adopted from human medicine. The same chemotherapeutics used in human lymphoma for example are also active in canine lymphoma (e.g., vincristine, cyclophosphamide, doxorubicin, mitoxantrone, cytarabine and methotrexate), and drugs that are ineffective in human lymphomas are also inactive in canines (i.e., gemcitabine, cisplatin and carboplatin) [5].

One of the biggest advantages of companion animals as models for human cancer is the spontaneous development of tumors in the presence of an intact immune system. Immune cells pose a significant barrier to the development of cancer, and undergo changes themselves, as cancer cells co-opt the immune response [76,77]. As a result, the tumor influences innate as well as adaptive immune cells to become regulatory, rather than tumoricidal [78,79]. This interplay results in the selection of tumor cells that are invisible to anti-tumor T-cell-mediated destruction, and is central for tumor editing and immune evasion. Additionally, similar to humans, dogs with advanced cancer exhibit intrinsic T-cell defects as well as T-cell exhaustion [80]. Immune cell interplay with cancer cells is a JAK-STAT3/5 affair, as detailed in the special issue in several articles [24,25,81].

Like all other model systems, companion animals have strengths as well as weaknesses, both practically and conceptually. The biggest hurdle, when working with cats and dogs in research is the paucity of investigational tools. Many antibodies and recombinant products that are available for humans and mice do not show cross-species reactivity. However, the sequencing of the canine genome and the development of genome editing via the CRISPR/Cas9 technology has relieved many of the restrictions [82]. It is now possible to introduced genetic alterations with high efficiency into any desired locus, facilitating for example the visualization of specific cell types via the expression of marker proteins, or targeted deletion of single genes [83,84]. Conceptually, the lack of standardized housing conditions, making it difficult to control for variables, has been held against non-rodent models. However, as pointed out here, this represents an advantage when it comes to closely mimicking the translational aspects in tumor biology.

The potential of companion animals as biological models between the basic mechanistic work that is possible in the mouse and translation to humans is demonstrated by 1. The high conservation at the genomic level, 2. The involvement of similar genetic and environmental risk factors, 3. The successful use of canine cancer as biological models for the early development of bone marrow transplantation protocols and 4. canine trials for the development of drug level and exposure durations. This sets an ideal stage to combine new perspectives of targeted therapies and specific molecular inhibitors in the field of comparative oncology for the benefit of human as well as veterinary medicine (Figure 1). Furthermore, this approach is not limited to cancer but applicable to any comparative condition including infectious or inflammatory diseases like inflammatory bowel disease or pre-malignant conditions like adenoma formation or clonal hematopoiesis.

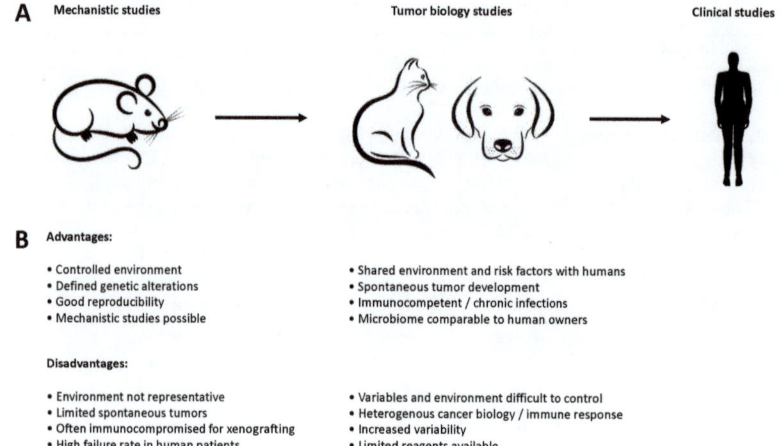

Figure 1. Advantages and disadvantages of different models during drug discovery. (**A**) Companion animals can be used as an intermediate step between the mechanistic work in murine models and clinical studies in humans, particularly with regard to comparative aspects of tumor biology. (**B**) Advantages and disadvantages of the individual models for translation into human clinical studies.

4. Relevance and Conservation of the JAK-STAT Signaling Pathway

JAK-STAT proteins constitute an evolutionary conserved signaling pathway [85]. Ligand binding of receptors leads to the activation of JAK kinases and STAT proteins, inducing transcription. The family of JAK kinases consists of four members, JAK1-3 and TYK2, and there are seven highly homologous STAT proteins, STAT1–4, STAT5a and STAT5b and STAT6 [22,86]. For details on the mechanism see also other reviews in this issue [87,88]. As such, this pathway provides a remarkably elegant and straightforward mechanism to transduce signals from receptors to the nucleus.

All family members show the same structural organization, i.e., an N-terminal domain required for oligomerization of dimers into tetramers, a coiled-coil domain, a DNA-binding domain, a linker domain, a Src homology 2 (Sh2) domain for dimerization and a C-terminal transactivation domain (Figure 2). Functionally, STAT2, STAT4 and STAT6 regulate specific immune cell responses, whereas STAT1, STAT3 and STAT5 have diverse physiological roles. STAT1 is mostly involved in immunity, host defense against pathogens and cell death, stimulating the transcription of pro-inflammatory and anti-proliferative genes like caspases, *NOS2*, *MDM2*, *CDKN1A* and *CDKN1B*. On the contrary, STAT3 and STAT5 are mostly involved in cell proliferation and prevention of apoptosis, activating the transcription of genes like *CCND1*, *BIRC5*, *c-MYC*, *VEGF*, *MCL1*, *BCL2L1* and *BCL2*. Additionally, STAT3 can also be found in mitochondria, where it supports RAS-dependent malignant transformation via sustained altered glycolytic and oxidative phosphorylation [89,90]. Given their roles in the stimulation of cellular proliferation, the prevention of apoptosis and the stimulation of metabolism, STAT5, and even more so STAT3, are activated in nearly 70% of solid and hematological human tumors [91–93].

Silencing or inhibition of STAT3 or STAT5 signaling impairs tumor growth and survival in murine and human studies, while only slightly affecting normal differentiated cells [94–97]. These findings lead to the concept of STAT3 and STAT5 constituting a "signaling bottleneck" situation for tumor cells, making them attractive targets for inhibition [98]. However, caution has to be exerted with regard to tissue-specificity, as tumor-suppressive functions have been ascribed to STAT3 in neuronal, hepatic and colorectal tumors and to STAT5 in breast cancer [99,100].

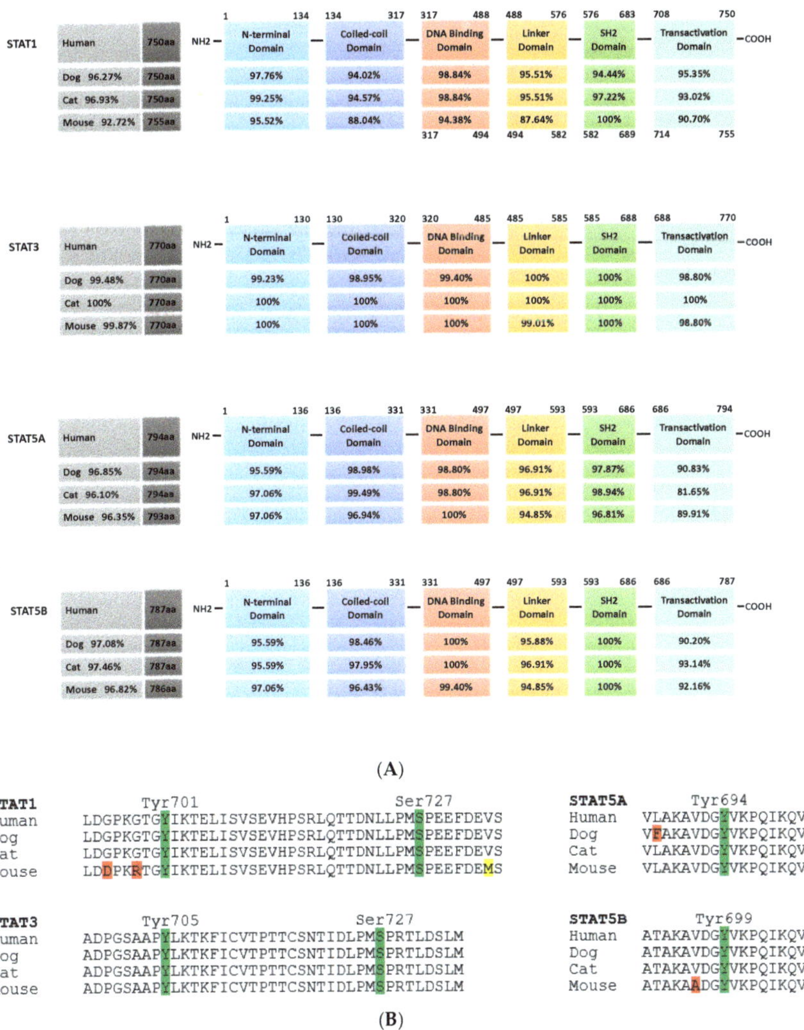

(A)

(B)

Figure 2. Cross-species conservation of STAT protein domains. (**A**) STAT1, STAT3, STAT5a and STAT5b from dog, cat and mouse are analyzed for their overall homology compared to the respective human protein (grey boxes, left). In the schematic representation of STAT protein domains, the amino acid positions are indicated above. All proteins share the same domain positions, except for murine STAT1, which has a five amino acid insertion in the DNA binding domain (numbers below the scheme indicate the aa position in this case). Percentages in the domain boxes of dog, cat and mouse STAT proteins show the homology of each domain to the human counterpart. Analyses were carried out using ClustalX. (**B**) Comparison of key phosphorylation sites in the transactivation domain of STAT1, STAT3, STAT5a and STAT5b from dog, cat and mouse to the human sequence. Amino acid sequence is shown, with phosphorylation sites in green and position indicated; positive amino acid exchanges (conserving protein function) are indicated in yellow, other exchanges in red. (STAT1: human NP_009330.1, dog XP_848353.1, cat XP_006935505.1, mouse NP_001192242.1; STAT3: human NP_644805.1, dog XP_005624514.1, cat XP_003996930.1, mouse NP_998824.1; STAT5a: human NP_001275647.1, dog XP_548091.2, cat XP_023099834.1, mouse NP_001157534.1; STAT5b: human NP_036580.2, dog XP_548092.1, cat XP_023100377.1, mouse NP_035619.3).

Several different ways of inhibiting STAT signaling are possible. Upstream of STAT proteins, JAK kinases are mutated in a broad range of diseases from severe combined immunodeficiency to various forms of cancer, including JAK1 in acute myeloid leukemia, JAK2 in myeloproliferative diseases and JAK3 in different leukemias and lymphomas, and inhibitors against JAK kinases are already approved by the US Food and Drug Administration (FDA) for clinical use [27]. Interestingly, different layers of negative regulators of JAK-STAT signaling are present such as suppressor of cytokine signaling (SOCS), protein inhibitor of activated STAT (PIAS) and protein tyrosine phosphatases, arguing for the necessity of a tightly controlled down-regulation of this signaling pathway [101].

Due to the broad activation, minor side-effects and the overall importance, major efforts by many laboratories and pharmaceutical companies are ongoing to develop inhibitors against STAT3 and STAT5. In both cases, all current inhibitors target one of three STAT motifs: the SH2 domain necessary for the interaction of phosphorylated monomers to form dimers, the N-terminal domain mediating the formation of tetramers from activated STAT dimers and the DNA-binding domain [102]. STAT3 and STAT5 from companion animals show more than 96% homology at the overall protein level to their human counterparts, with a particular high level of conservation of 98% to complete alignment in these three domains (Figure 2). This high level of conservation opens up the possibility to use pet animals as models for diseases in which the JAK-STAT signaling pathway is over-activated.

A good example for such a successful application is already established. Cytokine dysregulation has been implicated in allergic skin disease, particularly in atopic dermatitis in humans. T-helper cells type 2 (Th2) produce increased levels of IL4, IL5, IL10, IL13 and IL31, in addition to elevated production of IFNγ by T-helper cells type 1 (Th1), signals that all converge on the JAK-STAT signaling pathway [27,103–105]. Dermatological problems are the second most common reason for dogs to present to veterinary practices, frequently including allergic skin diseases like atopic dermatitis [106,107]. In the skin of atopic dogs, a cytokine profile can be found that resembles the human condition, and in an experimental model of canine allergic dermatitis, elevated transcripts of IL6, IL13 and IL18 and IFNγ were detected, supporting the idea that cytokine dysregulation plays a role in allergic skin disease [108–110]. The novel JAK inhibitor oclacitinib is most potent against JAK1, but also affects JAK2 and JAK3 at reduced efficiency, inhibiting the function of JAK1-dependent inflammatory cytokines [111]. Treatment of dogs suffering from atopic dermatitis results in a reduction of associated skin lesions and oclacitinib recently has been approved in the US and Europe for the treatment of allergic/atopic dermatitis [112–114]. Overall, inhibition of JAK1-dependent cytokines is an effective and novel way to treat canine allergic skin disease, proving the high similarity and cross-species conservation of JAK-STAT signaling.

5. Inhibition of STAT3 and STAT5 in Companion Animals: Current Status/Future Perspectives

Currently, STAT3 and STAT5 are studied in several cancer types of companion animals. Canine mammary cancer cells and diffuse large B cell lymphoma as well as feline oral squamous cell carcinoma and mammary tumors show activation of STAT3 or STAT5. First results indicate reduced proliferation and increased apoptosis upon inhibition of JAK1/2 or STAT3 respectively [115–121]. Two areas exemplify the potential of research and development concerning STAT proteins and human—canine comparative oncology best [122,123]. Osteosarcoma is the most frequent form of malignant bone disease in dogs, however, it is relatively rare in humans. The estimated incidence rate is at least 13.9/100,000 in canines and 1.02/100,000 in humans, affecting primarily children [124,125]. Such a high incidence rate provides a good opportunity to study a rare human disease using dogs as a preclinical model. Moreover, canine and human osteosarcoma share many key features like tumor location, early metastasis, development of chemotherapy-resistant metastases and altered expression or activation of several proteins [126].

Consistent activation of STAT3 occurs in a large subset of human and canine osteosarcoma and osteosarcoma cell lines, but not in normal osteoblasts. Down-regulation of STAT3 expression or activity reduces proliferation and induces apoptosis in human and canine cell lines [127–130]. Additionally, human and canine osteosarcoma possess overlapping transcriptional profiles, further supporting the concept that these diseases are similar at the molecular level [131]. Spontaneous canine osteosarcoma has been used for the development of novel therapeutics such as muramyl tripeptide, IGF-1R inhibitors and rapamycin [132]. However, despite aggressive treatment, 30–40% of children and >90% of dogs are still dying from disease, demonstrating the need for new therapeutic options. The similarities between human and canine osteosarcoma, together with the dependency on JAK-STAT signaling, particularly STAT3, make osteosarcoma one of the prime areas for comparative oncology studies, especially since a large number of STAT3 inhibitors are currently being developed.

Mast cell tumors arise from the uncontrolled proliferation of transformed mast cells mostly in skin, spreading primarily to spleen, liver and bone marrow [133]. While mast cell tumors are relatively rare in humans, they are frequent in dogs, accounting for 7–21% of cutaneous tumors [134,135]. Metastasized tumors in humans as well as in canines have a poor prognosis and short survival times [133,134,136]. Stem cell factor and its receptor c-KIT are essential for mast cell survival and inhibition of apoptosis, and gain-of-function mutations are present in human and canine mast cell tumors [137–142]. Accordingly, two inhibitors of c-KIT, masitinib and toceranib have been approved for use in c-KIT driven mast cell tumors in dogs. Both drugs are able to suppress tumor growth temporarily, however, relapses are high, indicating the need for the identification of further targets and therapeutic options [143,144].

STAT1, STAT3 and STAT5 are activated down-stream of mutant c-KIT, but only STAT5 is transcriptionally active in neoplastic mast cells [145,146]. Inhibition of JAK2 and STAT5 was recently discovered to inhibit proliferation and survival of canine mastocytoma cell lines, identifying JAK2/STAT5 signaling as a new potential target in mast cell tumors [147]. Therefore, an attractive option is to use the frequent occurrence of canine mast cell tumors to determine if inhibitors of STAT5 can be used alone or in combination with inhibitors of c-kit or other kinases as a new therapeutic option. Indeed, the potential of STAT5 inhibitors to overcome resistance to a multi-kinase inhibitor in neoplastic mast cells has been demonstrated [148]. Consequently, the next step is to determine the potential of new STAT5 inhibitors in vivo [149]. Both of these examples, osteosarcoma as well as mast cell tumors offer thereby the possibility of using new pharmaceutical in companion animals as a closer biological mimic of the human situation.

6. Conclusions

Taken together, cancer in dogs resembles cancer in humans in many ways, like its latency, clinical manifestation and metastatic behavior, its pathobiological characteristics like tumor heterogeneity, its genomic instability and pharmacogenomic signatures including chemoresistance and last but not least its multifactorial nature, including genetic and environmental risk factors [150]. The inability of murine cancer models to recapitulate certain aspects of human tumors is increasingly recognized and illuminates the huge potential of spontaneous canine and, to a lesser extent, feline cancer [151–153]. The JAK-STAT pathway is activated in the vast majority of solid and hematological tumors, and is necessary for tumor growth and prevention of apoptosis. The major efforts that are ongoing to develop inhibitors specifically for STAT3 and STAT5 can be extended to studies in companion animals, an option that is particularly attractive for rare human diseases occurring more frequently in dogs and/or cats.

Funding: This research received no external funding.

Acknowledgments: Open Access Funding by the University of Veterinary Medicine Vienna.

References

1.	Vail, D.M.; MacEwen, E.G. Spontaneously occurring tumors of companion animals as models for human cancer. *Cancer Invest.* **2000**, *18*, 781–792. [CrossRef]

2. Bonnett, B.N.; Egenvall, A.; Hedhammar, A.; Olson, P. Mortality in over 350,000 insured Swedish dogs from 1995-2000: I. Breed-, gender-, age- and cause-specific rates. *Acta Vet. Scand.* **2005**, *46*, 105–120. [CrossRef]

3. Egenvall, A.; Bonnett, B.N.; Hedhammar, A.; Olson, P. Mortality in over 350,000 insured Swedish dogs from 1995-2000: II. Breed-specific age and survival patterns and relative risk for causes of death. *Acta Vet. Scand.* **2005**, *46*, 121–136. [CrossRef] [PubMed]

4. Vascellari, M.; Baioni, E.; Ru, G.; Carminato, A.; Mutinelli, F. Animal tumour registry of two provinces in northern Italy: Incidence of spontaneous tumours in dogs and cats. *BMC Vet. Res.* **2009**, *5*, 39. [CrossRef] [PubMed]

5. Withrow, S.J.; Vail, D.M. *Withrow & MacEwen's Small Animal Clinical Oncology*, 5th ed.; Saunders Elsevier: Philadelphia, PA, USA, 2013.

6. Paoloni, M.; Khanna, C. Translation of new cancer treatments from pet dogs to humans. *Nat. Rev. Cancer* **2008**, *8*, 147–156. [CrossRef] [PubMed]

7. National Cancer Institute. Surveillance, Epidemiology and End Results (SEER) Program 2012–2016 Data. Available online: www.seer.cancer.gov/statfacts (accessed on 6 November 2019).

8. Vascellari, M.; Capello, K.; Carminato, A.; Zanardello, C.; Baioni, E.; Mutinelli, F. Incidence of mammary tumors in the canine population living in the Veneto region (Northeastern Italy): Risk factors and similarities to human breast cancer. *Prev. Vet. Med.* **2016**, *126*, 183–189. [CrossRef]

9. Hassan, B.B.; Elshafae, S.M.; Supsavhad, W.; Simmons, J.K.; Dirksen, W.P.; Sokkar, S.M.; Rosol, T.J. Feline Mammary Cancer. *Vet. Pathol.* **2017**, *54*, 32–43. [CrossRef]

10. Graf, R.; Pospischil, A.; Guscetti, F.; Meier, D.; Welle, M.; Dettwiler, M. Cutaneous tumors in Swiss dogs: Retrospective data from the Swiss canine cancer registry, 2008–2013. *Vet. Pathol.* **2018**, *55*, 809–820. [CrossRef]

11. Merlo, D.F.; Rossi, L.; Pellegrino, C.; Ceppi, M.; Cardellino, U.; Capurro, C.; Ratto, A.; Sambucco, P.L.; Sestito, V.; Tanara, G.; et al. Cancer incidence in pet dogs: Findings of the animal tumor registry of Genoa, Italy. *J. Vet. Intern. Med.* **2008**, *22*, 976–984. [CrossRef]

12. Lomas, A.; Leonardi-Bee, J.; Bath-Hextall, F. A systematic review of worldwide incidence of nonmelanoma skin cancer. *Br. J. Dermatol.* **2012**, *166*, 1069–1080. [CrossRef]

13. Pinello, K.C.; Niza-Ribeiro, J.; Fonseca, L.; de Matos, A.J. Incidence, characteristics and geographical distributions of canine and human non-Hodgkin's lymphoma in the Porto region (North West Portugal). *Vet. J.* **2019**, *245*, 70–76. [CrossRef] [PubMed]

14. Grover, S. Gastronintestinal lymphoma in cats. *Oncol. Compend.* **2005**, *27*, 741–751.

15. Egenvall, A.; Nodtvedt, A.; von Euler, H. Bone tumors in a population of 400 000 insured Swedish dogs up to 10 y of age: Incidence and survival. *Can. J. Vet. Res.* **2007**, *71*, 292–299. [PubMed]

16. Dimopoulou, M.; Kirpensteijn, J.; Moens, H.; Kik, M. Histologic prognosticators in feline osteosarcoma: A comparison with phenotypically similar canine osteosarcoma. *Vet. Surg.* **2008**, *37*, 466–471. [CrossRef] [PubMed]

17. Cheon, D.J.; Orsulic, S. Mouse models of cancer. *Annu. Rev. Pathol* **2011**, *6*, 95–119. [CrossRef] [PubMed]

18. Ghebranious, N.; Donehower, L.A. Mouse models in tumor suppression. *Oncogene* **1998**, *17*, 3385–3400. [CrossRef] [PubMed]

19. Fowlis, D.J.; Balmain, A. Oncogenes and tumour suppressor genes in transgenic mouse models of neoplasia. *Eur. J. Cancer* **1993**, *29A*, 638–645. [CrossRef]

20. Murphy, W.J. Of mice and men. *Biol. Blood Marrow Transplant.* **2013**, *19*, 1140–1141. [CrossRef]

21. Cekanova, M.; Rathore, K. Animal models and therapeutic molecular targets of cancer: Utility and limitations. *Drug Des. Devel. Ther.* **2014**, *8*, 1911–1921. [CrossRef]

22. Stark, G.R.; Darnell, J.E., Jr. The JAK-STAT pathway at twenty. *Immunity* **2012**, *36*, 503–514. [CrossRef]

23. Igelmann, S.; Neubauer, H.A.; Ferbeyre, G. STAT3 and STAT5 Activation in Solid Cancers. *Cancers* **2019**, *11*, 1428. [CrossRef] [PubMed]

24. Logotheti, S.; Putzer, B.M. STAT3 and STAT5 targeting for simultaneous management of melanoma and autoimmune diseases. *Cancers* **2019**, *11*, 1448. [CrossRef] [PubMed]

25. Rebe, C.; Ghiringhelli, F. STAT3, a master regulator of anti-tumor immune response. *Cancers* **2019**, *11*, 1280. [CrossRef] [PubMed]

26. Bromberg, J. Stat proteins and oncogenesis. *J. Clin. Invest.* **2002**, *109*, 1139–1142. [CrossRef]

27. O'Shea, J.J.; Schwartz, D.M.; Villarino, A.V.; Gadina, M.; McInnes, I.B.; Laurence, A. The JAK-STAT pathway: Impact on human disease and therapeutic intervention. *Annu. Rev. Med.* **2015**, *66*, 311–328. [CrossRef]

28. Fagard, R.; Metelev, V.; Souissi, I.; Baran-Marszak, F. STAT3 inhibitors for cancer therapy: Have all roads been explored? *JAKSTAT* **2013**, *2*, e22882. [CrossRef]

29. Rangarajan, A.; Weinberg, R.A. Opinion: Comparative biology of mouse versus human cells: Modelling human cancer in mice. *Nat. Rev. Cancer* **2003**, *3*, 952–959. [CrossRef]

30. Dobrowolski, P.; Fischer, M.; Naumann, R. Novel insights into the genetic background of genetically modified mice. *Transgenic Res.* **2018**, *27*, 265–275. [CrossRef]

31. Kilkenny, C.; Browne, W.J.; Cuthill, I.C.; Emerson, M.; Altman, D.G. Improving bioscience research reporting: The ARRIVE guidelines for reporting animal research. *PLoS Biol.* **2010**, *8*, e1000412. [CrossRef]

32. Bouchlaka, M.N.; Murphy, W.J. Impact of aging in cancer immunotherapy: The importance of using accurate preclinical models. *Oncoimmunology* **2013**, *2*, e27186. [CrossRef]

33. Hooper, L.V.; Littman, D.R.; Macpherson, A.J. Interactions between the microbiota and the immune system. *Science* **2012**, *336*, 1268–1273. [CrossRef] [PubMed]

34. Ivanov, I.I.; Honda, K. Intestinal commensal microbes as immune modulators. *Cell Host Microbe* **2012**, *12*, 496–508. [CrossRef] [PubMed]

35. Sivan, A.; Corrales, L.; Hubert, N.; Williams, J.B.; Aquino-Michaels, K.; Earley, Z.M.; Benyamin, F.W.; Lei, Y.M.; Jabri, B.; Alegre, M.L.; et al. Commensal bifidobacterium promotes antitumor immunity and facilitates anti-PD-L1 efficacy. *Science* **2015**, *350*, 1084–1089. [CrossRef] [PubMed]

36. Antonovic, L.; Martinez, M. Role of the cytochrome P450 enzyme system in veterinary pharmacokinetics: Where are we now? Where are we going? *Future Med. Chem.* **2011**, *3*, 855–879. [CrossRef] [PubMed]

37. Brekke, T.D.; Steele, K.A.; Mulley, J.F. Inbred or outbred? Genetic diversity in laboratory rodent colonies. *G3 (Bethesda)* **2018**, *8*, 679–686. [CrossRef] [PubMed]

38. Cutler, G.; Kassner, P.D. Copy number variation in the mouse genome: Implications for the mouse as a model organism for human disease. *Cytogenet. Genome Res.* **2008**, *123*, 297–306. [CrossRef]

39. Hartman, J.L.t.; Garvik, B.; Hartwell, L. Principles for the buffering of genetic variation. *Science* **2001**, *291*, 1001–1004. [CrossRef]

40. Srivastava, A.; Morgan, A.P.; Najarian, M.L.; Sarsani, V.K.; Sigmon, J.S.; Shorter, J.R.; Kashfeen, A.; McMullan, R.C.; Williams, L.H.; Giusti-Rodriguez, P.; et al. Genomes of the mouse collaborative cross. *Genetics* **2017**, *206*, 537–556. [CrossRef]

41. Dunn, G.P.; Bruce, A.T.; Ikeda, H.; Old, L.J.; Schreiber, R.D. Cancer immunoediting: From immunosurveillance to tumor escape. *Nat. Immunol.* **2002**, *3*, 991–998. [CrossRef]

42. Cespedes, M.V.; Casanova, I.; Parreno, M.; Mangues, R. Mouse models in oncogenesis and cancer therapy. *Clin. Transl. Oncol.* **2006**, *8*, 318–329. [CrossRef]

43. Holzapfel, B.M.; Wagner, F.; Thibaudeau, L.; Levesque, J.P.; Hutmacher, D.W. Concise review: Humanized models of tumor immunology in the 21st century: Convergence of cancer Research and tissue engineering. *Stem Cells* **2015**, *33*, 1696–1704. [CrossRef] [PubMed]

44. Richmond, A.; Su, Y. Mouse xenograft models vs. GEM models for human cancer therapeutics. *Dis Model. Mech.* **2008**, *1*, 78–82. [CrossRef] [PubMed]

45. Wald, A.; Corey, L. Persistence in the population: Epidemiology, transmission. In *Human Herpesviruses: Biology, Therapy, and Immunoprophylaxis*; Arvin, A., Campadelli-Fiume, G., Mocarski, E., Moore, P.S., Roizman, B., Whitley, R., Yamanishi, K., Eds.; Cambridge University Press: Cambridge, UK, 2007.

46. Smatti, M.K.; Al-Sadeq, D.W.; Ali, N.H.; Pintus, G.; Abou-Saleh, H.; Nasrallah, G.K. Epstein-Barr virus epidemiology, serology, and genetic variability of LMP-1 oncogene among healthy population: An update. *Front. Oncol.* **2018**, *8*, 211. [CrossRef] [PubMed]

47. Colugnati, F.A.; Staras, S.A.; Dollard, S.C.; Cannon, M.J. Incidence of cytomegalovirus infection among the general population and pregnant women in the United States. *BMC Infect. Dis* **2007**, *7*, 71. [CrossRef] [PubMed]

48. Day, C.P.; Merlino, G.; Van Dyke, T. Preclinical mouse cancer models: A maze of opportunities and challenges. *Cell* **2015**, *163*, 39–53. [CrossRef] [PubMed]

49. Hanahan, D.; Weinberg, R.A. Hallmarks of cancer: The next generation. *Cell* **2011**, *144*, 646–674. [CrossRef]

50. Willyard, C. The mice with human tumours: Growing pains for a popular cancer model. *Nature* **2018**, *560*, 156–157. [CrossRef]

51. Ireson, C.R.; Alavijeh, M.S.; Palmer, A.M.; Fowler, E.R.; Jones, H.J. The role of mouse tumour models in the discovery and development of anticancer drugs. *Br. J. Cancer* **2019**, *121*, 101–108. [CrossRef]

52. Ciociola, A.A.; Cohen, L.B.; Kulkarni, P.; Gastroenterology, F.D.-R.M.C.o.t.A.C.o. How drugs are developed and approved by the FDA: Current process and future directions. *Am. J. Gastroenterol.* **2014**, *109*, 620–623. [CrossRef]

53. Schiffman, J.D.; Breen, M. Comparative oncology: What dogs and other species can teach us about humans with cancer. *Philos. Trans. R Soc. Lond. B Biol. Sci.* **2015**, *370*. [CrossRef]

54. Hahn, K.A.; Bravo, L.; Adams, W.H.; Frazier, D.L. Naturally occurring tumors in dogs as comparative models for cancer therapy research. *In Vivo* **1994**, *8*, 133–143. [PubMed]

55. Reif, J.S.; Bruns, C.; Lower, K.S. Cancer of the nasal cavity and paranasal sinuses and exposure to environmental tobacco smoke in pet dogs. *Am. J. Epidemiol.* **1998**, *147*, 488–492. [CrossRef] [PubMed]

56. Reif, J.S.; Dunn, K.; Ogilvie, G.K.; Harris, C.K. Passive smoking and canine lung cancer risk. *Am. J. Epidemiol.* **1992**, *135*, 234–239. [CrossRef] [PubMed]

57. Laflamme, D.; Izquierdo, O.; Eirmann, L.; Binder, S. Myths and misperceptions about ingredients used in commercial pet foods. *Vet. Clin. North. Am. Small Anim. Pract.* **2014**, *44*, 689–698. [CrossRef]

58. Song, S.J.; Lauber, C.; Costello, E.K.; Lozupone, C.A.; Humphrey, G.; Berg-Lyons, D.; Caporaso, J.G.; Knights, D.; Clemente, J.C.; Nakielny, S.; et al. Cohabiting family members share microbiota with one another and with their dogs. *Elife* **2013**, *2*, e00458. [CrossRef]

59. Lindblad-Toh, K.; Wade, C.M.; Mikkelsen, T.S.; Karlsson, E.K.; Jaffe, D.B.; Kamal, M.; Clamp, M.; Chang, J.L.; Kulbokas, E.J., 3rd; Zody, M.C.; et al. Genome sequence, comparative analysis and haplotype structure of the domestic dog. *Nature* **2005**, *438*, 803–819. [CrossRef]

60. Vonholdt, B.M.; Pollinger, J.P.; Lohmueller, K.E.; Han, E.; Parker, H.G.; Quignon, P.; Degenhardt, J.D.; Boyko, A.R.; Earl, D.A.; Auton, A.; et al. Genome-wide SNP and haplotype analyses reveal a rich history underlying dog domestication. *Nature* **2010**, *464*, 898–902. [CrossRef]

61. Hoffman, M.M.; Birney, E. Estimating the neutral rate of nucleotide substitution using introns. *Mol. Biol Evol* **2007**, *24*, 522–531. [CrossRef]

62. Paoloni, M.C.; Khanna, C. Comparative oncology today. *Vet. Clin. North. Am. Small Anim Pract* **2007**, *37*, 1023–1032. [CrossRef]

63. Khanna, C.; Lindblad-Toh, K.; Vail, D.; London, C.; Bergman, P.; Barber, L.; Breen, M.; Kitchell, B.; McNeil, E.; Modiano, J.F.; et al. The dog as a cancer model. *Nat. Biotechnol.* **2006**, *24*, 1065–1066. [CrossRef]

64. Di Cerbo, A.; Palmieri, B.; De Vico, G.; Iannitti, T. Onco-epidemiology of domestic animals and targeted therapeutic attempts: Perspectives on human oncology. *J. Cancer Res. Clin. Oncol* **2014**, *140*, 1807–1814. [CrossRef] [PubMed]

65. Modiano, J.F.; Breen, M.; Burnett, R.C.; Parker, H.G.; Inusah, S.; Thomas, R.; Avery, P.R.; Lindblad-Toh, K.; Ostrander, E.A.; Cutter, G.C.; et al. Distinct B-cell and T-cell lymphoproliferative disease prevalence among dog breeds indicates heritable risk. *Cancer Res.* **2005**, *65*, 5654–5661. [CrossRef] [PubMed]

66. Nishiya, A.T.; Massoco, C.O.; Felizzola, C.R.; Perlmann, E.; Batschinski, K.; Tedardi, M.V.; Garcia, J.S.; Mendonca, P.P.; Teixeira, T.F.; Zaidan Dagli, M.L. Comparative aspects of canine melanoma. *Vet. Sci.* **2016**, *3*, 7. [CrossRef] [PubMed]

67. Dobson, J.M. Breed-predispositions to cancer in pedigree dogs. *ISRN Vet. Sci* **2013**, *2013*, 941275. [CrossRef] [PubMed]

68. Klopfenstein, M.; Howard, J.; Rossetti, M.; Geissbuhler, U. Life expectancy and causes of death in Bernese mountain dogs in Switzerland. *BMC Vet. Res.* **2016**, *12*, 153. [CrossRef] [PubMed]

69. Lewis, T.W.; Wiles, B.M.; Llewellyn-Zaidi, A.M.; Evans, K.M.; O'Neill, D.G. Longevity and mortality in Kennel Club registered dog breeds in the UK in 2014. *Canine Genet. Epidemiol.* **2018**, *5*, 10. [CrossRef] [PubMed]

70. Dobson, J.; Hoather, T.; McKinley, T.J.; Wood, J.L. Mortality in a cohort of flat-coated retrievers in the UK. *Vet. Comp. Oncol.* **2009**, *7*, 115–121. [CrossRef]

71. Zink, M.C.; Farhoody, P.; Elser, S.E.; Ruffini, L.D.; Gibbons, T.A.; Rieger, R.H. Evaluation of the risk and age of onset of cancer and behavioral disorders in gonadectomized Vizslas. *J. Am. Vet. Med. Assoc.* **2014**, *244*, 309–319. [CrossRef]

72. Henry, C.J.; Higginbotham, H.M. *Cancer Management in Small Animal Practice*; Saunders Elsevier: Philadelphia, PA, USA, 2009.

73. Fan, T.M.; Khanna, C. Comparative Aspects of Osteosarcoma Pathogenesis in Humans and Dogs. *Vet. Sci* **2015**, *2*, 210–230. [CrossRef]

74. Urfer, S.R.; Gaillard, C.; Steiger, A. Lifespan and disease predispositions in the Irish Wolfhound: A review. *Vet. Q.* **2007**, *29*, 102–111. [CrossRef]

75. Zandvliet, M. Canine lymphoma: A review. *Vet. Q.* **2016**, *36*, 76–104. [CrossRef] [PubMed]

76. Gajewski, T.F.; Schreiber, H.; Fu, Y.X. Innate and adaptive immune cells in the tumor microenvironment. *Nat. Immunol.* **2013**, *14*, 1014–1022. [CrossRef] [PubMed]

77. Ribas, A. Adaptive immune resistance: How cancer protects from immune attack. *Cancer Discov.* **2015**, *5*, 915–919. [CrossRef] [PubMed]

78. Goulart, M.R.; Pluhar, G.E.; Ohlfest, J.R. Identification of myeloid derived suppressor cells in dogs with naturally occurring cancer. *PLoS ONE* **2012**, *7*, e33274. [CrossRef]

79. Pinheiro, D.; Chang, Y.M.; Bryant, H.; Szladovits, B.; Dalessandri, T.; Davison, L.J.; Yallop, E.; Mills, E.; Leo, C.; Lara, A.; et al. Dissecting the regulatory microenvironment of a large animal model of non-Hodgkin lymphoma: Evidence of a negative prognostic impact of FOXP3+ T cells in canine B cell lymphoma. *PLoS ONE* **2014**, *9*, e105027. [CrossRef]

80. Coy, J.; Caldwell, A.; Chow, L.; Guth, A.; Dow, S. PD-1 expression by canine T cells and functional effects of PD-1 blockade. *Vet. Comp. Oncol.* **2017**, *15*, 1487–1502. [CrossRef]

81. Zerdes, I.; Wallerius, M.; Sifakis, E.G.; Wallmann, T.; Betts, S.; Bartish, M.; Tsesmetzis, N.; Tobin, N.P.; Coucoravas, C.; Bergh, J.; et al. STAT3 activity promotes programmed-death ligand 1 expression and suppresses immune responses in breast cancer. *Cancers* **2019**, *11*, 1479. [CrossRef]

82. Reardon, S. CRISPR gene-editing creates wave of exotic model organisms. *Nature* **2019**, *568*, 441–442. [CrossRef]

83. Simoff, I.; Karlgren, M.; Backlund, M.; Lindstrom, A.C.; Gaugaz, F.Z.; Matsson, P.; Artursson, P. Complete knockout of endogenous Mdr1 (Abcb1) in MDCK cells by CRISPR-Cas9. *J. Pharm. Sci.* **2016**, *105*, 1017–1021. [CrossRef]

84. Amoasii, L.; Hildyard, J.C.W.; Li, H.; Sanchez-Ortiz, E.; Mireault, A.; Caballero, D.; Harron, R.; Stathopoulou, T.R.; Massey, C.; Shelton, J.M.; et al. Gene editing restores dystrophin expression in a canine model of Duchenne muscular dystrophy. *Science* **2018**, *362*, 86–91. [CrossRef]

85. Liongue, C.; Ward, A.C. Evolution of the JAK-STAT pathway. *JAKSTAT* **2013**, *2*, e22756. [CrossRef] [PubMed]

86. Morris, R.; Kershaw, N.J.; Babon, J.J. The molecular details of cytokine signaling via the JAK/STAT pathway. *Protein Sci.* **2018**, *27*, 1984–2009. [CrossRef] [PubMed]

87. Polak, K.L.; Chernosky, N.M.; Smigiel, J.M.; Tamagno, I.; Jackson, M.W. Balancing STAT Activity as a Therapeutic Strategy. *Cancers* **2019**, *11*, 1716. [CrossRef] [PubMed]

88. Yue-Ting, K.L.; Ramaiyer, M.; E. Johnson, D.; R. Grandis, J. Targeting STAT3 in cancer with nucleotide therapeutics. *Cancers* **2019**, *11*, 1681. [CrossRef]

89. Gough, D.J.; Corlett, A.; Schlessinger, K.; Wegrzyn, J.; Larner, A.C.; Levy, D.E. Mitochondrial STAT3 supports Ras-dependent oncogenic transformation. *Science* **2009**, *324*, 1713–1716. [CrossRef]

90. Meier, J.A.; Larner, A.C. Toward a new STATe: The role of STATs in mitochondrial function. *Semin. Immunol.* **2014**, *26*, 20–28. [CrossRef]

91. Frank, D.A. STAT3 as a central mediator of neoplastic cellular transformation. *Cancer Lett.* **2007**, *251*, 199–210. [CrossRef]

92. Bar-Natan, M.; Nelson, E.A.; Xiang, M.; Frank, D.A. STAT signaling in the pathogenesis and treatment of myeloid malignancies. *JAKSTAT* **2012**, *1*, 55–64. [CrossRef]

93. Roeser, J.C.; Leach, S.D.; McAllister, F. Emerging strategies for cancer immunoprevention. *Oncogene* **2015**, *34*, 6029–6039. [CrossRef]

94. Yu, H.; Jove, R. The STATs of cancer–new molecular targets come of age. *Nat. Rev. Cancer* **2004**, *4*, 97–105. [CrossRef]

95. Sen, M.; Thomas, S.M.; Kim, S.; Yeh, J.I.; Ferris, R.L.; Johnson, J.T.; Duvvuri, U.; Lee, J.; Sahu, N.; Joyce, S.; et al. First-in-human trial of a STAT3 decoy oligonucleotide in head and neck tumors: Implications for cancer therapy. *Cancer Discov.* **2012**, *2*, 694–705. [CrossRef] [PubMed]

96. Park, J.S.; Kwok, S.K.; Lim, M.A.; Kim, E.K.; Ryu, J.G.; Kim, S.M.; Oh, H.J.; Ju, J.H.; Park, S.H.; Kim, H.Y.; et al. STA-21, a promising STAT-3 inhibitor that reciprocally regulates Th17 and Treg cells, inhibits osteoclastogenesis in mice and humans and alleviates autoimmune inflammation in an experimental model of rheumatoid arthritis. *Arthritis Rheumatol.* **2014**, *66*, 918–929. [CrossRef] [PubMed]

97. Kunigal, S.; Lakka, S.S.; Sodadasu, P.K.; Estes, N.; Rao, J.S. Stat3-siRNA induces Fas-mediated apoptosis in vitro and in vivo in breast cancer. *Int. J. Oncol.* **2009**, *34*, 1209–1220. [PubMed]

98. Darnell, J.E., Jr. Transcription factors as targets for cancer therapy. *Nat. Rev. Cancer* **2002**, *2*, 740–749. [CrossRef] [PubMed]

99. Ferbeyre, G.; Moriggl, R. The role of Stat5 transcription factors as tumor suppressors or oncogenes. *Biochim. Biophys. Acta* **2011**, *1815*, 104–114. [CrossRef]

100. Zhang, H.F.; Lai, R. STAT3 in Cancer-Friend or Foe? *Cancers* **2014**, *6*, 1408–1440. [CrossRef]

101. Starr, R.; Hilton, D.J. Negative regulation of the JAK/STAT pathway. *Bioessays* **1999**, *21*, 47–52. [CrossRef]

102. Wake, M.S.; Watson, C.J. STAT3 the oncogene - still eluding therapy? *FEBS J.* **2015**, *282*, 2600–2611. [CrossRef]

103. Carmi-Levy, I.; Homey, B.; Soumelis, V. A modular view of cytokine networks in atopic dermatitis. *Clin. Rev. Allergy Immunol.* **2011**, *41*, 245–253. [CrossRef]

104. Brandt, E.B.; Sivaprasad, U. Th2 Cytokines and Atopic Dermatitis. *J. Clin. Cell Immunol* **2011**, *2*. [CrossRef]

105. Ong, P.Y.; Leung, D.Y. Immune dysregulation in atopic dermatitis. *Curr. Allergy Asthma Rep.* **2006**, *6*, 384–389. [CrossRef] [PubMed]

106. Hill, P.B.; Lo, A.; Eden, C.A.; Huntley, S.; Morey, V.; Ramsey, S.; Richardson, C.; Smith, D.J.; Sutton, C.; Taylor, M.D.; et al. Survey of the prevalence, diagnosis and treatment of dermatological conditions in small animals in general practice. *Vet. Rec.* **2006**, *158*, 533–539. [CrossRef] [PubMed]

107. Olivry, T.; Bizikova, P. A systematic review of randomized controlled trials for prevention or treatment of atopic dermatitis in dogs: 2008-2011 update. *Vet. Dermatol.* **2013**, *24*, 97–117, e25–e26. [CrossRef] [PubMed]

108. Marsella, R.; Olivry, T.; Maeda, S. Cellular and cytokine kinetics after epicutaneous allergen challenge (atopy patch testing) with house dust mites in high-IgE beagles. *Vet. Dermatol.* **2006**, *17*, 111–120. [CrossRef] [PubMed]

109. Nuttall, T.J.; Knight, P.A.; McAleese, S.M.; Lamb, J.R.; Hill, P.B. Expression of Th1, Th2 and immunosuppressive cytokine gene transcripts in canine atopic dermatitis. *Clin. Exp. Allergy* **2002**, *32*, 789–795. [CrossRef] [PubMed]

110. Schlotter, Y.M.; Rutten, V.P.; Riemers, F.M.; Knol, E.F.; Willemse, T. Lesional skin in atopic dogs shows a mixed Type-1 and Type-2 immune responsiveness. *Vet. Immunol. Immunopathol.* **2011**, *143*, 20–26. [CrossRef]

111. Gonzales, A.J.; Bowman, J.W.; Fici, G.J.; Zhang, M.; Mann, D.W.; Mitton-Fry, M. Oclacitinib (APOQUEL((R))) is a novel Janus kinase inhibitor with activity against cytokines involved in allergy. *J. Vet. Pharmacol. Ther.* **2014**, *37*, 317–324. [CrossRef]

112. Cosgrove, S.B.; Wren, J.A.; Cleaver, D.M.; Walsh, K.F.; Follis, S.I.; King, V.I.; Tena, J.K.; Stegemann, M.R. A blinded, randomized, placebo-controlled trial of the efficacy and safety of the Janus kinase inhibitor oclacitinib (Apoquel(R)) in client-owned dogs with atopic dermatitis. *Vet. Dermatol.* **2013**, *24*, 587–597, e141–e142. [CrossRef]

113. Cosgrove, S.B.; Wren, J.A.; Cleaver, D.M.; Martin, D.D.; Walsh, K.F.; Harfst, J.A.; Follis, S.L.; King, V.L.; Boucher, J.F.; Stegemann, M.R. Efficacy and safety of oclacitinib for the control of pruritus and associated skin lesions in dogs with canine allergic dermatitis. *Vet. Dermatol.* **2013**, *24*, 479–e114. [CrossRef]

114. Collard, W.T.; Hummel, B.D.; Fielder, A.F.; King, V.L.; Boucher, J.F.; Mullins, M.A.; Malpas, P.B.; Stegemann, M.R. The pharmacokinetics of oclacitinib maleate, a Janus kinase inhibitor, in the dog. *J. Vet. Pharmacol Ther* **2014**, *37*, 279–285. [CrossRef]

115. Lu, Z.; Hong, C.C.; Jark, P.C.; Assumpcao, A.; Bollig, N.; Kong, G.; Pan, X. JAK1/2 Inhibitors AZD1480 and CYT387 inhibit canine B-Cell lymphoma growth by increasing apoptosis and disrupting cell proliferation. *J. Vet. Intern. Med.* **2017**, *31*, 1804–1815. [CrossRef] [PubMed]

116. Assumpcao, A.; Jark, P.C.; Hong, C.C.; Lu, Z.; Ruetten, H.M.; Heaton, C.M.; Pinkerton, M.E.; Pan, X. STAT3 Expression and Activity are Up-Regulated in Diffuse Large B Cell Lymphoma of Dogs. *J. Vet. Intern. Med.* **2018**, *32*, 361–369. [CrossRef] [PubMed]

117. Krol, M.; Pawlowski, K.M.; Dolka, I.; Musielak, O.; Majchrzak, K.; Mucha, J.; Motyl, T. Density of Gr1-positive myeloid precursor cells, p-STAT3 expression and gene expression pattern in canine mammary cancer metastasis. *Vet. Res. Commun.* **2011**, *35*, 409–423. [CrossRef] [PubMed]

118. Van Garderen, E.; Swennenhuis, J.F.; Hellmen, E.; Schalken, J.A. Growth hormone induces tyrosyl phosphorylation of the transcription factors Stat5a and Stat5b in CMT-U335 canine mammary tumor cells. *Domest. Anim. Endocrinol.* **2001**, *20*, 123–135. [CrossRef]

119. Brown, M.E.; Bear, M.D.; Rosol, T.J.; Premanandan, C.; Kisseberth, W.C.; London, C.A. Characterization of STAT3 expression, signaling and inhibition in feline oral squamous cell carcinoma. *BMC Vet. Res.* **2015**, *11*, 206. [CrossRef]

120. Petterino, C.; Podesta, G.; Ratto, A.; Drigo, M.; Pellegrino, C. Immunohistochemical study of phospho-Stat3-ser727 expression in feline mammary gland tumours. *Vet. Res. Commun.* **2007**, *31*, 173–184. [CrossRef]

121. Petterino, C.; Ratto, A.; Podesta, G.; Drigo, M.; Pellegrino, C. Immunohistochemical evaluation of STAT3-p-tyr705 expression in feline mammary gland tumours and correlation with histologic grade. *Res. Vet. Sci.* **2007**, *82*, 218–224. [CrossRef]

122. Willmann, M.; Hadzijusufovic, E.; Hermine, O.; Dacasto, M.; Marconato, L.; Bauer, K.; Peter, B.; Gamperl, S.; Eisenwort, G.; Jensen-Jarolim, E.; et al. Comparative oncology: The paradigmatic example of canine and human mast cell neoplasms. *Vet. Comp. Oncol.* **2019**, *17*, 1–10. [CrossRef]

123. Makielski, K.M.; Mills, L.J.; Sarver, A.L.; Henson, M.S.; Spector, L.G.; Naik, S.; Modiano, J.F. Risk factors for development of canine and human osteosarcoma: A comparative review. *Vet. Sci.* **2019**, *6*, 48. [CrossRef]

124. Rowell, J.L.; McCarthy, D.O.; Alvarez, C.E. Dog models of naturally occurring cancer. *Trends Mol. Med.* **2011**, *17*, 380–388. [CrossRef]

125. Mirabello, L.; Troisi, R.J.; Savage, S.A. Osteosarcoma incidence and survival rates from 1973 to 2004: Data from the surveillance, epidemiology, and end results program. *Cancer* **2009**, *115*, 1531–1543. [CrossRef] [PubMed]

126. Fenger, J.M.; London, C.A.; Kisseberth, W.C. Canine osteosarcoma: A naturally occurring disease to inform pediatric oncology. *ILAR J.* **2014**, *55*, 69–85. [CrossRef] [PubMed]

127. Fossey, S.L.; Liao, A.T.; McCleese, J.K.; Bear, M.D.; Lin, J.; Li, P.K.; Kisseberth, W.C.; London, C.A. Characterization of STAT3 activation and expression in canine and human osteosarcoma. *BMC Cancer* **2009**, *9*, 81. [CrossRef] [PubMed]

128. Fossey, S.L.; Bear, M.D.; Lin, J.; Li, C.; Schwartz, E.B.; Li, P.K.; Fuchs, J.R.; Fenger, J.; Kisseberth, W.C.; London, C.A. The novel curcumin analog FLLL32 decreases STAT3 DNA binding activity and expression, and induces apoptosis in osteosarcoma cell lines. *BMC Cancer* **2011**, *11*, 112. [CrossRef] [PubMed]

129. Couto, J.I.; Bear, M.D.; Lin, J.; Pennel, M.; Kulp, S.K.; Kisseberth, W.C.; London, C.A. Biologic activity of the novel small molecule STAT3 inhibitor LLL12 against canine osteosarcoma cell lines. *BMC Vet. Res.* **2012**, *8*, 244. [CrossRef] [PubMed]

130. Onimoe, G.I.; Liu, A.; Lin, L.; Wei, C.C.; Schwartz, E.B.; Bhasin, D.; Li, C.; Fuchs, J.R.; Li, P.K.; Houghton, P.; et al. Small molecules, LLL12 and FLLL32, inhibit STAT3 and exhibit potent growth suppressive activity in osteosarcoma cells and tumor growth in mice. *Invest. New Drugs* **2012**, *30*, 916–926. [CrossRef]

131. Paoloni, M.; Davis, S.; Lana, S.; Withrow, S.; Sangiorgi, L.; Picci, P.; Hewitt, S.; Triche, T.; Meltzer, P.; Khanna, C. Canine tumor cross-species genomics uncovers targets linked to osteosarcoma progression. *BMC Genomics* **2009**, *10*, 625. [CrossRef]

132. Rodriguez, C.O., Jr. Using canine osteosarcoma as a model to assess efficacy of novel therapies: Can old dogs teach us new tricks? *Adv. Exp. Med. Biol.* **2014**, *804*, 237–256. [CrossRef]

133. London, C.A.; Seguin, B. Mast cell tumors in the dog. *Vet. Clin. North. Am. Small Anim. Pract.* **2003**, *33*, 473–489. [CrossRef]

134. Misdorp, W. Mast cells and canine mast cell tumours. A review. *Vet. Q.* **2004**, *26*, 156–169. [CrossRef]

135. Shoop, S.J.; Marlow, S.; Church, D.B.; English, K.; McGreevy, P.D.; Stell, A.J.; Thomson, P.C.; O'Neill, D.G.; Brodbelt, D.C. Prevalence and risk factors for mast cell tumours in dogs in England. *Canine Genet. Epidemiol.* **2015**, *2*, 1. [CrossRef] [PubMed]

136. Lim, K.H.; Tefferi, A.; Lasho, T.L.; Finke, C.; Patnaik, M.; Butterfield, J.H.; McClure, R.F.; Li, C.Y.; Pardanani, A. Systemic mastocytosis in 342 consecutive adults: Survival studies and prognostic factors. *Blood* **2009**, *113*, 5727–5736. [CrossRef] [PubMed]

137. Letard, S.; Yang, Y.; Hanssens, K.; Palmerini, F.; Leventhal, P.S.; Guery, S.; Moussy, A.; Kinet, J.P.; Hermine, O.; Dubreuil, P. Gain-of-function mutations in the extracellular domain of KIT are common in canine mast cell tumors. *Mol. Cancer Res.* **2008**, *6*, 1137–1145. [CrossRef] [PubMed]

138. London, C.A.; Galli, S.J.; Yuuki, T.; Hu, Z.Q.; Helfand, S.C.; Geissler, E.N. Spontaneous canine mast cell tumors express tandem duplications in the proto-oncogene c-kit. *Exp. Hematol.* **1999**, *27*, 689–697. [CrossRef]

139. Nagata, H.; Worobec, A.S.; Oh, C.K.; Chowdhury, B.A.; Tannenbaum, S.; Suzuki, Y.; Metcalfe, D.D. Identification of a point mutation in the catalytic domain of the protooncogene c-kit in peripheral blood mononuclear cells of patients who have mastocytosis with an associated hematologic disorder. *Proc. Natl. Acad. Sci. USA* **1995**, *92*, 10560–10564. [CrossRef]

140. Metcalfe, D.D. Mast cells and mastocytosis. *Blood* **2008**, *112*, 946–956. [CrossRef]

141. Iemura, A.; Tsai, M.; Ando, A.; Wershil, B.K.; Galli, S.J. The c-kit ligand, stem cell factor, promotes mast cell survival by suppressing apoptosis. *Am. J. Pathol.* **1994**, *144*, 321–328.

142. Moller, C.; Alfredsson, J.; Engstrom, M.; Wootz, H.; Xiang, Z.; Lennartsson, J.; Jonsson, J.I.; Nilsson, G. Stem cell factor promotes mast cell survival via inactivation of FOXO3a-mediated transcriptional induction and MEK-regulated phosphorylation of the proapoptotic protein Bim. *Blood* **2005**, *106*, 1330–1336. [CrossRef]

143. Hahn, K.A.; Ogilvie, G.; Rusk, T.; Devauchelle, P.; Leblanc, A.; Legendre, A.; Powers, B.; Leventhal, P.S.; Kinet, J.P.; Palmerini, F.; et al. Masitinib is safe and effective for the treatment of canine mast cell tumors. *J. Vet. Intern. Med.* **2008**, *22*, 1301–1309. [CrossRef]

144. London, C.A.; Malpas, P.B.; Wood-Follis, S.L.; Boucher, J.F.; Rusk, A.W.; Rosenberg, M.P.; Henry, C.J.; Mitchener, K.L.; Klein, M.K.; Hintermeister, J.G.; et al. Multi-center, placebo-controlled, double-blind, randomized study of oral toceranib phosphate (SU11654), a receptor tyrosine kinase inhibitor, for the treatment of dogs with recurrent (either local or distant) mast cell tumor following surgical excision. *Clin. Cancer Res.* **2009**, *15*, 3856–3865. [CrossRef]

145. Chaix, A.; Lopez, S.; Voisset, E.; Gros, L.; Dubreuil, P.; De Sepulveda, P. Mechanisms of STAT protein activation by oncogenic KIT mutants in neoplastic mast cells. *J. Biol. Chem.* **2011**, *286*, 5956–5966. [CrossRef] [PubMed]

146. Baumgartner, C.; Cerny-Reiterer, S.; Sonneck, K.; Mayerhofer, M.; Gleixner, K.V.; Fritz, R.; Kerenyi, M.; Boudot, C.; Gouilleux, F.; Kornfeld, J.W.; et al. Expression of activated STAT5 in neoplastic mast cells in systemic mastocytosis: Subcellular distribution and role of the transforming oncoprotein KIT D816V. *Am. J. Pathol.* **2009**, *175*, 2416–2429. [CrossRef] [PubMed]

147. Keller, A.; Wingelhofer, B.; Peter, B.; Bauer, K.; Berger, D.; Gamperl, S.; Reifinger, M.; Cerny-Reiterer, S.; Moriggl, R.; Willmann, M.; et al. The JAK2/STAT5 signaling pathway as a potential therapeutic target in canine mastocytoma. *Vet. Comp. Oncol.* **2018**, *16*, 55–68. [CrossRef] [PubMed]

148. Peter, B.; Bibi, S.; Eisenwort, G.; Wingelhofer, B.; Berger, D.; Stefanzl, G.; Blatt, K.; Herrmann, H.; Hadzijusufovic, E.; Hoermann, G.; et al. Drug-induced inhibition of phosphorylation of STAT5 overrides drug resistance in neoplastic mast cells. *Leukemia* **2018**, *32*, 1016–1022. [CrossRef] [PubMed]

149. Page, B.D.; Khoury, H.; Laister, R.C.; Fletcher, S.; Vellozo, M.; Manzoli, A.; Yue, P.; Turkson, J.; Minden, M.D.; Gunning, P.T. Small molecule STAT5-SH2 domain inhibitors exhibit potent antileukemia activity. *J. Med. Chem.* **2012**, *55*, 1047–1055. [CrossRef]

150. Park, J.S.; Withers, S.S.; Modiano, J.F.; Kent, M.S.; Chen, M.; Luna, J.I.; Culp, W.T.N.; Sparger, E.E.; Rebhun, R.B.; Monjazeb, A.M.; et al. Canine cancer immunotherapy studies: Linking mouse and human. *J. Immunother. Cancer* **2016**, *4*, 97. [CrossRef]

151. Alvarez, C.E. Naturally occurring cancers in dogs: Insights for translational genetics and medicine. *ILAR J.* **2014**, *55*, 16–45. [CrossRef]

152. Furdos, I.; Fazekas, J.; Singer, J.; Jensen-Jarolim, E. Translating clinical trials from human to veterinary oncology and back. *J. Transl. Med.* **2015**, *13*, 265. [CrossRef]

153. Riccardo, F.; Aurisicchio, L.; Impellizeri, J.A.; Cavallo, F. The importance of comparative oncology in translational medicine. *Cancer Immunol. Immunother.* **2015**, *64*, 137–148. [CrossRef]

Review

Chaperoning STAT3/5 by Heat Shock Proteins: Interest of Their Targeting in Cancer Therapy

Gaëtan Jego [1,2,*,†], **François Hermetet** [1,2,†], **François Girodon** [1,2,3] and **Carmen Garrido** [1,2,4,*]

1   INSERM, LNC UMR1231, team HSP-Pathies, University of Bourgogne Franche-Comté, F-21000 Dijon, France; francois.hermetet@u-bourgogne.fr (F.H.); francois.girodon@chu-dijon.fr (F.G.)
2   UFR des Sciences de Santé, University of Burgundy and Franche-Comté, F-21000 Dijon, France
3   Haematology laboratory, Dijon University Hospital, F-21000 Dijon, France
4   Centre Georges François Leclerc, 21000 Dijon, France
*   Correspondence: cgarrido@u-bourgogne.fr (C.G.); gaetan.jego@u-bourgogne.fr (G.J.); Tel.: +33-3-8039-3345 (G.J.); Fax: +33-3-8039-3434 (C.G. & G.J.)
†   These authors contributed equally to this paper.

Received: 16 October 2019; Accepted: 13 December 2019; Published: 19 December 2019

Abstract: While cells from multicellular organisms are dependent upon exogenous signals for their survival, growth, and proliferation, commitment to a specific cell fate requires the correct folding and maturation of proteins, as well as the degradation of misfolded or aggregated proteins within the cell. This general control of protein quality involves the expression and the activity of molecular chaperones such as heat shock proteins (HSPs). HSPs, through their interaction with the STAT3/STAT5 transcription factor pathway, can be crucial both for the tumorigenic properties of cancer cells (cell proliferation, survival) and for the microenvironmental immune cell compartment (differentiation, activation, cytokine secretion) that contributes to immunosuppression, which, in turn, potentially promotes tumor progression. Understanding the contribution of chaperones such as HSP27, HSP70, HSP90, and HSP110 to the STAT3/5 signaling pathway has raised the possibility of targeting such HSPs to specifically restrain STAT3/5 oncogenic functions. In this review, we present how HSPs control STAT3 and STAT5 activation, and vice versa, how the STAT signaling pathways modulate HSP expression. We also discuss whether targeting HSPs is a valid therapeutic option and which HSP would be the best candidate for such a strategy.

Keywords: heat shock proteins; chaperones; stabilization; targeted therapy

1. Introduction to Heat Shock Proteins/Chaperones

Heat shock proteins (HSPs), also called stress proteins, are highly conserved molecular chaperones induced by a broad variety of exogenous or intracellular stresses, including chemotherapy. Based on their molecular weight, HSPs have been classified into five conserved families: HSP110 (also called HSPH), HSP90 (HSPC), HSP70 (HSPA), HSP60 (HSPD/E), and the small HSPs (HSPB). The expression of HSPs is mostly regulated by heat shock factor 1 (HSF1), which is able to translocate from the cytoplasm to the nucleus following stress to bind to a short, highly conserved DNA sequence known as a heat shock element (HSE) [1]. As cytoprotective proteins, HSPs participate in the correct folding, activity, transport, and stability of proteins [2], which are essential processes for cell survival. In physiological conditions, these proteins support neosynthesized proteins, favoring post-translational modification processes and protein folding. Otherwise, the functions attributed to them include the subcellular transport of their "client" proteins, or participation in certain signaling pathways. In response to stress, many partially denatured proteins accumulate and cluster together forming protein aggregates via the exposure of their hydrophobic residues. Some HSP proteins are then able to bind to these partially denatured proteins, thus preventing protein aggregation and favoring their correct folding. However, HSPs can also promote the elimination of these proteins by orienting them towards different degradation pathways, notably the ubiquitin–proteasome system, when correct folding is no longer possible. It has also been reported that HSP proteins are able to inhibit the intrinsic apoptotic (inhibition of apoptosome formation) and extrinsic (inhibition of signal transduction of death receptors) processes [3,4]. Interestingly, HSPs can also be secreted and act extracellularly via membrane receptors or within extracellular vesicles as a damage-associated molecular pattern and exhibit immune-cell dysregulation properties.

2. HSP Chaperones and Cancer

As cancer cells accumulate mutations and generate physiologically stressful conditions, they require a constitutively high level of HSPs for their survival and maintenance. In 2011, in order to simplify the complexity of this disease, researchers suggested that tumor development was organized around six essential alterations [5]. These major modifications include (i) self-sufficiency in growth signals; (ii) insensitivity to growth inhibition; (iii) tissue invasion and capacity to develop metastases; (iv) unlimited replication potential; (v) de novo angiogenesis; and (vi) inhibition of programmed cell death. Although the appearance of these changes is mainly linked to instability genomics, many studies have demonstrated the involvement of HSPs in these processes, indicating that these molecular chaperones have an oncogenic role. Comprehensive discussion about the oncogene-like functions of these different molecular chaperones and their participation in the progression of resistance to cancer treatment can be found in excellent recent reviews elsewhere [6,7]. Given the role of HSPs in cancer biology, these chaperones have also been suggested as potential therapeutic targets [3,8]. A number of these proteins have been correlated to cancer aggressiveness and/or cancer resistance to radiotherapy and adjuvant chemotherapy [9].

Targeting HSPs has emerged as a promising sensitization strategy in cancer therapy since HSPs have oncogene-like functions and mediate "non-oncogene addiction" of stressed tumor cells that must adapt to a hostile microenvironment. Except for one inhibitor of HSP27 (an antisense oligonucleotide in phase I/II) [10], all the HSP inhibitors used in clinical trials target HSP90 [11,12]. In this review, we mainly focused on the chaperones HSP90, HSP70, HSP110, and HSP27 and their regulation of protein misfolding and signaling in TYK2-STAT3/5 core cancer pathways, as well as the possibility of targeting such HSPs to specifically restrain STAT3/5 oncogenic functions. We also discuss the machinery behind the chaperones, which is becoming a major therapeutic target in cancer, and the emergence of promising HSP inhibitor-based drugs, which are currently being clinically tested or developed for cancer treatment (Table 1).

Table 1. Summary of the main strategies for HSP inhibition.

Inhibitor		Study Type	Cancer Model	Ref.
Name	**Nature/Structure**			
Target: HSP27				
Apatorsen (OGX-427)	2nd generation 2′-methoxyethyl-modified ASOs	in vitro/preclinical	Prostate, Ovary	[13,14]
		clinical trial (phase I)	CRPC, Breast, Ovary, Lung, Bladder	[10]
		in vitro/preclinical	Pancreatic, NSCLC	[15,16]
		clinical trial (phase II)	Stage IV non-squamous NSCLC	[17]
3-arylethynyltriazolyl ribonucleoside	ASOs	in vitro	Pancreatic	[18]
ASOs-Hsp27	ASOs	in vitro	Lymphoma	[19]
RP101 (Brivudine)	Uridine derivative and nucleoside analog	in vitro/preclinical/clinical	Pancreatic	[20]

Table 1. *Cont.*

Inhibitor		Study Type	Cancer Model	Ref.
Name	**Nature/Structure**			
Target: HSP70				
Pifithrin-μ (PFTμ, PES)	Drug-like small molecule	in vitro	AML, ALL, Primary AML blasts	[21,22]
VER-155008	ATP-derivative inhibitor	in vitro	Breast, Colon, Prostatic, Myeloma	[23–25]
A17/A8	Peptide aptamer	in vitro/preclinical	Cervix (HeLa cells), Melanoma	[26]
ADD70	Peptide aptamer	in vitro/preclinical	Rat colon carcinoma, Mouse melanoma	[27]
cmHsp70.1	Antibody	preclinical	Colorectal	[28]
Hsp70-peptide targeted NK based adoptive immunotherapy	A specific amino acid sequence (TKD) of Hsp70	clinical trials (phase I/II)	NSCLC (and colon cancer) patients with ex vivo Hsp70 peptide activated, autologous NK	[29]
Target: HSP90				
Radicicol	natural product isolated from the fungus *Monosporium bonorden*	in vitro	CML	[30]
17-AAG; 17-DMAG	Derivative of the antibiotic geldanamycin	in vitro/preclinical	Breast, Brain, Medulloblastoma	[31–33]
17-DMAG		in vitro	CLL	[34]
IPI-504 (retaspimycin)	Water-soluble derivate of 17-AAG	in vitro/preclinical	Breast, Pancreatic, Metastatic gastrointestinal stromal tumor	[35–39]
		in vitro/preclinical	NSCLC	[40]
IPI-504, AUY922 Ganetespib, Onalespib	-	clinical trials (phase I–III)	NSCLC Breast, Ovary, Colon	[41]
Novobiocin	Aminocoumarin antibiotic, produced by the actinomycete *Streptomyces nivens*	in vitro/preclinical	Leukemia, Prostate	[42–44]
Panaxynol	Natural pesticide and fatty alcohol	in vitro/preclinical	Lung	[45]
Ganetespib (STA-9090)	Synthetic, non-geldanamycin, small molecule inhibitor	preclinical	Thyroid	[46]
		in vitro	Breast	[47]
BIIB021 (CNF2024)	Orally available, fully synthetic purine scaffold, small molecule inhibitor	in vitro/preclinical	Blood malignancies, Solid tumors	[48]
PU-H71	Non-ansamycin, purine scaffold inhibitor	preclinical	mouse models of the MPN PV and ET	[49]
			MPN	[50]

Table 1. *Cont.*

Inhibitor		Study Type	Cancer Model	Ref.
Name	**Nature/Structure**			
NVP_AUY922 (AUY922)	Esorcinylic isoxazole amide, second-generation non-geldanamycin inhibitor	in vitro/preclinical	Gastric, Small cell lung, Thyroid	[51–55]
		in vitro	32D mouse hematopoietic cells expressing wild-type BCR-ABL (b3a2, 32Dp210) and mutant BCR-ABL imatinib-resistant cell lines	[56]
		in vitro/preclinical	Drug-resistant chronic myelogenous leukemia	[57]
		clinical trial (phase II)	Myeloproliferative neoplasms	[58]
		clinical trials (phase I/II)	EGFR-mutant lung cancer with acquired resistance to epidermal growth factor receptor tyrosine kinase inhibitors	[59]
AUY922, HSP990, PU-H71	-	in vitro/preclinical	Leukemia	[60]
Onalespib (AT13387)	second-generation, non-ansamycin inhibitor	in vitro	Transformed kidney cells, primary lung adenocarcinoma	[61]
		in vitro/preclinical	Melanoma	[62]
		in vitro/preclinical	NSCLC	[63]
		in vitro/preclinical	NSCLC	[64]
XL888	Orally available inhibitor with high selectivity for HSP90α and HSP90β	clinical trial (phase I)	Melanoma	[65]
SNX2112 SNX5422	Orally bioavailable, synthetic, small molecule inhibitors that competitively bind to HSP90α, HSP90β, Grp94 and Trap-1	in vitro/preclinical	Head and neck squamous cell carcinoma	[66]
			NSCLC	[67]
CUDC-305, Ganetespib CH5164840, WK88-1 17-DMAG	-	preclinical	NSCLC	[68–71]
Target: HSP110				
Foldamers 33 and 52	Protein–protein interaction inhibitors, based on pyridyl scaffolds mimicking α-helix	in vitro/preclinical	Colorectal	[72]

ALL: acute lymphoblastic leukemia; AML: acute myeloid leukemia; ASOs: antisense oligonucleotides; CLL: chronic lymphocytic leukemia; CML: chronic myelogenous leukemia; CRPC: castration-resistant prostate cancer; EGFR: epidermal growth factor receptor; ET: essential thrombocytosis; MPN: myeloproliferative neoplasm; NK: natural killer cells; NSCLC: non-small cell lung cancer; PES: 2-phenylethynesulfonamide; PV: polycythemia vera.

3. HSP90

3.1. HSP90 Structure and Functions

HSP90 (also known as HSPC) is one of the most abundant chaperones in eukaryotic cells in the absence of stress. HSP90 is critical for the operation of cellular machinery under physiological conditions through interactions with so-called "client" proteins. This is only achieved through the formation of a multimeric protein complex of cochaperones that binds to all three domains of HSP90. Hundreds of client proteins for HSP90 have been identified so far [73]. Many of these proteins are involved in essential cellular functions that promote cell growth, proliferation, cell survival, and immune responses. Most of these processes are also involved in cancer development. Three main groups of "client proteins" can be described for HSP90: first, the group of kinases represents the main group because HSP90 interacts with 60% of them [74]; second, the group of multiprotein complexes for which HSP90 promotes assembly [75]; and third, the group of ligands that HSP90 stabilizes with their receptors. It is difficult to identify new client proteins because HSP90 does not bind particular sequences. In contrast, the interaction seems to be based on the overall structural instability of the client proteins [74,76]. Among the client proteins, here we focus on the kinases and receptor tyrosine kinases involved in the STAT3/5 signaling pathway.

3.2. HSP90 and Nonfusion Protein Kinases

3.2.1. Jak Kinases

The mammalian family of Janus kinases (JAKs) is composed of 4 members: JAK1, JAK2, JAK3, and Tyrosine kinase 2 (Tyk2). This family is the main activator of STAT proteins. JAK regulation by HSP90 was discovered by studying the effect of HSP90 inhibitors on the type I and II Interferon (IFN) signaling. In several cell lines, this treatment suppressed the expression of multiple IFN-γ-induced genes and decreased IFN-γ-induced STAT1 phosphorylation on Tyr-701, required for dimerization, and on Ser-727, required for transcription factor activation. As JAK1/2 were known to be the protein kinases responsible for STAT1 phosphorylation, Shang et al. investigated the effect of HSP90 inhibitors on JAK1/2. They showed that HSP90 inhibition led to the proteasome-mediated degradation of JAK1/2. Further they showed that JAK1 interacted with HSP90 (and the CDC37 cochaperone [77]), and that both interactions were destabilized by HSP90 inhibitors [78]. As overactivation or constitutive JAK1/2 signaling promotes cell proliferation and survival in a variety of solid tumors and leukemia [79,80], this discovery paved the way for the identification of the critical role of HSP90 in the aberrant JAK/STAT signaling pathway. In particular, an activating point mutation in JAK2 (*JAK2*$^{\text{V617F}}$) was described as being highly frequent in chronic myeloproliferative neoplasms (MPN) that promote disease progression [81–84].

Despite this activating mutation, HSP90 inhibition in cell lines homozygous for *JAK2*$^{\text{V617F}}$ reduced total and phospho-JAK2, and subsequently cell viability [85]. In vivo experiments in a mouse model of MPN confirmed the efficacy of HSP90 targeting because treatment with the HSP90 inhibitor PU-H71 resulted in significant reductions in disease parameters and better chances of survival [49]. Furthermore, combined treatment that included HSP90 inhibitors and JAK2 inhibitors induced a greater depletion of the signaling proteins than a single inhibitor alone, and synergistically induced apoptosis in human primary CD34(+) MPN cells harboring JAK2$^{\text{V617F}}$ [50]. Therefore, HSP90 interaction with JAK2 is not altered by activated mutations, but instead could be used as a therapeutic target. This point is of great value, as mutations within the JAK2 kinase domain that confer resistance across a panel of JAK inhibitors have been described (G935R, Y931C, and E864K). Fortunately, genetic resistance to JAK2 enzymatic inhibitors can be overcome by HSP90 inhibitors, which still promote the degradation of both wild-type and mutant JAK2 [60]. Recently, results from a phase II clinical trial with the HSP90 inhibitor AUY922 (Novartis, transferred to Vernalis) have been published and have demonstrated a clinical response in five out of seven patients with MPN [58]. This response correlated with a reduction in overall levels of JAK2, pYSTAT3, and pYSTAT5. Unfortunately, most patients experienced severe adverse effects due to the toxicity of the inhibitor, a phenomenon that has already been observed with other HSP90 inhibitors.

3.2.2. Src Kinases

The members of the Src family of nonreceptor tyrosine kinases (Src, Fyn, Yes, Blk, Yrk, Rak, Fgr, Hck, Lck, Srm, and Lyn) are implicated in numerous important functions in eukaryotic cells. They control proliferation, survival, and differentiation, therefore playing a critical role in many cancer types [86]. Src members can activate STAT3 directly and synergize with JAK family tyrosine kinase action [87]. Among the family members, c-Src has been linked to cancer development [88]. The viral homolog of c-Src kinase, v-Src (from the *Rous sarcoma* virus), has a constitutive kinase activity and was the first discovered oncogene [89,90]. Both homologs bind to the HSP90/CDC37 complex but with striking differences. HSP90, which binds weakly and transiently to c-Src, binds strongly to v-Src, which appears to be its strongest client protein [91–93]. Accordingly, v-Src kinase activity depends strongly on HSP90 [74,94]. Recently, Boczek et al. provided more insight by determining the influence of HSP90 isoforms α and β on purified c-Src and v-Src activity. They have shown that HSP90 does not affect c-Src activity in vitro, whereas v-Src activity was increased two-fold when human HSP90β (but not HSP90α) was added to the experimental setting. HSP90β also stabilized v-Src at high temperatures when it would be inactive otherwise. Until recently, the mechanism behind this striking difference was unknown [76,95]. To solve this issue, Bolcek et al. generated an Src mutant that mimics the oncogenic v-Src kinase activity (c-src3MΔC). This mutant exhibited a more extended activation loop (A-loop) (usually present in an open form during wild-type Src active state to allow substrate binding). The A-loop from c-Src3MΔC is also less stable in comparison with the wild-type Src. Consequently, the c-Src3MΔC is conformationally uncontrolled, which enhances its interaction with HSP90 and suggests this could be a more general mechanism for the interaction between HSP90 and oncogenic kinases than the presence of a general client sequence motif. Indeed, HSP90 potentially interacts more strongly with structurally extended kinases, a frequent state observed upon activating mutations. Interestingly, a very similar mechanism needed to aid the initial folding of immature kinases such as c-Src, which is furtive is this case, governs the binding of HSP90 to conformationally unstable but mature kinases like v-Src. In this context, CDC37 appears to bind to parts of the unfolded kinase first (which might be considered as an independent kinase binding unit), partly unfolding it further before HSP90 clamps around the CDC37/kinase complex [96]. Other Src family members, like Lck^{Y505F} and HCK^{499F}, are probably stabilized by the same mechanism [97,98].

3.2.3. ACK1

Another nonreceptor tyrosine kinase, activated CDC42-associated kinase-1 (ACK1), catalyzes the phosphorylation of STAT1, STAT3, and STAT5. HSP90 interacts with ACK1 [99] and is necessary for the phosphorylation of STAT1 in transformed kidney cells and STAT3 in primary lung adenocarcinoma by ACK1 [61].

3.2.4. BRAF

The activated serine/threonine kinase BRAF mutant is a main driver of melanoma growth and progression [100] and is a HSP90 client protein [74,101]. Inhibition of HSP90 by AT13387 delays the emergence of resistance to BRAF inhibitors [62]. A recent phase I dose escalation clinical trial in melanoma has shown that another HSP90 inhibitor (XL888) in combination with a specific anti BRAF inhibitor (vemurafenib) has clinical activity in patients with advanced BRAF^{V600}-mutant melanoma, with a tolerable side effect profile [65].

3.3. HSP90 and Fusion Protein Kinases

3.3.1. BCR-ABL

Chronic myeloid leukemia (CML) is driven by the BCR-ABL fusion oncoprotein [102], which is involved, among other pathways, in the transcriptional regulation of STAT3 [103,104] and STAT5 [105,106]. In this context, the BCR-ABL/STAT3/STAT5 signaling pathway is mainly involved in tumor-initiating stem cell maintenance [107]. BCR-ABL is a HSP90 client protein that is destabilized by HSP90 inhibition, which leads to cell death [56]. In CML cells, BCR-ABL forms a high molecular weight network with JAK2, STAT3, and AKT. This network pushes disease progression, but could also be its Achilles' heel. Indeed, HSP90 directly binds to this signaling network, and its inhibition breaks the whole network apart [56]. As for other targeted therapies, resistance to BCR-ABL tyrosine kinase inhibitors can develop during the course of the treatment because of acquired *BCR-ABL* mutations. Hopefully, combination therapies involving HSP90 inhibitors and anti-JAK2 may overcome this resistance [57].

We are still unsure of how this mechanism of action can be extended to the interaction with other oncogenic mature kinases, but an important process for protein stabilization by HSP90 and CDC37 has been uncovered.

3.3.2. EML4-ALK

The echinoderm microtubule-associated protein-like 4-anaplastic lymphoma kinase (EML4-ALK) fusion gene is an oncogenic driver in about 5% of patients with non-small cell lung cancer (NSCLC). It is also an HSP90 client protein (one of the most sensitive), which is very rapidly degraded upon exposure to HSP90 inhibitors [40]. These results prompted the initiation of numerous clinical trials reviewed elsewhere [41,108]. Yet despite encouraging results, clinical response was weak, and so the development of HSP90 inhibitors was halted for NSCLC.

3.4. HSP90 and ErbB Family of Receptor Tyrosine Kinase (RTK)

STAT3 and STAT5 are also known to be phosphorylated by several receptor tyrosine kinases (RTK), such as the ErbB family, IGF-1R, or FGFR [109], most of which are HSP90 client proteins. Interestingly, they are also activated or captured for signal transduction by those RTK without phosphorylation [110]. Given their membrane localization, these RTK must go through a complex process of folding, maturation, and membrane insertion that requires significant chaperone cooperation. This is particularly true for mutated RTK, which is frequently observed in cancers. For instance, ErbB2 stability and maturation is regulated by its binding to HSP90 through its cytoplasmic tail [111] and is ATP dependent [112].

Accordingly, HSP90 inhibition leads to RTK destabilization and absence of STAT3 activation in different models of cancer [47,66,67]. Many drugs have been developed to inhibit mutated or rearranged RTK, but despite early success, most patients develop resistance and eventually relapse [113]. The strong HSP90/EGFR interaction has then been used to propose an alternative therapeutic strategy combining an HSP90 inhibitor with an EGFR inhibitor. Interestingly, this combination (with the EGFR inhibitor erlotinib) resulted in prolonged animal survival in nonmutated and erlotinib-resistant models [67,68,70,71,114,115].

3.5. STAT3/5 and HSP90

The STAT3/5 signaling pathway is also regulated downstream from the tyrosine kinases and RTK. Indeed, HSP90 is found within the cytosol, directly bound to dimers of STAT3 or STAT5 via its N-terminal regions [116]. However, in contrast to its role in TK or RTK folding and stabilization, HSP90 is not required for STAT3/5 maturation or total protein levels. They are therefore nonclassical HSP90 client proteins. The chaperone would rather change STAT conformation to ease the phosphorylation process and/or, once phosphorylated, maintain this active state for a prolonged period of time. Moulick et al. have suggested this pattern in chronic myeloid leukemia [69]. They showed that HSP90 directly binds to active pYSTAT5 (Tyr694), but not to inactive STAT5, and that pYSTAT5 acquires a conformation that is more susceptible to trypsine cleavage in the presence of HSP90. HSP90/STAT3 also protects pYSTAT3 from dephosphorylation by the phosphatase SHP-1 in gastric cancer cells. Luteolin (3,4,5,7-tetrahydroxyflavone), a natural flavonoid present in fruits and vegetables, inhibits STAT3 activation by disrupting the association of HSP90 to STAT3, which allows it to interact with SHP-1 [117].

In order to function as a transcription factor, STAT3/5 needs to translocate into the nucleus and form a stable interaction with DNA. In this context, as suggested by Longshaw et al., HSP90 appears to play a specific role in association with the cochaperone HOP [118]. They have shown that the depletion of HOP decreased the nuclear localization of STAT3. Although it is not yet clear how HSP90 promotes

STAT3/5 nuclear shuttling, it may involve the capacity of HSP90 to form molecular complexes with specific carriers, like importins alpha [119], that can transport STAT3 to the nucleus. This scheme would mimic what has been described for other molecular complexes implicating HSP90, such as the glucocorticoid receptor [120] or PKCZeta [121]. After entering the nucleus, HSP90 seems to promote STAT3/5 transcriptional activity as STAT3/5 interaction with promoters of target genes is enhanced by the presence of HSP90. Indeed, the STAT3/5 complexes and HSP90 have been shown to colocalize in MYC and in CCND2 promoters [69]. Furthermore, nuclear hormone receptors form multiprotein complexes with STAT3 and STAT5 [122,123], which together with HSPs could contribute to chromatin landscaping [124].

In conclusion, HSP90 appears to be a key chaperone for the STAT3/5 pathway. It operates at all levels of message transmission, from interaction with RTK in the cytoplasmic membrane, to the interaction with multiple kinases in the cytosol, and to favoring active STAT3/5 localization and binding of target genes the nucleus (Figure 1). This role is also central in the pathological overactivation of STAT signaling where HSP90 favors oncogenic proteins (Figure 1), promoting the development of several inhibitors for cancer treatment (Table 1, Figure 2). However so far, most clinical trials have yielded mixed results and frequent side effects that precluded the broad utilization of these treatments.

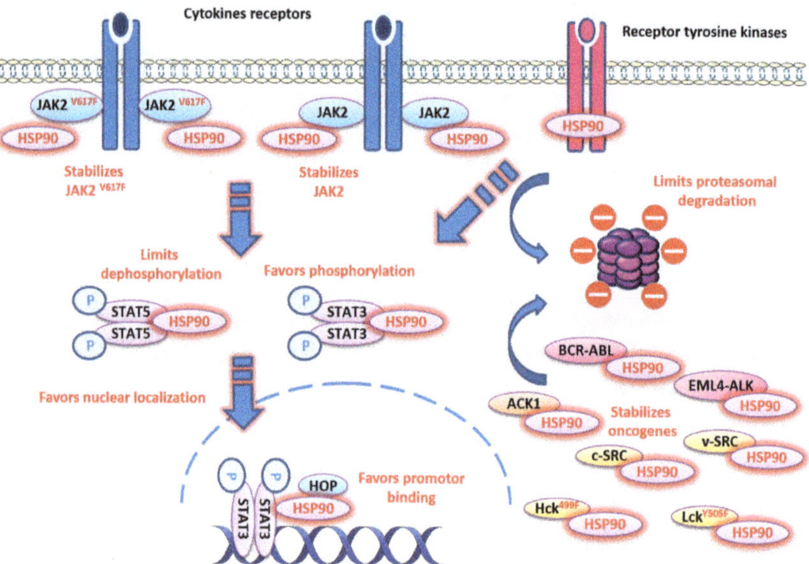

Figure 1. Localization and described functions of HSP90 within the STAT3 and STAT5 signaling pathways. HSP90 promotes those pathways through direct interaction with STAT3 or STAT5 dimers and favors their phosphorylation, nuclear localization, and promoter binding, but HSP90 also limits dephosphorylation and proteasomal degradation. Upstream of STAT3/STAT5 activation, HSP90 stabilizes several kinases, like JAK2, JAK2^{V617F}, c-Src, v-Src, ACK1, BCR-ABL, EML4-ALK, Lck^{Y505F}, and HCK^{499F}, and several receptor tyrosine kinases, such as the ErbB family.

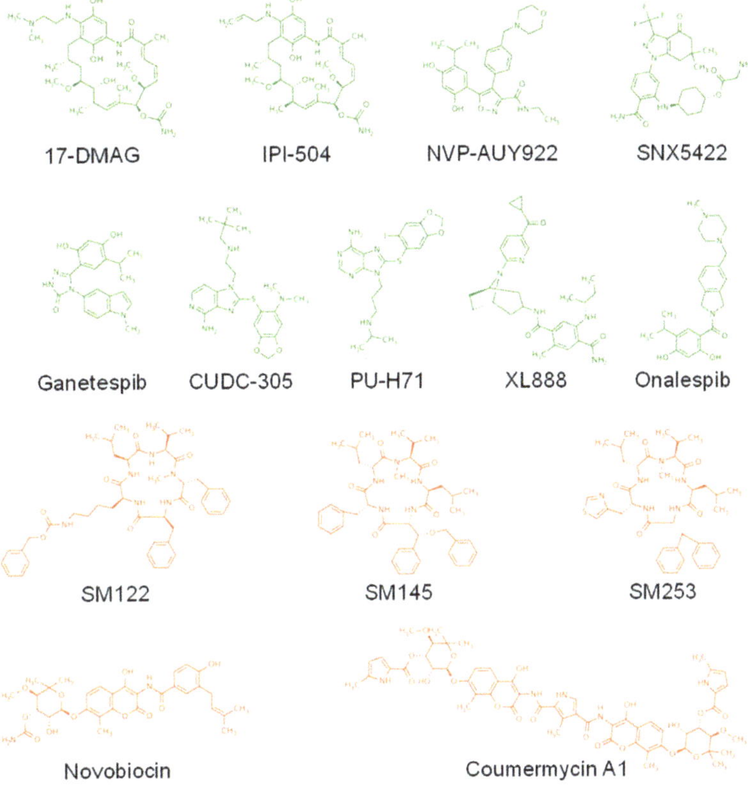

Figure 2. Structure of the main HSP90 inhibitors. The inhibitors targeting the ATP binding site at the N-terminus and the C-terminus of HSP90 are depicted in green and orange, respectively.

4. HSP27

4.1. HSP27 Structure

HSP27 (27 kDa), also known as HSPB1, belongs to the small HSP family. In contrast to other HSPs, it is an ATP-independent chaperone [125]. HSP27 shows a highly dynamic process of oligomerization that transforms proteins from dimers to large oligomers, which can culminate at 1000 kDa. This state of oligomerization dictates the affinity of HSP27 towards the proteins to be chaperoned, given that the multimer form is the most binding-competent state [126].

Four phosphorylation sites (S15, S78, S82 and T143) in the N-terminal domain regulate the assembly of oligomers [127]. Phosphorylation promotes the formation of small oligomers, while dephosphorylation promotes the formation of large oligomers [128]. Stressors such as anticancer agents, hydrogen peroxide, mitogens, inflammatory cytokines (TNF-α, IL-1b, etc.), and kinases (p38 MAPK, p70RSK, PKB, PKC, PKD and PKG) can promote HSP27 phosphorylation [129]. However, we have also shown that oligomerization can occur independently of phosphorylation through cell–cell contact, as observed in confluent cultures in vitro or solid tumors in vivo [130].

4.2. HSP27 Functions

HSP27 chaperone functions have been less described than higher molecular weight HSPs like HSP70 and HSP90. However, HSP27's main function as a chaperone is to stabilize denatured or aggregated proteins to bring them to a soluble and stable form [131,132].

HSP27 plays an important role and has been associated with poor prognosis in many cancers (for a recent review, see [127]). For instance, Rocchi et al. described the promotion of the STAT3 signaling pathway in prostate tumors [133]. They showed that HSP27 directly interacts with STAT3 and that total STAT3 levels correlated directly with HSP27 levels. Furthermore, the cytoprotective effect of HSP27 was attenuated by STAT3-reduced expression, underlying the importance of this pathway in prostate cancer [133]. HSP27 is also involved in the process of prostate cancer metastasis through the promotion of IL-6-mediated epithelial-to-mesenchymal transition (EMT), because reduced HSP27 expression decreases cell migration and invasion [134]. In the absence of HSP27, IL-6-induced phosphorylation of STAT3 is reduced, but not total STAT3 content. This inhibition leads naturally to reduced STAT3 nuclear localization and binding to the *TWIST* gene, which codes for a key EMT transcription factor.

Beside cancer development, HSP27 regulation of STAT3 has recently been implicated in placental implantation. HSP27 is indeed expressed during placenta formation and in the first two trimesters of pregnancy [135,136]. In particular, HSP27 is highly expressed during the differentiation of cytotrophoblast cells and extravillous trophoblast cells, and its silencing was found to significantly reduce total STAT3. Interestingly, the phosphorylation state of STAT3 was not altered in the absence of HSP27 in placental explants, suggesting a role in protein protection from proteasomal degradation [137]. This could be explained by the fact that STAT3 and STAT5 have a relatively low thermodynamic stability as isolated proteins and are thus more prone to aggregation, which would be limited by HSP27 [124]. Given the importance of STAT3 in embryonic development (STAT3 knock-out mice have a lethal embryonic phenotype) [138], this finding revealed the critical role of HSP27 in this process.

There has been little study of the state of HSP27 phosphorylation, and consequently the state of oligomerization, required for STAT3/5 binding, despite the fact that therapeutic targeting of specific kinases would logically impact HSP27 functions. We only know, from one study of prostate cancer, that IL-6 stimulation leads to HSP27 phosphorylation and correlates with the EMT, suggesting the phosphorylated form is required for STAT3 activation [134]. Other STAT family members are HSP27 client proteins, like STAT2 (a STAT family member involved in viral or interferon responses), which was also shown to be degraded upon HSP27 knockdown in Hela cells [139]. However, this process was reversed by proteasome inhibition. However, STAT3 and 5 were not or were only weakly reduced in this particular tumor cell line. Again, the discrepancy between tumoral contexts or cell lines could come from differences in HSP27 phosphorylation or oligomerization status.

As stated previously, HSP27/STAT3 interaction occurs also in nontumoral contexts. In normal liver cells under a high fat diet, the phosphorylated form of HSP27 stimulates autophagy and lipid droplet clearance through interaction with STAT3. In this particular situation, no STAT3 activation by HSP27 is described, but rather the disruption of STAT3/PKR complexes, facilitating PKR and eIF2α mediated autophagy [140]. These data suggest that dimers and multiprotein complexes can be displaced by the action of phospho-HSP27 on different binding partners, including STAT3, and this can also mediate critical cellular physiological processes.

Upstream from STAT3/5 activation, JAK2 plays a major role that can be modulated by HSP27 [141]. In the specific context of thrombopoietin- and JAK2$^{\text{V617F}}$-induced myelofibrosis (a chronic degenerative disorder of the hematopoietic system associated with the aberrant activation of the JAK/STAT pathway) [142], our team has recently shown that HSP27 interacts directly with JAK2/STAT5, stabilizing the complex. Neither total JAK2 nor STAT5 protein levels were affected, but we found that the state of phosphorylation of STAT5 (Tyr694) by JAK2 was HSP27 dependent. We demonstrated that HSP27, through interaction with STAT5, physically prevented its dephosphorylation by the phosphatase SHP2 in those cells.

In conclusion, HSP27 plays a very important role in STAT3/5 signaling, both in contexts of tumor development and others such as placenta development. In contrast with HSP90, whose functions depend on the client proteins, HSP27's different functions (protein stability, phosphorylation, disruption of complex of proteins, etc.) may rely on the phosphorylation and oligomerization status of the chaperone (Figure 3). The general vision is mainly dichotomous: on one side there are large phosphorylated oligomers, and on the other nonphosphorylated dimers. This simplistic description does not reflect the reality of cellular dynamics. Future studies will be needed to specify the proportion of each HSP27 oligomer within the cells and the associated function.

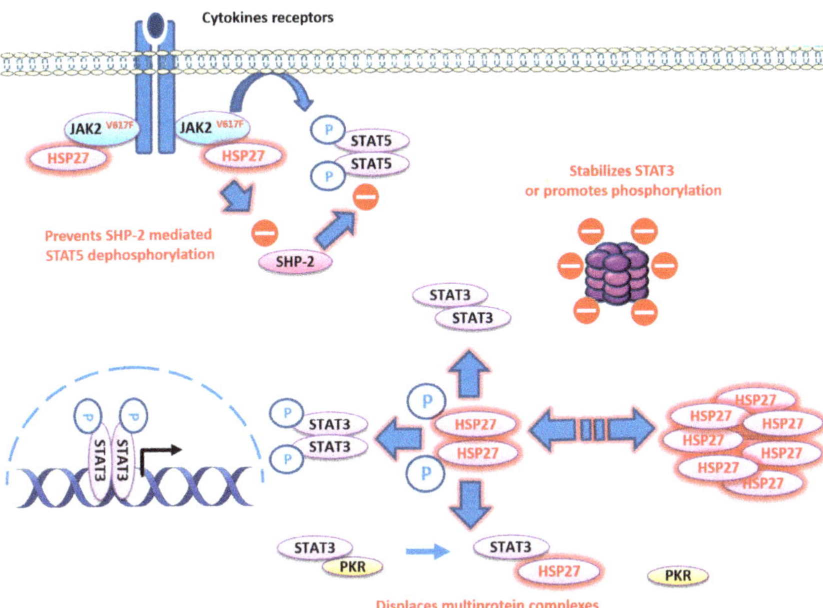

Figure 3. Localization and described functions of HSP27 within the STAT3 and STAT5 signaling pathways. HSP27 state of oligomerization varies dynamically to modulate its binding capacity to target proteins in a context-dependent way. HSP27 directly interacts with pYSTAT3 and total STAT3 to promote stabilization and phosphorylation. HSP27 directly binds to JAK2^{V617F}/STAT5 complexes to prevent STAT5 dephosphorylation by SHP-2 in MPN. HSP27 also displaces multiprotein complexes like the STAT3/PKR complex.

5. HSP110

5.1. HSP110 Structure

The 110 kDa heat shock protein (HSP110), also known as HSP105 or HSPH1, belongs to the members of the HSPH family. Although it appears to be distinct from other HSPs (HSP27, HSP40, HSP70, and HSP90) because of its molecular weight and the specificity of certain sequences, HSP110 is a member of the family of HSP70 proteins [143,144]. Until quite recently, HSP110 was considered as a mere nucleotide exchange factor of HSP70. However, it is now well established that HSP110 is able to act as an unfolding chaperone on its own using ATP hydrolysis to lead to conversion of stable misfolded polypeptide substrates into natively refolded products, even when HSP70 is not present [145]. In an ATP-independent manner, HSP110 also has the antiaggregating properties of unfolded or misfolded proteins.

5.2. HSP110 Functions

In contrast to other chaperones, like HSP90 or HSP70, little is known about the cellular and extracellular functions of HSP110. It is a ubiquitous and conserved chaperone with antiaggregation capabilities that act in synergy with the refolding activity of HSP70, which contributes to efficient protein homeostasis [145,146]. HSP110 expression is induced by a wide array of stress including hyperthermia, ethanol, oxidative stress, recovery from anoxia (i.e., reperfusion injury), some anticancer drugs, and inflammation. HSP110 is approximately four-fold more efficient at binding and stabilizing denatured protein substrates than HSP70 [147]. Due to its strong chaperone (or holder) function, HSP110 is a very good antigen carrier. It is therefore used in vaccine formulations [148] as a recombinant chaperone vaccine for antigen-targeted cancer immunotherapy. These vaccines have generated robust antigen-specific T lymphocyte responses in different preclinical cancer models [149]. Two forms of HSP110 exist: HSP110α and an alternatively spliced form called HSP110β, which contains 43 fewer amino acids [150]. While HSP110α is constitutively expressed in the cytoplasm of cells and can be induced in stressful conditions such as heat shock, HSP110β is strictly heat-inducible and specifically localized in the nucleus (Figure 4) [151]. Beyond the different localization of these two forms of HSP110, their differential roles remain unclear. Like other HSP proteins, the expression of HSP110 can be induced by a number of physical or chemical sources of stress and depends on the heat shock factor 1 (HSF1) transcription factor. Moreover, the presence STAT3 on the HSP110 promoter has been recently reported in humans, suggesting its regulation by STAT3 [152]. Conversely, HSP110β can induce HSP70 expression through STAT3 in mammalian cells (Figure 4) [153]. It is now clearly established that HSP110 favors several signaling pathways, including the Wnt/β-Catenin, MyD88/TLR, and STAT3 pathways [154,155].

Concerning the STAT signaling pathway, we have demonstrated that HSP110 directly binds to STAT3 and favors its phosphorylation (Tyr705) by JAK2 in the cytosol, thereby promoting cell proliferation (Figure 4) [156]. Colon cancer cells in which HSP110 has been shRNA-mediated and knocked down hardly proliferate, but proliferation is reactivated by the re-expression of HSP110 in these cells. Tumors from patients with high levels of HSP110 show high STAT3 phosphorylation levels and strong expression of proliferation markers [154,156]. Therefore, both the protein homeostasis function and the role of HSP110 on proliferative pathways may explain why this protein is linked to aggressive tumors. We suggest that HSP110 expression could be a surrogate prognostic marker and a potential therapeutic target, particularly for treatment of carcinomas, particularly colorectal cancer, for which there is strong evidence.

Given the emerging role of HSP110 in cancer and its role on STAT3 in particular, we selected two foldamers upon screening of a chemical library based on their ability to inhibit and block recombinant HSP110-mediated antiaggregation activity and to disrupt HSP110–STAT3 interaction [72]. These compounds, named 33 and 52 (Table 1, Figure 5), inhibit HSP110 chaperone function and colorectal cancer growth in vitro and in vivo [72]. Altogether these results confirm the interest of targeting HSP110, at least in colorectal cancers, and probably in other types of cancer, such as B-cell lymphoma.

Although HSPs are generally considered intracellular proteins, we now know that HSP110 can also be released to act extracellularly like HSP27, 70, and 90 [157–159]. The release of HSP110 from human intestinal epithelial cells has also been described, suggesting a role in the physiological process of epithelial renewal [160]. More recently, we have demonstrated that HSP110, like other HSPs, can be secreted by cancer cells and is abundantly observed in the cancer microenvironment [161]. Interestingly, extracellular HSP110 affects macrophage differentiation/polarization by favoring a protumor, anti-inflammatory profile and the formation of tumor-associated macrophages (TAMs), which are associated with immune suppression. Furthermore, we found a correlation between the level of extracellular HSP110 and the number of TAMs in patient biopsies [161], suggesting that the effect of extracellular HSP110 function on macrophages may also contribute to the poor outcomes that are associated with HSP110 expression.

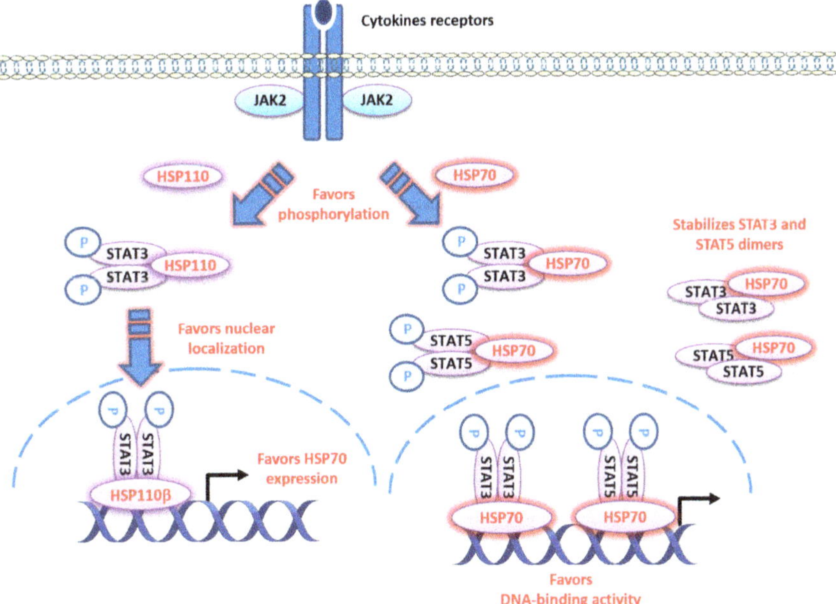

Figure 4. Localization and described functions of HSP110 and HSP70 within the STAT3 and STAT5 signaling pathways. HSP110 directly binds to STAT3 in the cytosol and favors its phosphorylation through JAK2, and, through this mechanism, participates in the promotion of cell proliferation. HSP110α and HSP110β localize to the cytoplasm and nucleus of cells, respectively. HSP110β induces the expression of HSP70 in mammalian cells. Overexpression of HSP110β stimulated the phosphorylation of STAT3 (Tyr705) and its translocation to the nucleus. STAT3 binds to the sequence of the HSP70 promoter at the level of a sequence (−206 to −187 base pair) whose mutation abrogated the activation of the HSP110β-mediated HSP70 promoter. HSP70 directly interacts with STAT3 and STAT5. It favors STAT3 phosphorylation and activity, and STAT5 levels, phosphorylation, and activity.

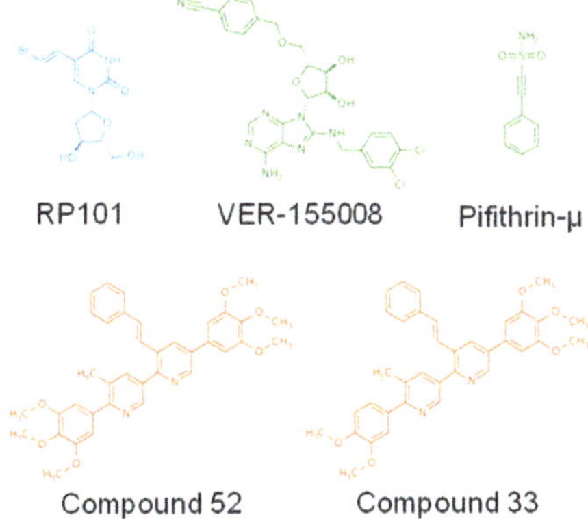

Figure 5. Structure of the main small-molecule inhibitors of HSP27 (blue), HSP70 (Green), and HSP110 (Orange). RP101 can inhibit HSP27 function via direct binding to Phe29 and Phe33. VER-155008 binds to the ATP-binding site at the N-terminus of HSP70. Pifithrin-μ inhibits specifically function of HSP70 via direct binding to its substrate binding domain. Compounds 52 and 33 bind to the ATP binding site at the N-terminus of HSP110.

6. HSP70

6.1. HSP70 Structure

Stress-inducible HSP70 (also called HSP72 or HSP1A, HSPA1B) is another member of the HSP superfamily that has emerged as a viable and very promising target for development of antitumor drugs to combat various forms of cancer [162–164]. As the most ubiquitous stress-inducible chaperone, HSP70 exhibits numerous chaperone functions that are critical for both the folding and proteasomal degradation of misfolded proteins. It thus participates in cellular protein quality control systems, leading to cell homeostasis and survival during stress conditions. The human HSP70 protein family consists of at least 13 members, including stress-induced HSP70 and the heat shock cognate protein 70 (HSC70) [165]. The chaperone mechanism of HSP70 has been extensively studied [166]. HSP70 associates with misfolded proteins in a manner controlled by its ATPase cycle and cochaperones such as HSP40, BAG family proteins, HIP, HOP, and HSPBP1 [165].

6.2. HSP70 Functions

Based on immunoprecipitation analysis from normal rat kidney interstitial fibroblast (NRK-49F) cells, HSP70 has been shown to directly interact with multiple STAT proteins, including STAT3/5 (Figure 4). This interaction increased when cells were stressed by exposure to advanced glycation end product [167]. Although the data are still limited, studies show that manipulating HSP70 expression or activity affects STAT protein activity within cells. HSP70 (HSP701A/B) knock down using siRNA further decreased constitutive STAT3 activity in an acute myeloid leukemia cell line (HEL) treated with arsenic trioxide and the HSP90 inhibitor 17-DMAG [168]. In addition, increasing HSP70 activity with administration of geranylgeranylacetone before and three days after intracerebral hemorrhage resulted in increased STAT3 and AKT phosphorylation and also cerebral levels of eNOS, which are collectively associated with preserved cerebral blood flow, decreased neuronal cell death, and improved functional recovery in rats [169]. Uchida et al. suggested that the mechanism for HSP70 induction through geranylgeranylacetone may be the result of geranylgeranylacetone-induced induction of protein kinase C [170].

It is worth noting that upregulation of HSP70 could stimulate cell proliferation through the control of tyrosine kinase functions. In cancer cells derived from chronic myeloid leukemia cells, BCR-ABL tyrosine kinase activity results in the phosphorylation and activation of AKT, and in the phosphorylation and DNA-binding activity of STAT5, which leads in turn to an increase in the expression of antiapoptotic protein Bcl-xL [171]. Interestingly, while increased expression of HSP70 results in the upregulation of STAT5 level and activity, inhibition of the BCR-ABL tyrosine kinase activity with imatinib or inhibition of PI-3K activity with wortmannin both result in a decrease in HSP70 expression and STAT5 activity. Thus, uncontrolled cell signaling may result in the transcription of HSP70, which in turn regulates the level and activity of STAT5 (Figure 4) [171]. Besides the intracellular role of HSP70 as a survival factor that promotes tumorigenesis [172], it is now known that HSP70 also regulates diverse immunoregulatory activities such as antigen cross presentation [173,174], dendritic cell maturation [175,176], and natural killer cell [177,178] and myeloid-derived suppressor cell [179] activities, by acting extracellularly as a cytokine [180,181]. We demonstrated that membrane-associated HSP70 is found extracellularly in tumor-derived exosomes and that it restrained tumor immune surveillance by promoting myeloid-derived suppressor cell functions in both mice and humans. Interestingly, tumor-derived exosomes harboring HSP70 were found to mediate the suppressive activity of the myeloid-derived suppressor cells via activation of STAT3 and ERK [179].

7. Regulation Mechanisms of HSF/HSPs by JAK/STAT Signaling: A Feedback Loop?

7.1. HSF and SOCS Regulation

The activation of the JAK family and the subsequent STAT signaling is regulated by the family of suppressors of cytokine signaling (SOCS) proteins. This system aims to protect organisms from permanent and/or overstimulation that could lead, for instance, to severe systemic inflammations mediated by IFN-γ signaling [182]. Conversely, a deficit in SOCS expression could play a pivotal role in the development and progression of cancers [183]. Expression of SOCS3, which has been shown to inhibit JAK1, JAK2, and TYK2 [184], is frequently reduced in cancer cells, thereby leading to a growth advantage. The effect of HSF1 and HSP on SOCS3 expression changes depending on the tissue and whether the context is normal or tumoral. In nontumoral microglia cells, the activation of HSF1 by paeoniflorin induced an indirect increase of SOCS3 expression mediated by HSP70 production and autocrine action [185]. Conversely, it was recently shown that HSF1 could directly bind to the SOCS3 promoter region and inhibit the transcriptional activity of its promoter [186]. This reinforces the role of HSF1 as a transcription factor for many genes not related to heat shock response [3]. SOCS3 repression of expression can also be mediated by HSPs, as seen in chronic lymphocytic leukemia. In this context, HSP90 inhibition by 17-DMAG induces the expression of SOCS3 through the activation of the p38 MAPK signaling [34]. This regulation is probably a specific to tumors, as normal B cells were not found to upregulate SOCS3 expression upon HSP90 inhibition, and their migration was not affected [34].

7.2. HSF/HSPs and JAK/STAT

The regulation of the STAT signaling pathway by HSPs has been largely described, but in a positive feedback loop STAT3/5 has been shown to regulate HSPs. Little is known about HSP regulation apart from the fundamental role of HSF. The STAT3/5 pathway modulates expression of HSP27, HSP70, HSP90, and HSP110 when faced with different stressful stimuli [187]. Among those, mild heat shock is a type of stress (that was used to first describe the function of HSPs) during which the induction of HSPs prevents cell death from intense heat shock that otherwise would have been lethal. This thermotolerance is also accompanied by STAT3 phosphorylation. Inhibition of STAT3 activation by STAT3 inhibitors AG490 or static partially suppresses thermotolerance and HSP110 expression, but not HSP70 and HSP27 [188]. This result underlines a particular role for HSP110 in the process, which is in line with its capacity to interact with α-tubulin. This mechanism could therefore protect microtubules from severe heat shock [189]. Besides STAT3, STAT1 is also involved in HSP70 and HSP90 regulation, particularly after IFN-γ activation [190,191] or under heat shock [192]. Together with HSF1, STAT1 is recruited to the first intron of the HSP90β gene to favor the recruitment of a chromatin-remodeling complex, leading to enhanced HSP90β expression.

Other types of stress, such as hypoxic stress (a common phenomenon in a majority of tumors), are strong inducers of HSP90α expression. In these conditions, STAT5 is one of the transcription factors that regulate HSP90α, and hypoxia increases the binding of STAT5 to the HSP90α promoter [193]. When the STAT signaling pathway is constitutively active in tumors, like in breast and colon cancer, STAT3 has also been shown to transcriptionally induce HSP27 expression [194]. Though less documented, this mechanism also exists in hematological malignancies like Burkitt's lymphoma, where pharmacological inhibition of STAT3 by AG490 downregulates HSP70 expression [195].

8. Conclusions

We have reviewed here the multiple roles of HSPs in the STAT3/5 activation network. From receptor tyrosine kinases to promoter binding of STAT3/5 target genes, HSPs appear to use several mechanisms to control this pathway. Among the various HSPs, HSP90 has a central role in the cellular machinery and is one of the most abundant cytosolic proteins. This role is clearly exposed when we describe both kinases and STAT proteins as HSP90 client proteins. The current literature points to the important role of HSP90/STAT3/STAT5 in cancer growth and the ability to thwart chemotherapy [3,196–198].

This discovery has prompted the development of drugs specifically targeting HSP90 and the subsequent initiation of clinical trials for cancer patients (Table 1, Figure 2). Almost all of the trials have reported clinical responses, confirming the relevance of this strategy. However, adverse side-effects were also frequent due to the inherent toxicity of HSP90 inhibitors. This could be explained by the number of client proteins that are simultaneously chaperoned by HSP90, which obviously endangers critical physiological processes in normal cells. A strategic question emerges from this observation: should we continue searching for more specific, less toxic drugs? Or should we limit studies to fundamental science to uncover new biological mechanisms? Other HSPs are gaining interest in the scientific community. In particular, HSP110 is a newly discovered player in the field of cancer aggressiveness that controls the STAT3 pathway in colon cancers and in B-cell lymphomas. Our team recently identified the first HSP110 inhibitors (Table 1, Figure 5) that block interaction with STAT3 and that effectively limit colon cancer progression in mouse models; we plan to bring these to the clinic within the next few years. No toxicity has been demonstrated so far, and we believe this new type of inhibitor is promising for cancers whose poor prognosis is associated with HSP110/pYSTAT3 expression. Both intra- and extracellular HSP70, which belong to a related HSP family, favor STAT3/5 phosphorylation by JAK2. This dual action is of particular therapeutic interest since targeting HSP70 would simultaneously blunt the macrophage-mediated immunosuppression and block the intratumoral growth signal. To reach the goal of bringing HSP70 inhibitors to the clinic, we have identified peptide aptamers that bind to the peptide-binding domain of HSP70. This has not been an easy task since HSP70, contrary to HSP90, is not a "druggable" protein. Despite this limitation, our HSP70 inhibitors are specific and have proven effective in xenograft models of colon cancer. Of course, further studies will be needed before clinical trials can be initiated. Finally, HSP27 has recently been found to play new roles in myeloproliferative neoplasms. In this disease, which is driven by the $JAK2^{V617F}$/STAT5 pathway, HSP27 inhibition destabilizes the protein complex and limits disease related myelofibrosis. Furthermore, in contrast to HSP90 inhibitors, the HSP27 inhibitor does not induce the compensatory expression of other HSPs that usually account for resistance to treatment. Targeting HSP27 would therefore be an alternative to the failed HSP90 therapy, which also targeted $JAK2^{V617F}$. An oligonucleotide antisense of second generation (OGX427) is currently under clinical evaluation (Table 1).

Furthermore, small molecules targeting STAT3/5 have been identified as enhancing protein degradation. The inhibition of STAT pathways is therefore likely highly amenable to HSP inhibition and presents a potential synergistic therapeutic strategy (Table 1).

In conclusion, we show in this review that the STAT3/5 pathways rely on multiple HSPs under physiological and pathological conditions. Targeting various members of the HSP family, alone or in combination, will probably improve the inhibition of this central pathway and should foster the development of new, more specific and less toxic HSP inhibitors to complete the existing therapeutic arsenal.

Author Contributions: G.J., F.H. co-wrote the paper, and designed and drew all the figures for this manuscript. C.G. proofread the manuscript, gave suggestions, and modified the language. F.G. contributed to the editing and critical reading of the manuscript. All authors have read and approved the final manuscript. All authors have read and agreed to the published version of the manuscript.

Funding: This research was funded by a French Government grant managed by the French National Research Agency under the program "Investissements d'Avenir" with reference ANR-11-LABX-0021 (LabEX LipSTIC), the Institut National du Cancer, the Agence Nationale de la Recherche, the Conseil Régional de Bourgogne, the European Union through the PO FEDER-FSE Bourgogne 2014/2020 programs, and the association "Châlon-sur-Saône Tulipes contre le cancer".

Acknowledgments: The authors thank S. Rankin (Dijon University Hospital) for proofreading the manuscript. F. Hermetet was supported by a fellowship from the University of Burgundy and Franche-Comté through the ISITE-BFC program. C Garrido's team is labeled by La Ligue Nationale contre le Cancer.

Conflicts of Interest: The authors declare no conflict of interest.

References

1. Sorger, P.K. Heat shock factor and the heat shock response. *Cell* **1991**, *65*, 363–366. [CrossRef]
2. Welch, W.J. Heat shock proteins functioning as molecular chaperones: Their roles in normal and stressed cells. *Philos. Trans. R. Soc. Lond. B Biol. Sci.* **1993**, *339*, 327–333. [PubMed]
3. Jego, G.; Hazoumé, A.; Seigneuric, R.; Garrido, C. Targeting heat shock proteins in cancer. *Cancer Lett.* **2013**, *332*, 275–285. [CrossRef] [PubMed]
4. Schmitt, E.; Gehrmann, M.; Brunet, M.; Multhoff, G.; Garrido, C. Intracellular and extracellular functions of heat shock proteins: Repercussions in cancer therapy. *J. Leukoc. Biol.* **2007**, *81*, 15–27. [CrossRef] [PubMed]
5. Hanahan, D.; Weinberg, R.A. Hallmarks of cancer: The next generation. *Cell* **2011**, *144*, 646–674. [CrossRef] [PubMed]
6. Calderwood, S.K.; Gong, J. Heat Shock Proteins Promote Cancer: It's a Protection Racket. *Trends BioChem. Sci.* **2016**, *41*, 311–323. [CrossRef] [PubMed]
7. Lianos, G.D.; Alexiou, G.A.; Mangano, A.; Mangano, A.; Rausei, S.; Boni, L.; Dionigi, G.; Roukos, D.H. The role of heat shock proteins in cancer. *Cancer Lett.* **2015**, *360*, 114–118. [CrossRef]
8. Wu, J.; Liu, T.; Rios, Z.; Mei, Q.; Lin, X.; Cao, S. Heat Shock Proteins and Cancer. *Trends Pharmacol. Sci.* **2017**, *38*, 226–256. [CrossRef]
9. Guttmann, D.M.; Koumenis, C. The heat shock proteins as targets for radiosensitization and chemosensitization in cancer. *Cancer Biol. Ther.* **2011**, *12*, 1023–1031. [CrossRef]
10. Chi, K.N.; Yu, E.Y.; Jacobs, C.; Bazov, J.; Kollmannsberger, C.; Higano, C.S.; Mukherjee, S.D.; Gleave, M.E.; Stewart, P.S.; Hotte, S.J. A phase I dose-escalation study of apatorsen (OGX-427), an antisense inhibitor targeting heat shock protein 27 (Hsp27), in patients with castration-resistant prostate cancer and oTher. advanced cancers. *Ann. Oncol.* **2016**, *27*, 1116–1122. [CrossRef]
11. Trepel, J.; Mollapour, M.; Giaccone, G.; Neckers, L. Targeting the dynamic HSP90 complex in cancer. *Nat. Rev. Cancer* **2010**, *10*, 537–549. [CrossRef] [PubMed]
12. Yuno, A.; Lee, M.J.; Lee, S.; Tomita, Y.; Rekhtman, D.; Moore, B.; Trepel, J.B. Clinical Evaluation and Biomarker Profiling of Hsp90 Inhibitors. *Methods Mol. Biol.* **2018**, *1709*, 423–441. [PubMed]
13. Rocchi, P.; So, A.; Kojima, S.; Signaevsky, M.; Beraldi, E.; Fazli, L.; Hurtado-Coll, A.; Yamanaka, K.; Gleave, M. Heat shock protein 27 increases after androgen ablation and plays a cytoprotective role in hormone-refractory prostate cancer. *Cancer Res.* **2004**, *64*, 6595–6602. [CrossRef] [PubMed]
14. Song, T.F.; Zhang, Z.F.; Liu, L.; Yang, T.; Jiang, J.; Li, P. Small interfering RNA-mediated silencing of heat shock protein 27 (HSP27) Increases chemosensitivity to paclitaxel by increasing production of reactive oxygen species in human ovarian cancer cells (HO8910). *J. Int. Med. Res.* **2009**, *37*, 1375–1388. [CrossRef]
15. Baylot, V.; Andrieu, C.; Katsogiannou, M.; Taieb, D.; Garcia, S.; Giusianom, S.; Acunzo, J.; Iovanna, J.; Gleave, M.; Garrido, C.; et al. OGX-427 inhibits tumor progression and enhances gemcitabine chemotherapy in pancreatic cancer. *Cell Death Dis.* **2011**, *2*, e221. [CrossRef]
16. Lelj-Garolla, B.; Kumano, M.; Beraldi, E.; Nappi, L.; Rocchi, P.; Ionescu, D.N.; Fazlim, L.; Zoubeidi, A.; Gleave, M.E. Hsp27 Inhibition with OGX-427 Sensitizes Non-Small Cell Lung Cancer Cells to Erlotinib and Chemotherapy. *Mol. Cancer Ther.* **2015**, *14*, 1107–1116. [CrossRef]
17. Spigel, D.R.; Shipley, D.L.; Waterhouse, D.M.; Jones, S.F.; Ward, P.J.; Shih, K.C.; Hemphill, B.; McCleod, M.; Whorf, R.C.; Page, R.D.; et al. A Randomized Double-Blinded Phase II Trial of Carboplatin and Pemetrexed with or without Apatorsen (OGX-427) in Patients with Previously Untreated Stage IV Non-Squamous-Non-Small-Cell Lung Cancer: The SPRUCE Trial. *Oncologist* **2019**, *24*, e1409–e1416. [CrossRef]
18. Xia, Y.; Liu, Y.; Wan, J.; Wang, M.; Rocchi, P.; Qu, F.; Iovanna, J.L.; Peng, L. Novel triazole ribonucleoside down-regulates heat shock protein 27 and induces potent anticancer activity on drug-resistant pancreatic cancer. *J. Med. Chem.* **2009**, *52*, 6083–6096. [CrossRef]
19. Chauhan, D.; Li, G.; Shringarpure, R.; Podar, K.; Ohtake, Y.; Hideshima, T.; Anderson, K.C. Blockade of Hsp27 overcomes Bortezomib/proteasome inhibitor PS-341 resistance in lymphoma cells. *Cancer Res.* **2003**, *63*, 6174–6177.
20. Heinrich, J.C.; Tuukkanen, A.; Schroeder, M.; Fahrig, T.; Fahrig, R. RP101 (brivudine) binds to heat shock protein HSP27 (HSPB1) and enhances survival in animals and pancreatic cancer patients. *J. Cancer Res. Clin. Oncol.* **2011**, *137*, 1349–1361. [CrossRef]

21. Kaiser, M.; Kühnl, A.; Reins, J.; Fischer, S.; Ortiz-Tanchez, J.; Schlee, C.; Mochmann, L.H.; Heesch, S.; Benlasfer, O.; Hofmann, W.K.; et al. Antileukemic activity of the HSP70 inhibitor pifithrin-μ in acute leukemia. *Blood Cancer J.* **2011**, *1*, e28. [CrossRef] [PubMed]

22. Goloudina, A.R.; Demidov, O.N.; Garrido, C. Inhibition of HSP70: A challenging anti-cancer strategy. *Cancer Lett.* **2012**, *325*, 117–124. [CrossRef] [PubMed]

23. Massey, A.J.; Williamson, D.S.; Browne, H.; Murray, J.B.; Dokurno, P.; Shaw, T.; Macias, A.T.; Daniels, Z.; Geoffroy, S.; Dopson, M.; et al. A novel small molecule inhibitor of Hsc70/Hsp70 potentiates Hsp90 inhibitor induced apoptosis in HCT116 colon carcinoma cells. *Cancer ChemoTher. Pharmacol.* **2010**, *66*, 535–545. [CrossRef] [PubMed]

24. Brünnert, D.; Langer, C.; Zimmermann, L.; Bargou, R.C.; Burchardt, M.; Chatterjee, M.; Stope, M.B. The heat shock protein 70 inhibitor VER155008 suppresses the expression of HSP27, HOP and HSP90β and the androgen receptor, induces apoptosis, and attenuates prostate cancer cell growth. *J. Cell BioChem.* **2020**, *121*, 407–417. [CrossRef] [PubMed]

25. Chatterjee, M.; Andrulis, M.; Stühmer, T.; Müller, E.; Hofmann, C.; Steinbrunn, T.; Heimberger, T.; Schraud, H.; Kressmann, S.; Einsele, H.; et al. The PI3K/Akt signaling pathway regulates the expression of Hsp70, which critically contributes to Hsp90-chaperone function and tumor cell survival in multiple myeloma. *Haematologica* **2013**, *98*, 1132–1141. [CrossRef] [PubMed]

26. Rérole, A.L.; Gobbo, J.; De Thonel, A.; Schmitt, E.; Pais de Barros, J.P.; Hammann, A.; Lanneau, D.; Fourmaux, E.; Demidov, O.N.; Micheau, O.; et al. Peptides and aptamers targeting HSP70: A novel approach for anticancer chemotherapy. *Cancer Res.* **2011**, *71*, 484–495. [CrossRef]

27. Schmitt, E.; Maingret, L.; Puig, P.E.; Rerole, A.L.; Ghiringhelli, F.; Hammann, A.; Solary, E.; Kroemer, G.; Garrido, C. Heat shock protein 70 neutralization exerts potent antitumor effects in animal models of colon cancer and melanoma. *Cancer Res.* **2006**, *66*, 4191–4197. [CrossRef]

28. Stangl, S.; Gehrmann, M.; Riegger, J.; Kuhs, K.; Riederer, I.; Sievert, W.; Hube, K.; Mocikat, R.; Dressel, R.; Kremmer, E.; et al. Targeting membrane heat-shock protein 70 (Hsp70) on tumors by cmHsp70.1 antibody. *Proc. Natl. Acad. Sci. USA* **2011**, *108*, 733–738. [CrossRef]

29. Krause, S.W.; Gastpar, R.; Andreesen, R.; Gross, C.; Ullrich, H.; Thonigs, G.; Pfister, K.; Multhoff, G. Treatment of colon and lung cancer patients with ex vivo heat shock protein 70-peptide-activated, autologous natural killer cells: A clinical phase i trial. *Clin. Cancer Res.* **2004**, *10*, 3699–3707. [CrossRef]

30. Shiotsu, Y.; Neckers, L.M.; Wortman, I.; An, W.G.; Schulte, T.W.; Soga, S.; Murakata, C.; Tamaoki, T.; Akinaga, S. Novel oxime derivatives of radicicol induce erythroid differentiation associated with preferential G (1) phase accumulation against chronic myelogenous leukemia cells through destabilization of Bcr-Abl with Hsp90 complex. *Blood* **2000**, *96*, 2284–2291. [CrossRef]

31. Palacios, C.; López-Pérez, A.I.; López-Rivas, A. Down-regulation of RIP expression by 17-dimethylaminoethylamino-17-demethoxygeldanamycin promotes TRAIL-induced apoptosis in breast tumor cells. *Cancer Lett.* **2010**, *287*, 207–215. [CrossRef] [PubMed]

32. Schaefer, S.; Svenstrup, T.H.; Guerra, B. The small-molecule kinase inhibitor D11 counteracts 17-AAG-mediated up-regulation of HSP70 in brain cancer cells. *PLoS ONE* **2017**, *12*, e0177706. [CrossRef] [PubMed]

33. Ayrault, O.; Godeny, M.D.; Dillon, C.; Zindy, F.; Fitzgerald, P.; Roussel, M.F.; Beere, H.M. Inhibition of Hsp90 via 17-DMAG induces apoptosis in a p53-dependent manner to prevent medulloblastoma. *Proc. Natl. Acad. Sci. USA* **2009**, *106*, 17037–17042. [CrossRef] [PubMed]

34. Chen, T.L.; Gupta, N.; Lehman, A.; Ruppert, A.S.; Yu, L.; Oakes, C.C.; Claus, R.; Plass, C.; Maddocks, K.J.; Andritsos, L.; et al. Hsp90 inhibition increases SOCS3 transcript and regulates migration and cell death in chronic lymphocytic leukemia. *Oncotarget* **2016**, *7*, 28684–28696. [CrossRef]

35. Hanson, B.E.; Vesole, D.H. Retaspimycin hydrochloride (IPI-504): A novel heat shock protein inhibitor as an anticancer agent. *Expert Opin. Investig. Drugs* **2009**, *18*, 1375–1383. [CrossRef]

36. Song, D.; Chaerkady, R.; Tan, A.C.; García-García, E.; Nalli, A.; Suárez-Gauthier, A.; López-Ríos, F.; Zhang, X.F.; Solomon, A.; Tong, J.; et al. Antitumor activity and molecular effects of the novel heat shock protein 90 inhibitor, IPI-504, in pancreatic cancer. *Mol. Cancer Ther.* **2008**, *7*, 3275–3284. [CrossRef]

37. Wagner, A.J.; Chugh, R.; Rosen, L.S.; Morgan, J.A.; George, S.; Gordon, M.; Dunbar, J.; Normant, E.; Grayzel, D.; Demetri, G.D.; et al. A phase I study of the HSP90 inhibitor retaspimycin hydrochloride (IPI-504) in patients with gastrointestinal stromal tumors or soft-tissue sarcomas. *Clin. Cancer Res.* **2013**, *19*, 6020–6029. [CrossRef]

38. Scaltriti, M.; Serra, V.; Normant, E.; Guzman, M.; Rodriguez, O.; Lim, A.R.; Slocum, K.L.; West, K.A.; Rodriguez, V.; Prudkin, L.; et al. Antitumor activity of the Hsp90 inhibitor IPI-504 in HER2-positive trastuzumab-resistant breast cancer. *Mol. Cancer Ther.* **2011**, *10*, 817–824. [CrossRef]

39. Floris, G.; Debiec-Rychter, M.; Wozniak, A.; Stefan, C.; Normant, E.; Faa, G.; Machiels, K.; Vanleeuw, U.; Sciot, R.; Schöffski, P. The heat shock protein 90 inhibitor IPI-504 induces KIT degradation, tumor shrinkage, and cell proliferation arrest in xenograft models of gastrointestinal stromal tumors. *Mol. Cancer Ther.* **2011**, *10*, 1897–1908. [CrossRef]

40. Normant, E.; Paez, G.; West, K.A.; Lim, A.R.; Slocum, K.L.; Tunkey, C.; McDougall, J.; Wylie, A.A.; Robison, K.; Caliri, K.; et al. The Hsp90 inhibitor IPI-504 rapidly lowers EML4-ALK levels and induces tumor regression in ALK-driven NSCLC models. *Oncogene* **2011**, *30*, 2581–2586. [CrossRef]

41. Hendriks, L.E.L.; Dingemans, A.C. Heat shock protein antagonists in early stage clinical trials for NSCLC. *Expert Opin. Investig. Drugs* **2017**, *26*, 541–550. [CrossRef] [PubMed]

42. Wu, L.X.; Xu, J.H.; Zhang, K.Z.; Lin, Q.; Huang, X.W.; Wen, C.X.; Chen, Y.Z. Disruption of the Bcr-Abl/Hsp90 protein complex: A possible mechanism to inhibit Bcr-Abl-positive human leukemic blasts by novobiocin. *Leukemia* **2008**, *22*, 1402–1409. [CrossRef] [PubMed]

43. Shelton, S.N.; Shawgo, M.E.; Matthews, S.B.; Lu, Y.; Donnelly, A.C.; Szabla, K.; Tanol, M.; Vielhauer, G.A.; Rajewski, R.A.; Matts, R.L.; et al. KU135, a novel novobiocin-derived C-terminal inhibitor of the 90-kDa heat shock protein, exerts potent antiproliferative effects in human leukemic cells. *Mol. Pharmacol.* **2009**, *76*, 1314–1322. [CrossRef] [PubMed]

44. Matthews, S.B.; Vielhauer, G.A.; Manthe, C.A.; Chaguturu, V.K.; Szabla, K.; Matts, R.L.; Donnelly, A.C.; Blagg, B.S.; Holzbeierlein, J.M. Characterization of a novel novobiocin analogue as a putative C-terminal inhibitor of heat shock protein 90 in prostate cancer cells. *Prostate* **2010**, *70*, 27–36. [CrossRef]

45. Le, H.T.; Vielhauer, G.A.; Manthe, C.A.; Chaguturu, V.K.; Szabla, K.; Matts, R.L.; Donnelly, A.C.; Blagg, B.S.; Holzbeierlein, J.M. Panaxynol, a natural Hsp90 inhibitor, effectively targets both lung cancer stem and non-stem cells. *Cancer Lett.* **2018**, *412*, 297–307. [CrossRef]

46. Lin, S.F.; Lin, J.D.; Hsueh, C.; Chou, T.C.; Yeh, C.N.; Chen, M.H.; Wong, R.J. Efficacy of an HSP90 inhibitor, ganetespib, in preclinical thyroid cancer models. *Oncotarget* **2017**, *8*, 41294–41304. [CrossRef]

47. Lee, H.; Saini, N.; Howard, E.W.; Parris, A.B.; Ma, Z.; Zhao, Q.; Zhao, M.; Liu, B.; Edgerton, S.M.; Thor, A.D.; et al. Ganetespib targets multiple levels of the receptor tyrosine kinase signaling cascade and preferentially inhibits ErbB2-overexpressing breast cancer cells. *Sci. Rep.* **2018**, *8*, 6829. [CrossRef]

48. He, W.; Hu, H. BIIB021, an Hsp90 inhibitor: A promising therapeutic strategy for blood malignancies (Review). *Oncol. Rep.* **2018**, *40*, 3–15. [CrossRef]

49. Marubayashi, S.; Koppikar, P.; Taldone, T.; Abdel-Wahab, O.; West, N.; Bhagwat, N.; Caldas-Lopes, E.; Ross, K.N.; Gönen, M.; Gozman, A.; et al. HSP90 is a therapeutic target in JAK2-dependent myeloproliferative neoplasms in mice and humans. *J. Clin. Investig.* **2010**, *120*, 3578–3593. [CrossRef]

50. Fiskus, W.; Verstovsek, S.; Manshouri, T.; Rao, R.; Balusu, R.; Venkannagari, S.; Rao, N.N.; Ha, K.; Smith, J.E.; Hembruff, S.L.; et al. Heat shock protein 90 inhibitor is synergistic with JAK2 inhibitor and overcomes resistance to JAK2-TKI in human myeloproliferative neoplasm cells. *Clin. Cancer Res.* **2011**, *17*, 7347–7358. [CrossRef]

51. Park, K.S.; Hong, Y.S.; Choi, J.; Yoon, S.; Kang, J.; Kim, D.; Lee, K.P.; Im, H.S.; Lee, C.H.; Seo, S.; et al. HSP90 inhibitor, AUY922, debilitates intrinsic and acquired lapatinib-resistant HER2-positive gastric cancer cells. *BMB Rep.* **2018**, *51*, 660–665. [CrossRef] [PubMed]

52. Yang, H.; Lee, M.H.; Park, I.; Jeon, H.; Choi, J.; Seo, S.; Kim, S.W.; Koh, G.Y.; Park, K.S.; Lee, D.H. HSP90 inhibitor (NVP-AUY922) enhances the anti-cancer effect of BCL-2 inhibitor (ABT-737) in small cell lung cancer expressing BCL-2. *Cancer Lett.* **2017**, *411*, 19–26. [CrossRef] [PubMed]

53. Kim, S.H.; Kang, J.G.; Kim, C.S.; Ihm, S.H.; Choi, M.G.; Yoo, H.J.; Lee, S.J. The dipeptidyl peptidase-IV inhibitor gemigliptin alone or in combination with NVP-AUY922 has a cytotoxic activity in thyroid carcinoma cells. *Tumour. Biol.* **2017**, *39*, 1010428317722068. [CrossRef] [PubMed]

54. Brough, P.A.; Aherne, W.; Barril, X.; Borgognoni, J.; Boxall, K.; Cansfield, J.E.; Cheung, K.M.; Collins, I.; Davies, N.G.; Drysdale, M.J.; et al. 4,5-diarylisoxazole Hsp90 chaperone inhibitors: Potential therapeutic agents for the treatment of cancer. *J. Med. Chem.* **2008**, *51*, 196–218. [CrossRef] [PubMed]

55. Liu, J.; Sun, W.; Dong, W.; Wang, Z.; Qin, Y.; Zhang, T.; Zhang, H. HSP90 inhibitor NVP-AUY922 induces cell apoptosis by disruption of the survivin in papillary thyroid carcinoma cells. *Biochem. Biophys. Res. Commun.* **2017**, *487*, 313–319. [CrossRef]

56. Tao, W.; Chakraborty, S.N.; Leng, X.; Ma, H.; Arlinghaus, R.B. HSP90 inhibitor AUY922 induces cell death by disruption of the Bcr-Abl, Jak2 and HSP90 signaling network complex in leukemia cells. *Genes Cancer* **2015**, *6*, 19–29.

57. Chakraborty, S.N.; Leng, X.; Perazzona, B.; Sun, X.; Lin, Y.H.; Arlinghaus, R.B. Combination of JAK2 and HSP90 inhibitors: An effective therapeutic option in drug-resistant chronic myelogenous leukemia. *Genes Cancer* **2016**, *7*, 201–208.

58. Hobbs, G.S.; Hanasoge Somasundara, A.V.; Kleppe, M.; Litvin, R.; Arcila, M.; Ahn, J.; McKenney, A.S.; Knapp, K.; Ptashkin, R.; Weinstein, H.; et al. Hsp90 inhibition disrupts JAK-STAT signaling and leads to reductions in splenomegaly in patients with myeloproliferative neoplasms. *Haematologica* **2018**, *103*, e5–e9. [CrossRef]

59. Johnson, M.L.; Yu, H.A.; Hart, E.M.; Weitner, B.B.; Rademaker, A.W.; Patel, J.D.; Kris, M.G.; Riely, G.J. Phase I/II Study of HSP90 Inhibitor AUY922 and Erlotinib for EGFR-Mutant Lung Cancer with Acquired Resistance to Epidermal Growth Factor Receptor Tyrosine Kinase Inhibitors. *J. Clin. Oncol.* **2015**, *33*, 1666–1673. [CrossRef]

60. Weigert, O.; Lane, A.A.; Bird, L.; Kopp, N.; Chapuy, B.; van Bodegom, D.; Toms, A.V.; Marubayashi, S.; Christie, A.L.; McKeown, M.; et al. Genetic resistance to JAK2 enzymatic inhibitors is overcome by HSP90 inhibition. *J. Exp. Med.* **2012**, *209*, 259–273. [CrossRef]

61. Mahendrarajah, N.; Borisova, M.E.; Reichardt, S.; Godmann, M.; Sellmer, A.; Mahboobi, S.; Haitel, A.; Schmid, K.; Kenner, L.; Heinzel, T.; et al. HSP90 is necessary for the ACK1-dependent phosphorylation of STAT1 and STAT3. *Cell Signal.* **2017**, *39*, 9–17. [CrossRef] [PubMed]

62. Smyth, T.; Paraiso, K.H.T.; Hearn, K.; Rodriguez-Lopez, A.M.; Munck, J.M.; Haarberg, H.E.; Sondak, V.K.; Thompson, N.T.; Azab, M.; Lyons, J.F.; et al. Inhibition of HSP90 by AT13387 delays the emergence of resistance to BRAF inhibitors and overcomes resistance to dual BRAF and MEK inhibition in melanoma models. *Mol. Cancer Ther.* **2014**, *13*, 2793–2804. [CrossRef] [PubMed]

63. Graham, B.; Curry, J.; Smyth, T.; Fazal, L.; Feltell, R.; Harada, I.; Coyle, J.; Williams, B.; Reule, M.; Angove, H.; et al. The heat shock protein 90 inhibitor, AT13387, displays a long duration of action in vitro and in vivo in non-small cell lung cancer. *Cancer Sci.* **2012**, *103*, 522–527. [CrossRef] [PubMed]

64. Courtin, A.; Smyth, T.; Hearn, K.; Saini, H.K.; Thompson, N.T.; Lyons, J.F.; Wallis, N.G. Emergence of resistance to tyrosine kinase inhibitors in non-small-cell lung cancer can be delayed by an upFront combination with the HSP90 inhibitor onalespib. *Br. J. Cancer* **2016**, *115*, 1069–1077. [CrossRef] [PubMed]

65. Eroglu, Z.; Chen, Y.A.; Gibney, G.T.; Weber, J.S.; Kudchadkar, R.R.; Khushalani, N.I.; Markowitz, J.; Brohl, A.S.; Tetteh, L.F.; Ramadan, H.; et al. Combined BRAF and HSP90 Inhibition in Patients with Unresectable. *Clin. Cancer Res.* **2018**, *24*, 5516–5524. [CrossRef] [PubMed]

66. Friedman, J.A.; Wise, S.C.; Hu, M.; Gouveia, C.; Vander Broek, R.; Freudlsperger, C.; Kannabiran, V.R.; Arun, P.; Mitchell, J.B.; Chen, Z.; et al. HSP90 Inhibitor SNX5422/2112 Targets the Dysregulated Signal and Transcription Factor Network and Malignant Phenotype of Head and Neck Squamous Cell Carcinoma. *Transl. Oncol.* **2013**, *6*, 429–441. [CrossRef] [PubMed]

67. Rice, J.W.; Veal, J.M.; Barabasz, A.; Foley, B.; Fadden, P.; Scott, A.; Huang, K.; Steed, P.; Hall, S. Targeting of multiple signaling pathways by the Hsp90 inhibitor SNX-2112 in EGFR resistance models as a single agent or in combination with erlotinib. *Oncol. Res.* **2009**, *18*, 229–242. [CrossRef]

68. Smith, D.L.; Acquaviva, J.; Sequeira, M.; Jimenez, J.P.; Zhang, C.; Sang, J.; Bates, R.C.; Proia, D.A. The HSP90 inhibitor ganetespib potentiates the antitumor activity of EGFR tyrosine kinase inhibition in mutant and wild-type non-small cell lung cancer. *Target. Oncol.* **2015**, *10*, 235–245. [CrossRef]

69. Moulick, K.; Ahn, J.H.; Zong, H.; Rodina, A.; Cerchietti, L.; Gomes DaGama, E.M.; Caldas-Lopes, E.; Beebe, K.; Perna, F.; Hatzi, K.; et al. Affinity-based proteomics reveal cancer-specific networks coordinated by Hsp90. *Nat. Chem. Biol.* **2011**, *7*, 818–826. [CrossRef]

70. Hong, Y.S.; Jang, W.J.; Chun, K.S.; Jeong, C.H. Hsp90 inhibition by WK88-1 potently suppresses the growth of gefitinib-resistant H1975 cells harboring the T790M mutation in EGFR. *Oncol. Rep.* **2014**, *31*, 2619–2624. [CrossRef]

71. Kobayashi, N.; Toyooka, S.; Soh, J.; Yamamoto, H.; Dote, H.; Kawasaki, K.; Otani, H.; Kubo, T.; Jida, M.; Ueno, T.; et al. The anti-proliferative effect of heat shock protein 90 inhibitor, 17-DMAG, on non-small-cell lung cancers being resistant to EGFR tyrosine kinase inhibitor. *Lung Cancer* **2012**, *75*, 161–166. [CrossRef] [PubMed]

72. Gozzi, G.J.; Gonzalez, D.; Boudesco, C.; Dias, A.M.M.; Gotthard, G.; Uyanik, B.; Dondaine, L.; Marcion, G.; Hermetet, F.; Denis, C.; et al. Selecting the first chemical molecule inhibitor of HSP110 for colorectal cancer therapy. *Cell Death Differ.* **2019**. [CrossRef] [PubMed]

73. McClellan, A.J.; Xia, Y.; Deutschbauer, A.M.; Davis, R.W.; Gerstein, M.; Frydman, J. Diverse cellular functions of the Hsp90 molecular chaperone uncovered using systems approaches. *Cell* **2007**, *131*, 121–135. [CrossRef] [PubMed]

74. Taipale, M.; Krykbaeva, I.; Koeva, M.; Kayatekin, C.; Westover, K.D.; Karras, G.I.; Lindquist, S. Quantitative analysis of HSP90-client interactions reveals principles of substrate recognition. *Cell* **2012**, *150*, 987–1001. [CrossRef]

75. Makhnevych, T.; Houry, W.A. The role of Hsp90 in protein complex assembly. *Biochim. Biophys. Acta* **2012**, *1823*, 674–682. [CrossRef] [PubMed]

76. Boczek, E.E.; Reefschläger, L.G.; Dehling, M.; Struller, T.J.; Häusler, E.; Seidl, A.; Kaila, V.R.; Buchner, J. Conformational processing of oncogenic v-Src kinase by the molecular chaperone Hsp90. *Proc. Natl. Acad. Sci. USA* **2015**, *112*, E3189–E3198. [CrossRef] [PubMed]

77. Gray, P.J.; Prince, T.; Chengm, J.; Stevenson, M.A.; Calderwood, S.K. Targeting the oncogene and kinome chaperone CDC37. *Nat. Rev. Cancer* **2008**, *8*, 491–495. [CrossRef]

78. Shang, L.; Tomasi, T.B. The heat shock protein 90-CDC37 chaperone complex is required for signaling by types I and II interferons. *J. Biol. Chem.* **2006**, *281*, 1876–1884. [CrossRef]

79. Peeters, P.; Wlodarska, I.; Baens, M.; Criel, A.; Selleslag, D.; Hagemeijer, A.; Van den Berghe, H.; Marynen, P. Fusion of ETV6 to MDS1/EVI1 as a result of t(3;12)(q26;p13) in myeloproliferative disorders. *Cancer Res.* **1997**, *57*, 564–569.

80. Lacronique, V.; Boureux, A.; Valle, V.D.; Poirel, H.; Quang, C.T.; Mauchauffé, M.; Berthou, C.; Lessard, M.; Berger, R.; Ghysdael, J.; et al. A TEL-JAK2 fusion protein with constitutive kinase activity in human leukemia. *Science* **1997**, *278*, 1309–1312. [CrossRef]

81. Baxter, E.J.; Scott, L.M.; Campbell, P.J.; East, C.; Fourouclas, N.; Swanton, S.; Vassiliou, G.S.; Bench, A.J.; Boyd, E.M.; Curtin, N.; et al. Acquired mutation of the tyrosine kinase JAK2 in human myeloproliferative disorders. *Lancet* **2005**, *365*, 1054–1061. [CrossRef]

82. Levine, R.L.; Loriaux, M.; Huntly, B.J.; Loh, M.L.; Beran, M.; Stoffregen, E.; Berger, R.; Clark, J.J.; Willis, S.G.; Nguyen, K.T.; et al. The JAK2V617F activating mutation occurs in chronic myelomonocytic leukemia and acute myeloid Leukemia but not in acute lymphoblastic leukemia or chronic lymphocytic leukemia. *Blood* **2005**, *106*, 3377–3379. [CrossRef] [PubMed]

83. James, C.; Ugo, V.; Le Couédic, J.P.; Staerk, J.; Delhommeau, F.; Lacout, C.; Garçon, L.; Raslova, H.; Berger, R.; Bennaceur-Griscelli, A.; et al. A unique clonal JAK2 mutation leading to constitutive signalling causes polycythaemia vera. *Nature* **2005**, *434*, 1144–1148. [CrossRef] [PubMed]

84. Jones, A.V.; Kreil, S.; Zoi, K.; Waghorn, K.; Curtis, C.; Zhang, L.; Score, J.; Seear, R.; Chase, A.J.; Grand, F.H.; et al. Widespread occurrence of the JAK2 V617F mutation in chronic myeloproliferative disorders. *Blood* **2005**, *106*, 2162–2168. [CrossRef]

85. Bareng, J.; Jilani, I.; Gorre, M.; Kantarjian, H.; Giles, F.; Hannah, A.; Albitar, M. A potential role for HSP90 inhibitors in the treatment of JAK2 mutant-positive diseases as demonstrated using quantitative flow cytometry. *Leuk. Lymphoma* **2007**, *48*, 2189–2195. [CrossRef]

86. Sen, B.; Johnson, F.M. Regulation of SRC family kinases in human cancers. *J. Signal. Transduct* **2011**, *2011*, 865819. [CrossRef]

87. Garcia, R.; Bowman, T.L.; Niu, G.; Yu, H.; Minton, S.; Muro-Cacho, C.A.; Cox, C.E.; Falcone, R.; Fairclough, R.; Parsons, S.; et al. Constitutive activation of Stat3 by the Src and JAK tyrosine kinases participates in growth regulation of human breast carcinoma cells. *Oncogene* **2001**, *20*, 2499–2513. [CrossRef]

88. Frame, M.C. Src in cancer: Deregulation and consequences for cell behaviour. *Biochim. Biophys. Acta* **2002**, *1602*, 114–130. [CrossRef]

89. Takeya, T.; Hanafusa, H. Structure and sequence of the cellular gene homologous to the RSV src gene and the mechanism for generating the transforming virus. *Cell* **1983**, *32*, 881–890. [CrossRef]

90. Stehelin, D.; Varmus, H.E.; Bishop, J.M.; Vogt, P.K. DNA related to the transforming gene(s) of avian sarcoma viruses is present in normal avian DNA. *Nature* **1976**, *260*, 170–173. [CrossRef]

91. Hunter, T.; Sefton, B.M. Transforming gene product of Rous sarcoma virus phosphorylates tyrosine. *Proc. Natl. Acad. Sci. USA* **1980**, *77*, 1311–1315. [CrossRef]

92. Xu, Y.; Lindquist, S. Heat-shock protein hsp90 governs the activity of pp60v-src kinase. *Proc. Natl. Acad. Sci. USA* **1993**, *90*, 7074–7078. [CrossRef]

93. Whitesell, L.; Mimnaugh, E.G.; De Costa, B.; Myers, C.E.; Neckers, L.M. Inhibition of heat shock protein HSP90-pp60v-src heteroprotein complex formation by benzoquinone ansamycins: Essential role for stress proteins in oncogenic transformation. *Proc. Natl. Acad. Sci. USA* **1994**, *91*, 8324–8328. [CrossRef]

94. Nathan, D.F.; Lindquist, S. Mutational analysis of Hsp90 function: Interactions with a steroid receptor and a protein kinase. *Mol. Cell Biol.* **1995**, *15*, 3917–3925. [CrossRef]

95. Luo, Q.; Boczek, E.E.; Wang, Q.; Buchner, J.; Kaila, V.R. Hsp90 dependence of a kinase is determined by its conformational landscape. *Sci. Rep.* **2017**, *7*, 43996. [CrossRef]

96. Verba, K.A.; Wang, R.Y.; Arakawa, A.; Liu, Y.; Shirouzu, M.; Yokoyama, S.; Agard, D.A. Atomic structure of Hsp90-Cdc37-Cdk4 reveals that Hsp90 traps and stabilizes an unfolded kinase. *Science* **2016**, *352*, 1542–1547. [CrossRef]

97. Giannini, A.; Bijlmakers, M.J. Regulation of the Src family kinase Lck by Hsp90 and ubiquitination. *Mol. Cell Biol.* **2004**, *24*, 5667–5676. [CrossRef]

98. Scholz, G.M.; Hartson, S.D.; Cartledge, K.; Volk, L.; Matts, R.L.; Dunn, A.R. The molecular chaperone Hsp90 is required for signal transduction by wild-type Hck and maintenance of its constitutively active counterpart. *Cell Growth Differ.* **2001**, *12*, 409–417.

99. Mahajan, N.P.; Whang, Y.E.; Mohler, J.L.; Earp, H.S. Activated tyrosine kinase Ack1 promotes prostate tumorigenesis: Role of Ack1 in polyubiquitination of tumor suppressor Wwox. *Cancer Res.* **2005**, *65*, 10514–10523. [CrossRef]

100. Karachaliou, N.; Pilotto, S.; Teixidó, C.; Viteri, S.; González-Cao, M.; Riso, A.; Morales-Espinosa, D.; Molina, M.A.; Chaib, I.; Santarpia, M.; et al. Melanoma: Oncogenic drivers and the immune system. *Ann. Transl. Med.* **2015**, *3*, 265.

101. Da Rocha Dias, S.; Friedlos, F.; Light, Y.; Springer, C.; Workman, P.; Marais, R. Activated B-RAF is an Hsp90 client protein that is targeted by the anticancer drug 17-allylamino-17-demethoxygeldanamycin. *Cancer Res.* **2005**, *65*, 10686–10691. [CrossRef]

102. Jabbour, E.; Kantarjian, H. Chronic myeloid leukemia: 2018 update on diagnosis, therapy and monitoring. *Am. J. Hematol.* **2018**, *93*, 442–459. [CrossRef]

103. Coppo, P.; Friedlos, F.; Light, Y.; Springer, C.; Workman, P.; Marais, R. BCR-ABL activates STAT3 via JAK and MEK pathways in human cells. *Br. J. Haematol.* **2006**, *134*, 171–179. [CrossRef]

104. Nair, R.R.; Tolentino, J.H.; Hazlehurst, L.A. Role of STAT3 in Transformation and Drug Resistance in CML. *Front. Oncol.* **2012**, *2*, 30. [CrossRef]

105. De Groot, R.P.; Raaijmakers, J.A.; Lammers, J.W.; Jove, R.; Koenderman, L. STAT5 activation by BCR-Abl contributes to transformation of K562 leukemia cells. *Blood* **1999**, *94*, 1108–1112. [CrossRef]

106. Sillaber, C.; Gesbert, F.; Frank, D.A.; Sattler, M.; Griffin, J.D. STAT5 activation contributes to growth and viability in Bcr/Abl-transformed cells. *Blood* **2000**, *95*, 2118–2125. [CrossRef]

107. Hoelbl, A.; Schuster, C.; Kovacic, B.; Zhu, B.; Wickre, M.; Hoelzl, M.A.; Fajmann, S.; Grebien, F.; Warsch, W.; Stengl, G.; et al. Stat5 is indispensable for the maintenance of bcr/abl-positive leukaemia. *EMBO Mol. Med.* **2010**, *2*, 98–110. [CrossRef]

108. Esfahani, K.; Cohen, V. HSP90 as a novel molecular target in non-small-cell lung cancer. *Lung Cancer (Auckl)* **2016**, *7*, 11–17.

109. Lemmon, M.A.; Schlessinger, J. Cell Signaling by receptor tyrosine kinases. *Cell* **2010**, *141*, 1117–1134. [CrossRef]

110. Kosack, L.; Wingelhofer, B.; Popa, A.; Orlova, A.; Agerer, B.; Vilagos, B.; Majek, P.; Parapatics, K.; Lercher, A.; Ringler, A.; et al. The ERBB-STAT3 Axis Drives Tasmanian Devil Facial Tumor Disease. *Cancer Cell* **2019**, *35*, 125–139.e9. [CrossRef]

111. Xu, W.; Mimnaugh, E.G.; Kim, J.S.; Trepel, J.B.; Neckers, L.M. Hsp90, not Grp94, regulates the intracellular trafficking and stability of nascent ErbB2. *Cell Stress Chaperones* **2002**, *7*, 91–96. [CrossRef]

112. Peng, X.; Guo, X.; Borkan, S.C.; Bharti, A.; Kuramochi, Y.; Calderwood, S.; Sawyer, D.B. Heat shock protein 90 stabilization of ErbB2 expression is disrupted by ATP depletion in myocytes. *J. Biol. Chem.* **2005**, *280*, 13148–13152. [CrossRef] [PubMed]

113. Wee, P.; Wang, Z. Epidermal Growth Factor Receptor Cell Proliferation Signaling Pathways. *Cancers (Basel)* **2017**, *9*, 52. [CrossRef]

114. Bao, R.; Lai, C.J.; Wang, D.G.; Qu, H.; Yin, L.; Zifcak, B.; Tao, X.; Wang, J.; Atoyan, R.; Samson, M.; et al. Targeting heat shock protein 90 with CUDC-305 overcomes erlotinib resistance in non-small cell lung cancer. *Mol. Cancer Ther.* **2009**, *8*, 3296–3306. [CrossRef]

115. Ono, N.; Yamazaki, T.; Tsukaguchi, T.; Fujii, T.; Sakata, K.; Suda, A.; Tsukuda, T.; Mio, T.; Ishii, N.; Kondoh, O.; et al. Enhanced antitumor activity of erlotinib in combination with the Hsp90 inhibitor CH5164840 against non-small-cell lung cancer. *Cancer Sci.* **2013**, *104*, 1346–1352. [CrossRef]

116. Sato, N.; Yamamoto, T.; Sekine, Y.; Yumioka, T.; Junicho, A.; Fuse, H.; Matsuda, T. Involvement of heat-shock protein 90 in the interleukin-6-mediated signaling pathway through STAT3. *Biochem. Biophys. Res. Commun.* **2003**, *300*, 847–852. [CrossRef]

117. Song, S.; Su, Z.; Xu, H.; Niu, M.; Chen, X.; Min, H.; Zhang, B.; Sun, G.; Xie, S.; Wang, H.; et al. Luteolin selectively kills STAT3 highly activated gastric cancer cells through enhancing the binding of STAT3 to SHP-1. *Cell Death Dis.* **2017**, *8*, e2612. [CrossRef]

118. Longshaw, V.M.; Baxter, M.; Prewitz, M.; Blatch, G.L. Knockdown of the co-chaperone Hop promotes extranuclear accumulation of Stat3 in mouse embryonic stem cells. *Eur. J. Cell Biol.* **2009**, *88*, 153–166. [CrossRef]

119. Liu, L.; McBride, K.M.; Reich, N.C. STAT3 nuclear import is independent of tyrosine phosphorylation and mediated by importin-alpha3. *Proc. Natl. Acad. Sci. USA* **2005**, *102*, 8150–8155. [CrossRef]

120. Echeverría, P.C.; Mazaira, G.; Erlejman, A.; Gomez-Sanchez, C.; Piwien Pilipuk, G.; Galigniana, M.D. Nuclear import of the glucocorticoid receptor-hsp90 complex through the nuclear pore complex is mediated by its interaction with Nup62 and importin beta. *Mol. Cell Biol.* **2009**, *29*, 4788–4797. [CrossRef]

121. Adwan, T.S.; Ohm, A.M.; Jones, D.N.; Humphries, M.J.; Reyland, M.E. Regulated binding of importin-α to protein kinase Cδ in response to apoptotic signals facilitates nuclear import. *J. Biol. Chem.* **2011**, *286*, 35716–35724. [CrossRef] [PubMed]

122. Langlais, D.; Couture, C.; Balsalobre, A.; Drouin, J. The Stat3/GR interaction code: Predictive value of direct/indirect DNA recruitment for transcription outcome. *Mol. Cell* **2012**, *47*, 38–49. [CrossRef] [PubMed]

123. Engblom, D.; Kornfeld, J.W.; Schwake, L.; Tronche, F.; Reimann, A.; Beug, H.; Hennighausen, L.; Moriggl, R.; Schütz, G. Direct glucocorticoid receptor-Stat5 interaction in hepatocytes controls body size and maturation-related gene expression. *Genes Dev.* **2007**, *21*, 1157–1162. [CrossRef] [PubMed]

124. Wingelhofer, B.; Neubauer, H.A.; Valent, P.; Han, X.; Constantinescu, S.N.; Gunning, P.T.; Müller, M.; Moriggl, R. Implications of STAT3 and STAT5 signaling on gene regulation and chromatin remodeling in hematopoietic cancer. *Leukemia* **2018**, *32*, 1713–1726. [CrossRef]

125. Ehrnsperger, M.; Gräber, S.; Gaestel, M.; Buchner, J. Binding of non-native protein to Hsp25 during heat shock creates a reservoir of folding intermediates for reactivation. *EMBO J.* **1997**, *16*, 221–229. [CrossRef]

126. Shashidharamurthy, R.; Koteiche, H.A.; Dong, J.; McHaourab, H.S. Mechanism of chaperone function in small heat shock proteins: Dissociation of the HSP27 oligomer is required for recognition and binding of destabilized T4 lysozyme. *J. Biol. Chem.* **2005**, *280*, 5281–5289. [CrossRef]

127. Choi, S.K.; Kam, H.; Kim, K.Y.; Park, S.I.; Lee, Y.S. Targeting Heat Shock Protein 27 in Cancer: A Druggable Target for Cancer Treatment? *Cancers (Basel)* **2019**, *11*, 1195. [CrossRef]

128. Jehle, S.; van Rossum, B.; Stout, J.R.; Noguchi, S.M.; Falber, K.; Rehbein, K.; Oschkinat, H.; Klevit, R.E.; Rajagopal, P. alphaB-crystallin: A hybrid solid-state/solution-state NMR investigation reveals structural aspects of the heterogeneous oligomer. *J. Mol. Biol.* **2009**, *385*, 1481–1497. [CrossRef]

129. Kostenko, S.; Moens, U. Heat shock protein 27 phosphorylation: Kinases, phosphatases, functions and pathology. *Cell Mol. Life Sci.* **2009**, *66*, 3289–3307. [CrossRef]

130. Bruey, J.M.; Paul, C.; Fromentin, A.; Hilpert, S.; Arrigo, A.P.; Solary, E.; Garrido, C. Differential regulation of HSP27 oligomerization in tumor cells grown in vitro and in vivo. *Oncogene* **2000**, *19*, 4855–4863. [CrossRef]

131. Mymrikov, E.V.; Daake, M.; Richter, B.; Haslbeck, M.; Buchner, J. The Chaperone Activity and Substrate Spectrum of Human Small Heat Shock Proteins. *J. Biol. Chem.* **2017**, *292*, 672–684. [CrossRef] [PubMed]

132. Ungelenk, S.; Moayed, F.; Ho, C.T.; Grousl, T.; Scharf, A.; Mashaghi, A.; Tans, S.; Mayer, M.P.; Mogk, A.; Bukau, B. Small heat shock proteins sequester misfolding proteins in near-native conformation for cellular protection and efficient refolding. *Nat. Commun.* **2016**, *7*, 13673. [CrossRef] [PubMed]

133. Rocchi, P.; Beraldi, E.; Ettinger, S.; Fazli, L.; Vessella, R.L.; Nelson, C.; Gleave, M. Increased Hsp27 after androgen ablation facilitates androgen-independent progression in prostate cancer via signal transducers and activators of transcription 3-mediated suppression of apoptosis. *Cancer Res.* **2005**, *65*, 11083–11093. [CrossRef] [PubMed]

134. Shiota, M.; Bishop, J.L.; Nip, K.M.; Zardan, A.; Takeuchi, A.; Cordonnier, T.; Beraldi, E.; Bazov, J.; Fazli, L.; Chi, K.; et al. Hsp27 regulates epithelial mesenchymal transition, metastasis, and circulating tumor cells in prostate cancer. *Cancer Res.* **2013**, *73*, 3109–3119. [CrossRef]

135. Shah, M.; Stanek, J.; Handwerger, S. Differential localization of heat shock proteins 90, 70, 60 and 27 in human decidua and placenta during pregnancy. *HistoChem. J.* **1998**, *30*, 509–518. [CrossRef]

136. Matalon, S.T.; Drucker, L.; Fishman, A.; Ornoy, A.; Lishner, M. The Role of heat shock protein 27 in extravillous trophoblast differentiation. *J. Cell BioChem.* **2008**, *103*, 719–729. [CrossRef]

137. Shochet, G.E.; Komemi, O.; Sadeh-Mestechkin, D.; Pomeranz, M.; Fishman, A.; Drucker, L.; Lishner, M.; Matalon, S.T. Heat shock protein-27 (HSP27) regulates STAT3 and eIF4G levels in first trimester human placenta. *J. Mol. Histol.* **2016**, *47*, 555–563. [CrossRef]

138. Suman, P.; Malhotra, S.S.; Gupta, S.K. LIF-STAT signaling and trophoblast biology. *JAKSTAT* **2013**, *2*, e25155. [CrossRef]

139. Gibert, B.; Eckel, B.; Fasquelle, L.; Moulin, M.; Bouhallier, F.; Gonin, V.; Mellier, G.; Simon, S.; Kretz-Remy, C.; Arrigo, A.P.; et al. Knock down of heat shock protein 27 (HspB1) induces degradation of several putative client proteins. *PLoS ONE* **2012**, *7*, e29719. [CrossRef]

140. Shen, L.; Qi, Z.; Zhu, Y.; Song, X.; Xuan, C.; Ben, P.; Lan, L.; Luo, L.; Yin, Z. Phosphorylated heat shock protein 27 promotes lipid clearance in hepatic cells through interacting with STAT3 and activating autophagy. *Cell Signal.* **2016**, *28*, 1086–1098. [CrossRef]

141. Sevin, M.; Kubovcakova, L.; Pernet, N.; Causse, S.; Vitte, F.; Villeval, J.L.; Lacout, C.; Cordonnier, M.; Rodrigues-Lima, F.; Chanteloup, G.; et al. HSP27 is a partner of JAK2-STAT5 and a potential therapeutic target in myelofibrosis. *Nat. Commun.* **2018**, *9*, 1431. [CrossRef] [PubMed]

142. Kleppe, M.; Kwak, M.; Koppikar, P.; Riester, M.; Keller, M.; Bastian, L.; Hricik, T.; Bhagwat, N.; McKenney, A.S.; Papalexi, E.; et al. JAK-STAT pathway activation in malignant and nonmalignant cells contributes to MPN pathogenesis and therapeutic response. *Cancer Discov.* **2015**, *5*, 316–331. [CrossRef] [PubMed]

143. Lee-Yoon, D.; Easton, D.; Murawski, M.; Burd, R.; Subjeck, J.R. Identification of a major subfamily of large hsp70-like proteins through the cloning of the mammalian 110-kDa heat shock protein. *J. Biol. Chem.* **1995**, *270*, 15725–15733. [CrossRef] [PubMed]

144. Oh, H.J.; Easton, D.; Murawski, M.; Kaneko, Y.; Subjeck, J.R. The chaperoning activity of hsp110. Identification of functional domains by use of targeted deletions. *J. Biol. Chem.* **1999**, *274*, 15712–15718. [CrossRef] [PubMed]

145. Mattoo, R.U.; Sharma, S.K.; Priya, S.; Finka, A.; Goloubinoff, P. Hsp110 is a bona fide chaperone using ATP to unfold stable misfolded polypeptides and reciprocally collaborate with Hsp70 to solubilize protein aggregates. *J. Biol. Chem.* **2013**, *288*, 21399–21411. [CrossRef] [PubMed]

146. Rampelt, H.; Kirstein-Miles, J.; Nillegoda, N.B.; Chi, K.; Scholz, S.R.; Morimoto, R.I.; Bukau, B. Metazoan Hsp70 machines use Hsp110 to power protein disaggregation. *EMBO J.* **2012**, *31*, 4221–4235. [CrossRef]

147. Wang, L.; Duke, L.; Zhang, P.S.; Arlinghaus, R.B.; Symmans, W.F.; Sahin, A.; Mendez, R.; Dai, J.L. Alternative splicing disrupts a nuclear localization signal in spleen tyrosine kinase that is required for invasion suppression in breast cancer. *Cancer Res.* **2003**, *63*, 4724–4730.

148. Manjili, M.H.; Henderson, R.; Wang, X.Y.; Chen, X.; Li, Y.; Repasky, E.; Kazim, L.; Subjeck, J.R. Development of a recombinant HSP110-HER-2/neu vaccine using the chaperoning properties of HSP110. *Cancer Res.* **2002**, *62*, 1737–1742.

149. Guo, C.; Subjeck, J.R.; Wang, X.Y. Creation of Recombinant Chaperone Vaccine Using Large Heat Shock Protein for Antigen-Targeted Cancer Immunotherapy. *Methods Mol. Biol.* **2018**, *1709*, 345–357.

150. Yasuda, K.; Nakai, A.; Hatayama, T.; Nagata, K. Cloning and expression of murine high molecular mass heat shock proteins, HSP105. *J. Biol. Chem.* **1995**, *270*, 29718–29723.

151. Saito, Y.; Yamagishi, N.; Hatayama, T. Nuclear localization mechanism of Hsp105beta and its possible function in mammalian cells. *J. BioChem.* **2009**, *145*, 185–191. [CrossRef] [PubMed]

152. Olszak, T.; Neves, J.F.; Dowds, C.M.; Baker, K.; Glickman, J.; Davidson, N.O.; Lin, C.S.; Jobin, C.; Brand, S.; Sotlar, K.; et al. Protective mucosal immunity mediated by epithelial CD1d and IL-10. *Nature* **2014**, *509*, 497–502. [CrossRef] [PubMed]

153. Yamagishi, N.; Fujii, H.; Saito, Y.; Hatayama, T. Hsp105beta upregulates hsp70 gene expression through signal transducer and activator of transcription-3. *FEBS J.* **2009**, *276*, 5870–5880. [CrossRef] [PubMed]

154. Yu, N.; Kakunda, M.; Pham, V.; Lill, J.R.; Du, P.; Wongchenko, M.; Yan, Y.; Firestein, R.; Huang, X. HSP105 recruits protein phosphatase 2A to dephosphorylate beta-catenin. *Mol. Cell Biol.* **2015**, *35*, 1390–1400. [CrossRef]

155. Boudesco, C.; Verhoeyen, E.; Martin, L.; Chassagne-Clement, C.; Salmi, L.; Mhaidly, R.; Pangault, C.; Fest, T.; Ramla, S.; Jardin, F.; et al. HSP110 sustains chronic NF-kappaB signaling in activated B-cell diffuse large B-cell lymphoma through MyD88 stabilization. *Blood* **2018**, *132*, 510–520. [CrossRef]

156. Berthenet, K.; Bokhari, A.; Lagrange, A.; Marcion, G.; Boudesco, C.; Causse, S.; De Thonel, A.; Svrcek, M.; Goloudina, A.R.; Dumont, S.; et al. HSP110 promotes colorectal cancer growth through STAT3 activation. *Oncogene* **2017**, *36*, 2328–2336. [CrossRef]

157. Ono, K.; Eguchi, T.; Sogawa, C.; Calderwood, S.K.; Futagawa, J.; Kasai, T.; Seno, M.; Okamoto, K.; Sasaki, A.; Kozaki, K.I. HSP-enriched properties of extracellular vesicles involve survival of metastatic oral cancer cells. *J. Cell BioChem.* **2018**, *119*, 7350–7362. [CrossRef]

158. Eguchi, T.; Sogawa, C.; Okusha, Y.; Uchibe, K.; Iinuma, R.; Ono, K.; Nakano, K.; Murakami, J.; Itoh, M.; Arai, K.; et al. Organoids with cancer stem cell-like properties secrete exosomes and HSP90 in a 3D nanoenvironment. *PLoS ONE* **2018**, *13*, e0191109. [CrossRef]

159. Taha, E.A.; Ono, K.; Eguchi, T. Roles of Extracellular HSPs as Biomarkers in Immune Surveillance and Immune Evasion. *Int. J. Mol. Sci.* **2019**, *20*, 4588. [CrossRef]

160. Colgan, S.P.; Pitman, R.S.; Nagaishi, T.; Mizoguchi, A.; Mizoguchi, E.; Mayer, L.F.; Shao, L.; Sartor, R.B.; Subjeck, J.R.; Blumberg, R.S. Intestinal heat shock protein 110 regulates expression of CD1d on intestinal epithelial cells. *J. Clin. Investig.* **2003**, *112*, 745–754. [CrossRef]

161. Berthenet, K.; Boudesco, C.; Collura, A.; Svrcek, M.; Richaud, S.; Hammann, A.; Causse, S.; Yousfi, N.; Wanherdrick, K.; Duplomb, L.; et al. Extracellular HSP110 skews macrophage polarization in colorectal cancer. *Oncoimmunology* **2016**, *5*, e1170264. [CrossRef] [PubMed]

162. Goetz, M.P.; Toft, D.; Reid, J.; Ames, M.; Stensgard, B.; Safgren, S.; Adjei, A.A.; Sloan, J.; Atherton, P.; Vasile, V.; et al. Phase I trial of 17-allylamino-17-demethoxygeldanamycin in patients with advanced cancer. *J. Clin. Oncol.* **2005**, *23*, 1078–1087. [CrossRef] [PubMed]

163. Boudesco, C.; Cause, S.; Jego, G.; Garrido, C. Hsp70: A Cancer Target Inside and Outside the Cell. *Methods Mol. Biol.* **2018**, *1709*, 371–396. [PubMed]

164. Eguchi, T.; Lang, B.J.; Murshid, A.; Prince, T.; Gong, J.; Calderwood, S.K. Regulatory Roles for Hsp70 in Cancer Incidence and Tumor Progression. In *Role of Molecular Chaperones in Structural Folding, Biological Functions, and Drug Interactions of Client Proteins*; Galigniana, M.D., Ed.; Bentham Science Publishers: Sharjah, UAE, 2018; Volume 1, pp. 1–22.

165. Liu, T.; Daniels, C.K.; Cao, S. Comprehensive review on the HSC70 functions, interactions with related molecules and involvement in clinical diseases and therapeutic potential. *Pharmacol. Ther.* **2012**, *136*, 354–374. [CrossRef] [PubMed]

166. Zhuravleva, A.; Gierasch, L.M. Allosteric signal transmission in the nucleotide-binding domain of 70-kDa heat shock protein (Hsp70) molecular chaperones. *Proc. Natl. Acad. Sci. USA* **2011**, *108*, 6987–6992. [CrossRef] [PubMed]

167. Chen, S.C.; Guh, J.Y.; Chen, H.C.; Yang, Y.L.; Huang, J.S.; Chuang, L.Y. Advanced glycation end-product-induced mitogenesis is dependent on Janus kinase 2-induced heat shock protein 70 in normal rat kidney interstitial fibroblast cells. *Transl. Res.* **2007**, *149*, 274–281. [CrossRef] [PubMed]

168. Ghoshal, S.; Rao, I.; Earp, J.C.; Jusko, W.J.; Wetzler, M. Down-regulation of heat shock protein 70 improves arsenic trioxide and 17-DMAG effects on constitutive signal transducer and activator of transcription 3 activity. *Cancer ChemoTher. Pharmacol.* **2010**, *66*, 681–689. [CrossRef]

169. Sinn, D.I.; Kim, S.J.; Chu, K.; Jung, K.H.; Lee, S.T.; Song, E.C.; Kim, J.M.; Park, D.K.; Kun Lee, S.; Kim, M.; et al. Valproic acid-mediated neuroprotection in intracerebral hemorrhage via histone deacetylase inhibition and transcriptional activation. *NeuroBiol. Dis.* **2007**, *26*, 464–472. [CrossRef]

170. Uchida, S.; Fujiki, M.; Nagai, Y.; Abe, T.; Kobayashi, H. Geranylgeranylacetone, a noninvasive heat shock protein inducer, induces protein kinase C and leads to neuroprotection against cerebral infarction in rats. *NeuroSci. Lett.* **2006**, *396*, 220–224. [CrossRef]

171. Guo, F.; Sigua, C.; Bali, P.; George, P.; Fiskus, W.; Scuto, A.; Annavarapu, S.; Mouttaki, A.; Sondarva, G.; Wei, S.; et al. Mechanistic role of heat shock protein 70 in Bcr-Abl-mediated resistance to apoptosis in human acute leukemia cells. *Blood* **2005**, *105*, 1246–1255. [CrossRef]

172. Ciocca, D.R.; Arrigo, A.P.; Calderwood, S.K. Heat shock proteins and heat shock factor 1 in carcinogenesis and tumor development: An update. *Arch. Toxicol.* **2013**, *87*, 19–48. [CrossRef] [PubMed]

173. Binder, R.J.; Srivastava, P.K. Peptides chaperoned by heat-shock proteins are a necessary and sufficient source of antigen in the cross-priming of CD8+ T cells. *Nat. Immunol.* **2005**, *6*, 593–599. [CrossRef] [PubMed]

174. Li, Z.; Menoret, A.; Srivastava, P. Roles of heat-shock proteins in antigen presentation and cross-presentation. *Curr. Opin. Immunol.* **2002**, *14*, 45–51. [CrossRef]

175. Srivastava, P. Interaction of heat shock proteins with peptides and antigen presenting cells: Chaperoning of the innate and adaptive immune responses. *Annu. Rev. Immunol.* **2002**, *20*, 395–425. [CrossRef] [PubMed]

176. Srivastava, P. Roles of heat-shock proteins in innate and adaptive immunity. *Nat. Rev. Immunol.* **2002**, *2*, 185–194. [CrossRef]

177. Elsner, L.; Muppala, V.; Gehrmann, M.; Lozano, J.; Malzahn, D.; Bickeböller, H.; Brunner, E.; Zientkowska, M.; Herrmann, T.; Walter, L.; et al. The heat shock protein HSP70 promotes mouse NK cell activity against tumors that express inducible NKG2D ligands. *J. Immunol.* **2007**, *179*, 5523–5533. [CrossRef]

178. Gross, C.; Hansch, D.; Gastpar, R.; Multhoff, G. Interaction of heat shock protein 70 peptide with NK cells involves the NK receptor CD94. *Biol. Chem.* **2003**, *384*, 267–279. [CrossRef]

179. Chalmin, F.; Ladoire, S.; Mignot, G.; Vincent, J.; Bruchard, M.; Remy-Martin, J.P.; Boireau, W.; Rouleau, A.; Simon, B.; Lanneau, D.; et al. Membrane-associated Hsp72 from tumor-derived exosomes mediates STAT3-dependent immunosuppressive function of mouse and human myeloid-derived suppressor cells. *J. Clin. Investig.* **2010**, *120*, 457–471.

180. Asea, A.; Kraeft, S.K.; Kurt-Jones, E.A.; Stevenson, M.A.; Chen, L.B.; Finberg, R.W.; Koo, G.C.; Calderwood, S.K. HSP70 stimulates cytokine production through a CD14-dependant pathway, demonstrating its dual role as a chaperone and cytokine. *Nat. Med.* **2000**, *6*, 435–442. [CrossRef]

181. Chen, T.; Guo, J.; Han, C.; Yang, M.; Cao, X. Heat shock protein 70, released from heat-stressed tumor cells, initiates antitumor immunity by inducing tumor cell chemokine production and activating dendritic cells via TLR4 pathway. *J. Immunol.* **2009**, *182*, 1449–1459. [CrossRef]

182. Starr, R.; Metcalf, D.; Elefanty, A.G.; Brysha, M.; Willson, T.A.; Nicola, N.A.; Hilton, D.J.; Alexander, W.S. Liver degeneration and lymphoid deficiencies in mice lacking suppressor of cytokine signaling-1. *Proc. Natl. Acad. Sci. USA* **1998**, *95*, 14395–14399. [CrossRef] [PubMed]

183. Jiang, M.; Zhang, W.W.; Liu, P.; Yu, W.; Liu, T.; Yu, J. Dysregulation of SOCS-Mediated Negative Feedback of Cytokine Signaling in Carcinogenesis and Its Significance in Cancer Treatment. *Front. Immunol.* **2017**, *8*, 70. [CrossRef]

184. Babon, J.J.; Kershaw, N.J.; Murphy, J.M.; Varghese, L.N.; Laktyushin, A.; Young, S.N.; Lucet, I.S.; Norton, R.S.; Nicola, N.A. Suppression of cytokine signaling by SOCS3: Characterization of the mode of inhibition and the basis of its specificity. *Immunity* **2012**, *36*, 239–250. [CrossRef] [PubMed]

185. Fan, Y.X.; Qian, C.; Liu, B.; Wang, C.; Liu, H.; Pan, X.; Teng, P.; Hu, L.; Zhang, G.; Han, Y.; et al. Induction of suppressor of cytokine signaling 3 via HSF-1-HSP70-TLR4 axis attenuates neuroinflammation and ameliorates postoperative pain. *Brain Behav. Immun.* **2018**, *68*, 111–122. [CrossRef] [PubMed]

186. Wang, G.; Xiao, G.; Liu, H.; Chen, G.; Wang, X.; Wen, P.; Li, T.; Wen, J.; Xiao, X. Heat Shock Factor 1 Inhibits the Expression of Suppressor of Cytokine Signaling 3 in Cerulein-Induced Acute Pancreatitis. *Shock* **2018**, *50*, 465–471. [CrossRef] [PubMed]

187. Stephanou, A.; Isenberg, D.A.; Nakajima, K.; Latchman, D.S. Signal transducer and activator of transcription-1 and heat shock factor-1 interact and activate the transcription of the Hsp-70 and Hsp-90beta gene promoters. *J. Biol. Chem.* **1999**, *274*, 1723–1728. [CrossRef]

188. Matozaki, M.; Saito, Y.; Yasutake, R.; Munira, S.; Kaibori, Y.; Yukawa, A.; Tada, M.; Nakayama, Y. Involvement of Stat3 phosphorylation in mild heat shock-induced thermotolerance. *Exp. Cell Res.* **2019**, *377*, 67–74. [CrossRef]

189. Saito, Y.; Yamagishi, N.; Ishihara, K.; Hatayama, T. Identification of alpha-tubulin as an hsp105alpha-binding protein by the yeast two-hybrid system. *Exp. Cell Res.* **2003**, *286*, 233–240. [CrossRef]

190. Fritchley, S.J.; Kirby, J.A.; Ali, S. The antagonism of interferon-gamma (IFN-gamma) by heparin: Examination of the blockade of class II MHC antigen and heat shock protein-70 expression. *Clin. Exp. Immunol.* **2000**, *120*, 247–252. [CrossRef]

191. Chen, X.S.; Zhang, Y.; Wang, J.S.; Li, X.Y.; Cheng, X.K.; Wu, N.H.; Shen, Y.F. Diverse effects of Stat1 on the regulation of hsp90alpha gene under heat shock. *J. Cell BioChem.* **2007**, *102*, 1059–1066. [CrossRef]

192. Cheng, M.B.; Zhang, Y.; Zhong, X.; Sutter, B.; Cao, C.Y.; Chen, X.S.; Cheng, X.K.; Xiao, L.; Shen, Y.F. Stat1 mediates an auto-regulation of hsp90beta gene in heat shock response. *Cell Signal.* **2010**, *22*, 1206–1213. [CrossRef] [PubMed]

193. Pak, S.H.; Joung, Y.H.; Park, J.H.; Lim, E.J.; Darvin, P.; Na, Y.M.; Hong, D.Y.; Lee, B.; Hwang, T.S.; Park, T.; et al. Hypoxia upregulates Hsp90α expression via STAT5b in cancer cells. *Int. J. Oncol.* **2012**, *41*, 161–168. [PubMed]

194. Song, H.; Ethier, S.P.; Dziubinski, M.L.; Lin, J. Stat3 modulates heat shock 27kDa protein expression in breast epithelial cells. *Biochem. Biophys. Res. Commun.* **2004**, *314*, 143–150. [CrossRef] [PubMed]

195. Xu, N.W.; Chen, Y.; Liu, W.; Chen, Y.J.; Fan, Z.M.; Liu, M.; Li, L.J. Inhibition of JAK2/STAT3 Signaling Pathway Suppresses Proliferation of Burkitt's Lymphoma Raji Cells via Cell Cycle Progression, Apoptosis, and Oxidative Stress by Modulating HSP70. *Med. Sci. Monit.* **2018**, *24*, 6255–6263. [CrossRef]

196. Krawczyk, Z.; Gogler-Pigłowska, A.; Sojka, D.R.; Scieglinska, D. The Role of Heat Shock Proteins in Cisplatin Resistance. *Anticancer Agents Med. Chem.* **2018**, *18*, 2093–2109. [CrossRef]

197. Chatterjee, S.; Burns, T.F. Targeting Heat Shock Proteins in Cancer: A Promising Therapeutic Approach. *Int. J. Mol. Sci.* **2017**, *18*, 1978. [CrossRef]

198. Saini, J.; Sharma, P.K. Clinical, Prognostic and Therapeutic Significance of Heat Shock Proteins in Cancer. *Curr. Drug Targets* **2018**, *19*, 1478–1490. [CrossRef]

Review

Targeting STAT3 in Cancer with Nucleotide Therapeutics

Yue-Ting K. Lau, Malini Ramaiyer, Daniel E. Johnson and Jennifer R. Grandis *

Department Otolaryngology—Head and Neck Surgery, University of California at San Francisco, 1450 3rd Street, Room HD268, Box 3111, San Francisco, CA 94143, USA; Kara.Lau@ucsf.edu (Y.-T.K.L.); malinisramaiyer@gmail.com (M.R.); daniel.johnson@ucsf.edu (D.E.J.)
* Correspondence: jennifer.grandis@ucsf.edu

Received: 3 October 2019; Accepted: 24 October 2019; Published: 29 October 2019

Abstract: Signal transducer and activator of transcription 3 (STAT3) plays a critical role in promoting the proliferation and survival of tumor cells. As a ubiquitously-expressed transcription factor, STAT3 has commonly been considered an "undruggable" target for therapy; thus, much research has focused on targeting upstream pathways to reduce the expression or phosphorylation/activation of STAT3 in tumor cells. Recently, however, novel approaches have been developed to directly inhibit STAT3 in human cancers, in the hope of reducing the survival and proliferation of tumor cells. Several of these agents are nucleic acid-based, including the antisense molecule AZD9150, CpG-coupled STAT3 siRNA, G-quartet oligodeoxynucleotides (GQ-ODNs), and STAT3 decoys. While the AZD9150 and CpG-STAT3 siRNA interfere with STAT3 expression, STAT3 decoys and GQ-ODNs target constitutively activated STAT3 and modulate its ability to bind to target genes. Both STAT3 decoy and AZD9150 have advanced to clinical testing in humans. Here we will review the current understanding of the structures, mechanisms, and potential clinical utilities of the nucleic acid-based STAT3 inhibitors.

Keywords: hedging; transaction costs; dynamic programming; risk management; post-decision state variable

---

1. Introduction

Signal transducer and activator of transcription 3 (STAT3) is a transcription factor that is overexpressed and/or hyperactivated in multiple human cancers, where it enhances tumor cell survival and invasion through transcription of anti-apoptotic and pro-proliferative genes [1]. STAT3 has also been shown to directly interact with mitochondrial DNA to contribute to Ras-dependent malignant transformation and cancer progression by amplifying electron transport chain function [2–6]. STAT3 overexpression has been significantly associated with poor overall survival rates in patients with solid tumors [7]. Aberrant phosphorylation of STAT3 on Tyrosine 705 or Serine 727 by upstream kinases results in hyperactivation of the STAT3 protein [8,9]. In addition, genome silencing of phosphatases that play a role in dephosphorylation/inactivation of STAT3, such as those encoded by *PTPR* genes [10], can also result in constitutive activation of STAT3 in cancer [11].

In addition to contributing to the proliferation and survival of tumor cells, STAT3 hyperactivation plays an important role in the resistance of tumors to conventional chemotherapy drugs, as well as molecular targeting agents [12]. Moreover, STAT3 promotes immunosuppression in the tumor microenvironment [13]. STAT3 activation in tumor cells leads to increased production of immunosuppressive cytokines, including interleukin-6 (IL-6), IL-10, vascular endothelial growth factor (VEGF), and transforming growth factor-b1 (TGF-β1) [14–17]. Cytokines and growth factors produced by tumor cells also commonly lead to STAT3 activation in tumor-infiltrating immune cells. The activation of STAT3 in infiltrating immune cells, in general, inhibits anti-tumor immunity. Specifically, STAT3 exhibits cell-autonomous inhibitory activities against cytotoxic T cells (CTLs), natural killer (NK) cells, and dendritic cells (DCs), while increasing the levels of immunosuppressive T regulatory (Treg) and myeloid-derived suppressor cells (MDSCs) [18].

Activation of STAT3 is known to occur via several general pathways (Figure 1). Ligand binding initiates engagement of receptor tyrosine kinases, such as the receptors for epidermal growth factor (EGF) or vascular endothelial growth factor (VEGF), or receptors that lack intrinsic tyrosine kinase activity, such as the IL-6 receptor/gp130 complex, and/or nonreceptor tyrosine kinases such as c-Src [19].

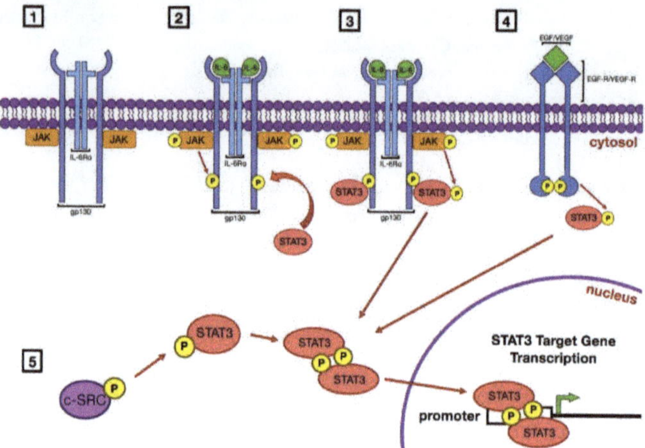

Figure 1. Pathways of signal transducer and activator of transcription 3 (STAT3) activation. Activation of STAT3 occurs via the initial phosphorylation of tyrosine 705 on the STAT3 molecule. This can occur in several ways: (1) Ligand binding activates Janus kinases (JAKs) that are associated with a receptor that lacks intrinsic tyrosine kinase activity, such as the interleukin-6 (IL-6) receptor/gp130 complex; (2) Activated JAKs phosphorylate the cytoplasmic region of the receptor molecule which then serves as a recruitment site for STAT3; (3) STAT3 is then phosphorylated by JAK. (4) Receptor tyrosine kinases (RTK) such as epidermal growth factor receptor (EGFR) or vascular endothelial growth factor receptor (VEGFR) have intrinsic kinase capabilities which directly phosphorylate STAT3 following ligand binding. (5) Receptor independent tyrosine kinases such as c–SRC can phosphorylate JAK without receptor activation. Once Y705 of STAT3 is phosphorylated the SH2 domain of each STAT3 molecule binds the phospho-tyrosine of another, resulting in dimerization of the two proteins. These homodimers are then able to translocate to the nucleus and bind the promoter regions of target genes and induce their transcription. Many of these target genes encode proteins that drive cellular proliferation and survival.

In the case of receptors lacking kinase activity, the binding of ligand leads to activation of receptor-associated Janus kinases (JAK). Activated JAKs phosphorylate the cytoplasmic region of the receptor molecule, which then serves as a docking site for STAT3. Recruitment of STAT3 then results in the direct phosphorylation of Tyrosine 705 by JAK. Phosphorylation of Serine 727 has also been shown to occur secondarily as a mechanism for maximal activation [20]. Receptor tyrosine kinases, including EGFR and VEGFR, harbor intrinsic kinase activity. Following ligand binding, the cytoplasmic regions of receptor tyrosine kinases are subjected to autophosphorylation, and these sites of phosphorylation then serve to recruit STAT3 which is subsequently phosphorylated/activated by the activated receptor.

Phosphorylation of Tyrosine 705 leads to homodimerization of STAT3 proteins, but STAT3 can also participate in alternative heterodimerization with STAT1α [21,22]. Canonical homodimerization occurs via SH2 recognition of the phospho-tyrosine residue in another STAT3 molecule and subsequent binding. STAT3 homodimers translocate to the nucleus where they bind to the promoter regions of STAT3 target genes, inducing the transcription of a broad number of genes whose products drive cellular proliferation and survival. STAT3 has also been demonstrated in an important role as a transcription modulator for mitochondrial respiration and oxidative metabolism [23–25].

1.1. Peptide and Small Molecule Inhibitors of STAT3

The critical role that overexpression and/or hyperactivation of STAT3 plays in the development of multiple cancers has spawned considerable effort to develop inhibitory molecules with potential for clinical application. Early efforts were focused on the development of peptide-based inhibitors. Additional pursuits have led to the identification of several small molecule inhibitors of STAT3. Furthermore, natural derivatives and natural models as lead scaffolds for molecular design have recently gained ground and demonstrated efficacy in STAT3 inhibition. In general, both peptidic inhibitors and small molecule inhibitors of STAT3 suffer from issues of low potency, poor cell penetrance, or undesirable nonspecific activities. However, a few recently discovered small molecule inhibitors are showing considerable promise and have reached the stage of advancing to clinical testing [26]. Hence, before describing nucleic acid-based inhibitors of STAT3 we will briefly review progress made in developing peptide and small molecule inhibitors.

1.2. Peptide Inhibitors

Critical to the function of STAT3 is the recognition of phosphotyrosine residues on activated cell surface receptors (for the purpose of binding to the receptor), or phosphotyrosine 705 on another STAT3 molecule (for the purpose of homodimerization). Recognition of these phosphotyrosine residues occurs via the STAT3 SH2 domain, which recognizes the consensus sequence PY*LKTK (where Y* represents phosphotyrosine). Early studies by Turkson et al. demonstrated that phosphorylated PY*LKTK peptide inhibited the DNA binding activity of STAT3 in nuclear extracts, but had no activity against STAT5 [27]. However, exceptionally high concentrations of the phosphorylated peptide were required to inhibit STAT3 activity in cells. In an effort to generate a more stable version of the PY*LKTK peptide, a peptidomimetic version named ISS-610 was developed and shown to have improved capacity for inhibiting STAT3 DNA binding activity in NIH3T3 cells (IC$_{50}$ = 42 μM) [28]. ISS-610 also demonstrated growth inhibitory activity against several different cancer cell lines characterized by hyperactivation of STAT3. Mandal et al. made additional modifications in an effort to prevent dephophorylation of the peptide, generating the agent PM-73G [29]. This novel drug disrupted STAT3 DNA binding activity in the nanomolar range and also slowed the growth of MDA-MB-468 tumor xenografts.

Despite advances made in generating peptide and peptidomimetic inhibitors of STAT3, further development is hindered by multiple factors. The potencies, stabilities, and cellular uptake of peptides and peptidomimetic compounds will need to be improved and specificities will need to be closely evaluated. Moreover, there is concern that peptides and peptidomimetics may stimulate an immune response, which could limit their effectiveness.

1.3. Small Molecule Inhibitors

Given the obstacles associated with developing clinically relevant peptide inhibitors, greater emphasis has been placed on identifying small molecule inhibitors of STAT3 [30,31]. A number of small-molecule STAT3 inhibitors have been discovered using either experimental screening strategies or virtual screening approaches. Among the best characterized small molecule inhibitors are STATTIC, OPB-31121, OPB-51602, OPB-111077, and C188-9. In addition, an alternative pathway in mitochondrial STAT3 inhibition has been described in the small molecule MDC-1112.

STATTIC was identified in a screen of 17,000 compounds that used a fluorescence polarization assay to detect compounds capable of dissociating phosphopeptide from the SH2 domain of STAT3 [29,32]. STATTIC inhibits STAT3 DNA binding activity and slows the growth of xenograft tumors representing breast cancer and head and neck squamous cell carcinoma [31]. Treatment with STATTIC has been shown to enhance the activities of chemotherapy and radiation against cancer cells in vitro [33].

OPB-31121, OPB-51602, and OPB-111077 are orally bioavailable inhibitors developed by Otsuka Pharmaceutical Company. Molecular and computational modeling studies indicate high affinity binding (Kd = 10 nM) of these compounds to the SH2 domain of STAT3 [34]. Preclinical studies with OPB-31121 have shown that the compound inhibits STAT3 DNA binding and enhances the activities of chemotherapy drugs in leukemia and gastric cancer models [34–36]. OPB-31121 also slows the growth of primary leukemia and SNU484 gastric cancer xenograft tumors in mice [35,36]. A Phase I trial of OPB-31121 revealed considerable dose-limiting toxicities at doses below those needed for STAT3 inhibition. Furthermore, limited anti-tumor activity was observed [37]. Similar to OPB-31121, OPB-51602 demonstrates growth inhibition against xenograft tumors in preclinical models [38]. Phase I testing of OPB-51602 demonstrated reduction of phosphorylated STAT3 in monocytic cells and regression of tumors in two patients with non-small cell lung cancer [39]. One of these patients exhibited complete regression of target lesions and a progression-free survival interval of 6.9 months, while the other responder showed a 41 percent decline in tumor burden. However, treatment with OPB-51602 for multiple cycles resulted in substantial toxicities in multiple patients, including diarrhea, nausea, lactic acidosis, and peripheral neuropathy, leading to necessary discontinuation of treatment [38,39]. Phase I evaluation of OPB-111077 has demonstrated greater tolerability, with only mild or moderate side effects. To date, only modest anti-tumor responses have been observed with the OPB-111077, and further clinical investigation is necessary.

The small molecule C188 was discovered via virtual screening of a large compound library (920,000 compounds) to identify small molecules that may bind to the STAT3 SH2 domain [40]. Evaluation of C188 in cell-free assays demonstrated its ability to abrogate binding of phosphopeptide to the SH2 domain. Chemical optimization of C188 yielded the compound C188-9, also called TTI-101, with the capacity to inhibit phosphopeptide/SH2 interactions in the low nanomolar range [41]. C188-9 is an orally bioavailable STAT3 inhibitor that exhibits in vitro and in vivo activity against preclinical models of non-small cell lung cancer [42], head and neck squamous cell carcinoma [41], liver cancer [43], and breast cancer [44]. Phase I evaluation of C188-9 in patients with solid tumors is currently underway.

MDC-1112, or V-P, is a valporic acid derivative synthesized by Medicon Pharmaceuticals Inc. which blocks the translocation of STAT3 into the mitochondria and promotes the production of reactive oxygen species, inducing cellular apoptosis [45]. In combination with cimetidine, MDC-1112 inhibited the growth of pancreatic cancer xenografts in mice by 60–70% [45] and glioblastoma multiforme xenografts by 78.2% [46].

Napabucasin (2-acetylfuro-1,4-naphthoquinone or BBI-608) is a small molecule in clinical development that has been reported to abrogate STAT3 signaling. A small Phase I trial in combination with paclitaxel for advanced/recurrent gastric cancer reported that the combination was well tolerated [47]. A Phase III trial investigating napabucassin in combination with nab-paclitaxel and gemcitabine for metastatic pancreatic cancer is ongoing [48].

1.4. Future Directions for Small Molecule STAT3 Inhibitors

Although current small molecule inhibitors of STAT3, particularly C188-9, hold potential promise, the success rate for generating clinically viable small molecule inhibitors is low. A number of factors contribute to this low rate of success, including poor pharmacokinetic properties, insufficient potency, and unacceptable levels of nonspecific activity leading to toxic side effects. In addition, acquired resistance to small molecule inhibitors frequently occurs via mutation of the binding site on the target protein. A new approach towards drug development using molecules called proteolysis-targeting chimeras (PROTACs) may help to overcome some of the issues related to nonspecific activity and acquired resistance [49,50]. In the PROTACs approach, a fusion molecule is generated in which a linker sequence is used to covalently connect a small molecule inhibitor to a molecule (eg., thalidomide) that can attract an intracellular E3 ubiquitin ligase. By bringing the E3 ligase in close proximity to the target protein (eg., STAT3), the inhibitor fusion molecule serves to facilitate ubiquitination and subsequent proteasomal degradation of the target protein [49,50]. This process is both rapid and catalytic, meaning the fusion inhibitor molecule is released after destruction of the target protein and can subsequently interact and facilitate destruction of additional target molecules. Because of this catalytic nature, low doses of the inhibitor fusion molecule can be used to achieve efficient removal of the target protein, minimizing the impact of nonspecific activities that would be seen when higher doses are required for target inhibition by a conventional small molecule inhibitor. Moreover, because elimination of the target protein occurs rapidly, the risk of developing acquired resistance via target protein mutation is minimized. The application of PROTACs to inhibition of STAT3 signaling is an exciting opportunity, although, to date, no STAT3 PROTACs inhibitors have been reported.

1.5. Natural Inhibitors

Of the currently available anti-cancer agents used in current practice, it was found that over 40% were based in or derived from natural products [51]. Because of the increasing development of resistance to chemotherapeutic drugs, novel strategies to generate treatments can be sought in bioprospecting and combinatorial biosynthesis from nature-based derivatives. Recent advances in plant-based derivatives that have shown efficacy in STAT3 inhibition include Erasin, bruceanitol, and Curcumin.

Erasin is a chromone-based STAT5 inhibitor that has hydrophobic substituents at the 6-position, resulting in STAT3 specific inhibition. Derived from the natural product classes of flavones and isoflavones, it was determined to decrease STAT3 Y705 phosphorylation and increase apoptosis in MDA-MB-231 breast cancer cells, as well as Erlotinib resistant non-small cell lung cancer (NSCLC) cells [52].

From Brucinea javanica, a chinese plant used to treat cancer, bruceanitol (BOL) was isolated and discovered to have potent anti-leukemic activity. BOL demonstrated the ability to suppress cell growth in a wide range of genetically varied colorectal cancer cell lines, all of which were equally sensitive to the compound [53]. Reductions in phosphorylated STAT3 and downstream target expression of Mcl-1, c-Myc, and Survivin was observed at BOL concentrations of 30 nM. In vivo experiments on mice with CRC xenografts reflected similar results with doses of 2 mg/kg and 4 mg/kg reducing p-STAT3 32% and 80%, and tumor volume 35% and 58% respectively [53].

Diferuloylmethane is a polyphenol derived from the plant Curcuma longa, known as Curcumin, and has been shown to downregulate the expression of downstream STAT3 expression targets such as BCL-2 and Bcl-xL. Curcumin was shown to downregulate the production of Survivin mRNA concordant with the reduction of p-STAT3 in pancreatic cancer cell lines [54].

Collectively these naturally-derived small molecules offer considerable potential in the inhibition of STAT3 hyperactivation in pre-clinical studies. Natural products serve as promising starting points for development of inhibitors, but represent a platform that is currently difficult to exploit because of the lack of knowledge surrounding lead structures for rational design.

1.6. Nucleic Acid-Based Agents to Inhibit Expression of STAT3

1.6.1. AZD9150

Inhibition of STAT3-mediated gene expression can be achieved by directly inhibiting STAT3 expression using antisense oligonucleotides that promote the destruction or inhibit the translation of STAT3 mRNA. These oligonucleotides are short, 12–25 nucleotide strands with sequences such as 5′-GCTCCAGCATCTGCTGCTTC-3′, designed to pair with complementary STAT3 mRNA sequence(s) [55,56]. Binding results in cleavage of the target via RNAse H, alteration of post-transcriptional RNA splicing, or arrest of translation, leading to downmodulation of STAT3 expression [55]. Several generations of antisense oligonucleotides have been developed and chemically modified to optimize stability and allow systemic administration [57].

Early versions of the STAT3 antisense molecule contained 2′-O-methyl or 2′-O-methoxyethyl moieties to prevent free-end degradation and were synthesized with phosphorothioate chemistry to provide further stability [58]. Expression of STAT3 and STAT3 response genes was reduced when prostate (DU145) [59,60], breast (SCK), and melanoma (B16) [61] cell lines were treated with STAT3 antisense oligonucleotides. Furthermore, in vivo experiments showed inhibition of STAT3-mediated tumorigenesis, angiogenesis, and tumor growth in xenograft mouse models of prostate and hepatocellular carcinoma [59,62]. Sensitivity to STAT3 antisense treatment was also demonstrated in androgen-resistant models of prostate cancer and in lung metastases arising from hepatocellular primary tumors [59,62].

The second-generation antisense oligonucleotide AZD9150 (ISIS 481464) is the only STAT3 antisense molecule to enter clinical trials. The ASO structure of AZD9150 was optimized by replacing the 2'-O-methoxyethyl groups with constrained ethyl modifications to increase stability and potency [63]. Preclinical studies with AZD9150 in lymphoma (KARPAS299 and SUP-M2) [63] and neuroblastoma (IMR 32) [64] cell lines with aberrantly activated STAT3 showed a decrease in the expression of STAT3 protein and downstream signaling targets. AZD9150 also demonstrated selective uptake and promoted impaired growth in hematopoietic myelodysplastic and leukemic stem cells [65]. Systemic administration of AZD9150 to immunodeficient mice harboring lymphoma, neuroblastoma, or non-small-cell lung cancer xenografts resulted in decreased STAT3 expression in tumor cells and reduction of tumor initiating potential following serial implantation of the AZD9150-treated tumors [63,64]. While established tumor growth in neuroblastoma xenografts was not inhibited, systemic administration of AZD9150 led to reversal of STAT3-mediated resistance to cisplatin, as indicated by a twofold decrease in the IC_{50} for cisplatin [64].

As a class, antisense oligonucleotides have demonstrated nonspecific immune system activation due to the presence of unmethylated CpG motifs that are recognized as pathogenic [66]. Toxicological effects of AZD9150 have been studied in cynomolgus monkeys and in mice, with key findings being transient prolongation of intrinsic pathway clotting times, elevation of serum transaminase levels, and accumulation in renal and liver tissue [67]. However, these effects occur in doses far greater than those used in clinical trials, with no signs of end organ damage in kidneys or liver until doses of 40 mg/kg or 70 mg/kg, respectively, are reached.

A sufficient therapeutic margin provided the foundation to carry out a phase I clinical trial (NCT01563302) in 30 patients with hematologic and solid malignancies refractory to at least one prior systemic therapy [68]. All responses following administration of AZD9150 at dose levels of 2 mg/kg and 3 mg/kg occurred in diffuse large B cell lymphoma (DLBCL) patients. Two DLBCL patients achieved complete response, two achieved a partial response, and one maintained stable disease for a response rate of 13% [69]. There was no significant difference in progression-free and overall survival between the dose levels. Notable adverse events included transaminitis, fatigue, thrombocytopenia, nausea, and anemia with thrombocytopenia being more closely related to high-grade events. Based on the observed toxicities, the recommended phase 2 trial dose was 3 mg/kg. Peripheral blood analyses in this trial suggested an impact of AZD9150 in selected immune cell populations [69].

The heterogeneity in the response of lymphoma patients, five of whom had DLBCL, to AZD9150 highlights the importance of exploring what factors predict the efficacy of treatment. STAT3 inhibition may increase the immunogenicity of malignant cells through cell-autonomous processes and prevent tumor microenvironment immunosuppression [70]. Immunogenic biomarkers could thus offer clues to treatment progress and success. This could also lay the foundation for examining how combination with immunotherapeutic agents may further increase therapeutic benefits.

1.6.2. CpG-coupled STAT3 siRNA

While STAT3 is known to be overexpressed in tumor cells, it is also dysregulated in tumor-associated myeloid cells, such as dendritic cells and macrophages [71]. This results in a loss of MHCII expression and the accumulation of inactive antigen presenting cells. In effect, there is reduced Th-1-mediated CD8+ cytotoxicity towards tumor cells [13], and an increased presence of myeloid-derived suppressor cells (MDSCs) and regulatory T-cells (Tregs) generating an immunosuppressive tumor microenvironment [72].

Selective targeting of STAT3 in tumor-associated myeloid-derived cells is possible with siRNA conjugation to a CpG TLR-9 ligand [73]. TLR9 is a cell-surface transmembrane receptor that is upregulated under conditions of cellular or environmental stress and is known to be expressed on myeloid-derived cells in the tumor microenvironment [74]. It is also upregulated in acute myeloid leukemia, multiple myeloma, and B-cell lymphoma, and activation has been shown to increase antigenicity of primary malignant B-cells and induce apoptosis [75]. When activated by CpG binding, TLR9 facilitates immunostimulatory signaling and the release of proinflammatory cytokines as well as the presentation of tumor-specific antigens [74]. However, STAT3 signaling abrogates CpG-activated immunostimulation.

Therefore, to generate a sufficient immune response, it is necessary to both stimulate TLR9 and deactivate STAT3 [76,77]. RNA interference or RNAi, is another mechanism by which STAT3 mRNA can be degraded, thereby silencing its expression. A segment of double-stranded RNA targeted to the STAT3 sequence is introduced to cells where it is metabolized to 20-21 base-pair fragments and incorporated into an RNA-induced silencing complex, RISC. The two strands of RNA are unwound, and this allows the antisense strand to bind STAT3 mRNA. RISC endonuclease activity can then cleave the target [78]. CpG-conjugated STAT3 siRNA is internalized into cells by endocytosis, where TLR9 binding facilitates the release of Dicer-uncoupled siRNA from the early endosome before acidification [79].

In vitro studies of CpG-conjugated STAT3 siRNA showed that more than 80% of mouse dendritic cells, macrophages, and B-cells were positive for uptake without transfection reagent within 60 minutes [73]. STAT3 gene silencing was achieved with maximal reduction in expression at high siRNA concentrations of 1 μM [73]. In mice bearing B16 melanoma, C4 melanoma, or CT26 colon xenograft tumors, peritumoral injection with CpG-Stat3 siRNA led to tumor regression and systemic administration reduced the number of B16 lung cancer metastases [73]. This was associated with enhanced CD8+ T-Cell recruitment and cytolytic activity in association with increased immunostimulatory cytokine and chemokine production [73,80].

Studies with murine models of AML have shown that intravenous delivery of CpG-STAT3 siRNA results in a 70-80%, CD8+ T-cell-mediated, reduction of leukemic cells in bone marrow, spleen, lymph nodes, and peripheral blood while simultaneously reducing Treg levels [81]. Combination intratumoral administration of CpG-STAT3 siRNA with radiotherapy has demonstrated complete rejection of A20 lymphoma tumors in mice and generates long-term protective immunity against the primary tumor [75]. Similarly, CpG-STAT3 siRNA inhibits tumor growth of androgen-independent prostate cancer, with a concomitant reduction of immunosuppressive MDSC levels in peritumoral lymph nodes [82].

1.7. Nucleic Acid-Based Agents that Act as Competitive Inhibitors of STAT3

1.7.1. G-Quartet Oligodeoxynucleotides (GQ-ODNs)

G-quartet oligodeoxynucleotides are macrocycles composed of four guanosine bases that, upon hydrogen-bonding, form a polyguanylate, tetrad-helical structure in the presence of a monovalent cation, usually potassium (Figure 2) [83,84]. In vivo, G-quartets are found in telomeric regions of chromosomes and in the transcriptional regulatory regions of some oncogenes [85]. Their therapeutic potential has been demonstrated as direct competitive inhibitors of HIV-1 integrase, blocking HIV-1 DNA integration into the host genome [86–88].

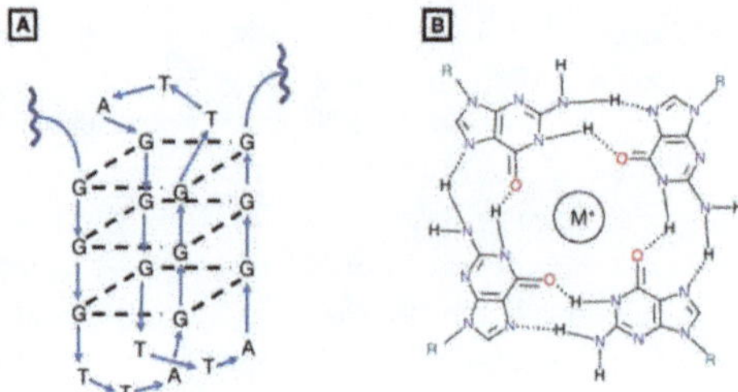

Figure 2. Structure of G-quartet inhibitor. (**A**) The G-quartet oligodeoxynucleotide is comprised of four guanosine macrocycles stacked on top of one another. To date, they have been used as competitive inhibitors of HIV-1 integrase but have demonstrated potential STAT3 dimerization inhibitors. With i.p. and i.v. injection in vivo mouse xenografts, G-quartets have been shown to reduce tumor growth in breast, prostate, and non-small cell lung cancers [85–87]. (**B**) An overhead view of a G-macrocycle demonstrates how hydrogen bonding generates a tetrad-helical structure with a monovalent cation at its core.

GQ-ODNs have been proposed as a class of unique, anti-cancer STAT3 inhibitors which directly destabilize the homodimerization of STAT3, and thus interfere with its DNA binding activity [89,90]. Computational analyses revealed a GQ-ODN interacts with residues Q643, Q644, N646, and N647 of the SH2 domain [91]. The folds that comprise the G-quartet intramolecular structure significantly occlude single-stranded endonuclease access to its phosphodiester linkages, resulting in an inability to cleave the oligonucleotide and in a long serum half-life [92]. However, because of their large size and charge, G-quartets cannot penetrate cell membranes and must be delivered via a polyethyleneimine (PEI) complex [93], or another delivery vehicle.

Preliminary in vitro assays demonstrated inhibition of IL-6-induced STAT3 activation in Hep2G cells incubated with the PEI-GQ-ODN T40214 (90% inhibition at 50 ng/mL). Preincubation at the same concentration inhibited the expression of anti-apoptotic mediators, with complete blockage of Bcl-XL mRNA upregulation and 50% blockage of Mcl-1 mRNA upregulation [89].

Treatment with GQ-ODN has demonstrated inhibition of STAT3 and tumor growth in in vivo models of head and neck squamous cell carcinoma (HNSCC) [94], breast cancer, prostate cancer [93], and non-small cell lung cancer (NSCLC) [95]. Inhibition of tumor growth was accompanied by a reduction in anti-apoptotic Bcl-2, Bcl-XL, Mcl-1, survivin, and VEGF in tumor tissue, and decreased expression of proliferation mediators cyclin D and c-Myc.

GQ-ODN selectivity for STAT3 is a potential consideration, given that a negative regulator of cell growth, STAT1 shares 50% sequence similarity with STAT3. Computational docking models indicate a two- to four-fold greater IC_{50} for STAT1 over STAT3 [89], and 3D analysis showed different residues involved in STAT1 and STAT3 surface interaction [96]. However, optimization of specificity for the therapeutic target is necessary for further development of clinical potential.

1.7.2. STAT3 Decoys

Once activated, STAT3 acts as a transcription factor, binding to a response element in the promoter regions of target genes to induce gene expression. Early investigation determined that STAT3 bound to a 15-base pair (bp) response element termed human serum-inducible element (hSIE) in the promoter region of the c-*fos* gene. We derived the decoy by systematically shortening the double stranded oligonucleotide to determine the smallest formulation that retained binding activity to STAT3 on gel shift assays. Figure 3A depicts the sequence of the SIE from the murine, feline, canine, and human c-*fos* genes, demonstrating nearly perfect conservation across these species. Subsequent studies identified optimal binding of STAT3 to a variant sequence of hSIE, specifically the sequence 5′-CATTTCCCGTAAATC-3′ (Figure 3A; [97]).

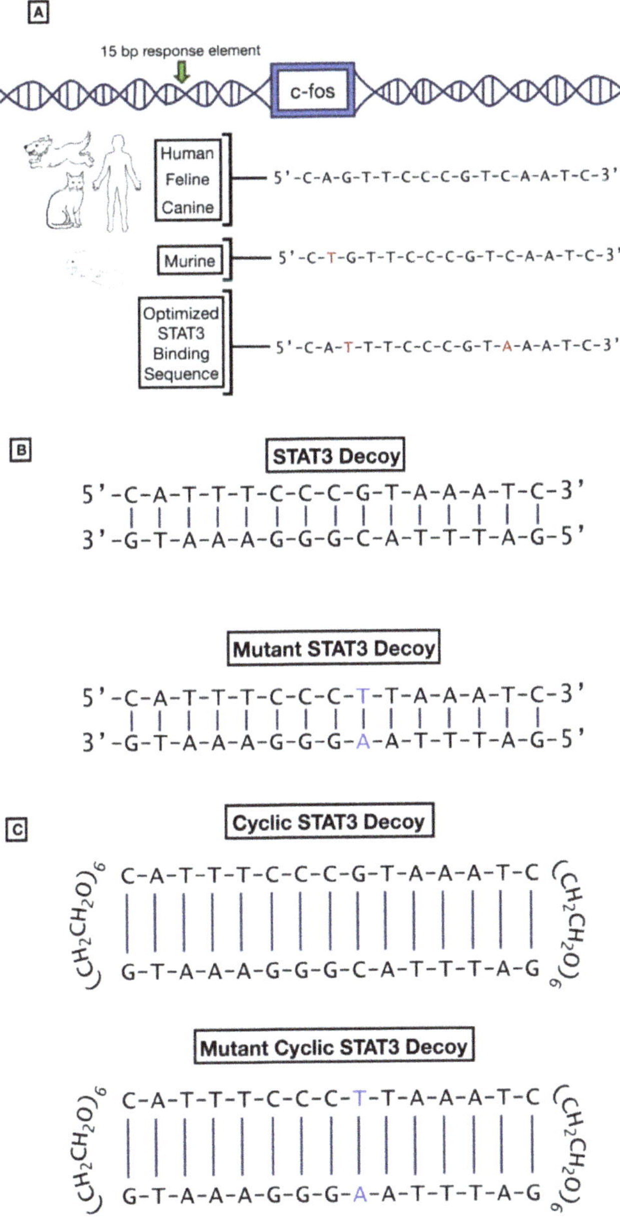

Figure 3. STAT3 response elements and STAT3 decoys. (**A**) The human serum-inducible element (hSIE) sequence is located upstream in the promoter region of the human *c-fos* gene and is almost perfectly conserved across human, feline, canine, and murine species. Sequence differences are shown in red. Below these is an optimized STAT3 binding sequence, with sequence modifications, shown in red, at positions 3 and 11 [52]. This allows the decoy to act as a direct competitive inhibitor of the DNA binding domain in the STAT3 molecule and thus inhibit expression of downstream anti-apoptotic and pro-proliferative signals. (**B**) Structure of the first-generation linear STAT3 decoy and mutant STAT3 decoy which is administered intratumorally in in vivo models. The location of the mutation in the mutant decoy is shown in blue. (**C**) Structure of the cyclic STAT3 decoy and mutant cyclic STAT3 decoy, with hexaethylene glycol linkages which is administered intravenously in in vivo models. The location of the mutation in the mutant decoy is shown in blue.

We synthesized the double-stranded STAT3 binding sequence (Figure 3B), the STAT3 decoy, and determined its effects when added to cells in culture [98]. As a negative control for binding specificity, a point mutant version of the decoy (termed mutant STAT3 decoy, or MT STAT3 decoy) was also synthesized and evaluated. In cell-free assays the STAT3 decoy, but not the MT STAT3 decoy, was confirmed to competitively inhibit binding of STAT3 protein to a radiolabeled decoy sequence. Fluorescently-tagged versions of both the STAT3 decoy and the MT STAT3 decoy were readily incorporated into the cytosol and nucleus of HNSCC cells and normal oral keratinocytes (NOKs) within 6 hours after treatment. Remarkably, the STAT3 decoy potently inhibited the growth of STAT3-dependent cancer cell lines, but not NOKs. The control molecule, MT STAT3, was largely ineffective against either the cancer cell lines or the NOKs.

The inhibition of cell growth resulting from treatment with STAT3 decoy was associated with induction of apoptosis and downregulation of the STAT3 target gene encoding Bcl-XL. Further investigation [99] revealed that the STAT3 decoy also inhibits the STAT1 protein, which is known to form heterodimers with STAT3. This finding was initially troubling, as STAT1 is known to have tumor suppressor activity. However, the expression or activation of STAT1 did not alter the apoptosis-inducing activity of the STAT3 decoy, indicating that the therapeutic activity of the decoy is independent of STAT1 signaling. Subsequent studies have determined that the STAT3 decoy can inhibit the growth of a broad variety of cancer cell lines, including cells representing melanoma and cancers of the bladder, brain, breast, colon, liver, lung, and ovary [98,100–103]. In addition, intratumoral injection of STAT3 decoy has been shown to inhibit the in vivo growth of breast, glioma, head and neck, lung, and ovarian xenograft tumors [98,100–103]. The anti-tumor effects of STAT3 decoy are associated with reduced tumor cell proliferation, induction of apoptosis, and reduced expression of STAT3 target genes, including the genes encoding Bcl-XL and cyclin D1 [101,104]. Based on these promising preclinical results, we conducted a Phase 0 clinical trial to assess the pharmacodynamic impact of the STAT3 decoy [105]. Intratumoral delivery of the decoy was found to downmodulate expression of Bcl-XL and cyclin D1 in the tumors of patients undergoing surgical resection of their head and neck cancer.

Fusion of the STAT3 decoy with CpG has enabled targeting of the decoy to TLR9-expressing leukemia and lymphoma cells [106,107]. Treatment of preclinical mouse models harboring leukemia or lymphoma with the CpG-STAT3 decoy resulted in potent in vivo growth inhibition of the malignant cells. Hence, the CpG-STAT3 decoy fusion molecule may provide an effective means for treating TLR9-expressing hematologic malignancies.

A major limitation of the first generation STAT3 decoy was the necessity for intratumoral injection. In preclinical studies, systemic delivery of STAT3 decoy failed to inhibit xenograft tumor growth [105], presumably due to rapid degradation of the molecule by nucleases in the blood. In an effort to produce a more stable molecule, we [105] generated a cyclic version of the decoy, using hexaethylene glycol linkages to cyclize the free ends (Figure 3C). The cyclic STAT3 decoy exhibited markedly enhanced thermal stability and a longer half-life in human serum (8 hours; [105,108]). Intravenous delivery of the cyclic STAT3 decoy has been shown to potently inhibit the growth of both non-small cell lung cancer and head and neck cancer xenograft tumors [105,109,110]. The cyclic STAT3 decoy was well tolerated in wild-type mice, with no apparent toxicity, even when delivered at a dose 20-fold higher than the maximal effective dose for growth inhibition of HNSCC xenograft tumors [110].

The therapeutic value of treatment with the STAT3 decoy, as with other STAT3 inhibitors, may be best realized in combination with other anticancer agents. In this regard, STAT3 decoy treatment has been shown to enhance the sensitivity of head and neck xenograft tumors to cisplatin [101,104]. Similarly, the STAT3 decoy heightens the response of head and neck cancer cells to bortezomib [111], ovarian cancer cells to paclitaxel [112], and leukemia cells to Adriamycin [113]. FDA IND-directed pharmacologic and toxicity studies are planned to enable a Phase I trial of the cyclic STAT3 decoy in cancer patients.

2. Conclusion

Hyperactivation of STAT3 in malignant cells and tumor-associated immune cells promotes cancer survival and growth through expression of anti-apoptotic and proliferative genes, as well as cytokines that generate an immunosuppressive tumor microenvironment. Understanding the role of STAT3 in cancer has led to promising strides in the development of specific nucleotide inhibitors (Figure 4). These anti-STAT3 agents can silence STAT3 expression or directly inhibit STAT3 DNA-binding ability and have shown promising in vivo results beyond proof-of-principle studies. However, optimization of specificity, potency, stability, and delivery of these nucleotide therapeutics will be important for enhancing their therapeutic benefits in the clinic.

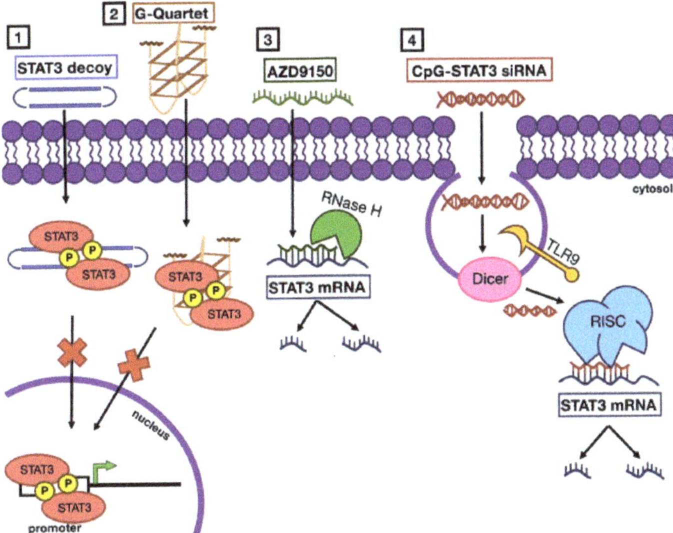

Figure 4. Mechanisms of STAT3 inhibition with nucleic acid-based agents. STAT3 inhibition with nucleic acid-based agents occurs via two main mechanisms: (1 and 2) STAT3 decoys or G-quartet deoxyoligonucleotides prevent STAT3 binding to promoter regions of target genes, and (3 and 4) antisense or siRNAs promote degradation of STAT3 mRNA.

3. Discussion

Currently, there are four classes of drugs that directly inhibit STAT3: SH2 domain inhibitors; DNA-binding domain inhibitors (eg., STAT3 decoys); N-terminal domain inhibitors; and STAT3 antisense and siRNA [114]. As a whole, translation to the clinical setting has been challenged by intracellular drug delivery, selectivity for the target, and minimization of systemic toxicity. More clinical success has been achieved with inhibition of upstream Janus kinases (JAKs) using Tofacitinib, developed for the treatment of rheumatoid arthritis [115], and Ruxolitinib, developed for the treatment of myelofibrosis [116]. Still, small molecule drugs often exhibit off-target effects and a lack of targeted potency; thus, the limitations of monotherapy should be considered.

The future of STAT3 inhibition is likely to be in combination therapy with currently existing or newly developed drugs. Blocking STAT3 could result in synergistic anti-tumor effects in combination with inhibition of EGFR or other tumorigenic dysregulated transcription factors [117,118]. STAT3 inhibition resulting in resensitization to immunotherapies could pave the way for more efficacious responses to checkpoint inhibitors or other immunotherapies [119–123]. Looking to the future, while the role of STAT3 has been established in cancer biology, identifying tumor biomarkers that can indicate patient sensitivity to STAT3 inhibition and expansion of target inhibition to other aberrant transcription factors will be important. Research efforts should continue to search for novel applications of STAT3 inhibition and continue to pursue clinical validation of efficacy.

Author Contributions: D.E.J. and J.R.G. are co-inventors of cyclic STAT3 decoy and have financial interests in STAT3 Therapeutics. STAT3 Therapeutics holds an interest in cyclic STAT3 decoy.

Funding: This work was supported by National Institutes of Health grants R35 CA231998 and P50 CA097190 (J.R.G.) and R01 DE024728 (D.E.J.).

Conflicts of Interest: The remaining authors declare no conflicts.

References

1. Huynh, J.; Chand, A.; Gough, D.; Ernst, M. Therapeutically Exploiting STAT3 Activity in Cancer—Using Tissue Repair as a Road Map. *Nat. Rev. Cancer* **2019**, *19*, 82–96. [CrossRef]

2. Gough, D.J.; Corlett, A.; Schlessinger, K.; Wegrzyn, J.; Larner, A.C.; Levy, D.E. Mitochondrial STAT3 Supports Ras-Dependent Oncogenic Transformation. *Science* **2009**, *324*, 1713–1716. [CrossRef]

3. Gough, D.J.; Koetz, L.; Levy, D.E. The MEK-ERK Pathway is Necessary for Serine Phosphorylation of Mitochondrial STAT3 and Ras-Mediated Transformation. *PLoS ONE* **2013**, *8*, e83395. [CrossRef]

4. Du, W.; Hong, J.; Wang, Y.-C.; Zhang, Y.-J.; Wang, P.; Su, W.-Y.; Lin, Y.-W.; Lu, R.; Zou, W.-P.; Xiong, H.; et al. Inhibition of JAK2/STAT3 Signalling Induces Colorectal Cancer Cell Apoptosis via Mitochondrial Pathway. *J. Cell. Mol. Med.* **2012**, *16*, 1878–1888. [CrossRef]

5. Yu, H.; Lee, H.; Herrmann, A.; Buettner, R.; Jove, R. Revisiting STAT3 Signalling in Cancer: New and Unexpected Biological Functions. *Nat. Rev. Cancer* **2014**, *14*, 736–746. [CrossRef]

6. Yang, R.; Rincon, M. Mitochondrial Stat3, the Need for Design Thinking. *Int. J. Biol. Sci.* **2016**, *12*, 532–544. [CrossRef]

7. Wu, P.; Wu, D.; Zhao, L.; Huang, L.; Shen, G.; Huang, J.; Chai, Y. Prognostic role of STAT3 in Solid Tumors: A Systematic Review and Meta-Analysis. *Oncotarget* **2016**, *7*, 19863–19883. [CrossRef]

8. Sakaguchi, M.; Oka, M.; Iwasaki, T.; Fukami, Y.; Nishigori, C. Role and Regulation of STAT3 Phosphorylation at Ser727 in Melanocytes and Melanoma Cells. *J. Investig. Dermatol.* **2012**, *132*, 1877–1885. [CrossRef]

9. Schuringa, J.-J.; Wierenga, A.T.J.; Kruijer, W.; Vellenga, E. Constitutive Stat3, Tyr705, and Ser727 Phosphorylation in Acute Myeloid Leukemia Cells Caused by the Autocrine Secretion of Interleukin-6. *Blood* **2000**, *95*, 3765–3770. [CrossRef]

10. Lui, V.W.Y.; Peyser, N.D.; Ng, P.K.-S.; Hritz, J.; Zeng, Y.; Lu, Y.; Li, H.; Wang, L.; Gilbert, B.R.; General, I.J.; et al. Frequent Mutation of Receptor Protein Tyrosine Phosphatases Provides a Mechanism for STAT3 Hyperactivation in Head and Neck Cancer. *Proc. Natl. Acad. Sci. USA* **2014**, *111*, 1114–1119. [CrossRef]

11. Geiger, J.L.; Grandis, J.R.; Bauman, J.E. The STAT3 Pathway as a Therapeutic Target in Head and Neck Cancer: Barriers and Innovations. *Oral Oncol.* **2016**, *56*, 84–92. [CrossRef]

12. Spitzner, M.; Ebner, R.; Wolff, H.A.; Michael Ghadimi, B.; Wienands, J.; Grade, M. STAT3: A Novel Molecular Mediator of Resistance to Chemoradiotherapy. *Cancers* **2014**, *6*, 1986–2011. [CrossRef]

13. Lee, H.; Pal, S.K.; Reckamp, K.; Figlin, R.A.; Yu, H. STAT3: A Target to Enhance Antitumor Immune Response. *Curr. Top. Microbiol. Immunol.* **2011**, *344*, 41–59.

14. Chen, M.-F.; Chen, P.-T.; Lu, M.S.; Lin, P.Y.; Chen, W.-C.; Lee, K.-D. IL-6 Expression Predicts Treatment Response and Outcome in Squamous Cell Carcinoma of the Esophagus. *Mol. Cancer* **2013**, *12*, 26. [CrossRef]

15. Wu, C.-T.; Chen, M.-F.; Chen, W.-C.; Hsieh, C.-C. The Role of IL-6 in the Radiation Response of Prostate Cancer. *Radiat. Oncol.* **2013**, *8*, 159. [CrossRef]

16. Cho, B.C.; Kim, S.M.; Solca, F.; Kim, J.-H. Abstract 1886: Activation of IL-6R/JAK1/STAT3 Signaling Induces de Novo Resistance to Irreversible EGFR Inhibitors in Non-Small Cell Lung Cancer with T790M Resistance Mutation. *Exp. Mol. Ther.* **2012**, *11*, 2254–2264.

17. Yu, H.; Kortylewski, M.; Pardoll, D. Crosstalk between Cancer and Immune Cells: Role of STAT3 in the Tumour Microenvironment. *Nat. Rev. Immunol.* **2007**, *7*, 41–51. [CrossRef]

18. Ferguson, S.D.; Srinivasan, V.M.; Heimberger, A.B. The Role of STAT3 in Tumor-Mediated Immune Suppression. *J. Neurooncol.* **2015**, *123*, 385–394. [CrossRef]

19. Rébé, C.; Végran, F.; Berger, H.; Ghiringhelli, F. STAT3 Activation: A Key Factor in Tumor Immunoescape. *JAKSTAT* **2013**, *2*, e23010. [CrossRef]

20. Wen, Z.; Zhong, Z.; Darnell, J.E., Jr. Maximal Activation of Transcription by Stat1 and Stat3 Requires both Tyrosine and Serine Phosphorylation. *Cell* **1995**, *82*, 241–250. [CrossRef]

21. Zhong, Z.; Wen, Z.; Darnell, J.E., Jr. Stat3: A STAT Family Member Activated by Tyrosine Phosphorylation in Response to Epidermal Growth Factor and Interleukin-6. *Science* **1994**, *264*, 95–98. [CrossRef]

22. Delgoffe, G.M.; Vignali, D.A.A. STAT Heterodimers in Immunity: A Mixed Message or a Unique Signal? *JAKSTAT* **2013**, *2*, e23060. [CrossRef]

23. Wegrzyn, J.; Potla, R.-J.; Chwae, Y.; Sepuri, N.B.V.; Zhang, Q.; Koeck, T.; Derecka, M.; Szczepanek, K.; Szelag, M.; Gornicka, A.; et al. Function of Mitochondrial Stat3 in Cellular Respiration. *Science* **2009**, *323*, 793–797. [CrossRef]

24. Carbognin, E.; Betto, R.M.; Soriano, M.E.; Smith, A.G.; Martello, G. Stat3 Promotes Mitochondrial Transcription and Oxidative Respiration during Maintenance and Induction of Naive Pluripotency. *EMBO J.* **2016**, *35*, 618–634. [CrossRef]

25. Liu, F.; Jia, L.; Farren, T.; Gribben, J.; Agrawal, S. 2.21 Autocrine Interleukin-6 Production Correlated with Survival of Chronic Lymphocytic Leukaemia Cells. *Clin. Lymphoma Myeloma Leuk.* **2011**, *11*, S172. [CrossRef]

26. Wake, M.S.; Watson, C.J. STAT3 the oncogene - still eluding therapy? *FEBS J.* **2015**, *282*, 2600–2611. [CrossRef]

27. Turkson, J.; Ryan, D.; Kim, J.S.; Zhang, Y.; Chen, Z.; Haura, E.; Laudano, A.; Sebti, S.; Hamilton, A.D.; Jove, R. Phosphotyrosyl Peptides Block Stat3-mediated DNA Binding Activity, Gene Regulation, and Cell Transformation. *J. Biol. Chem.* **2001**, *276*, 45443–45455. [CrossRef]

28. Turkson, J.; Kim, J.S.; Zhang, S.; Yuan, J.; Huang, M.; Glenn, M.; Haura, E.; Sebti, S.; Hamilton, A.D.; Jove, R. Novel Peptidomimetic Inhibitors of Signal Transducer and Activator of Transcription 3 Dimerization and Biological Activity. *Mol. Cancer Ther.* **2004**, *3*, 261–269.

29. Mandal, P.K.; Gao, F.; Lu, Z.; Ren, Z.; Ramesh, R.; Birtwistle, J.S.; Kaluarachchi, K.K.; Chen, X.; Bast, R.C., Jr.; Liao, W.S.; et al. Potent and Selective Phosphopeptide Mimetic Prodrugs Targeted to the Src Homology 2 (SH2) Domain of Signal Transducer and Activator of Transcription 3. *J. Med. Chem.* **2011**, *54*, 3549–3563. [CrossRef]

30. Schust, J.; Berg, T. A High-Throughput Fluorescence Polarization Assay for Signal Transducer and Activator of Transcription 3. *Anal. Biochem.* **2004**, *330*, 114–118. [CrossRef]

31. Schust, J.; Sperl, B.; Hollis, A.; Mayer, T.U.; Berg, T. Stattic: A Small-Molecule Inhibitor of STAT3 Activation and Dimerization. *Chem. Biol.* **2006**, *13*, 1235–1242. [CrossRef]

32. Auzenne, E.J.; Klostergaard, J.; Mandal, P.K.; Liao, W.S.; Lu, Z.; Gao, F.; Bast, R.C., Jr.; Robertson, F.M.; McMurray, J.S. A Phosphopeptide Mimetic Prodrug Targeting the SH2 Domain of Stat3 Inhibits Tumor Growth and Angiogenesis. *J. Exp. Ther. Oncol.* **2012**, *10*, 155–162.

33. Pan, Y.; Zhou, F.; Zhang, R.; Claret, F.X. Stat3 inhibitor Stattic Exhibits Potent Antitumor Activity and Induces Chemo- and Radio-Sensitivity in Nasopharyngeal Carcinoma. *PLoS ONE* **2013**, *8*, e54565. [CrossRef]

34. Brambilla, L.; Genini, D.; Laurini, E.; Merulla, J.; Perez, L.; Fermeglia, M.; Carbone, G.M.; Pricl, S.; Catapano, C.V. Hitting the right spot: Mechanism of action of OPB-31121, a novel and potent inhibitor of the Signal Transducer and Activator of Transcription 3 (STAT3). *Mol. Oncol.* **2015**, *9*, 1194–1206. [CrossRef]

35. Hayakawa, F.; Sugimoto, K.; Harada, Y.; Hashimoto, N.; Ohi, N.; Kurahashi, S.; Naoe, T. A novel STAT inhibitor, OPB-31121, has a Significant Antitumor Effect on Leukemia with STAT-Addictive Oncokinases. *Blood Cancer J.* **2013**, *3*, e166. [CrossRef]

36. Kim, M.-J.; Nam, H.-J.; Kim, H.-P.; Han, S.-W.; Im, S.-A.; Kim, T.-Y.; Oh, D.-Y.; Bang, Y.-J. OPB-31121, a Novel Small Molecular Inhibitor, Disrupts the JAK2/STAT3 Pathway and Exhibits an Antitumor Activity in Gastric Cancer Cells. *Cancer Lett.* **2013**, *335*, 145–152. [CrossRef]

37. Bendell, J.C.; Hong, D.S.; Burris, H.A., 3rd; Naing, A.; Jones, S.F.; Falchook, G.; Bricmont, P.; Elekes, A.; Rock, E.P.; Kurzrock, R. Phase 1, Open-Label, Dose-Escalation, and Pharmacokinetic Study of STAT3 Inhibitor OPB-31121 in Subjects with Advanced Solid Tumors. *Cancer Chemother. Pharmacol.* **2014**, *74*, 125–130. [CrossRef]

38. Ogura, M.; Uchida, T.; Terui, Y.; Hayakawa, F.; Kobayashi, Y.; Taniwaki, M.; Takamatsu, Y.; Naoe, T.; Tobinai, K.; Munakata, W.; et al. Phase I Study of OPB-51602, an Oral Inhibitor of Signal Transducer and Activator of Transcription 3, in Patients with Relapsed/Refractory Hematological Malignancies. *Cancer Sci.* **2015**, *106*, 896–901. [CrossRef]

39. Wong, A.L.; Soo, R.A.; Tan, D.S.; Lee, S.C.; Lim, J.S.; Marban, P.C.; Kong, L.R.; Lee, Y.J.; Wang, L.Z.; Thuya, W.L.; et al. Phase I and Biomarker Study of OPB-51602, a Novel Signal Transducer and Activator

of Transcription (STAT) 3 Inhibitor, in Patients with Refractory Solid Malignancies. *Ann. Oncol.* **2015**, *26*, 998–1005. [CrossRef]

40. Xu, X.; Kasembeli, M.M.; Jiang, X.; Tweardy, B.J.; Tweardy, D.J. Chemical Probes that Competitively and Selectively Inhibit Stat3 Activation. *PLoS ONE* **2009**, *4*, e4783. [CrossRef]

41. Bharadwaj, U.; Eckols, T.K.; Xu, X.; Kasembeli, M.M.; Chen, Y.; Adachi, M.; Song, Y.; Mo, Q.; Lai, S.Y.; Tweardy, D.J. Small-Molecule Inhibition of STAT3 in Radioresistant Head and Neck Squamous Cell Carcinoma. *Oncotarget* **2016**, *7*, 26307–26330. [CrossRef]

42. Lewis, K.M.; Bharadwaj, U.; Eckols, T.K.; Kolosov, M.; Kasembeli, M.M.; Fridley, C.; Siller, R.; Tweardy, D.J. Small-Molecule Targeting of Signal Transducer and Activator of Transcription (STAT) 3 to Treat Non-Small Cell Lung Cancer. *Lung Cancer* **2015**, *90*, 182–190. [CrossRef]

43. Jung, K.H.; Yoo, W.; Stevenson, H.L.; Deshpande, D.; Shen, H.; Gagea, M.; Yoo, S.-Y.; Wang, J.; Kris Eckols, T.; Bharadwaj, U.; et al. Multifunctional Effects of a Small-Molecule STAT3 Inhibitor on NASH and Hepatocellular Carcinoma in Mice. *Clin. Cancer Res.* **2017**, *23*, 5537–5546. [CrossRef]

44. Kettner, N.M.; Vijayaraghavan, S.; Durak, M.G.; Bui, T.; Kohansal, M.; Ha, M.J.; Liu, B.; Rao, X.; Wang, J.; Yi, M.; et al. Combined Inhibition of STAT3 and DNA Repair in Palbociclib-Resistant ER-Positive Breast Cancer. *Clin. Cancer Res.* **2019**, *25*, 3996–4013. [CrossRef]

45. Mackenzie, G.G.; Huang, L.; Alston, N.; Ouyang, N.; Vrankova, K.; Mattheolabakis, G.; Constantinides, P.P.; Rigas, B. Targeting Mitochondrial STAT3 with the Novel Phospho-Valproic Acid (MDC-1112) Inhibits Pancreatic Cancer Growth in Mice. *PLoS ONE* **2013**, *8*, e61532. [CrossRef]

46. Luo, D.; Fraga-Lauhirat, M.; Millings, J.; Ho, C.; Villarreal, E.M.; Fletchinger, T.C.; Bonfiglio, J.V.; Mata, L.; Nemesure, M.D.; Bartels, L.E.; et al. Phospho-Valproic Acid (MDC-1112) Suppresses Glioblastoma Growth in Preclinical Models Through the Inhibition of STAT3 Phosphorylation. *Carcinogenesis* **2019**. [CrossRef]

47. Shitara, K.; Yodo, Y.; Iino, S. A Phase I Study of Napabucasin Plus Paclitaxel for Japanese Patients with Advanced/Recurrent Gastric Cancer. *In Vivo* **2019**, *33*, 933–937. [CrossRef]

48. Sonbol, M.B.; Ahn, D.H.; Goldstein, D.; Okusaka, T.; Tabernero, J.; Macarulla, T.; Reni, M.; Li, C.-P.; O'Neil, B.; Van Cutsem, E.; et al. CanStem111P trial: A Phase III Study of Napabucasin Plus Nab-Paclitaxel with Gemcitabine. *Future Oncol.* **2019**, *15*, 1295–1302. [CrossRef]

49. Pettersson, M.; Crews, C.M. PROteolysis TArgeting Chimeras (PROTACs)—Past, Present and Future. *Drug Discov. Today Technol.* **2019**, *31*, 15–27. [CrossRef]

50. Paiva, S.-L.; Crews, C.M. Targeted Protein Degradation: Elements of PROTAC Design. *Curr. Opin. Chem. Biol.* **2019**, *50*, 111–119. [CrossRef]

51. Demain, A.L.; Vaishnav, P. Natural Products for Cancer Chemotherapy. *Microb. Biotechnol.* **2011**, *4*, 687–699. [CrossRef]

52. Lis, C.; Rubner, S.; Roatsch, M.; Berg, A.; Gilcrest, T.; Fu, D.; Nguyen, E.; Schmidt, A.-M.; Krautscheid, H.; Meiler, J.; et al. Development of Erasin: A Chromone-Based STAT3 Inhibitor which Induces Apoptosis in Erlotinib-Resistant Lung Cancer Cells. *Sci. Rep.* **2017**, *7*, 17390. [CrossRef]

53. Wei, N.; Li, J.; Fang, C.; Chang, J.; Xirou, V.; Syrigos, N.K.; Marks, B.J.; Chu, E.; Schmitz, J.C. Targeting Colon Cancer with the Novel STAT3 Inhibitor Bruceantinol. *Oncogene* **2019**, *38*, 1676–1687. [CrossRef]

54. Glienke, W.; Maute, L.; Wicht, J.; Bergmann, L. Curcumin Inhibits Constitutive STAT3 Phosphorylation in Human Pancreatic Cancer Cell Lines and Downregulation of Survivin/BIRC5 Gene Expression. *Cancer Investig.* **2010**, *28*, 166–171. [CrossRef]

55. Dean, N.M.; Bennett, C.F. Antisense Oligonucleotide-Based Therapeutics for Cancer. *Oncogene* **2003**, *22*, 9087–9096. [CrossRef]

56. Engelhard, H.H. Antisense Oligodeoxynucleotide Technology: Potential Use for the Treatment of Malignant Brain Tumors. *Cancer Control* **1998**, *5*, 163–170. [CrossRef]

57. Khvorova, A.; Watts, J.K. The Chemical Evolution of Oligonucleotide Therapies of Clinical Utility. *Nat. Biotechnol.* **2017**, *35*, 238–248. [CrossRef]

58. Shen, X.; Corey, D.R. Chemistry, Mechanism and Clinical Status of Antisense Oligonucleotides and Duplex RNAs. *Nucleic Acids Res.* **2018**, *46*, 1584–1600. [CrossRef]

59. Barton, B.E.; Karras, J.G.; Murphy, T.F.; Barton, A.; Huang, H.F.-S. Signal Transducer and Activator of Transcription 3 (STAT3) Activation in Prostate Cancer: Direct STAT3 Inhibition Induces Apoptosis in Prostate Cancer Lines. *Mol. Cancer Ther.* **2004**, *3*, 11–20.

60. Mora, L.B.; Buettner, R.; Seigne, J.; Diaz, J.; Ahmad, N.; Garcia, R.; Bowman, T.; Falcone, R.; Fairclough, R.; Cantor, A.; et al. Constitutive Activation of Stat3 in Human Prostate Tumors and Cell Lines: Direct Inhibition of Stat3 Signaling Induces Apoptosis of Prostate Cancer Cells. *Cancer Res.* **2002**, *62*, 6659–6666.

61. Niu, G.; Wright, K.L.; Huang, M.; Song, L.; Haura, E.; Turkson, J.; Zhang, S.; Wang, T.; Sinibaldi, D.; Coppola, D.; et al. Constitutive Stat3 Activity Up-Regulates VEGF Expression and Tumor Angiogenesis. *Oncogene* **2002**, *21*, 2000–2008. [CrossRef]

62. Li, W.-C.; Ye, S.-L.; Sun, R.-X.; Liu, Y.-K.; Tang, Z.-Y.; Kim, Y.; Karras, J.G.; Zhang, H. Inhibition of Growth and Metastasis of Human Hepatocellular Carcinoma by Antisense Oligonucleotide Targeting Signal Transducer and Activator of Transcription 3. *Clin. Cancer Res.* **2006**, *12*, 7140–7148. [CrossRef]

63. Hong, D.; Kurzrock, R.; Kim, Y.; Woessner, R.; Younes, A.; Nemunaitis, J.; Fowler, N.; Zhou, T.; Schmidt, J.; Jo, M.; et al. AZD9150, a Next-Generation Antisense Oligonucleotide Inhibitor of STAT3 with Early Evidence of Clinical Activity in Lymphoma and Lung Cancer. *Sci. Transl. Med.* **2015**, *7*, 314ra185. [CrossRef]

64. Odate, S.; Veschi, V.; Yan, S.; Lam, N.; Woessner, R.; Thiele, C.J. Inhibition of STAT3 with the Generation 2.5 Antisense Oligonucleotide, AZD9150, Decreases Neuroblastoma Tumorigenicity and Increases Chemosensitivity. *Clin. Cancer Res.* **2017**, *23*, 1771–1784. [CrossRef]

65. Shastri, A.; Choudhary, G.; Teixeira, M.; Gordon-Mitchell, S.; Ramachandra, N.; Bernard, L.; Bhattacharyya, S.; Lopez, R.; Pradhan, K.; Giricz, O.; et al. Antisense STAT3 Inhibitor Decreases Viability of Myelodysplastic and Leukemic Stem Cells. *J. Clin. Investig.* **2018**, *128*, 5479–5488. [CrossRef]

66. Barton, B.E.; Murphy, T.F.; Shu, P.; Huang, H.F.; Meyenhofer, M.; Barton, A. Novel Single-Stranded Oligonucleotides that Inhibit Signal Transducer and Activator of Transcription 3 Induce Apoptosis in Vitro and in Vivo in Prostate Cancer Cell Lines. *Mol. Cancer Ther.* **2004**, *3*, 1183–1191.

67. Burel, S.A.; Han, S.-R.; Lee, H.-S.; Norris, D.A.; Lee, B.-S.; Machemer, T.; Park, S.-Y.; Zhou, T.; He, G.; Kim, Y.; et al. Preclinical Evaluation of the Toxicological Effects of a Novel Constrained Ethyl Modified Antisense Compound Targeting Signal Transducer and Activator of Transcription 3 in Mice and Cynomolgus Monkeys. *Nucleic Acid Ther.* **2013**, *23*, 213–227. [CrossRef]

68. Phase 1/2, Open-Label, Dose-Escalation Study of IONIS-STAT3Rx, Administered to Patients with Advanced Cancers—Full Text View—ClinicalTrials.gov. Available online: https://clinicaltrials.gov/ct2/show/NCT01563302 (accessed on 10 July 2019).

69. Reilley, M.J.; McCoon, P.; Cook, C.; Lyne, P.; Kurzrock, R.; Kim, Y.; Woessner, R.; Younes, A.; Nemunaitis, J.; Fowler, N.; et al. STAT3 Antisense Oligonucleotide AZD9150 in a Subset of Patients with Heavily Pretreated Lymphoma: Results of a Phase 1b Trial. *J. Immunother. Cancer* **2018**, *6*, 119. [CrossRef]

70. Kroemer, G.; Galluzzi, L.; Zitvogel, L. STAT3 Inhibition for Cancer Therapy: Cell-Autonomous Effects Only? *Oncoimmunology* **2016**, *5*, e1126063. [CrossRef]

71. Kortylewski, M.; Moreira, D. Myeloid Cells as a Target for Oligonucleotide Therapeutics: Turning Obstacles into Opportunities. *Cancer Immunol. Immunother.* **2017**, *66*, 979–988. [CrossRef]

72. Hossain, D.M.S.; Pal, S.K.; Moreira, D.; Duttagupta, P.; Zhang, Q.; Won, H.; Jones, J.; D'Apuzzo, M.; Forman, S.; Kortylewski, M. TLR9-Targeted STAT3 Silencing Abrogates Immunosuppressive Activity of Myeloid-Derived Suppressor Cells from Prostate Cancer Patients. *Clin. Cancer Res.* **2015**, *21*, 3771–3782. [CrossRef]

73. Kortylewski, M.; Swiderski, P.; Herrmann, A.; Wang, L.; Kowolik, C.; Kujawski, M.; Lee, H.; Scuto, A.; Liu, Y.; Yang, C.; et al. In Vivo Delivery of siRNA to Immune Cells by Conjugation to a TLR9 Agonist Enhances Antitumor Immune Responses. *Nat. Biotechnol.* **2009**, *27*, 925–932. [CrossRef] [PubMed]

74. Kortylewski, M.; Kuo, Y.-H. Push and Release: TLR9 activation plus STAT3 Blockade for Systemic Antitumor Immunity. *Oncoimmunology* **2014**, *3*, e27441. [CrossRef] [PubMed]

75. Zhang, Q.; Hossain, D.M.S.; Nechaev, S.; Kozlowska, A.; Zhang, W.; Liu, Y.; Kowolik, C.M.; Swiderski, P.; Rossi, J.J.; Forman, S.; et al. TLR9-Mediated siRNA Delivery for Targeting of Normal and Malignant Human Hematopoietic Cells in Vivo. *Blood* **2013**, *121*, 1304–1315. [CrossRef] [PubMed]

76. Zhu, G.; Mei, L.; Vishwasrao, H.D.; Jacobson, O.; Wang, Z.; Liu, Y.; Yung, B.C.; Fu, X.; Jin, A.; Niu, G.; et al. Intertwining DNA-RNA Nanocapsules Loaded with Tumor Neoantigens as Synergistic Nanovaccines for Cancer Immunotherapy. *Nat. Commun.* **2017**, *8*, 1482. [CrossRef]

77. Kortylewski, M.; Kujawski, M.; Herrmann, A.; Yang, C.; Wang, L.; Liu, Y.; Salcedo, R.; Yu, H. Toll-like Receptor 9 Activation of Signal Transducer and Activator of Transcription 3 Constrains Its Agonist-Based Immunotherapy. *Cancer Res.* **2009**, *69*, 2497–2505. [CrossRef] [PubMed]

78. Setten, R.L.; Rossi, J.J.; Han, S.-P. The Current State and Future Directions of RNAi-Based Therapeutics. *Nat. Rev. Drug Discov.* **2019**, *18*, 421–446. [CrossRef]

79. Nechaev, S.; Gao, C.; Moreira, D.; Swiderski, P.; Jozwiak, A.; Kowolik, C.M.; Zhou, J.; Armstrong, B.; Raubitschek, A.; Rossi, J.J.; et al. Intracellular Processing of Immunostimulatory CpG-siRNA: Toll-like Receptor 9 Facilitates siRNA Dicing and Endosomal Escape. *J. Control. Release* **2013**, *170*, 307–315. [CrossRef]

80. Herrmann, A.; Kortylewski, M.; Kujawski, M.; Zhang, C.; Reckamp, K.; Armstrong, B.; Wang, L.; Kowolik, C.; Deng, J.; Figlin, R.; et al. Targeting Stat3 in the Myeloid Compartment Drastically Improves the in vivo Antitumor Functions of Adoptively Transferred T cells. *Cancer Res.* **2010**, *70*, 7455–7464. [CrossRef]

81. Hossain, D.M.S.; Dos Santos, C.; Zhang, Q.; Kozlowska, A.; Liu, H.; Gao, C.; Moreira, D.; Swiderski, P.; Jozwiak, A.; Kline, J.; et al. Leukemia Cell-Targeted STAT3 Silencing and TLR9 Triggering Generate Systemic Antitumor Immunity. *Blood* **2014**, *123*, 15–25. [CrossRef]

82. Moreira, D.; Zhang, Q.; Hossain, D.M.S.; Nechaev, S.; Li, H.; Kowolik, C.M.; D'Apuzzo, M.; Forman, S.; Jones, J.; Pal, S.K.; et al. TLR9 Signaling Through NF-κB/RELA and STAT3 Promotes Tumor-Propagating Potential of Prostate Cancer Cells. *Oncotarget* **2015**, *6*, 17302–17313. [CrossRef] [PubMed]

83. Davis, J.T. G-quartets 40 years later: From 5′-GMP to Molecular Biology and Supramolecular Chemistry. *Angew. Chem. Int. Ed. Engl.* **2004**, *43*, 668–698. [CrossRef] [PubMed]

84. Jing, N.; Gao, X.; Rando, R.F.; Hogan, M.E. Potassium-Induced Loop Conformational Transition of a Potent anti-HIV Oligonucleotide. *J. Biomol. Struct. Dyn.* **1997**, *15*, 573–585. [CrossRef] [PubMed]

85. McMicken, H.W.; Bates, P.J.; Chen, Y. Antiproliferative activity of G-Quartet-Containing Oligonucleotides Generated by a Novel Single-Stranded DNA Expression System. *Cancer Gene Ther.* **2003**, *10*, 867–869. [CrossRef] [PubMed]

86. Jing, N.; Marchand, C.; Guan, Y.; Liu, J.; Pallansch, L.; Lackman-Smith, C.; De Clercq, E.; Pommier, Y. Structure–Activity of Inhibition of HIV-1 Integrase and Virus Replication by G-quartet Oligonucleotides. *DNA Cell Biol.* **2001**, *20*, 499–508. [CrossRef]

87. Mazumder, A.; Neamati, N.; Ojwang, J.O.; Sunder, S.; Rando, R.F.; Pommier, Y. Inhibition of the human immunodeficiency virus type 1 integrase by guanosine quartet structures. *Biochemistry* **1996**, *35*, 13762–13771. [CrossRef]

88. Rando, R.F.; Ojwang, J.; Elbaggari, A.; Reyes, G.R.; Tinder, R.; McGrath, M.S.; Hogan, M.E. Suppression of Human Immunodeficiency Virus Type 1 Activity in Vitro by Oligonucleotides which Form Intramolecular Tetrads. *J. Biol. Chem.* **1995**, *270*, 1754–1760. [CrossRef]

89. Jing, N.; Tweardy, D.J. Targeting Stat3 in Cancer Therapy. *Anticancer Drugs* **2005**, *16*, 601–607. [CrossRef]

90. Sen, M.; Grandis, J.R. Nucleic Acid-Based Approaches to STAT Inhibition. *JAKSTAT* **2012**, *1*, 285–291. [CrossRef]

91. Zhu, Q.; Jing, N. Computational Study on Mechanism of G-Quartet Oligonucleotide T40214 Selectively Targeting Stat3. *J. Comput.-Aided Mol. Des.* **2007**, *21*, 641–648. [CrossRef]

92. Bishop, J.S.; Guy-Caffey, J.K.; Ojwang, J.O.; Smith, S.R.; Hogan, M.E.; Cossum, P.A.; Rando, R.F.; Chaudhary, N. Intramolecular G-quartet Motifs Confer Nuclease Resistance to a Potent anti-HIV Oligonucleotide. *J. Biol. Chem.* **1996**, *271*, 5698–5703. [CrossRef] [PubMed]

93. Weerasinghe, P.; Li, Y.; Guan, Y.; Zhang, R.; Tweardy, D.J.; Jing, N. T40214/PEI complex: A Potent Therapeutics for Prostate Cancer that Targets STAT3 Signaling. *Prostate* **2008**, *68*, 1430–1442. [CrossRef]

94. Jing, N.; Zhu, Q.; Yuan, P.; Li, Y.; Mao, L.; Tweardy, D.J. Targeting Signal Transducer and Activator of Transcription 3 with G-quartet Oligonucleotides: A Potential Novel Therapy for Head and Neck Cancer. *Mol. Cancer Ther.* **2006**, *5*, 279–286. [CrossRef] [PubMed]

95. Weerasinghe, P.; Garcia, G.E.; Zhu, Q.; Yuan, P.; Feng, L.; Mao, L.; Jing, N. Inhibition of Stat3 Activation and Tumor Growth Suppression of Non-Small Cell Lung Cancer by G-Quartet Oligonucleotides. *Int. J. Oncol.* **2007**, *31*, 129–136. [CrossRef] [PubMed]

96. Fagard, R.; Metelev, V.; Souissi, I.; Baran-Marszak, F. STAT3 Inhibitors for Cancer Therapy: Have All Roads Been Explored? *JAKSTAT* **2013**, *2*, e22882. [CrossRef]

97. Wagner, B.J.; Hayes, T.E.; Hoban, C.J.; Cochran, B.H. The SIF Binding Element Confers sis/PDGF Inducibility onto the c-fos Promoter. *EMBO J.* **1990**, *9*, 4477–4484. [CrossRef]

98. Leong, P.L.; Andrews, G.A.; Johnson, D.E.; Dyer, K.F.; Xi, S.; Mai, J.C.; Robbins, P.D.; Gadiparthi, S.; Burke, N.A.; Watkins, S.F.; et al. Targeted Inhibition of Stat3 with a Decoy Oligonucleotide Abrogates Head and Neck Cancer Cell Growth. *Proc. Natl. Acad. Sci. USA* **2003**, *100*, 4138–4143. [CrossRef]

99. Lui, V.W.Y.; Boehm, A.L.; Koppikar, P.; Leeman, R.J.; Johnson, D.; Ogagan, M.; Childs, E.; Freilino, M.; Grandis, J.R. Antiproliferative Mechanisms of a Transcription Factor Decoy Targeting Signal Transducer and Activator of Transcription (STAT) 3: The Role of STAT1. *Mol. Pharmacol.* **2007**, *71*, 1435–1443. [CrossRef]

100. Shen, J.; Li, R.; Li, G. Inhibitory Effects of Decoy-ODN Targeting Activated STAT3 on Human Glioma Growth in Vivo. *In Vivo* **2009**, *23*, 237–243.

101. Xi, S.; Gooding, W.E.; Grandis, J.R. In Vivo Antitumor Efficacy of STAT3 Blockade Using a Transcription Factor Decoy Approach: Implications for Cancer Therapy. *Oncogene* **2005**, *24*, 970–979. [CrossRef]

102. Zhang, X.; Liu, P.; Zhang, B.; Mao, H.; Shen, L.; Ma, Y. Inhibitory Effects of STAT3 Decoy Oligodeoxynucleotides on Human Epithelial Ovarian Cancer Cell Growth in Vivo. *Int. J. Mol. Med.* **2013**, *32*, 623–628. [CrossRef] [PubMed]

103. Zhang, X.; Zhang, J.; Wang, L.; Wei, H.; Tian, Z. Therapeutic Effects of STAT3 Decoy Oligodeoxynucleotide on Human Lung Cancer in Xenograft Mice. *BMC Cancer* **2007**, *7*, 149. [CrossRef] [PubMed]

104. Sen, M.; Joyce, S.; Panahandeh, M.; Li, C.; Thomas, S.M.; Maxwell, J.; Wang, L.; Gooding, W.E.; Johnson, D.E.; Grandis, J.R. Targeting Stat3 Abrogates EGFR Inhibitor Resistance in Cancer. *Clin. Cancer Res.* **2012**, *18*, 4986–4996. [CrossRef] [PubMed]

105. Sen, M.; Thomas, S.M.; Kim, S.; Yeh, J.I.; Ferris, R.L.; Johnson, J.T.; Duvvuri, U.; Lee, J.; Sahu, N.; Joyce, S.; et al. First-in-Human Trial of a STAT3 Decoy Oligonucleotide in Head and Neck Tumors: Implications for Cancer Therapy. *Cancer Discov.* **2012**, *2*, 694–705. [CrossRef]

106. Zhang, Q.; Hossain, D.M.S.; Duttagupta, P.; Moreira, D.; Zhao, X.; Won, H.; Buettner, R.; Nechaev, S.; Majka, M.; Zhang, B.; et al. Serum-Resistant CpG-STAT3 Decoy for Targeting Survival and Immune Checkpoint Signaling in Acute Myeloid Leukemia. *Blood* **2016**, *127*, 1687–1700. [CrossRef]

107. Zhao, X.; Zhang, Z.; Moreira, D.; Su, Y.-L.; Won, H.; Adamus, T.; Dong, Z.; Liang, Y.; Yin, H.H.; Swiderski, P.; et al. B Cell Lymphoma Immunotherapy Using TLR9-Targeted Oligonucleotide STAT3 Inhibitors. *Mol. Ther.* **2018**, *26*, 695–707. [CrossRef]

108. Lee, D.S.; O'Keefe, R.A.; Ha, P.K.; Grandis, J.R.; Johnson, D.E. Biochemical Properties of a Decoy Oligodeoxynucleotide Inhibitor of STAT3 Transcription Factor. *Int. J. Mol. Sci.* **2018**, *19*, 1608. [CrossRef]

109. Njatcha, C.; Farooqui, M.; Kornberg, A.; Johnson, D.E.; Grandis, J.R.; Siegfried, J.M. STAT3 Cyclic Decoy Demonstrates Robust Antitumor Effects in Non-Small Cell Lung Cancer. *Mol. Cancer Ther.* **2018**, *17*, 1917–1926. [CrossRef]

110. Sen, M.; Paul, K.; Freilino, M.L.; Li, H.; Li, C.; Johnson, D.E.; Wang, L.; Eiseman, J.; Grandis, J.R. Systemic Administration of a Cyclic Signal Transducer and Activator of Transcription 3 (STAT3) Decoy Oligonucleotide Inhibits Tumor Growth without Inducing Toxicological Effects. *Mol. Med.* **2014**, *20*, 46–56. [CrossRef]

111. Li, C.; Zang, Y.; Sen, M.; Leeman-Neill, R.J.; Man, D.S.K.; Grandis, J.R.; Johnson, D.E. Bortezomib up-Regulates Activated Signal Transducer and Activator of Transcription-3 and Synergizes with Inhibitors of Signal Transducer and Activator of Transcription-3 to Promote Head and Neck Squamous Cell Carcinoma Cell Death. *Mol. Cancer Ther.* **2009**, *8*, 2211–2220. [CrossRef]

112. Zhang, X.; Liu, P.; Zhang, B.; Wang, A.; Yang, M. Role of STAT3 Decoy Oligodeoxynucleotides on Cell Invasion and chemosensitivity in Human Epithelial Ovarian Cancer Cells. *Cancer Genet. Cytogenet.* **2010**, *197*, 46–53. [CrossRef] [PubMed]

113. Zhang, X.; Xiao, W.; Wang, L.; Tian, Z.; Zhang, J. Deactivation of Signal Transducer and Activator of Transcription 3 Reverses Chemotherapeutics Resistance of Leukemia Cells via Down-Regulating P-gp. *PLoS ONE* **2011**, *6*, e20965. [CrossRef] [PubMed]

114. Bharadwaj, U.; Kasembeli, M.M.; Tweardy, D.J. STAT3 Inhibitors in Cancer: A Comprehensive Update. In *Cancer Drug Discovery and Development*; Springer: Berlin/Heidelberg, Germany, 2016; pp. 95–161.

115. Hodge, J.A.; Kawabata, T.T.; Krishnaswami, S.; Clark, J.D.; Telliez, J.-B.; Dowty, M.E.; Menon, S.; Lamba, M.; Zwillich, S. The Mechanism of Action of Tofacitinib—An Oral Janus Kinase Inhibitor for the Treatment of Rheumatoid Arthritis. *Clin. Exp. Rheumatol.* **2016**, *34*, 318–328. [PubMed]

116. Zhou, T.; Georgeon, S.; Moser, R.; Moore, D.J.; Caflisch, A.; Hantschel, O. Specificity and Mechanism-of-Action of the JAK2 Tyrosine Kinase Inhibitors Ruxolitinib and SAR302503 (TG101348). *Leukemia* **2014**, *28*, 404–407. [CrossRef]

117. Eiring, A.M.; Page, B.D.G.; Kraft, I.L.; Mason, C.C.; Vellore, N.A.; Resetca, D.; Zabriskie, M.S.; Zhang, T.Y.; Khorashad, J.S.; Engar, A.J.; et al. Combined STAT3 and BCR-ABL1 Inhibition Induces Synthetic Lethality in Therapy-Resistant Chronic Myeloid Leukemia. *Leukemia* **2017**, *31*, 1253–1254. [CrossRef]

118. Wen, W.; Wu, J.; Liu, L.; Tian, Y.; Buettner, R.; Hsieh, M.-Y.; Horne, D.; Dellinger, T.H.; Han, E.S.; Jove, R.; et al. Synergistic Anti-Tumor Effect of Combined Inhibition of EGFR and JAK/STAT3 Pathways in Human Ovarian Cancer. *Mol. Cancer* **2015**, *14*, 100. [CrossRef]

119. Woessner, R.; McCoon, P.; Bell, K.; DuPont, R.; Collins, M.; Lawson, D.; Nadella, P.; Pablo, L.; Reimer, C.; Sah, V.; et al. Abstract A93: STAT3 Inhibition Enhances the Activity of Immune Checkpoint Inhibitors in Murine Syngeneic Tumor Models by Creating a More Immunogenic Tumor Microenvironment. *Tumor Microenviron.* **2015**, *3*. [CrossRef]

120. Gao, Y.; Li, Y.; Hsu, E.; Wang, Y.; Huang, J.; Brooks, E.; Li, C.J. Abstract LB-140: Inhibition of Cancer Stemness Sensitizes Colorectal Cancer to Immune Checkpoint Inhibitors. *Tumor Biol.* **2017**. [CrossRef]

121. Weber, R.; Fleming, V.; Hu, X.; Nagibin, V.; Groth, C.; Altevogt, P.; Utikal, J.; Umansky, V. Myeloid-Derived Suppressor Cells Hinder the Anti-Cancer Activity of Immune Checkpoint Inhibitors. *Front. Immunol.* **2018**, *9*, 1310. [CrossRef]

122. Johnson, D.E.; O'Keefe, R.A.; Grandis, J.R. Targeting the IL-6/JAK/STAT3 signalling axis in cancer. *Nat. Rev. Clin. Oncol.* **2018**, *15*, 234. [CrossRef]

123. Yang, H.; Yamazaki, T.; Pietrocola, F.; Zhou, H.; Zitvogel, L.; Ma, Y.; Kroemer, G. STAT3 Inhibition Enhances the Therapeutic Efficacy of Immunogenic Chemotherapy by Stimulating Type 1 Interferon Production by Cancer Cells. *Cancer Res.* **2015**, *75*, 3812–3822. [CrossRef] [PubMed]

Review

Nucleocytoplasmic Shuttling of STATs. A Target for Intervention?

Sabrina Ernst [1,2] **and Gerhard Müller-Newen** [1,*]

1   Institute of Biochemistry and Molecular Biology, RWTH Aachen University, 52074 Aachen, Germany; sabrina.ernst@rwth-aachen.de
2   Confocal Microscopy Facility, Interdisciplinary Center for Clinical Research IZKF, RWTH Aachen University, 52074 Aachen, Germany
*   Correspondence: mueller-newen@rwth-aachen.de

Received: 3 September 2019; Accepted: 13 November 2019; Published: 19 November 2019

Abstract: Signal transducer and activator of transcription (STAT) proteins are transcription factors that in the latent state are located predominantly in the cytoplasm. Activation of STATs through phosphorylation of a single tyrosine residue results in nuclear translocation. The requirement of tyrosine phosphorylation for nuclear accumulation is shared by all STAT family members but mechanisms of nuclear translocation vary between different STATs. These differences offer opportunities for specific intervention. To achieve this, the molecular mechanisms of nucleocytoplasmic shuttling of STATs need to be understood in more detail. In this review we will give an overview on the various aspects of nucleocytoplasmic shuttling of latent and activated STATs with a special focus on STAT3 and STAT5. Potential targets for cancer treatment will be identified and discussed.

Keywords: STAT3; STAT5; nuclear pore complex; nuclear transport receptors; nucleocytoplasmic shuttling; cancer; targeting

---

1. Aim and Scope

STAT (Signal transducer and activator of transcription) proteins can be seen as intracellular messengers that relay signals sensed at the plasma membrane to chromatin and genes in the nucleus. To achieve this, STATs must pass the nuclear envelope through nuclear pore complexes (NPCs). Thus, passage through the NPC is an essential step in the sequence of events from activation of STATs at cytokine receptors to DNA-binding and target gene induction. As detailed in the reviews and articles of this Special Issue of Cancers, deregulated activation of STAT3 and STAT5 contributes to various cancers in many ways. Thus, STAT3 and STAT5 proteins have emerged as promising therapeutic targets. Protein–protein interactions involved in nucleocytoplasmic transfer of STATs have not been exploited yet as molecular targets for intervention. Detailed knowledge of the involved molecules and mechanisms is an essential prerequisite for successful and specific targeting. In this review, we will first describe the general mechanisms involved in import and export of proteins in and out of the nucleus, concentrating on those which are most relevant for transcription factors. We will then focus on the mechanisms involved in nucleocytoplasmic shuttling of STAT3 and STAT5 and finally assess possible molecular targets for specific intervention.

2. General Mechanisms of Nucleocytoplasmic Transport of Proteins

2.1. The Nuclear Pore Complex

To enter or exit the nucleus, proteins must pass through the nuclear pore complex (NPC) [1]. NPCs are huge macromolecular assemblies (about 120 MDa in humans) made up of multiple copies of Nucleoporins (NUPs). More than 30 different NUPs have been identified that are built into the NPC as multiples of eight (8–64) resulting in the eight-fold rotational symmetry of the NPC [2]. The NPC can be seen as a channel that allows selective transfer of cargo and at the same time forms a soft barrier for free diffusion of macromolecules larger than about 30 kDa [3,4], preventing their access to the nucleus without permission. The barrier is formed by phenylalanine-glycine (FG)-repeats that protrude from certain NUPs into the lumen of the channel [5,6].

How exactly the FG-repeats form a selective permeability barrier is not completely understood and several models are currently being discussed [7,8]. One of the most prevalent is the selective phase model that relies on interactions between the FG-repeats creating a sieve-like meshwork with hydrogel-like properties, which would explain the observed mass exclusion limit [9]. Selectivity for cargo allowed to pass might result from phase separation that prevents passage of macromolecules that are unable to mix or interact with the selective phase made up by the FG-repeats [8].

The import/export pathways through the NPC involve soluble nuclear transport receptors (NTRs) that bind cargo in conjunction with the Ran-GTP/GDP cycle. NTRs can interact with FG-repeats [10] and facilitate passage of bound cargo through the NPC. According to the selective phase model, interaction of NTRs with the FG-repeats leads to local disturbance of the meshwork allowing the NTR/cargo complex to travel almost freely between cytoplasm and nucleoplasm [11]. The energy-consuming Ran-GTP/GDP cycle provides directionality of the transport through control of cargo to NTR binding which is differently regulated in nucleoplasm vs. cytoplasm [12,13].

NTRs, also known as Karyopherins, can be subdivided in Importins and Exportins, facilitating nuclear import and export, respectively. Biportins have also been described that support transport of cargo in both directions with imported and exported cargo being distinct [14].

2.2. Importins

The best characterized Karyopherin is Importin-β1 which either binds cargo directly or indirectly through interaction with adapters such as α-Importins or Snurportin-1 [15]. The transcription factors Snail1 [16,17] and SREBP2 [18] are among the cargoes directly bound by Importin-β1. Snurportin-1 is best known for its involvement in the nuclear import of spliceosomal snRNPs [19]. The interaction site of Importin-β1 with FG-repeats has been mapped [20,21] and is different from the well-defined interaction sites with cargoes and adapters [22]. This means that cargo-loaded Importin-β1 can interact with FG-Nucleoporins of the NPC and thereby facilitate passage of the Importin-β1/adapter/cargo complex.

The import pathway using α-Importins as adapters to connect cargo with Importin-β1 has been intensively studied and is now known as the classical import pathway [23]. Accordingly, the term classical nuclear localization signal (cNLS) refers to linear sequence motifs of cargoes that bind to α-Importins. The cNLS can be further subdivided into monopartite cNLS and bipartite cNLS. Monopartite cNLS consist of a short stretch of basic amino acid residues, the sequence PKKKRRV of the SV40 large T-antigen being the first identified [24]. The first bipartite cNLS has been found in the Nucleoplasmin protein of Xenopus laevis, consisting of two short stretches of basic amino acids, both essential for its function, separated by a few less relevant amino acids [25].

In humans, seven Importin-α isoforms have been identified: Importin-α1, -α3, -α4, -α5, -α6, -α7 [23] and the most recently discovered Importin-α8 [26]. All α-Importins are mainly built up of ten Armadillo (ARM)-repeats resulting in a structurally conserved solenoid protein domain [27]. The cNLS of cargo binds along a groove on the inner concave surface of α-Importins. Although the cNLS binding region is quite conserved, the α-Importins are specific for a set of cargo proteins with some considerable overlap. This specificity results in part from preferential binding of certain NLS but also from the three-dimensional context in which the NLS is presented by the cargo [28]. Apart from cNLS, so called non-classical or atypical NLS have been identified, which do not fit in the cNLS consensus motifs [29]. A short sequence termed Importin-β binding (IBB) domain precedes the solenoid domain of α-Importins. In the absence of cargo, the IBB folds back to the NLS binding site. Upon binding of cargo, the IBB is replaced by the cNLS of the cargo protein. The now exposed IBB binds to Importin-β1, facilitating the nuclear import of the ternary Importin-β1/Importin-α/cargo complex [30].

Transportin-1, also known as Importin-β2, also imports a broad spectrum of cargoes, including transcription factors and mRNA-binding proteins. Transportin-1 binds cargo directly through a broad range of loosely related NLS that are different from cNLS [31]. Symportin-1 has been identified as an adapter for nuclear import of some ribosomal proteins by Transportin-1 [32]. Transportin-3 recognizes arginine-serine (RS)-rich NLS found in proteins typically involved in mRNA metabolism such as the Splicing Factor 2 (SF2) [33]. Importin-13 is closely related to Transportin-3 but mediates both protein import and export. Therefore, by definition Importin-13 can be regarded as a Biportin. Some transcription factors are among the many cargoes identified so far. Cargo binding by Importin-13 relies on the recognition of folded domains rather than linear NLS [34,35].

Some import pathways have been identified that do not rely on β-Importins. Among those, Calmodulin-mediated nuclear import seems to be most relevant for transcription factors [36]. Finally, some proteins including transcription factors enter the nucleus independent of NTR or other carrier proteins, e.g., through direct interaction with NUPs [37].

2.3. Exportins

Crm1 (Chromosome region maintenance 1; in the systematic nomenclature of Karyopherins designated as Exportin-1 or Xpo1) is the most promiscuous Exportin, which mediates the export of about 1000 substrates, including various RNAs, ribonucleoproteins, and transcription factors [38]. Most protein cargo of Crm1 contains a leucine-rich nuclear export signal (NES) of 8–15 amino acids [39–41]. Like all β-Karyopherins, Crm1 is built up of HEAT-repeats (Huntingtin, Elongation factor 3, protein phosphatase 2A and TOR kinase). Each HEAT-repeat consists of two α-helices connected by loops of varying length. Similar to the ARM repeats of α-Importins, the HEAT-repeats of β-Karyopherins form a solenoid protein domain [42] with a slightly curved superhelical structure of high conformational flexibility. In all of the available structures of cargo bound to Crm1, the NES of the cargo fits into a hydrophobic groove between HEAT repeats 11 and 12 (out of 20) at the outer convex surface of Crm1 [43]. Binding at the outer convex surface of Crm1 allows accommodation of cargo of various sizes.

Other Exportins, such as Exportin-5, Exportin-t, and Exportin-2 (better known as CAS for Cellular Apoptosis Susceptibility), bind their cargo on the inner concave surface. Accordingly, these Exportins are more selective. Exportin-5 mediates export of several types of RNA but not protein; Exportin-t is even more restricted to the export of tRNAs [44]. CAS is specialized for the nuclear export of α-Importins, which is needed to maintain the classical import pathway [45]. Exportin-4, a distant member of the β-Karyopherin family, originally described to mediate nuclear export of the translation initiation factor 5A [46] and the transcription factor Smad3 [47], has more recently been described to be involved in the nuclear import of Sox transcription factors [48], thus acting as a Biportin. Another Biportin, Exportin-7 (also known as RanBP16) [49], has recently been described to act as a broad-spectrum Karyopherin with about 200 export and 30 import substrates [50].

2.4. Ran-GTP/GDP Cycle

In principle, through interaction with the FG-repeats, β-Karyopherins can pass the NPC in both directions [51]. Directionality of nucleocytoplasmic transport is achieved through coupling of cargo binding to β-Karyopherins with the energy-consuming Ran-GTP/GDP cycle [12,52]. Ran is a small G-protein of the Ras superfamily of GTPases that exists in a GTP- or GDP-bound state. The GTP-bound state results from replacement of bound GDP with GTP that requires the activity of a Ran-specific, chromatin-bound guanine-nucleotide exchange factor (RanGEF, also known as RCC1), which is predominantly located in the nucleus. Therefore, the Ran-GTP concentration in the nucleus is high. The GDP-bound state is generated through release of the β-phosphate of bound GTP by hydrolysis, which requires the activity of a ran-specific GTPase-activating protein (RanGAP). Through interaction with the Nucleoporin Ran-binding protein 2 (RanBP2, Nup358), RanGAP is located predominantly at the cytoplasmic surface of the NPC. Therefore, in the cytoplasm close to the nuclear envelope, the Ran-GTP concentration is low and the Ran-GDP concentration is high [53,54].

Cargo binding to Importins in the cytoplasm is Ran-independent. After passage of the Importin/cargo complex through the NPC, release of cargo in the nucleus is induced upon Ran-GTP binding. In contrast, Ran-GTP-binding to Exportins is required for cargo loading in the nucleus. After passage of the Exportin/cargo/Ran-GTP complex through the NPC, RanGAP-mediated GTP-hydrolysis results in release of RanGDP and cargo into the cytoplasm [55].

As mentioned above, NTRs facilitate nucleocytoplasmic transport through transient interactions with the hydrophobic FG-repeats within the channel of the NPC. In principle, other proteins with hydrophobic surface patches should to some extent also be able to disturb the FG-repeat meshwork, allowing them to shuttle between the cytoplasm and the nucleus in an NTR- and Ran-independent, passive manner. Indeed, in a recent study [56] it has been found that a continuum exists with gradual differences in the ability of proteins to pass the NPC in an active (i.e., NTR- and Ran-dependent) or passive manner. Depending on their surface properties, even large proteins can leak into the nucleus without involvement of NTRs.

3. STAT1—Using a Side Track for Nuclear Import

STAT1 is deeply involved in the antiviral response triggered by endogenously produced Interferons but also in the therapeutic responses to exogenously administered Interferons in cancer treatments and anti-viral therapies [57]. STAT1 is best known for its tumor suppressive activity in cancer but some tumor-promoting effects have also been documented [58]. IFNγ (type II Interferon) induces the activation of STAT1 through phosphorylation of Y701 by IFNγ receptor-associated Janus kinase 1 (JAK1) and JAK2, leading to dimerization and nuclear accumulation of the transcription factor [59]. Among all STAT family members, the mechanism of nuclear accumulation of the activated STAT1 dimer is best understood. Activated STAT1 interacts with Importin-α5 [60], but in an unconventional manner. Instead of employing the binding site for cNLS, which would involve ARM repeats 2–4 and 6–8 as the major and minor binding sites, respectively [22], STAT1 binds to the more C-terminally located ARM repeats of Importin-α5 [61] involving a critical tyrosine residue (Y476) located in ARM repeat 10 [62]. Accordingly, the activated STAT1 dimer does not expose a cNLS but a dimer-specific surface area to interact with Importin-α5. This surface has been termed dimer-specific NLS (dsNLS) and includes regions within the N-terminal domain (NTD) [63] and the DNA-binding domain (DBD; see Figure 1a for a scheme of the domain structure of STAT proteins; see Table 1 for an overview on putative NLS and NES sequences of STAT proteins; and see Figure 1b showing the corresponding putative NLS and NES motifs in the STAT1 dimer structure) [64]. Hence, Importin-α5 can be displaced from STAT1 by DNA-binding [65,66]. Most interestingly, this unique mode of interaction is exploited by the Ebola virus VP24 protein, that interacts with ARM repeats 8–10 of Importin-α5, preventing binding and nuclear translocation of activated STAT1 [67]. Interaction of Importin-α5 with cNLS cargo remains unaffected by VP24 binding. This example shows that, in principle, nuclear accumulation of a STAT transcription factor can be selectively blocked.

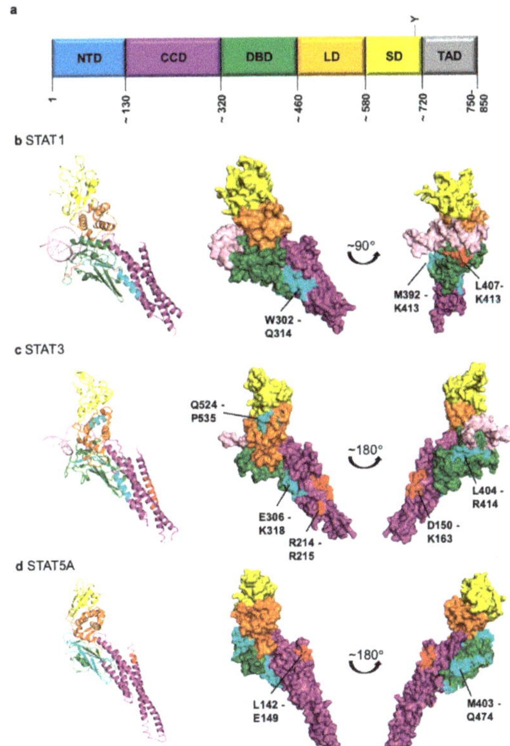

Figure 1. Structures of STAT proteins with putative nuclear localization signals (NLS) and nuclear export signals (NES) highlighted as listed in Table 1. In this context, "putative" means that the corresponding sequences do not fulfill classical NLS or NES functions but are required for nuclear transport or interaction with nuclear transport receptors. (**a**) General scheme of structural domains of STAT proteins (NTD, N-terminal domain; CCD coiled-coil domain; DBD, DNA-binding domain; LD, linker domain; SD, SH2 domain; TAD, transactivation domain). Numbers refer to amino acid positions. (**b–d**) Structures of individual STAT proteins as ribbon representations (left) or space-filling representations (middle and right). Domains are stained according to the coloring of the scheme in (**a**), DNA is shown in pink. Putative NLS and NES are marked in red and cyan, respectively. The corresponding references are listed in Table 1. PDB IDs: STAT1, 1BF5; STAT3, 1BG1; STAT5A, 1Y1U. Images of structures were generated with PyMOL.

Table 1. NTRs, NLS and NES involved in transport of STAT proteins.

	STAT1	STAT3	STAT5A	STAT5B
Importin	α5 [60,68], α7 [69]	α1 [70], α3 [70–72], α5 [69,70,72], α6 [71], α7 [69,72]	α3 [73]	n.d. [1]
Putative [2] NLS	L407-K413 [64,68] NTD involved [63,74]	D150-K163 [71] R214/R215 [69] NTD involved [75]	L142-E149 [76] intact CCD required [73]	n.d.
Exportin	Crm1 [66]	Crm1 [75]	Crm1 (not exclusively) [73]	Crm1 (not exclusively) [77]
Putative [2] NES	W302-Q314 [78] M392-K413 [66]	E306-K318 L404-R414 Q524-P535 [79]	L119/L133 M403-Q474 [73]	n.d.

[1] n.d., not determined; [2] in this context, "putative" means that the corresponding sequences do not fulfill classical NLS or NES functions but are required for nuclear transport or interaction with NTRs.

4. STAT3—Acting on Many Stages

4.1. Role of STAT3 in Cancer

STAT3 is involved in cancer in multiple ways, as detailed in several reviews and articles of this Special Issue of Cancers. Besides acting as an oncogene in a cell autonomous manner [80,81], activation of STAT3 in the tumor microenvironment contributes to metastasis, angiogenesis, cancer stem cell maintenance, and immune evasion [82]. Therefore, blocking STAT3 activity in cancer is of particular interest. However, it should also be noted that in some cancers STAT3 has tumor suppressive activities [83,84]. Several steps in the STAT3 activation pathway are potentially targetable and thus might be used to interfere with STAT3 activation. These include phosphorylation on Y705 by (oncogenic) tyrosine kinases, dimerization through phosphotyrosine/SH2 domain interactions, nuclear translocation, DNA-binding to GAS (γ-interferon activated sequence) elements, and interaction with cofactors. However, passage of the activated STAT3 dimer through the NPC has only rarely been considered as a potential molecular target for intervention [85,86]. This might be due to the fact that import mechanism for activated STAT3 and the interactions of STAT3 with NTRs have been analyzed in some detail but are not completely understood as outlined below.

4.2. Nuclear Import

Compared to STAT1, the mechanism of nuclear import of the activated STAT3 dimer is less well defined. Reports on interactions with α-Importins are contradictory in part. In one of the first studies, interactions of STAT3 with Importin-α1, -α3, and -α5 were detected [70]. Additionally, interactions of STAT3 with Importin-α5 and -α7, but not with Importin-α1, -α3, and -α4, were found [69]. In both reports, interactions were only seen upon activation of STAT3. Other groups found interactions with Importin-α3 and -α6 [71] and Importin-α3, -α5, and -α7 [72] being independent of activation of STAT3. In this context it is worth noting that expression of Importin-α6 is restricted to testis [87]. Therefore, Importin-α6 is less relevant for nuclear import of ubiquitously expressed transcription factors such as STAT3 and STAT5. Based on import assays with permeabilized cells, the MgcRacGAP protein bound to Rac has been suggested as an NLS-containing adapter for Importin-α-mediated nuclear import of STAT3 and STAT5 [88]. In another study using an MgcRacGAP inhibitor and siRNA mediated knockdown, the functional role of MgcRacGAP for nuclear import of STAT3 could not be substantiated [89]. These discrepancies might be the result of different experimental conditions and cellular systems used. For instance, subcellular distribution of STAT3 analyzed by immunofluorescence heavily depends on the applied fixation method [90]. Alternatively, the different findings indicate that STAT3 employs a different import system depending on the cytokines used for its activation, e.g., Interleukin-6 (IL-6) vs. Epidermal Growth Factor (EGF). Clearly, more work is required for better understanding.

Functional relevance of Importin-α5 in nuclear import of activated STAT3 has been demonstrated, as well as the involvement of Importin-β and Ran [70]. Moreover, it has been shown that the epitope on Importin-α5 employed in the interaction with STAT3 is different from the one used for the interaction with STAT1. A mutation in ARM repeat 10 of Importin-α5 blocks interaction with STAT1 [62] but does not affect the interaction with STAT3 [72]. Correspondingly, the epitopes used by STAT3 and STAT1 for interaction with Importin-α5 are also different. Amino acids in the DBD of STAT1 are involved in the recognition of Importin-α5 [65,66], connecting Importin-binding and nuclear presence with DNA-binding activity of STAT1 [91]. In STAT3, mutation of R214 and R215 located in the coiled-coil domain (CCD) abolishes binding of Importin-α5 and nuclear accumulation in response to stimulation [69]. Liu et al. [71] defined the sequence D150-K163 as being indispensable for nuclear import of STAT3, which is also located in the CCD. Nuclear accumulation of activated STAT3 does not require DNA-binding activity [72]. Common to STAT3 and STAT1 is the involvement of the N-terminal domain in stimulation-dependent nuclear import [63,74,75]. The essential role of the NTD for active nuclear import of STAT3 is further supported by the observation that the tumor suppressor ARHI

(A Ras Homologue member I/DIRAS3) blocks nuclear translocation of phosphorylated STAT3 via direct interaction with the NTD [92].

It seems that Importin-α5 is a major determinant for nuclear import of STAT3 and STAT1 but the molecular interfaces of the interactions are considerably different. This opens an avenue for specific interventions. However, to achieve a selective block of nuclear import of STAT3, the structural details of the interactions with NTRs must be characterized in more detail. It should be noted that no classical NLS or NES sequences have been identified in STAT3 in the sense of transferable functional motifs. The above-mentioned amino acids in the putative NLS and NES are most probably important for the integrity of domain structures or the part of epitopes that interact with NTRs in a non-conventional manner [75], as exemplified for STAT1 [62] (Figure 1c showing putative NLS and NES motifs in the activated STAT3 dimer structure).

4.3. Nuclear Export and Nucleocytoplasmic Shuttling

In the course of defining nuclear export signals of STAT3, it was found that pharmacological inhibition of Crm1-mediated nuclear export results in partial nuclear accumulation of STAT3 independent of cytokine stimulation [79]. From these and other studies it became evident that the subcellular distribution of latent STAT3 with high cytoplasmic and low nuclear concentration is the result of a steady-state of constitutive nuclear import and rapid export independent of phosphorylation at Y705 [93,94]. Determinants of stimulation-dependent nuclear import, such as the NTD and the R214/R215-motif, are not involved in basal nuclear import of latent STAT3 [75]. Unphosphorylated STAT1 has also been detected in the nucleus [95], and the mechanisms of nuclear import of latent and activated STAT1 were also found to be different [64]. Nuclear import of unphosphorylated STAT3 might be independent of NTRs and could involve direct interactions with FG-repeat NUPs of the NPC as observed for STAT1 [96]. From the three sequence motifs identified as putative NES in STAT3, at least the one containing L525/L528 is involved in constitutive nucleocytoplasmic shuttling [75,79].

It has been firmly established that STAT3 and STAT1 form homodimers in the absence of the activating tyrosine phosphorylation resulting in so called preformed dimers [97,98]. The phosphotyrosine-independent dimerization involves homotypic interactions of the NTDs [99]. However, basal nucleocytoplasmic shuttling of STAT3 does not require the NTD [75] and is not sensitive to a single point mutation (L78R) that prevents preformed dimer formation [100]. Not only is dimer formation not required for basal shuttling, but monomeric STAT3 shuttles even faster [75]. The increased shuttling rate of monomeric STAT3 could be attributed to the smaller size compared to the preformed dimer. However, a recent study suggests that hydrophobic surface patches that interact with FG-repeats of FG-Nups might be a stronger predictor for NTR-independent passage of the NPC than size [56]. Thus, accessible hydrophobic surfaces that are masked in the dimer may facilitate faster shuttling of monomeric STAT3 in an NTR-independent manner.

Since basal and stimulation-dependent nuclear import of STATs occur through different mechanisms, they could, in principle, be targeted independently of each other. However, the functional relevance of nuclear presence of unphosphorylated STAT3 and STAT1 (U-STATs) is not entirely clear. Nuclear U-STATs might be involved as cofactors in gene regulation independent of their DNA-binding activity [101].

5. STAT5—Leukemia and More

Important functional roles in hematopoiesis downstream of hematopoietic cytokines such as Erythropoietin and Thrombopoietin have been attributed to STAT5. Consequently, deregulated STAT5 signaling is prominently implicated in myeloproliferative diseases and leukemias [102]. The involvement of STAT5 in solid tumors is also well documented, both as an oncogene and a tumor suppressor [103], as detailed in several reviews and articles of this Special Issue of Cancers. In particular, in myeloproliferative neoplasms and leukemias, STAT5 has been identified as a promising therapeutic target.

Compared with the many reports on STAT3 and STAT1, only a handful of studies dedicated to the molecular mechanisms governing nucleocytoplasmic distribution of STAT5 exist. In one of the first studies [104], a sequence motif in the DNA-binding domain of STAT5B (V466-I469) was identified whose mutation prevented Growth Hormone-induced nuclear accumulation and DNA-binding. However, from this finding it cannot be concluded that DNA-binding per se is required for nuclear accumulation of activated STAT5B because the mutation could also interfere with binding of NTRs, which was not investigated in this study. We found that another mutation that affects DNA-binding of STAT5A does not impair nuclear accumulation upon Epo stimulation, in agreement with the DNA-binding independent nuclear accumulation of activated STAT3 [72]. In general, nuclear accumulation of STATs through retention by binding to nuclear structures seems to be of minor importance. If nuclear accumulation was dependent on retention of STATs on subnuclear structures, nuclear accumulation would be saturable, meaning that as soon as all binding sites are occupied, the remaining STAT molecules would not accumulate in the nucleus. This is not what is observed in the above-mentioned experiments with transfected cells: Even upon forced overexpression, STATs almost completely accumulate in the nucleus upon stimulation. This view is supported by the results of fluorescence recovery after photobleaching (FRAP) experiments that have been exemplarily performed on STAT1-GFP. Upon nuclear accumulation induced by Interferon treatment, STAT1-GFP freely diffuses through the nucleoplasm with the exclusion of nucleoli [105]. In this and another study [93], the involvement of the cytoskeleton in directed nuclear import of STAT1 and STAT3 was also excluded.

In one of the reports on binding of α-Importins to STAT3, binding to STAT5A and STAT5B was analyzed in parallel but no interaction could be detected [69]. This is in agreement with our own observations made with co-precipitation experiments from cellular lysates using Importins fused to GST (glutathione-S-transferase). Accordingly, no cNLS has been detected in STAT5. In another study, interaction of STAT5A with Importin-α3 has been shown [73], again supporting the notion that the outcomes of in vitro assays (mostly co-precipitations) for studying the interaction of STATs with NTRs are very sensitive to the experimental conditions used.

Similar to STAT3 and STAT1, constitutive nucleocytoplasmic shuttling of STAT5 independent of tyrosine phosphorylation has been detected [77]. A functional role of nuclear U-STAT5 has been established in megakaryocyte differentiation [106]. Basal and cytokine-induced import of STAT5A requires an intact CCD [73] and was abolished upon deletion of eight amino acids in the CCD (L142–E149, see Table 1 and Figure 1d showing putative NLS and NES motifs in the STAT5A monomer structure; a structure of the activated STAT5 dimer is not available) [76]. Interaction of STAT5A with Importin-α3 occurs in an unconventional manner involving the CCD [73]. As for STAT3 and STAT1, stimulation-dependent but not basal nuclear translocation requires the NTD of STAT5 [77]. In a study that tested all sequence motifs that might function as a cNES, a region between aa578 and aa675 in STAT5B comprising the SH2 domain was identified to be required for nuclear export [77]. Another region important for Crm1-mediated export was identified involving L119 and L133 in the NTD of STAT5A [73].

A discrepancy between nuclear localization of STAT5A and STAT5B was found in response to phosphorylation of the critical tyrosine residue (Y694 in STAT5A and Y699 in STAT5B) by Src family kinases (SFK). While phosphorylation of Y699 in STAT5B, as well as Y705 in STAT3, by Src leads to nuclear accumulation of the transcription factors, phosphorylation of Y694 in STAT5A by Src does not [107]. Later it was found that this observation is of some relevance in Bcr-Abl-positive chronic myeloid leukemia (CML), where STAT5 is predominantly localized in the cytoplasm despite being phosphorylated at the critical tyrosine residue [108]. Cytoplasmic retention of STAT5A in response to activation by Bcr-Abl was found to be mediated by SFK such as Src and Hck. Specific inhibition of SFKs resulted in nuclear accumulation of STAT5A, enhanced STAT5 target gene expression, and increased colony formation of CML cells [109]. SFK interfere specifically with dimerization of activated STAT5A and thereby prevent nuclear accumulation [110]. In another report, phosphorylation of S779 has been suggested as an additional requirement for the nuclear translocation of STAT5A in Bcr-Abl positive cells [111]. These intriguing examples show that even the highly homologous transcription factors STAT5A and STAT5B differ in some aspects of nucleocytoplasmic shuttling and these differences might be relevant for disease.

6. Perspectives for Therapeutic Interventions

Nucleocytoplasmic transport has been recognized as a target for cancer therapy [112,113] in part based on the observation that components of the nuclear transport machinery are differentially expressed in transformed cells [114]. These changes in expression can be relevant for disease. It has been shown that Crm1-mediated nuclear export contributes to drug resistance in multiple ways [115]. For example, Topoisomerase II inhibitors such as doxorubicin require nuclear Topoisomerase activity. As a drug resistance mechanism, Topoisomerase II can be exported from the nucleus in a Crm1-dependent manner [116]. Treatment with Crm1 inhibitors therefore sensitizes cancer cells to doxorubicin treatment [117]. Only recently, in two independent studies, was Crm1 convincingly identified as a synthetic lethality gene in different types of cancer [118,119]. Earlier clinical studies already showed that the highly efficient Crm1 inhibitor leptomycin B is not well tolerated by patients [120]. Less toxic small molecule inhibitors of Crm1 termed Selective Inhibitors of Nuclear Export (SINE), such as selinexor, are currently used in clinical trials in patients with hematological malignancies and solid tumors [112]. Most recently, selinexor in combination with dexamethasone has been approved for the treatment of refractory multiple myeloma. In this context, it has already been recognized that a more selective export inhibition could be more effective and reduce the serious side effects of treatment [115]. There are only a few inhibitors for some other NTRs available (Importin-β1, Importin-α/β1 heterodimers, and Transportin-1) which have not been explored in such detail yet [121].

For transcription factors that respond to extracellular cues, control of nucleocytoplasmic distribution contributes to efficient signal transduction. High cytoplasmic concentration of the latent transcription factor increases the sensitivity to stimulation, which usually occurs at the cytoplasmic face of membrane-bound receptors. After stimulation and activation of the transcription factor, high nuclear concentration increases the efficiency of DNA-binding and thus gene induction. However, the molecular mechanisms involved in the control of subcellular distribution vary considerably between different transcription factors. For instance, the NLS of NF-κB is masked by IκB [122]. Upon sensing of inflammatory mediators by membrane-bound receptors, IκB is phosphorylated and ubiquitinated resulting in its proteasomal degradation. The now exposed NLS facilitates α-Importin-mediated nuclear accumulation of NF-κB [123]. Likewise, the NLS of some nuclear receptors, such as the Glucocorticoid Receptor and Estrogen Receptor, are masked. Ligand binding to the receptors results in release of Hsp90 and exposition of the NLS [124]. Similar cytoplasmic interaction partners that would prevent nuclear translocation of latent STAT proteins have not been identified yet. Various other mechanisms involving SUMOylation, ubiquitination, mono-ADP-ribosylation, and acetylation have been described to control the nucleocytoplasmic distribution of transcription factors [22,125]. Beside the well-established Karyopherin-mediated transport, other non-conventional transport mechanisms exist [126]. Nucleocytoplasmic transport of β-Catenin is regulated independently of Ran and classical transport factors [127,128]. At the same time, β-Catenin acts as a transport receptor for its transcription factor partner LEF-1 [129]. The regulatory 14-3-3 proteins have also been suggested to be involved in such a piggy-back mechanism of nuclear transport [130,131], which has so far not been described for STAT proteins. Because of these various transport modes, specific interference should be feasible in principle.

In the case of STAT transcription factors, a non-classical NLS is generated through phosphorylation-induced dimerization (dsNLS) that is recognized by α-Importins or other yet uncharacterized factors. Because this mechanism is quite unique, a specific intervention through targeting the dsNLS/α-Importin interface should in general be feasible, as exemplified for STAT1 [67]. Extended surface areas are involved in the STAT/NTR interaction, including the NTD, DBD and CCD. The NTD seems to be a common denominator of nuclear import of activated STATs. Because the NTDs of different STATs vary in their sequence, a specific targeting should be possible, as exemplified by the specific interaction of ARHI with the NTD of STAT3, but not with the NTDs of STAT5 and STAT1 [92]. The impact of the ARHI/STAT3-NTD interaction on Importin-binding was further analyzed in a study applying homology modelling and molecular dynamics simulation [132]. Indeed, targeting of the NTD of STAT3 has been reported, however, with a focus on the role of the NTD as a tetramerization domain at enhancers and its function in gene induction [133]. The more recently solved structure of the NTD of STAT3 [134] should pave the way for a closer inspection of the surface areas involved in nuclear import by mutation of surface-exposed residues. For nuclear import of STAT5 and STAT3, data concerning use of the classical α-Importin/β1-Importin pathway are conflicting. Here, a fresh approach is needed to identify the NTRs involved, including α-Importins/Importin-β1-independent pathways mentioned in Section 2 of this review.

Reduced export of the imported STAT dimers also contributes to nuclear accumulation because export and redistribution requires dephosphorylation [135]. Blocking nuclear export of STATs as a therapeutic approach seems at first sight counterintuitive because it might trap the activated transcription factor in the nucleus where it could drive persistent oncogenic gene induction. However, STAT1, STAT3, and STAT5 are dephosphorylated in the nucleus by TC45, the 45 kDa isoform of T Cell Protein Tyrosine Phosphatase (TC-PTP) [136–138]. For STAT1 it has been shown that only the dephosphorylated form leaves the nucleus for reactivation at the receptor [91]. Exit out of the nucleus is required for reactivation of STATs [139]. Thus, STAT-mediated gene induction could be inhibited by trapping the dephosphorylated, transcriptionally inactive STATs (in the sense of canonical STAT signaling) in the nucleus through blockade of nuclear export. This has already been shown for STAT3 using the Crm1 inhibitor ratjadone A [140]. The more clinically advanced SINE have also been tested. These compounds inhibited STAT3-mediated Survivin expression in breast cancer models [141]. To interfere more specifically with export of STATs, the Exportins or Biportins involved must be characterized and the molecular interfaces of the STAT/NTR interaction mapped in detail.

Mechanistically, the export of dephosphorylated STATs most probably corresponds to the exit mechanisms involved in constitutive shuttling of latent STATs. Treatment of cells with specific inhibitors for the Exportin Crm1 often leads to only a partial nuclear accumulation of latent STATs. This means that either passive NTR-independent export occurs through direct hydrophobic interactions with FG-Nups or that other Exportins are involved. Specifically for STAT5A, a Crm1-independent export signal in the DBD has been mapped [73]. Redundancy in cargo recognition by NTRs in the sense that one cargo is transported by several NTRs is not unusual. As outlined in Section 2, broad-specificity Exportins or Biportins have been identified in recent years, worthy of analysis with respect to their involvement in the nuclear export of STAT proteins.

7. Conclusions

Taken together, effective inhibition of oncogenic STAT activity can in principle be achieved through specific blockade of both nuclear import of phosphorylated STAT dimers and export of dephosphorylated STATs out of the nucleus. However, this requires a deeper understanding of the protein–protein interactions involved in these processes. First, all NTRs facilitating nucleocytoplasmic shuttling need to be unambiguously identified for the individual STAT proteins. These studies should also include those NTRs that have been characterized in more detail in recent years. Then, the interactions between STATs and NTRs must be mapped in more detail using all structural data available. The ultimate goal would be to solve the structure of STAT/NTR complexes by X-ray crystallography or cryo-electron microscopy. Furthermore, the contribution of NTR-independent nuclear transport of STATs should be evaluated through assessment and mutation of surface patches as delineated by Frey et al. [56]. Based on solid structural and functional data, specific blockade of nucleocytoplasmic transport of individual STATs by tailor-made inhibitory molecules might be feasible.

Author Contributions: S.E. edited the manuscript, prepared figure and table and searched the literature, G.M.-N. drew the conceptual framework and wrote the manuscript.

Funding: This research was funded by the Interdisciplinary Center for Clinical Research (IZKF) Aachen within the Faculty of Medicine at RWTH Aachen University.

Acknowledgments: This work was supported by the Confocal Microscopy Facility, a core facility of the Interdisciplinary Center for Clinical Research (IZKF) Aachen within the Faculty of Medicine at RWTH Aachen University. The authors would like to thank Wolfram Antonin and Bernhard Lüscher for critical reading of the manuscript. The authors apologize to all colleagues whose contributions have not been cited.

Conflicts of Interest: The authors declare no conflict of interest.

References

1. Feldherr, C.M. The uptake and accumulation of proteins by the cell nucleus. *Bioessays* **1985**, *3*, 52–55. [CrossRef] [PubMed]
2. Beck, M.; Hurt, E. The nuclear pore complex: Understanding its function through structural insight. *Nat. Rev. Mol. Cell Biol.* **2017**, *18*, 73–89. [CrossRef] [PubMed]
3. Mohr, D.; Frey, S.; Fischer, T.; Guttler, T.; Gorlich, D. Characterisation of the passive permeability barrier of nuclear pore complexes. *EMBO J.* **2009**, *28*, 2541–2553. [CrossRef] [PubMed]
4. Timney, B.L.; Raveh, B.; Mironska, R.; Trivedi, J.M.; Kim, S.J.; Russel, D.; Wente, S.R.; Sali, A.; Rout, M.P. Simple rules for passive diffusion through the nuclear pore complex. *J. Cell Biol.* **2016**, *215*, 57–76. [CrossRef]
5. Frey, S.; Gorlich, D. A saturated FG-repeat hydrogel can reproduce the permeability properties of nuclear pore complexes. *Cell* **2007**, *130*, 512–523. [CrossRef]
6. Lim, R.Y.; Huang, N.P.; Koser, J.; Deng, J.; Lau, K.H.; Schwarz-Herion, K.; Fahrenkrog, B.; Aebi, U. Flexible phenylalanine-glycine nucleoporins as entropic barriers to nucleocytoplasmic transport. *Proc. Natl. Acad. Sci. USA* **2006**, *103*, 9512–9517. [CrossRef]
7. Rout, M.P.; Aitchison, J.D.; Magnasco, M.O.; Chait, B.T. Virtual gating and nuclear transport: The hole picture. *Trends Cell Biol.* **2003**, *13*, 622–628. [CrossRef]
8. Schmidt, H.B.; Gorlich, D. Transport selectivity of nuclear pores, phase separation, and membraneless organelles. *Trends Biochem. Sci.* **2016**, *41*, 46–61. [CrossRef]
9. Hulsmann, B.B.; Labokha, A.A.; Gorlich, D. The permeability of reconstituted nuclear pores provides direct evidence for the selective phase model. *Cell* **2012**, *150*, 738–751. [CrossRef]
10. Radu, A.; Moore, M.S.; Blobel, G. The peptide repeat domain of nucleoporin Nup98 functions as a docking site in transport across the nuclear pore complex. *Cell* **1995**, *81*, 215–222. [CrossRef]
11. Ribbeck, K.; Gorlich, D. Kinetic analysis of translocation through nuclear pore complexes. *EMBO J.* **2001**, *20*, 1320–1330. [CrossRef] [PubMed]
12. Rexach, M.; Blobel, G. Protein import into nuclei: Association and dissociation reactions involving transport substrate, transport factors, and nucleoporins. *Cell* **1995**, *83*, 683–692. [CrossRef]
13. Wente, S.R.; Rout, M.P. The nuclear pore complex and nuclear transport. *Cold Spring Harb Perspect. Biol.* **2010**, *2*, a000562. [CrossRef] [PubMed]
14. Aksu, M.; Trakhanov, S.; Vera Rodriguez, A.; Gorlich, D. Structural basis for the nuclear import and export functions of the biportin Pdr6/Kap122. *J. Cell Biol.* **2019**, *218*, 1839–1852. [CrossRef] [PubMed]
15. Harel, A.; Forbes, D.J. Importin beta: Conducting a much larger cellular symphony. *Mol. Cell* **2004**, *16*, 319–330. [CrossRef] [PubMed]
16. Choi, S.; Yamashita, E.; Yasuhara, N.; Song, J.; Son, S.Y.; Won, Y.H.; Hong, H.R.; Shin, Y.S.; Sekimoto, T.; Park, I.Y.; et al. Structural basis for the selective nuclear import of the C2H2 zinc-finger protein Snail by importin beta. *Acta Crystallogr. D Biol. Crystallogr.* **2014**, *70*, 1050–1060. [CrossRef]
17. Yamasaki, H.; Sekimoto, T.; Ohkubo, T.; Douchi, T.; Nagata, Y.; Ozawa, M.; Yoneda, Y. Zinc finger domain of Snail functions as a nuclear localization signal for importin beta-mediated nuclear import pathway. *Genes Cells* **2005**, *10*, 455–464. [CrossRef]
18. Lee, S.J.; Sekimoto, T.; Yamashita, E.; Nagoshi, E.; Nakagawa, A.; Imamoto, N.; Yoshimura, M.; Sakai, H.; Chong, K.T.; Tsukihara, T.; et al. The structure of importin-beta bound to SREBP-2: Nuclear import of a transcription factor. *Science* **2003**, *302*, 1571–1575. [CrossRef]
19. Huber, J.; Cronshagen, U.; Kadokura, M.; Marshallsay, C.; Wada, T.; Sekine, M.; Luhrmann, R. Snurportin1, an m3G-cap-specific nuclear import receptor with a novel domain structure. *EMBO J.* **1998**, *17*, 4114–4126. [CrossRef]

20. Bayliss, R.; Littlewood, T.; Strawn, L.A.; Wente, S.R.; Stewart, M. GLFG and FxFG nucleoporins bind to overlapping sites on importin-beta. *J. Biol. Chem.* **2002**, *277*, 50597–50606. [CrossRef]

21. Bednenko, J.; Cingolani, G.; Gerace, L. Importin beta contains a COOH-terminal nucleoporin binding region important for nuclear transport. *J. Cell Biol.* **2003**, *162*, 391–401. [CrossRef] [PubMed]

22. Christie, M.; Chang, C.W.; Rona, G.; Smith, K.M.; Stewart, A.G.; Takeda, A.A.; Fontes, M.R.; Stewart, M.; Vertessy, B.G.; Forwood, J.K.; et al. Structural biology and regulation of protein import into the nucleus. *J. Mol. Biol.* **2016**, *428*, 2060–2090. [CrossRef] [PubMed]

23. Pumroy, R.A.; Cingolani, G. Diversification of importin-alpha isoforms in cellular trafficking and disease states. *Biochem. J.* **2015**, *466*, 13–28. [CrossRef] [PubMed]

24. Kalderon, D.; Richardson, W.D.; Markham, A.F.; Smith, A.E. Sequence requirements for nuclear location of simian virus 40 large-T antigen. *Nature* **1984**, *311*, 33–38. [CrossRef]

25. Dingwall, C.; Robbins, J.; Dilworth, S.M.; Roberts, B.; Richardson, W.D. The nucleoplasmin nuclear location sequence is larger and more complex than that of SV-40 large T antigen. *J. Cell Biol.* **1988**, *107*, 841–849. [CrossRef] [PubMed]

26. Kelley, J.B.; Talley, A.M.; Spencer, A.; Gioeli, D.; Paschal, B.M. Karyopherin alpha7 (KPNA7), a divergent member of the importin alpha family of nuclear import receptors. *BMC Cell Biol.* **2010**, *11*, 63. [CrossRef]

27. Andrade, M.A.; Perez-Iratxeta, C.; Ponting, C.P. Protein repeats: Structures, functions, and evolution. *J. Struct. Biol.* **2001**, *134*, 117–131. [CrossRef]

28. Sankhala, R.S.; Lokareddy, R.K.; Begum, S.; Pumroy, R.A.; Gillilan, R.E.; Cingolani, G. Three-dimensional context rather than NLS amino acid sequence determines importin alpha subtype specificity for RCC1. *Nat. Commun.* **2017**, *8*, 979. [CrossRef]

29. Kosugi, S.; Hasebe, M.; Matsumura, N.; Takashima, H.; Miyamoto-Sato, E.; Tomita, M.; Yanagawa, H. Six classes of nuclear localization signals specific to different binding grooves of importin alpha. *J. Biol. Chem.* **2009**, *284*, 478–485. [CrossRef]

30. Cingolani, G.; Petosa, C.; Weis, K.; Muller, C.W. Structure of importin-beta bound to the IBB domain of importin-alpha. *Nature* **1999**, *399*, 221–229. [CrossRef]

31. Lee, B.J.; Cansizoglu, A.E.; Suel, K.E.; Louis, T.H.; Zhang, Z.; Chook, Y.M. Rules for nuclear localization sequence recognition by karyopherin beta 2. *Cell* **2006**, *126*, 543–558. [CrossRef] [PubMed]

32. Kressler, D.; Bange, G.; Ogawa, Y.; Stjepanovic, G.; Bradatsch, B.; Pratte, D.; Amlacher, S.; Strauss, D.; Yoneda, Y.; Katahira, J.; et al. Synchronizing nuclear import of ribosomal proteins with ribosome assembly. *Science* **2012**, *338*, 666–671. [CrossRef] [PubMed]

33. Lai, M.C.; Lin, R.I.; Tarn, W.Y. Transportin-SR2 mediates nuclear import of phosphorylated SR proteins. *Proc. Natl. Acad. Sci. USA* **2001**, *98*, 10154–10159. [CrossRef] [PubMed]

34. Baade, I.; Spillner, C.; Schmitt, K.; Valerius, O.; Kehlenbach, R.H. Extensive identification and in-depth validation of importin 13 cargoes. *Mol. Cell. Proteomics* **2018**, *17*, 1337–1353. [CrossRef] [PubMed]

35. Grunwald, M.; Lazzaretti, D.; Bono, F. Structural basis for the nuclear export activity of Importin13. *EMBO J.* **2013**, *32*, 899–913. [CrossRef]

36. Hanover, J.A.; Love, D.C.; Prinz, W.A. Calmodulin-driven nuclear entry: Trigger for sex determination and terminal differentiation. *J. Biol. Chem.* **2009**, *284*, 12593–12597. [CrossRef]

37. Sharma, M.; Jamieson, C.; Lui, C.; Henderson, B.R. Distinct hydrophobic "patches" in the N- and C-tails of beta-catenin contribute to nuclear transport. *Exp. Cell Res.* **2016**, *348*, 132–145. [CrossRef]

38. Kirli, K.; Karaca, S.; Dehne, H.J.; Samwer, M.; Pan, K.T.; Lenz, C.; Urlaub, H.; Gorlich, D. A deep proteomics perspective on CRM1-mediated nuclear export and nucleocytoplasmic partitioning. *eLife* **2015**, *4*, e11466. [CrossRef]

39. Fornerod, M.; Ohno, M.; Yoshida, M.; Mattaj, I.W. CRM1 is an export receptor for leucine-rich nuclear export signals. *Cell* **1997**, *90*, 1051–1060. [CrossRef]

40. Kosugi, S.; Hasebe, M.; Tomita, M.; Yanagawa, H. Nuclear export signal consensus sequences defined using a localization-based yeast selection system. *Traffic* **2008**, *9*, 2053–2062. [CrossRef]

41. Stade, K.; Ford, C.S.; Guthrie, C.; Weis, K. Exportin 1 (Crm1p) is an essential nuclear export factor. *Cell* **1997**, *90*, 1041–1050. [CrossRef]

42. Andrade, M.A.; Petosa, C.; O'Donoghue, S.I.; Muller, C.W.; Bork, P. Comparison of ARM and HEAT protein repeats. *J. Mol. Biol.* **2001**, *309*, 1–18. [CrossRef] [PubMed]

43. Matsuura, Y. Mechanistic insights from structural analyses of Ran-GTPase-Driven nuclear export of proteins and RNAs. *J. Mol. Biol.* **2016**, *428*, 2025–2039. [CrossRef] [PubMed]

44. Kohler, A.; Hurt, E. Exporting RNA from the nucleus to the cytoplasm. *Nat. Rev. Mol. Cell Biol.* **2007**, *8*, 761–773. [CrossRef]

45. Kutay, U.; Bischoff, F.R.; Kostka, S.; Kraft, R.; Gorlich, D. Export of importin alpha from the nucleus is mediated by a specific nuclear transport factor. *Cell* **1997**, *90*, 1061–1071. [CrossRef]

46. Lipowsky, G.; Bischoff, F.R.; Schwarzmaier, P.; Kraft, R.; Kostka, S.; Hartmann, E.; Kutay, U.; Gorlich, D. Exportin 4: A mediator of a novel nuclear export pathway in higher eukaryotes. *EMBO J.* **2000**, *19*, 4362–4371. [CrossRef]

47. Kurisaki, A.; Kurisaki, K.; Kowanetz, M.; Sugino, H.; Yoneda, Y.; Heldin, C.H.; Moustakas, A. The mechanism of nuclear export of Smad3 involves exportin 4 and Ran. *Mol. Cell. Biol.* **2006**, *26*, 1318–1332. [CrossRef]

48. Gontan, C.; Guttler, T.; Engelen, E.; Demmers, J.; Fornerod, M.; Grosveld, F.G.; Tibboel, D.; Gorlich, D.; Poot, R.A.; Rottier, R.J. Exportin 4 mediates a novel nuclear import pathway for Sox family transcription factors. *J. Cell Biol.* **2009**, *185*, 27–34. [CrossRef]

49. Mingot, J.M.; Bohnsack, M.T.; Jakle, U.; Gorlich, D. Exportin 7 defines a novel general nuclear export pathway. *EMBO J.* **2004**, *23*, 3227–3236. [CrossRef]

50. Aksu, M.; Pleiner, T.; Karaca, S.; Kappert, C.; Dehne, H.J.; Seibel, K.; Urlaub, H.; Bohnsack, M.T.; Gorlich, D. Xpo7 is a broad-spectrum exportin and a nuclear import receptor. *J. Cell Biol.* **2018**, *217*, 2329–2340. [CrossRef]

51. Nachury, M.V.; Weis, K. The direction of transport through the nuclear pore can be inverted. *Proc. Natl. Acad. Sci. USA* **1999**, *96*, 9622–9627. [CrossRef] [PubMed]

52. Kuersten, S.; Ohno, M.; Mattaj, I.W. Nucleocytoplasmic transport: Ran, beta and beyond. *Trends Cell Biol.* **2001**, *11*, 497–503. [CrossRef]

53. Macara, I.G. Transport into and out of the nucleus. *Microbiol. Mol. Biol. Rev.* **2001**, *65*, 570–594. [CrossRef] [PubMed]

54. Melchior, F.; Gerace, L. Two-way trafficking with Ran. *Trends Cell Biol.* **1998**, *8*, 175–179. [CrossRef]

55. Guttler, T.; Gorlich, D. Ran-dependent nuclear export mediators: A structural perspective. *EMBO J.* **2011**, *30*, 3457–3474. [CrossRef] [PubMed]

56. Frey, S.; Rees, R.; Schunemann, J.; Ng, S.C.; Funfgeld, K.; Huyton, T.; Gorlich, D. Surface properties determining passage rates of proteins through nuclear pores. *Cell* **2018**, *174*, 202–217. [CrossRef]

57. Parker, B.S.; Rautela, J.; Hertzog, P.J. Antitumour actions of interferons: Implications for cancer therapy. *Nat. Rev. Cancer* **2016**, *16*, 131–144. [CrossRef]

58. Meissl, K.; Macho-Maschler, S.; Muller, M.; Strobl, B. The good and the bad faces of STAT1 in solid tumours. *Cytokine* **2017**, *89*, 12–20. [CrossRef]

59. Green, D.S.; Young, H.A.; Valencia, J.C. Current prospects of type II interferon gamma signaling and autoimmunity. *J. Biol. Chem.* **2017**, *292*, 13925–13933. [CrossRef]

60. Sekimoto, T.; Imamoto, N.; Nakajima, K.; Hirano, T.; Yoneda, Y. Extracellular signal-dependent nuclear import of Stat1 is mediated by nuclear pore-targeting complex formation with NPI-1, but not Rch1. *EMBO J.* **1997**, *16*, 7067–7077. [CrossRef]

61. Melen, K.; Fagerlund, R.; Franke, J.; Kohler, M.; Kinnunen, L.; Julkunen, I. Importin alpha nuclear localization signal binding sites for STAT1, STAT2, and influenza A virus nucleoprotein. *J. Biol. Chem.* **2003**, *278*, 28193–28200. [CrossRef] [PubMed]

62. Nardozzi, J.; Wenta, N.; Yasuhara, N.; Vinkemeier, U.; Cingolani, G. Molecular basis for the recognition of phosphorylated STAT1 by importin alpha5. *J. Mol. Biol.* **2010**, *402*, 83–100. [CrossRef] [PubMed]

63. Meissner, T.; Krause, E.; Lodige, I.; Vinkemeier, U. Arginine methylation of STAT1: A reassessment. *Cell* **2004**, *119*, 587–589. [CrossRef] [PubMed]

64. Meyer, T.; Begitt, A.; Lodige, I.; van Rossum, M.; Vinkemeier, U. Constitutive and IFN-gamma-induced nuclear import of STAT1 proceed through independent pathways. *EMBO J.* **2002**, *21*, 344–354. [CrossRef]

65. Fagerlund, R.; Melen, K.; Kinnunen, L.; Julkunen, I. Arginine/lysine-rich nuclear localization signals mediate interactions between dimeric STATs and importin alpha 5. *J. Biol. Chem.* **2002**, *277*, 30072–30078. [CrossRef]

66. McBride, K.M.; McDonald, C.; Reich, N.C. Nuclear export signal located within theDNA-binding domain of the STAT1 transcription factor. *EMBO J.* **2000**, *19*, 6196–6206. [CrossRef]

67. Xu, W.; Edwards, M.R.; Borek, D.M.; Feagins, A.R.; Mittal, A.; Alinger, J.B.; Berry, K.N.; Yen, B.; Hamilton, J.; Brett, T.J.; et al. Ebola virus VP24 targets a unique NLS binding site on karyopherin alpha 5 to selectively compete with nuclear import of phosphorylated STAT1. *Cell Host Microbe* **2014**, *16*, 187–200. [CrossRef]

68. McBride, K.M.; Banninger, G.; McDonald, C.; Reich, N.C. Regulated nuclear import of the STAT1 transcription factor by direct binding of importin-alpha. *EMBO J.* **2002**, *21*, 1754–1763. [CrossRef]

69. Ma, J.; Cao, X. Regulation of Stat3 nuclear import by importin alpha5 and importin alpha7 via two different functional sequence elements. *Cell. Signal.* **2006**, *18*, 1117–1126. [CrossRef]

70. Ushijima, R.; Sakaguchi, N.; Kano, A.; Maruyama, A.; Miyamoto, Y.; Sekimoto, T.; Yoneda, Y.; Ogino, K.; Tachibana, T. Extracellular signal-dependent nuclear import of STAT3 is mediated by various importin alphas. *Biochem. Biophys. Res. Commun.* **2005**, *330*, 880–886. [CrossRef]

71. Liu, L.; McBride, K.M.; Reich, N.C. STAT3 nuclear import is independent of tyrosine phosphorylation and mediated by importin-alpha3. *Proc. Natl. Acad. Sci. USA* **2005**, *102*, 8150–8155. [CrossRef] [PubMed]

72. Martincuks, A.; Fahrenkamp, D.; Haan, S.; Herrmann, A.; Kuster, A.; Muller-Newen, G. Dissecting functions of the N-terminal domain and GAS-site recognition in STAT3 nuclear trafficking. *Cell. Signal.* **2016**, *28*, 810–825. [CrossRef] [PubMed]

73. Shin, H.Y.; Reich, N.C. Dynamic trafficking of STAT5 depends on an unconventional nuclear localization signal. *J. Cell Sci.* **2013**, *126*, 3333–3343. [CrossRef] [PubMed]

74. Strehlow, I.; Schindler, C. Amino-terminal signal transducer and activator of transcription (STAT) domains regulate nuclear translocation and STAT deactivation. *J. Biol. Chem.* **1998**, *273*, 28049–28056. [CrossRef]

75. Vogt, M.; Domoszlai, T.; Kleshchanok, D.; Lehmann, S.; Schmitt, A.; Poli, V.; Richtering, W.; Muller-Newen, G. The role of the N-terminal domain in dimerization and nucleocytoplasmic shuttling of latent STAT3. *J. Cell Sci.* **2011**, *124*, 900–909. [CrossRef]

76. Iyer, J.; Reich, N.C. Constitutive nuclear import of latent and activated STAT5a by its coiled coil domain. *FASEB J.* **2008**, *22*, 391–400. [CrossRef]

77. Zeng, R.; Aoki, Y.; Yoshida, M.; Arai, K.; Watanabe, S. Stat5B shuttles between cytoplasm and nucleus in a cytokine-dependent and -independent manner. *J. Immunol.* **2002**, *168*, 4567–4575. [CrossRef]

78. Begitt, A.; Meyer, T.; van Rossum, M.; Vinkemeier, U. Nucleocytoplasmic translocation of Stat1 is regulated by a leucine-rich export signal in the coiled-coil domain. *Proc. Natl. Acad. Sci. USA* **2000**, *97*, 10418–10423. [CrossRef]

79. Bhattacharya, S.; Schindler, C. Regulation of Stat3 nuclear export. *J. Clin. Investig.* **2003**, *111*, 553–559. [CrossRef]

80. Bromberg, J.F.; Wrzeszczynska, M.H.; Devgan, G.; Zhao, Y.; Pestell, R.G.; Albanese, C.; Darnell, J.E., Jr. Stat3 as an oncogene. *Cell* **1999**, *98*, 295–303. [CrossRef]

81. Pilati, C.; Zucman-Rossi, J. Mutations leading to constitutive active gp130/JAK1/STAT3 pathway. *Cytokine Growth Factor Rev.* **2015**, *26*, 499–506. [CrossRef] [PubMed]

82. Yu, H.; Lee, H.; Herrmann, A.; Buettner, R.; Jove, R. Revisiting STAT3 signalling in cancer: New and unexpected biological functions. *Nat. Rev. Cancer* **2014**, *14*, 736–746. [CrossRef] [PubMed]

83. Aigner, P.; Mizutani, T.; Horvath, J.; Eder, T.; Heber, S.; Lind, K.; Just, V.; Moll, H.P.; Yeroslaviz, A.; Fischer, M.J.M.; et al. STAT3beta is a tumor suppressor in acute myeloid leukemia. *Blood Adv.* **2019**, *3*, 1989–2002. [CrossRef]

84. Pencik, J.; Schlederer, M.; Gruber, W.; Unger, C.; Walker, S.M.; Chalaris, A.; Marie, I.J.; Hassler, M.R.; Javaheri, T.; Aksoy, O.; et al. STAT3 regulated ARF expression suppresses prostate cancer metastasis. *Nat. Commun.* **2015**, *6*, 7736. [CrossRef] [PubMed]

85. Fagard, R.; Metelev, V.; Souissi, I.; Baran-Marszak, F. STAT3 inhibitors for cancer therapy: Have all roads been explored? *JAKSTAT* **2013**, *2*, e22882. [CrossRef]

86. Meyer, T.; Vinkemeier, U. STAT nuclear translocation: Potential for pharmacological intervention. *Expert Opin. Ther. Targets* **2007**, *11*, 1355–1365. [CrossRef]

87. Kohler, M.; Ansieau, S.; Prehn, S.; Leutz, A.; Haller, H.; Hartmann, E. Cloning of two novel human importin-alpha subunits and analysis of the expression pattern of the importin-alpha protein family. *FEBS Lett.* **1997**, *417*, 104–108. [CrossRef]

88. Kawashima, T.; Bao, Y.C.; Minoshima, Y.; Nomura, Y.; Hatori, T.; Hori, T.; Fukagawa, T.; Fukada, T.; Takahashi, N.; Nosaka, T.; et al. A Rac GTPase-activating protein, MgcRacGAP, is a nuclear localizing signal-containing nuclear chaperone in the activation of STAT transcription factors. *Mol. Cell. Biol.* **2009**, *29*, 1796–1813. [CrossRef]

89. van Adrichem, A.J.; Wennerberg, K. MgcRacGAP inhibition stimulates JAK-dependent STAT3 activity. *FEBS Lett.* **2015**, *589*, 3859–3865. [CrossRef]

90. Martincuks, A.; Andryka, K.; Kuster, A.; Schmitz-Van de Leur, H.; Komorowski, M.; Muller-Newen, G. Nuclear translocation of STAT3 and NF-kappaB are independent of each other but NF-kappaB supports expression and activation of STAT3. *Cell. Signal.* **2017**, *32*, 36–47. [CrossRef]

91. Meyer, T.; Marg, A.; Lemke, P.; Wiesner, B.; Vinkemeier, U. DNA binding controls inactivation and nuclear accumulation of the transcription factor Stat1. *Genes Dev.* **2003**, *17*, 1992–2005. [CrossRef] [PubMed]

92. Nishimoto, A.; Yu, Y.; Lu, Z.; Mao, X.; Ren, Z.; Watowich, S.S.; Mills, G.B.; Liao, W.S.; Chen, X.; Bast, R.C., Jr.; et al. A Ras homologue member I directly inhibits signal transducers and activators of transcription 3 translocation and activity in human breast and ovarian cancer cells. *Cancer Res.* **2005**, *65*, 6701–6710. [CrossRef] [PubMed]

93. Mohr, A.; Chatain, N.; Domoszlai, T.; Rinis, N.; Sommerauer, M.; Vogt, M.; Muller-Newen, G. Dynamics and non-canonical aspects of JAK/STAT signalling. *Eur. J. Cell Biol.* **2012**, *91*, 524–532. [CrossRef] [PubMed]

94. Pranada, A.L.; Metz, S.; Herrmann, A.; Heinrich, P.C.; Muller-Newen, G. Real time analysis of STAT3 nucleocytoplasmic shuttling. *J. Biol. Chem.* **2004**, *279*, 15114–15123. [CrossRef] [PubMed]

95. Meyer, T.; Gavenis, K.; Vinkemeier, U. Cell type-specific and tyrosine phosphorylation-independent nuclear presence of STAT1 and STAT3. *Exp. Cell Res.* **2002**, *272*, 45–55. [CrossRef] [PubMed]

96. Marg, A.; Shan, Y.; Meyer, T.; Meissner, T.; Brandenburg, M.; Vinkemeier, U. Nucleocytoplasmic shuttling by nucleoporins Nup153 and Nup214 and CRM1-dependent nuclear export control the subcellular distribution of latent Stat1. *J. Cell Biol.* **2004**, *165*, 823–833. [CrossRef] [PubMed]

97. Braunstein, J.; Brutsaert, S.; Olson, R.; Schindler, C. STATs dimerize in the absence of phosphorylation. *J. Biol. Chem.* **2003**, *278*, 34133–34140. [CrossRef]

98. Haan, S.; Kortylewski, M.; Behrmann, I.; Muller-Esterl, W.; Heinrich, P.C.; Schaper, F. Cytoplasmic STAT proteins associate prior to activation. *Biochem. J.* **2000**, *345 Pt 3*, 417–421. [CrossRef]

99. Wenta, N.; Strauss, H.; Meyer, S.; Vinkemeier, U. Tyrosine phosphorylation regulates the partitioning of STAT1 between different dimer conformations. *Proc. Natl. Acad. Sci. USA* **2008**, *105*, 9238–9243. [CrossRef]

100. Domoszlai, T.; Martincuks, A.; Fahrenkamp, D.; Schmitz-Van de Leur, H.; Kuster, A.; Muller-Newen, G. Consequences of the disease-related L78R mutation for dimerization and activity of STAT3. *J. Cell Sci.* **2014**, *127*, 1899–1910. [CrossRef]

101. Majoros, A.; Platanitis, E.; Kernbauer-Holzl, E.; Rosebrock, F.; Muller, M.; Decker, T. Canonical and non-canonical aspects of JAK-STAT signaling: Lessons from interferons for cytokine responses. *Front. Immunol.* **2017**, *8*, 29. [CrossRef] [PubMed]

102. Wingelhofer, B.; Neubauer, H.A.; Valent, P.; Han, X.; Constantinescu, S.N.; Gunning, P.T.; Muller, M.; Moriggl, R. Implications of STAT3 and STAT5 signaling on gene regulation and chromatin remodeling in hematopoietic cancer. *Leukemia* **2018**, *32*, 1713–1726. [CrossRef] [PubMed]

103. Ferbeyre, G.; Moriggl, R. The role of Stat5 transcription factors as tumor suppressors or oncogenes. *Biochim. Biophys. Acta* **2011**, *1815*, 104–114. [CrossRef] [PubMed]

104. Herrington, J.; Rui, L.; Luo, G.; Yu-Lee, L.Y.; Carter-Su, C. A functional DNA binding domain is required for growth hormone-induced nuclear accumulation of Stat5B. *J. Biol. Chem.* **1999**, *274*, 5138–5145. [CrossRef] [PubMed]

105. Lillemeier, B.F.; Koster, M.; Kerr, I.M. STAT1 from the cell membrane to the DNA. *EMBO J.* **2001**, *20*, 2508–2517. [CrossRef]

106. Park, H.J.; Li, J.; Hannah, R.; Biddie, S.; Leal-Cervantes, A.I.; Kirschner, K.; Flores Santa Cruz, D.; Sexl, V.; Gottgens, B.; Green, A.R. Cytokine-induced megakaryocytic differentiation is regulated by genome-wide loss of a uSTAT transcriptional program. *EMBO J.* **2016**, *35*, 580–594. [CrossRef]

107. Kazansky, A.V.; Kabotyanski, E.B.; Wyszomierski, S.L.; Mancini, M.A.; Rosen, J.M. Differential effects of prolactin and src/abl kinases on the nuclear translocation of STAT5B and STAT5A. *J. Biol. Chem.* **1999**, *274*, 22484–22492. [CrossRef]

108. Harir, N.; Pecquet, C.; Kerenyi, M.; Sonneck, K.; Kovacic, B.; Nyga, R.; Brevet, M.; Dhennin, I.; Gouilleux-Gruart, V.; Beug, H.; et al. Constitutive activation of Stat5 promotes its cytoplasmic localization and association with PI3-kinase in myeloid leukemias. *Blood* **2007**, *109*, 1678–1686. [CrossRef]

109. Chatain, N.; Ziegler, P.; Fahrenkamp, D.; Jost, E.; Moriggl, R.; Schmitz-Van de Leur, H.; Muller-Newen, G. Src family kinases mediate cytoplasmic retention of activated STAT5 in BCR-ABL-positive cells. *Oncogene* **2013**, *32*, 3587–3597. [CrossRef]

110. Fahrenkamp, D.; de Leur, H.S.; Kuster, A.; Chatain, N.; Muller-Newen, G. Src family kinases interfere with dimerization of STAT5A through a phosphotyrosine-SH2 domain interaction. *Cell Commun. Signal.* **2015**, *13*, 10. [CrossRef]

111. Berger, A.; Hoelbl-Kovacic, A.; Bourgeais, J.; Hoefling, L.; Warsch, W.; Grundschober, E.; Uras, I.Z.; Menzl, I.; Putz, E.M.; Hoermann, G.; et al. PAK-dependent STAT5 serine phosphorylation is required for BCR-ABL-induced leukemogenesis. *Leukemia* **2014**, *28*, 629–641. [CrossRef] [PubMed]

112. Cagatay, T.; Chook, Y.M. Karyopherins in cancer. *Curr. Opin. Cell Biol.* **2018**, *52*, 30–42. [CrossRef] [PubMed]

113. Mahipal, A.; Malafa, M. Importins and exportins as therapeutic targets in cancer. *Pharmacol. Ther.* **2016**, *164*, 135–143. [CrossRef] [PubMed]

114. Poon, I.K.; Jans, D.A. Regulation of nuclear transport: Central role in development and transformation? *Traffic* **2005**, *6*, 173–186. [CrossRef]

115. Turner, J.G.; Dawson, J.; Cubitt, C.L.; Baz, R.; Sullivan, D.M. Inhibition of CRM1-dependent nuclear export sensitizes malignant cells to cytotoxic and targeted agents. *Semin. Cancer Biol.* **2014**, *27*, 62–73. [CrossRef]

116. Turner, J.G.; Engel, R.; Derderian, J.A.; Jove, R.; Sullivan, D.M. Human topoisomerase IIalpha nuclear export is mediated by two CRM-1-dependent nuclear export signals. *J. Cell Sci.* **2004**, *117*, 3061–3071. [CrossRef]

117. Turner, J.G.; Dawson, J.; Emmons, M.F.; Cubitt, C.L.; Kauffman, M.; Shacham, S.; Hazlehurst, L.A.; Sullivan, D.M. CRM1 inhibition sensitizes drug resistant human myeloma cells to topoisomerase II and proteasome inhibitors both In Vitro and Ex Vivo. *J. Cancer* **2013**, *4*, 614–625. [CrossRef]

118. Hong, A.L.; Tseng, Y.Y.; Cowley, G.S.; Jonas, O.; Cheah, J.H.; Kynnap, B.D.; Doshi, M.B.; Oh, C.; Meyer, S.C.; Church, A.J.; et al. Integrated genetic and pharmacologic interrogation of rare cancers. *Nat. Commun.* **2016**, *7*, 11987. [CrossRef]

119. Kim, J.; McMillan, E.; Kim, H.S.; Venkateswaran, N.; Makkar, G.; Rodriguez-Canales, J.; Villalobos, P.; Neggers, J.E.; Mendiratta, S.; Wei, S.; et al. XPO1-dependent nuclear export is a druggable vulnerability in KRAS-mutant lung cancer. *Nature* **2016**, *538*, 114–117. [CrossRef]

120. Newlands, E.S.; Rustin, G.J.; Brampton, M.H. Phase I trial of elactocin. *Br. J. Cancer* **1996**, *74*, 648–649. [CrossRef]

121. Stelma, T.; Chi, A.; van der Watt, P.J.; Verrico, A.; Lavia, P.; Leaner, V.D. Targeting nuclear transporters in cancer: Diagnostic, prognostic and therapeutic potential. *IUBMB Life* **2016**, *68*, 268–280. [CrossRef] [PubMed]

122. Beg, A.A.; Ruben, S.M.; Scheinman, R.I.; Haskill, S.; Rosen, C.A.; Baldwin, A.S., Jr. I kappa B interacts with the nuclear localization sequences of the subunits of NF-kappa B: A mechanism for cytoplasmic retention. *Genes Dev.* **1992**, *6*, 1899–1913. [CrossRef] [PubMed]

123. Fagerlund, R.; Kinnunen, L.; Kohler, M.; Julkunen, I.; Melen, K. NF-κB is transported into the nucleus by importin α3 and importin α4. *J. Biol. Chem.* **2005**, *280*, 15942–15951. [CrossRef] [PubMed]

124. Echeverria, P.C.; Picard, D. Molecular chaperones, essential partners of steroid hormone receptors for activity and mobility. *Biochim. Biophys. Acta* **2010**, *1803*, 641–649. [CrossRef] [PubMed]

125. Verheugd, P.; Forst, A.H.; Milke, L.; Herzog, N.; Feijs, K.L.; Kremmer, E.; Kleine, H.; Luscher, B. Regulation of NF-kappaB signalling by the mono-ADP-ribosyltransferase ARTD10. *Nat. Commun.* **2013**, *4*, 1683. [CrossRef] [PubMed]

126. Wagstaff, K.M.; Jans, D.A. Importins and beyond: Non-conventional nuclear transport mechanisms. *Traffic* **2009**, *10*, 1188–1198. [CrossRef]

127. Wiechens, N.; Fagotto, F. CRM1- and Ran-independent nuclear export of beta-catenin. *Curr. Biol.* **2001**, *11*, 18–27. [CrossRef]

128. Yokoya, F.; Imamoto, N.; Tachibana, T.; Yoneda, Y. Beta-catenin can be transported into the nucleus in a Ran-unassisted manner. *Mol. Biol. Cell* **1999**, *10*, 1119–1131. [CrossRef]

129. Asally, M.; Yoneda, Y. Beta-catenin can act as a nuclear import receptor for its partner transcription factor, lymphocyte enhancer factor-1 (lef-1). *Exp. Cell Res.* **2005**, *308*, 357–363. [CrossRef]

130. Brunet, A.; Kanai, F.; Stehn, J.; Xu, J.; Sarbassova, D.; Frangioni, J.V.; Dalal, S.N.; DeCaprio, J.A.; Greenberg, M.E.; Yaffe, M.B. 14-3-3 transits to the nucleus and participates in dynamic nucleocytoplasmic transport. *J. Cell Biol.* **2002**, *156*, 817–828. [CrossRef]

131. van Hemert, M.J.; Niemantsverdriet, M.; Schmidt, T.; Backendorf, C.; Spaink, H.P. Isoform-specific differences in rapid nucleocytoplasmic shuttling cause distinct subcellular distributions of 14-3-3 sigma and 14-3-3 zeta. *J. Cell Sci.* **2004**, *117*, 1411–1420. [CrossRef] [PubMed]

132. Muthu, K.; Panneerselvam, M.; Topno, N.S.; Jayaraman, M.; Ramadas, K. Structural perspective of ARHI mediated inhibition of STAT3 signaling: An insight into the inactive to active transition of ARHI and its interaction with STAT3 and importinbeta. *Cell. Signal.* **2015**, *27*, 739–755. [CrossRef] [PubMed]

133. Timofeeva, O.A.; Tarasova, N.I.; Zhang, X.; Chasovskikh, S.; Cheema, A.K.; Wang, H.; Brown, M.L.; Dritschilo, A. STAT3 suppresses transcription of proapoptotic genes in cancer cells with the involvement of its N-terminal domain. *Proc. Natl. Acad. Sci. USA* **2013**, *110*, 1267–1272. [CrossRef] [PubMed]

134. Hu, T.; Yeh, J.E.; Pinello, L.; Jacob, J.; Chakravarthy, S.; Yuan, G.C.; Chopra, R.; Frank, D.A. Impact of the N-Terminal Domain of STAT3 in STAT3-Dependent Transcriptional Activity. *Mol. Cell. Biol.* **2015**, *35*, 3284–3300. [CrossRef]

135. Haspel, R.L.; Darnell, J.E., Jr. A nuclear protein tyrosine phosphatase is required for the inactivation of Stat1. *Proc. Natl. Acad. Sci. USA* **1999**, *96*, 10188–10193. [CrossRef]

136. Aoki, N.; Matsuda, T. A nuclear protein tyrosine phosphatase TC-PTP is a potential negative regulator of the PRL-mediated signaling pathway: Dephosphorylation and deactivation of signal transducer and activator of transcription 5a and 5b by TC-PTP in nucleus. *Mol. Endocrinol.* **2002**, *16*, 58–69. [CrossRef]

137. ten Hoeve, J.; de Jesus Ibarra-Sanchez, M.; Fu, Y.; Zhu, W.; Tremblay, M.; David, M.; Shuai, K. Identification of a nuclear Stat1 protein tyrosine phosphatase. *Mol. Cell. Biol.* **2002**, *22*, 5662–5668. [CrossRef]

138. Yamamoto, T.; Sekine, Y.; Kashima, K.; Kubota, A.; Sato, N.; Aoki, N.; Matsuda, T. The nuclear isoform of protein-tyrosine phosphatase TC-PTP regulates interleukin-6-mediated signaling pathway through STAT3 dephosphorylation. *Biochem. Biophys. Res. Commun.* **2002**, *297*, 811–817. [CrossRef]

139. Lodige, I.; Marg, A.; Wiesner, B.; Malecova, B.; Oelgeschlager, T.; Vinkemeier, U. Nuclear export determines the cytokine sensitivity of STAT transcription factors. *J. Biol. Chem.* **2005**, *280*, 43087–43099. [CrossRef]

140. Herrmann, A.; Vogt, M.; Monnigmann, M.; Clahsen, T.; Sommer, U.; Haan, S.; Poli, V.; Heinrich, P.C.; Muller-Newen, G. Nucleocytoplasmic shuttling of persistently activated STAT3. *J. Cell Sci.* **2007**, *120*, 3249–3261. [CrossRef]

141. Cheng, Y.; Holloway, M.P.; Nguyen, K.; McCauley, D.; Landesman, Y.; Kauffman, M.G.; Shacham, S.; Altura, R.A. XPO1 (CRM1) inhibition represses STAT3 activation to drive a survivin-dependent oncogenic switch in triple-negative breast cancer. *Mol. Cancer Ther.* **2014**, *13*, 675–686. [CrossRef] [PubMed]

Review

TYK2: An Upstream Kinase of STATs in Cancer

Katharina Wöss, Natalija Simonović, Birgit Strobl, Sabine Macho-Maschler and Mathias Müller *

Institute of Animal Breeding and Genetics, University of Veterinary Medicine Vienna, A-1210 Vienna, Austria;
Katharina.Woess@vetmeduni.ac.at (K.W.); natalijabozovic@gmail.com (N.S.);
birgit.strobl@vetmeduni.ac.at (B.S.); sabine.macho-maschler@vetmeduni.ac.at (S.M.-M.)
* Correspondence: mathias.mueller@vetmeduni.ac.at

Received: 30 September 2019; Accepted: 2 November 2019; Published: 5 November 2019

Abstract: In this review we concentrate on the recent findings describing the oncogenic potential of the protein tyrosine kinase 2 (TYK2). The overview on the current understanding of TYK2 functions in cytokine responses and carcinogenesis focusses on the activation of the signal transducers and activators of transcription (STAT) 3 and 5. Insight gained from loss-of-function (LOF) gene-modified mice and human patients homozygous for *Tyk2/TYK2*-mutated alleles established the central role in immunological and inflammatory responses. For the description of physiological TYK2 structure/function relationships in cytokine signaling and of overarching molecular and pathologic properties in carcinogenesis, we mainly refer to the most recent reviews. Dysregulated TYK2 activation, aberrant TYK2 protein levels, and gain-of-function (GOF) TYK2 mutations are found in various cancers. We discuss the molecular consequences thereof and briefly describe the molecular means to counteract TYK2 activity under (patho-)physiological conditions by cellular effectors and by pharmacological intervention. For the role of TYK2 in tumor immune-surveillance we refer to the recent Special Issue of Cancers "JAK-STAT Signaling Pathway in Cancer".

Keywords: tyrosine kinase 2; JAK family of protein tyrosine kinases; signal transducer and activator of transcription; cytokine receptor signaling; gain-of-function mutation; tumorigenesis

1. TYK2-Mediated Cytokine Signaling and Activation of STAT3 and STAT5

TYK2 was the first identified member of a family of non-receptor kinases later termed Janus kinases (JAK), which additionally comprises JAK1-3 [1,2]. JAKs are associated with cytokine and growth factor receptors and activate STAT (STAT1-4, STAT5A, STAT5B, STAT6) family members [2,3]. JAKs share four functional domains (from N- to C-terminal): (i) a four-point-one, ezrin, radixin, moesin (FERM) homology domain; (ii) an atypical Src-homology 2 (SH2) domain, both facilitating protein-protein interactions (PPIs); (iii) a kinase-like or pseudokinase (JAK homology (JH) 2) domain negatively regulating the kinase activity; and (iv) a tyrosine kinase (JH1) domain which, upon conformational changes at ligand bound receptors, increases its catalytic activity by trans-/autophosphorylation of its activation loop [2,4].

To date, the requirement for TYK2 in signaling has been shown for numerous cytokines, including distinct interleukin (ILs) and interferons (IFNs), which comprise several subtypes (i.e., type I and III IFNs). The heterodimeric cytokine receptor complexes are composed of four distinct TYK2-associated receptor chains (IFNAR1, IL-12Rβ1, IL-10R2, and IL-13Rα1) and a respective second receptor chain associated either with JAK1 or JAK2, which serves as the signal transducing chain harboring STAT docking sites. Usually, these sites contain critical tyrosine residues that are phosphorylated by JAKs upon receptor complex activation (Figure 1). TYK2 also associates with the gp130 receptor chain, yet there is no evidence that gp130-utilizing cytokines rely on TYK2 for signal transduction [5,6]. Note that comprehensive reviews [2,7] provide lists of various other receptors utilizing TYK2-STAT signaling; however, TYK2-STAT activation/utilization is frequently only biochemically assessed by phosphorylation of critical tyrosine residues and cannot be put on a level with dissected downstream cellular activities. Here we review the cytokines which clearly transduce the TYK2 phosphorylation events into downstream physiological changes (Figure 1).

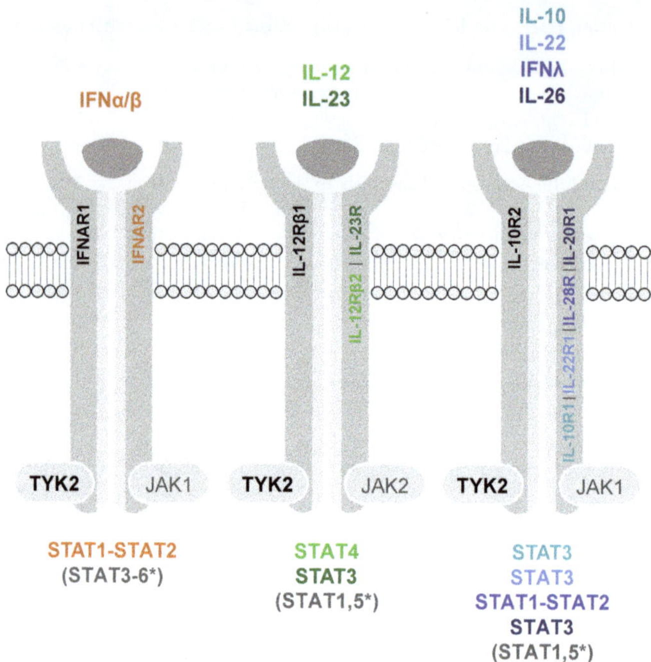

Figure 1. Cytokine receptor families signaling with the participation of TYK2 and JAK1 or JAK2. Cytokines are depicted only upon appearance in humans and mice and proof of TYK2 dependency. The color codes indicate the major STAT(s) activated by the respective cytokines. STAT1-STAT2 heterodimers combine with IFN regulatory factor (IRF) 9 and form the interferon-stimulated gene factor 3 (ISGF3) complex; * STAT activation is dependent on cell type or of less clear biological relevance.

The biological relevance for TYK2-dependent cytokines activating STAT3 is best established for the IL-10R2 utilizing IL-22 [8,9] and the IL-12Rβ1-utilizing IL-12 and IL-23 [10–12]. IL-22 is a central cytokine in tissue-barrier function, wound healing, and epithelial homeostasis and repair. Cancer promoting, as well as restraining, functions were described [13,14]. IL-23 is a key mediator of inflammation, bridges innate and adaptive immune responses, and is known to support tumorigenesis and metastasis [15,16]. IL-12 is central in promoting cell-mediated immunity to infection and cancer [12]. However, this anti-carcinogenic function can be counteracted by IL-12-STAT3-promoted production of pro-carcinogenic IL-23 [17]. While STAT3 is activated by type I and III IFN stimulation in various cell types, its biological functions in the IFN responses are less clear. Growing evidence suggests that STAT3 is a negative regulator of type I IFN activities, thereby providing a pro-viral and pro-survival cellular program [18]; there is, however, also a report on an opposite, i.e., anti-viral activity of STAT3 [19]. The role of TYK2 in IL-10 signaling through STAT3 is not entirely clear and may be cell type- or context-dependent [6]. The double-edged role of IL-10 in immunity and cancer is reviewed elsewhere [9,20]. IL-19, IL-20, IL-24, and IL-26 (absent in mice) constitute a subfamily within the IL-10 cytokine family and signal primarily through activation of STAT3 [9]. Activation of TYK2 at the respective receptors has not been formally shown but can be inferred from the receptor-chain composition. As this subfamily constitutes relatively recently discovered cytokines, cellular responses are still poorly defined, and we refer to recent publications and reviews for a potential cancer connection [21–23]. Lastly, without specification of the cytokines involved, TYK2 via STAT3 was reported to be crucial for the mediation of cell death in an auto-inflammatory context [24].

STAT5, in contrast, is not among the primarily activated STATs downstream of TYK2 (Figure 1) and occurs dependently on cell type and differentiation stage, in response to type I and III IFNs [25,26], IL-10R2-, and IL-12Rβ1-receptor family cytokines [9,12]. Neither a cytokine-TYK2-STAT5 axis nor its significance have been established under physiological conditions.

2. Aberrant Expression and/or Activity of TYK2 in Cancers

The JAK-STAT pathway is recognized as a core cancer pathway [27] and directly contributes to all hallmarks of cancer [28]. Oncogenic JAK activity can originate from aberrant JAK expression, deregulated upstream signals, GOF mutations, or generation of fusion proteins, as well as loss of negative feedback regulation [2,29–31]. Initially, cancer research focused on JAK1-3, while the TYK2 impact on disease was predominantly studied in inflammatory and (auto-)immune diseases [32,33]. Table 1 summarizes the literature on constitutive or hyperactivated TYK2, as well as GOF-mutated TYK2 and the resulting activation of STATs in cancers.

2.1. Aberrant TYK2 Levels

In vitro studies with overexpressed JAKs revealed that aberrant TYK2 levels lead to cellular transformation with constitutive phosphorylation of STAT3 [34]. An unusually high expression of TYK2 associated with or causative for carcinogenesis (reviewed [35]) was described for various cancer cell lines and samples from patients suffering from prostate [36,37], ovarian [38], cervical [39], and breast cancer [40,41], as well as malignant peripheral nerve sheath tumors (MPNST) [42,43]. Conflictingly, lowered TYK2 levels in tumor samples and sections (tumor cells and stroma) are generally considered to be an unfavorable prognostic marker (e.g., [44], www.proteinatlas.org). This is supported by a recently published meta-analysis of JAKs and STATs in hepatocellular carcinoma (HCC) patients, where normal or higher TYK2 levels correlated with longer survival and were found in healthy tissue [45]. The underlying reason for these conflicting reports may be attributed to the anti-proliferative/pro-apoptotic and/or tumor surveillance properties of TYK2 [5], as well as the undetermined tumor cell intrinsic and extrinsic state of TYK2. The important role of TYK2 in immune-surveillance is also in line with findings in patients who carry mutated TYK2 alleles which lead to loss of TYK2, lowered TYK2 levels [46], or expression of kinase-inactive TYK2 [47–49], and that show primarily immunodeficiencies. Nonetheless, proteomics suggested that low TYK2 facilitates local metastasis in breast cancer [50], and a comprehensive screen for protein tyrosine kinase variants in numerous cancer cell lines identified splice variants that render TYK2 inactive [51]. On a molecular mechanistic level, the cell intrinsic tumor-promoting consequences of low TYK2 or LOF of TYK2 currently remain elusive.

2.2. Aberrant Activation of TYK2

A comprehensive list of receptors (over-)expressed in various cancer types which allows us to deduce putative upstream signals involved in hyperactivation of TYK2 was compiled recently [7]. Primary hematological neoplasm (ALCL, anaplastic large cell lymphoma; T-ALL, T cell acute lymphoblastic leukemia) patient samples and cell lines were shown to be dependent on TYK2 activated by upstream IL-10 and/or IL-22 signals and established an upregulation of anti-apoptotic BCL2 family members via STAT1 and/or STAT3 [52,53]. A similar high TYK2-STAT1/3-BCL2 axis was found in MPNST [43]. Cytotoxic T-lymphocyte-associated antigen 4 (CTLA4, CD152) is mainly expressed on T cells and is a well-established immune checkpoint. CTLA4 signaling is initiated through binding to CD80 (B7-1) or CD86 (B7-2) on the surface of antigen-presenting cells (APCs). Ectopic expression of CTLA4 was found on diverse B-cell lymphoma. Mechanistically, it was established that CD86-CTLA4 engagement resulted in recruitment/activation of TYK2, which, in turn, led to a STAT3-driven tumor-promoting transcriptional program [54]. A STAT-independent involvement of activated TYK2 in fibroblast growth factor 2 (FGF-2) mediated escape from drug-induced death was reported for a sarcoma cell line [55].

2.3. TYK2 Mutations

Oncogenic JAK2 with the prominent JAK2^{V617F} mutation found in over 50% of myeloproliferative neoplasia (MPN) patients [56] is the paradigm for the understanding of structure/function relations of JAK activity [57–59] and for the general alertness of the cancer field for mutated JAK family members as potential oncogenes. TYK2 joined the club of GOF-mutated JAKs causative for patient hematopoietic malignancies only recently: In 2013, the first TYK2 GOF point mutations were found in T-ALL cell lines and characterized to have transforming capacity via STAT1 and a BCL2 family member [53]. With respect to biochemical studies, the first GOF mutation of TYK2 was V678F, which is the homologous mutation to JAK2^{V617F} [60,61]. Until now, this mutation was not found in patients. The only mutation reported in a public cancer genome database (www.stjude.cloud) for this residue is the V678L mutation, albeit with unknown structure/function consequences. Point mutations at the TYK2 locus are distributed throughout the whole gene body, with GOF mutations—similar to the other JAKs—primarily accumulating in the JH1 and JH2 domains ([2,5] and see public databases, e.g., Genomic Data Commons of the National Cancer Institute [62,63], Catalogue of Somatic Mutations in Cancer (COSMIC [64], and cBioPortal for Cancer Genomics [65,66]).

In addition to the somatic cancer cell mutations, two GOF TYK2 germline mutations (P760L and G761V) were found in pediatric patients developing several de novo leukemias. These mutations are located in the JH2 pseudokinase domain of TYK2 and are predicted to attenuate the negative regulation on the JH1 kinase domain, leading to constitutively activated TYK2 [67].

A prominent germline TYK2 mutation is P1104A/V, which was first found to be associated with solid and hematopoietic cancers [68,69] and later with immunological and inflammatory disorders (reviewed in [5]). While analyzing MPNST tumor samples, it was proposed that TYK2^{P1104A} is an unfavorable prognostic marker for the disease [42]. Notably, this study solely genotyped the somatic cancer cells and overlooked that this mutation impairs TYK2 catalytic activity; cellular signaling, however, is not completely abrogated, and the detected induction of BCL2 expression might favor an anti-apoptotic program [69,70]. Recent studies show that TYK2^{P1104A} is a LOF mutation, because patients homozygous for this allele are either susceptible to microbial infection or protected from autoimmune disease [47,49,71]. These mechanistic and phenotypic features of TYK2^{P1104A} were confirmed in independent mouse models [48,71].

2.4. TYK2 Fusion Proteins

Chromosomal rearrangements account for a number of driver kinase fusion genes in cancer [72–74]. The first fusion kinase involving a JAK was TEL-JAK2, consisting of a 3′ portion of JAK2 and a 5′ region of TEL, a member of the ETS transcription factor family [75]. This chromosomal translocation is found in T-ALL in patients [75] and transgenic mice expressing TEL-JAK2 develop T-cell leukemia [76]. In vitro studies with a TEL-TYK2 fusion showed constitutive activation of STAT1/3/5 and transforming capacities [77], albeit respective translocations have not yet been identified in patients. As observed for GOF-mutated JAKs, JAK2 kinase fusions occur most frequently compared to the other JAKs, which suggests that the JAK2 locus is a mutation and rearrangement hotspot [56,78,79]. The first leukemia patients carrying TYK2 fusion genes described were combinations of the TYK2 kinase domain and a part of the pseudokinase domain with 5′ portions of nucleophosmin (NPM) 1, polyadenylate binding protein (PABPC) 4, or the transcription factors MYB or NFκB2 [80–82]. Structurally and mechanistically, the TYK2 fusion proteins lack the negatively regulating function of the pseudokinase (JH2) domain leading to a GOF kinase activity and hyperactivity of STAT3 and depending on the cellularity also STAT1 and 5 (reviewed in [5,58,59]).

Subsequent analysis of patient samples and cell lines [83–87] and screening of cancer data sets revealed more than 50 chromosomal *TYK2* rearrangements found mostly in hematological, but also in solid cancers [88]. For the fusions, it is currently not known if they contribute as driver oncogenes to early tumorigenesis or are rather the result of genomic instability at later tumor stages [89]. Recently, chromothripsis was identified as a new type of chromosomal rearrangement during carcinogenesis. Based on a single chromosome-shattering event and DNA repair complex, intra- and interchromosomal rearrangements, such as fusion genes, are produced within a few cell cycles. If the fusion event(s) allow for growth or survival advantages, a cancer driver gene might be generated [90,91]. Chromothripsis was assigned to genomic alterations in childhood cancer [92], and mechanistically it is caused by defects in the nuclear envelope composition or formation and failures during mitosis [93]. It is tempting to speculate that the remarkably high number of described TYK2 fusions were—at least in part—generated through chromothripsis and thus might act as driver mutations.

3. Tumor-Promoting Activities of (Hyper-)Active TYK2

The molecular contribution of TYK2 signaling and known protein–protein interactions to the hallmarks of cancer were reviewed previously [5,28]. Here, we highlight the latest findings on the consequences of TYK2 hyperactivity in cancer cells.

3.1. TYK2 Activation of (Oncogenic) STAT Signaling

As shown in Figure 1, the heterodimeric cytokine receptors with engagement of TYK2 are capable of activating all STATs. Hyperactive, GOF-mutated TYK2 or TYK2 fusions in oncogenic settings preferentially lead to aberrant activation of STAT1, STAT3, and STAT5. The oncogenic potential of STAT3 and STAT5 was recognized early on and is well documented [94,95]. STAT1 was initially considered to exert tumor suppressor functions, and its oncogenic potential emerged more recently [96–98].

STAT1/3/5 were found hyperactivated in patient-tailored cell lines with activated TYK2 [53], as well as carrying somatic or germline TYK2 GOF mutations [53,67] or TYK2-NPM1 and -NFkB2 fusions [80,82]. In other tumor samples or experimental tissue culture settings, STAT3 only, or other dual combinations of activated STAT1/3/5, are described (see Table 1).

Interestingly, TYK2 does not only phosphorylate the major phosphorylation site Y705 in STAT3, but also Y640, which represses STAT3 activation [99]. This phosphorylation site in STAT3 is often mutated in cancers [100,101]. Neither the general (patho-)physiological impact nor the contribution to malignancies of this phosphorylation event are currently known.

Table 1. (Hyper-)active TYK2, GOF-, or LOF-mutated TYK2 and STAT activation in various cancers and cancer cell lines.

TYK2 Status	Disease	Activated STAT	Ref.
Activating somatic mutations (GOF)			
TYK2-G36D; -S47N	T-ALL	STAT1, STAT3	[53] (3) (2) (2 *)
TYK2-731I	T-ALL	STAT1, STAT3, STAT4	[53] (3) (2) (2 *)
TYK2-E957D	T-ALL	STAT1, STAT3, STAT5	[53] (3) (2) (2 *)
TYK2-R1027H	T-ALL	STAT1, STAT3	[53] (3) (2) (2 *)
TYK2-V678F	—	STAT3, STAT5	[61] (2 *)
Inactivating germline mutations (LOF)			
TYK2-P1104A	MPNST	n.d.	[42] (4 *) (1)
TYK2-P1104A	Breast-, colon-, stomach-cancer	n.d.	[68] (1)
TYK2-P1104V	AML	n.d.	[69] (5) (1) (2 *)
Activating germline mutations (GOF)			
TYK2-P760L	B-ALL	STAT1, STAT3, STAT5	[67] (3) (1) (2 *)
TYK2-G761V	T-ALL	STAT1, STAT3, STAT5	[67] (3) (1) (2 *)

Table 1. *Cont.*

TYK2 Status	Disease	Activated STAT	Ref.
Oncogenic fusion proteins (GOF)			
NPM1-TYK2	CD30-positive LPDs	STAT1, STAT3, STAT5	[82] (3) (1) (2) (2 *)
NFkB2-TYK2	ALCL	STAT1, STAT3, STAT5	[80] (3) (1) (2 *)
ELAVL1-TYK2	AML	STAT3, STAT5	[84] (2)
PABPC4-TYK2	ALCL	n.d.	[80] (1)
TEL-TYK2	—	STAT1, STAT3, STAT5	[77] (2 *)
MYB-TYK2	Ph-like ALL	n.d.	[81] (1)
High wildtype TYK2 levels			
TYK2 WT	T-ALL	STAT1, STAT3, STAT4, STAT5	[53] (1) (2) (2 *)
TYK2 WT	ALCL	STAT1, STAT3	[52] (4) (1) (2)
TYK2 WT	Hepatocarcinoma	STAT1, STAT3	[102] (3 *) (2 **)
TYK2 WT	MPNST	STAT1, STAT3	[43] (4) (?) (1) (2)
TYK2 WT	B-cell lymphoma	STAT3	[54] (3 *) (2)
TYK2 WT	Lung cancer	STAT3	[103] (3 *) (1) (2) (2 *)
TYK2 WT	Hepatocarcinoma	STAT3	[104] (3 *) (2 **)
TYK2 WT	Ovarian cancer	STAT3	[38] (3 *) (2)
TYK2 WT	Prostate cancer	n.d.	[36] (4) (1) (2)
TYK2 WT	Prostate cancer	n.d.	[37] (4) (1) (2)
TYK2 WT	Osteosarcoma	no	[55] (3 *) (2)
TYK2 WT	Breast cancer	n.d.	[40,41] (4) (1) (2)
TYK2 WT	Squamous cervical carcinoma	n.d.	[39] (4) (1)
TYK2 WT	MPNST	n.d.	[42] (4) (?) (1)
TYK2 WT	Lung cancer	STAT1	[105] (2 **)
Low wildtype TYK2 levels			
TYK2 WT	Breast cancer (metastatic)	n.d.	[50] (6) (1) (2)

— Unrelated to disease, in vitro findings in stable cell lines, (1) found in patient samples and primary material, (2) in vitro findings endogenous TYK2 expression, (2 *) in vitro findings exogenous TYK2 expression, (2 **) in vitro findings exogenous claudin expression, (3) phosphorylated mutated TYK2 protein, (3 *) phosphorylated wildtype TYK2 protein, (4) high levels of wildtype TYK2, (4 *) high levels of mutated TYK2, (5) reduced levels of phosphorylated mutated TYK2, (6) reduced levels of wildtype TYK2, and (?) not specified if wildtype or mutated TYK2. Please note that some references did not study the activation of all STATs and that not all described STATs in the table are active in all cell systems used. Ph = Philadelphia, and n.d. not determined.

3.2. TYK2 Stimulation of Tumor Cell Invasion

The families of tight junction proteins claudins (CLDNs) and of matrix metalloproteinases (MMPs) are central for the invasion of tumor cells and, in consequence, metastasis formation [106,107]. Recent studies show that, in liver and lung carcinoma, high levels of CLDN9/12/17 caused activation of TYK2 and STAT1/3 and promoted metastasis [102,104,105]. The promoters of various MMP genes harbor STAT binding sites, and many MMPs are transcriptionally activated through TYK2-associated cytokine receptors [108,109]. Gene-targeted mice revealed that TYK2 and STAT1 are required for expression of MMP2/9/14 under inflammatory conditions [110]. Biochemical studies showed that, dependent on context and inflammatory conditions, MMP1/3 induction involves STAT1 alone [108] or also STAT3 [111]. In a hematopoietic tumor TYK2-STAT3 induced MMP9 and tumor cell invasiveness [54] and in a solid tumor TYK2-STAT3 signaling induced MMP1 expression [103].

The urokinase-type plasminogen activator (uPA)/receptor (uPAR) system is central for a cascade of proteolytic events, including activation of MMPs, which allow for tumor cell migration and metastasis [112]. Signaling via uPAR involves TYK2 and PI3K [113], and, at the post-transcriptional level, TYK2 inhibits the accumulation of plasminogen activator inhibitor (PAI) 2 [114]. In prostate cancer, high levels of TYK2 correlate with invasion and metastasis [36,37]. In an ovarian cancer cell line pY-STAT3 co-localizes with TYK2 and JAK2 at focal adhesions, and hyperactive STAT3 was shown to promote cancer cell motility [38]. Without providing molecular details, a mouse model for aggressive lymphoma showed reduced tumor cell invasiveness upon loss of TYK2 [115]. In addition, without providing molecular insights, a siRNA screen assessing the role of the tyrosine kinome in metastasis formation identified TYK2 as a promoter of invadopodia, which are cellular structures characteristic for tumor cell migration [116,117]. Connexin43 (Cx43) is the most widely expressed member of a large family of transmembrane proteins involved in gap junction formation. Cx43 can be both pro- and anti-tumorigenic, e.g., by promoting invasion and metastasis and by acting as a tumor suppressor [118,119]. TYK2 was found to play a dual role in regulation of Cx43: On the one hand, TYK2 is capable of directly phosphorylating Cx43, thereby decreasing its stability; on the other hand, angiotensin II-activated TYK2 increased Cx43 levels in a STAT3-dependent manner [120]. This regulatory loop has not yet been studied in the context of carcinogenesis. Furthermore, knockdown of TYK2 reduced migration of breast cancer cell lines [50].

3.3. TYK2 Prevention of Apoptosis

IFNs in general are capable of promoting apoptosis of cancer cells [121]; hence, provided that IFN stimulus and responsiveness in the tumor is given, TYK2 acts tumor suppressive. Tumor cells are able to resist cell death by upregulation of anti-apoptotic BCL-2 family members [122,123]. TYK2 was shown to drive either in a STAT1- and/or a STAT3-dependent manner or in a STAT-independent but ERK1/2-dependent manner high expression of BCL-2 [43,53,55] or its family members BCL-2L1 [54] and MCL1 [52,55]. In contrast, an in vitro study demonstrated that TYK2 physically interacts with SIVA-1 and promotes SIVA-1 mediated apoptosis, as well as inhibits BCL-2 [124].

3.4. TYK2 Crosstalk to Oncogenes and Proto-Oncogenic Pathways

In a mouse model of ALCL, as well as in patient cells, TYK2 showed co-operativity with the oncogenic fusion kinase NPM-ALK [52]. In contrast, no co-operation of TYK2 with mutated FLT3-ITD or JAK2^{V617F} in MPN mouse models was found [125,126]. The latter is consistent with the observation that, in JAK2^{V617F} MPN patients (see below) resistant to pharmacological JAK2 inhibition, only JAK1, and not TYK2, leads to heterodimeric STAT activation, despite both kinases show equal tyrosine phosphorylation at the activating loop [127]. This is to be expected, since, in contrast to the other JAKs, loss of TYK2 at heterodimeric JAK-associated cytokine receptors leads only to a partial impairment in signaling [5,6], and, as experimentally described for the IFNAR receptor, TYK2 is the subordinated JAK at cytokine receptors [128,129].

Early biochemical studies suggest that, upon type I IFN treatment, TYK2 interacts with various proto-oncogenes, including the guanine nucleotide exchange factor 1 VAV, the E3 ubiquitin-protein ligase C-CBL, and the SRC family tyrosine kinases FYN and LYN [130–134]. The importance of these PPIs for tumorigenesis is currently unknown. In cancer samples or cell lines, TYK2 was found to cooperate with other oncogenic effectors and pathways, such as the RAF/ERK [53,55,61], MAPKs [135], PIM1/2 [84], and PI3K/AKT/mTOR pathway [36,53,61]. Reported solely in the context of skin inflammation is the TYK2-STAT3 requirement for expression of IκBζ (encoded by *NFKBIZ*) [136]; however, emerging reports suggest cell-intrinsic oncogenic, as well as tumor-suppressive, functions of IκBζ [137].

The mapped and predicted PPIs of TYK2 based on proteomics [138,139] and next generation sequencing (NGS) are accessible at various open-source databases (for a review, see [140]). The TYK2 kinase domain and a STAT3-based reporter system were used to establish the first mammalian two hybrid kinase substrate sensor (KISS) screening platform [141,142]. These databases and the screening approaches should be systematically exploited to further define and fine tune the TYK2 interactome in health and disease.

4. Deactivation and Stabilization of TYK2 under (Patho-)Physiological Conditions

JAK activity is counter regulated by molecule-intrinsic events, such as post-translational modifications (PTMs) and the inhibitory function of the pseudokinase domain [143] as well as by extrinsic inhibitory regulators, such as suppressor of cytokine signaling (SOCS) proteins and protein tyrosine phosphatases (PTPs) [144].

Databases [145,146] provide curated PTMs, but with the exception of the well described activating phosphotyrosines, there is still a lack of information on the properties of JAKs that are defined by PTMs. For TYK2, ubiquitination and phosphorylation are detected at multiple residues and discussed in the context of stability/decay (PhosphoSitePlus®, [146]), albeit the (patho-) physiological relevance is unknown.

SOCS proteins are encoded by STAT target genes and are negative feedback inhibitors of JAK signaling. SOCS1 and 3 are the most potent JAK inhibitors because, in addition to recruitment of JAKs to E3 ubiquitination/degradation mediated by all SOCS family members, they also harbor a kinase inhibitory region (KIR), which efficiently shuts down JAK activity by binding to the JH1 domain [147]. Activated JAKs and cytokine receptor chains are dephosphorylated by multiple PTPs [148]. The current literature regarding deactivation of TYK2 by SOCS1/3, the PTPs PTB1B and SHP1, as well as the global impact of SOCS and PTP family members in cancer are reviewed elsewhere [5,149–151].

In vitro studies showed that in hematopoietic tumor cells the PTP SHP1 suppresses growth via accelerating the TYK2 protein degradation [152]. In lung cancer cells, overexpression of the E3 ubiquitin ligase seven-in-absentia-2 (SIAH2) accelerates the proteasomal degradation of TYK2, thereby attenuating STAT3 signaling [103].

HSP90 is a chaperone supporting folding, stability, and function of many client proteins, including JAKs and STATs [153–155]. Cancer cells frequently use HSP90 to stabilize and/or increase the function of numerous oncogenes, and HSP90 inhibitors have been studied as anticancer drugs for more than two decades [156,157]. Physical interaction of HSP90 with TYK2 was demonstrated in cancer cell lines and confirmed in a proteome-wide assessment of the HSP90 interactome [158,159]. HSP90 inhibitor treatments in various tumor settings showed beneficial effects by reducing the activity of TYK2 or its fusion proteins [158,160,161].

An emerging field is the involvement of noncoding RNAs in the regulation of the JAK-STAT pathway in carcinogenesis [162–164]. Recently, the long noncoding RNA (lncRNA) MEG3 in concert with a microRNA (miR-147) was reported to modulate JAK-STAT signaling in chronic myeloid leukemia (CML). Interestingly, the lncRNA was found to physically interact with TYK2, JAK2, and STAT3, thereby diminishing the activity level of STAT3 (and STAT5) [165].

5. Pharmaceutical TYK2 Inhibition

The first selective JAK inhibitor (JAKinib) to be tested in humans was tofacitinib, which potently inhibits JAK3 and JAK1, and, to a lesser extent JAK2, and has little effect on TYK2 [166]. Historically, JAKinibs were developed as immunosuppressive drugs for the clinical use in organ transplants and autoimmune diseases [167]. The success story of ruxolitinib, a JAK2 and JAK1 inhibitor which was the first JAKinib approved for treatment of a hematopoietic malignancy, pushed the perception of JAKinibs as anticancer drugs [168,169]. For insight in development and clinical use, as well as side effects of JAKinibs, we refer to the most recent reviews [170–173].

TYK2inibs are mainly envisaged as therapeutics for treatment of autoimmune and inflammatory diseases [33,174], in which JAKinib selectivity is currently considered not to be of utmost importance [175]. As for the other JAKinibs, the first generation TYK2inibs are directed to the JH1 domain and compete with ATP in binding to the enzymatic pocket. These inhibitors are potent in inhibiting wildtype (overexpressed) TYK2, mutated (hyperactive) TYK2, and TYK2 fusion proteins harboring the JH1 domain. Since the JAKs show high homology in the JH1 domain, it is hard to develop ATP-competing inhibitors with high selectivity for one particular JAK family member [170,172]. A next-generation inhibitor of TYK2 is directed against the JH2 domain and recently passed the phase II clinical trial for psoriasis treatment [176]. A comprehensive report on the high selectivity and the biological effects of this TYK2inib in mouse models, as well as its efficacy in human cells collected from autoimmune patients, was recently published [177]. JH2-specific TYK2inibs are currently further improved, and additional compounds are being developed [178–181]. The only TYK2inib reported and successfully tested to block TYK2 activity in an oncogenic setting is a JH1-specific TYK2inib [135].

Notably, JH2 domain inhibitors might not be working for treatment of diseases driven by TYK2 fusion genes missing parts of the JH2 domain.

6. Conclusions and Future Perspectives

Since the discovery of TYK2 and the JAK-STAT signaling paradigm in the early 1990s, enormous progress has been made in the structural and functional understanding of the linear JAK-STAT axis and the crosstalk of JAKs or STATs to other signaling hubs, as well as the cell type-specific contributions of JAKs and STATs in health and disease. The striking phenotypical similarities between mouse models deficient for TYK2 or engineered to express kinase-inactive TYK2 and human patients carrying the respective germline mutations established TYK2 as a fundamental component in both innate and adaptive immunity. The (patho-)physiological and molecular pathway similarities of TYK2 in human and mice allow for highly informative comparative biomedical studies and efficient translation of basic molecular insights into clinical applications. The use of TYK2inibs in the treatment of immunological and inflammatory diseases is within reach [182] and is also attractive for malignancies with the involvement of hyperactivated TYK2. The role of TYK2 and GOF-mutated TYK2 upstream of oncogenic STAT3—and, less frequently, STAT1—is established, while, up to now, no mechanistic evidence for an oncogenic TYK2-STAT5 axis is given. Mouse models as genetic mimics of kinase-inhibited TYK2 exist [48,71,183,184] and are currently exploited to further dissect the kinase-dependent from the scaffolding functions of TYK2.

In a short-term perspective, work should concentrate on the use of refined TYK2 mouse models that allow studying the kinase-independent and cell type-specific functions, in order to fully in vivo assess TYK2inibs with respect to their benefits and unwanted side effects. Mouse models to study the consequences of aberrant high TYK2 and GOF-mutated TYK2 are underway (K. Wöss, T. Rülicke et al., unpublished). For pharmacological intervention with oncogenic TYK2, TYK2inibs with the highest possible selectivity are required, and efforts should focus on the further development and in vivo testing of these next-generation TYK2inibs.

In a long-term perspective, the further understanding of the TYK2 function requires the in-depth elucidation of the PTMs and the interactome of TYK2 under spatiotemporal conditions. Additionally, computational modelling and structure predictions (e.g., [185]) should complement the attempts to determine the holo-crystal structure of TYK2 and to use high-resolution imaging (e.g., [186]) to gain insight into the structural features of full-length wildtype and mutated TYK2, as well as its conformation bound to various cytokine receptors.

Author Contributions: K.W., S.M.-M., and M.M. designed the draft of the review and performed the literature search. K.W. and S.M.-M. compiled the table, and N.S. designed the figure. S.M.-M., M.M., N.S., and B.S. provided the final version of the manuscript.

Funding: This work was funded by the Austrian Science Fund FWF DK W1212, SFB F6101 and F6106, and DocFund DOC32-B28. We are thankful to Tanja Bulat for critically reading the manuscript.

Conflicts of Interest: The authors declare no conflicts of interest.

References

1. Krolewski, J.J.; Lee, R.; Eddy, R.; Shows, T.B.; Dalla-Favera, R. Identification and chromosomal mapping of new human tyrosine kinase genes. *Oncogene* **1990**, *5*, 277–282. [PubMed]

2. Hammaren, H.M.; Virtanen, A.T.; Raivola, J.; Silvennoinen, O. The regulation of JAKs in cytokine signaling and its breakdown in disease. *Cytokine* **2019**, *118*, 48–63. [CrossRef] [PubMed]

3. Stark, G.R.; Darnell, J.E. The JAK-STAT Pathway at Twenty. *Immunity* **2012**, *36*, 503–514. [CrossRef] [PubMed]

4. Ferrao, R.; Lupardus, P.J. The Janus Kinase (JAK) FERM and SH2 Domains: Bringing Specificity to JAK-Receptor Interactions. *Front. Endocrinol.* **2017**, *8*, 71. [CrossRef]

5. Leitner, N.R.; Witalisz-Siepracka, A.; Strobl, B.; Muller, M. Tyrosine kinase 2-Surveillant of tumours and bona fide oncogene. *Cytokine* **2017**, *89*, 209–218. [CrossRef]

6. Strobl, B.; Stoiber, D.; Sexl, V.; Mueller, M. Tyrosine kinase 2 (TYK2) in cytokine signalling and host immunity. *Front. Biosci.* **2011**, *16*, 3214–3232. [CrossRef]

7. Bousoik, E.; Montazeri Aliabadi, H. "Do We Know Jack" About JAK? A Closer Look at JAK/STAT Signaling Pathway. *Front. Oncol.* **2018**, *8*, 287. [CrossRef]

8. Dudakov, J.A.; Hanash, A.M.; van den Brink, M.R. Interleukin-22: Immunobiology and pathology. *Annu. Rev. Immunol.* **2015**, *33*, 747–785. [CrossRef]

9. Ouyang, W.; O'Garra, A. IL-10 Family Cytokines IL-10 and IL-22: From Basic Science to Clinical Translation. *Immunity* **2019**, *50*, 871–891. [CrossRef]

10. Kastelein, R.A.; Hunter, C.A.; Cua, D.J. Discovery and biology of IL-23 and IL-27: Related but functionally distinct regulators of inflammation. *Annu. Rev. Immunol.* **2007**, *25*, 221–242. [CrossRef]

11. Rutz, S.; Wang, X.; Ouyang, W. The IL-20 subfamily of cytokines–from host defence to tissue homeostasis. *Nat. Rev. Immunol.* **2014**, *14*, 783–795. [CrossRef] [PubMed]

12. Tait Wojno, E.D.; Hunter, C.A.; Stumhofer, J.S. The Immunobiology of the Interleukin-12 Family: Room for Discovery. *Immunity* **2019**, *50*, 851–870. [CrossRef] [PubMed]

13. Hernandez, P.; Gronke, K.; Diefenbach, A. A catch-22: Interleukin-22 and cancer. *Eur. J. Immunol.* **2018**, *48*, 15–31. [CrossRef] [PubMed]

14. Lim, C.; Savan, R. The role of the IL-22/IL-22R1 axis in cancer. *Cytokine Growth Factor Rev.* **2014**, *25*, 257–271. [CrossRef] [PubMed]

15. Huynh, J.; Chand, A.; Gough, D.; Ernst, M. Therapeutically exploiting STAT3 activity in cancer—Using tissue repair as a road map. *Nat. Rev. Cancer* **2019**, *19*, 82–96. [CrossRef]

16. Yan, J.M.; Smyth, M.J.; Teng, M.W.L. Interleukin (IL)-12 and IL-23 and Their Conflicting Roles in Cancer. *Cold Spring Harb. Perspect. Biol.* **2018**, *10*, a028530. [CrossRef]

17. Kortylewski, M.; Xin, H.; Kujawski, M.; Lee, H.; Liu, Y.; Harris, T.; Drake, C.; Pardoll, D.; Yu, H. Regulation of the IL-23 and IL-12 balance by Stat3 signaling in the tumor microenvironment. *Cancer Cell* **2009**, *15*, 114–123. [CrossRef]

18. Tsai, M.H.; Pai, L.M.; Lee, C.K. Fine-Tuning of Type I Interferon Response by STAT3. *Front. Immunol.* **2019**, *10*, 1448. [CrossRef]

19. Mahony, R.; Gargan, S.; Roberts, K.L.; Bourke, N.; Keating, S.E.; Bowie, A.G.; O'Farrelly, C.; Stevenson, N.J. A novel anti-viral role for STAT3 in IFN-alpha signalling responses. *Cell. Mol. Life Sci.* **2017**, *74*, 1755–1764. [CrossRef]

20. Mannino, M.H.; Zhu, Z.; Xiao, H.; Bai, Q.; Wakefield, M.R.; Fang, Y. The paradoxical role of IL-10 in immunity and cancer. *Cancer Lett.* **2015**, *367*, 103–107. [CrossRef]

21. Chen, Y.Y.; Li, C.F.; Yeh, C.H.; Chang, M.S.; Hsing, C.H. Interleukin-19 in breast cancer. *Clin. Dev. Immunol.* **2013**, *2013*, 294320. [CrossRef] [PubMed]

22. Niess, J.H.; Hruz, P.; Kaymak, T. The Interleukin-20 Cytokines in Intestinal Diseases. *Front. Immunol.* **2018**, *9*, 1373. [CrossRef] [PubMed]

23. You, W.; Tang, Q.; Zhang, C.; Wu, J.; Gu, C.; Wu, Z.; Li, X. IL-26 promotes the proliferation and survival of human gastric cancer cells by regulating the balance of STAT1 and STAT3 activation. *PLoS ONE* **2013**, *8*, e63588. [CrossRef] [PubMed]

24. Wan, J.; Fu, A.K.; Ip, F.C.; Ng, H.K.; Hugon, J.; Page, G.; Wang, J.H.; Lai, K.O.; Wu, Z.; Ip, N.Y. Tyk2/STAT3 signaling mediates beta-amyloid-induced neuronal cell death: Implications in Alzheimer's disease. *J. Neurosci.* **2010**, *30*, 6873–6881. [CrossRef] [PubMed]

25. Kotenko, S.V. IFN-lambdas. *Curr. Opin. Immunol.* **2011**, *23*, 583–590. [CrossRef] [PubMed]

26. Schreiber, G. The molecular basis for differential type I interferon signaling. *J. Biol. Chem.* **2017**, *292*, 7285–7294. [CrossRef] [PubMed]

27. Vogelstein, B.; Papadopoulos, N.; Velculescu, V.E.; Zhou, S.; Diaz, L.A., Jr.; Kinzler, K.W. Cancer genome landscapes. *Science* **2013**, *339*, 1546–1558. [CrossRef]

28. Hanahan, D.; Weinberg, R.A. Hallmarks of cancer: The next generation. *Cell* **2011**, *144*, 646–674. [CrossRef]

29. Chen, E.; Staudt, L.M.; Green, A.R. Janus kinase deregulation in leukemia and lymphoma. *Immunity* **2012**, *36*, 529–541. [CrossRef]

30. Groner, B.; von Manstein, V. Jak Stat signaling and cancer: Opportunities, benefits and side effects of targeted inhibition. *Mol. Cell. Endocrinol.* **2017**, *451*, 1–14. [CrossRef]

31. Thomas, S.J.; Snowden, J.A.; Zeidler, M.P.; Danson, S.J. The role of JAK/STAT signalling in the pathogenesis, prognosis and treatment of solid tumours. *Br. J. Cancer* **2015**, *113*, 365–371. [CrossRef] [PubMed]

32. O'Shea, J.J.; Holland, S.M.; Staudt, L.M. JAKs and STATs in immunity, immunodeficiency, and cancer. *N. Engl. J. Med.* **2013**, *368*, 161–170. [CrossRef] [PubMed]

33. Schwartz, D.M.; Kanno, Y.; Villarino, A.; Ward, M.; Gadina, M.; O'Shea, J.J. JAK inhibition as a therapeutic strategy for immune and inflammatory diseases. *Nat. Rev. Drug Discov.* **2017**. [CrossRef] [PubMed]

34. Knoops, L.; Hornakova, T.; Royer, Y.; Constantinescu, S.N.; Renauld, J.C. JAK kinases overexpression promotes in vitro cell transformation. *Oncogene* **2008**, *27*, 1511–1519. [CrossRef] [PubMed]

35. Ubel, C.; Mousset, S.; Trufa, D.; Sirbu, H.; Finotto, S. Establishing the role of tyrosine kinase 2 in cancer. *Oncoimmunology* **2013**, *2*, e22840. [CrossRef]

36. Ide, H.; Nakagawa, T.; Terado, Y.; Kamiyama, Y.; Muto, S.; Horie, S. Tyk2 expression and its signaling enhances the invasiveness of prostate cancer cells. *Biochem. Biophys. Res. Commun.* **2008**, *369*, 292–296. [CrossRef] [PubMed]

37. Santos, J.; Mesquita, D.; Barros-Silva, J.D.; Jeronimo, C.; Henrique, R.; Morais, A.; Paulo, P.; Teixeira, M.R. Uncovering potential downstream targets of oncogenic GRPR overexpression in prostate carcinomas harboring ETS rearrangements. *Oncoscience* **2015**, *2*, 497–507. [CrossRef]

38. Silver, D.L.; Naora, H.; Liu, J.; Cheng, W.; Montell, D.J. Activated signal transducer and activator of transcription (STAT) 3: Localization in focal adhesions and function in ovarian cancer cell motility. *Cancer Res.* **2004**, *64*, 3550–3558. [CrossRef] [PubMed]

39. Zhu, X.; Lv, J.; Yu, L.; Zhu, X.; Wu, J.; Zou, S.; Jiang, S. Proteomic identification of differentially-expressed proteins in squamous cervical cancer. *Gynecol. Oncol.* **2009**, *112*, 248–256. [CrossRef] [PubMed]

40. Christy, J.; Priyadharshini, L. Differential expression analysis of JAK/STAT pathway related genes in breast cancer. *Meta Gene* **2018**, *16*, 122–129. [CrossRef]

41. Song, X.C.; Fu, G.; Yang, X.; Jiang, Z.; Wang, Y.; Zhou, G.W. Protein expression profiling of breast cancer cells by dissociable antibody microarray (DAMA) staining. *Mol. Cell. Proteom.* **2008**, *7*, 163–169. [CrossRef] [PubMed]

42. Hirbe, A.C.; Kaushal, M.; Sharma, M.K.; Dahiya, S.; Pekmezci, M.; Perry, A.; Gutmann, D.H. Clinical genomic profiling identifies TYK2 mutation and overexpression in patients with neurofibromatosis type 1-associated malignant peripheral nerve sheath tumors. *Cancer* **2017**, *123*, 1194–1201. [CrossRef] [PubMed]

43. Qin, W.J.; Godec, A.; Zhang, X.C.; Zhu, C.G.; Shao, J.Y.; Tao, Y.; Bu, X.Z.; Hirlbe, A.C. TYK2 promotes malignant peripheral nerve sheath tumor progression through inhibition of cell death. *Cancer Med.* **2019**, *8*, 5232–5241. [CrossRef] [PubMed]

44. Uhlen, M.; Zhang, C.; Lee, S.; Sjostedt, E.; Fagerberg, L.; Bidkhori, G.; Benfeitas, R.; Arif, M.; Liu, Z.; Edfors, F.; et al. A pathology atlas of the human cancer transcriptome. *Science* **2017**, *357*, eaan2507. [CrossRef]

45. Wang, X.; Liao, X.; Yu, T.; Gong, Y.; Zhang, L.; Huang, J.; Yang, C.; Han, C.; Yu, L.; Zhu, G.; et al. Analysis of clinical significance and prospective molecular mechanism of main elements of the JAK/STAT pathway in hepatocellular carcinoma. *Int. J. Oncol.* **2019**, *55*, 805–822. [CrossRef]

46. Nemoto, M.; Hattori, H.; Maeda, N.; Akita, N.; Muramatsu, H.; Moritani, S.; Kawasaki, T.; Maejima, M.; Ode, H.; Hachiya, A.; et al. Compound heterozygous TYK2 mutations underlie primary immunodeficiency with T-cell lymphopenia. *Sci. Rep.* **2018**, *8*, 6956. [CrossRef]

47. Boisson-Dupuis, S.; Ramirez-Alejo, N.; Li, Z.; Patin, E.; Rao, G.; Kerner, G.; Lim, C.K.; Krementsov, D.N.; Hernandez, N.; Ma, C.S.; et al. Tuberculosis and impaired IL-23-dependent IFN-gamma immunity in humans homozygous for a common TYK2 missense variant. *Sci. Immunol.* **2018**, *3*, eaau8714. [CrossRef]

48. Dendrou, C.A.; Cortes, A.; Shipman, L.; Evans, H.G.; Attfield, K.E.; Jostins, L.; Barber, T.; Kaur, G.; Kuttikkatte, S.B.; Leach, O.A.; et al. Resolving TYK2 locus genotype-to-phenotype differences in autoimmunity. *Sci. Transl. Med.* **2016**, *8*, 363ra149. [CrossRef]

49. Kerner, G.; Ramirez-Alejo, N.; Seeleuthner, Y.; Yang, R.; Ogishi, M.; Cobat, A.; Patin, E.; Quintana-Murci, L.; Boisson-Dupuis, S.; Casanova, J.L.; et al. Homozygosity for TYK2 P1104A underlies tuberculosis in about 1% of patients in a cohort of European ancestry. *Proc. Natl. Acad. Sci. USA* **2019**, *116*, 10430–10434. [CrossRef]

50. Sang, Q.X.; Man, Y.G.; Sung, Y.M.; Khamis, Z.I.; Zhang, L.; Lee, M.H.; Byers, S.W.; Sahab, Z.J. Non-receptor tyrosine kinase 2 reaches its lowest expression levels in human breast cancer during regional nodal metastasis. *Clin. Exp. Metastasis* **2012**, *29*, 143–153. [CrossRef]

51. Ruhe, J.E.; Streit, S.; Hart, S.; Wong, C.H.; Specht, K.; Knyazev, P.; Knyazeva, T.; Tay, L.S.; Loo, H.L.; Foo, P.; et al. Genetic alterations in the tyrosine kinase transcriptome of human cancer cell lines. *Cancer Res.* **2007**, *67*, 11368–11376. [CrossRef]

52. Prutsch, N.; Gurnhofer, E.; Suske, T.; Liang, H.C.; Schlederer, M.; Roos, S.; Wu, L.C.; Simonitsch-Klupp, I.; Alvarez-Hernandez, A.; Kornauth, C.; et al. Dependency on the TYK2/STAT1/MCL1 axis in anaplastic large cell lymphoma. *Leukemia* **2019**, *33*, 696–709. [CrossRef] [PubMed]

53. Sanda, T.; Tyner, J.W.; Gutierrez, A.; Ngo, V.N.; Glover, J.; Chang, B.H.; Yost, A.; Ma, W.; Fleischman, A.G.; Zhou, W.; et al. TYK2-STAT1-BCL2 pathway dependence in T-cell acute lymphoblastic leukemia. *Cancer Discov.* **2013**, *3*, 564–577. [CrossRef] [PubMed]

54. Herrmann, A.; Lahtz, C.; Nagao, T.; Song, J.Y.; Chan, W.C.; Lee, H.; Yue, C.; Look, T.; Mulfarth, R.; Li, W.; et al. CTLA4 Promotes Tyk2-STAT3-Dependent B-cell Oncogenicity. *Cancer Res.* **2017**, *77*, 5118–5128. [CrossRef] [PubMed]

55. Carmo, C.R.; Lyons-Lewis, J.; Seckl, M.J.; Costa-Pereira, A.P. A Novel Requirement for Janus Kinases as Mediators of Drug Resistance Induced by Fibroblast Growth Factor-2 in Human Cancer Cells. *PLoS ONE* **2011**, *6*, e19861. [CrossRef]

56. Vainchenker, W.; Constantinescu, S.N. JAK/STAT signaling in hematological malignancies. *Oncogene* **2013**, *32*, 2601–2613. [CrossRef]

57. Hubbard, S.R. Mechanistic Insights into Regulation of JAK2 Tyrosine Kinase. *Front. Endocrinol.* **2017**, *8*, 361. [CrossRef]

58. Morris, R.; Kershaw, N.J.; Babon, J.J. The molecular details of cytokine signaling via the JAK/STAT pathway. *Protein Sci.* **2018**, *27*, 1984–2009. [CrossRef]

59. Silvennoinen, O.; Hubbard, S.R. Molecular insights into regulation of JAK2 in myeloproliferative neoplasms. *Blood* **2015**, *125*, 3388–3392. [CrossRef]

60. Gakovic, M.; Ragimbeau, J.; Francois, V.; Constantinescu, S.N.; Pellegrini, S. The Stat3-activating Tyk2 V678F mutant does not up-regulate signaling through the type I interferon receptor but confers ligand hypersensitivity to a homodimeric receptor. *J. Biol. Chem.* **2008**, *283*, 18522–18529. [CrossRef]

61. Staerk, J.; Kallin, A.; Demoulin, J.B.; Vainchenker, W.; Constantinescu, S.N. JAK1 and Tyk2 activation by the homologous polycythemia vera JAK2 V617F mutation: Cross-talk with IGF1 receptor. *J. Biol. Chem.* **2005**, *280*, 41893–41899. [CrossRef] [PubMed]

62. Grossman, R.L.; Heath, A.P.; Ferretti, V.; Varmus, H.E.; Lowy, D.R.; Kibbe, W.A.; Staudt, L.M. Toward a Shared Vision for Cancer Genomic Data. *N. Engl. Med.* **2016**, *375*, 1109–1112. [CrossRef] [PubMed]

63. Jensen, M.A.; Ferretti, V.; Grossman, R.L.; Staudt, L.M. The NCI Genomic Data Commons as an engine for precision medicine. *Blood* **2017**, *130*, 453–459. [CrossRef] [PubMed]

64. Tate, J.G.; Bamford, S.; Jubb, H.C.; Sondka, Z.; Beare, D.M.; Bindal, N.; Boutselakis, H.; Cole, C.G.; Creatore, C.; Dawson, E.; et al. COSMIC: The Catalogue of Somatic Mutations in Cancer. *Nucleic Acids Res.* **2019**, *47*, D941–D947. [CrossRef]

65. Cerami, E.; Gao, J.; Dogrusoz, U.; Gross, B.E.; Sumer, S.O.; Aksoy, B.A.; Jacobsen, A.; Byrne, C.J.; Heuer, M.L.; Larsson, E.; et al. The cBio cancer genomics portal: An open platform for exploring multidimensional cancer genomics data. *Cancer Discov.* **2012**, *2*, 401–404. [CrossRef]

66. Gao, J.; Aksoy, B.A.; Dogrusoz, U.; Dresdner, G.; Gross, B.; Sumer, S.O.; Sun, Y.; Jacobsen, A.; Sinha, R.; Larsson, E.; et al. Integrative analysis of complex cancer genomics and clinical profiles using the cBioPortal. *Sci. Signal* **2013**, *6*, pl1. [CrossRef]

67. Waanders, E.; Scheijen, B.; Jongmans, M.C.; Venselaar, H.; van Reijmersdal, S.V.; van Dijk, A.H.; Pastorczak, A.; Weren, R.D.; van der Schoot, C.E.; van de Vorst, M.; et al. Germline activating TYK2 mutations in pediatric patients with two primary acute lymphoblastic leukemia occurrences. *Leukemia* **2017**, *31*, 821–828. [CrossRef]

68. Kaminker, J.S.; Zhang, Y.; Waugh, A.; Haverty, P.M.; Peters, B.; Sebisanovic, D.; Stinson, J.; Forrest, W.F.; Bazan, J.F.; Seshagiri, S.; et al. Distinguishing cancer-associated missense mutations from common polymorphisms. *Cancer Res.* **2007**, *67*, 465–473. [CrossRef]

69. Tomasson, M.H.; Xiang, Z.; Walgren, R.; Zhao, Y.; Kasai, Y.; Miner, T.; Ries, R.E.; Lubman, O.; Fremont, D.H.; McLellan, M.D.; et al. Somatic mutations and germline sequence variants in the expressed tyrosine kinase genes of patients with de novo acute myeloid leukemia. *Blood* **2008**, *111*, 4797–4808. [CrossRef]

70. Li, Z.; Gakovic, M.; Ragimbeau, J.; Eloranta, M.L.; Ronnblom, L.; Michel, F.; Pellegrini, S. Two rare disease-associated Tyk2 variants are catalytically impaired but signaling competent. *J. Immunol.* **2013**, *190*, 2335–2344. [CrossRef]

71. Gorman, J.A.; Hundhausen, C.; Kinsman, M.; Arkatkar, T.; Allenspach, E.J.; Clough, C.; West, S.E.; Thomas, K.; Eken, A.; Khim, S.; et al. The TYK2-P1104A Autoimmune Protective Variant Limits Coordinate Signals Required to Generate Specialized T Cell Subsets. *Front. Immunol.* **2019**, *10*, 44. [CrossRef] [PubMed]

72. Gao, Q.S.; Liang, W.W.; Foltz, S.M.; Mutharasu, G.; Jayasinghe, R.G.; Cao, S.; Liao, W.W.; Reynolds, S.M.; Wyczalkowski, M.A.; Yao, L.J.; et al. Driver Fusions and Their Implications in the Development and Treatment of Human Cancers. *Cell Rep.* **2018**, *23*, 227–238. [CrossRef] [PubMed]

73. Medves, S.; Demoulin, J.B. Tyrosine kinase gene fusions in cancer: Translating mechanisms into targeted therapies. *J. Cell. Mol. Med.* **2012**, *16*, 237–248. [CrossRef] [PubMed]

74. Stransky, N.; Cerami, E.; Schalm, S.; Kim, J.L.; Lengauer, C. The landscape of kinase fusions in cancer. *Nat. Commun.* **2014**, *5*, 4846. [CrossRef] [PubMed]

75. Lacronique, V.; Boureux, A.; Valle, V.D.; Poirel, H.; Quang, C.T.; Mauchauffe, M.; Berthou, C.; Lessard, M.; Berger, R.; Ghysdael, J.; et al. A TEL-JAK2 fusion protein with constitutive kinase activity in human leukemia. *Science* **1997**, *278*, 1309–1312. [CrossRef]

76. Carron, C.; Cormier, F.; Janin, A.; Lacronique, V.; Giovannini, M.; Daniel, M.T.; Bernard, O.; Ghysdael, J. TEL-JAK2 transgenic mice develop T-cell leukemia. *Blood* **2000**, *95*, 3891–3899. [CrossRef]

77. Lacronique, V.; Boureux, A.; Monni, R.; Dumon, S.; Mauchauffe, M.; Mayeux, P.; Gouilleux, F.; Berger, R.; Gisselbrecht, S.; Ghysdael, J.; et al. Transforming properties of chimeric TEL-JAK proteins in Ba/F3 cells. *Blood* **2000**, *95*, 2076–2083. [CrossRef]

78. Ho, K.; Valdez, F.; Garcia, R.; Tirado, C.A. JAK2 Translocations in hematological malignancies: Review of the literature. *J. Assoc. Genet. Technol.* **2010**, *36*, 107–109.

79. Levavi, H.; Tripodi, J.; Marcellino, B.; Mascarenhas, J.; Jones, A.V.; Cross, N.C.P.; Gruenstein, D.; Najfeld, V. A Novel t(1;9)(p36;p24.1) JAK2 Translocation and Review of the Literature. *Acta Haematol.* **2019**, *142*, 105–112. [CrossRef]

80. Crescenzo, R.; Abate, F.; Lasorsa, E.; Tabbo, F.; Gaudiano, M.; Chiesa, N.; Di Giacomo, F.; Spaccarotella, E.; Barbarossa, L.; Ercole, E.; et al. Convergent mutations and kinase fusions lead to oncogenic STAT3 activation in anaplastic large cell lymphoma. *Cancer Cell* **2015**, *27*, 516–532. [CrossRef]

81. Roberts, K.G.; Li, Y.; Payne-Turner, D.; Harvey, R.C.; Yang, Y.L.; Pei, D.; McCastlain, K.; Ding, L.; Lu, C.; Song, G.; et al. Targetable kinase-activating lesions in Ph-like acute lymphoblastic leukemia. *N. Engl. J. Med.* **2014**, *371*, 1005–1015. [CrossRef] [PubMed]

82. Velusamy, T.; Kiel, M.J.; Sahasrabuddhe, A.A.; Rolland, D.; Dixon, C.A.; Bailey, N.G.; Betz, B.L.; Brown, N.A.; Hristov, A.C.; Wilcox, R.A.; et al. A novel recurrent NPM1-TYK2 gene fusion in cutaneous CD30-positive lymphoproliferative disorders. *Blood* **2014**, *124*, 3768–3771. [CrossRef] [PubMed]

83. Roberts, K.G.; Gu, Z.; Payne-Turner, D.; McCastlain, K.; Harvey, R.C.; Chen, I.M.; Pei, D.; Iacobucci, I.; Valentine, M.; Pounds, S.B.; et al. High Frequency and Poor Outcome of Philadelphia Chromosome-Like Acute Lymphoblastic Leukemia in Adults. *J. Clin. Oncol.* **2017**, *35*, 394–401. [CrossRef] [PubMed]

84. Tron, A.E.; Keeton, E.K.; Ye, M.; Casas-Selves, M.; Chen, H.; Dillman, K.S.; Gale, R.E.; Stengel, C.; Zinda, M.; Linch, D.C.; et al. Next-generation sequencing identifies a novel ELAVL1-TYK2 fusion gene in MOLM-16, an AML cell line highly sensitive to the PIM kinase inhibitor AZD1208. *Leuk. Lymphoma* **2016**, *57*, 2927–2929. [CrossRef]

85. Gu, Z.; Churchman, M.; Roberts, K.; Li, Y.; Liu, Y.; Harvey, R.C.; McCastlain, K.; Reshmi, S.C.; Payne-Turner, D.; Iacobucci, I.; et al. Genomic analyses identify recurrent MEF2D fusions in acute lymphoblastic leukaemia. *Nat. Commun.* **2016**, *7*, 13331. [CrossRef]

86. Roberts, K.G.; Yang, Y.L.; Payne-Turner, D.; Lin, W.; Files, J.K.; Dickerson, K.; Gu, Z.; Taunton, J.; Janke, L.J.; Chen, T.; et al. Oncogenic role and therapeutic targeting of ABL-class and JAK-STAT activating kinase alterations in Ph-like ALL. *Blood Adv.* **2017**, *1*, 1657–1671. [CrossRef]

87. Prasad, A.; Rabionet, R.; Espinet, B.; Zapata, L.; Puiggros, A.; Melero, C.; Puig, A.; Sarria-Trujillo, Y.; Ossowski, S.; Garcia-Muret, M.P.; et al. Identification of gene mutations and fusion genes in patients with Sezary Syndrome. *J. Investig. Dermatol.* **2016**, *136*, 1490–1499. [CrossRef]

88. Kim, P.; Zhou, X. FusionGDB: Fusion gene annotation DataBase. *Nucleic Acids Res.* **2019**, *47*, D994–D1004. [CrossRef]

89. Negrini, S.; Gorgoulis, V.G.; Halazonetis, T.D. Genomic instability—An evolving hallmark of cancer. *Nat. Rev. Mol. Cell Biol.* **2010**, *11*, 220–228. [CrossRef]

90. Ly, P.; Cleveland, D.W. Rebuilding Chromosomes After Catastrophe: Emerging Mechanisms of Chromothripsis. *Trends Cell Biol.* **2017**, *27*, 917–930. [CrossRef]

91. Rode, A.; Maass, K.K.; Willmund, K.V.; Lichter, P.; Ernst, A. Chromothripsis in cancer cells: An update. *Int. J. Cancer* **2016**, *138*, 2322–2333. [CrossRef] [PubMed]

92. Grobner, S.N.; Worst, B.C.; Weischenfeldt, J.; Buchhalter, I.; Kleinheinz, K.; Rudneva, V.A.; Johann, P.D.; Balasubramanian, G.P.; Segura-Wang, M.; Brabetz, S.; et al. The landscape of genomic alterations across childhood cancers. *Nature* **2018**, *555*, 321–327. [CrossRef] [PubMed]

93. Liu, S.; Kwon, M.; Mannino, M.; Yang, N.; Renda, F.; Khodjakov, A.; Pellman, D. Nuclear envelope assembly defects link mitotic errors to chromothripsis. *Nature* **2018**, *561*, 551–555. [CrossRef] [PubMed]

94. Bromberg, J. Stat proteins and oncogenesis. *J. Clin. Investig.* **2002**, *109*, 1139–1142. [CrossRef]

95. Wingelhofer, B.; Neubauer, H.A.; Valent, P.; Han, X.; Constantinescu, S.N.; Gunning, P.T.; Muller, M.; Moriggl, R. Implications of STAT3 and STAT5 signaling on gene regulation and chromatin remodeling in hematopoietic cancer. *Leukemia* **2018**, *32*, 1713–1726. [CrossRef]

96. Avalle, L.; Pensa, S.; Regis, G.; Novelli, F.; Poli, V. STAT1 and STAT3 in tumorigenesis: A matter of balance. *JAKSTAT* **2012**, *1*, 65–72. [CrossRef]

97. Meissl, K.; Macho-Maschler, S.; Muller, M.; Strobl, B. The good and the bad faces of STAT1 in solid tumours. *Cytokine* **2017**, *89*, 12–20. [CrossRef]

98. Zhang, Y.; Liu, Z. STAT1 in cancer: Friend or foe? *Discov. Med.* **2017**, *24*, 19–29.

99. Mori, R.; Wauman, J.; Icardi, L.; Van der Heyden, J.; De Cauwer, L.; Peelman, F.; De Bosscher, K.; Tavernier, J. TYK2-induced phosphorylation of Y640 suppresses STAT3 transcriptional activity. *Sci. Rep.* **2017**, *7*, 15919. [CrossRef]

100. Koskela, H.L.; Eldfors, S.; Ellonen, P.; van Adrichem, A.J.; Kuusanmaki, H.; Andersson, E.I.; Lagstrom, S.; Clemente, M.J.; Olson, T.; Jalkanen, S.E.; et al. Somatic STAT3 mutations in large granular lymphocytic leukemia. *N. Engl. J. Med.* **2012**, *366*, 1905–1913. [CrossRef] [PubMed]

101. Pilati, C.; Amessou, M.; Bihl, M.P.; Balabaud, C.; Nhieu, J.T.; Paradis, V.; Nault, J.C.; Izard, T.; Bioulac-Sage, P.; Couchy, G.; et al. Somatic mutations activating STAT3 in human inflammatory hepatocellular adenomas. *J. Exp. Med.* **2011**, *208*, 1359–1366. [CrossRef] [PubMed]

102. Sun, L.M.; Feng, L.S.; Cui, J.W. Increased expression of claudin-17 promotes a malignant phenotype in hepatocyte via Tyk2/Stat3 signaling and is associated with poor prognosis in patients with hepatocellular carcinoma. *Diagn. Pathol.* **2018**, *13*, 72. [CrossRef] [PubMed]

103. Muller, S.; Chen, Y.; Ginter, T.; Schafer, C.; Buchwald, M.; Schmitz, L.M.; Klitzsch, J.; Schutz, A.; Haitel, A.; Schmid, K.; et al. SIAH2 antagonizes TYK2-STAT3 signaling in lung carcinoma cells. *Oncotarget* **2014**, *5*, 3184–3196. [CrossRef]

104. Liu, H.; Wang, M.; Liang, N.; Guan, L. Claudin-9 enhances the metastatic potential of hepatocytes via Tyk2/Stat3 signaling. *Turk. J. Gastroenterol.* **2019**, *30*, 722–731. [CrossRef] [PubMed]

105. Sun, L.M.; Feng, L.S.; Cui, J.W. Increased expression of claudin-12 promotes the metastatic phenotype of human bronchial epithelial cells and is associated with poor prognosis in lung squamous cell carcinoma. *Exp. Ther. Med.* **2019**, *17*, 165–174. [CrossRef] [PubMed]

106. Cathcart, J.; Pulkoski-Gross, A.; Cao, J. Targeting matrix metalloproteinases in cancer: Bringing new life to old ideas. *Genes Dis.* **2015**, *2*, 26–34. [CrossRef] [PubMed]

107. Tabaries, S.; Siegel, P.M. The role of claudins in cancer metastasis. *Oncogene* **2017**, *36*, 1176–1190. [CrossRef]

108. Cutler, S.J.; Doecke, J.D.; Ghazawi, I.; Yang, J.; Griffiths, L.R.; Spring, K.J.; Ralph, S.J.; Mellick, A.S. Novel STAT binding elements mediate IL-6 regulation of MMP-1 and MMP-3. *Sci. Rep.* **2017**, *7*, 8526. [CrossRef]

109. Fanjul-Fernandez, M.; Folgueras, A.R.; Cabrera, S.; Lopez-Otin, C. Matrix metalloproteinases: Evolution, gene regulation and functional analysis in mouse models. *Biochim. Biophys. Acta* **2010**, *1803*, 3–19. [CrossRef]

110. Costantino, G.; Egerbacher, M.; Kolbe, T.; Karaghiosoff, M.; Strobl, B.; Vogl, C.; Helmreich, M.; Muller, M. Tyk2 and signal transducer and activator of transcription 1 contribute to intestinal I/R injury. *Shock* **2008**, *29*, 238–244. [CrossRef]

111. Araki, Y.; Tsuzuki Wada, T.; Aizaki, Y.; Sato, K.; Yokota, K.; Fujimoto, K.; Kim, Y.T.; Oda, H.; Kurokawa, R.; Mimura, T. Histone Methylation and STAT-3 Differentially Regulate Interleukin-6-Induced Matrix Metalloproteinase Gene Activation in Rheumatoid Arthritis Synovial Fibroblasts. *Arthritis Rheumatol.* **2016**, *68*, 1111–1123. [CrossRef] [PubMed]

112. Mahmood, N.; Mihalcioiu, C.; Rabbani, S.A. Multifaceted Role of the Urokinase-Type Plasminogen Activator (uPA) and Its Receptor (uPAR): Diagnostic, Prognostic, and Therapeutic Applications. *Front. Oncol.* **2018**, *8*, 24. [CrossRef] [PubMed]

113. Kusch, A.; Tkachuk, S.; Haller, H.; Dietz, R.; Gulba, D.C.; Lipp, M.; Dumler, I. Urokinase stimulates human vascular smooth muscle cell migration via a phosphatidylinositol 3-kinase-Tyk2 interaction. *J. Biol. Chem.* **2000**, *275*, 39466–39473. [CrossRef] [PubMed]

114. Radwan, M.; Miller, I.; Grunert, T.; Marchetti-Deschmann, M.; Vogl, C.; O'Donoghue, N.; Dunn, M.J.; Kolbe, T.; Allmaier, G.; Gemeiner, M.; et al. The impact of tyrosine kinase 2 (Tyk2) on the proteome of murine macrophages and their response to lipopolysaccharide (LPS). *Proteomics* **2008**, *8*, 3469–3485. [CrossRef] [PubMed]

115. Schuster, C.; Muller, M.; Freissmuth, M.; Sexl, V.; Stoiber, D. Commentary on H. Ide. Tyk2 expression and its signaling enhances the invasiveness of prostate cancer cells. *Biochem. Biophys. Res. Commun.* **2008**, *366*, 869–870. [CrossRef]

116. Murphy, D.A.; Courtneidge, S.A. The 'ins' and 'outs' of podosomes and invadopodia: Characteristics, formation and function. *Nat. Rev. Mol. Cell Biol.* **2011**, *12*, 413–426. [CrossRef]

117. Revach, O.Y.; Sandler, O.; Samuels, Y.; Geiger, B. Cross-Talk between Receptor Tyrosine Kinases AXL and ERBB3 Regulates Invadopodia Formation in Melanoma Cells. *Cancer Res.* **2019**, *79*, 2634–2648. [CrossRef]

118. Aasen, T.; Leithe, E.; Graham, S.V.; Kameritsch, P.; Mayán, M.D.; Mesnil, M.; Pogoda, K.; Tabernero, A. Connexins in cancer: Bridging the gap to the clinic. *Oncogene* **2019**, *38*, 4429–4451. [CrossRef]

119. Bonacquisti, E.E.; Nguyen, J. Connexin 43 (Cx43) in cancer: Implications for therapeutic approaches via gap junctions. *Cancer Lett.* **2019**, *442*, 439–444. [CrossRef]

120. Li, H.; Spagnol, G.; Zheng, L.; Stauch, K.L.; Sorgen, P.L. Regulation of Connexin43 Function and Expression by Tyrosine Kinase 2. *J. Biol. Chem.* **2016**, *291*, 15867–15880. [CrossRef]

121. Kotredes, K.P.; Gamero, A.M. Interferons as inducers of apoptosis in malignant cells. *J. Interferon Cytokine Res.* **2013**, *33*, 162–170. [CrossRef] [PubMed]

122. Kale, J.; Osterlund, E.J.; Andrews, D.W. BCL-2 family proteins: Changing partners in the dance towards death. *Cell Death Differ.* **2018**, *25*, 65–80. [CrossRef] [PubMed]

123. Singh, R.; Letai, A.; Sarosiek, K. Regulation of apoptosis in health and disease: The balancing act of BCL-2 family proteins. *Nat. Rev. Mol. Cell Biol.* **2019**, *20*, 175–193. [CrossRef] [PubMed]

124. Shimoda, H.K.; Shide, K.; Kameda, T.; Matsunaga, T.; Shimoda, K. Tyrosine kinase 2 interacts with the proapoptotic protein Siva-1 and augments its apoptotic functions. *Biochem. Biophys. Res. Commun.* **2010**, *400*, 252–257. [CrossRef] [PubMed]

125. Nakajima, H.; Shibata, F.; Kumagai, H.; Shimoda, K.; Kitamura, T. Tyk2 is dispensable for induction of myeloproliferative disease by mutant FLT3. *Int. J. Hematol.* **2006**, *84*, 54–59. [CrossRef]

126. Yamaji, T.; Shide, K.; Kameda, T.; Sekine, M.; Kamiunten, A.; Hidaka, T.; Kubuki, Y.; Shimoda, H.; Abe, H.; Miike, T.; et al. Loss of Tyrosine Kinase 2 Does Not Affect the Severity of Jak2V617F-induced Murine Myeloproliferative Neoplasm. *Anticancer Res.* **2017**, *37*, 3841–3847. [CrossRef]

127. Koppikar, P.; Bhagwat, N.; Kilpivaara, O.; Manshouri, T.; Adli, M.; Hricik, T.; Liu, F.; Saunders, L.M.; Mullally, A.; Abdel-Wahab, O.; et al. Heterodimeric JAK-STAT activation as a mechanism of persistence to JAK2 inhibitor therapy. *Nature* **2012**, *489*, 155–159. [CrossRef]

128. Kohlhuber, F.; Rogers, N.C.; Watling, D.; Feng, J.; Guschin, D.; Briscoe, J.; Witthuhn, B.A.; Kotenko, S.V.; Pestka, S.; Stark, G.R.; et al. A JAK1/JAK2 chimera can sustain alpha and gamma interferon responses. *Mol. Cell. Biol.* **1997**, *17*, 695–706. [CrossRef]

129. Briscoe, J.; Rogers, N.C.; Witthuhn, B.A.; Watling, D.; Harpur, A.G.; Wilks, A.; Stark, G.R.; Ihle, J.N.; Kerr, I.M. Kinase-negative mutants of JAK1 can sustain interferon-gamma-inducible gene expression but not an antiviral state. *EMBO J.* **1996**, *15*, 799–809. [CrossRef]

130. Adam, L.; Bandyopadhyay, D.; Kumar, R. Interferon-alpha signaling promotes nucleus-to-cytoplasmic redistribution of p95Vav, and formation of a multisubunit complex involving Vav, Ku80, and Tyk2. *Biochem. Biophys. Res. Commun.* **2000**, *267*, 692–696. [CrossRef]

131. Uddin, S.; Gardziola, C.; Dangat, A.; Yi, T.; Platanias, L.C. Interaction of the c-cbl proto-oncogene product with the Tyk-2 protein tyrosine kinase. *Biochem. Biophys. Res. Commun.* **1996**, *225*, 833–838. [CrossRef] [PubMed]

132. Uddin, S.; Grumbach, I.M.; Yi, T.; Colamonici, O.R.; Platanias, L.C. Interferon alpha activates the tyrosine kinase Lyn in haemopoietic cells. *Br. J. Haematol.* **1998**, *101*, 446–449. [CrossRef] [PubMed]

133. Uddin, S.; Sher, D.A.; Alsayed, Y.; Pons, S.; Colamonici, O.R.; Fish, E.N.; White, M.F.; Platanias, L.C. Interaction of p59(fyn) with interferon-activated Jak kinases. *Biochem. Biophys. Res. Commun.* **1997**, *235*, 83–88. [CrossRef] [PubMed]

134. Uddin, S.; Sweet, M.; Colamonici, O.R.; Krolewski, J.J.; Platanias, L.C. The vav proto-oncogene product (p95vav) interacts with the Tyk-2 protein tyrosine kinase. *FEBS Lett.* **1997**, *403*, 31–34. [CrossRef]

135. Akahane, K.; Li, Z.; Etchin, J.; Berezovskaya, A.; Gjini, E.; Masse, C.E.; Miao, W.; Rocnik, J.; Kapeller, R.; Greenwood, J.R.; et al. Anti-leukaemic activity of the TYK2 selective inhibitor NDI-031301 in T-cell acute lymphoblastic leukaemia. *Br. J. Haematol.* **2017**, *177*, 271–282. [CrossRef] [PubMed]

136. Muromoto, R.; Tawa, K.; Ohgakiuchi, Y.; Sato, A.; Saino, Y.; Hirashima, K.; Minoguchi, H.; Kitai, Y.; Kashiwakura, J.I.; Shimoda, K.; et al. IkappaB-zeta Expression Requires Both TYK2/STAT3 Activity and IL-17-Regulated mRNA Stabilization. *Immunohorizons* **2019**, *3*, 172–185. [CrossRef] [PubMed]

137. Willems, M.; Dubois, N.; Musumeci, L.; Bours, V.; Robe, P.A. IkappaBzeta: An emerging player in cancer. *Oncotarget* **2016**, *7*, 66310–66322. [CrossRef]

138. Luck, K.; Sheynkman, G.M.; Zhang, I.; Vidal, M. Proteome-Scale Human Interactomics. *Trends Biochem. Sci.* **2017**, *42*, 342–354. [CrossRef]

139. Rolland, T.; Tasan, M.; Charloteaux, B.; Pevzner, S.J.; Zhong, Q.; Sahni, N.; Yi, S.; Lemmens, I.; Fontanillo, C.; Mosca, R.; et al. A proteome-scale map of the human interactome network. *Cell* **2014**, *159*, 1212–1226. [CrossRef]

140. Szklarczyk, D.; Jensen, L.J. Protein-protein interaction databases. *Methods Mol. Biol.* **2015**, *1278*, 39–56. [CrossRef]

141. Lievens, S.; Gerlo, S.; Lemmens, I.; De Clercq, D.J.; Risseeuw, M.D.; Vanderroost, N.; De Smet, A.S.; Ruyssinck, E.; Chevet, E.; Van Calenbergh, S.; et al. Kinase Substrate Sensor (KISS), a mammalian in situ protein interaction sensor. *Mol. Cell. Proteom.* **2014**, *13*, 3332–3342. [CrossRef] [PubMed]

142. Masschaele, D.; Gerlo, S.; Lemmens, I.; Lievens, S.; Tavernier, J. KISS: A Mammalian Two-Hybrid Method for In Situ Analysis of Protein-Protein Interactions. *Methods Mol. Biol.* **2018**, *1794*, 269–278. [CrossRef] [PubMed]

143. Babon, J.J.; Lucet, I.S.; Murphy, J.M.; Nicola, N.A.; Varghese, L.N. The molecular regulation of Janus kinase (JAK) activation. *Biochem. J.* **2014**, *462*, 1–13. [CrossRef] [PubMed]

144. Murray, P.J. The JAK-STAT signaling pathway: Input and output integration. *J. Immunol.* **2007**, *178*, 2623–2629. [CrossRef] [PubMed]

145. Chen, C.; Huang, H.; Wu, C.H. Protein Bioinformatics Databases and Resources. *Methods Mol. Biol.* **2017**, *1558*, 3–39. [CrossRef] [PubMed]

146. Hornbeck, P.V.; Kornhauser, J.M.; Latham, V.; Murray, B.; Nandhikonda, V.; Nord, A.; Skrzypek, E.; Wheeler, T.; Zhang, B.; Gnad, F. 15 years of PhosphoSitePlus(R): Integrating post-translationally modified sites, disease variants and isoforms. *Nucleic Acids Res.* **2019**, *47*, D433–D441. [CrossRef]

147. Kershaw, N.J.; Murphy, J.M.; Lucet, I.S.; Nicola, N.A.; Babon, J.J. Regulation of Janus kinases by SOCS proteins. *Biochem. Soc. Trans.* **2013**, *41*, 1042–1047. [CrossRef]

148. Xu, D.; Qu, C.K. Protein tyrosine phosphatases in the JAK/STAT pathway. *Front. Biosci.* **2008**, *13*, 4925–4932. [CrossRef]

149. Bollu, L.R.; Mazumdar, A.; Savage, M.I.; Brown, P.H. Molecular Pathways: Targeting Protein Tyrosine Phosphatases in Cancer. *Clin. Cancer Res.* **2017**, *23*, 2136–2142. [CrossRef]

150. Inagaki-Ohara, K.; Kondo, T.; Ito, M.; Yoshimura, A. SOCS, inflammation, and cancer. *JAKSTAT* **2013**, *2*, e24053. [CrossRef]

151. Jiang, M.; Zhang, W.W.; Liu, P.; Yu, W.; Liu, T.; Yu, J. Dysregulation of SOCS-Mediated Negative Feedback of Cytokine Signaling in Carcinogenesis and Its Significance in Cancer Treatment. *Front. Immunol.* **2017**, *8*, 70. [CrossRef] [PubMed]

152. Wu, C.; Guan, Q.; Wang, Y.; Zhao, Z.J.; Zhou, G.W. SHP-1 suppresses cancer cell growth by promoting degradation of JAK kinases. *J. Cell. Biochem.* **2003**, *90*, 1026–1037. [CrossRef] [PubMed]

153. Bocchini, C.E.; Kasembeli, M.M.; Roh, S.H.; Tweardy, D.J. Contribution of chaperones to STAT pathway signaling. *JAKSTAT* **2014**, *3*, e970459. [CrossRef] [PubMed]
154. Prodromou, C. Mechanisms of Hsp90 regulation. *Biochem. J.* **2016**, *473*, 2439–2452. [CrossRef] [PubMed]
155. Taipale, M.; Jarosz, D.F.; Lindquist, S. HSP90 at the hub of protein homeostasis: Emerging mechanistic insights. *Nat. Rev. Mol. Cell Biol.* **2010**, *11*, 515–528. [CrossRef] [PubMed]
156. Jaeger, A.M.; Whitesell, L. HSP90: Enabler of Cancer Adaptation. *Annu. Rev. Cancer Biol.* **2019**, *3*, 275–297. [CrossRef]
157. Trepel, J.; Mollapour, M.; Giaccone, G.; Neckers, L. Targeting the dynamic HSP90 complex in cancer. *Nat. Rev. Cancer* **2010**, *10*, 537–549. [CrossRef]
158. Caldas-Lopes, E.; Cerchietti, L.; Ahn, J.H.; Clement, C.C.; Robles, A.I.; Rodina, A.; Moulick, K.; Taldone, T.; Gozman, A.; Guo, Y.; et al. Hsp90 inhibitor PU-H71, a multimodal inhibitor of malignancy, induces complete responses in triple-negative breast cancer models. *Proc. Natl. Acad. Sci. USA* **2009**, *106*, 8368–8373. [CrossRef]
159. Taipale, M.; Krykbaeva, I.; Koeva, M.; Kayatekin, C.; Westover, K.D.; Karras, G.I.; Lindquist, S. Quantitative analysis of HSP90-client interactions reveals principles of substrate recognition. *Cell* **2012**, *150*, 987–1001. [CrossRef]
160. Akahane, K.; Sanda, T.; Mansour, M.R.; Radimerski, T.; DeAngelo, D.J.; Weinstock, D.M.; Look, A.T. HSP90 inhibition leads to degradation of the TYK2 kinase and apoptotic cell death in T-cell acute lymphoblastic leukemia. *Leukemia* **2016**, *30*, 219–228. [CrossRef]
161. Schoof, N.; von Bonin, F.; Trumper, L.; Kube, D. HSP90 is essential for Jak-STAT signaling in classical Hodgkin Lymphoma cells. *Cell Commun. Sig.* **2009**, *7*, 17. [CrossRef] [PubMed]
162. Witte, S.; Muljo, S.A. Integrating non-coding RNAs in JAK-STAT regulatory networks. *JAKSTAT* **2014**, *3*, e28055. [CrossRef] [PubMed]
163. Mullany, L.E.; Herrick, J.S.; Sakoda, L.C.; Samowitz, W.; Stevens, J.R.; Wolff, R.K.; Slattery, M.L. MicroRNA-messenger RNA interactions involving JAK-STAT signaling genes in colorectal cancer. *Genes Cancer* **2018**, *9*, 232–246. [CrossRef] [PubMed]
164. Pencik, J.; Pham, H.T.; Schmoellerl, J.; Javaheri, T.; Schlederer, M.; Culig, Z.; Merkel, O.; Moriggl, R.; Grebien, F.; Kenner, L. JAK-STAT signaling in cancer: From cytokines to non-coding genome. *Cytokine* **2016**, *87*, 26–36. [CrossRef]
165. Li, Z.Y.; Yang, L.; Liu, X.J.; Wang, X.Z.; Pan, Y.X.; Luo, J.M. The Long Noncoding RNA MEG3 and its Target miR-147 Regulate JAK/STAT Pathway in Advanced Chronic Myeloid Leukemia. *EBioMedicine* **2018**, *34*, 61–75. [CrossRef]
166. Ghoreschi, K.; Jesson, M.I.; Li, X.; Lee, J.L.; Ghosh, S.; Alsup, J.W.; Warner, J.D.; Tanaka, M.; Steward-Tharp, S.M.; Gadina, M.; et al. Modulation of innate and adaptive immune responses by tofacitinib (CP-690,550). *J. Immunol.* **2011**, *186*, 4234–4243. [CrossRef]
167. Changelian, P.S.; Flanagan, M.E.; Ball, D.J.; Kent, C.R.; Magnuson, K.S.; Martin, W.H.; Rizzuti, B.J.; Sawyer, P.S.; Perry, B.D.; Brissette, W.H.; et al. Prevention of organ allograft rejection by a specific Janus kinase 3 inhibitor. *Science* **2003**, *302*, 875–878. [CrossRef]
168. Harrison, C.; Kiladjian, J.J.; Al-Ali, H.K.; Gisslinger, H.; Waltzman, R.; Stalbovskaya, V.; McQuitty, M.; Hunter, D.S.; Levy, R.; Knoops, L.; et al. JAK inhibition with ruxolitinib versus best available therapy for myelofibrosis. *N. Engl. J. Med.* **2012**, *366*, 787–798. [CrossRef]
169. Verstovsek, S.; Mesa, R.A.; Gotlib, J.; Levy, R.S.; Gupta, V.; DiPersio, J.F.; Catalano, J.V.; Deininger, M.; Miller, C.; Silver, R.T.; et al. A double-blind, placebo-controlled trial of ruxolitinib for myelofibrosis. *N. Engl. J. Med.* **2012**, *366*, 799–807. [CrossRef]
170. Gadina, M.; Johnson, C.; Schwartz, D.; Bonelli, M.; Hasni, S.; Kanno, Y.; Changelian, P.; Laurence, A.; O'Shea, J.J. Translational and clinical advances in JAK-STAT biology: The present and future of jakinibs. *J. Leukoc. Biol.* **2018**, *104*, 499–514. [CrossRef]
171. Gadina, M.; Le, M.T.; Schwartz, D.M.; Silvennoinen, O.; Nakayamada, S.; Yamaoka, K.; O'Shea, J.J. Janus kinases to jakinibs: From basic insights to clinical practice. *Rheumatology* **2019**, *58*, 4–16. [CrossRef] [PubMed]
172. Vainchenker, W.; Leroy, E.; Gilles, L.; Marty, C.; Plo, I.; Constantinescu, S.N. JAK inhibitors for the treatment of myeloproliferative neoplasms and other disorders. *F1000Res* **2018**, *7*, 82. [CrossRef] [PubMed]
173. Virtanen, A.T.; Haikarainen, T.; Raivola, J.; Silvennoinen, O. Selective JAKinibs: Prospects in Inflammatory and Autoimmune Diseases. *Biodrugs* **2019**, *33*, 15–32. [CrossRef] [PubMed]

174. He, X.R.; Chen, X.B.; Zhang, H.C.; Xie, T.; Ye, X.Y. Selective Tyk2 inhibitors as potential therapeutic agents: A patent review (2015-2018). *Expert. Opin. Ther. Pat.* **2019**, *29*, 137–149. [CrossRef] [PubMed]
175. Danese, S.; Argollo, M.; Le Berre, C.; Peyrin-Biroulet, L. JAK selectivity for inflammatory bowel disease treatment: Does it clinically matter? *Gut* **2019**, *68*, 1893–1899. [CrossRef] [PubMed]
176. Papp, K.; Gordon, K.; Thaci, D.; Morita, A.; Gooderham, M.; Foley, P.; Girgis, I.G.; Kundu, S.; Banerjee, S. Phase 2 Trial of Selective Tyrosine Kinase 2 Inhibition in Psoriasis. *N. Engl. J. Med.* **2018**, *379*, 1313–1321. [CrossRef]
177. Burke, J.R.; Cheng, L.; Gillooly, K.M.; Strnad, J.; Zupa-Fernandez, A.; Catlett, I.M.; Zhang, Y.; Heimrich, E.M.; McIntyre, K.W.; Cunningham, M.D.; et al. Autoimmune pathways in mice and humans are blocked by pharmacological stabilization of the TYK2 pseudokinase domain. *Sci. Transl. Med.* **2019**, *11*, eaaw1736. [CrossRef]
178. Liu, C.; Lin, J.; Moslin, R.; Tokarski, J.S.; Muckelbauer, J.; Chang, C.; Tredup, J.; Xie, D.; Park, H.; Li, P.; et al. Identification of Imidazo[1,2-b]pyridazine Derivatives as Potent, Selective, and Orally Active Tyk2 JH2 Inhibitors. *ACS Med. Chem. Lett.* **2019**, *10*, 383–388. [CrossRef]
179. Moslin, R.; Gardner, D.; Santella, J.; Zhang, Y.; Duncia, J.V.; Liu, C.; Lin, J.; Tokarski, J.S.; Strnad, J.; Pedicord, D.; et al. Identification of imidazo[1,2-b] pyridazine TYK2 pseudokinase ligands as potent and selective allosteric inhibitors of TYK2 signalling. *Medchemcomm* **2017**, *8*, 700–712. [CrossRef]
180. Moslin, R.; Zhang, Y.; Wrobleski, S.T.; Lin, S.; Mertzman, M.; Spergel, S.; Tokarski, J.S.; Strnad, J.; Gillooly, K.; McIntyre, K.W.; et al. Identification of N-Methyl Nicotinamide and N-Methyl Pyridazine-3-Carboxamide Pseudokinase Domain Ligands as Highly Selective Allosteric Inhibitors of Tyrosine Kinase 2 (TYK2). *J. Med. Chem.* **2019**, *62*, 8953–8972. [CrossRef]
181. Wrobleski, S.T.; Moslin, R.; Lin, S.; Zhang, Y.; Spergel, S.; Kempson, J.; Tokarski, J.S.; Strnad, J.; Zupa-Fernandez, A.; Cheng, L.; et al. Highly Selective Inhibition of Tyrosine Kinase 2 (TYK2) for the Treatment of Autoimmune Diseases: Discovery of the Allosteric Inhibitor BMS-986165. *J. Med. Chem.* **2019**, *62*, 8973–8995. [CrossRef] [PubMed]
182. Villanueva, M.T. TYK2 inhibition shows promise. *Nat. Rev. Drug Discov.* **2019**, *18*, 668. [CrossRef] [PubMed]
183. Prchal-Murphy, M.; Semper, C.; Lassnig, C.; Wallner, B.; Gausterer, C.; Teppner-Klymiuk, I.; Kobolak, J.; Muller, S.; Kolbe, T.; Karaghiosoff, M.; et al. TYK2 kinase activity is required for functional type I interferon responses in vivo. *PLoS ONE* **2012**, *7*, e39141. [CrossRef] [PubMed]
184. Raje, V.; Derecka, M.; Cantwell, M.; Meier, J.; Szczepanek, K.; Sisler, J.D.; Strobl, B.; Gamero, A.; Harris, T.E.; Larner, A.C. Kinase Inactive Tyrosine Kinase (Tyk2) Supports Differentiation of Brown Fat Cells. *Endocrinology* **2017**, *158*, 148–157. [CrossRef] [PubMed]
185. Lesgidou, N.; Eliopoulos, E.; Goulielmos, G.N.; Vlassi, M. Insights on the alteration of functionality of a tyrosine kinase 2 variant: A molecular dynamics study. *Bioinformatics* **2018**, *34*, i781–i786. [CrossRef] [PubMed]
186. Lupardus, P.J.; Skiniotis, G.; Rice, A.J.; Thomas, C.; Fischer, S.; Walz, T.; Garcia, K.C. Structural snapshots of full-length Jak1, a transmembrane gp130/IL-6/IL-6Ralpha cytokine receptor complex, and the receptor-Jak1 holocomplex. *Structure* **2011**, *19*, 45–55. [CrossRef] [PubMed]

Article

mTOR and STAT3 Pathway Hyper-Activation is Associated with Elevated Interleukin-6 Levels in Patients with Shwachman-Diamond Syndrome: Further Evidence of Lymphoid Lineage Impairment

Antonio Vella [1,†], Elisabetta D'Aversa [2,†], Martina Api [3], Giulia Breveglieri [2], Marisole Allegri [3], Alice Giacomazzi [1], Elena Marinelli Busilacchi [4], Benedetta Fabrizzi [3], Tiziana Cestari [1], Claudio Sorio [5], Gloria Bedini [6], Giovanna D'Amico [6], Vincenzo Bronte [1], Antonella Poloni [4], Antonio Benedetti [7], Chiara Bovo [8], Seth J. Corey [9], Monica Borgatti [2,10,‡], Marco Cipolli [11,‡] and Valentino Bezzerri [3,*,‡]

1 Unit of Immunology, Azienda Ospedaliera Universitaria Integrata, 37134 Verona, Italy; antonio.vella@univr.it (A.V.); alice.giacomazzi@univr.it (A.G.); tiziana.cestari@univr.it (T.C.); vincenzo.bronte@univr.it (V.B.)
2 Department of Life Sciences and Biotechnology, University of Ferrara, 44100 Ferrara, Italy; elisabetta.daversa@unife.it (E.D.); giulia.breveglieri@unife.it (G.B.); monica.borgatti@unife.it (M.B.)
3 Cystic Fibrosis Center, Azienda Ospedaliero Universitaria Ospedali Riuniti, 60126 Ancona, Italy; martina.api@ospedaliriuniti.marche.it (M.A.); marisole.allegri@ospedaliriuniti.marche.it (M.A.); benedetta.fabrizzi@ospedaliriuniti.marche.it (B.F.)
4 Hematology Clinic, Università Politecnica delle Marche -AOU Ospedali Riuniti, 60126 Ancona, Italy; e.busilacchi@staff.univpm.it (E.M.B.); Antonella.poloni@ospedaliriuniti.marche.it (A.P.)
5 Department of Medicine, University of Verona, 37134 Verona, Italy; claudio.sorio@univr.it
6 Immunology and Cell Therapy Unit, Tettamanti Research Center, University of Milano-Bicocca, 20900 Monza, Italy; gloriabedini85@gmail.com (G.B.); giovanna.damico@asst-monza.it (G.D.)
7 Department of Gastroenterology and Hepatology, Università Politecnica delle Marche, 60126 Ancona, Italy; antonio.benedetti@ospedaliriuniti.marche.it
8 Hospital Health Direction, Azienda Ospedaliera Universitaria Integrata, 37126 Verona, Italy; Direzione.sanitaria@aovr.veneto.it
9 Department of Pediatric Hematology/Oncology and Stem Cell Transplantation, Cleveland Clinic, Cleveland, OH 44195, USA; coreys2@ccf.org
10 Biotechnology Center, University of Ferrara, 44100 Ferrara, Italy
11 Cystic Fibrosis Center, Azienda Ospedaliera Universitaria Integrata, 37126 Verona, Italy; marco.cipolli@aovr.veneto.it
* Correspondence: valentino.bezzerri@ospedaliriuniti.marche.it
† These authors contributed equally to this work.
‡ Shared seniorship.

Received: 11 February 2020; Accepted: 3 March 2020; Published: 5 March 2020

Abstract: Shwachman–Diamond syndrome (SDS) is a rare inherited bone marrow failure syndrome, resulting in neutropenia and a risk of myeloid neoplasia. A mutation in a ribosome maturation factor accounts for almost all of the cases. Lymphoid involvement in SDS has not been well characterized. We recently reported that lymphocyte subpopulations are reduced in SDS patients. We have also shown that the mTOR-STAT3 pathway is hyper-activated in SDS myeloid cell populations. Here we show that mTOR-STAT3 signaling is markedly upregulated in the lymphoid compartment of SDS patients. Furthermore, our data reveal elevated IL-6 levels in cellular supernatants obtained from lymphoblasts, bone marrow mononuclear and mesenchymal stromal cells, and plasma samples obtained from a cohort of 10 patients. Of note, everolimus-mediated inhibition of mTOR signaling is associated with basal state of phosphorylated STAT3. Finally, inhibition of mTOR-STAT3 pathway activation leads to normalization of IL-6 expression in SDS cells. Altogether, our data strengthen the hypothesis that SDS affects both lymphoid and myeloid blood compartment and suggest everolimus as a potential therapeutic agent to reduce excessive mTOR-STAT3 activation in SDS.

Keywords: STAT3; mTOR; Bone Marrow Failure Syndromes; lymphocytes

---

1. Introduction

Shwachman–Diamond syndrome (SDS) is one of the most common inherited bone marrow failure syndromes (IBMFS), occurring in almost 1 out of 75,000 live births [1]. SDS results from biallelic mutations in the Shwachman–Bodian–Diamond syndrome gene (*SBDS*), which encode the SBDS protein. SBDS protein cooperates with its partner elongation factor-like GTPase 1 (EFL1) to catalyze the release of the ribosomal anti-association factor eIF6, facilitating the assembly of the functional 80S ribosome [2–4]. The IBMFS are also cancer predisposition syndromes, in particular myelodysplastic syndrome (MDS) and acute myeloid leukemia (AML). In the general population, MDS has an incidence ranging from 2–12 cases per 100,000 people, which increases as individuals age [5]. Patients with SDS demonstrate a risk of evolution to MDS of 8.1% and 36% at 10 and 30 years, respectively [6]. A recent genomic analysis of 1514 patients with MDS who underwent a stem cell transplant showed that 4% of the young adult patients had undiagnosed compound heterozygous mutations in *SBDS*, suggesting that SDS prevalence among MDS/AML patients may be underestimated [7]. AML derives from dysregulated proliferation and accumulation of immature myeloid progenitor cells into the bone marrow and peripheral blood, which finally leads to a severe impairment of the hematopoietic system. Acute leukemias rapidly disseminate after initial inception, escaping the anti-leukemic immunity process. Regulatory T cells play a key role in the maintenance of immune tolerance, which acts as a regulator of the tumor immunity [8]. CD4/CD8 double negative (DN) T cells have gained prominence among T regulatory cell subsets engaged in immunosurveillance. DN T cells are mature T cells representing almost 3–5% of the total peripheral T cell population [9]. Most human and murine T cells express and rearrange the α and β chains of the T cell receptor (TCR) and are recognized as TCR$\alpha\beta$ T cells, whereas a small part of T cells do express the γ and δ chains, which are mostly DN T cells [9]. Interestingly, DN T cells showed anti-leukemic activity and synergy with conventional chemotherapies both in vitro and in patient-derived xenograft models of AML [10,11]. In a mouse model of AML, the leukemic cells promoted T cell tolerance with suppression of anti-tumor CD8$^+$ T cells [12]. Failure of T cell-mediated anti-cancer immune response is associated with disease progression and poor outcome in MDS and AML.

IL-6/JAK/STAT3 signaling axis plays a key role in leukemogenesis [13]. The genes encoding the kinase protein JAK2 have indeed often been mutated in myeloproliferative disorders, leading to constitutive hyper-activation of its downstream effector, the transcription factor STAT3 [14]. Of note, STAT3 hyper-activation has been found in tumor-infiltrating leukocytes, in which STAT3 orchestrates the crosstalk between cancer and immune cells [15]. Furthermore, it has been reported that STAT3 inhibition in the myeloid compartment remarkably induces the anti-tumor capabilities of T cells and promotes their expansion in vivo [16,17]. STAT3 activation in immune cells is indeed associated with suppression of anti-tumor immunity. STAT3 excessive activation may be triggered by elevated levels of IL-6 present in the serum or released within the tumor microenvironment [14]. IL-6 may act in an autocrine or paracrine manner. IL-6 binds to its IL-6 receptor (IL-6R), localized onto the plasma membrane (membrane bound (mb)IL-6R), which is physically associated with the gp130 protein. This process is recognized as classical IL-6 signaling and leads to gp130 homodimerization, resulting in the activation of the IL-6 receptor complex [18]. In addition, IL-6 binds to the small extracellular secretory soluble IL-6 receptor (sIL-6R), which is generated by alternative splicing of IL-6R gene or by metalloproteinase-dependent cleavage of mbIL-6R. The sIL-6R mediates JAK-STAT3 activation in gp130 positive cells, which do not express mbIL-6R through a process termed IL-6 trans-signaling [19]. The

IL-6 trans-signaling pathway has been reported in murine hematopoietic stem cells [20,21]. Elevated levels of IL-6 have been found in adult bone marrow niche of patients with AML [22]. Serum IL-6 levels were found to be significantly increased in pediatric patients with AML [23]. Moreover, increased IL-6 serum levels are associated with poor prognosis in several types of cancer, including AML [23]. Interestingly, *IL-6* gene expression can be regulated by STAT3 itself, resulting in a feedforward autocrine feedback loop [14].

STAT3 has been reported as a direct substrate for the mammalian target of rapamycin (mTOR), which induces STAT3 S727 phosphorylation [24,25]. In addition, the mTOR-inhibitor rapamycin inhibits STAT3 S727 phosphorylation [24]. Moreover, we have previously shown that mTOR can promote STAT3 phosphorylation both at residue tyrosine 705 and serine 727 in SDS leukocytes [26]. Previous studies have reported that relapse of AML is associated with the gain of additional mutations in the mTOR gene, often due to the cytotoxic chemotherapy received by patients [27,28]. Inhibition of the mTOR pathway using rapamycin or other analogue molecules, including everolimus (RAD001) as anti-leukemic agent, has shown potent anti-cancer capabilities both in vitro and in vivo [29,30].

To date, no pharmacological therapy has been developed for IBMFS-related MDS or AML, and allogeneic hematopoietic stem cell transplantation remains the unique option in these cases. Unfortunately, its efficacy is limited by the morbidity and mortality associated with graft-versus-host disease.

Here we show further analysis of the mTOR-STAT3 axis in an extended panel of lymphocytic populations including CD4$^+$, CD8$^+$, T cells, DN T cells, γδT cells, and Natural Killer cells (NK). Moreover, we have checked the expression of IL-6 in different cellular and clinical SDS models, including lymphoblastoid cell lines (LCL), bone marrow mononuclear hematopoietic progenitors (BM-MNC) and mesenchymal stromal cells (BM-MSC), and plasma obtained from an enlarged cohort of 31 patients with SDS (Table 1). Our data indicate that everolimus can restore a normal level of mTOR and STAT3 activation in primary SDS lymphocytes. Importantly, mTOR-STAT3 inhibition was paralleled by a downregulation of IL-6 expression in hematopoietic SDS cells. Our results suggest the existence of a mTOR-STAT3-IL-6 loop of activation in hematopoietic SDS cells, which may affect both myeloid and lymphoid compartment, thus contributing to malignant transformation over the time. Taken together, these data strengthen the hypothesis of the involvement of lymphoid lineage in SDS and suggest everolimus or new rapalogs as potential therapeutic agents in SDS patients.

Table 1. Clinical data and genetics of patients with SDS recruited in this study.

UPN	Gender	Age	Genotype	PMN (Cell/mm^3)	Phenotype	Cytogenetics
1	M	27	258+2T>C/183–184TA>CT	1460	PI, FTT, recurrent infections, HbF > 2%, bone malformation, thrombocytopenia	46, XY, i(7)(q10)
6	M	27	258+2T>C/101A>T	3515	PI, FTT, recurrent infections, bone malformation, thrombocytopenia, cognitive impairment	46, XY, del(20)q
13	M	19	258+2T>C/183–184TA>CT+258+2T>C	1130	PI, FTT, recurrent infections, bone malformation, thrombocytopenia, anemia, cognitive impairment	46, XY, del(20)q
26	M	16	258+2T>C/183–184TA>CT	58	PI, FTT, bone malformation, thrombocytopenia	46, XY
33	F	8	258+2T>C/183–184TA>CT	1100	PI, FTT, bone malformation, thrombocytopenia	46, XX
35	F	15	258+2T>C/183–184TA>CT	970	PI, FTT, thrombocytopenia, anemia	46, XX
37	F	10	258+2T>C/183–184TA>CT	1280	PI, FTT, recurrent infections	46, XX, i(7)(q10)
43	M	22	258+2T>C/258+2T>C+533–549+403del	970	PI, FTT, bone malformation, thrombocytopenia, cognitive impairment	46, XY
47	M	12	258+2T>C/183–184TA>CT	550	PI, FTT, recurrent infections, bone malformation, thrombocytopenia	46, XY

Table 1. *Cont.*

UPN	Gender	Age	Genotype	PMN (Cell/mm^3)	Phenotype	Cytogenetics
52	M	10	258+2T>C/183–184TA>CT+258+2T>C	1070	PI, FTT, bone malformation	46, XY
56	F	15	258+2T>C/183–184TA>CT	1840	PI, FTT, recurrent infections, HbF > 2%, bone malformation, thrombocytopenia	46, XX
57	F	40	258+2T>C/G63C	500	PI, FTT, HbF > 2%, bone malformation, thrombocytopenia, cognitive impairment	46, XX
58	M	12	258+2T>C/183–184TA>CT	390	PI, FTT, HbF > 2%, bone malformation, thrombocytopenia, anemia	46, XY
63	M	15	258+2T>C/183–184TA>CT+258+2T>C	536	PI, FTT, recurrent infections, HbF > 2%, bone malformation, thrombocytopenia	46, XY
65	M	19	258+2T>C/258+2T>C	1390	PI, recurrent infections, bone malformation, thrombocytopenia, cognitive impairment	46, XY, del(20)q
66	M	23	258+2T>C/183–184TA>CT	1340	PI, bone malformation, thrombocytopenia, cognitive impairment	46, XY
67	M	8	258+2T>C/183–184TA>CT	500	PI, FTT, HbF > 2%, bone malformation	46, XY
68	M	22	258+2T>C/183–184TA>CT+258+2T>C	600	PI, FTT, recurrent infections, bone malformation, thrombocytopenia, cognitive impairment	46, XY
69	F	8	258+2T>C/183–184TA>CT	770	PI, FTT, HbF > 2%, anemia, cognitive impairment	46, XX
72	M	28	258+2T>C/183–184TA>CT	380	PI, FTT, recurrent infections, HbF > 2%, bone malformation, thrombocytopenia, anemia, cognitive impairment	46, XY
73	F	7	258+2T>C/183–184TA>CT	520	PI, FTT, HbF > 2%, thrombocytopenia	46, XX
74	M	9	258+2T>C/183–184TA>CT	1430	PI, FTT, HbF > 2%, cognitive impairment	46, XY
75	F	7	258+2T>C/183–184TA>CT	1000	PI, FTT, HbF > 2%, bone malformation, thrombocytopenia, cognitive impairment	46, XX
80	M	7	258+2T>C/183–184TA>CT	680	PI, FTT, recurrent infections, bone malformation, anemia, cognitive impairment	46, XY
82	M	16	258+2T>C/183–184TA>CT	300	PI, FTT, recurrent infections, bone malformation, thrombocytopenia, anemia, cognitive impairment	46, XY
87	M	18	258+2T>C/183–184TA>CT	880	PI, FTT, recurrent infections, bone malformation, cognitive impairment	46, XY
91	M	4	258+2T>C/183–184TA>CT+258+2T>C	1050	PI, FTT, bone malformation	46, XY
94	F	19	258+2T>C/352A>G	2420	PI, HbF > 2%, thrombocytopenia	46, XX
104	M	10	258+2T>C/183–184TA>CT	500	PI, FTT, recurrent infections, HbF > 2%, bone malformation, thrombocytopenia	46, XY
106	M	36	258+2T>C/183–184TA>CT	1210	PI, FTT, bone malformation, anemia	46, XY
108	M	17	258+2T>C/183–184TA>CT	970	PS, FTT, bone malformation, anemia	46, XY

UPN, unique patient number; PI, pancreas insufficiency; PS, pancreas sufficiency; FTT, failure to thrive; HbF, fetal hemoglobin.

2. Results

2.1. mTOR-STAT3 Pathway is Hyper-Activated Also in SDS Lymphocyte Subsets and Everolimus Can Reduce This Process In Vitro

The JAK/STAT3 pathway regulates T cell cytotoxic gene expression, proliferation, and survival. STAT3 inhibition in the myeloid compartment displays a remarkable induction of the T cell anti-tumor capabilities and promotes their expansion in vivo [16,17]. Thus, STAT3 activation in immune cells is associated with suppression of anti-tumor immunity. The protein kinase mTOR acts as an activator of the STAT3 pathway [24,25,31]. Accordingly, we recently reported that mTOR can activate STAT3 pathway in B cells, neutrophils and monocytes from SDS patients [26]. To assess the activation of mTOR-STAT3 pathway in SDS lymphoid compartment, we determined the phosphorylation of mTOR (S2448) and STAT3 (Y705 and S727) in CD4+ and CD8+ T cells, DN T cells, γδT cells, and NK cells isolated from peripheral blood from seven patients with SDS. In SDS patients, all cell subsets except NK displayed significantly elevated levels of phospho-mTOR (Figure 1) and -STAT3 (Figure 2) compared to age-matched healthy donors. Then, we tested the effect of the clinically-approved rapamycin analog, everolimus (RAD001) on mTOR, and STAT3 phosphorylation in SDS patient-derived T cell subsets. Results show that everolimus restores normal levels of phosphorylation of mTOR (Figure 1) and STAT3 (Figure 2), confirming the existence of an mTOR-STAT3 axis activation in the lymphoid compartment of SDS patients. To determine whether upregulation *STAT3* gene expression eventually exists in SDS cells along with hyper-phosphorylation, we measured and correlated *STAT3* transcript expression with protein levels in LCL and primary PBMC isolated from several SDS patients with different genotypes, compared to age-matched healthy controls. Our studies showed that *STAT3* expression is elevated in lymphocytes obtained from SDS patients compared to control subjects (Figures 3 and 4, Figures S1 and S2).

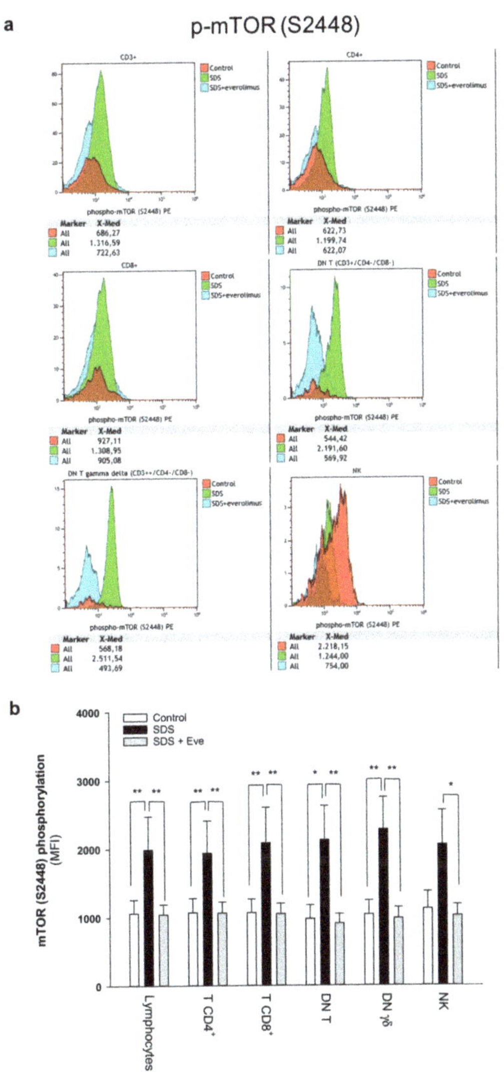

Figure 1. Phospho-flow analysis of mTOR S2448 in a panel of primary lymphocyte subsets. (**a**) Representative experiments conducted on peripheral blood samples obtained from patients with

SDS. Red histograms represent age-matched healthy control cells; green histograms represent SDS lymphocytes; blue histograms represent SDS lymphocytes upon everolimus (350 nM) treatment. (**b**) Data are mean ± SEM of seven experiments conducted on SDS lymphocytes obtained from seven SDS patients (UPN37, UPN58, UPN69, UPN73, UPN87, UPN106, UPN108). Statistics: normal distribution was tested by the Shapiro–Wilk test. Subsequently, the Mann–Whitney Rank Sum Test was calculated. * $p < 0.05$; ** $p < 0.01$.

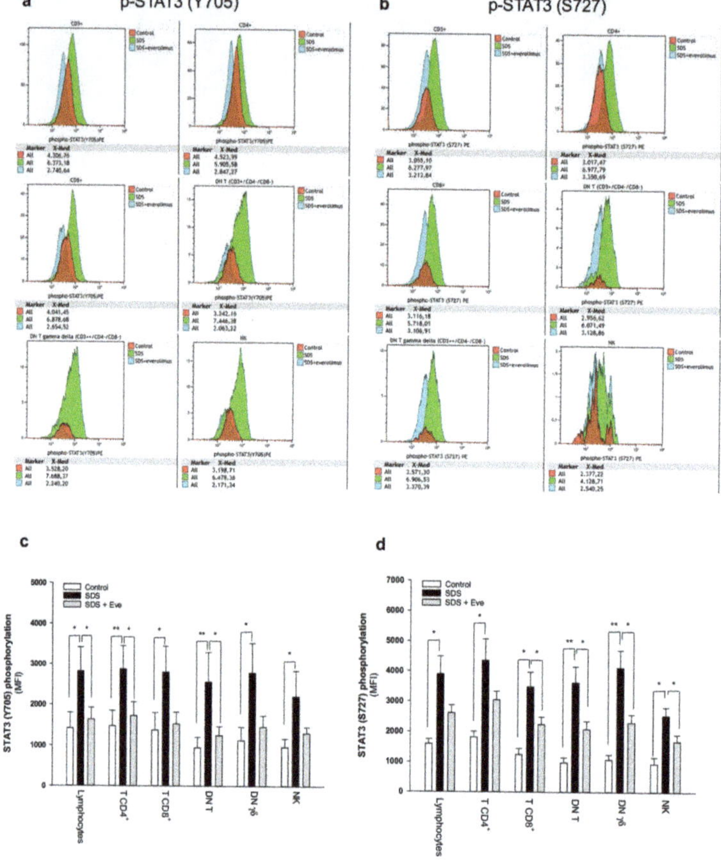

Figure 2. Phospho-flow analysis of STAT3 in a panel of primary lymphocyte subsets. (**a**) Representative analysis of phospho-STAT3 Y705 in lymphocyte subsets. (**b**) Representative analysis of phospho-STAT3 S727 in lymphocyte subsets. Red histograms represent age-matched healthy control cells; green histograms represent SDS patient-derived lymphocytes; blue histograms represent SDS lymphocytes upon everolimus (350nM) treatment. Data are mean ± SEM of seven experiments conducted on SDS patient-derived lymphocytes obtained from seven SDS patients (UPN37, UPN58, UPN69, UPN73, UPN87, UPN106, UPN108). Statistical Student's *t* test for paired data has been calculated. * $p < 0.05$; ** $p < 0.01$. (**c**) STAT3 (Y705) Median Fluorescence Intensity as measured by phospho-flow assays. (**d**) STAT3 (S727) Median Fluorescence Intensity as measured by phospho-flow assays. Data are mean ± SEM of seven experiments conducted on SDS lymphocytes obtained from seven SDS patients (UPN37, UPN58, UPN69, UPN73, UPN87, UPN106, UPN108). Statistics: Normal distribution was tested by the Shapiro–Wilk test. Subsequently, the Mann–Whitney Rank Sum Test was calculated. * $p < 0.05$; ** $p < 0.01$.

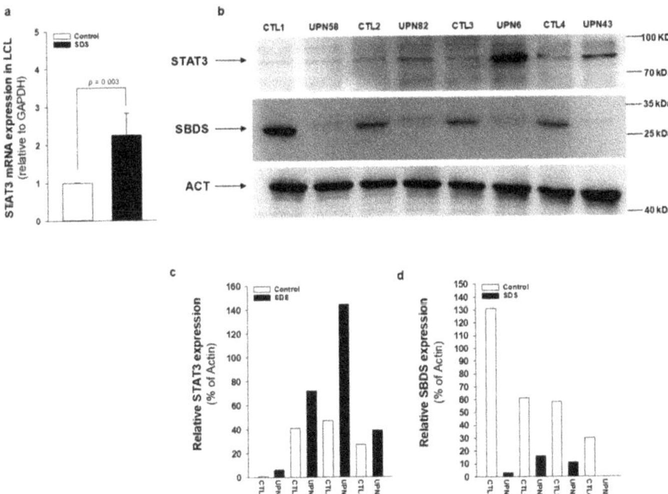

Figure 3. *STAT3* transcript and protein expression is upregulated in SDS patient-derived LCL. (**a**) *STAT3* mRNA expression in LCL isolated from UPN6, UPN43, UPN58, UPN82 (black bar), and from age-matched controls (white bar), measured by qRT-PCR. Data are mean ± SEM of four experiments performed in duplicate. (**b**) STAT3 protein level was measured in LCL (UPN6, UPN43, UPN58, UPN82) by Western blot analysis. (**c**,**d**) Densitometric analysis of Western blots showed in panel (**b**). Statistics: Normal distribution was tested by the Shapiro–Wilk test. Subsequently, the Mann–Whitney Rank Sum Test was calculated and reported within histograms.

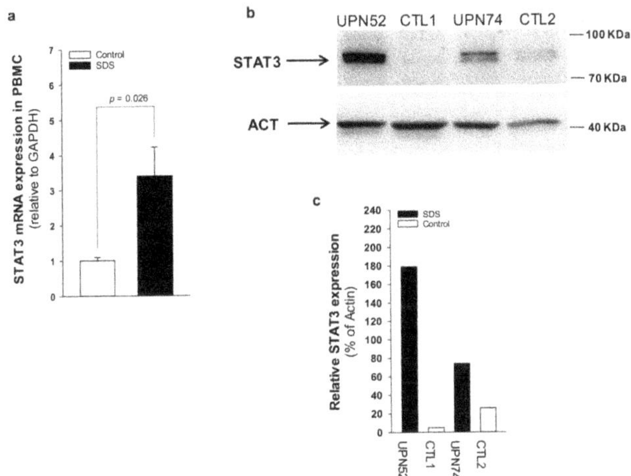

Figure 4. *STAT3* transcript and protein expression is upregulated in SDS patient-derived primary PBMC. (**a**) *STAT3* mRNA expression in primary PBMC isolated from UPN26, UPN69, UPN73, UPN87, UPN94, UPN106, and UPN108 (black bar), and from age-matched controls (white bar), measured by qRT-PCR. Data are mean ± SEM of seven experiments. (**b**) STAT3 protein level was measured in PBMC (UPN52 and UPN74) by Western blot analysis. (**c**) Densitometric analysis of Western blots showed in panel **b**. Statistics: Normal distribution was tested by the Shapiro–Wilk test. Subsequently, the Mann-Whitney Rank Sum Test was calculated and reported within histograms.

2.2. IL-6 Expression Is Upregulated in SDS

Plasma levels of IL-6 are generally close to the detection limit (1 pg/mL) in healthy individuals but significantly increase during the inflammatory process and cancers [32]. IL-6 is a major activator of JAK-STAT3 signaling, and *IL-6* transcript expression is upregulated by STAT3 activation, generating an autocrine/paracrine loop of activation [14]. Since mTOR-STAT3 axis is upregulated in SDS patient-derived myeloid cells [26], we sought to find out whether hyper-activation of this pathway is associated with *IL-6* over-expression in non-myeloid SDS cell models. Both LCL and BM-MSC obtained from patients with SDS displayed significantly upregulated IL-6 release into cell culture supernatants compared with age-matched healthy controls (Figure 5a,b). In particular, IL-6 levels were significantly elevated in primary SDS BM-MSC, which released ~8 ng/mL in 2×10^5 cells in our experimental conditions (Figure 5b). Plasma samples obtained from the peripheral blood collected from an expanded cohort of 21 patients with SDS showed significantly increased levels of IL-6 (mean 3.66 ± 4.58) compared to aged-matched controls (mean 1.19 ± 1.89), consistent with *in vitro* results (Figure 5c). Since SDS BM-MSC showed an impressive upregulation of *IL-6* expression, we sought to find out whether IL-6 were further concentrated in plasma derived from the bone marrow of patients. IL-6 levels in bone marrow plasma were even more elevated (mean 4.75 ± 3.82) than those found in peripheral blood (mean 3.04 ± 2.05) obtained in parallel, from the same patients (Figure 5d).

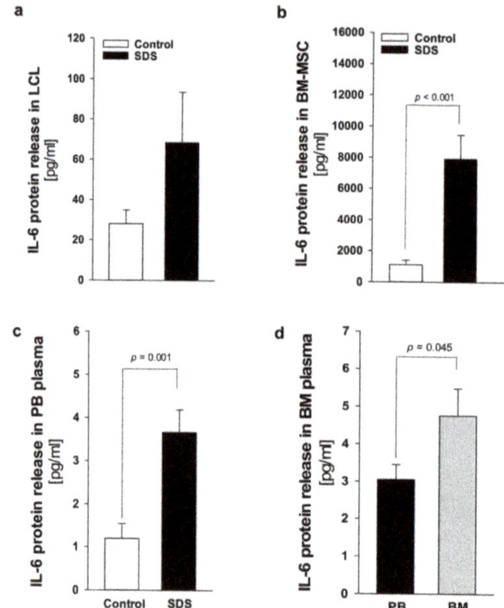

Figure 5. IL-6 release is elevated in SDS specimens compared to age-matched donor controls. (**a**) Measurement of IL-6 released in supernatants collected from 1×10^6 LCL after 48h. Data are mean ± SEM of 10 experiments conducted on LCL obtained from UPN24, UPN26, UPN58, UPN68, UPN75, and UPN106. (**b**) IL-6 released in supernatants collected from 2×10^5 primary BM-MSC after 48 h. (**c**) IL-6 concentration in peripheral blood (PB) plasma samples obtained from 21 patients with SDS (UPN1, 13, 26, 37, 47, 52, 56, 57, 58, 63, 65, 66, 69, 72, 73, 74, 87, 94, 104, 106, 108) compared with age-matched plasma controls. (**d**) IL-6 concentration in bone marrow (BM) plasma samples obtained from eight patients with SDS (UPN 47, 56, 65, 74, 87, 94, 106, 108) compared to PB plasma samples obtained from the same patients. Normal distribution was tested by the Shapiro–Wilk test. The Mann–Whitney Rank Sum Test was calculated and reported within histograms.

2.3. Patients with SDS ShowReduced Levels of Soluble IL-6 Receptor

IL-6 signaling cascade occurs by classical activation through mbIL-6R or trans-signaling via the soluble sIL-6R [19]. The latter mechanism allows cells that do not express mbIL-6R, but do express gp130, to be responsive to IL-6. Undifferentiated MSC lack the expression of IL-6 receptor although they normally express gp130 [33]. Thus, trans-signaling should be required to activate these cells upon IL-6 stimulation. Nevertheless, BM-MSC can constitutively release large quantities of IL-6 [34]. In this work, we quantified sIL-6R in plasma samples obtained from 21 patients with SDS. In healthy individuals, plasma levels of sIL-6R range between 50–70 ng/mL [35]. Of note, we found that the soluble receptor release is reduced in SDS patients (44.3 ± 15.4 pg/µL) compared to age-matched healthy donors (71.6 ± 14.7 pg/µL) (Figure 6a). To verify whether the bone marrow compartment, which showed increased levels of IL-6, exhibits also higher levels of sIL-6R, we compared sIL-6R expression in peripheral blood plasma with the expression found in bone marrow plasma obtained in parallel from the same patients. However, peripheral blood and bone marrow plasma showed comparable levels of sIL-6R expression (Figure 6b).

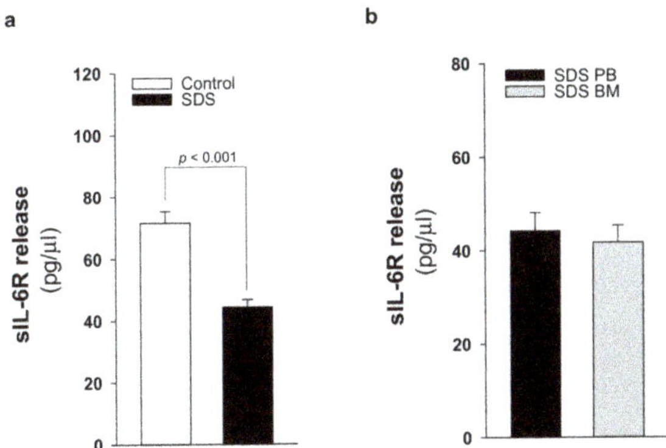

Figure 6. sIL6-R release is reduced in patients with SDS. (**a**) sIL-6R protein release was quantified by ELISA in PB plasma samples obtained from 21 patients with SDS (UPN 1, 13, 26, 37, 47, 52, 56, 57, 58, 63, 65, 66, 69, 72, 73, 74, 87, 94, 104, 106, 108) compared with age-matched plasma controls. (**b**) sIL-6R concentration in BM plasma samples obtained from eight patients with SDS (UPN 47, 56, 65, 74, 87, 94, 106, 108) compared with PB plasma samples obtained from the same patients. Normal distribution was tested by the Shapiro–Wilk test. The Mann–Whitney Rank Sum Test was calculated and reported within histograms.

2.4. Elevated IL-6 Gene Expressionin Hematopoietic CellsIs Primarily Driven by mTOR-STAT3 Pathway in SDS

Fifty to eighty percent of patients with AML present a constitutive activation of the mTOR pathway, showing significantly shorter disease-free and overall survival rates compared with patients without constitutive activation [36,37]. In the last decade, the development of new rapamycin (sirolimus) analogs showing improved pharmacokinetic profile, such as the clinically approved rapalog everolimus, have given rise to great interest for anti-cancer therapy [29]. We previously reported that rapamycin-dependent mTOR inhibition leads to normal levels of phosphorylation of STAT3 in SDS cells [26]. Here we show that everolimus (350 nM) and STAT3 inhibitor stattic (7.5 µM) significantly reduce *IL-6* mRNA expression in LCL and in primary BM-MNC obtained from patients with SDS (Figure 7a,c). Decreased *IL-6* mRNA expression is paralleled by a reduction of IL-6 release in culture supernatants upon everolimus and stattic treatments both in LCL (-46.6% and -68%, respectively) and BM-MNC (-34.6% and -66%, respectively) (Figure 7b,d). In order to validate these data, we also knocked-down the expression of *mTOR* and *STAT3* in SDS cells using a short interference (si)RNA strategy. To this aim, we transfected two different siRNA sequences for each target gene and a negative control sequence (scrambled), which was previously validated [29], using siRNA-specific liposomes as a vector. We significantly knocked-down *mTOR* and *STAT3* gene expression in SDS LCL (Figure 8a,b). Consistently with pharmacological inhibition, both *mTOR* and *STAT3* gene silencing lead to a strong inhibition of *IL-6* expression in SDS cells (Figure 8c,d). In particular, knock-down of the *STAT3* gene resulted in a statistically significant reduction of *IL-6* expression in terms of mRNA (-85%) and protein release (-51%). However, no effect of everolimus nor stattic on IL-6 release was reported in BM-MSC obtained from four SDS patients (Figure 7f), and *IL-6* mRNA transcription was surprisingly increased (two-fold increase) upon stattic treatment in these cells (Figure 7e).

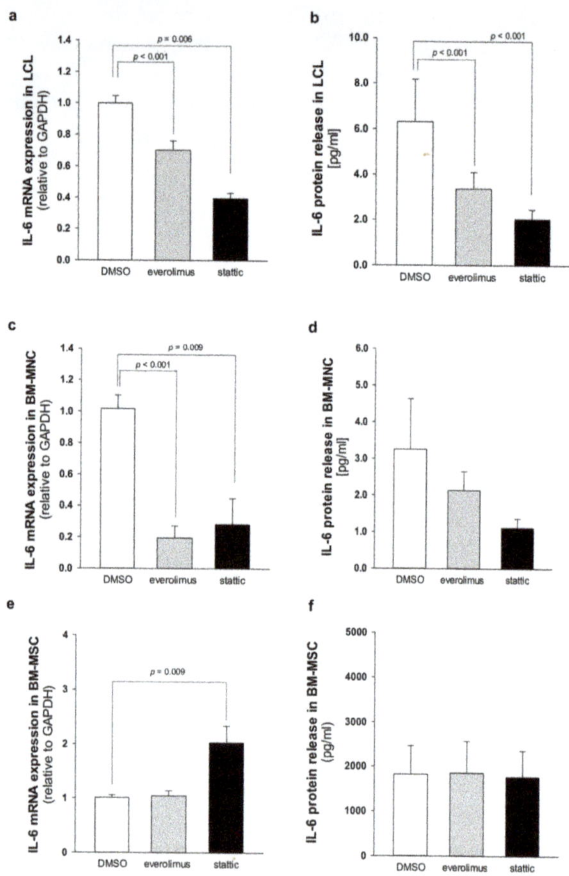

Figure 7. Everolimus and stattic inhibit *IL-6* expression in SDS patient-derived hematopoietic cells. (**a**) *IL-6* transcript expression in LCL incubated in the absence (DMSO) or in the presence of 350 nM everolimus or 7.5 μM stattic for 24 h was quantified by qRT-PCR. Data are mean ± SEM of six experiments performed in duplicate from three affected individuals (UPN58, UPN75, and UPN106). (**b**) IL-6 release in supernatants collected from LCL incubated in the absence (DMSO) or in the presence of 350 nM everolimus or 7.5 μM stattic for 24 h, as measured by Bio-plex assay. Data are mean ± SEM of six experiments conducted as reported in panel a. (**c**) *IL-6* mRNA expression in primary BM-MNC incubated in the absence (DMSO) or in the presence of 350 nM everolimus or 7.5 μM stattic for 24 h was

quantified by qRT-PCR. Data are mean ± SEM of four experiments performed in duplicate from four affected individuals (UPN74, UPN80, UPN94, and UPN106). (**d**) IL-6 release in supernatants collected from BM-MNC as measured by Bio-plex assay. Data are mean ± SEM of four experiments conducted as reported in panel **c**. (**e**) IL-6 transcript expression in BM-MSC incubated in the absence (DMSO) or in the presence of 350 nM everolimus or 7.5 μM stattic for 24 h was quantified by qRT-PCR. Data are mean ± SEM of four experiments performed in duplicate from four affected individuals (UPN33, UPN35, UPN67 and UPN91). (**f**) IL-6 release in supernatants collected from BM-MSC (UPN33, UPN35, UPN67 and UPN91) as measured by Bio-plex assay. Data are mean ± SEM of four experiments conducted as reported in panel **e**. Statistics: Normal distribution was tested by the Shapiro–Wilk test, and the Student's t test for paired data has been calculated and reported within histograms, accordingly.

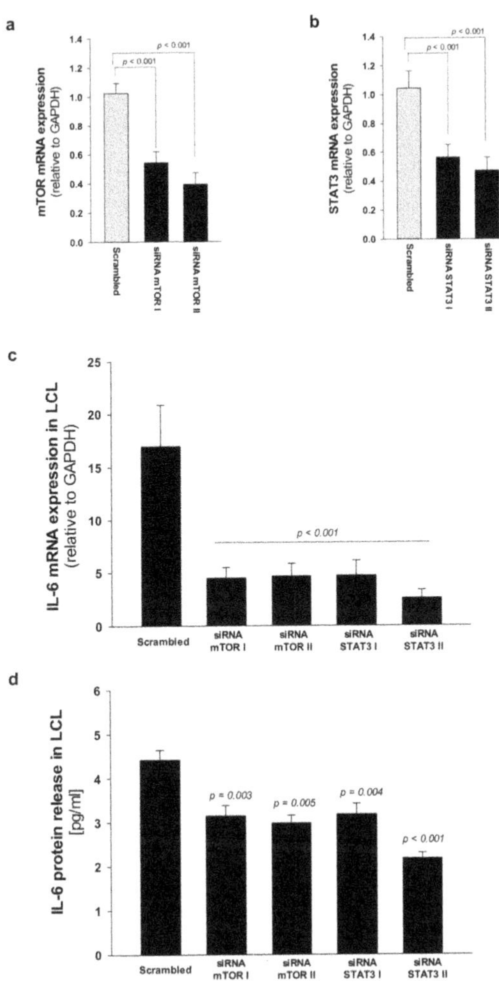

Figure 8. *STAT3* and *mTOR* gene silencing inhibit *IL-6* expression in LCL from SDS patients. (**a**) Reduced expression of *mTOR* mRNA after siRNA-mediated gene silencing. LCL obtained from UPN58, UPN75

and UPN106 were cultured with two different siRNA sequences against target genes, or scrambled sequence (negative control for 48 h). Data are mean ± SEM of four experiments performed in duplicate. (**b**) Reduced expression of STAT3 mRNA after siRNA-mediated gene silencing in LCL, as performed in panel a. Data are mean ± SEM of four experiments performed in duplicate. (**c**) *IL-6* mRNA expression in LCL treated as indicated in panels (**a**) and (**b**). (**d**) IL-6 release was measured by Bio-plex assay in supenatants obtained from LCL treated as indicated in **a**. Statistics: Normal distribution was tested by the Shapiro–Wilk test. Subsequently, the Mann–Whitney Rank Sum Test was calculated and reported within histograms.

3. Discussion

Although it has been widely reported that SDS mainly involves the neutrophil lineage, a number of patients suffer from anemia, thrombocytopenia or pancytopenia (Table 1). Because the bone marrow is often hypocellular, lymphoid and stromal cells may contribute to reduced blood cell formation and myelodysplasia. We recently described a severe deficit of T cells, in particular of DN T subpopulation in SDS patients [38]. Here we show that T cell subpopulations isolated from SDS patients display also hyper-activation of mTOR-STAT3 pathway. In addition, *STAT3* transcript and protein expression are markedly increased in PBMC and LCL obtained from SDS patients, confirming the involvement of the STAT3 pathway in lymphoid lineages. Given the key role of STAT3 in reducing T regulatory cell accumulation [39], our results could partially explain the reduced number of DN T cells observed in SDS patients [38]. Furthermore, STAT3 activation in lymphocytes is associated with T cell impaired functions [17] and reduced anti-tumor activity [16,40]. Accordingly, impaired functions of T cells have been previously described in SDS [41]. Thus, STAT3 pathway upregulation could lead to harmful consequences both on myeloid differentiation in the bone marrow and on innate and adaptive immunosurveillance mechanisms in SDS.

IL-6 is a major activator of STAT3, and *IL-6* transcript expression is itself a target of STAT3 [14]. Thus, we measured *IL-6* expression both at mRNA and protein levels in several SDS cell types. We found that *IL-6* expression is elevated in LCL, primary BM-MNC and BM-MSC, compared to age-matched healthy controls. BM-MSC produce large amounts of IL-6 even in unstimulated and undifferentiated conditions [34]. Accordingly, BM-MSC obtained from SDS patients released huge amounts of IL-6 protein into the supernatants (~8 ng/mL, from 2×10^5 cells), that is an amount comparable with the dose of IL-6 commonly used to stimulate in vitro these cells (10 ng/mL). Interestingly, that concentration of IL-6 has been also reported as the driving force leading to a further increase of mTOR-STAT3 activation in SDS leukocytes [26], suggesting the existence of a feedforward autocrine/paracrine feedback loop between STAT3 and IL-6 in SDS bone marrow. In order to verify this hypothesis in clinical samples, we measured peripheral blood plasma levels of IL-6 in a cohort of 21 patients with SDS. We found that IL-6 levels in peripheral blood plasma are elevated in SDS patients. Then, we measured IL-6 released into bone marrow plasma and we found even more elevated cytokine levels. Since BM-MNC do release very low amounts of IL-6 compared to BM-MSC, we speculate that IL-6 accumulation into the bone marrow environment is mainly due to the contribution from the stromal compartment. The MSC compartment is involved in AML development, contributing to disease initiation both in animal models and in patients [42–44]. For instance, studies on BM-MSC obtained from patients with MDS or AML reported altered expression of several cytokines and other soluble pro-inflammatory mediators [45]. Thus, increased interleukin-6 levels in bone marrow of patients with SDS, together with mTOR-STAT3 axis hyper-activation in hematopoietic cells, may clarify the reason why these subjects are prone to develop AML. We previously reported that rapamycin (sirolimus) treatment reduced the phosphorylation of both mTOR (S2448) and STAT3 (Y705 and S727) in cell lines as well as in primary neutrophils, monocytes and B cells isolated from patients with SDS. Here, we show that the clinically approved rapalog, everolimus, can restore normal level of activation of both mTOR and STAT3 in SDS lymphocyte populations. Importantly, in this study we show that everolimus and a commercially available STAT3 chemical inhibitor, namely stattic, can significantly inhibit IL-6 release in SDS LCL and BM-MNC. In the light of these findings, novel clinically approved inhibitors of mTOR and STAT3 might be helpful in reducing pro-leukemic pathways in SDS hematopoietic cells. STAT3 inhibitors are being evaluated as chemotherapeutic agents in leukemias, due to their strong pro-apoptotic activity [13]. Although we used 7.5 µM stattic, which was a previously reported dose that can inhibit STAT3 without affecting lymphoid cell viability in vitro [46], we observed a stattic-dependent induction of late apoptosis in LCL (Figure S3). We cannot exclude the fact that the effect observed in this case may partially be due to pro-apoptotic activity. However, since everolimus reduced both STAT3 activation and *IL-6* expression in SDS LCL and BM-MNC without inducing pro-apoptotic processes (Figure S3), we can assume that mTOR-STAT3 pathway inhibition might be useful in reducing the excessive cytokine release. Both everolimus and stattic did not reduce the huge release of IL-6 from undifferentiated SDS BM-MSC, which remains one of the main sources of IL-6 within the bone marrow compartment. This finding suggests other regulatory pathways of *IL-6* gene expression exist in BM-MSC.

IL-6 can trigger JAK-STAT3 signaling activation through the direct binding to mbIL-6R or via the soluble sIL-6R (IL-6 trans-signaling), as result of alternative splicing or protease cleavage of mbIL-6R [19]. In order to evaluate whether the mTOR-STAT3-IL6 loop is generated by increased IL-6 trans-signaling in SDS, we measured sIL-6R levels in plasma samples obtained from 21 patients. However, results indicate that sIL-6R expression is reduced in SDS patients compared to age-matched healthy donors, thus suggesting that this loop is mainly generated via the classical IL-6 signaling in hematopoietic cells. Interestingly, ex vivo undifferentiated BM-MSC lack the expression of mbIL-6R, even if they normally do express the gp130 protein [33]. Besides, only a few cell types do express mbIL-6R and therefore are able to respond to IL-6 generating the IL-6-STAT3-IL6 loop of activation. Among these cells, major players are represented by macrophages, neutrophils, T-cells, and hepatocytes. On the contrary, gp130 is generally ubiquitously expressed [47,48]. It has been suggested that the lack of mbIL-6R is a regulatory mechanism of BM-MSC that inhibits the IL-6-dependent chondrogenic/osteogenic differentiation, thus maintaining the stemness [33]. During osteogenic differentiation, the expression of IL-6R indeed increases in BM-MSC, allowing the autocrine/paracrine activation of the IL-6-STAT3 signaling pathway [49]. Thus, reduced sIL-6R levels in plasma suggest that the impaired ossification reported in SDS [50] might be partially due to decreased IL-6 trans-signaling in BM-MSC. The lack of IL-6 trans-signaling might therefore also partially explain why everolimus and stattic cannot reduce *IL-6* expression in BM-MSC. Our results indicate that BM-MSC are unable to activate the autocrine/paracrine feedforward loop of mTOR-STAT3-IL6.

4. Materials and Methods

4.1. Human Subjects

Human samples were obtained and analyzed in accordance with the declaration of Helsinki, after written consent. All protocols were approved by the Ethics Committee of San Gerardo Hospital (Monza, Italy), approval No. 504 4/9/2012, Azienda Ospedaliera Universitaria Integrata (Verona, Italy), approval No. 658 CESC, and Azienda Ospedaliero Universitaria Ospedali Riuniti (Ancona, Italy), approval No. CERM 2018-82.

4.2. Plasma Isolation

Bone marrow and peripheral blood specimens were collected into EDTA-containing tubes from patients. Specimens were collected during a bone marrow harvest from healthy donors serving as donors for a related HLA-matched transplant, as permitted by the local hospital ethics committee and after informed consent was obtained. Peripheral blood and bone marrow plasma samples were prepared by centrifugation for 10 min at 600× *g* at 4 °C. To obtain platelet-poor plasma fractions, another centrifugation was performed for 15 min at 1800× *g*. Plasma specimens were then stored at −80 °C until analysis.

4.3. Cell Cultures

LCL, PBMC, BM-MNC, and BM-MSC were obtained from peripheral blood or bone marrow samples from31 patients with Shwachman–Diamond Syndrome. All patients enrolled in this study were diagnosed as affected by SDS on the basis of clinical and genetics criteria, and they were excluded if MDS/AML were present (Table 1). None of the patients underwent granulocyte colony-stimulating factor (G-CSF, filgrastim) nor steroids therapies. PBMC and BM-MNC were separated by stratification on Ficoll-Paque PLUS (density 1.077 g/mL, GE Healthcare, Waukesha, WI, USA) gradient and washed twice with PBS (Euroclone, Milan, Italy). Cells were seeded in 6-wells cell culture plates at 1×10^6 BM-MNC/well in 1 mL RPMI-1640 Medium (Gibco, Waltham, MA, USA) supplemented with 10% FBS (Sigma-Aldrich, St. Louis, MO) and incubated in the presence or absence of 350 nM everolimus (Sigma-Aldrich) or 7.5 µM stattic (Selleckchem, Houston, TX, USA) at 37 °C in humidified atmosphere, with 5% CO_2 for 24 h. Treated cell pellets were collected by centrifugation and the supernatant isolated and stored at −80 °C. BM-MSC were isolated after seeding 1.6×10^5 BM-MNC cells/cm^2 from fresh bone marrow, in Dulbecco's Modified Eagle Medium low glucose (Gibco) supplemented with 10% of FBS, 1% L-glutamine and 1% penicillin/streptomycin. BM-MNC were incubated at 37 °C in humidified atmosphere, with 5% CO_2 for 24 h. Subsequently, non-adherent cells were removed by washing with PBS and culture medium was finally replaced.BM-MSC were seeded at 4×10^3 cell/cm^2 and incubated at 37 °C in humidified atmosphere, with 5% CO_2 since they reached 70–80% confluence. 1×10^5 MSC were seeded in a 6-well plate and incubated for 48 h at 37 °C in humidified atmosphere, with 5% CO_2. Eventually, treated cells were collected and the supernatant isolated and stored at −80 °C. To obtain LCL, B cells were seeded at a density of 3×10^6 cells in 12-well cell culture plates, incubated with 3 mL RPMI-1640 supplemented with 10% FBS, and infected for 18 h with Epstein–Barr virus (EBV) released from marmoset blood leukocytes B95.8 virus-producer cell lines as previously described [26]. 1×10^6 LCL were seeded in 6 well-cell culture plate in RPMI-1640 supplemented with 10% FBS and incubated for 48 h at 37 °C in humidified atmosphere, with 5% CO_2. Eventually,1×10^6 LCL were seeded in 6 well-cell culture plate in RPMI-1640 supplemented with 10% FBS, in the presence or absence of 350 nM everolimus or 7.5 µM stattic at 37 °C in humidified atmosphere, with 5% CO_2 for 24 h. Treated cell pellets were collected by centrifugation and the supernatant isolated and stored at −80 °C. LCL, BM-MNC and MSC cultured in medium containing DMSO (Sigma, dilution 1:10,000) were used as negative control.

4.4. Flow Cytometry

Plasma was separated from peripheral blood samples derived from SDS patients or healthy subjects by centrifugation at 1500× *g* for 10 min. Red blood cells were lysed in 40 mL of solution containing 0.89% (*w/v*) NH_4Cl, 0.10% (*w/v*) $KHCO_3$ and 200 µM EDTA. Leukocytes were cultured in 6-well plates containing RPMI-1640 supplemented with 10% freshly prepared, heat-inactivated human plasma, in the presence or absence (DMSO) of 350 nM everolimus for 1h. Fifty to one-hundred-100 microliter aliquots of blood cells were incubated for 15 min at room temperature with combinations of the following fluorochrome-conjugated monoclonal antibodies: CD3-APC750, CD4-PC7, CD8-ECD, CD16-Pacific Blue, CD19-Chrome-Orange, CD45-APC700, and CD56-PC5 (Beckman-Coulter). Cells were gently centrifuged (600× *g*) for 10 min, washed with ice-cold PBS, fixed, and permeabilized with Intracellular Fixation and Permeabilization Buffer Set (eBioscience, San Diego, CA, USA), following the manufacturer's recommendations. Subsequently, cells were washed with flow buffer and stained with anti-pS727-STAT3-PE, anti-Y705-STAT3-PE, anti-p-S2448-mTOR-PE or isotype control-PE conjugated antibodies for 30 min. Data acquisition was performed by a 10 color, three laser Navios flow cytometer (Beckman Coulter, Indianapolis, IN, USA). Analysis of the acquired data was performed by the "Navios" or Kaluza software, version 1.3 (Beckman Coulter, Indianapolis, IN, USA). In order to define different subsets of lymphocytes, gating strategy and data filtering were established as follows. Cell debris were excluded using a dot plot for morphological parameters (FS, SS). Lymphocytes were gated into the side scatter (SS low) and CD45 positive area. Within the lymphocyte compartment, CD3$^+$

events were further gated into CD4$^+$ and CD8$^+$ T cells. NK cells were instead identified as CD3/CD19 double-negative events (CD56$^+$ and/or CD16$^+$). Double-negative T cells were gated by plotting T cells (CD3$^+$) in CD4 versus CD8 dot plots. Since it has been reported that TCR $\gamma\delta^+$T cells are recognizable by the bright expression on their membrane of the CD3 molecule [51], CD3 bright DN T cells (DN $\gamma\delta$) were gated by plotting CD3 bright positive area into CD4 versus CD8 dot plots. Flow cytometry was performed on at least 25,000 events. The total white blood cell count (WBC) was determined by Hematology Analyzer XN- 9000 (Sysmex Europe GmbH, Norderstedt, Germany).

4.5. qRT-PCR

Total RNA from LCL and MNC was isolated using High Pure RNA Isolation Kit (Roche, Mannheim, Germany) following the manufacturer's instructions. RNA was eluted in 50 µL of RNAse free water. Final RNA concentration was determined using NanoDrop 2000 spectrophotometer (Thermo Fisher Scientific, Foster City, CA, USA) and then stored at −80 °C until use. A total amount of 500 ng of RNA was reverse transcribed to cDNA using a High Capacity cDNA Reverse Transcription Kit with random primers (Applied Biosystems by Thermo Fisher Scientific) for 120 min at 37 °C and 5 min at 85 °C in a final reaction volume of 20 µL. For real-time qPCR analysis, 5 Ml of cDNA were used for each reaction to quantify the relative gene expression. The cDNA was then amplified for 45 PCR cycles using Platinum SYBR Green qPCR Super Mix-UDG (Invitrogen by Thermo Fisher Scientific) in a 20 µL reaction using the Rotor-Gene 6000 cycler (Qiagen, Hilden, Germany). In order to perform the PCR reaction, QuantiTect Primer Assays (Qiagen) for IL-6 (Hs_IL6_1_SG, NM_000600), STAT3 (Hs_STAT3_1_SG, NM_003150), mTOR (Hs_MTOR_1_SG, NM_004958), and GAPDH (HS_GAPDH_1_SG, NM_001256799) were purchased. Each target gene expression was normalized with GAPDH gene expression and relative quantification was calculated by the ΔCt method.

4.6. IL-6 and sIL-6R Detection

Bio-Plex Pro Human IL-6 Assay (Bio-Rad Laboratories, Philadelphia, PA, USA) was used to measure IL-6 protein released into cell supernatants and plasma samples (sample volume 50 µL). The assays were performed using the Bio-Plex Suspension Array System, with the Bio-Rad 96-well plate reader. Data were analyzed by the Bio-Plex Manager software (Bio-Rad Laboratories). The in-vitro quantitative determination of soluble IL-6 Receptor (sIL-6R) in plasma was performed by using the Human IL-6R/CD126 ELISA Kit (Origene, Rockville, MD, USA), according to the manufacturer's instructions. Plasma samples were assayed after 1:200 dilution.

4.7. Western Blot

A total of 30 µg of cell extract was denatured for 5 min at 95 °C in 4× Laemmli Sample Buffer, (277.8 mM Tris-HCl, pH 6.8, 44.4% glycerol, 4.4% LDS, 0.02% bromophenol blue) (Bio-Rad Laboratoires) supplemented with 355 mM 2-mercaptoethanol. The samples were loaded on 7.5% polyacrylamide SDS-PAGE gel in Tris-glycine Buffer (25 mM Tris, 192 mM glycine, and 0.1% SDS) using tag protein ladder (Spectra Multicolor Broad Range Protein Ladder, Thermo Scientific, Waltham, MA, USA) to determine molecular weight. The electrotransfer was performed into nitrocellulose membrane (iBlot Gel Transfer Stacks Nitrocellulose, Thermo Fisher) at 20V using iBlot Dry Blotting System (Invitrogen, Waltham, MA, USA) for 10 min. The membranes were blocked in BSA 5% for 90 min at room temperature and washed with TBS (5 mM Tris-HCl pH 7.6, 150 mM NaCl) supplemented with 0.05% tween-20 (TBS/T) for 15 min. Subsequently, membranes were probed with primary anti-human STAT3 rabbit polyclonal antibody (SAB2104912 Sigma-Aldrich, Missouri, USA, dilution 1:500) in primary antibody dilution buffer (TBS/T with 5% BSA) and incubated overnight at 4 °C. After washes, membranes were incubated with mouse anti-rabbit IgG- Horseradish Peroxidase-Coupled secondary antibody (Sigma-Aldrich, dilution 1:2000) diluted TBS/T for 90 min at room temperature. Immunocomplexes were visualized using chemiluminescence (ECL Plus Western Blotting Substrate, Pierce, Thermo Scientific), incubating ECL for 5 min at room temperature. Band intensity was calculated by scanning video densitometry using the Chemi Doc imaging system (UVP, LCC, Upland, CA, USA). Blots were re-probed with monoclonal β-Actin-Peroxidase clone AC-15 antibody (Sigma-Aldrich, dilution1:5000) in TBS/T for 90 min.

4.8. Gene Silencing

Both *mTOR* and *STAT3* genes were knocked-down by siRNA. The lipid-based agent for reverse transfection, siPORT NeoFX (Thermo Fisher), was used according to the manufacturer's instructions. LCL derived from healthy donors and SDS patients were transiently transfected with two different specific siRNA sequences designed against mTOR and STAT3 genes or with a scrambled sequence as control. The siRNA molecules were complexed with cationic liposomes, siPORT Neo-FX (Thermo Fisher). Briefly, siPORT Neo-FX (1 μL/well) was complexed with siRNA or Scrambled oligos (40 nM each) in 250 μL RPMI-1640 medium supplemented with 0.5% FBS. 1×10^5 LCL were incubated in 24-wells plates for 48 h at 37 °C. Knock-down of *mTOR* and *STAT3* gene expression was determined by Real Time qRT-PCR.

4.9. Statistical Analysis

Normal distribution was tested in each experiment using the Shapiro–Wilk test. Based on that evaluation, independent group determination was tested using Mann–Whitney test, while Student's *t*-test was used in case of paired data. A *p*-value < 0.05 was considered statistically significant. The statistical software SigmaPlot (Systat Software Inc., San Jose, CA, USA) was used.

5. Conclusions

STAT3 acts as a double-edged sword in SDS cells, as this pathways controls myeloid progenitor growth and proliferation [52] and promotes leukemogenesis. Extending our findings on mTOR-STAT3 signaling dysregulation in myeloid lineage [26], we now show that constitutive activation of mTOR-STAT3 axis occurs in the lymphoid compartment of SDS patients. An autocrine or paracrine feedforward loop of STAT3-IL6 exists in hematopoietic SDS cells. Since the loss of *SBDS* expression in healthy donor-derived cells is sufficient to reproduce the hyper-activation of mTOR-STAT3 signaling [26], we assume that alteration of the mTOR-STAT3-IL6 axis could be used by SDS cells as a survival mechanism that induces cell proliferation and myeloid growth in bone marrow, thus trying to escape from incoming neutropenia/aplasia processes.

Supplementary Materials: The following are available online at http://www.mdpi.com/2072-6694/12/3/597/s1, Figure S1: Whole blots showing all the bands with all molecular weight markers on the Western blot reported in Figure 3b, Figure S2: Whole blots showing all the bands with all molecular weight markers on the Western blot reported in Figure 4b, Figure S3: Apoptosis assay in LCL.

Author Contributions: Conceptualization, V.B. (Valentino Bezzerri), M.B. and M.C.; methodology, A.V. and V.B.(Valentino Bezzerri); formal analysis, A.V., E.D., M.B. and V.B. (Valentino Bezzerri); investigation, E.D., M.A. (Martina Api), M.A. (Marisole Allegri), G.B. (Giulia Breveglieri), A.G., E.M.B., T.C. and G.B. (Gloria Bedini).; data curation, V.B.(Valentino Bezzerri),A.V., S.J.C. and M.B.; writing—original draft preparation, V.B. (Valentino Bezzerri); writing—review and editing, S.J.C., M.B. and M.C.; supervision, V.B. (Valentino Bezzerri), M.B., M.C.; project administration, V.B. (Valentino Bezzerri), G.B. (Giulia Breveglieri). and M.B.; resources, B.F., V.B. (Vincenzo Bronte)., M.C., C.S., A.B., A.P., C.B. and G.D.; Funding acquisition, V.B. (Valentino Bezzerri), G.B. (Giulia Breveglieri), M.B and M.C. All authors have read and agreed to the published version of the manuscript.

Funding: This research was funded by the Italian Ministry of Health, grant number GR-2016-02363570 to V.B. (Valentino Bezzerri), G.B. (Giulia Breveglieri), M.B. and M.C., the Associazione Italiana Sindrome di Shwachman-Diamond (AISS), grant number 9/2017 to V.B. (Valentino Bezzerri) and M.C, and the Rotaract Distretto 2060 association, Italy (district funding to V.N. (Valentino Bezzerri)).

Acknowledgments: The authors are grateful to Anastasia D'Antuono and Federica Masseria (Azienda Ospedaliero Universitaria Ospedali Riuniti, Ancona) for excellent patient care, to Emily Pintani for SDS Italian Registry data management, to Francesco Pasquali (Università degli Studi dell'Insubria) for cytogenetics of SDS patients, and to Laura Donnini and Silvia Duranti (Azienda Ospedaliero Universitaria Ospedali Riuniti, Ancona) for the admirable administrative support.

Conflicts of Interest: The authors declare no conflict of interest.

References

1. Dror, Y.; Donadieu, J.; Koglmeier, J.; Dodge, J.; Toiviainen-Salo, S.; Makitie, O.; Kerr, E.; Zeidler, C.; Shimamura, A.; Shah, N.; et al. Draft consensus guidelines for diagnosis and treatment of shwachman-diamond syndrome. *Ann. N Y Acad. Sci.* **2011**, *1242*, 40–55. [CrossRef]

2. Kargas, V.; Castro-Hartmann, P.; Escudero-Urquijo, N.; Dent, K.; Hilcenko, C.; Sailer, C.; Zisser, G.; Marques-Carvalho, M.J.; Pellegrino, S.; Wawiórka, L.; et al. Mechanism of completion of peptidyltransferase centre assembly in eukaryotes. *Elife* **2019**, *8*, e44904. [CrossRef] [PubMed]

3. Weis, F.; Giudice, E.; Churcher, M.; Jin, L.; Hilcenko, C.; Wong, C.C.; Traynor, D.; Kay, R.R.; Warren, A.J. Mechanism of eif6 release from the nascent 60s ribosomal subunit. *Nat. Struct. Mol. Biol.* **2015**, *22*, 914–919. [CrossRef]

4. Finch, A.J.; Hilcenko, C.; Basse, N.; Drynan, L.F.; Goyenechea, B.; Menne, T.F.; González Fernández, A.; Simpson, P.; D'Santos, C.S.; Arends, M.J.; et al. Uncoupling of gtp hydrolysis from eif6 release on the ribosome causes shwachman-diamond syndrome. *Genes Dev.* **2011**, *25*, 917–929. [CrossRef] [PubMed]

5. Corey, S.J.; Minden, M.D.; Barber, D.L.; Kantarjian, H.; Wang, J.C.; Schimmer, A.D. Myelodysplastic syndromes: The complexity of stem-cell diseases. *Nat. Rev. Cancer* **2007**, *7*, 118–129. [CrossRef] [PubMed]

6. Donadieu, J.; Leblanc, T.; Bader Meunier, B.; Barkaoui, M.; Fenneteau, O.; Bertrand, Y.; Maier-Redelsperger, M.; Micheau, M.; Stephan, J.L.; Phillipe, N.; et al. Analysis of risk factors for myelodysplasias, leukemias and death from infection among patients with congenital neutropenia. Experience of the french severe chronic neutropenia study group. *Haematologica* **2005**, *90*, 45–53. [PubMed]

7. Lindsley, R.C.; Saber, W.; Mar, B.G.; Redd, R.; Wang, T.; Haagenson, M.D.; Grauman, P.V.; Hu, Z.H.; Spellman, S.R.; Lee, S.J.; et al. Prognostic mutations in myelodysplastic syndrome after stem-cell transplantation. *N. Engl. J. Med.* **2017**, *376*, 536–547. [CrossRef]

8. Curran, E.K.; Godfrey, J.; Kline, J. Mechanisms of immune tolerance in leukemia and lymphoma. *Trends Immunol.* **2017**, *38*, 513–525. [CrossRef]

9. D'Acquisto, F.; Crompton, T. Cd3+cd4-cd8- (double negative) t cells: Saviours or villains of the immune response? *Biochem. Pharmacol.* **2011**, *82*, 333–340. [CrossRef]

10. Chen, B.; Lee, J.B.; Kang, H.; Minden, M.D.; Zhang, L. Targeting chemotherapy-resistant leukemia by combining dnt cellular therapy with conventional chemotherapy. *J. Exp. Clin. Cancer Res.* **2018**, *37*, 88. [CrossRef]

11. Lee, J.; Minden, M.D.; Chen, W.C.; Streck, E.; Chen, B.; Kang, H.; Arruda, A.; Ly, D.; Der, S.D.; Kang, S.; et al. Allogeneic human double negative t cells as a novel immunotherapy for acute myeloid leukemia and its underlying mechanisms. *Clin. Cancer Res.* **2018**, *24*, 370–382. [CrossRef] [PubMed]

12. Zhang, L.; Chen, X.; Liu, X.; Kline, D.E.; Teague, R.M.; Gajewski, T.F.; Kline, J. Cd40 ligation reverses t cell tolerance in acute myeloid leukemia. *J. Clin. Investig.* **2013**, *123*, 1999–2010. [CrossRef] [PubMed]

13. O'Shea, J.J.; Holland, S.M.; Staudt, L.M. Jaks and stats in immunity, immunodeficiency, and cancer. *N. Engl. J. Med.* **2013**, *368*, 161–170. [CrossRef] [PubMed]

14. Johnson, D.E.; O'Keefe, R.A.; Grandis, J.R. Targeting the il-6/jak/stat3 signalling axis in cancer. *Nat. Rev. Clin. Oncol.* **2018**, *15*, 234–248. [CrossRef] [PubMed]

15. Yu, H.; Kortylewski, M.; Pardoll, D. Crosstalk between cancer and immune cells: Role of stat3 in the tumour microenvironment. *Nat. Rev. Immunol.* **2007**, *7*, 41–51. [CrossRef] [PubMed]

16. Herrmann, A.; Kortylewski, M.; Kujawski, M.; Zhang, C.; Reckamp, K.; Armstrong, B.; Wang, L.; Kowolik, C.; Deng, J.; Figlin, R.; et al. Targeting stat3 in the myeloid compartment drastically improves the in vivo antitumor functions of adoptively transferred t cells. *Cancer Res.* **2010**, *70*, 7455–7464. [CrossRef]

17. Kujawski, M.; Zhang, C.; Herrmann, A.; Reckamp, K.; Scuto, A.; Jensen, M.; Deng, J.; Forman, S.; Figlin, R.; Yu, H. Targeting stat3 in adoptively transferred t cells promotes their in vivo expansion and antitumor effects. *Cancer Res.* **2010**, *70*, 9599–9610. [CrossRef]

18. Kumari, N.; Dwarakanath, B.S.; Das, A.; Bhatt, A.N. Role of interleukin-6 in cancer progression and therapeutic resistance. *Tumour Biol.* **2016**, *37*, 11553–11572. [CrossRef]

19. Chalaris, A.; Garbers, C.; Rabe, B.; Rose-John, S.; Scheller, J. The soluble interleukin 6 receptor: Generation and role in inflammation and cancer. *Eur. J. Cell Biol.* **2011**, *90*, 484–494. [CrossRef]

20. Audet, J.; Miller, C.L.; Rose-John, S.; Piret, J.M.; Eaves, C.J. Distinct role of gp130 activation in promoting self-renewal divisions by mitogenically stimulated murine hematopoietic stem cells. *Proc. Natl. Acad. Sci. USA* **2001**, *98*, 1757–1762. [CrossRef]

21. Walker, F.; Zhang, H.H.; Matthews, V.; Weinstock, J.; Nice, E.C.; Ernst, M.; Rose-John, S.; Burgess, A.W. Il6/sil6r complex contributes to emergency granulopoietic responses in g-csf- and gm-csf-deficient mice. *Blood* **2008**, *111*, 3978–3985. [CrossRef] [PubMed]

22. Han, Y.; Ye, A.; Bi, L.; Wu, J.; Yu, K.; Zhang, S. Th17 cells and interleukin-17 increase with poor prognosis in patients with acute myeloid leukemia. *Cancer Sci.* **2014**, *105*, 933–942. [CrossRef] [PubMed]

23. Sanchez-Correa, B.; Bergua, J.M.; Campos, C.; Gayoso, I.; Arcos, M.J.; Bañas, H.; Morgado, S.; Casado, J.G.; Solana, R.; Tarazona, R. Cytokine profiles in acute myeloid leukemia patients at diagnosis: Survival is inversely correlated with il-6 and directly correlated with il-10 levels. *Cytokine* **2013**, *61*, 885–891. [CrossRef] [PubMed]

24. Yokogami, K.; Wakisaka, S.; Avruch, J.; Reeves, S.A. Serine phosphorylation and maximal activation of stat3 during cntf signaling is mediated by the rapamycin target mtor. *Curr. Biol.* **2000**, *10*, 47–50. [CrossRef]

25. Dodd, K.M.; Yang, J.; Shen, M.H.; Sampson, J.R.; Tee, A.R. Mtorc1 drives hif-1α and vegf-a signalling via multiple mechanisms involving 4e-bp1, s6k1 and stat3. *Oncogene* **2015**, *34*, 2239–2250. [CrossRef]

26. Bezzerri, V.; Vella, A.; Calcaterra, E.; Finotti, A.; Gasparello, J.; Gambari, R.; Assael, B.M.; Cipolli, M.; Sorio, C. New insights into the shwachman-diamond syndrome-related haematological disorder: Hyper-activation of mtor and stat3 in leukocytes. *Sci. Rep.* **2016**, *6*, 33165. [CrossRef]

27. Chapuis, N.; Tamburini, J.; Green, A.S.; Willems, L.; Bardet, V.; Park, S.; Lacombe, C.; Mayeux, P.; Bouscary, D. Perspectives on inhibiting mtor as a future treatment strategy for hematological malignancies. *Leukemia* **2010**, *24*, 1686–1699. [CrossRef]

28. Hoshii, T.; Matsuda, S.; Hirao, A. Pleiotropic roles of mtor complexes in haemato-lymphopoiesis and leukemogenesis. *J. Biochem.* **2014**, *156*, 73–83. [CrossRef]

29. Porta, C.; Paglino, C.; Mosca, A. Targeting pi3k/akt/mtor signaling in cancer. *Front. Oncol.* **2014**, *4*, 64. [CrossRef]

30. Witzig, T.E.; Reeder, C.; Han, J.J.; LaPlant, B.; Stenson, M.; Tun, H.W.; Macon, W.; Ansell, S.M.; Habermann, T.M.; Inwards, D.J.; et al. The mtorc1 inhibitor everolimus has antitumor activity in vitro and produces tumor responses in patients with relapsed t-cell lymphoma. *Blood* **2015**, *126*, 328–335. [CrossRef]

31. Cai, Y.; Xue, F.; Qin, H.; Chen, X.; Liu, N.; Fleming, C.; Hu, X.; Zhang, H.G.; Chen, F.; Zheng, J.; et al. Differential roles of the mtor-stat3 signaling in dermal γδ t cell effector function in skin inflammation. *Cell Rep.* **2019**, *27*, 3034–3048. [CrossRef] [PubMed]

32. Rose-John, S. Il-6 trans-signaling via the soluble il-6 receptor: Importance for the pro-inflammatory activities of il-6. *Int. J. Biol. Sci.* **2012**, *8*, 1237–1247. [CrossRef] [PubMed]

33. Erices, A.; Conget, P.; Rojas, C.; Minguell, J.J. Gp130 activation by soluble interleukin-6 receptor/interleukin-6 enhances osteoblastic differentiation of human bone marrow-derived mesenchymal stem cells. *Exp. Cell Res.* **2002**, *280*, 24–32. [CrossRef] [PubMed]

34. Pricola, K.L.; Kuhn, N.Z.; Haleem-Smith, H.; Song, Y.; Tuan, R.S. Interleukin-6 maintains bone marrow-derived mesenchymal stem cell stemness by an erk1/2-dependent mechanism. *J. Cell Biochem.* **2009**, *108*, 577–588. [CrossRef]

35. Calabrese, L.H.; Rose-John, S. Il-6 biology: Implications for clinical targeting in rheumatic disease. *Nat. Rev. Rheumatol.* **2014**, *10*, 720–727. [CrossRef]

36. Altman, J.K.; Sassano, A.; Platanias, L.C. Targeting mtor for the treatment of aml. New agents and new directions. *Oncotarget* **2011**, *2*, 510–517. [CrossRef]

37. Dinner, S.; Platanias, L.C. Targeting the mtor pathway in leukemia. *J. Cell Biochem.* **2016**, *117*, 1745–1752. [CrossRef]

38. Bezzerri, V.; Vella, A.; Gennaro, G.D.; Ortolani, R.; Nicolis, E.; Cesaro, S.; Fabrizzi, B.; Bronte, V.; Corey, S.J.; Cipolli, M. Peripheral blood immunophenotyping in a large cohort of patients with shwachman-diamond syndrome. *Pediatr. Blood Cancer* **2019**, *66*, e27597. [CrossRef]

39. Durant, L.; Watford, W.T.; Ramos, H.L.; Laurence, A.; Vahedi, G.; Wei, L.; Takahashi, H.; Sun, H.W.; Kanno, Y.; Powrie, F.; et al. Diverse targets of the transcription factor stat3 contribute to t cell pathogenicity and homeostasis. *Immunity* **2010**, *32*, 605–615. [CrossRef]

40. Hossain, D.M.; Dos Santos, C.; Zhang, Q.; Kozlowska, A.; Liu, H.; Gao, C.; Moreira, D.; Swiderski, P.; Jozwiak, A.; Kline, J.; et al. Leukemia cell-targeted stat3 silencing and tlr9 triggering generate systemic antitumor immunity. *Blood* **2014**, *123*, 15–25. [CrossRef]

41. Dror, Y.; Ginzberg, H.; Dalal, I.; Cherepanov, V.; Downey, G.; Durie, P.; Roifman, C.M.; Freedman, M.H. Immune function in patients with shwachman-diamond syndrome. *Br. J. Haematol.* **2001**, *114*, 712–717. [CrossRef]

42. Raaijmakers, M.H.; Mukherjee, S.; Guo, S.; Zhang, S.; Kobayashi, T.; Schoonmaker, J.A.; Ebert, B.L.; Al-Shahrour, F.; Hasserjian, R.P.; Scadden, E.O.; et al. Bone progenitor dysfunction induces myelodysplasia and secondary leukaemia. *Nature* **2010**, *464*, 852–857. [CrossRef]

43. Santamaría, C.; Muntión, S.; Rosón, B.; Blanco, B.; López-Villar, O.; Carrancio, S.; Sánchez-Guijo, F.M.; Díez-Campelo, M.; Alvarez-Fernández, S.; Sarasquete, M.E.; et al. Impaired expression of dicer, drosha, sbds and some micrornas in mesenchymal stromal cells from myelodysplastic syndrome patients. *Haematologica* **2012**, *97*, 1218–1224. [CrossRef] [PubMed]

44. Geyh, S.; Oz, S.; Cadeddu, R.P.; Fröbel, J.; Brückner, B.; Kündgen, A.; Fenk, R.; Bruns, I.; Zilkens, C.; Hermsen, D.; et al. Insufficient stromal support in mds results from molecular and functional deficits of mesenchymal stromal cells. *Leukemia* **2013**, *27*, 1841–1851. [CrossRef]

45. Schroeder, T.; Geyh, S.; Germing, U.; Haas, R. Mesenchymal stromal cells in myeloid malignancies. *Blood Res.* **2016**, *51*, 225–232. [CrossRef] [PubMed]

46. Severin, F.; Frezzato, F.; Visentin, A.; Martini, V.; Trimarco, V.; Carraro, S.; Tibaldi, E.; Brunati, A.M.; Piazza, F.; Semenzato, G.; et al. In chronic lymphocytic leukemia the jak2/stat3 pathway is constitutively activated and its inhibition leads to cll cell death unaffected by the protective bone marrow microenvironment. *Cancers* **2019**, *11*, 1939. [CrossRef] [PubMed]

47. Oberg, H.H.; Wesch, D.; Grüssel, S.; Rose-John, S.; Kabelitz, D. Differential expression of cd126 and cd130 mediates different stat-3 phosphorylation in cd4+cd25- and cd25high regulatory t cells. *Int. Immunol.* **2006**, *18*, 555–563. [CrossRef]

48. Scheller, J.; Rose-John, S. Interleukin-6 and its receptor: From bench to bedside. *Med. Microbiol. Immunol.* **2006**, *195*, 173–183. [CrossRef]

49. Kondo, M.; Yamaoka, K.; Sakata, K.; Sonomoto, K.; Lin, L.; Nakano, K.; Tanaka, Y. Contribution of the interleukin-6/stat-3 signaling pathway to chondrogenic differentiation of human mesenchymal stem cells. *Arthritis Rheumatol.* **2015**, *67*, 1250–1260. [CrossRef]

50. Bezzerri, V.; Cipolli, M. Shwachman-diamond syndrome: Molecular mechanisms and current perspectives. *Mol. Diagn. Ther.* **2018**, *23*, 281–290. [CrossRef]

51. Lambert, C.; Genin, C. Cd3 bright lymphocyte population reveal gammadelta t cells. *Cytom. B Clin. Cytom.* **2004**, *61*, 45–53. [CrossRef] [PubMed]

52. Zhang, H.; Nguyen-Jackson, H.; Panopoulos, A.D.; Li, H.S.; Murray, P.J.; Watowich, S.S. Stat3 controls myeloid progenitor growth during emergency granulopoiesis. *Blood* **2010**, *116*, 2462–2471. [CrossRef] [PubMed]

MDPI

St. Alban-Anlage 66

4052 Basel

Switzerland

Tel. +41 61 683 77 34

Fax +41 61 302 89 18

www.mdpi.com

Cancers Editorial Office

E-mail: cancers@mdpi.com

www.mdpi.com/journal/cancers